STUDENT'S SOLUTIONS MANUAL
SINGLE VARIABLE

WILLIAM ARDIS
Collin County Community College

THOMAS' CALCULUS
TWELFTH EDITION

BASED ON THE ORIGINAL WORK BY

George B. Thomas, Jr.
Massachusetts Institute of Technology

AS REVISED BY

Maurice D. Weir
Naval Postgraduate School

Joel Hass
University of California, Davis

Addison-Wesley
is an imprint of

Reproduced by Pearson Addison-Wesley from electronic files supplied by the author.

Publishing as Pearson Addison-Wesley, 75 Arlington Street, Boston, MA 02116.

ISBN-13: 978-0-321-60070-7
ISBN-10: 0-321-60070-3

6 BRR 12 11 10

Addison-Wesley
is an imprint of

www.pearsonhighered.com

PREFACE TO THE STUDENT

The Student's Solutions Manual contains the solutions to all of the odd-numbered exercise in the 12th Edition of THOMAS' CALCULUS by Maurice Weir and Joel Hass, excluding the Computer Algebra System (CAS) exercises. We have worked each solution to ensure that it

- conforms exactly to the methods, procedures and steps presented in the text
- is mathematically correct
- includes all of the steps necessary so you can follow the logical argument and algebra
- includes a graph or figure whenever called for by the exercise, or if needed to help with the explanation
- is formatted in an appropriate style to aid in its understanding

How to use a solution's manual

- solve the assigned problem yourself
- if you get stuck along the way, refer to the solution in the manual as an aid but continue to solve the problem on your own
- if you cannot continue, reread the textbook section, or work through that section in the Student Study Guide, or consult your instructor
- if your answer is correct by your solution procedure seems to differ from the one in the manual, and you are unsure your method is correct, consult your instructor
- if your answer is incorrect and you cannot find your error, consult your instructor

For more information about other resources available with Thomas' Calculus, visit http://pearsonhighered.com.

TABLE OF CONTENTS

11 Parametric Equations and Polar Coordinates 321

CHAPTER 1 FUNCTIONS

1.1 FUNCTIONS AND THEIR GRAPHS

1. domain $= (-\infty, \infty)$; range $= [1, \infty)$

3. domain $= [-2, \infty)$; y in range and $y = \sqrt{5x + 10} \geq 0 \Rightarrow$ y can be any positive real number $\Rightarrow$ range $= [0, \infty)$.

5. domain $= (-\infty, 3) \cup (3, \infty)$; y in range and $y = \frac{4}{3-t}$, now if $t < 3 \Rightarrow 3 - t > 0 \Rightarrow \frac{4}{3-t} > 0$, or if $t > 3$ $\Rightarrow 3 - t < 0 \Rightarrow \frac{4}{3-t} < 0 \Rightarrow$ y can be any nonzero real number $\Rightarrow$ range $= (-\infty, 0) \cup (0, \infty)$.

7. (a) Not the graph of a function of x since it fails the vertical line test.
   (b) Is the graph of a function of x since any vertical line intersects the graph at most once.

9. base $= x$; $(\text{height})^2 + \left(\frac{x}{2}\right)^2 = x^2 \Rightarrow$ height $= \frac{\sqrt{3}}{2}x$; area is $a(x) = \frac{1}{2}(\text{base})(\text{height}) = \frac{1}{2}(x)\left(\frac{\sqrt{3}}{2}x\right) = \frac{\sqrt{3}}{4}x^2$; perimeter is $p(x) = x + x + x = 3x$.

11. Let D = diagonal length of a face of the cube and ℓ = the length of an edge. Then $\ell^2 + D^2 = d^2$ and $D^2 = 2\ell^2 \Rightarrow 3\ell^2 = d^2 \Rightarrow \ell = \frac{d}{\sqrt{3}}$. The surface area is $6\ell^2 = \frac{6d^2}{3} = 2d^2$ and the volume is $\ell^3 = \left(\frac{d^2}{3}\right)^{3/2} = \frac{d^3}{3\sqrt{3}}$.

13. $2x + 4y = 5 \Rightarrow y = -\frac{1}{2}x + \frac{5}{4}$; $L = \sqrt{(x-0)^2 + (y-0)^2} = \sqrt{x^2 + (-\frac{1}{2}x + \frac{5}{4})^2} = \sqrt{x^2 + \frac{1}{4}x^2 - \frac{5}{4}x + \frac{25}{16}}$ $= \sqrt{\frac{5}{4}x^2 - \frac{5}{4}x + \frac{25}{16}} = \sqrt{\frac{20x^2 - 20x + 25}{16}} = \frac{\sqrt{20x^2 - 20x + 25}}{4}$

15. The domain is $(-\infty, \infty)$.

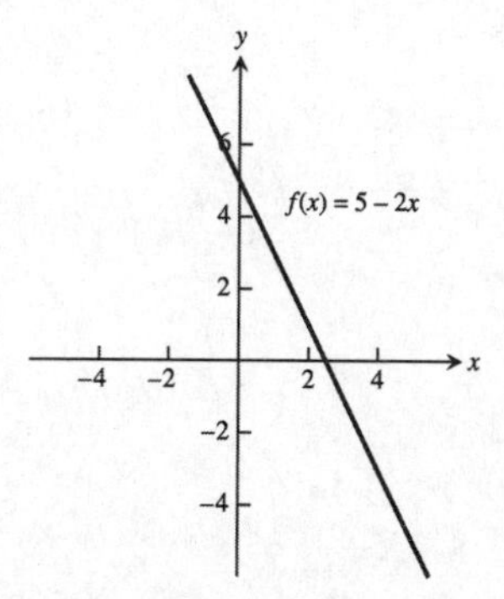

17. The domain is $(-\infty, \infty)$.

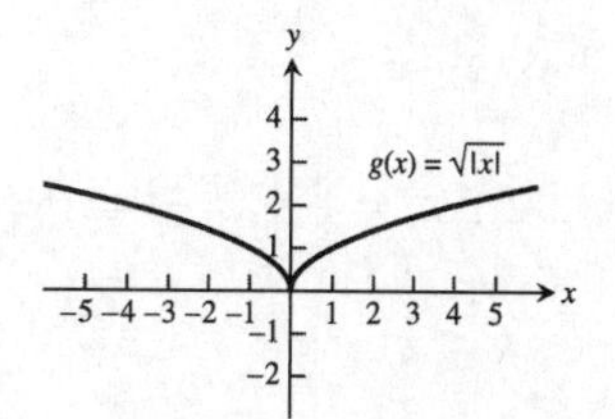

19. The domain is $(-\infty, 0) \cup (0, \infty)$.

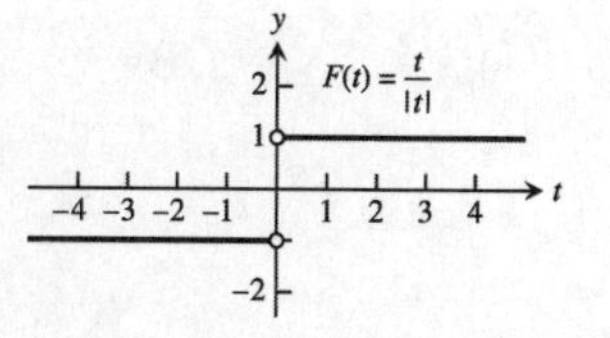

21. The domain is $(-\infty, -5) \cup (-5, -3] \cup [3, 5) \cup (5, \infty)$

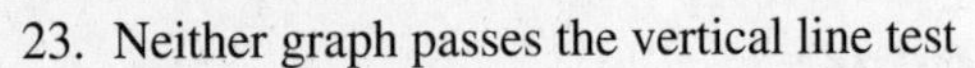

23. Neither graph passes the vertical line test

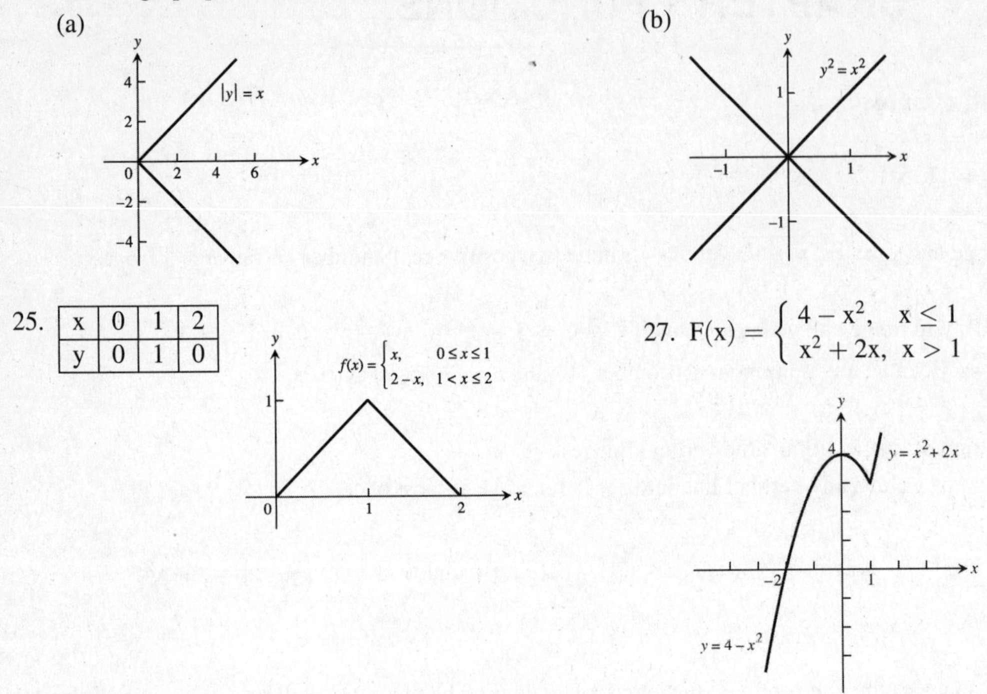

25.

x	0	1	2
y	0	1	0

27. $F(x) = \begin{cases} 4 - x^2, & x \le 1 \\ x^2 + 2x, & x > 1 \end{cases}$

29. (a) Line through (0, 0) and (1, 1): $y = x$; Line through (1, 1) and (2, 0): $y = -x + 2$

$$f(x) = \begin{cases} x, & 0 \le x \le 1 \\ -x + 2, & 1 < x \le 2 \end{cases}$$

(b) $f(x) = \begin{cases} 2, & 0 \le x < 1 \\ 0, & 1 \le x < 2 \\ 2, & 2 \le x < 3 \\ 0, & 3 \le x \le 4 \end{cases}$

31. (a) Line through (−1, 1) and (0, 0): $y = -x$

Line through (0, 1) and (1, 1): $y = 1$

Line through (1, 1) and (3, 0): $m = \frac{0-1}{3-1} = \frac{-1}{2} = -\frac{1}{2}$, so $y = -\frac{1}{2}(x - 1) + 1 = -\frac{1}{2}x + \frac{3}{2}$

$$f(x) = \begin{cases} -x & -1 \le x < 0 \\ 1 & 0 < x \le 1 \\ -\frac{1}{2}x + \frac{3}{2} & 1 < x < 3 \end{cases}$$

(b) Line through (−2, −1) and (0, 0): $y = \frac{1}{2}x$

Line through (0, 2) and (1, 0): $y = -2x + 2$

Line through (1, −1) and (3, −1): $y = -1$

$$f(x) = \begin{cases} \frac{1}{2}x & -2 \le x \le 0 \\ -2x + 2 & 0 < x \le 1 \\ -1 & 1 < x \le 3 \end{cases}$$

33. (a) $\lfloor x \rfloor = 0$ for $x \in [0, 1)$ (b) $\lceil x \rceil = 0$ for $x \in (-1, 0]$

35. For any real number x, $n \le x \le n + 1$, where n is an integer. Now: $n \le x \le n + 1 \Rightarrow -(n + 1) \le -x \le -n$. By definition: $\lceil -x \rceil = -n$ and $\lfloor x \rfloor = n \Rightarrow -\lfloor x \rfloor = -n$. So $\lceil -x \rceil = -\lfloor x \rfloor$ for all $x \in \Re$.

37. Symmetric about the origin
Dec: $-\infty < x < \infty$
Inc: nowhere

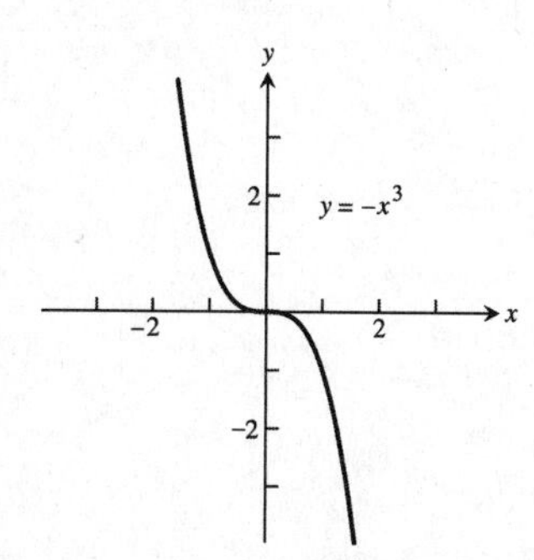

39. Symmetric about the origin
Dec: nowhere
Inc: $-\infty < x < 0$
$0 < x < \infty$

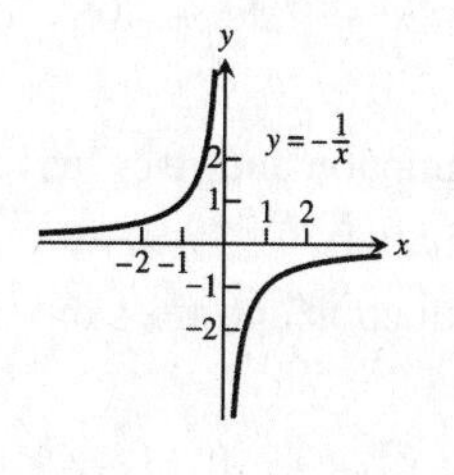

41. Symmetric about the y-axis
Dec: $-\infty < x \leq 0$
Inc: $0 < x < \infty$

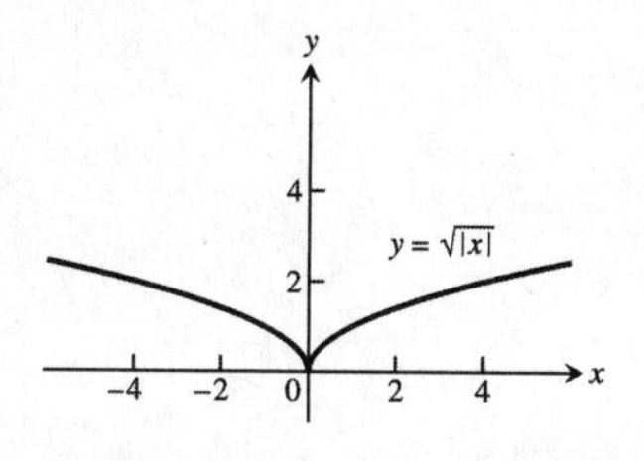

43. Symmetric about the origin
Dec: nowhere
Inc: $-\infty < x < \infty$

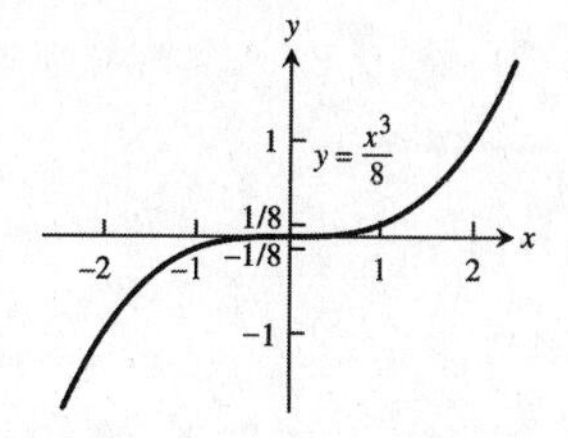

45. No symmetry
Dec: $0 \leq x < \infty$
Inc: nowhere

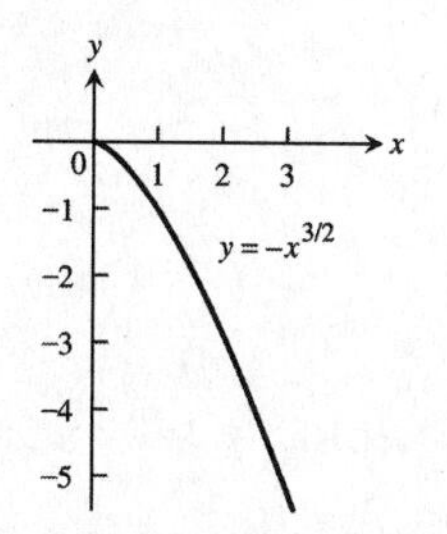

47. Since a horizontal line not through the origin is symmetric with respect to the y-axis, but not with respect to the origin, the function is even.

49. Since $f(x) = x^2 + 1 = (-x)^2 + 1 = -f(x)$. The function is even.

51. Since $g(x) = x^3 + x$, $g(-x) = -x^3 - x = -(x^3 + x) = -g(x)$. So the function is odd.

53. $g(x) = \frac{1}{x^2 - 1} = \frac{1}{(-x)^2 - 1} = g(-x)$. Thus the function is even.

55. $h(t) = \frac{1}{t-1}$; $h(-t) = \frac{1}{-t-1}$; $-h(t) = \frac{1}{1-t}$. Since $h(t) \neq -h(t)$ and $h(t) \neq h(-t)$, the function is neither even nor odd.

57. $h(t) = 2t + 1$, $h(-t) = -2t + 1$. So $h(t) \neq h(-t)$. $-h(t) = -2t - 1$, so $h(t) \neq -h(t)$. The function is neither even nor odd.

59. $s = kt \Rightarrow 25 = k(75) \Rightarrow k = \frac{1}{3} \Rightarrow s = \frac{1}{3}t;\ 60 = \frac{1}{3}t \Rightarrow t = 180$

61. $r = \frac{k}{s} \Rightarrow 6 = \frac{k}{4} \Rightarrow k = 24 \Rightarrow r = \frac{24}{s};\ 10 = \frac{24}{s} \Rightarrow s = \frac{12}{5}$

63. $v = f(x) = x(14 - 2x)(22 - 2x) = 4x^3 - 72x^2 + 308x;\ 0 < x < 7.$

65. (a) Graph h because it is an even function and rises less rapidly than does Graph g.
(b) Graph f because it is an odd function.
(c) Graph g because it is an even function and rises more rapidly than does Graph h.

67. (a) From the graph, $\frac{x}{2} > 1 + \frac{4}{x} \Rightarrow x \in (-2, 0) \cup (4, \infty)$
(b) $\frac{x}{2} > 1 + \frac{4}{x} \Rightarrow \frac{x}{2} - 1 - \frac{4}{x} > 0$
$x > 0$: $\frac{x}{2} - 1 - \frac{4}{x} > 0 \Rightarrow \frac{x^2-2x-8}{2x} > 0 \Rightarrow \frac{(x-4)(x+2)}{2x} > 0$
$\Rightarrow x > 4$ since x is positive;
$x < 0$: $\frac{x}{2} - 1 - \frac{4}{x} > 0 \Rightarrow \frac{x^2-2x-8}{2x} < 0 \Rightarrow \frac{(x-4)(x+2)}{2x} < 0$
$\Rightarrow x < -2$ since x is negative;
sign of $(x - 4)(x + 2)$
$+ \quad - \quad +$
$-2 \quad 4$
Solution interval: $(-2, 0) \cup (4, \infty)$

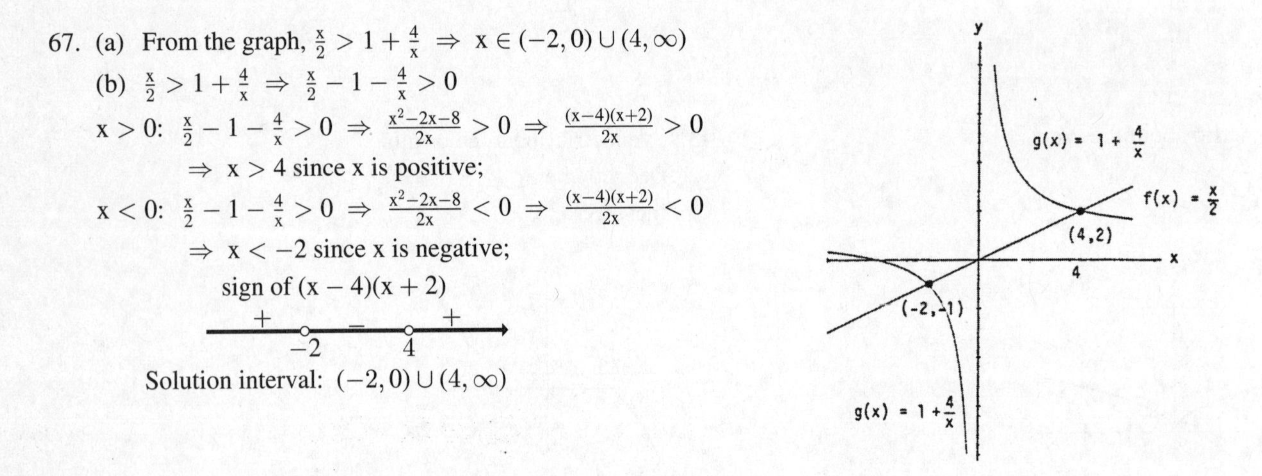

69. A curve symmetric about the x-axis will not pass the vertical line test because the points (x, y) and (x, −y) lie on the same vertical line. The graph of the function $y = f(x) = 0$ is the x-axis, a horizontal line for which there is a single y-value, 0, for any x.

71. $x^2 + x^2 = h^2 \Rightarrow x = \frac{h}{\sqrt{2}} = \frac{\sqrt{2}h}{2}$; cost $= 5(2x) + 10h \Rightarrow C(h) = 10\left(\frac{\sqrt{2}h}{2}\right) + 10h = 5h\left(\sqrt{2} + 2\right)$

1.2 COMBINING FUNCTIONS; SHIFTING AND SCALING GRAPHS

1. $D_f: -\infty < x < \infty$, $D_g: x \ge 1 \Rightarrow D_{f+g} = D_{fg}: x \ge 1$. $R_f: -\infty < y < \infty$, $R_g: y \ge 0$, $R_{f+g}: y \ge 1$, $R_{fg}: y \ge 0$

3. $D_f: -\infty < x < \infty$, $D_g: -\infty < x < \infty$, $D_{f/g}: -\infty < x < \infty$, $D_{g/f}: -\infty < x < \infty$, $R_f: y = 2$, $R_g: y \ge 1$, $R_{f/g}: 0 < y \le 2$, $R_{g/f}: \frac{1}{2} \le y < \infty$

5. (a) 2 (b) 22 (c) $x^2 + 2$
(d) $(x + 5)^2 - 3 = x^2 + 10x + 22$ (e) 5 (f) -2
(g) $x + 10$ (h) $(x^2 - 3)^2 - 3 = x^4 - 6x^2 + 6$

7. $(f \circ g \circ h)(x) = f(g(h(x))) = f(g(4 - x)) = f(3(4 - x)) = f(12 - 3x) = (12 - 3x) + 1 = 13 - 3x$

9. $(f \circ g \circ h)(x) = f(g(h(x))) = f\left(g\left(\frac{1}{x}\right)\right) = f\left(\frac{1}{\frac{1}{x}+4}\right) = f\left(\frac{x}{1+4x}\right) = \sqrt{\frac{x}{1+4x} + 1} = \sqrt{\frac{5x+1}{1+4x}}$

11. (a) $(f \circ g)(x)$ (b) $(j \circ g)(x)$ (c) $(g \circ g)(x)$
(d) $(j \circ j)(x)$ (e) $(g \circ h \circ f)(x)$ (f) $(h \circ j \circ f)(x)$

13.

	$g(x)$	$f(x)$	$(f \circ g)(x)$
(a)	$x - 7$	$\sqrt{x}$	$\sqrt{x - 7}$
(b)	$x + 2$	$3x$	$3(x + 2) = 3x + 6$
(c)	x^2	$\sqrt{x - 5}$	$\sqrt{x^2 - 5}$
(d)	$\frac{x}{x-1}$	$\frac{x}{x-1}$	$\frac{\frac{x}{x-1}}{\frac{x}{x-1} - 1} = \frac{x}{x-(x-1)} = x$
(e)	$\frac{1}{x-1}$	$1 + \frac{1}{x}$	x
(f)	$\frac{1}{x}$	$\frac{1}{x}$	x

15. (a) $f(g(-1)) = f(1) = 1$ (b) $g(f(0)) = g(-2) = 2$ (c) $f(f(-1)) = f(0) = -2$
(d) $g(g(2)) = g(0) = 0$ (e) $g(f(-2)) = g(1) = -1$ (f) $f(g(1)) = f(-1) = 0$

17. (a) $(f \circ g)(x) = f(g(x)) = \sqrt{\frac{1}{x} + 1} = \sqrt{\frac{1+x}{x}}$
$(g \circ f)(x) = g(f(x)) = \frac{1}{\sqrt{x+1}}$
(b) Domain $(f \circ g)$: $(-\infty, -1] \cup (0, \infty)$, domain $(g \circ f)$: $(-1, \infty)$
(c) Range $(f \circ g)$: $(1, \infty)$, range $(g \circ f)$: $(0, \infty)$

19. $(f \circ g)(x) = x \Rightarrow f(g(x)) = x \Rightarrow \frac{g(x)}{g(x)-2} = x \Rightarrow g(x) = (g(x) - 2)x = x \cdot g(x) - 2x$
$\Rightarrow g(x) - x \cdot g(x) = -2x \Rightarrow g(x) = -\frac{2x}{1-x} = \frac{2x}{x-1}$

21. (a) $y = -(x + 7)^2$ (b) $y = -(x - 4)^2$

23. (a) Position 4 (b) Position 1 (c) Position 2 (d) Position 3

25.

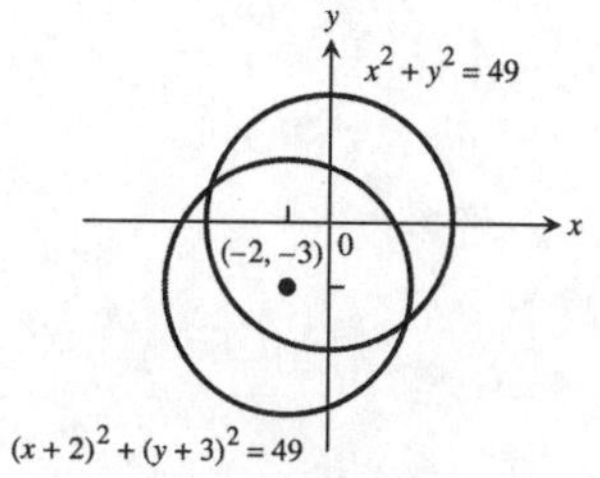

$x^2 + y^2 = 49$; $(-2, -3)$; $(x + 2)^2 + (y + 3)^2 = 49$

27.

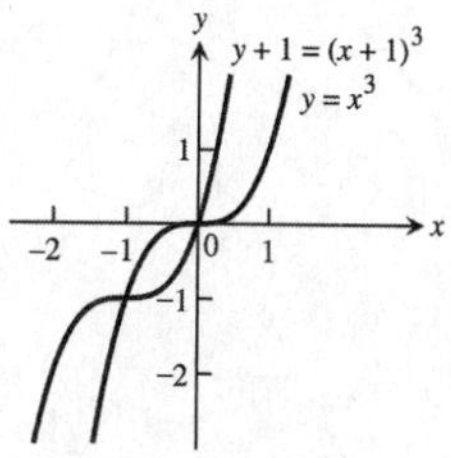

$y + 1 = (x + 1)^3$; $y = x^3$

29.

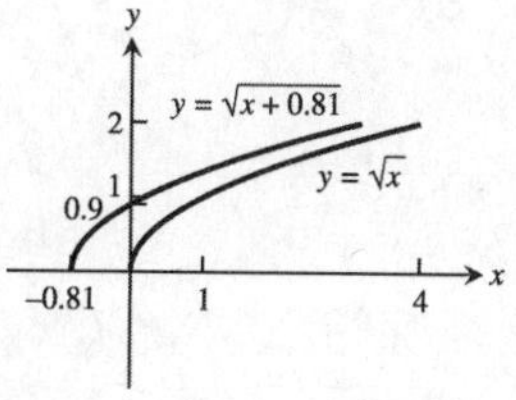

$y = \sqrt{x + 0.81}$; $y = \sqrt{x}$; 0.9; −0.81

31.

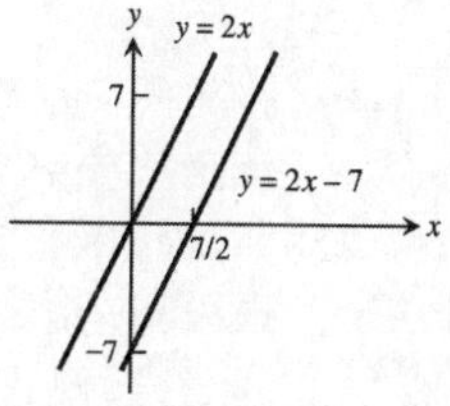

$y = 2x$; $y = 2x - 7$; 7; 7/2; −7

33.

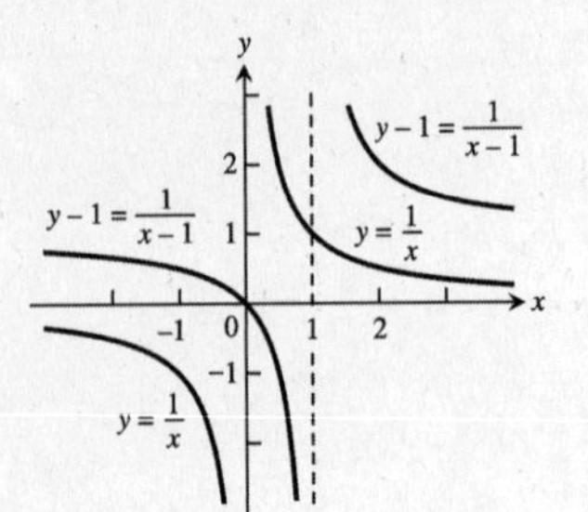

35.

$y = \sqrt{x+4}$

37.

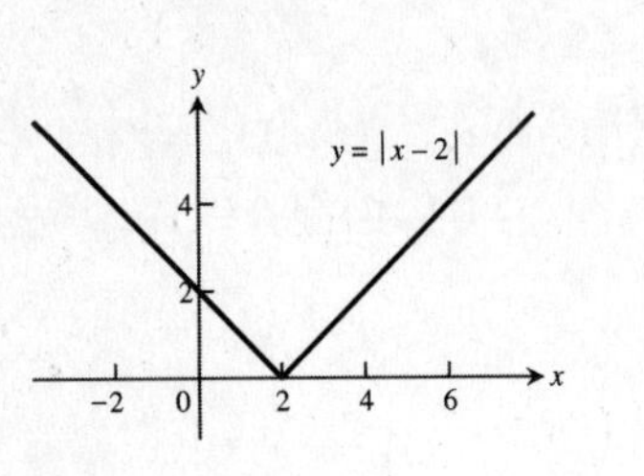

39.

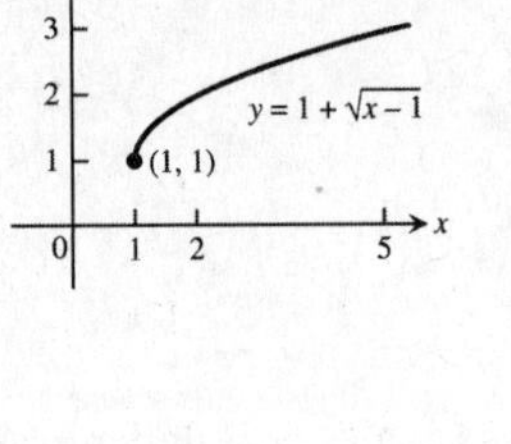

41.

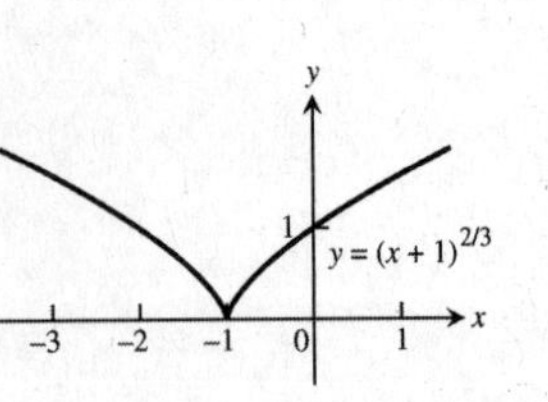

43.

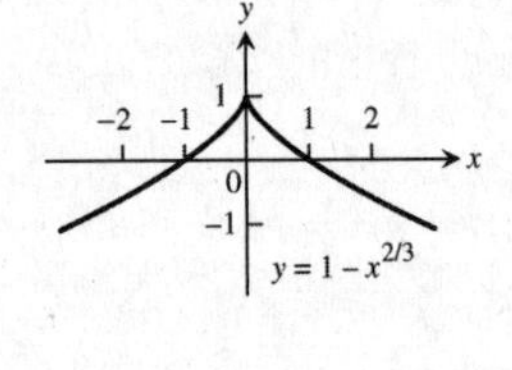

45.

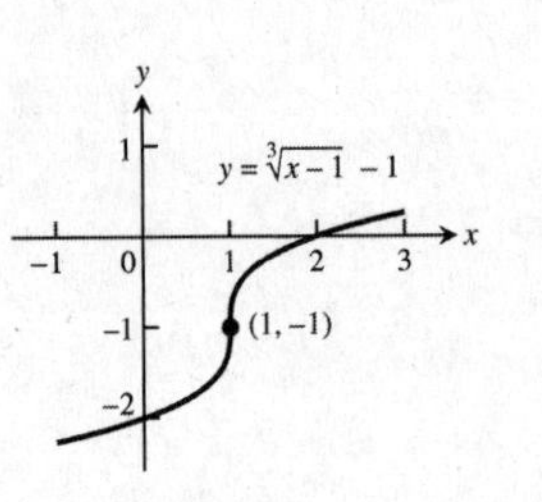

47.

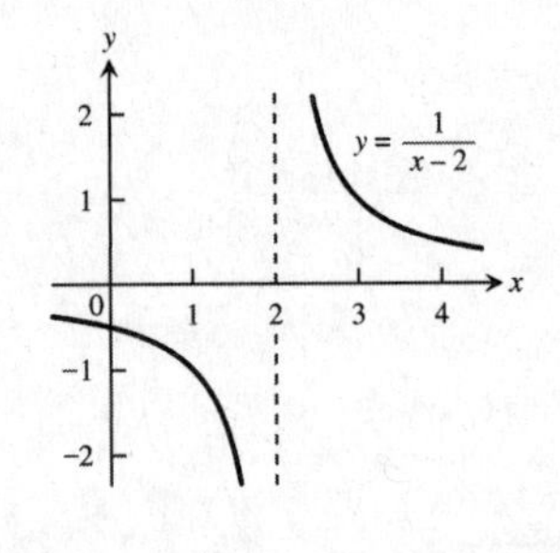

49.

$y = \frac{1}{x} + 2$

51.

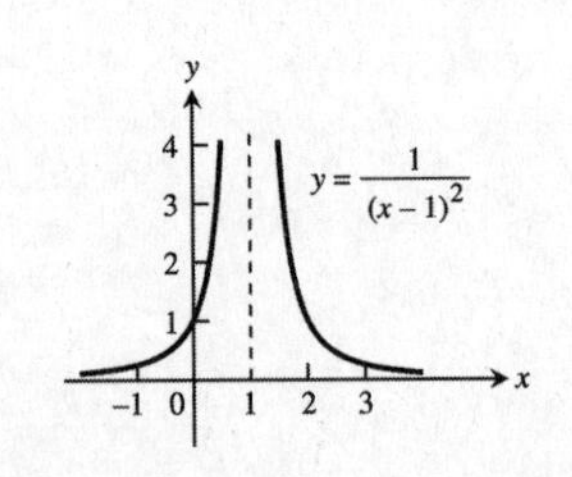

53.

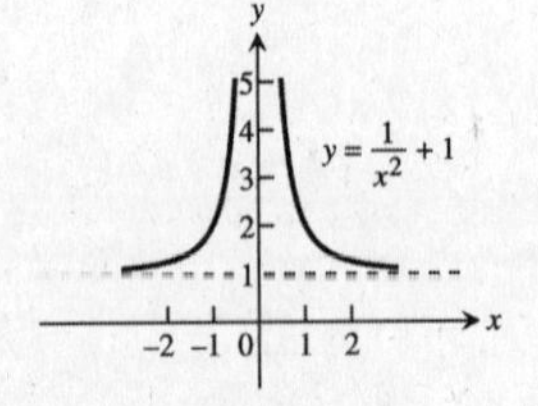

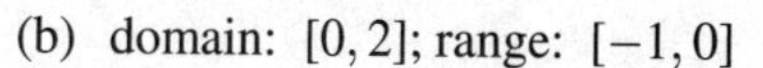

55. (a) domain: $[0, 2]$; range: $[2, 3]$

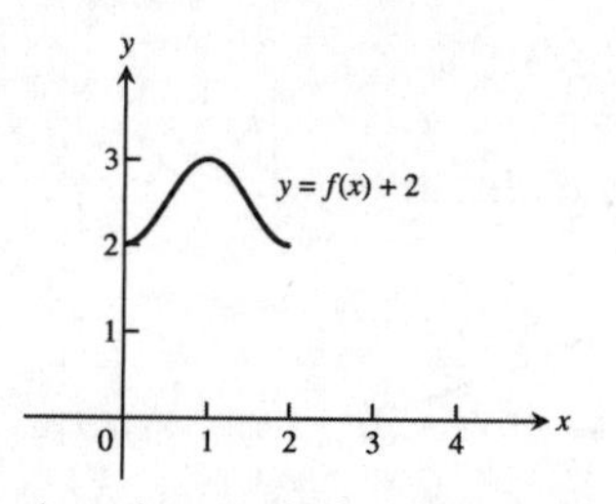

(b) domain: $[0, 2]$; range: $[-1, 0]$

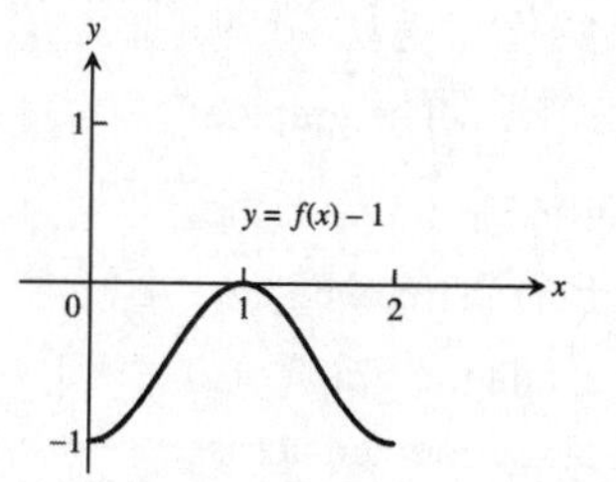

(c) domain: $[0, 2]$; range: $[0, 2]$

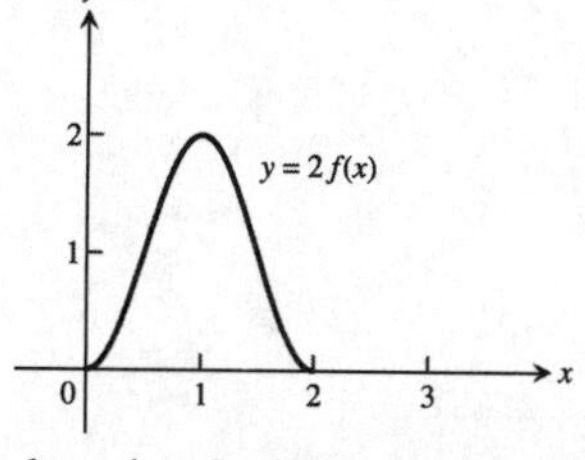

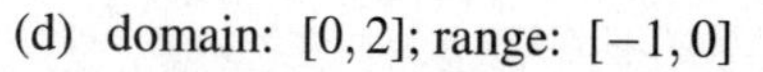

(d) domain: $[0, 2]$; range: $[-1, 0]$

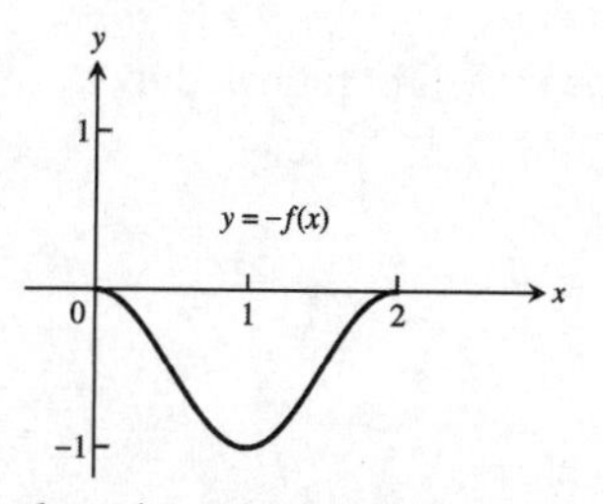

(e) domain: $[-2, 0]$; range: $[0, 1]$

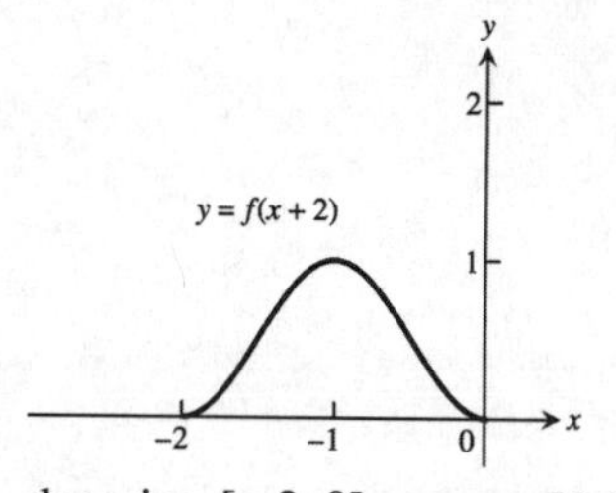

(f) domain: $[1, 3]$; range: $[0, 1]$

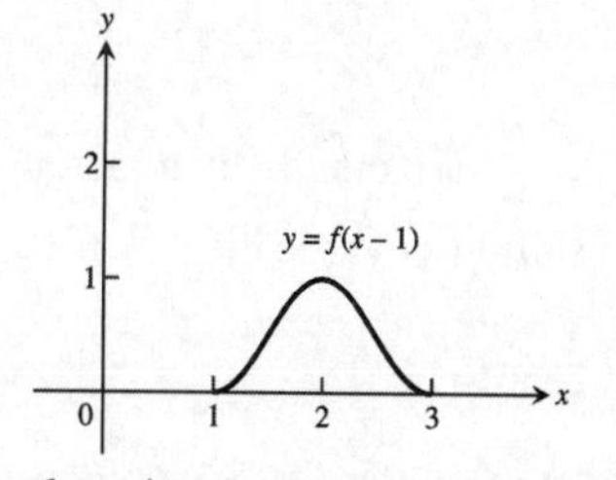

(g) domain: $[-2, 0]$; range: $[0, 1]$

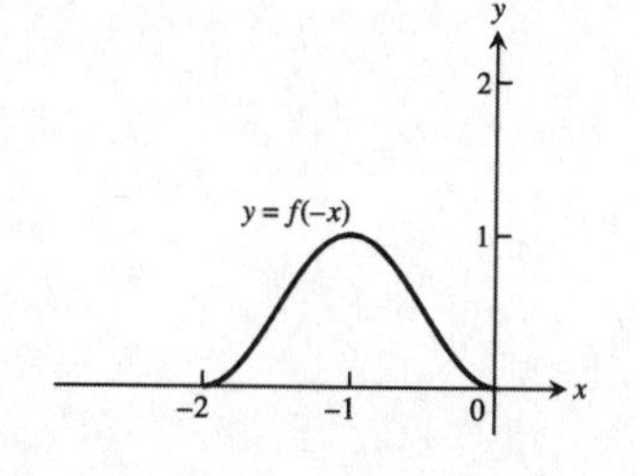

(h) domain: $[-1, 1]$; range: $[0, 1]$

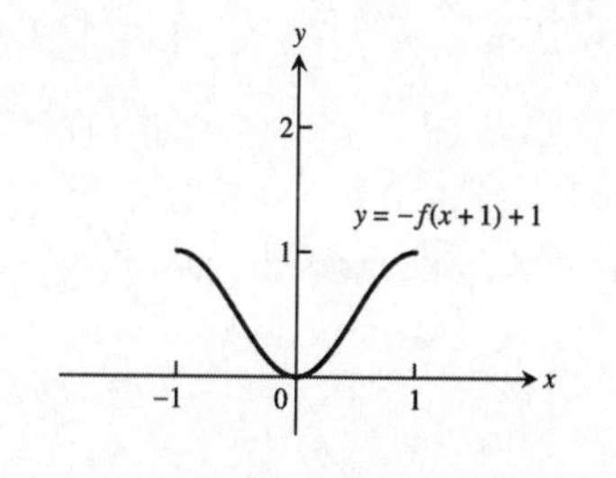

57. $y = 3x^2 - 3$

59. $y = \frac{1}{2}\left(1 + \frac{1}{x^2}\right) = \frac{1}{2} + \frac{1}{2x^2}$

61. $y = \sqrt{4x + 1}$

63. $y = \sqrt{4 - \left(\frac{x}{2}\right)^2} = \frac{1}{2}\sqrt{16 - x^2}$

65. $y = 1 - (3x)^3 = 1 - 27x^3$

67. Let $y = -\sqrt{2x+1} = f(x)$ and let $g(x) = x^{1/2}$, $h(x) = \left(x + \frac{1}{2}\right)^{1/2}$, $i(x) = \sqrt{2}\left(x + \frac{1}{2}\right)^{1/2}$, and $j(x) = -\left[\sqrt{2}\left(x + \frac{1}{2}\right)^{1/2}\right] = f(x)$. The graph of h(x) is the graph of g(x) shifted left $\frac{1}{2}$ unit; the graph of i(x) is the graph of h(x) stretched vertically by a factor of $\sqrt{2}$; and the graph of $j(x) = f(x)$ is the graph of i(x) reflected across the x-axis.

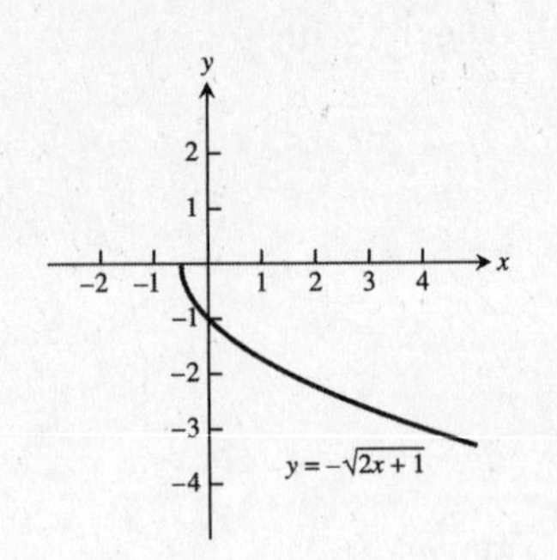

69. $y = f(x) = x^3$. Shift f(x) one unit right followed by a shift two units up to get $g(x) = (x-1)^3 + 2$.

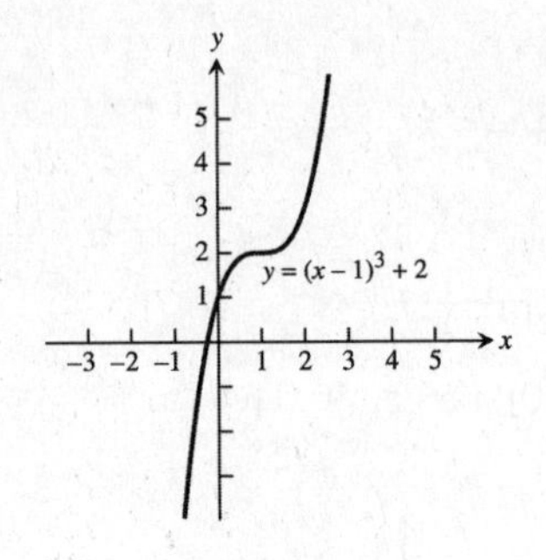

71. Compress the graph of $f(x) = \frac{1}{x}$ horizontally by a factor of 2 to get $g(x) = \frac{1}{2x}$. Then shift g(x) vertically down 1 unit to get $h(x) = \frac{1}{2x} - 1$.

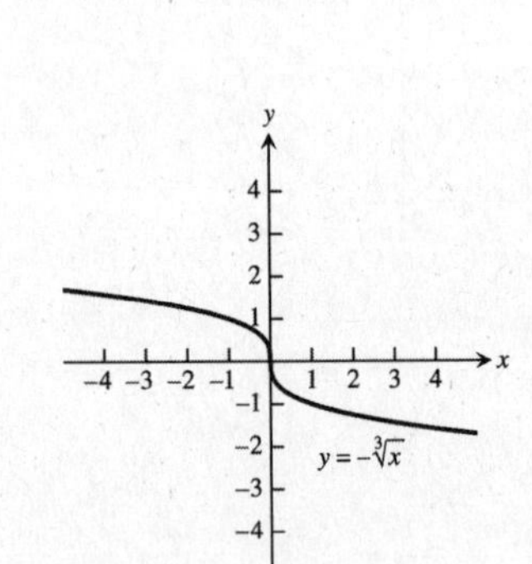

73. Reflect the graph of $y = f(x) = \sqrt[3]{x}$ across the x-axis to get $g(x) = -\sqrt[3]{x}$.

75.

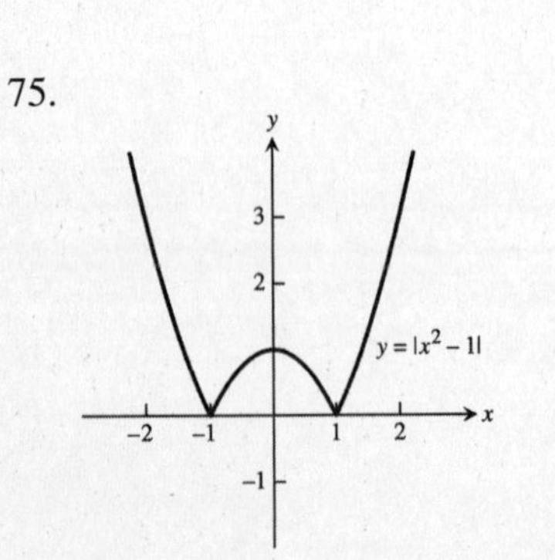

77. $9x^2 + 25y^2 = 225 \Rightarrow \frac{x^2}{5^2} + \frac{y^2}{3^2} = 1$

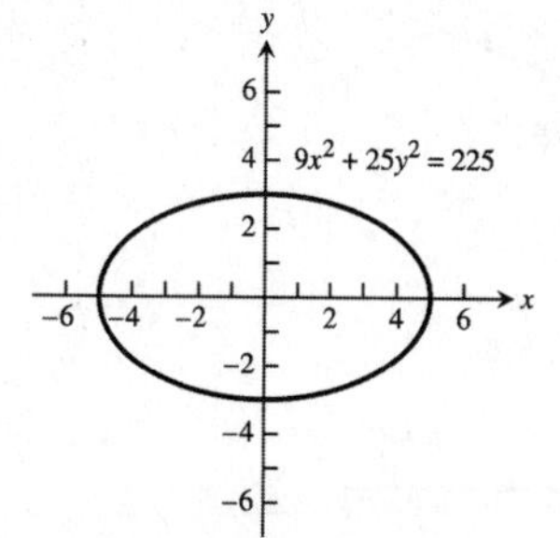

79. $3x^2 + (y-2)^2 = 3 \Rightarrow \frac{x^2}{1^2} + \frac{(y-2)^2}{(\sqrt{3})^2} = 1$

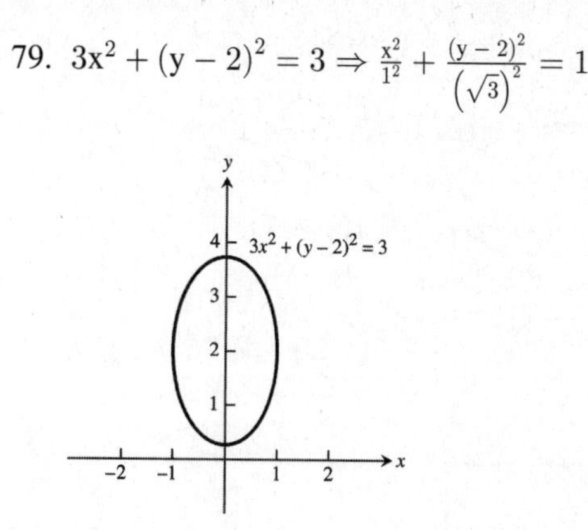

81. $3(x-1)^2 + 2(y+2)^2 = 6 \Rightarrow \frac{(x-1)^2}{(\sqrt{2})^2} + \frac{[y-(-2)]^2}{(\sqrt{3})^2} = 1$

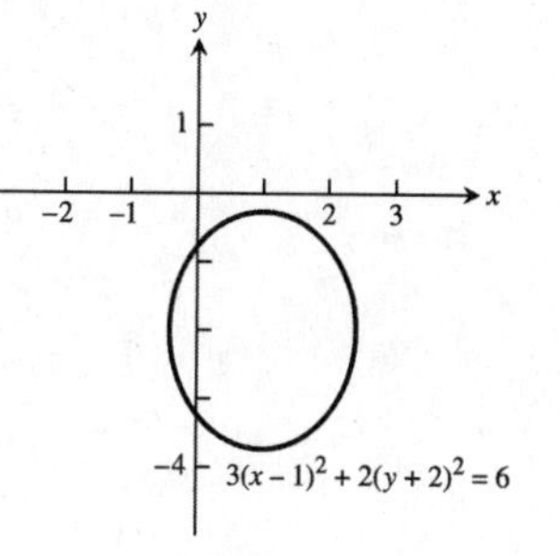

83. $\frac{x^2}{16} + \frac{y^2}{9} = 1$ has its center at $(0, 0)$. Shiftinig 4 units left and 3 units up gives the center at $(h, k) = (-4, 3)$. So the equation is $\frac{[x-(-4)]^2}{4^2} + \frac{(y-3)^2}{3^2} = 1$ $\Rightarrow \frac{(x+4)^2}{4^2} + \frac{(y-3)^2}{3^2} = 1$. Center, C, is $(-4, 3)$, and major axis, $\overline{AB}$, is the segment from $(-8, 3)$ to $(0, 3)$.

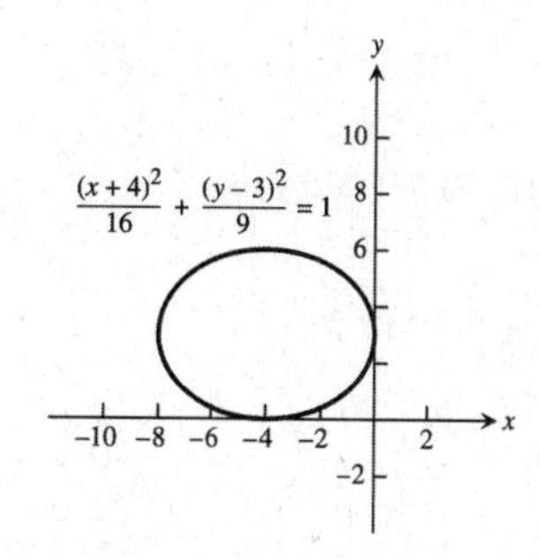

85. (a) $(fg)(-x) = f(-x)g(-x) = f(x)(-g(x)) = -(fg)(x)$, odd

(b) $\left(\frac{f}{g}\right)(-x) = \frac{f(-x)}{g(-x)} = \frac{f(x)}{-g(x)} = -\left(\frac{f}{g}\right)(x)$, odd

(c) $\left(\frac{g}{f}\right)(-x) = \frac{g(-x)}{f(-x)} = \frac{-g(x)}{f(x)} = -\left(\frac{g}{f}\right)(x)$, odd

(d) $f^2(-x) = f(-x)f(-x) = f(x)f(x) = f^2(x)$, even

(e) $g^2(-x) = (g(-x))^2 = (-g(x))^2 = g^2(x)$, even

(f) $(f \circ g)(-x) = f(g(-x)) = f(-g(x)) = f(g(x)) = (f \circ g)(x)$, even

(g) $(g \circ f)(-x) = g(f(-x)) = g(f(x)) = (g \circ f)(x)$, even

(h) $(f \circ f)(-x) = f(f(-x)) = f(f(x)) = (f \circ f)(x)$, even

(i) $(g \circ g)(-x) = g(g(-x)) = g(-g(x)) = -g(g(x)) = -(g \circ g)(x)$, odd

87. (a) (b) (c) (d)

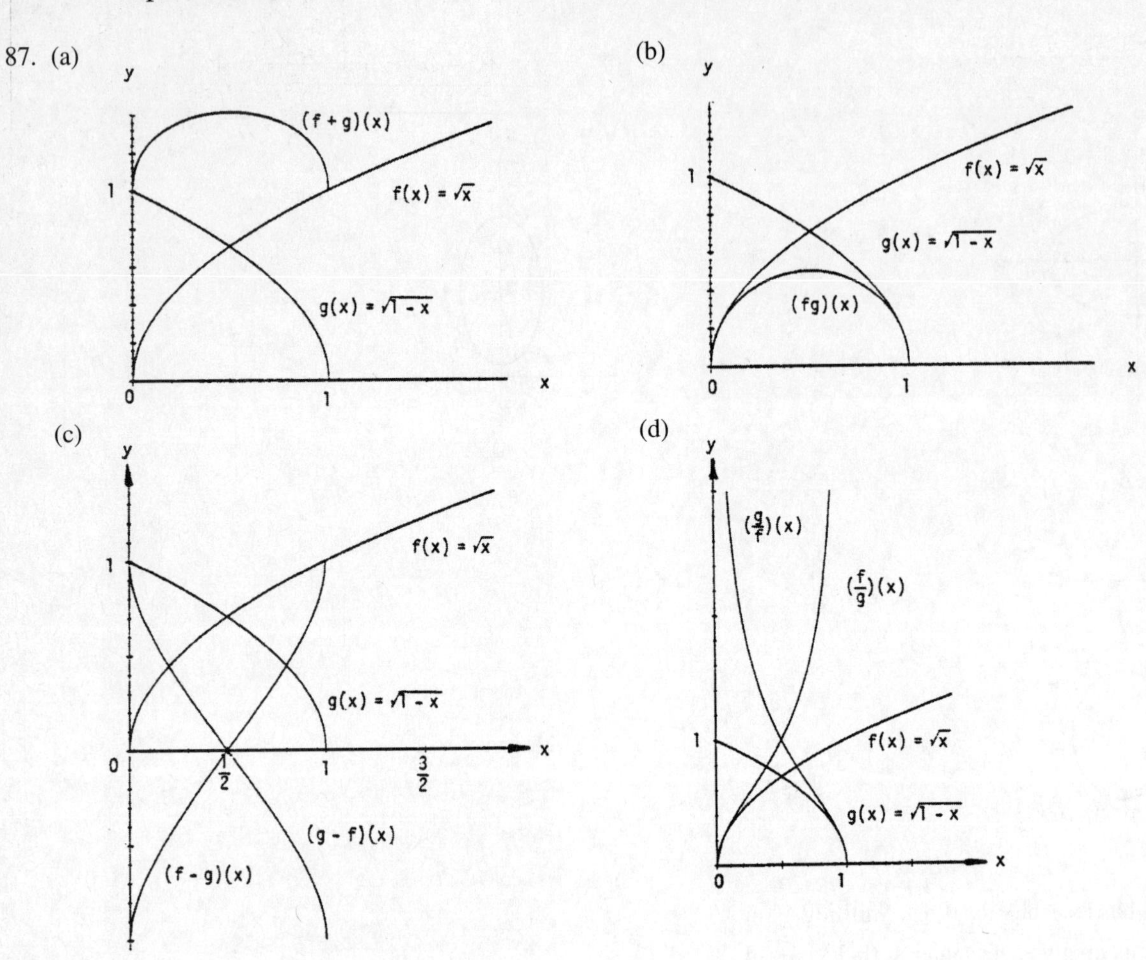

1.3 TRIGONOMETRIC FUNCTIONS

1. (a) $s = r\theta = (10)\left(\frac{4\pi}{5}\right) = 8\pi$ m (b) $s = r\theta = (10)(110°)\left(\frac{\pi}{180°}\right) = \frac{110\pi}{18} = \frac{55\pi}{9}$ m

3. $\theta = 80° \Rightarrow \theta = 80°\left(\frac{\pi}{180°}\right) = \frac{4\pi}{9} \Rightarrow s = (6)\left(\frac{4\pi}{9}\right) = 8.4$ in. (since the diameter = 12 in. $\Rightarrow$ radius = 6 in.)

5.

θ	$-\pi$	$-\frac{2\pi}{3}$	0	$\frac{\pi}{2}$	$\frac{3\pi}{4}$
$\sin\theta$	0	$-\frac{\sqrt{3}}{2}$	0	1	$\frac{1}{\sqrt{2}}$
$\cos\theta$	-1	$-\frac{1}{2}$	1	0	$-\frac{1}{\sqrt{2}}$
$\tan\theta$	0	$\sqrt{3}$	0	und.	-1
$\cot\theta$	und.	$\frac{1}{\sqrt{3}}$	und.	0	-1
$\sec\theta$	-1	-2	1	und.	$-\sqrt{2}$
$\csc\theta$	und.	$-\frac{2}{\sqrt{3}}$	und.	1	$\sqrt{2}$

7. $\cos x = -\frac{4}{5}$, $\tan x = -\frac{3}{4}$

9. $\sin x = -\frac{\sqrt{8}}{3}$, $\tan x = -\sqrt{8}$

11. $\sin x = -\frac{1}{\sqrt{5}}$, $\cos x = -\frac{2}{\sqrt{5}}$

13.

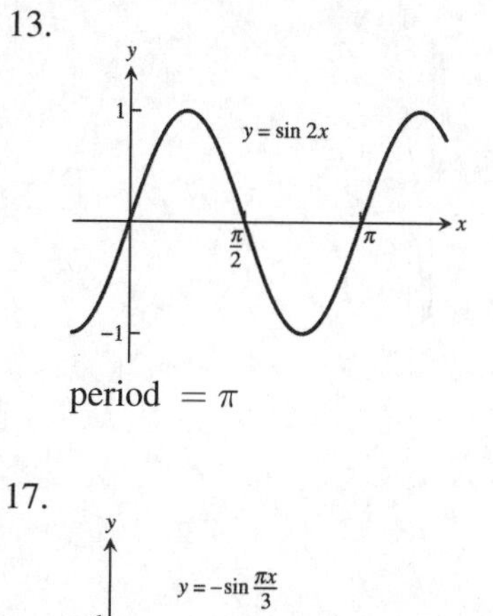

period $= \pi$

15.

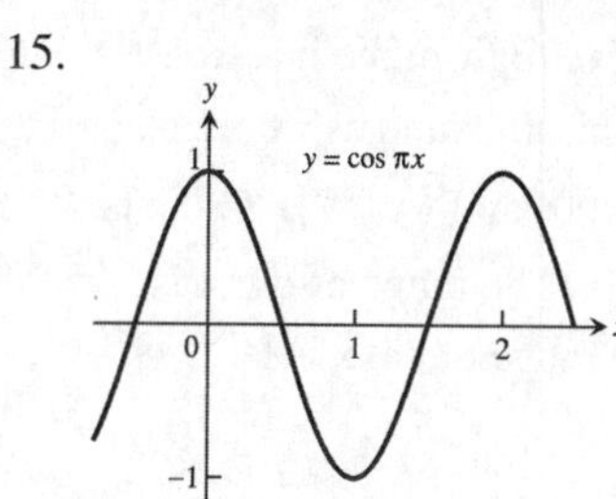

period $= 2$

17.

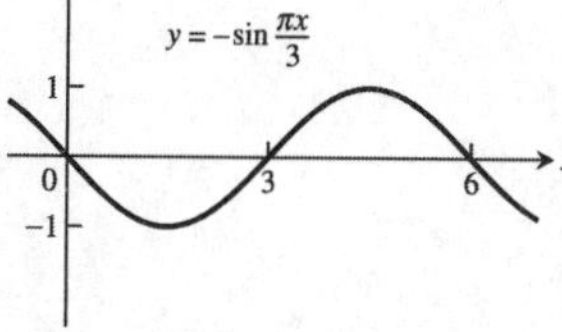

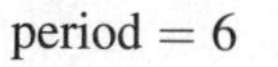
period $= 6$

19.

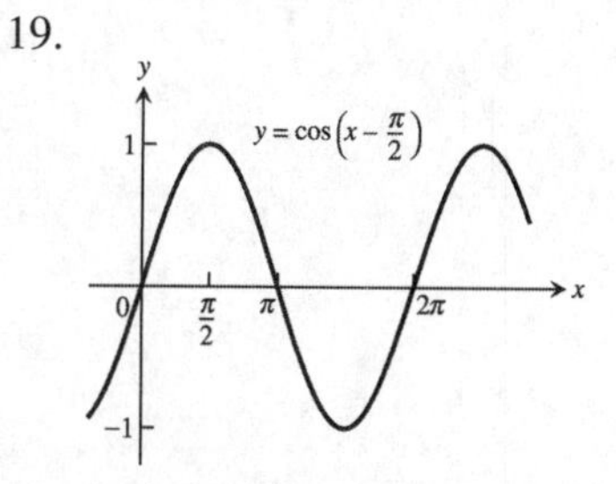

period $= 2\pi$

21.

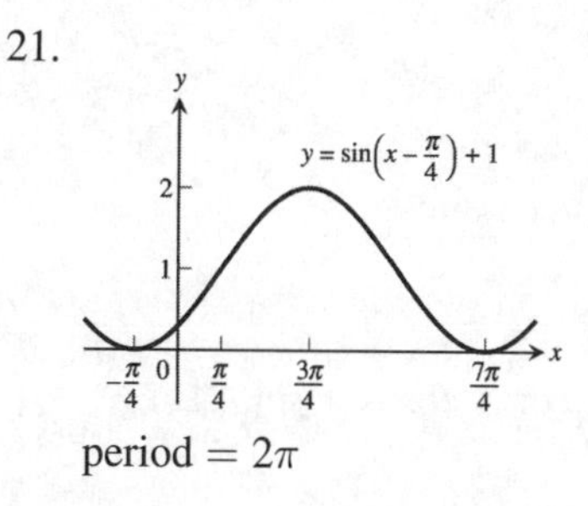

period $= 2\pi$

23. period $= \frac{\pi}{2}$, symmetric about the origin

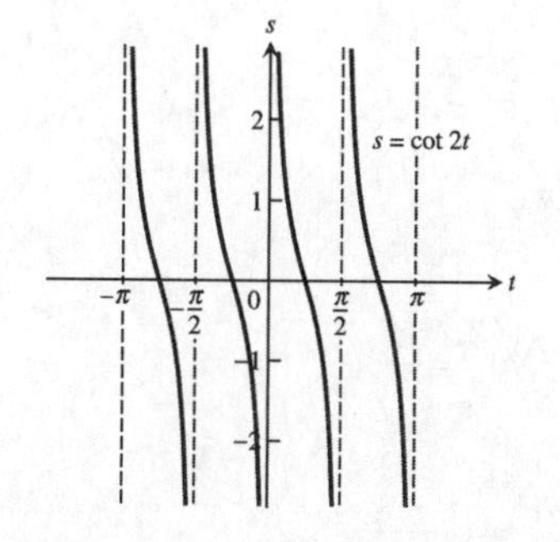

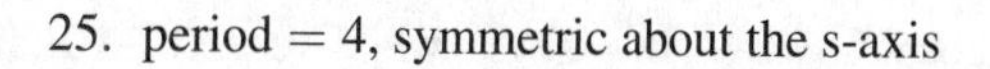
25. period $= 4$, symmetric about the s-axis

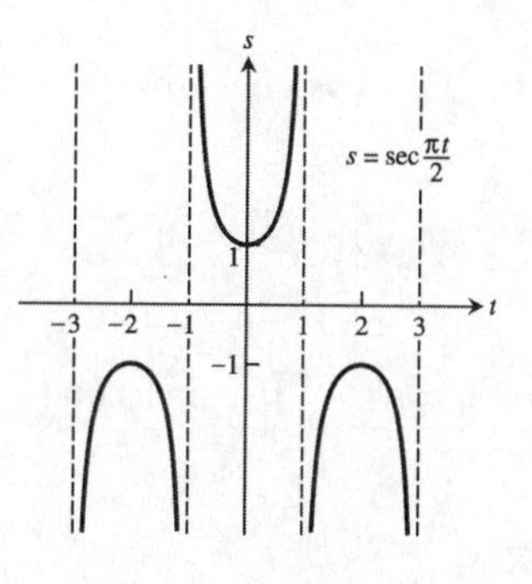

27. (a) Cos x and sec x are positive for x in the interval $\left(-\frac{\pi}{2}, \frac{\pi}{2}\right)$; and cos x and sec x are negative for x in the intervals $\left(-\frac{3\pi}{2}, -\frac{\pi}{2}\right)$ and $\left(\frac{\pi}{2}, \frac{3\pi}{2}\right)$. Sec x is undefined when cos x is 0. The range of sec x is $(-\infty, -1] \cup [1, \infty)$; the range of cos x is $[-1, 1]$.

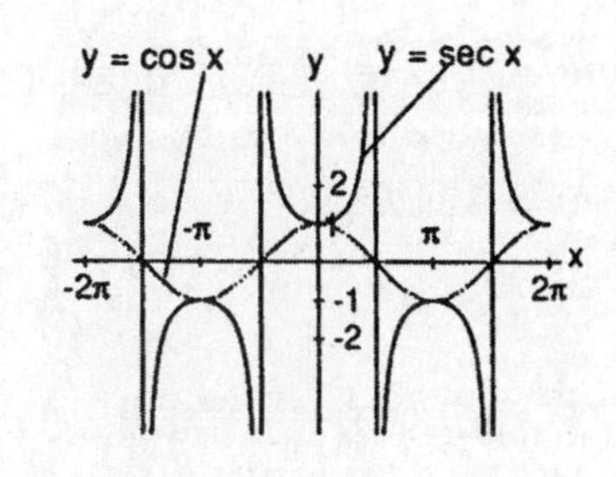

(b) Sin x and csc x are positive for x in the intervals $\left(-\frac{3\pi}{2}, -\pi\right)$ and $(0, \pi)$; and sin x and csc x are negative for x in the intervals $(-\pi, 0)$ and $\left(\pi, \frac{3\pi}{2}\right)$. Csc x is undefined when sin x is 0. The range of csc x is $(-\infty, -1] \cup [1, \infty)$; the range of sin x is $[-1, 1]$.

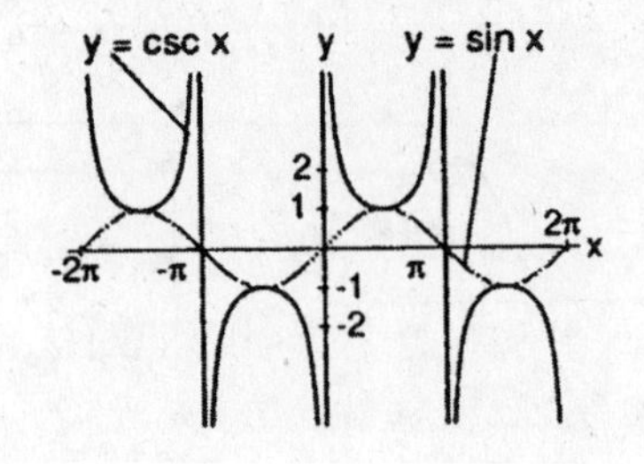

29. D: $-\infty < x < \infty$; R: $y = -1, 0, 1$

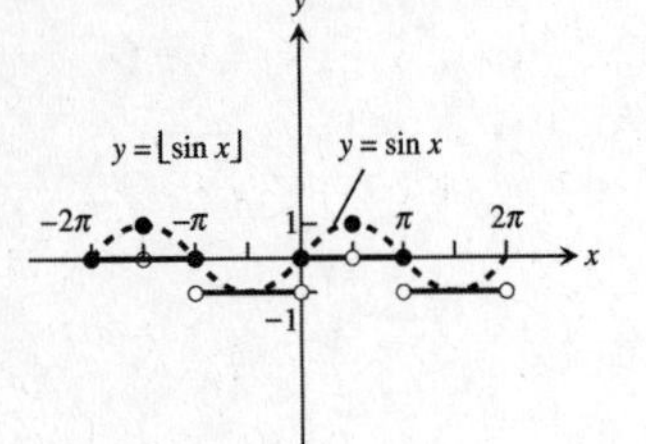

31. $\cos\left(x - \frac{\pi}{2}\right) = \cos x \cos\left(-\frac{\pi}{2}\right) - \sin x \sin\left(-\frac{\pi}{2}\right) = (\cos x)(0) - (\sin x)(-1) = \sin x$

33. $\sin\left(x + \frac{\pi}{2}\right) = \sin x \cos\left(\frac{\pi}{2}\right) + \cos x \sin\left(\frac{\pi}{2}\right) = (\sin x)(0) + (\cos x)(1) = \cos x$

35. $\cos(A - B) = \cos(A + (-B)) = \cos A \cos(-B) - \sin A \sin(-B) = \cos A \cos B - \sin A(-\sin B)$
$= \cos A \cos B + \sin A \sin B$

37. If $B = A$, $A - B = 0 \Rightarrow \cos(A - B) = \cos 0 = 1$. Also $\cos(A - B) = \cos(A - A) = \cos A \cos A + \sin A \sin A$ $= \cos^2 A + \sin^2 A$. Therefore, $\cos^2 A + \sin^2 A = 1$.

39. $\cos(\pi + x) = \cos\pi \cos x - \sin\pi \sin x = (-1)(\cos x) - (0)(\sin x) = -\cos x$

41. $\sin\left(\frac{3\pi}{2} - x\right) = \sin\left(\frac{3\pi}{2}\right)\cos(-x) + \cos\left(\frac{3\pi}{2}\right)\sin(-x) = (-1)(\cos x) + (0)(\sin(-x)) = -\cos x$

43. $\sin\frac{7\pi}{12} = \sin\left(\frac{\pi}{4} + \frac{\pi}{3}\right) = \sin\frac{\pi}{4}\cos\frac{\pi}{3} + \cos\frac{\pi}{4}\sin\frac{\pi}{3} = \left(\frac{\sqrt{2}}{2}\right)\left(\frac{1}{2}\right) + \left(\frac{\sqrt{2}}{2}\right)\left(\frac{\sqrt{3}}{2}\right) = \frac{\sqrt{6}+\sqrt{2}}{4}$

45. $\cos\frac{\pi}{12} = \cos\left(\frac{\pi}{3} - \frac{\pi}{4}\right) = \cos\frac{\pi}{3}\cos\left(-\frac{\pi}{4}\right) - \sin\frac{\pi}{3}\sin\left(-\frac{\pi}{4}\right) = \left(\frac{1}{2}\right)\left(\frac{\sqrt{2}}{2}\right) - \left(\frac{\sqrt{3}}{2}\right)\left(-\frac{\sqrt{2}}{2}\right) = \frac{1+\sqrt{3}}{2\sqrt{2}}$

47. $\cos^2\frac{\pi}{8} = \frac{1+\cos\left(\frac{2\pi}{8}\right)}{2} = \frac{1+\frac{\sqrt{2}}{2}}{2} = \frac{2+\sqrt{2}}{4}$

49. $\sin^2\frac{\pi}{12} = \frac{1-\cos\left(\frac{2\pi}{12}\right)}{2} = \frac{1-\frac{\sqrt{3}}{2}}{2} = \frac{2-\sqrt{3}}{4}$

51. $\sin^2\theta = \frac{3}{4} \Rightarrow \sin\theta = \pm\frac{\sqrt{3}}{2} \Rightarrow \theta = \frac{\pi}{3}, \frac{2\pi}{3}, \frac{4\pi}{3}, \frac{5\pi}{3}$

53. $\sin 2\theta - \cos\theta = 0 \Rightarrow 2\sin\theta\cos\theta - \cos\theta = 0 \Rightarrow \cos\theta(2\sin\theta - 1) = 0 \Rightarrow \cos\theta = 0$ or $2\sin\theta - 1 = 0 \Rightarrow \cos\theta = 0$ or $\sin\theta = \frac{1}{2} \Rightarrow \theta = \frac{\pi}{2}, \frac{3\pi}{2}$, or $\theta = \frac{\pi}{6}, \frac{5\pi}{6} \Rightarrow \theta = \frac{\pi}{6}, \frac{\pi}{2}, \frac{5\pi}{6}, \frac{3\pi}{2}$

55. $\tan(A + B) = \frac{\sin(A+B)}{\cos(A+B)} = \frac{\sin A\cos B + \cos A\sin B}{\cos A\cos B - \sin A\sin B} = \frac{\frac{\sin A\cos B}{\cos A\cos B} + \frac{\cos A\sin B}{\cos A\cos B}}{\frac{\cos A\cos B}{\cos A\cos B} - \frac{\sin A\sin B}{\cos A\cos B}} = \frac{\tan A+\tan B}{1-\tan A\tan B}$

57. According to the figure in the text, we have the following: By the law of cosines, $c^2 = a^2 + b^2 - 2ab\cos\theta$ $= 1^2 + 1^2 - 2\cos(A - B) = 2 - 2\cos(A - B)$. By distance formula, $c^2 = (\cos A - \cos B)^2 + (\sin A - \sin B)^2$

$= \cos^2 A - 2\cos A\cos B + \cos^2 B + \sin^2 A - 2\sin A\sin B + \sin^2 B = 2 - 2(\cos A\cos B + \sin A\sin B)$. Thus $c^2 = 2 - 2\cos(A - B) = 2 - 2(\cos A\cos B + \sin A\sin B) \Rightarrow \cos(A - B) = \cos A\cos B + \sin A\sin B$.

59. $c^2 = a^2 + b^2 - 2ab\cos C = 2^2 + 3^2 - 2(2)(3)\cos(60°) = 4 + 9 - 12\cos(60°) = 13 - 12\left(\frac{1}{2}\right) = 7$. Thus, $c = \sqrt{7} \approx 2.65$.

61. From the figures in the text, we see that $\sin B = \frac{h}{c}$. If C is an acute angle, then $\sin C = \frac{h}{b}$. On the other hand, if C is obtuse (as in the figure on the right), then $\sin C = \sin(\pi - C) = \frac{h}{b}$. Thus, in either case, $h = b\sin C = c\sin B \Rightarrow ah = ab\sin C = ac\sin B$.

By the law of cosines, $\cos C = \frac{a^2+b^2-c^2}{2ab}$ and $\cos B = \frac{a^2+c^2-b^2}{2ac}$. Moreover, since the sum of the interior angles of a triangle is π, we have $\sin A = \sin(\pi - (B + C)) = \sin(B + C) = \sin B\cos C + \cos B\sin C$
$= \left(\frac{h}{c}\right)\left[\frac{a^2+b^2-c^2}{2ab}\right] + \left[\frac{a^2+c^2-b^2}{2ac}\right]\left(\frac{h}{b}\right) = \left(\frac{h}{2abc}\right)(2a^2 + b^2 - c^2 + c^2 - b^2) = \frac{ah}{bc} \Rightarrow ah = bc\sin A$.

Combining our results we have $ah = ab\sin C$, $ah = ac\sin B$, and $ah = bc\sin A$. Dividing by abc gives $\frac{h}{bc} = \underbrace{\frac{\sin A}{a} = \frac{\sin C}{c} = \frac{\sin B}{b}}_{\text{law of sines}}$.

63. From the figure at the right and the law of cosines,
$b^2 = a^2 + 2^2 - 2(2a)\cos B$
$= a^2 + 4 - 4a\left(\frac{1}{2}\right) = a^2 - 2a + 4$.
Applying the law of sines to the figure, $\frac{\sin A}{a} = \frac{\sin B}{b}$
$\Rightarrow \frac{\sqrt{2}/2}{a} = \frac{\sqrt{3}/2}{b} \Rightarrow b = \sqrt{\frac{3}{2}}\,a$. Thus, combining results,
$a^2 - 2a + 4 = b^2 = \frac{3}{2}a^2 \Rightarrow 0 = \frac{1}{2}a^2 + 2a - 4$
$\Rightarrow 0 = a^2 + 4a - 8$. From the quadratic formula and the fact that $a > 0$, we have
$a = \frac{-4+\sqrt{4^2-4(1)(-8)}}{2} = \frac{4\sqrt{3}-4}{2} \simeq 1.464$.

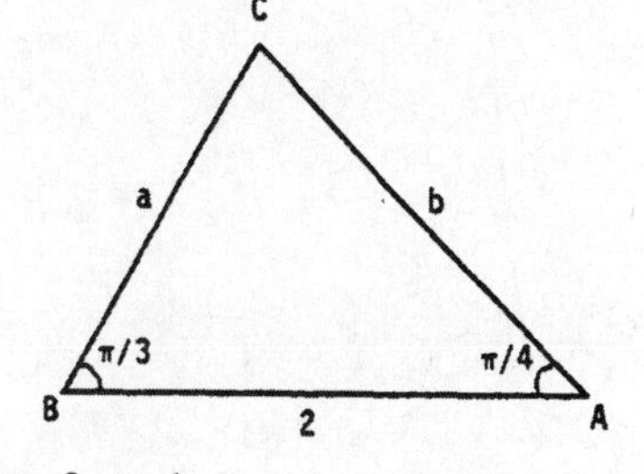

65. $A = 2, B = 2\pi, C = -\pi, D = -1$

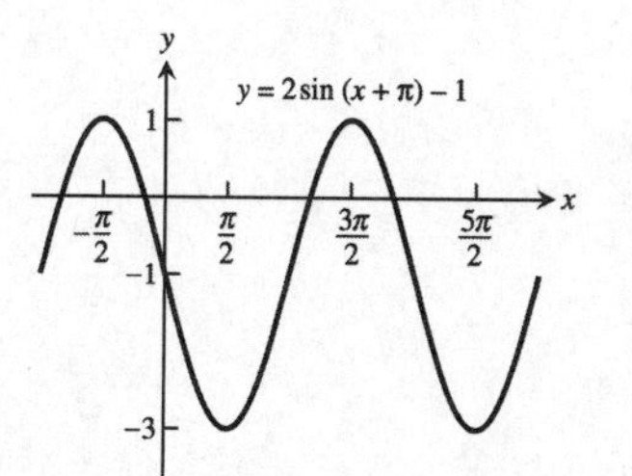

67. $A = -\frac{2}{\pi}, B = 4, C = 0, D = \frac{1}{\pi}$

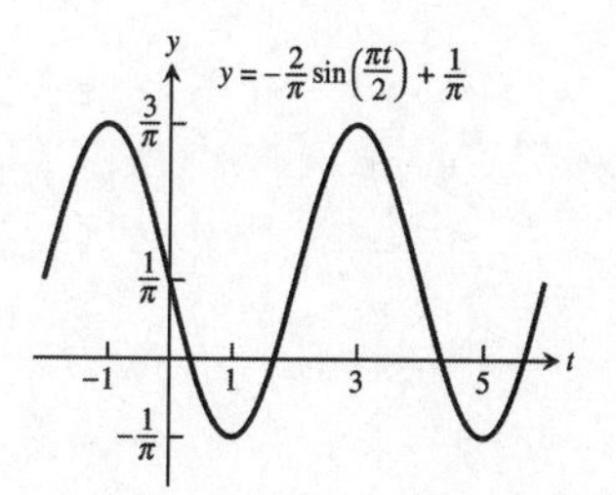

1.4 GRAPHING WITH CALCULATORS AND COMPUTERS

1-4. The most appropriate viewing window displays the maxima, minima, intercepts, and end behavior of the graphs and has little unused space.

1. d.

3. d.

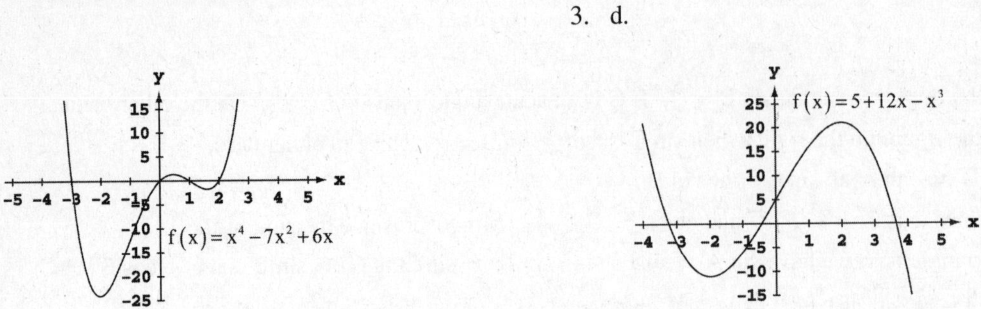

5-30. For any display there are many appropriate display widows. The graphs given as answers in Exercises 5–30 are not unique in appearance.

5. [−2, 5] by [−15, 40]

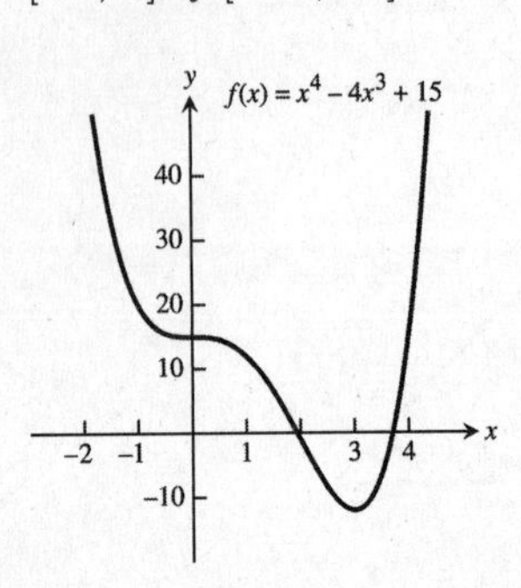

7. [−2, 6] by [−250, 50]

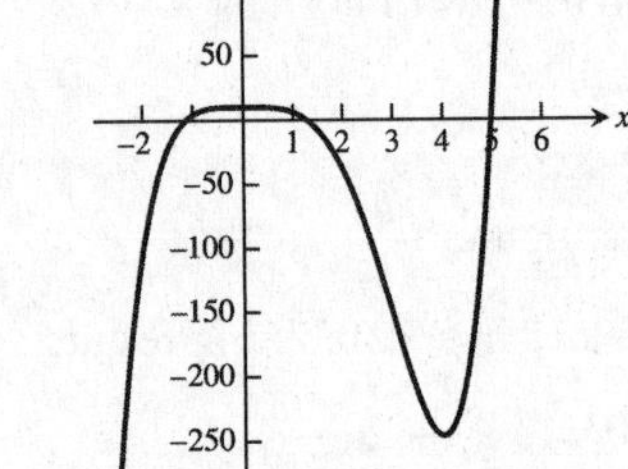
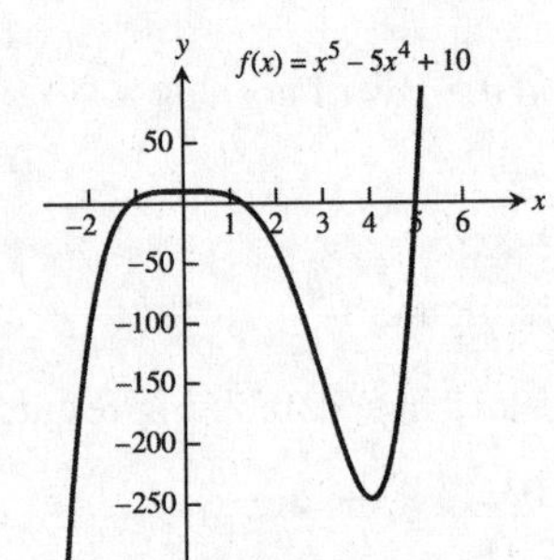

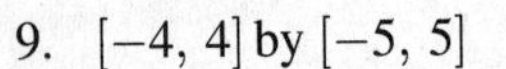
9. [−4, 4] by [−5, 5]

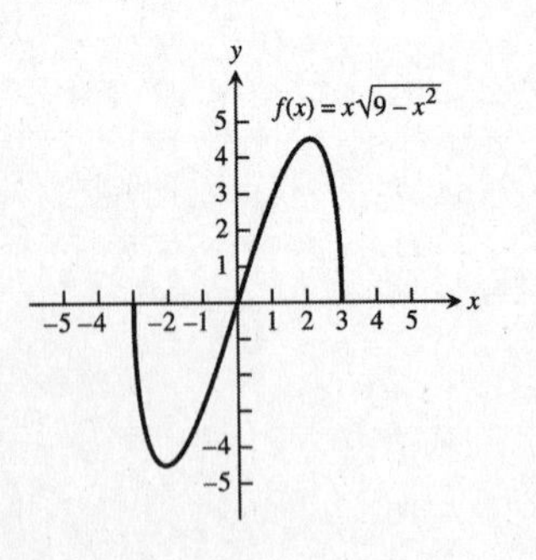

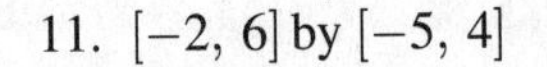
11. [−2, 6] by [−5, 4]

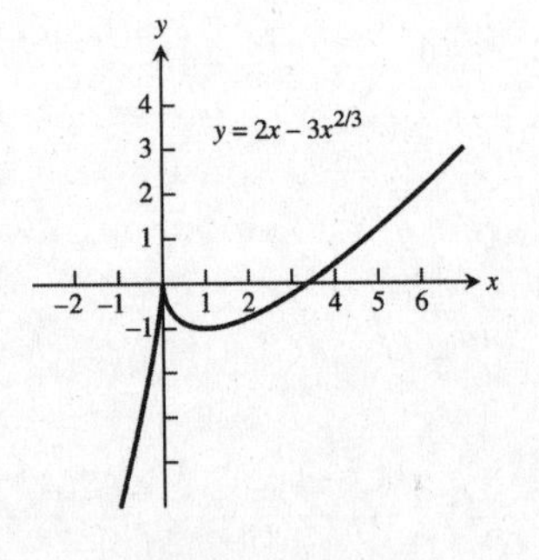

13. [−1, 6] by [−1, 4]

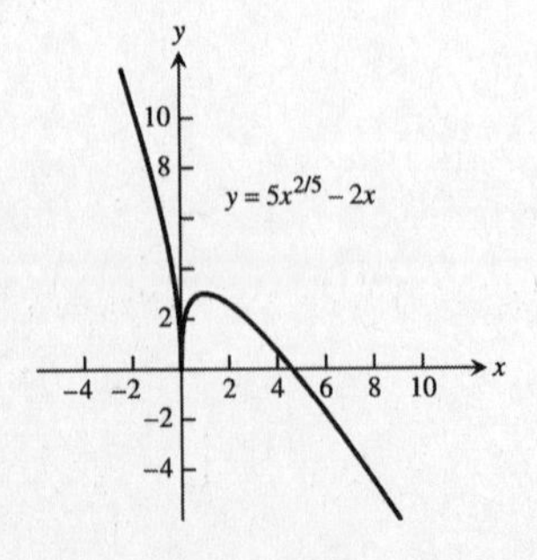

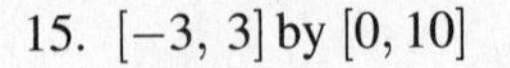
15. [−3, 3] by [0, 10]

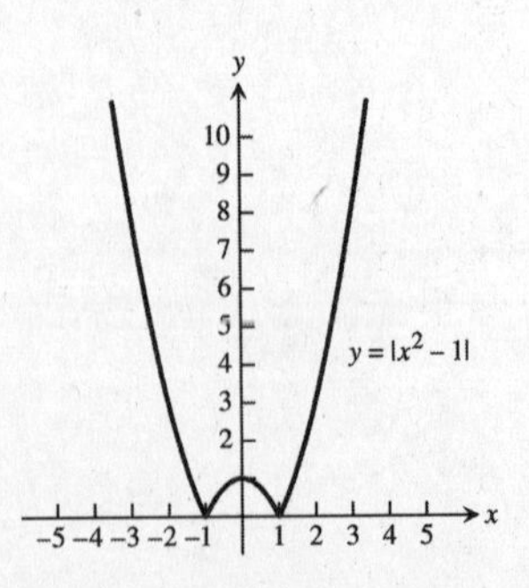

17. $[-5, 1]$ by $[-5, 5]$

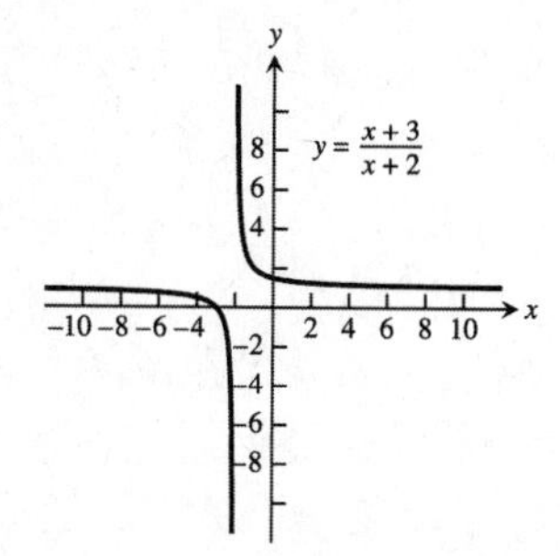

19. $[-4, 4]$ by $[0, 3]$

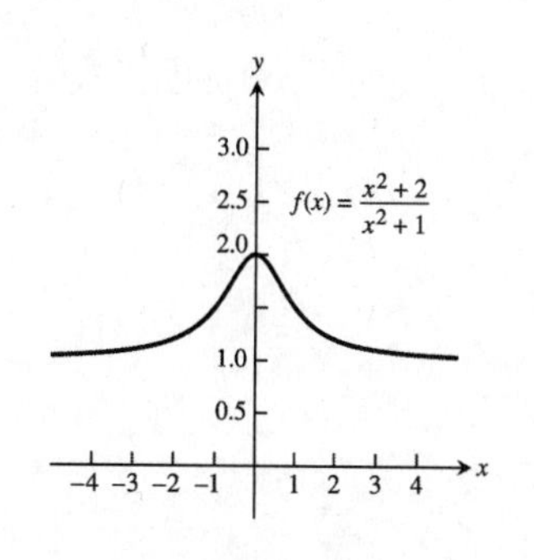

21. $[-10, 10]$ by $[-6, 6]$

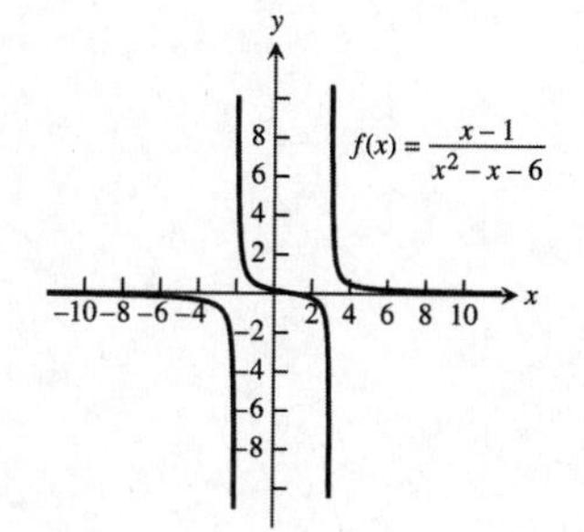

23. $[-6, 10]$ by $[-6, 6]$

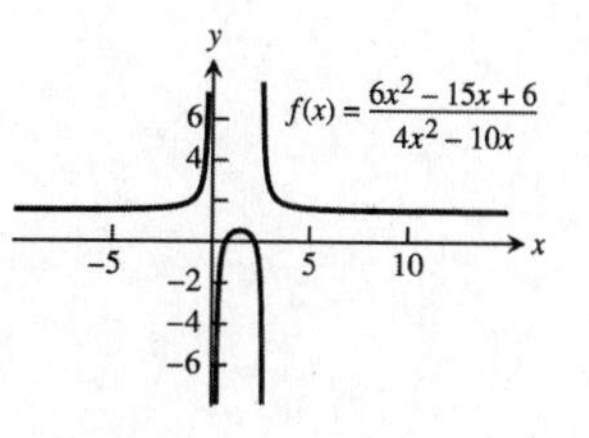

25. $[-0.03, 0.03]$ by $[-1.25, 1.25]$

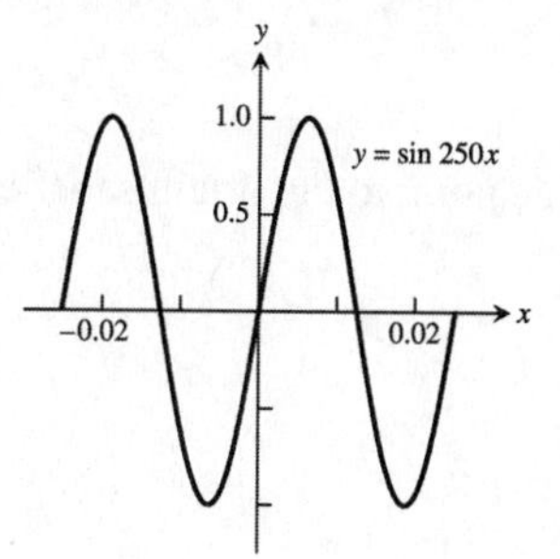

27. $[-300, 300]$ by $[-1.25, 1.25]$

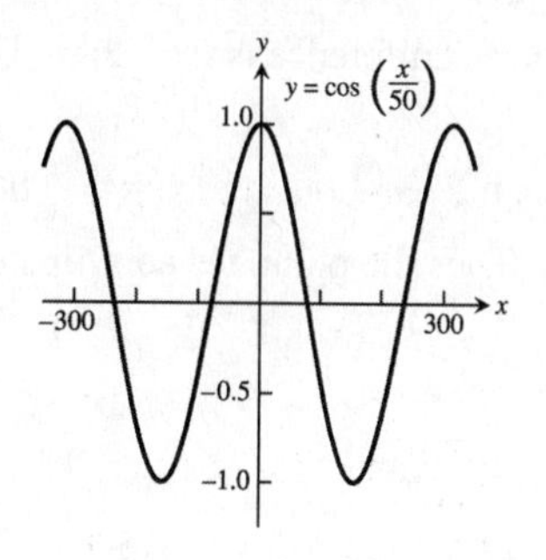

29. $[-0.25, 0.25]$ by $[-0.3, 0.3]$

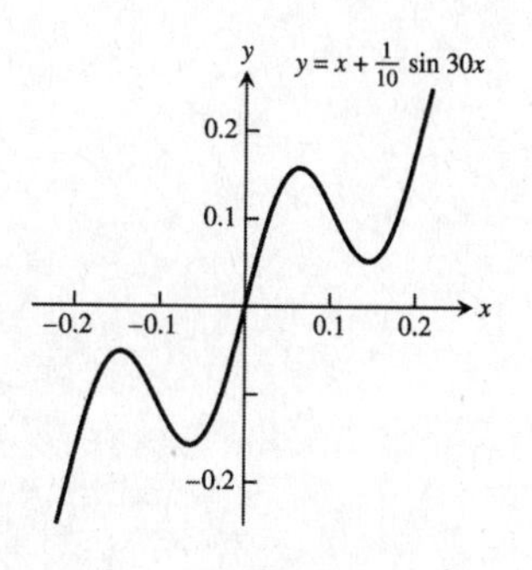

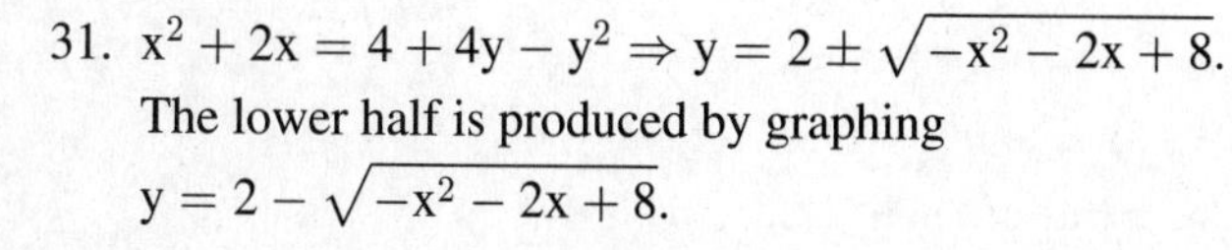
31. $x^2 + 2x = 4 + 4y - y^2 \Rightarrow y = 2 \pm \sqrt{-x^2 - 2x + 8}$.
The lower half is produced by graphing
$y = 2 - \sqrt{-x^2 - 2x + 8}$.

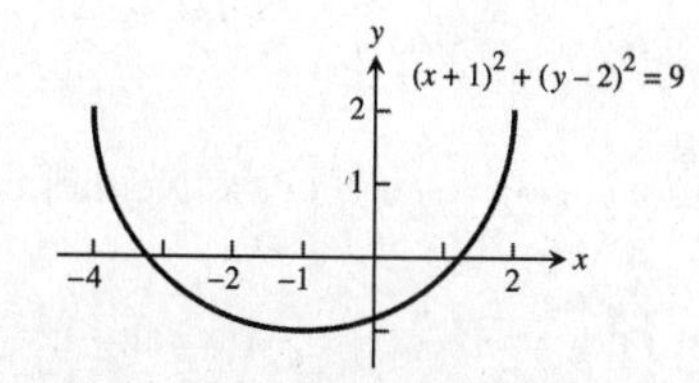

33.

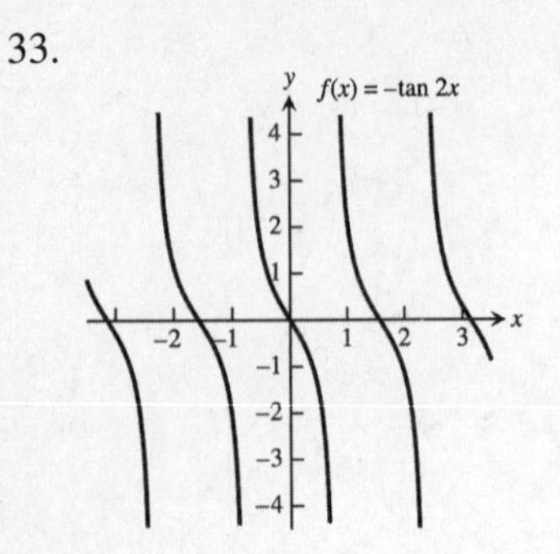

35.

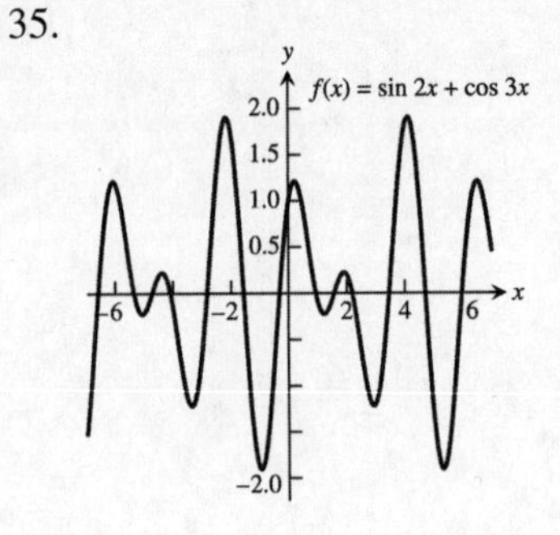

37.

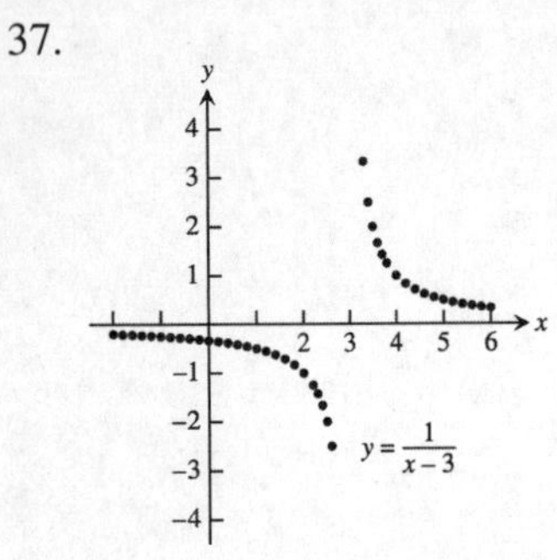

39.

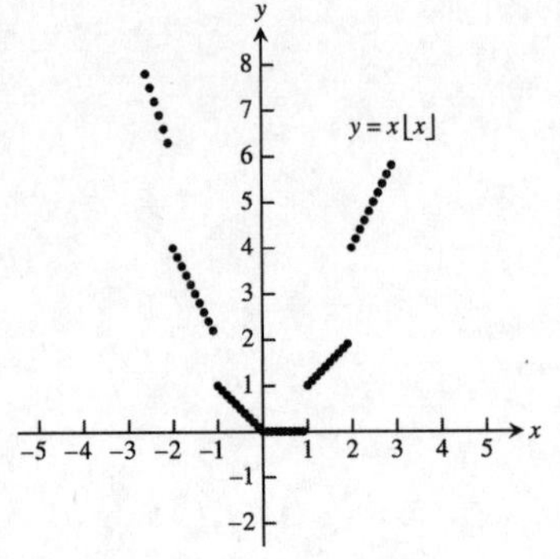

CHAPTER 1 PRACTICE EXERCISES

1. The area is $A = \pi r^2$ and the circumference is $C = 2\pi r$. Thus, $r = \frac{C}{2\pi} \Rightarrow A = \pi\left(\frac{C}{2\pi}\right)^2 = \frac{C^2}{4\pi}$.

3. The coordinates of a point on the parabola are (x, x^2). The angle of inclination θ joining this point to the origin satisfies the equation $\tan\theta = \frac{x^2}{x} = x$. Thus the point has coordinates $(x, x^2) = (\tan\theta, \tan^2\theta)$.

5.

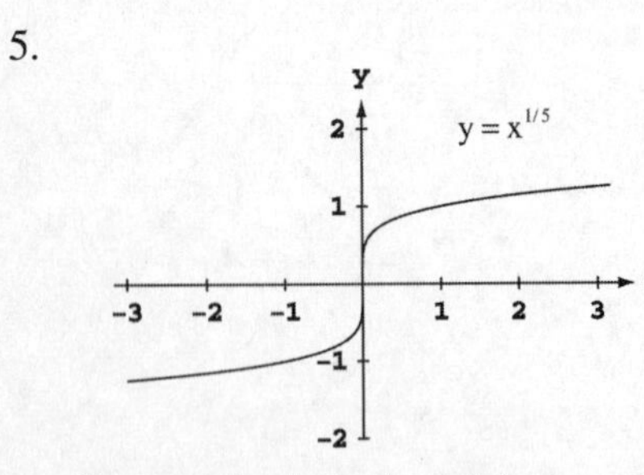

Symmetric about the origin.

7.

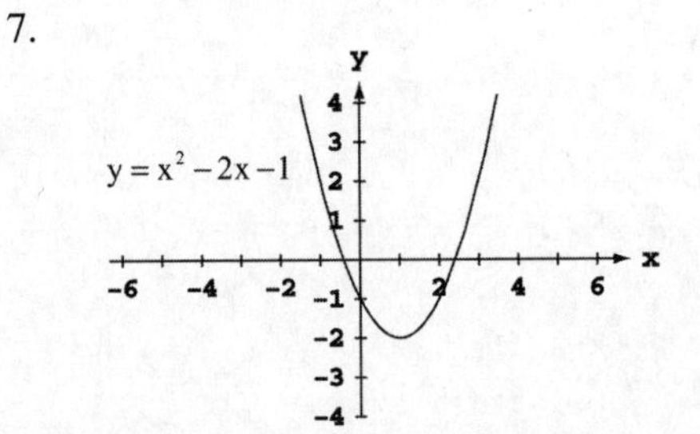

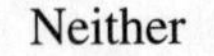
Neither

9. $y(-x) = (-x)^2 + 1 = x^2 + 1 = y(x)$. Even.

11. $y(-x) = 1 - \cos(-x) = 1 - \cos x = y(x)$. Even.

13. $y(-x) = \frac{(-x)^4+1}{(-x)^3-2(-x)} = \frac{x^4+1}{-x^3+2x} = -\frac{x^4+1}{x^3-2x} = -y(x)$. Odd.

15. $y(-x) = -x + \cos(-x) = -x + \cos x$. Neither even nor odd.

17. Since f and g are odd $\Rightarrow f(-x) = -f(x)$ and $g(-x) = -g(x)$.
    (a) $(f \cdot g)(-x) = f(-x)g(-x) = [-f(x)][-g(x)] = f(x)g(x) = (f \cdot g)(x) \Rightarrow f \cdot g$ is even

(b) $f^3(-x) = f(-x)f(-x)f(-x) = [-f(x)][-f(x)][-f(x)] = -f(x) \cdot f(x) \cdot f(x) = -f^3(x) \Rightarrow f^3$ is odd.
(c) $f(\sin(-x)) = f(-\sin(x)) = -f(\sin(x)) \Rightarrow f(\sin(x))$ is odd.
(d) $g(\sec(-x)) = g(\sec(x)) \Rightarrow g(\sec(x))$ is even.
(e) $|g(-x)| = |-g(x)| = |g(x)| \Rightarrow |g|$ is even.

19. (a) The function is defined for all values of x, so the domain is $(-\infty, \infty)$.
(b) Since $|x|$ attains all nonnegative values, the range is $[-2, \infty)$.

21. (a) Since the square root requires $16 - x^2 \ge 0$, the domain is $[-4, 4]$.
(b) For values of x in the domain, $0 \le 16 - x^2 \le 16$, so $0 \le \sqrt{16 - x^2} \le 4$. The range is $[0, 4]$.

23. (a) The function is defined for all values of x, so the domain is $(-\infty, \infty)$.
(b) Since $2e^{-x}$ attains all positive values, the range is $(-3, \infty)$.

25. (a) The function is defined for all values of x, so the domain is $(-\infty, \infty)$.
(b) The sine function attains values from -1 to 1, so $-2 \le 2\sin(3x + \pi) \le 2$ and hence $-3 \le 2\sin(3x + \pi) - 1 \le 1$. The range is $[-3, 1]$.

27. (a) The logarithm requires $x - 3 > 0$, so the domain is $(3, \infty)$.
(b) The logarithm attains all real values, so the range is $(-\infty, \infty)$.

29. (a) Increasing because volume increases as radius increases
(b) Neither, since the greatest integer function is composed of horizontal (constant) line segments
(c) Decreasing because as the height increases, the atmospheric pressure decreases.
(d) Increasing because the kinetic (motion) energy increases as the particles velocity increases.

31. (a) The function is defined for $-4 \le x \le 4$, so the domain is $[-4, 4]$.
(b) The function is equivalent to $y = \sqrt{|x|}$, $-4 \le x \le 4$, which attains values from 0 to 2 for x in the domain. The range is $[0, 2]$.

33. First piece: Line through (0, 1) and (1, 0). $m = \frac{0-1}{1-0} = \frac{-1}{1} = -1 \Rightarrow y = -x + 1 = 1 - x$
Second piece: Line through (1, 1) and (2, 0). $m = \frac{0-1}{2-1} = \frac{-1}{1} = -1 \Rightarrow y = -(x - 1) + 1 = -x + 2 = 2 - x$

$$f(x) = \begin{cases} 1 - x, & 0 \le x < 1 \\ 2 - x, & 1 \le x \le 2 \end{cases}$$

35. (a) $(f \circ g)(-1) = f(g(-1)) = f\left(\frac{1}{\sqrt{-1+2}}\right) = f(1) = \frac{1}{1} = 1$
(b) $(g \circ f)(2) = g(f(2)) = g\left(\frac{1}{2}\right) = \frac{1}{\sqrt{\frac{1}{2}+2}} = \frac{1}{\sqrt{2.5}}$ or $\sqrt{\frac{2}{5}}$
(c) $(f \circ f)(x) = f(f(x)) = f\left(\frac{1}{x}\right) = \frac{1}{1/x} = x,\ x \ne 0$
(d) $(g \circ g)(x) = g(g(x)) = g\left(\frac{1}{\sqrt{x+2}}\right) = \frac{1}{\sqrt{\frac{1}{\sqrt{x+2}}+2}} = \frac{\sqrt[4]{x+2}}{\sqrt{1+2\sqrt{x+2}}}$

37. (a) $(f \circ g)(x) = f(g(x)) = f\left(\sqrt{x+2}\right) = 2 - \left(\sqrt{x+2}\right)^2 = -x,\ x \ge -2.$
$(g \circ f)(x) = f(g(x)) = g(2 - x^2) = \sqrt{(2 - x^2) + 2} = \sqrt{4 - x^2}$
(b) Domain of $f \circ g$: $[-2, \infty)$.
Domain of $g \circ f$: $[-2, 2]$.
(c) Range of $f \circ g$: $(-\infty, 2]$.
Range of $g \circ f$: $[0, 2]$.

39. $y = f(x)$ $\qquad$ $y = (f \circ f)(x)$

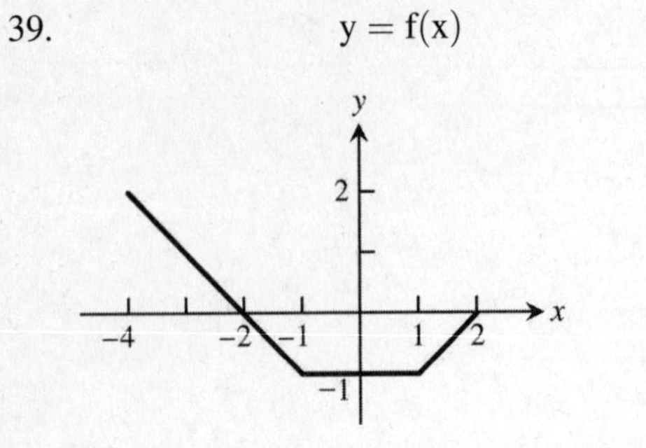

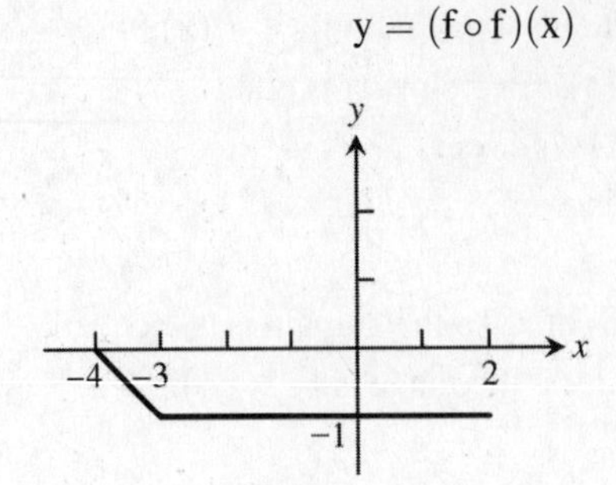

41.

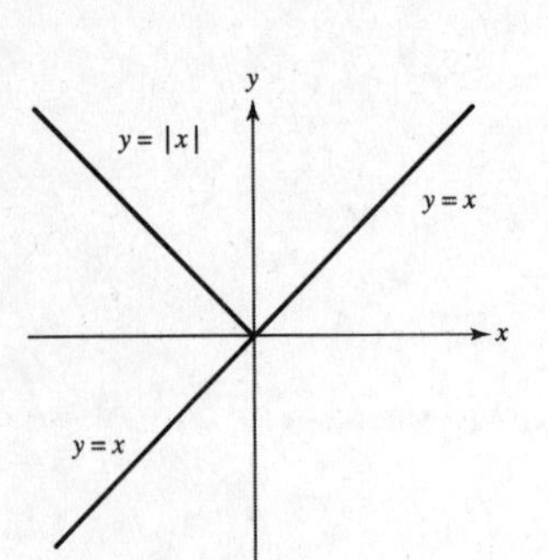

The graph of $f_2(x) = f_1(|x|)$ is the same as the graph of $f_1(x)$ to the right of the y-axis. The graph of $f_2(x)$ to the left of the y-axis is the reflection of $y = f_1(x)$, $x \geq 0$ across the y-axis.

43.

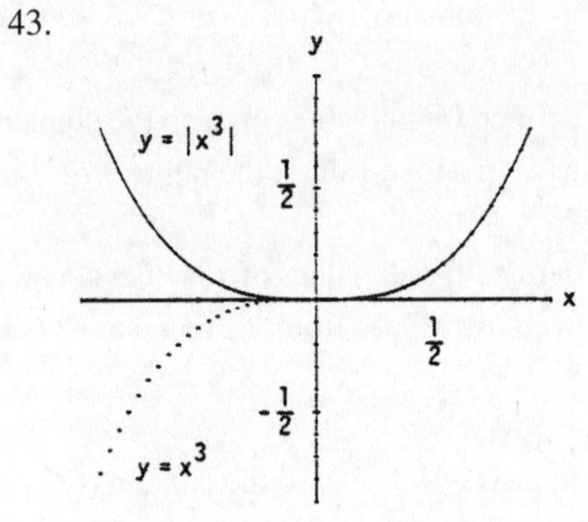

Whenever $g_1(x)$ is positive, the graph of $y = g_2(x) = |g_1(x)|$ is the same as the graph of $y = g_1(x)$. When $g_1(x)$ is negative, the graph of $y = g_2(x)$ is the reflection of the graph of $y = g_1(x)$ across the x-axis.

45.

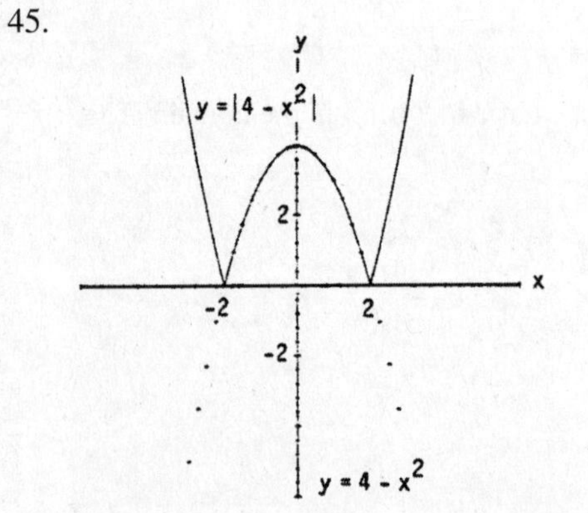

Whenever $g_1(x)$ is positive, the graph of $y = g_2(x) = |g_1(x)|$ is the same as the graph of $y = g_1(x)$. When $g_1(x)$ is negative, the graph of $y = g_2(x)$ is the reflection of the graph of $y = g_1(x)$ across the x-axis.

47.

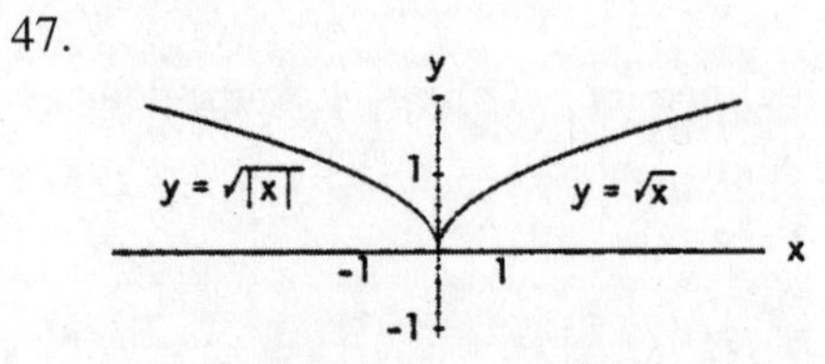

The graph of $f_2(x) = f_1(|x|)$ is the same as the graph of $f_1(x)$ to the right of the y-axis. The graph of $f_2(x)$ to the left of the y-axis is the reflection of $y = f_1(x)$, $x \geq 0$ across the y-axis.

49. (a) $y = g(x - 3) + \frac{1}{2}$ $\qquad$ (b) $y = g\left(x + \frac{2}{3}\right) - 2$

(c) $y = g(-x)$ $\qquad$ (d) $y = -g(x)$

(e) $y = 5 \cdot g(x)$ $\qquad$ (f) $y = g(5x)$

51. Reflection of the grpah of $y = \sqrt{x}$ about the x-axis followed by a horizontal compression by a factor of $\frac{1}{2}$ then a shift left 2 units.

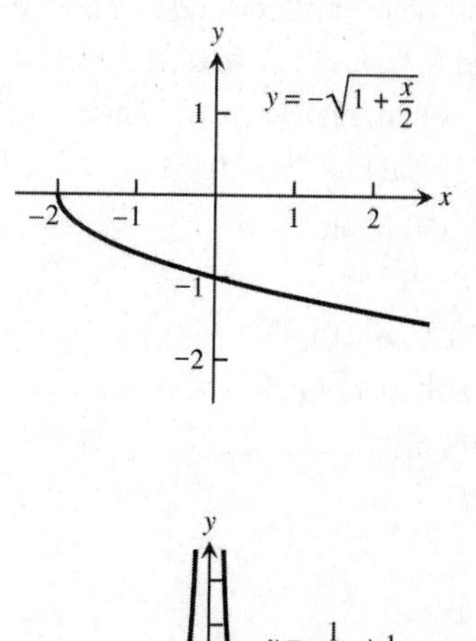

53. Vertical compression of the graph of $y = \frac{1}{x^2}$ by a factor of 2, then shift the graph up 1 unit.

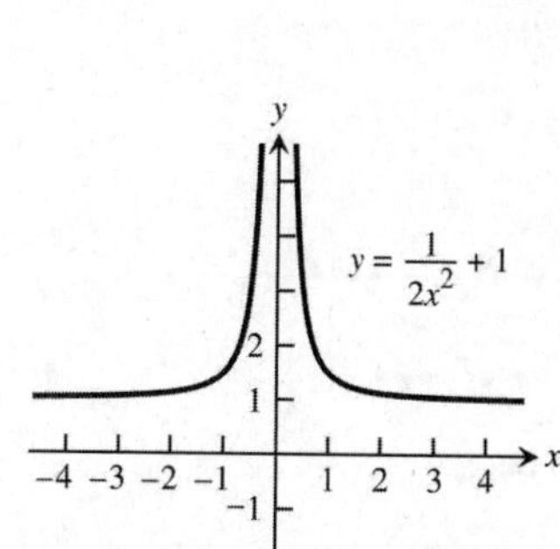

55.

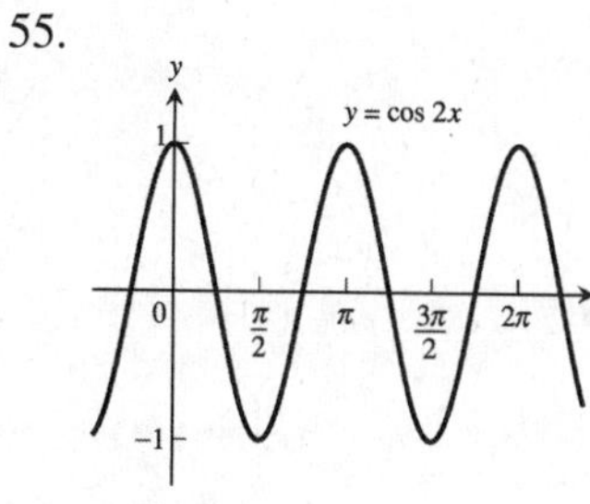

period $= \pi$

57.

y
$y = \sin \pi x$
1
1
2
x
-1

period $= 2$

59.

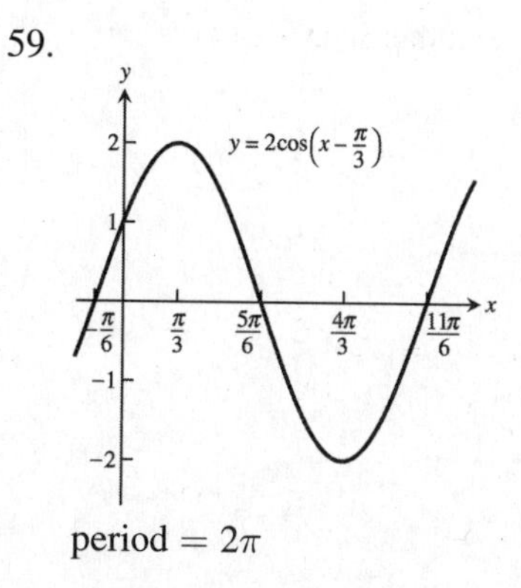

period $= 2\pi$

61. (a) $\sin B = \sin \frac{\pi}{3} = \frac{b}{c} = \frac{b}{2} \Rightarrow b = 2 \sin \frac{\pi}{3} = 2\left(\frac{\sqrt{3}}{2}\right) = \sqrt{3}$. By the theorem of Pythagoras, $a^2 + b^2 = c^2 \Rightarrow a = \sqrt{c^2 - b^2} = \sqrt{4 - 3} = 1$.

(b) $\sin B = \sin \frac{\pi}{3} = \frac{b}{c} = \frac{2}{c} \Rightarrow c = \frac{2}{\sin \frac{\pi}{3}} = \frac{2}{\left(\frac{\sqrt{3}}{2}\right)} = \frac{4}{\sqrt{3}}$. Thus, $a = \sqrt{c^2 - b^2} = \sqrt{\left(\frac{4}{\sqrt{3}}\right)^2 - (2)^2} = \sqrt{\frac{4}{3}} = \frac{2}{\sqrt{3}}$.

63. (a) $\tan B = \frac{b}{a} \Rightarrow a = \frac{b}{\tan B}$

(b) $\sin A = \frac{a}{c} \Rightarrow c = \frac{a}{\sin A}$

65. Let h = height of vertical pole, and let b and c denote the distances of points B and C from the base of the pole, measured along the flatground, respectively. Then, $\tan 50° = \frac{h}{c}$, $\tan 35° = \frac{h}{b}$, and $b - c = 10$.
Thus, $h = c \tan 50°$ and $h = b \tan 35° = (c + 10) \tan 35°$
$\Rightarrow c \tan 50° = (c + 10) \tan 35°$
$\Rightarrow c(\tan 50° - \tan 35°) = 10 \tan 35°$
$\Rightarrow c = \frac{10 \tan 35°}{\tan 50° - \tan 35°} \Rightarrow h = c \tan 50°$
$= \frac{10 \tan 35° \tan 50°}{\tan 50° - \tan 35°} \approx 16.98$ m.

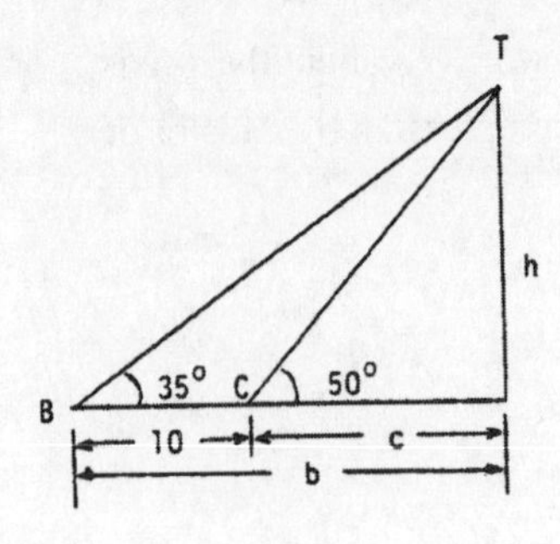

67. (a)

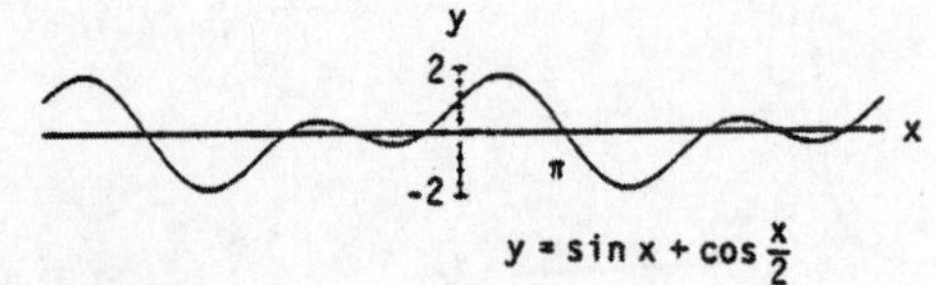

(b) The period appears to be 4π.

(c) $f(x + 4\pi) = \sin(x + 4\pi) + \cos\left(\frac{x+4\pi}{2}\right) = \sin(x + 2\pi) + \cos\left(\frac{x}{2} + 2\pi\right) = \sin x + \cos\frac{x}{2}$ since the period of sine and cosine is 2π. Thus, f(x) has period 4π.

CHAPTER 1 ADDITIONAL AND ADVANCED EXERCISES

1. There are (infinitely) many such function pairs. For example, $f(x) = 3x$ and $g(x) = 4x$ satisfy $f(g(x)) = f(4x) = 3(4x) = 12x = 4(3x) = g(3x) = g(f(x))$.

3. If f is odd and defined at x, then $f(-x) = -f(x)$. Thus $g(-x) = f(-x) - 2 = -f(x) - 2$ whereas $-g(x) = -(f(x) - 2) = -f(x) + 2$. Then g cannot be odd because $g(-x) = -g(x) \Rightarrow -f(x) - 2 = -f(x) + 2$ $\Rightarrow 4 = 0$, which is a contradiction. Also, g(x) is not even unless $f(x) = 0$ for all x. On the other hand, if f is even, then $g(x) = f(x) - 2$ is also even: $g(-x) = f(-x) - 2 = f(x) - 2 = g(x)$.

5. For (x, y) in the 1st quadrant, $|x| + |y| = 1 + x$ $\Leftrightarrow x + y = 1 + x \Leftrightarrow y = 1$. For (x, y) in the 2nd quadrant, $|x| + |y| = x + 1 \Leftrightarrow -x + y = x + 1$ $\Leftrightarrow y = 2x + 1$. In the 3rd quadrant, $|x| + |y| = x + 1$ $\Leftrightarrow -x - y = x + 1 \Leftrightarrow y = -2x - 1$. In the 4th quadrant, $|x| + |y| = x + 1 \Leftrightarrow x + (-y) = x + 1$ $\Leftrightarrow y = -1$. The graph is given at the right.

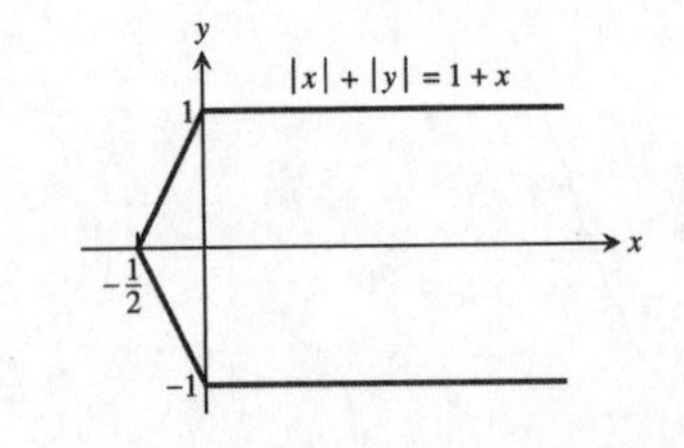

7. (a) $\sin^2 x + \cos^2 x = 1 \Rightarrow \sin^2 x = 1 - \cos^2 x = (1 - \cos x)(1 + \cos x) \Rightarrow (1 - \cos x) = \frac{\sin^2 x}{1+\cos x}$
$\Rightarrow \frac{1-\cos x}{\sin x} = \frac{\sin x}{1+\cos x}$

(b) Using the definition of the tangent function and the double angle formulas, we have

$$\tan^2\left(\frac{x}{2}\right) = \frac{\sin^2\left(\frac{x}{2}\right)}{\cos^2\left(\frac{x}{2}\right)} = \frac{\frac{1-\cos\left(2\left(\frac{x}{2}\right)\right)}{2}}{\frac{1+\cos\left(2\left(\frac{x}{2}\right)\right)}{2}} = \frac{1-\cos x}{1+\cos x}.$$

9. As in the proof of the law of sines of Section 1.3, Exercise 61, $ah = bc \sin A = ab \sin C = ac \sin B$ $\Rightarrow$ the area of ABC $= \frac{1}{2}$(base)(height) $= \frac{1}{2}ah = \frac{1}{2}bc \sin A = \frac{1}{2}ab \sin C = \frac{1}{2}ac \sin B$.

11. If f is even and odd, then $f(-x) = -f(x)$ and $f(-x) = f(x) \Rightarrow f(x) = -f(x)$ for all x in the domain of f. Thus $2f(x) = 0 \Rightarrow f(x) = 0$.

13. $y = ax^2 + bx + c = a\left(x^2 + \frac{b}{a}x + \frac{b^2}{4a^2}\right) - \frac{b^2}{4a} + c = a\left(x + \frac{b}{2a}\right)^2 - \frac{b^2}{4a} + c$

    (a) If $a > 0$ the graph is a parabola that opens upward. Increasing a causes a vertical stretching and a shift of the vertex toward the y-axis and upward. If $a < 0$ the graph is a parabola that opens downward. Decreasing a causes a vertical stretching and a shift of the vertex toward the y-axis and downward.

    (b) If $a > 0$ the graph is a parabola that opens upward. If also $b > 0$, then increasing b causes a shift of the graph downward to the left; if $b < 0$, then decreasing b causes a shift of the graph downward and to the right.

    If $a < 0$ the graph is a parabola that opens downward. If $b > 0$, increasing b shifts the graph upward to the right. If $b < 0$, decreasing b shifts the graph upward to the left.

    (c) Changing c (for fixed a and b) by Δc shifts the graph upward Δc units if $\Delta c > 0$, and downward $-\Delta c$ units if $\Delta c < 0$.

15. Each of the triangles pictured has the same base $b = v\Delta t = v(1 \text{ sec})$. Moreover, the height of each triangle is the same value h. Thus $\frac{1}{2}$(base)(height) $= \frac{1}{2}bh = A_1 = A_2 = A_3 = \ldots$. In conclusion, the object sweeps out equal areas in each one second interval.

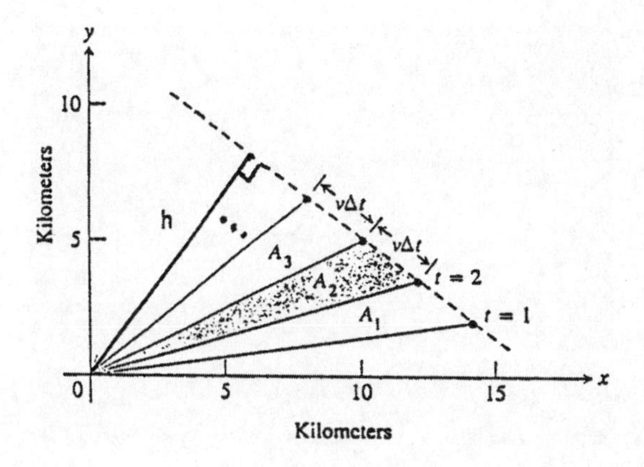

17. From the figure we see that $0 \le \theta \le \frac{\pi}{2}$ and $AB = AD = 1$. From trigonometry we have the following: $\sin\theta = \frac{EB}{AB} = EB$, $\cos\theta = \frac{AE}{AB} = AE$, $\tan\theta = \frac{CD}{AD} = CD$, and $\tan\theta = \frac{EB}{AE} = \frac{\sin\theta}{\cos\theta}$. We can see that:
area $\triangle AEB <$ area sector $\widehat{DB} <$ area $\triangle ADC \Rightarrow \frac{1}{2}(AE)(EB) < \frac{1}{2}(AD)^2\theta < \frac{1}{2}(AD)(CD)$
$\Rightarrow \frac{1}{2}\sin\theta\cos\theta < \frac{1}{2}(1)^2\theta < \frac{1}{2}(1)(\tan\theta) \Rightarrow \frac{1}{2}\sin\theta\cos\theta < \frac{1}{2}\theta < \frac{1}{2}\frac{\sin\theta}{\cos\theta}$

NOTES:

CHAPTER 2 LIMITS AND CONTINUITY

2.1 RATES OF CHANGE AND TANGENTS TO CURVES

1. (a) $\frac{\Delta f}{\Delta x} = \frac{f(3) - f(2)}{3 - 2} = \frac{28 - 9}{1} = 19$ (b) $\frac{\Delta f}{\Delta x} = \frac{f(1) - f(-1)}{1 - (-1)} = \frac{2 - 0}{2} = 1$

3. (a) $\frac{\Delta h}{\Delta t} = \frac{h\left(\frac{3\pi}{4}\right) - h\left(\frac{\pi}{4}\right)}{\frac{3\pi}{4} - \frac{\pi}{4}} = \frac{-1 - 1}{\frac{\pi}{2}} = -\frac{4}{\pi}$ (b) $\frac{\Delta h}{\Delta t} = \frac{h\left(\frac{\pi}{2}\right) - h\left(\frac{\pi}{6}\right)}{\frac{\pi}{2} - \frac{\pi}{6}} = \frac{0 - \sqrt{3}}{\frac{\pi}{3}} = \frac{-3\sqrt{3}}{\pi}$

5. $\frac{\Delta R}{\Delta \theta} = \frac{R(2) - R(0)}{2 - 0} = \frac{\sqrt{8+1} - \sqrt{1}}{2} = \frac{3 - 1}{2} = 1$

7. (a) $\frac{\Delta y}{\Delta x} = \frac{((2+h)^2 - 3) - (2^2 - 3)}{h} = \frac{4 + 4h + h^2 - 3 - 1}{h} = \frac{4h + h^2}{h} = 4 + h$. As $h \to 0$, $4 + h \to 4 \Rightarrow$ at $P(2, 1)$ the slope is 4.
   (b) $y - 1 = 4(x - 2) \Rightarrow y - 1 = 4x - 8 \Rightarrow y = 4x - 7$

9. (a) $\frac{\Delta y}{\Delta x} = \frac{((2+h)^2 - 2(2+h) - 3) - (2^2 - 2(2) - 3)}{h} = \frac{4 + 4h + h^2 - 4 - 2h - 3 - (-3)}{h} = \frac{2h + h^2}{h} = 2 + h$. As $h \to 0$, $2 + h \to 2 \Rightarrow$ at $P(2, -3)$ the slope is 2.
   (b) $y - (-3) = 2(x - 2) \Rightarrow y + 3 = 2x - 4 \Rightarrow y = 2x - 7$.

11. (a) $\frac{\Delta y}{\Delta x} = \frac{(2+h)^3 - 2^3}{h} = \frac{8 + 12h + 4h^2 + h^3 - 8}{h} = \frac{12h + 4h^2 + h^3}{h} = 12 + 4h + h^2$. As $h \to 0$, $12 + 4h + h^2 \to 12$, $\Rightarrow$ at $P(2, 8)$ the slope is 12.
   (b) $y - 8 = 12(x - 2) \Rightarrow y - 8 = 12x - 24 \Rightarrow y = 12x - 16$.

13. (a) $\frac{\Delta y}{\Delta x} = \frac{(1+h)^3 - 12(1+h) - (1^3 - 12(1))}{h} = \frac{1 + 3h + 3h^2 + h^3 - 12 - 12h - (-11)}{h} = \frac{-9h + 3h^2 + h^3}{h} = -9 + 3h + h^2$. As $h \to 0$, $-9 + 3h + h^2 \to -9 \Rightarrow$ at $P(1, -11)$ the slope is -9.
   (b) $y - (-11) = (-9)(x - 1) \Rightarrow y + 11 = -9x + 9 \Rightarrow y = -9x - 2$.

15. (a)

Q	Slope of PQ $= \frac{\Delta p}{\Delta t}$
$Q_1(10, 225)$	$\frac{650 - 225}{20 - 10} = 42.5$ m/sec
$Q_2(14, 375)$	$\frac{650 - 375}{20 - 14} = 45.83$ m/sec
$Q_3(16.5, 475)$	$\frac{650 - 475}{20 - 16.5} = 50.00$ m/sec
$Q_4(18, 550)$	$\frac{650 - 550}{20 - 18} = 50.00$ m/sec

   (b) At $t = 20$, the sportscar was traveling approximately 50 m/sec or 180 km/h.

17. (a)

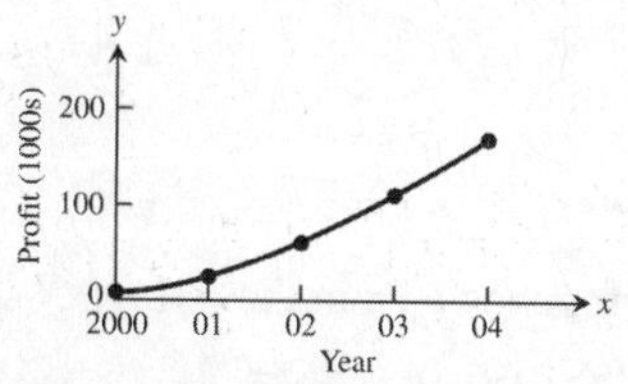

   (b) $\frac{\Delta p}{\Delta t} = \frac{174 - 62}{2004 - 2002} = \frac{112}{2} = 56$ thousand dollars per year
   (c) The average rate of change from 2001 to 2002 is $\frac{\Delta p}{\Delta t} = \frac{62 - 27}{20022 - 2001} = 35$ thousand dollars per year.
   The average rate of change from 2002 to 2003 is $\frac{\Delta p}{\Delta t} = \frac{111 - 62}{2003 - 2002} = 49$ thousand dollars per year.
   So, the rate at which profits were changing in 2002 is approximatley $\frac{1}{2}(35 + 49) = 42$ thousand dollars per year.

19. (a) $\frac{\Delta g}{\Delta x} = \frac{g(2)-g(1)}{2-1} = \frac{\sqrt{2}-1}{2-1} \approx 0.414213$ $\frac{\Delta g}{\Delta x} = \frac{g(1.5)-g(1)}{1.5-1} = \frac{\sqrt{1.5}-1}{0.5} \approx 0.449489$

$\frac{\Delta g}{\Delta x} = \frac{g(1+h)-g(1)}{(1+h)-1} = \frac{\sqrt{1+h}-1}{h}$

(b) $g(x) = \sqrt{x}$

$1+h$	1.1	1.01	1.001	1.0001	1.00001	1.000001
$\sqrt{1+h}$	1.04880	1.004987	1.0004998	1.0000499	1.000005	1.0000005
$\left(\sqrt{1+h}-1\right)/h$	0.4880	0.4987	0.4998	0.499	0.5	0.5

(c) The rate of change of g(x) at x = 1 is 0.5.

(d) The calculator gives $\lim\limits_{h \to 0} \frac{\sqrt{1+h}-1}{h} = \frac{1}{2}$.

NOTE: Answers will vary in Exercise 21.

21. (a) $[0, 1]$: $\frac{\Delta s}{\Delta t} = \frac{15-0}{1-0} = 15$ mph; $[1, 2.5]$: $\frac{\Delta s}{\Delta t} = \frac{20-15}{2.5-1} = \frac{10}{3}$ mph; $[2.5, 3.5]$: $\frac{\Delta s}{\Delta t} = \frac{30-20}{3.5-2.5} = 10$ mph

(b) At $P\left(\frac{1}{2}, 7.5\right)$: Since the portion of the graph from t = 0 to t = 1 is nearly linear, the instantaneous rate of change will be almost the same as the average rate of change, thus the instantaneous speed at $t = \frac{1}{2}$ is $\frac{15-7.5}{1-0.5} = 15$ mi/hr.

At P(2, 20): Since the portion of the graph from t = 2 to t = 2.5 is nearly linear, the instantaneous rate of change will be nearly the same as the average rate of change, thus $v = \frac{20-20}{2.5-2} = 0$ mi/hr. For values of t less than 2, we have

Q	Slope of PQ $= \frac{\Delta s}{\Delta t}$
$Q_1(1, 15)$	$\frac{15-20}{1-2} = 5$ mi/hr
$Q_2(1.5, 19)$	$\frac{19-20}{1.5-2} = 2$ mi/hr
$Q_3(1.9, 19.9)$	$\frac{19.9-20}{1.9-2} = 1$ mi/hr

Thus, it appears that the instantaneous speed at t = 2 is 0 mi/hr.

At P(3, 22):

Q	Slope of PQ $= \frac{\Delta s}{\Delta t}$
$Q_1(4, 35)$	$\frac{35-22}{4-3} = 13$ mi/hr
$Q_2(3.5, 30)$	$\frac{30-22}{3.5-3} = 16$ mi/hr
$Q_3(3.1, 23)$	$\frac{23-22}{3.1-3} = 10$ mi/hr

Q	Slope of PQ $= \frac{\Delta s}{\Delta t}$
$Q_1(2, 20)$	$\frac{20-22}{2-3} = 2$ mi/hr
$Q_2(2.5, 20)$	$\frac{20-22}{2.5-3} = 4$ mi/hr
$Q_3(2.9, 21.6)$	$\frac{21.6-22}{2.9-3} = 4$ mi/hr

Thus, it appears that the instantaneous speed at t = 3 is about 7 mi/hr.

(c) It appears that the curve is increasing the fastest at t = 3.5. Thus for P(3.5, 30)

Q	Slope of PQ $= \frac{\Delta s}{\Delta t}$
$Q_1(4, 35)$	$\frac{35-30}{4-3.5} = 10$ mi/hr
$Q_2(3.75, 34)$	$\frac{34-30}{3.75-3.5} = 16$ mi/hr
$Q_3(3.6, 32)$	$\frac{32-30}{3.6-3.5} = 20$ mi/hr

Q	Slope of PQ $= \frac{\Delta s}{\Delta t}$
$Q_1(3, 22)$	$\frac{22-30}{3-3.5} = 16$ mi/hr
$Q_2(3.25, 25)$	$\frac{25-30}{3.25-3.5} = 20$ mi/hr
$Q_3(3.4, 28)$	$\frac{28-30}{3.4-3.5} = 20$ mi/hr

Thus, it appears that the instantaneous speed at t = 3.5 is about 20 mi/hr.

2.2 LIMIT OF A FUNCTION AND LIMIT LAWS

1. (a) Does not exist. As x approaches 1 from the right, g(x) approaches 0. As x approaches 1 from the left, g(x) approaches 1. There is no single number L that all the values g(x) get arbitrarily close to as $x \to 1$.
(b) 1 (c) 0 (d) 0.5

3. (a) True (b) True (c) False
(d) False (e) False (f) True
(g) True

5. $\lim_{x \to 0} \frac{x}{|x|}$ does not exist because $\frac{x}{|x|} = \frac{x}{x} = 1$ if $x > 0$ and $\frac{x}{|x|} = \frac{x}{-x} = -1$ if $x < 0$. As x approaches 0 from the left, $\frac{x}{|x|}$ approaches -1. As x approaches 0 from the right, $\frac{x}{|x|}$ approaches 1. There is no single number L that all the function values get arbitrarily close to as $x \to 0$.

7. Nothing can be said about f(x) because the existence of a limit as $x \to x_0$ does not depend on how the function is defined at x_0. In order for a limit to exist, f(x) must be arbitrarily close to a single real number L when x is close enough to x_0. That is, the existence of a limit depends on the values of f(x) for x near x_0, not on the definition of f(x) at x_0 itself.

9. No, the definition does not require that f be defined at $x = 1$ in order for a limiting value to exist there. If f(1) is defined, it can be any real number, so we can conclude nothing about f(1) from $\lim_{x \to 1} f(x) = 5$.

11. $\lim_{x \to -7} (2x + 5) = 2(-7) + 5 = -14 + 5 = -9$

13. $\lim_{t \to 6} 8(t - 5)(t - 7) = 8(6 - 5)(6 - 7) = -8$

15. $\lim_{x \to 2} \frac{x+3}{x+6} = \frac{2+3}{2+6} = \frac{5}{8}$

17. $\lim_{x \to -1} 3(2x - 1)^2 = 3(2(-1) - 1)^2 = 3(-3)^2 = 27$

19. $\lim_{y \to -3} (5 - y)^{4/3} = [5 - (-3)]^{4/3} = (8)^{4/3} = \left((8)^{1/3}\right)^4 = 2^4 = 16$

21. $\lim_{h \to 0} \frac{3}{\sqrt{3h+1}+1} = \frac{3}{\sqrt{3(0)+1}+1} = \frac{3}{\sqrt{1}+1} = \frac{3}{2}$

23. $\lim_{x \to 5} \frac{x-5}{x^2-25} = \lim_{x \to 5} \frac{x-5}{(x+5)(x-5)} = \lim_{x \to 5} \frac{1}{x+5} = \frac{1}{5+5} = \frac{1}{10}$

25. $\lim_{x \to -5} \frac{x^2+3x-10}{x+5} = \lim_{x \to -5} \frac{(x+5)(x-2)}{x+5} = \lim_{x \to -5} (x - 2) = -5 - 2 = -7$

27. $\lim_{t \to 1} \frac{t^2+t-2}{t^2-1} = \lim_{t \to 1} \frac{(t+2)(t-1)}{(t-1)(t+1)} = \lim_{t \to 1} \frac{t+2}{t+1} = \frac{1+2}{1+1} = \frac{3}{2}$

29. $\lim_{x \to -2} \frac{-2x-4}{x^3+2x^2} = \lim_{x \to -2} \frac{-2(x+2)}{x^2(x+2)} = \lim_{x \to -2} \frac{-2}{x^2} = \frac{-2}{4} = -\frac{1}{2}$

31. $\lim_{x \to 1} \frac{\frac{1}{x}-1}{x-1} = \lim_{x \to 1} \frac{\frac{1-x}{x}}{x-1} = \lim_{x \to 1} \left(\frac{1-x}{x} \cdot \frac{1}{x-1}\right) = \lim_{x \to 1} -\frac{1}{x} = -1$

33. $\lim_{u \to 1} \frac{u^4-1}{u^3-1} = \lim_{u \to 1} \frac{(u^2+1)(u+1)(u-1)}{(u^2+u+1)(u-1)} = \lim_{u \to 1} \frac{(u^2+1)(u+1)}{u^2+u+1} = \frac{(1+1)(1+1)}{1+1+1} = \frac{4}{3}$

35. $\lim_{x \to 9} \frac{\sqrt{x}-3}{x-9} = \lim_{x \to 9} \frac{\sqrt{x}-3}{(\sqrt{x}-3)(\sqrt{x}+3)} = \lim_{x \to 9} \frac{1}{\sqrt{x}+3} = \frac{1}{\sqrt{9}+3} = \frac{1}{6}$

37. $\lim_{x \to 1} \frac{x-1}{\sqrt{x+3}-2} = \lim_{x \to 1} \frac{(x-1)(\sqrt{x+3}+2)}{(\sqrt{x+3}-2)(\sqrt{x+3}+2)} = \lim_{x \to 1} \frac{(x-1)(\sqrt{x+3}+2)}{(x+3)-4} = \lim_{x \to 1} \left(\sqrt{x+3}+2\right)$
$= \sqrt{4}+2 = 4$

39. $\lim_{x \to 2} \frac{\sqrt{x^2+12}-4}{x-2} = \lim_{x \to 2} \frac{\left(\sqrt{x^2+12}-4\right)\left(\sqrt{x^2+12}+4\right)}{(x-2)\left(\sqrt{x^2+12}+4\right)} = \lim_{x \to 2} \frac{(x^2+12)-16}{(x-2)\left(\sqrt{x^2+12}+4\right)}$
$= \lim_{x \to 2} \frac{(x-2)(x+2)}{(x-2)\left(\sqrt{x^2+12}+4\right)} = \lim_{x \to 2} \frac{x+2}{\sqrt{x^2+12}+4} = \frac{4}{\sqrt{16}+4} = \frac{1}{2}$

41. $\lim\limits_{x \to -3} \frac{2-\sqrt{x^2-5}}{x+3} = \lim\limits_{x \to -3} \frac{\left(2-\sqrt{x^2-5}\right)\left(2+\sqrt{x^2-5}\right)}{(x+3)\left(2+\sqrt{x^2-5}\right)} = \lim\limits_{x \to -3} \frac{4-(x^2-5)}{(x+3)\left(2+\sqrt{x^2-5}\right)}$

$= \lim\limits_{x \to -3} \frac{9-x^2}{(x+3)\left(2+\sqrt{x^2-5}\right)} = \lim\limits_{x \to -3} \frac{(3-x)(3+x)}{(x+3)\left(2+\sqrt{x^2-5}\right)} = \lim\limits_{x \to -3} \frac{3-x}{2+\sqrt{x^2-5}} = \frac{6}{2+\sqrt{4}} = \frac{3}{2}$

43. $\lim\limits_{x \to 0} (2\sin x - 1) = 2\sin 0 - 1 = 0 - 1 = -1$

45. $\lim\limits_{x \to 0} \sec x = \lim\limits_{x \to 0} \frac{1}{\cos x} = \frac{1}{\cos 0} = \frac{1}{1} = 1$

47. $\lim\limits_{x \to 0} \frac{1+x+\sin x}{3\cos x} = \frac{1+0+\sin 0}{3\cos 0} = \frac{1+0+0}{3} = \frac{1}{3}$

49. $\lim\limits_{x \to -\pi} \sqrt{x+4}\cos(x+\pi) = \lim\limits_{x \to -\pi} \sqrt{x+4} \cdot \lim\limits_{x \to -\pi} \cos(x+\pi) = \sqrt{-\pi+4} \cdot \cos 0 = \sqrt{4-\pi} \cdot 1 = \sqrt{4-\pi}$

51. (a) quotient rule
(b) difference and power rules
(c) sum and constant multiple rules

53. (a) $\lim\limits_{x \to c} f(x)\,g(x) = \left[\lim\limits_{x \to c} f(x)\right]\left[\lim\limits_{x \to c} g(x)\right] = (5)(-2) = -10$

(b) $\lim\limits_{x \to c} 2f(x)\,g(x) = 2\left[\lim\limits_{x \to c} f(x)\right]\left[\lim\limits_{x \to c} g(x)\right] = 2(5)(-2) = -20$

(c) $\lim\limits_{x \to c} [f(x) + 3g(x)] = \lim\limits_{x \to c} f(x) + 3\lim\limits_{x \to c} g(x) = 5 + 3(-2) = -1$

(d) $\lim\limits_{x \to c} \frac{f(x)}{f(x)-g(x)} = \frac{\lim_{x \to c} f(x)}{\lim_{x \to c} f(x) - \lim_{x \to c} g(x)} = \frac{5}{5-(-2)} = \frac{5}{7}$

55. (a) $\lim\limits_{x \to b} [f(x) + g(x)] = \lim\limits_{x \to b} f(x) + \lim\limits_{x \to b} g(x) = 7 + (-3) = 4$

(b) $\lim\limits_{x \to b} f(x) \cdot g(x) = \left[\lim\limits_{x \to b} f(x)\right]\left[\lim\limits_{x \to b} g(x)\right] = (7)(-3) = -21$

(c) $\lim\limits_{x \to b} 4g(x) = \left[\lim\limits_{x \to b} 4\right]\left[\lim\limits_{x \to b} g(x)\right] = (4)(-3) = -12$

(d) $\lim\limits_{x \to b} f(x)/g(x) = \lim\limits_{x \to b} f(x) / \lim\limits_{x \to b} g(x) = \frac{7}{-3} = -\frac{7}{3}$

57. $\lim\limits_{h \to 0} \frac{(1+h)^2 - 1^2}{h} = \lim\limits_{h \to 0} \frac{1+2h+h^2-1}{h} = \lim\limits_{h \to 0} \frac{h(2+h)}{h} = \lim\limits_{h \to 0} (2+h) = 2$

59. $\lim\limits_{h \to 0} \frac{[3(2+h)-4]-[3(2)-4]}{h} = \lim\limits_{h \to 0} \frac{3h}{h} = 3$

61. $\lim\limits_{h \to 0} \frac{\sqrt{7+h}-\sqrt{7}}{h} = \lim\limits_{h \to 0} \frac{\left(\sqrt{7+h}-\sqrt{7}\right)\left(\sqrt{7+h}+\sqrt{7}\right)}{h\left(\sqrt{7+h}+\sqrt{7}\right)} = \lim\limits_{h \to 0} \frac{(7+h)-7}{h\left(\sqrt{7+h}+\sqrt{7}\right)}$

$= \lim\limits_{h \to 0} \frac{h}{h\left(\sqrt{7+h}+\sqrt{7}\right)} = \lim\limits_{h \to 0} \frac{1}{\sqrt{7+h}+\sqrt{7}} = \frac{1}{2\sqrt{7}}$

63. $\lim\limits_{x \to 0} \sqrt{5-2x^2} = \sqrt{5-2(0)^2} = \sqrt{5}$ and $\lim\limits_{x \to 0} \sqrt{5-x^2} = \sqrt{5-(0)^2} = \sqrt{5}$; by the sandwich theorem,

$\lim\limits_{x \to 0} f(x) = \sqrt{5}$

65. (a) $\lim\limits_{x \to 0} \left(1 - \frac{x^2}{6}\right) = 1 - \frac{0}{6} = 1$ and $\lim\limits_{x \to 0} 1 = 1$; by the sandwich theorem, $\lim\limits_{x \to 0} \frac{x \sin x}{2 - 2\cos x} = 1$

(b) For $x \neq 0$, $y = (x \sin x)/(2 - 2\cos x)$ lies between the other two graphs in the figure, and the graphs converge as $x \to 0$.

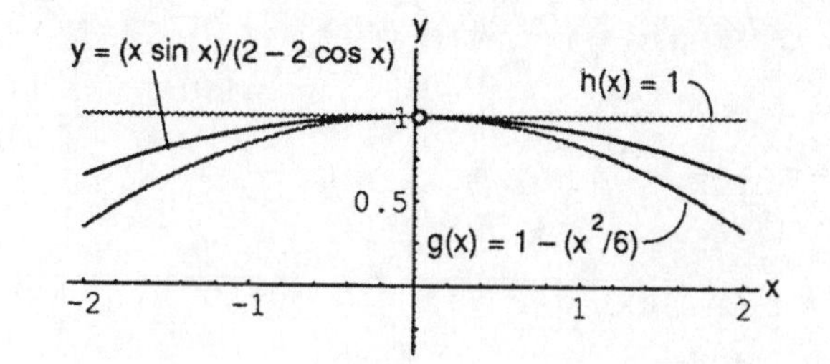

67. (a) $f(x) = (x^2 - 9)/(x + 3)$

x	−3.1	−3.01	−3.001	−3.0001	−3.00001	−3.000001
f(x)	−6.1	−6.01	−6.001	−6.0001	−6.00001	−6.000001

x	−2.9	−2.99	−2.999	−2.9999	−2.99999	−2.999999
f(x)	−5.9	−5.99	−5.999	−5.9999	−5.99999	−5.999999

The estimate is $\lim_{x \to -3} f(x) = -6$.

(b)

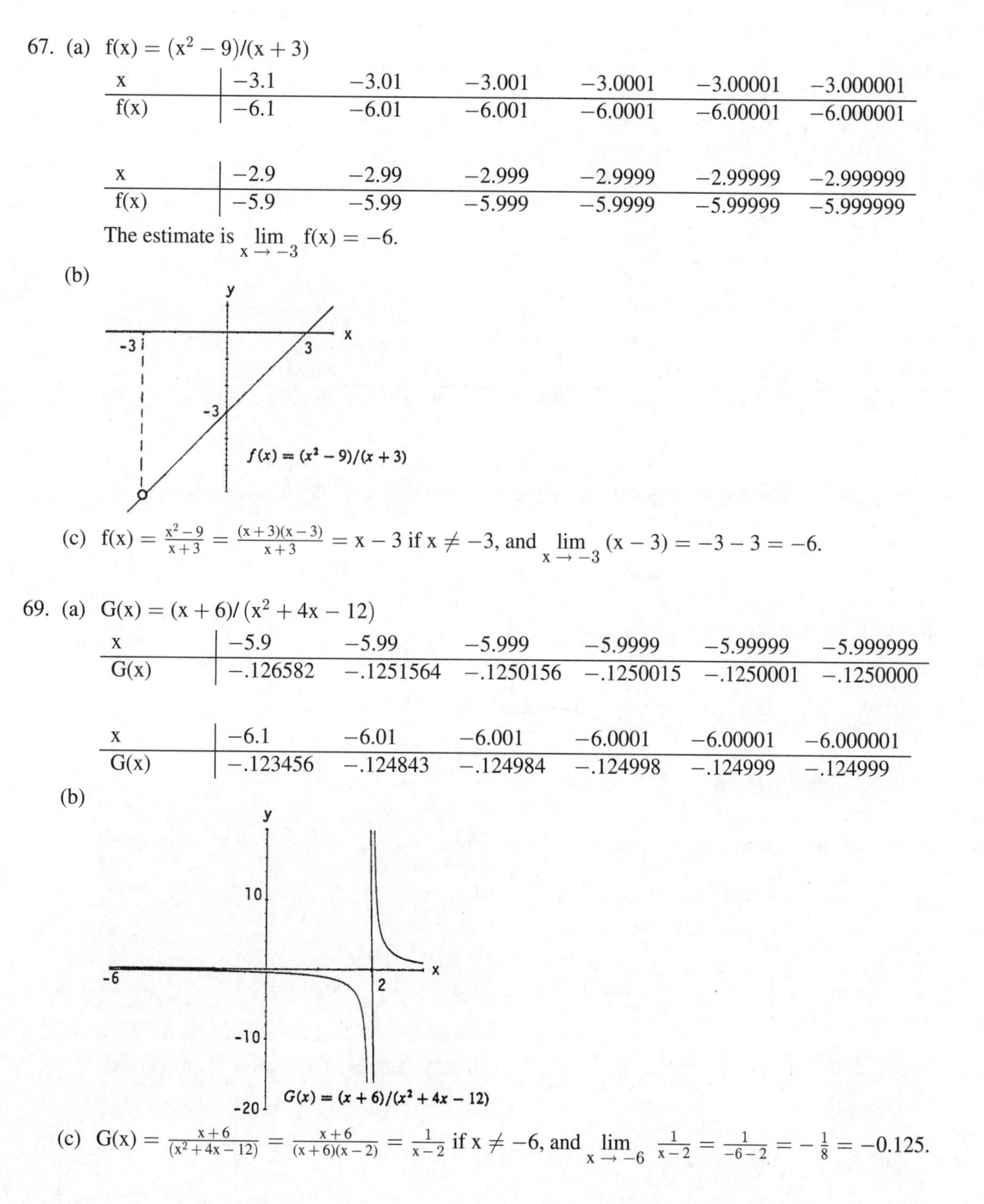

(c) $f(x) = \frac{x^2-9}{x+3} = \frac{(x+3)(x-3)}{x+3} = x - 3$ if $x \neq -3$, and $\lim_{x \to -3} (x - 3) = -3 - 3 = -6$.

69. (a) $G(x) = (x + 6)/(x^2 + 4x - 12)$

x	−5.9	−5.99	−5.999	−5.9999	−5.99999	−5.999999
G(x)	−.126582	−.1251564	−.1250156	−.1250015	−.1250001	−.1250000

x	−6.1	−6.01	−6.001	−6.0001	−6.00001	−6.000001
G(x)	−.123456	−.124843	−.124984	−.124998	−.124999	−.124999

(b)

(c) $G(x) = \frac{x+6}{(x^2+4x-12)} = \frac{x+6}{(x+6)(x-2)} = \frac{1}{x-2}$ if $x \neq -6$, and $\lim_{x \to -6} \frac{1}{x-2} = \frac{1}{-6-2} = -\frac{1}{8} = -0.125$.

71. (a) $f(x) = (x^2 - 1)/(|x| - 1)$

x	−1.1	−1.01	−1.001	−1.0001	−1.00001	−1.000001
f(x)	2.1	2.01	2.001	2.0001	2.00001	2.000001

x	−.9	−.99	−.999	−.9999	−.99999	−.999999
f(x)	1.9	1.99	1.999	1.9999	1.99999	1.999999

(b)

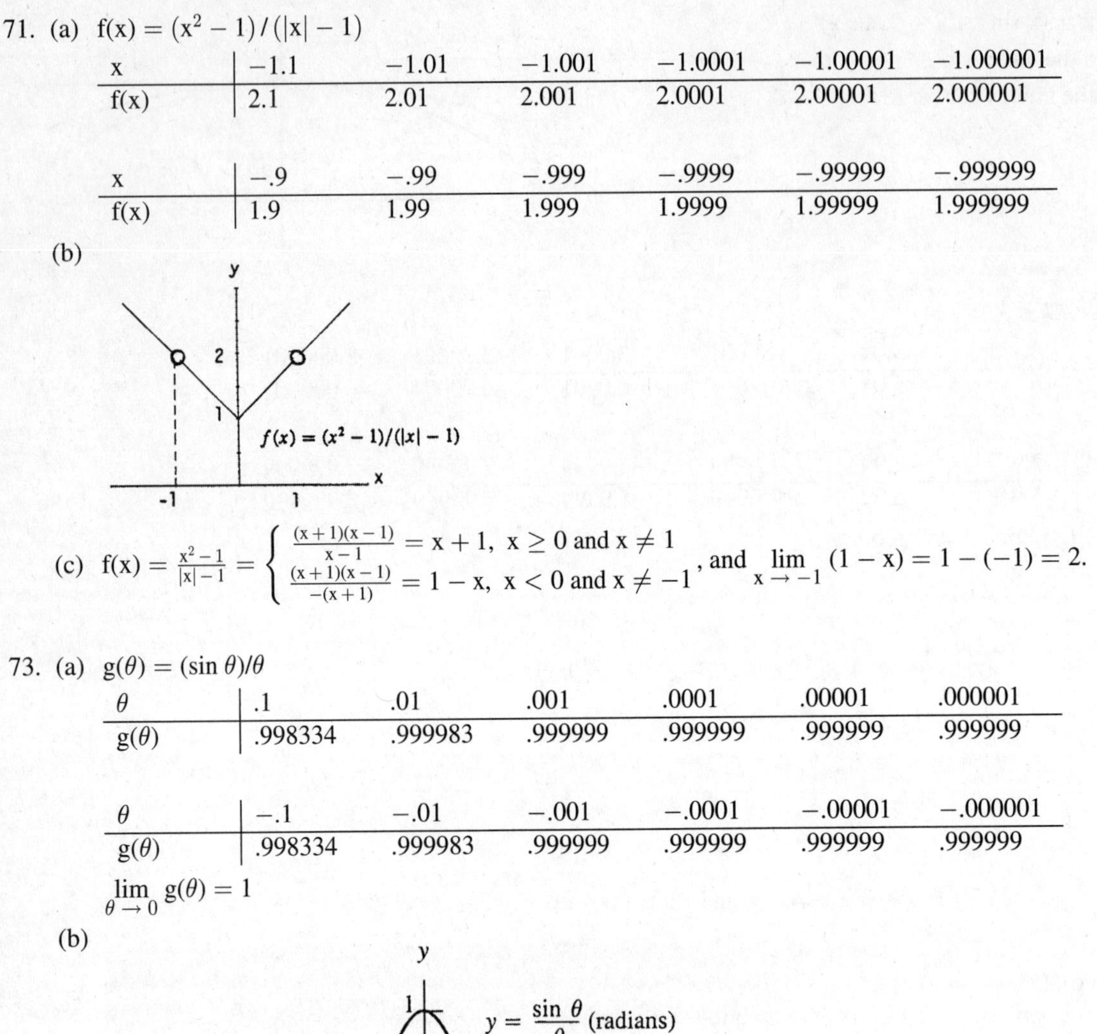

(c) $f(x) = \frac{x^2-1}{|x|-1} = \begin{cases} \frac{(x+1)(x-1)}{x-1} = x+1, & x \ge 0 \text{ and } x \ne 1 \\ \frac{(x+1)(x-1)}{-(x+1)} = 1 - x, & x < 0 \text{ and } x \ne -1 \end{cases}$, and $\lim_{x \to -1} (1 - x) = 1 - (-1) = 2.$

73. (a) $g(\theta) = (\sin\theta)/\theta$

θ	.1	.01	.001	.0001	.00001	.000001
$g(\theta)$	.998334	.999983	.999999	.999999	.999999	.999999

θ	−.1	−.01	−.001	−.0001	−.00001	−.000001
$g(\theta)$	.998334	.999983	.999999	.999999	.999999	.999999

$\lim_{\theta \to 0} g(\theta) = 1$

(b)

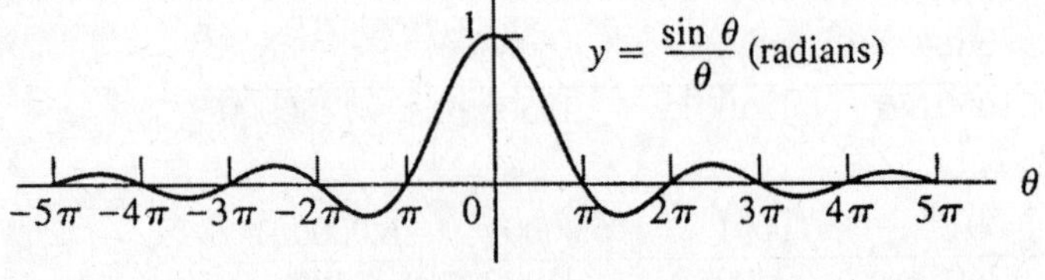

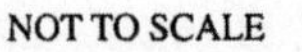

75. $\lim_{x \to c} f(x)$ exists at those points c where $\lim_{x \to c} x^4 = \lim_{x \to c} x^2$. Thus, $c^4 = c^2 \Rightarrow c^2(1 - c^2) = 0 \Rightarrow c = 0, 1,$ or -1.

Moreover, $\lim_{x \to 0} f(x) = \lim_{x \to 0} x^2 = 0$ and $\lim_{x \to -1} f(x) = \lim_{x \to 1} f(x) = 1$.

77. $1 = \lim_{x \to 4} \frac{f(x)-5}{x-2} = \frac{\lim_{x\to4} f(x) - \lim_{x\to4} 5}{\lim_{x\to4} x - \lim_{x\to4} 2} = \frac{\lim_{x\to4} f(x) - 5}{4-2} \Rightarrow \lim_{x \to 4} f(x) - 5 = 2(1) \Rightarrow \lim_{x \to 4} f(x) = 2 + 5 = 7.$

79. (a) $0 = 3 \cdot 0 = \left[\lim_{x \to 2} \frac{f(x)-5}{x-2}\right]\left[\lim_{x \to 2} (x-2)\right] = \lim_{x \to 2} \left[\left(\frac{f(x)-5}{x-2}\right)(x-2)\right] = \lim_{x \to 2} [f(x) - 5] = \lim_{x \to 2} f(x) - 5$

$\Rightarrow \lim_{x \to 2} f(x) - 5.$

(b) $0 = 4 \cdot 0 = \left[\lim_{x \to 2} \frac{f(x)-5}{x-2}\right]\left[\lim_{x \to 2} (x-2)\right] \Rightarrow \lim_{x \to 2} f(x) = 5$ as in part (a).

81. (a) $\lim_{x \to 0} x \sin \frac{1}{x} = 0$

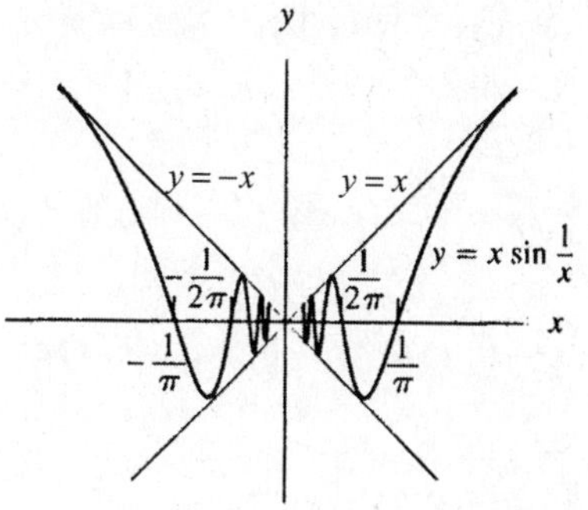

(b) $-1 \le \sin \frac{1}{x} \le 1$ for $x \neq 0$:

$x > 0 \Rightarrow -x \le x \sin \frac{1}{x} \le x \Rightarrow \lim_{x \to 0} x \sin \frac{1}{x} = 0$ by the sandwich theorem;

$x < 0 \Rightarrow -x \ge x \sin \frac{1}{x} \ge x \Rightarrow \lim_{x \to 0} x \sin \frac{1}{x} = 0$ by the sandwich theorem.

2.3 THE PRECISE DEFINITION OF A LIMIT

1. 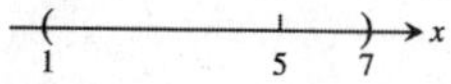

Step 1: $|x - 5| < \delta \Rightarrow -\delta < x - 5 < \delta \Rightarrow -\delta + 5 < x < \delta + 5$

Step 2: $\delta + 5 = 7 \Rightarrow \delta = 2$, or $-\delta + 5 = 1 \Rightarrow \delta = 4$.

The value of δ which assures $|x - 5| < \delta \Rightarrow 1 < x < 7$ is the smaller value, $\delta = 2$.

3. 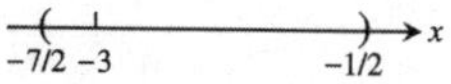

Step 1: $|x - (-3)| < \delta \Rightarrow -\delta < x + 3 < \delta \Rightarrow -\delta - 3 < x < \delta - 3$

Step 2: $-\delta - 3 = -\frac{7}{2} \Rightarrow \delta = \frac{1}{2}$, or $\delta - 3 = -\frac{1}{2} \Rightarrow \delta = \frac{5}{2}$.

The value of δ which assures $|x - (-3)| < \delta \Rightarrow -\frac{7}{2} < x < -\frac{1}{2}$ is the smaller value, $\delta = \frac{1}{2}$.

5. 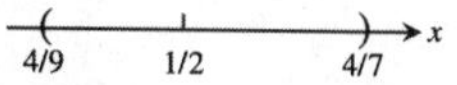

Step 1: $\left|x - \frac{1}{2}\right| < \delta \Rightarrow -\delta < x - \frac{1}{2} < \delta \Rightarrow -\delta + \frac{1}{2} < x < \delta + \frac{1}{2}$

Step 2: $-\delta + \frac{1}{2} = \frac{4}{9} \Rightarrow \delta = \frac{1}{18}$, or $\delta + \frac{1}{2} = \frac{4}{7} \Rightarrow \delta = \frac{1}{14}$.

The value of δ which assures $\left|x - \frac{1}{2}\right| < \delta \Rightarrow \frac{4}{9} < x < \frac{4}{7}$ is the smaller value, $\delta = \frac{1}{18}$.

7. Step 1: $|x - 5| < \delta \Rightarrow -\delta < x - 5 < \delta \Rightarrow -\delta + 5 < x < \delta + 5$

Step 2: From the graph, $-\delta + 5 = 4.9 \Rightarrow \delta = 0.1$, or $\delta + 5 = 5.1 \Rightarrow \delta = 0.1$; thus $\delta = 0.1$ in either case.

9. Step 1: $|x - 1| < \delta \Rightarrow -\delta < x - 1 < \delta \Rightarrow -\delta + 1 < x < \delta + 1$

Step 2: From the graph, $-\delta + 1 = \frac{9}{16} \Rightarrow \delta = \frac{7}{16}$, or $\delta + 1 = \frac{25}{16} \Rightarrow \delta = \frac{9}{16}$; thus $\delta = \frac{7}{16}$.

11. Step 1: $|x - 2| < \delta \Rightarrow -\delta < x - 2 < \delta \Rightarrow -\delta + 2 < x < \delta + 2$

Step 2: From the graph, $-\delta + 2 = \sqrt{3} \Rightarrow \delta = 2 - \sqrt{3} \approx 0.2679$, or $\delta + 2 = \sqrt{5} \Rightarrow \delta = \sqrt{5} - 2 \approx 0.2361$; thus $\delta = \sqrt{5} - 2$.

13. Step 1: $|x - (-1)| < \delta \Rightarrow -\delta < x + 1 < \delta \Rightarrow -\delta - 1 < x < \delta - 1$

Step 2: From the graph, $-\delta - 1 = -\frac{16}{9} \Rightarrow \delta = \frac{7}{9} \approx 0.77$, or $\delta - 1 = -\frac{16}{25} \Rightarrow \frac{9}{25} = 0.36$; thus $\delta = \frac{9}{25} = 0.36$.

15. Step 1: $|(x+1)-5| < 0.01 \Rightarrow |x-4| < 0.01 \Rightarrow -0.01 < x-4 < 0.01 \Rightarrow 3.99 < x < 4.01$
Step 2: $|x-4| < \delta \Rightarrow -\delta < x-4 < \delta \Rightarrow -\delta+4 < x < \delta+4 \Rightarrow \delta = 0.01.$

17. Step 1: $\left|\sqrt{x+1}-1\right| < 0.1 \Rightarrow -0.1 < \sqrt{x+1}-1 < 0.1 \Rightarrow 0.9 < \sqrt{x+1} < 1.1 \Rightarrow 0.81 < x+1 < 1.21$
$\Rightarrow -0.19 < x < 0.21$
Step 2: $|x-0| < \delta \Rightarrow -\delta < x < \delta$. Then, $-\delta = -0.19 \Rightarrow \delta = 0.19$ or $\delta = 0.21$; thus, $\delta = 0.19$.

19. Step 1: $\left|\sqrt{19-x}-3\right| < 1 \Rightarrow -1 < \sqrt{19-x}-3 < 1 \Rightarrow 2 < \sqrt{19-x} < 4 \Rightarrow 4 < 19-x < 16$
$\Rightarrow -4 > x-19 > -16 \Rightarrow 15 > x > 3$ or $3 < x < 15$
Step 2: $|x-10| < \delta \Rightarrow -\delta < x-10 < \delta \Rightarrow -\delta+10 < x < \delta+10.$
Then $-\delta+10 = 3 \Rightarrow \delta = 7$, or $\delta+10 = 15 \Rightarrow \delta = 5$; thus $\delta = 5$.

21. Step 1: $\left|\frac{1}{x}-\frac{1}{4}\right| < 0.05 \Rightarrow -0.05 < \frac{1}{x}-\frac{1}{4} < 0.05 \Rightarrow 0.2 < \frac{1}{x} < 0.3 \Rightarrow \frac{10}{2} > x > \frac{10}{3}$ or $\frac{10}{3} < x < 5.$
Step 2: $|x-4| < \delta \Rightarrow -\delta < x-4 < \delta \Rightarrow -\delta+4 < x < \delta+4.$
Then $-\delta+4 = \frac{10}{3}$ or $\delta = \frac{2}{3}$, or $\delta+4 = 5$ or $\delta = 1$; thus $\delta = \frac{2}{3}$.

23. Step 1: $|x^2-4| < 0.5 \Rightarrow -0.5 < x^2-4 < 0.5 \Rightarrow 3.5 < x^2 < 4.5 \Rightarrow \sqrt{3.5} < |x| < \sqrt{4.5} \Rightarrow -\sqrt{4.5} < x < -\sqrt{3.5},$
for x near -2.
Step 2: $|x-(-2)| < \delta \Rightarrow -\delta < x+2 < \delta \Rightarrow -\delta-2 < x < \delta-2.$
Then $-\delta-2 = -\sqrt{4.5} \Rightarrow \delta = \sqrt{4.5}-2 \approx 0.1213$, or $\delta-2 = -\sqrt{3.5} \Rightarrow \delta = 2-\sqrt{3.5} \approx 0.1292$;
thus $\delta = \sqrt{4.5}-2 \approx 0.12$.

25. Step 1: $|(x^2-5)-11| < 1 \Rightarrow |x^2-16| < 1 \Rightarrow -1 < x^2-16 < 1 \Rightarrow 15 < x^2 < 17 \Rightarrow \sqrt{15} < x < \sqrt{17}.$
Step 2: $|x-4| < \delta \Rightarrow -\delta < x-4 < \delta \Rightarrow -\delta+4 < x < \delta+4.$
Then $-\delta+4 = \sqrt{15} \Rightarrow \delta = 4-\sqrt{15} \approx 0.1270$, or $\delta+4 = \sqrt{17} \Rightarrow \delta = \sqrt{17}-4 \approx 0.1231$;
thus $\delta = \sqrt{17}-4 \approx 0.12$.

27. Step 1: $|mx-2m| < 0.03 \Rightarrow -0.03 < mx-2m < 0.03 \Rightarrow -0.03+2m < mx < 0.03+2m \Rightarrow$
$2-\frac{0.03}{m} < x < 2+\frac{0.03}{m}.$
Step 2: $|x-2| < \delta \Rightarrow -\delta < x-2 < \delta \Rightarrow -\delta+2 < x < \delta+2.$
Then $-\delta+2 = 2-\frac{0.03}{m} \Rightarrow \delta = \frac{0.03}{m}$, or $\delta+2 = 2+\frac{0.03}{m} \Rightarrow \delta = \frac{0.03}{m}$. In either case, $\delta = \frac{0.03}{m}$.

29. Step 1: $\left|(mx+b)-\left(\frac{m}{2}+b\right)\right| < c \Rightarrow -c < mx-\frac{m}{2} < c \Rightarrow -c+\frac{m}{2} < mx < c+\frac{m}{2} \Rightarrow \frac{1}{2}-\frac{c}{m} < x < \frac{1}{2}+\frac{c}{m}.$
Step 2: $\left|x-\frac{1}{2}\right| < \delta \Rightarrow -\delta < x-\frac{1}{2} < \delta \Rightarrow -\delta+\frac{1}{2} < x < \delta+\frac{1}{2}.$
Then $-\delta+\frac{1}{2} = \frac{1}{2}-\frac{c}{m} \Rightarrow \delta = \frac{c}{m}$, or $\delta+\frac{1}{2} = \frac{1}{2}+\frac{c}{m} \Rightarrow \delta = \frac{c}{m}$. In either case, $\delta = \frac{c}{m}$.

31. $\lim_{x\to 3}(3-2x) = 3-2(3) = -3$
Step 1: $|(3-2x)-(-3)| < 0.02 \Rightarrow -0.02 < 6-2x < 0.02 \Rightarrow -6.02 < -2x < -5.98 \Rightarrow 3.01 > x > 2.99$ or
$2.99 < x < 3.01.$
Step 2: $0 < |x-3| < \delta \Rightarrow -\delta < x-3 < \delta \Rightarrow -\delta+3 < x < \delta+3$
Then $-\delta+3 = 2.99 \Rightarrow \delta = 0.01$, or $\delta+3 = 3.01 \Rightarrow \delta = 0.01$; thus $\delta = 0.01$.

33. $\lim_{x\to 2}\frac{x^2-4}{x-2} = \lim_{x\to 2}\frac{(x+2)(x-2)}{(x-2)} = \lim_{x\to 2}(x+2) = 2+2 = 4,\ x \neq 2$
Step 1: $\left|\left(\frac{x^2-4}{x-2}\right)-4\right| < 0.05 \Rightarrow -0.05 < \frac{(x+2)(x-2)}{(x-2)}-4 < 0.05 \Rightarrow 3.95 < x+2 < 4.05,\ x \neq 2$
$\Rightarrow 1.95 < x < 2.05,\ x \neq 2.$

Step 2: $|x-2|<\delta \Rightarrow -\delta<x-2<\delta \Rightarrow -\delta+2<x<\delta+2$.
Then $-\delta+2=1.95 \Rightarrow \delta=0.05$, or $\delta+2=2.05 \Rightarrow \delta=0.05$; thus $\delta=0.05$.

35. $\lim\limits_{x\to -3}\sqrt{1-5x}=\sqrt{1-5(-3)}=\sqrt{16}=4$

Step 1: $\left|\sqrt{1-5x}-4\right|<0.5 \Rightarrow -0.5<\sqrt{1-5x}-4<0.5 \Rightarrow 3.5<\sqrt{1-5x}<4.5 \Rightarrow 12.25<1-5x<20.25$
$\Rightarrow 11.25<-5x<19.25 \Rightarrow -3.85<x<-2.25$.

Step 2: $|x-(-3)|<\delta \Rightarrow -\delta<x+3<\delta \Rightarrow -\delta-3<x<\delta-3$.
Then $-\delta-3=-3.85 \Rightarrow \delta=0.85$, or $\delta-3=-2.25 \Rightarrow 0.75$; thus $\delta=0.75$.

37. Step 1: $|(9-x)-5|<\epsilon \Rightarrow -\epsilon<4-x<\epsilon \Rightarrow -\epsilon-4<-x<\epsilon-4 \Rightarrow \epsilon+4>x>4-\epsilon \Rightarrow 4-\epsilon<x<4+\epsilon$.

Step 2: $|x-4|<\delta \Rightarrow -\delta<x-4<\delta \Rightarrow -\delta+4<x<\delta+4$.
Then $-\delta+4=-\epsilon+4 \Rightarrow \delta=\epsilon$, or $\delta+4=\epsilon+4 \Rightarrow \delta=\epsilon$. Thus choose $\delta=\epsilon$.

39. Step 1: $\left|\sqrt{x-5}-2\right|<\epsilon \Rightarrow -\epsilon<\sqrt{x-5}-2<\epsilon \Rightarrow 2-\epsilon<\sqrt{x-5}<2+\epsilon \Rightarrow (2-\epsilon)^2<x-5<(2+\epsilon)^2$
$\Rightarrow (2-\epsilon)^2+5<x<(2+\epsilon)^2+5$.

Step 2: $|x-9|<\delta \Rightarrow -\delta<x-9<\delta \Rightarrow -\delta+9<x<\delta+9$.
Then $-\delta+9=\epsilon^2-4\epsilon+9 \Rightarrow \delta=4\epsilon-\epsilon^2$, or $\delta+9=\epsilon^2+4\epsilon+9 \Rightarrow \delta=4\epsilon+\epsilon^2$. Thus choose the smaller distance, $\delta=4\epsilon-\epsilon^2$.

41. Step 1: For $x\neq 1$, $|x^2-1|<\epsilon \Rightarrow -\epsilon<x^2-1<\epsilon \Rightarrow 1-\epsilon<x^2<1+\epsilon \Rightarrow \sqrt{1-\epsilon}<|x|<\sqrt{1+\epsilon}$
$\Rightarrow \sqrt{1-\epsilon}<x<\sqrt{1+\epsilon}$ near $x=1$.

Step 2: $|x-1|<\delta \Rightarrow -\delta<x-1<\delta \Rightarrow -\delta+1<x<\delta+1$.
Then $-\delta+1=\sqrt{1-\epsilon} \Rightarrow \delta=1-\sqrt{1-\epsilon}$, or $\delta+1=\sqrt{1+\epsilon} \Rightarrow \delta=\sqrt{1+\epsilon}-1$. Choose $\delta=\min\left\{1-\sqrt{1-\epsilon},\sqrt{1+\epsilon}-1\right\}$, that is, the smaller of the two distances.

43. Step 1: $\left|\frac{1}{x}-1\right|<\epsilon \Rightarrow -\epsilon<\frac{1}{x}-1<\epsilon \Rightarrow 1-\epsilon<\frac{1}{x}<1+\epsilon \Rightarrow \frac{1}{1+\epsilon}<x<\frac{1}{1-\epsilon}$.

Step 2: $|x-1|<\delta \Rightarrow -\delta<x-1<\delta \Rightarrow 1-\delta<x<1+\delta$.
Then $1-\delta=\frac{1}{1+\epsilon} \Rightarrow \delta=1-\frac{1}{1+\epsilon}=\frac{\epsilon}{1+\epsilon}$, or $1+\delta=\frac{1}{1-\epsilon} \Rightarrow \delta=\frac{1}{1-\epsilon}-1=\frac{\epsilon}{1-\epsilon}$.
Choose $\delta=\frac{\epsilon}{1+\epsilon}$, the smaller of the two distances.

45. Step 1: $\left|\left(\frac{x^2-9}{x+3}\right)-(-6)\right|<\epsilon \Rightarrow -\epsilon<(x-3)+6<\epsilon,\ x\neq -3 \Rightarrow -\epsilon<x+3<\epsilon \Rightarrow -\epsilon-3<x<\epsilon-3$.

Step 2: $|x-(-3)|<\delta \Rightarrow -\delta<x+3<\delta \Rightarrow -\delta-3<x<\delta-3$.
Then $-\delta-3=-\epsilon-3 \Rightarrow \delta=\epsilon$, or $\delta-3=\epsilon-3 \Rightarrow \delta=\epsilon$. Choose $\delta=\epsilon$.

47. Step 1: $x<1$: $|(4-2x)-2|<\epsilon \Rightarrow 0<2-2x<\epsilon$ since $x<1$. Thus, $1-\frac{\epsilon}{2}<x<0$;
$x\geq 1$: $|(6x-4)-2|<\epsilon \Rightarrow 0\leq 6x-6<\epsilon$ since $x\geq 1$. Thus, $1\leq x<1+\frac{\epsilon}{6}$.

Step 2: $|x-1|<\delta \Rightarrow -\delta<x-1<\delta \Rightarrow 1-\delta<x<1+\delta$.
Then $1-\delta=1-\frac{\epsilon}{2} \Rightarrow \delta=\frac{\epsilon}{2}$, or $1+\delta=1+\frac{\epsilon}{6} \Rightarrow \delta=\frac{\epsilon}{6}$. Choose $\delta=\frac{\epsilon}{6}$.

49. By the figure, $-x\leq x\sin\frac{1}{x}\leq x$ for all $x>0$ and $-x\geq x\sin\frac{1}{x}\geq x$ for $x<0$. Since $\lim\limits_{x\to 0}(-x)=\lim\limits_{x\to 0}x=0$, then by the sandwich theorem, in either case, $\lim\limits_{x\to 0}x\sin\frac{1}{x}=0$.

51. As x approaches the value 0, the values of g(x) approach k. Thus for every number $\epsilon>0$, there exists a $\delta>0$ such that $0<|x-0|<\delta \Rightarrow |g(x)-k|<\epsilon$.

53. Let $f(x) = x^2$. The function values do get closer to -1 as x approaches 0, but $\lim_{x \to 0} f(x) = 0$, not -1. The function $f(x) = x^2$ never gets arbitrarily close to -1 for x near 0.

55. $|A - 9| \le 0.01 \Rightarrow -0.01 \le \pi\left(\frac{x}{2}\right)^2 - 9 \le 0.01 \Rightarrow 8.99 \le \frac{\pi x^2}{4} \le 9.01 \Rightarrow \frac{4}{\pi}(8.99) \le x^2 \le \frac{4}{\pi}(9.01)$ $\Rightarrow 2\sqrt{\frac{8.99}{\pi}} \le x \le 2\sqrt{\frac{9.01}{\pi}}$ or $3.384 \le x \le 3.387$. To be safe, the left endpoint was rounded up and the right endpoint was rounded down.

57. (a) $-\delta < x - 1 < 0 \Rightarrow 1 - \delta < x < 1 \Rightarrow f(x) = x$. Then $|f(x) - 2| = |x - 2| = 2 - x > 2 - 1 = 1$. That is, $|f(x) - 2| \ge 1 \ge \frac{1}{2}$ no matter how small δ is taken when $1 - \delta < x < 1 \Rightarrow \lim_{x \to 1} f(x) \ne 2$.
(b) $0 < x - 1 < \delta \Rightarrow 1 < x < 1 + \delta \Rightarrow f(x) = x + 1$. Then $|f(x) - 1| = |(x + 1) - 1| = |x| = x > 1$. That is, $|f(x) - 1| \ge 1$ no matter how small δ is taken when $1 < x < 1 + \delta \Rightarrow \lim_{x \to 1} f(x) \ne 1$.
(c) $-\delta < x - 1 < 0 \Rightarrow 1 - \delta < x < 1 \Rightarrow f(x) = x$. Then $|f(x) - 1.5| = |x - 1.5| = 1.5 - x > 1.5 - 1 = 0.5$. Also, $0 < x - 1 < \delta \Rightarrow 1 < x < 1 + \delta \Rightarrow f(x) = x + 1$. Then $|f(x) - 1.5| = |(x + 1) - 1.5| = |x - 0.5|$ $= x - 0.5 > 1 - 0.5 = 0.5$. Thus, no matter how small δ is taken, there exists a value of x such that $-\delta < x - 1 < \delta$ but $|f(x) - 1.5| \ge \frac{1}{2} \Rightarrow \lim_{x \to 1} f(x) \ne 1.5$.

59. (a) For $3 - \delta < x < 3 \Rightarrow f(x) > 4.8 \Rightarrow |f(x) - 4| \ge 0.8$. Thus for $\epsilon < 0.8$, $|f(x) - 4| \ge \epsilon$ whenever $3 - \delta < x < 3$ no matter how small we choose $\delta > 0 \Rightarrow \lim_{x \to 3} f(x) \ne 4$.
(b) For $3 < x < 3 + \delta \Rightarrow f(x) < 3 \Rightarrow |f(x) - 4.8| \ge 1.8$. Thus for $\epsilon < 1.8$, $|f(x) - 4.8| \ge \epsilon$ whenever $3 < x < 3 + \delta$ no matter how small we choose $\delta > 0 \Rightarrow \lim_{x \to 3} f(x) \ne 4.8$.
(c) For $3 - \delta < x < 3 \Rightarrow f(x) > 4.8 \Rightarrow |f(x) - 3| \ge 1.8$. Again, for $\epsilon < 1.8$, $|f(x) - 3| \ge \epsilon$ whenever $3 - \delta < x < 3$ no matter how small we choose $\delta > 0 \Rightarrow \lim_{x \to 3} f(x) \ne 3$.

2.4 ONE-SIDED LIMITS

1. (a) True (b) True (c) False (d) True
(e) True (f) True (g) False (h) False
(i) False (j) False (k) True (l) False

3. (a) $\lim_{x \to 2^+} f(x) = \frac{2}{2} + 1 = 2$, $\lim_{x \to 2^-} f(x) = 3 - 2 = 1$
(b) No, $\lim_{x \to 2} f(x)$ does not exist because $\lim_{x \to 2^+} f(x) \ne \lim_{x \to 2^-} f(x)$
(c) $\lim_{x \to 4^-} f(x) = \frac{4}{2} + 1 = 3$, $\lim_{x \to 4^+} f(x) = \frac{4}{2} + 1 = 3$
(d) Yes, $\lim_{x \to 4} f(x) = 3$ because $3 = \lim_{x \to 4^-} f(x) = \lim_{x \to 4^+} f(x)$

5. (a) No, $\lim_{x \to 0^+} f(x)$ does not exist since $\sin\left(\frac{1}{x}\right)$ does not approach any single value as x approaches 0
(b) $\lim_{x \to 0^-} f(x) = \lim_{x \to 0^-} 0 = 0$
(c) $\lim_{x \to 0} f(x)$ does not exist because $\lim_{x \to 0^+} f(x)$ does not exist

7. (a)

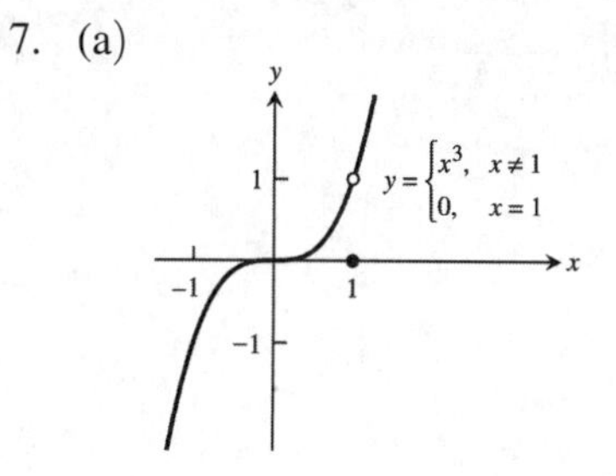

(b) $\lim\limits_{x\to 1^-} f(x) = 1 = \lim\limits_{x\to 1^+} f(x)$

(c) Yes, $\lim\limits_{x\to 1} f(x) = 1$ since the right-hand and left-hand limits exist and equal 1

9. (a) domain: $0 \le x \le 2$
range: $0 < y \le 1$ and $y = 2$

(b) $\lim\limits_{x\to c} f(x)$ exists for c belonging to $(0,1)\cup(1,2)$

(c) $x = 2$

(d) $x = 0$

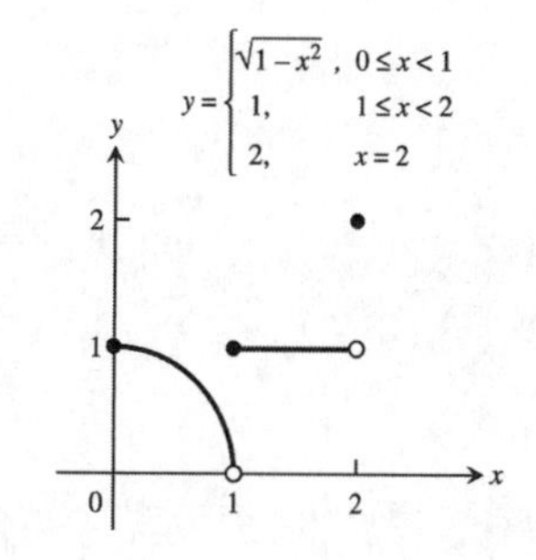

11. $\lim\limits_{x\to -0.5^-} \sqrt{\frac{x+2}{x-1}} = \sqrt{\frac{-0.5+2}{-0.5+1}} = \sqrt{\frac{3/2}{1/2}} = \sqrt{3}$

13. $\lim\limits_{x\to -2^+} \left(\frac{x}{x+1}\right)\left(\frac{2x+5}{x^2+x}\right) = \left(\frac{-2}{-2+1}\right)\left(\frac{2(-2)+5}{(-2)^2+(-2)}\right) = (2)\left(\frac{1}{2}\right) = 1$

15. $\lim\limits_{h\to 0^+} \frac{\sqrt{h^2+4h+5}-\sqrt{5}}{h} = \lim\limits_{h\to 0^+} \left(\frac{\sqrt{h^2+4h+5}-\sqrt{5}}{h}\right)\left(\frac{\sqrt{h^2+4h+5}+\sqrt{5}}{\sqrt{h^2+4h+5}+\sqrt{5}}\right)$
$= \lim\limits_{h\to 0^+} \frac{(h^2+4h+5)-5}{h\left(\sqrt{h^2+4h+5}+\sqrt{5}\right)} = \lim\limits_{h\to 0^+} \frac{h(h+4)}{h\left(\sqrt{h^2+4h+5}+\sqrt{5}\right)} = \frac{0+4}{\sqrt{5}+\sqrt{5}} = \frac{2}{\sqrt{5}}$

17. (a) $\lim\limits_{x\to -2^+} (x+3)\frac{|x+2|}{x+2} = \lim\limits_{x\to -2^+} (x+3)\frac{(x+2)}{(x+2)}$ $\quad(|x+2| = (x+2)$ for $x > -2)$
$= \lim\limits_{x\to -2^+} (x+3) = ((-2)+3) = 1$

(b) $\lim\limits_{x\to -2^-} (x+3)\frac{|x+2|}{x+2} = \lim\limits_{x\to -2^-} (x+3)\left[\frac{-(x+2)}{(x+2)}\right]$ $\quad(|x+2| = -(x+2)$ for $x < -2)$
$= \lim\limits_{x\to -2^-} (x+3)(-1) = -(-2+3) = -1$

19. (a) $\lim\limits_{\theta\to 3^+} \frac{\lfloor\theta\rfloor}{\theta} = \frac{3}{3} = 1$ (b) $\lim\limits_{\theta\to 3^-} \frac{\lfloor\theta\rfloor}{\theta} = \frac{2}{3}$

21. $\lim\limits_{\theta\to 0} \frac{\sin\sqrt{2}\theta}{\sqrt{2}\theta} = \lim\limits_{x\to 0} \frac{\sin x}{x} = 1$ (where $x = \sqrt{2}\theta$)

23. $\lim\limits_{y\to 0} \frac{\sin 3y}{4y} = \frac{1}{4}\lim\limits_{y\to 0} \frac{3\sin 3y}{3y} = \frac{3}{4}\lim\limits_{y\to 0} \frac{\sin 3y}{3y} = \frac{3}{4}\lim\limits_{\theta\to 0} \frac{\sin\theta}{\theta} = \frac{3}{4}$ (where $\theta = 3y$)

25. $\lim\limits_{x\to 0} \frac{\tan 2x}{x} = \lim\limits_{x\to 0} \frac{\left(\frac{\sin 2x}{\cos 2x}\right)}{x} = \lim\limits_{x\to 0} \frac{\sin 2x}{x\cos 2x} = \left(\lim\limits_{x\to 0} \frac{1}{\cos 2x}\right)\left(\lim\limits_{x\to 0} \frac{2\sin 2x}{2x}\right) = 1\cdot 2 = 2$

27. $\lim\limits_{x\to 0} \frac{x\csc 2x}{\cos 5x} = \lim\limits_{x\to 0} \left(\frac{x}{\sin 2x}\cdot\frac{1}{\cos 5x}\right) = \left(\frac{1}{2}\lim\limits_{x\to 0} \frac{2x}{\sin 2x}\right)\left(\lim\limits_{x\to 0} \frac{1}{\cos 5x}\right) = \left(\frac{1}{2}\cdot 1\right)(1) = \frac{1}{2}$

29. $\lim\limits_{x\to 0} \frac{x+x\cos x}{\sin x\cos x} = \lim\limits_{x\to 0} \left(\frac{x}{\sin x\cos x} + \frac{x\cos x}{\sin x\cos x}\right) = \lim\limits_{x\to 0} \left(\frac{x}{\sin x}\cdot\frac{1}{\cos x}\right) + \lim\limits_{x\to 0} \frac{x}{\sin x}$
$= \lim\limits_{x\to 0} \left(\frac{1}{\frac{\sin x}{x}}\right)\cdot\lim\limits_{x\to 0} \left(\frac{1}{\cos x}\right) + \lim\limits_{x\to 0} \left(\frac{1}{\frac{\sin x}{x}}\right) = (1)(1) + 1 = 2$

31. $\lim\limits_{\theta \to 0} \frac{1-\cos\theta}{\sin 2\theta} = \lim\limits_{\theta \to 0} \frac{(1-\cos\theta)(1+\cos\theta)}{(2\sin\theta\cos\theta)(1+\cos\theta)} = \lim\limits_{\theta \to 0} \frac{1-\cos^2\theta}{(2\sin\theta\cos\theta)(1+\cos\theta)} = \lim\limits_{\theta \to 0} \frac{\sin^2\theta}{(2\sin\theta\cos\theta)(1+\cos\theta)}$
$= \lim\limits_{\theta \to 0} \frac{\sin\theta}{(2\cos\theta)(1+\cos\theta)} = \frac{0}{(2)(2)} = 0$

33. $\lim\limits_{t \to 0} \frac{\sin(1-\cos t)}{1-\cos t} = \lim\limits_{\theta \to 0} \frac{\sin\theta}{\theta} = 1$ since $\theta = 1 - \cos t \to 0$ as $t \to 0$

35. $\lim\limits_{\theta \to 0} \frac{\sin\theta}{\sin 2\theta} = \lim\limits_{\theta \to 0} \left(\frac{\sin\theta}{\sin 2\theta} \cdot \frac{2\theta}{2\theta}\right) = \frac{1}{2} \lim\limits_{\theta \to 0} \left(\frac{\sin\theta}{\theta} \cdot \frac{2\theta}{\sin 2\theta}\right) = \frac{1}{2} \cdot 1 \cdot 1 = \frac{1}{2}$

37. $\lim\limits_{\theta \to 0} \theta\cos\theta = 0 \cdot 1 = 0$

39. $\lim\limits_{x \to 0} \frac{\tan 3x}{\sin 8x} = \lim\limits_{x \to 0} \left(\frac{\sin 3x}{\cos 3x} \cdot \frac{1}{\sin 8x}\right) = \lim\limits_{x \to 0} \left(\frac{\sin 3x}{\cos 3x} \cdot \frac{1}{\sin 8x} \cdot \frac{8x}{3x} \cdot \frac{3}{8}\right)$
$= \frac{3}{8} \lim\limits_{x \to 0} \left(\frac{1}{\cos 3x}\right)\left(\frac{\sin 3x}{3x}\right)\left(\frac{8x}{\sin 8x}\right) = \frac{3}{8} \cdot 1 \cdot 1 \cdot 1 = \frac{3}{8}$

41. $\lim\limits_{\theta \to 0} \frac{\tan\theta}{\theta^2 \cot 3\theta} = \lim\limits_{\theta \to 0} \frac{\frac{\sin\theta}{\cos\theta}}{\theta^2 \frac{\cos 3\theta}{\sin 3\theta}} = \lim\limits_{\theta \to 0} \frac{\sin\theta \sin 3\theta}{\theta^2 \cos\theta \cos 3\theta} = \lim\limits_{\theta \to 0} \left(\frac{\sin\theta}{\theta}\right)\left(\frac{\sin 3\theta}{3\theta}\right)\left(\frac{3}{\cos\theta\cos 3\theta}\right) = (1)(1)\left(\frac{3}{1 \cdot 1}\right) = 3$

43. Yes. If $\lim\limits_{x \to a^+} f(x) = L = \lim\limits_{x \to a^-} f(x)$, then $\lim\limits_{x \to a} f(x) = L$. If $\lim\limits_{x \to a^+} f(x) \neq \lim\limits_{x \to a^-} f(x)$, then $\lim\limits_{x \to a} f(x)$ does not exist.

45. If f is an odd function of x, then $f(-x) = -f(x)$. Given $\lim\limits_{x \to 0^+} f(x) = 3$, then $\lim\limits_{x \to 0^-} f(x) = -3$.

47. $I = (5, 5+\delta) \Rightarrow 5 < x < 5+\delta$. Also, $\sqrt{x-5} < \epsilon \Rightarrow x - 5 < \epsilon^2 \Rightarrow x < 5 + \epsilon^2$. Choose $\delta = \epsilon^2$
$\Rightarrow \lim\limits_{x \to 5^+} \sqrt{x-5} = 0$.

49. As $x \to 0^-$ the number x is always negative. Thus, $\left|\frac{x}{|x|} - (-1)\right| < \epsilon \Rightarrow \left|\frac{x}{-x} + 1\right| < \epsilon \Rightarrow 0 < \epsilon$ which is always true independent of the value of x. Hence we can choose any $\delta > 0$ with $-\delta < x < 0 \Rightarrow \lim\limits_{x \to 0^-} \frac{x}{|x|} = -1$.

51. (a) $\lim\limits_{x \to 400^+} \lfloor x \rfloor = 400$. Just observe that if $400 < x < 401$, then $\lfloor x \rfloor = 400$. Thus if we choose $\delta = 1$, we have for any number $\epsilon > 0$ that $400 < x < 400 + \delta \Rightarrow |\lfloor x \rfloor - 400| = |400 - 400| = 0 < \epsilon$.
(b) $\lim\limits_{x \to 400^-} \lfloor x \rfloor = 399$. Just observe that if $399 < x < 400$ then $\lfloor x \rfloor = 399$. Thus if we choose $\delta = 1$, we have for any number $\epsilon > 0$ that $400 - \delta < x < 400 \Rightarrow |\lfloor x \rfloor - 399| = |399 - 399| = 0 < \epsilon$.
(c) Since $\lim\limits_{x \to 400^+} \lfloor x \rfloor \neq \lim\limits_{x \to 400^-} \lfloor x \rfloor$ we conclude that $\lim\limits_{x \to 400} \lfloor x \rfloor$ does not exist.

2.5 CONTINUITY

1. No, discontinuous at $x = 2$, not defined at $x = 2$

3. Continuous on $[-1, 3]$

5. (a) Yes
(b) Yes, $\lim\limits_{x \to -1^+} f(x) = 0$
(c) Yes
(d) Yes

7. (a) No
(b) No

9. $f(2) = 0$, since $\lim\limits_{x \to 2^-} f(x) = -2(2) + 4 = 0 = \lim\limits_{x \to 2^+} f(x)$

11. Nonremovable discontinuity at $x = 1$ because $\lim_{x \to 1} f(x)$ fails to exist ($\lim_{x \to 1^-} f(x) = 1$ and $\lim_{x \to 1^+} f(x) = 0$).
Removable discontinuity at $x = 0$ by assigning the number $\lim_{x \to 0} f(x) = 0$ to be the value of $f(0)$ rather than $f(0) = 1$.

13. Discontinuous only when $x - 2 = 0 \Rightarrow x = 2$

15. Discontinuous only when $x^2 - 4x + 3 = 0 \Rightarrow (x - 3)(x - 1) = 0 \Rightarrow x = 3$ or $x = 1$

17. Continuous everywhere. ($|x - 1| + \sin x$ defined for all x; limits exist and are equal to function values.)

19. Discontinuous only at $x = 0$

21. Discontinuous when 2x is an integer multiple of π, i.e., $2x = n\pi$, n an integer $\Rightarrow x = \frac{n\pi}{2}$, n an integer, but continuous at all other x.

23. Discontinuous at odd integer multiples of $\frac{\pi}{2}$, i.e., $x = (2n - 1)\frac{\pi}{2}$, n an integer, but continuous at all other x.

25. Discontinuous when $2x + 3 < 0$ or $x < -\frac{3}{2} \Rightarrow$ continuous on the interval $\left[-\frac{3}{2}, \infty\right)$.

27. Continuous everywhere: $(2x - 1)^{1/3}$ is defined for all x; limits exist and are equal to function values.

29. Continuous everywhere since $\lim_{x \to 3} \frac{x^2 - x - 6}{x - 3} = \lim_{x \to 3} \frac{(x - 3)(x + 2)}{x - 3} = \lim_{x \to 3} (x + 2) = 5 = g(3)$

31. $\lim_{x \to \pi} \sin(x - \sin x) = \sin(\pi - \sin \pi) = \sin(\pi - 0) = \sin \pi = 0$, and function continuous at $x = \pi$.

33. $\lim_{y \to 1} \sec(y \sec^2 y - \tan^2 y - 1) = \lim_{y \to 1} \sec(y \sec^2 y - \sec^2 y) = \lim_{y \to 1} \sec((y - 1)\sec^2 y) = \sec((1 - 1)\sec^2 1)$
$= \sec 0 = 1$, and function continuous at $y = 1$.

35. $\lim_{t \to 0} \cos\left[\frac{\pi}{\sqrt{19 - 3 \sec 2t}}\right] = \cos\left[\frac{\pi}{\sqrt{19 - 3 \sec 0}}\right] = \cos \frac{\pi}{\sqrt{16}} = \cos \frac{\pi}{4} = \frac{\sqrt{2}}{2}$, and function continuous at $t = 0$.

37. $g(x) = \frac{x^2 - 9}{x - 3} = \frac{(x + 3)(x - 3)}{(x - 3)} = x + 3, x \neq 3 \Rightarrow g(3) = \lim_{x \to 3} (x + 3) = 6$

39. $f(s) = \frac{s^3 - 1}{s^2 - 1} = \frac{(s^2 + s + 1)(s - 1)}{(s + 1)(s - 1)} = \frac{s^2 + s + 1}{s + 1}, s \neq 1 \Rightarrow f(1) = \lim_{s \to 1} \left(\frac{s^2 + s + 1}{s + 1}\right) = \frac{3}{2}$

41. As defined, $\lim_{x \to 3^-} f(x) = (3)^2 - 1 = 8$ and $\lim_{x \to 3^+} (2a)(3) = 6a$. For f(x) to be continuous we must have $6a = 8 \Rightarrow a = \frac{4}{3}$.

43. As defined, $\lim_{x \to 2^-} f(x) = 12$ and $\lim_{x \to 2^+} f(x) = a^2(2) - 2a = 2a^2 - 2a$. For f(x) to be continuous we must have $12 = 2a^2 - 2a \Rightarrow a = 3$ or $a = -2$.

45. As defined, $\lim_{x \to -1^-} f(x) = -2$ and $\lim_{x \to -1^+} f(x) = a(-1) + b = -a + b$, and $\lim_{x \to 1^-} f(x) = a(1) + b = a + b$ and $\lim_{x \to 1^+} f(x) = 3$. For f(x) to be continuous we must have $-2 = -a + b$ and $a + b = 3 \Rightarrow a = \frac{5}{2}$ and $b = \frac{1}{2}$.

47. The function can be extended: $f(0) \approx 2.3$.

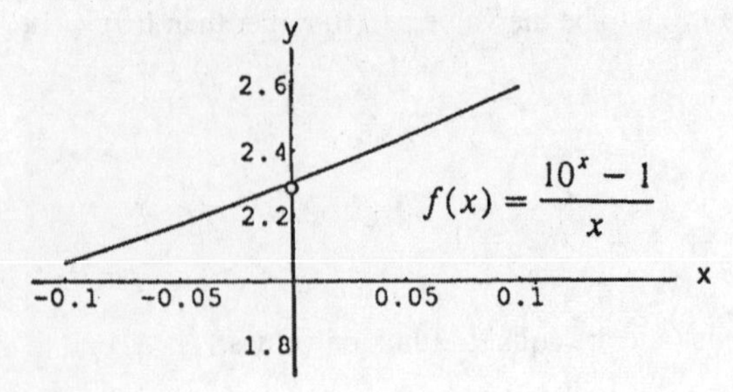

49. The function cannot be extended to be continuous at $x = 0$. If $f(0) = 1$, it will be continuous from the right. Or if $f(0) = -1$, it will be continuous from the left.

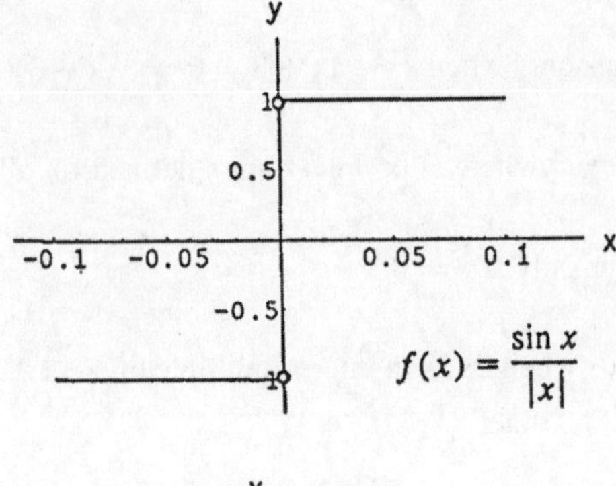

51. $f(x)$ is continuous on $[0, 1]$ and $f(0) < 0$, $f(1) > 0$ $\Rightarrow$ by the Intermediate Value Theorem $f(x)$ takes on every value between $f(0)$ and $f(1)$ $\Rightarrow$ the equation $f(x) = 0$ has at least one solution between $x = 0$ and $x = 1$.

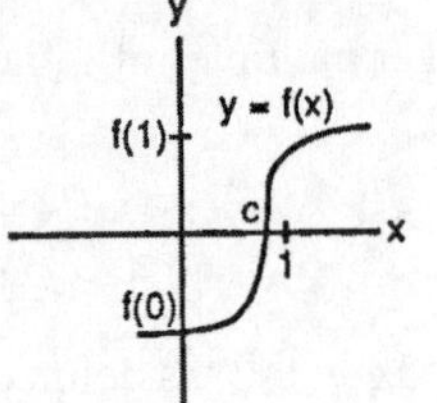

53. Let $f(x) = x^3 - 15x + 1$, which is continuous on $[-4, 4]$. Then $f(-4) = -3$, $f(-1) = 15$, $f(1) = -13$, and $f(4) = 5$. By the Intermediate Value Theorem, $f(x) = 0$ for some x in each of the intervals $-4 < x < -1$, $-1 < x < 1$, and $1 < x < 4$. That is, $x^3 - 15x + 1 = 0$ has three solutions in $[-4, 4]$. Since a polynomial of degree 3 can have at most 3 solutions, these are the only solutions.

55. Answers may vary. Note that f is continuous for every value of x.
 (a) $f(0) = 10$, $f(1) = 1^3 - 8(1) + 10 = 3$. Since $3 < \pi < 10$, by the Intermediate Value Theorem, there exists a c so that $0 < c < 1$ and $f(c) = \pi$.
 (b) $f(0) = 10$, $f(-4) = (-4)^3 - 8(-4) + 10 = -22$. Since $-22 < -\sqrt{3} < 10$, by the Intermediate Value Theorem, there exists a c so that $-4 < c < 0$ and $f(c) = -\sqrt{3}$.
 (c) $f(0) = 10$, $f(1000) = (1000)^3 - 8(1000) + 10 = 999{,}992{,}010$. Since $10 < 5{,}000{,}000 < 999{,}992{,}010$, by the Intermediate Value Theorem, there exists a c so that $0 < c < 1000$ and $f(c) = 5{,}000{,}000$.

57. Answers may vary. For example, $f(x) = \frac{\sin(x-2)}{x-2}$ is discontinuous at $x = 2$ because it is not defined there. However, the discontinuity can be removed because f has a limit (namely 1) as $x \to 2$.

59. (a) Suppose x_0 is rational $\Rightarrow f(x_0) = 1$. Choose $\epsilon = \frac{1}{2}$. For any $\delta > 0$ there is an irrational number x (actually infinitely many) in the interval $(x_0 - \delta, x_0 + \delta) \Rightarrow f(x) = 0$. Then $0 < |x - x_0| < \delta$ but $|f(x) - f(x_0)| = 1 > \frac{1}{2} = \epsilon$, so $\lim_{x \to x_0} f(x)$ fails to exist $\Rightarrow$ f is discontinuous at x_0 rational.
 On the other hand, x_0 irrational $\Rightarrow f(x_0) = 0$ and there is a rational number x in $(x_0 - \delta, x_0 + \delta) \Rightarrow f(x) = 1$. Again $\lim_{x \to x_0} f(x)$ fails to exist $\Rightarrow$ f is discontinuous at x_0 irrational. That is, f is discontinuous at every point.
 (b) f is neither right-continuous nor left-continuous at any point x_0 because in every interval $(x_0 - \delta, x_0)$ or $(x_0, x_0 + \delta)$ there exist both rational and irrational real numbers. Thus neither limits $\lim_{x \to x_0^-} f(x)$ and $\lim_{x \to x_0^+} f(x)$ exist by the same arguments used in part (a).

61. No. For instance, if $f(x) = 0$, $g(x) = \lceil x \rceil$, then $h(x) = 0(\lceil x \rceil) = 0$ is continuous at $x = 0$ and $g(x)$ is not.

63. Yes, because of the Intermediate Value Theorem. If f(a) and f(b) did have different signs then f would have to equal zero at some point between a and b since f is continuous on [a, b].

65. If $f(0) = 0$ or $f(1) = 1$, we are done (i.e., $c = 0$ or $c = 1$ in those cases). Then let $f(0) = a > 0$ and $f(1) = b < 1$ because $0 \le f(x) \le 1$. Define $g(x) = f(x) - x \Rightarrow$ g is continuous on [0, 1]. Moreover, $g(0) = f(0) - 0 = a > 0$ and $g(1) = f(1) - 1 = b - 1 < 0 \Rightarrow$ by the Intermediate Value Theorem there is a number c in (0, 1) such that $g(c) = 0 \Rightarrow f(c) - c = 0$ or $f(c) = c$.

67. By Exercises 52 in Section 2.3, we have $\lim_{x \to c} f(x) = L \Leftrightarrow \lim_{h \to 0} f(c+h) = L$.
Thus, f(x) is continuous at $x = c \Leftrightarrow \lim_{x \to c} f(x) = f(c) \Leftrightarrow \lim_{h \to 0} f(c+h) = f(c)$.

69. $x \approx 1.8794, -1.5321, -0.3473$

71. $x \approx 1.7549$

73. $x \approx 3.5156$

75. $x \approx 0.7391$

2.6 LIMITS INVOLVING INFINITY; ASYMPTOTES OF GRAPHS

1. (a) $\lim_{x \to 2} f(x) = 0$
(b) $\lim_{x \to -3^+} f(x) = -2$
(c) $\lim_{x \to -3^-} f(x) = 2$
(d) $\lim_{x \to -3} f(x) =$ does not exist
(e) $\lim_{x \to 0^+} f(x) = -1$
(f) $\lim_{x \to 0^-} f(x) = +\infty$
(g) $\lim_{x \to 0} f(x) =$ does not exist
(h) $\lim_{x \to \infty} f(x) = 1$
(i) $\lim_{x \to -\infty} f(x) = 0$

Note: In these exercises we use the result $\lim_{x \to \pm\infty} \frac{1}{x^{m/n}} = 0$ whenever $\frac{m}{n} > 0$. This result follows immediately from Theorem 8 and the power rule in Theorem 1: $\lim_{x \to \pm\infty} \left(\frac{1}{x^{m/n}}\right) = \lim_{x \to \pm\infty} \left(\frac{1}{x}\right)^{m/n} = \left(\lim_{x \to \pm\infty} \frac{1}{x}\right)^{m/n} = 0^{m/n} = 0$.

3. (a) -3
(b) -3

5. (a) $\frac{1}{2}$
(b) $\frac{1}{2}$

7. (a) $-\frac{5}{3}$
(b) $-\frac{5}{3}$

9. $-\frac{1}{x} \le \frac{\sin 2x}{x} \le \frac{1}{x} \Rightarrow \lim_{x \to \infty} \frac{\sin 2x}{x} = 0$ by the Sandwich Theorem

11. $\lim_{t \to \infty} \frac{2-t+\sin t}{t+\cos t} = \lim_{t \to \infty} \frac{\frac{2}{t} - 1 + \left(\frac{\sin t}{t}\right)}{1 + \left(\frac{\cos t}{t}\right)} = \frac{0-1+0}{1+0} = -1$

13. (a) $\lim_{x \to \infty} \frac{2x+3}{5x+7} = \lim_{x \to \infty} \frac{2+\frac{3}{x}}{5+\frac{7}{x}} = \frac{2}{5}$
(b) $\frac{2}{5}$ (same process as part (a))

15. (a) $\lim_{x \to \infty} \frac{x+1}{x^2+3} = \lim_{x \to \infty} \frac{\frac{1}{x}+\frac{1}{x^2}}{1+\frac{3}{x^2}} = 0$
(b) 0 (same process as part (a))

17. (a) $\lim_{x \to \infty} \frac{7x^3}{x^3 - 3x^2 + 6x} = \lim_{x \to \infty} \frac{7}{1 - \frac{3}{x} + \frac{6}{x^2}} = 7$
(b) 7 (same process as part (a))

19. (a) $\lim_{x\to\infty} \frac{10x^5+x^4+31}{x^6} = \lim_{x\to\infty} \frac{\frac{10}{x}+\frac{1}{x^2}+\frac{31}{x^6}}{1} = 0$ (b) 0 (same process as part (a))

21. (a) $\lim_{x\to\infty} \frac{-2x^3-2x+3}{3x^3+3x^2-5x} = \lim_{x\to\infty} \frac{-2-\frac{2}{x^2}+\frac{3}{x^3}}{3+\frac{3}{x}-\frac{5}{x^2}} = -\frac{2}{3}$

(b) $-\frac{2}{3}$ (same process as part (a))

23. $\lim_{x\to\infty} \sqrt{\frac{8x^2-3}{2x^2+x}} = \lim_{x\to\infty} \sqrt{\frac{8-\frac{3}{x^2}}{2+\frac{1}{x}}} = \sqrt{\lim_{x\to\infty} \frac{8-\frac{3}{x^2}}{2+\frac{1}{x}}} = \sqrt{\frac{8-0}{2+0}} = \sqrt{4} = 2$

25. $\lim_{x\to-\infty} \left(\frac{1-x^3}{x^2-7x}\right)^5 = \lim_{x\to-\infty} \left(\frac{\frac{1}{x^2}-x}{1-\frac{7}{x}}\right)^5 = \left(\lim_{x\to-\infty} \frac{\frac{1}{x^2}-x}{1-\frac{7}{x}}\right)^5 = \left(\frac{0+\infty}{1-0}\right)^5 = \infty$

27. $\lim_{x\to\infty} \frac{2\sqrt{x}+x^{-1}}{3x-7} = \lim_{x\to\infty} \frac{\left(\frac{2}{x^{1/2}}\right)+\left(\frac{1}{x^2}\right)}{3-\frac{7}{x}} = 0$

29. $\lim_{x\to-\infty} \frac{\sqrt[3]{x}-\sqrt[5]{x}}{\sqrt[3]{x}+\sqrt[5]{x}} = \lim_{x\to-\infty} \frac{1-x^{(1/5)-(1/3)}}{1+x^{(1/5)-(1/3)}} = \lim_{x\to-\infty} \frac{1-\left(\frac{1}{x^{2/15}}\right)}{1+\left(\frac{1}{x^{2/15}}\right)} = 1$

31. $\lim_{x\to\infty} \frac{2x^{5/3}-x^{1/3}+7}{x^{8/5}+3x+\sqrt{x}} = \lim_{x\to\infty} \frac{2x^{1/15}-\frac{1}{x^{19/15}}+\frac{7}{x^{8/5}}}{1+\frac{3}{x^{3/5}}+\frac{1}{x^{11/10}}} = \infty$

33. $\lim_{x\to\infty} \frac{\sqrt{x^2+1}}{x+1} = \lim_{x\to\infty} \frac{\sqrt{x^2+1}/\sqrt{x^2}}{(x+1)/\sqrt{x^2}} = \lim_{x\to\infty} \frac{\sqrt{(x^2+1)/x^2}}{(x+1)/x} = \lim_{x\to\infty} \frac{\sqrt{1+1/x^2}}{(1+1/x)} = \frac{\sqrt{1+0}}{(1+0)} = 1$

35. $\lim_{x\to\infty} \frac{x-3}{\sqrt{4x^2+25}} = \lim_{x\to\infty} \frac{(x-3)/\sqrt{x^2}}{\sqrt{4x^2+25}/\sqrt{x^2}} = \lim_{x\to\infty} \frac{(x-3)/x}{\sqrt{(4x^2+25)/x^2}} = \lim_{x\to\infty} \frac{(1-3/x)}{\sqrt{4+25/x^2}} = \frac{(1-0)}{\sqrt{4+0}} = \frac{1}{2}$

37. $\lim_{x\to 0^+} \frac{1}{3x} = \infty$ $\left(\frac{\text{positive}}{\text{positive}}\right)$

39. $\lim_{x\to 2^-} \frac{3}{x-2} = -\infty$ $\left(\frac{\text{positive}}{\text{negative}}\right)$

41. $\lim_{x\to -8^+} \frac{2x}{x+8} = -\infty$ $\left(\frac{\text{negative}}{\text{positive}}\right)$

43. $\lim_{x\to 7} \frac{4}{(x-7)^2} = \infty$ $\left(\frac{\text{positive}}{\text{positive}}\right)$

45. (a) $\lim_{x\to 0^+} \frac{2}{3x^{1/3}} = \infty$ (b) $\lim_{x\to 0^-} \frac{2}{3x^{1/3}} = -\infty$

47. $\lim_{x\to 0} \frac{4}{x^{2/5}} = \lim_{x\to 0} \frac{4}{(x^{1/5})^2} = \infty$

49. $\lim_{x\to \left(\frac{\pi}{2}\right)^-} \tan x = \infty$

51. $\lim_{\theta\to 0^-} (1+\csc\theta) = -\infty$

53. (a) $\lim_{x\to 2^+} \frac{1}{x^2-4} = \lim_{x\to 2^+} \frac{1}{(x+2)(x-2)} = \infty$ $\left(\frac{1}{\text{positive}\cdot\text{positive}}\right)$

(b) $\lim_{x\to 2^-} \frac{1}{x^2-4} = \lim_{x\to 2^-} \frac{1}{(x+2)(x-2)} = -\infty$ $\left(\frac{1}{\text{positive}\cdot\text{negative}}\right)$

(c) $\lim_{x\to -2^+} \frac{1}{x^2-4} = \lim_{x\to -2^+} \frac{1}{(x+2)(x-2)} = -\infty$ $\left(\frac{1}{\text{positive}\cdot\text{negative}}\right)$

(d) $\lim_{x\to -2^-} \frac{1}{x^2-4} = \lim_{x\to -2^-} \frac{1}{(x+2)(x-2)} = \infty$ $\left(\frac{1}{\text{negative}\cdot\text{negative}}\right)$

55. (a) $\lim_{x\to 0^+} \frac{x^2}{2} - \frac{1}{x} = 0 + \lim_{x\to 0^+} \frac{1}{-x} = -\infty$ $\left(\frac{1}{\text{negative}}\right)$

(b) $\lim_{x\to 0^-} \frac{x^2}{2} - \frac{1}{x} = 0 + \lim_{x\to 0^-} \frac{1}{-x} = \infty$ $\left(\frac{1}{\text{positive}}\right)$

(c) $\lim\limits_{x \to \sqrt[3]{2}} \frac{x^2}{2} - \frac{1}{x} = \frac{2^{2/3}}{2} - \frac{1}{2^{1/3}} = 2^{-1/3} - 2^{-1/3} = 0$

(d) $\lim\limits_{x \to -1} \frac{x^2}{2} - \frac{1}{x} = \frac{1}{2} - \left(\frac{1}{-1}\right) = \frac{3}{2}$

57. (a) $\lim\limits_{x \to 0^+} \frac{x^2-3x+2}{x^3-2x^2} = \lim\limits_{x \to 0^+} \frac{(x-2)(x-1)}{x^2(x-2)} = -\infty$ $\left(\frac{\text{negative}\cdot\text{negative}}{\text{positive}\cdot\text{negative}}\right)$

(b) $\lim\limits_{x \to 2^+} \frac{x^2-3x+2}{x^3-2x^2} = \lim\limits_{x \to 2^+} \frac{(x-2)(x-1)}{x^2(x-2)} = \lim\limits_{x \to 2^+} \frac{x-1}{x^2} = \frac{1}{4}, x \neq 2$

(c) $\lim\limits_{x \to 2^-} \frac{x^2-3x+2}{x^3-2x^2} = \lim\limits_{x \to 2^-} \frac{(x-2)(x-1)}{x^2(x-2)} = \lim\limits_{x \to 2^-} \frac{x-1}{x^2} = \frac{1}{4}, x \neq 2$

(d) $\lim\limits_{x \to 2} \frac{x^2-3x+2}{x^3-2x^2} = \lim\limits_{x \to 2} \frac{(x-2)(x-1)}{x^2(x-2)} = \lim\limits_{x \to 2} \frac{x-1}{x^2} = \frac{1}{4}, x \neq 2$

(e) $\lim\limits_{x \to 0} \frac{x^2-3x+2}{x^3-2x^2} = \lim\limits_{x \to 0} \frac{(x-2)(x-1)}{x^2(x-2)} = -\infty$ $\left(\frac{\text{negative}\cdot\text{negative}}{\text{positive}\cdot\text{negative}}\right)$

59. (a) $\lim\limits_{t \to 0^+} \left[2 - \frac{3}{t^{1/3}}\right] = -\infty$

(b) $\lim\limits_{t \to 0^-} \left[2 - \frac{3}{t^{1/3}}\right] = \infty$

61. (a) $\lim\limits_{x \to 0^+} \left[\frac{1}{x^{2/3}} + \frac{2}{(x-1)^{2/3}}\right] = \infty$

(b) $\lim\limits_{x \to 0^-} \left[\frac{1}{x^{2/3}} + \frac{2}{(x-1)^{2/3}}\right] = \infty$

(c) $\lim\limits_{x \to 1^+} \left[\frac{1}{x^{2/3}} + \frac{2}{(x-1)^{2/3}}\right] = \infty$

(d) $\lim\limits_{x \to 1^-} \left[\frac{1}{x^{2/3}} + \frac{2}{(x-1)^{2/3}}\right] = \infty$

63. $y = \frac{1}{x-1}$

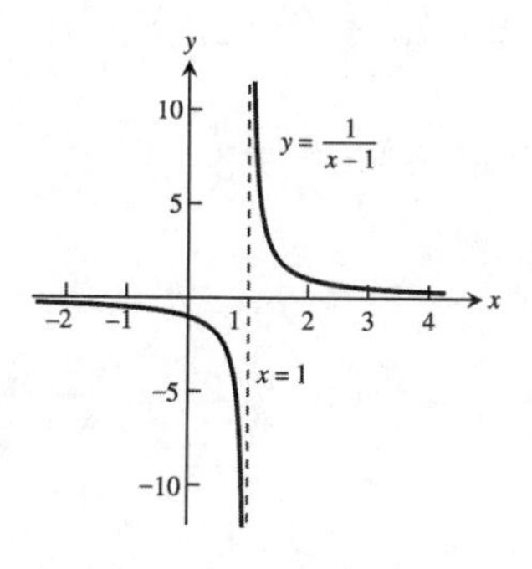

65. $y = \frac{1}{2x+4}$

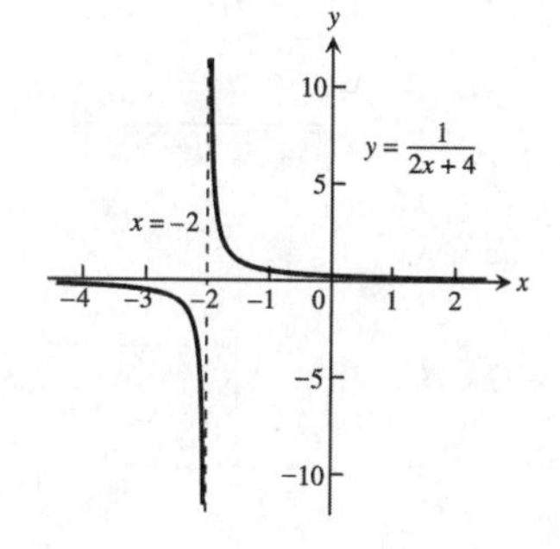

67. $y = \frac{x+3}{x+2} = 1 + \frac{1}{x+2}$

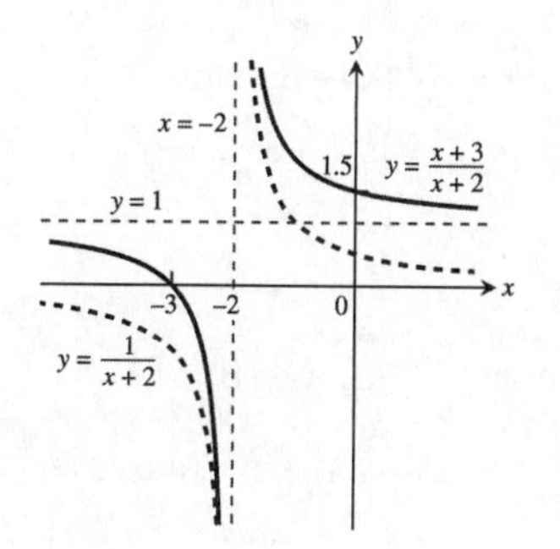

69. Here is one possibility.

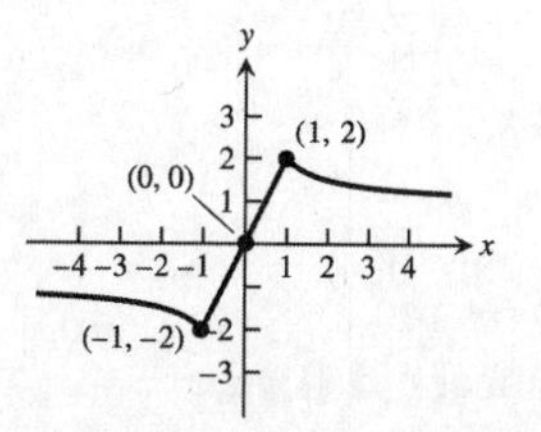

71. Here is one possibility.

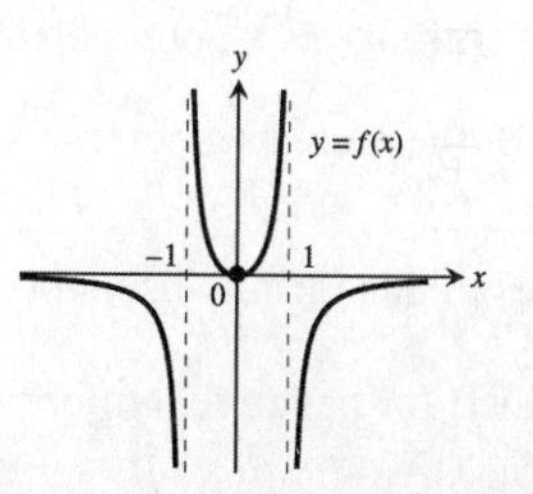

73. Here is one possibility.

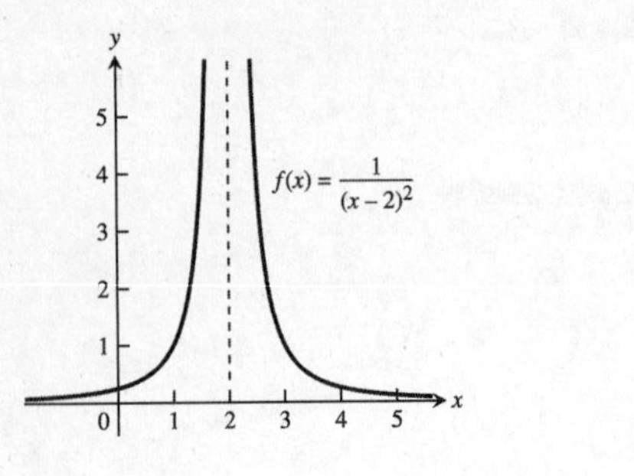

75. Here is one possibility.

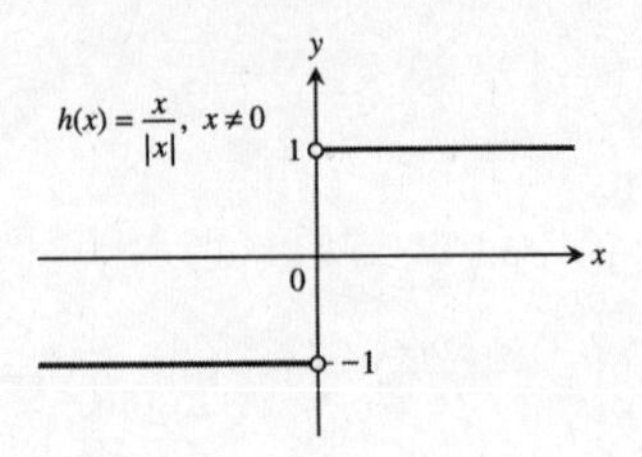

77. Yes. If $\lim_{x \to \infty} \frac{f(x)}{g(x)} = 2$ then the ratio of the polynomials' leading coefficients is 2, so $\lim_{x \to -\infty} \frac{f(x)}{g(x)} = 2$ as well.

79. At most 1 horizontal asymptote: If $\lim_{x \to \infty} \frac{f(x)}{g(x)} = L$, then the ratio of the polynomials' leading coefficients is L, so $\lim_{x \to -\infty} \frac{f(x)}{g(x)} = L$ as well.

81. $\lim_{x \to \infty} \left(\sqrt{x^2+25} - \sqrt{x^2-1}\right) = \lim_{x \to \infty} \left[\sqrt{x^2+25} - \sqrt{x^2-1}\right] \cdot \left[\frac{\sqrt{x^2+25}+\sqrt{x^2-1}}{\sqrt{x^2+25}+\sqrt{x^2-1}}\right] = \lim_{x \to \infty} \frac{(x^2+25)-(x^2-1)}{\sqrt{x^2+25}+\sqrt{x^2-1}}$

$= \lim_{x \to \infty} \frac{26}{\sqrt{x^2+25}+\sqrt{x^2-1}} = \lim_{x \to \infty} \frac{\frac{26}{x}}{\sqrt{1+\frac{25}{x^2}}+\sqrt{1-\frac{1}{x^2}}} = \frac{0}{1+1} = 0$

83. $\lim_{x \to -\infty} \left(2x + \sqrt{4x^2+3x-2}\right) = \lim_{x \to -\infty} \left[2x + \sqrt{4x^2+3x-2}\right] \cdot \left[\frac{2x - \sqrt{4x^2+3x-2}}{2x - \sqrt{4x^2+3x-2}}\right] = \lim_{x \to -\infty} \frac{(4x^2)-(4x^2+3x-2)}{2x - \sqrt{4x^2+3x-2}}$

$= \lim_{x \to -\infty} \frac{-3x+2}{2x - \sqrt{4x^2+3x-2}} = \lim_{x \to -\infty} \frac{\frac{-3x+2}{\sqrt{x^2}}}{\frac{2x}{\sqrt{x^2}} - \sqrt{4+\frac{3}{x}-\frac{2}{x^2}}} = \lim_{x \to -\infty} \frac{\frac{-3x+2}{-x}}{\frac{2x}{-x} - \sqrt{4+\frac{3}{x}-\frac{2}{x^2}}} = \lim_{x \to -\infty} \frac{3-\frac{2}{x}}{-2-\sqrt{4+\frac{3}{x}-\frac{2}{x^2}}}$

$= \frac{3-0}{-2-2} = -\frac{3}{4}$

85. $\lim_{x \to \infty} \left(\sqrt{x^2+3x} - \sqrt{x^2-2x}\right) = \lim_{x \to \infty} \left[\sqrt{x^2+3x} - \sqrt{x^2-2x}\right] \cdot \left[\frac{\sqrt{x^2+3x}+\sqrt{x^2-2x}}{\sqrt{x^2+3x}+\sqrt{x^2-2x}}\right] = \lim_{x \to \infty} \frac{(x^2+3x)-(x^2-2x)}{\sqrt{x^2+3x}+\sqrt{x^2-2x}}$

$= \lim_{x \to \infty} \frac{5x}{\sqrt{x^2+3x}+\sqrt{x^2-2x}} = \lim_{x \to \infty} \frac{5}{\sqrt{1+\frac{3}{x}}+\sqrt{1-\frac{2}{x}}} = \frac{5}{1+1} = \frac{5}{2}$

87. For any $\epsilon > 0$, take $N = 1$. Then for all $x > N$ we have that $|f(x) - k| = |k - k| = 0 < \epsilon$.

89. For every real number $-B < 0$, we must find a $\delta > 0$ such that for all x, $0 < |x - 0| < \delta \Rightarrow \frac{-1}{x^2} < -B$. Now, $-\frac{1}{x^2} < -B < 0 \Leftrightarrow \frac{1}{x^2} > B > 0 \Leftrightarrow x^2 < \frac{1}{B} \Leftrightarrow |x| < \frac{1}{\sqrt{B}}$. Choose $\delta = \frac{1}{\sqrt{B}}$, then $0 < |x| < \delta \Rightarrow |x| < \frac{1}{\sqrt{B}}$ $\Rightarrow \frac{-1}{x^2} < -B$ so that $\lim_{x \to 0} -\frac{1}{x^2} = -\infty$.

91. For every real number $-B < 0$, we must find a $\delta > 0$ such that for all x, $0 < |x - 3| < \delta \Rightarrow \frac{-2}{(x-3)^2} < -B$. Now, $\frac{-2}{(x-3)^2} < -B < 0 \Leftrightarrow \frac{2}{(x-3)^2} > B > 0 \Leftrightarrow \frac{(x-3)^2}{2} < \frac{1}{B} \Leftrightarrow (x-3)^2 < \frac{2}{B} \Leftrightarrow 0 < |x-3| < \sqrt{\frac{2}{B}}$. Choose $\delta = \sqrt{\frac{2}{B}}$, then $0 < |x - 3| < \delta \Rightarrow \frac{-2}{(x-3)^2} < -B < 0$ so that $\lim_{x \to 3} \frac{-2}{(x-3)^2} = -\infty$.

93. (a) We say that $f(x)$ approaches infinity as x approaches x_0 from the left, and write $\lim_{x \to x_0^-} f(x) = \infty$, if for every positive number B, there exists a corresponding number $\delta > 0$ such that for all x, $x_0 - \delta < x < x_0 \Rightarrow f(x) > B$.

(b) We say that $f(x)$ approaches minus infinity as x approaches x_0 from the right, and write $\lim_{x \to x_0^+} f(x) = -\infty$, if for every positive number B (or negative number $-B$) there exists a corresponding number $\delta > 0$ such that for all x, $x_0 < x < x_0 + \delta \Rightarrow f(x) < -B$.

(c) We say that f(x) approaches minus infinity as x approaches x_0 from the left, and write $\lim_{x \to x_0^-} f(x) = -\infty$, if for every positive number B (or negative number $-B$) there exists a corresponding number $\delta > 0$ such that for all x, $x_0 - \delta < x < x_0 \Rightarrow f(x) < -B$.

95. For $B > 0$, $\frac{1}{x} < -B < 0 \Leftrightarrow -\frac{1}{x} > B > 0 \Leftrightarrow -x < \frac{1}{B} \Leftrightarrow -\frac{1}{B} < x$. Choose $\delta = \frac{1}{B}$. Then $-\delta < x < 0$ $\Rightarrow -\frac{1}{B} < x \Rightarrow \frac{1}{x} < -B$ so that $\lim_{x \to 0^-} \frac{1}{x} = -\infty$.

97. For $B > 0$, $\frac{1}{x-2} > B \Leftrightarrow 0 < x - 2 < \frac{1}{B}$. Choose $\delta = \frac{1}{B}$. Then $2 < x < 2 + \delta \Rightarrow 0 < x - 2 < \delta \Rightarrow 0 < x - 2 < \frac{1}{B}$ $\Rightarrow \frac{1}{x-2} > B > 0$ so that $\lim_{x \to 2^+} \frac{1}{x-2} = \infty$.

99. $y = \frac{x^2}{x-1} = x + 1 + \frac{1}{x-1}$

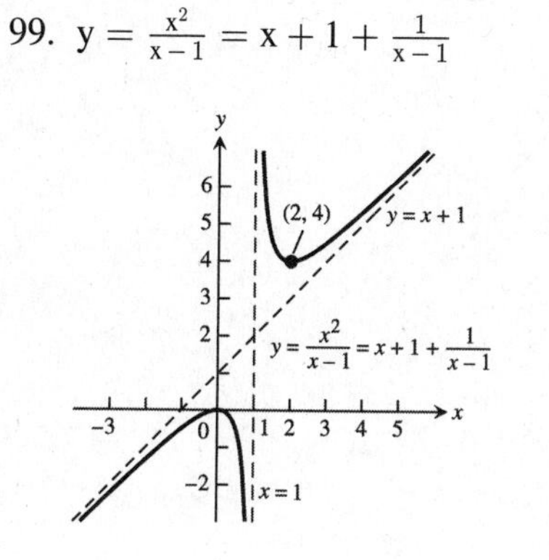

101. $y = \frac{x^2-4}{x-1} = x + 1 - \frac{3}{x-1}$

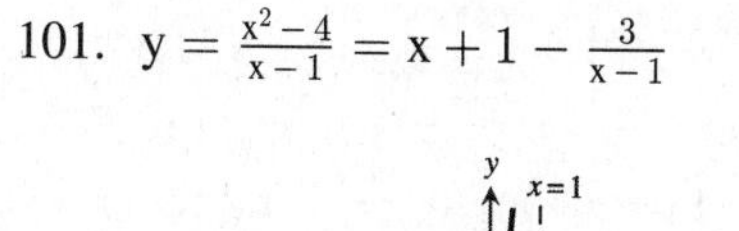

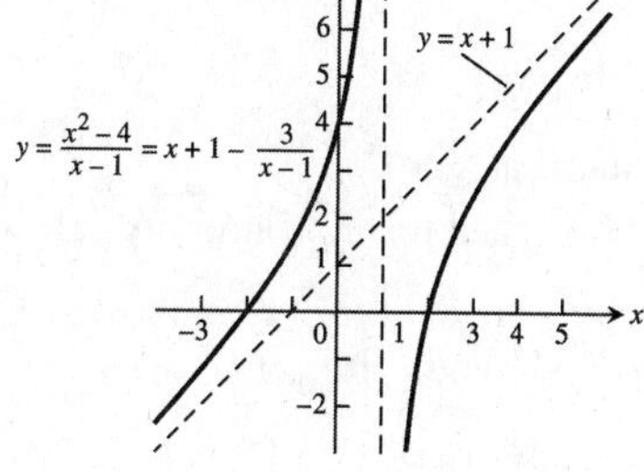

103. $y = \frac{x^2-1}{x} = x - \frac{1}{x}$

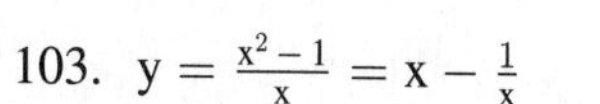

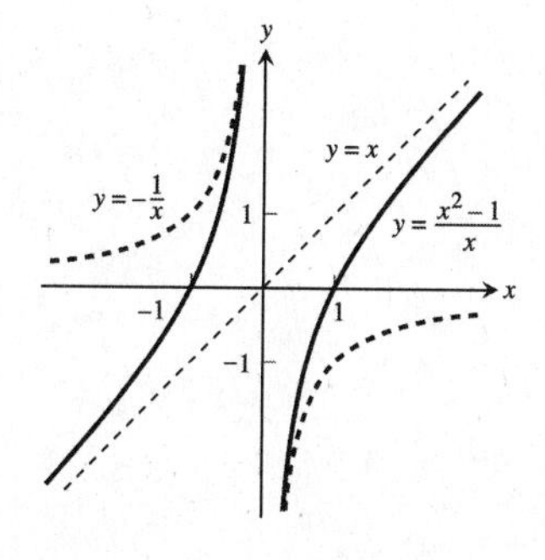

105. $y = \frac{x}{\sqrt{4-x^2}}$

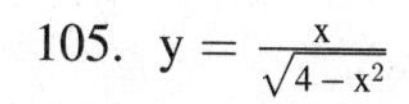

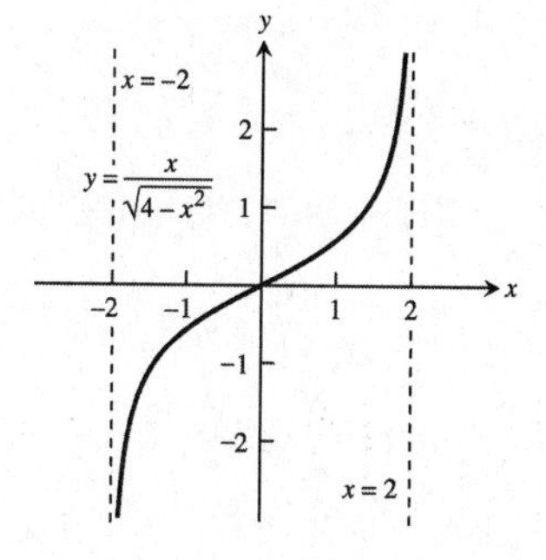

107. $y = x^{2/3} + \frac{1}{x^{1/3}}$

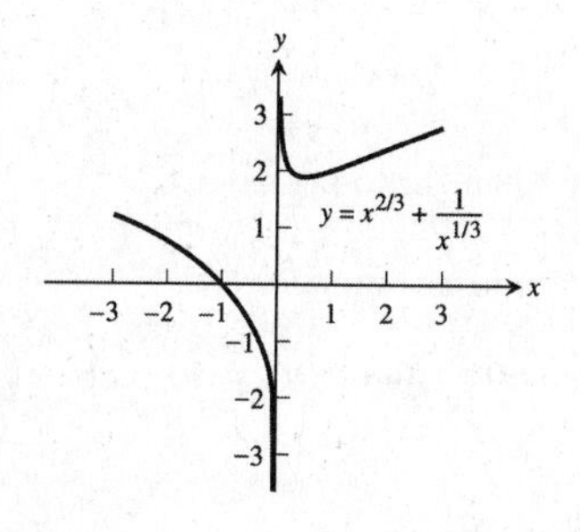

109. (a) $y \to \infty$ (see accompanying graph)
(b) $y \to \infty$ (see accompanying graph)
(c) cusps at $x = \pm 1$ (see accompanying graph)

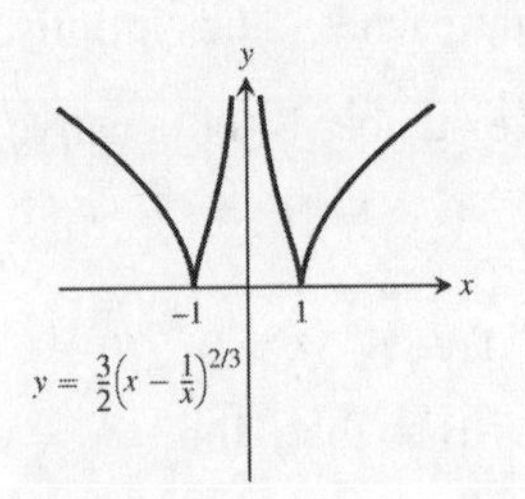

CHAPTER 2 PRACTICE EXERCISES

1. At $x = -1$: $\lim\limits_{x \to -1^-} f(x) = \lim\limits_{x \to -1^+} f(x) = 1$
$\Rightarrow \lim\limits_{x \to -1} f(x) = 1 = f(-1)$
$\Rightarrow$ f is continuous at $x = -1$.
At $x = 0$: $\lim\limits_{x \to 0^-} f(x) = \lim\limits_{x \to 0^+} f(x) = 0 \Rightarrow \lim\limits_{x \to 0} f(x) = 0$.
But $f(0) = 1 \neq \lim\limits_{x \to 0} f(x)$
$\Rightarrow$ f is discontinuous at $x = 0$.
If we define $f(0) = 0$, then the discontinuity at $x = 0$ is removable.
At $x = 1$: $\lim\limits_{x \to 1^-} f(x) = -1$ and $\lim\limits_{x \to 1^+} f(x) = 1$
$\Rightarrow \lim\limits_{x \to 1} f(x)$ does not exist
$\Rightarrow$ f is discontinuous at $x = 1$.

3. (a) $\lim\limits_{t \to t_0} (3f(t)) = 3 \lim\limits_{t \to t_0} f(t) = 3(-7) = -21$
(b) $\lim\limits_{t \to t_0} (f(t))^2 = \left(\lim\limits_{t \to t_0} f(t)\right)^2 = (-7)^2 = 49$
(c) $\lim\limits_{t \to t_0} (f(t) \cdot g(t)) = \lim\limits_{t \to t_0} f(t) \cdot \lim\limits_{t \to t_0} g(t) = (-7)(0) = 0$
(d) $\lim\limits_{t \to t_0} \frac{f(t)}{g(t)-7} = \frac{\lim_{t \to t_0} f(t)}{\lim_{t \to t_0} (g(t)-7)} = \frac{\lim_{t \to t_0} f(t)}{\lim_{t \to t_0} g(t) - \lim_{t \to t_0} 7} = \frac{-7}{0-7} = 1$
(e) $\lim\limits_{t \to t_0} \cos(g(t)) = \cos\left(\lim\limits_{t \to t_0} g(t)\right) = \cos 0 = 1$
(f) $\lim\limits_{t \to t_0} |f(t)| = \left|\lim\limits_{t \to t_0} f(t)\right| = |-7| = 7$
(g) $\lim\limits_{t \to t_0} (f(t) + g(t)) = \lim\limits_{t \to t_0} f(t) + \lim\limits_{t \to t_0} g(t) = -7 + 0 = -7$
(h) $\lim\limits_{t \to t_0} \left(\frac{1}{f(t)}\right) = \frac{1}{\lim_{t \to t_0} f(t)} = \frac{1}{-7} = -\frac{1}{7}$

5. Since $\lim\limits_{x \to 0} x = 0$ we must have that $\lim\limits_{x \to 0} (4 - g(x)) = 0$. Otherwise, if $\lim\limits_{x \to 0} (4 - g(x))$ is a finite positive number, we would have $\lim\limits_{x \to 0^-} \left[\frac{4-g(x)}{x}\right] = -\infty$ and $\lim\limits_{x \to 0^+} \left[\frac{4-g(x)}{x}\right] = \infty$ so the limit could not equal 1 as $x \to 0$. Similar reasoning holds if $\lim\limits_{x \to 0} (4 - g(x))$ is a finite negative number. We conclude that $\lim\limits_{x \to 0} g(x) = 4$.

7. (a) $\lim\limits_{x \to c} f(x) = \lim\limits_{x \to c} x^{1/3} = c^{1/3} = f(c)$ for every real number c $\Rightarrow$ f is continuous on $(-\infty, \infty)$.
(b) $\lim\limits_{x \to c} g(x) = \lim\limits_{x \to c} x^{3/4} = c^{3/4} = g(c)$ for every nonnegative real number c $\Rightarrow$ g is continuous on $[0, \infty)$.
(c) $\lim\limits_{x \to c} h(x) = \lim\limits_{x \to c} x^{-2/3} = \frac{1}{c^{2/3}} = h(c)$ for every nonzero real number c $\Rightarrow$ h is continuous on $(-\infty, 0)$ and $(-\infty, \infty)$.
(d) $\lim\limits_{x \to c} k(x) = \lim\limits_{x \to c} x^{-1/6} = \frac{1}{c^{1/6}} = k(c)$ for every positive real number c $\Rightarrow$ k is continuous on $(0, \infty)$

9. (a) $\lim\limits_{x\to 0} \frac{x^2-4x+4}{x^3+5x^2-14x} = \lim\limits_{x\to 0} \frac{(x-2)(x-2)}{x(x+7)(x-2)} = \lim\limits_{x\to 0} \frac{x-2}{x(x+7)}$, $x \neq 2$; the limit does not exist because $\lim\limits_{x\to 0^-} \frac{x-2}{x(x+7)} = \infty$ and $\lim\limits_{x\to 0^+} \frac{x-2}{x(x+7)} = -\infty$

(b) $\lim\limits_{x\to 2} \frac{x^2-4x+4}{x^3+5x^2-14x} = \lim\limits_{x\to 2} \frac{(x-2)(x-2)}{x(x+7)(x-2)} = \lim\limits_{x\to 2} \frac{x-2}{x(x+7)}$, $x \neq 2$, and $\lim\limits_{x\to 2} \frac{x-2}{x(x+7)} = \frac{0}{2(9)} = 0$

11. $\lim\limits_{x\to 1} \frac{1-\sqrt{x}}{1-x} = \lim\limits_{x\to 1} \frac{1-\sqrt{x}}{(1-\sqrt{x})(1+\sqrt{x})} = \lim\limits_{x\to 1} \frac{1}{1+\sqrt{x}} = \frac{1}{2}$

13. $\lim\limits_{h\to 0} \frac{(x+h)^2-x^2}{h} = \lim\limits_{h\to 0} \frac{(x^2+2hx+h^2)-x^2}{h} = \lim\limits_{h\to 0} (2x+h) = 2x$

15. $\lim\limits_{x\to 0} \frac{\frac{1}{2+x}-\frac{1}{2}}{x} = \lim\limits_{x\to 0} \frac{2-(2+x)}{2x(2+x)} = \lim\limits_{x\to 0} \frac{-1}{4+2x} = -\frac{1}{4}$

17. $\lim\limits_{x\to 1} \frac{x^{1/3}-1}{\sqrt{x}-1} = \lim\limits_{x\to 1} \frac{(x^{1/3}-1)}{(\sqrt{x}-1)} \cdot \frac{(x^{2/3}+x^{1/3}+1)(\sqrt{x}+1)}{(\sqrt{x}+1)(x^{2/3}+x^{1/3}+1)} = \lim\limits_{x\to 1} \frac{(x-1)(\sqrt{x}+1)}{(x-1)(x^{2/3}+x^{1/3}+1)} = \lim\limits_{x\to 1} \frac{\sqrt{x}+1}{x^{2/3}+x^{1/3}+1}$
$= \frac{1+1}{1+1+1} = \frac{2}{3}$

19. $\lim\limits_{x\to 0} \frac{\tan 2x}{\tan \pi x} = \lim\limits_{x\to 0} \frac{\sin 2x}{\cos 2x} \cdot \frac{\cos \pi x}{\sin \pi x} = \lim\limits_{x\to 0} \left(\frac{\sin 2x}{2x}\right)\left(\frac{\cos \pi x}{\cos 2x}\right)\left(\frac{\pi x}{\sin \pi x}\right)\left(\frac{2x}{\pi x}\right) = 1 \cdot 1 \cdot 1 \cdot \frac{2}{\pi} = \frac{2}{\pi}$

21. $\lim\limits_{x\to \pi} \sin\left(\frac{x}{2} + \sin x\right) = \sin\left(\frac{\pi}{2} + \sin \pi\right) = \sin\left(\frac{\pi}{2}\right) = 1$

23. $\lim\limits_{x\to 0} \frac{8x}{3\sin x - x} = \lim\limits_{x\to 0} \frac{8}{3\frac{\sin x}{x} - 1} = \frac{8}{3(1)-1} = 4$

25. $\lim\limits_{x\to 0^+} [4\,g(x)]^{1/3} = 2 \Rightarrow \left[\lim\limits_{x\to 0^+} 4\,g(x)\right]^{1/3} = 2 \Rightarrow \lim\limits_{x\to 0^+} 4\,g(x) = 8$, since $2^3 = 8$. Then $\lim\limits_{x\to 0^+} g(x) = 2$.

27. $\lim\limits_{x\to 1} \frac{3x^2+1}{g(x)} = \infty \Rightarrow \lim\limits_{x\to 1} g(x) = 0$ since $\lim\limits_{x\to 1} (3x^2+1) = 4$

29. At $x = -1$: $\lim\limits_{x\to -1^-} f(x) = \lim\limits_{x\to -1^-} \frac{x(x^2-1)}{|x^2-1|}$
$= \lim\limits_{x\to -1^-} \frac{x(x^2-1)}{x^2-1} = \lim\limits_{x\to -1^-} x = -1$, and
$\lim\limits_{x\to -1^+} f(x) = \lim\limits_{x\to -1^+} \frac{x(x^2-1)}{|x^2-1|} = \lim\limits_{x\to -1^+} \frac{x(x^2-1)}{-(x^2-1)}$
$= \lim\limits_{x\to -1} (-x) = -(-1) = 1$. Since
$\lim\limits_{x\to -1^-} f(x) \neq \lim\limits_{x\to -1^+} f(x)$
$\Rightarrow \lim\limits_{x\to -1} f(x)$ does not exist, the function f cannot be extended to a continuous function at $x = -1$.

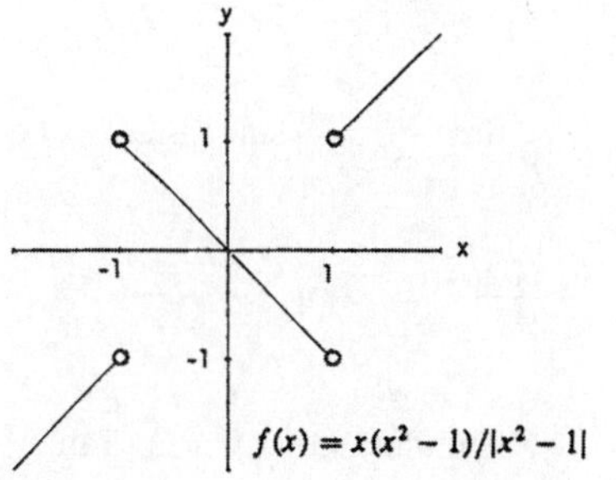

$f(x) = x(x^2-1)/|x^2-1|$

At $x = 1$: $\lim\limits_{x\to 1^-} f(x) = \lim\limits_{x\to 1^-} \frac{x(x^2-1)}{|x^2-1|} = \lim\limits_{x\to 1^-} \frac{x(x^2-1)}{-(x^2-1)} = \lim\limits_{x\to 1^-} (-x) = -1$, and
$\lim\limits_{x\to 1^+} f(x) = \lim\limits_{x\to 1^+} \frac{x(x^2-1)}{|x^2-1|} = \lim\limits_{x\to 1^+} \frac{x(x^2-1)}{x^2-1} = \lim\limits_{x\to 1^+} x = 1$. Again $\lim\limits_{x\to 1} f(x)$ does not exist so f cannot be extended to a continuous function at $x = 1$ either.

31. Yes, f does have a continuous extension to a = 1:
define $f(1) = \lim\limits_{x \to 1} \frac{x-1}{x-\sqrt[4]{x}} = \frac{4}{3}$.

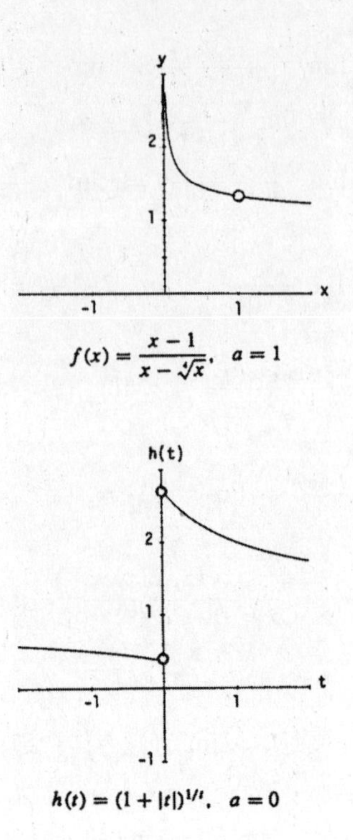

$f(x) = \frac{x-1}{x-\sqrt[4]{x}}, \quad a = 1$

33. From the graph we see that $\lim\limits_{t \to 0^-} h(t) \neq \lim\limits_{t \to 0^+} h(t)$
so h cannot be extended to a continuous function at a = 0.

$h(t) = (1+|t|)^{1/t}, \quad a = 0$

35. (a) $f(-1) = -1$ and $f(2) = 5 \Rightarrow$ f has a root between -1 and 2 by the Intermediate Value Theorem.
(b), (c) root is 1.32471795724

37. $\lim\limits_{x \to \infty} \frac{2x+3}{5x+7} = \lim\limits_{x \to \infty} \frac{2+\frac{3}{x}}{5+\frac{7}{x}} = \frac{2+0}{5+0} = \frac{2}{5}$

39. $\lim\limits_{x \to -\infty} \frac{x^2-4x+8}{3x^3} = \lim\limits_{x \to -\infty} \left(\frac{1}{3x} - \frac{4}{3x^2} + \frac{8}{3x^3}\right) = 0 - 0 + 0 = 0$

41. $\lim\limits_{x \to -\infty} \frac{x^2-7x}{x+1} = \lim\limits_{x \to -\infty} \frac{x-7}{1+\frac{1}{x}} = -\infty$

43. $\lim\limits_{x \to \infty} \frac{\sin x}{\lfloor x \rfloor} \le \lim\limits_{x \to \infty} \frac{1}{\lfloor x \rfloor} = 0$ since int $x \to \infty$ as $x \to \infty \Rightarrow \lim\limits_{x \to \infty} \frac{\sin x}{\lfloor x \rfloor} = 0$.

45. $\lim\limits_{x \to \infty} \frac{x + \sin x + 2\sqrt{x}}{x + \sin x} = \lim\limits_{x \to \infty} \frac{1 + \frac{\sin x}{x} + \frac{2}{\sqrt{x}}}{1 + \frac{\sin x}{x}} = \frac{1+0+0}{1+0} = 1$

47. (a) $y = \frac{x^2+4}{x-3}$ is undefined at $x = 3$: $\lim\limits_{x \to 3^-} \frac{x^2+4}{x-3} = -\infty$ and $\lim\limits_{x \to 3^+} \frac{x^2+4}{x-3} = +\infty$, thus $x = 3$ is a vertical asymptote.
(b) $y = \frac{x^2-x-2}{x^2-2x+1}$ is undefined at $x = 1$: $\lim\limits_{x \to 1^-} \frac{x^2-x-2}{x^2-2x+1} = -\infty$ and $\lim\limits_{x \to 1^+} \frac{x^2-x-2}{x^2-2x+1} = -\infty$, thus $x = 1$ is a vertical asymptote.
(c) $y = \frac{x^2+x-6}{x^2+2x-8}$ is undefined at $x = 2$ and -4: $\lim\limits_{x \to 2} \frac{x^2+x-6}{x^2+2x-8} = \lim\limits_{x \to 2} \frac{x+3}{x+4} = \frac{5}{6}$; $\lim\limits_{x \to -4^-} \frac{x^2+x-6}{x^2+2x-8} = \lim\limits_{x \to -4^-} \frac{x+3}{x+4} = \infty$
$\lim\limits_{x \to -4^+} \frac{x^2+x-6}{x^2+2x-8} = \lim\limits_{x \to -4^+} \frac{x+3}{x+4} = -\infty$. Thus $x = -4$ is a vertical asymptote.

CHAPTER 2 ADDITIONAL AND ADVANCED EXERCISES

1. (a)

x	0.1	0.01	0.001	0.0001	0.00001
x^x	0.7943	0.9550	0.9931	0.9991	0.9999

Apparently, $\lim\limits_{x \to 0^+} x^x = 1$

(b)

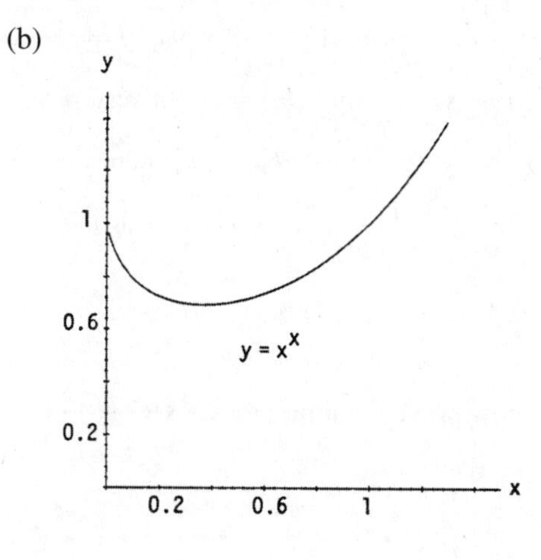

3. $\lim\limits_{v \to c^-} L = \lim\limits_{v \to c^-} L_0 \sqrt{1 - \frac{v^2}{c^2}} = L_0 \sqrt{1 - \frac{\lim\limits_{v \to c^-} v^2}{c^2}} = L_0 \sqrt{1 - \frac{c^2}{c^2}} = 0$

The left-hand limit was needed because the function L is undefined if $v > c$ (the rocket cannot move faster than the speed of light).

5. $|10 + (t - 70) \times 10^{-4} - 10| < 0.0005 \Rightarrow |(t - 70) \times 10^{-4}| < 0.0005 \Rightarrow -0.0005 < (t - 70) \times 10^{-4} < 0.0005$
$\Rightarrow -5 < t - 70 < 5 \Rightarrow 65° < t < 75° \Rightarrow$ Within 5° F.

7. Show $\lim\limits_{x \to 1} f(x) = \lim\limits_{x \to 1} (x^2 - 7) = -6 = f(1)$.

Step 1: $|(x^2 - 7) + 6| < \epsilon \Rightarrow -\epsilon < x^2 - 1 < \epsilon \Rightarrow 1 - \epsilon < x^2 < 1 + \epsilon \Rightarrow \sqrt{1 - \epsilon} < x < \sqrt{1 + \epsilon}$.

Step 2: $|x - 1| < \delta \Rightarrow -\delta < x - 1 < \delta \Rightarrow -\delta + 1 < x < \delta + 1$.

Then $-\delta + 1 = \sqrt{1 - \epsilon}$ or $\delta + 1 = \sqrt{1 + \epsilon}$. Choose $\delta = \min\left\{1 - \sqrt{1 - \epsilon}, \sqrt{1 + \epsilon} - 1\right\}$, then $0 < |x - 1| < \delta \Rightarrow |(x^2 - 7) - 6| < \epsilon$ and $\lim\limits_{x \to 1} f(x) = -6$. By the continuity test, f(x) is continuous at $x = 1$.

9. Show $\lim\limits_{x \to 2} h(x) = \lim\limits_{x \to 2} \sqrt{2x - 3} = 1 = h(2)$.

Step 1: $\left|\sqrt{2x - 3} - 1\right| < \epsilon \Rightarrow -\epsilon < \sqrt{2x - 3} - 1 < \epsilon \Rightarrow 1 - \epsilon < \sqrt{2x - 3} < 1 + \epsilon \Rightarrow \frac{(1-\epsilon)^2 + 3}{2} < x < \frac{(1+\epsilon)^2 + 3}{2}$.

Step 2: $|x - 2| < \delta \Rightarrow -\delta < x - 2 < \delta$ or $-\delta + 2 < x < \delta + 2$.

Then $-\delta + 2 = \frac{(1-\epsilon)^2 + 3}{2} \Rightarrow \delta = 2 - \frac{(1-\epsilon)^2 + 3}{2} = \frac{1 - (1-\epsilon)^2}{2} = \epsilon - \frac{\epsilon^2}{2}$, or $\delta + 2 = \frac{(1+\epsilon)^2 + 3}{2}$
$\Rightarrow \delta = \frac{(1+\epsilon)^2 + 3}{2} - 2 = \frac{(1+\epsilon)^2 - 1}{2} = \epsilon + \frac{\epsilon^2}{2}$. Choose $\delta = \epsilon - \frac{\epsilon^2}{2}$, the smaller of the two values. Then,
$0 < |x - 2| < \delta \Rightarrow \left|\sqrt{2x - 3} - 1\right| < \epsilon$, so $\lim\limits_{x \to 2} \sqrt{2x - 3} = 1$. By the continuity test, h(x) is continuous at $x = 2$.

11. Suppose L_1 and L_2 are two different limits. Without loss of generality assume $L_2 > L_1$. Let $\epsilon = \frac{1}{3}(L_2 - L_1)$.
Since $\lim\limits_{x \to x_0} f(x) = L_1$ there is a $\delta_1 > 0$ such that $0 < |x - x_0| < \delta_1 \Rightarrow |f(x) - L_1| < \epsilon \Rightarrow -\epsilon < f(x) - L_1 < \epsilon$
$\Rightarrow -\frac{1}{3}(L_2 - L_1) + L_1 < f(x) < \frac{1}{3}(L_2 - L_1) + L_1 \Rightarrow 4L_1 - L_2 < 3f(x) < 2L_1 + L_2$. Likewise, $\lim\limits_{x \to x_0} f(x) = L_2$ so there is a δ_2 such that $0 < |x - x_0| < \delta_2 \Rightarrow |f(x) - L_2| < \epsilon \Rightarrow -\epsilon < f(x) - L_2 < \epsilon$
$\Rightarrow -\frac{1}{3}(L_2 - L_1) + L_2 < f(x) < \frac{1}{3}(L_2 - L_1) + L_2 \Rightarrow 2L_2 + L_1 < 3f(x) < 4L_2 - L_1$
$\Rightarrow L_1 - 4L_2 < -3f(x) < -2L_2 - L_1$. If $\delta = \min\{\delta_1, \delta_2\}$ both inequalities must hold for $0 < |x - x_0| < \delta$:
$\left.\begin{array}{l} 4L_1 - L_2 < 3f(x) < 2L_1 + L_2 \\ L_1 - 4L_2 < -3f(x) < -2L_2 - L_1 \end{array}\right\} \Rightarrow 5(L_1 - L_2) < 0 < L_1 - L_2$. That is, $L_1 - L_2 < 0$ and $L_1 - L_2 > 0$,
a contradiction.

13. (a) Since $x \to 0^+, 0 < x^3 < x < 1 \Rightarrow (x^3 - x) \to 0^- \Rightarrow \lim_{x \to 0^+} f(x^3 - x) = \lim_{y \to 0^-} f(y) = B$ where $y = x^3 - x$.

(b) Since $x \to 0^-, -1 < x < x^3 < 0 \Rightarrow (x^3 - x) \to 0^+ \Rightarrow \lim_{x \to 0^-} f(x^3 - x) = \lim_{y \to 0^+} f(y) = A$ where $y = x^3 - x$.

(c) Since $x \to 0^+, 0 < x^4 < x^2 < 1 \Rightarrow (x^2 - x^4) \to 0^+ \Rightarrow \lim_{x \to 0^+} f(x^2 - x^4) = \lim_{y \to 0^+} f(y) = A$ where $y = x^2 - x^4$.

(d) Since $x \to 0^-, -1 < x < 0 \Rightarrow 0 < x^4 < x^2 < 1 \Rightarrow (x^2 - x^4) \to 0^+ \Rightarrow \lim_{x \to 0^+} f(x^2 - x^4) = A$ as in part (c).

15. Show $\lim_{x \to -1} f(x) = \lim_{x \to -1} \frac{x^2-1}{x+1} = \lim_{x \to -1} \frac{(x+1)(x-1)}{(x+1)} = -2, x \neq -1$.

Define the continuous extension of f(x) as $F(x) = \begin{cases} \frac{x^2-1}{x+1}, & x \neq -1 \\ -2, & x = -1 \end{cases}$. We now prove the limit of f(x) as $x \to -1$ exists and has the correct value.

Step 1: $\left|\frac{x^2-1}{x+1} - (-2)\right| < \epsilon \Rightarrow -\epsilon < \frac{(x+1)(x-1)}{(x+1)} + 2 < \epsilon \Rightarrow -\epsilon < (x-1) + 2 < \epsilon, x \neq -1 \Rightarrow -\epsilon - 1 < x < \epsilon - 1$.

Step 2: $|x - (-1)| < \delta \Rightarrow -\delta < x + 1 < \delta \Rightarrow -\delta - 1 < x < \delta - 1$.

Then $-\delta - 1 = -\epsilon - 1 \Rightarrow \delta = \epsilon$, or $\delta - 1 = \epsilon - 1 \Rightarrow \delta = \epsilon$. Choose $\delta = \epsilon$. Then $0 < |x - (-1)| < \delta$ $\Rightarrow \left|\frac{x^2-1}{x+1} - (-2)\right| < \epsilon \Rightarrow \lim_{x \to -1} F(x) = -2$. Since the conditions of the continuity test are met by F(x), then f(x) has a continuous extension to F(x) at $x = -1$.

17. (a) Let $\epsilon > 0$ be given. If x is rational, then $f(x) = x \Rightarrow |f(x) - 0| = |x - 0| < \epsilon \Leftrightarrow |x - 0| < \epsilon$; i.e., choose $\delta = \epsilon$. Then $|x - 0| < \delta \Rightarrow |f(x) - 0| < \epsilon$ for x rational. If x is irrational, then $f(x) = 0 \Rightarrow |f(x) - 0| < \epsilon$ $\Leftrightarrow 0 < \epsilon$ which is true no matter how close irrational x is to 0, so again we can choose $\delta = \epsilon$. In either case, given $\epsilon > 0$ there is a $\delta = \epsilon > 0$ such that $0 < |x - 0| < \delta \Rightarrow |f(x) - 0| < \epsilon$. Therefore, f is continuous at $x = 0$.

(b) Choose $x = c > 0$. Then within any interval $(c - \delta, c + \delta)$ there are both rational and irrational numbers. If c is rational, pick $\epsilon = \frac{c}{2}$. No matter how small we choose $\delta > 0$ there is an irrational number x in $(c - \delta, c + \delta) \Rightarrow |f(x) - f(c)| = |0 - c| = c > \frac{c}{2} = \epsilon$. That is, f is not continuous at any rational $c > 0$. On the other hand, suppose c is irrational $\Rightarrow f(c) = 0$. Again pick $\epsilon = \frac{c}{2}$. No matter how small we choose $\delta > 0$ there is a rational number x in $(c - \delta, c + \delta)$ with $|x - c| < \frac{c}{2} = \epsilon \Leftrightarrow \frac{c}{2} < x < \frac{3c}{2}$. Then $|f(x) - f(c)| = |x - 0|$ $= |x| > \frac{c}{2} = \epsilon \Rightarrow$ f is not continuous at any irrational $c > 0$.

If $x = c < 0$, repeat the argument picking $\epsilon = \frac{|c|}{2} = \frac{-c}{2}$. Therefore f fails to be continuous at any nonzero value $x = c$.

19. Yes. Let R be the radius of the equator (earth) and suppose at a fixed instant of time we label noon as the zero point, 0, on the equator $\Rightarrow 0 + \pi R$ represents the midnight point (at the same exact time). Suppose x_1 is a point on the equator "just after" noon $\Rightarrow x_1 + \pi R$ is simultaneously "just after" midnight. It seems reasonable that the temperature T at a point just after noon is hotter than it would be at the diametrically opposite point just after midnight: That is, $T(x_1) - T(x_1 + \pi R) > 0$. At exactly the same moment in time pick x_2 to be a point just before midnight $\Rightarrow x_2 + \pi R$ is just before noon. Then $T(x_2) - T(x_2 + \pi R) < 0$. Assuming the temperature function T is continuous along the equator (which is reasonable), the Intermediate Value Theorem says there is a point c between 0 (noon) and πR (simultaneously midnight) such that $T(c) - T(c + \pi R) = 0$; i.e., there is always a pair of antipodal points on the earth's equator where the temperatures are the same.

21. (a) At $x = 0$: $\lim_{a \to 0} r_+(a) = \lim_{a \to 0} \frac{-1+\sqrt{1+a}}{a} = \lim_{a \to 0} \left(\frac{-1+\sqrt{1+a}}{a}\right)\left(\frac{-1-\sqrt{1+a}}{-1-\sqrt{1+a}}\right)$

$= \lim_{a \to 0} \frac{1-(1+a)}{a(-1-\sqrt{1+a})} = \frac{-1}{-1-\sqrt{1+0}} = \frac{1}{2}$

At $x = -1$: $\lim_{a \to -1^+} r_+(a) = \lim_{a \to -1^+} \frac{1-(1+a)}{a(-1-\sqrt{1+a})} = \lim_{a \to -1} \frac{-a}{a(-1-\sqrt{1+a})} = \frac{-1}{-1-\sqrt{0}} = 1$

(b) At $x=0$: $\lim_{a\to 0^-} r_-(a) = \lim_{a\to 0^-} \frac{-1-\sqrt{1+a}}{a} = \lim_{a\to 0^-} \left(\frac{-1-\sqrt{1+a}}{a}\right)\left(\frac{-1+\sqrt{1+a}}{-1+\sqrt{1+a}}\right)$
$= \lim_{a\to 0^-} \frac{1-(1+a)}{a(-1+\sqrt{1+a})} = \lim_{a\to 0^-} \frac{-a}{a(-1+\sqrt{1+a})} = \lim_{a\to 0^-} \frac{-1}{-1+\sqrt{1+a}} = \infty$ (because the denominator is always negative); $\lim_{a\to 0^+} r_-(a) = \lim_{a\to 0^+} \frac{-1}{-1+\sqrt{1+a}} = -\infty$ (because the denominator is always positive). Therefore, $\lim_{a\to 0} r_-(a)$ does not exist.

At $x=-1$: $\lim_{a\to -1^+} r_-(a) = \lim_{a\to -1^+} \frac{-1-\sqrt{1+a}}{a} = \lim_{a\to -1^+} \frac{-1}{-1+\sqrt{1+a}} = 1$

(c)

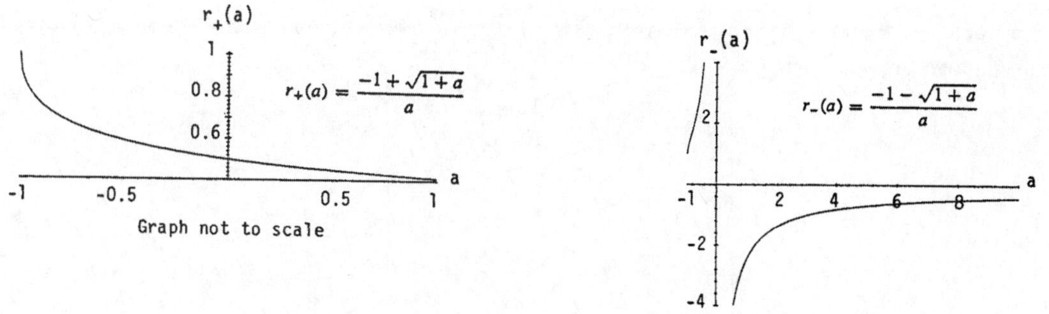

(d)

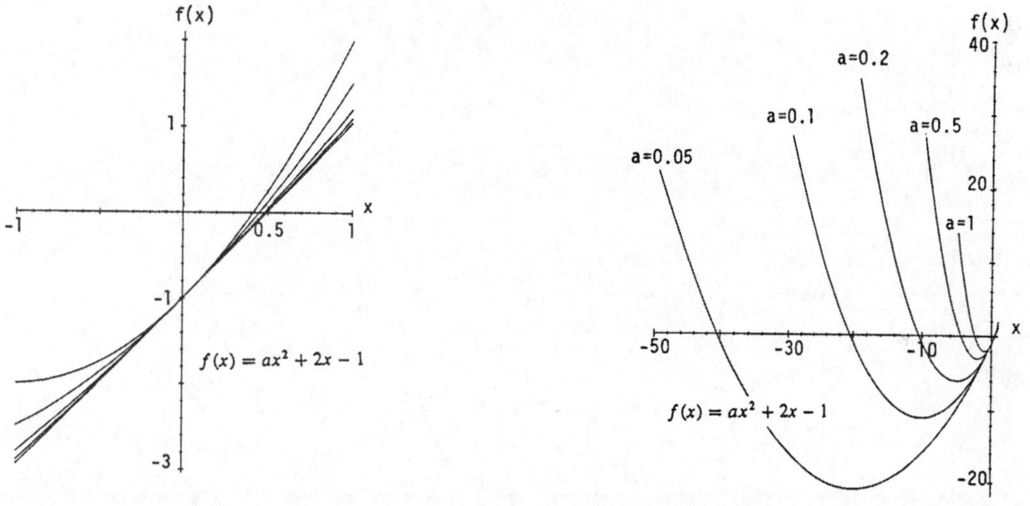

23. (a) The function f is bounded on D if $f(x) \ge M$ and $f(x) \le N$ for all x in D. This means $M \le f(x) \le N$ for all x in D. Choose B to be $\max\{|M|, |N|\}$. Then $|f(x)| \le B$. On the other hand, if $|f(x)| \le B$, then $-B \le f(x) \le B \Rightarrow f(x) \ge -B$ and $f(x) \le B \Rightarrow f(x)$ is bounded on D with $N = B$ an upper bound and $M = -B$ a lower bound.

(b) Assume $f(x) \le N$ for all x and that $L > N$. Let $\epsilon = \frac{L-N}{2}$. Since $\lim_{x\to x_0} f(x) = L$ there is a $\delta > 0$ such that $0 < |x - x_0| < \delta \Rightarrow |f(x) - L| < \epsilon \Leftrightarrow L - \epsilon < f(x) < L + \epsilon \Leftrightarrow L - \frac{L-N}{2} < f(x) < L + \frac{L-N}{2}$ $\Leftrightarrow \frac{L+N}{2} < f(x) < \frac{3L-N}{2}$. But $L > N \Rightarrow \frac{L+N}{2} > N \Rightarrow N < f(x)$ contrary to the boundedness assumption $f(x) \le N$. This contradiction proves $L \le N$.

(c) Assume $M \le f(x)$ for all x and that $L < M$. Let $\epsilon = \frac{M-L}{2}$. As in part (b), $0 < |x - x_0| < \delta$ $\Rightarrow L - \frac{M-L}{2} < f(x) < L + \frac{M-L}{2} \Leftrightarrow \frac{3L-M}{2} < f(x) < \frac{M+L}{2} < M$, a contradiction.

25. $\lim_{x\to 0} \frac{\sin(1-\cos x)}{x} = \lim_{x\to 0} \frac{\sin(1-\cos x)}{1-\cos x} \cdot \frac{1-\cos x}{x} \cdot \frac{1+\cos x}{1+\cos x} = \lim_{x\to 0} \frac{\sin(1-\cos x)}{1-\cos x} \cdot \lim_{x\to 0} \frac{1-\cos^2 x}{x(1+\cos x)} = 1 \cdot \lim_{x\to 0} \frac{\sin^2 x}{x(1+\cos x)}$
$= \lim_{x\to 0} \frac{\sin x}{x} \cdot \frac{\sin x}{1+\cos x} = 1 \cdot \left(\frac{0}{2}\right) = 0.$

27. $\lim_{x \to 0} \frac{\sin(\sin x)}{x} = \lim_{x \to 0} \frac{\sin(\sin x)}{\sin x} \cdot \frac{\sin x}{x} = \lim_{x \to 0} \frac{\sin(\sin x)}{\sin x} \cdot \lim_{x \to 0} \frac{\sin x}{x} = 1 \cdot 1 = 1.$

29. $\lim_{x \to 2} \frac{\sin(x^2 - 4)}{x - 2} = \lim_{x \to 2} \frac{\sin(x^2 - 4)}{x^2 - 4} \cdot (x + 2) = \lim_{x \to 2} \frac{\sin(x^2 - 4)}{x^2 - 4} \cdot \lim_{x \to 2} (x + 2) = 1 \cdot 4 = 4$

31. Since the highest power of x in the numerator is 1 more than the highest power of x in the denominator, there is an oblique asymptote. $y = \frac{2x^{3/2} + 2x - 3}{\sqrt{x} + 1} = 2x - \frac{3}{\sqrt{x} + 1}$, thus the oblique asymptote is $y = 2x$.

33. As $x \to \pm\infty$, $x^2 + 1 \to x^2 \Rightarrow \sqrt{x^2 + 1} \to \sqrt{x^2}$; as $x \to -\infty$, $\sqrt{x^2} = -x$, and as $x \to +\infty$, $\sqrt{x^2} = x$; thus the oblique asymptotes are $y = x$ and $y = -x$.

CHAPTER 3 DIFFERENTIATION

3.1 TANGENTS AND THE DERIVATIVE AT A POINT

1. P_1: $m_1 = 1$, P_2: $m_2 = 5$

3. P_1: $m_1 = \frac{5}{2}$, P_2: $m_2 = -\frac{1}{2}$

5. $m = \lim\limits_{h \to 0} \frac{[4-(-1+h)^2]-(4-(-1)^2)}{h}$
$= \lim\limits_{h \to 0} \frac{-(1-2h+h^2)+1}{h} = \lim\limits_{h \to 0} \frac{h(2-h)}{h} = 2;$
at $(-1, 3)$: $y = 3 + 2(x - (-1)) \Rightarrow y = 2x + 5$, tangent line

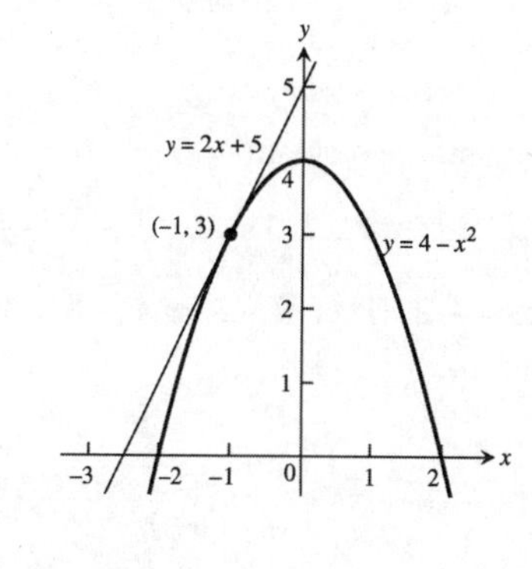

7. $m = \lim\limits_{h \to 0} \frac{2\sqrt{1+h}-2\sqrt{1}}{h} = \lim\limits_{h \to 0} \frac{2\sqrt{1+h}-2}{h} \cdot \frac{2\sqrt{1+h}+2}{2\sqrt{1+h}+2}$
$= \lim\limits_{h \to 0} \frac{4(1+h)-4}{2h\left(\sqrt{1+h}+1\right)} = \lim\limits_{h \to 0} \frac{2}{\sqrt{1+h}+1} = 1;$
at $(1, 2)$: $y = 2 + 1(x - 1) \Rightarrow y = x + 1$, tangent line

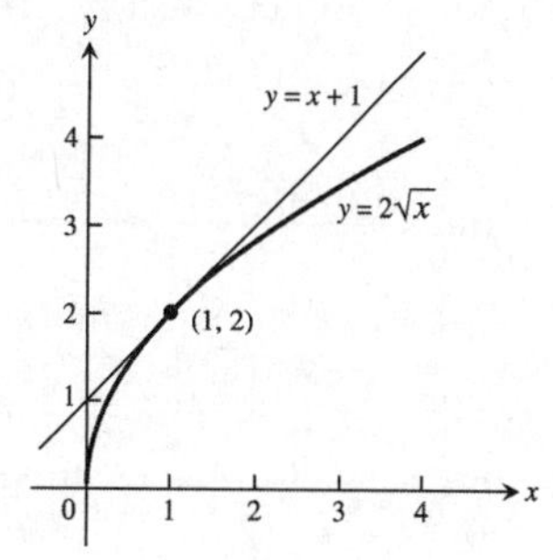

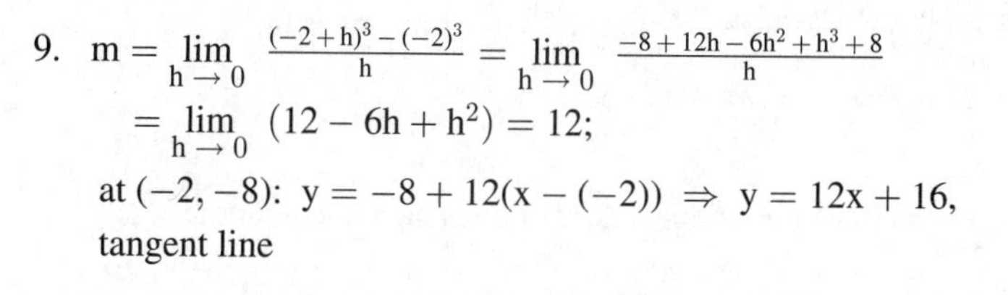

9. $m = \lim\limits_{h \to 0} \frac{(-2+h)^3-(-2)^3}{h} = \lim\limits_{h \to 0} \frac{-8+12h-6h^2+h^3+8}{h}$
$= \lim\limits_{h \to 0} (12 - 6h + h^2) = 12;$
at $(-2, -8)$: $y = -8 + 12(x - (-2)) \Rightarrow y = 12x + 16$, tangent line

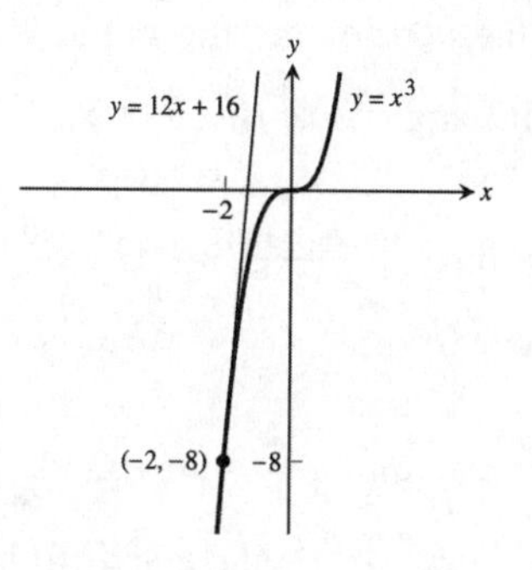

11. $m = \lim\limits_{h \to 0} \frac{[(2+h)^2+1]-5}{h} = \lim\limits_{h \to 0} \frac{(5+4h+h^2)-5}{h} = \lim\limits_{h \to 0} \frac{h(4+h)}{h} = 4;$
at $(2, 5)$: $y - 5 = 4(x - 2)$, tangent line

13. $m = \lim\limits_{h \to 0} \frac{\frac{3+h}{(3+h)-2}-3}{h} = \lim\limits_{h \to 0} \frac{(3+h)-3(h+1)}{h(h+1)} = \lim\limits_{h \to 0} \frac{-2h}{h(h+1)} = -2;$
at $(3, 3)$: $y - 3 = -2(x - 3)$, tangent line

15. $m = \lim\limits_{h \to 0} \frac{(2+h)^3-8}{h} = \lim\limits_{h \to 0} \frac{(8+12h+6h^2+h^3)-8}{h} = \lim\limits_{h \to 0} \frac{h(12+6h+h^2)}{h} = 12;$
at $(2, 8)$: $y - 8 = 12(t - 2)$, tangent line

17. $m = \lim_{h \to 0} \frac{\sqrt{4+h}-2}{h} = \lim_{h \to 0} \frac{\sqrt{4+h}-2}{h} \cdot \frac{\sqrt{4+h}+2}{\sqrt{4+h}+2} = \lim_{h \to 0} \frac{(4+h)-4}{h\left(\sqrt{4+h}+2\right)} = \lim_{h \to 0} \frac{h}{h\left(\sqrt{4+h}+2\right)} = \frac{1}{\sqrt{4}+2}$
$= \frac{1}{4}$; at (4, 2): $y - 2 = \frac{1}{4}(x-4)$, tangent line

19. At $x = -1, y = 5 \Rightarrow m = \lim_{h \to 0} \frac{5(-1+h)^2-5}{h} = \lim_{h \to 0} \frac{5(1-2h+h^2)-5}{h} = \lim_{h \to 0} \frac{5h(-2+h)}{h} = -10$, slope

21. At $x = 3, y = \frac{1}{2} \Rightarrow m = \lim_{h \to 0} \frac{\frac{1}{(3+h)-1} - \frac{1}{2}}{h} = \lim_{h \to 0} \frac{2-(2+h)}{2h(2+h)} = \lim_{h \to 0} \frac{-h}{2h(2+h)} = -\frac{1}{4}$, slope

23. At a horizontal tangent the slope $m = 0 \Rightarrow 0 = m = \lim_{h \to 0} \frac{[(x+h)^2+4(x+h)-1]-(x^2+4x-1)}{h}$
$= \lim_{h \to 0} \frac{(x^2+2xh+h^2+4x+4h-1)-(x^2+4x-1)}{h} = \lim_{h \to 0} \frac{(2xh+h^2+4h)}{h} = \lim_{h \to 0}(2x+h+4) = 2x+4$;
$2x+4 = 0 \Rightarrow x = -2$. Then $f(-2) = 4-8-1 = -5 \Rightarrow (-2, -5)$ is the point on the graph where there is a horizontal tangent.

25. $-1 = m = \lim_{h \to 0} \frac{\frac{1}{(x+h)-1} - \frac{1}{x-1}}{h} = \lim_{h \to 0} \frac{(x-1)-(x+h-1)}{h(x-1)(x+h-1)} = \lim_{h \to 0} \frac{-h}{h(x-1)(x+h-1)} = -\frac{1}{(x-1)^2}$
$\Rightarrow (x-1)^2 = 1 \Rightarrow x^2 - 2x = 0 \Rightarrow x(x-2) = 0 \Rightarrow x = 0$ or $x = 2$. If $x = 0$, then $y = -1$ and $m = -1$
$\Rightarrow y = -1-(x-0) = -(x+1)$. If $x = 2$, then $y = 1$ and $m = -1 \Rightarrow y = 1-(x-2) = -(x-3)$.

27. $\lim_{h \to 0} \frac{f(2+h)-f(2)}{h} = \lim_{h \to 0} \frac{(100-4.9(2+h)^2)-(100-4.9(2)^2)}{h} = \lim_{h \to 0} \frac{-4.9(4+4h+h^2)+4.9(4)}{h}$
$= \lim_{h \to 0}(-19.6-4.9h) = -19.6$. The minus sign indicates the object is falling <u>downward</u> at a speed of 19.6 m/sec.

29. $\lim_{h \to 0} \frac{f(3+h)-f(3)}{h} = \lim_{h \to 0} \frac{\pi(3+h)^2-\pi(3)^2}{h} = \lim_{h \to 0} \frac{\pi[9+6h+h^2-9]}{h} = \lim_{h \to 0} \pi(6+h) = 6\pi$

31. At (x_0, mx_0+b) the slope of the tangent line is $\lim_{h \to 0} \frac{(m(x_0+h)+b)-(mx_0+b)}{(x_0+h)-x_0} = \lim_{h \to 0} \frac{mh}{h} = \lim_{h \to 0} m = m$.
The equation of the tangent line is $y-(mx_0+b) = m(x-x_0) \Rightarrow y = mx+b$.

33. Slope at origin $= \lim_{h \to 0} \frac{f(0+h)-f(0)}{h} = \lim_{h \to 0} \frac{h^2 \sin\left(\frac{1}{h}\right)}{h} = \lim_{h \to 0} h \sin\left(\frac{1}{h}\right) = 0 \Rightarrow$ yes, f(x) does have a tangent at the origin with slope 0.

35. $\lim_{h \to 0^-} \frac{f(0+h)-f(0)}{h} = \lim_{h \to 0^-} \frac{-1-0}{h} = \infty$, and $\lim_{h \to 0^+} \frac{f(0+h)-f(0)}{h} = \lim_{h \to 0^+} \frac{1-0}{h} = \infty$. Therefore,
$\lim_{h \to 0} \frac{f(0+h)-f(0)}{h} = \infty \Rightarrow$ yes, the graph of f has a vertical tangent at the origin.

37. (a) The graph appears to have a cusp at $x = 0$.

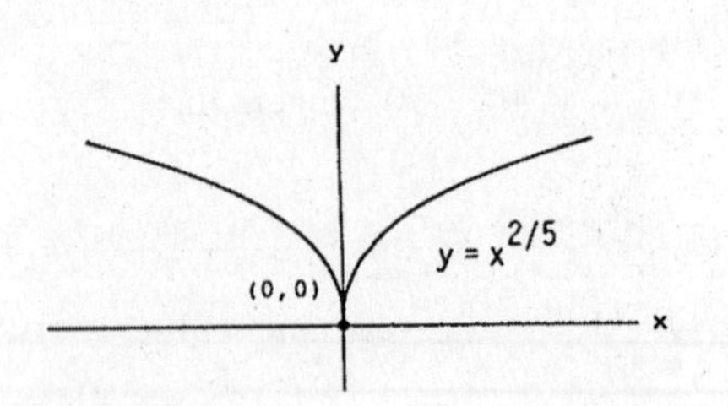

(b) $\lim_{h \to 0^-} \frac{f(0+h)-f(0)}{h} = \lim_{h \to 0^-} \frac{h^{2/5}-0}{h} = \lim_{h \to 0^-} \frac{1}{h^{3/5}} = -\infty$ and $\lim_{h \to 0^+} \frac{1}{h^{3/5}} = \infty \Rightarrow$ limit does not exist
$\Rightarrow$ the graph of $y = x^{2/5}$ does not have a vertical tangent at $x = 0$.

39. (a) The graph appears to have a vertical tangent at $x = 0$.

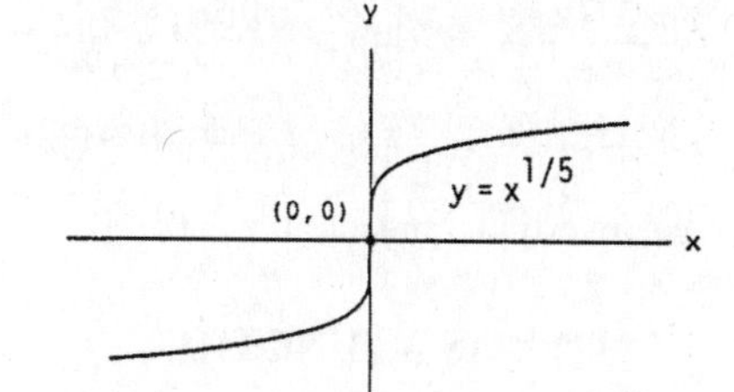

(b) $\lim\limits_{h \to 0} \frac{f(0+h)-f(0)}{h} = \lim\limits_{h \to 0} \frac{h^{1/5}-0}{h} = \lim\limits_{h \to 0} \frac{1}{h^{4/5}} = \infty \Rightarrow y = x^{1/5}$ has a vertical tangent at $x = 0$.

41. (a) The graph appears to have a cusp at $x = 0$.

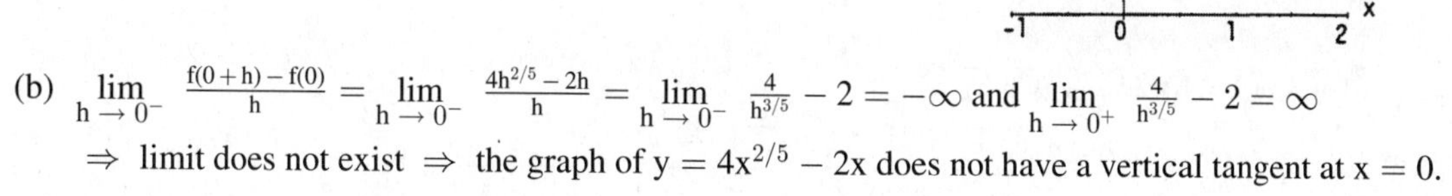

(b) $\lim\limits_{h \to 0^-} \frac{f(0+h)-f(0)}{h} = \lim\limits_{h \to 0^-} \frac{4h^{2/5}-2h}{h} = \lim\limits_{h \to 0^-} \frac{4}{h^{3/5}} - 2 = -\infty$ and $\lim\limits_{h \to 0^+} \frac{4}{h^{3/5}} - 2 = \infty$

$\Rightarrow$ limit does not exist $\Rightarrow$ the graph of $y = 4x^{2/5} - 2x$ does not have a vertical tangent at $x = 0$.

43. (a) The graph appears to have a vertical tangent at $x = 1$ and a cusp at $x = 0$.

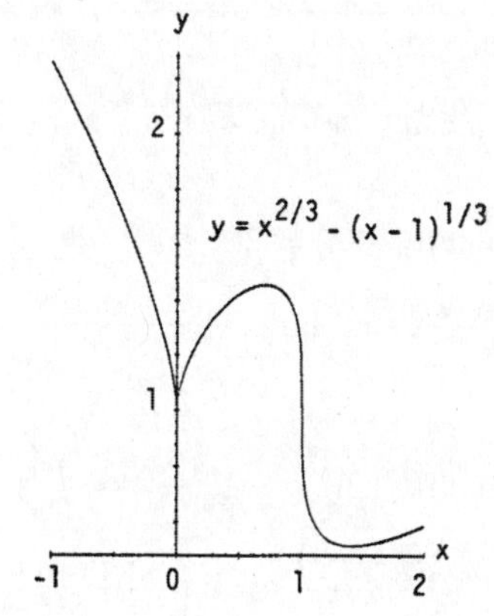

(b) $x = 1$: $\lim\limits_{h \to 0} \frac{(1+h)^{2/3}-(1+h-1)^{1/3}-1}{h} = \lim\limits_{h \to 0} \frac{(1+h)^{2/3}-h^{1/3}-1}{h} = -\infty$

$\Rightarrow y = x^{2/3} - (x-1)^{1/3}$ has a vertical tangent at $x = 1$;

$x = 0$: $\lim\limits_{h \to 0} \frac{f(0+h)-f(0)}{h} = \lim\limits_{h \to 0} \frac{h^{2/3}-(h-1)^{1/3}-(-1)^{1/3}}{h} = \lim\limits_{h \to 0} \left[\frac{1}{h^{1/3}} - \frac{(h-1)^{1/3}}{h} + \frac{1}{h}\right]$

does not exist $\Rightarrow y = x^{2/3} - (x-1)^{1/3}$ does not have a vertical tangent at $x = 0$.

45. (a) The graph appears to have a vertical tangent at $x = 0$.

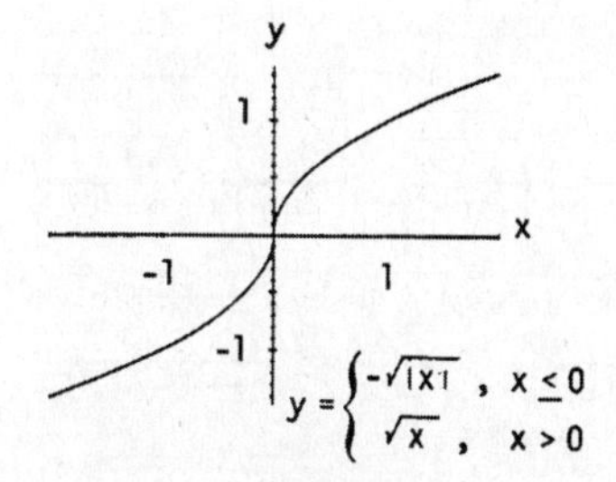

(b) $\lim\limits_{h \to 0^+} \frac{f(0+h)-f(0)}{h} = \lim\limits_{x \to 0^+} \frac{\sqrt{h}-0}{h} = \lim\limits_{h \to 0} \frac{1}{\sqrt{h}} = \infty;$

$\lim\limits_{h \to 0^-} \frac{f(0+h)-f(0)}{h} = \lim\limits_{h \to 0^-} \frac{-\sqrt{|h|}-0}{h} = \lim\limits_{h \to 0^-} \frac{-\sqrt{|h|}}{-|h|} = \lim\limits_{h \to 0^-} \frac{1}{\sqrt{|h|}} = \infty$

$\Rightarrow$ y has a vertical tangent at $x = 0$.

3.2 THE DERIVATIVE AS A FUNCTION

1. Step 1: $f(x) = 4 - x^2$ and $f(x+h) = 4 - (x+h)^2$

Step 2: $\frac{f(x+h)-f(x)}{h} = \frac{[4-(x+h)^2]-(4-x^2)}{h} = \frac{(4-x^2-2xh-h^2)-4+x^2}{h} = \frac{-2xh-h^2}{h} = \frac{h(-2x-h)}{h}$

$= -2x - h$

Step 3: $f'(x) = \lim\limits_{h \to 0} (-2x-h) = -2x$; $f'(-3) = 6$, $f'(0) = 0$, $f'(1) = -2$

3. Step 1: $g(t) = \frac{1}{t^2}$ and $g(t+h) = \frac{1}{(t+h)^2}$

Step 2: $\frac{g(t+h)-g(t)}{h} = \frac{\frac{1}{(t+h)^2} - \frac{1}{t^2}}{h} = \frac{\left(\frac{t^2-(t+h)^2}{(t+h)^2 \cdot t^2}\right)}{h} = \frac{t^2-(t^2+2th+h^2)}{(t+h)^2 \cdot t^2 \cdot h} = \frac{-2th-h^2}{(t+h)^2 t^2 h}$

$= \frac{h(-2t-h)}{(t+h)^2 t^2 h} = \frac{-2t-h}{(t+h)^2 t^2}$

Step 3: $g'(t) = \lim\limits_{h \to 0} \frac{-2t-h}{(t+h)^2 t^2} = \frac{-2t}{t^2 \cdot t^2} = \frac{-2}{t^3}$; $g'(-1) = 2$, $g'(2) = -\frac{1}{4}$, $g'\left(\sqrt{3}\right) = -\frac{2}{3\sqrt{3}}$

5. Step 1: $p(\theta) = \sqrt{3\theta}$ and $p(\theta+h) = \sqrt{3(\theta+h)}$

Step 2: $\frac{p(\theta+h)-p(\theta)}{h} = \frac{\sqrt{3(\theta+h)}-\sqrt{3\theta}}{h} = \frac{\left(\sqrt{3\theta+3h}-\sqrt{3\theta}\right)}{h} \cdot \frac{\left(\sqrt{3\theta+3h}+\sqrt{3\theta}\right)}{\left(\sqrt{3\theta+3h}+\sqrt{3\theta}\right)} = \frac{(3\theta+3h)-3\theta}{h\left(\sqrt{3\theta+3h}+\sqrt{3\theta}\right)}$

$= \frac{3h}{h\left(\sqrt{3\theta+3h}+\sqrt{3\theta}\right)} = \frac{3}{\sqrt{3\theta+3h}+\sqrt{3\theta}}$

Step 3: $p'(\theta) = \lim\limits_{h \to 0} \frac{3}{\sqrt{3\theta+3h}+\sqrt{3\theta}} = \frac{3}{\sqrt{3\theta}+\sqrt{3\theta}} = \frac{3}{2\sqrt{3\theta}}$; $p'(1) = \frac{3}{2\sqrt{3}}$, $p'(3) = \frac{1}{2}$, $p'\left(\frac{2}{3}\right) = \frac{3}{2\sqrt{2}}$

7. $y = f(x) = 2x^3$ and $f(x+h) = 2(x+h)^3 \Rightarrow \frac{dy}{dx} = \lim\limits_{h \to 0} \frac{2(x+h)^3 - 2x^3}{h} = \lim\limits_{h \to 0} \frac{2\left(x^3+3x^2h+3xh^2+h^3\right)-2x^3}{h}$

$= \lim\limits_{h \to 0} \frac{6x^2h+6xh^2+2h^3}{h} = \lim\limits_{h \to 0} \frac{h\left(6x^2+6xh+2h^2\right)}{h} = \lim\limits_{h \to 0} \left(6x^2+6xh+2h^2\right) = 6x^2$

9. $s = r(t) = \frac{t}{2t+1}$ and $r(t+h) = \frac{t+h}{2(t+h)+1} \Rightarrow \frac{ds}{dt} = \lim\limits_{h \to 0} \frac{\left(\frac{t+h}{2(t+h)+1}\right) - \left(\frac{t}{2t+1}\right)}{h}$

$= \lim\limits_{h \to 0} \frac{\left(\frac{(t+h)(2t+1)-t(2t+2h+1)}{(2t+2h+1)(2t+1)}\right)}{h} = \lim\limits_{h \to 0} \frac{(t+h)(2t+1)-t(2t+2h+1)}{(2t+2h+1)(2t+1)h}$

$= \lim\limits_{h \to 0} \frac{2t^2+t+2ht+h-2t^2-2ht-t}{(2t+2h+1)(2t+1)h} = \lim\limits_{h \to 0} \frac{h}{(2t+2h+1)(2t+1)h} = \lim\limits_{h \to 0} \frac{1}{(2t+2h+1)(2t+1)}$

$= \frac{1}{(2t+1)(2t+1)} = \frac{1}{(2t+1)^2}$

11. $p = f(q) = \frac{1}{\sqrt{q+1}}$ and $f(q+h) = \frac{1}{\sqrt{(q+h)+1}} \Rightarrow \frac{dp}{dq} = \lim\limits_{h \to 0} \frac{\left(\frac{1}{\sqrt{(q+h)+1}}\right) - \left(\frac{1}{\sqrt{q+1}}\right)}{h}$

$= \lim\limits_{h \to 0} \frac{\left(\frac{\sqrt{q+1}-\sqrt{q+h+1}}{\sqrt{q+h+1}\sqrt{q+1}}\right)}{h} = \lim\limits_{h \to 0} \frac{\sqrt{q+1}-\sqrt{q+h+1}}{h\sqrt{q+h+1}\sqrt{q+1}}$

$= \lim\limits_{h \to 0} \frac{\left(\sqrt{q+1}-\sqrt{q+h+1}\right)}{h\sqrt{q+h+1}\sqrt{q+1}} \cdot \frac{\left(\sqrt{q+1}+\sqrt{q+h+1}\right)}{\left(\sqrt{q+1}+\sqrt{q+h+1}\right)} = \lim\limits_{h \to 0} \frac{(q+1)-(q+h+1)}{h\sqrt{q+h+1}\sqrt{q+1}\left(\sqrt{q+1}+\sqrt{q+h+1}\right)}$

$= \lim\limits_{h \to 0} \frac{-h}{h\sqrt{q+h+1}\sqrt{q+1}\left(\sqrt{q+1}+\sqrt{q+h+1}\right)} = \lim\limits_{h \to 0} \frac{-1}{\sqrt{q+h+1}\sqrt{q+1}\left(\sqrt{q+1}+\sqrt{q+h+1}\right)}$

$= \frac{-1}{\sqrt{q+1}\sqrt{q+1}\left(\sqrt{q+1}+\sqrt{q+1}\right)} = \frac{-1}{2(q+1)\sqrt{q+1}}$

13. $f(x) = x + \frac{9}{x}$ and $f(x+h) = (x+h) + \frac{9}{(x+h)} \Rightarrow \frac{f(x+h)-f(x)}{h} = \frac{\left[(x+h)+\frac{9}{(x+h)}\right]-\left[x+\frac{9}{x}\right]}{h}$
$= \frac{x(x+h)^2+9x-x^2(x+h)-9(x+h)}{x(x+h)h} = \frac{x^3+2x^2h+xh^2+9x-x^3-x^2h-9x-9h}{x(x+h)h} = \frac{x^2h+xh^2-9h}{x(x+h)h}$
$= \frac{h(x^2+xh-9)}{x(x+h)h} = \frac{x^2+xh-9}{x(x+h)}$; $f'(x) = \lim_{h\to 0} \frac{x^2+xh-9}{x(x+h)} = \frac{x^2-9}{x^2} = 1 - \frac{9}{x^2}$; $m = f'(-3) = 0$

15. $\frac{ds}{dt} = \lim_{h\to 0} \frac{[(t+h)^3-(t+h)^2]-(t^3-t^2)}{h} = \lim_{h\to 0} \frac{(t^3+3t^2h+3th^2+h^3)-(t^2+2th+h^2)-t^3+t^2}{h}$
$= \lim_{h\to 0} \frac{3t^2h+3th^2+h^3-2th-h^2}{h} = \lim_{h\to 0} \frac{h(3t^2+3th+h^2-2t-h)}{h} = \lim_{h\to 0} (3t^2+3th+h^2-2t-h)$
$= 3t^2 - 2t$; $m = \left.\frac{ds}{dt}\right|_{t=-1} = 5$

17. $f(x) = \frac{8}{\sqrt{x-2}}$ and $f(x+h) = \frac{8}{\sqrt{(x+h)-2}} \Rightarrow \frac{f(x+h)-f(x)}{h} = \frac{\frac{8}{\sqrt{(x+h)-2}}-\frac{8}{\sqrt{x-2}}}{h}$
$= \frac{8\left(\sqrt{x-2}-\sqrt{x+h-2}\right)}{h\sqrt{x+h-2}\sqrt{x-2}} \cdot \frac{\left(\sqrt{x-2}+\sqrt{x+h-2}\right)}{\left(\sqrt{x-2}+\sqrt{x+h-2}\right)} = \frac{8[(x-2)-(x+h-2)]}{h\sqrt{x+h-2}\sqrt{x-2}\left(\sqrt{x-2}+\sqrt{x+h-2}\right)}$
$= \frac{-8h}{h\sqrt{x+h-2}\sqrt{x-2}\left(\sqrt{x-2}+\sqrt{x+h-2}\right)} \Rightarrow f'(x) = \lim_{h\to 0} \frac{-8}{\sqrt{x+h-2}\sqrt{x-2}\left(\sqrt{x-2}+\sqrt{x+h-2}\right)}$
$= \frac{-8}{\sqrt{x-2}\sqrt{x-2}\left(\sqrt{x-2}+\sqrt{x-2}\right)} = \frac{-4}{(x-2)\sqrt{x-2}}$; $m = f'(6) = \frac{-4}{4\sqrt{4}} = -\frac{1}{2} \Rightarrow$ the equation of the tangent line at $(6, 4)$ is $y - 4 = -\frac{1}{2}(x-6) \Rightarrow y = -\frac{1}{2}x + 3 + 4 \Rightarrow y = -\frac{1}{2}x + 7$.

19. $s = f(t) = 1 - 3t^2$ and $f(t+h) = 1 - 3(t+h)^2 = 1 - 3t^2 - 6th - 3h^2 \Rightarrow \frac{ds}{dt} = \lim_{h\to 0} \frac{f(t+h)-f(t)}{h}$
$= \lim_{h\to 0} \frac{(1-3t^2-6th-3h^2)-(1-3t^2)}{h} = \lim_{h\to 0} (-6t-3h) = -6t \Rightarrow \left.\frac{ds}{dt}\right|_{t=-1} = 6$

21. $r = f(\theta) = \frac{2}{\sqrt{4-\theta}}$ and $f(\theta+h) = \frac{2}{\sqrt{4-(\theta+h)}} \Rightarrow \frac{dr}{d\theta} = \lim_{h\to 0} \frac{f(\theta+h)-f(\theta)}{h} = \lim_{h\to 0} \frac{\frac{2}{\sqrt{4-\theta-h}}-\frac{2}{\sqrt{4-\theta}}}{h}$
$= \lim_{h\to 0} \frac{2\sqrt{4-\theta}-2\sqrt{4-\theta-h}}{h\sqrt{4-\theta}\sqrt{4-\theta-h}} = \lim_{h\to 0} \frac{2\sqrt{4-\theta}-2\sqrt{4-\theta-h}}{h\sqrt{4-\theta}\sqrt{4-\theta-h}} \cdot \frac{\left(2\sqrt{4-\theta}+2\sqrt{4-\theta-h}\right)}{\left(2\sqrt{4-\theta}+2\sqrt{4-\theta-h}\right)}$
$= \lim_{h\to 0} \frac{4(4-\theta)-4(4-\theta-h)}{2h\sqrt{4-\theta}\sqrt{4-\theta-h}\left(\sqrt{4-\theta}+\sqrt{4-\theta-h}\right)} = \lim_{h\to 0} \frac{2}{\sqrt{4-\theta}\sqrt{4-\theta-h}\left(\sqrt{4-\theta}+\sqrt{4-\theta-h}\right)}$
$= \frac{2}{(4-\theta)\left(2\sqrt{4-\theta}\right)} = \frac{1}{(4-\theta)\sqrt{4-\theta}} \Rightarrow \left.\frac{dr}{d\theta}\right|_{\theta=0} = \frac{1}{8}$

23. $f'(x) = \lim_{z\to x} \frac{f(z)-f(x)}{z-x} = \lim_{z\to x} \frac{\frac{1}{z+2}-\frac{1}{x+2}}{z-x} = \lim_{z\to x} \frac{(x+2)-(z+2)}{(z-x)(z+2)(x+2)} = \lim_{z\to x} \frac{x-z}{(z-x)(z+2)(x+2)} = \lim_{z\to x} \frac{-1}{(z+2)(x+2)} = \frac{-1}{(x+2)^2}$

25. $g'(x) = \lim_{z\to x} \frac{g(z)-g(x)}{z-x} = \lim_{z\to x} \frac{\frac{z}{z-1}-\frac{x}{x-1}}{z-x} = \lim_{z\to x} \frac{z(x-1)-x(z-1)}{(z-x)(z-1)(x-1)} = \lim_{z\to x} \frac{-z+x}{(z-x)(z-1)(x-1)} = \lim_{z\to x} \frac{-1}{(z-1)(x-1)} = \frac{-1}{(x-1)^2}$

27. Note that as x increases, the slope of the tangent line to the curve is first negative, then zero (when $x = 0$), then positive $\Rightarrow$ the slope is always increasing which matches (b).

29. $f_3(x)$ is an oscillating function like the cosine. Everywhere that the graph of f_3 has a horizontal tangent we expect f_3' to be zero, and (d) matches this condition.

31. (a) f' is not defined at $x = 0, 1, 4$. At these points, the left-hand and right-hand derivatives do not agree. For example, $\lim_{x\to 0^-} \frac{f(x)-f(0)}{x-0}$ = slope of line joining $(-4, 0)$ and $(0, 2) = \frac{1}{2}$ but $\lim_{x\to 0^+} \frac{f(x)-f(0)}{x-0}$ = slope of line joining $(0, 2)$ and $(1, -2) = -4$. Since these values are not equal, $f'(0) = \lim_{x\to 0} \frac{f(x)-f(0)}{x-0}$ does not exist.

(b)

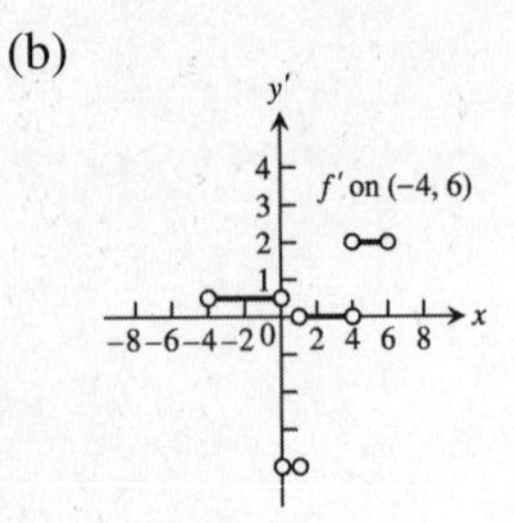

33.

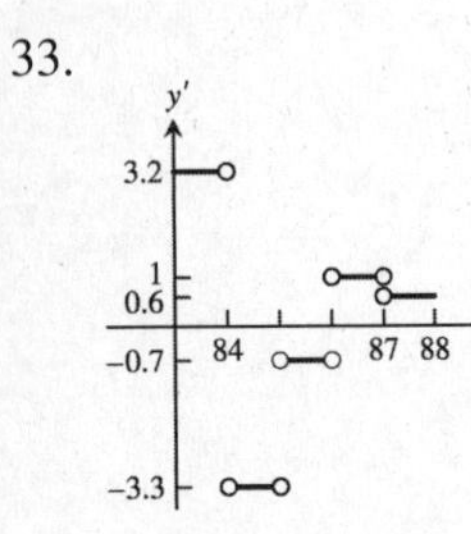

35. Answers may vary. In each case, draw a tangent line and estimate its slope.

(a) i) slope $\approx 1.54 \Rightarrow \frac{dT}{dt} \approx 1.54° \frac{F}{hr}$ ii) slope $\approx 2.86 \Rightarrow \frac{dT}{dt} \approx 2.86° \frac{F}{hr}$

iii) slope $\approx 0 \Rightarrow \frac{dT}{dt} \approx 0° \frac{F}{hr}$ iv) slope $\approx -3.75 \Rightarrow \frac{dT}{dt} \approx -3.75° \frac{F}{hr}$

(b) The tangent with the steepest positive slope appears to occur at $t = 6 \Rightarrow 12$ p.m. and slope $\approx 7.27 \Rightarrow \frac{dT}{dt} \approx 7.27° \frac{F}{hr}$. The tangent with the steepest negative slope appears to occur at $t = 12 \Rightarrow 6$ p.m. and slope $\approx -8.00 \Rightarrow \frac{dT}{dt} \approx -8.00° \frac{F}{hr}$

(c)

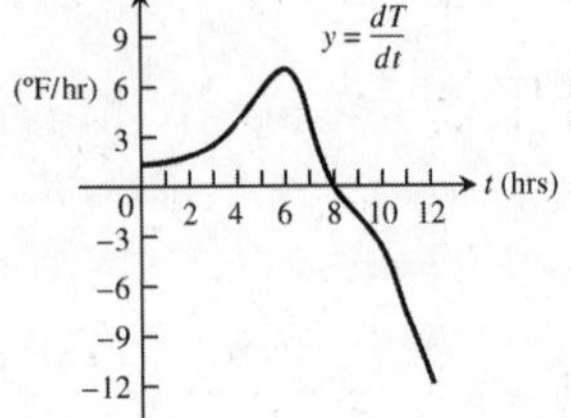

37. Left-hand derivative: For $h < 0$, $f(0 + h) = f(h) = h^2$ (using $y = x^2$ curve) $\Rightarrow \lim_{h \to 0^-} \frac{f(0+h) - f(0)}{h}$
$= \lim_{h \to 0^-} \frac{h^2 - 0}{h} = \lim_{h \to 0^-} h = 0$;

Right-hand derivative: For $h > 0$, $f(0 + h) = f(h) = h$ (using $y = x$ curve) $\Rightarrow \lim_{h \to 0^+} \frac{f(0+h) - f(0)}{h}$
$= \lim_{h \to 0^+} \frac{h - 0}{h} = \lim_{h \to 0^+} 1 = 1$;

Then $\lim_{h \to 0^-} \frac{f(0+h) - f(0)}{h} \neq \lim_{h \to 0^+} \frac{f(0+h) - f(0)}{h} \Rightarrow$ the derivative $f'(0)$ does not exist.

39. Left-hand derivative: When $h < 0$, $1 + h < 1 \Rightarrow f(1 + h) = \sqrt{1 + h} \Rightarrow \lim_{h \to 0^-} \frac{f(1+h) - f(1)}{h}$
$= \lim_{h \to 0^-} \frac{\sqrt{1+h} - 1}{h} = \lim_{h \to 0^-} \frac{\left(\sqrt{1+h} - 1\right)}{h} \cdot \frac{\left(\sqrt{1+h} + 1\right)}{\left(\sqrt{1+h} + 1\right)} = \lim_{h \to 0^-} \frac{(1+h) - 1}{h\left(\sqrt{1+h} + 1\right)} = \lim_{h \to 0^-} \frac{1}{\sqrt{1+h} + 1} = \frac{1}{2}$;

Right-hand derivative: When $h > 0$, $1 + h > 1 \Rightarrow f(1 + h) = 2(1 + h) - 1 = 2h + 1 \Rightarrow \lim_{h \to 0^+} \frac{f(1+h) - f(1)}{h}$
$= \lim_{h \to 0^+} \frac{(2h + 1) - 1}{h} = \lim_{h \to 0^+} 2 = 2$;

Then $\lim_{h \to 0^-} \frac{f(1+h) - f(1)}{h} \neq \lim_{h \to 0^+} \frac{f(1+h) - f(1)}{h} \Rightarrow$ the derivative $f'(1)$ does not exist.

41. f is not continuous at $x = 0$ since $\lim_{x \to 0} f(x) =$ does not exist and $f(0) = -1$

43. (a) The function is differentiable on its domain $-3 \le x \le 2$ (it is smooth)
    (b) none
    (c) none

45. (a) The function is differentiable on $-3 \le x < 0$ and $0 < x \le 3$
    (b) none
    (c) The function is neither continuous nor differentiable at $x = 0$ since $\lim\limits_{x \to 0^-} f(x) \ne \lim\limits_{x \to 0^+} f(x)$

47. (a) f is differentiable on $-1 \le x < 0$ and $0 < x \le 2$
    (b) f is continuous but not differentiable at $x = 0$: $\lim\limits_{x \to 0} f(x) = 0$ exists but there is a cusp at $x = 0$, so $f'(0) = \lim\limits_{h \to 0} \frac{f(0+h) - f(0)}{h}$ does not exist
    (c) none

49. (a) $f'(x) = \lim\limits_{h \to 0} \frac{f(x+h) - f(x)}{h} = \lim\limits_{h \to 0} \frac{-(x+h)^2 - (-x^2)}{h} = \lim\limits_{h \to 0} \frac{-x^2 - 2xh - h^2 + x^2}{h} = \lim\limits_{h \to 0} (-2x - h) = -2x$
    (b)

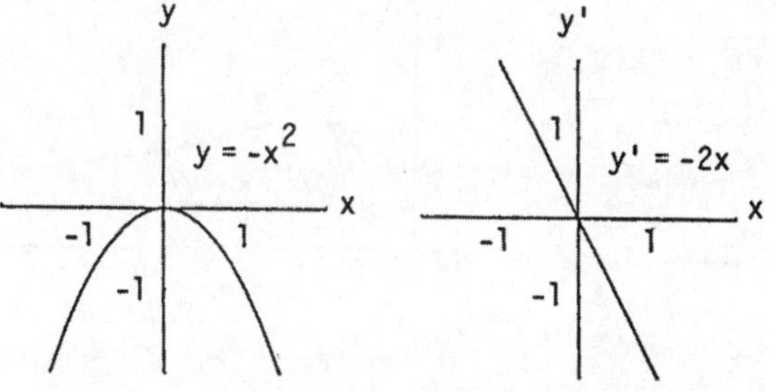

    (c) $y' = -2x$ is positive for $x < 0$, y' is zero when $x = 0$, y' is negative when $x > 0$
    (d) $y = -x^2$ is increasing for $-\infty < x < 0$ and decreasing for $0 < x < \infty$; the function is increasing on intervals where $y' > 0$ and decreasing on intervals where $y' < 0$

51. (a) Using the alternate formula for calculating derivatives: $f'(x) = \lim\limits_{z \to x} \frac{f(z) - f(x)}{z - x} = \lim\limits_{z \to x} \frac{\left(\frac{z^3}{3} - \frac{x^3}{3}\right)}{z - x}$
    $= \lim\limits_{z \to x} \frac{z^3 - x^3}{3(z - x)} = \lim\limits_{z \to x} \frac{(z - x)(z^2 + zx + x^2)}{3(z - x)} = \lim\limits_{z \to x} \frac{z^2 + zx + x^2}{3} = x^2 \Rightarrow f'(x) = x^2$
    (b)

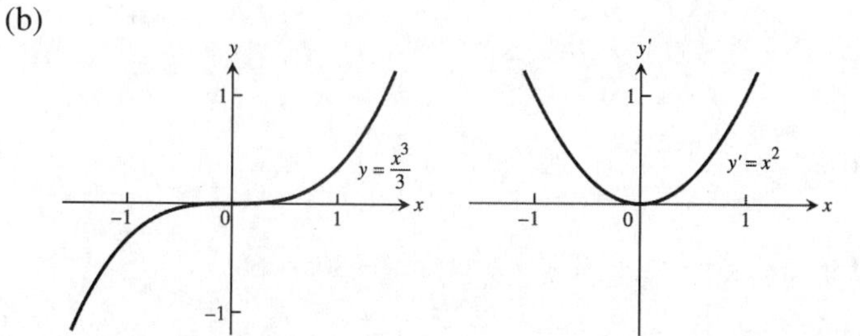

    (c) y' is positive for all $x \ne 0$, and $y' = 0$ when $x = 0$; y' is never negative
    (d) $y = \frac{x^3}{3}$ is increasing for all $x \ne 0$ (the graph is horizontal at $x = 0$) because y is increasing where $y' > 0$; y is never decreasing

53. $y' = \lim\limits_{h \to 0} \frac{(2(x+h)^2 - 13(x+h) + 5) - (2x^2 - 13x + 5)}{h} = \lim\limits_{h \to 0} \frac{2x^2 + 4xh + 2h^2 - 13x - 13h + 5 - 2x^2 + 13x - 5}{h}$
    $= \lim\limits_{h \to 0} \frac{4xh + 2h^2 - 13h}{h} = \lim\limits_{h \to 0} (4x + 2h - 13) = 4x - 13$, slope at x. The slope is -1 when $4x - 13 = -1$
    $\Rightarrow 4x = 12 \Rightarrow x = 3 \Rightarrow y = 2 \cdot 3^2 - 13 \cdot 3 + 5 = -16$. Thus the tangent line is $y + 16 = (-1)(x - 3)$
    $\Rightarrow y = -x - 13$ and the point of tangency is $(3, -16)$.

55. Yes; the derivative of $-f$ is $-f'$ so that $f'(x_0)$ exists $\Rightarrow -f'(x_0)$ exists as well.

57. Yes, $\lim_{t\to 0} \frac{g(t)}{h(t)}$ can exist but it need not equal zero. For example, let $g(t) = mt$ and $h(t) = t$. Then $g(0) = h(0)$ $= 0$, but $\lim_{t\to 0} \frac{g(t)}{h(t)} = \lim_{t\to 0} \frac{mt}{t} = \lim_{t\to 0} m = m$, which need not be zero.

59. The graphs are shown below for $h = 1, 0.5, 0.1$. The function $y = \frac{1}{2\sqrt{x}}$ is the derivative of the function $y = \sqrt{x}$ so that $\frac{1}{2\sqrt{x}} = \lim_{h\to 0} \frac{\sqrt{x+h}-\sqrt{x}}{h}$. The graphs reveal that $y = \frac{\sqrt{x+h}-\sqrt{x}}{h}$ gets closer to $y = \frac{1}{2\sqrt{x}}$ as h gets smaller and smaller.

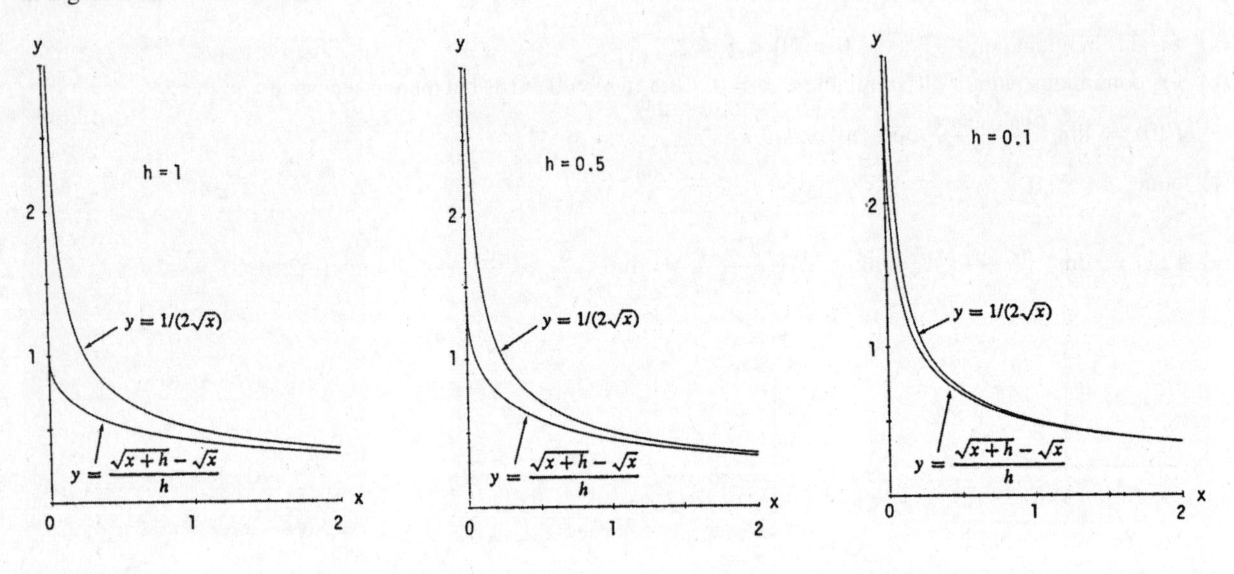

61. The graphs are the same. So we know that for $f(x) = |x|$, we have $f'(x) = \frac{|x|}{x}$.

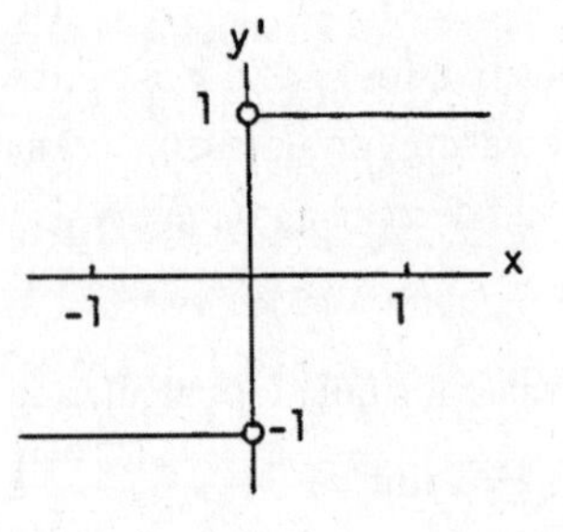

3.3 DIFFERENTIATION RULES

1. $y = -x^2 + 3 \Rightarrow \frac{dy}{dx} = \frac{d}{dx}(-x^2) + \frac{d}{dx}(3) = -2x + 0 = -2x \Rightarrow \frac{d^2y}{dx^2} = -2$

3. $s = 5t^3 - 3t^5 \Rightarrow \frac{ds}{dt} = \frac{d}{dt}(5t^3) - \frac{d}{dt}(3t^5) = 15t^2 - 15t^4 \Rightarrow \frac{d^2s}{dt^2} = \frac{d}{dt}(15t^2) - \frac{d}{dt}(15t^4) = 30t - 60t^3$

5. $y = \frac{4}{3}x^3 - x \Rightarrow \frac{dy}{dx} = 4x^2 - 1 \Rightarrow \frac{d^2y}{dx^2} = 8x$

7. $w = 3z^{-2} - z^{-1} \Rightarrow \frac{dw}{dz} = -6z^{-3} + z^{-2} = \frac{-6}{z^3} + \frac{1}{z^2} \Rightarrow \frac{d^2w}{dz^2} = 18z^{-4} - 2z^{-3} = \frac{18}{z^4} - \frac{2}{z^3}$

9. $y = 6x^2 - 10x - 5x^{-2} \Rightarrow \frac{dy}{dx} = 12x - 10 + 10x^{-3} = 12x - 10 + \frac{10}{x^3} \Rightarrow \frac{d^2y}{dx^2} = 12 - 0 - 30x^{-4} = 12 - \frac{30}{x^4}$

11. $r = \frac{1}{3}s^{-2} - \frac{5}{2}s^{-1} \Rightarrow \frac{dr}{ds} = -\frac{2}{3}s^{-3} + \frac{5}{2}s^{-2} = \frac{-2}{3s^3} + \frac{5}{2s^2} \Rightarrow \frac{d^2r}{ds^2} = 2s^{-4} - 5s^{-3} = \frac{2}{s^4} - \frac{5}{s^3}$

13. (a) $y = (3 - x^2)(x^3 - x + 1) \Rightarrow y' = (3 - x^2) \cdot \frac{d}{dx}(x^3 - x + 1) + (x^3 - x + 1) \cdot \frac{d}{dx}(3 - x^2)$
$= (3 - x^2)(3x^2 - 1) + (x^3 - x + 1)(-2x) = -5x^4 + 12x^2 - 2x - 3$

(b) $y = -x^5 + 4x^3 - x^2 - 3x + 3 \Rightarrow y' = -5x^4 + 12x^2 - 2x - 3$

15. (a) $y = (x^2+1)\left(x+5+\frac{1}{x}\right) \Rightarrow y' = (x^2+1)\cdot\frac{d}{dx}\left(x+5+\frac{1}{x}\right) + \left(x+5+\frac{1}{x}\right)\cdot\frac{d}{dx}(x^2+1)$
$= (x^2+1)(1-x^{-2}) + (x+5+x^{-1})(2x) = (x^2-1+1-x^{-2}) + (2x^2+10x+2) = 3x^2+10x+2-\frac{1}{x^2}$
(b) $y = x^3+5x^2+2x+5+\frac{1}{x} \Rightarrow y' = 3x^2+10x+2-\frac{1}{x^2}$

17. $y = \frac{2x+5}{3x-2}$; use the quotient rule: $u = 2x+5$ and $v = 3x-2 \Rightarrow u' = 2$ and $v' = 3 \Rightarrow y' = \frac{vu'-uv'}{v^2}$
$= \frac{(3x-2)(2)-(2x+5)(3)}{(3x-2)^2} = \frac{6x-4-6x-15}{(3x-2)^2} = \frac{-19}{(3x-2)^2}$

19. $g(x) = \frac{x^2-4}{x+0.5}$; use the quotient rule: $u = x^2-4$ and $v = x+0.5 \Rightarrow u' = 2x$ and $v' = 1 \Rightarrow g'(x) = \frac{vu'-uv'}{v^2}$
$= \frac{(x+0.5)(2x)-(x^2-4)(1)}{(x+0.5)^2} = \frac{2x^2+x-x^2+4}{(x+0.5)^2} = \frac{x^2+x+4}{(x+0.5)^2}$

21. $v = (1-t)(1+t^2)^{-1} = \frac{1-t}{1+t^2} \Rightarrow \frac{dv}{dt} = \frac{(1+t^2)(-1)-(1-t)(2t)}{(1+t^2)^2} = \frac{-1-t^2-2t+2t^2}{(1+t^2)^2} = \frac{t^2-2t-1}{(1+t^2)^2}$

23. $f(s) = \frac{\sqrt{s}-1}{\sqrt{s}+1} \Rightarrow f'(s) = \frac{(\sqrt{s}+1)\left(\frac{1}{2\sqrt{s}}\right)-(\sqrt{s}-1)\left(\frac{1}{2\sqrt{s}}\right)}{(\sqrt{s}+1)^2} = \frac{(\sqrt{s}+1)-(\sqrt{s}-1)}{2\sqrt{s}(\sqrt{s}+1)^2} = \frac{1}{\sqrt{s}(\sqrt{s}+1)^2}$
NOTE: $\frac{d}{ds}(\sqrt{s}) = \frac{1}{2\sqrt{s}}$ from Example 2 in Section 3.2

25. $v = \frac{1+x-4\sqrt{x}}{x} \Rightarrow v' = \frac{x\left(1-\frac{2}{\sqrt{x}}\right)-(1+x-4\sqrt{x})}{x^2} = \frac{2\sqrt{x}-1}{x^2}$

27. $y = \frac{1}{(x^2-1)(x^2+x+1)}$; use the quotient rule: $u = 1$ and $v = (x^2-1)(x^2+x+1) \Rightarrow u' = 0$ and
$v' = (x^2-1)(2x+1)+(x^2+x+1)(2x) = 2x^3+x^2-2x-1+2x^3+2x^2+2x = 4x^3+3x^2-1$
$\Rightarrow \frac{dy}{dx} = \frac{vu'-uv'}{v^2} = \frac{0-1(4x^3+3x^2-1)}{(x^2-1)^2(x^2+x+1)^2} = \frac{-4x^3-3x^2+1}{(x^2-1)^2(x^2+x+1)^2}$

29. $y = \frac{1}{2}x^4 - \frac{3}{2}x^2 - x \Rightarrow y' = 2x^3-3x-1 \Rightarrow y'' = 6x^2-3 \Rightarrow y''' = 12x \Rightarrow y^{(4)} = 12 \Rightarrow y^{(n)} = 0$ for all $n \ge 5$

31. $y = (x-1)(x^2+3x-5) = x^3+2x^2-8x+5 \Rightarrow y' = 3x^2+4x-8 \Rightarrow y'' = 6x+4 \Rightarrow y''' = 6 \Rightarrow y^{(n)} = 0$ for all $n \ge 4$

33. $y = \frac{x^3+7}{x} = x^2+7x^{-1} \Rightarrow \frac{dy}{dx} = 2x-7x^{-2} = 2x-\frac{7}{x^2} \Rightarrow \frac{d^2y}{dx^2} = 2+14x^{-3} = 2+\frac{14}{x^3}$

35. $r = \frac{(\theta-1)(\theta^2+\theta+1)}{\theta^3} = \frac{\theta^3-1}{\theta^3} = 1-\frac{1}{\theta^3} = 1-\theta^{-3} \Rightarrow \frac{dr}{d\theta} = 0+3\theta^{-4} = 3\theta^{-4} = \frac{3}{\theta^4} \Rightarrow \frac{d^2r}{d\theta^2} = -12\theta^{-5} = \frac{-12}{\theta^5}$

37. $w = \left(\frac{1+3z}{3z}\right)(3-z) = \left(\frac{1}{3}z^{-1}+1\right)(3-z) = z^{-1}-\frac{1}{3}+3-z = z^{-1}+\frac{8}{3}-z \Rightarrow \frac{dw}{dz} = -z^{-2}+0-1 = -z^{-2}-1$
$= \frac{-1}{z^2}-1 \Rightarrow \frac{d^2w}{dz^2} = 2z^{-3}-0 = 2z^{-3} = \frac{2}{z^3}$

39. $p = \left(\frac{q^2+3}{12q}\right)\left(\frac{q^4-1}{q^3}\right) = \frac{q^6-q^2+3q^4-3}{12q^4} = \frac{1}{12}q^2-\frac{1}{12}q^{-2}+\frac{1}{4}-\frac{1}{4}q^{-4} \Rightarrow \frac{dp}{dq} = \frac{1}{6}q+\frac{1}{6}q^{-3}+q^{-5} = \frac{1}{6}q+\frac{1}{6q^3}+\frac{1}{q^5}$
$\Rightarrow \frac{d^2p}{dq^2} = \frac{1}{6}-\frac{1}{2}q^{-4}-5q^{-6} = \frac{1}{6}-\frac{1}{2q^4}-\frac{5}{q^6}$

41. $u(0) = 5,\ u'(0) = -3,\ v(0) = -1,\ v'(0) = 2$
(a) $\frac{d}{dx}(uv) = uv'+vu' \Rightarrow \frac{d}{dx}(uv)\big|_{x=0} = u(0)v'(0)+v(0)u'(0) = 5\cdot 2+(-1)(-3) = 13$
(b) $\frac{d}{dx}\left(\frac{u}{v}\right) = \frac{vu'-uv'}{v^2} \Rightarrow \frac{d}{dx}\left(\frac{u}{v}\right)\big|_{x=0} = \frac{v(0)u'(0)-u(0)v'(0)}{(v(0))^2} = \frac{(-1)(-3)-(5)(2)}{(-1)^2} = -7$
(c) $\frac{d}{dx}\left(\frac{v}{u}\right) = \frac{uv'-vu'}{u^2} \Rightarrow \frac{d}{dx}\left(\frac{v}{u}\right)\big|_{x=0} = \frac{u(0)v'(0)-v(0)u'(0)}{(u(0))^2} = \frac{(5)(2)-(-1)(-3)}{(5)^2} = \frac{7}{25}$

(d) $\frac{d}{dx}(7v - 2u) = 7v' - 2u' \Rightarrow \frac{d}{dx}(7v - 2u)\big|_{x=0} = 7v'(0) - 2u'(0) = 7 \cdot 2 - 2(-3) = 20$

43. $y = x^3 - 4x + 1$. Note that $(2, 1)$ is on the curve: $1 = 2^3 - 4(2) + 1$

(a) Slope of the tangent at (x, y) is $y' = 3x^2 - 4 \Rightarrow$ slope of the tangent at $(2, 1)$ is $y'(2) = 3(2)^2 - 4 = 8$. Thus the slope of the line perpendicular to the tangent at $(2, 1)$ is $-\frac{1}{8} \Rightarrow$ the equation of the line perpendicular to the tangent line at $(2, 1)$ is $y - 1 = -\frac{1}{8}(x - 2)$ or $y = -\frac{x}{8} + \frac{5}{4}$.

(b) The slope of the curve at x is $m = 3x^2 - 4$ and the smallest value for m is -4 when $x = 0$ and $y = 1$.

(c) We want the slope of the curve to be $8 \Rightarrow y' = 8 \Rightarrow 3x^2 - 4 = 8 \Rightarrow 3x^2 = 12 \Rightarrow x^2 = 4 \Rightarrow x = \pm 2$. When $x = 2$, $y = 1$ and the tangent line has equation $y - 1 = 8(x - 2)$ or $y = 8x - 15$; when $x = -2$, $y = (-2)^3 - 4(-2) + 1 = 1$, and the tangent line has equation $y - 1 = 8(x + 2)$ or $y = 8x + 17$.

45. $y = \frac{4x}{x^2+1} \Rightarrow \frac{dy}{dx} = \frac{(x^2+1)(4) - (4x)(2x)}{(x^2+1)^2} = \frac{4x^2+4-8x^2}{(x^2+1)^2} = \frac{4(-x^2+1)}{(x^2+1)^2}$. When $x = 0$, $y = 0$ and $y' = \frac{4(0+1)}{1} = 4$, so the tangent to the curve at $(0, 0)$ is the line $y = 4x$. When $x = 1$, $y = 2 \Rightarrow y' = 0$, so the tangent to the curve at $(1, 2)$ is the line $y = 2$.

47. $y = ax^2 + bx + c$ passes through $(0, 0) \Rightarrow 0 = a(0) + b(0) + c \Rightarrow c = 0$; $y = ax^2 + bx$ passes through $(1, 2)$ $\Rightarrow 2 = a + b$; $y' = 2ax + b$ and since the curve is tangent to $y = x$ at the origin, its slope is 1 at $x = 0$ $\Rightarrow y' = 1$ when $x = 0 \Rightarrow 1 = 2a(0) + b \Rightarrow b = 1$. Then $a + b = 2 \Rightarrow a = 1$. In summary $a = b = 1$ and $c = 0$ so the curve is $y = x^2 + x$.

49. $y = 8x + 5 \Rightarrow m = 8$; $f(x) = 3x^2 - 4x \Rightarrow f'(x) = 6x - 4$; $6x - 4 = 8 \Rightarrow x = 2 \Rightarrow f(2) = 3(2)^2 - 4(2) = 4 \Rightarrow (2, 4)$

51. $y = 2x + 3 \Rightarrow m = 2 \Rightarrow m_\perp = -\frac{1}{2}$; $y = \frac{x}{x-2} \Rightarrow y' = \frac{(x-2)(1) - x(1)}{(x-2)^2} = \frac{-2}{(x-2)^2}$; $\frac{-2}{(x-2)^2} = -\frac{1}{2} \Rightarrow 4 = (x-2)^2$ $\Rightarrow \pm 2 = x - 2 \Rightarrow x = 4$ or $x = 0 \Rightarrow$ if $x = 4$, $y = \frac{4}{4-2} = 2$, and if $x = 0$, $y = \frac{0}{0-2} = 0 \Rightarrow (4, 2)$ or $(0, 0)$.

53. (a) $y = x^3 - x \Rightarrow y' = 3x^2 - 1$. When $x = -1$, $y = 0$ and $y' = 2 \Rightarrow$ the tangent line to the curve at $(-1, 0)$ is $y = 2(x + 1)$ or $y = 2x + 2$.

(b)

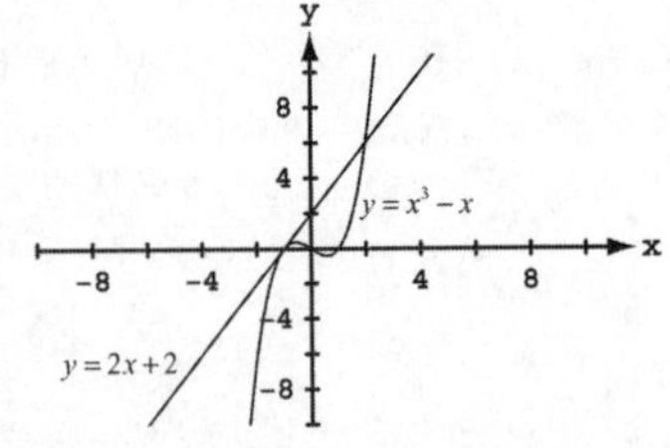

(c) $\left.\begin{array}{l} y = x^3 - x \\ y = 2x + 2 \end{array}\right\} \Rightarrow x^3 - x = 2x + 2 \Rightarrow x^3 - 3x - 2 = (x - 2)(x + 1)^2 = 0 \Rightarrow x = 2$ or $x = -1$. Since $y = 2(2) + 2 = 6$; the other intersection point is $(2, 6)$

55. $\lim\limits_{x \to 1} \frac{x^{50} - 1}{x - 1} = 50x^{49}\Big|_{x=1} = 50(1)^{49} = 50$

57. $g'(x) = \begin{cases} 2x - 3 & x > 0 \\ a & x < 0 \end{cases}$, since g is differentiable at $x = 0 \Rightarrow \lim\limits_{x \to 0^+}(2x - 3) = -3$ and $\lim\limits_{x \to 0^-} a = a \Rightarrow a = -3$

59. $P(x) = a_n x^n + a_{n-1}x^{n-1} + \cdots + a_2 x^2 + a_1 x + a_0 \Rightarrow P'(x) = na_n x^{n-1} + (n-1)a_{n-1}x^{n-2} + \cdots + 2a_2 x + a_1$

61. Let c be a constant $\Rightarrow \frac{dc}{dx} = 0 \Rightarrow \frac{d}{dx}(u \cdot c) = u \cdot \frac{dc}{dx} + c \cdot \frac{du}{dx} = u \cdot 0 + c\,\frac{du}{dx} = c\,\frac{du}{dx}$. Thus when one of the functions is a constant, the Product Rule is just the Constant Multiple Rule $\Rightarrow$ the Constant Multiple Rule is a special case of the Product Rule.

63. (a) $\frac{d}{dx}(uvw) = \frac{d}{dx}((uv) \cdot w) = (uv)\frac{dw}{dx} + w \cdot \frac{d}{dx}(uv) = uv\frac{dw}{dx} + w\left(u\frac{dv}{dx} + v\frac{du}{dx}\right) = uv\frac{dw}{dx} + wu\frac{dv}{dx} + wv\frac{du}{dx}$
$= uvw' + uv'w + u'vw$

(b) $\frac{d}{dx}(u_1u_2u_3u_4) = \frac{d}{dx}((u_1u_2u_3)\,u_4) = (u_1u_2u_3)\frac{du_4}{dx} + u_4\frac{d}{dx}(u_1u_2u_3) \Rightarrow \frac{d}{dx}(u_1u_2u_3u_4)$
$= u_1u_2u_3\frac{du_4}{dx} + u_4\left(u_1u_2\frac{du_3}{dx} + u_3u_1\frac{du_2}{dx} + u_3u_2\frac{du_1}{dx}\right)$ (using (a) above)
$\Rightarrow \frac{d}{dx}(u_1u_2u_3u_4) = u_1u_2u_3\frac{du_4}{dx} + u_1u_2u_4\frac{du_3}{dx} + u_1u_3u_4\frac{du_2}{dx} + u_2u_3u_4\frac{du_1}{dx}$
$= u_1u_2u_3u_4' + u_1u_2u_3'u_4 + u_1u_2'u_3u_4 + u_1'u_2u_3u_4$

(c) Generalizing (a) and (b) above, $\frac{d}{dx}(u_1 \cdots u_n) = u_1u_2\cdots u_{n-1}u_n' + u_1u_2\cdots u_{n-2}u_{n-1}'u_n + \ldots + u_1'u_2\cdots u_n$

65. $P = \frac{nRT}{V-nb} - \frac{an^2}{V^2}$. We are holding T constant, and a, b, n, R are also constant so their derivatives are zero
$\Rightarrow \frac{dP}{dV} = \frac{(V-nb)\cdot 0 - (nRT)(1)}{(V-nb)^2} - \frac{V^2(0) - (an^2)(2V)}{(V^2)^2} = \frac{-nRT}{(V-nb)^2} + \frac{2an^2}{V^3}$

3.4 THE DERIVATIVE AS A RATE OF CHANGE

1. $s = t^2 - 3t + 2,\ 0 \le t \le 2$

(a) displacement $= \Delta s = s(2) - s(0) = 0\text{m} - 2\text{m} = -2$ m, $v_{av} = \frac{\Delta s}{\Delta t} = \frac{-2}{2} = -1$ m/sec

(b) $v = \frac{ds}{dt} = 2t - 3 \Rightarrow |v(0)| = |-3| = 3$ m/sec and $|v(2)| = 1$ m/sec;
$a = \frac{d^2s}{dt^2} = 2 \Rightarrow a(0) = 2\ \text{m/sec}^2$ and $a(2) = 2\ \text{m/sec}^2$

(c) $v = 0 \Rightarrow 2t - 3 = 0 \Rightarrow t = \frac{3}{2}$. v is negative in the interval $0 < t < \frac{3}{2}$ and v is positive when $\frac{3}{2} < t < 2 \Rightarrow$ the body changes direction at $t = \frac{3}{2}$.

3. $s = -t^3 + 3t^2 - 3t,\ 0 \le t \le 3$

(a) displacement $= \Delta s = s(3) - s(0) = -9$ m, $v_{av} = \frac{\Delta s}{\Delta t} = \frac{-9}{3} = -3$ m/sec

(b) $v = \frac{ds}{dt} = -3t^2 + 6t - 3 \Rightarrow |v(0)| = |-3| = 3$ m/sec and $|v(3)| = |-12| = 12$ m/sec; $a = \frac{d^2s}{dt^2} = -6t + 6$
$\Rightarrow a(0) = 6\ \text{m/sec}^2$ and $a(3) = -12\ \text{m/sec}^2$

(c) $v = 0 \Rightarrow -3t^2 + 6t - 3 = 0 \Rightarrow t^2 - 2t + 1 = 0 \Rightarrow (t-1)^2 = 0 \Rightarrow t = 1$. For all other values of t in the interval the velocity v is negative (the graph of $v = -3t^2 + 6t - 3$ is a parabola with vertex at $t = 1$ which opens downward $\Rightarrow$ the body never changes direction).

5. $s = \frac{25}{t^2} - \frac{5}{t},\ 1 \le t \le 5$

(a) $\Delta s = s(5) - s(1) = -20$ m, $v_{av} = \frac{-20}{4} = -5$ m/sec

(b) $v = \frac{-50}{t^3} + \frac{5}{t^2} \Rightarrow |v(1)| = 45$ m/sec and $|v(5)| = \frac{1}{5}$ m/sec; $a = \frac{150}{t^4} - \frac{10}{t^3} \Rightarrow a(1) = 140\ \text{m/sec}^2$ and $a(5) = \frac{4}{25}\ \text{m/sec}^2$

(c) $v = 0 \Rightarrow \frac{-50+5t}{t^3} = 0 \Rightarrow -50 + 5t = 0 \Rightarrow t = 10 \Rightarrow$ the body does not change direction in the interval

7. $s = t^3 - 6t^2 + 9t$ and let the positive direction be to the right on the s-axis.

(a) $v = 3t^2 - 12t + 9$ so that $v = 0 \Rightarrow t^2 - 4t + 3 = (t-3)(t-1) = 0 \Rightarrow t = 1$ or 3; $a = 6t - 12 \Rightarrow a(1) = -6\ \text{m/sec}^2$ and $a(3) = 6\ \text{m/sec}^2$. Thus the body is motionless but being accelerated left when $t = 1$, and motionless but being accelerated right when $t = 3$.

(b) $a = 0 \Rightarrow 6t - 12 = 0 \Rightarrow t = 2$ with speed $|v(2)| = |12 - 24 + 9| = 3$ m/sec

(c) The body moves to the right or forward on $0 \le t < 1$, and to the left or backward on $1 < t < 2$. The positions are $s(0) = 0$, $s(1) = 4$ and $s(2) = 2 \Rightarrow$ total distance $= |s(1) - s(0)| + |s(2) - s(1)| = |4| + |-2| = 6$ m.

9. $s_m = 1.86t^2 \Rightarrow v_m = 3.72t$ and solving $3.72t = 27.8 \Rightarrow t \approx 7.5$ sec on Mars; $s_j = 11.44t^2 \Rightarrow v_j = 22.88t$ and solving $22.88t = 27.8 \Rightarrow t \approx 1.2$ sec on Jupiter.

11. $s = 15t - \frac{1}{2} g_s t^2 \Rightarrow v = 15 - g_s t$ so that $v = 0 \Rightarrow 15 - g_s t = 0 \Rightarrow g_s = \frac{15}{t}$. Therefore $g_s = \frac{15}{20} = \frac{3}{4} = 0.75$ m/sec^2

13. (a) $s = 179 - 16t^2 \Rightarrow v = -32t \Rightarrow \text{speed} = |v| = 32t$ ft/sec and $a = -32$ ft/sec^2

(b) $s = 0 \Rightarrow 179 - 16t^2 = 0 \Rightarrow t = \sqrt{\frac{179}{16}} \approx 3.3$ sec

(c) When $t = \sqrt{\frac{179}{16}}$, $v = -32\sqrt{\frac{179}{16}} = -8\sqrt{179} \approx -107.0$ ft/sec

15. (a) at 2 and 7 seconds

(b) between 3 and 6 seconds: $3 \le t \le 6$

(c)

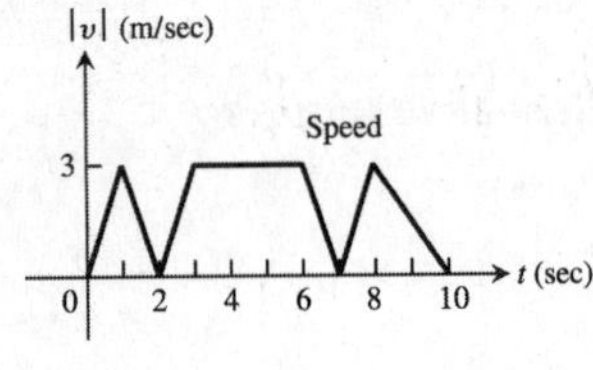

(d)

17. (a) 190 ft/sec

(b) 2 sec

(c) at 8 sec, 0 ft/sec

(d) 10.8 sec, 90 ft/sec

(e) From $t = 8$ until $t = 10.8$ sec, a total of 2.8 sec

(f) Greatest acceleration happens 2 sec after launch

(g) From $t = 2$ to $t = 10.8$ sec; during this period, $a = \frac{v(10.8) - v(2)}{10.8 - 2} \approx -32$ ft/sec^2

19. $s = 490t^2 \Rightarrow v = 980t \Rightarrow a = 980$

(a) Solving $160 = 490t^2 \Rightarrow t = \frac{4}{7}$ sec. The average velocity was $\frac{s(4/7) - s(0)}{4/7} = 280$ cm/sec.

(b) At the 160 cm mark the balls are falling at $v(4/7) = 560$ cm/sec. The acceleration at the 160 cm mark was 980 cm/sec^2.

(c) The light was flashing at a rate of $\frac{17}{4/7} = 29.75$ flashes per second.

21. C = position, A = velocity, and B = acceleration. Neither A nor C can be the derivative of B because B's derivative is constant. Graph C cannot be the derivative of A either, because A has some negative slopes while C has only positive values. So, C (being the derivative of neither A nor B) must be the graph of position. Curve C has both positive and negative slopes, so its derivative, the velocity, must be A and not B. That leaves B for acceleration.

23. (a) $c(100) = 11{,}000 \Rightarrow c_{av} = \frac{11{,}000}{100} = \$110$

(b) $c(x) = 2000 + 100x - .1x^2 \Rightarrow c'(x) = 100 - .2x$. Marginal cost $= c'(x) \Rightarrow$ the marginal cost of producing 100 machines is $c'(100) = \$80$

(c) The cost of producing the 101st machine is $c(101) - c(100) = 100 - \frac{201}{10} = \$79.90$

25. $b(t) = 10^6 + 10^4 t - 10^3 t^2 \Rightarrow b'(t) = 10^4 - (2)(10^3 t) = 10^3(10 - 2t)$

(a) $b'(0) = 10^4$ bacteria/hr

(b) $b'(5) = 0$ bacteria/hr

(c) $b'(10) = -10^4$ bacteria/hr

27. (a) $y = 6\left(1 - \frac{t}{12}\right)^2 = 6\left(1 - \frac{t}{6} + \frac{t^2}{144}\right) \Rightarrow \frac{dy}{dt} = \frac{t}{12} - 1$

(b) The largest value of $\frac{dy}{dt}$ is 0 m/h when $t = 12$ and the fluid level is falling the slowest at that time. The smallest value of $\frac{dy}{dt}$ is -1 m/h, when $t = 0$, and the fluid level is falling the fastest at that time.

(c) In this situation, $\frac{dy}{dt} \le 0 \Rightarrow$ the graph of y is always decreasing. As $\frac{dy}{dt}$ increases in value, the slope of the graph of y increases from -1 to 0 over the interval $0 \le t \le 12$.

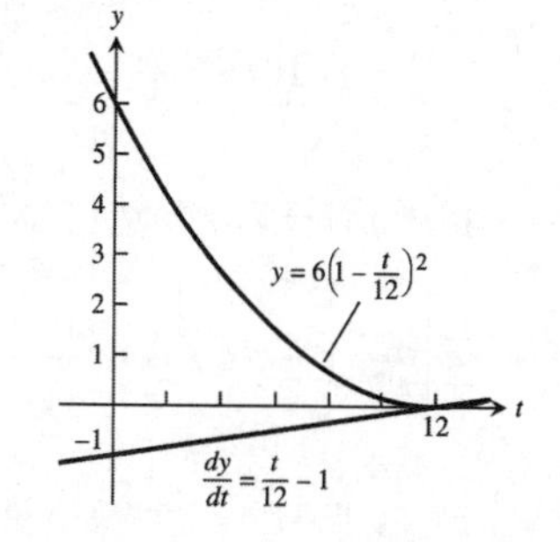

29. 200 km/hr $= 55\frac{5}{9}$ m/sec $= \frac{500}{9}$ m/sec, and $D = \frac{10}{9}t^2 \Rightarrow V = \frac{20}{9}t$. Thus $V = \frac{500}{9} \Rightarrow \frac{20}{9}t = \frac{500}{9} \Rightarrow t = 25$ sec. When $t = 25$, $D = \frac{10}{9}(25)^2 = \frac{6250}{9}$ m

31.

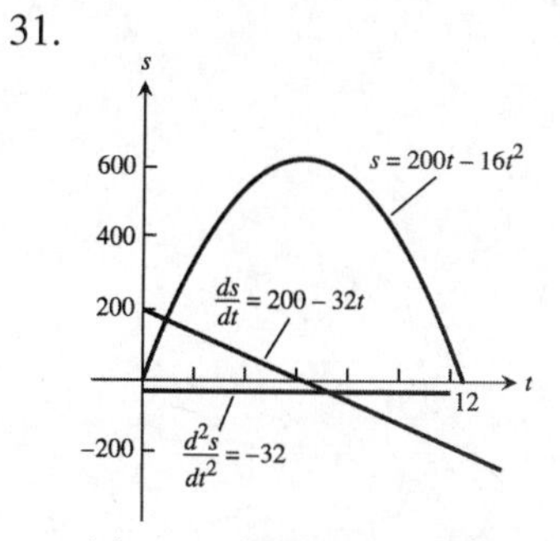

(a) $v = 0$ when $t = 6.25$ sec

(b) $v > 0$ when $0 \le t < 6.25 \Rightarrow$ body moves right (up); $v < 0$ when $6.25 < t \le 12.5 \Rightarrow$ body moves left (down)

(c) body changes direction at $t = 6.25$ sec

(d) body speeds up on $(6.25, 12.5]$ and slows down on $[0, 6.25)$

(e) The body is moving fastest at the endpoints $t = 0$ and $t = 12.5$ when it is traveling 200 ft/sec. It's moving slowest at $t = 6.25$ when the speed is 0.

(f) When $t = 6.25$ the body is $s = 625$ m from the origin and farthest away.

33.

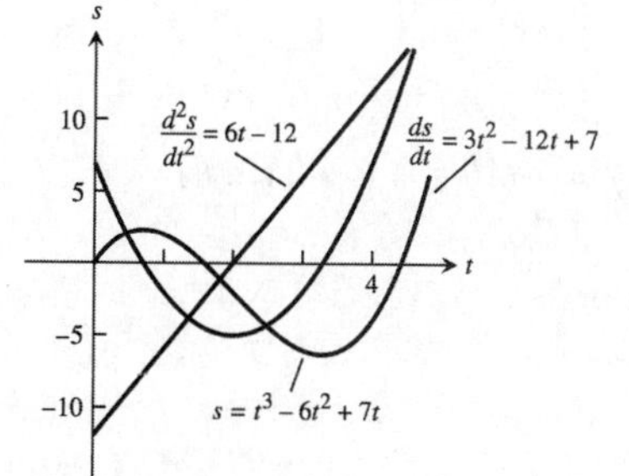

(a) $v = 0$ when $t = \frac{6 \pm \sqrt{15}}{3}$ sec

(b) $v < 0$ when $\frac{6-\sqrt{15}}{3} < t < \frac{6+\sqrt{15}}{3} \Rightarrow$ body moves left (down); $v > 0$ when $0 \le t < \frac{6-\sqrt{15}}{3}$ or $\frac{6+\sqrt{15}}{3} < t \le 4$ $\Rightarrow$ body moves right (up)

(c) body changes direction at $t = \frac{6 \pm \sqrt{15}}{3}$ sec

(d) body speeds up on $\left(\frac{6-\sqrt{15}}{3}, 2\right) \cup \left(\frac{6+\sqrt{15}}{3}, 4\right]$ and slows down on $\left[0, \frac{6-\sqrt{15}}{3}\right) \cup \left(2, \frac{6+\sqrt{15}}{3}\right)$.

(e) The body is moving fastest at $t = 0$ and $t = 4$ when it is moving 7 units/sec and slowest at $t = \frac{6 \pm \sqrt{15}}{3}$ sec

(f) When $t = \frac{6+\sqrt{15}}{3}$ the body is at position $s \approx -6.303$ units and farthest from the origin.

3.5 DERIVATIVES OF TRIGONOMETRIC FUNCTIONS

1. $y = -10x + 3\cos x \Rightarrow \frac{dy}{dx} = -10 + 3\frac{d}{dx}(\cos x) = -10 - 3\sin x$

3. $y = x^2\cos x \Rightarrow \frac{dy}{dx} = x^2(-\sin x) + 2x\cos x = -x^2\sin x + 2x\cos x$

5. $y = \csc x - 4\sqrt{x} + 7 \Rightarrow \frac{dy}{dx} = -\csc x\cot x - \frac{4}{2\sqrt{x}} + 0 = -\csc x\cot x - \frac{2}{\sqrt{x}}$

7. $f(x) = \sin x\tan x \Rightarrow f'(x) = \sin x\sec^2 x + \cos x\tan x = \sin x\sec^2 x + \cos x\frac{\sin x}{\cos x} = \sin x(\sec^2 x + 1)$

9. $y = (\sec x + \tan x)(\sec x - \tan x) \Rightarrow \frac{dy}{dx} = (\sec x + \tan x)\frac{d}{dx}(\sec x - \tan x) + (\sec x - \tan x)\frac{d}{dx}(\sec x + \tan x)$
$= (\sec x + \tan x)(\sec x\tan x - \sec^2 x) + (\sec x - \tan x)(\sec x\tan x + \sec^2 x)$
$= (\sec^2 x\tan x + \sec x\tan^2 x - \sec^3 x - \sec^2 x\tan x) + (\sec^2 x\tan x - \sec x\tan^2 x + \sec^3 x - \tan x\sec^2 x) = 0.$
$\left(\text{Note also that } y = \sec^2 x - \tan^2 x = (\tan^2 x + 1) - \tan^2 x = 1 \Rightarrow \frac{dy}{dx} = 0.\right)$

11. $y = \frac{\cot x}{1+\cot x} \Rightarrow \frac{dy}{dx} = \frac{(1+\cot x)\frac{d}{dx}(\cot x) - (\cot x)\frac{d}{dx}(1+\cot x)}{(1+\cot x)^2} = \frac{(1+\cot x)(-\csc^2 x) - (\cot x)(-\csc^2 x)}{(1+\cot x)^2}$
$= \frac{-\csc^2 x - \csc^2 x\cot x + \csc^2 x\cot x}{(1+\cot x)^2} = \frac{-\csc^2 x}{(1+\cot x)^2}$

13. $y = \frac{4}{\cos x} + \frac{1}{\tan x} = 4\sec x + \cot x \Rightarrow \frac{dy}{dx} = 4\sec x\tan x - \csc^2 x$

15. $y = x^2\sin x + 2x\cos x - 2\sin x \Rightarrow \frac{dy}{dx} = (x^2\cos x + (\sin x)(2x)) + ((2x)(-\sin x) + (\cos x)(2)) - 2\cos x$
$= x^2\cos x + 2x\sin x - 2x\sin x + 2\cos x - 2\cos x = x^2\cos x$

17. $f(x) = x^3\sin x\cos x \Rightarrow f'(x) = x^3\sin x(-\sin x) + x^3\cos x(\cos x) + 3x^2\sin x\cos x = -x^3\sin^2 x + x^3\cos^2 x + 3x^2\sin x\cos x$

19. $s = \tan t - t \Rightarrow \frac{ds}{dt} = \sec^2 t - 1$

21. $s = \frac{1+\csc t}{1-\csc t} \Rightarrow \frac{ds}{dt} = \frac{(1-\csc t)(-\csc t\cot t) - (1+\csc t)(\csc t\cot t)}{(1-\csc t)^2}$
$= \frac{-\csc t\cot t + \csc^2 t\cot t - \csc t\cot t - \csc^2 t\cot t}{(1-\csc t)^2} = \frac{-2\csc t\cot t}{(1-\csc t)^2}$

23. $r = 4 - \theta^2\sin\theta \Rightarrow \frac{dr}{d\theta} = -\left(\theta^2\frac{d}{d\theta}(\sin\theta) + (\sin\theta)(2\theta)\right) = -(\theta^2\cos\theta + 2\theta\sin\theta) = -\theta(\theta\cos\theta + 2\sin\theta)$

25. $r = \sec\theta\csc\theta \Rightarrow \frac{dr}{d\theta} = (\sec\theta)(-\csc\theta\cot\theta) + (\csc\theta)(\sec\theta\tan\theta)$
$= \left(\frac{-1}{\cos\theta}\right)\left(\frac{1}{\sin\theta}\right)\left(\frac{\cos\theta}{\sin\theta}\right) + \left(\frac{1}{\sin\theta}\right)\left(\frac{1}{\cos\theta}\right)\left(\frac{\sin\theta}{\cos\theta}\right) = \frac{-1}{\sin^2\theta} + \frac{1}{\cos^2\theta} = \sec^2\theta - \csc^2\theta$

27. $p = 5 + \frac{1}{\cot q} = 5 + \tan q \Rightarrow \frac{dp}{dq} = \sec^2 q$

29. $p = \frac{\sin q + \cos q}{\cos q} \Rightarrow \frac{dp}{dq} = \frac{(\cos q)(\cos q - \sin q) - (\sin q + \cos q)(-\sin q)}{\cos^2 q}$
$= \frac{\cos^2 q - \cos q\sin q + \sin^2 q + \cos q\sin q}{\cos^2 q} = \frac{1}{\cos^2 q} = \sec^2 q$

31. $p = \frac{q \sin q}{q^2 - 1} \Rightarrow \frac{dp}{dq} = \frac{(q^2-1)(q\cos q + \sin q(1)) - (q \sin q)(2q)}{(q^2-1)^2} = \frac{q^3 \cos q + q^2 \sin q - q\cos q - \sin q - 2q^2 \sin q}{(q^2-1)^2}$
$= \frac{q^3 \cos q - q^2 \sin q - q \cos q - \sin q}{(q^2-1)^2}$

33. (a) $y = \csc x \Rightarrow y' = -\csc x \cot x \Rightarrow y'' = -((\csc x)(-\csc^2 x) + (\cot x)(-\csc x \cot x)) = \csc^3 x + \csc x \cot^2 x$
$= (\csc x)(\csc^2 x + \cot^2 x) = (\csc x)(\csc^2 x + \csc^2 x - 1) = 2\csc^3 x - \csc x$
(b) $y = \sec x \Rightarrow y' = \sec x \tan x \Rightarrow y'' = (\sec x)(\sec^2 x) + (\tan x)(\sec x \tan x) = \sec^3 x + \sec x \tan^2 x$
$= (\sec x)(\sec^2 x + \tan^2 x) = (\sec x)(\sec^2 x + \sec^2 x - 1) = 2\sec^3 x - \sec x$

35. $y = \sin x \Rightarrow y' = \cos x \Rightarrow$ slope of tangent at $x = -\pi$ is $y'(-\pi) = \cos(-\pi) = -1$; slope of tangent at $x = 0$ is $y'(0) = \cos(0) = 1$; and slope of tangent at $x = \frac{3\pi}{2}$ is $y'\left(\frac{3\pi}{2}\right) = \cos \frac{3\pi}{2}$ $= 0$. The tangent at $(-\pi, 0)$ is $y - 0 = -1(x + \pi)$, or $y = -x - \pi$; the tangent at $(0, 0)$ is $y - 0 = 1(x - 0)$, or $y = x$; and the tangent at $\left(\frac{3\pi}{2}, -1\right)$ is $y = -1$.

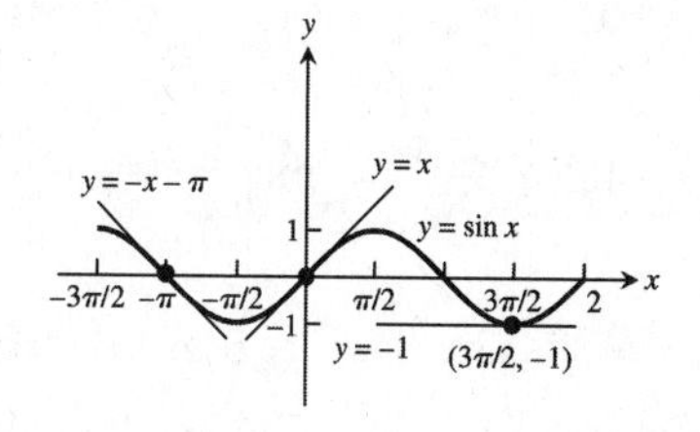

37. $y = \sec x \Rightarrow y' = \sec x \tan x \Rightarrow$ slope of tangent at $x = -\frac{\pi}{3}$ is $\sec\left(-\frac{\pi}{3}\right)\tan\left(-\frac{\pi}{3}\right) = -2\sqrt{3}$; slope of tangent at $x = \frac{\pi}{4}$ is $\sec\left(\frac{\pi}{4}\right)\tan\left(\frac{\pi}{4}\right) = \sqrt{2}$. The tangent at the point $\left(-\frac{\pi}{3}, \sec\left(-\frac{\pi}{3}\right)\right) = \left(-\frac{\pi}{3}, 2\right)$ is $y - 2 = -2\sqrt{3}\left(x + \frac{\pi}{3}\right)$; the tangent at the point $\left(\frac{\pi}{4}, \sec\left(\frac{\pi}{4}\right)\right) = \left(\frac{\pi}{4}, \sqrt{2}\right)$ is $y - \sqrt{2}$ $= \sqrt{2}\left(x - \frac{\pi}{4}\right)$.

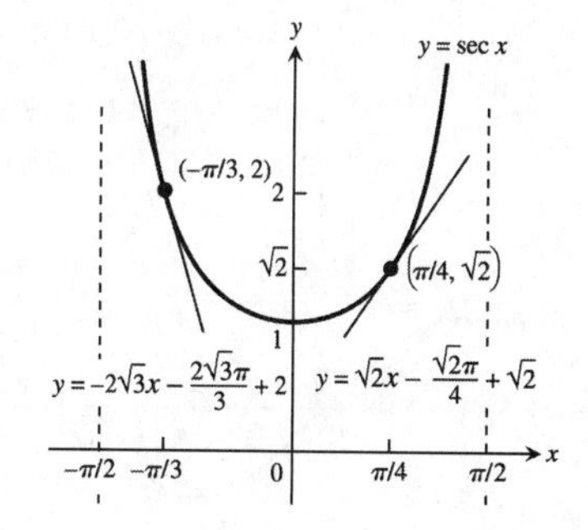

39. Yes, $y = x + \sin x \Rightarrow y' = 1 + \cos x$; horizontal tangent occurs where $1 + \cos x = 0 \Rightarrow \cos x = -1 \Rightarrow x = \pi$

41. No, $y = x - \cot x \Rightarrow y' = 1 + \csc^2 x$; horizontal tangent occurs where $1 + \csc^2 x = 0 \Rightarrow \csc^2 x = -1$. But there are no x-values for which $\csc^2 x = -1$.

43. We want all points on the curve where the tangent line has slope 2. Thus, $y = \tan x \Rightarrow y' = \sec^2 x$ so that $y' = 2 \Rightarrow \sec^2 x = 2 \Rightarrow \sec x = \pm\sqrt{2}$ $\Rightarrow x = \pm\frac{\pi}{4}$. Then the tangent line at $\left(\frac{\pi}{4}, 1\right)$ has equation $y - 1 = 2\left(x - \frac{\pi}{4}\right)$; the tangent line at $\left(-\frac{\pi}{4}, -1\right)$ has equation $y + 1 = 2\left(x + \frac{\pi}{4}\right)$.

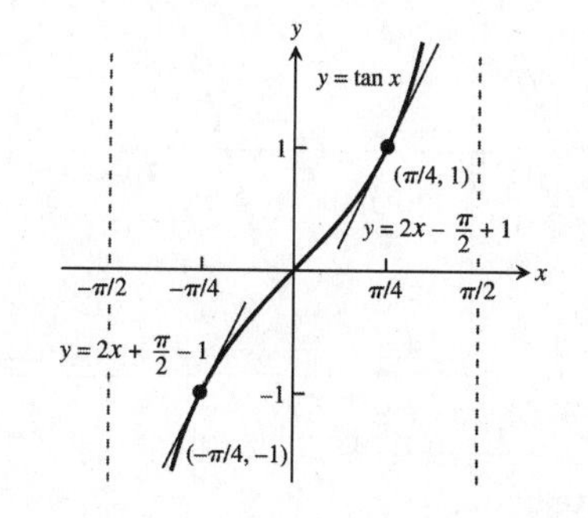

45. $y = 4 + \cot x - 2\csc x \Rightarrow y' = -\csc^2 x + 2\csc x \cot x = -\left(\frac{1}{\sin x}\right)\left(\frac{1 - 2\cos x}{\sin x}\right)$
(a) When $x = \frac{\pi}{2}$, then $y' = -1$; the tangent line is $y = -x + \frac{\pi}{2} + 2$.
(b) To find the location of the horizontal tangent set $y' = 0 \Rightarrow 1 - 2\cos x = 0 \Rightarrow x = \frac{\pi}{3}$ radians. When $x = \frac{\pi}{3}$, then $y = 4 - \sqrt{3}$ is the horizontal tangent.

47. $\lim_{x \to 2} \sin\left(\frac{1}{x} - \frac{1}{2}\right) = \sin\left(\frac{1}{2} - \frac{1}{2}\right) = \sin 0 = 0$

49. $\lim\limits_{\theta \to \frac{\pi}{6}} \frac{\sin\theta - \frac{1}{2}}{\theta - \frac{\pi}{6}} = \frac{d}{d\theta}(\sin\theta)\Big|_{\theta=\frac{\pi}{6}} = \cos\theta\Big|_{\theta=\frac{\pi}{6}} = \cos\left(\frac{\pi}{6}\right) = \frac{\sqrt{3}}{2}$

51. $\lim\limits_{x \to 0} \sec\left[\cos x + \pi \tan\left(\frac{\pi}{4 \sec x}\right) - 1\right] = \sec\left[1 + \pi \tan\left(\frac{\pi}{4 \sec 0}\right) - 1\right] = = \sec\left[\pi \tan\left(\frac{\pi}{4}\right)\right] = \sec \pi = -1$

53. $\lim\limits_{t \to 0} \tan\left(1 - \frac{\sin t}{t}\right) = \tan\left(1 - \lim\limits_{t \to 0} \frac{\sin t}{t}\right) = \tan(1 - 1) = 0$

55. $s = 2 - 2\sin t \Rightarrow v = \frac{ds}{dt} = -2\cos t \Rightarrow a = \frac{dv}{dt} = 2\sin t \Rightarrow j = \frac{da}{dt} = 2\cos t$. Therefore, velocity $= v\left(\frac{\pi}{4}\right)$ $= -\sqrt{2}$ m/sec; speed $= \left|v\left(\frac{\pi}{4}\right)\right| = \sqrt{2}$ m/sec; acceleration $= a\left(\frac{\pi}{4}\right) = \sqrt{2}$ m/sec^2; jerk $= j\left(\frac{\pi}{4}\right) = \sqrt{2}$ m/sec^3.

57. $\lim\limits_{x \to 0} f(x) = \lim\limits_{x \to 0} \frac{\sin^2 3x}{x^2} = \lim\limits_{x \to 0} 9\left(\frac{\sin 3x}{3x}\right)\left(\frac{\sin 3x}{3x}\right) = 9$ so that f is continuous at $x = 0 \Rightarrow \lim\limits_{x \to 0} f(x) = f(0) \Rightarrow 9 = c$.

59. $\frac{d^{999}}{dx^{999}}(\cos x) = \sin x$ because $\frac{d^4}{dx^4}(\cos x) = \cos x \Rightarrow$ the derivative of cos x any number of times that is a multiple of 4 is cos x. Thus, dividing 999 by 4 gives $999 = 249 \cdot 4 + 3 \Rightarrow \frac{d^{999}}{dx^{999}}(\cos x)$ $= \frac{d^3}{dx^3}\left[\frac{d^{249 \cdot 4}}{dx^{249 \cdot 4}}(\cos x)\right] = \frac{d^3}{dx^3}(\cos x) = \sin x$.

61. (a) $t = 0 \to x = 10\cos(0) = 10\,\text{cm}; t = \frac{\pi}{3} \to x = 10\cos\left(\frac{\pi}{3}\right) = 5\,\text{cm}; t = \frac{3\pi}{4} \to x = 10\cos\left(\frac{3\pi}{4}\right) = -5\sqrt{2}\,\text{cm}$

(b) $t = 0 \to v = -10\sin(0) = 0\,\frac{\text{cm}}{\text{sec}}; t = \frac{\pi}{3} \to v = -10\sin\left(\frac{\pi}{3}\right) = -5\sqrt{3}\,\frac{\text{cm}}{\text{sec}}; t = \frac{3\pi}{4} \to v = -10\sin\left(\frac{3\pi}{4}\right) = -5\sqrt{2}\,\frac{\text{cm}}{\text{sec}}$

63.

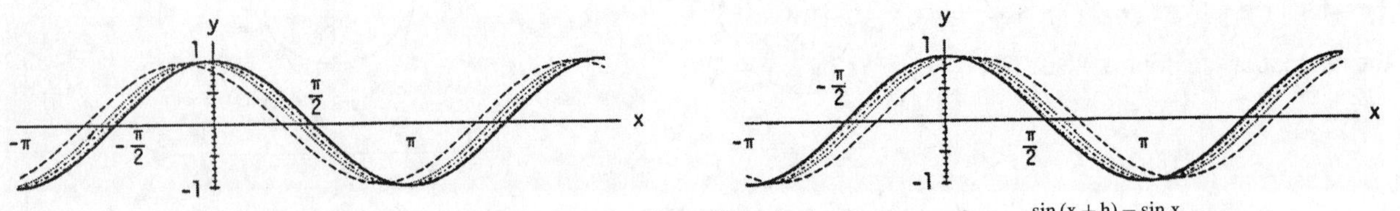

As h takes on the values of 1, 0.5, 0.3 and 0.1 the corresponding dashed curves of $y = \frac{\sin(x+h) - \sin x}{h}$ get closer and closer to the black curve $y = \cos x$ because $\frac{d}{dx}(\sin x) = \lim\limits_{h \to 0} \frac{\sin(x+h) - \sin x}{h} = \cos x$. The same is true as h takes on the values of $-1, -0.5, -0.3$ and -0.1.

65. (a)

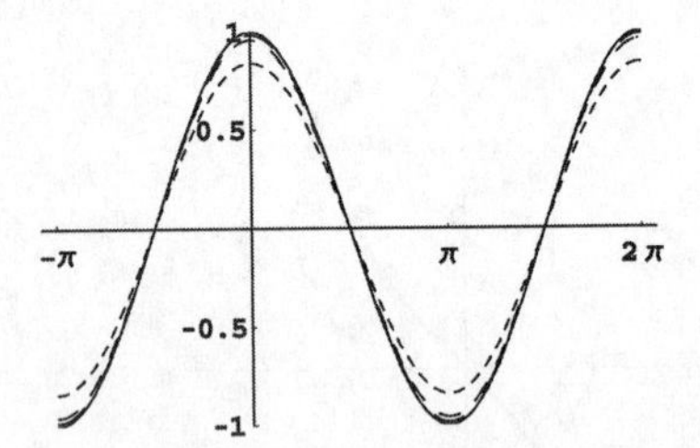

The dashed curves of $y = \frac{\sin(x+h) - \sin(x-h)}{2h}$ are closer to the black curve $y = \cos x$ than the corresponding dashed curves in Exercise 63 illustrating that the centered difference quotient is a better approximation of the derivative of this function.

(b)

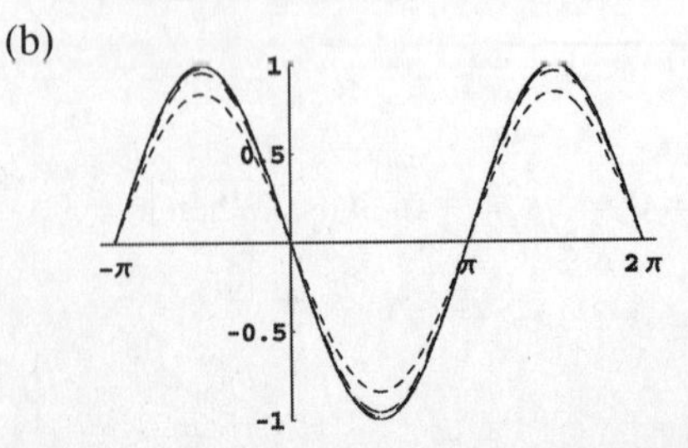

The dashed curves of $y = \frac{\cos(x+h) - \cos(x-h)}{2h}$ are closer to the black curve $y = -\sin x$ than the corresponding dashed curves in Exercise 64 illustrating that the centered difference quotient is a better approximation of the derivative of this function.

67. $y = \tan x \Rightarrow y' = \sec^2 x$, so the smallest value $y' = \sec^2 x$ takes on is $y' = 1$ when $x = 0$; y' has no maximum value since $\sec^2 x$ has no largest value on $\left(-\frac{\pi}{2}, \frac{\pi}{2}\right)$; y' is never negative since $\sec^2 x \geq 1$.

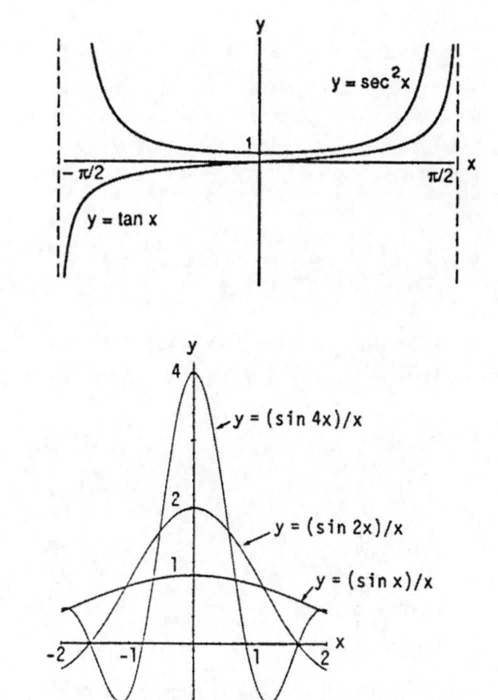

69. $y = \frac{\sin x}{x}$ appears to cross the y-axis at $y = 1$, since $\lim\limits_{x \to 0} \frac{\sin x}{x} = 1$; $y = \frac{\sin 2x}{x}$ appears to cross the y-axis at $y = 2$, since $\lim\limits_{x \to 0} \frac{\sin 2x}{x} = 2$; $y = \frac{\sin 4x}{x}$ appears to cross the y-axis at $y = 4$, since $\lim\limits_{x \to 0} \frac{\sin 4x}{x} = 4$.
However, none of these graphs actually cross the y-axis since $x = 0$ is not in the domain of the functions. Also, $\lim\limits_{x \to 0} \frac{\sin 5x}{x} = 5$, $\lim\limits_{x \to 0} \frac{\sin(-3x)}{x} = -3$, and $\lim\limits_{x \to 0} \frac{\sin kx}{x} = k \Rightarrow$ the graphs of $y = \frac{\sin 5x}{x}$, $y = \frac{\sin(-3x)}{x}$, and $y = \frac{\sin kx}{x}$ approach 5, -3, and k, respectively, as $x \to 0$. However, the graphs do not actually cross the y-axis.

3.6 THE CHAIN RULE

1. $f(u) = 6u - 9 \Rightarrow f'(u) = 6 \Rightarrow f'(g(x)) = 6$; $g(x) = \frac{1}{2}x^4 \Rightarrow g'(x) = 2x^3$; therefore $\frac{dy}{dx} = f'(g(x))g'(x) = 6 \cdot 2x^3 = 12x^3$

3. $f(u) = \sin u \Rightarrow f'(u) = \cos u \Rightarrow f'(g(x)) = \cos(3x+1)$; $g(x) = 3x + 1 \Rightarrow g'(x) = 3$; therefore $\frac{dy}{dx} = f'(g(x))g'(x)$ $= (\cos(3x+1))(3) = 3\cos(3x+1)$

5. $f(u) = \cos u \Rightarrow f'(u) = -\sin u \Rightarrow f'(g(x)) = -\sin(\sin x)$; $g(x) = \sin x \Rightarrow g'(x) = \cos x$; therefore $\frac{dy}{dx} = f'(g(x))g'(x) = -(\sin(\sin x))\cos x$

7. $f(u) = \tan u \Rightarrow f'(u) = \sec^2 u \Rightarrow f'(g(x)) = \sec^2(10x - 5)$; $g(x) = 10x - 5 \Rightarrow g'(x) = 10$; therefore $\frac{dy}{dx} = f'(g(x))g'(x) = (\sec^2(10x-5))(10) = 10\sec^2(10x-5)$

9. With $u = (2x+1)$, $y = u^5$: $\frac{dy}{dx} = \frac{dy}{du}\frac{du}{dx} = 5u^4 \cdot 2 = 10(2x+1)^4$

11. With $u = \left(1 - \frac{x}{7}\right)$, $y = u^{-7}$: $\frac{dy}{dx} = \frac{dy}{du}\frac{du}{dx} = -7u^{-8} \cdot \left(-\frac{1}{7}\right) = \left(1 - \frac{x}{7}\right)^{-8}$

13. With $u = \left(\frac{x^2}{8} + x - \frac{1}{x}\right)$, $y = u^4$: $\frac{dy}{dx} = \frac{dy}{du}\frac{du}{dx} = 4u^3 \cdot \left(\frac{x}{4} + 1 + \frac{1}{x^2}\right) = 4\left(\frac{x^2}{8} + x - \frac{1}{x}\right)^3 \left(\frac{x}{4} + 1 + \frac{1}{x^2}\right)$

15. With $u = \tan x$, $y = \sec u$: $\frac{dy}{dx} = \frac{dy}{du}\frac{du}{dx} = (\sec u \tan u)(\sec^2 x) = (\sec(\tan x)\tan(\tan x))\sec^2 x$

17. With $u = \sin x$, $y = u^3$: $\frac{dy}{dx} = \frac{dy}{du}\frac{du}{dx} = 3u^2 \cos x = 3(\sin^2 x)(\cos x)$

19. $p = \sqrt{3-t} = (3-t)^{1/2} \Rightarrow \frac{dp}{dt} = \frac{1}{2}(3-t)^{-1/2} \cdot \frac{d}{dt}(3-t) = -\frac{1}{2}(3-t)^{-1/2} = \frac{-1}{2\sqrt{3-t}}$

21. $s = \frac{4}{3\pi}\sin 3t + \frac{4}{5\pi}\cos 5t \Rightarrow \frac{ds}{dt} = \frac{4}{3\pi}\cos 3t \cdot \frac{d}{dt}(3t) + \frac{4}{5\pi}(-\sin 5t) \cdot \frac{d}{dt}(5t) = \frac{4}{\pi}\cos 3t - \frac{4}{\pi}\sin 5t$
$= \frac{4}{\pi}(\cos 3t - \sin 5t)$

23. $r = (\csc\theta + \cot\theta)^{-1} \Rightarrow \frac{dr}{d\theta} = -(\csc\theta + \cot\theta)^{-2}\frac{d}{d\theta}(\csc\theta + \cot\theta) = \frac{\csc\theta\cot\theta + \csc^2\theta}{(\csc\theta + \cot\theta)^2} = \frac{\csc\theta(\cot\theta + \csc\theta)}{(\csc\theta + \cot\theta)^2} = \frac{\csc\theta}{\csc\theta + \cot\theta}$

25. $y = x^2\sin^4 x + x\cos^{-2} x \Rightarrow \frac{dy}{dx} = x^2\frac{d}{dx}(\sin^4 x) + \sin^4 x \cdot \frac{d}{dx}(x^2) + x\frac{d}{dx}(\cos^{-2} x) + \cos^{-2} x \cdot \frac{d}{dx}(x)$
$= x^2\left(4\sin^3 x\frac{d}{dx}(\sin x)\right) + 2x\sin^4 x + x\left(-2\cos^{-3} x \cdot \frac{d}{dx}(\cos x)\right) + \cos^{-2} x$
$= x^2(4\sin^3 x\cos x) + 2x\sin^4 x + x((-2\cos^{-3} x)(-\sin x)) + \cos^{-2} x$
$= 4x^2\sin^3 x\cos x + 2x\sin^4 x + 2x\sin x\cos^{-3} x + \cos^{-2} x$

27. $y = \frac{1}{21}(3x-2)^7 + \left(4 - \frac{1}{2x^2}\right)^{-1} \Rightarrow \frac{dy}{dx} = \frac{7}{21}(3x-2)^6 \cdot \frac{d}{dx}(3x-2) + (-1)\left(4 - \frac{1}{2x^2}\right)^{-2} \cdot \frac{d}{dx}\left(4 - \frac{1}{2x^2}\right)$
$= \frac{7}{21}(3x-2)^6 \cdot 3 + (-1)\left(4 - \frac{1}{2x^2}\right)^{-2}\left(\frac{1}{x^3}\right) = (3x-2)^6 - \frac{1}{x^3\left(4 - \frac{1}{2x^2}\right)^2}$

29. $y = (4x+3)^4(x+1)^{-3} \Rightarrow \frac{dy}{dx} = (4x+3)^4(-3)(x+1)^{-4} \cdot \frac{d}{dx}(x+1) + (x+1)^{-3}(4)(4x+3)^3 \cdot \frac{d}{dx}(4x+3)$
$= (4x+3)^4(-3)(x+1)^{-4}(1) + (x+1)^{-3}(4)(4x+3)^3(4) = -3(4x+3)^4(x+1)^{-4} + 16(4x+3)^3(x+1)^{-3}$
$= \frac{(4x+3)^3}{(x+1)^4}[-3(4x+3) + 16(x+1)] = \frac{(4x+3)^3(4x+7)}{(x+1)^4}$

31. $h(x) = x\tan\left(2\sqrt{x}\right) + 7 \Rightarrow h'(x) = x\frac{d}{dx}\left(\tan\left(2x^{1/2}\right)\right) + \tan\left(2x^{1/2}\right) \cdot \frac{d}{dx}(x) + 0$
$= x\sec^2\left(2x^{1/2}\right) \cdot \frac{d}{dx}\left(2x^{1/2}\right) + \tan\left(2x^{1/2}\right) = x\sec^2\left(2\sqrt{x}\right) \cdot \frac{1}{\sqrt{x}} + \tan\left(2\sqrt{x}\right) = \sqrt{x}\sec^2\left(2\sqrt{x}\right) + \tan\left(2\sqrt{x}\right)$

33. $f(x) = \sqrt{7 + x\sec x} \Rightarrow f'(x) = \frac{1}{2}(7 + x\sec x)^{-1/2}(x \cdot (\sec x\tan x) + (\sec x) \cdot 1) = \frac{x\sec x\tan x + \sec x}{2\sqrt{7+x\sec x}}$

35. $f(\theta) = \left(\frac{\sin\theta}{1+\cos\theta}\right)^2 \Rightarrow f'(\theta) = 2\left(\frac{\sin\theta}{1+\cos\theta}\right) \cdot \frac{d}{d\theta}\left(\frac{\sin\theta}{1+\cos\theta}\right) = \frac{2\sin\theta}{1+\cos\theta} \cdot \frac{(1+\cos\theta)(\cos\theta) - (\sin\theta)(-\sin\theta)}{(1+\cos\theta)^2}$
$= \frac{(2\sin\theta)(\cos\theta + \cos^2\theta + \sin^2\theta)}{(1+\cos\theta)^3} = \frac{(2\sin\theta)(\cos\theta + 1)}{(1+\cos\theta)^3} = \frac{2\sin\theta}{(1+\cos\theta)^2}$

37. $r = \sin(\theta^2)\cos(2\theta) \Rightarrow \frac{dr}{d\theta} = \sin(\theta^2)(-\sin 2\theta)\frac{d}{d\theta}(2\theta) + \cos(2\theta)(\cos(\theta^2)) \cdot \frac{d}{d\theta}(\theta^2)$
$= \sin(\theta^2)(-\sin 2\theta)(2) + (\cos 2\theta)(\cos(\theta^2))(2\theta) = -2\sin(\theta^2)\sin(2\theta) + 2\theta\cos(2\theta)\cos(\theta^2)$

39. $q = \sin\left(\frac{t}{\sqrt{t+1}}\right) \Rightarrow \frac{dq}{dt} = \cos\left(\frac{t}{\sqrt{t+1}}\right) \cdot \frac{d}{dt}\left(\frac{t}{\sqrt{t+1}}\right) = \cos\left(\frac{t}{\sqrt{t+1}}\right) \cdot \frac{\sqrt{t+1}(1) - t \cdot \frac{d}{dt}\left(\sqrt{t+1}\right)}{\left(\sqrt{t+1}\right)^2}$
$= \cos\left(\frac{t}{\sqrt{t+1}}\right) \cdot \frac{\sqrt{t+1} - \frac{t}{2\sqrt{t+1}}}{t+1} = \cos\left(\frac{t}{\sqrt{t+1}}\right)\left(\frac{2(t+1) - t}{2(t+1)^{3/2}}\right) = \left(\frac{t+2}{2(t+1)^{3/2}}\right)\cos\left(\frac{t}{\sqrt{t+1}}\right)$

41. $y = \sin^2(\pi t - 2) \Rightarrow \frac{dy}{dt} = 2\sin(\pi t - 2) \cdot \frac{d}{dt}\sin(\pi t - 2) = 2\sin(\pi t - 2) \cdot \cos(\pi t - 2) \cdot \frac{d}{dt}(\pi t - 2)$
$= 2\pi\sin(\pi t - 2)\cos(\pi t - 2)$

43. $y = (1 + \cos 2t)^{-4} \Rightarrow \frac{dy}{dt} = -4(1 + \cos 2t)^{-5} \cdot \frac{d}{dt}(1 + \cos 2t) = -4(1 + \cos 2t)^{-5}(-\sin 2t) \cdot \frac{d}{dt}(2t) = \frac{8\sin 2t}{(1 + \cos 2t)^5}$

45. $y = (t\tan t)^{10} \Rightarrow \frac{dy}{dt} = 10(t\tan t)^9(t \cdot \sec^2 t + 1 \cdot \tan t) = 10t^9\tan^9 t(t\sec^2 t + \tan t) = 10t^{10}\tan^9 t\sec^2 t + 10t^9\tan^{10} t$

47. $y=\left(\frac{t^2}{t^3-4t}\right)^3 \Rightarrow \frac{dy}{dt}=3\left(\frac{t^2}{t^3-4t}\right)^2\cdot\frac{(t^3-4t)(2t)-t^2(3t^2-4)}{(t^3-4t)^2}=\frac{3t^4}{(t^3-4t)^2}\cdot\frac{2t^4-8t^2-3t^4+4t^2}{(t^3-4t)^2}=\frac{3t^4(-t^4-4t^2)}{t^4(t^2-4)^4}=\frac{-3t^2(t^2+4)}{(t^2-4)^4}$

49. $y=\sin(\cos(2t-5)) \Rightarrow \frac{dy}{dt}=\cos(\cos(2t-5))\cdot\frac{d}{dt}\cos(2t-5)=\cos(\cos(2t-5))\cdot(-\sin(2t-5))\cdot\frac{d}{dt}(2t-5)$
$=-2\cos(\cos(2t-5))(\sin(2t-5))$

51. $y=\left[1+\tan^4\left(\frac{t}{12}\right)\right]^3 \Rightarrow \frac{dy}{dt}=3\left[1+\tan^4\left(\frac{t}{12}\right)\right]^2\cdot\frac{d}{dt}\left[1+\tan^4\left(\frac{t}{12}\right)\right]=3\left[1+\tan^4\left(\frac{t}{12}\right)\right]^2\left[4\tan^3\left(\frac{t}{12}\right)\cdot\frac{d}{dt}\tan\left(\frac{t}{12}\right)\right]$
$=12\left[1+\tan^4\left(\frac{t}{12}\right)\right]^2\left[\tan^3\left(\frac{t}{12}\right)\sec^2\left(\frac{t}{12}\right)\cdot\frac{1}{12}\right]=\left[1+\tan^4\left(\frac{t}{12}\right)\right]^2\left[\tan^3\left(\frac{t}{12}\right)\sec^2\left(\frac{t}{12}\right)\right]$

53. $y=(1+\cos(t^2))^{1/2} \Rightarrow \frac{dy}{dt}=\frac{1}{2}(1+\cos(t^2))^{-1/2}\cdot\frac{d}{dt}(1+\cos(t^2))=\frac{1}{2}(1+\cos(t^2))^{-1/2}\left(-\sin(t^2)\cdot\frac{d}{dt}(t^2)\right)$
$=-\frac{1}{2}(1+\cos(t^2))^{-1/2}(\sin(t^2))\cdot 2t=-\frac{t\sin(t^2)}{\sqrt{1+\cos(t^2)}}$

55. $y=\tan^2(\sin^3 t) \Rightarrow \frac{dy}{dt}=2\tan(\sin^3 t)\cdot\sec^2(\sin^3 t)\cdot(3\sin^2 t\cdot(\cos t))=6\tan(\sin^3 t)\sec^2(\sin^3 t)\sin^2 t\cos t$

57. $y=3t(2t^2-5)^4 \Rightarrow \frac{dy}{dt}=3t\cdot 4(2t^2-5)^3(4t)+3\cdot(2t^2-5)^4=3(2t^2-5)^3\left[16t^2+2t^2-5\right]=3(2t^2-5)^3(18t^2-5)$

59. $y=\left(1+\frac{1}{x}\right)^3 \Rightarrow y'=3\left(1+\frac{1}{x}\right)^2\left(-\frac{1}{x^2}\right)=-\frac{3}{x^2}\left(1+\frac{1}{x}\right)^2 \Rightarrow y''=\left(-\frac{3}{x^2}\right)\cdot\frac{d}{dx}\left(1+\frac{1}{x}\right)^2-\left(1+\frac{1}{x}\right)^2\cdot\frac{d}{dx}\left(\frac{3}{x^2}\right)$
$=\left(-\frac{3}{x^2}\right)\left(2\left(1+\frac{1}{x}\right)\left(-\frac{1}{x^2}\right)\right)+\left(\frac{6}{x^3}\right)\left(1+\frac{1}{x}\right)^2=\frac{6}{x^4}\left(1+\frac{1}{x}\right)+\frac{6}{x^3}\left(1+\frac{1}{x}\right)^2=\frac{6}{x^3}\left(1+\frac{1}{x}\right)\left(\frac{1}{x}+1+\frac{1}{x}\right)$
$=\frac{6}{x^3}\left(1+\frac{1}{x}\right)\left(1+\frac{2}{x}\right)$

61. $y=\frac{1}{9}\cot(3x-1) \Rightarrow y'=-\frac{1}{9}\csc^2(3x-1)(3)=-\frac{1}{3}\csc^2(3x-1) \Rightarrow y''=\left(-\frac{2}{3}\right)\left(\csc(3x-1)\cdot\frac{d}{dx}\csc(3x-1)\right)$
$=-\frac{2}{3}\csc(3x-1)\left(-\csc(3x-1)\cot(3x-1)\cdot\frac{d}{dx}(3x-1)\right)=2\csc^2(3x-1)\cot(3x-1)$

63. $y=x(2x+1)^4 \Rightarrow y'=x\cdot 4(2x+1)^3(2)+1\cdot(2x+1)^4=(2x+1)^3(8x+(2x+1))=(2x+1)^3(10x+1)$
$\Rightarrow y''=(2x+1)^3(10)+3(2x+1)^2(2)(10x+1)=2(2x+1)^2(5(2x+1)+3(10x+1))=2(2x+1)^2(40x+8)$
$=16(2x+1)^2(5x+1)$

65. $g(x)=\sqrt{x} \Rightarrow g'(x)=\frac{1}{2\sqrt{x}} \Rightarrow g(1)=1$ and $g'(1)=\frac{1}{2}$; $f(u)=u^5+1 \Rightarrow f'(u)=5u^4 \Rightarrow f'(g(1))=f'(1)=5$;
therefore, $(f\circ g)'(1)=f'(g(1))\cdot g'(1)=5\cdot\frac{1}{2}=\frac{5}{2}$

67. $g(x)=5\sqrt{x} \Rightarrow g'(x)=\frac{5}{2\sqrt{x}} \Rightarrow g(1)=5$ and $g'(1)=\frac{5}{2}$; $f(u)=\cot\left(\frac{\pi u}{10}\right) \Rightarrow f'(u)=-\csc^2\left(\frac{\pi u}{10}\right)\left(\frac{\pi}{10}\right)=\frac{-\pi}{10}\csc^2\left(\frac{\pi u}{10}\right)$
$\Rightarrow f'(g(1))=f'(5)=-\frac{\pi}{10}\csc^2\left(\frac{\pi}{2}\right)=-\frac{\pi}{10}$; therefore, $(f\circ g)'(1)=f'(g(1))g'(1)=-\frac{\pi}{10}\cdot\frac{5}{2}=-\frac{\pi}{4}$

69. $g(x)=10x^2+x+1 \Rightarrow g'(x)=20x+1 \Rightarrow g(0)=1$ and $g'(0)=1$; $f(u)=\frac{2u}{u^2+1} \Rightarrow f'(u)=\frac{(u^2+1)(2)-(2u)(2u)}{(u^2+1)^2}$
$=\frac{-2u^2+2}{(u^2+1)^2} \Rightarrow f'(g(0))=f'(1)=0$; therefore, $(f\circ g)'(0)=f'(g(0))g'(0)=0\cdot 1=0$

71. $y=f(g(x)),\ f'(3)=-1,\ g'(2)=5,\ g(2)=3 \Rightarrow y'=f'(g(x))g'(x) \Rightarrow y'\big|_{x=2}=f'(g(2))g'(2)=f'(3)\cdot 5$
$=(-1)\cdot 5=-5$

73. (a) $y=2f(x) \Rightarrow \frac{dy}{dx}=2f'(x) \Rightarrow \frac{dy}{dx}\Big|_{x=2}=2f'(2)=2\left(\frac{1}{3}\right)=\frac{2}{3}$

(b) $y=f(x)+g(x) \Rightarrow \frac{dy}{dx}=f'(x)+g'(x) \Rightarrow \frac{dy}{dx}\Big|_{x=3}=f'(3)+g'(3)=2\pi+5$

(c) $y=f(x)\cdot g(x) \Rightarrow \frac{dy}{dx}=f(x)g'(x)+g(x)f'(x) \Rightarrow \frac{dy}{dx}\Big|_{x=3}=f(3)g'(3)+g(3)f'(3)=3\cdot 5+(-4)(2\pi)=15-8\pi$

(d) $y = \frac{f(x)}{g(x)} \Rightarrow \frac{dy}{dx} = \frac{g(x)f'(x) - f(x)g'(x)}{[g(x)]^2} \Rightarrow \left.\frac{dy}{dx}\right|_{x=2} = \frac{g(2)f'(2) - f(2)g'(2)}{[g(2)]^2} = \frac{(2)\left(\frac{1}{3}\right) - (8)(-3)}{2^2} = \frac{37}{6}$

(e) $y = f(g(x)) \Rightarrow \frac{dy}{dx} = f'(g(x))g'(x) \Rightarrow \left.\frac{dy}{dx}\right|_{x=2} = f'(g(2))g'(2) = f'(2)(-3) = \frac{1}{3}(-3) = -1$

(f) $y = (f(x))^{1/2} \Rightarrow \frac{dy}{dx} = \frac{1}{2}(f(x))^{-1/2} \cdot f'(x) = \frac{f'(x)}{2\sqrt{f(x)}} \Rightarrow \left.\frac{dy}{dx}\right|_{x=2} = \frac{f'(2)}{2\sqrt{f(2)}} = \frac{\left(\frac{1}{3}\right)}{2\sqrt{8}} = \frac{1}{6\sqrt{8}} = \frac{1}{12\sqrt{2}} = \frac{\sqrt{2}}{24}$

(g) $y = (g(x))^{-2} \Rightarrow \frac{dy}{dx} = -2(g(x))^{-3} \cdot g'(x) \Rightarrow \left.\frac{dy}{dx}\right|_{x=3} = -2(g(3))^{-3}g'(3) = -2(-4)^{-3} \cdot 5 = \frac{5}{32}$

(h) $y = \left((f(x))^2 + (g(x))^2\right)^{1/2} \Rightarrow \frac{dy}{dx} = \frac{1}{2}\left((f(x))^2 + (g(x))^2\right)^{-1/2}(2f(x) \cdot f'(x) + 2g(x) \cdot g'(x))$

$\Rightarrow \left.\frac{dy}{dx}\right|_{x=2} = \frac{1}{2}\left((f(2))^2 + (g(2))^2\right)^{-1/2}(2f(2)f'(2) + 2g(2)g'(2)) = \frac{1}{2}\left(8^2 + 2^2\right)^{-1/2}\left(2 \cdot 8 \cdot \frac{1}{3} + 2 \cdot 2 \cdot (-3)\right) = -\frac{5}{3\sqrt{17}}$

75. $\frac{ds}{dt} = \frac{ds}{d\theta} \cdot \frac{d\theta}{dt}$: $s = \cos\theta \Rightarrow \frac{ds}{d\theta} = -\sin\theta \Rightarrow \left.\frac{ds}{d\theta}\right|_{\theta = \frac{3\pi}{2}} = -\sin\left(\frac{3\pi}{2}\right) = 1$ so that $\frac{ds}{dt} = \frac{ds}{d\theta} \cdot \frac{d\theta}{dt} = 1 \cdot 5 = 5$

77. With $y = x$, we should get $\frac{dy}{dx} = 1$ for both (a) and (b):

(a) $y = \frac{u}{5} + 7 \Rightarrow \frac{dy}{du} = \frac{1}{5}$; $u = 5x - 35 \Rightarrow \frac{du}{dx} = 5$; therefore, $\frac{dy}{dx} = \frac{dy}{du} \cdot \frac{du}{dx} = \frac{1}{5} \cdot 5 = 1$, as expected

(b) $y = 1 + \frac{1}{u} \Rightarrow \frac{dy}{du} = -\frac{1}{u^2}$; $u = (x-1)^{-1} \Rightarrow \frac{du}{dx} = -(x-1)^{-2}(1) = \frac{-1}{(x-1)^2}$; therefore $\frac{dy}{dx} = \frac{dy}{du} \cdot \frac{du}{dx}$

$= \frac{-1}{u^2} \cdot \frac{-1}{(x-1)^2} = \frac{-1}{((x-1)^{-1})^2} \cdot \frac{-1}{(x-1)^2} = (x-1)^2 \cdot \frac{1}{(x-1)^2} = 1$, again as expected

79. $y = \left(\frac{x-1}{x+1}\right)^2$ and $x = 0 \Rightarrow y = \left(\frac{0-1}{0+1}\right)^2 = (-1)^2 = 1$. $y' = 2\left(\frac{x-1}{x+1}\right) \cdot \frac{(x+1)\cdot 1 - (x-1)\cdot 1}{(x+1)^2} = 2\frac{(x-1)}{(x+1)}\frac{2}{(x+1)^2} = \frac{4(x-1)}{(x+1)^3}$

$\left.y'\right|_{x=0} = \frac{4(0-1)}{(0+1)^3} = \frac{-4}{1^3} = -4 \Rightarrow y - 1 = -4(x-0) \Rightarrow y = -4x + 1$

81. $y = 2\tan\left(\frac{\pi x}{4}\right) \Rightarrow \frac{dy}{dx} = \left(2\sec^2\frac{\pi x}{4}\right)\left(\frac{\pi}{4}\right) = \frac{\pi}{2}\sec^2\frac{\pi x}{4}$

(a) $\left.\frac{dy}{dx}\right|_{x=1} = \frac{\pi}{2}\sec^2\left(\frac{\pi}{4}\right) = \pi \Rightarrow$ slope of tangent is 2; thus, $y(1) = 2\tan\left(\frac{\pi}{4}\right) = 2$ and $y'(1) = \pi \Rightarrow$ tangent line is given by $y - 2 = \pi(x-1) \Rightarrow y = \pi x + 2 - \pi$

(b) $y' = \frac{\pi}{2}\sec^2\left(\frac{\pi x}{4}\right)$ and the smallest value the secant function can have in $-2 < x < 2$ is $1 \Rightarrow$ the minimum value of y' is $\frac{\pi}{2}$ and that occurs when $\frac{\pi}{2} = \frac{\pi}{2}\sec^2\left(\frac{\pi x}{4}\right) \Rightarrow 1 = \sec^2\left(\frac{\pi x}{4}\right) \Rightarrow \pm 1 = \sec\left(\frac{\pi x}{4}\right) \Rightarrow x = 0$.

83. $s = A\cos(2\pi bt) \Rightarrow v = \frac{ds}{dt} = -A\sin(2\pi bt)(2\pi b) = -2\pi bA\sin(2\pi bt)$. If we replace b with 2b to double the frequency, the velocity formula gives $v = -4\pi bA\sin(4\pi bt) \Rightarrow$ doubling the frequency causes the velocity to double. Also $v = -2\pi bA\sin(2\pi bt) \Rightarrow a = \frac{dv}{dt} = -4\pi^2b^2A\cos(2\pi bt)$. If we replace b with 2b in the acceleration formula, we get $a = -16\pi^2b^2A\cos(4\pi bt) \Rightarrow$ doubling the frequency causes the acceleration to quadruple. Finally, $a = -4\pi^2b^2A\cos(2\pi bt) \Rightarrow j = \frac{da}{dt} = 8\pi^3b^3A\sin(2\pi bt)$. If we replace b with 2b in the jerk formula, we get $j = 64\pi^3b^3A\sin(4\pi bt) \Rightarrow$ doubling the frequency multiplies the jerk by a factor of 8.

85. $s = (1+4t)^{1/2} \Rightarrow v = \frac{ds}{dt} = \frac{1}{2}(1+4t)^{-1/2}(4) = 2(1+4t)^{-1/2} \Rightarrow v(6) = 2(1+4\cdot 6)^{-1/2} = \frac{2}{5}$ m/sec;
$v = 2(1+4t)^{-1/2} \Rightarrow a = \frac{dv}{dt} = -\frac{1}{2} \cdot 2(1+4t)^{-3/2}(4) = -4(1+4t)^{-3/2} \Rightarrow a(6) = -4(1+4\cdot 6)^{-3/2} = -\frac{4}{125}$ m/sec²

87. v proportional to $\frac{1}{\sqrt{s}} \Rightarrow v = \frac{k}{\sqrt{s}}$ for some constant $k \Rightarrow \frac{dv}{ds} = -\frac{k}{2s^{3/2}}$. Thus, $a = \frac{dv}{dt} = \frac{dv}{ds} \cdot \frac{ds}{dt} = \frac{dv}{ds} \cdot v$
$= -\frac{k}{2s^{3/2}} \cdot \frac{k}{\sqrt{s}} = -\frac{k^2}{2}\left(\frac{1}{s^2}\right) \Rightarrow$ acceleration is a constant times $\frac{1}{s^2}$ so a is inversely proportional to s^2.

89. $T = 2\pi\sqrt{\frac{L}{g}} \Rightarrow \frac{dT}{dL} = 2\pi \cdot \frac{1}{2\sqrt{\frac{L}{g}}} \cdot \frac{1}{g} = \frac{\pi}{g\sqrt{\frac{L}{g}}} = \frac{\pi}{\sqrt{gL}}$. Therefore, $\frac{dT}{du} = \frac{dT}{dL} \cdot \frac{dL}{du} = \frac{\pi}{\sqrt{gL}} \cdot kL = \frac{\pi k\sqrt{L}}{\sqrt{g}} = \frac{1}{2} \cdot 2\pi k\sqrt{\frac{L}{g}}$
$= \frac{kT}{2}$, as required.

91. As $h \to 0$, the graph of $y = \frac{\sin 2(x+h) - \sin 2x}{h}$ approaches the graph of $y = 2\cos 2x$ because $\lim\limits_{h \to 0} \frac{\sin 2(x+h) - \sin 2x}{h} = \frac{d}{dx}(\sin 2x) = 2\cos 2x$.

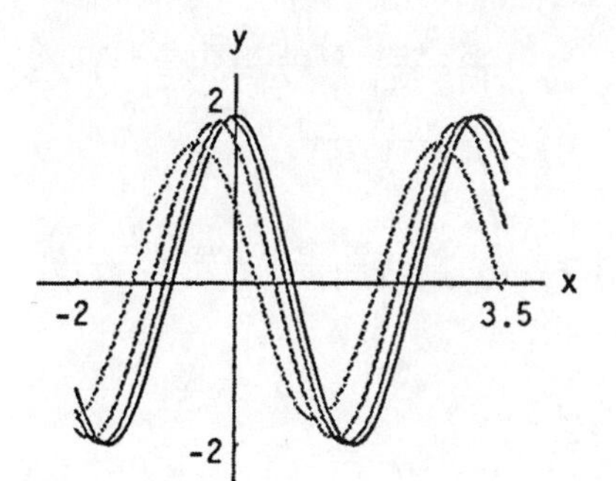

93. (a)

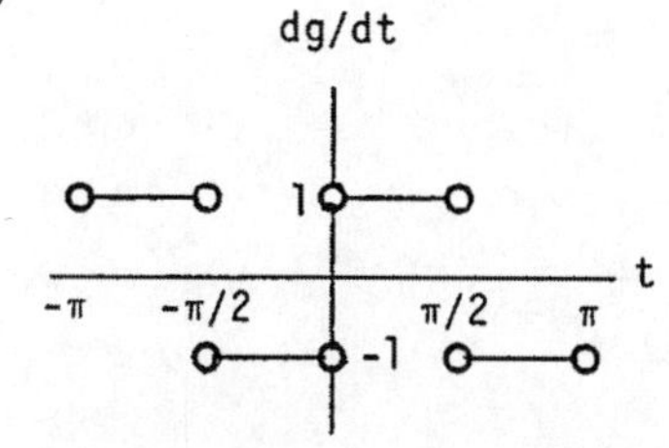

(b) $\frac{df}{dt} = 1.27324 \sin 2t + 0.42444 \sin 6t + 0.2546 \sin 10t + 0.18186 \sin 14t$

(c) The curve of $y = \frac{df}{dt}$ approximates $y = \frac{dg}{dt}$ the best when t is not $-\pi, -\frac{\pi}{2}, 0, \frac{\pi}{2}$, nor π.

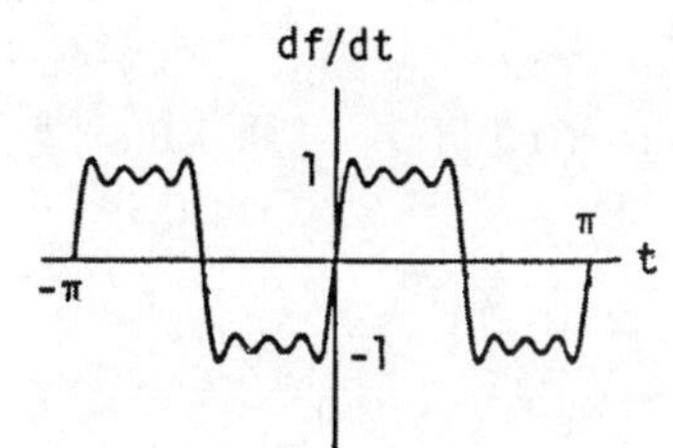

3.7 IMPLICIT DIFFERENTIATION

1. $x^2y + xy^2 = 6$:

Step 1: $\left(x^2 \frac{dy}{dx} + y \cdot 2x\right) + \left(x \cdot 2y \frac{dy}{dx} + y^2 \cdot 1\right) = 0$

Step 2: $x^2 \frac{dy}{dx} + 2xy \frac{dy}{dx} = -2xy - y^2$

Step 3: $\frac{dy}{dx}(x^2 + 2xy) = -2xy - y^2$

Step 4: $\frac{dy}{dx} = \frac{-2xy - y^2}{x^2 + 2xy}$

3. $2xy + y^2 = x + y$:

Step 1: $\left(2x \frac{dy}{dx} + 2y\right) + 2y \frac{dy}{dx} = 1 + \frac{dy}{dx}$

Step 2: $2x \frac{dy}{dx} + 2y \frac{dy}{dx} - \frac{dy}{dx} = 1 - 2y$

Step 3: $\frac{dy}{dx}(2x + 2y - 1) = 1 - 2y$

Step 4: $\frac{dy}{dx} = \frac{1 - 2y}{2x + 2y - 1}$

5. $x^2(x - y)^2 = x^2 - y^2$:

Step 1: $x^2\left[2(x - y)\left(1 - \frac{dy}{dx}\right)\right] + (x - y)^2(2x) = 2x - 2y \frac{dy}{dx}$

Step 2: $-2x^2(x - y) \frac{dy}{dx} + 2y \frac{dy}{dx} = 2x - 2x^2(x - y) - 2x(x - y)^2$

Step 3: $\frac{dy}{dx}\left[-2x^2(x - y) + 2y\right] = 2x\left[1 - x(x - y) - (x - y)^2\right]$

Step 4: $\frac{dy}{dx} = \frac{2x[1 - x(x-y) - (x-y)^2]}{-2x^2(x-y) + 2y} = \frac{x[1 - x(x-y) - (x-y)^2]}{y - x^2(x-y)} = \frac{x(1 - x^2 + xy - x^2 + 2xy - y^2)}{x^2y - x^3 + y}$

$= \frac{x - 2x^3 + 3x^2y - xy^2}{x^2y - x^3 + y}$

7. $y^2 = \frac{x-1}{x+1} \Rightarrow 2y\frac{dy}{dx} = \frac{(x+1)-(x-1)}{(x+1)^2} = \frac{2}{(x+1)^2} \Rightarrow \frac{dy}{dx} = \frac{1}{y(x+1)^2}$

9. $x = \tan y \Rightarrow 1 = (\sec^2 y)\frac{dy}{dx} \Rightarrow \frac{dy}{dx} = \frac{1}{\sec^2 y} = \cos^2 y$

11. $x + \tan(xy) = 0 \Rightarrow 1 + [\sec^2(xy)]\left(y + x\frac{dy}{dx}\right) = 0 \Rightarrow x\sec^2(xy)\frac{dy}{dx} = -1 - y\sec^2(xy) \Rightarrow \frac{dy}{dx} = \frac{-1 - y\sec^2(xy)}{x\sec^2(xy)}$

$= \frac{-1}{x\sec^2(xy)} - \frac{y}{x} = \frac{-\cos^2(xy)}{x} - \frac{y}{x} = \frac{-\cos^2(xy) - y}{x}$

13. $y\sin\left(\frac{1}{y}\right) = 1 - xy \Rightarrow y\left[\cos\left(\frac{1}{y}\right)\cdot(-1)\frac{1}{y^2}\cdot\frac{dy}{dx}\right] + \sin\left(\frac{1}{y}\right)\cdot\frac{dy}{dx} = -x\frac{dy}{dx} - y \Rightarrow$

$\frac{dy}{dx}\left[-\frac{1}{y}\cos\left(\frac{1}{y}\right) + \sin\left(\frac{1}{y}\right) + x\right] = -y \Rightarrow \frac{dy}{dx} = \frac{-y}{-\frac{1}{y}\cos\left(\frac{1}{y}\right) + \sin\left(\frac{1}{y}\right) + x} = \frac{-y^2}{y\sin\left(\frac{1}{y}\right) - \cos\left(\frac{1}{y}\right) + xy}$

15. $\theta^{1/2} + r^{1/2} = 1 \Rightarrow \frac{1}{2}\theta^{-1/2} + \frac{1}{2}r^{-1/2}\cdot\frac{dr}{d\theta} = 0 \Rightarrow \frac{dr}{d\theta}\left[\frac{1}{2\sqrt{r}}\right] = \frac{-1}{2\sqrt{\theta}} \Rightarrow \frac{dr}{d\theta} = -\frac{2\sqrt{r}}{2\sqrt{\theta}} = -\frac{\sqrt{r}}{\sqrt{\theta}}$

17. $\sin(r\theta) = \frac{1}{2} \Rightarrow [\cos(r\theta)]\left(r + \theta\frac{dr}{d\theta}\right) = 0 \Rightarrow \frac{dr}{d\theta}[\theta\cos(r\theta)] = -r\cos(r\theta) \Rightarrow \frac{dr}{d\theta} = \frac{-r\cos(r\theta)}{\theta\cos(r\theta)} = -\frac{r}{\theta},\ \cos(r\theta) \neq 0$

19. $x^2 + y^2 = 1 \Rightarrow 2x + 2yy' = 0 \Rightarrow 2yy' = -2x \Rightarrow \frac{dy}{dx} = y' = -\frac{x}{y}$; now to find $\frac{d^2y}{dx^2}$, $\frac{d}{dx}(y') = \frac{d}{dx}\left(-\frac{x}{y}\right)$

$\Rightarrow y'' = \frac{y(-1) + xy'}{y^2} = \frac{-y + x\left(-\frac{x}{y}\right)}{y^2}$ since $y' = -\frac{x}{y} \Rightarrow \frac{d^2y}{dx^2} = y'' = \frac{-y^2 - x^2}{y^3} = \frac{-y^2 - (1 - y^2)}{y^3} = \frac{-1}{y^3}$

21. $y^2 = x^2 + 2x \Rightarrow 2yy' = 2x + 2 \Rightarrow y' = \frac{2x+2}{2y} = \frac{x+1}{y}$; then $y'' = \frac{y - (x+1)y'}{y^2} = \frac{y - (x+1)\left(\frac{x+1}{y}\right)}{y^2}$

$\Rightarrow \frac{d^2y}{dx^2} = y'' = \frac{y^2 - (x+1)^2}{y^3}$

23. $2\sqrt{y} = x - y \Rightarrow y^{-1/2}y' = 1 - y' \Rightarrow y'\left(y^{-1/2} + 1\right) = 1 \Rightarrow \frac{dy}{dx} = y' = \frac{1}{y^{-1/2} + 1} = \frac{\sqrt{y}}{\sqrt{y} + 1}$; we can differentiate the equation $y'\left(y^{-1/2} + 1\right) = 1$ again to find y'': $y'\left(-\frac{1}{2}y^{-3/2}y'\right) + \left(y^{-1/2} + 1\right)y'' = 0$

$\Rightarrow \left(y^{-1/2} + 1\right)y'' = \frac{1}{2}[y']^2 y^{-3/2} \Rightarrow \frac{d^2y}{dx^2} = y'' = \frac{\frac{1}{2}\left(\frac{1}{y^{-1/2}+1}\right)^2 y^{-3/2}}{\left(y^{-1/2} + 1\right)} = \frac{1}{2y^{3/2}\left(y^{-1/2} + 1\right)^3} = \frac{1}{2\left(1 + \sqrt{y}\right)^3}$

25. $x^3 + y^3 = 16 \Rightarrow 3x^2 + 3y^2y' = 0 \Rightarrow 3y^2y' = -3x^2 \Rightarrow y' = -\frac{x^2}{y^2}$; we differentiate $y^2y' = -x^2$ to find y'':

$y^2y'' + y'[2y\cdot y'] = -2x \Rightarrow y^2y'' = -2x - 2y[y']^2 \Rightarrow y'' = \frac{-2x - 2y\left(-\frac{x^2}{y^2}\right)^2}{y^2} = \frac{-2x - \frac{2x^4}{y^3}}{y^2}$

$= \frac{-2xy^3 - 2x^4}{y^5} \Rightarrow \left.\frac{d^2y}{dx^2}\right|_{(2,2)} = \frac{-32 - 32}{32} = -2$

27. $y^2 + x^2 = y^4 - 2x$ at $(-2, 1)$ and $(-2, -1) \Rightarrow 2y\frac{dy}{dx} + 2x = 4y^3\frac{dy}{dx} - 2 \Rightarrow 2y\frac{dy}{dx} - 4y^3\frac{dy}{dx} = -2 - 2x$

$\Rightarrow \frac{dy}{dx}\left(2y - 4y^3\right) = -2 - 2x \Rightarrow \frac{dy}{dx} = \frac{x+1}{2y^3 - y} \Rightarrow \left.\frac{dy}{dx}\right|_{(-2,1)} = -1$ and $\left.\frac{dy}{dx}\right|_{(-2,-1)} = 1$

29. $x^2 + xy - y^2 = 1 \Rightarrow 2x + y + xy' - 2yy' = 0 \Rightarrow (x - 2y)y' = -2x - y \Rightarrow y' = \frac{2x + y}{2y - x}$;

(a) the slope of the tangent line $m = y'|_{(2,3)} = \frac{7}{4} \Rightarrow$ the tangent line is $y - 3 = \frac{7}{4}(x - 2) \Rightarrow y = \frac{7}{4}x - \frac{1}{2}$

(b) the normal line is $y - 3 = -\frac{4}{7}(x - 2) \Rightarrow y = -\frac{4}{7}x + \frac{29}{7}$

31. $x^2y^2 = 9 \Rightarrow 2xy^2 + 2x^2yy' = 0 \Rightarrow x^2yy' = -xy^2 \Rightarrow y' = -\frac{y}{x}$;

(a) the slope of the tangent line $m = y'|_{(-1,3)} = -\frac{y}{x}\Big|_{(-1,3)} = 3 \Rightarrow$ the tangent line is $y - 3 = 3(x+1) \Rightarrow y = 3x + 6$

(b) the normal line is $y - 3 = -\frac{1}{3}(x+1) \Rightarrow y = -\frac{1}{3}x + \frac{8}{3}$

33. $6x^2 + 3xy + 2y^2 + 17y - 6 = 0 \Rightarrow 12x + 3y + 3xy' + 4yy' + 17y' = 0 \Rightarrow y'(3x + 4y + 17) = -12x - 3y$ $\Rightarrow y' = \frac{-12x-3y}{3x+4y+17}$;

(a) the slope of the tangent line $m = y'|_{(-1,0)} = \frac{-12x-3y}{3x+4y+17}\Big|_{(-1,0)} = \frac{6}{7} \Rightarrow$ the tangent line is $y - 0 = \frac{6}{7}(x+1)$ $\Rightarrow y = \frac{6}{7}x + \frac{6}{7}$

(b) the normal line is $y - 0 = -\frac{7}{6}(x+1) \Rightarrow y = -\frac{7}{6}x - \frac{7}{6}$

35. $2xy + \pi \sin y = 2\pi \Rightarrow 2xy' + 2y + \pi(\cos y)y' = 0 \Rightarrow y'(2x + \pi \cos y) = -2y \Rightarrow y' = \frac{-2y}{2x + \pi \cos y}$;

(a) the slope of the tangent line $m = y'|_{(1,\frac{\pi}{2})} = \frac{-2y}{2x+\pi\cos y}\Big|_{(1,\frac{\pi}{2})} = -\frac{\pi}{2} \Rightarrow$ the tangent line is $y - \frac{\pi}{2} = -\frac{\pi}{2}(x-1) \Rightarrow y = -\frac{\pi}{2}x + \pi$

(b) the normal line is $y - \frac{\pi}{2} = \frac{2}{\pi}(x-1) \Rightarrow y = \frac{2}{\pi}x - \frac{2}{\pi} + \frac{\pi}{2}$

37. $y = 2\sin(\pi x - y) \Rightarrow y' = 2[\cos(\pi x - y)] \cdot (\pi - y') \Rightarrow y'[1 + 2\cos(\pi x - y)] = 2\pi\cos(\pi x - y) \Rightarrow y' = \frac{2\pi\cos(\pi x - y)}{1 + 2\cos(\pi x - y)}$;

(a) the slope of the tangent line $m = y'|_{(1,0)} = \frac{2\pi\cos(\pi x - y)}{1+2\cos(\pi x - y)}\Big|_{(1,0)} = 2\pi \Rightarrow$ the tangent line is $y - 0 = 2\pi(x-1) \Rightarrow y = 2\pi x - 2\pi$

(b) the normal line is $y - 0 = -\frac{1}{2\pi}(x-1) \Rightarrow y = -\frac{x}{2\pi} + \frac{1}{2\pi}$

39. Solving $x^2 + xy + y^2 = 7$ and $y = 0 \Rightarrow x^2 = 7 \Rightarrow x = \pm\sqrt{7} \Rightarrow \left(-\sqrt{7}, 0\right)$ and $\left(\sqrt{7}, 0\right)$ are the points where the curve crosses the x-axis. Now $x^2 + xy + y^2 = 7 \Rightarrow 2x + y + xy' + 2yy' = 0 \Rightarrow (x + 2y)y' = -2x - y$ $\Rightarrow y' = -\frac{2x+y}{x+2y} \Rightarrow m = -\frac{2x+y}{x+2y} \Rightarrow$ the slope at $\left(-\sqrt{7}, 0\right)$ is $m = -\frac{-2\sqrt{7}}{-\sqrt{7}} = -2$ and the slope at $\left(\sqrt{7}, 0\right)$ is $m = -\frac{2\sqrt{7}}{\sqrt{7}} = -2$. Since the slope is -2 in each case, the corresponding tangents must be parallel.

41. $y^4 = y^2 - x^2 \Rightarrow 4y^3y' = 2yy' - 2x \Rightarrow 2(2y^3 - y)y' = -2x \Rightarrow y' = \frac{x}{y - 2y^3}$; the slope of the tangent line at $\left(\frac{\sqrt{3}}{4}, \frac{\sqrt{3}}{2}\right)$ is $\frac{x}{y-2y^3}\Big|_{\left(\frac{\sqrt{3}}{4}, \frac{\sqrt{3}}{2}\right)} = \frac{\frac{\sqrt{3}}{4}}{\frac{\sqrt{3}}{2} - \frac{6\sqrt{3}}{8}} = \frac{\frac{1}{4}}{\frac{1}{2} - \frac{3}{4}} = \frac{1}{2-3} = -1$; the slope of the tangent line at $\left(\frac{\sqrt{3}}{4}, \frac{1}{2}\right)$ is $\frac{x}{y-2y^3}\Big|_{\left(\frac{\sqrt{3}}{4}, \frac{1}{2}\right)} = \frac{\frac{\sqrt{3}}{4}}{\frac{1}{2} - \frac{2}{8}} = \frac{2\sqrt{3}}{4-2} = \sqrt{3}$

43. $y^4 - 4y^2 = x^4 - 9x^2 \Rightarrow 4y^3y' - 8yy' = 4x^3 - 18x \Rightarrow y'(4y^3 - 8y) = 4x^3 - 18x \Rightarrow y' = \frac{4x^3 - 18x}{4y^3 - 8y} = \frac{2x^3 - 9x}{2y^3 - 4y}$ $= \frac{x(2x^2-9)}{y(2y^2-4)} = m$; $(-3, 2)$: $m = \frac{(-3)(18-9)}{2(8-4)} = -\frac{27}{8}$; $(-3, -2)$: $m = \frac{27}{8}$; $(3, 2)$: $m = \frac{27}{8}$; $(3, -2)$: $m = -\frac{27}{8}$

45. $x^2 + 2xy - 3y^2 = 0 \Rightarrow 2x + 2xy' + 2y - 6yy' = 0 \Rightarrow y'(2x - 6y) = -2x - 2y \Rightarrow y' = \frac{x+y}{3y-x} \Rightarrow$ the slope of the tangent line $m = y'|_{(1,1)} = \frac{x+y}{3y-x}\Big|_{(1,1)} = 1 \Rightarrow$ the equation of the normal line at $(1, 1)$ is $y - 1 = -1(x - 1) \Rightarrow y = -x + 2$. To find where the normal line intersects the curve we substitute into its equation: $x^2 + 2x(2-x) - 3(2-x)^2 = 0$ $\Rightarrow x^2 + 4x - 2x^2 - 3(4 - 4x + x^2) = 0 \Rightarrow -4x^2 + 16x - 12 = 0 \Rightarrow x^2 - 4x + 3 = 0 \Rightarrow (x-3)(x-1) = 0$ $\Rightarrow x = 3$ and $y = -x + 2 = -1$. Therefore, the normal to the curve at $(1, 1)$ intersects the curve at the point $(3, -1)$. Note that it also intersects the curve at $(1, 1)$.

47. $y^2 = x \Rightarrow \frac{dy}{dx} = \frac{1}{2y}$. If a normal is drawn from $(a, 0)$ to (x_1, y_1) on the curve its slope satisfies $\frac{y_1 - 0}{x_1 - a} = -2y_1$ $\Rightarrow y_1 = -2y_1(x_1 - a)$ or $a = x_1 + \frac{1}{2}$. Since $x_1 \geq 0$ on the curve, we must have that $a \geq \frac{1}{2}$. By symmetry, the two points on the parabola are $(x_1, \sqrt{x_1})$ and $(x_1, -\sqrt{x_1})$. For the normal to be perpendicular, $\left(\frac{\sqrt{x_1}}{x_1 - a}\right)\left(\frac{\sqrt{x_1}}{a - x_1}\right) = -1$ $\Rightarrow \frac{x_1}{(a - x_1)^2} = 1 \Rightarrow x_1 = (a - x_1)^2 \Rightarrow x_1 = \left(x_1 + \frac{1}{2} - x_1\right)^2 \Rightarrow x_1 = \frac{1}{4}$ and $y_1 = \pm\frac{1}{2}$. Therefore, $\left(\frac{1}{4}, \pm\frac{1}{2}\right)$ and $a = \frac{3}{4}$.

49. (a) $x^2 + y^2 = 4, x^2 = 3y^2 \Rightarrow (3y^2) + y^2 = 4 \Rightarrow y^2 = 1 \Rightarrow y = \pm 1$. If $y = 1 \Rightarrow x^2 + (1)^2 = 4 \Rightarrow x^2 = 3$ $\Rightarrow x = \pm\sqrt{3}$. If $y = -1 \Rightarrow x^2 + (-1)^2 = 4 \Rightarrow x^2 = 3 \Rightarrow x = \pm\sqrt{3}$.
$x^2 + y^2 = 4 \Rightarrow 2x + 2y\frac{dy}{dx} = 0 \Rightarrow m_1 = \frac{dy}{dx} = -\frac{x}{y}$ and $x^2 = 3y^2 \Rightarrow 2x = 6y\frac{dy}{dx} \Rightarrow m_2 = \frac{dy}{dx} = \frac{x}{3y}$
At $\left(\sqrt{3}, 1\right)$: $m_1 = \frac{dy}{dx} = -\frac{\sqrt{3}}{1} = -\sqrt{3}$ and $m_2 = \frac{dy}{dx} = \frac{\sqrt{3}}{3(1)} = \frac{\sqrt{3}}{3} \Rightarrow m_1 \cdot m_2 = \left(-\sqrt{3}\right)\left(\frac{\sqrt{3}}{3}\right) = -1$
At $\left(\sqrt{3}, -1\right)$: $m_1 = \frac{dy}{dx} = -\frac{\sqrt{3}}{(-1)} = \sqrt{3}$ and $m_2 = \frac{dy}{dx} = \frac{\sqrt{3}}{3(-1)} = -\frac{\sqrt{3}}{3} \Rightarrow m_1 \cdot m_2 = \left(\sqrt{3}\right)\left(-\frac{\sqrt{3}}{3}\right) = -1$
At $\left(-\sqrt{3}, 1\right)$: $m_1 = \frac{dy}{dx} = -\frac{\left(-\sqrt{3}\right)}{1} = \sqrt{3}$ and $m_2 = \frac{dy}{dx} = \frac{-\sqrt{3}}{3(1)} = -\frac{\sqrt{3}}{3} \Rightarrow m_1 \cdot m_2 = \left(\sqrt{3}\right)\left(-\frac{\sqrt{3}}{3}\right) = -1$
At $\left(-\sqrt{3}, -1\right)$: $m_1 = \frac{dy}{dx} = -\frac{\left(-\sqrt{3}\right)}{(-1)} = -\sqrt{3}$ and $m_2 = \frac{dy}{dx} = \frac{\left(-\sqrt{3}\right)}{3(-1)} = \frac{\sqrt{3}}{3} \Rightarrow m_1 \cdot m_2 = \left(-\sqrt{3}\right)\left(\frac{\sqrt{3}}{3}\right) = -1$

(b) $x = 1 - y^2, x = \frac{1}{3}y^2 \Rightarrow \left(\frac{1}{3}y^2\right) = 1 - y^2 \Rightarrow y^2 = \frac{3}{4} \Rightarrow y = \pm\frac{\sqrt{3}}{2}$. If $y = \frac{\sqrt{3}}{2} \Rightarrow x = 1 - \left(\frac{\sqrt{3}}{2}\right)^2 = \frac{1}{4}$. If $y = -\frac{\sqrt{3}}{2} \Rightarrow x = 1 - \left(-\frac{\sqrt{3}}{2}\right)^2 = \frac{1}{4}$. $x = 1 - y^2 \Rightarrow 1 = -2y\frac{dy}{dx} \Rightarrow m_1 = \frac{dy}{dx} = -\frac{1}{2y}$ and $x = \frac{1}{3}y^2$ $\Rightarrow 1 = \frac{2}{3}y\frac{dy}{dx} \Rightarrow m_2 = \frac{dy}{dx} = \frac{3}{2y}$
At $\left(\frac{1}{4}, \frac{\sqrt{3}}{2}\right)$: $m_1 = \frac{dy}{dx} = -\frac{1}{2\left(\sqrt{3}/2\right)} = -\frac{1}{\sqrt{3}}$ and $m_2 = \frac{dy}{dx} = \frac{3}{2\left(\sqrt{3}/2\right)} = \frac{3}{\sqrt{3}} \Rightarrow m_1 \cdot m_2 = \left(-\frac{1}{\sqrt{3}}\right)\left(\frac{3}{\sqrt{3}}\right) = -1$
At $\left(\frac{1}{4}, -\frac{\sqrt{3}}{2}\right)$: $m_1 = \frac{dy}{dx} = -\frac{1}{2\left(-\sqrt{3}/2\right)} = \frac{1}{\sqrt{3}}$ and $m_2 = \frac{dy}{dx} = \frac{3}{2\left(-\sqrt{3}/2\right)} = -\frac{3}{\sqrt{3}} \Rightarrow m_1 \cdot m_2 = \left(\frac{1}{\sqrt{3}}\right)\left(-\frac{3}{\sqrt{3}}\right) = -1$

51. $xy^3 + x^2y = 6 \Rightarrow x\left(3y^2\frac{dy}{dx}\right) + y^3 + x^2\frac{dy}{dx} + 2xy = 0 \Rightarrow \frac{dy}{dx}(3xy^2 + x^2) = -y^3 - 2xy \Rightarrow \frac{dy}{dx} = \frac{-y^3 - 2xy}{3xy^2 + x^2}$ $= -\frac{y^3 + 2xy}{3xy^2 + x^2}$; also, $xy^3 + x^2y = 6 \Rightarrow x(3y^2) + y^3\frac{dx}{dy} + x^2 + y\left(2x\frac{dx}{dy}\right) = 0 \Rightarrow \frac{dx}{dy}(y^3 + 2xy) = -3xy^2 - x^2$ $\Rightarrow \frac{dx}{dy} = -\frac{3xy^2 + x^2}{y^3 + 2xy}$; thus $\frac{dx}{dy}$ appears to equal $\frac{1}{\frac{dy}{dx}}$. The two different treatments view the graphs as functions symmetric across the line $y = x$, so their slopes are reciprocals of one another at the corresponding points (a, b) and (b, a).

3.8 RELATED RATES

1. $A = \pi r^2 \Rightarrow \frac{dA}{dt} = 2\pi r\frac{dr}{dt}$

3. $y = 5x, \frac{dx}{dt} = 2 \Rightarrow \frac{dy}{dt} = 5\frac{dx}{dt} \Rightarrow \frac{dy}{dt} = 5(2) = 10$

5. $y = x^2, \frac{dx}{dt} = 3 \Rightarrow \frac{dy}{dt} = 2x\frac{dx}{dt}$; when $x = -1 \Rightarrow \frac{dy}{dt} = 2(-1)(3) = -6$

7. $x^2 + y^2 = 25, \frac{dx}{dt} = -2 \Rightarrow 2x\frac{dx}{dt} + 2y\frac{dy}{dt} = 0$; when $x = 3$ and $y = -4 \Rightarrow 2(3)(-2) + 2(-4)\frac{dy}{dt} = 0 \Rightarrow \frac{dy}{dt} = -\frac{3}{2}$

9. $L = \sqrt{x^2 + y^2}, \frac{dx}{dt} = -1, \frac{dy}{dt} = 3 \Rightarrow \frac{dL}{dt} = \frac{1}{2\sqrt{x^2 + y^2}}\left(2x\frac{dx}{dt} + 2y\frac{dy}{dt}\right) = \frac{x\frac{dx}{dt} + y\frac{dy}{dt}}{\sqrt{x^2 + y^2}}$; when $x = 5$ and $y = 12$ $\Rightarrow \frac{dL}{dt} = \frac{(5)(-1) + (12)(3)}{\sqrt{(5)^2 + (12)^2}} = \frac{31}{13}$

11. (a) $S = 6x^2$, $\frac{dx}{dt} = -5\,\frac{m}{min} \Rightarrow \frac{dS}{dt} = 12x\,\frac{dx}{dt}$; when $x = 3 \Rightarrow \frac{dS}{dt} = 12(3)(-5) = -180\,\frac{m^2}{min}$
(b) $V = x^3$, $\frac{dx}{dt} = -5\,\frac{m}{min} \Rightarrow \frac{dV}{dt} = 3x^2\,\frac{dx}{dt}$; when $x = 3 \Rightarrow \frac{dV}{dt} = 3(3)^2(-5) = -135\,\frac{m^3}{min}$

13. (a) $V = \pi r^2 h \Rightarrow \frac{dV}{dt} = \pi r^2\,\frac{dh}{dt}$
(b) $V = \pi r^2 h \Rightarrow \frac{dV}{dt} = 2\pi rh\,\frac{dr}{dt}$
(c) $V = \pi r^2 h \Rightarrow \frac{dV}{dt} = \pi r^2\,\frac{dh}{dt} + 2\pi rh\,\frac{dr}{dt}$

15. (a) $\frac{dV}{dt} = 1$ volt/sec
(b) $\frac{dI}{dt} = -\frac{1}{3}$ amp/sec
(c) $\frac{dV}{dt} = R\left(\frac{dI}{dt}\right) + I\left(\frac{dR}{dt}\right) \Rightarrow \frac{dR}{dt} = \frac{1}{I}\left(\frac{dV}{dt} - R\,\frac{dI}{dt}\right) \Rightarrow \frac{dR}{dt} = \frac{1}{I}\left(\frac{dV}{dt} - \frac{V}{I}\,\frac{dI}{dt}\right)$
(d) $\frac{dR}{dt} = \frac{1}{2}\left[1 - \frac{12}{2}\left(-\frac{1}{3}\right)\right] = \left(\frac{1}{2}\right)(3) = \frac{3}{2}$ ohms/sec, R is increasing

17. (a) $s = \sqrt{x^2 + y^2} = (x^2 + y^2)^{1/2} \Rightarrow \frac{ds}{dt} = \frac{x}{\sqrt{x^2+y^2}}\,\frac{dx}{dt}$
(b) $s = \sqrt{x^2 + y^2} = (x^2 + y^2)^{1/2} \Rightarrow \frac{ds}{dt} = \frac{x}{\sqrt{x^2+y^2}}\,\frac{dx}{dt} + \frac{y}{\sqrt{x^2+y^2}}\,\frac{dy}{dt}$
(c) $s = \sqrt{x^2 + y^2} \Rightarrow s^2 = x^2 + y^2 \Rightarrow 2s\,\frac{ds}{dt} = 2x\,\frac{dx}{dt} + 2y\,\frac{dy}{dt} \Rightarrow 2s \cdot 0 = 2x\,\frac{dx}{dt} + 2y\,\frac{dy}{dt} \Rightarrow \frac{dx}{dt} = -\frac{y}{x}\,\frac{dy}{dt}$

19. (a) $A = \frac{1}{2}ab\sin\theta \Rightarrow \frac{dA}{dt} = \frac{1}{2}ab\cos\theta\,\frac{d\theta}{dt}$
(b) $A = \frac{1}{2}ab\sin\theta \Rightarrow \frac{dA}{dt} = \frac{1}{2}ab\cos\theta\,\frac{d\theta}{dt} + \frac{1}{2}b\sin\theta\,\frac{da}{dt}$
(c) $A = \frac{1}{2}ab\sin\theta \Rightarrow \frac{dA}{dt} = \frac{1}{2}ab\cos\theta\,\frac{d\theta}{dt} + \frac{1}{2}b\sin\theta\,\frac{da}{dt} + \frac{1}{2}a\sin\theta\,\frac{db}{dt}$

21. Given $\frac{d\ell}{dt} = -2$ cm/sec, $\frac{dw}{dt} = 2$ cm/sec, $\ell = 12$ cm and $w = 5$ cm.
(a) $A = \ell w \Rightarrow \frac{dA}{dt} = \ell\,\frac{dw}{dt} + w\,\frac{d\ell}{dt} \Rightarrow \frac{dA}{dt} = 12(2) + 5(-2) = 14\ \text{cm}^2/\text{sec}$, increasing
(b) $P = 2\ell + 2w \Rightarrow \frac{dP}{dt} = 2\,\frac{d\ell}{dt} + 2\,\frac{dw}{dt} = 2(-2) + 2(2) = 0$ cm/sec, constant
(c) $D = \sqrt{w^2 + \ell^2} = (w^2 + \ell^2)^{1/2} \Rightarrow \frac{dD}{dt} = \frac{1}{2}(w^2 + \ell^2)^{-1/2}\left(2w\,\frac{dw}{dt} + 2\ell\,\frac{d\ell}{dt}\right) \Rightarrow \frac{dD}{dt} = \frac{w\,\frac{dw}{dt} + \ell\,\frac{d\ell}{dt}}{\sqrt{w^2+\ell^2}}$
$= \frac{(5)(2) + (12)(-2)}{\sqrt{25+144}} = -\frac{14}{13}$ cm/sec, decreasing

23. Given: $\frac{dx}{dt} = 5$ ft/sec, the ladder is 13 ft long, and $x = 12$, $y = 5$ at the instant of time
(a) Since $x^2 + y^2 = 169 \Rightarrow \frac{dy}{dt} = -\frac{x}{y}\,\frac{dx}{dt} = -\left(\frac{12}{5}\right)(5) = -12$ ft/sec, the ladder is sliding down the wall
(b) The area of the triangle formed by the ladder and walls is $A = \frac{1}{2}xy \Rightarrow \frac{dA}{dt} = \left(\frac{1}{2}\right)\left(x\,\frac{dy}{dt} + y\,\frac{dx}{dt}\right)$. The area is changing at $\frac{1}{2}[12(-12) + 5(5)] = -\frac{119}{2} = -59.5\ \text{ft}^2/\text{sec}$.
(c) $\cos\theta = \frac{x}{13} \Rightarrow -\sin\theta\,\frac{d\theta}{dt} = \frac{1}{13}\cdot\frac{dx}{dt} \Rightarrow \frac{d\theta}{dt} = -\frac{1}{13\sin\theta}\cdot\frac{dx}{dt} = -\left(\frac{1}{5}\right)(5) = -1$ rad/sec

25. Let s represent the distance between the girl and the kite and x represents the horizontal distance between the girl and kite $\Rightarrow s^2 = (300)^2 + x^2 \Rightarrow \frac{ds}{dt} = \frac{x}{s}\,\frac{dx}{dt} = \frac{400(25)}{500} = 20$ ft/sec.

27. $V = \frac{1}{3}\pi r^2 h$, $h = \frac{3}{8}(2r) = \frac{3r}{4} \Rightarrow r = \frac{4h}{3} \Rightarrow V = \frac{1}{3}\pi\left(\frac{4h}{3}\right)^2 h = \frac{16\pi h^3}{27} \Rightarrow \frac{dV}{dt} = \frac{16\pi h^2}{9}\,\frac{dh}{dt}$
(a) $\left.\frac{dh}{dt}\right|_{h=4} = \left(\frac{9}{16\pi 4^2}\right)(10) = \frac{90}{256\pi} \approx 0.1119$ m/sec $= 11.19$ cm/sec
(b) $r = \frac{4h}{3} \Rightarrow \frac{dr}{dt} = \frac{4}{3}\,\frac{dh}{dt} = \frac{4}{3}\left(\frac{90}{256\pi}\right) = \frac{15}{32\pi} \approx 0.1492$ m/sec $= 14.92$ cm/sec

29. (a) $V = \frac{\pi}{3}y^2(3R - y) \Rightarrow \frac{dV}{dt} = \frac{\pi}{3}[2y(3R - y) + y^2(-1)]\,\frac{dy}{dt} \Rightarrow \frac{dy}{dt} = \left[\frac{\pi}{3}(6Ry - 3y^2)\right]^{-1}\frac{dV}{dt} \Rightarrow$ at $R = 13$ and $y = 8$ we have $\frac{dy}{dt} = \frac{1}{144\pi}(-6) = \frac{-1}{24\pi}$ m/min
(b) The hemisphere is on the circle $r^2 + (13 - y)^2 = 169 \Rightarrow r = \sqrt{26y - y^2}$ m
(c) $r = (26y - y^2)^{1/2} \Rightarrow \frac{dr}{dt} = \frac{1}{2}(26y - y^2)^{-1/2}(26 - 2y)\,\frac{dy}{dt} \Rightarrow \frac{dr}{dt} = \frac{13 - y}{\sqrt{26y - y^2}}\,\frac{dy}{dt} \Rightarrow \left.\frac{dr}{dt}\right|_{y=8} = \frac{13 - 8}{\sqrt{26\cdot 8 - 64}}\left(\frac{-1}{24\pi}\right)$
$= \frac{-5}{288\pi}$ m/min

31. If $V = \frac{4}{3}\pi r^3$, r = 5, and $\frac{dV}{dt} = 100\pi$ ft³/min, then $\frac{dV}{dt} = 4\pi r^2 \frac{dr}{dt} \Rightarrow \frac{dr}{dt} = 1$ ft/min. Then $S = 4\pi r^2 \Rightarrow \frac{dS}{dt} = 8\pi r \frac{dr}{dt} = 8\pi(5)(1) = 40\pi$ ft²/min, the rate at which the surface area is increasing.

33. Let s represent the distance between the bicycle and balloon, h the height of the balloon and x the horizontal distance between the balloon and the bicycle. The relationship between the variables is $s^2 = h^2 + x^2$ $\Rightarrow \frac{ds}{dt} = \frac{1}{s}\left(h\frac{dh}{dt} + x\frac{dx}{dt}\right) \Rightarrow \frac{ds}{dt} = \frac{1}{85}[68(1) + 51(17)] = 11$ ft/sec.

35. $y = QD^{-1} \Rightarrow \frac{dy}{dt} = D^{-1}\frac{dQ}{dt} - QD^{-2}\frac{dD}{dt} = \frac{1}{41}(0) - \frac{233}{(41)^2}(-2) = \frac{466}{1681}$ L/min $\Rightarrow$ increasing about 0.2772 L/min

37. The distance from the origin is $s = \sqrt{x^2 + y^2}$ and we wish to find $\left.\frac{ds}{dt}\right|_{(5,12)} = \frac{1}{2}(x^2 + y^2)^{-1/2}\left.\left(2x\frac{dx}{dt} + 2y\frac{dy}{dt}\right)\right|_{(5,12)}$ $= \frac{(5)(-1) + (12)(-5)}{\sqrt{25 + 144}} = -5$ m/sec

39. Let $s = 16t^2$ represent the distance the ball has fallen, h the distance between the ball and the ground, and I the distance between the shadow and the point directly beneath the ball. Accordingly, $s + h = 50$ and since the triangle LOQ and triangle PRQ are similar we have $I = \frac{30h}{50 - h} \Rightarrow h = 50 - 16t^2$ and $I = \frac{30(50 - 16t^2)}{50 - (50 - 16t^2)} = \frac{1500}{16t^2} - 30 \Rightarrow \frac{dI}{dt} = -\frac{1500}{8t^3}$ $\Rightarrow \left.\frac{dI}{dt}\right|_{t=\frac{1}{2}} = -1500$ ft/sec.

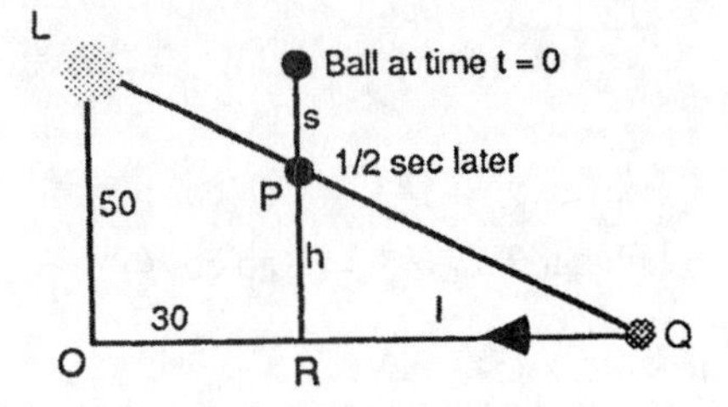

41. The volume of the ice is $V = \frac{4}{3}\pi r^3 - \frac{4}{3}\pi 4^3 \Rightarrow \frac{dV}{dt} = 4\pi r^2 \frac{dr}{dt} \Rightarrow \left.\frac{dr}{dt}\right|_{r=6} = \frac{-5}{72\pi}$ in./min when $\frac{dV}{dt} = -10$ in³/min, the thickness of the ice is decreasing at $\frac{5}{72\pi}$ in/min. The surface area is $S = 4\pi r^2 \Rightarrow \frac{dS}{dt} = 8\pi r \frac{dr}{dt} \Rightarrow \left.\frac{dS}{dt}\right|_{r=6} = 48\pi\left(\frac{-5}{72\pi}\right)$ $= -\frac{10}{3}$ in²/min, the outer surface area of the ice is decreasing at $\frac{10}{3}$ in²/min.

43. Let x represent distance of the player from second base and s the distance to third base. Then $\frac{dx}{dt} = -16$ ft/sec

(a) $s^2 = x^2 + 8100 \Rightarrow 2s\frac{ds}{dt} = 2x\frac{dx}{dt} \Rightarrow \frac{ds}{dt} = \frac{x}{s}\frac{dx}{dt}$. When the player is 30 ft from first base, x = 60 $\Rightarrow s = 30\sqrt{13}$ and $\frac{ds}{dt} = \frac{60}{30\sqrt{13}}(-16) = \frac{-32}{\sqrt{13}} \approx -8.875$ ft/sec

(b) $\sin\theta_1 = \frac{90}{s} \Rightarrow \cos\theta_1 \frac{d\theta_1}{dt} = -\frac{90}{s^2}\cdot\frac{ds}{dt} \Rightarrow \frac{d\theta_1}{dt} = -\frac{90}{s^2\cos\theta_1}\cdot\frac{ds}{dt} = -\frac{90}{s\cdot x}\cdot\frac{ds}{dt}$. Therefore, x = 60 and $s = 30\sqrt{13}$ $\Rightarrow \frac{d\theta_1}{dt} = -\frac{90}{\left(30\sqrt{13}\right)(60)}\cdot\left(\frac{-32}{\sqrt{13}}\right) = \frac{8}{65}$ rad/sec; $\cos\theta_2 = \frac{90}{s} \Rightarrow -\sin\theta_2\frac{d\theta_2}{dt} = -\frac{90}{s^2}\cdot\frac{ds}{dt} \Rightarrow \frac{d\theta_2}{dt} = \frac{90}{s^2\sin\theta_2}\cdot\frac{ds}{dt}$ $= \frac{90}{s\cdot x}\cdot\frac{ds}{dt}$. Therefore, x = 60 and $s = 30\sqrt{13} \Rightarrow \frac{d\theta_2}{dt} = \frac{90}{\left(30\sqrt{13}\right)(60)}\cdot\left(\frac{-32}{\sqrt{13}}\right) = -\frac{8}{65}$ rad/sec.

(c) $\frac{d\theta_1}{dt} = -\frac{90}{s^2\cos\theta_1}\cdot\frac{ds}{dt} = -\frac{90}{\left(s^2\cdot\frac{x}{s}\right)}\cdot\left(\frac{x}{s}\right)\cdot\left(\frac{dx}{dt}\right) = \left(-\frac{90}{s^2}\right)\left(\frac{dx}{dt}\right) = \left(-\frac{90}{x^2 + 8100}\right)\frac{dx}{dt} \Rightarrow \lim_{x\to 0}\frac{d\theta_1}{dt}$ $= \lim_{x\to 0}\left(-\frac{90}{x^2 + 8100}\right)(-15) = \frac{1}{6}$ rad/sec; $\frac{d\theta_2}{dt} = \frac{90}{s^2\sin\theta_2}\cdot\frac{ds}{dt} = \left(\frac{90}{s^2\cdot\frac{x}{s}}\right)\left(\frac{x}{s}\right)\left(\frac{dx}{dt}\right) = \left(\frac{90}{s^2}\right)\left(\frac{dx}{dt}\right)$ $= \left(\frac{90}{x^2 + 8100}\right)\frac{dx}{dt} \Rightarrow \lim_{x\to 0}\frac{d\theta_2}{dt} = -\frac{1}{6}$ rad/sec

3.9 LINEARIZATION AND DIFFERENTIALS

1. $f(x) = x^3 - 2x + 3 \Rightarrow f'(x) = 3x^2 - 2 \Rightarrow L(x) = f'(2)(x - 2) + f(2) = 10(x - 2) + 7 \Rightarrow L(x) = 10x - 13$ at x = 2

3. $f(x) = x + \frac{1}{x} \Rightarrow f'(x) = 1 - x^{-2} \Rightarrow L(x) = f(1) + f'(1)(x - 1) = 2 + 0(x - 1) = 2$

5. $f(x) = \tan x \Rightarrow f'(x) = \sec^2 x \Rightarrow L(x) = f(\pi) + f'(\pi)(x - \pi) = 0 + 1(x - \pi) = x - \pi$

7. $f(x) = x^2 + 2x \Rightarrow f'(x) = 2x + 2 \Rightarrow L(x) = f'(0)(x - 0) + f(0) = 2(x - 0) + 0 \Rightarrow L(x) = 2x$ at $x = 0$

9. $f(x) = 2x^2 + 4x - 3 \Rightarrow f'(x) = 4x + 4 \Rightarrow L(x) = f'(-1)(x + 1) + f(-1) = 0(x + 1) + (-5) \Rightarrow L(x) = -5$ at $x = -1$

11. $f(x) = \sqrt[3]{x} = x^{1/3} \Rightarrow f'(x) = \left(\frac{1}{3}\right) x^{-2/3} \Rightarrow L(x) = f'(8)(x - 8) + f(8) = \frac{1}{12}(x - 8) + 2 \Rightarrow L(x) = \frac{1}{12}x + \frac{4}{3}$ at $x = 8$

13. $f'(x) = k(1 + x)^{k-1}$. We have $f(0) = 1$ and $f'(0) = k$. $L(x) = f(0) + f'(0)(x - 0) = 1 + k(x - 0) = 1 + kx$

15. (a) $(1.0002)^{50} = (1 + 0.0002)^{50} \approx 1 + 50(0.0002) = 1 + .01 = 1.01$
(b) $\sqrt[3]{1.009} = (1 + 0.009)^{1/3} \approx 1 + \left(\frac{1}{3}\right)(0.009) = 1 + 0.003 = 1.003$

17. $y = x^3 - 3\sqrt{x} = x^3 - 3x^{1/2} \Rightarrow dy = \left(3x^2 - \frac{3}{2}x^{-1/2}\right) dx \Rightarrow dy = \left(3x^2 - \frac{3}{2\sqrt{x}}\right) dx$

19. $y = \frac{2x}{1+x^2} \Rightarrow dy = \left(\frac{(2)(1+x^2) - (2x)(2x)}{(1+x^2)^2}\right) dx = \frac{2 - 2x^2}{(1+x^2)^2}\, dx$

21. $2y^{3/2} + xy - x = 0 \Rightarrow 3y^{1/2}\,dy + y\,dx + x\,dy - dx = 0 \Rightarrow \left(3y^{1/2} + x\right) dy = (1 - y)\,dx \Rightarrow dy = \frac{1-y}{3\sqrt{y}+x}\, dx$

23. $y = \sin\left(5\sqrt{x}\right) = \sin\left(5x^{1/2}\right) \Rightarrow dy = \left(\cos\left(5x^{1/2}\right)\right)\left(\frac{5}{2}x^{-1/2}\right) dx \Rightarrow dy = \frac{5\cos\left(5\sqrt{x}\right)}{2\sqrt{x}}\, dx$

25. $y = 4\tan\left(\frac{x^3}{3}\right) \Rightarrow dy = 4\left(\sec^2\left(\frac{x^3}{3}\right)\right)(x^2)\,dx \Rightarrow dy = 4x^2\sec^2\left(\frac{x^3}{3}\right) dx$

27. $y = 3\csc\left(1 - 2\sqrt{x}\right) = 3\csc\left(1 - 2x^{1/2}\right) \Rightarrow dy = 3\left(-\csc\left(1 - 2x^{1/2}\right)\right)\cot\left(1 - 2x^{1/2}\right)\left(-x^{-1/2}\right) dx$
$\Rightarrow dy = \frac{3}{\sqrt{x}}\csc\left(1 - 2\sqrt{x}\right)\cot\left(1 - 2\sqrt{x}\right) dx$

29. $f(x) = x^2 + 2x,\ x_0 = 1,\ dx = 0.1 \Rightarrow f'(x) = 2x + 2$
(a) $\Delta f = f(x_0 + dx) - f(x_0) = f(1.1) - f(1) = 3.41 - 3 = 0.41$
(b) $df = f'(x_0)\,dx = [2(1) + 2](0.1) = 0.4$
(c) $|\Delta f - df| = |0.41 - 0.4| = 0.01$

31. $f(x) = x^3 - x,\ x_0 = 1,\ dx = 0.1 \Rightarrow f'(x) = 3x^2 - 1$
(a) $\Delta f = f(x_0 + dx) - f(x_0) = f(1.1) - f(1) = .231$
(b) $df = f'(x_0)\,dx = [3(1)^2 - 1](.1) = .2$
(c) $|\Delta f - df| = |.231 - .2| = .031$

33. $f(x) = x^{-1},\ x_0 = 0.5,\ dx = 0.1 \Rightarrow f'(x) = -x^{-2}$
(a) $\Delta f = f(x_0 + dx) - f(x_0) = f(.6) - f(.5) = -\frac{1}{3}$
(b) $df = f'(x_0)\,dx = (-4)\left(\frac{1}{10}\right) = -\frac{2}{5}$
(c) $|\Delta f - df| = \left|-\frac{1}{3} + \frac{2}{5}\right| = \frac{1}{15}$

35. $V = \frac{4}{3}\pi r^3 \Rightarrow dV = 4\pi r_0^2\, dr$

37. $S = 6x^2 \Rightarrow dS = 12x_0\, dx$

39. $V = \pi r^2 h$, height constant $\Rightarrow dV = 2\pi r_0 h\, dr$

41. Given $r = 2$ m, $dr = .02$ m
   (a) $A = \pi r^2 \Rightarrow dA = 2\pi r\, dr = 2\pi(2)(.02) = .08\pi\ m^2$
   (b) $\left(\frac{.08\pi}{4\pi}\right)(100\%) = 2\%$

43. The volume of a cylinder is $V = \pi r^2 h$. When h is held fixed, we have $\frac{dV}{dr} = 2\pi rh$, and so $dV = 2\pi rh\, dr$. For $h = 30$ in., $r = 6$ in., and $dr = 0.5$ in., the volume of the material in the shell is approximately $dV = 2\pi rh\, dr = 2\pi(6)(30)(0.5)$ $= 180\pi \approx 565.5\ in^3$.

45. The percentage error in the radius is $\frac{\left(\frac{dr}{dt}\right)}{r} \times 100 \le 2\%$.
   (a) Since $C = 2\pi r \Rightarrow \frac{dC}{dt} = 2\pi \frac{dr}{dt}$. The percentage error in calculating the circle's circumference is $\frac{\left(\frac{dC}{dt}\right)}{C} \times 100$ $= \frac{\left(2\pi \frac{dr}{dt}\right)}{2\pi r} \times 100 = \frac{\left(\frac{dr}{dt}\right)}{r} \times 100 \le 2\%$.
   (b) Since $A = \pi r^2 \Rightarrow \frac{dA}{dt} = 2\pi r \frac{dr}{dt}$. The percentage error in calculating the circle's area is given by $\frac{\left(\frac{dA}{dt}\right)}{A} \times 100$ $= \frac{\left(2\pi r \frac{dr}{dt}\right)}{\pi r^2} \times 100 = 2\frac{\left(\frac{dr}{dt}\right)}{r} \times 100 \le 2(2\%) = 4\%$.

47. $V = \pi h^3 \Rightarrow dV = 3\pi h^2\, dh$; recall that $\Delta V \approx dV$. Then $|\Delta V| \le (1\%)(V) = \frac{(1)(\pi h^3)}{100} \Rightarrow |dV| \le \frac{(1)(\pi h^3)}{100}$ $\Rightarrow |3\pi h^2\, dh| \le \frac{(1)(\pi h^3)}{100} \Rightarrow |dh| \le \frac{1}{300} h = \left(\frac{1}{3}\%\right) h$. Therefore the greatest tolerated error in the measurement of h is $\frac{1}{3}\%$.

49. Given $D = 100$ cm, $dD = 1$ cm, $V = \frac{4}{3}\pi\left(\frac{D}{2}\right)^3 = \frac{\pi D^3}{6} \Rightarrow dV = \frac{\pi}{2} D^2\, dD = \frac{\pi}{2}(100)^2(1) = \frac{10^4\pi}{2}$. Then $\frac{dV}{V}(100\%)$ $= \left[\frac{\frac{10^4\pi}{2}}{\frac{10^6\pi}{6}}\right](10^2\%) = \left[\frac{\frac{10^6\pi}{2}}{\frac{10^6\pi}{6}}\right]\% = 3\%$

51. $W = a + \frac{b}{g} = a + bg^{-1} \Rightarrow dW = -bg^{-2}\, dg = -\frac{b\, dg}{g^2} \Rightarrow \frac{dW_{moon}}{dW_{earth}} = \frac{\left(-\frac{b\, dg}{(5.2)^2}\right)}{\left(-\frac{b\, dg}{(32)^2}\right)} = \left(\frac{32}{5.2}\right)^2 = 37.87$, so a change of gravity on the moon has about 38 times the effect that a change of the same magnitude has on Earth.

53. $E(x) = f(x) - g(x) \Rightarrow E(x) = f(x) - m(x-a) - c$. Then $E(a) = 0 \Rightarrow f(a) - m(a-a) - c = 0 \Rightarrow c = f(a)$. Next we calculate m: $\lim_{x \to a} \frac{E(x)}{x-a} = 0 \Rightarrow \lim_{x \to a} \frac{f(x) - m(x-a) - c}{x-a} = 0 \Rightarrow \lim_{x \to a}\left[\frac{f(x) - f(a)}{x-a} - m\right] = 0$ (since $c = f(a)$) $\Rightarrow f'(a) - m = 0 \Rightarrow m = f'(a)$. Therefore, $g(x) = m(x-a) + c = f'(a)(x-a) + f(a)$ is the linear approximation, as claimed.

CHAPTER 3 PRACTICE EXERCISES

1. $y = x^5 - 0.125x^2 + 0.25x \Rightarrow \frac{dy}{dx} = 5x^4 - 0.25x + 0.25$

3. $y = x^3 - 3(x^2 + \pi^2) \Rightarrow \frac{dy}{dx} = 3x^2 - 3(2x + 0) = 3x^2 - 6x = 3x(x-2)$

5. $y = (x+1)^2(x^2 + 2x) \Rightarrow \frac{dy}{dx} = (x+1)^2(2x+2) + (x^2+2x)(2(x+1)) = 2(x+1)\left[(x+1)^2 + x(x+2)\right]$ $= 2(x+1)(2x^2 + 4x + 1)$

7. $y = (\theta^2 + \sec\theta + 1)^3 \Rightarrow \frac{dy}{d\theta} = 3(\theta^2 + \sec\theta + 1)^2(2\theta + \sec\theta\tan\theta)$

9. $s = \frac{\sqrt{t}}{1+\sqrt{t}} \Rightarrow \frac{ds}{dt} = \frac{(1+\sqrt{t})\cdot\frac{1}{2\sqrt{t}} - \sqrt{t}\left(\frac{1}{2\sqrt{t}}\right)}{(1+\sqrt{t})^2} = \frac{(1+\sqrt{t}) - \sqrt{t}}{2\sqrt{t}(1+\sqrt{t})^2} = \frac{1}{2\sqrt{t}(1+\sqrt{t})^2}$

11. $y = 2\tan^2 x - \sec^2 x \Rightarrow \frac{dy}{dx} = (4\tan x)(\sec^2 x) - (2\sec x)(\sec x \tan x) = 2\sec^2 x \tan x$

13. $s = \cos^4(1-2t) \Rightarrow \frac{ds}{dt} = 4\cos^3(1-2t)(-\sin(1-2t))(-2) = 8\cos^3(1-2t)\sin(1-2t)$

15. $s = (\sec t + \tan t)^5 \Rightarrow \frac{ds}{dt} = 5(\sec t + \tan t)^4(\sec t \tan t + \sec^2 t) = 5(\sec t)(\sec t + \tan t)^5$

17. $r = \sqrt{2\theta \sin\theta} = (2\theta\sin\theta)^{1/2} \Rightarrow \frac{dr}{d\theta} = \frac{1}{2}(2\theta\sin\theta)^{-1/2}(2\theta\cos\theta + 2\sin\theta) = \frac{\theta\cos\theta + \sin\theta}{\sqrt{2\theta\sin\theta}}$

19. $r = \sin\sqrt{2\theta} = \sin(2\theta)^{1/2} \Rightarrow \frac{dr}{d\theta} = \cos(2\theta)^{1/2}\left(\frac{1}{2}(2\theta)^{-1/2}(2)\right) = \frac{\cos\sqrt{2\theta}}{\sqrt{2\theta}}$

21. $y = \frac{1}{2}x^2 \csc\frac{2}{x} \Rightarrow \frac{dy}{dx} = \frac{1}{2}x^2\left(-\csc\frac{2}{x}\cot\frac{2}{x}\right)\left(\frac{-2}{x^2}\right) + \left(\csc\frac{2}{x}\right)\left(\frac{1}{2}\cdot 2x\right) = \csc\frac{2}{x}\cot\frac{2}{x} + x\csc\frac{2}{x}$

23. $y = x^{-1/2}\sec(2x)^2 \Rightarrow \frac{dy}{dx} = x^{-1/2}\sec(2x)^2\tan(2x)^2(2(2x)\cdot 2) + \sec(2x)^2\left(-\frac{1}{2}x^{-3/2}\right)$
$= 8x^{1/2}\sec(2x)^2\tan(2x)^2 - \frac{1}{2}x^{-3/2}\sec(2x)^2 = \frac{1}{2}x^{1/2}\sec(2x)^2\left[16\tan(2x)^2 - x^{-2}\right]$ or $\frac{1}{2x^{3/2}}\sec(2x)^2\left[16x^2\tan(2x)^2 - 1\right]$

25. $y = 5\cot x^2 \Rightarrow \frac{dy}{dx} = 5(-\csc^2 x^2)(2x) = -10x\csc^2(x^2)$

27. $y = x^2\sin^2(2x^2) \Rightarrow \frac{dy}{dx} = x^2(2\sin(2x^2))(\cos(2x^2))(4x) + \sin^2(2x^2)(2x) = 8x^3\sin(2x^2)\cos(2x^2) + 2x\sin^2(2x^2)$

29. $s = \left(\frac{4t}{t+1}\right)^{-2} \Rightarrow \frac{ds}{dt} = -2\left(\frac{4t}{t+1}\right)^{-3}\left(\frac{(t+1)(4) - (4t)(1)}{(t+1)^2}\right) = -2\left(\frac{4t}{t+1}\right)^{-3}\frac{4}{(t+1)^2} = -\frac{(t+1)}{8t^3}$

31. $y = \left(\frac{\sqrt{x}}{x+1}\right)^2 \Rightarrow \frac{dy}{dx} = 2\left(\frac{\sqrt{x}}{x+1}\right)\cdot\frac{(x+1)\left(\frac{1}{2\sqrt{x}}\right) - (\sqrt{x})(1)}{(x+1)^2} = \frac{(x+1) - 2x}{(x+1)^3} = \frac{1-x}{(x+1)^3}$

33. $y = \sqrt{\frac{x^2+x}{x^2}} = \left(1 + \frac{1}{x}\right)^{1/2} \Rightarrow \frac{dy}{dx} = \frac{1}{2}\left(1 + \frac{1}{x}\right)^{-1/2}\left(-\frac{1}{x^2}\right) = -\frac{1}{2x^2\sqrt{1+\frac{1}{x}}}$

35. $r = \left(\frac{\sin\theta}{\cos\theta - 1}\right)^2 \Rightarrow \frac{dr}{d\theta} = 2\left(\frac{\sin\theta}{\cos\theta-1}\right)\left[\frac{(\cos\theta-1)(\cos\theta) - (\sin\theta)(-\sin\theta)}{(\cos\theta-1)^2}\right] = 2\left(\frac{\sin\theta}{\cos\theta-1}\right)\left(\frac{\cos^2\theta - \cos\theta + \sin^2\theta}{(\cos\theta-1)^2}\right)$
$= \frac{(2\sin\theta)(1-\cos\theta)}{(\cos\theta-1)^3} = \frac{-2\sin\theta}{(\cos\theta-1)^2}$

37. $y = (2x+1)\sqrt{2x+1} = (2x+1)^{3/2} \Rightarrow \frac{dy}{dx} = \frac{3}{2}(2x+1)^{1/2}(2) = 3\sqrt{2x+1}$

39. $y = 3(5x^2 + \sin 2x)^{-3/2} \Rightarrow \frac{dy}{dx} = 3\left(-\frac{3}{2}\right)(5x^2 + \sin 2x)^{-5/2}[10x + (\cos 2x)(2)] = \frac{-9(5x + \cos 2x)}{(5x^2 + \sin 2x)^{5/2}}$

41. $xy + 2x + 3y = 1 \Rightarrow (xy' + y) + 2 + 3y' = 0 \Rightarrow xy' + 3y' = -2 - y \Rightarrow y'(x+3) = -2 - y \Rightarrow y' = -\frac{y+2}{x+3}$

43. $x^3 + 4xy - 3y^{4/3} = 2x \Rightarrow 3x^2 + \left(4x\frac{dy}{dx} + 4y\right) - 4y^{1/3}\frac{dy}{dx} = 2 \Rightarrow 4x\frac{dy}{dx} - 4y^{1/3}\frac{dy}{dx} = 2 - 3x^2 - 4y$
$\Rightarrow \frac{dy}{dx}\left(4x - 4y^{1/3}\right) = 2 - 3x^2 - 4y \Rightarrow \frac{dy}{dx} = \frac{2 - 3x^2 - 4y}{4x - 4y^{1/3}}$

45. $(xy)^{1/2} = 1 \Rightarrow \frac{1}{2}(xy)^{-1/2}\left(x\frac{dy}{dx} + y\right) = 0 \Rightarrow x^{1/2}y^{-1/2}\frac{dy}{dx} = -x^{-1/2}y^{1/2} \Rightarrow \frac{dy}{dx} = -x^{-1}y \Rightarrow \frac{dy}{dx} = -\frac{y}{x}$

47. $y^2 = \frac{x}{x+1} \Rightarrow 2y\frac{dy}{dx} = \frac{(x+1)(1) - (x)(1)}{(x+1)^2} \Rightarrow \frac{dy}{dx} = \frac{1}{2y(x+1)^2}$

49. $p^3 + 4pq - 3q^2 = 2 \Rightarrow 3p^2 \frac{dp}{dq} + 4\left(p + q\frac{dp}{dq}\right) - 6q = 0 \Rightarrow 3p^2 \frac{dp}{dq} + 4q\frac{dp}{dq} = 6q - 4p \Rightarrow \frac{dp}{dq}(3p^2 + 4q) = 6q - 4p$
$\Rightarrow \frac{dp}{dq} = \frac{6q - 4p}{3p^2 + 4q}$

51. $r\cos 2s + \sin^2 s = \pi \Rightarrow r(-\sin 2s)(2) + (\cos 2s)\left(\frac{dr}{ds}\right) + 2\sin s\cos s = 0 \Rightarrow \frac{dr}{ds}(\cos 2s) = 2r\sin 2s - 2\sin s\cos s$
$\Rightarrow \frac{dr}{ds} = \frac{2r\sin 2s - \sin 2s}{\cos 2s} = \frac{(2r-1)(\sin 2s)}{\cos 2s} = (2r-1)(\tan 2s)$

53. (a) $x^3 + y^3 = 1 \Rightarrow 3x^2 + 3y^2\frac{dy}{dx} = 0 \Rightarrow \frac{dy}{dx} = -\frac{x^2}{y^2} \Rightarrow \frac{d^2y}{dx^2} = \frac{y^2(-2x) - (-x^2)\left(2y\frac{dy}{dx}\right)}{y^4}$
$\Rightarrow \frac{d^2y}{dx^2} = \frac{-2xy^2 + (2yx^2)\left(-\frac{x^2}{y^2}\right)}{y^4} = \frac{-2xy^2 - \frac{2x^4}{y}}{y^4} = \frac{-2xy^3 - 2x^4}{y^5}$
(b) $y^2 = 1 - \frac{2}{x} \Rightarrow 2y\frac{dy}{dx} = \frac{2}{x^2} \Rightarrow \frac{dy}{dx} = \frac{1}{yx^2} \Rightarrow \frac{dy}{dx} = (yx^2)^{-1} \Rightarrow \frac{d^2y}{dx^2} = -(yx^2)^{-2}\left[y(2x) + x^2\frac{dy}{dx}\right]$
$\Rightarrow \frac{d^2y}{dx^2} = \frac{-2xy - x^2\left(\frac{1}{yx^2}\right)}{y^2x^4} = \frac{-2xy^2 - 1}{y^3x^4}$

55. (a) Let $h(x) = 6f(x) - g(x) \Rightarrow h'(x) = 6f'(x) - g'(x) \Rightarrow h'(1) = 6f'(1) - g'(1) = 6\left(\frac{1}{2}\right) - (-4) = 7$
(b) Let $h(x) = f(x)g^2(x) \Rightarrow h'(x) = f(x)(2g(x))g'(x) + g^2(x)f'(x) \Rightarrow h'(0) = 2f(0)g(0)g'(0) + g^2(0)f'(0)$
$= 2(1)(1)\left(\frac{1}{2}\right) + (1)^2(-3) = -2$
(c) Let $h(x) = \frac{f(x)}{g(x)+1} \Rightarrow h'(x) = \frac{(g(x)+1)f'(x) - f(x)g'(x)}{(g(x)+1)^2} \Rightarrow h'(1) = \frac{(g(1)+1)f'(1) - f(1)g'(1)}{(g(1)+1)^2} = \frac{(5+1)\left(\frac{1}{2}\right) - 3(-4)}{(5+1)^2} = \frac{5}{12}$
(d) Let $h(x) = f(g(x)) \Rightarrow h'(x) = f'(g(x))g'(x) \Rightarrow h'(0) = f'(g(0))g'(0) = f'(1)\left(\frac{1}{2}\right) = \left(\frac{1}{2}\right)\left(\frac{1}{2}\right) = \frac{1}{4}$
(e) Let $h(x) = g(f(x)) \Rightarrow h'(x) = g'(f(x))f'(x) \Rightarrow h'(0) = g'(f(0))f'(0) = g'(1)f'(0) = (-4)(-3) = 12$
(f) Let $h(x) = (x + f(x))^{3/2} \Rightarrow h'(x) = \frac{3}{2}(x + f(x))^{1/2}(1 + f'(x)) \Rightarrow h'(1) = \frac{3}{2}(1 + f(1))^{1/2}(1 + f'(1))$
$= \frac{3}{2}(1+3)^{1/2}\left(1 + \frac{1}{2}\right) = \frac{9}{2}$
(g) Let $h(x) = f(x + g(x)) \Rightarrow h'(x) = f'(x + g(x))(1 + g'(x)) \Rightarrow h'(0) = f'(g(0))(1 + g'(0))$
$= f'(1)\left(1 + \frac{1}{2}\right) = \left(\frac{1}{2}\right)\left(\frac{3}{2}\right) = \frac{3}{4}$

57. $x = t^2 + \pi \Rightarrow \frac{dx}{dt} = 2t;\ y = 3\sin 2x \Rightarrow \frac{dy}{dx} = 3(\cos 2x)(2) = 6\cos 2x = 6\cos(2t^2 + 2\pi) = 6\cos(2t^2)$; thus,
$\frac{dy}{dt} = \frac{dy}{dx}\cdot\frac{dx}{dt} = 6\cos(2t^2)\cdot 2t \Rightarrow \left.\frac{dy}{dt}\right|_{t=0} = 6\cos(0)\cdot 0 = 0$

59. $r = 8\sin\left(s + \frac{\pi}{6}\right) \Rightarrow \frac{dr}{ds} = 8\cos\left(s + \frac{\pi}{6}\right);\ w = \sin\left(\sqrt{r} - 2\right) \Rightarrow \frac{dw}{dr} = \cos\left(\sqrt{r} - 2\right)\left(\frac{1}{2\sqrt{r}}\right)$
$= \frac{\cos\sqrt{8\sin\left(s+\frac{\pi}{6}\right)} - 2}{2\sqrt{8\sin\left(s+\frac{\pi}{6}\right)}}$; thus, $\frac{dw}{ds} = \frac{dw}{dr}\cdot\frac{dr}{ds} = \frac{\cos\left(\sqrt{8\sin\left(s+\frac{\pi}{6}\right)} - 2\right)}{2\sqrt{8\sin\left(s+\frac{\pi}{6}\right)}}\cdot\left[8\cos\left(s + \frac{\pi}{6}\right)\right]$
$\Rightarrow \left.\frac{dw}{ds}\right|_{s=0} = \frac{\cos\left(\sqrt{8\sin\left(\frac{\pi}{6}\right)} - 2\right)\cdot 8\cos\left(\frac{\pi}{6}\right)}{2\sqrt{8\sin\left(\frac{\pi}{6}\right)}} = \frac{(\cos 0)(8)\left(\frac{\sqrt{3}}{2}\right)}{2\sqrt{4}} = \sqrt{3}$

61. $y^3 + y = 2\cos x \Rightarrow 3y^2\frac{dy}{dx} + \frac{dy}{dx} = -2\sin x \Rightarrow \frac{dy}{dx}(3y^2 + 1) = -2\sin x \Rightarrow \frac{dy}{dx} = \frac{-2\sin x}{3y^2+1} \Rightarrow \left.\frac{dy}{dx}\right|_{(0,1)}$
$= \frac{-2\sin(0)}{3+1} = 0;\ \frac{d^2y}{dx^2} = \frac{(3y^2+1)(-2\cos x) - (-2\sin x)\left(6y\frac{dy}{dx}\right)}{(3y^2+1)^2}$
$\Rightarrow \left.\frac{d^2y}{dx^2}\right|_{(0,1)} = \frac{(3+1)(-2\cos 0) - (-2\sin 0)(6\cdot 0)}{(3+1)^2} = -\frac{1}{2}$

63. $f(t) = \frac{1}{2t+1}$ and $f(t+h) = \frac{1}{2(t+h)+1} \Rightarrow \frac{f(t+h) - f(t)}{h} = \frac{\frac{1}{2(t+h)+1} - \frac{1}{2t+1}}{h} = \frac{2t + 1 - (2t + 2h + 1)}{(2t + 2h + 1)(2t + 1)h}$
$= \frac{-2h}{(2t+2h+1)(2t+1)h} = \frac{-2}{(2t+2h+1)(2t+1)} \Rightarrow f'(t) = \lim_{h\to 0}\frac{f(t+h) - f(t)}{h} = \lim_{h\to 0}\frac{-2}{(2t+2h+1)(2t+1)}$
$= \frac{-2}{(2t+1)^2}$

65. (a)

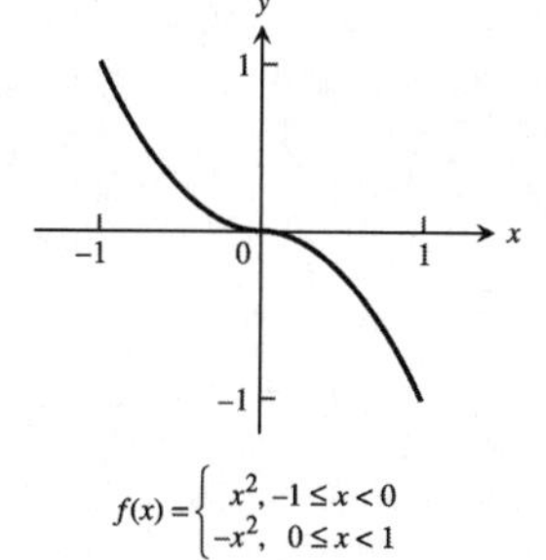

(b) $\lim_{x \to 0^-} f(x) = \lim_{x \to 0^-} x^2 = 0$ and $\lim_{x \to 0^+} f(x) = \lim_{x \to 0^+} -x^2 = 0 \Rightarrow \lim_{x \to 0} f(x) = 0$. Since $\lim_{x \to 0} f(x) = 0 = f(0)$ it follows that f is continuous at $x = 0$.

(c) $\lim_{x \to 0^-} f'(x) = \lim_{x \to 0^-} (2x) = 0$ and $\lim_{x \to 0^+} f'(x) = \lim_{x \to 0^+} (-2x) = 0 \Rightarrow \lim_{x \to 0} f'(x) = 0$. Since this limit exists, it follows that f is differentiable at $x = 0$.

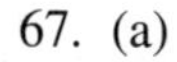
67. (a)

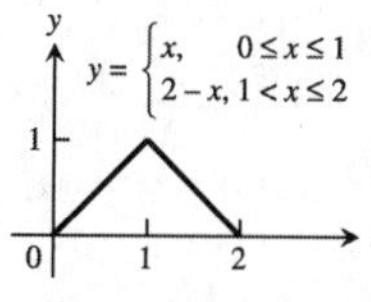

(b) $\lim_{x \to 1^-} f(x) = \lim_{x \to 1^-} x = 1$ and $\lim_{x \to 1^+} f(x) = \lim_{x \to 1^+} (2 - x) = 1 \Rightarrow \lim_{x \to 1} f(x) = 1$. Since $\lim_{x \to 1} f(x) = 1 = f(1)$, it follows that f is continuous at $x = 1$.

(c) $\lim_{x \to 1^-} f'(x) = \lim_{x \to 1^-} 1 = 1$ and $\lim_{x \to 1^+} f'(x) = \lim_{x \to 1^+} -1 = -1 \Rightarrow \lim_{x \to 1^-} f'(x) \neq \lim_{x \to 1^+} f'(x)$, so $\lim_{x \to 1} f'(x)$ does not exist $\Rightarrow$ f is not differentiable at $x = 1$.

69. $y = \frac{x}{2} + \frac{1}{2x-4} = \frac{1}{2}x + (2x-4)^{-1} \Rightarrow \frac{dy}{dx} = \frac{1}{2} - 2(2x-4)^{-2}$; the slope of the tangent is $-\frac{3}{2} \Rightarrow -\frac{3}{2} = \frac{1}{2} - 2(2x-4)^{-2}$ $\Rightarrow -2 = -2(2x-4)^{-2} \Rightarrow 1 = \frac{1}{(2x-4)^2} \Rightarrow (2x-4)^2 = 1 \Rightarrow 4x^2 - 16x + 16 = 1 \Rightarrow 4x^2 - 16x + 15 = 0$ $\Rightarrow (2x-5)(2x-3) = 0 \Rightarrow x = \frac{5}{2}$ or $x = \frac{3}{2} \Rightarrow \left(\frac{5}{2}, \frac{9}{4}\right)$ and $\left(\frac{3}{2}, -\frac{1}{4}\right)$ are points on the curve where the slope is $-\frac{3}{2}$.

71. $y = 2x^3 - 3x^2 - 12x + 20 \Rightarrow \frac{dy}{dx} = 6x^2 - 6x - 12$; the tangent is parallel to the x-axis when $\frac{dy}{dx} = 0$ $\Rightarrow 6x^2 - 6x - 12 = 0 \Rightarrow x^2 - x - 2 = 0 \Rightarrow (x-2)(x+1) = 0 \Rightarrow x = 2$ or $x = -1 \Rightarrow (2, 0)$ and $(-1, 27)$ are points on the curve where the tangent is parallel to the x-axis.

73. $y = 2x^3 - 3x^2 - 12x + 20 \Rightarrow \frac{dy}{dx} = 6x^2 - 6x - 12$

(a) The tangent is perpendicular to the line $y = 1 - \frac{x}{24}$ when $\frac{dy}{dx} = -\left(\frac{1}{-\left(\frac{1}{24}\right)}\right) = 24$; $6x^2 - 6x - 12 = 24$ $\Rightarrow x^2 - x - 2 = 4 \Rightarrow x^2 - x - 6 = 0 \Rightarrow (x-3)(x+2) = 0 \Rightarrow x = -2$ or $x = 3 \Rightarrow (-2, 16)$ and $(3, 11)$ are points where the tangent is perpendicular to $y = 1 - \frac{x}{24}$.

(b) The tangent is parallel to the line $y = \sqrt{2} - 12x$ when $\frac{dy}{dx} = 12 \Rightarrow 6x^2 - 6x - 12 = -12 \Rightarrow x^2 - x = 0$ $\Rightarrow x(x-1) = 0 \Rightarrow x = 0$ or $x = 1 \Rightarrow (0, 20)$ and $(1, 7)$ are points where the tangent is parallel to $y = \sqrt{2} - 12x$.

75. $y = \tan x, -\frac{\pi}{2} < x < \frac{\pi}{2} \Rightarrow \frac{dy}{dx} = \sec^2 x$; now the slope of $y = -\frac{x}{2}$ is $-\frac{1}{2} \Rightarrow$ the normal line is parallel to $y = -\frac{x}{2}$ when $\frac{dy}{dx} = 2$. Thus, $\sec^2 x = 2 \Rightarrow \frac{1}{\cos^2 x} = 2 \Rightarrow \cos^2 x = \frac{1}{2} \Rightarrow \cos x = \frac{\pm 1}{\sqrt{2}} \Rightarrow x = -\frac{\pi}{4}$ and $x = \frac{\pi}{4}$ for $-\frac{\pi}{2} < x < \frac{\pi}{2} \Rightarrow \left(-\frac{\pi}{4}, -1\right)$ and $\left(\frac{\pi}{4}, 1\right)$ are points where the normal is parallel to $y = -\frac{x}{2}$.

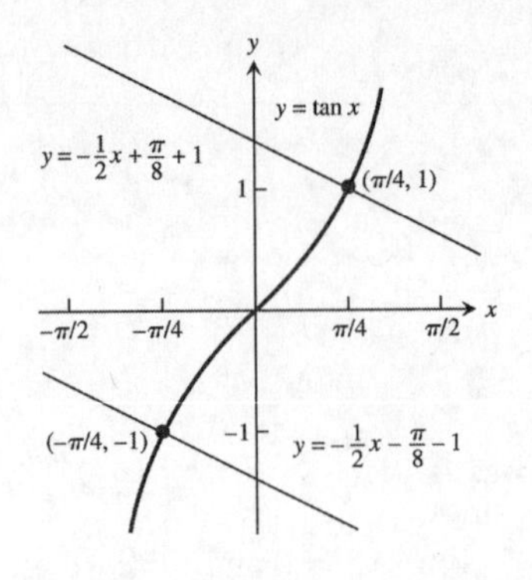

77. $y = x^2 + C \Rightarrow \frac{dy}{dx} = 2x$ and $y = x \Rightarrow \frac{dy}{dx} = 1$; the parabola is tangent to $y = x$ when $2x = 1 \Rightarrow x = \frac{1}{2} \Rightarrow y = \frac{1}{2}$; thus, $\frac{1}{2} = \left(\frac{1}{2}\right)^2 + C \Rightarrow C = \frac{1}{4}$

79. The line through $(0, 3)$ and $(5, -2)$ has slope $m = \frac{3-(-2)}{0-5} = -1 \Rightarrow$ the line through $(0, 3)$ and $(5, -2)$ is $y = -x + 3$; $y = \frac{c}{x+1} \Rightarrow \frac{dy}{dx} = \frac{-c}{(x+1)^2}$, so the curve is tangent to $y = -x + 3 \Rightarrow \frac{dy}{dx} = -1 = \frac{-c}{(x+1)^2}$ $\Rightarrow (x+1)^2 = c, x \neq -1$. Moreover, $y = \frac{c}{x+1}$ intersects $y = -x + 3 \Rightarrow \frac{c}{x+1} = -x + 3, x \neq -1$ $\Rightarrow c = (x+1)(-x+3), x \neq -1$. Thus $c = c \Rightarrow (x+1)^2 = (x+1)(-x+3) \Rightarrow (x+1)[x+1-(-x+3)]$ $= 0, x \neq -1 \Rightarrow (x+1)(2x-2) = 0 \Rightarrow x = 1$ (since $x \neq -1$) $\Rightarrow c = 4$.

81. $x^2 + 2y^2 = 9 \Rightarrow 2x + 4y \frac{dy}{dx} = 0 \Rightarrow \frac{dy}{dx} = -\frac{x}{2y} \Rightarrow \left.\frac{dy}{dx}\right|_{(1,2)} = -\frac{1}{4} \Rightarrow$ the tangent line is $y = 2 - \frac{1}{4}(x-1)$ $= -\frac{1}{4}x + \frac{9}{4}$ and the normal line is $y = 2 + 4(x-1) = 4x - 2$.

83. $xy + 2x - 5y = 2 \Rightarrow \left(x \frac{dy}{dx} + y\right) + 2 - 5 \frac{dy}{dx} = 0 \Rightarrow \frac{dy}{dx}(x-5) = -y - 2 \Rightarrow \frac{dy}{dx} = \frac{-y-2}{x-5} \Rightarrow \left.\frac{dy}{dx}\right|_{(3,2)} = 2$ $\Rightarrow$ the tangent line is $y = 2 + 2(x-3) = 2x - 4$ and the normal line is $y = 2 + \frac{-1}{2}(x-3) = -\frac{1}{2}x + \frac{7}{2}$.

85. $x + \sqrt{xy} = 6 \Rightarrow 1 + \frac{1}{2\sqrt{xy}}\left(x \frac{dy}{dx} + y\right) = 0 \Rightarrow x \frac{dy}{dx} + y = -2\sqrt{xy} \Rightarrow \frac{dy}{dx} = \frac{-2\sqrt{xy} - y}{x} \Rightarrow \left.\frac{dy}{dx}\right|_{(4,1)} = \frac{-5}{4}$ $\Rightarrow$ the tangent line is $y = 1 - \frac{5}{4}(x-4) = -\frac{5}{4}x + 6$ and the normal line is $y = 1 + \frac{4}{5}(x-4) = \frac{4}{5}x - \frac{11}{5}$.

87. $x^3y^3 + y^2 = x + y \Rightarrow \left[x^3\left(3y^2 \frac{dy}{dx}\right) + y^3(3x^2)\right] + 2y \frac{dy}{dx} = 1 + \frac{dy}{dx} \Rightarrow 3x^3y^2 \frac{dy}{dx} + 2y \frac{dy}{dx} - \frac{dy}{dx} = 1 - 3x^2y^3$ $\Rightarrow \frac{dy}{dx}(3x^3y^2 + 2y - 1) = 1 - 3x^2y^3 \Rightarrow \frac{dy}{dx} = \frac{1-3x^2y^3}{3x^3y^2+2y-1} \Rightarrow \left.\frac{dy}{dx}\right|_{(1,1)} = -\frac{2}{4}$, but $\left.\frac{dy}{dx}\right|_{(1,-1)}$ is undefined. Therefore, the curve has slope $-\frac{1}{2}$ at $(1, 1)$ but the slope is undefined at $(1, -1)$.

89. B = graph of f, A = graph of f′. Curve B cannot be the derivative of A because A has only negative slopes while some of B's values are positive.

91.

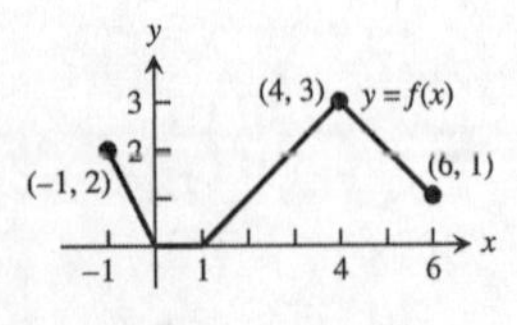

93. (a) 0, 0 (b) largest 1700, smallest about 1400

95. $\lim\limits_{x \to 0} \frac{\sin x}{2x^2 - x} = \lim\limits_{x \to 0} \left[\left(\frac{\sin x}{x}\right) \cdot \frac{1}{(2x-1)}\right] = (1)\left(\frac{1}{-1}\right) = -1$

97. $\lim\limits_{r \to 0} \frac{\sin r}{\tan 2r} = \lim\limits_{r \to 0} \left(\frac{\sin r}{r} \cdot \frac{2r}{\tan 2r} \cdot \frac{1}{2}\right) = \left(\frac{1}{2}\right)(1) \lim\limits_{r \to 0} \frac{\cos 2r}{\left(\frac{\sin 2r}{2r}\right)} = \left(\frac{1}{2}\right)(1)\left(\frac{1}{1}\right) = \frac{1}{2}$

99. $\lim\limits_{\theta \to \left(\frac{\pi}{2}\right)^-} \frac{4\tan^2\theta + \tan\theta + 1}{\tan^2\theta + 5} = \lim\limits_{\theta \to \left(\frac{\pi}{2}\right)^-} \frac{\left(4 + \frac{1}{\tan\theta} + \frac{1}{\tan^2\theta}\right)}{\left(1 + \frac{5}{\tan^2\theta}\right)} = \frac{(4+0+0)}{(1+0)} = 4$

101. $\lim\limits_{x \to 0} \frac{x \sin x}{2 - 2\cos x} = \lim\limits_{x \to 0} \frac{x \sin x}{2(1 - \cos x)} = \lim\limits_{x \to 0} \frac{x \sin x}{2\left(2\sin^2\left(\frac{x}{2}\right)\right)} = \lim\limits_{x \to 0} \left[\frac{\frac{x}{2} \cdot \frac{x}{2}}{\sin^2\left(\frac{x}{2}\right)} \cdot \frac{\sin x}{x}\right] = \lim\limits_{x \to 0} \left[\frac{\left(\frac{x}{2}\right)}{\sin\left(\frac{x}{2}\right)} \cdot \frac{\left(\frac{x}{2}\right)}{\sin\left(\frac{x}{2}\right)} \cdot \frac{\sin x}{x}\right]$ $= (1)(1)(1) = 1$

103. $\lim\limits_{x \to 0} \frac{\tan x}{x} = \lim\limits_{x \to 0} \left(\frac{1}{\cos x} \cdot \frac{\sin x}{x}\right) = 1$; let $\theta = \tan x \Rightarrow \theta \to 0$ as $x \to 0 \Rightarrow \lim\limits_{x \to 0} g(x) = \lim\limits_{x \to 0} \frac{\tan(\tan x)}{\tan x}$ $= \lim\limits_{\theta \to 0} \frac{\tan\theta}{\theta} = 1$. Therefore, to make g continuous at the origin, define $g(0) = 1$.

105. (a) $S = 2\pi r^2 + 2\pi rh$ and h constant $\Rightarrow \frac{dS}{dt} = 4\pi r \frac{dr}{dt} + 2\pi h \frac{dr}{dt} = (4\pi r + 2\pi h)\frac{dr}{dt}$

(b) $S = 2\pi r^2 + 2\pi rh$ and r constant $\Rightarrow \frac{dS}{dt} = 2\pi r \frac{dh}{dt}$

(c) $S = 2\pi r^2 + 2\pi rh \Rightarrow \frac{dS}{dt} = 4\pi r \frac{dr}{dt} + 2\pi\left(r\frac{dh}{dt} + h\frac{dr}{dt}\right) = (4\pi r + 2\pi h)\frac{dr}{dt} + 2\pi r \frac{dh}{dt}$

(d) S constant $\Rightarrow \frac{dS}{dt} = 0 \Rightarrow 0 = (4\pi r + 2\pi h)\frac{dr}{dt} + 2\pi r\frac{dh}{dt} \Rightarrow (2r + h)\frac{dr}{dt} = -r\frac{dh}{dt} \Rightarrow \frac{dr}{dt} = \frac{-r}{2r+h}\frac{dh}{dt}$

107. $A = \pi r^2 \Rightarrow \frac{dA}{dt} = 2\pi r \frac{dr}{dt}$; so $r = 10$ and $\frac{dr}{dt} = -\frac{2}{\pi}$ m/sec $\Rightarrow \frac{dA}{dt} = (2\pi)(10)\left(-\frac{2}{\pi}\right) = -40\ \text{m}^2/\text{sec}$

109. $\frac{dR_1}{dt} = -1$ ohm/sec, $\frac{dR_2}{dt} = 0.5$ ohm/sec; and $\frac{1}{R} = \frac{1}{R_1} + \frac{1}{R_2} \Rightarrow \frac{-1}{R^2}\frac{dR}{dt} = \frac{-1}{R_1^2}\frac{dR_1}{dt} - \frac{1}{R_2^2}\frac{dR_2}{dt}$. Also, $R_1 = 75$ ohms and $R_2 = 50$ ohms $\Rightarrow \frac{1}{R} = \frac{1}{75} + \frac{1}{50} \Rightarrow R = 30$ ohms. Therefore, from the derivative equation, $\frac{-1}{(30)^2}\frac{dR}{dt} = \frac{-1}{(75)^2}(-1) - \frac{1}{(50)^2}(0.5) = \left(\frac{1}{5625} - \frac{1}{5000}\right) \Rightarrow \frac{dR}{dt} = (-900)\left(\frac{5000-5625}{5625 \cdot 5000}\right) = \frac{9(625)}{50(5625)} = \frac{1}{50} = 0.02$ ohm/sec.

111. Given $\frac{dx}{dt} = 10$ m/sec and $\frac{dy}{dt} = 5$ m/sec, let D be the distance from the origin $\Rightarrow D^2 = x^2 + y^2 \Rightarrow 2D\frac{dD}{dt}$ $= 2x\frac{dx}{dt} + 2y\frac{dy}{dt} \Rightarrow D\frac{dD}{dt} = x\frac{dx}{dt} + y\frac{dy}{dt}$. When $(x, y) = (3, -4)$, $D = \sqrt{3^2 + (-4)^2} = 5$ and $5\frac{dD}{dt} = (3)(10) + (-4)(5) \Rightarrow \frac{dD}{dt} = \frac{10}{5} = 2$. Therefore, the particle is moving away from the origin at 2 m/sec (because the distance D is increasing).

113. (a) From the diagram we have $\frac{10}{h} = \frac{4}{r} \Rightarrow r = \frac{2}{5}h$.

(b) $V = \frac{1}{3}\pi r^2 h = \frac{1}{3}\pi\left(\frac{2}{5}h\right)^2 h = \frac{4\pi h^3}{75} \Rightarrow \frac{dV}{dt} = \frac{4\pi h^2}{25}\frac{dh}{dt}$, so $\frac{dV}{dt} = -5$ and $h = 6 \Rightarrow \frac{dh}{dt} = -\frac{125}{144\pi}$ ft/min.

115. (a) From the sketch in the text, $\frac{d\theta}{dt} = -0.6$ rad/sec and $x = \tan\theta$. Also $x = \tan\theta \Rightarrow \frac{dx}{dt} = \sec^2\theta\frac{d\theta}{dt}$; at point A, $x = 0$ $\Rightarrow \theta = 0 \Rightarrow \frac{dx}{dt} = (\sec^2 0)(-0.6) = -0.6$. Therefore the speed of the light is $0.6 = \frac{3}{5}$ km/sec when it reaches point A.

(b) $\frac{(3/5)\text{ rad}}{\text{sec}} \cdot \frac{1\text{ rev}}{2\pi\text{ rad}} \cdot \frac{60\text{ sec}}{\text{min}} = \frac{18}{\pi}$ revs/min

117. (a) If $f(x) = \tan x$ and $x = -\frac{\pi}{4}$, then $f'(x) = \sec^2 x$, $f\left(-\frac{\pi}{4}\right) = -1$ and $f'\left(-\frac{\pi}{4}\right) = 2$. The linearization of $f(x)$ is $L(x) = 2\left(x + \frac{\pi}{4}\right) + (-1) = 2x + \frac{\pi-2}{2}$.

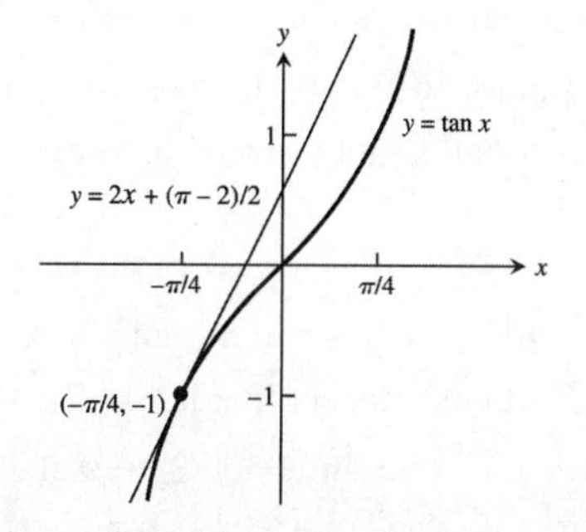

(b) If $f(x) = \sec x$ and $x = -\frac{\pi}{4}$, then $f'(x) = \sec x \tan x$, $f\left(-\frac{\pi}{4}\right) = \sqrt{2}$ and $f'\left(-\frac{\pi}{4}\right) = -\sqrt{2}$. The linearization of $f(x)$ is $L(x) = -\sqrt{2}\left(x + \frac{\pi}{4}\right) + \sqrt{2}$ $= -\sqrt{2}x + \frac{\sqrt{2}(4-\pi)}{4}$.

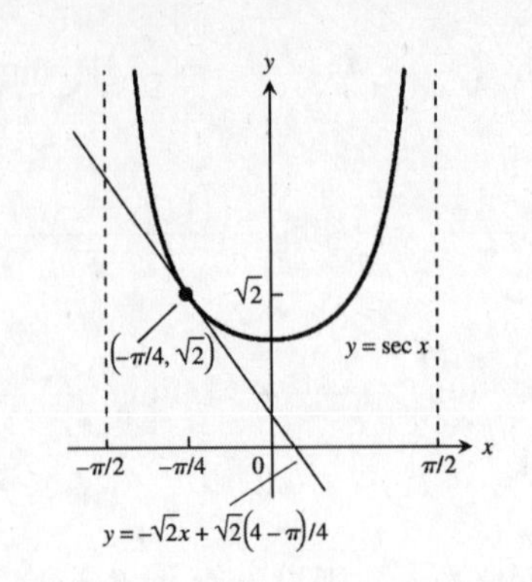

119. $f(x) = \sqrt{x+1} + \sin x - 0.5 = (x+1)^{1/2} + \sin x - 0.5 \Rightarrow f'(x) = \left(\frac{1}{2}\right)(x+1)^{-1/2} + \cos x$ $\Rightarrow L(x) = f'(0)(x-0) + f(0) = 1.5(x-0) + 0.5 \Rightarrow L(x) = 1.5x + 0.5$, the linearization of $f(x)$.

121. $S = \pi r\sqrt{r^2 + h^2}$, r constant $\Rightarrow dS = \pi r \cdot \frac{1}{2}(r^2 + h^2)^{-1/2} 2h\, dh = \frac{\pi r h}{\sqrt{r^2 + h^2}} dh$. Height changes from h_0 to $h_0 + dh$ $\Rightarrow dS = \frac{\pi r h_0 (dh)}{\sqrt{r^2 + h_0^2}}$

123. $C = 2\pi r \Rightarrow r = \frac{C}{2\pi}$, $S = 4\pi r^2 = \frac{C^2}{\pi}$, and $V = \frac{4}{3}\pi r^3 = \frac{C^3}{6\pi^2}$. It also follows that $dr = \frac{1}{2\pi} dC$, $dS = \frac{2C}{\pi} dC$ and $dV = \frac{C^2}{2\pi^2} dC$. Recall that $C = 10$ cm and $dC = 0.4$ cm.

(a) $dr = \frac{0.4}{2\pi} = \frac{0.2}{\pi}$ cm $\Rightarrow \left(\frac{dr}{r}\right)(100\%) = \left(\frac{0.2}{\pi}\right)\left(\frac{2\pi}{10}\right)(100\%) = (.04)(100\%) = 4\%$

(b) $dS = \frac{20}{\pi}(0.4) = \frac{8}{\pi}$ cm $\Rightarrow \left(\frac{dS}{S}\right)(100\%) = \left(\frac{8}{\pi}\right)\left(\frac{\pi}{100}\right)(100\%) = 8\%$

(c) $dV = \frac{10^2}{2\pi^2}(0.4) = \frac{20}{\pi^2}$ cm $\Rightarrow \left(\frac{dV}{V}\right)(100\%) = \left(\frac{20}{\pi^2}\right)\left(\frac{6\pi^2}{1000}\right)(100\%) = 12\%$

CHAPTER 3 ADDITIONAL AND ADVANCED EXERCISES

1. (a) $\sin 2\theta = 2\sin\theta\cos\theta \Rightarrow \frac{d}{d\theta}(\sin 2\theta) = \frac{d}{d\theta}(2\sin\theta\cos\theta) \Rightarrow 2\cos 2\theta = 2[(\sin\theta)(-\sin\theta) + (\cos\theta)(\cos\theta)]$ $\Rightarrow \cos 2\theta = \cos^2\theta - \sin^2\theta$

(b) $\cos 2\theta = \cos^2\theta - \sin^2\theta \Rightarrow \frac{d}{d\theta}(\cos 2\theta) = \frac{d}{d\theta}(\cos^2\theta - \sin^2\theta) \Rightarrow -2\sin 2\theta = (2\cos\theta)(-\sin\theta) - (2\sin\theta)(\cos\theta)$ $\Rightarrow \sin 2\theta = \cos\theta\sin\theta + \sin\theta\cos\theta \Rightarrow \sin 2\theta = 2\sin\theta\cos\theta$

3. (a) $f(x) = \cos x \Rightarrow f'(x) = -\sin x \Rightarrow f''(x) = -\cos x$, and $g(x) = a + bx + cx^2 \Rightarrow g'(x) = b + 2cx \Rightarrow g''(x) = 2c$; also, $f(0) = g(0) \Rightarrow \cos(0) = a \Rightarrow a = 1$; $f'(0) = g'(0) \Rightarrow -\sin(0) = b \Rightarrow b = 0$; $f''(0) = g''(0) \Rightarrow -\cos(0) = 2c$ $\Rightarrow c = -\frac{1}{2}$. Therefore, $g(x) = 1 - \frac{1}{2}x^2$.

(b) $f(x) = \sin(x+a) \Rightarrow f'(x) = \cos(x+a)$, and $g(x) = b\sin x + c\cos x \Rightarrow g'(x) = b\cos x - c\sin x$; also, $f(0) = g(0)$ $\Rightarrow \sin(a) = b\sin(0) + c\cos(0) \Rightarrow c = \sin a$; $f'(0) = g'(0) \Rightarrow \cos(a) = b\cos(0) - c\sin(0) \Rightarrow b = \cos a$. Therefore, $g(x) = \sin x\cos a + \cos x\sin a$.

(c) When $f(x) = \cos x$, $f'''(x) = \sin x$ and $f^{(4)}(x) = \cos x$; when $g(x) = 1 - \frac{1}{2}x^2$, $g'''(x) = 0$ and $g^{(4)}(x) = 0$. Thus $f'''(0) = 0 = g'''(0)$ so the third derivatives agree at $x = 0$. However, the fourth derivatives do not agree since $f^{(4)}(0) = 1$ but $g^{(4)}(0) = 0$. In case (b), when $f(x) = \sin(x+a)$ and $g(x) = \sin x\cos a + \cos x\sin a$, notice that $f(x) = g(x)$ for all x, not just $x = 0$. Since this is an identity, we have $f^{(n)}(x) = g^{(n)}(x)$ for any x and any positive integer n.

5. If the circle $(x-h)^2 + (y-k)^2 = a^2$ and $y = x^2 + 1$ are tangent at $(1,2)$, then the slope of this tangent is $m = 2x\big|_{(1,2)} = 2$ and the tangent line is $y = 2x$. The line containing (h,k) and $(1,2)$ is perpendicular to $y = 2x \Rightarrow \frac{k-2}{h-1} = -\frac{1}{2} \Rightarrow h = 5 - 2k \Rightarrow$ the location of the center is $(5-2k, k)$. Also, $(x-h)^2 + (y-k)^2 = a^2$ $\Rightarrow x - h + (y-k)y' = 0 \Rightarrow 1 + (y')^2 + (y-k)y'' = 0 \Rightarrow y'' = \frac{1+(y')^2}{k-y}$. At the point $(1,2)$ we know

$y' = 2$ from the tangent line and that $y'' = 2$ from the parabola. Since the second derivatives are equal at $(1, 2)$ we obtain $2 = \frac{1+(2)^2}{k-2} \Rightarrow k = \frac{9}{2}$. Then $h = 5 - 2k = -4 \Rightarrow$ the circle is $(x+4)^2 + \left(y - \frac{9}{2}\right)^2 = a^2$. Since $(1, 2)$ lies on the circle we have that $a = \frac{5\sqrt{5}}{2}$.

7. (a) $y = uv \Rightarrow \frac{dy}{dt} = \frac{du}{dt} v + u \frac{dv}{dt} = (0.04u)v + u(0.05v) = 0.09uv = 0.09y \Rightarrow$ the rate of growth of the total production is 9% per year.
   (b) If $\frac{du}{dt} = -0.02u$ and $\frac{dv}{dt} = 0.03v$, then $\frac{dy}{dt} = (-0.02u)v + (0.03v)u = 0.01uv = 0.01y$, increasing at 1% per year.

9. Answers will vary. Here is one possibility.

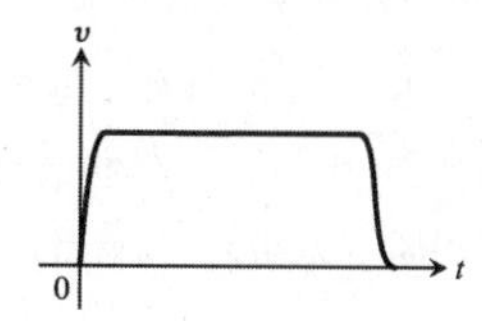

11. (a) $s(t) = 64t - 16t^2 \Rightarrow v(t) = \frac{ds}{dt} = 64 - 32t = 32(2 - t)$. The maximum height is reached when $v(t) = 0$ $\Rightarrow t = 2$ sec. The velocity when it leaves the hand is $v(0) = 64$ ft/sec.
   (b) $s(t) = 64t - 2.6t^2 \Rightarrow v(t) = \frac{ds}{dt} = 64 - 5.2t$. The maximum height is reached when $v(t) = 0 \Rightarrow t \approx 12.31$ sec. The maximum height is about $s(12.31) = 393.85$ ft.

13. $m\left(v^2 - v_0^2\right) = k\left(x_0^2 - x^2\right) \Rightarrow m\left(2v \frac{dv}{dt}\right) = k\left(-2x \frac{dx}{dt}\right) \Rightarrow m \frac{dv}{dt} = k\left(-\frac{2x}{2v}\right) \frac{dx}{dt} \Rightarrow m \frac{dv}{dt} = -kx\left(\frac{1}{v}\right) \frac{dx}{dt}$. Then substituting $\frac{dx}{dt} = v \Rightarrow m \frac{dv}{dt} = -kx$, as claimed.

15. (a) To be continuous at $x = \pi$ requires that $\lim_{x \to \pi^-} \sin x = \lim_{x \to \pi^+} (mx + b) \Rightarrow 0 = m\pi + b \Rightarrow m = -\frac{b}{\pi}$;
   (b) If $y' = \begin{cases} \cos x, & x < \pi \\ m, & x \geq \pi \end{cases}$ is differentiable at $x = \pi$, then $\lim_{x \to \pi^-} \cos x = m \Rightarrow m = -1$ and $b = \pi$.

17. (a) For all a, b and for all $x \neq 2$, f is differentiable at x. Next, f differentiable at $x = 2 \Rightarrow$ f continuous at $x = 2$ $\Rightarrow \lim_{x \to 2^-} f(x) = f(2) \Rightarrow 2a = 4a - 2b + 3 \Rightarrow 2a - 2b + 3 = 0$. Also, f differentiable at $x \neq 2$ $\Rightarrow f'(x) = \begin{cases} a, & x < 2 \\ 2ax - b, & x > 2 \end{cases}$. In order that $f'(2)$ exist we must have $a = 2a(2) - b \Rightarrow a = 4a - b \Rightarrow 3a = b$. Then $2a - 2b + 3 = 0$ and $3a = b \Rightarrow a = \frac{3}{4}$ and $b = \frac{9}{4}$.
   (b) For $x < 2$, the graph of f is a straight line having a slope of $\frac{3}{4}$ and passing through the origin; for $x \geq 2$, the graph of f is a parabola. At $x = 2$, the value of the y-coordinate on the parabola is $\frac{3}{2}$ which matches the y-coordinate of the point on the straight line at $x = 2$. In addition, the slope of the parabola at the match up point is $\frac{3}{4}$ which is equal to the slope of the straight line. Therefore, since the graph is differentiable at the match up point, the graph is smooth there.

19. f odd $\Rightarrow f(-x) = -f(x) \Rightarrow \frac{d}{dx}(f(-x)) = \frac{d}{dx}(-f(x)) \Rightarrow f'(-x)(-1) = -f'(x) \Rightarrow f'(-x) = f'(x) \Rightarrow f'$ is even.

21. Let $h(x) = (fg)(x) = f(x)\,g(x) \Rightarrow h'(x) = \lim_{x \to x_0} \frac{h(x) - h(x_0)}{x - x_0} = \lim_{x \to x_0} \frac{f(x)\,g(x) - f(x_0)\,g(x_0)}{x - x_0}$
$= \lim_{x \to x_0} \frac{f(x)\,g(x) - f(x)\,g(x_0) + f(x)\,g(x_0) - f(x_0)\,g(x_0)}{x - x_0} = \lim_{x \to x_0} \left[f(x)\left[\frac{g(x) - g(x_0)}{x - x_0}\right]\right] + \lim_{x \to x_0} \left[g(x_0)\left[\frac{f(x) - f(x_0)}{x - x_0}\right]\right]$
$= f(x_0) \lim_{x \to x_0} \left[\frac{g(x) - g(x_0)}{x - x_0}\right] + g(x_0)\,f'(x_0) = 0 \cdot \lim_{x \to x_0} \left[\frac{g(x) - g(x_0)}{x - x_0}\right] + g(x_0)\,f'(x_0) = g(x_0)\,f'(x_0)$, if g is continuous at x_0. Therefore $(fg)(x)$ is differentiable at x_0 if $f(x_0) = 0$, and $(fg)'(x_0) = g(x_0)\,f'(x_0)$.

23. If $f(x) = x$ and $g(x) = x \sin\left(\frac{1}{x}\right)$, then $x^2 \sin\left(\frac{1}{x}\right)$ is differentiable at $x = 0$ because $f'(0) = 1$, $f(0) = 0$ and $\lim_{x \to 0} x \sin\left(\frac{1}{x}\right) = \lim_{x \to 0} \frac{\sin\left(\frac{1}{x}\right)}{\frac{1}{x}} = \lim_{t \to \infty} \frac{\sin t}{t} = 0$ (so g is continuous at $x = 0$). In fact, from Exercise 21, $h'(0) = g(0)\, f'(0) = 0$. However, for $x \neq 0$, $h'(x) = \left[x^2 \cos\left(\frac{1}{x}\right)\right]\left(-\frac{1}{x^2}\right) + 2x \sin\left(\frac{1}{x}\right)$. But $\lim_{x \to 0} h'(x) = \lim_{x \to 0} \left[-\cos\left(\frac{1}{x}\right) + 2x \sin\left(\frac{1}{x}\right)\right]$ does not exist because $\cos\left(\frac{1}{x}\right)$ has no limit as $x \to 0$. Therefore, the derivative is not continuous at $x = 0$ because it has no limit there.

25. Step 1: The formula holds for $n = 2$ (a single product) since $y = u_1 u_2 \Rightarrow \frac{dy}{dx} = \frac{du_1}{dx} u_2 + u_1 \frac{du_2}{dx}$.

Step 2: Assume the formula holds for $n = k$:

$$y = u_1 u_2 \cdots u_k \Rightarrow \frac{dy}{dx} = \frac{du_1}{dx} u_2 u_3 \cdots u_k + u_1 \frac{du_2}{dx} u_3 \cdots u_k + \ldots + u_1 u_2 \cdots u_{k-1} \frac{du_k}{dx}.$$

If $y = u_1 u_2 \cdots u_k u_{k+1} = (u_1 u_2 \cdots u_k)\, u_{k+1}$, then $\frac{dy}{dx} = \frac{d(u_1 u_2 \cdots u_k)}{dx} u_{k+1} + u_1 u_2 \cdots u_k \frac{du_{k+1}}{dx}$

$= \left(\frac{du_1}{dx} u_2 u_3 \cdots u_k + u_1 \frac{du_2}{dx} u_3 \cdots u_k + \cdots + u_1 u_2 \cdots u_{k-1} \frac{du_k}{dx}\right) u_{k+1} + u_1 u_2 \cdots u_k \frac{du_{k+1}}{dx}$

$= \frac{du_1}{dx} u_2 u_3 \cdots u_{k+1} + u_1 \frac{du_2}{dx} u_3 \cdots u_{k+1} + \cdots + u_1 u_2 \cdots u_{k-1} \frac{du_k}{dx} u_{k+1} + u_1 u_2 \cdots u_k \frac{du_{k+1}}{dx}$.

Thus the original formula holds for $n = (k+1)$ whenever it holds for $n = k$.

27. (a) $T^2 = \frac{4\pi^2 L}{g} \Rightarrow L = \frac{T^2 g}{4\pi^2} \Rightarrow L = \frac{(1\text{ sec}^2)(32.2\text{ ft/sec}^2)}{4\pi^2} \Rightarrow L \approx 0.8156$ ft

(b) $T^2 = \frac{4\pi^2 L}{g} \Rightarrow T = \frac{2\pi}{\sqrt{g}}\sqrt{L}$; $dT = \frac{2\pi}{\sqrt{g}} \cdot \frac{1}{2\sqrt{L}} dL = \frac{\pi}{\sqrt{Lg}} dL$; $dT = \frac{\pi}{\sqrt{(0.8156\text{ ft})(32.2\text{ ft/sec}^2)}}(0.01\text{ ft}) \approx 0.00613$ sec.

(c) Since there are 86,400 sec in a day, we have (0.00613 sec)(86,400 sec/day) $\approx$ 529.6 sec/day, or 8.83 min/day; the clock will lose about 8.83 min/day.

CHAPTER 4 APPLICATIONS OF DERIVATIVES

4.1 EXTREME VALUES OF FUNCTIONS

1. An absolute minimum at $x = c_2$, an absolute maximum at $x = b$. Theorem 1 guarantees the existence of such extreme values because h is continuous on [a, b].

3. No absolute minimum. An absolute maximum at $x = c$. Since the function's domain is an open interval, the function does not satisfy the hypotheses of Theorem 1 and need not have absolute extreme values.

5. An absolute minimum at $x = a$ and an absolute maximum at $x = c$. Note that $y = g(x)$ is not continuous but still has extrema. When the hypothesis of Theorem 1 is satisfied then extrema are guaranteed, but when the hypothesis is not satisfied, absolute extrema may or may not occur.

7. Local minimum at $(-1, 0)$, local maximum at $(1, 0)$

9. Maximum at $(0, 5)$. Note that there is no minimum since the endpoint $(2, 0)$ is excluded from the graph.

11. Graph (c), since this the only graph that has positive slope at c.

13. Graph (d), since this is the only graph representing a funtion that is differentiable at b but not at a.

15. f has an absolute min at $x = 0$ but does not have an absolute max. Since the interval on which f is defined, $-1 < x < 2$, is an open interval, we do not meet the conditions of Theorem 1.

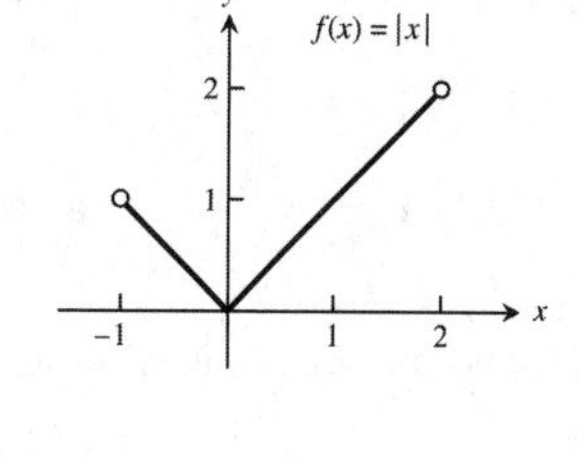

17. f has an absolute max at $x = 2$ but does not have an absolute min. Since the function is not continuous at $x = 1$, we do not meet the conditions of Theorem 1.

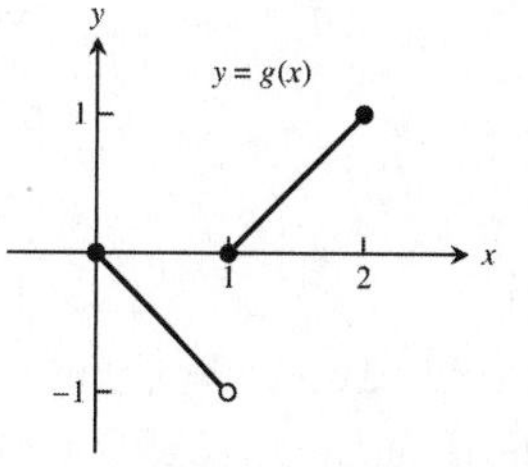

19. f has an absolute max at $x = \frac{\pi}{2}$ and an absolute min at $x = \frac{3\pi}{2}$. Since the interval on which f is defined, $0 < x < 2\pi$, is an open interval, we do not meet the conditions of Theorem 1.

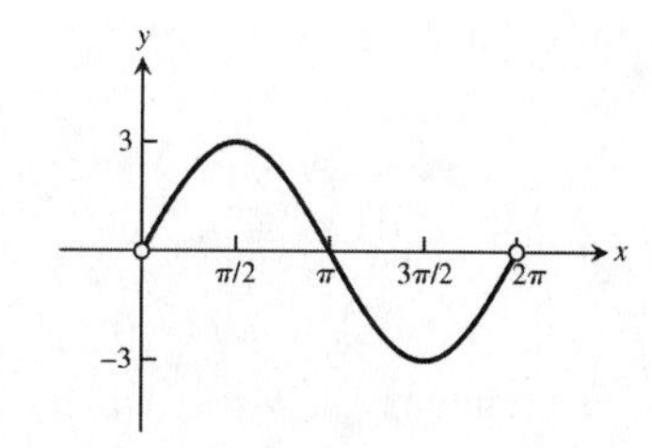

21. $f(x) = \frac{2}{3}x - 5 \Rightarrow f'(x) = \frac{2}{3} \Rightarrow$ no critical points; $f(-2) = -\frac{19}{3}$, $f(3) = -3 \Rightarrow$ the absolute maximum is -3 at $x = 3$ and the absolute minimum is $-\frac{19}{3}$ at $x = -2$

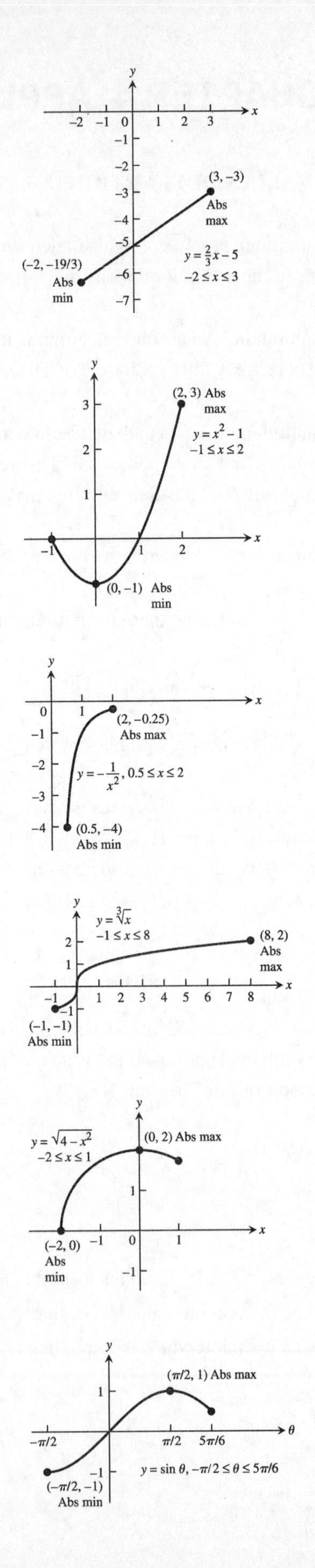

23. $f(x) = x^2 - 1 \Rightarrow f'(x) = 2x \Rightarrow$ a critical point at $x = 0$; $f(-1) = 0$, $f(0) = -1$, $f(2) = 3 \Rightarrow$ the absolute maximum is 3 at $x = 2$ and the absolute minimum is -1 at $x = 0$

25. $F(x) = -\frac{1}{x^2} = -x^{-2} \Rightarrow F'(x) = 2x^{-3} = \frac{2}{x^3}$, however $x = 0$ is not a critical point since 0 is not in the domain; $F(0.5) = -4$, $F(2) = -0.25 \Rightarrow$ the absolute maximum is -0.25 at $x = 2$ and the absolute minimum is -4 at $x = 0.5$

27. $h(x) = \sqrt[3]{x} = x^{1/3} \Rightarrow h'(x) = \frac{1}{3}x^{-2/3} \Rightarrow$ a critical point at $x = 0$; $h(-1) = -1$, $h(0) = 0$, $h(8) = 2 \Rightarrow$ the absolute maximum is 2 at $x = 8$ and the absolute minimum is -1 at $x = -1$

29. $g(x) = \sqrt{4 - x^2} = (4 - x^2)^{1/2}$
$\Rightarrow g'(x) = \frac{1}{2}(4 - x^2)^{-1/2}(-2x) = \frac{-x}{\sqrt{4 - x^2}}$
$\Rightarrow$ critical points at $x = -2$ and $x = 0$, but not at $x = 2$ because 2 is not in the domain; $g(-2) = 0$, $g(0) = 2$, $g(1) = \sqrt{3} \Rightarrow$ the absolute maximum is 2 at $x = 0$ and the absolute minimum is 0 at $x = -2$

31. $f(\theta) = \sin\theta \Rightarrow f'(\theta) = \cos\theta \Rightarrow \theta = \frac{\pi}{2}$ is a critical point, but $\theta = \frac{-\pi}{2}$ is not a critical point because $\frac{-\pi}{2}$ is not interior to the domain; $f\left(\frac{-\pi}{2}\right) = -1$, $f\left(\frac{\pi}{2}\right) = 1$, $f\left(\frac{5\pi}{6}\right) = \frac{1}{2}$
$\Rightarrow$ the absolute maximum is 1 at $\theta = \frac{\pi}{2}$ and the absolute minimum is -1 at $\theta = \frac{-\pi}{2}$

33. $g(x) = \csc x \Rightarrow g'(x) = -(\csc x)(\cot x) \Rightarrow$ a critical point at $x = \frac{\pi}{2}$; $g\left(\frac{\pi}{3}\right) = \frac{2}{\sqrt{3}}$, $g\left(\frac{\pi}{2}\right) = 1$, $g\left(\frac{2\pi}{3}\right) = \frac{2}{\sqrt{3}} \Rightarrow$ the absolute maximum is $\frac{2}{\sqrt{3}}$ at $x = \frac{\pi}{3}$ and $x = \frac{2\pi}{3}$, and the absolute minimum is 1 at $x = \frac{\pi}{2}$

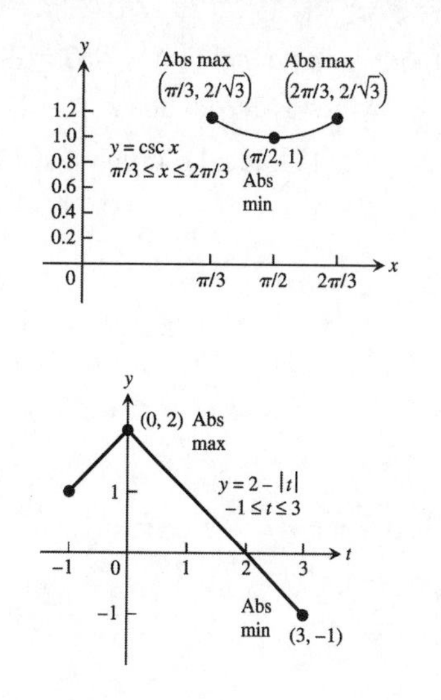

35. $f(t) = 2 - |t| = 2 - \sqrt{t^2} = 2 - (t^2)^{1/2}$
$\Rightarrow f'(t) = -\frac{1}{2}(t^2)^{-1/2}(2t) = -\frac{t}{\sqrt{t^2}} = -\frac{t}{|t|}$
$\Rightarrow$ a critical point at $t = 0$; $f(-1) = 1$,
$f(0) = 2$, $f(3) = -1 \Rightarrow$ the absolute maximum is 2 at $t = 0$ and the absolute minimum is -1 at $t = 3$

37. $f(x) = x^{4/3} \Rightarrow f'(x) = \frac{4}{3}x^{1/3} \Rightarrow$ a critical point at $x = 0$; $f(-1) = 1$, $f(0) = 0$, $f(8) = 16 \Rightarrow$ the absolute maximum is 16 at $x = 8$ and the absolute minimum is 0 at $x = 0$

39. $g(\theta) = \theta^{3/5} \Rightarrow g'(\theta) = \frac{3}{5}\theta^{-2/5} \Rightarrow$ a critical point at $\theta = 0$; $g(-32) = -8$, $g(0) = 0$, $g(1) = 1 \Rightarrow$ the absolute maximum is 1 at $\theta = 1$ and the absolute minimum is -8 at $\theta = -32$

41. $y = x^2 - 6x + 7 \Rightarrow y' = 2x - 6 \Rightarrow 2x - 6 = 0 \Rightarrow x = 3$. The critical point is $x = 3$.

43. $f(x) = x(4-x)^3 \Rightarrow f'(x) = x\left[3(4-x)^2(-1)\right] + (4-x)^3 = (4-x)^2\left[-3x + (4-x)\right] = (4-x)^2(4-4x)$
$= 4(4-x)^2(1-x) \Rightarrow 4(4-x)^2(1-x) = 0 \Rightarrow x = 1$ or $x = 4$. The critical points are $x = 1$ and $x = 4$.

45. $y = x^2 + \frac{2}{x} \Rightarrow y' = 2x - \frac{2}{x^2} = \frac{2x^3 - 2}{x^2} \Rightarrow \frac{2x^3 - 2}{x^2} = 0 \Rightarrow 2x^3 - 2 = 0 \Rightarrow x = 1$; $\frac{2x^3 - 2}{x^2} =$ undefined $\Rightarrow x^2 = 0 \Rightarrow x = 0$.
The domain of the function is $(-\infty, 0) \cup (0, \infty)$, thus $x = 0$ is not in the domain, so the only critical point is $x = 1$.

47. $y = x^2 - 32\sqrt{x} \Rightarrow y' = 2x - \frac{16}{\sqrt{x}} = \frac{2x^{3/2} - 16}{\sqrt{x}} \Rightarrow \frac{2x^{3/2} - 16}{\sqrt{x}} = 0 \Rightarrow 2x^{3/2} - 16 = 0 \Rightarrow x = 4$; $\frac{2x^{3/2} - 16}{\sqrt{x}} =$ undefined
$\Rightarrow \sqrt{x} = 0 \Rightarrow x = 0$. The critical points are $x = 4$ and $x = 0$.

49. Minimum value is 1 at $x = 2$.

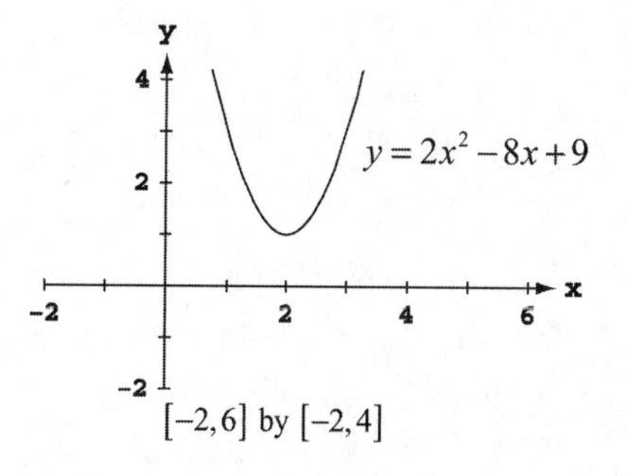

51. To find the exact values, note that that $y' = 3x^2 + 2x - 8$ $= (3x - 4)(x + 2)$, which is zero when $x = -2$ or $x = \frac{4}{3}$. Local maximum at $(-2, 17)$; local minimum at $\left(\frac{4}{3}, -\frac{41}{27}\right)$

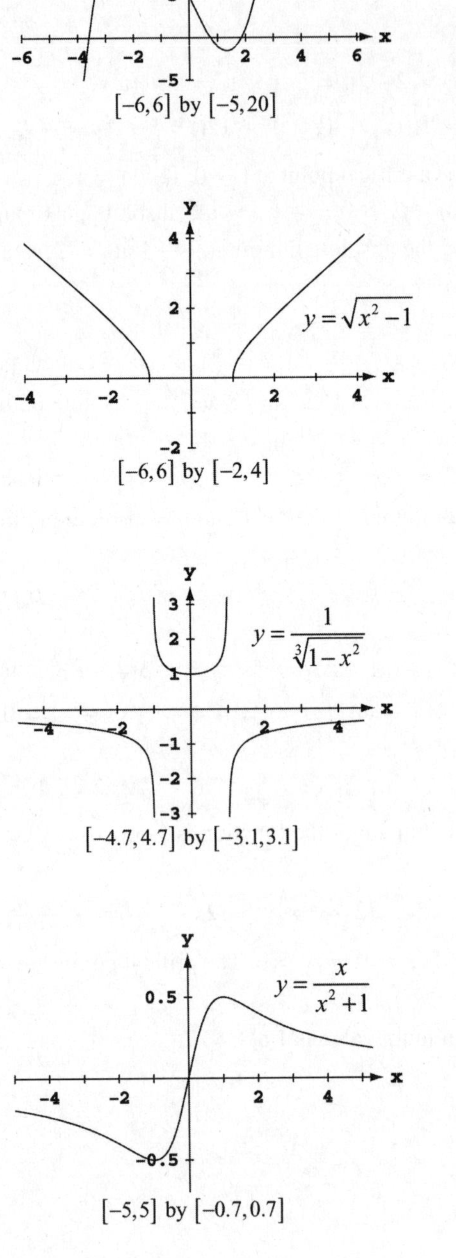

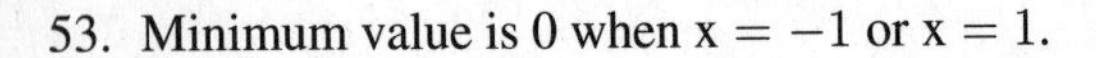

53. Minimum value is 0 when $x = -1$ or $x = 1$.

55. The actual graph of the function has asymptotes at $x = \pm 1$, so there are no extrema near these values. (This is an example of grapher failure.) There is a local minimum at $(0, 1)$.

57. Maximum value is $\frac{1}{2}$ at $x = 1$; minimum value is $-\frac{1}{2}$ as $x = -1$.

59. $y' = x^{2/3}(1) + \frac{2}{3}x^{-1/3}(x+2) = \frac{5x+4}{3\sqrt[3]{x}}$

crit. pt.	derivative	extremum	value
$x = -\frac{4}{5}$	0	local max	$\frac{12}{25}10^{1/3} = 1.034$
$x = 0$	undefined	local min	0

$y = x^{2/3}(x+2)$

[−4, 4] by [−3, 3]

61. $y' = x\frac{1}{2\sqrt{4-x^2}}(-2x) + (1)\sqrt{4-x^2}$

$= \frac{-x^2+(4-x^2)}{\sqrt{4-x^2}} = \frac{4-2x^2}{\sqrt{4-x^2}}$

crit. pt.	derivative	extremum	value
$x = -2$	undefined	local max	0
$x = -\sqrt{2}$	0	minimum	−2
$x = \sqrt{2}$	0	maximum	2
$x = 2$	undefined	local min	0

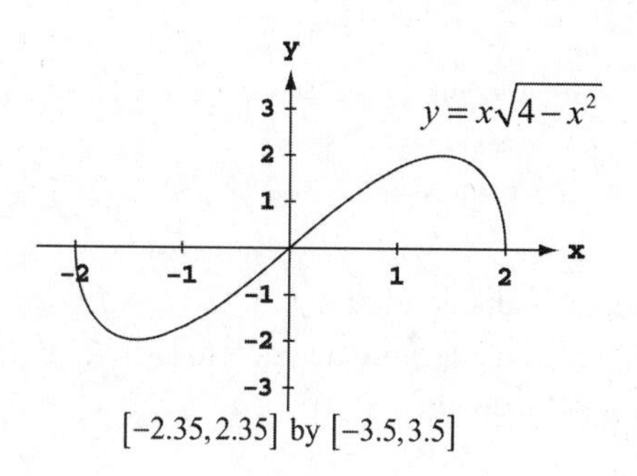

[−2.35, 2.35] by [−3.5, 3.5]

63. $y' = \begin{cases} -2, & x < 1 \\ 1, & x > 1 \end{cases}$

crit. pt.	derivative	extremum	value
$x = 1$	undefined	minimum	2

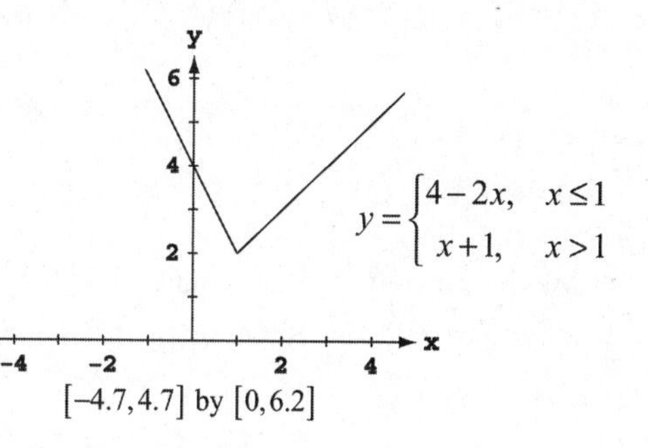

[−4.7, 4.7] by [0, 6.2]

65. $y' = \begin{cases} -2x-2, & x < 1 \\ -2x+6, & x > 1 \end{cases}$

crit. pt.	derivative	extremum	value
$x = -1$	0	maximum	5
$x = 1$	undefined	local min	1
$x = 3$	0	maximum	5

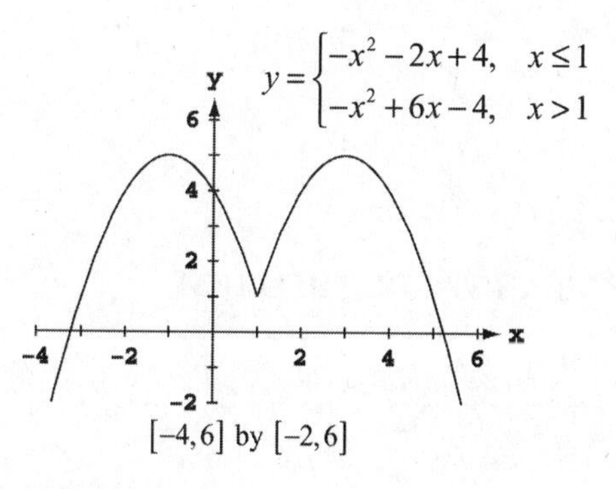

[−4, 6] by [−2, 6]

67. (a) No, since $f'(x) = \frac{2}{3}(x-2)^{-1/3}$, which is undefined at $x = 2$.

(b) The derivative is defined and nonzero for all $x \neq 2$. Also, $f(2) = 0$ and $f(x) > 0$ for all $x \neq 2$.

(c) No, $f(x)$ need not have a global maximum because its domain is all real numbers. Any restriction of f to a closed interval of the form [a, b] would have both a maximum value and minimum value on the interval.

(d) The answers are the same as (a) and (b) with 2 replaced by a.

69. Yes, since $f(x) = |x| = \sqrt{x^2} = (x^2)^{1/2} \Rightarrow f'(x) = \frac{1}{2}(x^2)^{-1/2}(2x) = \frac{x}{(x^2)^{1/2}} = \frac{x}{|x|}$ is not defined at $x = 0$. Thus it is not required that f' be zero at a local extreme point since f' may be undefined there.

71. If $g(c)$ is a local minimum value of g, then $g(x) \geq g(c)$ for all x in some open interval (a, b) containing c. Since g is odd, $g(-x) = -g(x) \leq -g(c) = g(-c)$ for all $-x$ in the open interval $(-b, -a)$ containing $-c$. That is, g assumes a local maximum at the point $-c$. This is also clear from the graph of g because the graph of an odd function is symmetric about the origin.

73. (a) $V(x) = 160x - 52x^2 + 4x^3$
$V'(x) = 160 - 104x + 12x^2 = 4(x-2)(3x-20)$
The only critical point in the interval $(0, 5)$ is at $x = 2$. The maximum value of $V(x)$ is 144 at $x = 2$.
(b) The largest possible volume of the box is 144 cubic units, and it occurs when $x = 2$ units.

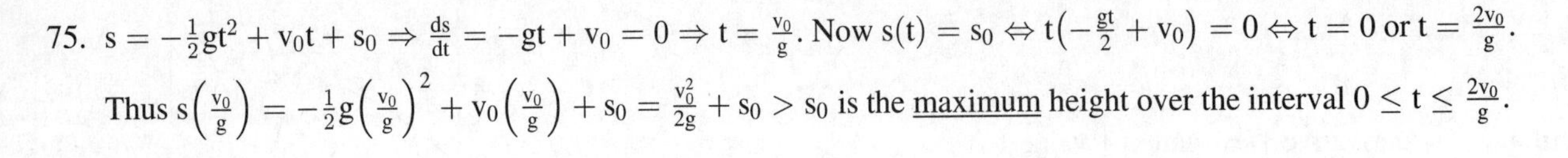

75. $s = -\frac{1}{2}gt^2 + v_0t + s_0 \Rightarrow \frac{ds}{dt} = -gt + v_0 = 0 \Rightarrow t = \frac{v_0}{g}$. Now $s(t) = s_0 \Leftrightarrow t\left(-\frac{gt}{2} + v_0\right) = 0 \Leftrightarrow t = 0$ or $t = \frac{2v_0}{g}$.
Thus $s\left(\frac{v_0}{g}\right) = -\frac{1}{2}g\left(\frac{v_0}{g}\right)^2 + v_0\left(\frac{v_0}{g}\right) + s_0 = \frac{v_0^2}{2g} + s_0 > s_0$ is the maximum height over the interval $0 \leq t \leq \frac{2v_0}{g}$.

77. Maximum value is 11 at $x = 5$;
minimum value is 5 on the interval $[-3, 2]$;
local maximum at $(-5, 9)$

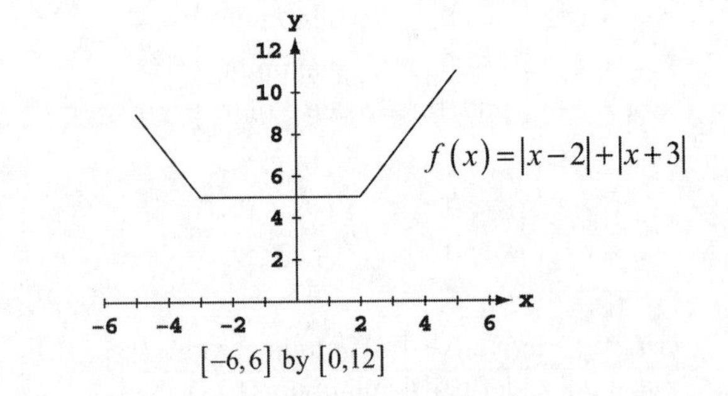

$[-6,6]$ by $[0,12]$

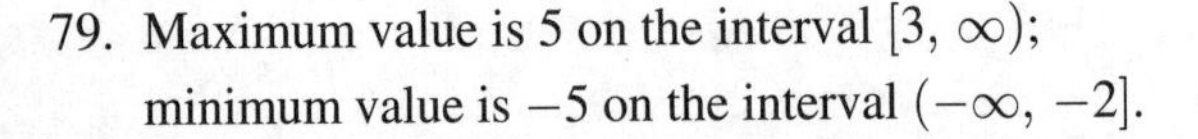

79. Maximum value is 5 on the interval $[3, \infty)$;
minimum value is -5 on the interval $(-\infty, -2]$.

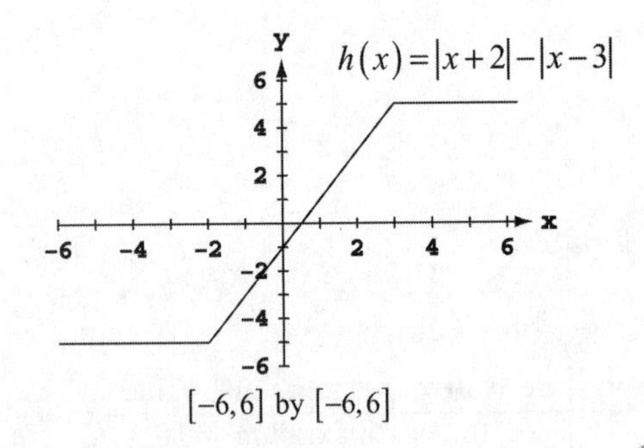

$[-6,6]$ by $[-6,6]$

4.2 THE MEAN VALUE THEOREM

1. When $f(x) = x^2 + 2x - 1$ for $0 \leq x \leq 1$, then $\frac{f(1)-f(0)}{1-0} = f'(c) \Rightarrow 3 = 2c + 2 \Rightarrow c = \frac{1}{2}$.

3. When $f(x) = x + \frac{1}{x}$ for $\frac{1}{2} \leq x \leq 2$, then $\frac{f(2)-f(1/2)}{2-1/2} = f'(c) \Rightarrow 0 = 1 - \frac{1}{c^2} \Rightarrow c = 1$.

5. When $f(x) = x^3 - x^2$ for $-1 \leq x \leq 2$, then $\frac{f(2)-f(-1)}{2-(-1)} = f'(c) \Rightarrow 2 = 3c^2 - 2c \Rightarrow c = \frac{1 \pm \sqrt{7}}{3}$.
$\frac{1+\sqrt{7}}{3} \approx 1.22$ and $\frac{1-\sqrt{7}}{3} \approx -0.549$ are both in the interval $-1 \leq x \leq 2$.

7. Does not; $f(x)$ is not differentiable at $x = 0$ in $(-1, 8)$.

9. Does; f(x) is continuous for every point of [0, 1] and differentiable for every point in (0, 1).

11. Does not; f is not differentiable at $x = -1$ in $(-2, 0)$.

13. Since f(x) is not continuous on $0 \le x \le 1$, Rolle's Theorem does not apply: $\lim_{x \to 1^-} f(x) = \lim_{x \to 1^-} x = 1 \ne 0 = f(1)$.

15. (a) i
ii
iii
iv

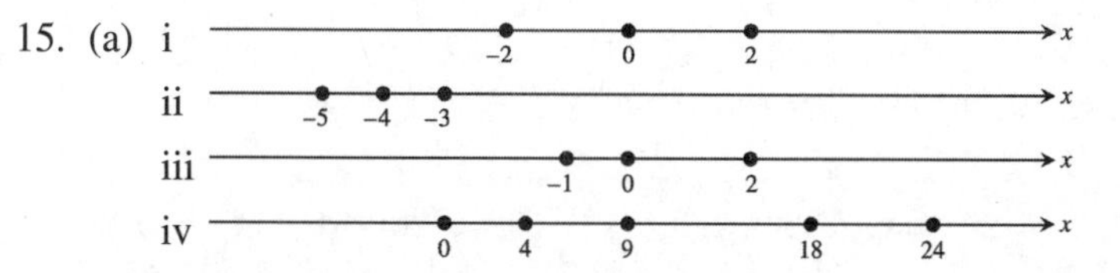

(b) Let r_1 and r_2 be zeros of the polynomial $P(x) = x^n + a_{n-1}x^{n-1} + \ldots + a_1x + a_0$, then $P(r_1) = P(r_2) = 0$. Since polynomials are everywhere continuous and differentiable, by Rolle's Theorem $P'(r) = 0$ for some r between r_1 and r_2, where $P'(x) = nx^{n-1} + (n-1)\,a_{n-1}x^{n-2} + \ldots + a_1$.

17. Since f'' exists throughout [a, b] the derivative function f' is continuous there. If f' has more than one zero in [a, b], say $f'(r_1) = f'(r_2) = 0$ for $r_1 \ne r_2$, then by Rolle's Theorem there is a c between r_1 and r_2 such that $f''(c) = 0$, contrary to $f'' > 0$ throughout [a, b]. Therefore f' has at most one zero in [a, b]. The same argument holds if $f'' < 0$ throughout [a, b].

19. With $f(-2) = 11 > 0$ and $f(-1) = -1 < 0$ we conclude from the Intermediate Value Theorem that $f(x) = x^4 + 3x + 1$ has at least one zero between -2 and -1. Then $-2 < x < -1 \Rightarrow -8 < x^3 < -1 \Rightarrow -32 < 4x^3 < -4$ $\Rightarrow -29 < 4x^3 + 3 < -1 \Rightarrow f'(x) < 0$ for $-2 < x < -1 \Rightarrow$ f(x) is decreasing on $[-2, -1] \Rightarrow f(x) = 0$ has exactly one solution in the interval $(-2, -1)$.

21. $g(t) = \sqrt{t} + \sqrt{t+1} - 4 \Rightarrow g'(t) = \frac{1}{2\sqrt{t}} + \frac{1}{2\sqrt{t+1}} > 0 \Rightarrow$ g(t) is increasing for t in $(0, \infty)$; $g(3) = \sqrt{3} - 2 < 0$ and $g(15) = \sqrt{15} > 0 \Rightarrow$ g(t) has exactly one zero in $(0, \infty)$.

23. $r(\theta) = \theta + \sin^2\left(\frac{\theta}{3}\right) - 8 \Rightarrow r'(\theta) = 1 + \frac{2}{3}\sin\left(\frac{\theta}{3}\right)\cos\left(\frac{\theta}{3}\right) = 1 + \frac{1}{3}\sin\left(\frac{2\theta}{3}\right) > 0$ on $(-\infty, \infty) \Rightarrow r(\theta)$ is increasing on $(-\infty, \infty)$; $r(0) = -8$ and $r(8) = \sin^2\left(\frac{8}{3}\right) > 0 \Rightarrow r(\theta)$ has exactly one zero in $(-\infty, \infty)$.

25. $r(\theta) = \sec\theta - \frac{1}{\theta^3} + 5 \Rightarrow r'(\theta) = (\sec\theta)(\tan\theta) + \frac{3}{\theta^4} > 0$ on $\left(0, \frac{\pi}{2}\right) \Rightarrow r(\theta)$ is increasing on $\left(0, \frac{\pi}{2}\right)$; $r(0.1) \approx -994$ and $r(1.57) \approx 1260.5 \Rightarrow r(\theta)$ has exactly one zero in $\left(0, \frac{\pi}{2}\right)$.

27. By Corollary 1, $f'(x) = 0$ for all $x \Rightarrow f(x) = C$, where C is a constant. Since $f(-1) = 3$ we have $C = 3 \Rightarrow f(x) = 3$ for all x.

29. $g(x) = x^2 \Rightarrow g'(x) = 2x = f'(x)$ for all x. By Corollary 2, $f(x) = g(x) + C$.
(a) $f(0) = 0 \Rightarrow 0 = g(0) + C = 0 + C \Rightarrow C = 0 \Rightarrow f(x) = x^2 \Rightarrow f(2) = 4$
(b) $f(1) = 0 \Rightarrow 0 = g(1) + C = 1 + C \Rightarrow C = -1 \Rightarrow f(x) = x^2 - 1 \Rightarrow f(2) = 3$
(c) $f(-2) = 3 \Rightarrow 3 = g(-2) + C \Rightarrow 3 = 4 + C \Rightarrow C = -1 \Rightarrow f(x) = x^2 - 1 \Rightarrow f(2) = 3$

31. (a) $y = \frac{x^2}{2} + C$ (b) $y = \frac{x^3}{3} + C$ (c) $y = \frac{x^4}{4} + C$

33. (a) $y' = -x^{-2} \Rightarrow y = \frac{1}{x} + C$ (b) $y = x + \frac{1}{x} + C$ (c) $y = 5x - \frac{1}{x} + C$

35. (a) $y = -\frac{1}{2}\cos 2t + C$ (b) $y = 2\sin\frac{t}{2} + C$
(c) $y = -\frac{1}{2}\cos 2t + 2\sin\frac{t}{2} + C$

37. $f(x) = x^2 - x + C; 0 = f(0) = 0^2 - 0 + C \Rightarrow C = 0 \Rightarrow f(x) = x^2 - x$

39. $r(\theta) = 8\theta + \cot\theta + C; 0 = r\left(\frac{\pi}{4}\right) = 8\left(\frac{\pi}{4}\right) + \cot\left(\frac{\pi}{4}\right) + C \Rightarrow 0 = 2\pi + 1 + C \Rightarrow C = -2\pi - 1$
$\Rightarrow r(\theta) = 8\theta + \cot\theta - 2\pi - 1$

41. $v = \frac{ds}{dt} = 9.8t + 5 \Rightarrow s = 4.9t^2 + 5t + C$; at $s = 10$ and $t = 0$ we have $C = 10 \Rightarrow s = 4.9t^2 + 5t + 10$

43. $v = \frac{ds}{dt} = \sin(\pi t) \Rightarrow s = -\frac{1}{\pi}\cos(\pi t) + C$; at $s = 0$ and $t = 0$ we have $C = \frac{1}{\pi} \Rightarrow s = \frac{1-\cos(\pi t)}{\pi}$

45. $a = 32 \Rightarrow v = 32t + C_1$; at $v = 20$ and $t = 0$ we have $C_1 = 20 \Rightarrow v = 32t + 20 \Rightarrow s = 16t^2 + 20t + C_2$; at $s = 5$ and $t = 0$ we have $C_2 = 5 \Rightarrow s = 16t^2 + 20t + 5$

47. $a = -4\sin(2t) \Rightarrow v = 2\cos(2t) + C_1$; at $v = 2$ and $t = 0$ we have $C_1 = 0 \Rightarrow v = 2\cos(2t) \Rightarrow s = \sin(2t) + C_2$; at $s = -3$ and $t = 0$ we have $C_2 = -3 \Rightarrow s = \sin(2t) - 3$

49. If T(t) is the temperature of the thermometer at time t, then $T(0) = -19°$ C and $T(14) = 100°$ C. From the Mean Value Theorem there exists a $0 < t_0 < 14$ such that $\frac{T(14)-T(0)}{14-0} = 8.5°$ C/sec $= T'(t_0)$, the rate at which the temperature was changing at $t = t_0$ as measured by the rising mercury on the thermometer.

51. Because its average speed was approximately 7.667 knots, and by the Mean Value Theorem, it must have been going that speed at least once during the trip.

53. Let d(t) represent the distance the automobile traveled in time t. The average speed over $0 \le t \le 2$ is $\frac{d(2)-d(0)}{2-0}$. The Mean Value Theorem says that for some $0 < t_0 < 2$, $d'(t_0) = \frac{d(2)-d(0)}{2-0}$. The value $d'(t_0)$ is the speed of the automobile at time t_0 (which is read on the speedometer).

55. The conclusion of the Mean Value Theorem yields $\frac{\frac{1}{b}-\frac{1}{a}}{b-a} = -\frac{1}{c^2} \Rightarrow c^2\left(\frac{a-b}{ab}\right) = a - b \Rightarrow c = \sqrt{ab}$.

57. $f'(x) = [\cos x \sin(x+2) + \sin x \cos(x+2)] - 2\sin(x+1)\cos(x+1) = \sin(x + x + 2) - \sin 2(x+1)$
$= \sin(2x+2) - \sin(2x+2) = 0$. Therefore, the function has the constant value $f(0) = -\sin^2 1 \approx -0.7081$ which explains why the graph is a horizontal line.

59. f(x) must be zero at least once between a and b by the Intermediate Value Theorem. Now suppose that f(x) is zero twice between a and b. Then by the Mean Value Theorem, $f'(x)$ would have to be zero at least once between the two zeros of f(x), but this can't be true since we are given that $f'(x) \neq 0$ on this interval. Therefore, f(x) is zero once and only once between a and b.

61. $f'(x) \le 1$ for $1 \le x \le 4 \Rightarrow f(x)$ is differentiable on $1 \le x \le 4 \Rightarrow f$ is continuous on $1 \le x \le 4 \Rightarrow f$ satisfies the conditions of the Mean Value Theorem $\Rightarrow \frac{f(4)-f(1)}{4-1} = f'(c)$ for some c in $1 < x < 4 \Rightarrow f'(c) \le 1 \Rightarrow \frac{f(4)-f(1)}{3} \le 1$
$\Rightarrow f(4) - f(1) \le 3$

63. Let $f(t) = \cos t$ and consider the interval $[0, x]$ where x is a real number. f is continuous on $[0, x]$ and f is differentiable on $(0, x)$ since $f'(t) = -\sin t \Rightarrow f$ satisfies the conditions of the Mean Value Theorem $\Rightarrow \frac{f(x)-f(0)}{x-(0)} = f'(c)$ for some c in $[0, x] \Rightarrow \frac{\cos x - 1}{x} = -\sin c$. Since $-1 \le \sin c \le 1 \Rightarrow -1 \le -\sin c \le 1 \Rightarrow -1 \le \frac{\cos x - 1}{x} \le 1$. If $x > 0$, $-1 \le \frac{\cos x - 1}{x} \le 1$
$\Rightarrow -x \le \cos x - 1 \le x \Rightarrow |\cos x - 1| \le x = |x|$. If $x < 0$, $-1 \le \frac{\cos x - 1}{x} \le 1 \Rightarrow -x \ge \cos x - 1 \ge x$

$\Rightarrow x \le \cos x - 1 \le -x \Rightarrow -(-x) \le \cos x - 1 \le -x \Rightarrow |\cos x - 1| \le -x = |x|$. Thus, in both cases, we have $|\cos x - 1| \le |x|$. If $x = 0$, then $|\cos 0 - 1| = |1 - 1| = |0| \le |0|$, thus $|\cos x - 1| \le |x|$ is true for all x.

65. Yes. By Corollary 2 we have $f(x) = g(x) + c$ since $f'(x) = g'(x)$. If the graphs start at the same point $x = a$, then $f(a) = g(a) \Rightarrow c = 0 \Rightarrow f(x) = g(x)$.

67. By the Mean Value Theorem we have $\frac{f(b) - f(a)}{b - a} = f'(c)$ for some point c between a and b. Since $b - a > 0$ and $f(b) < f(a)$, we have $f(b) - f(a) < 0 \Rightarrow f'(c) < 0$.

69. $f'(x) = (1 + x^4 \cos x)^{-1} \Rightarrow f''(x) = -(1 + x^4 \cos x)^{-2}(4x^3 \cos x - x^4 \sin x)$
$= -x^3(1 + x^4 \cos x)^{-2}(4 \cos x - x \sin x) < 0$ for $0 \le x \le 0.1 \Rightarrow f'(x)$ is decreasing when $0 \le x \le 0.1$
$\Rightarrow \min f' \approx 0.9999$ and $\max f' = 1$. Now we have $0.9999 \le \frac{f(0.1) - 1}{0.1} \le 1 \Rightarrow 0.09999 \le f(0.1) - 1 \le 0.1$
$\Rightarrow 1.09999 \le f(0.1) \le 1.1$.

71. (a) Suppose $x < 1$, then by the Mean Value Theorem $\frac{f(x) - f(1)}{x - 1} < 0 \Rightarrow f(x) > f(1)$. Suppose $x > 1$, then by the Mean Value Theorem $\frac{f(x) - f(1)}{x - 1} > 0 \Rightarrow f(x) > f(1)$. Therefore $f(x) \ge 1$ for all x since $f(1) = 1$.
(b) Yes. From part (a), $\lim_{x \to 1^-} \frac{f(x) - f(1)}{x - 1} \le 0$ and $\lim_{x \to 1^+} \frac{f(x) - f(1)}{x - 1} \ge 0$. Since $f'(1)$ exists, these two one-sided limits are equal and have the value $f'(1) \Rightarrow f'(1) \le 0$ and $f'(1) \ge 0 \Rightarrow f'(1) = 0$.

4.3 MONOTONIC FUNCTIONS AND THE FIRST DERIVATIVE TEST

1. (a) $f'(x) = x(x - 1) \Rightarrow$ critical points at 0 and 1
(b) $f' = +++\underset{0}{|}---\underset{1}{|}+++ \Rightarrow$ increasing on $(-\infty, 0)$ and $(1, \infty)$, decreasing on $(0, 1)$
(c) Local maximum at $x = 0$ and a local minimum at $x = 1$

3. (a) $f'(x) = (x - 1)^2(x + 2) \Rightarrow$ critical points at -2 and 1
(b) $f' = ---\underset{-2}{|}+++\underset{1}{|}+++ \Rightarrow$ increasing on $(-2, 1)$ and $(1, \infty)$, decreasing on $(-\infty, -2)$
(c) No local maximum and a local minimum at $x = -2$

5. (a) $f'(x) = (x - 1)(x + 2)(x - 3) \Rightarrow$ critical points at -2, 1 and 3
(b) $f' = ---\underset{-2}{|}+++\underset{1}{|}---\underset{3}{|}+++ \Rightarrow$ increasing on $(-2, 1)$ and $(3, \infty)$, decreasing on $(-\infty, -2)$ and $(1, 3)$
(c) Local maximum at $x = 1$, local minima at $x = -2$ and $x = 3$

7. (a) $f'(x) = \frac{x^2(x - 1)}{(x + 2)} \Rightarrow$ critical points at $x = 0$, $x = 1$ and $x = -2$
(b) $f' = +++\underset{-2}{)(}---\underset{0}{|}---\underset{1}{|}+++ \Rightarrow$ increasing on $(-\infty, -2)$ and $(1, \infty)$, decreasing on $(-2, 0)$ and $(0, 1)$
(c) Local minimum at $x = 1$

9. (a) $f'(x) = 1 - \frac{4}{x^2} = \frac{x^2 - 4}{x^2} \Rightarrow$ critical points at $x = -2$, $x = 2$ and $x = 0$.
(b) $f' = +++\underset{-2}{|}---\underset{0}{)(}---\underset{2}{|}+++ \Rightarrow$ increasing on $(-\infty, -2)$ and $(2, \infty)$, decreasing on $(-2, 0)$ and $(0, 2)$
(c) Local maximum at $x = -2$, local minimum at $x = 2$

11. (a) $f'(x) = x^{-1/3}(x + 2) \Rightarrow$ critical points at $x = -2$ and $x = 0$
(b) $f' = +++\underset{-2}{|}---\underset{0}{)(}+++ \Rightarrow$ increasing on $(-\infty, -2)$ and $(0, \infty)$, decreasing on $(-2, 0)$

(c) Local maximum at $x = -2$, local minimum at $x = 0$

13. (a) $f'(x) = (\sin x - 1)(2\cos x + 1), 0 \le x \le 2\pi \Rightarrow$ critical points at $x = \frac{\pi}{2}, x = \frac{2\pi}{3}$, and $x = \frac{4\pi}{3}$

(b) $f' = \underset{0}{[}---\underset{\frac{\pi}{2}}{|}---\underset{\frac{2\pi}{3}}{|}+++\underset{\frac{4\pi}{3}}{|}---\underset{2\pi}{]} \Rightarrow$ increasing on $\left(\frac{2\pi}{3}, \frac{4\pi}{3}\right)$, decreasing on $\left(0, \frac{\pi}{2}\right)$, $\left(\frac{\pi}{2}, \frac{2\pi}{3}\right)$ and $\left(\frac{4\pi}{3}, 2\pi\right)$

(c) Local maximum at $x = \frac{4\pi}{3}$ and $x = 0$, local minimum at $x = \frac{2\pi}{3}$ and $x = 2\pi$

15. (a) Increasing on $(-2, 0)$ and $(2, 4)$, decreasing on $(-4, -2)$ and $(0, 2)$

(b) Absolute maximum at $(-4, 2)$, local maximum at $(0, 1)$ and $(4, -1)$; Absolute minimum at $(2, -3)$, local minimum at $(-2, 0)$

17. (a) Increasing on $(-4, -1)$, $(0.5, 2)$, and $(2, 4)$, decreasing on $(-1, 0.5)$

(b) Absolute maximum at $(4, 3)$, local maximum at $(-1, 2)$ and $(2, 1)$; No absolute minimum, local minimum at $(-4, -1)$and $(0.5, -1)$

19. (a) $g(t) = -t^2 - 3t + 3 \Rightarrow g'(t) = -2t - 3 \Rightarrow$ a critical point at $t = -\frac{3}{2}$; $g' = +++\underset{-3/2}{|}---$, increasing on $\left(-\infty, -\frac{3}{2}\right)$, decreasing on $\left(-\frac{3}{2}, \infty\right)$

(b) local maximum value of $g\left(-\frac{3}{2}\right) = \frac{21}{4}$ at $t = -\frac{3}{2}$, absolute maximum is $\frac{21}{4}$ at $t = -\frac{3}{2}$

21. (a) $h(x) = -x^3 + 2x^2 \Rightarrow h'(x) = -3x^2 + 4x = x(4 - 3x) \Rightarrow$ critical points at $x = 0, \frac{4}{3}$
$\Rightarrow h' = ---\underset{0}{|}+++\underset{4/3}{|}---$, increasing on $\left(0, \frac{4}{3}\right)$, decreasing on $(-\infty, 0)$ and $\left(\frac{4}{3}, \infty\right)$

(b) local maximum value of $h\left(\frac{4}{3}\right) = \frac{32}{27}$ at $x = \frac{4}{3}$; local minimum value of $h(0) = 0$ at $x = 0$, no absolute extrema

23. (a) $f(\theta) = 3\theta^2 - 4\theta^3 \Rightarrow f'(\theta) = 6\theta - 12\theta^2 = 6\theta(1 - 2\theta) \Rightarrow$ critical points at $\theta = 0, \frac{1}{2} \Rightarrow f' = ---\underset{0}{|}+++\underset{1/2}{|}---$, increasing on $\left(0, \frac{1}{2}\right)$, decreasing on $(-\infty, 0)$ and $\left(\frac{1}{2}, \infty\right)$

(b) a local maximum is $f\left(\frac{1}{2}\right) = \frac{1}{4}$ at $\theta = \frac{1}{2}$, a local minimum is $f(0) = 0$ at $\theta = 0$, no absolute extrema

25. (a) $f(r) = 3r^3 + 16r \Rightarrow f'(r) = 9r^2 + 16 \Rightarrow$ no critical points $\Rightarrow f' = +++++$, increasing on $(-\infty, \infty)$, never decreasing

(b) no local extrema, no absolute extrema

27. (a) $f(x) = x^4 - 8x^2 + 16 \Rightarrow f'(x) = 4x^3 - 16x = 4x(x + 2)(x - 2) \Rightarrow$ critical points at $x = 0$ and $x = \pm 2$
$\Rightarrow f' = ---\underset{-2}{|}+++\underset{0}{|}---\underset{2}{|}+++$, increasing on $(-2, 0)$ and $(2, \infty)$, decreasing on $(-\infty, -2)$ and $(0, 2)$

(b) a local maximum is $f(0) = 16$ at $x = 0$, local minima are $f(\pm 2) = 0$ at $x = \pm 2$, no absolute maximum; absolute minimum is 0 at $x = \pm 2$

29. (a) $H(t) = \frac{3}{2}t^4 - t^6 \Rightarrow H'(t) = 6t^3 - 6t^5 = 6t^3(1 + t)(1 - t) \Rightarrow$ critical points at $t = 0, \pm 1$
$\Rightarrow H' = +++\underset{-1}{|}---\underset{0}{|}+++\underset{1}{|}---$, increasing on $(-\infty, -1)$ and $(0, 1)$, decreasing on $(-1, 0)$ and $(1, \infty)$

(b) the local maxima are $H(-1) = \frac{1}{2}$ at $t = -1$ and $H(1) = \frac{1}{2}$ at $t = 1$, the local minimum is $H(0) = 0$ at $t = 0$, absolute maximum is $\frac{1}{2}$ at $t = \pm 1$; no absolute minimum

31. (a) $f(x) = x - 6\sqrt{x - 1} \Rightarrow f'(x) = 1 - \frac{3}{\sqrt{x - 1}} = \frac{\sqrt{x-1} - 3}{\sqrt{x - 1}} \Rightarrow$ critical points at $x = 1$ and $x = 10$
$\Rightarrow f' = \underset{1}{(}---\underset{10}{|}+++$, increasing on $(10, \infty)$, decreasing on $(1, 10)$

(b) a local minimum is $f(10) = -8$, a local and absolute maximum is $f(1) = 1$, absolute minimum of -8 at $x = 10$

33. (a) $g(x) = x\sqrt{8 - x^2} = x(8 - x^2)^{1/2} \Rightarrow g'(x) = (8 - x^2)^{1/2} + x\left(\frac{1}{2}\right)(8 - x^2)^{-1/2}(-2x) = \frac{2(2 - x)(2 + x)}{\sqrt{(2\sqrt{2} - x)(2\sqrt{2} + x)}}$

$\Rightarrow$ critical points at $x = \pm 2, \pm 2\sqrt{2} \Rightarrow g' = (\underset{-2\sqrt{2}}{} \; --- \underset{-2}{|} +++ \underset{2}{|} --- \underset{2\sqrt{2}}{)}$, increasing on $(-2, 2)$, decreasing on $\left(-2\sqrt{2}, -2\right)$ and $\left(2, 2\sqrt{2}\right)$

(b) local maxima are $g(2) = 4$ at $x = 2$ and $g\left(-2\sqrt{2}\right) = 0$ at $x = -2\sqrt{2}$, local minima are $g(-2) = -4$ at $x = -2$ and $g\left(2\sqrt{2}\right) = 0$ at $x = 2\sqrt{2}$, absolute maximum is 4 at $x = 2$; absolute minimum is -4 at $x = -2$

35. (a) $f(x) = \frac{x^2 - 3}{x - 2} \Rightarrow f'(x) = \frac{2x(x - 2) - (x^2 - 3)(1)}{(x - 2)^2} = \frac{(x - 3)(x - 1)}{(x - 2)^2} \Rightarrow$ critical points at $x = 1, 3$

$\Rightarrow f' = +++ \underset{1}{|} --- \underset{2}{)(} --- \underset{3}{|} +++$, increasing on $(-\infty, 1)$ and $(3, \infty)$, decreasing on $(1, 2)$ and $(2, 3)$, discontinuous at $x = 2$

(b) a local maximum is $f(1) = 2$ at $x = 1$, a local minimum is $f(3) = 6$ at $x = 3$, no absolute extrema

37. (a) $f(x) = x^{1/3}(x + 8) = x^{4/3} + 8x^{1/3} \Rightarrow f'(x) = \frac{4}{3}x^{1/3} + \frac{8}{3}x^{-2/3} = \frac{4(x + 2)}{3x^{2/3}} \Rightarrow$ critical points at $x = 0, -2$

$\Rightarrow f' = --- \underset{-2}{|} +++ \underset{0}{)(} +++$, increasing on $(-2, 0) \cup (0, \infty)$, decreasing on $(-\infty, -2)$

(b) no local maximum, a local minimum is $f(-2) = -6\sqrt[3]{2} \approx -7.56$ at $x = -2$, no absolute maximum; absolute minimum is $-6\sqrt[3]{2}$ at $x = -2$

39. (a) $h(x) = x^{1/3}(x^2 - 4) = x^{7/3} - 4x^{1/3} \Rightarrow h'(x) = \frac{7}{3}x^{4/3} - \frac{4}{3}x^{-2/3} = \frac{\left(\sqrt{7}x + 2\right)\left(\sqrt{7}x - 2\right)}{3\sqrt[3]{x^2}} \Rightarrow$ critical points at $x = 0, \frac{\pm 2}{\sqrt{7}} \Rightarrow h' = +++ \underset{-2/\sqrt{7}}{|} --- \underset{0}{)(} --- \underset{2/\sqrt{7}}{|} +++$, increasing on $\left(-\infty, \frac{-2}{\sqrt{7}}\right)$ and $\left(\frac{2}{\sqrt{7}}, \infty\right)$, decreasing on $\left(\frac{-2}{\sqrt{7}}, 0\right)$ and $\left(0, \frac{2}{\sqrt{7}}\right)$

(b) local maximum is $h\left(\frac{-2}{\sqrt{7}}\right) = \frac{24\sqrt[3]{2}}{7^{7/6}} \approx 3.12$ at $x = \frac{-2}{\sqrt{7}}$, the local minimum is $h\left(\frac{2}{\sqrt{7}}\right) = -\frac{24\sqrt[3]{2}}{7^{7/6}} \approx -3.12$, no absolute extrema

41. (a) $f(x) = 2x - x^2 \Rightarrow f'(x) = 2 - 2x \Rightarrow$ a critical point at $x = 1 \Rightarrow f' = +++ \underset{1}{|} --- \underset{2}{]}$ and $f(1) = 1$ and $f(2) = 0$

a local maximum is 1 at $x = 1$, a local minimum is 0 at $x = 2$.

(b) There is an absolute maximum of 1 at $x = 1$; no absolute minimum.

(c)

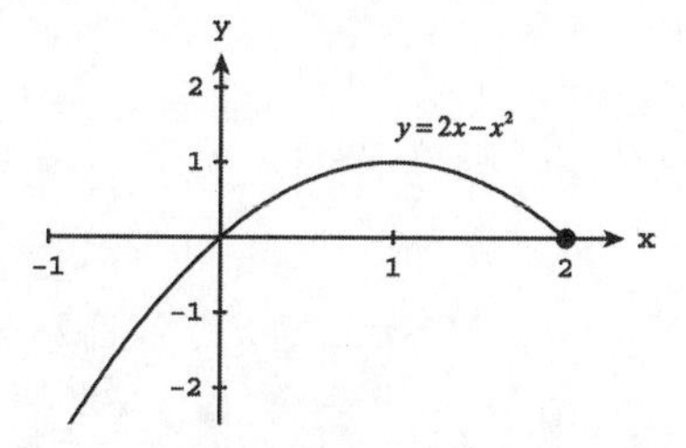

43. (a) $g(x) = x^2 - 4x + 4 \Rightarrow g'(x) = 2x - 4 = 2(x - 2) \Rightarrow$ a critical point at $x = 2 \Rightarrow g' = \underset{1}{[} --- \underset{2}{|} +++$ and $g(1) = 1, g(2) = 0 \Rightarrow$ a local maximum is 1 at $x = 1$, a local minimum is $g(2) = 0$ at $x = 2$

(b) no absolute maximum; absolute minimum is 0 at $x = 2$

(c)

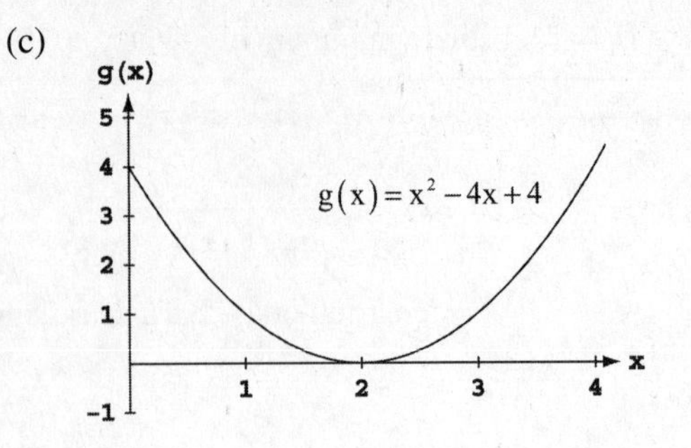

45. (a) $f(t) = 12t - t^3 \Rightarrow f'(t) = 12 - 3t^2 = 3(2 + t)(2 - t) \Rightarrow$ critical points at $t = \pm 2 \Rightarrow f' = \underset{-3}{[}\, - - - \underset{-2}{|}\, + + + \underset{2}{|}\, - - -$ and $f(-3) = -9$, $f(-2) = -16$, $f(2) = 16 \Rightarrow$ local maxima are -9 at $t = -3$ and 16 at $t = 2$, a local minimum is -16 at $t = -2$

(b) absolute maximum is 16 at $t = 2$; no absolute minimum

(c)

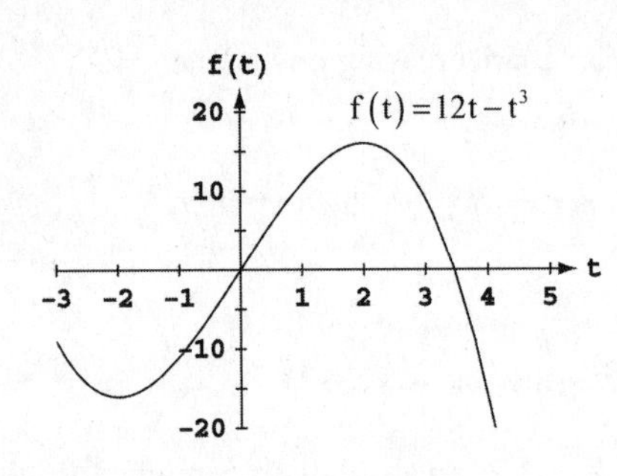

47. (a) $h(x) = \frac{x^3}{3} - 2x^2 + 4x \Rightarrow h'(x) = x^2 - 4x + 4 = (x - 2)^2 \Rightarrow$ a critical point at $x = 2 \Rightarrow h' = \underset{0}{[}\, + + + \underset{2}{|}\, + + +$ and $h(0) = 0 \Rightarrow$ no local maximum, a local minimum is 0 at $x = 0$

(b) no absolute maximum; absolute minimum is 0 at $x = 0$

(c)

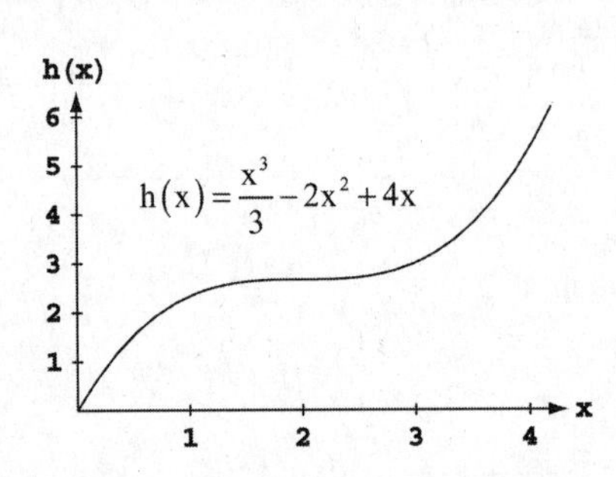

49. (a) $f(x) = \sqrt{25 - x^2} \Rightarrow f'(x) = \frac{-x}{\sqrt{25 - x^2}} \Rightarrow$ critical points at $x = 0$, $x = -5$, and $x = 5$ $\Rightarrow f' = \underset{-5}{(}\, + + + \underset{0}{|}\, - - - \underset{5}{)}$, $f(-5) = 0$, $f(0) = 5$, $f(5) = 0 \Rightarrow$ local maximum is 5 at $x = 0$; local minimum of 0 at $x = -5$ and $x = 5$

(b) absolute maximum is 5 at $x = 0$; absolute minimum of 0 at $x = -5$ and $x = 5$

(c)

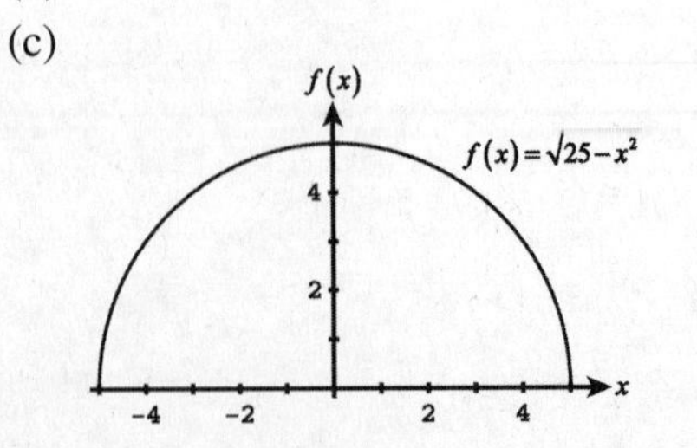

51. (a) $g(x) = \frac{x-2}{x^2-1}, 0 \le x < 1 \Rightarrow g'(x) = \frac{-x^2+4x-1}{(x^2-1)^2} \Rightarrow$ only critical point in $0 \le x < 1$ is $x = 2 - \sqrt{3} \approx 0.268$

$\Rightarrow g' = \underset{0}{[}\; --- \underset{0.268}{|} +++ \underset{1}{)}$, $g\left(2-\sqrt{3}\right) = \frac{\sqrt{3}}{4\sqrt{3}-6} \approx 1.866 \Rightarrow$ local minimum of $\frac{\sqrt{3}}{4\sqrt{3}-6}$ at $x = 2-\sqrt{3}$, local maximum at $x = 0$.

(b) absolute minimum of $\frac{\sqrt{3}}{4\sqrt{3}-6}$ at $x = 2-\sqrt{3}$, no absolute maximum

(c)

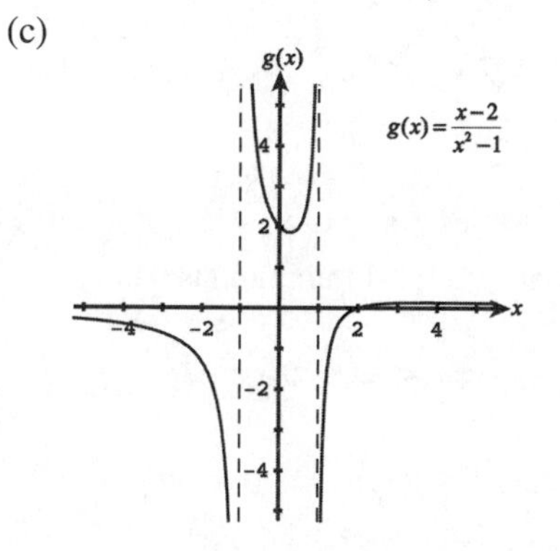

53. (a) $f(x) = \sin 2x, 0 \le x \le \pi \Rightarrow f'(x) = 2\cos 2x, f'(x) = 0 \Rightarrow \cos 2x = 0 \Rightarrow$ critical points are $x = \frac{\pi}{4}$ and $x = \frac{3\pi}{4}$

$\Rightarrow f' = \underset{0}{[}+++\underset{\frac{\pi}{4}}{|}---\underset{\frac{3\pi}{4}}{|}+++\underset{\pi}{]}$, $f(0) = 0, f\left(\frac{\pi}{4}\right) = 1, f\left(\frac{3\pi}{4}\right) = -1, f(\pi) = 0 \Rightarrow$ local maxima are 1 at $x = \frac{\pi}{4}$ and 0 at $x = \pi$, and local minima are -1 at $x = \frac{3\pi}{4}$ and 0 at $x = 0$.

(b) The graph of f rises when $f' > 0$, falls when $f' < 0$, and has local extreme values where $f' = 0$. The function f has a local minimum value at $x = 0$ and $x = \frac{3\pi}{4}$, where the values of f' change from negative to positive. The function f has a local maximum value at $x = \pi$ and $x = \frac{\pi}{4}$, where the values of f'change from positive to negative.

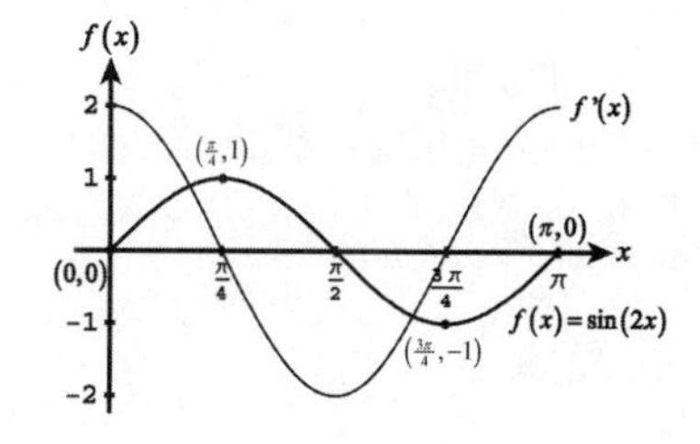

55. (a) $f(x) = \sqrt{3}\cos x + \sin x, 0 \le x \le 2\pi \Rightarrow f'(x) = -\sqrt{3}\sin x + \cos x, f'(x) = 0 \Rightarrow \tan x = \frac{1}{\sqrt{3}} \Rightarrow$ critical points are $x = \frac{\pi}{6}$ and $x = \frac{7\pi}{6} \Rightarrow f' = \underset{0}{[}+++\underset{\frac{\pi}{6}}{|}---\underset{\frac{7\pi}{6}}{|}+++\underset{2\pi}{]}$, $f(0) = \sqrt{3}, f\left(\frac{\pi}{6}\right) = 2, f\left(\frac{7\pi}{6}\right) = -2, f(2\pi) = \sqrt{3} \Rightarrow$ local maxima are 2 at $x = \frac{\pi}{6}$ and $\sqrt{3}$ at $x = 2\pi$, and local minima are -2 at $x = \frac{7\pi}{6}$ and $\sqrt{3}$ at $x = 0$.

(b) The graph of f rises when $f' > 0$, falls when $f' < 0$, and has local extreme values where $f' = 0$. The function f has a local minimum value at $x = 0$ and $x = \frac{7\pi}{6}$, where the values of f' change from negative to positive. The function f has a local maximum value at $x = 2\pi$ and $x = \frac{\pi}{6}$, where the values of f'change from positive to negative.

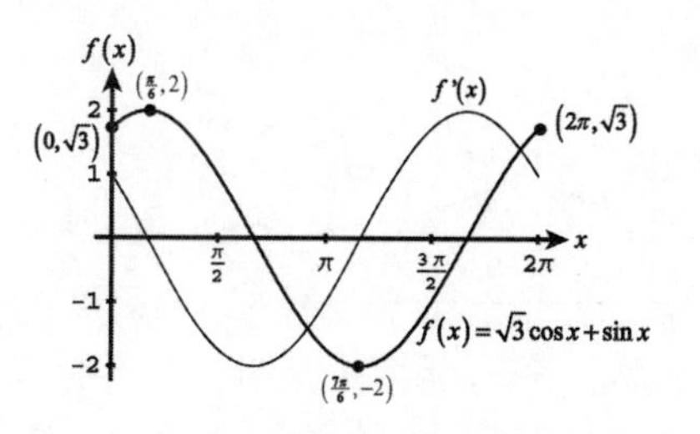

57. (a) $f(x) = \frac{x}{2} - 2\sin\left(\frac{x}{2}\right) \Rightarrow f'(x) = \frac{1}{2} - \cos\left(\frac{x}{2}\right), f'(x) = 0 \Rightarrow \cos\left(\frac{x}{2}\right) = \frac{1}{2} \Rightarrow$ a critical point at $x = \frac{2\pi}{3}$

$\Rightarrow f' = \underset{0}{[}---\underset{2\pi/3}{|}+++\underset{2\pi}{]}$ and $f(0) = 0, f\left(\frac{2\pi}{3}\right) = \frac{\pi}{3} - \sqrt{3}, f(2\pi) = \pi \Rightarrow$ local maxima are 0 at $x = 0$ and π at $x = 2\pi$, a local minimum is $\frac{\pi}{3} - \sqrt{3}$ at $x = \frac{2\pi}{3}$

(b) The graph of f rises when $f' > 0$, falls when $f' < 0$, and has a local minimum value at the point where f' changes from negative to positive.

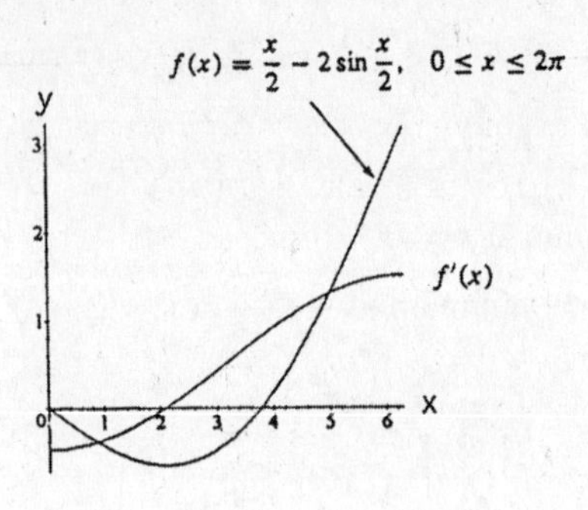

59. (a) $f(x) = \csc^2 x - 2\cot x \Rightarrow f'(x) = 2(\csc x)(-\csc x)(\cot x) - 2(-\csc^2 x) = -2(\csc^2 x)(\cot x - 1) \Rightarrow$ a critical point at $x = \frac{\pi}{4} \Rightarrow f' = \underset{0}{(}\ ---\ \underset{\pi/4}{|}\ +++\ \underset{\pi}{)}$ and $f\left(\frac{\pi}{4}\right) = 0 \Rightarrow$ no local maximum, a local minimum is 0 at $x = \frac{\pi}{4}$

(b) The graph of f rises when $f' > 0$, falls when $f' < 0$, and has a local minimum value at the point where $f' = 0$ and the values of f' change from negative to positive. The graph of f steepens as $f'(x) \to \pm\infty$.

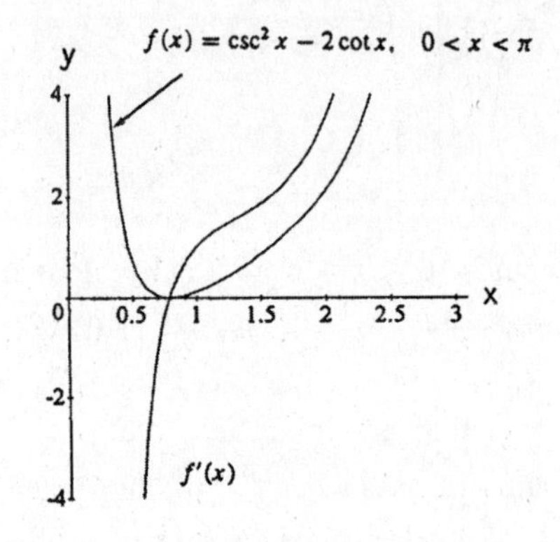

61. $h(\theta) = 3\cos\left(\frac{\theta}{2}\right) \Rightarrow h'(\theta) = -\frac{3}{2}\sin\left(\frac{\theta}{2}\right) \Rightarrow h' = \underset{0}{[}\ ---\ \underset{2\pi}{]}$, $(0, 3)$ and $(2\pi, -3) \Rightarrow$ a local maximum is 3 at $\theta = 0$, a local minimum is -3 at $\theta = 2\pi$

63. (a) (b) (c) (d)

65. (a) (b)

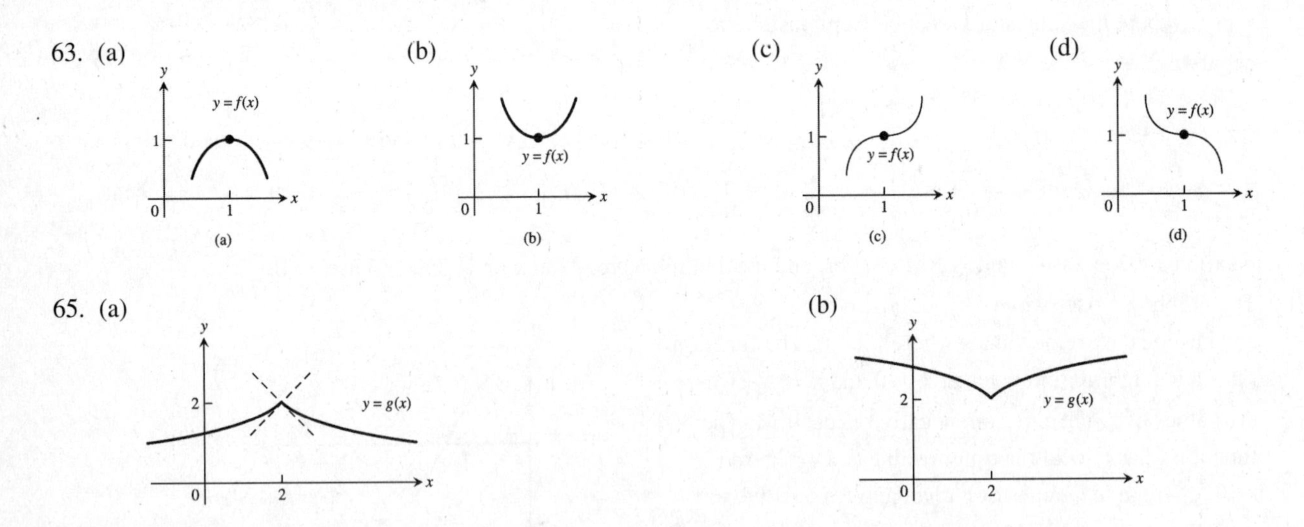

67. The function $f(x) = x\sin\left(\frac{1}{x}\right)$ has an infinite number of local maxima and minima. The function $\sin x$ has the following properties: a) it is continuous on $(-\infty, \infty)$; b) it is periodic; and c) its range is $[-1, 1]$. Also, for $a > 0$, the function $\frac{1}{x}$ has a range of $(-\infty, -a] \cup [a, \infty)$ on $\left[-\frac{1}{a}, \frac{1}{a}\right]$. In particular, if $a = 1$, then $\frac{1}{x} \le -1$ or $\frac{1}{x} \ge 1$ when x is in $[-1, 1]$. This means $\sin\left(\frac{1}{x}\right)$ takes on the values of 1 and -1 infinitely many times in times on the interval $[-1, 1]$, which occur when $\frac{1}{x} = \pm\frac{\pi}{2}, \pm\frac{3\pi}{2}, \pm\frac{5\pi}{2}, \ldots \Rightarrow x = \pm\frac{2}{\pi}, \pm\frac{2}{3\pi}, \pm\frac{2}{5\pi}, \ldots$. Thus $\sin\left(\frac{1}{x}\right)$ has infinitely many local maxima and minima in the interval $[-1, 1]$. On the interval $[0, 1]$, $-1 \le \sin\left(\frac{1}{x}\right) \le 1$ and since $x > 0$ we have $-x \le x\sin\left(\frac{1}{x}\right) \le x$. On the interval $[-1, 0]$, $-1 \le \sin\left(\frac{1}{x}\right) \le 1$ and since $x < 0$ we have $-x \ge x\sin\left(\frac{1}{x}\right) \ge x$. Thus f(x) is bounded by the lines $y = x$

and $y = -x$. Since $\sin\left(\frac{1}{x}\right)$ oscillates between 1 and -1 infinitely many times on $[-1, 1]$ then f will oscillate between $y = x$ and $y = -x$ infinitely many times. Thus f has infinitely many local maxima and minima. We can see from the graph (and verify later in Chapter 7) that $\lim\limits_{x\to\infty} x\sin\left(\frac{1}{x}\right) = 1$ and $\lim\limits_{x\to-\infty} x\sin\left(\frac{1}{x}\right) = 1$. The graph of f does not have any absolute maxima., but it does have two absolute minima.

69. $f(x) = ax^2 + bx \Rightarrow f'(x) = 2ax + b$, $f(1) = 2 \Rightarrow a + b = 2$, $f'(1) = 0 \Rightarrow 2a + b = 0 \Rightarrow a = -2, b = 4$
$\Rightarrow f(x) = -2x^2 + 4x$

4.4 CONCAVITY AND CURVE SKETCHING

1. $y = \frac{x^3}{3} - \frac{x^2}{2} - 2x + \frac{1}{3} \Rightarrow y' = x^2 - x - 2 = (x-2)(x+1) \Rightarrow y'' = 2x - 1 = 2\left(x - \frac{1}{2}\right)$. The graph is rising on $(-\infty, -1)$ and $(2, \infty)$, falling on $(-1, 2)$, concave up on $\left(\frac{1}{2}, \infty\right)$ and concave down on $\left(-\infty, \frac{1}{2}\right)$. Consequently, a local maximum is $\frac{3}{2}$ at $x = -1$, a local minimum is -3 at $x = 2$, and $\left(\frac{1}{2}, -\frac{3}{4}\right)$ is a point of inflection.

3. $y = \frac{3}{4}\left(x^2 - 1\right)^{2/3} \Rightarrow y' = \left(\frac{3}{4}\right)\left(\frac{2}{3}\right)\left(x^2-1\right)^{-1/3}(2x) = x\left(x^2-1\right)^{-1/3}$, $y' = \underset{-1}{---)}\ \underset{0}{(+++|}\ \underset{1}{---)}(+++$
$\Rightarrow$ the graph is rising on $(-1, 0)$ and $(1, \infty)$, falling on $(-\infty, -1)$ and $(0, 1)$ $\Rightarrow$ a local maximum is $\frac{3}{4}$ at $x = 0$, local minima are 0 at $x = \pm 1$; $y'' = \left(x^2-1\right)^{-1/3} + (x)\left(-\frac{1}{3}\right)\left(x^2-1\right)^{-4/3}(2x) = \frac{x^2-3}{3\sqrt[3]{\left(x^2-1\right)^4}}$,
$y'' = \underset{-\sqrt{3}}{+++|}\ \underset{-1}{---)}\ \underset{1}{(---)}\underset{\sqrt{3}}{(---|}\ +++$ $\Rightarrow$ the graph is concave up on $\left(-\infty, -\sqrt{3}\right)$ and $\left(\sqrt{3}, \infty\right)$, concave down on $\left(-\sqrt{3}, \sqrt{3}\right)$ $\Rightarrow$ points of inflection at $\left(\pm\sqrt{3}, \frac{3\sqrt[3]{4}}{4}\right)$

5. $y = x + \sin 2x \Rightarrow y' = 1 + 2\cos 2x$, $y' = \underset{-2\pi/3}{[---|}\underset{-\pi/3}{}\ \underset{\pi/3}{+++|}\ \underset{2\pi/3}{---]}$ $\Rightarrow$ the graph is rising on $\left(-\frac{\pi}{3}, \frac{\pi}{3}\right)$, falling on $\left(-\frac{2\pi}{3}, -\frac{\pi}{3}\right)$ and $\left(\frac{\pi}{3}, \frac{2\pi}{3}\right)$ $\Rightarrow$ local maxima are $-\frac{2\pi}{3} + \frac{\sqrt{3}}{2}$ at $x = -\frac{2\pi}{3}$ and $\frac{\pi}{3} + \frac{\sqrt{3}}{2}$ at $x = \frac{\pi}{3}$, local minima are $-\frac{\pi}{3} - \frac{\sqrt{3}}{2}$ at $x = -\frac{\pi}{3}$ and $\frac{2\pi}{3} - \frac{\sqrt{3}}{2}$ at $x = \frac{2\pi}{3}$; $y'' = -4\sin 2x$, $y'' = \underset{-2\pi/3}{[}\ \underset{-\pi/2}{---|}\ \underset{0}{+++|}\ \underset{\pi/2}{---|}\ \underset{2\pi/3}{+++]}$ $\Rightarrow$ the graph is concave up on $\left(-\frac{\pi}{2}, 0\right)$ and $\left(\frac{\pi}{2}, \frac{2\pi}{3}\right)$, concave down on $\left(-\frac{2\pi}{3}, -\frac{\pi}{2}\right)$ and $\left(0, \frac{\pi}{2}\right)$ $\Rightarrow$ points of inflection at $\left(-\frac{\pi}{2}, -\frac{\pi}{2}\right)$, $(0, 0)$, and $\left(\frac{\pi}{2}, \frac{\pi}{2}\right)$

7. If $x \geq 0$, $\sin|x| = \sin x$ and if $x < 0$, $\sin|x| = \sin(-x)$ $= -\sin x$. From the sketch the graph is rising on $\left(-\frac{3\pi}{2}, -\frac{\pi}{2}\right)$, $\left(0, \frac{\pi}{2}\right)$ and $\left(\frac{3\pi}{2}, 2\pi\right)$, falling on $\left(-2\pi, -\frac{3\pi}{2}\right)$, $\left(-\frac{\pi}{2}, 0\right)$ and $\left(\frac{\pi}{2}, \frac{3\pi}{2}\right)$; local minima are -1 at $x = \pm\frac{3\pi}{2}$ and 0 at $x = 0$; local maxima are 1 at $x = \pm\frac{\pi}{2}$ and 0 at $x = \pm 2\pi$; concave up on $(-2\pi, -\pi)$ and $(\pi, 2\pi)$, and concave down on $(-\pi, 0)$ and $(0, \pi)$ $\Rightarrow$ points of inflection are $(-\pi, 0)$ and $(\pi, 0)$

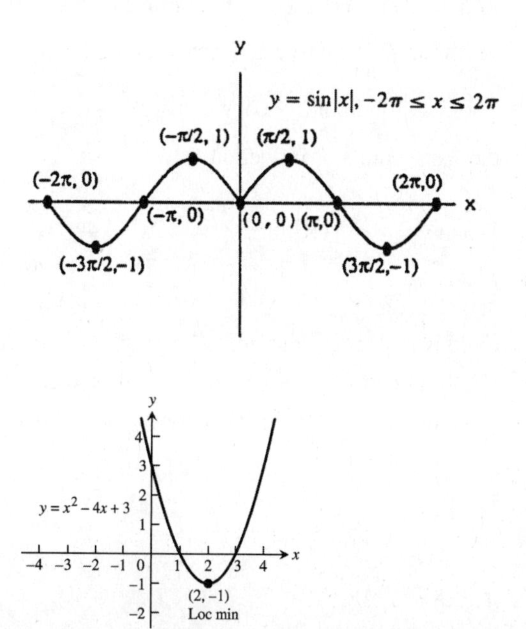

9. When $y = x^2 - 4x + 3$, then $y' = 2x - 4 = 2(x - 2)$ and $y'' = 2$. The curve rises on $(2, \infty)$ and falls on $(-\infty, 2)$. At $x = 2$ there is a minimum. Since $y'' > 0$, the curve is concave up for all x.

11. When $y = x^3 - 3x + 3$, then $y' = 3x^2 - 3 = 3(x - 1)(x + 1)$ and $y'' = 6x$. The curve rises on $(-\infty, -1) \cup (1, \infty)$ and falls on $(-1, 1)$. At $x = -1$ there is a local maximum and at $x = 1$ a local minimum. The curve is concave down on $(-\infty, 0)$ and concave up on $(0, \infty)$. There is a point of inflection at $x = 0$.

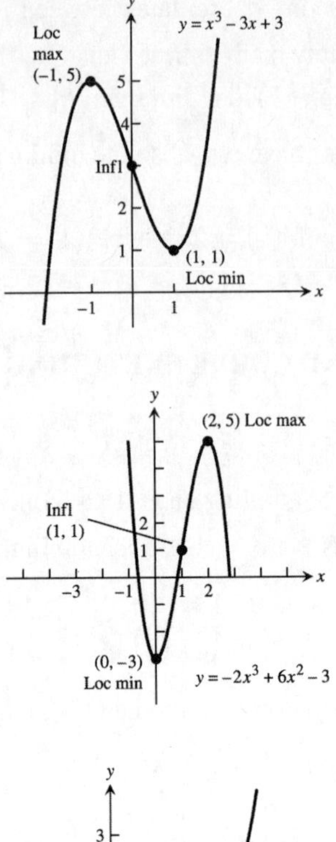

13. When $y = -2x^3 + 6x^2 - 3$, then $y' = -6x^2 + 12x = -6x(x - 2)$ and $y'' = -12x + 12 = -12(x - 1)$. The curve rises on $(0, 2)$ and falls on $(-\infty, 0)$ and $(2, \infty)$. At $x = 0$ there is a local minimum and at $x = 2$ a local maximum. The curve is concave up on $(-\infty, 1)$ and concave down on $(1, \infty)$. At $x = 1$ there is a point of inflection.

15. When $y = (x - 2)^3 + 1$, then $y' = 3(x - 2)^2$ and $y'' = 6(x - 2)$. The curve never falls and there are no local extrema. The curve is concave down on $(-\infty, 2)$ and concave up on $(2, \infty)$. At $x = 2$ there is a point of inflection.

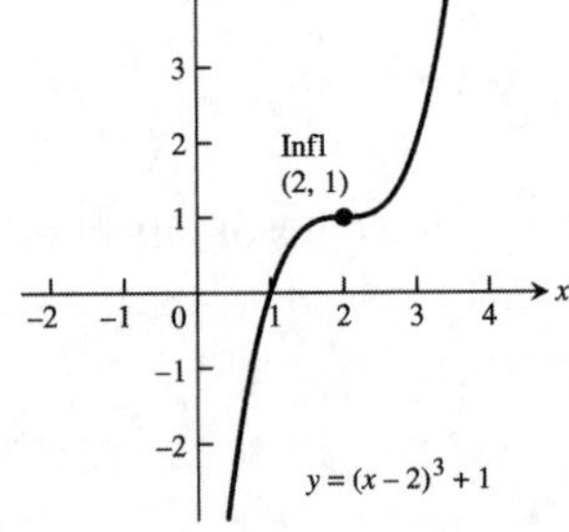

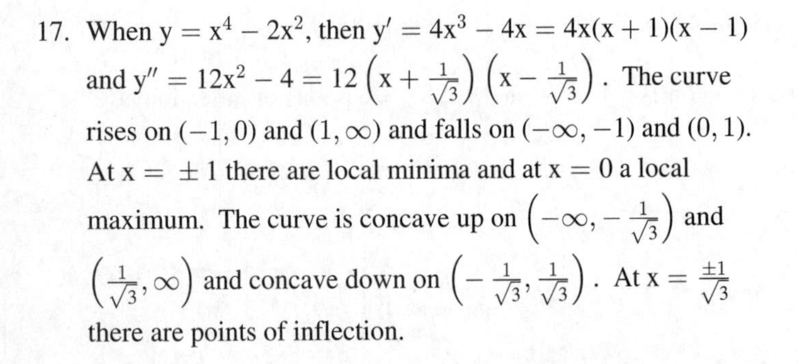

17. When $y = x^4 - 2x^2$, then $y' = 4x^3 - 4x = 4x(x + 1)(x - 1)$ and $y'' = 12x^2 - 4 = 12\left(x + \frac{1}{\sqrt{3}}\right)\left(x - \frac{1}{\sqrt{3}}\right)$. The curve rises on $(-1, 0)$ and $(1, \infty)$ and falls on $(-\infty, -1)$ and $(0, 1)$. At $x = \pm 1$ there are local minima and at $x = 0$ a local maximum. The curve is concave up on $\left(-\infty, -\frac{1}{\sqrt{3}}\right)$ and $\left(\frac{1}{\sqrt{3}}, \infty\right)$ and concave down on $\left(-\frac{1}{\sqrt{3}}, \frac{1}{\sqrt{3}}\right)$. At $x = \frac{\pm 1}{\sqrt{3}}$ there are points of inflection.

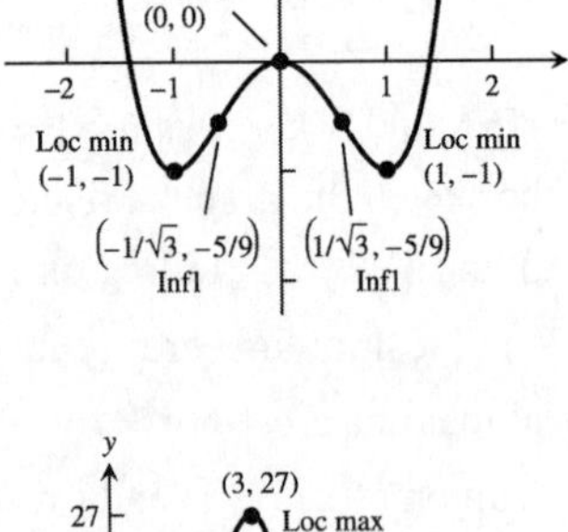

19. When $y = 4x^3 - x^4$, then $y' = 12x^2 - 4x^3 = 4x^2(3 - x)$ and $y'' = 24x - 12x^2 = 12x(2 - x)$. The curve rises on $(-\infty, 3)$ and falls on $(3, \infty)$. At $x = 3$ there is a local maximum, but there is no local minimum. The graph is concave up on $(0, 2)$ and concave down on $(-\infty, 0)$ and $(2, \infty)$. There are inflection points at $x = 0$ and $x = 2$.

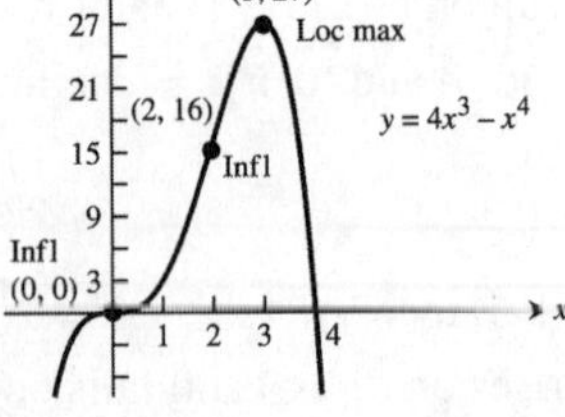

21. When $y = x^5 - 5x^4$, then $y' = 5x^4 - 20x^3 = 5x^3(x-4)$ and $y'' = 20x^3 - 60x^2 = 20x^2(x-3)$. The curve rises on $(-\infty, 0)$ and $(4, \infty)$, and falls on $(0, 4)$. There is a local maximum at $x = 0$, and a local minimum at $x = 4$. The curve is concave down on $(-\infty, 3)$ and concave up on $(3, \infty)$. At $x = 3$ there is a point of inflection.

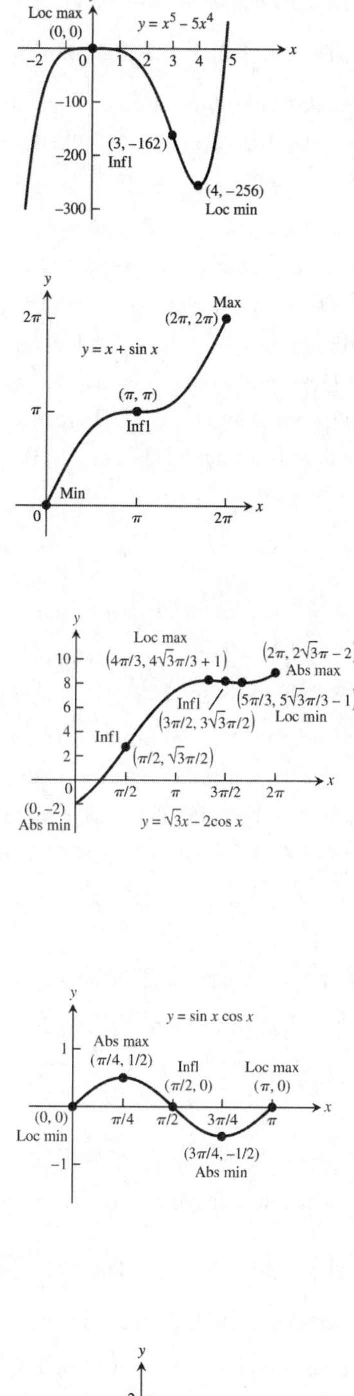

23. When $y = x + \sin x$, then $y' = 1 + \cos x$ and $y'' = -\sin x$. The curve rises on $(0, 2\pi)$. At $x = 0$ there is a local and absolute minimum and at $x = 2\pi$ there is a local and absolute maximum. The curve is concave down on $(0, \pi)$ and concave up on $(\pi, 2\pi)$. At $x = \pi$ there is a point of inflection.

25. When $y = \sqrt{3}x - 2\cos x$, then $y' = \sqrt{3} + 2\sin x$ and $y'' = 2\cos x$. The curve is increasing on $\left(0, \frac{4\pi}{3}\right)$ and $\left(\frac{5\pi}{3}, 2\pi\right)$, and decreasing on $\left(\frac{4\pi}{3}, \frac{5\pi}{3}\right)$. At $x = 0$ there is a local and absolute minimum, at $x = \frac{4\pi}{3}$ there is a local maximum, at $x = \frac{5\pi}{3}$ there is a local minimum, and and at $x = 2\pi$ there is a local and absolute maximum. The curve is concave up on $\left(0, \frac{\pi}{2}\right)$ and $\left(\frac{3\pi}{2}, 2\pi\right)$, and is concave down on$\left(\frac{\pi}{2}, \frac{3\pi}{2}\right)$. At $x = \frac{\pi}{2}$ and $x = \frac{3\pi}{2}$ there are points of inflection.

27. When $y = \sin x \cos x$, then $y' = -\sin^2 x + \cos^2 x = \cos 2x$ and $y'' = -2\sin 2x$. The curve is increasing on $\left(0, \frac{\pi}{4}\right)$ and $\left(\frac{3\pi}{4}, \pi\right)$, and decreasing on $\left(\frac{\pi}{4}, \frac{3\pi}{4}\right)$. At $x = 0$ there is a local minimum, at $x = \frac{\pi}{4}$ there is a local and absolute maximum, at $x = \frac{3\pi}{4}$ there is a local and absolute minimum, and at $x = \pi$ there is a local maximum. The curve is concave down on $\left(0, \frac{\pi}{2}\right)$, and is concave up on$\left(\frac{\pi}{2}, \pi\right)$. At $x = \frac{\pi}{2}$ there is a point of inflection.

29. When $y = x^{1/5}$, then $y' = \frac{1}{5}x^{-4/5}$ and $y'' = -\frac{4}{25}x^{-9/5}$. The curve rises on $(-\infty, \infty)$ and there are no extrema. The curve is concave up on $(-\infty, 0)$ and concave down on $(0, \infty)$. At $x = 0$ there is a point of inflection.

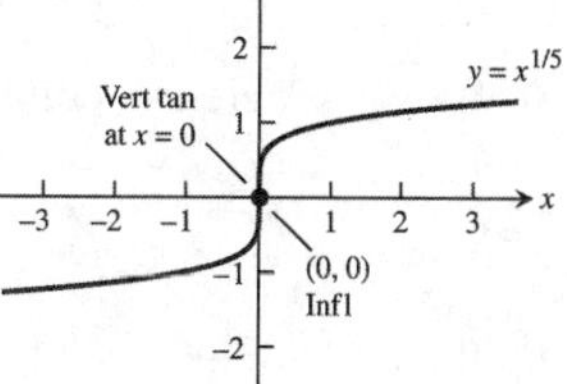

31. When $y = \frac{x}{\sqrt{x^2+1}}$, then $y' = \frac{1}{(x^2+1)^{3/2}}$ and $y'' = \frac{-3x}{(x^2+1)^{5/2}}$. The curve is increasing on $(-\infty, \infty)$. There are no local or absolute extrema. The curve is concave up on $(-\infty, 0)$ and concave down on $(0, \infty)$. At $x = 0$ there is a point of inflection.

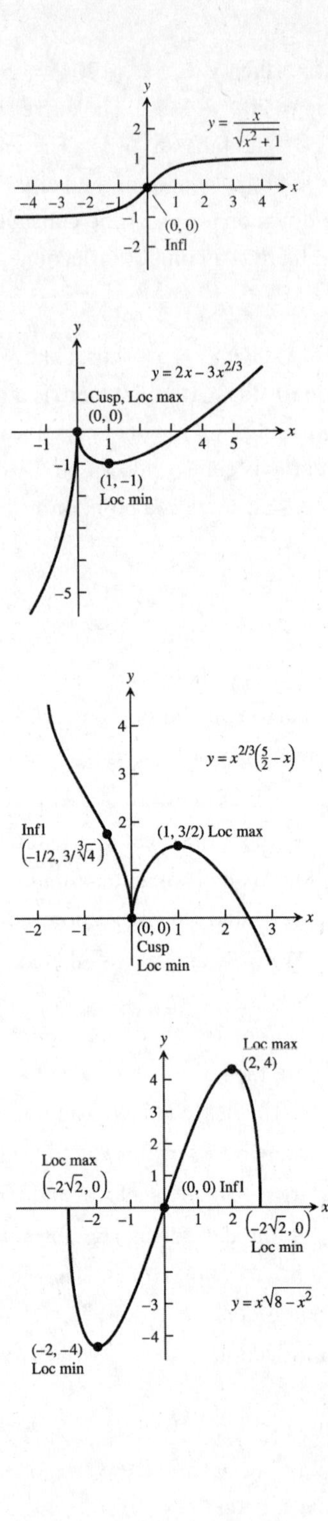

33. When $y = 2x - 3x^{2/3}$, then $y' = 2 - 2x^{-1/3}$ and $y'' = \frac{2}{3}x^{-4/3}$. The curve is rising on $(-\infty, 0)$ and $(1, \infty)$, and falling on $(0, 1)$. There is a local maximum at $x = 0$ and a local minimum at $x = 1$. The curve is concave up on $(-\infty, 0)$ and $(0, \infty)$. There are no points of inflection, but a cusp exists at $x = 0$.

35. When $y = x^{2/3}\left(\frac{5}{2} - x\right) = \frac{5}{2}x^{2/3} - x^{5/3}$, then $y' = \frac{5}{3}x^{-1/3} - \frac{5}{3}x^{2/3} = \frac{5}{3}x^{-1/3}(1 - x)$ and $y'' = -\frac{5}{9}x^{-4/3} - \frac{10}{9}x^{-1/3} = -\frac{5}{9}x^{-4/3}(1 + 2x)$. The curve is rising on $(0, 1)$ and falling on $(-\infty, 0)$ and $(1, \infty)$. There is a local minimum at $x = 0$ and a local maximum at $x = 1$. The curve is concave up on $\left(-\infty, -\frac{1}{2}\right)$ and concave down on $\left(-\frac{1}{2}, 0\right)$ and $(0, \infty)$. There is a point of inflection at $x = -\frac{1}{2}$ and a cusp at $x = 0$.

37. When $y = x\sqrt{8 - x^2} = x\,(8 - x^2)^{1/2}$, then

$$y' = (8 - x^2)^{1/2} + (x)\left(\tfrac{1}{2}\right)(8 - x^2)^{-1/2}(-2x)$$
$$= (8 - x^2)^{-1/2}(8 - 2x^2) = \frac{2(2-x)(2+x)}{\sqrt{\left(2\sqrt{2}+x\right)\left(2\sqrt{2}-x\right)}}$$ and

$$y'' = \left(-\tfrac{1}{2}\right)(8 - x^2)^{-\frac{3}{2}}(-2x)(8 - 2x^2) + (8 - x^2)^{-\frac{1}{2}}(-4x)$$
$$= \frac{2x\,(x^2 - 12)}{\sqrt{(8 - x^2)^3}}$$. The curve is rising on $(-2, 2)$, and falling on $\left(-2\sqrt{2}, -2\right)$ and $\left(2, 2\sqrt{2}\right)$. There are local minima $x = -2$ and $x = 2\sqrt{2}$, and local maxima at $x = -2\sqrt{2}$ and $x = 2$. The curve is concave up on $\left(-2\sqrt{2}, 0\right)$ and concave down on $\left(0, 2\sqrt{2}\right)$. There is a point of inflection at $x = 0$.

39. When $y = \sqrt{16 - x^2}$, then $y' = \frac{-x}{\sqrt{16 - x^2}}$ and $y'' = \frac{-16}{(16 - x^2)^{3/2}}$. The curve is rising on $(-4, 0)$ and falling on $(0, 4)$. There is a local and absolute maximum at $x = 0$ and local and absolute minima at $x = -4$ and $x = 4$. The curve is concave down on $(-4, 4)$. There are no points of inflection.

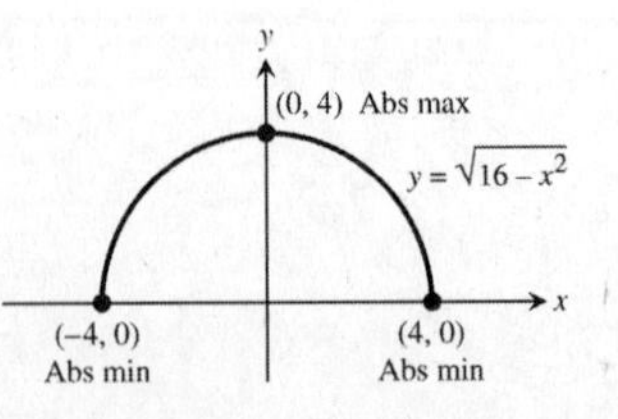

41. When $y = \frac{x^2-3}{x-2}$, then $y' = \frac{2x(x-2)-(x^2-3)(1)}{(x-2)^2}$

$= \frac{(x-3)(x-1)}{(x-2)^2}$ and

$y'' = \frac{(2x-4)(x-2)^2-(x^2-4x+3)2(x-2)}{(x-2)^4} = \frac{2}{(x-2)^3}$.

The curve is rising on $(-\infty, 1)$ and $(3, \infty)$, and falling on $(1, 2)$ and $(2, 3)$. There is a local maximum at $x = 1$ and a local minimum at $x = 3$. The curve is concave down on $(-\infty, 2)$ and concave up on $(2, \infty)$. There are no points of inflection because $x = 2$ is not in the domain.

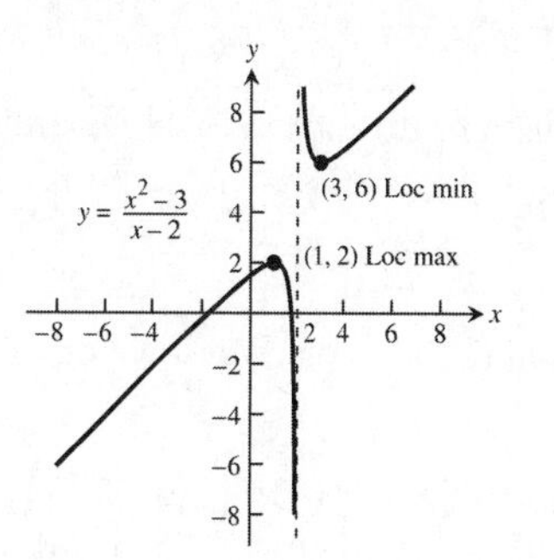

43. When $y = \frac{8x}{x^2+4}$, then $y' = \frac{-8(x^2-4)}{(x^2+4)^2}$ and

$y'' = \frac{16x(x^2-12)}{(x^2+4)^3}$. The curve is fallng on $(-\infty, -2)$ and $(2, \infty)$, and is rising on $(-2, 2)$. There is a local and absolute minimum at $x = -2$, and a local and absolute maximum at $x = 2$. The curve is concave down on $\left(-\infty, -2\sqrt{3}\right)$ and $\left(0, 2\sqrt{3}\right)$, and concave up on $\left(-2\sqrt{3}, 0\right)$ and $\left(2\sqrt{3}, \infty\right)$. There are points of inflection at $x = -2\sqrt{3}$, $x = 0$, and $x = 2\sqrt{3}$. $y = 0$ is a horizontal asymptote.

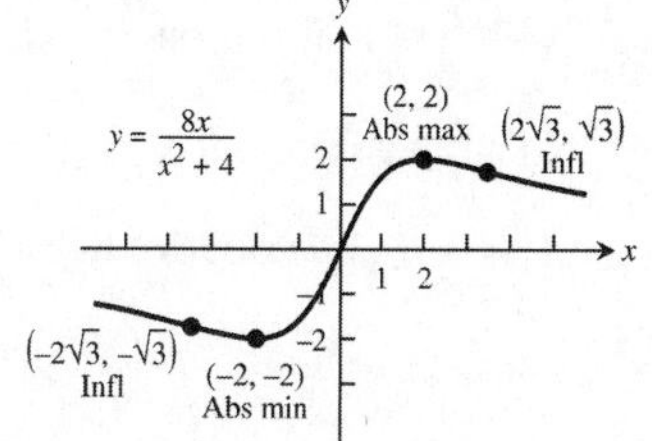

45. When $y = |x^2-1| = \begin{cases} x^2-1, & |x| \ge 1 \\ 1-x^2, & |x| < 1 \end{cases}$, then

$y' = \begin{cases} 2x, & |x| > 1 \\ -2x, & |x| < 1 \end{cases}$ and $y'' = \begin{cases} 2, & |x| > 1 \\ -2, & |x| < 1 \end{cases}$. The curve rises on $(-1, 0)$ and $(1, \infty)$ and falls on $(-\infty, -1)$ and $(0, 1)$. There is a local maximum at $x = 0$ and local minima at $x = \pm 1$. The curve is concave up on $(-\infty, -1)$ and $(1, \infty)$, and concave down on $(-1, 1)$. There are no points of inflection because y is not differentiable at $x = \pm 1$ (so there is no tangent line at those points).

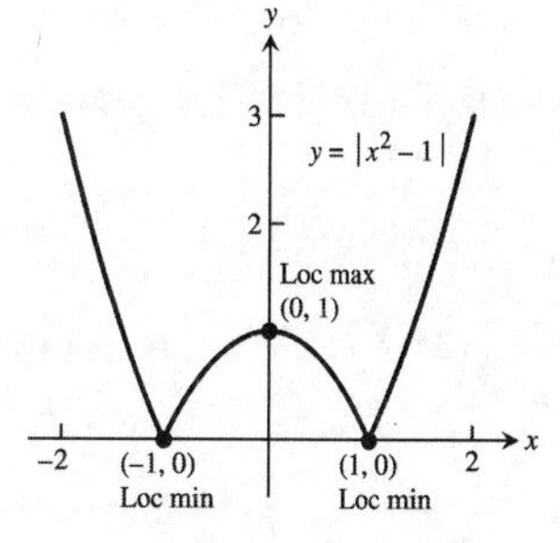

47. When $y = \sqrt{|x|} = \begin{cases} \sqrt{x}, & x \ge 0 \\ \sqrt{-x}, & x < 0 \end{cases}$, then

$y' = \begin{cases} \frac{1}{2\sqrt{x}}, & x > 0 \\ \frac{-1}{2\sqrt{-x}}, & x < 0 \end{cases}$ and $y'' = \begin{cases} \frac{-x^{-3/2}}{4}, & x > 0 \\ \frac{-(-x)^{-3/2}}{4}, & x < 0 \end{cases}$.

Since $\lim_{x \to 0^-} y' = -\infty$ and $\lim_{x \to 0^+} y' = \infty$ there is a cusp at $x = 0$. There is a local minimum at $x = 0$, but no local maximum. The curve is concave down on $(-\infty, 0)$ and $(0, \infty)$. There are no points of inflection.

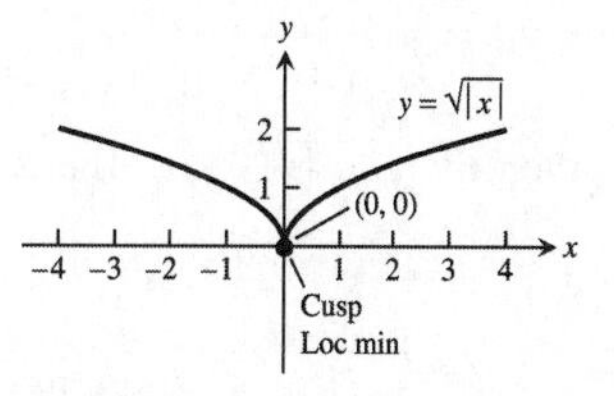

49. $y' = 2 + x - x^2 = (1+x)(2-x)$, $y' = ---\,|\,+++\,|\,---$ (sign changes at -1 and 2)

$\Rightarrow$ rising on $(-1, 2)$, falling on $(-\infty, -1)$ and $(2, \infty)$

$\Rightarrow$ there is a local maximum at $x = 2$ and a local minimum at $x = -1$; $y'' = 1 - 2x$, $y'' = +++\,|\,---$ (sign change at $1/2$)

$\Rightarrow$ concave up on $\left(-\infty, \frac{1}{2}\right)$, concave down on $\left(\frac{1}{2}, \infty\right)$ $\Rightarrow$ a point of inflection at $x = \frac{1}{2}$

Loc max $x = 2$; Infl $x = \frac{1}{2}$; Loc min $x = -1$

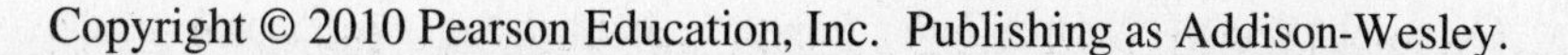

51. $y' = x(x-3)^2$, $y' = \underset{0}{---|}\underset{3}{+++|}+++ \Rightarrow$ rising on $(0, \infty)$, falling on $(-\infty, 0) \Rightarrow$ no local maximum, but there is a local minimum at $x = 0$; $y'' = (x-3)^2 + x(2)(x-3)$ $= 3(x-3)(x-1)$, $y'' = \underset{1}{+++|}\underset{3}{---|}+++ \Rightarrow$ concave up on $(-\infty, 1)$ and $(3, \infty)$, concave down on $(1, 3) \Rightarrow$ points of inflection at $x = 1$ and $x = 3$

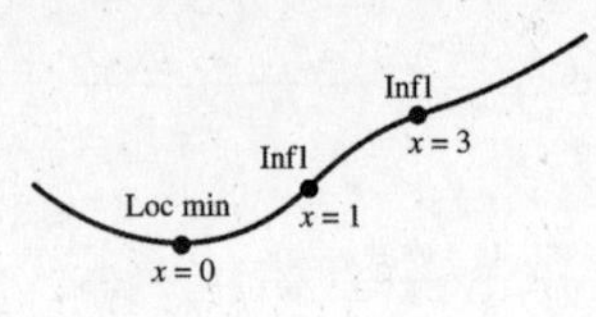

53. $y' = x(x^2 - 12) = x\left(x - 2\sqrt{3}\right)\left(x + 2\sqrt{3}\right)$,
$y' = \underset{-2\sqrt{3}}{---|}\underset{0}{+++|}\underset{2\sqrt{3}}{---|}+++ \Rightarrow$ rising on $\left(-2\sqrt{3}, 0\right)$ and $\left(2\sqrt{3}, \infty\right)$, falling on $\left(-\infty, -2\sqrt{3}\right)$ and $\left(0, 2\sqrt{3}\right) \Rightarrow$ a local maximum at $x = 0$, local minima at $x = \pm 2\sqrt{3}$; $y'' = 1\,(x^2 - 12) + x(2x) = 3(x-2)(x+2)$,
$y'' = \underset{-2}{+++|}\underset{2}{---|}+++ \Rightarrow$ concave up on $(-\infty, -2)$ and $(2, \infty)$, concave down on $(-2, 2) \Rightarrow$ points of inflection at $x = \pm 2$

55. $y' = (8x - 5x^2)(4-x)^2 = x(8-5x)(4-x)^2$,
$y' = \underset{0}{---|}\underset{8/5}{+++|}\underset{4}{---|}--- \Rightarrow$ rising on $\left(0, \frac{8}{5}\right)$, falling on $(-\infty, 0)$ and $\left(\frac{8}{5}, \infty\right) \Rightarrow$ a local maximum at $x = \frac{8}{5}$, a local minimum at $x = 0$;
$y'' = (8 - 10x)(4-x)^2 + (8x - 5x^2)(2)(4-x)(-1) = 4(4-x)(5x^2 - 16x + 8)$,
$y'' = \underset{\frac{8-2\sqrt{6}}{5}}{+++|}\underset{\frac{8+2\sqrt{6}}{5}}{---|}\underset{4}{+++|}--- \Rightarrow$ concave up on $\left(-\infty, \frac{8-2\sqrt{6}}{5}\right)$ and $\left(\frac{8+2\sqrt{6}}{5}, 4\right)$, concave down on $\left(\frac{8-2\sqrt{6}}{5}, \frac{8+2\sqrt{6}}{5}\right)$ and $(4, \infty) \Rightarrow$ points of inflection at $x = \frac{8 \pm 2\sqrt{6}}{5}$ and $x = 4$

57. $y' = \sec^2 x$, $y' = \underset{-\pi/2}{(}\,+++\,\underset{\pi/2}{)} \Rightarrow$ rising on $\left(-\frac{\pi}{2}, \frac{\pi}{2}\right)$, never falling $\Rightarrow$ no local extrema; $y'' = 2(\sec x)(\sec x)(\tan x)$ $= 2(\sec^2 x)(\tan x)$, $y'' = \underset{-\pi/2}{(}\,\underset{0}{---|}+++\,\underset{\pi/2}{)} \Rightarrow$ concave up on $\left(0, \frac{\pi}{2}\right)$, concave down on $\left(-\frac{\pi}{2}, 0\right)$, 0 is a point of inflection.

59. $y' = \cot\frac{\theta}{2}$, $y' = \underset{0}{(}\underset{\pi}{+++|}---\underset{2\pi}{)} \Rightarrow$ rising on $(0, \pi)$, falling on $(\pi, 2\pi) \Rightarrow$ a local maximum at $\theta = \pi$, no local minimum; $y'' = -\frac{1}{2}\csc^2\frac{\theta}{2}$, $y'' = \underset{0}{(}---\underset{2\pi}{)} \Rightarrow$ never concave up, concave down on $(0, 2\pi) \Rightarrow$ no points of inflection

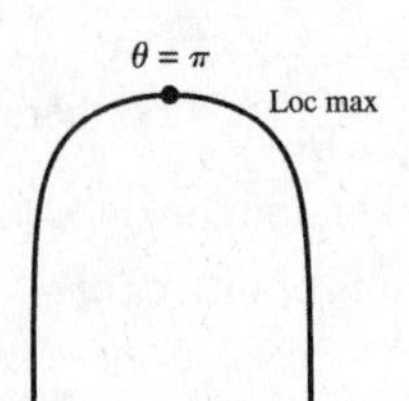

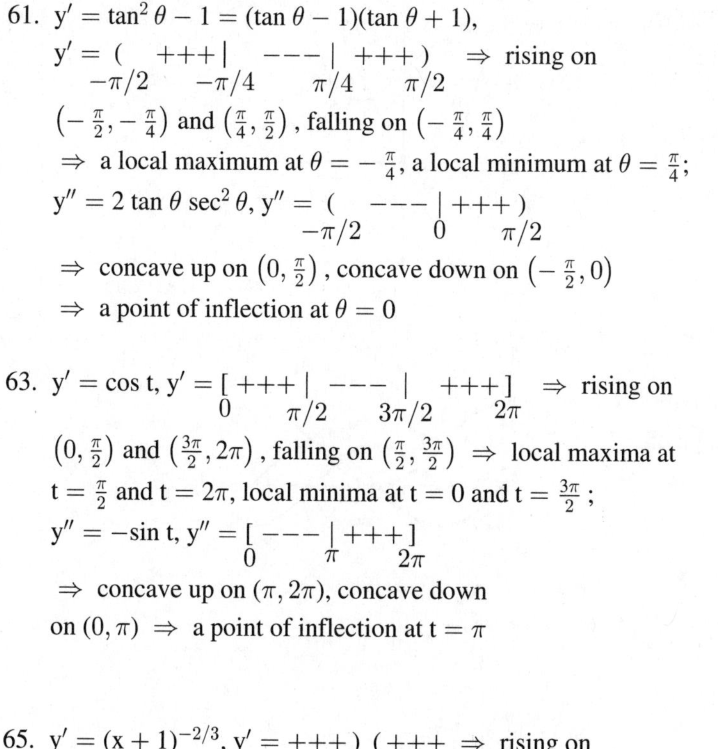

61. $y' = \tan^2\theta - 1 = (\tan\theta - 1)(\tan\theta + 1)$,
$y' = \underset{-\pi/2}{(}\ \ +++\underset{-\pi/4}{|}\ \ ---\underset{\pi/4}{|}\ +++\underset{\pi/2}{)}$ $\Rightarrow$ rising on
$\left(-\frac{\pi}{2}, -\frac{\pi}{4}\right)$ and $\left(\frac{\pi}{4}, \frac{\pi}{2}\right)$, falling on $\left(-\frac{\pi}{4}, \frac{\pi}{4}\right)$
$\Rightarrow$ a local maximum at $\theta = -\frac{\pi}{4}$, a local minimum at $\theta = \frac{\pi}{4}$;
$y'' = 2\tan\theta\sec^2\theta$, $y'' = \underset{-\pi/2}{(}\ \ ---\underset{0}{|}\ +++\underset{\pi/2}{)}$
$\Rightarrow$ concave up on $\left(0, \frac{\pi}{2}\right)$, concave down on $\left(-\frac{\pi}{2}, 0\right)$
$\Rightarrow$ a point of inflection at $\theta = 0$

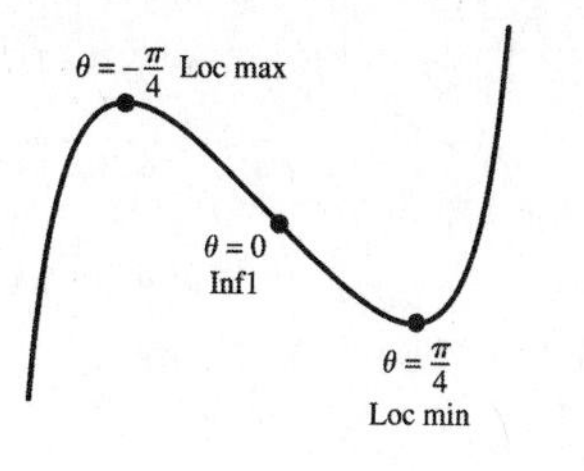

63. $y' = \cos t$, $y' = \underset{0}{[}\ +++\underset{\pi/2}{|}\ ---\underset{3\pi/2}{|}\ +++\underset{2\pi}{]}$ $\Rightarrow$ rising on
$\left(0, \frac{\pi}{2}\right)$ and $\left(\frac{3\pi}{2}, 2\pi\right)$, falling on $\left(\frac{\pi}{2}, \frac{3\pi}{2}\right)$ $\Rightarrow$ local maxima at
$t = \frac{\pi}{2}$ and $t = 2\pi$, local minima at $t = 0$ and $t = \frac{3\pi}{2}$;
$y'' = -\sin t$, $y'' = \underset{0}{[}\ ---\underset{\pi}{|}\ +++\underset{2\pi}{]}$
$\Rightarrow$ concave up on $(\pi, 2\pi)$, concave down
on $(0, \pi)$ $\Rightarrow$ a point of inflection at $t = \pi$

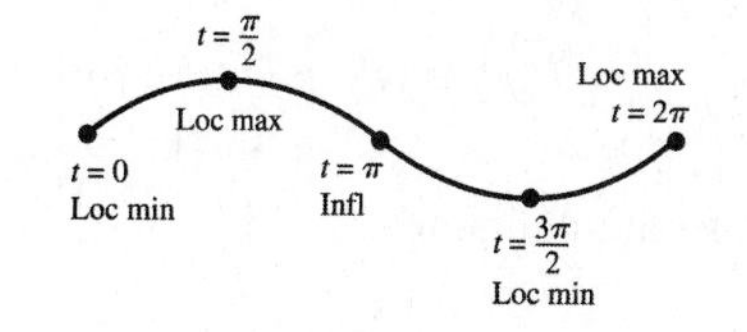

65. $y' = (x+1)^{-2/3}$, $y' = +++\underset{-1}{)\,(}+++$ $\Rightarrow$ rising on
$(-\infty, \infty)$, never falling $\Rightarrow$ no local extrema;
$y'' = -\frac{2}{3}(x+1)^{-5/3}$, $y'' = +++\underset{-1}{)\,(}---$
$\Rightarrow$ concave up on $(-\infty, -1)$, concave down on $(-1, \infty)$
$\Rightarrow$ a point of inflection and vertical tangent at $x = -1$

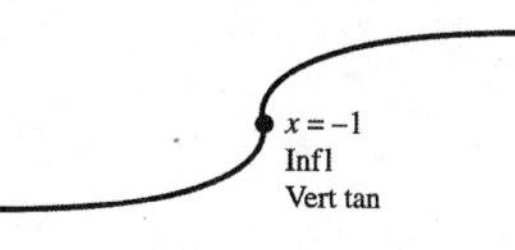

67. $y' = x^{-2/3}(x-1)$, $y' = ---\underset{0}{)\,(}---\underset{1}{|}+++$ $\Rightarrow$ rising on
$(1, \infty)$, falling on $(-\infty, 1)$ $\Rightarrow$ no local maximum, but a
local minimum at $x = 1$; $y'' = \frac{1}{3}x^{-2/3} + \frac{2}{3}x^{-5/3}$
$= \frac{1}{3}x^{-5/3}(x+2)$, $y'' = +++\underset{-2}{|}---\underset{0}{)\,(}+++$
$\Rightarrow$ concave up on $(-\infty, -2)$ and $(0, \infty)$, concave down on
$(-2, 0)$ $\Rightarrow$ points of inflection at $x = -2$ and $x = 0$, and a
vertical tangent at $x = 0$

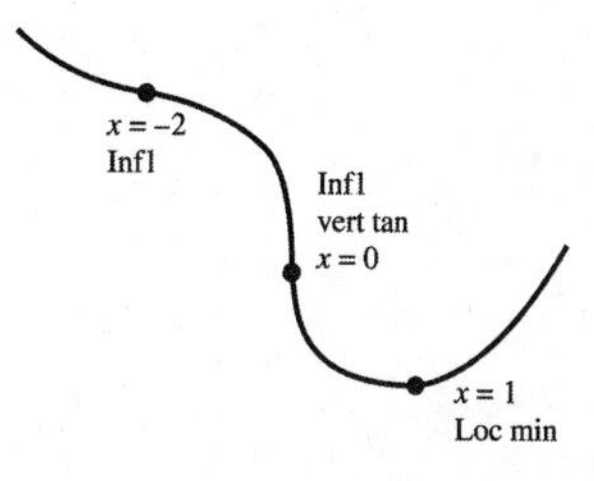

69. $y' = \begin{cases} -2x, & x \le 0 \\ 2x, & x > 0 \end{cases}$, $y' = +++\underset{0}{|}+++$ $\Rightarrow$ rising on
$(-\infty, \infty) \Rightarrow$ no local extrema; $y'' = \begin{cases} -2, & x < 0 \\ 2, & x > 0 \end{cases}$,
$y'' = ---\underset{0}{)\,(}+++$ $\Rightarrow$ concave up on $(0, \infty)$, concave
down on $(-\infty, 0)$ $\Rightarrow$ a point of inflection at $x = 0$

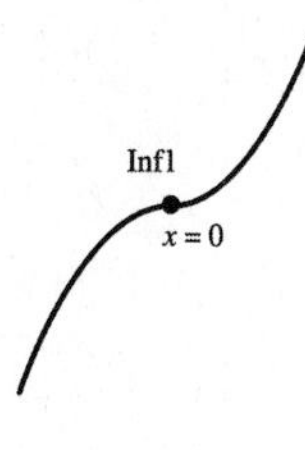

71. The graph of $y = f''(x) \Rightarrow$ the graph of $y = f(x)$ is concave up on $(0, \infty)$, concave down on $(-\infty, 0) \Rightarrow$ a point of inflection at $x = 0$; the graph of $y = f'(x)$ $\Rightarrow y' = +++ \mid --- \mid +++ \Rightarrow$ the graph $y = f(x)$ has both a local maximum and a local minimum

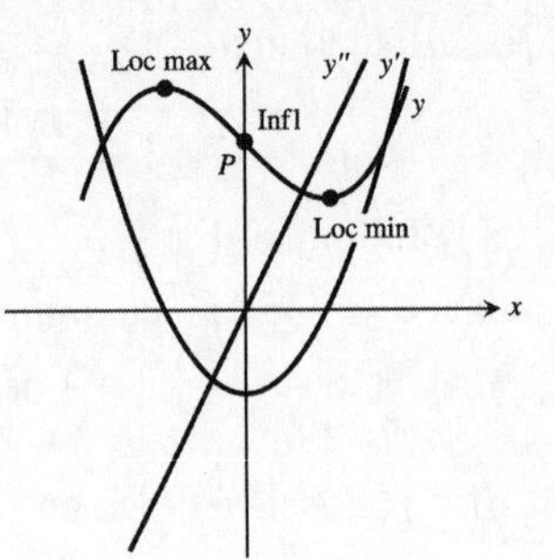

73. The graph of $y = f''(x) \Rightarrow y'' = --- \mid +++ \mid ---$ $\Rightarrow$ the graph of $y = f(x)$ has two points of inflection, the graph of $y = f'(x) \Rightarrow y' = --- \mid +++ \Rightarrow$ the graph of $y = f(x)$ has a local minimum

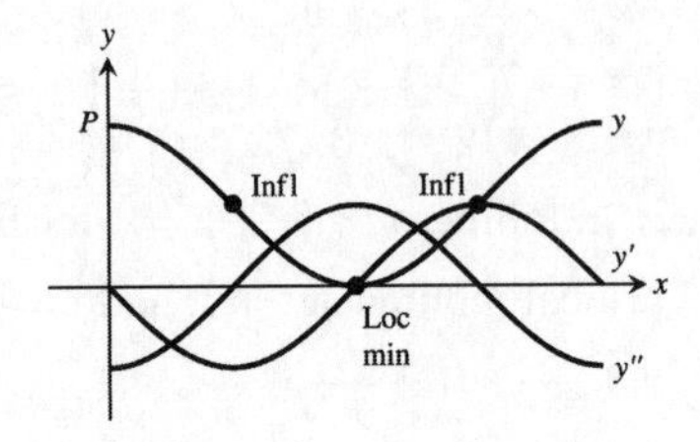

75. $y = \frac{2x^2+x-1}{x^2-1}$

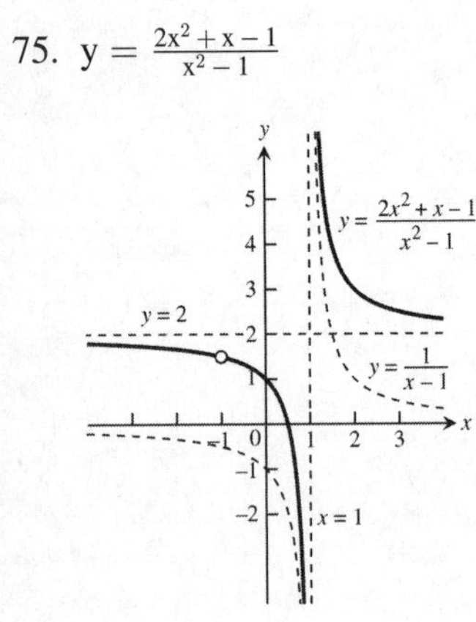

77. $y = \frac{x^4+1}{x^2} = x^2 + \frac{1}{x^2}$

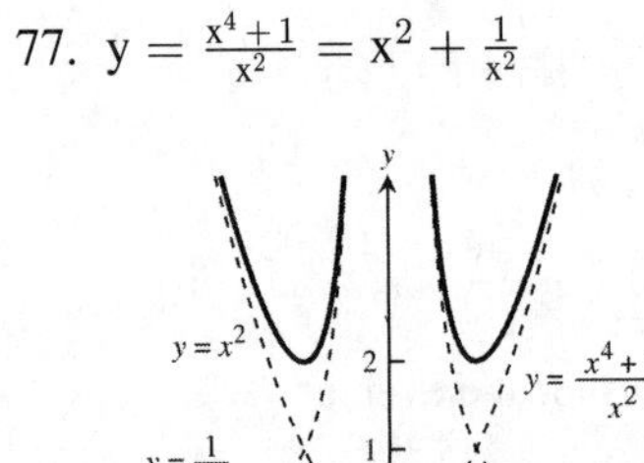
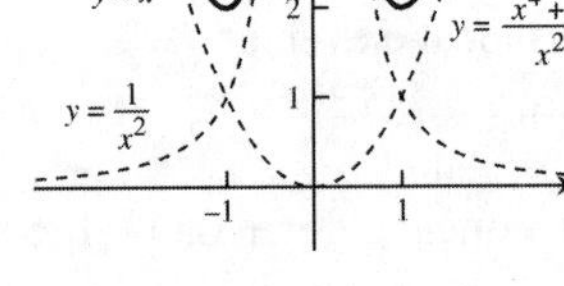

79. $y = \frac{1}{x^2-1}$

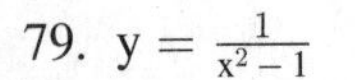

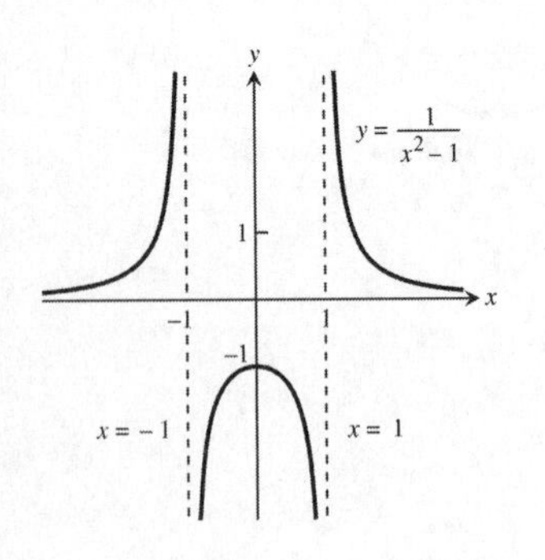

81. $y = -\frac{x^2-2}{x^2-1} = -1 + \frac{1}{x^2-1}$

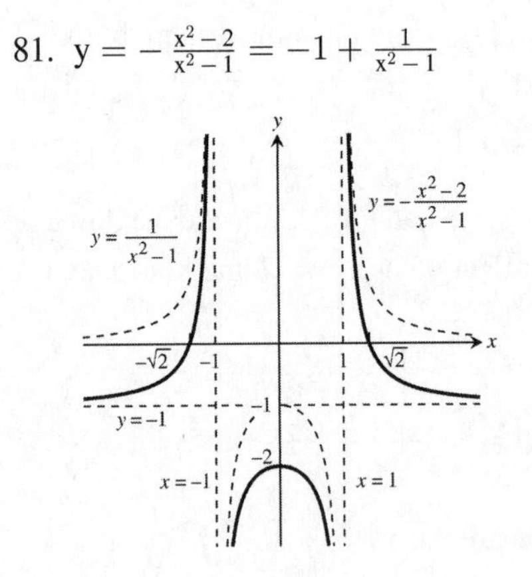

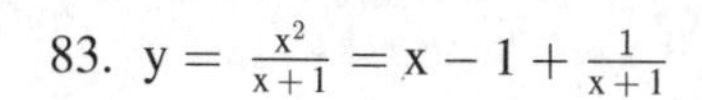

83. $y = \frac{x^2}{x+1} = x - 1 + \frac{1}{x+1}$

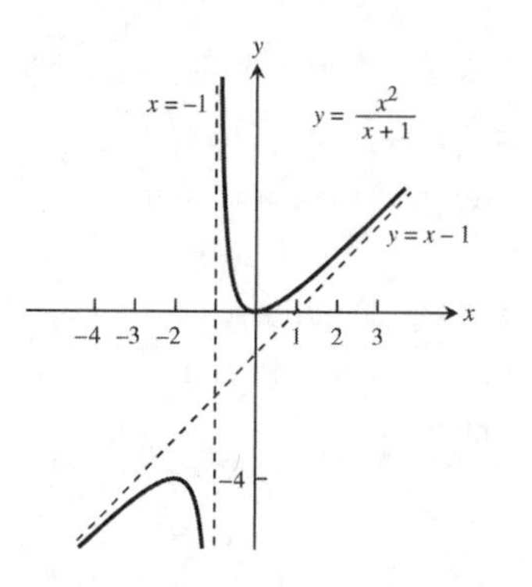

85. $y = \frac{x^2-x+1}{x-1} = x + \frac{1}{x-1}$

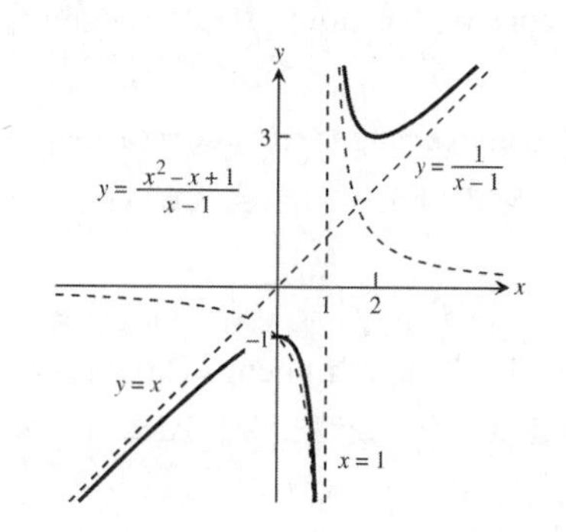

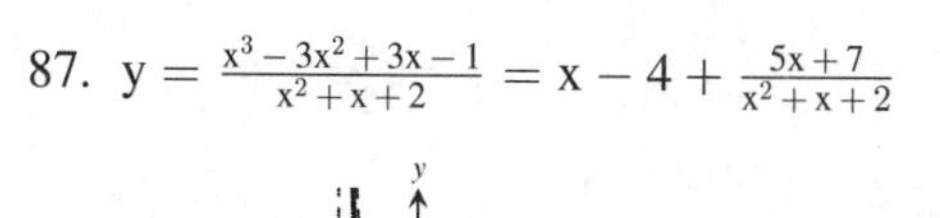

87. $y = \frac{x^3-3x^2+3x-1}{x^2+x+2} = x - 4 + \frac{5x+7}{x^2+x+2}$

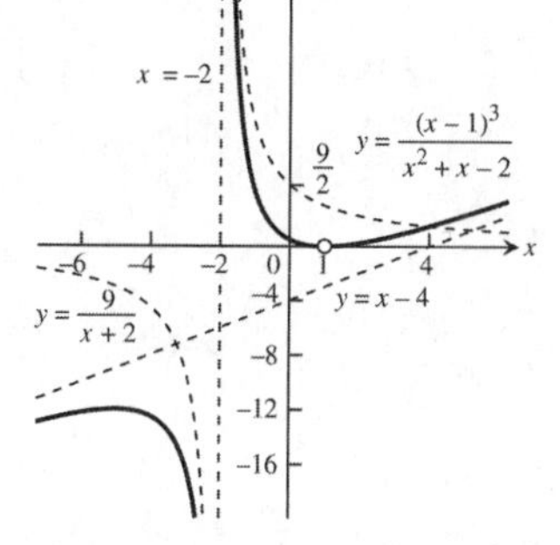

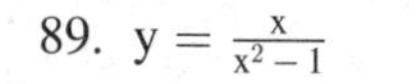

89. $y = \frac{x}{x^2-1}$

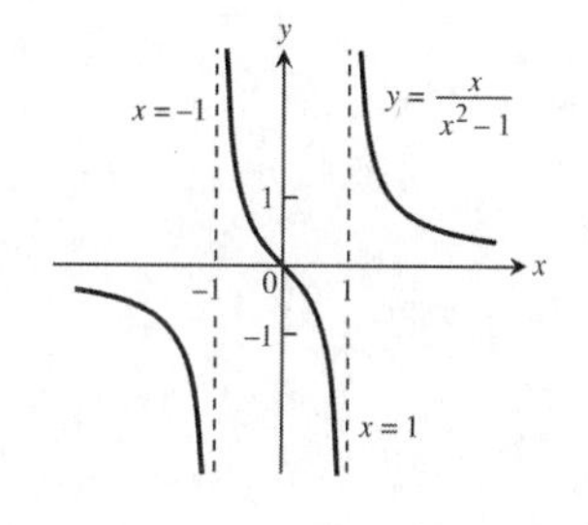

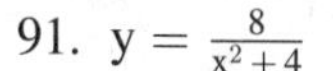

91. $y = \frac{8}{x^2+4}$

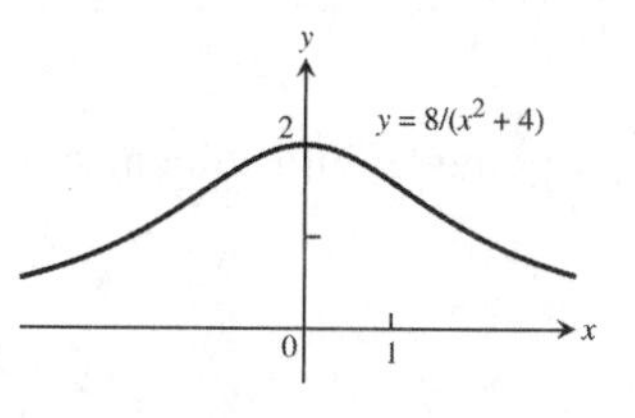

93.

Point	y'	y''
P	−	+
Q	+	0
R	+	−
S	0	−
T	−	−

95.

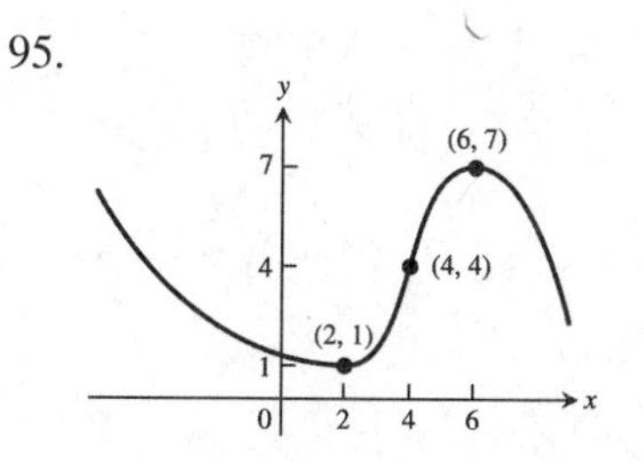

97. Graphs printed in color can shift during a press run, so your values may differ somewhat from those given here.

(a) The body is moving away from the origin when |displacement| is increasing as t increases, $0 < t < 2$ and $6 < t < 9.5$; the body is moving toward the origin when |displacement| is decreasing as t increases, $2 < t < 6$ and $9.5 < t < 15$

(b) The velocity will be zero when the slope of the tangent line for $y = s(t)$ is horizontal. The velocity is zero when t is approximately 2, 6, or 9.5 sec.

(c) The acceleration will be zero at those values of t where the curve $y = s(t)$ has points of inflection. The acceleration is zero when t is approximately 4, 7.5, or 12.5 sec.

(d) The acceleration is positive when the concavity is up, $4 < t < 7.5$ and $12.5 < t < 15$; the acceleration is negative when the concavity is down, $0 < t < 4$ and $7.5 < t < 12.5$

99. The marginal cost is $\frac{dc}{dx}$ which changes from decreasing to increasing when its derivative $\frac{d^2c}{dx^2}$ is zero. This is a point of inflection of the cost curve and occurs when the production level x is approximately 60 thousand units.

101. When $y' = (x-1)^2(x-2)$, then $y'' = 2(x-1)(x-2) + (x-1)^2$. The curve falls on $(-\infty, 2)$ and rises on $(2, \infty)$. At $x = 2$ there is a local minimum. There is no local maximum. The curve is concave upward on $(-\infty, 1)$ and $\left(\frac{5}{3}, \infty\right)$, and concave downward on $\left(1, \frac{5}{3}\right)$. At $x = 1$ or $x = \frac{5}{3}$ there are inflection points.

103. The graph must be concave down for $x > 0$ because $f''(x) = -\frac{1}{x^2} < 0$.

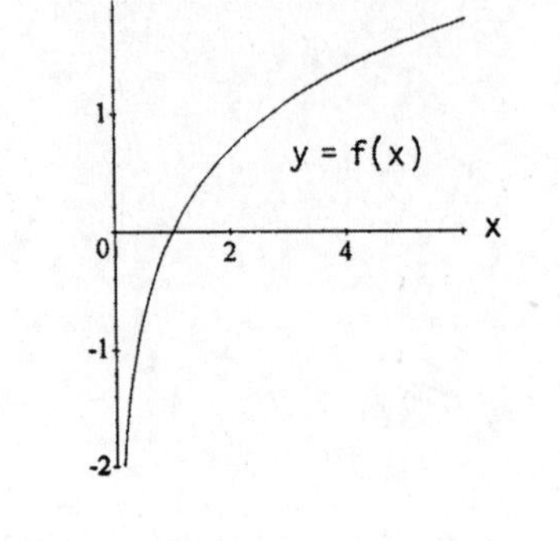

105. The curve will have a point of inflection at $x = 1$ if 1 is a solution of $y'' = 0$; $y = x^3 + bx^2 + cx + d$ $\Rightarrow y' = 3x^2 + 2bx + c \Rightarrow y'' = 6x + 2b$ and $6(1) + 2b = 0 \Rightarrow b = -3$.

107. A quadratic curve never has an inflection point. If $y = ax^2 + bx + c$ where $a \neq 0$, then $y' = 2ax + b$ and $y'' = 2a$. Since 2a is a constant, it is not possible for y'' to change signs.

109. $y'' = (x+1)(x-2)$, when $y'' = 0 \Rightarrow x = -1$ or $x = 2$; $y'' = +++\underset{-1}{|}---\underset{2}{|}+++ \Rightarrow$ points of inflection at $x = -1$ and $x = 2$

111. $y = ax^3 + bx^2 + cx \Rightarrow y' = 3ax^2 + 2bx + c$ and $y'' = 6ax + 2b$; local maximum at $x = 3$ $\Rightarrow 3a(3)^2 + 2b(3) + c = 0 \Rightarrow 27a + 6b + c = 0$; local mimimum at $x = -1 \Rightarrow 3a(-1)^2 + 2b(-1) + c = 0$ $\Rightarrow 3a - 2b + c = 0$; point of inflection at $(1, 11) \Rightarrow a(1)^3 + b(1)^2 + c(1) = 11 \Rightarrow a + b + c = 11$ and $6a(1) + 2b = 0 \Rightarrow 6a + 2b = 0$. Solving $27a + 6b + c = 0$, $3a - 2b + c = 0$, $a + b + c = 11$, and $6a + 2b = 0$ $\Rightarrow a = -1, b = 3$, and $c = 9 \Rightarrow y = -x^3 + 3x^2 + 9x$

113. If $y = x^5 - 5x^4 - 240$, then $y' = 5x^3(x-4)$ and $y'' = 20x^2(x-3)$. The zeros of y' are extrema, and there is a point of inflection at $x = 3$.

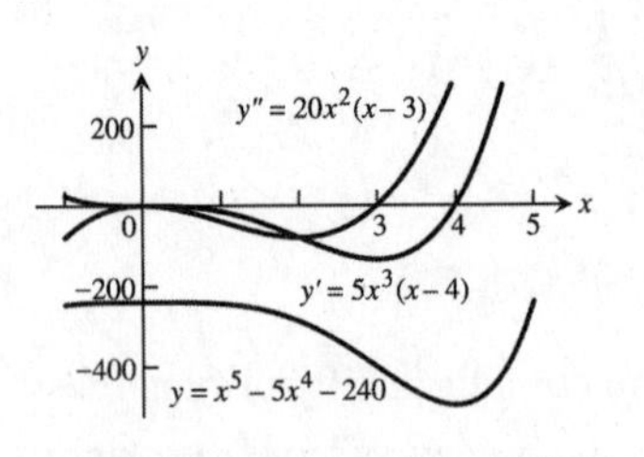

115. If $y = \frac{4}{5}x^5 + 16x^2 - 25$, then $y' = 4x(x^3 + 8)$ and $y'' = 16(x^3 + 2)$. The zeros of y' and y'' are extrema and points of inflection, respectively.

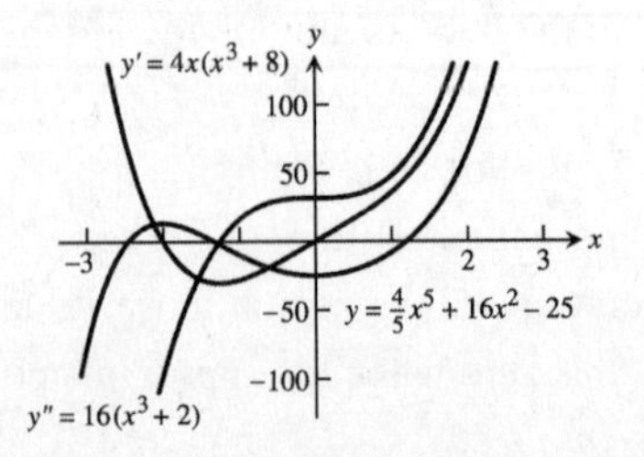

117. The graph of f falls where $f' < 0$, rises where $f' > 0$, and has horizontal tangents where $f' = 0$. It has local minima at points where f' changes from negative to positive and local maxima where f' changes from positive to negative. The graph of f is concave down where $f'' < 0$ and concave up where $f'' > 0$. It has an inflection point each time f'' changes sign, provided a tangent line exists there.

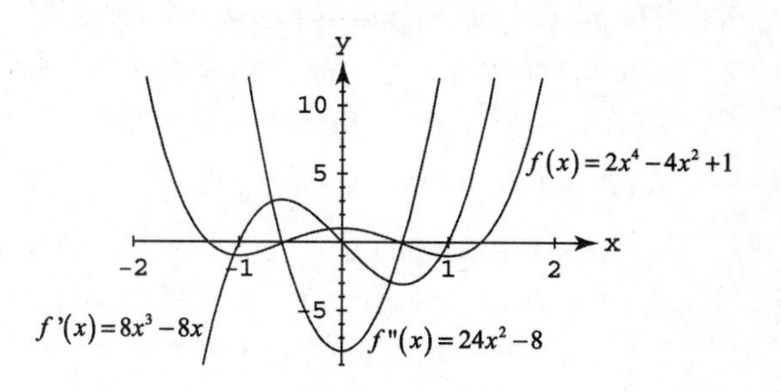

4.5 APPLIED OPTIMIZATION

1. Let ℓ and w represent the length and width of the rectangle, respectively. With an area of 16 in.^{2}, we have that $(\ell)(w) = 16 \Rightarrow w = 16\ell^{-1} \Rightarrow$ the perimeter is $P = 2\ell + 2w = 2\ell + 32\ell^{-1}$ and $P'(\ell) = 2 - \frac{32}{\ell^2} = \frac{2(\ell^2-16)}{\ell^2}$. Solving $P'(\ell) = 0 \Rightarrow \frac{2(\ell+4)(\ell-4)}{\ell^2} = 0 \Rightarrow \ell = -4, 4$. Since $\ell > 0$ for the length of a rectangle, ℓ must be 4 and $w = 4 \Rightarrow$ the perimeter is 16 in., a minimum since $P''(\ell) = \frac{16}{\ell^3} > 0$.

3. (a) The line containing point P also contains the points (0, 1) and (1, 0) $\Rightarrow$ the line containing P is $y = 1 - x$ $\Rightarrow$ a general point on that line is $(x, 1 - x)$.
   (b) The area $A(x) = 2x(1 - x)$, where $0 \le x \le 1$.
   (c) When $A(x) = 2x - 2x^2$, then $A'(x) = 0 \Rightarrow 2 - 4x = 0 \Rightarrow x = \frac{1}{2}$. Since $A(0) = 0$ and $A(1) = 0$, we conclude that $A\left(\frac{1}{2}\right) = \frac{1}{2}$ sq units is the largest area. The dimensions are 1 unit by $\frac{1}{2}$ unit.

5. The volume of the box is $V(x) = x(15 - 2x)(8 - 2x)$ $= 120x - 46x^2 + 4x^3$, where $0 \le x \le 4$. Solving $V'(x) = 0$ $\Rightarrow 120 - 92x + 12x^2 = 4(6 - x)(5 - 3x) = 0 \Rightarrow x = \frac{5}{3}$ or 6, but 6 is not in the domain. Since $V(0) = V(4) = 0$, $V\left(\frac{5}{3}\right) = \frac{2450}{27} \approx 91$ in^3 must be the maximum volume of the box with dimensions $\frac{14}{3} \times \frac{35}{3} \times \frac{5}{3}$ inches.

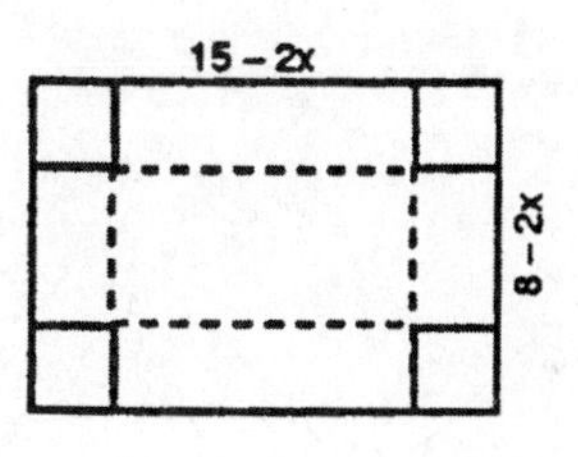

7. The area is $A(x) = x(800 - 2x)$, where $0 \le x \le 400$. Solving $A'(x) = 800 - 4x = 0 \Rightarrow x = 200$. With $A(0) = A(400) = 0$, the maximum area is $A(200) = 80{,}000$ m^2. The dimensions are 200 m by 400 m.

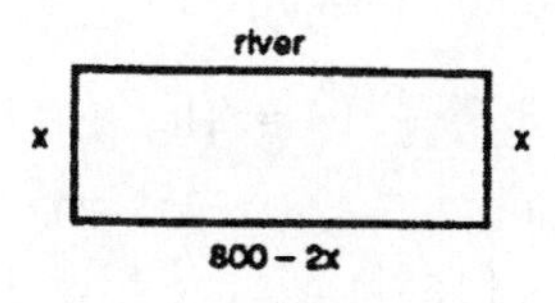

9. (a) We minimize the weight = tS where S is the surface area, and t is the thickness of the steel walls of the tank. The surface area is $S = x^2 + 4xy$ where x is the length of a side of the square base of the tank, and y is its depth. The volume of the tank must be 500ft^3 $\Rightarrow y = \frac{500}{x^2}$. Therefore, the weight of the tank is $w(x) = t\left(x^2 + \frac{2000}{x}\right)$. Treating the thickness as a constant gives $w'(x) = t\left(2x - \frac{2000}{x^2}\right)$. The critical value is at $x = 10$. Since $w''(10) = t\left(2 + \frac{4000}{10^3}\right) > 0$, there is a minimum at $x = 10$. Therefore, the optimum dimensions of the tank are 10 ft on the base edges and 5 ft deep.
   (b) Minimizing the surface area of the tank minimizes its weight for a given wall thickness. The thickness of the steel walls would likely be determined by other considerations such as structural requirements.

11. The area of the printing is $(y - 4)(x - 8) = 50$. Consequently, $y = \left(\frac{50}{x-8}\right) + 4$. The area of the paper is $A(x) = x\left(\frac{50}{x-8} + 4\right)$, where $8 < x$. Then $A'(x) = \left(\frac{50}{x-8} + 4\right) - x\left(\frac{50}{(x-8)^2}\right) = \frac{4(x-8)^2 - 400}{(x-8)^2} = 0$ $\Rightarrow$ the critical points are -2 and 18, but -2 is not in the domain. Thus $A''(18) > 0 \Rightarrow$ at $x = 18$ we have a minimum Therefore the dimensions 18 by 9 inches minimize the amount minimize the amount of paper.

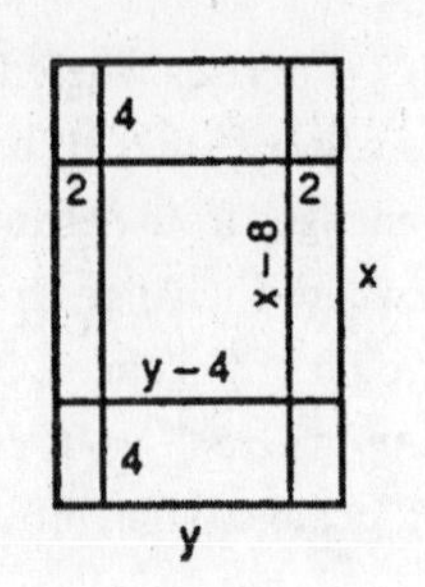

13. The area of the triangle is $A(\theta) = \frac{ab \sin\theta}{2}$, where $0 < \theta < \pi$. Solving $A'(\theta) = 0 \Rightarrow \frac{ab\cos\theta}{2} = 0 \Rightarrow \theta = \frac{\pi}{2}$. Since $A''(\theta) = -\frac{ab\sin\theta}{2} \Rightarrow A''\left(\frac{\pi}{2}\right) < 0$, there is a maximum at $\theta = \frac{\pi}{2}$.

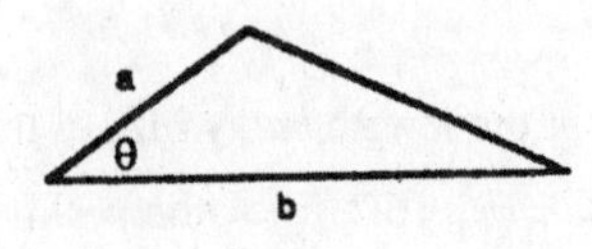

15. With a volume of 1000 cm and $V = \pi r^2 h$, then $h = \frac{1000}{\pi r^2}$. The amount of aluminum used per can is $A = 8r^2 + 2\pi rh = 8r^2 + \frac{2000}{r}$. Then $A'(r) = 16r - \frac{2000}{r^2} = 0 \Rightarrow \frac{8r^3 - 1000}{r^2} = 0 \Rightarrow$ the critical points are 0 and 5, but $r = 0$ results in no can. Since $A''(r) = 16 + \frac{1000}{r^3} > 0$ we have a minimum at $r = 5 \Rightarrow h = \frac{40}{\pi}$ and $h:r = 8:\pi$.

17. (a) The "sides" of the suitcase will measure $24 - 2x$ in. by $18 - 2x$ in. and will be $2x$ in. apart, so the volume formula is $V(x) = 2x(24 - 2x)(18 - 2x) = 8x^3 - 168x^2 + 862x$.

(b) We require $x > 0$, $2x < 18$, and $2x < 12$. Combining these requirements, the domain is the interval $(0, 9)$.

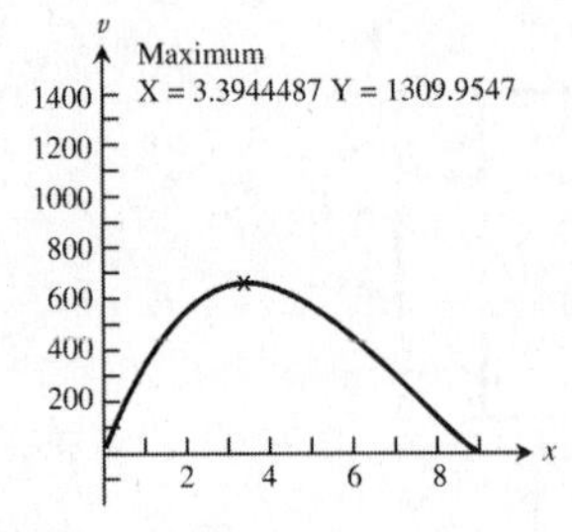

(c) The maximum volume is approximately 1309.95 in.^{3} when $x \approx 3.39$ in.

(d) $V'(x) = 24x^2 - 336x + 864 = 24(x^2 - 14x + 36)$. The critical point is at $x = \frac{14 \pm \sqrt{(-14)^2 - 4(1)(36)}}{2(1)} = \frac{14 \pm \sqrt{52}}{2}$ $= 7 \pm \sqrt{13}$, that is, $x \approx 3.39$ or $x \approx 10.61$. We discard the larger value because it is not in the domain. Since $V''(x) = 24(2x - 14)$ which is negative when $x \approx 3.39$, the critical point corresponds to the maximum volume. The maximum value occurs at $x = 7 - \sqrt{13} \approx 3.39$, which confirms the results in (c).

(e) $8x^3 - 168x^2 + 862x = 1120 \Rightarrow 8(x^3 - 21x^2 + 108x - 140) = 0 \Rightarrow 8(x-2)(x-5)(x-14) = 0$. Since 14 is not in the fomain, the possible values of x are $x = 2$ in. or $x = 5$ in.

(f) The dimensions of the resulting box are $2x$ in., $(24 - 2x)$ in., and $(18 - 2x)$. Each of these measurements must be positive, so that gives the domain of $(0, 9)$.

19. Let the radius of the cylinder be r cm, $0 < r < 10$. Then the height is $2\sqrt{100 - r^2}$ and the volume is $V(r) = 2\pi r^2\sqrt{100 - r^2}$ cm^3. Then, $V'(r) = 2\pi r^2\left(\frac{1}{\sqrt{100 - r^2}}\right)(-2r) + \left(2\pi\sqrt{100 - r^2}\right)(2r)$ $= \frac{-2\pi r^3 + 4\pi r(100 - r^2)}{\sqrt{100 - r^2}} = \frac{2\pi r(200 - 3r^2)}{\sqrt{100 - r^2}}$. The critical point for $0 < r < 10$ occurs at $r = \sqrt{\frac{200}{3}} = 10\sqrt{\frac{2}{3}}$. Since $V'(r) > 0$ for $0 < r < 10\sqrt{\frac{2}{3}}$ and $V'(r) < 0$ for $10\sqrt{\frac{2}{3}} < r < 10$, the critical point corresponds to the maximum volume. The dimensions are $r = 10\sqrt{\frac{2}{3}} \approx 8.16$ cm and $h = \frac{20}{\sqrt{3}} \approx 11.55$ cm, and the volume is $\frac{4000\pi}{3\sqrt{3}} \approx 2418.40$ cm^3.

21. (a) From the diagram we have $3h + 2w = 108$ and $V = h^2 w \Rightarrow V(h) = h^2\left(54 - \frac{3}{2}h\right) = 54h^2 - \frac{3}{2}h^3$. Then $V'(h) = 108h - \frac{9}{2}h^2 = \frac{9}{2}h(24 - h) = 0$ $\Rightarrow h = 0$ or $h = 24$, but $h = 0$ results in no box. Since $V''(h) = 108 - 9h < 0$ at $h = 24$, we have a maximum volume at $h = 24$ and $w = 54 - \frac{3}{2}h = 18$.

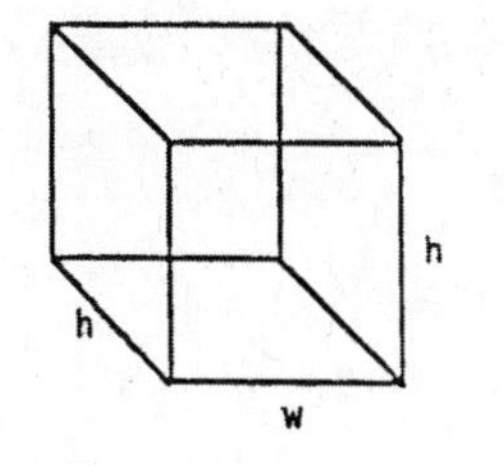

(b)

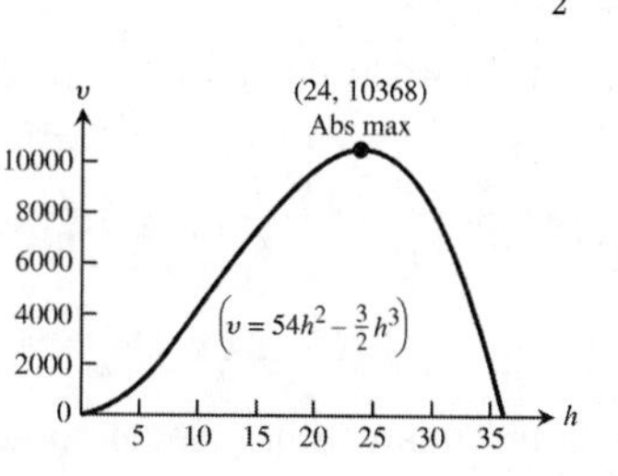

23. The fixed volume is $V = \pi r^2 h + \frac{2}{3}\pi r^3 \Rightarrow h = \frac{V}{\pi r^2} - \frac{2r}{3}$, where h is the height of the cylinder and r is the radius of the hemisphere. To minimize the cost we must minimize surface area of the cylinder added to twice the surface area of the hemisphere. Thus, we minimize $C = 2\pi rh + 4\pi r^2 = 2\pi r\left(\frac{V}{\pi r^2} - \frac{2r}{3}\right) + 4\pi r^2 = \frac{2V}{r} + \frac{8}{3}\pi r^2$. Then $\frac{dC}{dr} = -\frac{2V}{r^2} + \frac{16}{3}\pi r = 0 \Rightarrow V = \frac{8}{3}\pi r^3 \Rightarrow r = \left(\frac{3V}{8\pi}\right)^{1/3}$. From the volume equation, $h = \frac{V}{\pi r^2} - \frac{2r}{3}$ $= \frac{4V^{1/3}}{\pi^{1/3}\cdot 3^{2/3}} - \frac{2\cdot 3^{1/3}\cdot V^{1/3}}{3\cdot 2\cdot \pi^{1/3}} = \frac{3^{1/3}\cdot 2\cdot 4\cdot V^{1/3} - 2\cdot 3^{1/3}\cdot V^{1/3}}{3\cdot 2\cdot \pi^{1/3}} = \left(\frac{3V}{\pi}\right)^{1/3}$. Since $\frac{d^2C}{dr^2} = \frac{4V}{r^3} + \frac{16}{3}\pi > 0$, these dimensions do minimize the cost.

25. (a) From the diagram we have: $\overline{AP} = x$, $\overline{RA} = \sqrt{L - x^2}$, $\overline{PB} = 8.5 - x$, $\overline{CH} = \overline{DR} = 11 - \overline{RA} = 11 - \sqrt{L - x^2}$, $\overline{QB} = \sqrt{x^2 - (8.5 - x)^2}$, $\overline{HQ} = 11 - \overline{CH} - \overline{QB}$ $= 11 - \left[11 - \sqrt{L - x^2} + \sqrt{x^2 - (8.5 - x)^2}\right]$ $= \sqrt{L - x^2} - \sqrt{x^2 - (8.5 - x)^2}$, $\overline{RQ}^2 = \overline{RH}^2 + \overline{HQ}^2$ $= (8.5)^2 + \left(\sqrt{L - x^2} - \sqrt{x^2 - (8.5 - x)^2}\right)^2$. It follows that $\overline{RP}^2 = \overline{PQ}^2 + \overline{RQ}^2 \Rightarrow L^2 = x^2 + \left(\sqrt{L^2 - x^2} - \sqrt{x^2 - (x - 8.5)^2}\right)^2 + (8.5)^2$

$\Rightarrow L^2 = x^2 + L^2 - x^2 - 2\sqrt{L^2 - x^2}\,\sqrt{17x - (8.5)^2} + 17x - (8.5)^2 + (8.5)^2$

$\Rightarrow 17^2 x^2 = 4\left(L^2 - x^2\right)\left(17x - (8.5)^2\right) \Rightarrow L^2 = x^2 + \frac{17^2 x^2}{4\left[17x - (8.5)^2\right]} = \frac{17x^3}{17x - (8.5)^2} = \frac{17x^3}{17x - \left(\frac{17}{2}\right)^2}$

$= \frac{4x^3}{4x - 17} = \frac{2x^3}{2x - 8.5}$.

(b) If $f(x) = \frac{4x^3}{4x-17}$ is minimized, then L^2 is minimized. Now $f'(x) = \frac{4x^2(8x - 51)}{(4x - 17)^2} \Rightarrow f'(x) < 0$ when $x < \frac{51}{8}$ and $f'(x) > 0$ when $x > \frac{51}{8}$. Thus L^2 is minimized when $x = \frac{51}{8}$.

(c) When $x = \frac{51}{8}$, then $L \approx 11.0$ in.

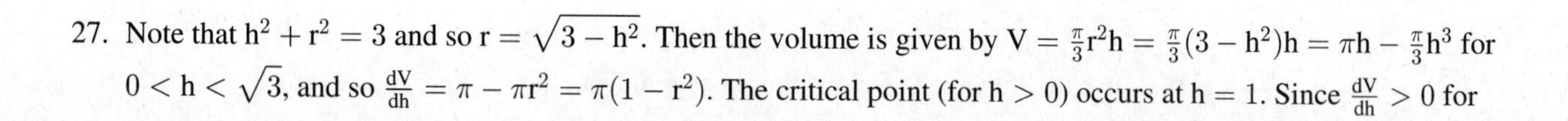

27. Note that $h^2 + r^2 = 3$ and so $r = \sqrt{3 - h^2}$. Then the volume is given by $V = \frac{\pi}{3}r^2 h = \frac{\pi}{3}(3 - h^2)h = \pi h - \frac{\pi}{3}h^3$ for $0 < h < \sqrt{3}$, and so $\frac{dV}{dh} = \pi - \pi r^2 = \pi(1 - r^2)$. The critical point (for $h > 0$) occurs at $h = 1$. Since $\frac{dV}{dh} > 0$ for

$0 < h < 1$, and $\frac{dV}{dh} < 0$ for $1 < h < \sqrt{3}$, the critical point corresponds to the maximum volume. The cone of greatest volume has radius $\sqrt{2}$ m, height 1m, and volume $\frac{2\pi}{3}$ m^3.

29. Let $S(x) = x + \frac{1}{x}, x > 0 \Rightarrow S'(x) = 1 - \frac{1}{x^2} = \frac{x^2-1}{x^2}$. $S'(x) = 0 \Rightarrow \frac{x^2-1}{x^2} = 0 \Rightarrow x^2 - 1 = 0 \Rightarrow x = \pm 1$. Since $x > 0$, we only consider $x = 1$. $S''(x) = \frac{2}{x^3} \Rightarrow S''(1) = \frac{2}{1^3} > 0 \Rightarrow$ local minimum when $x = 1$

31. The length of the wire b = perimeter of the triangle + circumference of the circle. Let x = length of a side of the equilateral triangle $\Rightarrow P = 3x$, and let r = radius of the circle $\Rightarrow C = 2\pi r$. Thus $b = 3x + 2\pi r \Rightarrow r = \frac{b-3x}{2\pi}$. The area of the circle is πr^2 and the area of an equilateral triangle whose sides are x is $\frac{1}{2}(x)\left(\frac{\sqrt{3}}{2}x\right) = \frac{\sqrt{3}}{4}x^2$. Thus, the total area is given by $A = \frac{\sqrt{3}}{4}x^2 + \pi r^2 = \frac{\sqrt{3}}{4}x^2 + \pi\left(\frac{b-3x}{2\pi}\right)^2 = \frac{\sqrt{3}}{4}x^2 + \frac{(b-3x)^2}{4\pi} \Rightarrow A' = \frac{\sqrt{3}}{2}x - \frac{3}{2\pi}(b-3x) = \frac{\sqrt{3}}{2}x - \frac{3b}{2\pi} + \frac{9}{2\pi}x$
$A' = 0 \Rightarrow \frac{\sqrt{3}}{2}x - \frac{3b}{2\pi} + \frac{9}{2\pi}x = 0 \Rightarrow x = \frac{3b}{\sqrt{3}\pi+9}$. $A'' = \frac{\sqrt{3}}{2} + \frac{9}{2\pi} > 0 \Rightarrow$ local minimum at the critical point.
$P = 3\left(\frac{3b}{\sqrt{3}\pi+9}\right) = \frac{9b}{\sqrt{3}\pi+9}$ m is the length of the trianglular segment and $C = 2\pi\left(\frac{b-3x}{2\pi}\right) = b - 3x$
$= b - \frac{9b}{\sqrt{3}\pi+9} = \frac{\sqrt{3}\pi b}{\sqrt{3}\pi+9}$ m is the length of the circular segment.

33. Let $(x, y) = \left(x, \frac{4}{3}x\right)$ be the coordinates of the corner that intersects the line. Then base $= 3 - x$ and height $= y = \frac{4}{3}x$, thus the area of therectangle is given by $A = (3-x)\left(\frac{4}{3}x\right) = 4x - \frac{4}{3}x^2, 0 \le x \le 3$. $A' = 4 - \frac{8}{3}x$, $A' = 0 \Rightarrow x = \frac{3}{2}$. $A'' = -\frac{4}{3}$ $\Rightarrow A''\left(\frac{3}{2}\right) < 0 \Rightarrow$ local maximum at the critical point. The base $= 3 - \frac{3}{2} = \frac{3}{2}$ and the height $= \frac{4}{3}\left(\frac{3}{2}\right) = 2$.

35. (a) $f(x) = x^2 + \frac{a}{x} \Rightarrow f'(x) = x^{-2}(2x^3 - a)$, so that $f'(x) = 0$ when $x = 2$ implies $a = 16$
(b) $f(x) = x^2 + \frac{a}{x} \Rightarrow f''(x) = 2x^{-3}(x^3 + a)$, so that $f''(x) = 0$ when $x = 1$ implies $a = -1$

37. (a) $s(t) = -16t^2 + 96t + 112 \Rightarrow v(t) = s'(t) = -32t + 96$. At $t = 0$, the velocity is $v(0) = 96$ ft/sec.
(b) The maximum height ocurs when $v(t) = 0$, when $t = 3$. The maximum height is $s(3) = 256$ ft and it occurs at $t = 3$ sec.
(c) Note that $s(t) = -16t^2 + 96t + 112 = -16(t+1)(t-7)$, so $s = 0$ at $t = -1$ or $t = 7$. Choosing the positive value of t, the velocity when $s = 0$ is $v(7) = -128$ ft/sec.

39. $\frac{8}{x} = \frac{h}{x+27} \Rightarrow h = 8 + \frac{216}{x}$ and $L(x) = \sqrt{h^2 + (x+27)^2}$
$= \sqrt{\left(8 + \frac{216}{x}\right)^2 + (x+27)^2}$ when $x \ge 0$. Note that $L(x)$ is minimized when $f(x) = \left(8 + \frac{216}{x}\right)^2 + (x+27)^2$ is minimized. If $f'(x) = 0$, then
$2\left(8 + \frac{216}{x}\right)\left(-\frac{216}{x^2}\right) + 2(x+27) = 0$
$\Rightarrow (x+27)\left(1 - \frac{1728}{x^3}\right) = 0 \Rightarrow x = -27$ (not acceptable since distance is never negative or $x = 12$. Then $L(12) = \sqrt{2197} \approx 46.87$ ft.

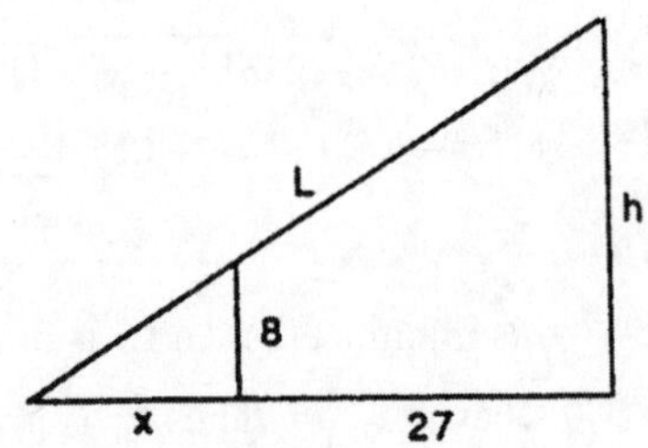

41. $I = \frac{k}{d^2}$, let x = distance the point is from the stronger light source $\Rightarrow 6 - x =$ distance the point is from the other light source. The intensity of illumination at the point from the stronger light is $I_1 = \frac{k_1}{x^2}$, and intensity of illumination at the point from the weaker light is $I_2 = \frac{k_2}{(6-x)^2}$. Since the intensity of the first light is eight times the intensity of the second light $\Rightarrow k_1 = 8k_2$. $\Rightarrow I_1 = \frac{8k_2}{x^2}$. The total intensity is given by $I = I_1 + I_2 = \frac{8k_2}{x^2} + \frac{k_2}{(6-x)^2} \Rightarrow I' = -\frac{16k_2}{x^3} + \frac{2k_2}{(6-x)^3}$
$= \frac{-16(6-x)^3k_2 + 2x^3k_2}{x^3(6-x)^3}$ and $I' = 0 \Rightarrow \frac{-16(6-x)^3k_2 + 2x^3k_2}{x^3(6-x)^3} = 0 \Rightarrow -16(6-x)^3k_2 + 2x^3k_2 = 0 \Rightarrow x = 4$ m. $I'' = \frac{48k_2}{x^4} + \frac{6k_2}{(6-x)^4}$
$\Rightarrow I''(4) = \frac{48k_2}{4^4} + \frac{6k_2}{(6-4)^4} > 0 \Rightarrow$ local minimum. The point should be 4 m from the stronger light source.

43. (a) From the diagram we have $d^2 = 4r^2 - w^2$. The strength of the beam is $S = kwd^2 = kw\left(4r^2 - w^2\right)$. When $r = 6$, then $S = 144kw - kw^3$. Also, $S'(w) = 144k - 3kw^2 = 3k\left(48 - w^2\right)$ so $S'(w) = 0 \Rightarrow w = \pm 4\sqrt{3}$; $S''\left(4\sqrt{3}\right) < 0$ and $-4\sqrt{3}$ is not acceptable. Therefore $S\left(4\sqrt{3}\right)$ is the maximum strength. The dimensions of the strongest beam are $4\sqrt{3}$ by $4\sqrt{6}$ inches.

(b) (c)

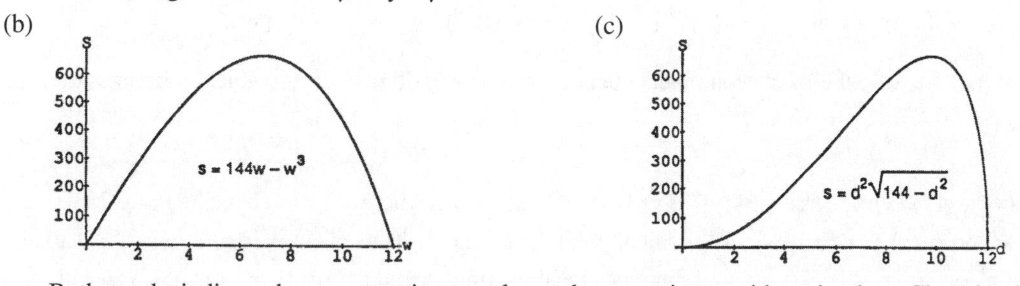

Both graphs indicate the same maximum value and are consistent with each other. Changing k does not change the dimensions that give the strongest beam (i.e., do not change the values of w and d that produce the strongest beam).

45. (a) $s = 10\cos(\pi t) \Rightarrow v = -10\pi\sin(\pi t) \Rightarrow$ speed $= |10\pi\sin(\pi t)| = 10\pi|\sin(\pi t)| \Rightarrow$ the maximum speed is $10\pi \approx 31.42$ cm/sec since the maximum value of $|\sin(\pi t)|$ is 1; the cart is moving the fastest at $t = 0.5$ sec, 1.5 sec, 2.5 sec and 3.5 sec when $|\sin(\pi t)|$ is 1. At these times the distance is $s = 10\cos\left(\frac{\pi}{2}\right) = 0$ cm and $a = -10\pi^2\cos(\pi t) \Rightarrow |a| = 10\pi^2|\cos(\pi t)| \Rightarrow |a| = 0 \text{ cm/sec}^2$

(b) $|a| = 10\pi^2|\cos(\pi t)|$ is greatest at $t = 0.0$ sec, 1.0 sec, 2.0 sec, 3.0 sec and 4.0 sec, and at these times the magnitude of the cart's position is $|s| = 10$ cm from the rest position and the speed is 0 cm/sec.

47. (a) $s = \sqrt{(12 - 12t)^2 + (8t)^2} = \left((12 - 12t)^2 + 64t^2\right)^{1/2}$

(b) $\frac{ds}{dt} = \frac{1}{2}\left((12 - 12t)^2 + 64t^2\right)^{-1/2}[2(12 - 12t)(-12) + 128t] = \frac{208t - 144}{\sqrt{(12 - 12t)^2 + 64t^2}} \Rightarrow \left.\frac{ds}{dt}\right|_{t=0} = -12$ knots and $\left.\frac{ds}{dt}\right|_{t=1} = 8$ knots

(c) The graph indicates that the ships did not see each other because $s(t) > 5$ for all values of t.

(d) The graph supports the conclusions in parts (b) and (c).

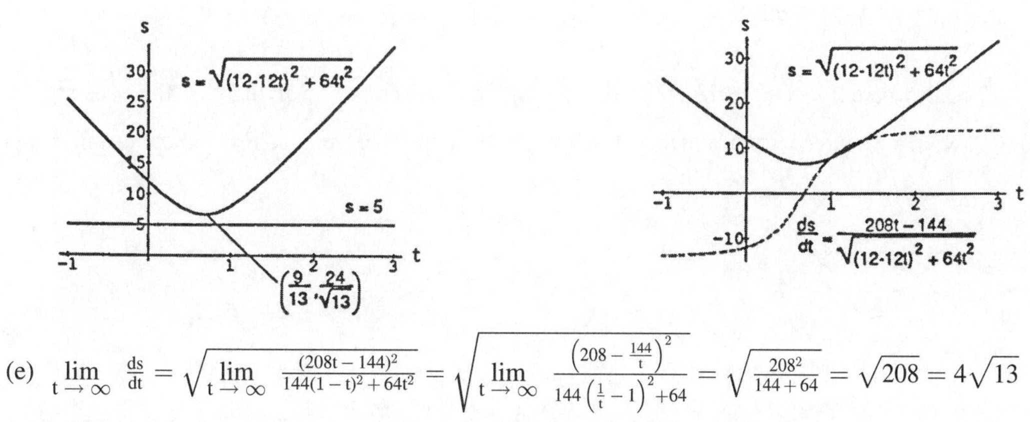

(e) $\lim\limits_{t \to \infty} \frac{ds}{dt} = \sqrt{\lim\limits_{t \to \infty} \frac{(208t - 144)^2}{144(1 - t)^2 + 64t^2}} = \sqrt{\lim\limits_{t \to \infty} \frac{\left(208 - \frac{144}{t}\right)^2}{144\left(\frac{1}{t} - 1\right)^2 + 64}} = \sqrt{\frac{208^2}{144 + 64}} = \sqrt{208} = 4\sqrt{13}$ which equals the square root of the sums of the squares of the individual speeds.

49. If $v = kax - kx^2$, then $v' = ka - 2kx$ and $v'' = -2k$, so $v' = 0 \Rightarrow x = \frac{a}{2}$. At $x = \frac{a}{2}$ there is a maximum since $v''\left(\frac{a}{2}\right) = -2k < 0$. The maximum value of v is $\frac{ka^2}{4}$.

51. The profit is $p = nx - nc = n(x - c) = \left[a(x - c)^{-1} + b(100 - x)\right](x - c) = a + b(100 - x)(x - c)$ $= a + (bc + 100b)x - 100bc - bx^2$. Then $p'(x) = bc + 100b - 2bx$ and $p''(x) = -2b$. Solving $p'(x) = 0 \Rightarrow x = \frac{c}{2} + 50$. At $x = \frac{c}{2} + 50$ there is a maximum profit since $p''(x) = -2b < 0$ for all x.

53. (a) $A(q) = kmq^{-1} + cm + \frac{h}{2}q$, where $q > 0 \Rightarrow A'(q) = -kmq^{-2} + \frac{h}{2} = \frac{hq^2 - 2km}{2q^2}$ and $A''(q) = 2kmq^{-3}$. The critical points are $-\sqrt{\frac{2km}{h}}$, 0, and $\sqrt{\frac{2km}{h}}$, but only $\sqrt{\frac{2km}{h}}$ is in the domain. Then $A''\left(\sqrt{\frac{2km}{h}}\right) > 0 \Rightarrow$ at $q = \sqrt{\frac{2km}{h}}$ there is a minimum average weekly cost.

(b) $A(q) = \frac{(k+bq)m}{q} + cm + \frac{h}{2}q = kmq^{-1} + bm + cm + \frac{h}{2}q$, where $q > 0 \Rightarrow A'(q) = 0$ at $q = \sqrt{\frac{2km}{h}}$ as in (a). Also $A''(q) = 2kmq^{-3} > 0$ so the most economical quantity to order is still $q = \sqrt{\frac{2km}{h}}$ which minimizes the average weekly cost.

55. The profit $p(x) = r(x) - c(x) = 6x - (x^3 - 6x^2 + 15x) = -x^3 + 6x^2 - 9x$, where $x \geq 0$. Then $p'(x) = -3x^2 + 12x - 9 = -3(x-3)(x-1)$ and $p''(x) = -6x + 12$. The critical points are 1 and 3. Thus $p''(1) = 6 > 0 \Rightarrow$ at $x = 1$ there is a local minimum, and $p''(3) = -6 < 0 \Rightarrow$ at $x = 3$ there is a local maximum. But $p(3) = 0 \Rightarrow$ the best you can do is break even.

57. Let x = the length of a side of the square base of the box and h = the height of the box. $V = x^2h = 48 \Rightarrow h = \frac{48}{x^2}$. The total cost is given by $C = 6 \cdot x^2 + 4(4 \cdot x h) = 6x^2 + 16x\left(\frac{48}{x^2}\right) = 6x^2 + \frac{768}{x}$, $x > 0 \Rightarrow C' = 12x - \frac{768}{x^2} = \frac{12x^3 - 768}{x^2}$
$C' = 0 \Rightarrow \frac{12x^3 - 768}{x^2} = 0 \Rightarrow 12x^3 - 768 = 0 \Rightarrow x = 4$; $C'' = 12 + \frac{1536}{x^2} \Rightarrow C''(4) = 12 + \frac{1536}{4^2} > 0 \Rightarrow$ local minimum.
$x = 4 \Rightarrow h = \frac{48}{4^2} = 3$ and $C(4) = 6(4)^2 + \frac{768}{4} = 288 \Rightarrow$ the box is 4 ft × 4 ft × 3 ft, with a minimum cost of \$288

59. We have $\frac{dR}{dM} = CM - M^2$. Solving $\frac{d^2R}{dM^2} = C - 2M = 0 \Rightarrow M = \frac{C}{2}$. Also, $\frac{d^3R}{dM^3} = -2 < 0 \Rightarrow$ at $M = \frac{C}{2}$ there is a maximum.

61. If $x > 0$, then $(x-1)^2 \geq 0 \Rightarrow x^2 + 1 \geq 2x \Rightarrow \frac{x^2+1}{x} \geq 2$. In particular if a, b, c and d are positive integers, then $\left(\frac{a^2+1}{a}\right)\left(\frac{b^2+1}{b}\right)\left(\frac{c^2+1}{c}\right)\left(\frac{d^2+1}{d}\right) \geq 16$.

63. At $x = c$, the tangents to the curves are parallel. Justification: The vertical distance between the curves is $D(x) = f(x) - g(x)$, so $D'(x) = f'(x) - g'(x)$. The maximum value of D will occur at a point c where $D' = 0$. At such a point, $f'(c) - g'(c) = 0$, or $f'(c) = g'(c)$.

65. (a) If $y = \cot x - \sqrt{2}\csc x$ where $0 < x < \pi$, then $y' = (\csc x)\left(\sqrt{2}\cot x - \csc x\right)$. Solving $y' = 0 \Rightarrow \cos x = \frac{1}{\sqrt{2}}$ $\Rightarrow x = \frac{\pi}{4}$. For $0 < x < \frac{\pi}{4}$ we have $y' > 0$, and $y' < 0$ when $\frac{\pi}{4} < x < \pi$. Therefore, at $x = \frac{\pi}{4}$ there is a maximum value of $y = -1$.

(b)

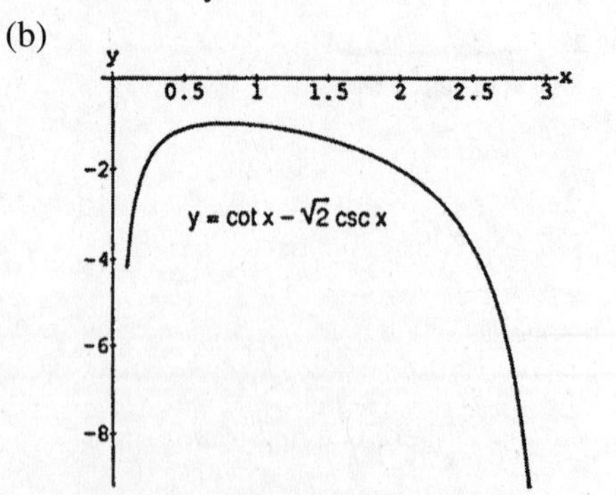

The graph confirms the findings in (a).

67. (a) The square of the distance is $D(x) = \left(x - \frac{3}{2}\right)^2 + \left(\sqrt{x} + 0\right)^2 = x^2 - 2x + \frac{9}{4}$, so $D'(x) = 2x - 2$ and the critical point occurs at $x = 1$. Since $D'(x) < 0$ for $x < 1$ and $D'(x) > 0$ for $x > 1$, the critical point corresponds to the minimum distance. The minimum distance is $\sqrt{D(1)} = \frac{\sqrt{5}}{2}$.

(b)

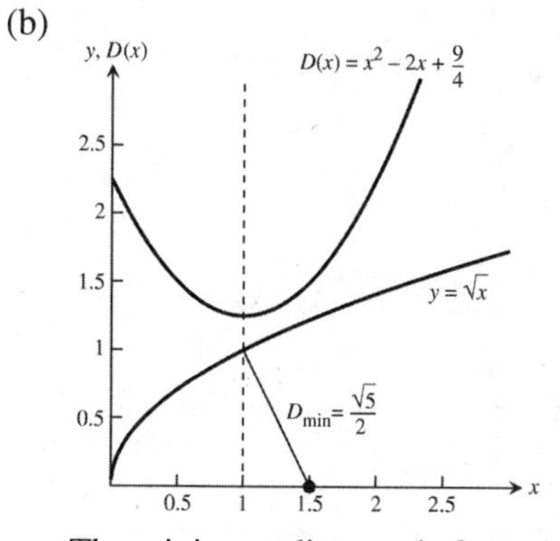

The minimum distance is from the point $\left(\frac{3}{2}, 0\right)$ to the point $(1, 1)$ on the graph of $y = \sqrt{x}$, and this occurs at the value $x = 1$ where $D(x)$, the distance squared, has its minimum value.

4.6 NEWTON'S METHOD

1. $y = x^2 + x - 1 \Rightarrow y' = 2x + 1 \Rightarrow x_{n+1} = x_n - \frac{x_n^2 + x_n - 1}{2x_n + 1}$; $x_0 = 1 \Rightarrow x_1 = 1 - \frac{1+1-1}{2+1} = \frac{2}{3}$
$\Rightarrow x_2 = \frac{2}{3} - \frac{\frac{4}{9} + \frac{2}{3} - 1}{\frac{4}{3} + 1} \Rightarrow x_2 = \frac{2}{3} - \frac{4+6-9}{12+9} = \frac{2}{3} - \frac{1}{21} = \frac{13}{21} \approx .61905$; $x_0 = -1 \Rightarrow x_1 = 1 - \frac{1-1-1}{-2+1} = -2$
$\Rightarrow x_2 = -2 - \frac{4-2-1}{-4+1} = -\frac{5}{3} \approx -1.66667$

3. $y = x^4 + x - 3 \Rightarrow y' = 4x^3 + 1 \Rightarrow x_{n+1} = x_n - \frac{x_n^4 + x_n - 3}{4x_n^3 + 1}$; $x_0 = 1 \Rightarrow x_1 = 1 - \frac{1+1-3}{4+1} = \frac{6}{5}$
$\Rightarrow x_2 = \frac{6}{5} - \frac{\frac{1296}{625} + \frac{6}{5} - 3}{\frac{864}{125} + 1} = \frac{6}{5} - \frac{1296 + 750 - 1875}{4320 + 625} = \frac{6}{5} - \frac{171}{4945} = \frac{5763}{4945} \approx 1.16542$; $x_0 = -1 \Rightarrow x_1 = -1 - \frac{1-1-3}{-4+1}$
$= -2 \Rightarrow x_2 = -2 - \frac{16-2-3}{-32+1} = -2 + \frac{11}{31} = -\frac{51}{31} \approx -1.64516$

5. $y = x^4 - 2 \Rightarrow y' = 4x^3 \Rightarrow x_{n+1} = x_n - \frac{x_n^4 - 2}{4x_n^3}$; $x_0 = 1 \Rightarrow x_1 = 1 - \frac{1-2}{4} = \frac{5}{4} \Rightarrow x_2 = \frac{5}{4} - \frac{\frac{625}{256} - 2}{\frac{125}{16}} = \frac{5}{4} - \frac{625-512}{2000}$
$= \frac{5}{4} - \frac{113}{2000} = \frac{2500-113}{2000} = \frac{2387}{2000} \approx 1.1935$

7. $f(x_0) = 0$ and $f'(x_0) \neq 0 \Rightarrow x_{n+1} = x_n - \frac{f(x_n)}{f'(x_n)}$ gives $x_1 = x_0 \Rightarrow x_2 = x_0 \Rightarrow x_n = x_0$ for all $n \geq 0$. That is, all of the approximations in Newton's method will be the root of $f(x) = 0$.

9. If $x_0 = h > 0 \Rightarrow x_1 = x_0 - \frac{f(x_0)}{f'(x_0)} = h - \frac{f(h)}{f'(h)}$
$= h - \frac{\sqrt{h}}{\left(\frac{1}{2\sqrt{h}}\right)} = h - \left(\sqrt{h}\right)\left(2\sqrt{h}\right) = -h$;
if $x_0 = -h < 0 \Rightarrow x_1 = x_0 - \frac{f(x_0)}{f'(x_0)} = -h - \frac{f(-h)}{f'(-h)}$
$= -h - \frac{\sqrt{h}}{\left(\frac{-1}{2\sqrt{h}}\right)} = -h + \left(\sqrt{h}\right)\left(2\sqrt{h}\right) = h.$

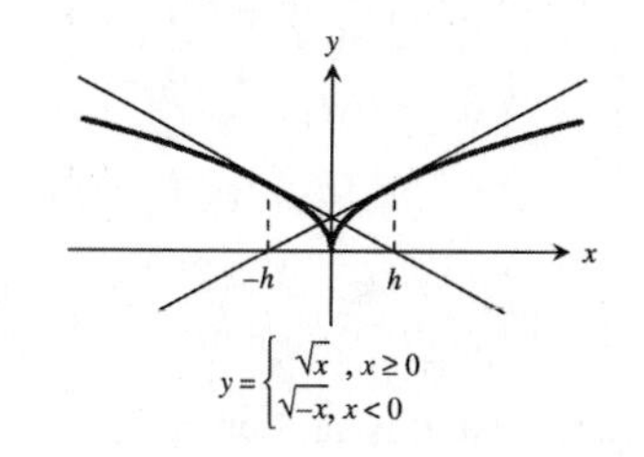

11. i) is equivalent to solving $x^3 - 3x - 1 = 0$.
ii) is equivalent to solving $x^3 - 3x - 1 = 0$.
iii) is equivalent to solving $x^3 - 3x - 1 = 0$.
iv) is equivalent to solving $x^3 - 3x - 1 = 0$.
All four equations are equivalent.

13. $f(x) = \tan x - 2x \Rightarrow f'(x) = \sec^2 x - 2 \Rightarrow x_{n+1} = x_n - \frac{\tan(x_n) - 2x_n}{\sec^2(x_n)}$; $x_0 = 1 \Rightarrow x_1 = 1.2920445$
$\Rightarrow x_2 = 1.155327774 \Rightarrow x_{16} = x_{17} = 1.165561185$

15. (a) The graph of $f(x) = \sin 3x - 0.99 + x^2$ in the window $-2 \le x \le 2, -2 \le y \le 3$ suggests three roots. However, when you zoom in on the x-axis near $x = 1.2$, you can see that the graph lies above the axis there. There are only two roots, one near $x = -1$, the other near $x = 0.4$.

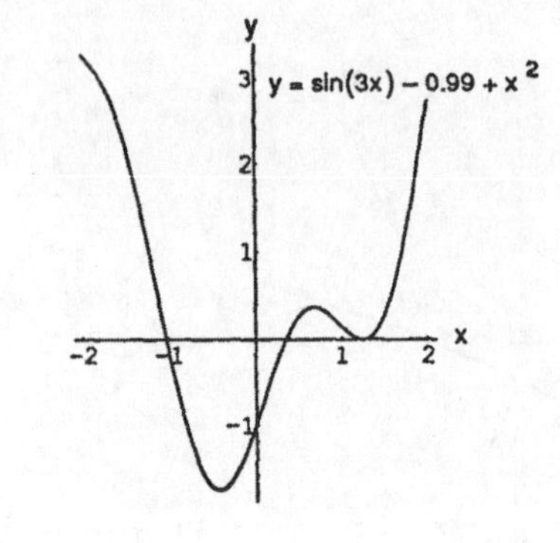

(b) $f(x) = \sin 3x - 0.99 + x^2 \Rightarrow f'(x) = 3\cos 3x + 2x$
$\Rightarrow x_{n+1} = x_n - \frac{\sin(3x_n) - 0.99 + x_n^2}{3\cos(3x_n) + 2x_n}$ and the solutions are approximately 0.35003501505249 and -1.0261731615301

17. $f(x) = 2x^4 - 4x^2 + 1 \Rightarrow f'(x) = 8x^3 - 8x \Rightarrow x_{n+1} = x_n - \frac{2x_n^4 - 4x_n^2 + 1}{8x_n^3 - 8x_n}$; if $x_0 = -2$, then $x_6 = -1.30656296$; if $x_0 = -0.5$, then $x_3 = -0.5411961$; the roots are approximately ± 0.5411961 and ± 1.30656296 because f(x) is an even function.

19. From the graph we let $x_0 = 0.5$ and $f(x) = \cos x - 2x$
$\Rightarrow x_{n+1} = x_n - \frac{\cos(x_n) - 2x_n}{-\sin(x_n) - 2} \Rightarrow x_1 = .45063$
$\Rightarrow x_2 = .45018 \Rightarrow$ at $x \approx 0.45$ we have $\cos x = 2x$.

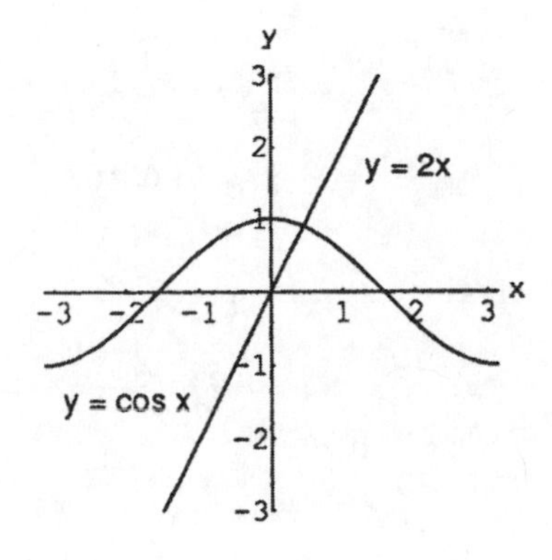

21. The x-coordinate of the point of intersection of $y = x^2(x + 1)$ and $y = \frac{1}{x}$ is the solution of $x^2(x + 1) = \frac{1}{x}$
$\Rightarrow x^3 + x^2 - \frac{1}{x} = 0 \Rightarrow$ The x-coordinate is the root of $f(x) = x^3 + x^2 - \frac{1}{x} \Rightarrow f'(x) = 3x^2 + 2x + \frac{1}{x^2}$. Let $x_0 = 1$
$\Rightarrow x_{n+1} = x_n - \frac{x_n^3 + x_n^2 - \frac{1}{x_n}}{3x_n^2 + 2x_n + \frac{1}{x_n^2}} \Rightarrow x_1 = 0.83333 \Rightarrow x_2 = 0.81924 \Rightarrow x_3 = 0.81917 \Rightarrow x_7 = 0.81917 \Rightarrow r \approx 0.8192$

23. If $f(x) = x^3 + 2x - 4$, then $f(1) = -1 < 0$ and $f(2) = 8 > 0 \Rightarrow$ by the Intermediate Value Theorem the equation $x^3 + 2x - 4 = 0$ has a solution between 1 and 2. Consequently, $f'(x) = 3x^2 + 2$ and $x_{n+1} = x_n - \frac{x_n^3 + 2x_n - 4}{3x_n^2 + 2}$.
Then $x_0 = 1 \Rightarrow x_1 = 1.2 \Rightarrow x_2 = 1.17975 \Rightarrow x_3 = 1.179509 \Rightarrow x_4 = 1.1795090 \Rightarrow$ the root is approximately 1.17951.

25. $f(x) = 4x^4 - 4x^2 \Rightarrow f'(x) = 16x^3 - 8x \Rightarrow x_{i+1} = x_i - \frac{f(x_i)}{f'(x_i)} = x_i - \frac{x_i^3 - x_i}{4x_i^2 - 2}$. Iterations are performed using the procedure in problem 13 in this section.
(a) For $x_0 = -2$ or $x_0 = -0.8$, $x_i \to -1$ as i gets large.
(b) For $x_0 = -0.5$ or $x_0 = 0.25$, $x_i \to 0$ as i gets large.
(c) For $x_0 = 0.8$ or $x_0 = 2$, $x_i \to 1$ as i gets large.

(d) (If your calculator has a CAS, put it in exact mode, otherwise approximate the radicals with a decimal value.) For $x_0 = -\frac{\sqrt{21}}{7}$ or $x_0 = -\frac{\sqrt{21}}{7}$, Newton's method does not converge. The values of x_i alternate between $x_0 = -\frac{\sqrt{21}}{7}$ or $x_0 = -\frac{\sqrt{21}}{7}$ as i increases.

27. $f(x) = (x-1)^{40} \Rightarrow f'(x) = 40(x-1)^{39} \Rightarrow x_{n+1} = x_n - \frac{(x_n-1)^{40}}{40(x_n-1)^{39}} = \frac{39x_n+1}{40}$. With $x_0 = 2$, our computer gave $x_{87} = x_{88} = x_{89} = \cdots = x_{200} = 1.11051$, coming within 0.11051 of the root $x = 1$.

4.7 ANTIDERIVATIVES

1. (a) x^2 (b) $\frac{x^3}{3}$ (c) $\frac{x^3}{3} - x^2 + x$

3. (a) x^{-3} (b) $-\frac{x^{-3}}{3}$ (c) $-\frac{x^{-3}}{3} + x^2 + 3x$

5. (a) $\frac{-1}{x}$ (b) $\frac{-5}{x}$ (c) $2x + \frac{5}{x}$

7. (a) $\sqrt{x^3}$ (b) $\sqrt{x}$ (c) $\frac{2}{3}\sqrt{x^3} + 2\sqrt{x}$

9. (a) $x^{2/3}$ (b) $x^{1/3}$ (c) $x^{-1/3}$

11. (a) $\cos(\pi x)$ (b) $-3\cos x$ (c) $\frac{-\cos(\pi x)}{\pi} + \cos(3x)$

13. (a) $\tan x$ (b) $2\tan\left(\frac{x}{3}\right)$ (c) $-\frac{2}{3}\tan\left(\frac{3x}{2}\right)$

15. (a) $-\csc x$ (b) $\frac{1}{5}\csc(5x)$ (c) $2\csc\left(\frac{\pi x}{2}\right)$

17. $\int (x+1)\,dx = \frac{x^2}{2} + x + C$

19. $\int \left(3t^2 + \frac{t}{2}\right) dt = t^3 + \frac{t^2}{4} + C$

21. $\int (2x^3 - 5x + 7)\,dx = \frac{1}{2}x^4 - \frac{5}{2}x^2 + 7x + C$

23. $\int \left(\frac{1}{x^2} - x^2 - \frac{1}{3}\right) dx = \int \left(x^{-2} - x^2 - \frac{1}{3}\right) dx = \frac{x^{-1}}{-1} - \frac{x^3}{3} - \frac{1}{3}x + C = -\frac{1}{x} - \frac{x^3}{3} - \frac{x}{3} + C$

25. $\int x^{-1/3}\,dx = \frac{x^{2/3}}{\frac{2}{3}} + C = \frac{3}{2}x^{2/3} + C$

27. $\int \left(\sqrt{x} + \sqrt[3]{x}\right) dx = \int \left(x^{1/2} + x^{1/3}\right) dx = \frac{x^{3/2}}{\frac{3}{2}} + \frac{x^{4/3}}{\frac{4}{3}} + C = \frac{2}{3}x^{3/2} + \frac{3}{4}x^{4/3} + C$

29. $\int \left(8y - \frac{2}{y^{1/4}}\right) dy = \int \left(8y - 2y^{-1/4}\right) dy = \frac{8y^2}{2} - 2\left(\frac{y^{3/4}}{\frac{3}{4}}\right) + C = 4y^2 - \frac{8}{3}y^{3/4} + C$

31. $\int 2x\left(1 - x^{-3}\right) dx = \int \left(2x - 2x^{-2}\right) dx = \frac{2x^2}{2} - 2\left(\frac{x^{-1}}{-1}\right) + C = x^2 + \frac{2}{x} + C$

33. $\int \frac{t\sqrt{t}+\sqrt{t}}{t^2}\,dt = \int \left(\frac{t^{3/2}}{t^2} + \frac{t^{1/2}}{t^2}\right) dt = \int \left(t^{-1/2} + t^{-3/2}\right) dt = \frac{t^{1/2}}{\frac{1}{2}} + \left(\frac{t^{-1/2}}{-\frac{1}{2}}\right) + C = 2\sqrt{t} - \frac{2}{\sqrt{t}} + C$

35. $\int -2\cos t\,dt = -2\sin t + C$

37. $\int 7\sin\frac{\theta}{3}\,d\theta = -21\cos\frac{\theta}{3} + C$

39. $\int -3\csc^2 x\, dx = 3\cot x + C$

41. $\int \frac{\csc\theta\cot\theta}{2}\, d\theta = -\frac{1}{2}\csc\theta + C$

43. $\int (4\sec x\tan x - 2\sec^2 x)\, dx = 4\sec x - 2\tan x + C$

45. $\int (\sin 2x - \csc^2 x)\, dx = -\frac{1}{2}\cos 2x + \cot x + C$

47. $\int \frac{1+\cos 4t}{2}\, dt = \int \left(\frac{1}{2} + \frac{1}{2}\cos 4t\right) dt = \frac{1}{2}t + \frac{1}{2}\left(\frac{\sin 4t}{4}\right) + C = \frac{t}{2} + \frac{\sin 4t}{8} + C$

49. $\int (1+\tan^2\theta)\, d\theta = \int \sec^2\theta\, d\theta = \tan\theta + C$

51. $\int \cot^2 x\, dx = \int (\csc^2 x - 1)\, dx = -\cot x - x + C$

53. $\int \cos\theta(\tan\theta + \sec\theta)\, d\theta = \int (\sin\theta + 1)\, d\theta = -\cos\theta + \theta + C$

55. $\frac{d}{dx}\left(\frac{(7x-2)^4}{28} + C\right) = \frac{4(7x-2)^3(7)}{28} = (7x-2)^3$

57. $\frac{d}{dx}\left(\frac{1}{5}\tan(5x-1) + C\right) = \frac{1}{5}(\sec^2(5x-1))(5) = \sec^2(5x-1)$

59. $\frac{d}{dx}\left(\frac{-1}{x+1} + C\right) = (-1)(-1)(x+1)^{-2} = \frac{1}{(x+1)^2}$

61. (a) Wrong: $\frac{d}{dx}\left(\frac{x^2}{2}\sin x + C\right) = \frac{2x}{2}\sin x + \frac{x^2}{2}\cos x = x\sin x + \frac{x^2}{2}\cos x \neq x\sin x$
(b) Wrong: $\frac{d}{dx}(-x\cos x + C) = -\cos x + x\sin x \neq x\sin x$
(c) Right: $\frac{d}{dx}(-x\cos x + \sin x + C) = -\cos x + x\sin x + \cos x = x\sin x$

63. (a) Wrong: $\frac{d}{dx}\left(\frac{(2x+1)^3}{3} + C\right) = \frac{3(2x+1)^2(2)}{3} = 2(2x+1)^2 \neq (2x+1)^2$
(b) Wrong: $\frac{d}{dx}((2x+1)^3 + C) = 3(2x+1)^2(2) = 6(2x+1)^2 \neq 3(2x+1)^2$
(c) Right: $\frac{d}{dx}((2x+1)^3 + C) = 6(2x+1)^2$

65. Right: $\frac{d}{dx}\left(\left(\frac{x+3}{x-2}\right)^3 + C\right) = 3\left(\frac{x+3}{x-2}\right)^2 \frac{(x-2)\cdot 1 - (x+3)\cdot 1}{(x-2)^2} = 3\frac{(x+3)^2}{(x-2)^2}\frac{-5}{(x-2)^2} = \frac{-15(x+3)^2}{(x-2)^4}$

67. Graph (b), because $\frac{dy}{dx} = 2x \Rightarrow y = x^2 + C$. Then $y(1) = 4 \Rightarrow C = 3$.

69. $\frac{dy}{dx} = 2x - 7 \Rightarrow y = x^2 - 7x + C$; at $x = 2$ and $y = 0$ we have $0 = 2^2 - 7(2) + C \Rightarrow C = 10 \Rightarrow y = x^2 - 7x + 10$

71. $\frac{dy}{dx} = \frac{1}{x^2} + x = x^{-2} + x \Rightarrow y = -x^{-1} + \frac{x^2}{2} + C$; at $x = 2$ and $y = 1$ we have $1 = -2^{-1} + \frac{2^2}{2} + C \Rightarrow C = -\frac{1}{2}$
$\Rightarrow y = -x^{-1} + \frac{x^2}{2} - \frac{1}{2}$ or $y = -\frac{1}{x} + \frac{x^2}{2} - \frac{1}{2}$

73. $\frac{dy}{dx} = 3x^{-2/3} \Rightarrow y = \frac{3x^{1/3}}{\frac{1}{3}} + C = 9$; at $x = 9x^{1/3} + C$; at $x = -1$ and $y = -5$ we have $-5 = 9(-1)^{1/3} + C \Rightarrow C = 4$
$\Rightarrow y = 9x^{1/3} + 4$

75. $\frac{ds}{dt} = 1 + \cos t \Rightarrow s = t + \sin t + C$; at $t = 0$ and $s = 4$ we have $4 = 0 + \sin 0 + C \Rightarrow C = 4 \Rightarrow s = t + \sin t + 4$

77. $\frac{dr}{d\theta} = -\pi\sin\pi\theta \Rightarrow r = \cos(\pi\theta) + C$; at $r = 0$ and $\theta = 0$ we have $0 = \cos(\pi 0) + C \Rightarrow C = -1 \Rightarrow r = \cos(\pi\theta) - 1$

79. $\frac{dv}{dt} = \frac{1}{2}\sec t \tan t \Rightarrow v = \frac{1}{2}\sec t + C$; at $v = 1$ and $t = 0$ we have $1 = \frac{1}{2}\sec(0) + C \Rightarrow C = \frac{1}{2} \Rightarrow v = \frac{1}{2}\sec t + \frac{1}{2}$

81. $\frac{d^2y}{dx^2} = 2 - 6x \Rightarrow \frac{dy}{dx} = 2x - 3x^2 + C_1$; at $\frac{dy}{dx} = 4$ and $x = 0$ we have $4 = 2(0) - 3(0)^2 + C_1 \Rightarrow C_1 = 4$
$\Rightarrow \frac{dy}{dx} = 2x - 3x^2 + 4 \Rightarrow y = x^2 - x^3 + 4x + C_2$; at $y = 1$ and $x = 0$ we have $1 = 0^2 - 0^3 + 4(0) + C_2 \Rightarrow C_2 = 1$
$\Rightarrow y = x^2 - x^3 + 4x + 1$

83. $\frac{d^2r}{dt^2} = \frac{2}{t^3} = 2t^{-3} \Rightarrow \frac{dr}{dt} = -t^{-2} + C_1$; at $\frac{dr}{dt} = 1$ and $t = 1$ we have $1 = -(1)^{-2} + C_1 \Rightarrow C_1 = 2 \Rightarrow \frac{dr}{dt} = -t^{-2} + 2$
$\Rightarrow r = t^{-1} + 2t + C_2$; at $r = 1$ and $t = 1$ we have $1 = 1^{-1} + 2(1) + C_2 \Rightarrow C_2 = -2 \Rightarrow r = t^{-1} + 2t - 2$ or
$r = \frac{1}{t} + 2t - 2$

85. $\frac{d^3y}{dx^3} = 6 \Rightarrow \frac{d^2y}{dx^2} = 6x + C_1$; at $\frac{d^2y}{dx^2} = -8$ and $x = 0$ we have $-8 = 6(0) + C_1 \Rightarrow C_1 = -8 \Rightarrow \frac{d^2y}{dx^2} = 6x - 8$
$\Rightarrow \frac{dy}{dx} = 3x^2 - 8x + C_2$; at $\frac{dy}{dx} = 0$ and $x = 0$ we have $0 = 3(0)^2 - 8(0) + C_2 \Rightarrow C_2 = 0 \Rightarrow \frac{dy}{dx} = 3x^2 - 8x$
$\Rightarrow y = x^3 - 4x^2 + C_3$; at $y = 5$ and $x = 0$ we have $5 = 0^3 - 4(0)^2 + C_3 \Rightarrow C_3 = 5 \Rightarrow y = x^3 - 4x^2 + 5$

87. $y^{(4)} = -\sin t + \cos t \Rightarrow y''' = \cos t + \sin t + C_1$; at $y''' = 7$ and $t = 0$ we have $7 = \cos(0) + \sin(0) + C_1 \Rightarrow C_1 = 6$
$\Rightarrow y''' = \cos t + \sin t + 6 \Rightarrow y'' = \sin t - \cos t + 6t + C_2$; at $y'' = -1$ and $t = 0$ we have
$-1 = \sin(0) - \cos(0) + 6(0) + C_2 \Rightarrow C_2 = 0 \Rightarrow y'' = \sin t - \cos t + 6t \Rightarrow y' = -\cos t - \sin t + 3t^2 + C_3$; at
$y' = -1$ and $t = 0$ we have $-1 = -\cos(0) - \sin(0) + 3(0)^2 + C_3 \Rightarrow C_3 = 0 \Rightarrow y' = -\cos t - \sin t + 3t^2$
$\Rightarrow y = -\sin t + \cos t + t^3 + C_4$; at $y = 0$ and $t = 0$ we have $0 = -\sin(0) + \cos(0) + 0^3 + C_4 \Rightarrow C_4 = -1$
$\Rightarrow y = -\sin t + \cos t + t^3 - 1$

89. $m = y' = 3\sqrt{x} = 3x^{1/2} \Rightarrow y = 2x^{3/2} + C$; at $(9, 4)$ we have $4 = 2(9)^{3/2} + C \Rightarrow C = -50 \Rightarrow y = 2x^{3/2} - 50$

91. $\frac{dy}{dx} = 1 - \frac{4}{3}x^{1/3} \Rightarrow y = \int \left(1 - \frac{4}{3}x^{1/3}\right) dx = x - x^{4/3} + C$; at $(1, 0.5)$ on the curve we have $0.5 = 1 - 1^{4/3} + C$
$\Rightarrow C = 0.5 \Rightarrow y = x - x^{4/3} + \frac{1}{2}$

93. $\frac{dy}{dx} = \sin x - \cos x \Rightarrow y = \int (\sin x - \cos x)\, dx = -\cos x - \sin x + C$; at $(-\pi, -1)$ on the curve we have
$-1 = -\cos(-\pi) - \sin(-\pi) + C \Rightarrow C = -2 \Rightarrow y = -\cos x - \sin x - 2$

95. (a) $\frac{ds}{dt} = 9.8t - 3 \Rightarrow s = 4.9t^2 - 3t + C$; (i) at $s = 5$ and $t = 0$ we have $C = 5 \Rightarrow s = 4.9t^2 - 3t + 5$;
displacement $= s(3) - s(1) = ((4.9)(9) - 9 + 5) - (4.9 - 3 + 5) = 33.2$ units; (ii) at $s = -2$ and $t = 0$ we have
$C = -2 \Rightarrow s = 4.9t^2 - 3t - 2$; displacement $= s(3) - s(1) = ((4.9)(9) - 9 - 2) - (4.9 - 3 - 2) = 33.2$ units;
(iii) at $s = s_0$ and $t = 0$ we have $C = s_0 \Rightarrow s = 4.9t^2 - 3t + s_0$; displacement $= s(3) - s(1)$
$= ((4.9)(9) - 9 + s_0) - (4.9 - 3 + s_0) = 33.2$ units

(b) True. Given an antiderivative f(t) of the velocity function, we know that the body's position function is $s = f(t) + C$ for some constant C. Therefore, the displacement from $t = a$ to $t = b$ is $(f(b) + C) - (f(a) + C)$ $= f(b) - f(a)$. Thus we can find the displacement from any antiderivative f as the numerical difference $f(b) - f(a)$ without knowing the exact values of C and s.

97. Step 1: $\frac{d^2s}{dt^2} = -k \Rightarrow \frac{ds}{dt} = -kt + C_1$; at $\frac{ds}{dt} = 88$ and $t = 0$ we have $C_1 = 88 \Rightarrow \frac{ds}{dt} = -kt + 88 \Rightarrow$
$s = -k\left(\frac{t^2}{2}\right) + 88t + C_2$; at $s = 0$ and $t = 0$ we have $C_2 = 0 \Rightarrow s = -\frac{kt^2}{2} + 88t$

Step 2: $\frac{ds}{dt} = 0 \Rightarrow 0 = -kt + 88 \Rightarrow t = \frac{88}{k}$

Step 3: $242 = \frac{-k\left(\frac{88}{k}\right)^2}{2} + 88\left(\frac{88}{k}\right) \Rightarrow 242 = -\frac{(88)^2}{2k} + \frac{(88)^2}{k} \Rightarrow 242 = \frac{(88)^2}{2k} \Rightarrow k = 16$

99. (a) $v = \int a\,dt = \int \left(15t^{1/2} - 3t^{-1/2}\right) dt = 10t^{3/2} - 6t^{1/2} + C$; $\frac{ds}{dt}(1) = 4 \Rightarrow 4 = 10(1)^{3/2} - 6(1)^{1/2} + C \Rightarrow C = 0$
$\Rightarrow v = 10t^{3/2} - 6t^{1/2}$

(b) $s = \int v\,dt = \int \left(10t^{3/2} - 6t^{1/2}\right) dt = 4t^{5/2} - 4t^{3/2} + C$; $s(1) = 0 \Rightarrow 0 = 4(1)^{5/2} - 4(1)^{3/2} + C \Rightarrow C = 0$
$\Rightarrow s = 4t^{5/2} - 4t^{3/2}$

101. $\frac{d^2s}{dt^2} = a \Rightarrow \frac{ds}{dt} = \int a\,dt = at + C$; $\frac{ds}{dt} = v_0$ when $t = 0 \Rightarrow C = v_0 \Rightarrow \frac{ds}{dt} = at + v_0 \Rightarrow s = \frac{at^2}{2} + v_0t + C_1$; $s = s_0$ when $t = 0 \Rightarrow s_0 = \frac{a(0)^2}{2} + v_0(0) + C_1 \Rightarrow C_1 = s_0 \Rightarrow s = \frac{at^2}{2} + v_0t + s_0$

CHAPTER 4 PRACTICE EXERCISES

1. No, since $f(x) = x^3 + 2x + \tan x \Rightarrow f'(x) = 3x^2 + 2 + \sec^2 x > 0 \Rightarrow f(x)$ is always increasing on its domain

3. No absolute minimum because $\lim_{x \to \infty} (7 + x)(11 - 3x)^{1/3} = -\infty$. Next $f'(x) = (11 - 3x)^{1/3} - (7 + x)(11 - 3x)^{-2/3} = \frac{(11 - 3x) - (7 + x)}{(11 - 3x)^{2/3}} = \frac{4(1 - x)}{(11 - 3x)^{2/3}} \Rightarrow x = 1$ and $x = \frac{11}{3}$ are critical points. Since $f' > 0$ if $x < 1$ and $f' < 0$ if $x > 1$, $f(1) = 16$ is the absolute maximum.

5. Yes, because at each point of $[0, 1)$ except $x = 0$, the function's value is a local minimum value as well as a local maximum value. At $x = 0$ the function's value, 0, is not a local minimum value because each open interval around $x = 0$ on the x-axis contains points to the left of 0 where f equals -1.

7. No, because the interval $0 < x < 1$ fails to be closed. The Extreme Value Theorem says that if the function is continuous throughout a finite closed interval $a \le x \le b$ then the existence of absolute extrema is guaranteed on that interval.

9. (a) There appear to be local minima at $x = -1.75$ and 1.8. Points of inflection are indicated at approximately $x = 0$ and $x = \pm 1$.

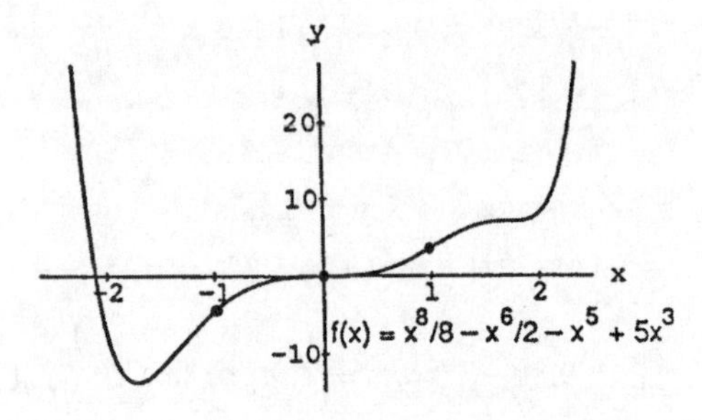

(b) $f'(x) = x^7 - 3x^5 - 5x^4 + 15x^2 = x^2(x^2 - 3)(x^3 - 5)$. The pattern $y' = - - - \underset{-\sqrt{3}}{|} + + + \underset{0}{|} + + + \underset{\sqrt[3]{5}}{|} - - - \underset{\sqrt{3}}{|} + + +$ indicates a local maximum at $x = \sqrt[3]{5}$ and local minima at $x = \pm\sqrt{3}$.

(c)

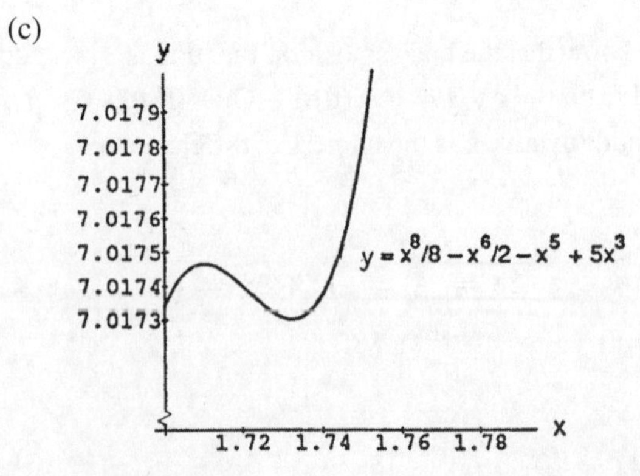

11. (a) $g(t) = \sin^2 t - 3t \Rightarrow g'(t) = 2 \sin t \cos t - 3 = \sin(2t) - 3 \Rightarrow g' < 0 \Rightarrow g(t)$ is always falling and hence must decrease on every interval in its domain.

(b) One, since $\sin^2 t - 3t - 5 = 0$ and $\sin^2 t - 3t = 5$ have the same solutions: $f(t) = \sin^2 t - 3t - 5$ has the same derivative as $g(t)$ in part (a) and is always decreasing with $f(-3) > 0$ and $f(0) < 0$. The Intermediate Value Theorem guarantees the continuous function f has a root in $[-3, 0]$.

13. (a) $f(x) = x^4 + 2x^2 - 2 \Rightarrow f'(x) = 4x^3 + 4x$. Since $f(0) = -2 < 0$, $f(1) = 1 > 0$ and $f'(x) \geq 0$ for $0 \leq x \leq 1$, we may conclude from the Intermediate Value Theorem that $f(x)$ has exactly one solution when $0 \leq x \leq 1$.

(b) $x^2 = \frac{-2 \pm \sqrt{4+8}}{2} > 0 \Rightarrow x^2 = \sqrt{3} - 1$ and $x \geq 0 \Rightarrow x \approx \sqrt{.7320508076} \approx .8555996772$

15. Let $V(t)$ represent the volume of the water in the reservoir at time t, in minutes, let $V(0) = a_0$ be the initial amount and $V(1440) = a_0 + (1400)(43{,}560)(7.48)$ gallons be the amount of water contained in the reservoir after the rain, where 24 hr = 1440 min. Assume that $V(t)$ is continuous on $[0, 1440]$ and differentiable on $(0, 1440)$. The Mean Value Theorem says that for some t_0 in $(0, 1440)$ we have $V'(t_0) = \frac{V(1440) - V(0)}{1440 - 0} = \frac{a_0 + (1400)(43{,}560)(7.48) - a_0}{1440} = \frac{456{,}160{,}320 \text{ gal}}{1440 \text{ min}}$ $= 316{,}778$ gal/min. Therefore at t_0 the reservoir's volume was increasing at a rate in excess of 225,000 gal/min.

17. No, $\frac{x}{x+1} = 1 + \frac{-1}{x+1} \Rightarrow \frac{x}{x+1}$ differs from $\frac{-1}{x+1}$ by the constant 1. Both functions have the same derivative $\frac{d}{dx}\left(\frac{x}{x+1}\right) = \frac{(x+1) - x(1)}{(x+1)^2} = \frac{1}{(x+1)^2} = \frac{d}{dx}\left(\frac{-1}{x+1}\right)$.

19. The global minimum value of $\frac{1}{2}$ occurs at $x = 2$.

21. (a) $t = 0, 6, 12$ (b) $t = 3, 9$ (c) $6 < t < 12$ (d) $0 < t < 6$, $12 < t < 14$

23.

25.

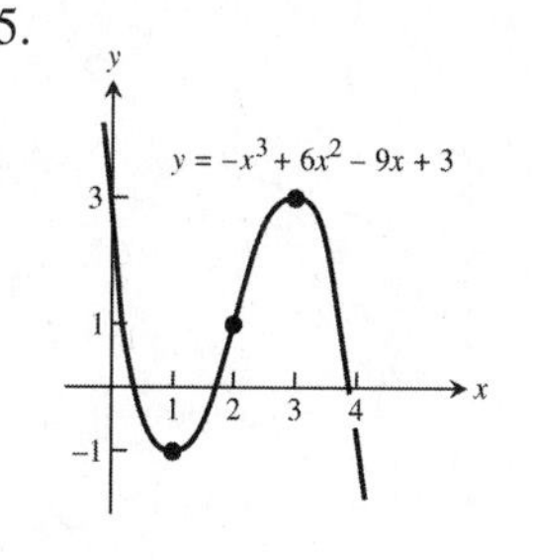

27.

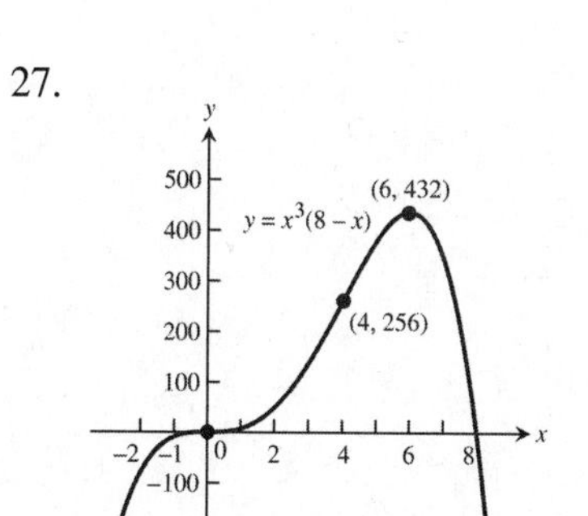

29.

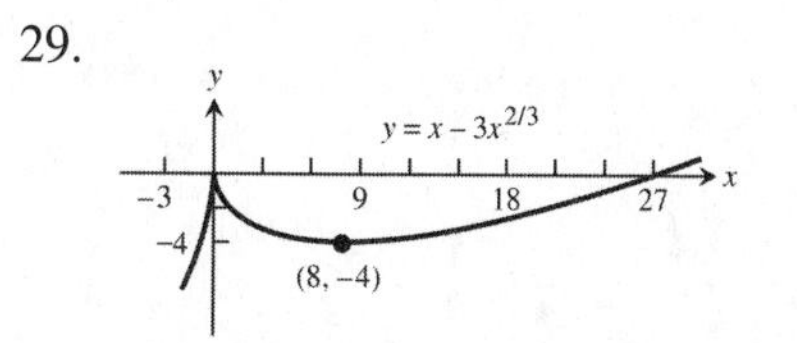

31.

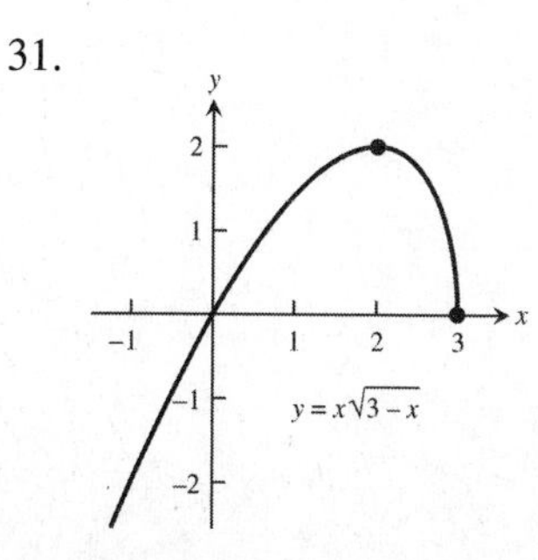

33. (a) $y' = 16 - x^2 \Rightarrow y' = \underset{-4}{---|}\ \underset{4}{+++|}\ ---$ $\Rightarrow$ the curve is rising on $(-4, 4)$, falling on $(-\infty, -4)$ and $(4, \infty)$ $\Rightarrow$ a local maximum at $x = 4$ and a local minimum at $x = -4$; $y'' = -2x \Rightarrow y'' = \underset{0}{+++|}\ ---$ $\Rightarrow$ the curve is concave up on $(-\infty, 0)$, concave down on $(0, \infty)$ $\Rightarrow$ a point of inflection at $x = 0$

(b)

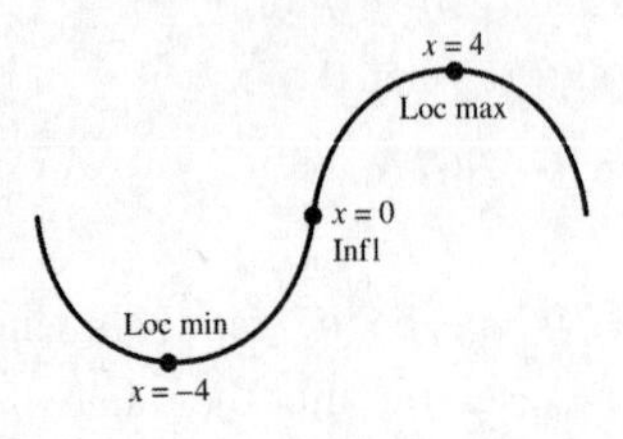

35. (a) $y' = 6x(x+1)(x-2) = 6x^3 - 6x^2 - 12x \Rightarrow y' = \underset{-1}{---|}\ \underset{0}{+++|}\ \underset{2}{---|}\ +++$ $\Rightarrow$ the graph is rising on $(-1, 0)$ and $(2, \infty)$, falling on $(-\infty, -1)$ and $(0, 2)$ $\Rightarrow$ a local maximum at $x = 0$, local minima at $x = -1$ and $x = 2$; $y'' = 18x^2 - 12x - 12 = 6(3x^2 - 2x - 2) = 6\left(x - \frac{1-\sqrt{7}}{3}\right)\left(x - \frac{1+\sqrt{7}}{3}\right) \Rightarrow$ $y'' = \underset{\frac{1-\sqrt{7}}{3}}{+++|}\ \underset{\frac{1+\sqrt{7}}{3}}{---|}\ +++$ $\Rightarrow$ the curve is concave up on $\left(-\infty, \frac{1-\sqrt{7}}{3}\right)$ and $\left(\frac{1+\sqrt{7}}{3}, \infty\right)$, concave down on $\left(\frac{1-\sqrt{7}}{3}, \frac{1+\sqrt{7}}{3}\right)$ $\Rightarrow$ points of inflection at $x = \frac{1\pm\sqrt{7}}{3}$

(b)

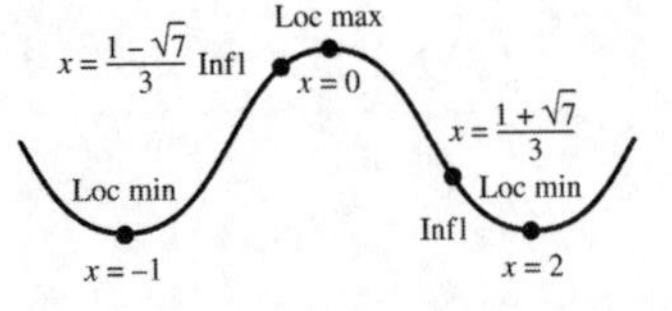

37. (a) $y' = x^4 - 2x^2 = x^2(x^2 - 2) \Rightarrow y' = \underset{-\sqrt{2}}{+++|}\ \underset{0}{---|}\ \underset{\sqrt{2}}{---|}\ +++$ $\Rightarrow$ the curve is rising on $\left(-\infty, -\sqrt{2}\right)$ and $\left(\sqrt{2}, \infty\right)$, falling on $\left(-\sqrt{2}, \sqrt{2}\right)$ $\Rightarrow$ a local maximum at $x = -\sqrt{2}$ and a local minimum at $x = \sqrt{2}$; $y'' = 4x^3 - 4x = 4x(x-1)(x+1) \Rightarrow y'' = \underset{-1}{---|}\ \underset{0}{+++|}\ \underset{1}{---|}\ +++$ $\Rightarrow$ concave up on $(-1, 0)$ and $(1, \infty)$, concave down on $(-\infty, -1)$ and $(0, 1)$ $\Rightarrow$ points of inflection at $x = 0$ and $x = \pm 1$

(b)

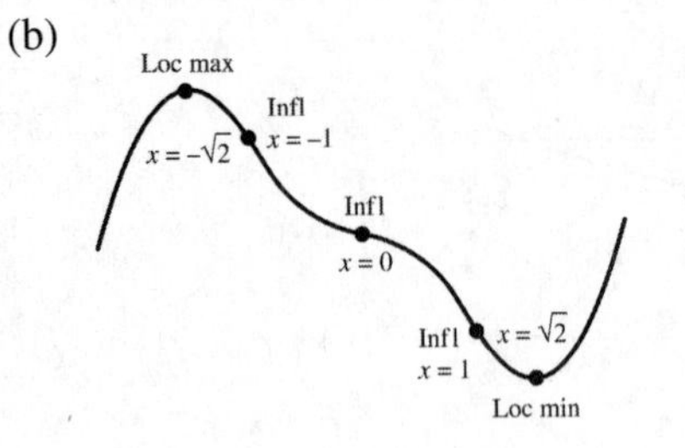

39. The values of the first derivative indicate that the curve is rising on $(0, \infty)$ and falling on $(-\infty, 0)$. The slope of the curve approaches $-\infty$ as $x \to 0^-$, and approaches ∞ as $x \to 0^+$ and $x \to 1$. The curve should therefore have a cusp and local minimum at $x = 0$, and a vertical tangent at $x = 1$.

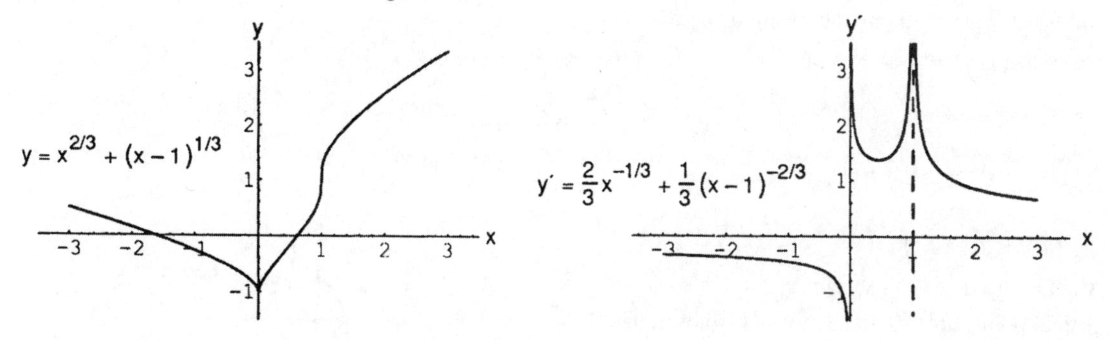

41. The values of the first derivative indicate that the curve is always rising. The slope of the curve approaches ∞ as $x \to 0$ and as $x \to 1$, indicating vertical tangents at both $x = 0$ and $x = 1$.

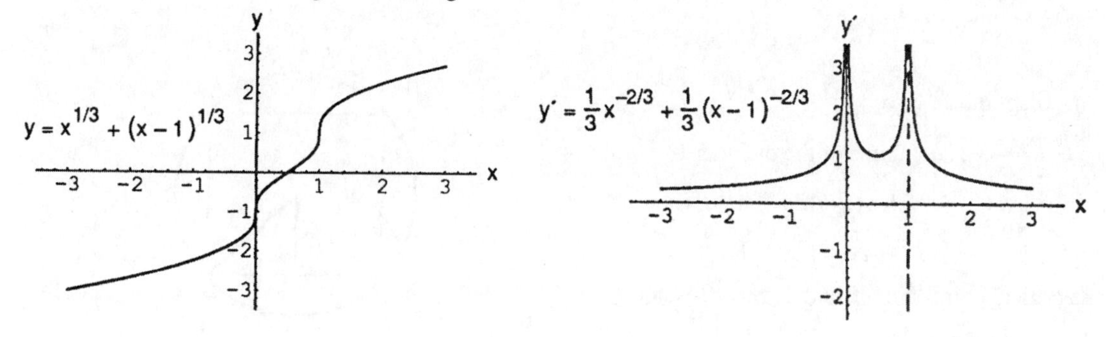

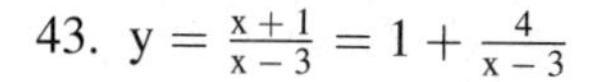

43. $y = \frac{x+1}{x-3} = 1 + \frac{4}{x-3}$

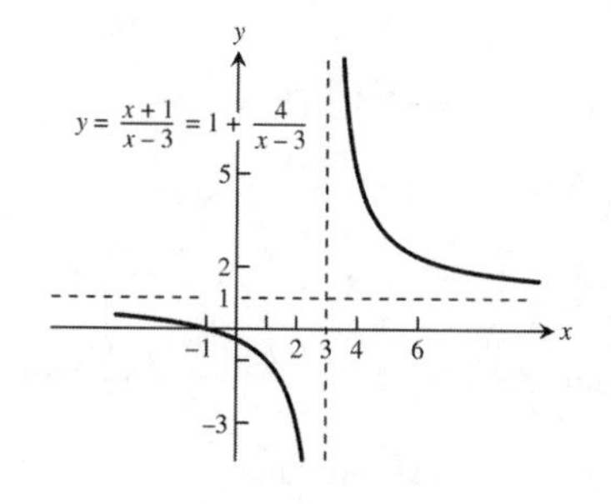

45. $y = \frac{x^2+1}{x} = x + \frac{1}{x}$

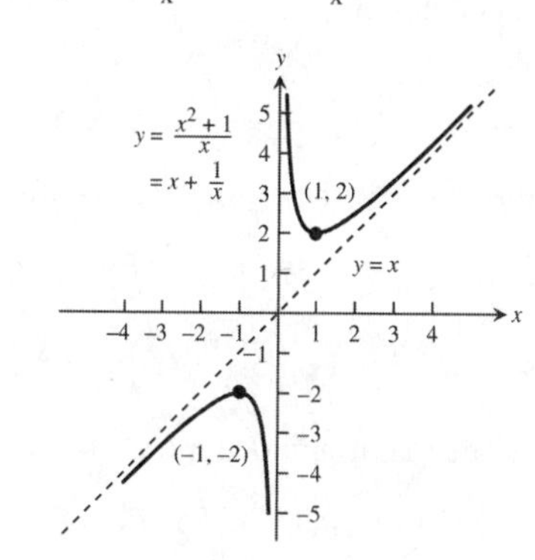

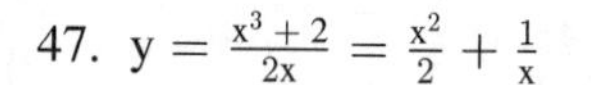

47. $y = \frac{x^3+2}{2x} = \frac{x^2}{2} + \frac{1}{x}$

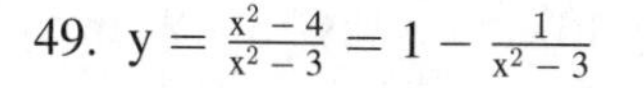

49. $y = \frac{x^2-4}{x^2-3} = 1 - \frac{1}{x^2-3}$

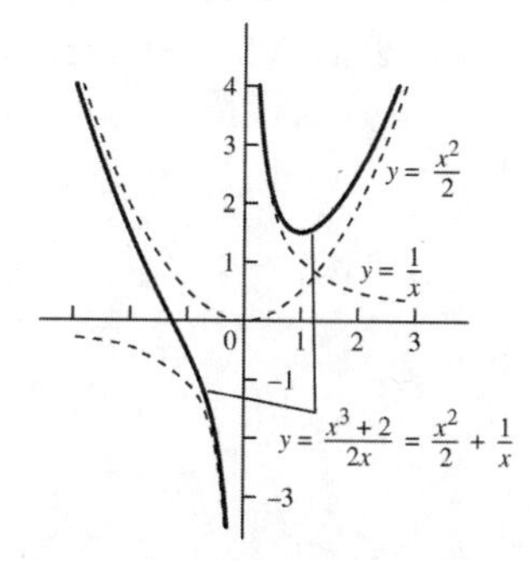

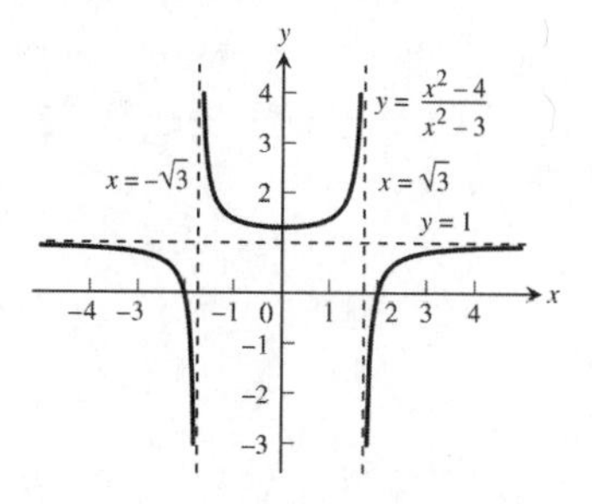

51. (a) Maximize $f(x) = \sqrt{x} - \sqrt{36 - x} = x^{1/2} - (36 - x)^{1/2}$ where $0 \le x \le 36$

$\Rightarrow f'(x) = \frac{1}{2} x^{-1/2} - \frac{1}{2}(36 - x)^{-1/2}(-1) = \frac{\sqrt{36 - x} + \sqrt{x}}{2\sqrt{x}\sqrt{36 - x}} \Rightarrow$ derivative fails to exist at 0 and 36; $f(0) = -6$,

and $f(36) = 6 \Rightarrow$ the numbers are 0 and 36

(b) Maximize $g(x) = \sqrt{x} + \sqrt{36 - x} = x^{1/2} + (36 - x)^{1/2}$ where $0 \le x \le 36$

$\Rightarrow g'(x) = \frac{1}{2} x^{-1/2} + \frac{1}{2}(36 - x)^{-1/2}(-1) = \frac{\sqrt{36 - x} - \sqrt{x}}{2\sqrt{x}\sqrt{36 - x}} \Rightarrow$ critical points at 0, 18 and 36; $g(0) = 6$,

$g(18) = 2\sqrt{18} = 6\sqrt{2}$ and $g(36) = 6 \Rightarrow$ the numbers are 18 and 18

53. $A(x) = \frac{1}{2}(2x)(27 - x^2)$ for $0 \le x \le \sqrt{27}$
$\Rightarrow A'(x) = 3(3 + x)(3 - x)$ and $A''(x) = -6x$.
The critical points are -3 and 3, but -3 is not in the domain. Since $A''(3) = -18 < 0$ and $A\left(\sqrt{27}\right) = 0$, the maximum occurs at $x = 3 \Rightarrow$ the largest area is $A(3) = 54$ sq units.

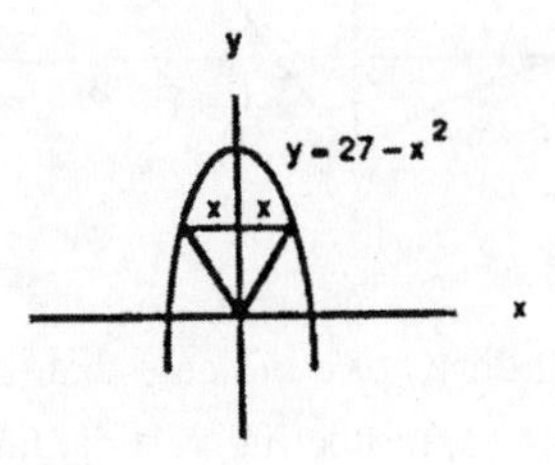

55. From the diagram we have $\left(\frac{h}{2}\right)^2 + r^2 = \left(\sqrt{3}\right)^2$
$\Rightarrow r^2 = \frac{12-h^2}{4}$. The volume of the cylinder is
$V = \pi r^2 h = \pi\left(\frac{12 - h^2}{4}\right) h = \frac{\pi}{4}(12h - h^3)$, where
$0 \le h \le 2\sqrt{3}$. Then $V'(h) = \frac{3\pi}{4}(2 + h)(2 - h)$
$\Rightarrow$ the critical points are -2 and 2, but -2 is not in the domain. At $h = 2$ there is a maximum since $V''(2) = -3\pi < 0$. The dimensions of the largest cylinder are radius $= \sqrt{2}$ and height $= 2$.

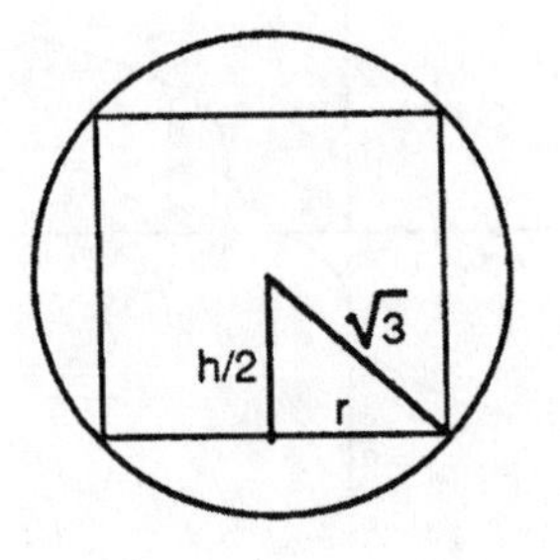

57. The profit $P = 2px + py = 2px + p\left(\frac{40 - 10x}{5 - x}\right)$, where p is the profit on grade B tires and $0 \le x \le 4$. Thus $P'(x) = \frac{2p}{(5 - x)^2}(x^2 - 10x + 20) \Rightarrow$ the critical points are $\left(5 - \sqrt{5}\right)$, 5, and $\left(5 + \sqrt{5}\right)$, but only $\left(5 - \sqrt{5}\right)$ is in the domain. Now $P'(x) > 0$ for $0 < x < \left(5 - \sqrt{5}\right)$ and $P'(x) < 0$ for $\left(5 - \sqrt{5}\right) < x < 4 \Rightarrow$ at $x = \left(5 - \sqrt{5}\right)$ there is a local maximum. Also $P(0) = 8p$, $P\left(5 - \sqrt{5}\right) = 4p\left(5 - \sqrt{5}\right) \approx 11p$, and $P(4) = 8p \Rightarrow$ at $x = \left(5 - \sqrt{5}\right)$ there is an absolute maximum. The maximum occurs when $x = \left(5 - \sqrt{5}\right)$ and $y = 2\left(5 - \sqrt{5}\right)$, the units are hundreds of tires, i.e., $x \approx 276$ tires and $y \approx 553$ tires.

59. The dimensions will be x in. by $10 - 2x$ in. by $16 - 2x$ in., so $V(x) = x(10 - 2x)(16 - 2x) = 4x^3 - 52x^2 + 160x$ for $0 < x < 5$. Then $V'(x) = 12x^2 - 104x + 160 = 4(x - 2)(3x - 20)$, so the critical point in the correct domain is $x = 2$. This critical point corresponds to the maximum possible volume because $V'(x) > 0$ for $0 < x < 2$ and $V'(x) < 0$ for $2 < x < 5$. The box of largest volume has a height of 2 in. and a base measuring 6 in. by 12 in., and its volume is 144 in.^{3}
Graphical support:

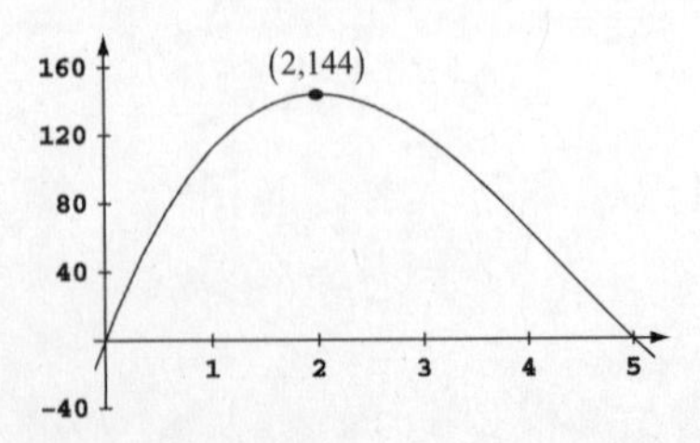

61. $g(x) = 3x - x^3 + 4 \Rightarrow g(2) = 2 > 0$ and $g(3) = -14 < 0 \Rightarrow g(x) = 0$ in the interval $[2, 3]$ by the Intermediate Value Theorem. Then $g'(x) = 3 - 3x^2 \Rightarrow x_{n+1} = x_n - \frac{3x_n - x_n^3 + 4}{3-3x_n^2}$; $x_0 = 2 \Rightarrow x_1 = 2.\overline{22} \Rightarrow x_2 = 2.196215$, and so forth to $x_5 = 2.195823345$.

63. $\int (x^3 + 5x - 7)\,dx = \frac{x^4}{4} + \frac{5x^2}{2} - 7x + C$

65. $\int \left(3\sqrt{t} + \frac{4}{t^2}\right) dt = \int \left(3t^{1/2} + 4t^{-2}\right) dt = \frac{3t^{3/2}}{\left(\frac{3}{2}\right)} + \frac{4t^{-1}}{-1} + C = 2t^{3/2} - \frac{4}{t} + C$

67. Let $u = r + 5 \Rightarrow du = dr$
$\int \frac{dr}{(r+5)^2} = \int \frac{du}{u^2} = \int u^{-2}\,du = \frac{u^{-1}}{-1} + C = -u^{-1} + C = -\frac{1}{(r+5)} + C$

69. Let $u = \theta^2 + 1 \Rightarrow du = 2\theta\,d\theta \Rightarrow \frac{1}{2}\,du = \theta\,d\theta$
$\int 3\theta\sqrt{\theta^2 + 1}\,d\theta = \int \sqrt{u}\left(\frac{3}{2}\,du\right) = \frac{3}{2}\int u^{1/2}\,du = \frac{3}{2}\left(\frac{u^{3/2}}{\frac{3}{2}}\right) + C = u^{3/2} + C = \left(\theta^2 + 1\right)^{3/2} + C$

71. Let $u = 1 + x^4 \Rightarrow du = 4x^3\,dx \Rightarrow \frac{1}{4}\,du = x^3\,dx$
$\int x^3\left(1 + x^4\right)^{-1/4} dx = \int u^{-1/4}\left(\frac{1}{4}\,du\right) = \frac{1}{4}\int u^{-1/4}\,du = \frac{1}{4}\left(\frac{u^{3/4}}{\frac{3}{4}}\right) + C = \frac{1}{3}u^{3/4} + C = \frac{1}{3}\left(1 + x^4\right)^{3/4} + C$

73. Let $u = \frac{s}{10} \Rightarrow du = \frac{1}{10}\,ds \Rightarrow 10\,du = ds$
$\int \sec^2 \frac{s}{10}\,ds = \int (\sec^2 u)(10\,du) = 10\int \sec^2 u\,du = 10\tan u + C = 10\tan\frac{s}{10} + C$

75. Let $u = \sqrt{2}\,\theta \Rightarrow du = \sqrt{2}\,d\theta \Rightarrow \frac{1}{\sqrt{2}}\,du = d\theta$
$\int \csc\sqrt{2}\theta\cot\sqrt{2}\theta\,d\theta = \int (\csc u\cot u)\left(\frac{1}{\sqrt{2}}\,du\right) = \frac{1}{\sqrt{2}}(-\csc u) + C = -\frac{1}{\sqrt{2}}\csc\sqrt{2}\theta + C$

77. Let $u = \frac{x}{4} \Rightarrow du = \frac{1}{4}\,dx \Rightarrow 4\,du = dx$
$\int \sin^2\frac{x}{4}\,dx = \int (\sin^2 u)(4\,du) = \int 4\left(\frac{1 - \cos 2u}{2}\right) du = 2\int (1 - \cos 2u)\,du = 2\left(u - \frac{\sin 2u}{2}\right) + C$
$= 2u - \sin 2u + C = 2\left(\frac{x}{4}\right) - \sin 2\left(\frac{x}{4}\right) + C = \frac{x}{2} - \sin\frac{x}{2} + C$

79. $y = \int \frac{x^2+1}{x^2}\,dx = \int \left(1 + x^{-2}\right) dx = x - x^{-1} + C = x - \frac{1}{x} + C$; $y = -1$ when $x = 1 \Rightarrow 1 - \frac{1}{1} + C = -1$
$\Rightarrow C = -1 \Rightarrow y = x - \frac{1}{x} - 1$

81. $\frac{dr}{dt} = \int \left(15\sqrt{t} + \frac{3}{\sqrt{t}}\right) dt = \int \left(15t^{1/2} + 3t^{-1/2}\right) dt = 10t^{3/2} + 6t^{1/2} + C$; $\frac{dr}{dt} = 8$ when $t = 1$
$\Rightarrow 10(1)^{3/2} + 6(1)^{1/2} + C = 8 \Rightarrow C = -8$. Thus $\frac{dr}{dt} = 10t^{3/2} + 6t^{1/2} - 8 \Rightarrow r = \int \left(10t^{3/2} + 6t^{1/2} - 8\right) dt$
$= 4t^{5/2} + 4t^{3/2} - 8t + C$; $r = 0$ when $t = 1 \Rightarrow 4(1)^{5/2} + 4(1)^{3/2} - 8(1) + C_1 = 0 \Rightarrow C_1 = 0$. Therefore, $r = 4t^{5/2} + 4t^{3/2} - 8t$

CHAPTER 4 ADDITIONAL AND ADVANCED EXERCISES

1. If M and m are the maximum and minimum values, respectively, then $m \le f(x) \le M$ for all $x \in I$. If $m = M$ then f is constant on I.

3. On an open interval the extreme values of a continuous function (if any) must occur at an interior critical point. On a half-open interval the extreme values of a continuous function may be at a critical point or at the closed endpoint. Extreme values occur only where $f' = 0$, f' does not exist, or at the endpoints of the interval. Thus the extreme points will not be at the ends of an open interval.

5. (a) If $y' = 6(x+1)(x-2)^2$, then $y' < 0$ for $x < -1$ and $y' > 0$ for $x > -1$. The sign pattern is $f' = \underset{-1}{---|} \underset{2}{+++|}+++ \Rightarrow$ f has a local minimum at $x = -1$. Also $y'' = 6(x-2)^2 + 12(x+1)(x-2)$ $= 6(x-2)(3x) \Rightarrow y'' > 0$ for $x < 0$ or $x > 2$, while $y'' < 0$ for $0 < x < 2$. Therefore f has points of inflection at $x = 0$ and $x = 2$. There is no local maximum.
   (b) If $y' = 6x(x+1)(x-2)$, then $y' < 0$ for $x < -1$ and $0 < x < 2$; $y' > 0$ for $-1 < x < 0$ and $x > 2$. The sign sign pattern is $y' = \underset{-1}{---|} \underset{0}{+++|} \underset{2}{---|}+++$. Therefore f has a local maximum at $x = 0$ and local minima at $x = -1$ and $x = 2$. Also, $y'' = 18\left[x - \left(\frac{1-\sqrt{7}}{3}\right)\right]\left[x - \left(\frac{1+\sqrt{7}}{3}\right)\right]$, so $y'' < 0$ for $\frac{1-\sqrt{7}}{3} < x < \frac{1+\sqrt{7}}{3}$ and $y'' > 0$ for all other x $\Rightarrow$ f has points of inflection at $x = \frac{1 \pm \sqrt{7}}{3}$.

7. If f is continuous on $[a, c)$ and $f'(x) \le 0$ on $[a, c)$, then by the Mean Value Theorem for all $x \in [a, c)$ we have $\frac{f(c) - f(x)}{c - x} \le 0 \Rightarrow f(c) - f(x) \le 0 \Rightarrow f(x) \ge f(c)$. Also if f is continuous on $(c, b]$ and $f'(x) \ge 0$ on $(c, b]$, then for all $x \in (c, b]$ we have $\frac{f(x) - f(c)}{x - c} \ge 0 \Rightarrow f(x) - f(c) \ge 0 \Rightarrow f(x) \ge f(c)$. Therefore $f(x) \ge f(c)$ for all $x \in [a, b]$.

9. No. Corollary 1 requires that $f'(x) = 0$ for <u>all</u> x in some interval I, not $f'(x) = 0$ at a single point in I.

11. From (ii), $f(-1) = \frac{-1+a}{b-c+2} = 0 \Rightarrow a = 1$; from (iii), either $1 = \lim_{x \to \infty} f(x)$ or $1 = \lim_{x \to -\infty} f(x)$. In either case, $\lim_{x \to \pm\infty} f(x) = \lim_{x \to \pm\infty} \frac{x+1}{bx^2+cx+2} = \lim_{x \to \pm\infty} \frac{1+\frac{1}{x}}{bx+c+\frac{2}{x}} = 1 \Rightarrow b = 0$ and $c = 1$. For if $b = 1$, then $\lim_{x \to \pm\infty} \frac{1+\frac{1}{x}}{x+c+\frac{2}{x}} = 0$ and if $c = 0$, then $\lim_{x \to \pm\infty} \frac{1+\frac{1}{x}}{bx+\frac{2}{x}} = \lim_{x \to \pm\infty} \frac{1+\frac{1}{x}}{\frac{2}{x}} = \pm\infty$. Thus $a = 1$, $b = 0$, and $c = 1$.

13. The area of the ΔABC is $A(x) = \frac{1}{2}(2)\sqrt{1-x^2} = (1-x^2)^{1/2}$, where $0 \le x \le 1$. Thus $A'(x) = \frac{-x}{\sqrt{1-x^2}} \Rightarrow 0$ and ± 1 are critical points. Also $A(\pm 1) = 0$ so $A(0) = 1$ is the maximum. When $x = 0$ the ΔABC is isosceles since $AC = BC = \sqrt{2}$.

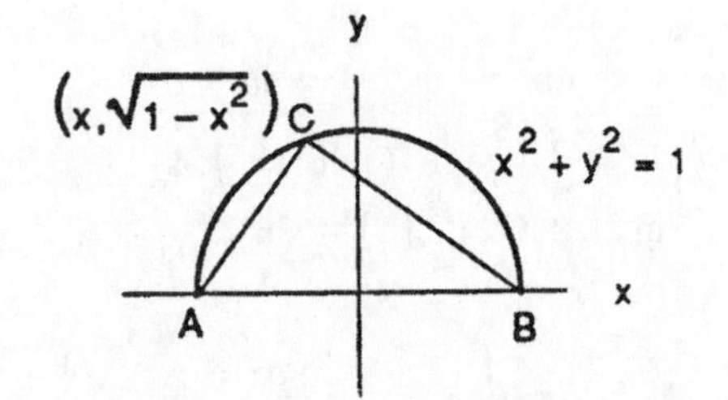

15. The time it would take the water to hit the ground from height y is $\sqrt{\frac{2y}{g}}$, where g is the acceleration of gravity. The product of time and exit velocity (rate) yields the distance the water travels:
$D(y) = \sqrt{\frac{2y}{g}}\sqrt{64(h-y)} = 8\sqrt{\frac{2}{g}}(hy - y^2)^{1/2}, 0 \le y \le h \Rightarrow D'(y) = -4\sqrt{\frac{2}{g}}(hy-y^2)^{-1/2}(h-2y) \Rightarrow 0, \frac{h}{2}$ and h are critical points. Now $D(0) = 0$, $D\left(\frac{h}{2}\right) = 8\sqrt{\frac{2}{g}}\left(h\left(\frac{h}{2}\right) - \left(\frac{h}{2}\right)^2\right)^{1/2} = 4h\sqrt{\frac{2}{g}}$ and $D(h) = 0 \Rightarrow$ the best place to drill the hole is at $y = \frac{h}{2}$.

17. The surface area of the cylinder is $S = 2\pi r^2 + 2\pi rh$. From the diagram we have $\frac{r}{R} = \frac{H-h}{H} \Rightarrow h = \frac{RH - rH}{R}$ and
$S(r) = 2\pi r(r+h) = 2\pi r\left(r + H - r\frac{H}{R}\right)$
$= 2\pi\left(1 - \frac{H}{R}\right)r^2 + 2\pi Hr$, where $0 \le r \le R$.

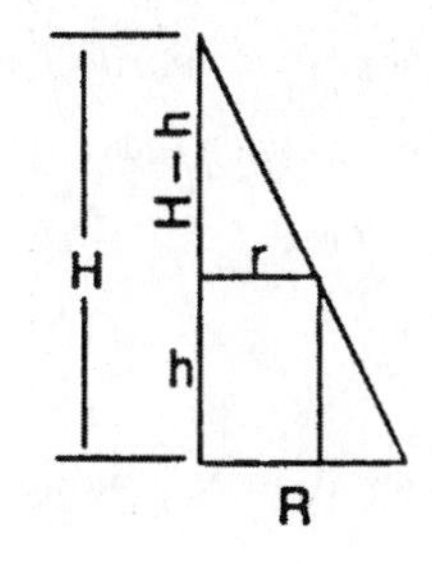

Case 1: $H < R \Rightarrow S(r)$ is a quadratic equation containing the origin and concave upward $\Rightarrow S(r)$ is maximum at $r = R$.

Case 2: $H = R \Rightarrow S(r)$ is a linear equation containing the origin with a positive slope $\Rightarrow S(r)$ is maximum at $r = R$.

Case 3: $H > R \Rightarrow S(r)$ is a quadratic equation containing the origin and concave downward. Then $\frac{dS}{dr} = 4\pi\left(1 - \frac{H}{R}\right)r + 2\pi H$ and $\frac{dS}{dr} = 0 \Rightarrow 4\pi\left(1 - \frac{H}{R}\right)r + 2\pi H = 0 \Rightarrow r = \frac{RH}{2(H-R)}$. For simplification we let $r^* = \frac{RH}{2(H-R)}$.

(a) If $R < H < 2R$, then $0 > H - 2R \Rightarrow H > 2(H-R) \Rightarrow r^* = \frac{RH}{2(H-R)} > R$. Therefore, the maximum occurs at the right endpoint R of the interval $0 \le r \le R$ because S(r) is an increasing function of r.

(b) If $H = 2R$, then $r^* = \frac{2R^2}{2R} = R \Rightarrow S(r)$ is maximum at $r = R$.

(c) If $H > 2R$, then $2R + H < 2H \Rightarrow H < 2(H-R) \Rightarrow \frac{H}{2(H-R)} < 1 \Rightarrow \frac{RH}{2(H-R)} < R \Rightarrow r^* < R$. Therefore, S(r) is a maximum at $r = r^* = \frac{RH}{2(H-R)}$.

<u>Conclusion</u>: If $H \in (0, 2R]$, then the maximum surface area is at $r = R$. If $H \in (2R, \infty)$, then the maximum is at $r = r^* = \frac{RH}{2(H-R)}$.

19. (a) The profit function is $P(x) = (c - ex)x - (a + bx) = -ex^2 + (c-b)x - a$. $P'(x) = -2ex + c - b = 0 \Rightarrow x = \frac{c-b}{2e}$. $P''(x) = -2e < 0$ if $e > 0$ so that the profit function is maximized at $x = \frac{c-b}{2e}$.

(b) The price therefore that corresponds to a production level yeilding a maximum profit is $p\Big|_{x=\frac{c-b}{2e}} = c - e\left(\frac{c-b}{2e}\right) = \frac{c+b}{2}$ dollars.

(c) The weekly profit at this production level is $P(x) = -e\left(\frac{c-b}{2e}\right)^2 + (c-b)\left(\frac{c-b}{2e}\right) - a = \frac{(c-b)^2}{4e} - a$.

(d) The tax increases cost to the new profit function is $F(x) = (c - ex)x - (a + bx + tx) = -ex^2 + (c - b - t)x - a$. Now $F'(x) = -2ex + c - b - t = 0$ when $x = \frac{t+b-c}{-2e} = \frac{c-b-t}{2e}$. Since $F''(x) = -2e < 0$ if $e > 0$, F is maximized when $x = \frac{c-b-t}{2e}$ units per week. Thus the price per unit is $p = c - e\left(\frac{c-b-t}{2e}\right) = \frac{c+b+t}{2}$ dollars. Thus, such a tax increases the cost per unit by $\frac{c+b+t}{2} - \frac{c+b}{2} = \frac{t}{2}$ dollars if units are priced to maximize profit.

21. $x_1 = x_0 - \frac{f(x_0)}{f'(x_0)} = x_0 - \frac{x_0^q - a}{qx_0^{q-1}} = \frac{qx_0^q - x_0^q + a}{qx_0^{q-1}} = \frac{x_0^q(q-1) + a}{qx_0^{q-1}} = x_0\left(\frac{q-1}{q}\right) + \frac{a}{x_0^{q-1}}\left(\frac{1}{q}\right)$ so that x_1 is a weighted average of x_0 and $\frac{a}{x_0^{q-1}}$ with weights $m_0 = \frac{q-1}{q}$ and $m_1 = \frac{1}{q}$.

In the case where $x_0 = \frac{a}{x_0^{q-1}}$ we have $x_0^q = a$ and $x_1 = \frac{a}{x_0^{q-1}}\left(\frac{q-1}{q}\right) + \frac{a}{x_0^{q-1}}\left(\frac{1}{q}\right) = \frac{a}{x_0^{q-1}}\left(\frac{q-1}{q} + \frac{1}{q}\right) = \frac{a}{x_0^{q-1}}$.

23. (a) $a(t) = s''(t) = -k \ (k > 0) \Rightarrow s'(t) = -kt + C_1$, where $s'(0) = 88 \Rightarrow C_1 = 88 \Rightarrow s'(t) = -kt + 88$. So $s(t) = \frac{-kt^2}{2} + 88t + C_2$ where $s(0) = 0 \Rightarrow C_2 = 0$ so $s(t) = \frac{-kt^2}{2} + 88t$. Now $s(t) = 100$ when $\frac{-kt^2}{2} + 88t = 100$. Solving for t we obtain $t = \frac{88 \pm \sqrt{88^2 - 200k}}{k}$. At such t we want $s'(t) = 0$, thus $-k\left(\frac{88 + \sqrt{88^2 - 200k}}{k}\right) + 88 = 0$ or $-k\left(\frac{88 - \sqrt{88^2 - 200k}}{k}\right) + 88 = 0$. In either case we obtain $88^2 - 200k = 0$ so that $k = \frac{88^2}{200} \approx 38.72$ ft/sec^2.

(b) The initial condition that $s'(0) = 44$ ft/sec implies that $s'(t) = -kt + 44$ and $s(t) = \frac{-kt^2}{2} + 44t$ where k is as above. The car is stopped at a time t such that $s'(t) = -kt + 44 = 0 \Rightarrow t = \frac{44}{k}$. At this time the car has traveled a distance $s\left(\frac{44}{k}\right) = \frac{-k}{2}\left(\frac{44}{k}\right)^2 + 44\left(\frac{44}{k}\right) = \frac{44^2}{2k} = \frac{968}{k} = 968\left(\frac{200}{88^2}\right) = 25$ feet. Thus halving the initial velocity quarters stopping distance.

25. Yes. The curve $y = x$ satisfies all three conditions since $\frac{dy}{dx} = 1$ everywhere, when $x = 0$, $y = 0$, and $\frac{d^2y}{dx^2} = 0$ everywhere.

27. $s''(t) = a = -t^2 \Rightarrow v = s'(t) = \frac{-t^3}{3} + C$. We seek $v_0 = s'(0) = C$. We know that $s(t^*) = b$ for some t^* and s is at a maximum for this t^*. Since $s(t) = \frac{-t^4}{12} + Ct + k$ and $s(0) = 0$ we have that $s(t) = \frac{-t^4}{12} + Ct$ and also $s'(t^*) = 0$ so that $t^* = (3C)^{1/3}$. So $\frac{\left[-(3C)^{1/3}\right]^4}{12} + C(3C)^{1/3} = b \Rightarrow (3C)^{1/3}\left(C - \frac{3C}{12}\right) = b \Rightarrow (3C)^{1/3}\left(\frac{3C}{4}\right) = b \Rightarrow 3^{1/3}C^{4/3} = \frac{4b}{3}$ $\Rightarrow C = \frac{(4b)^{3/4}}{3}$. Thus $v_0 = s'(0) = \frac{(4b)^{3/4}}{3} = \frac{2\sqrt{2}}{3}b^{3/4}$.

29. The graph of $f(x) = ax^2 + bx + c$ with $a > 0$ is a parabola opening upwards. Thus $f(x) \geq 0$ for all x if $f(x) = 0$ for at most one real value of x. The solutions to $f(x) = 0$ are, by the quadratic equation $\frac{-2b \pm \sqrt{(2b)^2 - 4ac}}{2a}$. Thus we require $(2b)^2 - 4ac \leq 0 \Rightarrow b^2 - ac \leq 0$.

CHAPTER 5 INTEGRATION

5.1 AREA AND ESTIMATING WITH FINITE SUMS

1. $f(x) = x^2$

Since f is increasing on $[0, 1]$, we use left endpoints to obtain lower sums and right endpoints to obtain upper sums.

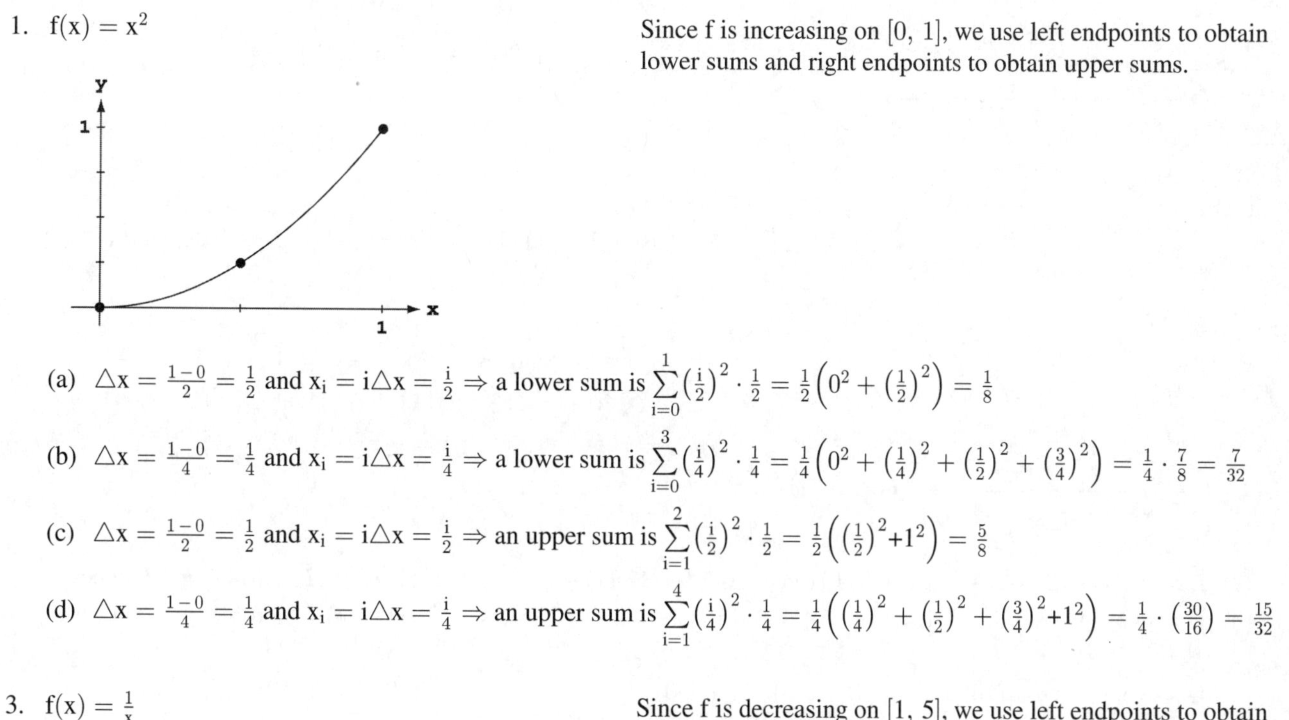

(a) $\triangle x = \frac{1-0}{2} = \frac{1}{2}$ and $x_i = i\triangle x = \frac{i}{2} \Rightarrow$ a lower sum is $\sum_{i=0}^{1} \left(\frac{i}{2}\right)^2 \cdot \frac{1}{2} = \frac{1}{2}\left(0^2 + \left(\frac{1}{2}\right)^2\right) = \frac{1}{8}$

(b) $\triangle x = \frac{1-0}{4} = \frac{1}{4}$ and $x_i = i\triangle x = \frac{i}{4} \Rightarrow$ a lower sum is $\sum_{i=0}^{3} \left(\frac{i}{4}\right)^2 \cdot \frac{1}{4} = \frac{1}{4}\left(0^2 + \left(\frac{1}{4}\right)^2 + \left(\frac{1}{2}\right)^2 + \left(\frac{3}{4}\right)^2\right) = \frac{1}{4} \cdot \frac{7}{8} = \frac{7}{32}$

(c) $\triangle x = \frac{1-0}{2} = \frac{1}{2}$ and $x_i = i\triangle x = \frac{i}{2} \Rightarrow$ an upper sum is $\sum_{i=1}^{2} \left(\frac{i}{2}\right)^2 \cdot \frac{1}{2} = \frac{1}{2}\left(\left(\frac{1}{2}\right)^2 + 1^2\right) = \frac{5}{8}$

(d) $\triangle x = \frac{1-0}{4} = \frac{1}{4}$ and $x_i = i\triangle x = \frac{i}{4} \Rightarrow$ an upper sum is $\sum_{i=1}^{4} \left(\frac{i}{4}\right)^2 \cdot \frac{1}{4} = \frac{1}{4}\left(\left(\frac{1}{4}\right)^2 + \left(\frac{1}{2}\right)^2 + \left(\frac{3}{4}\right)^2 + 1^2\right) = \frac{1}{4} \cdot \left(\frac{30}{16}\right) = \frac{15}{32}$

3. $f(x) = \frac{1}{x}$

Since f is decreasing on $[1, 5]$, we use left endpoints to obtain upper sums and right endpoints to obtain lower sums.

(a) $\triangle x = \frac{5-1}{2} = 2$ and $x_i = 1 + i\triangle x = 1 + 2i \Rightarrow$ a lower sum is $\sum_{i=1}^{2} \frac{1}{x_i} \cdot 2 = 2\left(\frac{1}{3} + \frac{1}{5}\right) = \frac{16}{15}$

(b) $\triangle x = \frac{5-1}{4} = 1$ and $x_i = 1 + i\triangle x = 1 + i \Rightarrow$ a lower sum is $\sum_{i=1}^{4} \frac{1}{x_i} \cdot 1 = 1\left(\frac{1}{2} + \frac{1}{3} + \frac{1}{4} + \frac{1}{5}\right) = \frac{77}{60}$

(c) $\triangle x = \frac{5-1}{2} = 2$ and $x_i = 1 + i\triangle x = 1 + 2i \Rightarrow$ an upper sum is $\sum_{i=0}^{1} \frac{1}{x_i} \cdot 2 = 2\left(1 + \frac{1}{3}\right) = \frac{8}{3}$

(d) $\triangle x = \frac{5-1}{4} = 1$ and $x_i = 1 + i\triangle x = 1 + i \Rightarrow$ an upper sum is $\sum_{i=0}^{3} \frac{1}{x_i} \cdot 1 = 1\left(1 + \frac{1}{2} + \frac{1}{3} + \frac{1}{4}\right) = \frac{25}{12}$

5. $f(x) = x^2$

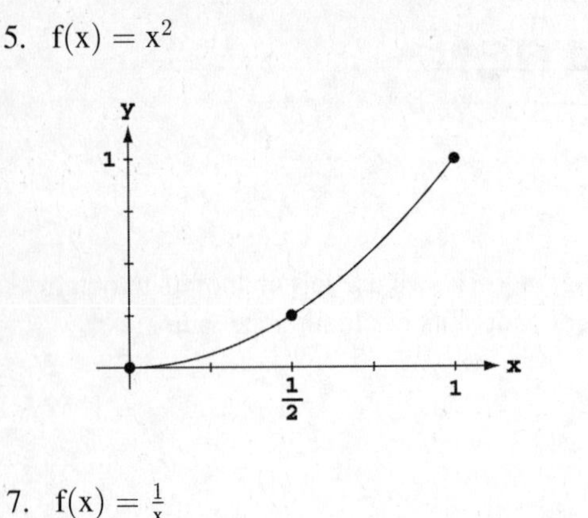

Using 2 rectangles $\Rightarrow \triangle x = \frac{1-0}{2} = \frac{1}{2} \Rightarrow \frac{1}{2}\left(f\left(\frac{1}{4}\right) + f\left(\frac{3}{4}\right)\right)$
$= \frac{1}{2}\left(\left(\frac{1}{4}\right)^2 + \left(\frac{3}{4}\right)^2\right) = \frac{10}{32} = \frac{5}{16}$

Using 4 rectangles $\Rightarrow \triangle x = \frac{1-0}{4} = \frac{1}{4}$
$\Rightarrow \frac{1}{4}\left(f\left(\frac{1}{8}\right) + f\left(\frac{3}{8}\right) + f\left(\frac{5}{8}\right) + f\left(\frac{7}{8}\right)\right)$
$= \frac{1}{4}\left(\left(\frac{1}{8}\right)^2 + \left(\frac{3}{8}\right)^2 + \left(\frac{5}{8}\right)^2 + \left(\frac{7}{8}\right)^2\right) = \frac{21}{64}$

7. $f(x) = \frac{1}{x}$

Using 2 rectangles $\Rightarrow \triangle x = \frac{5-1}{2} = 2 \Rightarrow 2(f(2) + f(4))$
$= 2\left(\frac{1}{2} + \frac{1}{4}\right) = \frac{3}{2}$

Using 4 rectangles $\Rightarrow \triangle x = \frac{5-1}{4} = 1$
$\Rightarrow 1\left(f\left(\frac{3}{2}\right) + f\left(\frac{5}{2}\right) + f\left(\frac{7}{2}\right) + f\left(\frac{9}{2}\right)\right)$
$= 1\left(\frac{2}{3} + \frac{2}{5} + \frac{2}{7} + \frac{2}{9}\right) = \frac{1488}{3 \cdot 5 \cdot 7 \cdot 9} = \frac{496}{5 \cdot 7 \cdot 9} = \frac{496}{315}$

9. (a) $D \approx (0)(1) + (12)(1) + (22)(1) + (10)(1) + (5)(1) + (13)(1) + (11)(1) + (6)(1) + (2)(1) + (6)(1) = 87$ inches
(b) $D \approx (12)(1) + (22)(1) + (10)(1) + (5)(1) + (13)(1) + (11)(1) + (6)(1) + (2)(1) + (6)(1) + (0)(1) = 87$ inches

11. (a) $D \approx (0)(10) + (44)(10) + (15)(10) + (35)(10) + (30)(10) + (44)(10) + (35)(10) + (15)(10) + (22)(10)$ $+ (35)(10) + (44)(10) + (30)(10) = 3490$ feet ≈ 0.66 miles
(b) $D \approx (44)(10) + (15)(10) + (35)(10) + (30)(10) + (44)(10) + (35)(10) + (15)(10) + (22)(10) + (35)(10)$ $+ (44)(10) + (30)(10) + (35)(10) = 3840$ feet ≈ 0.73 miles

13. (a) Because the acceleration is decreasing, an upper estimate is obtained using left end-points in summing acceleration $\cdot \Delta t$. Thus, $\Delta t = 1$ and speed $\approx [32.00 + 19.41 + 11.77 + 7.14 + 4.33](1) = 74.65$ ft/sec
(b) Using right end-points we obtain a lower estimate: speed $\approx [19.41 + 11.77 + 7.14 + 4.33 + 2.63](1)$ $= 45.28$ ft/sec
(c) Upper estimates for the speed at each second are:

t	0	1	2	3	4	5
v	0	32.00	51.41	63.18	70.32	74.65

Thus, the distance fallen when $t = 3$ seconds is $s \approx [32.00 + 51.41 + 63.18](1) = 146.59$ ft.

15. Partition $[0, 2]$ into the four subintervals $[0, 0.5]$, $[0.5, 1]$, $[1, 1.5]$, and $[1.5, 2]$. The midpoints of these subintervals are $m_1 = 0.25$, $m_2 = 0.75$, $m_3 = 1.25$, and $m_4 = 1.75$. The heights of the four approximating rectangles are $f(m_1) = (0.25)^3 = \frac{1}{64}$, $f(m_2) = (0.75)^3 = \frac{27}{64}$, $f(m_3) = (1.25)^3 = \frac{125}{64}$, and $f(m_4) = (1.75)^3 = \frac{343}{64}$
Notice that the average value is approximated by $\frac{1}{2}\left[\left(\frac{1}{4}\right)^3\left(\frac{1}{2}\right) + \left(\frac{3}{4}\right)^3\left(\frac{1}{2}\right) + \left(\frac{5}{4}\right)^3\left(\frac{1}{2}\right) + \left(\frac{7}{4}\right)^3\left(\frac{1}{2}\right)\right] = \frac{31}{16}$
$= \frac{1}{\text{length of }[0,2]} \cdot \begin{bmatrix}\text{approximate area under} \\ \text{curve } f(x) = x^3\end{bmatrix}$. We use this observation in solving the next several exercises.

17. Partition $[0, 2]$ into the four subintervals $[0, 0.5]$, $[0.5, 1]$, $[1, 1.5]$, and $[1.5, 2]$. The midpoints of the subintervals are $m_1 = 0.25$, $m_2 = 0.75$, $m_3 = 1.25$, and $m_4 = 1.75$. The heights of the four approximating rectangles are $f(m_1) = \frac{1}{2} + \sin^2\frac{\pi}{4} = \frac{1}{2} + \frac{1}{2} = 1$, $f(m_2) = \frac{1}{2} + \sin^2\frac{3\pi}{4} = \frac{1}{2} + \frac{1}{2} = 1$, $f(m_3) = \frac{1}{2} + \sin^2\frac{5\pi}{4} = \frac{1}{2} + \left(-\frac{1}{\sqrt{2}}\right)^2$

$= \frac{1}{2} + \frac{1}{2} = 1$, and $f(m_4) = \frac{1}{2} + \sin^2 \frac{7\pi}{4} = \frac{1}{2} + \left(-\frac{1}{\sqrt{2}}\right)^2 = 1$. The width of each rectangle is $\Delta x = \frac{1}{2}$. Thus, Area $\approx (1+1+1+1)\left(\frac{1}{2}\right) = 2 \Rightarrow$ average value $\approx \frac{\text{area}}{\text{length of } [0,2]} = \frac{2}{2} = 1$.

19. Since the leakage is increasing, an upper estimate uses right endpoints and a lower estimate uses left endpoints:
    (a) upper estimate $= (70)(1) + (97)(1) + (136)(1) + (190)(1) + (265)(1) = 758$ gal,
    lower estimate $= (50)(1) + (70)(1) + (97)(1) + (136)(1) + (190)(1) = 543$ gal.
    (b) upper estimate $= (70 + 97 + 136 + 190 + 265 + 369 + 516 + 720) = 2363$ gal,
    lower estimate $= (50 + 70 + 97 + 136 + 190 + 265 + 369 + 516) = 1693$ gal.
    (c) worst case: $2363 + 720t = 25{,}000 \Rightarrow t \approx 31.4$ hrs;
    best case: $1693 + 720t = 25{,}000 \Rightarrow t \approx 32.4$ hrs

21. (a) The diagonal of the square has length 2, so the side length is $\sqrt{2}$. Area $= \left(\sqrt{2}\right)^2 = 2$
    (b) Think of the octagon as a collection of 16 right triangles with a hypotenuse of length 1 and an acute angle measuring $\frac{2\pi}{16} = \frac{\pi}{8}$.
    Area $= 16\left(\frac{1}{2}\right)\left(\sin \frac{\pi}{8}\right)\left(\cos \frac{\pi}{8}\right) = 4 \sin \frac{\pi}{4} = 2\sqrt{2} \approx 2.828$
    (c) Think of the 16-gon as a collection of 32 right triangles with a hypotenuse of length 1 and an acute angle measuring $\frac{2\pi}{32} = \frac{\pi}{16}$.
    Area $= 32\left(\frac{1}{2}\right)\left(\sin \frac{\pi}{16}\right)\left(\cos \frac{\pi}{16}\right) = 8 \sin \frac{\pi}{8} = 2\sqrt{2} \approx 3.061$
    (d) Each area is less than the area of the circle, π. As n increases, the area approaches π.

5.2 SIGMA NOTATION AND LIMITS OF FINITE SUMS

1. $\sum_{k=1}^{2} \frac{6k}{k+1} = \frac{6(1)}{1+1} + \frac{6(2)}{2+1} = \frac{6}{2} + \frac{12}{3} = 7$

3. $\sum_{k=1}^{4} \cos k\pi = \cos(1\pi) + \cos(2\pi) + \cos(3\pi) + \cos(4\pi) = -1 + 1 - 1 + 1 = 0$

5. $\sum_{k=1}^{3} (-1)^{k+1} \sin \frac{\pi}{k} = (-1)^{1+1} \sin \frac{\pi}{1} + (-1)^{2+1} \sin \frac{\pi}{2} + (-1)^{3+1} \sin \frac{\pi}{3} = 0 - 1 + \frac{\sqrt{3}}{2} = \frac{\sqrt{3}-2}{2}$

7. (a) $\sum_{k=1}^{6} 2^{k-1} = 2^{1-1} + 2^{2-1} + 2^{3-1} + 2^{4-1} + 2^{5-1} + 2^{6-1} = 1 + 2 + 4 + 8 + 16 + 32$
    (b) $\sum_{k=0}^{5} 2^{k} = 2^0 + 2^1 + 2^2 + 2^3 + 2^4 + 2^5 = 1 + 2 + 4 + 8 + 16 + 32$
    (c) $\sum_{k=-1}^{4} 2^{k+1} = 2^{-1+1} + 2^{0+1} + 2^{1+1} + 2^{2+1} + 2^{3+1} + 2^{4+1} = 1 + 2 + 4 + 8 + 16 + 32$
    All of them represent $1 + 2 + 4 + 8 + 16 + 32$

9. (a) $\sum_{k=2}^{4} \frac{(-1)^{k-1}}{k-1} = \frac{(-1)^{2-1}}{2-1} + \frac{(-1)^{3-1}}{3-1} + \frac{(-1)^{4-1}}{4-1} = -1 + \frac{1}{2} - \frac{1}{3}$
    (b) $\sum_{k=0}^{2} \frac{(-1)^{k}}{k+1} = \frac{(-1)^0}{0+1} + \frac{(-1)^1}{1+1} + \frac{(-1)^2}{2+1} = 1 - \frac{1}{2} + \frac{1}{3}$
    (c) $\sum_{k=-1}^{1} \frac{(-1)^{k}}{k+2} = \frac{(-1)^{-1}}{-1+2} + \frac{(-1)^0}{0+2} + \frac{(-1)^1}{1+2} = -1 + \frac{1}{2} - \frac{1}{3}$
    (a) and (c) are equivalent; (b) is not equivalent to the other two.

11. $\sum_{k=1}^{6} k$

13. $\sum_{k=1}^{4} \frac{1}{2^k}$

15. $\sum_{k=1}^{5} (-1)^{k+1} \frac{1}{k}$

17. (a) $\sum_{k=1}^{n} 3a_k = 3\sum_{k=1}^{n} a_k = 3(-5) = -15$

(b) $\sum_{k=1}^{n} \frac{b_k}{6} = \frac{1}{6}\sum_{k=1}^{n} b_k = \frac{1}{6}(6) = 1$

(c) $\sum_{k=1}^{n} (a_k + b_k) = \sum_{k=1}^{n} a_k + \sum_{k=1}^{n} b_k = -5 + 6 = 1$

(d) $\sum_{k=1}^{n} (a_k - b_k) = \sum_{k=1}^{n} a_k - \sum_{k=1}^{n} b_k = -5 - 6 = -11$

(e) $\sum_{k=1}^{n} (b_k - 2a_k) = \sum_{k=1}^{n} b_k - 2\sum_{k=1}^{n} a_k = 6 - 2(-5) = 16$

19. (a) $\sum_{k=1}^{10} k = \frac{10(10+1)}{2} = 55$

(b) $\sum_{k=1}^{10} k^2 = \frac{10(10+1)(2(10)+1)}{6} = 385$

(c) $\sum_{k=1}^{10} k^3 = \left[\frac{10(10+1)}{2}\right]^2 = 55^2 = 3025$

21. $\sum_{k=1}^{7} -2k = -2\sum_{k=1}^{7} k = -2\left(\frac{7(7+1)}{2}\right) = -56$

23. $\sum_{k=1}^{6} (3 - k^2) = \sum_{k=1}^{6} 3 - \sum_{k=1}^{6} k^2 = 3(6) - \frac{6(6+1)(2(6)+1)}{6} = -73$

25. $\sum_{k=1}^{5} k(3k+5) = \sum_{k=1}^{5} (3k^2 + 5k) = 3\sum_{k=1}^{5} k^2 + 5\sum_{k=1}^{5} k = 3\left(\frac{5(5+1)(2(5)+1)}{6}\right) + 5\left(\frac{5(5+1)}{2}\right) = 240$

27. $\sum_{k=1}^{5} \frac{k^3}{225} + \left(\sum_{k=1}^{5} k\right)^3 = \frac{1}{225}\sum_{k=1}^{5} k^3 + \left(\sum_{k=1}^{5} k\right)^3 = \frac{1}{225}\left(\frac{5(5+1)}{2}\right)^2 + \left(\frac{5(5+1)}{2}\right)^3 = 3376$

29. (a) $\sum_{k=1}^{7} 3 = 3(7) = 21$

(b) $\sum_{k=1}^{500} 7 = 7(500) = 3500$

(c) Let $j = k - 2 \Rightarrow k = j + 2$; if $k = 3 \Rightarrow j = 1$ and if $k = 264 \Rightarrow j = 262 \Rightarrow \sum_{k=3}^{264} 10 = \sum_{j=1}^{262} 10 = 10(262) = 2620$

31. (a) $\sum_{k=1}^{n} 4 = 4n$

(b) $\sum_{k=1}^{n} c = cn$

(c) $\sum_{k=1}^{n} (k-1) = \sum_{k=1}^{n} k - \sum_{k=1}^{n} 1 = \frac{n(n+1)}{2} - n = \frac{n^2 - n}{2}$

33. (a) (b) (c)

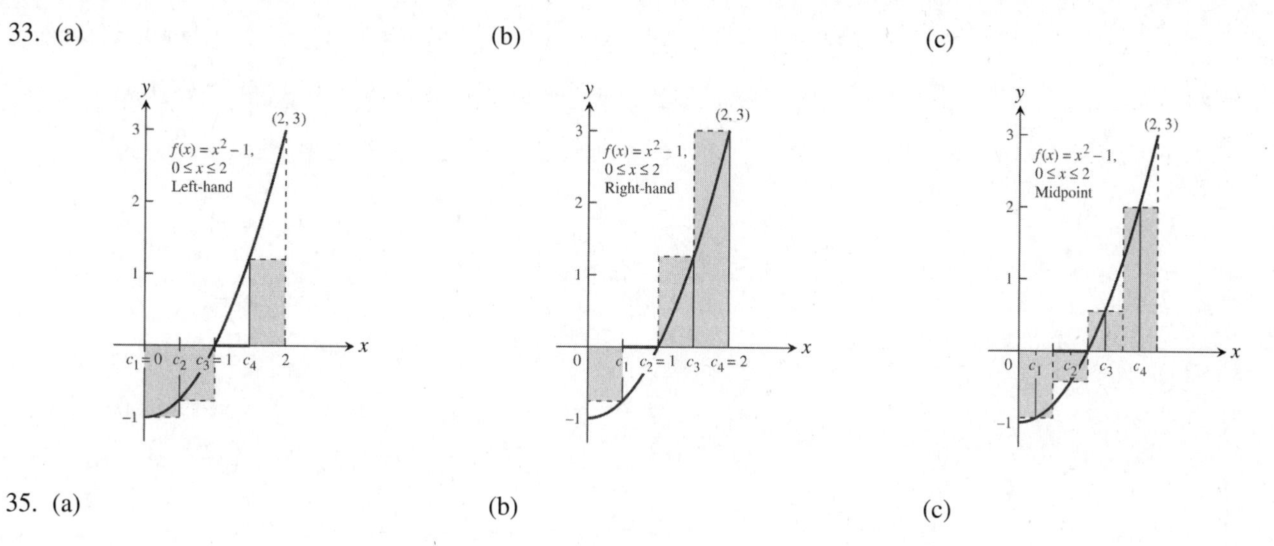

35. (a) (b) (c)

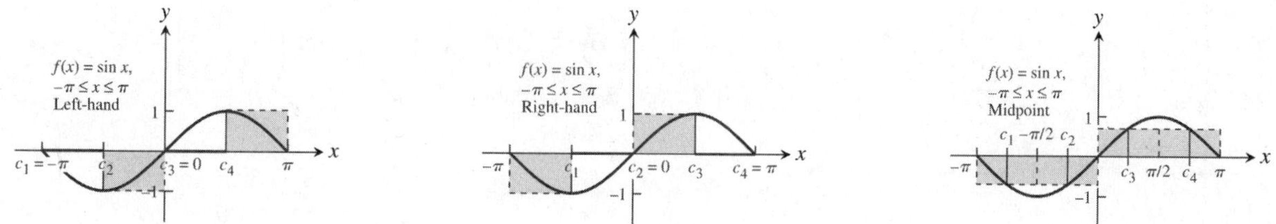

37. $|x_1 - x_0| = |1.2 - 0| = 1.2$, $|x_2 - x_1| = |1.5 - 1.2| = 0.3$, $|x_3 - x_2| = |2.3 - 1.5| = 0.8$, $|x_4 - x_3| = |2.6 - 2.3| = 0.3$, and $|x_5 - x_4| = |3 - 2.6| = 0.4$; the largest is $\|P\| = 1.2$.

39. $f(x) = 1 - x^2$

Let $\triangle x = \frac{1-0}{n} = \frac{1}{n}$ and $c_i = i\triangle x = \frac{i}{n}$. The right-hand sum is $\sum_{i=1}^{n}(1 - c_i^2)\frac{1}{n} = \frac{1}{n}\sum_{i=1}^{n}\left(1 - \left(\frac{i}{n}\right)^2\right) = \frac{1}{n^3}\sum_{i=1}^{n}(n^2 - i^2)$
$= \frac{n^3}{n^3} - \frac{1}{n^3}\sum_{i=1}^{n} i^2 = 1 - \frac{n(n+1)(2n+1)}{6n^3} = 1 - \frac{2n^3+3n^2+n}{6n^3}$
$= 1 - \frac{2+\frac{3}{n}+\frac{1}{n^2}}{6}$. Thus, $\lim_{n\to\infty}\sum_{i=1}^{n}(1 - c_i^2)\frac{1}{n}$
$= \lim_{n\to\infty}\left(1 - \frac{2-\frac{3}{n}+\frac{1}{n^2}}{6}\right) = 1 - \frac{1}{3} = \frac{2}{3}$

41. $f(x) = x^2 + 1$

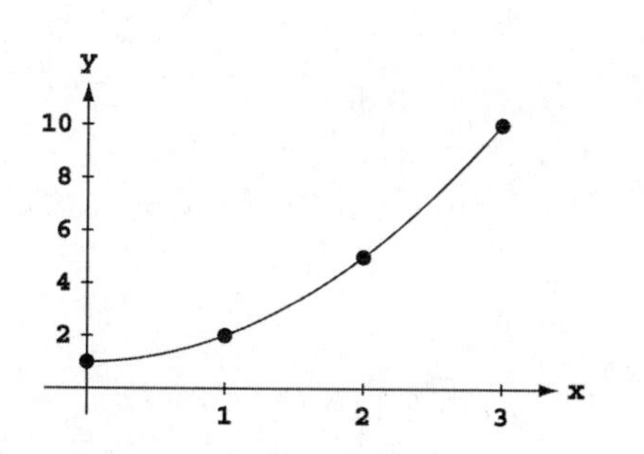

Let $\triangle x = \frac{3-0}{n} = \frac{3}{n}$ and $c_i = i\triangle x = \frac{3i}{n}$. The right-hand sum is $\sum_{i=1}^{n}(c_i^2 + 1)\frac{3}{n} = \sum_{i=1}^{n}\left(\left(\frac{3i}{n}\right)^2 + 1\right)\frac{3}{n} = \frac{3}{n}\sum_{i=1}^{n}\left(\frac{9i^2}{n^2} + 1\right)$
$= \frac{27}{n}\sum_{i=1}^{n} i^2 + \frac{3}{n}\cdot n = \frac{27}{n^3}\left(\frac{n(n+1)(2n+1)}{6}\right) + 3$
$= \frac{9(2n^3+3n^2+n)}{2n^3} + 3 = \frac{18+\frac{27}{n}+\frac{9}{n^2}}{2} + 3$. Thus,
$\lim_{n\to\infty}\sum_{i=1}^{n}(c_i^2 + 1)\frac{3}{n} = \lim_{n\to\infty}\left(\frac{18+\frac{27}{n}+\frac{9}{n^2}}{2} + 3\right) = 9 + 3 = 12.$

43. $f(x) = x + x^2 = x(1+x)$

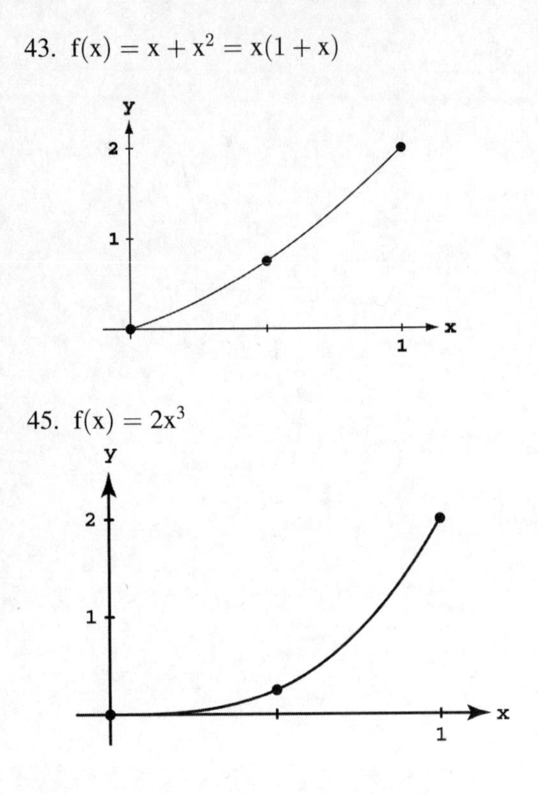

Let $\triangle x = \frac{1-0}{n} = \frac{1}{n}$ and $c_i = i\triangle x = \frac{i}{n}$. The right-hand sum is
$\sum_{i=1}^{n}(c_i + c_i^2)\frac{1}{n} = \sum_{i=1}^{n}\left(\frac{i}{n} + \left(\frac{i}{n}\right)^2\right)\frac{1}{n} = \frac{1}{n^2}\sum_{i=1}^{n} i + \frac{1}{n^3}\sum_{i=1}^{n} i^2$
$= \frac{1}{n^2}\left(\frac{n(n+1)}{2}\right) + \frac{1}{n^3}\left(\frac{n(n+1)(2n+1)}{6}\right) = \frac{n^2+n}{2n^2} + \frac{2n^3+3n^2+n}{6n^3}$
$= \frac{1+\frac{1}{n}}{2} + \frac{2+\frac{3}{n}+\frac{1}{n^2}}{6}$. Thus, $\lim_{n\to\infty} \sum_{i=1}^{n}(c_i + c_i^2)\frac{1}{n}$
$= \lim_{n\to\infty}\left[\left(\frac{1+\frac{1}{n}}{2}\right) + \left(\frac{2+\frac{3}{n}+\frac{1}{n^2}}{6}\right)\right] = \frac{1}{2} + \frac{2}{6} = \frac{5}{6}$.

45. $f(x) = 2x^3$

Let $\triangle x = \frac{1-0}{n} = \frac{1}{n}$ and $c_i = i\triangle x = \frac{i}{n}$. The right-hand sum is
$\sum_{i=1}^{n}(2c_i^3)\frac{1}{n} = \sum_{i=1}^{n}\left(2\left(\frac{i}{n}\right)^3\right)\frac{1}{n} = \frac{2}{n^4}\sum_{i=1}^{n} i^3 = \frac{2}{n^4}\left(\frac{n(n+1)}{2}\right)^2$
$= \frac{2n^2(n^2+2n+1)}{4n^4} = \frac{n^2+2n+1}{2n^2} = \frac{1+\frac{2}{n}+\frac{1}{n^2}}{2}$.
Thus, $\lim_{n\to\infty} \sum_{i=1}^{n}(2c_i^3)\frac{1}{n} = \lim_{n\to\infty}\left[\frac{1+\frac{2}{n}+\frac{1}{n^2}}{2}\right] = \frac{1}{2}$.

5.3 THE DEFINITE INTEGRAL

1. $\int_0^2 x^2\,dx$

3. $\int_{-7}^{5} (x^2 - 3x)\,dx$

5. $\int_2^3 \frac{1}{1-x}\,dx$

7. $\int_{-\pi/4}^{0} (\sec x)\,dx$

9. (a) $\int_2^2 g(x)\,dx = 0$

(b) $\int_5^1 g(x)\,dx = -\int_1^5 g(x)\,dx = -8$

(c) $\int_1^2 3f(x)\,dx = 3\int_1^2 f(x)\,dx = 3(-4) = -12$

(d) $\int_2^5 f(x)\,dx = \int_1^5 f(x)\,dx - \int_1^2 f(x)\,dx = 6 - (-4) = 10$

(e) $\int_1^5 [f(x) - g(x)]\,dx = \int_1^5 f(x)\,dx - \int_1^5 g(x)\,dx = 6 - 8 = -2$

(f) $\int_1^5 [4f(x) - g(x)]\,dx = 4\int_1^5 f(x)\,dx - \int_1^5 g(x)\,dx = 4(6) - 8 = 16$

11. (a) $\int_1^2 f(u)\,du = \int_1^2 f(x)\,dx = 5$

(b) $\int_1^2 \sqrt{3}\,f(z)\,dz = \sqrt{3}\int_1^2 f(z)\,dz = 5\sqrt{3}$

(c) $\int_2^1 f(t)\,dt = -\int_1^2 f(t)\,dt = -5$

(d) $\int_1^2 [-f(x)]\,dx = -\int_1^2 f(x)\,dx = -5$

13. (a) $\int_3^4 f(z)\,dz = \int_0^4 f(z)\,dz - \int_0^3 f(z)\,dz = 7 - 3 = 4$

(b) $\int_4^3 f(t)\,dt = -\int_3^4 f(t)\,dt = -4$

15. The area of the trapezoid is $A = \frac{1}{2}(B + b)h$
$= \frac{1}{2}(5 + 2)(6) = 21 \Rightarrow \int_{-2}^{4} \left(\frac{x}{2} + 3\right) dx$
$= 21$ square units

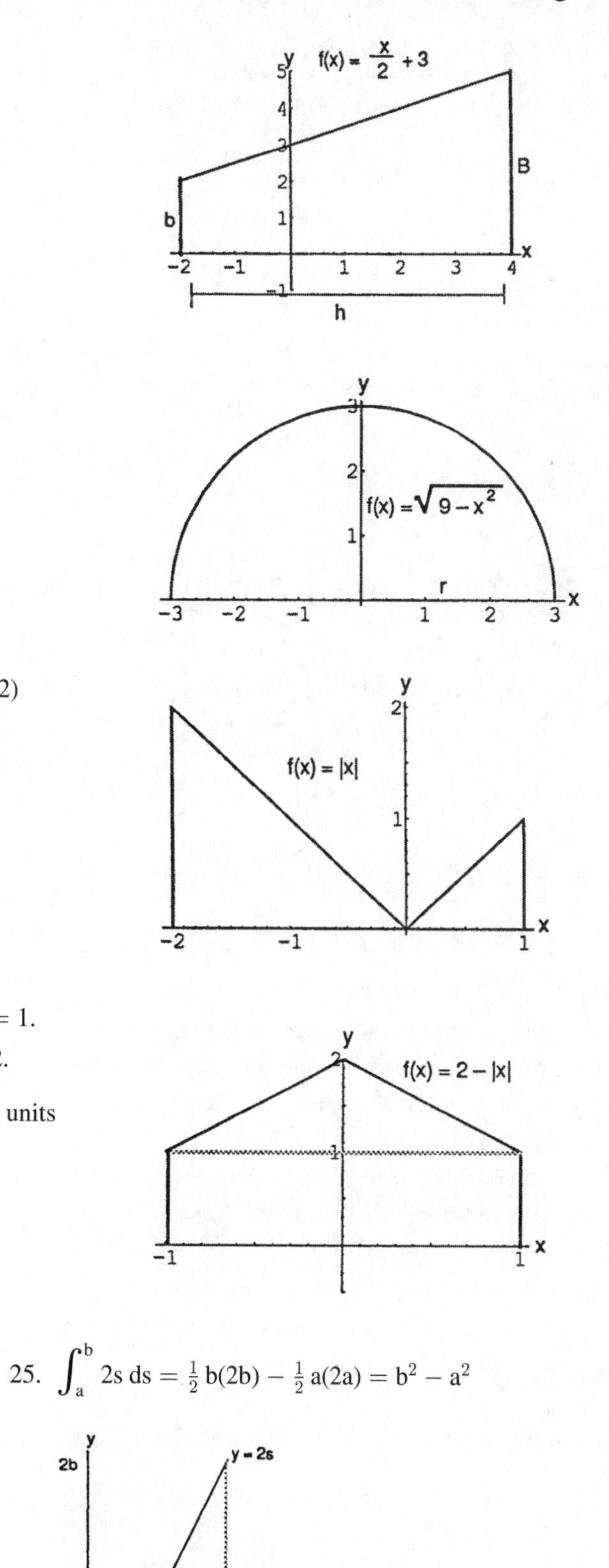

17. The area of the semicircle is $A = \frac{1}{2}\pi r^2 = \frac{1}{2}\pi(3)^2$
$= \frac{9}{2}\pi \Rightarrow \int_{-3}^{3} \sqrt{9 - x^2}\, dx = \frac{9}{2}\pi$ square units

19. The area of the triangle on the left is $A = \frac{1}{2}bh = \frac{1}{2}(2)(2)$
$= 2$. The area of the triangle on the right is $A = \frac{1}{2}bh$
$= \frac{1}{2}(1)(1) = \frac{1}{2}$. Then, the total area is 2.5
$\Rightarrow \int_{-2}^{1} |x|\, dx = 2.5$ square units

21. The area of the triangular peak is $A = \frac{1}{2}bh = \frac{1}{2}(2)(1) = 1$.
The area of the rectangular base is $S = \ell w = (2)(1) = 2$.
Then the total area is $3 \Rightarrow \int_{-1}^{1} (2 - |x|)\, dx = 3$ square units

23. $\int_0^b \frac{x}{2}\, dx = \frac{1}{2}(b)\left(\frac{b}{2}\right) = \frac{b^2}{4}$

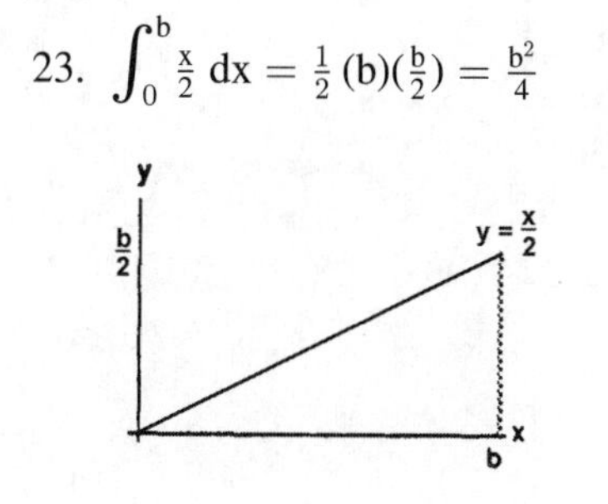

25. $\int_a^b 2s\, ds = \frac{1}{2}b(2b) - \frac{1}{2}a(2a) = b^2 - a^2$

27. (a) $\int_{-2}^{2} \sqrt{4 - x^2}\,dx = \frac{1}{2}\left[\pi(2)^2\right] = 2\pi$ (b) $\int_{0}^{2} \sqrt{4 - x^2}\,dx = \frac{1}{4}\left[\pi(2)^2\right] = \pi$

29. $\int_{1}^{\sqrt{2}} x\,dx = \frac{\left(\sqrt{2}\right)^2}{2} - \frac{(1)^2}{2} = \frac{1}{2}$

31. $\int_{\pi}^{2\pi} \theta\,d\theta = \frac{(2\pi)^2}{2} - \frac{\pi^2}{2} = \frac{3\pi^2}{2}$

33. $\int_{0}^{\sqrt[3]{7}} x^2\,dx = \frac{\left(\sqrt[3]{7}\right)^3}{3} = \frac{7}{3}$

35. $\int_{0}^{1/2} t^2\,dt = \frac{\left(\frac{1}{2}\right)^3}{3} = \frac{1}{24}$

37. $\int_{a}^{2a} x\,dx = \frac{(2a)^2}{2} - \frac{a^2}{2} = \frac{3a^2}{2}$

39. $\int_{0}^{\sqrt[3]{b}} x^2\,dx = \frac{\left(\sqrt[3]{b}\right)^3}{3} = \frac{b}{3}$

41. $\int_{3}^{1} 7\,dx = 7(1 - 3) = -14$

43. $\int_{0}^{2} (2t - 3)\,dt = 2\int_{1}^{1} t\,dt - \int_{0}^{2} 3\,dt = 2\left[\frac{2^2}{2} - \frac{0^2}{2}\right] - 3(2 - 0) = 4 - 6 = -2$

45. $\int_{2}^{1} \left(1 + \frac{z}{2}\right) dz = \int_{2}^{1} 1\,dz + \int_{2}^{1} \frac{z}{2}\,dz = \int_{2}^{1} 1\,dz - \frac{1}{2}\int_{1}^{2} z\,dz = 1[1 - 2] - \frac{1}{2}\left[\frac{2^2}{2} - \frac{1^2}{2}\right] = -1 - \frac{1}{2}\left(\frac{3}{2}\right) = -\frac{7}{4}$

47. $\int_{1}^{2} 3u^2\,du = 3\int_{1}^{2} u^2\,du = 3\left[\int_{0}^{2} u^2\,du - \int_{0}^{1} u^2\,du\right] = 3\left(\left[\frac{2^3}{3} - \frac{0^3}{3}\right] - \left[\frac{1^3}{3} - \frac{0^3}{3}\right]\right) = 3\left[\frac{2^3}{3} - \frac{1^3}{3}\right] = 3\left(\frac{7}{3}\right) = 7$

49. $\int_{0}^{2} (3x^2 + x - 5)\,dx = 3\int_{0}^{2} x^2\,dx + \int_{0}^{2} x\,dx - \int_{0}^{2} 5\,dx = 3\left[\frac{2^3}{3} - \frac{0^3}{3}\right] + \left[\frac{2^2}{2} - \frac{0^2}{2}\right] - 5[2 - 0] = (8 + 2) - 10 = 0$

51. Let $\Delta x = \frac{b-0}{n} = \frac{b}{n}$ and let $x_0 = 0$, $x_1 = \Delta x$, $x_2 = 2\Delta x, \ldots, x_{n-1} = (n-1)\Delta x$, $x_n = n\Delta x = b$. Let the c_k's be the right end-points of the subintervals $\Rightarrow c_1 = x_1$, $c_2 = x_2$, and so on. The rectangles defined have areas:

$f(c_1)\,\Delta x = f(\Delta x)\,\Delta x = 3(\Delta x)^2\,\Delta x = 3(\Delta x)^3$

$f(c_2)\,\Delta x = f(2\Delta x)\,\Delta x = 3(2\Delta x)^2\,\Delta x = 3(2)^2(\Delta x)^3$

$f(c_3)\,\Delta x = f(3\Delta x)\,\Delta x = 3(3\Delta x)^2\,\Delta x = 3(3)^2(\Delta x)^3$

$\vdots$

$f(c_n)\,\Delta x = f(n\Delta x)\,\Delta x = 3(n\Delta x)^2\,\Delta x = 3(n)^2(\Delta x)^3$

Then $S_n = \sum_{k=1}^{n} f(c_k)\,\Delta x = \sum_{k=1}^{n} 3k^2(\Delta x)^3$

$= 3(\Delta x)^3 \sum_{k=1}^{n} k^2 = 3\left(\frac{b^3}{n^3}\right)\left(\frac{n(n+1)(2n+1)}{6}\right)$

$= \frac{b^3}{2}\left(2 + \frac{3}{n} + \frac{1}{n^2}\right) \Rightarrow \int_{0}^{b} 3x^2\,dx = \lim_{n \to \infty} \frac{b^3}{2}\left(2 + \frac{3}{n} + \frac{1}{n^2}\right) = b^3.$

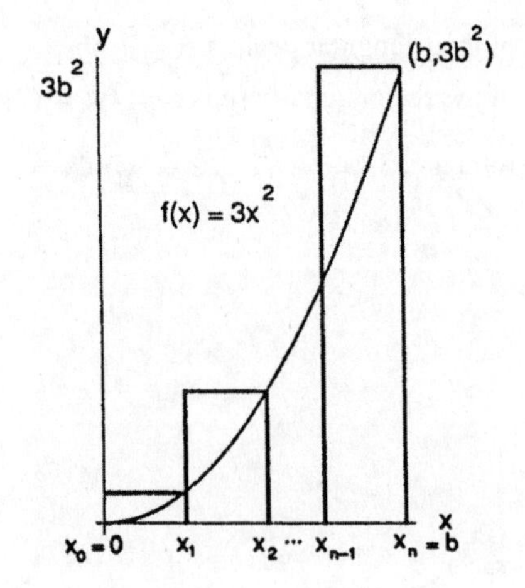

53. Let $\Delta x = \frac{b-0}{n} = \frac{b}{n}$ and let $x_0 = 0$, $x_1 = \Delta x$, $x_2 = 2\Delta x, \ldots, x_{n-1} = (n-1)\Delta x$, $x_n = n\Delta x = b$. Let the c_k's be the right end-points of the subintervals $\Rightarrow c_1 = x_1$, $c_2 = x_2$, and so on. The rectangles defined have areas:

$$f(c_1)\,\Delta x = f(\Delta x)\,\Delta x = 2(\Delta x)(\Delta x) = 2(\Delta x)^2$$
$$f(c_2)\,\Delta x = f(2\Delta x)\,\Delta x = 2(2\Delta x)(\Delta x) = 2(2)(\Delta x)^2$$
$$f(c_3)\,\Delta x = f(3\Delta x)\,\Delta x = 2(3\Delta x)(\Delta x) = 2(3)(\Delta x)^2$$
$$\vdots$$
$$f(c_n)\,\Delta x = f(n\Delta x)\,\Delta x = 2(n\Delta x)(\Delta x) = 2(n)(\Delta x)^2$$

Then $S_n = \sum_{k=1}^{n} f(c_k)\,\Delta x = \sum_{k=1}^{n} 2k(\Delta x)^2$

$= 2(\Delta x)^2 \sum_{k=1}^{n} k = 2\left(\frac{b^2}{n^2}\right)\left(\frac{n(n+1)}{2}\right)$

$= b^2\left(1+\frac{1}{n}\right) \Rightarrow \int_0^b 2x\,dx = \lim_{n\to\infty} b^2\left(1+\frac{1}{n}\right) = b^2.$

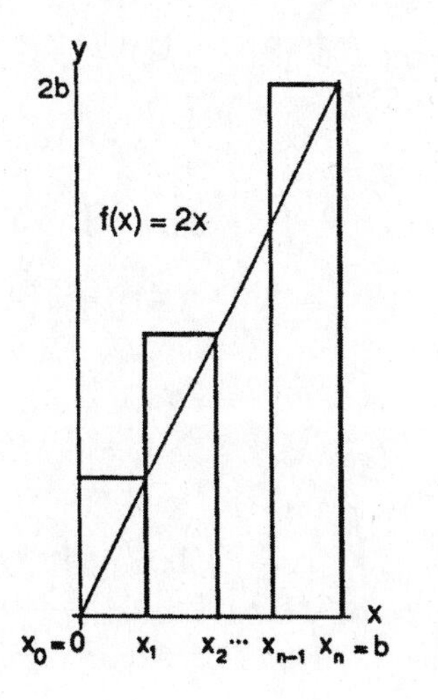

55. $\text{av}(f) = \left(\frac{1}{\sqrt{3}-0}\right)\int_0^{\sqrt{3}} (x^2-1)\,dx$

$= \frac{1}{\sqrt{3}}\int_0^{\sqrt{3}} x^2\,dx - \frac{1}{\sqrt{3}}\int_0^{\sqrt{3}} 1\,dx$

$= \frac{1}{\sqrt{3}}\left(\frac{(\sqrt{3})^3}{3}\right) - \frac{1}{\sqrt{3}}\left(\sqrt{3}-0\right) = 1 - 1 = 0.$

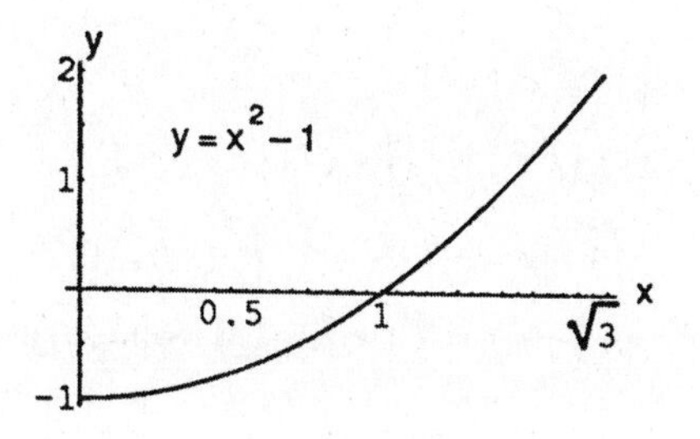

57. $\text{av}(f) = \left(\frac{1}{1-0}\right)\int_0^1 (-3x^2-1)\,dx =$

$= -3\int_0^1 x^2\,dx - \int_0^1 1\,dx = -3\left(\frac{1^3}{3}\right) - (1-0)$

$= -2.$

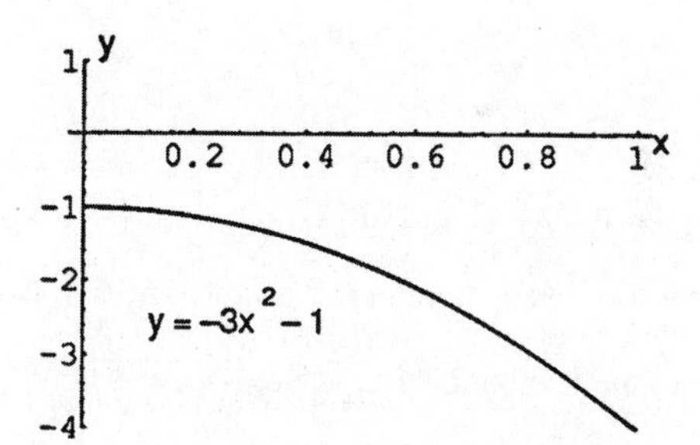

59. $\text{av}(f) = \left(\frac{1}{3-0}\right)\int_0^3 (t-1)^2\,dt$

$= \frac{1}{3}\int_0^3 t^2\,dt - \frac{2}{3}\int_0^3 t\,dt + \frac{1}{3}\int_0^3 1\,dt$

$= \frac{1}{3}\left(\frac{3^3}{3}\right) - \frac{2}{3}\left(\frac{3^2}{2} - \frac{0^2}{2}\right) + \frac{1}{3}(3-0) = 1.$

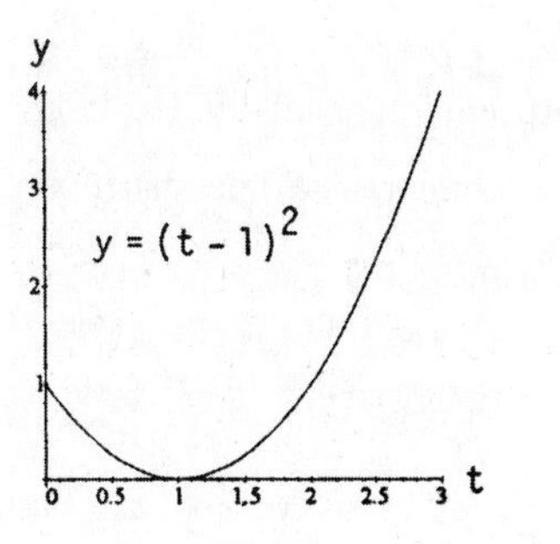

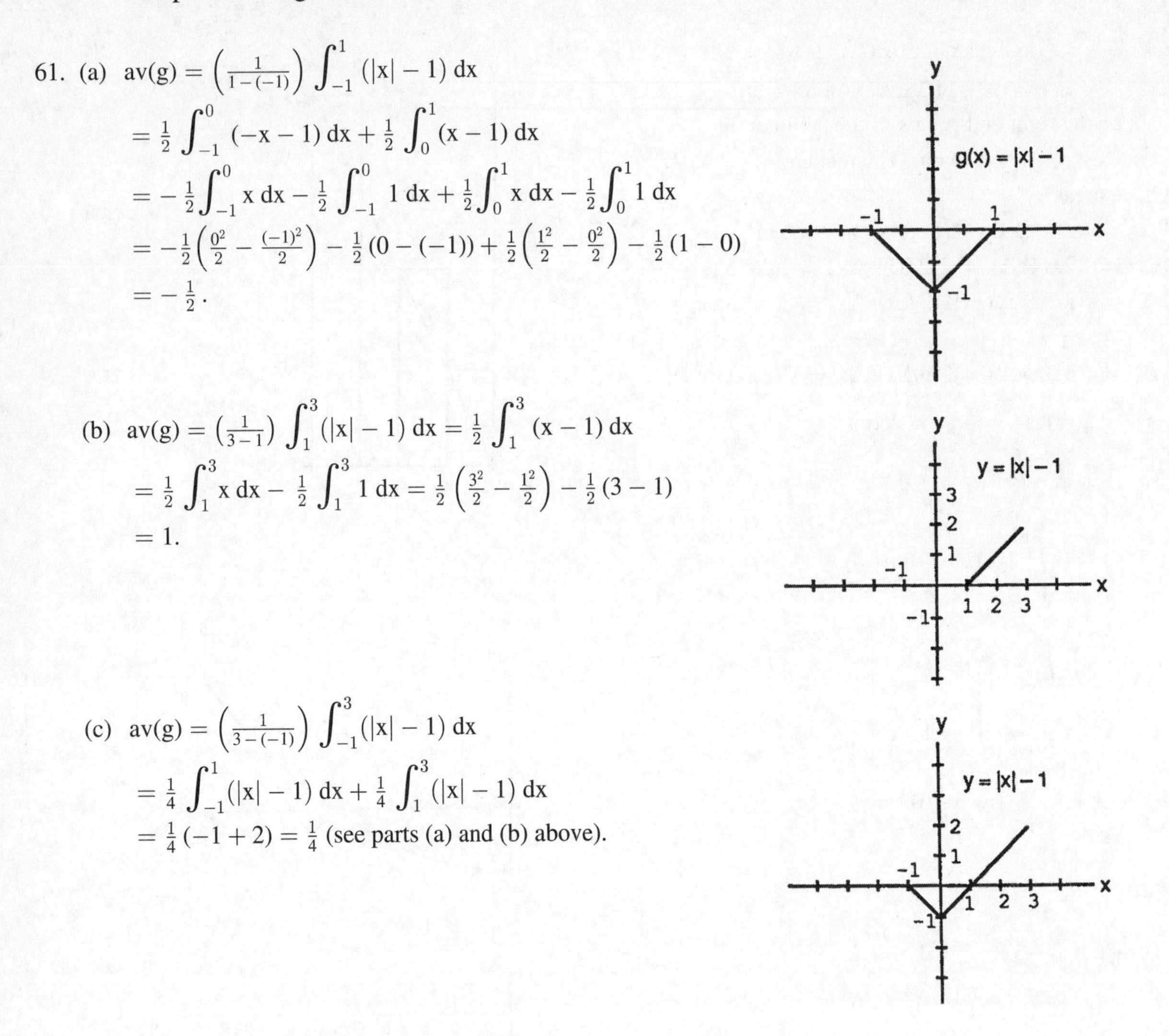

61. (a) $\text{av}(g) = \left(\frac{1}{1-(-1)}\right)\int_{-1}^{1}(|x|-1)\,dx$

$= \frac{1}{2}\int_{-1}^{0}(-x-1)\,dx + \frac{1}{2}\int_{0}^{1}(x-1)\,dx$

$= -\frac{1}{2}\int_{-1}^{0}x\,dx - \frac{1}{2}\int_{-1}^{0}1\,dx + \frac{1}{2}\int_{0}^{1}x\,dx - \frac{1}{2}\int_{0}^{1}1\,dx$

$= -\frac{1}{2}\left(\frac{0^2}{2} - \frac{(-1)^2}{2}\right) - \frac{1}{2}(0-(-1)) + \frac{1}{2}\left(\frac{1^2}{2} - \frac{0^2}{2}\right) - \frac{1}{2}(1-0)$

$= -\frac{1}{2}.$

(b) $\text{av}(g) = \left(\frac{1}{3-1}\right)\int_{1}^{3}(|x|-1)\,dx = \frac{1}{2}\int_{1}^{3}(x-1)\,dx$

$= \frac{1}{2}\int_{1}^{3}x\,dx - \frac{1}{2}\int_{1}^{3}1\,dx = \frac{1}{2}\left(\frac{3^2}{2} - \frac{1^2}{2}\right) - \frac{1}{2}(3-1)$

$= 1.$

(c) $\text{av}(g) = \left(\frac{1}{3-(-1)}\right)\int_{-1}^{3}(|x|-1)\,dx$

$= \frac{1}{4}\int_{-1}^{1}(|x|-1)\,dx + \frac{1}{4}\int_{1}^{3}(|x|-1)\,dx$

$= \frac{1}{4}(-1+2) = \frac{1}{4}$ (see parts (a) and (b) above).

63. Consider the partition P that subdivides the interval $[a, b]$ into n subintervals of width $\Delta x = \frac{b-a}{n}$ and let c_k be the right endpoint of each subinterval. So the partition is $P = \left\{a, a + \frac{b-a}{n}, a + \frac{2(b-a)}{n}, \ldots, a + \frac{n(b-a)}{n}\right\}$ and $c_k = a + \frac{k(b-a)}{n}$. We get the Riemann sum $\sum_{k=1}^{n} f(c_k)\Delta x = \sum_{k=1}^{n} c \cdot \frac{b-a}{n} = \frac{c(b-a)}{n}\sum_{k=1}^{n} 1 = \frac{c(b-a)}{n} \cdot n = c(b-a)$. As $n \to \infty$ and $\|P\| \to 0$ this expression remains $c(b-a)$. Thus, $\int_a^b c\,dx = c(b-a)$.

65. Consider the partition P that subdivides the interval $[a, b]$ into n subintervals of width $\Delta x = \frac{b-a}{n}$ and let c_k be the right endpoint of each subinterval. So the partition is $P = \left\{a, a + \frac{b-a}{n}, a + \frac{2(b-a)}{n}, \ldots, a + \frac{n(b-a)}{n}\right\}$ and $c_k = a + \frac{k(b-a)}{n}$. We get the Riemann sum $\sum_{k=1}^{n} f(c_k)\Delta x = \sum_{k=1}^{n} c_k^2\left(\frac{b-a}{n}\right) = \frac{b-a}{n}\sum_{k=1}^{n}\left(a + \frac{k(b-a)}{n}\right)^2 = \frac{b-a}{n}\sum_{k=1}^{n}\left(a^2 + \frac{2ak(b-a)}{n} + \frac{k^2(b-a)^2}{n^2}\right)$

$= \frac{b-a}{n}\left(\sum_{k=1}^{n}a^2 + \frac{2a(b-a)}{n}\sum_{k=1}^{n}k + \frac{(b-a)^2}{n^2}\sum_{k=1}^{n}k^2\right) = \frac{b-a}{n} \cdot na^2 + \frac{2a(b-a)^2}{n^2} \cdot \frac{n(n+1)}{2} + \frac{(b-a)^3}{n^3} \cdot \frac{n(n+1)(2n+1)}{6}$

$= (b-a)a^2 + a(b-a)^2 \cdot \frac{n+1}{n} + \frac{(b-a)^3}{6} \cdot \frac{(n+1)(2n+1)}{n^2} = (b-a)a^2 + a(b-a)^2 \cdot \frac{1+\frac{1}{n}}{1} + \frac{(b-a)^3}{6} \cdot \frac{2+\frac{3}{n}+\frac{1}{n^2}}{1}$

As $n \to \infty$ and $\|P\| \to 0$ this expression has value $(b-a)a^2 + a(b-a)^2 \cdot 1 + \frac{(b-a)^3}{6} \cdot 2$

$= ba^2 - a^3 + ab^2 - 2a^2b + a^3 + \frac{1}{3}(b^3 - 3b^2a + 3ba^2 - a^3) = \frac{b^3}{3} - \frac{a^3}{3}$. Thus, $\int_a^b x^2\,dx = \frac{b^3}{3} - \frac{a^3}{3}$.

67. Consider the partition P that subdivides the interval $[-1, 2]$ into n subintervals of width $\triangle x = \frac{2-(-1)}{n} = \frac{3}{n}$ and let c_k be the right endpoint of each subinterval. So the partition is $P = \{-1, -1 + \frac{3}{n}, -1 + 2 \cdot \frac{3}{n}, \ldots, -1 + n \cdot \frac{3}{n} = 2\}$ and $c_k = -1 + k \cdot \frac{3}{n} = -1 + \frac{3k}{n}$. We get the Riemann sum $\sum_{k=1}^{n} f(c_k)\triangle x = \sum_{k=1}^{n}\left(3\left(-1 + \frac{3k}{n}\right)^2 - 2\left(-1 + \frac{3k}{n}\right) + 1\right) \cdot \frac{3}{n}$
$= \frac{3}{n}\sum_{k=1}^{n}\left(3 - \frac{18k}{n} + \frac{27k^2}{n^2} + 2 - \frac{6k}{n} + 1\right) = \frac{18}{n}\sum_{k=1}^{n} 1 - \frac{72}{n^2}\sum_{k=1}^{n} k + \frac{81}{n^3}\sum_{k=1}^{n} k^2 = \frac{18}{n} \cdot n - \frac{72}{n^2} \cdot \frac{n(n+1)}{2} + \frac{81}{n^3} \cdot \frac{n(n+1)(2n+1)}{6}$
$= 18 - \frac{36(n+1)}{n} + \frac{27(n+1)(2n+1)}{2n^2}$. As $n \to \infty$ and $\|P\| \to 0$ this expression has value $18 - 36 + 27 = 9$. Thus, $\int_{-1}^{2}(3x^2 - 2x + 1)dx = 9$.

69. Consider the partition P that subdivides the interval $[a, b]$ into n subintervals of width $\triangle x = \frac{b-a}{n}$ and let c_k be the right endpoint of each subinterval. So the partition is $P = \{a, a + \frac{b-a}{n}, a + \frac{2(b-a)}{n}, \ldots, a + \frac{n(b-a)}{n} = b\}$ and $c_k = a + \frac{k(b-a)}{n}$. We get the Riemann sum $\sum_{k=1}^{n} f(c_k)\triangle x = \sum_{k=1}^{n} c_k^3\left(\frac{b-a}{n}\right) = \frac{b-a}{n}\sum_{k=1}^{n}\left(a + \frac{k(b-a)}{n}\right)^3$
$= \frac{b-a}{n}\sum_{k=1}^{n}\left(a^3 + \frac{3a^2k(b-a)}{n} + \frac{3ak^2(b-a)^2}{n^2} + \frac{k^3(b-a)^3}{n^3}\right) = \frac{b-a}{n}\left(\sum_{k=1}^{n} a^3 + \frac{3a^2(b-a)}{n}\sum_{k=1}^{n} k + \frac{3a(b-a)^2}{n^2}\sum_{k=1}^{n} k^2 + \frac{(b-a)^3}{n^3}\sum_{k=1}^{n} k^3\right)$
$= \frac{b-a}{n} \cdot na^3 + \frac{3a^2(b-a)^2}{n^2} \cdot \frac{n(n+1)}{2} + \frac{3a(b-a)^3}{n^3} \cdot \frac{n(n+1)(2n+1)}{6} + \frac{(b-a)^4}{n^4} \cdot \left(\frac{n(n+1)}{2}\right)^2$
$= (b-a)a^3 + \frac{3a^2(b-a)^2}{2} \cdot \frac{n+1}{n} + \frac{a(b-a)^3}{2} \cdot \frac{(n+1)(2n+1)}{n^2} + \frac{(b-a)^4}{4} \cdot \frac{(n+1)^2}{n^2}$
$= (b-a)a^3 + \frac{3a^2(b-a)^2}{2} \cdot \frac{1+\frac{1}{n}}{1} + \frac{a(b-a)^3}{2} \cdot \frac{2+\frac{3}{n}+\frac{1}{n^2}}{1} + \frac{(b-a)^4}{4} \cdot \frac{1+\frac{2}{n}+\frac{1}{n^2}}{1}$. As $n \to \infty$ and $\|P\| \to 0$ this expression has value $(b-a)a^3 + \frac{3a^2(b-a)^2}{2} + a(b-a)^3 + \frac{(b-a)^4}{4} = \frac{b^4}{4} - \frac{a^4}{4}$. Thus, $\int_a^b x^3 dx = \frac{b^4}{4} - \frac{a^4}{4}$.

71. To find where $x - x^2 \ge 0$, let $x - x^2 = 0 \Rightarrow x(1-x) = 0 \Rightarrow x = 0$ or $x = 1$. If $0 < x < 1$, then $0 < x - x^2 \Rightarrow a = 0$ and $b = 1$ maximize the integral.

73. $f(x) = \frac{1}{1+x^2}$ is decreasing on $[0, 1] \Rightarrow$ maximum value of f occurs at $0 \Rightarrow \max f = f(0) = 1$; minimum value of f occurs at $1 \Rightarrow \min f = f(1) = \frac{1}{1+1^2} = \frac{1}{2}$. Therefore, $(1-0)\min f \le \int_0^1 \frac{1}{1+x^2}\,dx \le (1-0)\max f \Rightarrow \frac{1}{2} \le \int_0^1 \frac{1}{1+x^2}\,dx \le 1$. That is, an upper bound $= 1$ and a lower bound $= \frac{1}{2}$.

75. $-1 \le \sin(x^2) \le 1$ for all $x \Rightarrow (1-0)(-1) \le \int_0^1 \sin(x^2)\,dx \le (1-0)(1)$ or $\int_0^1 \sin x^2\,dx \le 1 \Rightarrow \int_0^1 \sin x^2\,dx$ cannot equal 2.

77. If $f(x) \ge 0$ on $[a, b]$, then $\min f \ge 0$ and $\max f \ge 0$ on $[a, b]$. Now, $(b-a)\min f \le \int_a^b f(x)\,dx \le (b-a)\max f$. Then $b \ge a \Rightarrow b - a \ge 0 \Rightarrow (b-a)\min f \ge 0 \Rightarrow \int_a^b f(x)\,dx \ge 0$.

79. $\sin x \le x$ for $x \ge 0 \Rightarrow \sin x - x \le 0$ for $x \ge 0 \Rightarrow \int_0^1 (\sin x - x)\,dx \le 0$ (see Exercise 78) $\Rightarrow \int_0^1 \sin x\,dx - \int_0^1 x\,dx \le 0$
$\Rightarrow \int_0^1 \sin x\,dx \le \int_0^1 x\,dx \Rightarrow \int_0^1 \sin x\,dx \le \left(\frac{1^2}{2} - \frac{0^2}{2}\right) \Rightarrow \int_0^1 \sin x\,dx \le \frac{1}{2}$. Thus an upper bound is $\frac{1}{2}$.

81. Yes, for the following reasons: $\text{av}(f) = \frac{1}{b-a}\int_a^b f(x)\,dx$ is a constant K. Thus $\int_a^b \text{av}(f)\,dx = \int_a^b K\,dx = K(b-a)$
$\Rightarrow \int_a^b \text{av}(f)\,dx = (b-a)K = (b-a) \cdot \frac{1}{b-a}\int_a^b f(x)\,dx = \int_a^b f(x)\,dx$.

83. (a) $U = \max_1 \Delta x + \max_2 \Delta x + \ldots + \max_n \Delta x$ where $\max_1 = f(x_1)$, $\max_2 = f(x_2), \ldots, \max_n = f(x_n)$ since f is increasing on [a, b]; $L = \min_1 \Delta x + \min_2 \Delta x + \ldots + \min_n \Delta x$ where $\min_1 = f(x_0)$, $\min_2 = f(x_1), \ldots,$ $\min_n = f(x_{n-1})$ since f is increasing on [a, b]. Therefore

$$U - L = (\max_1 - \min_1)\Delta x + (\max_2 - \min_2)\Delta x + \ldots + (\max_n - \min_n)\Delta x$$
$$= (f(x_1) - f(x_0))\Delta x + (f(x_2) - f(x_1))\Delta x + \ldots + (f(x_n) - f(x_{n-1}))\Delta x = (f(x_n) - f(x_0))\Delta x = (f(b) - f(a))\Delta x.$$

(b) $U = \max_1 \Delta x_1 + \max_2 \Delta x_2 + \ldots + \max_n \Delta x_n$ where $\max_1 = f(x_1)$, $\max_2 = f(x_2), \ldots, \max_n = f(x_n)$ since f is increasing on[a, b]; $L = \min_1 \Delta x_1 + \min_2 \Delta x_2 + \ldots + \min_n \Delta x_n$ where $\min_1 = f(x_0)$, $\min_2 = f(x_1), \ldots, \min_n = f(x_{n-1})$ since f is increasing on [a, b]. Therefore

$$U - L = (\max_1 - \min_1)\Delta x_1 + (\max_2 - \min_2)\Delta x_2 + \ldots + (\max_n - \min_n)\Delta x_n$$
$$= (f(x_1) - f(x_0))\Delta x_1 + (f(x_2) - f(x_1))\Delta x_2 + \ldots + (f(x_n) - f(x_{n-1}))\Delta x_n$$
$$\le (f(x_1) - f(x_0))\Delta x_{max} + (f(x_2) - f(x_1))\Delta x_{max} + \ldots + (f(x_n) - f(x_{n-1}))\Delta x_{max}.$$ Then

$U - L \le (f(x_n) - f(x_0))\Delta x_{max} = (f(b) - f(a))\Delta x_{max} = |f(b) - f(a)|\,\Delta x_{max}$ since $f(b) \ge f(a)$. Thus $\lim\limits_{\|P\| \to 0} (U - L) = \lim\limits_{\|P\| \to 0} (f(b) - f(a))\Delta x_{max} = 0$, since $\Delta x_{max} = \|P\|$.

85. (a) Partition $\left[0, \frac{\pi}{2}\right]$ into n subintervals, each of length $\Delta x = \frac{\pi}{2n}$ with points $x_0 = 0$, $x_1 = \Delta x$, $x_2 = 2\Delta x, \ldots, x_n = n\Delta x = \frac{\pi}{2}$. Since sin x is increasing on $\left[0, \frac{\pi}{2}\right]$, the upper sum U is the sum of the areas of the circumscribed rectangles of areas $f(x_1)\Delta x = (\sin \Delta x)\Delta x$, $f(x_2)\Delta x = (\sin 2\Delta x)\Delta x, \ldots, f(x_n)\Delta x = (\sin n\Delta x)\Delta x$. Then $U = (\sin \Delta x + \sin 2\Delta x + \ldots + \sin n\Delta x)\Delta x = \left[\frac{\cos \frac{\Delta x}{2} - \cos\left(\left(n + \frac{1}{2}\right)\Delta x\right)}{2 \sin \frac{\Delta x}{2}}\right]\Delta x$

$$= \left[\frac{\cos \frac{\pi}{4n} - \cos\left(\left(n + \frac{1}{2}\right)\frac{\pi}{2n}\right)}{2 \sin \frac{\pi}{4n}}\right]\left(\frac{\pi}{2n}\right) = \frac{\pi\left(\cos \frac{\pi}{4n} - \cos\left(\frac{\pi}{2} + \frac{\pi}{4n}\right)\right)}{4n \sin \frac{\pi}{4n}} = \frac{\cos \frac{\pi}{4n} - \cos\left(\frac{\pi}{2} + \frac{\pi}{4n}\right)}{\left(\frac{\sin \frac{\pi}{4n}}{\frac{\pi}{4n}}\right)}$$

(b) The area is $\int_0^{\pi/2} \sin x\, dx = \lim\limits_{n \to \infty} \frac{\cos \frac{\pi}{4n} - \cos\left(\frac{\pi}{2} + \frac{\pi}{4n}\right)}{\left(\frac{\sin \frac{\pi}{4n}}{\frac{\pi}{4n}}\right)} = \frac{1 - \cos \frac{\pi}{2}}{1} = 1.$

87. By Exercise 86, $U - L = \sum\limits_{i=1}^{n} \Delta x_i \cdot M_i - \sum\limits_{i=1}^{n} \Delta x_i \cdot m_i$ where $M_i = \max\{f(x) \text{ on the ith subinterval}\}$ and $m_i = \min\{f(x) \text{ on the ith subinterval}\}$. Thus $U - L = \sum\limits_{i=1}^{n} (M_i - m_i)\Delta x_i < \sum\limits_{i=1}^{n} \epsilon \cdot \Delta x_i$ provided $\Delta x_i < \delta$ for each $i = 1, \ldots, n$. Since $\sum\limits_{i=1}^{n} \epsilon \cdot \Delta x_i = \epsilon \sum\limits_{i=1}^{n} \Delta x_i = \epsilon(b - a)$ the result, $U - L < \epsilon(b - a)$ follows.

5.4 THE FUNDAMENTAL THEOREM OF CALCULUS

1. $\int_{-2}^{0} (2x + 5)\, dx = \left[x^2 + 5x\right]_{-2}^{0} = \left(0^2 + 5(0)\right) - \left((-2)^2 + 5(-2)\right) = 6$

3. $\int_0^2 x(x - 3)\, dx = \int_0^2 (x^2 - 3x)\, dx = \left[\frac{x^3}{3} - \frac{3x^2}{2}\right]_0^2 = \left(\frac{(2)^3}{3} - \frac{3(2)^2}{2}\right) - \left(\frac{(0)^3}{3} - \frac{3(0)^2}{2}\right) = -\frac{10}{3}$

5. $\int_0^4 \left(3x - \frac{x^3}{4}\right) dx = \left[\frac{3x^2}{2} - \frac{x^4}{16}\right]_0^4 = \left(\frac{3(4)^2}{2} - \frac{4^4}{16}\right) - \left(\frac{3(0)^2}{2} - \frac{(0)^4}{16}\right) = 8$

7. $\int_0^1 \left(x^2 + \sqrt{x}\right) dx = \left[\frac{x^3}{3} + \frac{2}{3}x^{3/2}\right]_0^1 = \left(\frac{1}{3} + \frac{2}{3}\right) - 0 = 1$

9. $\int_0^{\pi/3} 2 \sec^2 x\, dx = \left[2 \tan x\right]_0^{\pi/3} = \left(2 \tan\left(\frac{\pi}{3}\right)\right) - (2 \tan 0) = 2\sqrt{3} - 0 = 2\sqrt{3}$

11. $\int_{\pi/4}^{3\pi/4} \csc\theta \cot\theta\, d\theta = \left[-\csc\theta\right]_{\pi/4}^{3\pi/4} = \left(-\csc\left(\frac{3\pi}{4}\right)\right) - \left(-\csc\left(\frac{\pi}{4}\right)\right) = -\sqrt{2} - \left(-\sqrt{2}\right) = 0$

13. $\int_{\pi/2}^{0} \frac{1+\cos 2t}{2}\,dt = \int_{\pi/2}^{0} \left(\frac{1}{2} + \frac{1}{2}\cos 2t\right) dt = \left[\frac{1}{2}t + \frac{1}{4}\sin 2t\right]_{\pi/2}^{0} = \left(\frac{1}{2}(0) + \frac{1}{4}\sin 2(0)\right) - \left(\frac{1}{2}\left(\frac{\pi}{2}\right) + \frac{1}{4}\sin 2\left(\frac{\pi}{2}\right)\right) = -\frac{\pi}{4}$

15. $\int_{0}^{\pi/4} \tan^2 x\,dx = \int_{0}^{\pi/4} (\sec^2 x - 1)dx = [\tan x - x]_0^{\pi/4} = \left(\tan\left(\frac{\pi}{4}\right) - \frac{\pi}{4}\right) - (\tan(0) - 0) = 1 - \frac{\pi}{4}$

17. $\int_{0}^{\pi/8} \sin 2x\,dx = \left[-\frac{1}{2}\cos 2x\right]_0^{\pi/8} = \left(-\frac{1}{2}\cos 2\left(\frac{\pi}{8}\right)\right) - \left(-\frac{1}{2}\cos 2(0)\right) = \frac{2-\sqrt{2}}{4}$

19. $\int_{1}^{-1} (r+1)^2\,dr = \int_{1}^{-1} (r^2 + 2r + 1)\,dr = \left[\frac{r^3}{3} + r^2 + r\right]_1^{-1} = \left(\frac{(-1)^3}{3} + (-1)^2 + (-1)\right) - \left(\frac{1^3}{3} + 1^2 + 1\right) = -\frac{8}{3}$

21. $\int_{\sqrt{2}}^{1} \left(\frac{u^7}{2} - \frac{1}{u^5}\right) du = \int_{\sqrt{2}}^{1} \left(\frac{u^7}{2} - u^{-5}\right) du = \left[\frac{u^8}{16} + \frac{1}{4u^4}\right]_{\sqrt{2}}^{1} = \left(\frac{1^8}{16} + \frac{1}{4(1)^4}\right) - \left(\frac{\left(\sqrt{2}\right)^8}{16} + \frac{1}{4\left(\sqrt{2}\right)^4}\right) = -\frac{3}{4}$

23. $\int_{1}^{\sqrt{2}} \frac{s^2+\sqrt{s}}{s^2}\,ds = \int_{1}^{\sqrt{2}} \left(1 + s^{-3/2}\right) ds = \left[s - \frac{2}{\sqrt{s}}\right]_1^{\sqrt{2}} = \left(\sqrt{2} - \frac{2}{\sqrt{\sqrt{2}}}\right) - \left(1 - \frac{2}{\sqrt{1}}\right) = \sqrt{2} - 2^{3/4} + 1$
$= \sqrt{2} - \sqrt[4]{8} + 1$

25. $\int_{\pi/2}^{\pi} \frac{\sin 2x}{2\sin x}\,dx = \int_{\pi/2}^{\pi} \frac{2\sin x\cos x}{2\sin x}\,dx = \int_{\pi/2}^{\pi} \cos x\,dx = \left[\sin x\right]_{\pi/2}^{\pi} = (\sin(\pi)) - \left(\sin\left(\frac{\pi}{2}\right)\right) = -1$

27. $\int_{-4}^{4} |x|\,dx = \int_{-4}^{0} |x|\,dx + \int_{0}^{4} |x|\,dx = -\int_{-4}^{0} x\,dx + \int_{0}^{4} x\,dx = \left[-\frac{x^2}{2}\right]_{-4}^{0} + \left[\frac{x^2}{2}\right]_0^4 = \left(-\frac{0^2}{2} + \frac{(-4)^2}{2}\right) + \left(\frac{4^2}{2} - \frac{0^2}{2}\right) = 16$

29. (a) $\int_{0}^{\sqrt{x}} \cos t\,dt = [\sin t]_0^{\sqrt{x}} = \sin\sqrt{x} - \sin 0 = \sin\sqrt{x} \Rightarrow \frac{d}{dx}\left(\int_{0}^{\sqrt{x}} \cos t\,dt\right) = \frac{d}{dx}\left(\sin\sqrt{x}\right) = \cos\sqrt{x}\left(\frac{1}{2}x^{-1/2}\right)$
$= \frac{\cos\sqrt{x}}{2\sqrt{x}}$

(b) $\frac{d}{dx}\left(\int_{0}^{\sqrt{x}} \cos t\,dt\right) = \left(\cos\sqrt{x}\right)\left(\frac{d}{dx}\left(\sqrt{x}\right)\right) = \left(\cos\sqrt{x}\right)\left(\frac{1}{2}x^{-1/2}\right) = \frac{\cos\sqrt{x}}{2\sqrt{x}}$

31. (a) $\int_{0}^{t^4} \sqrt{u}\,du = \int_{0}^{t^4} u^{1/2}\,du = \left[\frac{2}{3}u^{3/2}\right]_0^{t^4} = \frac{2}{3}\left(t^4\right)^{3/2} - 0 = \frac{2}{3}t^6 \Rightarrow \frac{d}{dt}\left(\int_{0}^{t^4} \sqrt{u}\,du\right) = \frac{d}{dt}\left(\frac{2}{3}t^6\right) = 4t^5$

(b) $\frac{d}{dt}\left(\int_{0}^{t^4} \sqrt{u}\,du\right) = \sqrt{t^4}\left(\frac{d}{dt}\left(t^4\right)\right) = t^2\left(4t^3\right) = 4t^5$

33. $y = \int_{0}^{x} \sqrt{1+t^2}\,dt \Rightarrow \frac{dy}{dx} = \sqrt{1+x^2}$

35. $y = \int_{\sqrt{x}}^{0} \sin t^2\,dt = -\int_{0}^{\sqrt{x}} \sin t^2\,dt \Rightarrow \frac{dy}{dx} = -\left(\sin\left(\sqrt{x}\right)^2\right)\left(\frac{d}{dx}\left(\sqrt{x}\right)\right) = -(\sin x)\left(\frac{1}{2}x^{-1/2}\right) = -\frac{\sin x}{2\sqrt{x}}$

37. $y = \int_{-1}^{x} \frac{t^2}{t^2+4}\,dt - \int_{3}^{x} \frac{t^2}{t^2+4}\,dt \Rightarrow \frac{dy}{dx} = \frac{x^2}{x^2+4} - \frac{x^2}{x^2+4} = 0$

39. $y = \int_{0}^{\sin x} \frac{dt}{\sqrt{1-t^2}},\ |x| < \frac{\pi}{2} \Rightarrow \frac{dy}{dx} = \frac{1}{\sqrt{1-\sin^2 x}}\left(\frac{d}{dx}(\sin x)\right) = \frac{1}{\sqrt{\cos^2 x}}(\cos x) = \frac{\cos x}{|\cos x|} = \frac{\cos x}{\cos x} = 1$ since $|x| < \frac{\pi}{2}$

41. $-x^2 - 2x = 0 \Rightarrow -x(x+2) = 0 \Rightarrow x = 0$ or $x = -2$; Area

$= -\int_{-3}^{-2} (-x^2 - 2x)dx + \int_{-2}^{0} (-x^2 - 2x)dx - \int_{0}^{2} (-x^2 - 2x)dx$

$= -\left[-\frac{x^3}{3} - x^2\right]_{-3}^{-2} + \left[-\frac{x^3}{3} - x^2\right]_{-2}^{0} - \left[-\frac{x^3}{3} - x^2\right]_{0}^{2}$

$= -\left(\left(-\frac{(-2)^3}{3} - (-2)^2\right) - \left(-\frac{(-3)^3}{3} - (-3)^2\right)\right)$

$+ \left(\left(-\frac{0^3}{3} - 0^2\right) - \left(-\frac{(-2)^3}{3} - (-2)^2\right)\right)$

$- \left(\left(-\frac{2^3}{3} - 2^2\right) - \left(-\frac{0^3}{3} - 0^2\right)\right) = \frac{28}{3}$

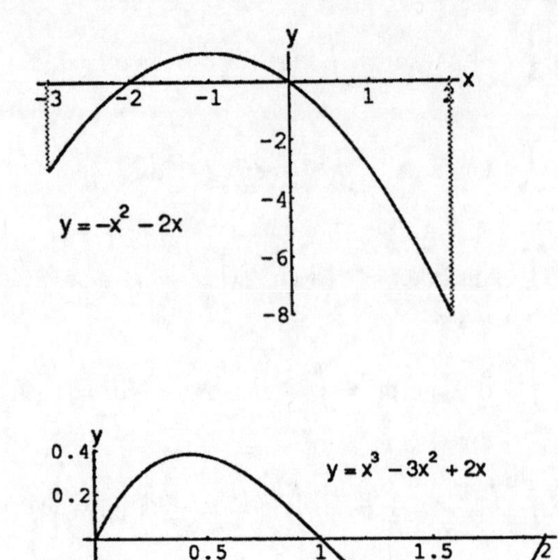

43. $x^3 - 3x^2 + 2x = 0 \Rightarrow x(x^2 - 3x + 2) = 0$

$\Rightarrow x(x-2)(x-1) = 0 \Rightarrow x = 0, 1,$ or 2;

Area $= \int_{0}^{1} (x^3 - 3x^2 + 2x)dx - \int_{1}^{2} (x^3 - 3x^2 + 2x)dx$

$= \left[\frac{x^4}{4} - x^3 + x^2\right]_0^1 - \left[\frac{x^4}{4} - x^3 + x^2\right]_1^2$

$= \left(\frac{1^4}{4} - 1^3 + 1^2\right) - \left(\frac{0^4}{4} - 0^3 + 0^2\right)$

$- \left[\left(\frac{2^4}{4} - 2^3 + 2^2\right) - \left(\frac{1^4}{4} - 1^3 + 1^2\right)\right] = \frac{1}{2}$

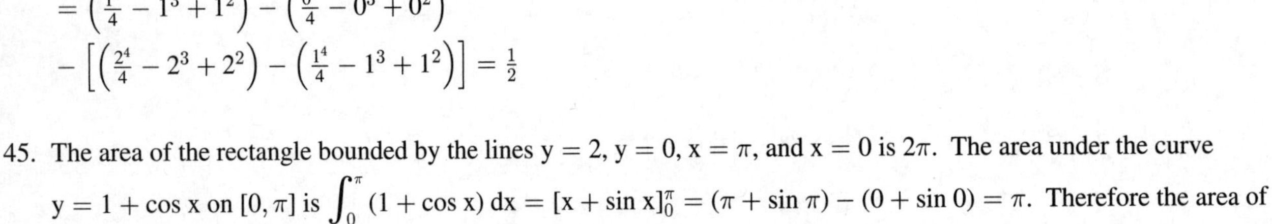

45. The area of the rectangle bounded by the lines $y = 2$, $y = 0$, $x = \pi$, and $x = 0$ is 2π. The area under the curve $y = 1 + \cos x$ on $[0, \pi]$ is $\int_0^{\pi} (1 + \cos x)\, dx = [x + \sin x]_0^{\pi} = (\pi + \sin \pi) - (0 + \sin 0) = \pi$. Therefore the area of the shaded region is $2\pi - \pi = \pi$.

47. On $\left[-\frac{\pi}{4}, 0\right]$: The area of the rectangle bounded by the lines $y = \sqrt{2}$, $y = 0$, $\theta = 0$, and $\theta = -\frac{\pi}{4}$ is $\sqrt{2}\left(\frac{\pi}{4}\right)$ $= \frac{\pi\sqrt{2}}{4}$. The area between the curve $y = \sec\theta\tan\theta$ and $y = 0$ is $-\int_{-\pi/4}^{0} \sec\theta\tan\theta\, d\theta = [-\sec\theta]_{-\pi/4}^{0}$ $= (-\sec 0) - \left(-\sec\left(-\frac{\pi}{4}\right)\right) = \sqrt{2} - 1$. Therefore the area of the shaded region on $\left[-\frac{\pi}{4}, 0\right]$ is $\frac{\pi\sqrt{2}}{4} + \left(\sqrt{2} - 1\right)$. On $\left[0, \frac{\pi}{4}\right]$: The area of the rectangle bounded by $\theta = \frac{\pi}{4}$, $\theta = 0$, $y = \sqrt{2}$, and $y = 0$ is $\sqrt{2}\left(\frac{\pi}{4}\right) = \frac{\pi\sqrt{2}}{4}$. The area under the curve $y = \sec\theta\tan\theta$ is $\int_0^{\pi/4} \sec\theta\tan\theta\, d\theta = [\sec\theta]_0^{\pi/4} = \sec\frac{\pi}{4} - \sec 0 = \sqrt{2} - 1$. Therefore the area of the shaded region on $\left[0, \frac{\pi}{4}\right]$ is $\frac{\pi\sqrt{2}}{4} - \left(\sqrt{2} - 1\right)$. Thus, the area of the total shaded region is $\left(\frac{\pi\sqrt{2}}{4} + \sqrt{2} - 1\right) + \left(\frac{\pi\sqrt{2}}{4} - \sqrt{2} + 1\right) = \frac{\pi\sqrt{2}}{2}$.

49. $y = \int_{\pi}^{x} \frac{1}{t}\, dt - 3 \Rightarrow \frac{dy}{dx} = \frac{1}{x}$ and $y(\pi) = \int_{\pi}^{\pi} \frac{1}{t}\, dt - 3 = 0 - 3 = -3 \Rightarrow$ (d) is a solution to this problem.

51. $y = \int_0^x \sec t\, dt + 4 \Rightarrow \frac{dy}{dx} = \sec x$ and $y(0) = \int_0^0 \sec t\, dt + 4 = 0 + 4 = 4 \Rightarrow$ (b) is a solution to this problem.

53. $y = \int_2^x \sec t\, dt + 3$

55. Area $= \int_{-b/2}^{b/2} \left(h - \left(\frac{4h}{b^2}\right)x^2\right) dx = \left[hx - \frac{4hx^3}{3b^2}\right]_{-b/2}^{b/2}$

$= \left(h\left(\frac{b}{2}\right) - \frac{4h\left(\frac{b}{2}\right)^3}{3b^2}\right) - \left(h\left(-\frac{b}{2}\right) - \frac{4h\left(-\frac{b}{2}\right)^3}{3b^2}\right)$

$= \left(\frac{bh}{2} - \frac{bh}{6}\right) - \left(-\frac{bh}{2} + \frac{bh}{6}\right) = bh - \frac{bh}{3} = \frac{2}{3}\,bh$

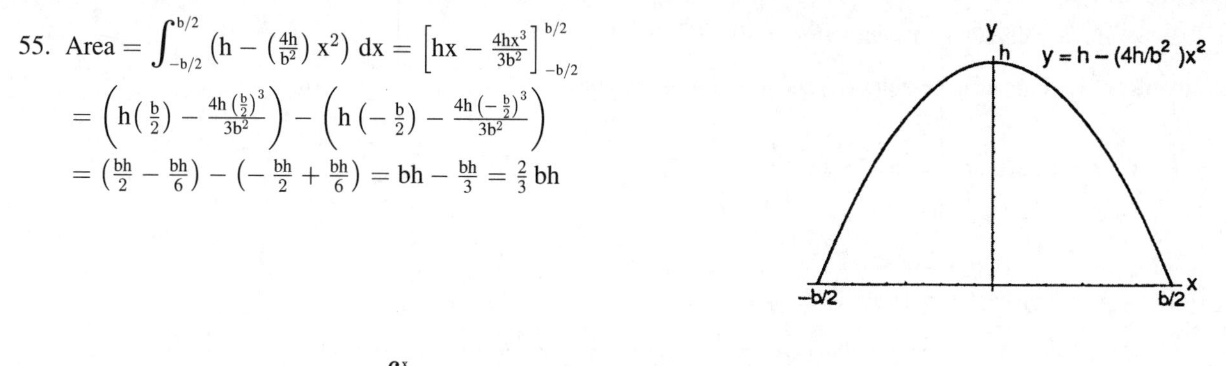

57. $\frac{dc}{dx} = \frac{1}{2\sqrt{x}} = \frac{1}{2}x^{-1/2} \Rightarrow c = \int_0^x \frac{1}{2}t^{-1/2}dt = \left[t^{1/2}\right]_0^x = \sqrt{x}$; $c(100) - c(1) = \sqrt{100} - \sqrt{1} = \$9.00$

59. (a) $t = 0 \Rightarrow T = 85 - 3\sqrt{25 - 0} = 70°$ F; $t = 16 \Rightarrow T = 85 - 3\sqrt{25 - 16} = 76°$ F;
$t = 25 \Rightarrow T = 85 - 3\sqrt{25 - 25} = 85°$ F

(b) average temperatuve $= \frac{1}{25-0}\int_0^{25}\left(85 - 3\sqrt{25 - t}\right) dt = \frac{1}{25}\left[85t + 2(25 - t)^{3/2}\right]_0^{25}$

$= \frac{1}{25}\left(85(25) + 2(25 - 25)^{3/2}\right) - \frac{1}{25}\left(85(0) + 2(25 - 0)^{3/2}\right) = 75°$ F

61. $\int_1^x f(t)\,dt = x^2 - 2x + 1 \Rightarrow f(x) = \frac{d}{dx}\int_1^x f(t)\,dt = \frac{d}{dx}(x^2 - 2x + 1) = 2x - 2$

63. $f(x) = 2 - \int_2^{x+1} \frac{9}{1+t}\,dt \Rightarrow f'(x) = -\frac{9}{1+(x+1)} = \frac{-9}{x+2} \Rightarrow f'(1) = -3$; $f(1) = 2 - \int_2^{1+1} \frac{9}{1+t}\,dt = 2 - 0 = 2$;
$L(x) = -3(x - 1) + f(1) = -3(x - 1) + 2 = -3x + 5$

65. (a) True: since f is continuous, g is differentiable by Part 1 of the Fundamental Theorem of Calculus.
(b) True: g is continuous because it is differentiable.
(c) True, since $g'(1) = f(1) = 0$.
(d) False, since $g''(1) = f'(1) > 0$.
(e) True, since $g'(1) = 0$ and $g''(1) = f'(1) > 0$.
(f) False: $g''(x) = f'(x) > 0$, so g'' never changes sign.
(g) True, since $g'(1) = f(1) = 0$ and $g'(x) = f(x)$ is an increasing function of x (because $f'(x) > 0$).

5.5 INDEFINTE INTEGRALS AND THE SUBSTITUTION RULE

1. Let $u = 2x + 4 \Rightarrow du = 2\,dx \Rightarrow \frac{1}{2}\,du = dx$
$\int 2(2x + 4)^5 dx = \int 2u^5\,\frac{1}{2}\,du = \int u^5\,du = \frac{1}{6}\,u^6 + C = \frac{1}{6}(2x + 4)^6 + C$

3. Let $u = x^2 + 5 \Rightarrow du = 2x\,dx \Rightarrow \frac{1}{2}\,du = x\,dx$
$\int 2x(x^2 + 5)^{-4} dx = \int 2u^{-4}\,\frac{1}{2}du = \int u^{-4}\,du = -\frac{1}{3}\,u^{-3} + C = -\frac{1}{3}(x^2 + 5)^{-3} + C$

5. Let $u = 3x^2 + 4x \Rightarrow du = (6x + 4)dx = 2(3x + 2)\,dx \Rightarrow \frac{1}{2}\,du = (3x + 2)\,dx$
$\int (3x + 2)(3x^2 + 4x)^4 dx = \int u^4\,\frac{1}{2}du = \frac{1}{2}\int u^4\,du = \frac{1}{10}\,u^5 + C = \frac{1}{10}\,(3x^2 + 4x)^5 + C$

7. Let $u = 3x \Rightarrow du = 3\,dx \Rightarrow \frac{1}{3}\,du = dx$
$\int \sin 3x\,dx = \int \frac{1}{3}\sin u\,du = -\frac{1}{3}\cos u + C = -\frac{1}{3}\cos 3x + C$

9. Let $u = 2t \Rightarrow du = 2\,dt \Rightarrow \frac{1}{2}\,du = dt$
$\int \sec 2t \tan 2t\,dt = \int \frac{1}{2} \sec u \tan u\,du = \frac{1}{2} \sec u + C = \frac{1}{2} \sec 2t + C$

11. Let $u = 1 - r^3 \Rightarrow du = -3r^2\,dr \Rightarrow -3\,du = 9r^2\,dr$
$\int \frac{9r^2\,dr}{\sqrt{1-r^3}} = \int -3u^{-1/2}\,du = -3(2)u^{1/2} + C = -6\left(1 - r^3\right)^{1/2} + C$

13. Let $u = x^{3/2} - 1 \Rightarrow du = \frac{3}{2} x^{1/2}\,dx \Rightarrow \frac{2}{3}\,du = \sqrt{x}\,dx$
$\int \sqrt{x} \sin^2\left(x^{3/2} - 1\right) dx = \int \frac{2}{3} \sin^2 u\,du = \frac{2}{3}\left(\frac{u}{2} - \frac{1}{4} \sin 2u\right) + C = \frac{1}{3}\left(x^{3/2} - 1\right) - \frac{1}{6} \sin\left(2x^{3/2} - 2\right) + C$

15. (a) Let $u = \cot 2\theta \Rightarrow du = -2 \csc^2 2\theta\,d\theta \Rightarrow -\frac{1}{2}\,du = \csc^2 2\theta\,d\theta$
$\int \csc^2 2\theta \cot 2\theta\,d\theta = -\int \frac{1}{2} u\,du = -\frac{1}{2}\left(\frac{u^2}{2}\right) + C = -\frac{u^2}{4} + C = -\frac{1}{4} \cot^2 2\theta + C$
(b) Let $u = \csc 2\theta \Rightarrow du = -2 \csc 2\theta \cot 2\theta\,d\theta \Rightarrow -\frac{1}{2}\,du = \csc 2\theta \cot 2\theta\,d\theta$
$\int \csc^2 2\theta \cot 2\theta\,d\theta = \int -\frac{1}{2} u\,du = -\frac{1}{2}\left(\frac{u^2}{2}\right) + C = -\frac{u^2}{4} + C = -\frac{1}{4} \csc^2 2\theta + C$

17. Let $u = 3 - 2s \Rightarrow du = -2\,ds \Rightarrow -\frac{1}{2}\,du = ds$
$\int \sqrt{3 - 2s}\,ds = \int \sqrt{u}\left(-\frac{1}{2}\,du\right) = -\frac{1}{2} \int u^{1/2}\,du = \left(-\frac{1}{2}\right)\left(\frac{2}{3} u^{3/2}\right) + C = -\frac{1}{3}(3 - 2s)^{3/2} + C$

19. Let $u = 1 - \theta^2 \Rightarrow du = -2\theta\,d\theta \Rightarrow -\frac{1}{2}\,du = \theta\,d\theta$
$\int \theta \sqrt[4]{1 - \theta^2}\,d\theta = \int \sqrt[4]{u}\left(-\frac{1}{2}\,du\right) = -\frac{1}{2} \int u^{1/4}\,du = \left(-\frac{1}{2}\right)\left(\frac{4}{5} u^{5/4}\right) + C = -\frac{2}{5}\left(1 - \theta^2\right)^{5/4} + C$

21. Let $u = 1 + \sqrt{x} \Rightarrow du = \frac{1}{2\sqrt{x}}\,dx \Rightarrow 2\,du = \frac{1}{\sqrt{x}}\,dx$
$\int \frac{1}{\sqrt{x}\left(1+\sqrt{x}\right)^2}\,dx = \int \frac{2\,du}{u^2} = -\frac{2}{u} + C = \frac{-2}{1+\sqrt{x}} + C$

23. Let $u = 3x + 2 \Rightarrow du = 3\,dx \Rightarrow \frac{1}{3}\,du = dx$
$\int \sec^2(3x + 2)\,dx = \int \left(\sec^2 u\right)\left(\frac{1}{3}\,du\right) = \frac{1}{3} \int \sec^2 u\,du = \frac{1}{3} \tan u + C = \frac{1}{3} \tan(3x + 2) + C$

25. Let $u = \sin\left(\frac{x}{3}\right) \Rightarrow du = \frac{1}{3} \cos\left(\frac{x}{3}\right) dx \Rightarrow 3\,du = \cos\left(\frac{x}{3}\right) dx$
$\int \sin^5\left(\frac{x}{3}\right) \cos\left(\frac{x}{3}\right) dx = \int u^5 (3\,du) = 3\left(\frac{1}{6} u^6\right) + C = \frac{1}{2} \sin^6\left(\frac{x}{3}\right) + C$

27. Let $u = \frac{r^3}{18} - 1 \Rightarrow du = \frac{r^2}{6}\,dr \Rightarrow 6\,du = r^2\,dr$
$\int r^2\left(\frac{r^3}{18} - 1\right)^5 dr = \int u^5 (6\,du) = 6 \int u^5\,du = 6\left(\frac{u^6}{6}\right) + C = \left(\frac{r^3}{18} - 1\right)^6 + C$

29. Let $u = x^{3/2} + 1 \Rightarrow du = \frac{3}{2} x^{1/2}\,dx \Rightarrow \frac{2}{3}\,du = x^{1/2}\,dx$
$\int x^{1/2} \sin\left(x^{3/2} + 1\right) dx = \int (\sin u)\left(\frac{2}{3}\,du\right) = \frac{2}{3} \int \sin u\,du = \frac{2}{3}(-\cos u) + C = -\frac{2}{3} \cos\left(x^{3/2} + 1\right) + C$

31. Let $u = \cos(2t + 1) \Rightarrow du = -2 \sin(2t + 1)\,dt \Rightarrow -\frac{1}{2}\,du = \sin(2t + 1)\,dt$
$\int \frac{\sin(2t+1)}{\cos^2(2t+1)}\,dt = \int -\frac{1}{2} \frac{du}{u^2} = \frac{1}{2u} + C = \frac{1}{2\cos(2t+1)} + C$

33. Let $u = \frac{1}{t} - 1 = t^{-1} - 1 \Rightarrow du = -t^{-2}\,dt \Rightarrow -du = \frac{1}{t^2}\,dt$
$\int \frac{1}{t^2} \cos\left(\frac{1}{t} - 1\right) dt = \int (\cos u)(-du) = -\int \cos u\,du = -\sin u + C = -\sin\left(\frac{1}{t} - 1\right) + C$

35. Let $u = \sin \frac{1}{\theta} \Rightarrow du = \left(\cos \frac{1}{\theta}\right)\left(-\frac{1}{\theta^2}\right) d\theta \Rightarrow -du = \frac{1}{\theta^2} \cos \frac{1}{\theta}\, d\theta$

$\int \frac{1}{\theta^2} \sin \frac{1}{\theta} \cos \frac{1}{\theta}\, d\theta = \int -u\, du = -\frac{1}{2} u^2 + C = -\frac{1}{2} \sin^2 \frac{1}{\theta} + C$

37. Let $u = 1 + t^4 \Rightarrow du = 4t^3\, dt \Rightarrow \frac{1}{4}\, du = t^3\, dt$

$\int t^3 \left(1 + t^4\right)^3 dt = \int u^3 \left(\frac{1}{4}\, du\right) = \frac{1}{4}\left(\frac{1}{4} u^4\right) + C = \frac{1}{16}\left(1 + t^4\right)^4 + C$

39. Let $u = 2 - \frac{1}{x} \Rightarrow du = \frac{1}{x^2}\, dx$

$\int \frac{1}{x^2}\sqrt{2 - \frac{1}{x}}\, dx = \int \sqrt{u}\, du = \int u^{1/2}\, du = \frac{2}{3} u^{3/2} + C = \frac{2}{3}\left(2 - \frac{1}{x}\right)^{3/2} + C$

41. Let $u = 1 - \frac{3}{x^3} \Rightarrow du = \frac{9}{x^4}\, dx \Rightarrow \frac{1}{9}\, du = \frac{1}{x^4}\, dx$

$\int \sqrt{\frac{x^3 - 3}{x^{11}}}\, dx = \int \frac{1}{x^4}\sqrt{\frac{x^3 - 3}{x^3}}\, dx = \int \frac{1}{x^4}\sqrt{1 - \frac{3}{x^3}}\, dx = \int \sqrt{u}\, \frac{1}{9}\, du = \frac{1}{9}\int u^{1/2}\, du = \frac{2}{27} u^{3/2} + C = \frac{2}{27}\left(1 - \frac{3}{x^3}\right)^{3/2} + C$

43. Let $u = x - 1$. Then $du = dx$ and $x = u + 1$. Thus $\int x(x - 1)^{10}\, dx = \int (u + 1)u^{10}\, du = \int \left(u^{11} + u^{10}\right) du$
$= \frac{1}{12}u^{12} + \frac{1}{11}u^{11} + C = \frac{1}{12}(x - 1)^{12} + \frac{1}{11}(x - 1)^{11} + C$

45. Let $u = 1 - x$. Then $du = -1\, dx$ and $(-1)\, du = dx$ and $x = 1 - u$. Thus $\int (x + 1)^2 (1 - x)^5 dx$
$= \int (2 - u)^2 u^5\, (-1)\, du = \int \left(-u^7 + 4u^6 - 4u^5\right) du = -\frac{1}{8}u^8 + \frac{4}{7}u^7 - \frac{2}{3}u^6 + C$
$= -\frac{1}{8}(1 - x)^8 + \frac{4}{7}(1 - x)^7 - \frac{2}{3}(1 - x)^6 + C$

47. Let $u = x^2 + 1$. Then $du = 2x\, dx$ and $\frac{1}{2}du = x\, dx$ and $x^2 = u - 1$. Thus $\int x^3\sqrt{x^2 + 1}\, dx = \int (u - 1)\frac{1}{2}\sqrt{u}\, du$
$= \frac{1}{2}\int \left(u^{3/2} - u^{1/2}\right) du = \frac{1}{2}\left[\frac{2}{5}u^{5/2} - \frac{2}{3}u^{3/2}\right] + C = \frac{1}{5}u^{5/2} - \frac{1}{3}u^{3/2} + C = \frac{1}{5}\left(x^2 + 1\right)^{5/2} - \frac{1}{3}\left(x^2 + 1\right)^{3/2} + C$

49. Let $u = x^2 - 4 \Rightarrow du = 2x\, dx$ and $\frac{1}{2}\, du = x\, dx$. Thus $\int \frac{x}{\left(x^2 - 4\right)^3}\, dx = \int \left(x^2 - 4\right)^{-3} x\, dx = \int u^{-3}\frac{1}{2}\, du = \frac{1}{2}\int u^{-3}\, du$
$= -\frac{1}{4}u^{-2} + C = -\frac{1}{4}\left(x^2 - 4\right)^{-2} + C$

51. (a) Let $u = \tan x \Rightarrow du = \sec^2 x\, dx$; $v = u^3 \Rightarrow dv = 3u^2\, du \Rightarrow 6\, dv = 18u^2\, du$; $w = 2 + v \Rightarrow dw = dv$

$\int \frac{18 \tan^2 x \sec^2 x}{\left(2 + \tan^3 x\right)^2}\, dx = \int \frac{18u^2}{\left(2 + u^3\right)^2}\, du = \int \frac{6\, dv}{(2 + v)^2} = \int \frac{6\, dw}{w^2} = 6\int w^{-2}\, dw = -6w^{-1} + C = -\frac{6}{2 + v} + C$
$= -\frac{6}{2 + u^3} + C = -\frac{6}{2 + \tan^3 x} + C$

(b) Let $u = \tan^3 x \Rightarrow du = 3 \tan^2 x \sec^2 x\, dx \Rightarrow 6\, du = 18 \tan^2 x \sec^2 x\, dx$; $v = 2 + u \Rightarrow dv = du$

$\int \frac{18 \tan^2 x \sec^2 x}{\left(2 + \tan^3 x\right)^2}\, dx = \int \frac{6\, du}{(2 + u)^2} = \int \frac{6\, dv}{v^2} = -\frac{6}{v} + C = -\frac{6}{2 + u} + C = -\frac{6}{2 + \tan^3 x} + C$

(c) Let $u = 2 + \tan^3 x \Rightarrow du = 3 \tan^2 x \sec^2 x\, dx \Rightarrow 6\, du = 18 \tan^2 x \sec^2 x\, dx$

$\int \frac{18 \tan^2 x \sec^2 x}{\left(2 + \tan^3 x\right)^2}\, dx = \int \frac{6\, du}{u^2} = -\frac{6}{u} + C = -\frac{6}{2 + \tan^3 x} + C$

53. Let $u = 3(2r - 1)^2 + 6 \Rightarrow du = 6(2r - 1)(2)\, dr \Rightarrow \frac{1}{12}\, du = (2r - 1)\, dr$; $v = \sqrt{u} \Rightarrow dv = \frac{1}{2\sqrt{u}}\, du \Rightarrow \frac{1}{6}\, dv = \frac{1}{12\sqrt{u}}\, du$

$\int \frac{(2r - 1) \cos \sqrt{3(2r - 1)^2 + 6}}{\sqrt{3(2r - 1)^2 + 6}}\, dr = \int \left(\frac{\cos \sqrt{u}}{\sqrt{u}}\right)\left(\frac{1}{12}\, du\right) = \int (\cos v)\left(\frac{1}{6}\, dv\right) = \frac{1}{6} \sin v + C = \frac{1}{6} \sin \sqrt{u} + C$
$= \frac{1}{6} \sin \sqrt{3(2r - 1)^2 + 6} + C$

55. Let $u = 3t^2 - 1 \Rightarrow du = 6t\,dt \Rightarrow 2\,du = 12t\,dt$

$s = \int 12t\,(3t^2-1)^3\,dt = \int u^3\,(2\,du) = 2\left(\frac{1}{4}u^4\right) + C = \frac{1}{2}u^4 + C = \frac{1}{2}(3t^2-1)^4 + C;$

$s = 3$ when $t = 1 \Rightarrow 3 = \frac{1}{2}(3-1)^4 + C \Rightarrow 3 = 8 + C \Rightarrow C = -5 \Rightarrow s = \frac{1}{2}(3t^2-1)^4 - 5$

57. Let $u = t + \frac{\pi}{12} \Rightarrow du = dt$

$s = \int 8\sin^2\left(t + \frac{\pi}{12}\right)dt = \int 8\sin^2 u\,du = 8\left(\frac{u}{2} - \frac{1}{4}\sin 2u\right) + C = 4\left(t + \frac{\pi}{12}\right) - 2\sin\left(2t + \frac{\pi}{6}\right) + C;$

$s = 8$ when $t = 0 \Rightarrow 8 = 4\left(\frac{\pi}{12}\right) - 2\sin\left(\frac{\pi}{6}\right) + C \Rightarrow C = 8 - \frac{\pi}{3} + 1 = 9 - \frac{\pi}{3}$

$\Rightarrow s = 4\left(t + \frac{\pi}{12}\right) - 2\sin\left(2t + \frac{\pi}{6}\right) + 9 - \frac{\pi}{3} = 4t - 2\sin\left(2t + \frac{\pi}{6}\right) + 9$

59. Let $u = 2t - \frac{\pi}{2} \Rightarrow du = 2\,dt \Rightarrow -2\,du = -4\,dt$

$\frac{ds}{dt} = \int -4\sin\left(2t - \frac{\pi}{2}\right)dt = \int (\sin u)(-2\,du) = 2\cos u + C_1 = 2\cos\left(2t - \frac{\pi}{2}\right) + C_1;$

at $t = 0$ and $\frac{ds}{dt} = 100$ we have $100 = 2\cos\left(-\frac{\pi}{2}\right) + C_1 \Rightarrow C_1 = 100 \Rightarrow \frac{ds}{dt} = 2\cos\left(2t - \frac{\pi}{2}\right) + 100$

$\Rightarrow s = \int \left(2\cos\left(2t - \frac{\pi}{2}\right) + 100\right)dt = \int (\cos u + 50)\,du = \sin u + 50u + C_2 = \sin\left(2t - \frac{\pi}{2}\right) + 50\left(2t - \frac{\pi}{2}\right) + C_2;$

at $t = 0$ and $s = 0$ we have $0 = \sin\left(-\frac{\pi}{2}\right) + 50\left(-\frac{\pi}{2}\right) + C_2 \Rightarrow C_2 = 1 + 25\pi$

$\Rightarrow s = \sin\left(2t - \frac{\pi}{2}\right) + 100t - 25\pi + (1 + 25\pi) \Rightarrow s = \sin\left(2t - \frac{\pi}{2}\right) + 100t + 1$

61. Let $u = 2t \Rightarrow du = 2\,dt \Rightarrow 3\,du = 6\,dt$

$s = \int 6\sin 2t\,dt = \int (\sin u)(3\,du) = -3\cos u + C = -3\cos 2t + C;$

at $t = 0$ and $s = 0$ we have $0 = -3\cos 0 + C \Rightarrow C = 3 \Rightarrow s = 3 - 3\cos 2t \Rightarrow s\left(\frac{\pi}{2}\right) = 3 - 3\cos(\pi) = 6$ m

63. All three integrations are correct. In each case, the derivative of the function on the right is the integrand on the left, and each formula has an arbitrary constant for generating the remaining antiderivatives. Moreover, $\sin^2 x + C_1 = 1 - \cos^2 x + C_1 \Rightarrow C_2 = 1 + C_1$; also $-\cos^2 x + C_2 = -\frac{\cos 2x}{2} - \frac{1}{2} + C_2 \Rightarrow C_3 = C_2 - \frac{1}{2} = C_1 + \frac{1}{2}$.

5.6 SUBSTITUTION AND AREA BETWEEN CURVES

1. (a) Let $u = y + 1 \Rightarrow du = dy$; $y = 0 \Rightarrow u = 1$, $y = 3 \Rightarrow u = 4$

$\int_0^3 \sqrt{y+1}\,dy = \int_1^4 u^{1/2}\,du = \left[\frac{2}{3}u^{3/2}\right]_1^4 = \left(\frac{2}{3}\right)(4)^{3/2} - \left(\frac{2}{3}\right)(1)^{3/2} = \left(\frac{2}{3}\right)(8) - \left(\frac{2}{3}\right)(1) = \frac{14}{3}$

(b) Use the same substitution for u as in part (a); $y = -1 \Rightarrow u = 0$, $y = 0 \Rightarrow u = 1$

$\int_{-1}^0 \sqrt{y+1}\,dy = \int_0^1 u^{1/2}\,du = \left[\frac{2}{3}u^{3/2}\right]_0^1 = \left(\frac{2}{3}\right)(1)^{3/2} - 0 = \frac{2}{3}$

3. (a) Let $u = \tan x \Rightarrow du = \sec^2 x\,dx$; $x = 0 \Rightarrow u = 0$, $x = \frac{\pi}{4} \Rightarrow u = 1$

$\int_0^{\pi/4} \tan x \sec^2 x\,dx = \int_0^1 u\,du = \left[\frac{u^2}{2}\right]_0^1 = \frac{1^2}{2} - 0 = \frac{1}{2}$

(b) Use the same substitution as in part (a); $x = -\frac{\pi}{4} \Rightarrow u = -1$, $x = 0 \Rightarrow u = 0$

$\int_{-\pi/4}^0 \tan x \sec^2 x\,dx = \int_{-1}^0 u\,du = \left[\frac{u^2}{2}\right]_{-1}^0 = 0 - \frac{1}{2} = -\frac{1}{2}$

5. (a) $u = 1 + t^4 \Rightarrow du = 4t^3\,dt \Rightarrow \frac{1}{4}du = t^3\,dt$; $t = 0 \Rightarrow u = 1$, $t = 1 \Rightarrow u = 2$

$\int_0^1 t^3(1+t^4)^3\,dt = \int_1^2 \frac{1}{4}u^3\,du = \left[\frac{u^4}{16}\right]_1^2 = \frac{2^4}{16} - \frac{1^4}{16} = \frac{15}{16}$

(b) Use the same substitution as in part (a); $t = -1 \Rightarrow u = 2$, $t = 1 \Rightarrow u = 2$

$\int_{-1}^1 t^3(1+t^4)^3\,dt = \int_2^2 \frac{1}{4}u^3\,du = 0$

7. (a) Let $u = 4 + r^2 \Rightarrow du = 2r\,dr \Rightarrow \frac{1}{2}\,du = r\,dr$; $r = -1 \Rightarrow u = 5$, $r = 1 \Rightarrow u = 5$

$$\int_{-1}^{1} \frac{5r}{(4+r^2)^2}\,dr = 5\int_5^5 \tfrac{1}{2}u^{-2}\,du = 0$$

(b) Use the same substitution as in part (a); $r = 0 \Rightarrow u = 4$, $r = 1 \Rightarrow u = 5$

$$\int_0^1 \frac{5r}{(4+r^2)^2}\,dr = 5\int_4^5 \tfrac{1}{2}u^{-2}\,du = 5\left[-\tfrac{1}{2}u^{-1}\right]_4^5 = 5\left(-\tfrac{1}{2}(5)^{-1}\right) - 5\left(-\tfrac{1}{2}(4)^{-1}\right) = \tfrac{1}{8}$$

9. (a) Let $u = x^2 + 1 \Rightarrow du = 2x\,dx \Rightarrow 2\,du = 4x\,dx$; $x = 0 \Rightarrow u = 1$, $x = \sqrt{3} \Rightarrow u = 4$

$$\int_0^{\sqrt{3}} \frac{4x}{\sqrt{x^2+1}}\,dx = \int_1^4 \frac{2}{\sqrt{u}}\,du = \int_1^4 2u^{-1/2}\,du = \left[4u^{1/2}\right]_1^4 = 4(4)^{1/2} - 4(1)^{1/2} = 4$$

(b) Use the same substitution as in part (a); $x = -\sqrt{3} \Rightarrow u = 4$, $x = \sqrt{3} \Rightarrow u = 4$

$$\int_{-\sqrt{3}}^{\sqrt{3}} \frac{4x}{\sqrt{x^2+1}}\,dx = \int_4^4 \frac{2}{\sqrt{u}}\,du = 0$$

11. (a) Let $u = 1 - \cos 3t \Rightarrow du = 3\sin 3t\,dt \Rightarrow \frac{1}{3}\,du = \sin 3t\,dt$; $t = 0 \Rightarrow u = 0$, $t = \frac{\pi}{6} \Rightarrow u = 1 - \cos\frac{\pi}{2} = 1$

$$\int_0^{\pi/6} (1 - \cos 3t)\sin 3t\,dt = \int_0^1 \tfrac{1}{3}u\,du = \left[\tfrac{1}{3}\left(\tfrac{u^2}{2}\right)\right]_0^1 = \tfrac{1}{6}(1)^2 - \tfrac{1}{6}(0)^2 = \tfrac{1}{6}$$

(b) Use the same substitution as in part (a); $t = \frac{\pi}{6} \Rightarrow u = 1$, $t = \frac{\pi}{3} \Rightarrow u = 1 - \cos\pi = 2$

$$\int_{\pi/6}^{\pi/3} (1 - \cos 3t)\sin 3t\,dt = \int_1^2 \tfrac{1}{3}u\,du = \left[\tfrac{1}{3}\left(\tfrac{u^2}{2}\right)\right]_1^2 = \tfrac{1}{6}(2)^2 - \tfrac{1}{6}(1)^2 = \tfrac{1}{2}$$

13. (a) Let $u = 4 + 3\sin z \Rightarrow du = 3\cos z\,dz \Rightarrow \frac{1}{3}\,du = \cos z\,dz$; $z = 0 \Rightarrow u = 4$, $z = 2\pi \Rightarrow u = 4$

$$\int_0^{2\pi} \frac{\cos z}{\sqrt{4+3\sin z}}\,dz = \int_4^4 \frac{1}{\sqrt{u}}\left(\tfrac{1}{3}\,du\right) = 0$$

(b) Use the same substitution as in part (a); $z = -\pi \Rightarrow u = 4 + 3\sin(-\pi) = 4$, $z = \pi \Rightarrow u = 4$

$$\int_{-\pi}^{\pi} \frac{\cos z}{\sqrt{4+3\sin z}}\,dz = \int_4^4 \frac{1}{\sqrt{u}}\left(\tfrac{1}{3}\,du\right) = 0$$

15. Let $u = t^5 + 2t \Rightarrow du = (5t^4 + 2)\,dt$; $t = 0 \Rightarrow u = 0$, $t = 1 \Rightarrow u = 3$

$$\int_0^1 \sqrt{t^5 + 2t}\,(5t^4 + 2)\,dt = \int_0^3 u^{1/2}\,du = \left[\tfrac{2}{3}u^{3/2}\right]_0^3 = \tfrac{2}{3}(3)^{3/2} - \tfrac{2}{3}(0)^{3/2} = 2\sqrt{3}$$

17. Let $u = \cos 2\theta \Rightarrow du = -2\sin 2\theta\,d\theta \Rightarrow -\frac{1}{2}\,du = \sin 2\theta\,d\theta$; $\theta = 0 \Rightarrow u = 1$, $\theta = \frac{\pi}{6} \Rightarrow u = \cos 2\left(\frac{\pi}{6}\right) = \frac{1}{2}$

$$\int_0^{\pi/6} \cos^{-3} 2\theta \sin 2\theta\,d\theta = \int_1^{1/2} u^{-3}\left(-\tfrac{1}{2}\,du\right) = -\tfrac{1}{2}\int_1^{1/2} u^{-3}\,du = \left[-\tfrac{1}{2}\left(\tfrac{u^{-2}}{-2}\right)\right]_1^{1/2} = \frac{1}{4\left(\frac{1}{2}\right)^2} - \frac{1}{4(1)^2} = \tfrac{3}{4}$$

19. Let $u = 5 - 4\cos t \Rightarrow du = 4\sin t\,dt \Rightarrow \frac{1}{4}\,du = \sin t\,dt$; $t = 0 \Rightarrow u = 5 - 4\cos 0 = 1$, $t = \pi \Rightarrow u = 5 - 4\cos\pi = 9$

$$\int_0^{\pi} 5(5 - 4\cos t)^{1/4}\sin t\,dt = \int_1^9 5u^{1/4}\left(\tfrac{1}{4}\,du\right) = \tfrac{5}{4}\int_1^9 u^{1/4}\,du = \left[\tfrac{5}{4}\left(\tfrac{4}{5}u^{5/4}\right)\right]_1^9 = 9^{5/4} - 1 = 3^{5/2} - 1$$

21. Let $u = 4y - y^2 + 4y^3 + 1 \Rightarrow du = (4 - 2y + 12y^2)\,dy$; $y = 0 \Rightarrow u = 1$, $y = 1 \Rightarrow u = 4(1) - (1)^2 + 4(1)^3 + 1 = 8$

$$\int_0^1 (4y - y^2 + 4y^3 + 1)^{-2/3}(12y^2 - 2y + 4)\,dy = \int_1^8 u^{-2/3}\,du = \left[3u^{1/3}\right]_1^8 = 3(8)^{1/3} - 3(1)^{1/3} = 3$$

23. Let $u = \theta^{3/2} \Rightarrow du = \frac{3}{2}\theta^{1/2}\,d\theta \Rightarrow \frac{2}{3}\,du = \sqrt{\theta}\,d\theta$; $\theta = 0 \Rightarrow u = 0$, $\theta = \sqrt[3]{\pi^2} \Rightarrow u = \pi$

$$\int_0^{\sqrt[3]{\pi^2}} \sqrt{\theta}\cos^2\left(\theta^{3/2}\right)d\theta = \int_0^{\pi} \cos^2 u\left(\tfrac{2}{3}\,du\right) = \left[\tfrac{2}{3}\left(\tfrac{u}{2} + \tfrac{1}{4}\sin 2u\right)\right]_0^{\pi} = \tfrac{2}{3}\left(\tfrac{\pi}{2} + \tfrac{1}{4}\sin 2\pi\right) - \tfrac{2}{3}(0) = \tfrac{\pi}{3}$$

25. Let $u = 4 - x^2 \Rightarrow du = -2x\,dx \Rightarrow -\frac{1}{2}\,du = x\,dx$; $x = -2 \Rightarrow u = 0$, $x = 0 \Rightarrow u = 4$, $x = 2 \Rightarrow u = 0$

$$A = -\int_{-2}^{0} x\sqrt{4-x^2}\,dx + \int_0^2 x\sqrt{4-x^2}\,dx = -\int_0^4 -\tfrac{1}{2}u^{1/2}\,du + \int_4^0 -\tfrac{1}{2}u^{1/2}\,du = 2\int_0^4 \tfrac{1}{2}u^{1/2}\,du = \int_0^4 u^{1/2}\,du$$
$$= \left[\tfrac{2}{3}u^{3/2}\right]_0^4 = \tfrac{2}{3}(4)^{3/2} - \tfrac{2}{3}(0)^{3/2} = \tfrac{16}{3}$$

27. Let $u = 1 + \cos x \Rightarrow du = -\sin x\,dx \Rightarrow -du = \sin x\,dx$; $x = -\pi \Rightarrow u = 1 + \cos(-\pi) = 0$, $x = 0 \Rightarrow u = 1 + \cos 0 = 2$

$$A = -\int_{-\pi}^{0} 3(\sin x)\sqrt{1+\cos x}\,dx = -\int_0^2 3u^{1/2}(-du) = 3\int_0^2 u^{1/2}\,du = \left[2u^{3/2}\right]_0^2 = 2(2)^{3/2} - 2(0)^{3/2} = 2^{5/2}$$

29. For the sketch given, $a = 0$, $b = \pi$; $f(x) - g(x) = 1 - \cos^2 x = \sin^2 x = \frac{1-\cos 2x}{2}$;

$$A = \int_0^{\pi} \tfrac{(1-\cos 2x)}{2}\,dx = \tfrac{1}{2}\int_0^{\pi}(1-\cos 2x)\,dx = \tfrac{1}{2}\left[x - \tfrac{\sin 2x}{2}\right]_0^{\pi} = \tfrac{1}{2}[(\pi - 0) - (0 - 0)] = \tfrac{\pi}{2}$$

31. For the sketch given, $a = -2$, $b = 2$; $f(x) - g(x) = 2x^2 - (x^4 - 2x^2) = 4x^2 - x^4$;

$$A = \int_{-2}^{2}(4x^2 - x^4)\,dx = \left[\tfrac{4x^3}{3} - \tfrac{x^5}{5}\right]_{-2}^{2} = \left(\tfrac{32}{3} - \tfrac{32}{5}\right) - \left[-\tfrac{32}{3} - \left(-\tfrac{32}{5}\right)\right] = \tfrac{64}{3} - \tfrac{64}{5} = \tfrac{320-192}{15} = \tfrac{128}{15}$$

33. For the sketch given, $c = 0$, $d = 1$; $f(y) - g(y) = (12y^2 - 12y^3) - (2y^2 - 2y) = 10y^2 - 12y^3 + 2y$;

$$A = \int_0^1 (10y^2 - 12y^3 + 2y)\,dy = \int_0^1 10y^2\,dy - \int_0^1 12y^3\,dy + \int_0^1 2y\,dy = \left[\tfrac{10}{3}y^3\right]_0^1 - \left[\tfrac{12}{4}y^4\right]_0^1 + \left[\tfrac{2}{2}y^2\right]_0^1$$
$$= \left(\tfrac{10}{3} - 0\right) - (3 - 0) + (1 - 0) = \tfrac{4}{3}$$

35. We want the area between the line $y = 1$, $0 \le x \le 2$, and the curve $y = \frac{x^2}{4}$, *minus* the area of a triangle (formed by $y = x$ and $y = 1$) with base 1 and height 1. Thus, $A = \int_0^2\left(1 - \frac{x^2}{4}\right)dx - \frac{1}{2}(1)(1) = \left[x - \frac{x^3}{12}\right]_0^2 - \frac{1}{2}$

$$= \left(2 - \tfrac{8}{12}\right) - \tfrac{1}{2} = 2 - \tfrac{2}{3} - \tfrac{1}{2} = \tfrac{5}{6}$$

37. AREA = A1 + A2

A1: For the sketch given, $a = -3$ and we find b by solving the equations $y = x^2 - 4$ and $y = -x^2 - 2x$ simultaneously for x: $x^2 - 4 = -x^2 - 2x \Rightarrow 2x^2 + 2x - 4 = 0 \Rightarrow 2(x+2)(x-1) \Rightarrow x = -2$ or $x = 1$ so $b = -2$: $f(x) - g(x) = (x^2 - 4) - (-x^2 - 2x) = 2x^2 + 2x - 4 \Rightarrow A1 = \int_{-3}^{-2}(2x^2 + 2x - 4)\,dx$

$$= \left[\tfrac{2x^3}{3} + \tfrac{2x^2}{2} - 4x\right]_{-3}^{-2} = \left(-\tfrac{16}{3} + 4 + 8\right) - (-18 + 9 + 12) = 9 - \tfrac{16}{3} = \tfrac{11}{3};$$

A2: For the sketch given, $a = -2$ and $b = 1$: $f(x) - g(x) = (-x^2 - 2x) - (x^2 - 4) = -2x^2 - 2x + 4$

$$\Rightarrow A2 = -\int_{-2}^{1}(2x^2 + 2x - 4)\,dx = -\left[\tfrac{2x^3}{3} + x^2 - 4x\right]_{-2}^{1} = -\left(\tfrac{2}{3} + 1 - 4\right) + \left(-\tfrac{16}{3} + 4 + 8\right)$$
$$= -\tfrac{2}{3} - 1 + 4 - \tfrac{16}{3} + 4 + 8 = 9;$$

Therefore, AREA = $A1 + A2 = \frac{11}{3} + 9 = \frac{38}{3}$

39. AREA = A1 + A2 + A3

A1: For the sketch given, $a = -2$ and $b = -1$: $f(x) - g(x) = (-x + 2) - (4 - x^2) = x^2 - x - 2$

$$\Rightarrow A1 = \int_{-2}^{-1}(x^2 - x - 2)\,dx = \left[\tfrac{x^3}{3} - \tfrac{x^2}{2} - 2x\right]_{-2}^{-1} = \left(-\tfrac{1}{3} - \tfrac{1}{2} + 2\right) - \left(-\tfrac{8}{3} - \tfrac{4}{2} + 4\right) = \tfrac{7}{3} - \tfrac{1}{2} = \tfrac{14-3}{6} = \tfrac{11}{6};$$

A2: For the sketch given, $a = -1$ and $b = 2$: $f(x) - g(x) = (4 - x^2) - (-x + 2) = -(x^2 - x - 2)$

$$\Rightarrow A2 = -\int_{-1}^{2}(x^2 - x - 2)\,dx = -\left[\tfrac{x^3}{3} - \tfrac{x^2}{2} - 2x\right]_{-1}^{2} = -\left(\tfrac{8}{3} - \tfrac{4}{2} - 4\right) + \left(-\tfrac{1}{3} - \tfrac{1}{2} + 2\right) = -3 + 8 - \tfrac{1}{2} = \tfrac{9}{2};$$

A3: For the sketch given, $a = 2$ and $b = 3$: $f(x) - g(x) = (-x + 2) - (4 - x^2) = x^2 - x - 2$

$$\Rightarrow A3 = \int_2^3 (x^2 - x - 2)\,dx = \left[\tfrac{x^3}{3} - \tfrac{x^2}{2} - 2x\right]_2^3 = \left(\tfrac{27}{3} - \tfrac{9}{2} - 6\right) - \left(\tfrac{8}{3} - \tfrac{4}{2} - 4\right) = 9 - \tfrac{9}{2} - \tfrac{8}{3};$$

Therefore, AREA = $A1 + A2 + A3 = \frac{11}{6} + \frac{9}{2} + \left(9 - \frac{9}{2} - \frac{8}{3}\right) = 9 - \frac{5}{6} = \frac{49}{6}$

41. $a = -2$, $b = 2$;
$f(x) - g(x) = 2 - (x^2 - 2) = 4 - x^2$
$\Rightarrow A = \int_{-2}^{2} (4 - x^2)dx = \left[4x - \frac{x^3}{3}\right]_{-2}^{2} = \left(8 - \frac{8}{3}\right) - \left(-8 + \frac{8}{3}\right)$
$= 2 \cdot \left(\frac{24}{3} - \frac{8}{3}\right) = \frac{32}{3}$

43. $a = 0$, $b = 2$;
$f(x) - g(x) = 8x - x^4 \Rightarrow A = \int_0^2 (8x - x^4)\, dx$
$= \left[\frac{8x^2}{2} - \frac{x^5}{5}\right]_0^2 = 16 - \frac{32}{5} = \frac{80 - 32}{5} = \frac{48}{5}$

45. Limits of integration: $x^2 = -x^2 + 4x \Rightarrow 2x^2 - 4x = 0$
$\Rightarrow 2x(x - 2) = 0 \Rightarrow a = 0$ and $b = 2$;
$f(x) - g(x) = (-x^2 + 4x) - x^2 = -2x^2 + 4x$
$\Rightarrow A = \int_0^2 (-2x^2 + 4x)\, dx = \left[\frac{-2x^3}{3} + \frac{4x^2}{2}\right]_0^2$
$= -\frac{16}{3} + \frac{16}{2} = \frac{-32+48}{6} = \frac{8}{3}$

47. Limits of integration: $x^4 - 4x^2 + 4 = x^2$
$\Rightarrow x^4 - 5x^2 + 4 = 0 \Rightarrow (x^2 - 4)(x^2 - 1) = 0$
$\Rightarrow (x + 2)(x - 2)(x + 1)(x - 1) = 0 \Rightarrow x = -2, -1, 1, 2$;
$f(x) - g(x) = (x^4 - 4x^2 + 4) - x^2 = x^4 - 5x^2 + 4$ and
$g(x) - f(x) = x^2 - (x^4 - 4x^2 + 4) = -x^4 + 5x^2 - 4$
$\Rightarrow A = \int_{-2}^{-1} (-x^4 + 5x^2 - 4)dx + \int_{-1}^{1} (x^4 - 5x^2 + 4)dx$
$+ \int_1^2 (-x^4 + 5x^2 - 4)dx$
$= \left[-\frac{x^5}{5} + \frac{5x^3}{3} - 4x\right]_{-2}^{-1} + \left[\frac{x^5}{5} - \frac{5x^3}{3} + 4x\right]_{-1}^{1} + \left[\frac{-x^5}{5} + \frac{5x^3}{3} - 4x\right]_1^2$
$= \left(\frac{1}{5} - \frac{5}{3} + 4\right) - \left(\frac{32}{5} - \frac{40}{3} + 8\right) + \left(\frac{1}{5} - \frac{5}{3} + 4\right) - \left(-\frac{1}{5} + \frac{5}{3} - 4\right) + \left(-\frac{32}{5} + \frac{40}{3} - 8\right) - \left(-\frac{1}{5} + \frac{5}{3} - 4\right)$
$= -\frac{60}{5} + \frac{60}{3} = \frac{300-180}{15} = 8$

49. Limits of integration: $y = \sqrt{|x|} = \begin{cases} \sqrt{-x}, & x \le 0 \\ \sqrt{x}, & x \ge 0 \end{cases}$ and
$5y = x + 6$ or $y = \frac{x}{5} + \frac{6}{5}$; for $x \le 0$: $\sqrt{-x} = \frac{x}{5} + \frac{6}{5}$
$\Rightarrow 5\sqrt{-x} = x + 6 \Rightarrow 25(-x) = x^2 + 12x + 36$
$\Rightarrow x^2 + 37x + 36 = 0 \Rightarrow (x + 1)(x + 36) = 0$
$\Rightarrow x = -1, -36$ (but $x = -36$ is not a solution);
for $x \ge 0$: $5\sqrt{x} = x + 6 \Rightarrow 25x = x^2 + 12x + 36$
$\Rightarrow x^2 - 13x + 36 = 0 \Rightarrow (x - 4)(x - 9) = 0$
$\Rightarrow x = 4, 9$; there are three intersection points and

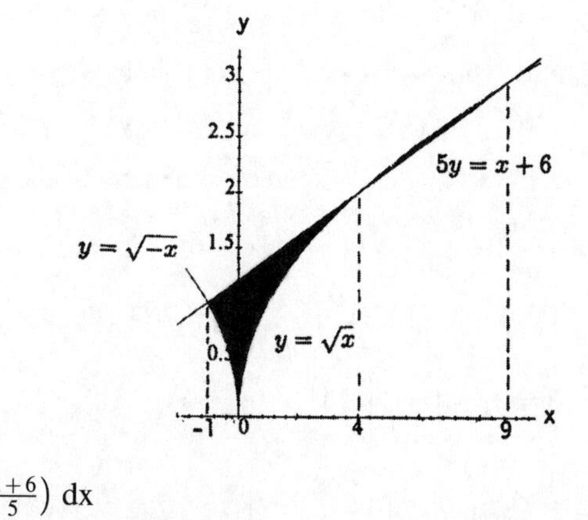

$A = \int_{-1}^{0} \left(\frac{x+6}{5} - \sqrt{-x}\right)dx + \int_0^4 \left(\frac{x+6}{5} - \sqrt{x}\right)dx + \int_4^9 \left(\sqrt{x} - \frac{x+6}{5}\right) dx$

$= \left[\frac{(x+6)^2}{10} + \frac{2}{3}(-x)^{3/2}\right]_{-1}^{0} + \left[\frac{(x+6)^2}{10} - \frac{2}{3}x^{3/2}\right]_{0}^{4} + \left[\frac{2}{3}x^{3/2} - \frac{(x+6)^2}{10}\right]_{4}^{9}$

$= \left(\frac{36}{10} - \frac{25}{10} - \frac{2}{3}\right) + \left(\frac{100}{10} - \frac{2}{3}\cdot 4^{3/2} - \frac{36}{10} + 0\right) + \left(\frac{2}{3}\cdot 9^{3/2} - \frac{225}{10} - \frac{2}{3}\cdot 4^{3/2} + \frac{100}{10}\right) = -\frac{50}{10} + \frac{20}{3} = \frac{5}{3}$

51. Limits of integration: $c = 0$ and $d = 3$;

$f(y) - g(y) = 2y^2 - 0 = 2y^2$

$\Rightarrow A = \int_0^3 2y^2\,dy = \left[\frac{2y^3}{3}\right]_0^3 = 2\cdot 9 = 18$

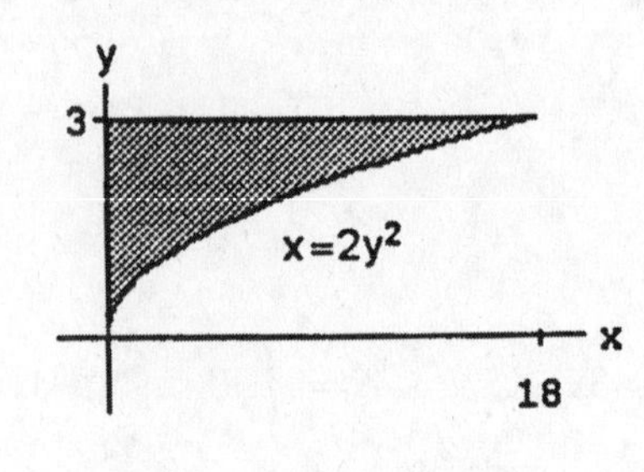

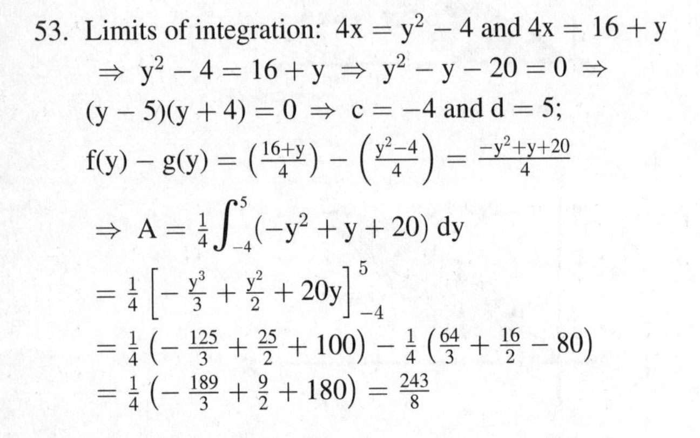

53. Limits of integration: $4x = y^2 - 4$ and $4x = 16 + y$

$\Rightarrow y^2 - 4 = 16 + y \Rightarrow y^2 - y - 20 = 0 \Rightarrow$

$(y-5)(y+4) = 0 \Rightarrow c = -4$ and $d = 5$;

$f(y) - g(y) = \left(\frac{16+y}{4}\right) - \left(\frac{y^2-4}{4}\right) = \frac{-y^2+y+20}{4}$

$\Rightarrow A = \frac{1}{4}\int_{-4}^{5}(-y^2 + y + 20)\,dy$

$= \frac{1}{4}\left[-\frac{y^3}{3} + \frac{y^2}{2} + 20y\right]_{-4}^{5}$

$= \frac{1}{4}\left(-\frac{125}{3} + \frac{25}{2} + 100\right) - \frac{1}{4}\left(\frac{64}{3} + \frac{16}{2} - 80\right)$

$= \frac{1}{4}\left(-\frac{189}{3} + \frac{9}{2} + 180\right) = \frac{243}{8}$

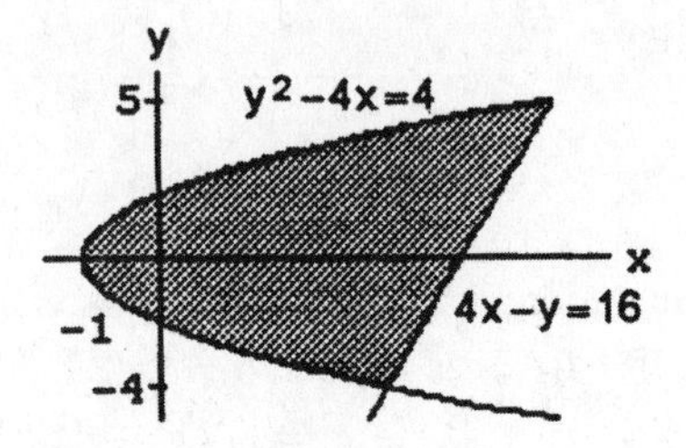

55. Limits of integration: $x = y^2 - y$ and $x = 2y^2 - 2y - 6$

$\Rightarrow y^2 - y = 2y^2 - 2y - 6 \Rightarrow y^2 - y - 6 = 0$

$\Rightarrow (y-3)(y+2) = 0 \Rightarrow c = -2$ and $d = 3$;

$f(y) - g(y) = (y^2 - y) - (2y^2 - 2y - 6) = -y^2 + y + 6$

$\Rightarrow A = \int_{-2}^{3}(-y^2 + y + 6)\,dy = \left[-\frac{y^3}{3} + \frac{1}{2}y^2 + 6y\right]_{-2}^{3}$

$= \left(-9 + \frac{9}{2} + 18\right) - \left(\frac{8}{3} + 2 - 12\right) = \frac{125}{6}$

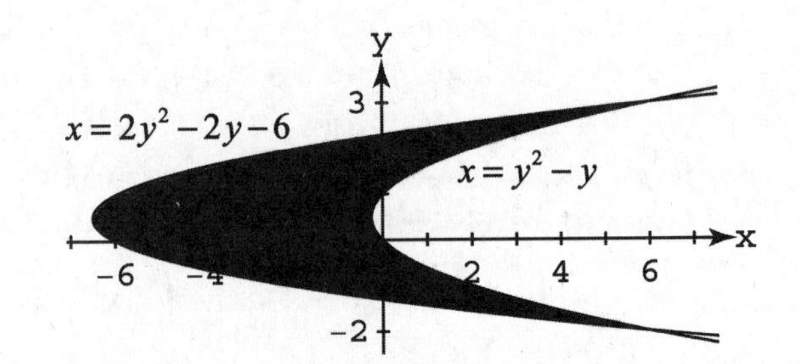

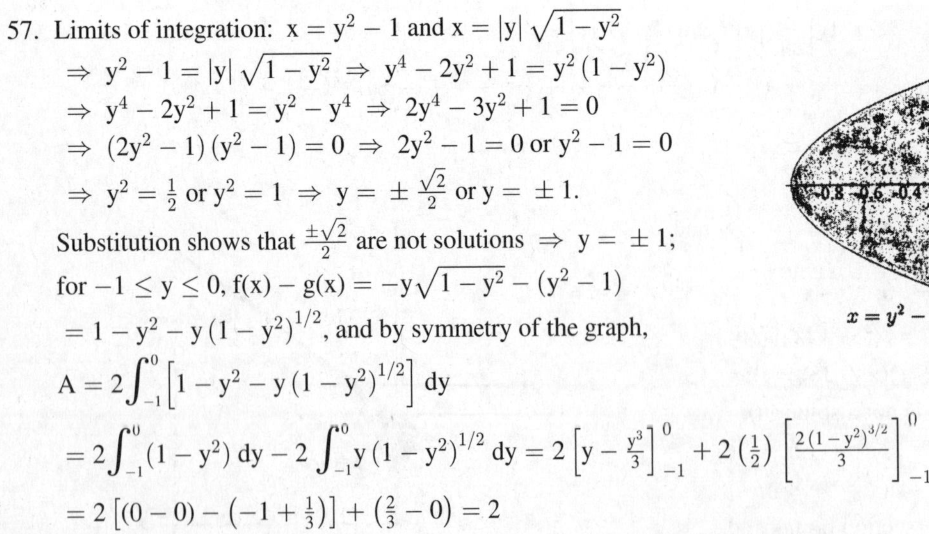

57. Limits of integration: $x = y^2 - 1$ and $x = |y|\sqrt{1-y^2}$

$\Rightarrow y^2 - 1 = |y|\sqrt{1-y^2} \Rightarrow y^4 - 2y^2 + 1 = y^2(1-y^2)$

$\Rightarrow y^4 - 2y^2 + 1 = y^2 - y^4 \Rightarrow 2y^4 - 3y^2 + 1 = 0$

$\Rightarrow (2y^2 - 1)(y^2 - 1) = 0 \Rightarrow 2y^2 - 1 = 0$ or $y^2 - 1 = 0$

$\Rightarrow y^2 = \frac{1}{2}$ or $y^2 = 1 \Rightarrow y = \pm\frac{\sqrt{2}}{2}$ or $y = \pm 1$.

Substitution shows that $\frac{\pm\sqrt{2}}{2}$ are not solutions $\Rightarrow y = \pm 1$;

for $-1 \le y \le 0$, $f(x) - g(x) = -y\sqrt{1-y^2} - (y^2 - 1)$

$= 1 - y^2 - y(1-y^2)^{1/2}$, and by symmetry of the graph,

$A = 2\int_{-1}^{0}\left[1 - y^2 - y(1-y^2)^{1/2}\right]dy$

$= 2\int_{-1}^{0}(1-y^2)\,dy - 2\int_{-1}^{0} y(1-y^2)^{1/2}\,dy = 2\left[y - \frac{y^3}{3}\right]_{-1}^{0} + 2\left(\frac{1}{2}\right)\left[\frac{2(1-y^2)^{3/2}}{3}\right]_{-1}^{0}$

$= 2\left[(0-0) - \left(-1 + \frac{1}{3}\right)\right] + \left(\frac{2}{3} - 0\right) = 2$

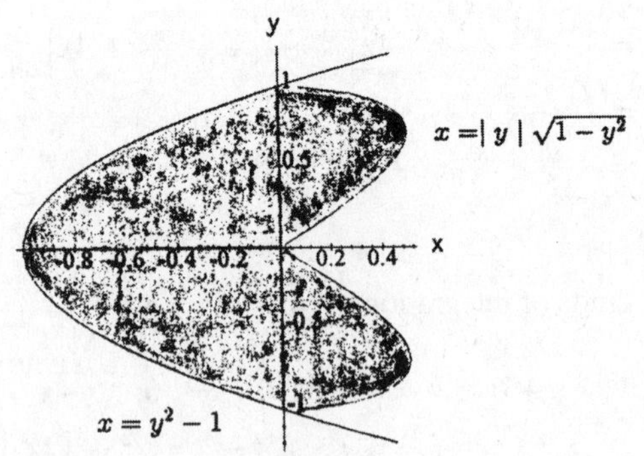

59. Limits of integration: $y = -4x^2 + 4$ and $y = x^4 - 1$
$\Rightarrow x^4 - 1 = -4x^2 + 4 \Rightarrow x^4 + 4x^2 - 5 = 0$
$\Rightarrow (x^2 + 5)(x - 1)(x + 1) = 0 \Rightarrow a = -1$ and $b = 1$;
$f(x) - g(x) = -4x^2 + 4 - x^4 + 1 = -4x^2 - x^4 + 5$
$\Rightarrow A = \int_{-1}^{1} (-4x^2 - x^4 + 5)\, dx = \left[-\frac{4x^3}{3} - \frac{x^5}{5} + 5x\right]_{-1}^{1}$
$= \left(-\frac{4}{3} - \frac{1}{5} + 5\right) - \left(\frac{4}{3} + \frac{1}{5} - 5\right) = 2\left(-\frac{4}{3} - \frac{1}{5} + 5\right) = \frac{104}{15}$

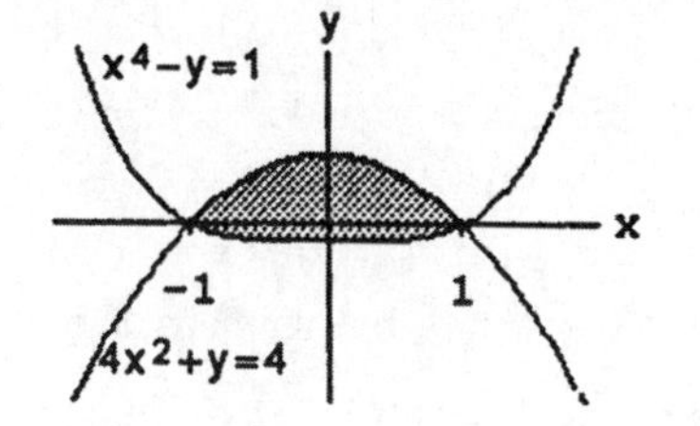

61. Limits of integration: $x = 4 - 4y^2$ and $x = 1 - y^4$
$\Rightarrow 4 - 4y^2 = 1 - y^4 \Rightarrow y^4 - 4y^2 + 3 = 0$
$\Rightarrow \left(y - \sqrt{3}\right)\left(y + \sqrt{3}\right)(y - 1)(y + 1) = 0 \Rightarrow c = -1$
and $d = 1$ since $x \ge 0$; $f(y) - g(y) = (4 - 4y^2) - (1 - y^4)$
$= 3 - 4y^2 + y^4 \Rightarrow A = \int_{-1}^{1} (3 - 4y^2 + y^4)\, dy$
$= \left[3y - \frac{4y^3}{3} + \frac{y^5}{5}\right]_{-1}^{1} = 2\left(3 - \frac{4}{3} + \frac{1}{5}\right) = \frac{56}{15}$

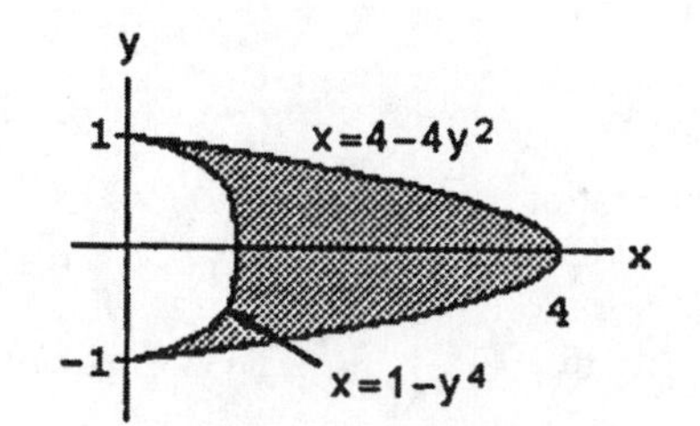

63. $a = 0, b = \pi$; $f(x) - g(x) = 2 \sin x - \sin 2x$
$\Rightarrow A = \int_{0}^{\pi} (2 \sin x - \sin 2x)\, dx = \left[-2 \cos x + \frac{\cos 2x}{2}\right]_{0}^{\pi}$
$= \left[-2(-1) + \frac{1}{2}\right] - \left(-2 \cdot 1 + \frac{1}{2}\right) = 4$

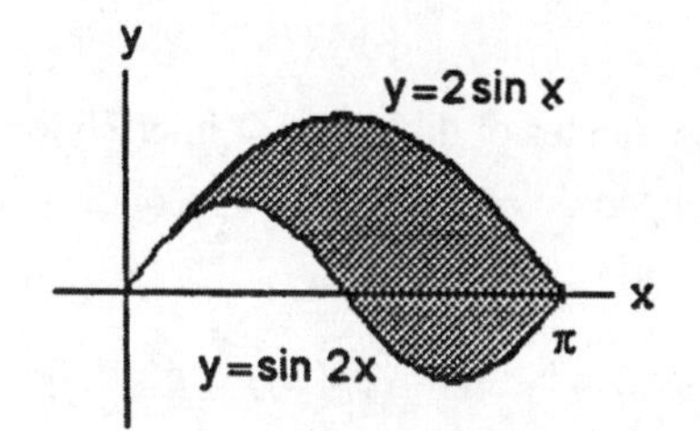

65. $a = -1, b = 1$; $f(x) - g(x) = (1 - x^2) - \cos\left(\frac{\pi x}{2}\right)$
$\Rightarrow A = \int_{-1}^{1} \left[1 - x^2 - \cos\left(\frac{\pi x}{2}\right)\right] dx = \left[x - \frac{x^3}{3} - \frac{2}{\pi} \sin\left(\frac{\pi x}{2}\right)\right]_{-1}^{1}$
$= \left(1 - \frac{1}{3} - \frac{2}{\pi}\right) - \left(-1 + \frac{1}{3} + \frac{2}{\pi}\right) = 2\left(\frac{2}{3} - \frac{2}{\pi}\right) = \frac{4}{3} - \frac{4}{\pi}$

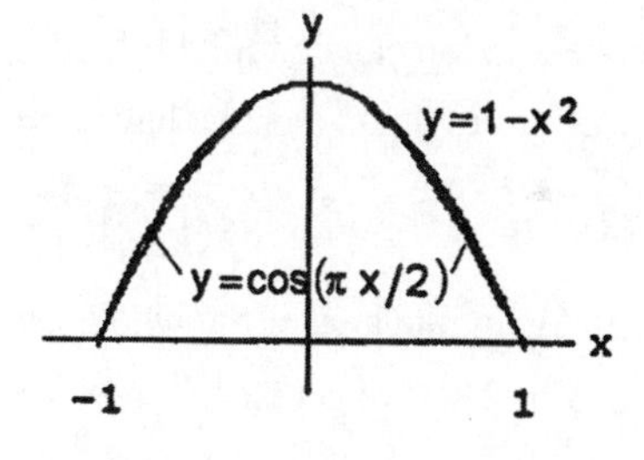

67. $a = -\frac{\pi}{4}, b = \frac{\pi}{4}$; $f(x) - g(x) = \sec^2 x - \tan^2 x$
$\Rightarrow A = \int_{-\pi/4}^{\pi/4} (\sec^2 x - \tan^2 x)\, dx$
$= \int_{-\pi/4}^{\pi/4} [\sec^2 x - (\sec^2 x - 1)]\, dx$
$= \int_{-\pi/4}^{\pi/4} 1 \cdot dx = [x]_{-\pi/4}^{\pi/4} = \frac{\pi}{4} - \left(-\frac{\pi}{4}\right) = \frac{\pi}{2}$

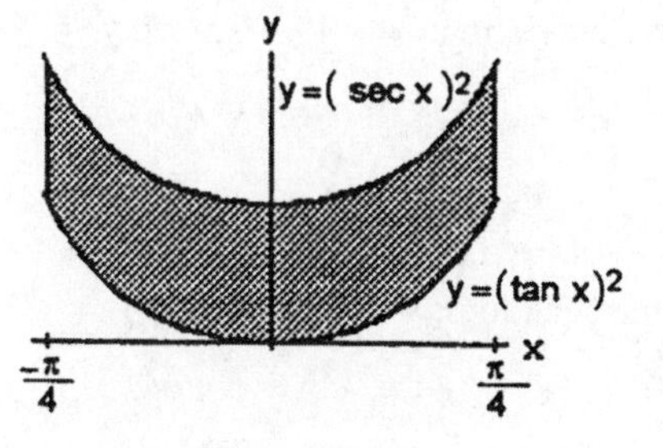

69. $c = 0, d = \frac{\pi}{2}$; $f(y) - g(y) = 3 \sin y \sqrt{\cos y} - 0 = 3 \sin y \sqrt{\cos y}$
$\Rightarrow A = 3 \int_{0}^{\pi/2} \sin y \sqrt{\cos y}\, dy = -3\left[\frac{2}{3} (\cos y)^{3/2}\right]_{0}^{\pi/2}$
$= -2(0 - 1) = 2$

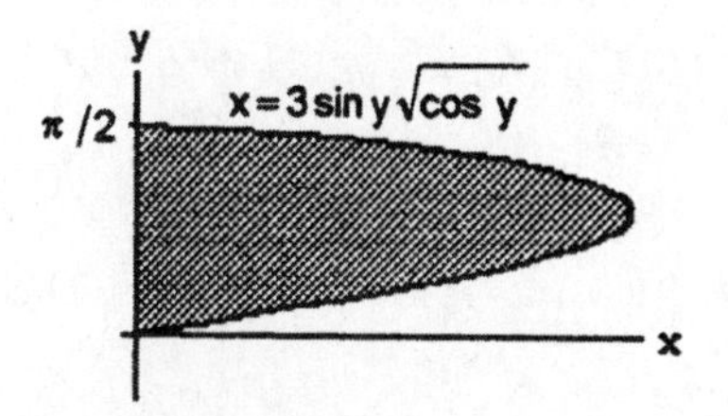

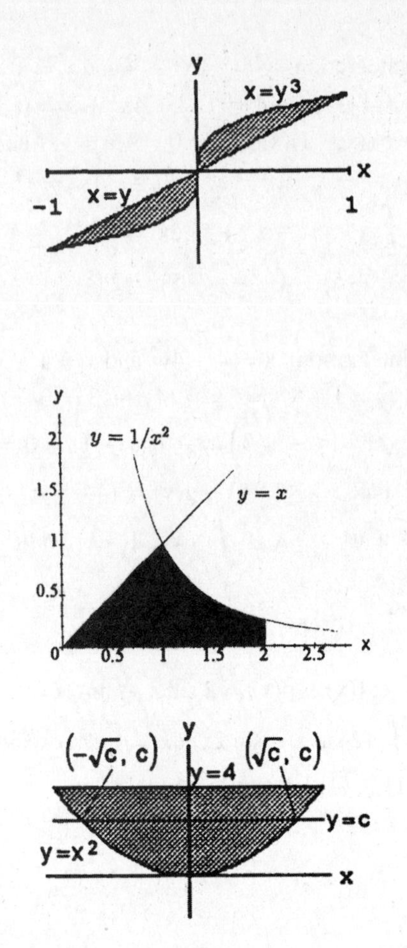

71. $A = A_1 + A_2$
Limits of integration: $x = y^3$ and $x = y \Rightarrow y = y^3$
$\Rightarrow y^3 - y = 0 \Rightarrow y(y-1)(y+1) = 0 \Rightarrow c_1 = -1, d_1 = 0$
and $c_2 = 0, d_2 = 1$; $f_1(y) - g_1(y) = y^3 - y$ and
$f_2(y) - g_2(y) = y - y^3 \Rightarrow$ by symmetry about the origin,
$A_1 + A_2 = 2A_2 \Rightarrow A = 2\int_0^1 (y - y^3)\, dy = 2\left[\frac{y^2}{2} - \frac{y^4}{4}\right]_0^1$
$= 2\left(\frac{1}{2} - \frac{1}{4}\right) = \frac{1}{2}$

73. $A = A_1 + A_2$
Limits of integration: $y = x$ and $y = \frac{1}{x^2} \Rightarrow x = \frac{1}{x^2}, x \neq 0$
$\Rightarrow x^3 = 1 \Rightarrow x = 1$, $f_1(x) - g_1(x) = x - 0 = x$
$\Rightarrow A_1 = \int_0^1 x\, dx = \left[\frac{x^2}{2}\right]_0^1 = \frac{1}{2}$; $f_2(x) - g_2(x) = \frac{1}{x^2} - 0$
$= x^{-2} \Rightarrow A_2 = \int_1^2 x^{-2}\, dx = \left[\frac{-1}{x}\right]_1^2 = -\frac{1}{2} + 1 = \frac{1}{2}$;
$A = A_1 + A_2 = \frac{1}{2} + \frac{1}{2} = 1$

75. (a) The coordinates of the points of intersection of the
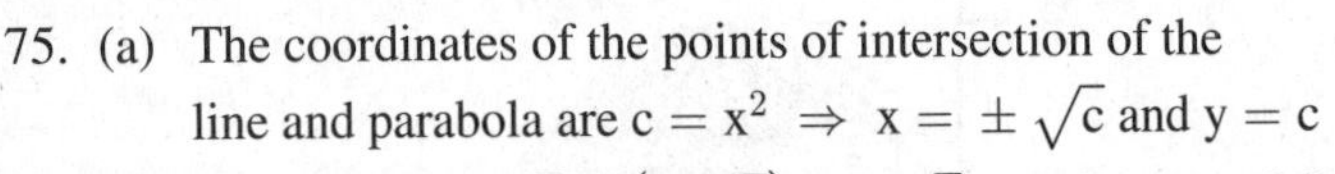
line and parabola are $c = x^2 \Rightarrow x = \pm\sqrt{c}$ and $y = c$
(b) $f(y) - g(y) = \sqrt{y} - (-\sqrt{y}) = 2\sqrt{y} \Rightarrow$ the area of the
lower section is, $A_L = \int_0^c [f(y) - g(y)]\, dy$
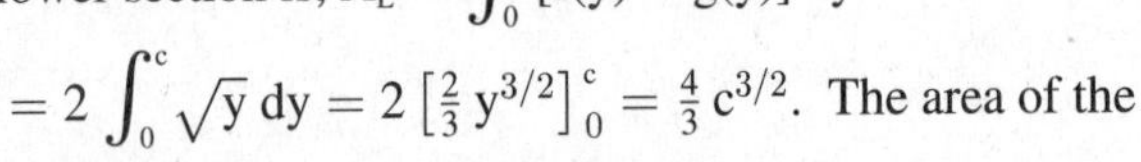
$= 2\int_0^c \sqrt{y}\, dy = 2\left[\frac{2}{3}y^{3/2}\right]_0^c = \frac{4}{3}c^{3/2}$. The area of the
entire shaded region can be found by setting $c = 4$: $A = \left(\frac{4}{3}\right)4^{3/2} = \frac{4 \cdot 8}{3} = \frac{32}{3}$. Since we want c to divide the region into subsections of equal area we have $A = 2A_L \Rightarrow \frac{32}{3} = 2\left(\frac{4}{3}c^{3/2}\right) \Rightarrow c = 4^{2/3}$
(c) $f(x) - g(x) = c - x^2 \Rightarrow A_L = \int_{-\sqrt{c}}^{\sqrt{c}} [f(x) - g(x)]\, dx = \int_{-\sqrt{c}}^{\sqrt{c}} (c - x^2)\, dx = \left[cx - \frac{x^3}{3}\right]_{-\sqrt{c}}^{\sqrt{c}} = 2\left[c^{3/2} - \frac{c^{3/2}}{3}\right]$
$= \frac{4}{3}c^{3/2}$. Again, the area of the whole shaded region can be found by setting $c = 4 \Rightarrow A = \frac{32}{3}$. From the condition $A = 2A_L$, we get $\frac{4}{3}c^{3/2} = \frac{32}{3} \Rightarrow c = 4^{2/3}$ as in part (b).

77. Limits of integration: $y = 1 + \sqrt{x}$ and $y = \frac{2}{\sqrt{x}}$
$\Rightarrow 1 + \sqrt{x} = \frac{2}{\sqrt{x}}, x \neq 0 \Rightarrow \sqrt{x} + x = 2 \Rightarrow x = (2 - x)^2$
$\Rightarrow x = 4 - 4x + x^2 \Rightarrow x^2 - 5x + 4 = 0$
$\Rightarrow (x-4)(x-1) = 0 \Rightarrow x = 1, 4$ (but $x = 4$ does not
satisfy the equation); $y = \frac{2}{\sqrt{x}}$ and $y = \frac{x}{4} \Rightarrow \frac{2}{\sqrt{x}} = \frac{x}{4}$
$\Rightarrow 8 = x\sqrt{x} \Rightarrow 64 = x^3 \Rightarrow x = 4.$
Therefore, AREA $= A_1 + A_2$: $f_1(x) - g_1(x) = \left(1 + x^{1/2}\right) - \frac{x}{4}$
$\Rightarrow A_1 = \int_0^1 \left(1 + x^{1/2} - \frac{x}{4}\right) dx = \left[x + \frac{2}{3}x^{3/2} - \frac{x^2}{8}\right]_0^1$
$= \left(1 + \frac{2}{3} - \frac{1}{8}\right) - 0 = \frac{37}{24}$; $f_2(x) - g_2(x) = 2x^{-1/2} - \frac{x}{4} \Rightarrow A_2 = \int_1^4 \left(2x^{-1/2} - \frac{x}{4}\right) dx = \left[4x^{1/2} - \frac{x^2}{8}\right]_1^4$
$= \left(4 \cdot 2 - \frac{16}{8}\right) - \left(4 - \frac{1}{8}\right) = 4 - \frac{15}{8} = \frac{17}{8}$; Therefore, AREA $= A_1 + A_2 = \frac{37}{24} + \frac{17}{8} = \frac{37+51}{24} = \frac{88}{24} = \frac{11}{3}$

79. Area between parabola and $y = a^2$: $A = 2\int_0^a (a^2 - x^2)\, dx = 2\left[a^2x - \frac{1}{3}x^3\right]_0^a = 2\left(a^3 - \frac{a^3}{3}\right) - 0 = \frac{4a^3}{3}$;
Area of triangle AOC: $\frac{1}{2}(2a)(a^2) = a^3$; limit of ratio $= \lim_{a \to 0^+} \frac{a^3}{\left(\frac{4a^3}{3}\right)} = \frac{3}{4}$ which is independent of a.

81. The lower boundary of the region is the line through the points $(z, 1 - z^2)$ and $\left(z + 1, 1 - (z + 1)^2\right)$. The equation of this line is $y - (1 - z^2) = \frac{(1-(z+1)^2) - (1-z^2)}{z+1-z}(x - 1) = -(2z + 1)(x - 1) \Rightarrow y = -(2z + 1)x + (z^2 + z + 1)$.

The area of theregion is given by $\int_z^{z+1} ((1 - x^2) - (-(2z + 1)x + (z^2 + z + 1)))dy$

$= \int_z^{z+1} (-x^2 + (2z + 1)x - z^2 - z)dy = \left[-\frac{1}{3}x^3 + \frac{1}{2}(2z + 1)x^2 - (z^2 + z)x\right]_z^{z+1}$

$= \left(-\frac{1}{3}(z+1)^3 + \frac{1}{2}(2z + 1)(z + 1)^2 - (z^2 + z)(z + 1)\right) - \left(-\frac{1}{3}z^3 + \frac{1}{2}(2z + 1)z^2 - (z^2 + z)z\right) = \frac{1}{6}$. No matter where we choose z, the area of the region bounded by $y = 1 - x^2$ and the line through the points $(z, 1 - z^2)$ and $\left(z + 1, 1 - (z + 1)^2\right)$ is always $\frac{1}{6}$.

83. Let $u = 2x \Rightarrow du = 2\,dx \Rightarrow \frac{1}{2}\,du = dx$; $x = 1 \Rightarrow u = 2$, $x = 3 \Rightarrow u = 6$

$\int_1^3 \frac{\sin 2x}{x}\,dx = \int_2^6 \frac{\sin u}{\left(\frac{u}{2}\right)}\left(\frac{1}{2}\,du\right) = \int_2^6 \frac{\sin u}{u}\,du = [F(u)]_2^6 = F(6) - F(2)$

85. (a) Let $u = -x \Rightarrow du = -\,dx$; $x = -1 \Rightarrow u = 1$, $x = 0 \Rightarrow u = 0$

f odd $\Rightarrow f(-x) = -f(x)$. Then $\int_{-1}^0 f(x)\,dx = \int_1^0 f(-u)\,(-\,du) = \int_1^0 -f(u)\,(-\,du) = \int_1^0 f(u)\,du = -\int_0^1 f(u)\,du$ $= -3$

(b) Let $u = -x \Rightarrow du = -\,dx$; $x = -1 \Rightarrow u = 1$, $x = 0 \Rightarrow u = 0$

f even $\Rightarrow f(-x) = f(x)$. Then $\int_{-1}^0 f(x)\,dx = \int_1^0 f(-u)\,(-\,du) = -\int_1^0 f(u)\,du = \int_0^1 f(u)\,du = 3$

87. Let $u = a - x \Rightarrow du = -\,dx$; $x = 0 \Rightarrow u = a$, $x = a \Rightarrow u = 0$

$I = \int_0^a \frac{f(x)\,dx}{f(x)+f(a-x)} = \int_a^0 \frac{f(a-u)}{f(a-u)+f(u)}(-\,du) = \int_0^a \frac{f(a-u)\,du}{f(u)+f(a-u)} = \int_0^a \frac{f(a-x)\,dx}{f(x)+f(a-x)}$

$\Rightarrow I + I = \int_0^a \frac{f(x)\,dx}{f(x)+f(a-x)} + \int_0^a \frac{f(a-x)\,dx}{f(x)+f(a-x)} = \int_0^a \frac{f(x)+f(a-x)}{f(x)+f(a-x)}\,dx = \int_0^a dx = [x]_0^a = a - 0 = a$.

Therefore, $2I = a \Rightarrow I = \frac{a}{2}$.

89. Let $u = x + c \Rightarrow du = dx$; $x = a - c \Rightarrow u = a$, $x = b - c \Rightarrow u = b$

$\int_{a-c}^{b-c} f(x + c)\,dx = \int_a^b f(u)\,du = \int_a^b f(x)\,dx$

CHAPTER 5 PRACTICE EXERCISES

1. (a) Each time subinterval is of length $\Delta t = 0.4$ sec. The distance traveled over each subinterval, using the midpoint rule, is $\Delta h = \frac{1}{2}(v_i + v_{i+1})\Delta t$, where v_i is the velocity at the left endpoint and v_{i+1} the velocity at the right endpoint of the subinterval. We then add Δh to the height attained so far at the left endpoint v_i to arrive at the height associated with velocity v_{i+1} at the right endpoint. Using this methodology we build the following table based on the figure in the text:

t (sec)	0	0.4	0.8	1.2	1.6	2.0	2.4	2.8	3.2	3.6	4.0	4.4	4.8	5.2	5.6	6.0
v (fps)	0	10	25	55	100	190	180	165	150	140	130	115	105	90	76	65
h (ft)	0	2	9	25	56	114	188	257	320	378	432	481	525	564	592	620.2

t (sec)	6.4	6.8	7.2	7.6	8.0
v (fps)	50	37	25	12	0
h (ft)	643.2	660.6	672	679.4	681.8

NOTE: Your table values may vary slightly from ours depending on the v-values you read from the graph. Remember that some shifting of the graph occurs in the printing process.

The total height attained is about 680 ft.

(b) The graph is based on the table in part (a).

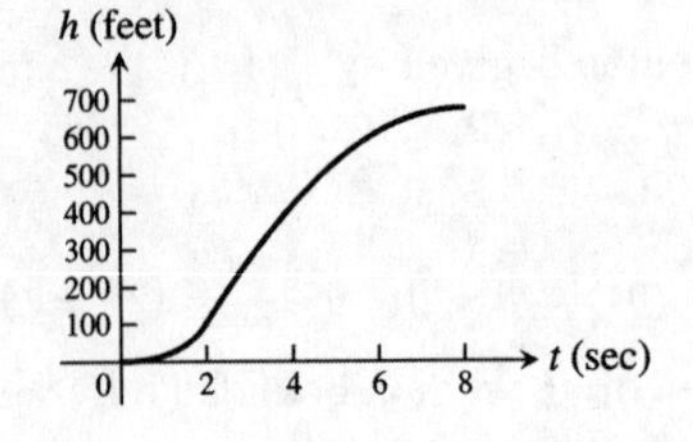

3. (a) $\sum_{k=1}^{10} \frac{a_k}{4} = \frac{1}{4} \sum_{k=1}^{10} a_k = \frac{1}{4}(-2) = -\frac{1}{2}$ (b) $\sum_{k=1}^{10} (b_k - 3a_k) = \sum_{k=1}^{10} b_k - 3 \sum_{k=1}^{10} a_k = 25 - 3(-2) = 31$

(c) $\sum_{k=1}^{10} (a_k + b_k - 1) = \sum_{k=1}^{10} a_k + \sum_{k=1}^{10} b_k - \sum_{k=1}^{10} 1 = -2 + 25 - (1)(10) = 13$

(d) $\sum_{k=1}^{10} \left(\frac{5}{2} - b_k\right) = \sum_{k=1}^{10} \frac{5}{2} - \sum_{k=1}^{10} b_k = \frac{5}{2}(10) - 25 = 0$

5. Let $u = 2x - 1 \Rightarrow du = 2\,dx \Rightarrow \frac{1}{2}\,du = dx$; $x = 1 \Rightarrow u = 1$, $x = 5 \Rightarrow u = 9$

$\int_1^5 (2x-1)^{-1/2}\,dx = \int_1^9 u^{-1/2}\left(\frac{1}{2}\,du\right) = \left[u^{1/2}\right]_1^9 = 3 - 1 = 2$

7. Let $u = \frac{x}{2} \Rightarrow 2\,du = dx$; $x = -\pi \Rightarrow u = -\frac{\pi}{2}$, $x = 0 \Rightarrow u = 0$

$\int_{-\pi}^{0} \cos\left(\frac{x}{2}\right) dx = \int_{-\pi/2}^{0} (\cos u)(2\,du) = [2 \sin u]_{-\pi/2}^{0} = 2 \sin 0 - 2 \sin\left(-\frac{\pi}{2}\right) = 2(0 - (-1)) = 2$

9. (a) $\int_{-2}^{2} f(x)\,dx = \frac{1}{3}\int_{-2}^{2} 3 f(x)\,dx = \frac{1}{3}(12) = 4$ (b) $\int_{2}^{5} f(x)\,dx = \int_{-2}^{5} f(x)\,dx - \int_{-2}^{2} f(x)\,dx = 6 - 4 = 2$

(c) $\int_{5}^{-2} g(x)\,dx = -\int_{-2}^{5} g(x)\,dx = -2$ (d) $\int_{-2}^{5} (-\pi\, g(x))\,dx = -\pi \int_{-2}^{5} g(x)\,dx = -\pi(2) = -2\pi$

(e) $\int_{-2}^{5} \left(\frac{f(x)+g(x)}{5}\right) dx = \frac{1}{5}\int_{-2}^{5} f(x)\,dx + \frac{1}{5}\int_{-2}^{5} g(x)\,dx = \frac{1}{5}(6) + \frac{1}{5}(2) = \frac{8}{5}$

11. $x^2 - 4x + 3 = 0 \Rightarrow (x-3)(x-1) = 0 \Rightarrow x = 3$ or $x = 1$;

$\text{Area} = \int_0^1 (x^2 - 4x + 3)\,dx - \int_1^3 (x^2 - 4x + 3)\,dx$

$= \left[\frac{x^3}{3} - 2x^2 + 3x\right]_0^1 - \left[\frac{x^3}{3} - 2x^2 + 3x\right]_1^3$

$= \left[\left(\frac{1^3}{3} - 2(1)^2 + 3(1)\right) - 0\right]$

$\quad - \left[\left(\frac{3^3}{3} - 2(3)^2 + 3(3)\right) - \left(\frac{1^3}{3} - 2(1)^2 + 3(1)\right)\right]$

$= \left(\frac{1}{3} + 1\right) - \left[0 - \left(\frac{1}{3} + 1\right)\right] = \frac{8}{3}$

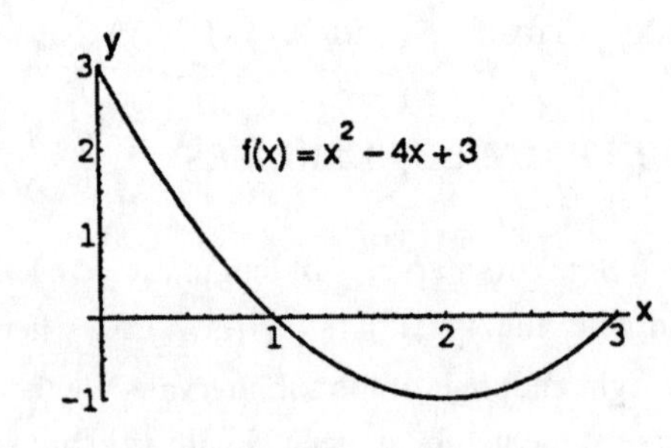

13. $5 - 5x^{2/3} = 0 \Rightarrow 1 - x^{2/3} = 0 \Rightarrow x = \pm 1$;

$\text{Area} = \int_{-1}^{1} (5 - 5x^{2/3})\,dx - \int_{1}^{8} (5 - 5x^{2/3})\,dx$

$= \left[5x - 3x^{5/3}\right]_{-1}^{1} - \left[5x - 3x^{5/3}\right]_{1}^{8}$

$= \left[(5(1) - 3(1)^{5/3}) - (5(-1) - 3(-1)^{5/3})\right]$

$\quad - \left[(5(8) - 3(8)^{5/3}) - (5(1) - 3(1)^{5/3})\right]$

$= [2 - (-2)] - [(40 - 96) - 2] = 62$

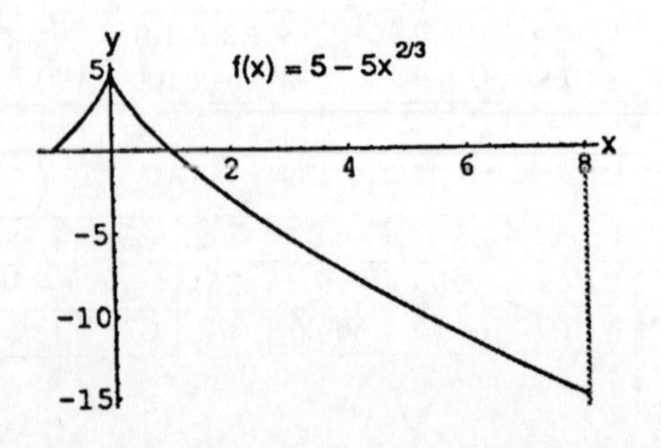

15. $f(x) = x$, $g(x) = \frac{1}{x^2}$, $a = 1$, $b = 2 \Rightarrow A = \int_a^b [f(x) - g(x)]\,dx$
$= \int_1^2 \left(x - \frac{1}{x^2}\right) dx = \left[\frac{x^2}{2} + \frac{1}{x}\right]_1^2 = \left(\frac{4}{2} + \frac{1}{2}\right) - \left(\frac{1}{2} + 1\right) = 1$

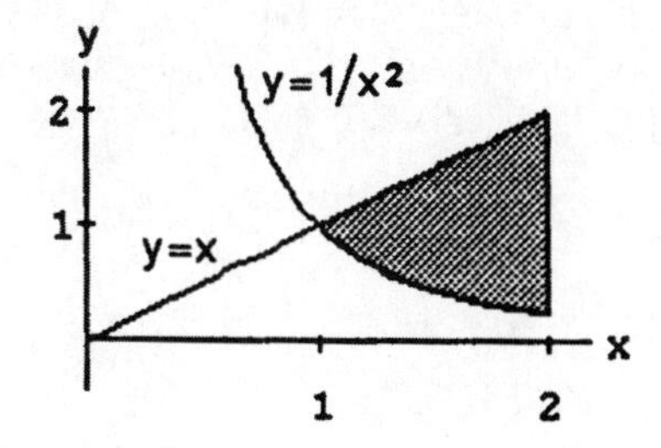

17. $f(x) = \left(1 - \sqrt{x}\right)^2$, $g(x) = 0$, $a = 0$, $b = 1 \Rightarrow A = \int_a^b [f(x) - g(x)]\,dx = \int_0^1 \left(1 - \sqrt{x}\right)^2 dx = \int_0^1 \left(1 - 2\sqrt{x} + x\right) dx$
$= \int_0^1 \left(1 - 2x^{1/2} + x\right) dx = \left[x - \frac{4}{3}x^{3/2} + \frac{x^2}{2}\right]_0^1 = 1 - \frac{4}{3} + \frac{1}{2} = \frac{1}{6}(6 - 8 + 3) = \frac{1}{6}$

19. $f(y) = 2y^2$, $g(y) = 0$, $c = 0$, $d = 3$
$\Rightarrow A = \int_c^d [f(y) - g(y)]\,dy = \int_0^3 (2y^2 - 0)\,dy$
$= 2\int_0^3 y^2\,dy = \frac{2}{3}\left[y^3\right]_0^3 = 18$

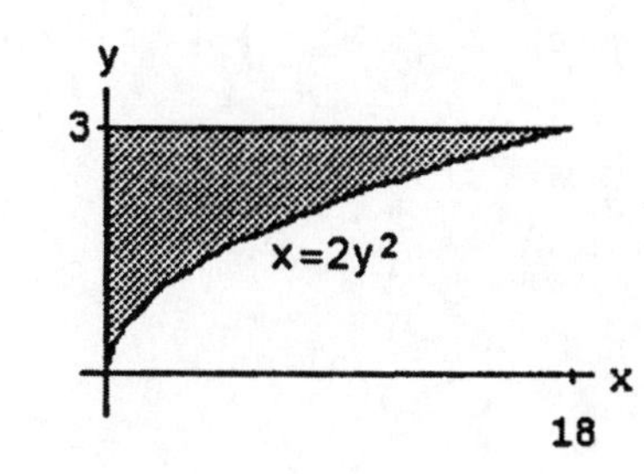

21. Let us find the intersection points: $\frac{y^2}{4} = \frac{y+2}{4}$
$\Rightarrow y^2 - y - 2 = 0 \Rightarrow (y-2)(y+1) = 0 \Rightarrow y = -1$
or $y = 2 \Rightarrow c = -1, d = 2$; $f(y) = \frac{y+2}{4}$, $g(y) = \frac{y^2}{4}$
$\Rightarrow A = \int_c^d [f(y) - g(y)]\,dy = \int_{-1}^2 \left(\frac{y+2}{4} - \frac{y^2}{4}\right) dy$
$= \frac{1}{4}\int_{-1}^2 (y + 2 - y^2)\,dy = \frac{1}{4}\left[\frac{y^2}{2} + 2y - \frac{y^3}{3}\right]_{-1}^2$
$= \frac{1}{4}\left[\left(\frac{4}{2} + 4 - \frac{8}{3}\right) - \left(\frac{1}{2} - 2 + \frac{1}{3}\right)\right] = \frac{9}{8}$

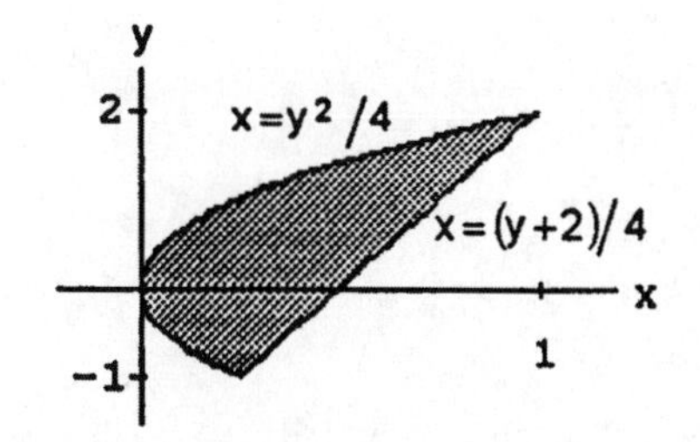

23. $f(x) = x$, $g(x) = \sin x$, $a = 0$, $b = \frac{\pi}{4}$
$\Rightarrow A = \int_a^b [f(x) - g(x)]\,dx = \int_0^{\pi/4} (x - \sin x)\,dx$
$= \left[\frac{x^2}{2} + \cos x\right]_0^{\pi/4} = \left(\frac{\pi^2}{32} + \frac{\sqrt{2}}{2}\right) - 1$

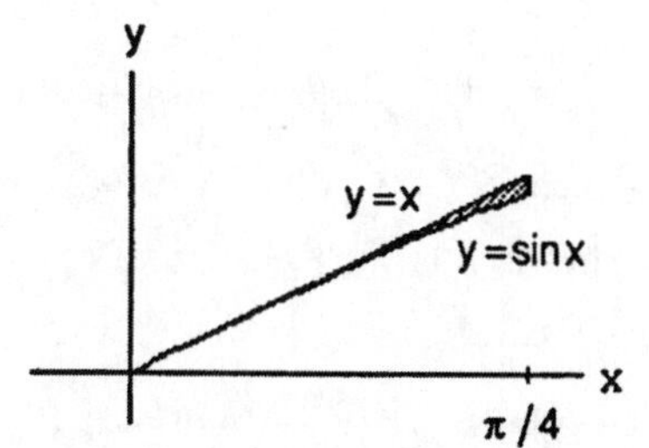

25. $a = 0$, $b = \pi$, $f(x) - g(x) = 2\sin x - \sin 2x$
$\Rightarrow A = \int_0^\pi (2\sin x - \sin 2x)\,dx = \left[-2\cos x + \frac{\cos 2x}{2}\right]_0^\pi$
$= \left[-2\cdot(-1) + \frac{1}{2}\right] - \left(-2\cdot 1 + \frac{1}{2}\right) = 4$

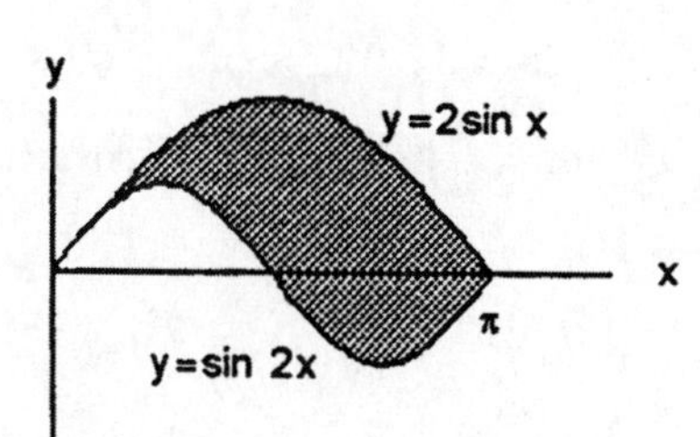

27. $f(y) = \sqrt{y}$, $g(y) = 2 - y$, $c = 1$, $d = 2$

$\Rightarrow A = \int_c^d [f(y) - g(y)]\, dy = \int_1^2 \left[\sqrt{y} - (2 - y)\right] dy$

$= \int_1^2 \left(\sqrt{y} - 2 + y\right) dy = \left[\frac{2}{3} y^{3/2} - 2y + \frac{y^2}{2}\right]_1^2$

$= \left(\frac{4}{3}\sqrt{2} - 4 + 2\right) - \left(\frac{2}{3} - 2 + \frac{1}{2}\right) = \frac{4}{3}\sqrt{2} - \frac{7}{6} = \frac{8\sqrt{2}-7}{6}$

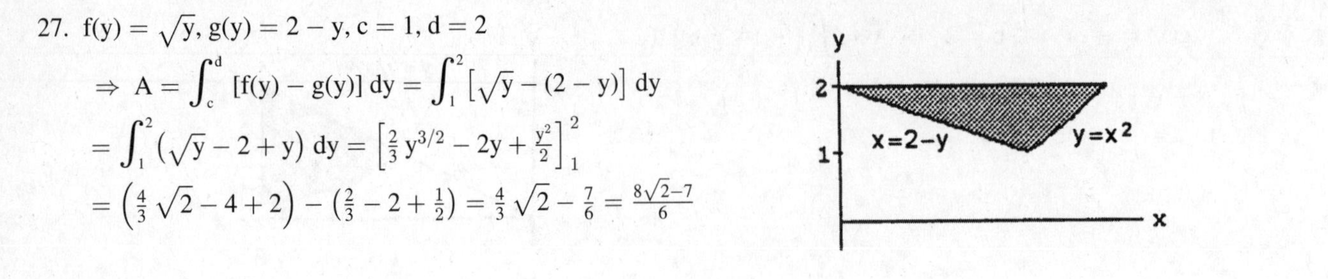

29. $f(x) = x^3 - 3x^2 = x^2(x - 3) \Rightarrow f'(x) = 3x^2 - 6x = 3x(x - 2) \Rightarrow f' = +++\underset{0}{|}----\underset{2}{|}+++$

$\Rightarrow f(0) = 0$ is a maximum and $f(2) = -4$ is a minimum. $A = -\int_0^3 (x^3 - 3x^2)\, dx = -\left[\frac{x^4}{4} - x^3\right]_0^3 = -\left(\frac{81}{4} - 27\right) = \frac{27}{4}$

31. The area above the x-axis is $A_1 = \int_0^1 \left(y^{2/3} - y\right) dy$

$= \left[\frac{3y^{5/3}}{5} - \frac{y^2}{2}\right]_0^1 = \frac{1}{10}$; the area below the x-axis is

$A_2 = \int_{-1}^0 \left(y^{2/3} - y\right) dy = \left[\frac{3y^{5/3}}{5} - \frac{y^2}{2}\right]_{-1}^0 = \frac{11}{10}$

$\Rightarrow$ the total area is $A_1 + A_2 = \frac{6}{5}$

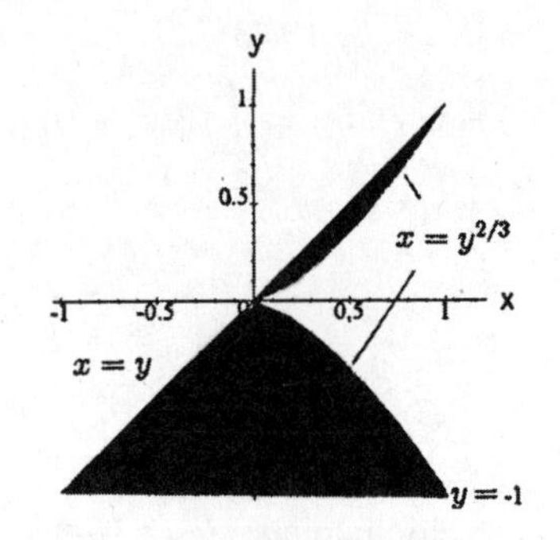

33. $y = x^2 + \int_1^x \frac{1}{t}\, dt \Rightarrow \frac{dy}{dx} = 2x + \frac{1}{x} \Rightarrow \frac{d^2y}{dx^2} = 2 - \frac{1}{x^2}$; $y(1) = 1 + \int_1^1 \frac{1}{t}\, dt = 1$ and $y'(1) = 2 + 1 = 3$

35. $y = \int_5^x \frac{\sin t}{t}\, dt - 3 \Rightarrow \frac{dy}{dx} = \frac{\sin x}{x}$; $x = 5 \Rightarrow y = \int_5^5 \frac{\sin t}{t}\, dt - 3 = -3$

37. Let $u = \cos x \Rightarrow du = -\sin x\, dx \Rightarrow -du = \sin x\, dx$

$\int 2(\cos x)^{-1/2} \sin x\, dx = \int 2u^{-1/2}(-du) = -2\int u^{-1/2}\, du = -2\left(\frac{u^{1/2}}{\frac{1}{2}}\right) + C = -4u^{1/2} + C = -4(\cos x)^{1/2} + C$

39. Let $u = 2\theta + 1 \Rightarrow du = 2\, d\theta \Rightarrow \frac{1}{2}\, du = d\theta$

$\int [2\theta + 1 + 2\cos(2\theta + 1)]\, d\theta = \int (u + 2\cos u)\left(\frac{1}{2}\, du\right) = \frac{u^2}{4} + \sin u + C_1 = \frac{(2\theta+1)^2}{4} + \sin(2\theta + 1) + C_1$

$= \theta^2 + \theta + \sin(2\theta + 1) + C$, where $C = C_1 + \frac{1}{4}$ is still an arbitrary constant

41. $\int \left(t - \frac{2}{t}\right)\left(t + \frac{2}{t}\right) dt = \int \left(t^2 - \frac{4}{t^2}\right) dt = \int \left(t^2 - 4t^{-2}\right) dt = \frac{t^3}{3} - 4\left(\frac{t^{-1}}{-1}\right) + C = \frac{t^3}{3} + \frac{4}{t} + C$

43. Let $u = 2t^{3/2} \Rightarrow du = 3\sqrt{t}\, dt \Rightarrow \frac{1}{3}du = \sqrt{t}\, dt$

$\int \sqrt{t} \sin\left(2t^{3/2}\right) dt = \frac{1}{3}\int \sin u\, du = -\frac{1}{3}\cos u + C = -\frac{1}{3}\cos\left(2t^{3/2}\right) + C$

45. $\int_{-1}^1 (3x^2 - 4x + 7)\, dx = \left[x^3 - 2x^2 + 7x\right]_{-1}^1 = [1^3 - 2(1)^2 + 7(1)] - [(-1)^3 - 2(-1)^2 + 7(-1)] = 6 - (-10) = 16$

47. $\int_1^2 \frac{4}{v^2}\, dv = \int_1^2 4v^{-2}\, dv = \left[-4v^{-1}\right]_1^2 = \left(\frac{-4}{2}\right) - \left(\frac{-4}{1}\right) = 2$

49. $\int_1^4 \frac{dt}{t\sqrt{t}} = \int_1^4 \frac{dt}{t^{3/2}} = \int_1^4 t^{-3/2}\, dt = \left[-2t^{-1/2}\right]_1^4 = \frac{-2}{\sqrt{4}} - \frac{(-2)}{\sqrt{1}} = 1$

51. Let $u = 2x + 1 \Rightarrow du = 2\,dx \Rightarrow 18\,du = 36\,dx$; $x = 0 \Rightarrow u = 1$, $x = 1 \Rightarrow u = 3$

$$\int_0^1 \frac{36\,dx}{(2x+1)^3} = \int_1^3 18u^{-3}\,du = \left[\frac{18u^{-2}}{-2}\right]_1^3 = \left[\frac{-9}{u^2}\right]_1^3 = \left(\frac{-9}{3^2}\right) - \left(\frac{-9}{1^2}\right) = 8$$

53. Let $u = 1 - x^{2/3} \Rightarrow du = -\frac{2}{3}x^{-1/3}\,dx \Rightarrow -\frac{3}{2}\,du = x^{-1/3}\,dx$; $x = \frac{1}{8} \Rightarrow u = 1 - \left(\frac{1}{8}\right)^{2/3} = \frac{3}{4}$, $x = 1 \Rightarrow u = 1 - 1^{2/3} = 0$

$$\int_{1/8}^1 x^{-1/3}\left(1 - x^{2/3}\right)^{3/2} dx = \int_{3/4}^0 u^{3/2}\left(-\frac{3}{2}\,du\right) = \left[\left(-\frac{3}{2}\right)\left(\frac{u^{5/2}}{\frac{5}{2}}\right)\right]_{3/4}^0 = \left[-\frac{3}{5}u^{5/2}\right]_{3/4}^0 = -\frac{3}{5}(0)^{5/2} - \left(-\frac{3}{5}\right)\left(\frac{3}{4}\right)^{5/2}$$
$$= \frac{27\sqrt{3}}{160}$$

55. Let $u = 5r \Rightarrow du = 5\,dr \Rightarrow \frac{1}{5}\,du = dr$; $r = 0 \Rightarrow u = 0$, $r = \pi \Rightarrow u = 5\pi$

$$\int_0^\pi \sin^2 5r\,dr = \int_0^{5\pi} (\sin^2 u)\left(\frac{1}{5}\,du\right) = \frac{1}{5}\left[\frac{u}{2} - \frac{\sin 2u}{4}\right]_0^{5\pi} = \left(\frac{\pi}{2} - \frac{\sin 10\pi}{20}\right) - \left(0 - \frac{\sin 0}{20}\right) = \frac{\pi}{2}$$

57. $$\int_0^{\pi/3} \sec^2\theta\,d\theta = [\tan\theta]_0^{\pi/3} = \tan\frac{\pi}{3} - \tan 0 = \sqrt{3}$$

59. Let $u = \frac{x}{6} \Rightarrow du = \frac{1}{6}\,dx \Rightarrow 6\,du = dx$; $x = \pi \Rightarrow u = \frac{\pi}{6}$, $x = 3\pi \Rightarrow u = \frac{\pi}{2}$

$$\int_\pi^{3\pi} \cot^2\frac{x}{6}\,dx = \int_{\pi/6}^{\pi/2} 6\cot^2 u\,du = 6\int_{\pi/6}^{\pi/2} (\csc^2 u - 1)\,du = [6(-\cot u - u)]_{\pi/6}^{\pi/2} = 6\left(-\cot\frac{\pi}{2} - \frac{\pi}{2}\right) - 6\left(-\cot\frac{\pi}{6} - \frac{\pi}{6}\right)$$
$$= 6\sqrt{3} - 2\pi$$

61. $$\int_{-\pi/3}^0 \sec x\tan x\,dx = [\sec x]_{-\pi/3}^0 = \sec 0 - \sec\left(-\frac{\pi}{3}\right) = 1 - 2 = -1$$

63. Let $u = \sin x \Rightarrow du = \cos x\,dx$; $x = 0 \Rightarrow u = 0$, $x = \frac{\pi}{2} \Rightarrow u = 1$

$$\int_0^{\pi/2} 5(\sin x)^{3/2}\cos x\,dx = \int_0^1 5u^{3/2}\,du = \left[5\left(\frac{2}{5}\right)u^{5/2}\right]_0^1 = \left[2u^{5/2}\right]_0^1 = 2(1)^{5/2} - 2(0)^{5/2} = 2$$

65. Let $u = \sin 3x \Rightarrow du = 3\cos 3x\,dx \Rightarrow \frac{1}{3}\,du = \cos 3x\,dx$; $x = -\frac{\pi}{2} \Rightarrow u = \sin\left(-\frac{3\pi}{2}\right) = 1$, $x = \frac{\pi}{2} \Rightarrow u = \sin\left(\frac{3\pi}{2}\right) = -1$

$$\int_{-\pi/2}^{\pi/2} 15\sin^4 3x\cos 3x\,dx = \int_1^{-1} 15u^4\left(\frac{1}{3}\,du\right) = \int_1^{-1} 5u^4\,du = \left[u^5\right]_1^{-1} = (-1)^5 - (1)^5 = -2$$

67. Let $u = 1 + 3\sin^2 x \Rightarrow du = 6\sin x\cos x\,dx \Rightarrow \frac{1}{2}\,du = 3\sin x\cos x\,dx$; $x = 0 \Rightarrow u = 1$, $x = \frac{\pi}{2} \Rightarrow u = 1 + 3\sin^2\frac{\pi}{2} = 4$

$$\int_0^{\pi/2} \frac{3\sin x\cos x}{\sqrt{1 + 3\sin^2 x}}\,dx = \int_1^4 \frac{1}{\sqrt{u}}\left(\frac{1}{2}\,du\right) = \int_1^4 \frac{1}{2}u^{-1/2}\,du = \left[\frac{1}{2}\left(\frac{u^{1/2}}{\frac{1}{2}}\right)\right]_1^4 = \left[u^{1/2}\right]_1^4 = 4^{1/2} - 1^{1/2} = 1$$

69. Let $u = \sec\theta \Rightarrow du = \sec\theta\tan\theta\,d\theta$; $\theta = 0 \Rightarrow u = \sec 0 = 1$, $\theta = \frac{\pi}{3} \Rightarrow u = \sec\frac{\pi}{3} = 2$

$$\int_0^{\pi/3} \frac{\tan\theta}{\sqrt{2\sec\theta}}\,d\theta = \int_0^{\pi/3} \frac{\sec\theta\tan\theta}{\sec\theta\sqrt{2\sec\theta}}\,d\theta = \int_0^{\pi/3} \frac{\sec\theta\tan\theta}{\sqrt{2}(\sec\theta)^{3/2}}\,d\theta = \int_1^2 \frac{1}{\sqrt{2}u^{3/2}}\,du = \frac{1}{\sqrt{2}}\int_1^2 u^{-3/2}\,du$$
$$= \frac{1}{\sqrt{2}}\left[\frac{u^{-1/2}}{\left(-\frac{1}{2}\right)}\right]_1^2 = \left[-\frac{2}{\sqrt{2u}}\right]_1^2 = -\frac{2}{\sqrt{2(2)}} - \left(-\frac{2}{\sqrt{2(1)}}\right) = \sqrt{2} - 1$$

71. (a) $$\text{av}(f) = \frac{1}{1-(-1)}\int_{-1}^1 (mx + b)\,dx = \frac{1}{2}\left[\frac{mx^2}{2} + bx\right]_{-1}^1 = \frac{1}{2}\left[\left(\frac{m(1)^2}{2} + b(1)\right) - \left(\frac{m(-1)^2}{2} + b(-1)\right)\right] = \frac{1}{2}(2b) = b$$

(b) $$\text{av}(f) = \frac{1}{k-(-k)}\int_{-k}^k (mx + b)\,dx = \frac{1}{2k}\left[\frac{mx^2}{2} + bx\right]_{-k}^k = \frac{1}{2k}\left[\left(\frac{m(k)^2}{2} + b(k)\right) - \left(\frac{m(-k)^2}{2} + b(-k)\right)\right] = \frac{1}{2k}(2bk) = b$$

73. $f'_{av} = \frac{1}{b-a}\int_a^b f'(x)\,dx = \frac{1}{b-a}[f(x)]_a^b = \frac{1}{b-a}[f(b) - f(a)] = \frac{f(b) - f(a)}{b - a}$ so the average value of f' over $[a, b]$ is the slope of the secant line joining the points $(a, f(a))$ and $(b, f(b))$, which is the average rate of change of f over $[a, b]$.

75. We want to evaluate

$$\frac{1}{365-0}\int_0^{365} f(x)\,dx = \frac{1}{365}\int_0^{365}\left(37\sin\left[\frac{2\pi}{365}(x-101)\right]+25\right)dx = \frac{37}{365}\int_0^{365}\sin\left[\frac{2\pi}{365}(x-101)\right]dx + \frac{25}{365}\int_0^{365}dx$$

Notice that the period of $y = \sin\left[\frac{2\pi}{365}(x-101)\right]$ is $\frac{2\pi}{\frac{2\pi}{365}} = 365$ and that we are integrating this function over an iterval of length 365. Thus the value of $\frac{37}{365}\int_0^{365}\sin\left[\frac{2\pi}{365}(x-101)\right]dx + \frac{25}{365}\int_0^{365}dx$ is $\frac{37}{365}\cdot 0 + \frac{25}{365}\cdot 365 = 25$.

77. $\frac{dy}{dx} = \sqrt{2+\cos^3 x}$

79. $\frac{dy}{dx} = \frac{d}{dx}\left(-\int_1^x \frac{6}{3+t^4}dt\right) = -\frac{6}{3+x^4}$

81. Yes. The function f, being differentiable on [a, b], is then continuous on [a, b]. The Fundamental Theorem of Calculus says that every continuous function on [a, b] is the derivative of a function on [a, b].

83. $y = \int_x^1 \sqrt{1+t^2}\,dt = -\int_1^x \sqrt{1+t^2}\,dt \Rightarrow \frac{dy}{dx} = \frac{d}{dx}\left[-\int_1^x \sqrt{1+t^2}\,dt\right] = -\frac{d}{dx}\left[\int_1^x \sqrt{1+t^2}\,dt\right] = -\sqrt{1+x^2}$

85. We estimate the area A using midpoints of the vertical intervals, and we will estimate the width of the parking lot on each interval by averaging the widths at top and bottom. This gives the estimate
$A \approx 15\cdot\left(\frac{0+36}{2} + \frac{36+54}{2} + \frac{54+51}{2} + \frac{51+49.5}{2} + \frac{49.5+54}{2} + \frac{54+64.4}{2} + \frac{64.4+67.5}{2} + \frac{67.5+42}{2}\right)$
$A \approx 5961\ \text{ft}^2$. The cost is Area $\cdot$ (\$2.10/ft²) $\approx$ (5961 ft²) (\$2.10/ft²) = \$12,518.10 $\Rightarrow$ the job cannot be done for \$11,000.

CHAPTER 5 ADDITIONAL AND ADVANCED EXERCISES

1. (a) Yes, because $\int_0^1 f(x)\,dx = \frac{1}{7}\int_0^1 7f(x)\,dx = \frac{1}{7}(7) = 1$

   (b) No. For example, $\int_0^1 8x\,dx = \left[4x^2\right]_0^1 = 4$, but $\int_0^1 \sqrt{8x}\,dx = \left[2\sqrt{2}\left(\frac{x^{3/2}}{\frac{3}{2}}\right)\right]_0^1 = \frac{4\sqrt{2}}{3}\left(1^{3/2}-0^{3/2}\right) = \frac{4\sqrt{2}}{3} \neq \sqrt{4}$

3. $y = \frac{1}{a}\int_0^x f(t)\sin a(x-t)\,dt = \frac{1}{a}\int_0^x f(t)\sin ax\cos at\,dt - \frac{1}{a}\int_0^x f(t)\cos ax\sin at\,dt$

   $= \frac{\sin ax}{a}\int_0^x f(t)\cos at\,dt - \frac{\cos ax}{a}\int_0^x f(t)\sin at\,dt \Rightarrow \frac{dy}{dx} = \cos ax\left(\int_0^x f(t)\cos at\,dt\right)$

   $+ \frac{\sin ax}{a}\left(\frac{d}{dx}\int_0^x f(t)\cos at\,dt\right) + \sin ax\int_0^x f(t)\sin at\,dt - \frac{\cos ax}{a}\left(\frac{d}{dx}\int_0^x f(t)\sin at\,dt\right)$

   $= \cos ax\int_0^x f(t)\cos at\,dt + \frac{\sin ax}{a}(f(x)\cos ax) + \sin ax\int_0^x f(t)\sin at\,dt - \frac{\cos ax}{a}(f(x)\sin ax)$

   $\Rightarrow \frac{dy}{dx} = \cos ax\int_0^x f(t)\cos at\,dt + \sin ax\int_0^x f(t)\sin at\,dt$. Next,

   $\frac{d^2y}{dx^2} = -a\sin ax\int_0^x f(t)\cos at\,dt + (\cos ax)\left(\frac{d}{dx}\int_0^x f(t)\cos at\,dt\right) + a\cos ax\int_0^x f(t)\sin at\,dt$

   $+ (\sin ax)\left(\frac{d}{dx}\int_0^x f(t)\sin at\,dt\right) = -a\sin ax\int_0^x f(t)\cos at\,dt + (\cos ax)f(x)\cos ax$

   $+ a\cos ax\int_0^x f(t)\sin at\,dt + (\sin ax)f(x)\sin ax = -a\sin ax\int_0^x f(t)\cos at\,dt + a\cos ax\int_0^x f(t)\sin at\,dt + f(x)$.

   Therefore, $y'' + a^2y = a\cos ax\int_0^x f(t)\sin at\,dt - a\sin ax\int_0^x f(t)\cos at\,dt + f(x)$

   $+ a^2\left(\frac{\sin ax}{a}\int_0^x f(t)\cos at\,dt - \frac{\cos ax}{a}\int_0^x f(t)\sin at\,dt\right) = f(x)$. Note also that $y'(0) = y(0) = 0$.

5. (a) $\int_0^{x^2} f(t)\,dt = x\cos\pi x \Rightarrow \frac{d}{dx}\int_0^{x^2} f(t)\,dt = \cos\pi x - \pi x\sin\pi x \Rightarrow f(x^2)(2x) = \cos\pi x - \pi x\sin\pi x$
$\Rightarrow f(x^2) = \frac{\cos\pi x - \pi x\sin\pi x}{2x}$. Thus, $x = 2 \Rightarrow f(4) = \frac{\cos 2\pi - 2\pi\sin 2\pi}{4} = \frac{1}{4}$

(b) $\int_0^{f(x)} t^2\,dt = \left[\frac{t^3}{3}\right]_0^{f(x)} = \frac{1}{3}(f(x))^3 \Rightarrow \frac{1}{3}(f(x))^3 = x\cos\pi x \Rightarrow (f(x))^3 = 3x\cos\pi x \Rightarrow f(x) = \sqrt[3]{3x\cos\pi x}$
$\Rightarrow f(4) = \sqrt[3]{3(4)\cos 4\pi} = \sqrt[3]{12}$

7. $\int_1^b f(x)\,dx = \sqrt{b^2+1} - \sqrt{2} \Rightarrow f(b) = \frac{d}{db}\int_1^b f(x)\,dx = \frac{1}{2}(b^2+1)^{-1/2}(2b) = \frac{b}{\sqrt{b^2+1}} \Rightarrow f(x) = \frac{x}{\sqrt{x^2+1}}$

9. $\frac{dy}{dx} = 3x^2 + 2 \Rightarrow y = \int(3x^2+2)\,dx = x^3 + 2x + C$. Then $(1,-1)$ on the curve $\Rightarrow 1^3 + 2(1) + C = -1 \Rightarrow C = -4$
$\Rightarrow y = x^3 + 2x - 4$

11. $\int_{-8}^3 f(x)\,dx = \int_{-8}^0 x^{2/3}\,dx + \int_0^3 -4\,dx$
$= \left[\frac{3}{5}x^{5/3}\right]_{-8}^0 + [-4x]_0^3$
$= \left(0 - \frac{3}{5}(-8)^{5/3}\right) + (-4(3) - 0) = \frac{96}{5} - 12$
$= \frac{36}{5}$

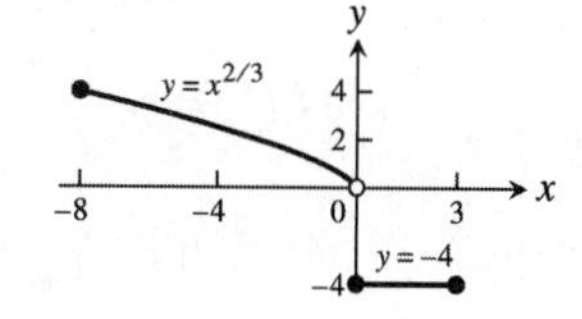

13. $\int_0^2 g(t)\,dt = \int_0^1 t\,dt + \int_1^2 \sin\pi t\,dt$
$= \left[\frac{t^2}{2}\right]_0^1 + \left[-\frac{1}{\pi}\cos\pi t\right]_1^2$
$= \left(\frac{1}{2} - 0\right) + \left[-\frac{1}{\pi}\cos 2\pi - \left(-\frac{1}{\pi}\cos\pi\right)\right]$
$= \frac{1}{2} - \frac{2}{\pi}$

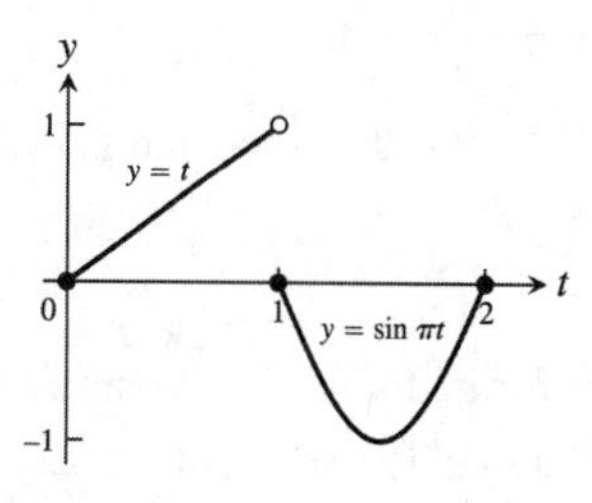

15. $\int_{-2}^2 f(x)\,dx = \int_{-2}^{-1} dx + \int_{-1}^1 (1-x^2)\,dx + \int_1^2 2\,dx$
$= [x]_{-2}^{-1} + \left[x - \frac{x^3}{3}\right]_{-1}^1 + [2x]_1^2$
$= (-1-(-2)) + \left[\left(1 - \frac{1^3}{3}\right) - \left(-1 - \frac{(-1)^3}{3}\right)\right] + [2(2) - 2(1)]$
$= 1 + \frac{2}{3} - \left(-\frac{2}{3}\right) + 4 - 2 = \frac{13}{3}$

17. Ave. value $= \frac{1}{b-a}\int_a^b f(x)\,dx = \frac{1}{2-0}\int_0^2 f(x)\,dx = \frac{1}{2}\left[\int_0^1 x\,dx + \int_1^2 (x-1)\,dx\right] = \frac{1}{2}\left[\frac{x^2}{2}\right]_0^1 + \frac{1}{2}\left[\frac{x^2}{2} - x\right]_1^2$
$= \frac{1}{2}\left[\left(\frac{1^2}{2} - 0\right) + \left(\frac{2^2}{2} - 2\right) - \left(\frac{1^2}{2} - 1\right)\right] = \frac{1}{2}$

19. Let $f(x) = x^5$ on $[0, 1]$. Partition $[0, 1]$ into n subintervals with $\Delta x = \frac{1-0}{n} = \frac{1}{n}$. Then $\frac{1}{n}, \frac{2}{n}, \ldots, \frac{n}{n}$ are the right-hand endpoints of the subintervals. Since f is increasing on $[0, 1]$, $U = \sum_{j=1}^{\infty}\left(\frac{j}{n}\right)^5\left(\frac{1}{n}\right)$ is the upper sum for $f(x) = x^5$ on $[0, 1] \Rightarrow \lim_{n\to\infty}\sum_{j=1}^{\infty}\left(\frac{j}{n}\right)^5\left(\frac{1}{n}\right) = \lim_{n\to\infty}\frac{1}{n}\left[\left(\frac{1}{n}\right)^5 + \left(\frac{2}{n}\right)^5 + \ldots + \left(\frac{n}{n}\right)^5\right] = \lim_{n\to\infty}\left[\frac{1^5 + 2^5 + \ldots + n^5}{n^6}\right]$
$= \int_0^1 x^5\,dx = \left[\frac{x^6}{6}\right]_0^1 = \frac{1}{6}$

21. Let $y = f(x)$ on $[0, 1]$. Partition $[0, 1]$ into n subintervals with $\Delta x = \frac{1-0}{n} = \frac{1}{n}$. Then $\frac{1}{n}, \frac{2}{n}, \ldots, \frac{n}{n}$ are the right-hand endpoints of the subintervals. Since f is continuous on $[0, 1]$, $\sum_{j=1}^{\infty} f\left(\frac{j}{n}\right)\left(\frac{1}{n}\right)$ is a Riemann sum of $y = f(x)$ on $[0, 1] \Rightarrow \lim_{n\to\infty} \sum_{j=1}^{\infty} f\left(\frac{j}{n}\right)\left(\frac{1}{n}\right) = \lim_{n\to\infty} \frac{1}{n}\left[f\left(\frac{1}{n}\right) + f\left(\frac{2}{n}\right) + \ldots + f\left(\frac{n}{n}\right)\right] = \int_0^1 f(x)\,dx$

23. (a) Let the polygon be inscribed in a circle of radius r. If we draw a radius from the center of the circle (and the polygon) to each vertex of the polygon, we have n isosceles triangles formed (the equal sides are equal to r, the radius of the circle) and a vertex angle of θ_n where $\theta_n = \frac{2\pi}{n}$. The area of each triangle is $A_n = \frac{1}{2}r^2 \sin\theta_n \Rightarrow$ the area of the polygon is $A = nA_n = \frac{nr^2}{2}\sin\theta_n = \frac{nr^2}{2}\sin\frac{2\pi}{n}$.

(b) $\lim_{n\to\infty} A = \lim_{n\to\infty} \frac{nr^2}{2}\sin\frac{2\pi}{n} = \lim_{n\to\infty} \frac{n\pi r^2}{2\pi}\sin\frac{2\pi}{n} = \lim_{n\to\infty} (\pi r^2)\frac{\sin\left(\frac{2\pi}{n}\right)}{\left(\frac{2\pi}{n}\right)} = (\pi r^2)\lim_{2\pi/n\to 0}\frac{\sin\left(\frac{2\pi}{n}\right)}{\left(\frac{2\pi}{n}\right)} = \pi r^2$

25. (a) $g(1) = \int_1^1 f(t)\,dt = 0$

(b) $g(3) = \int_1^3 f(t)\,dt = -\frac{1}{2}(2)(1) = -1$

(c) $g(-1) = \int_1^{-1} f(t)\,dt = -\int_{-1}^1 f(t)\,dt = -\frac{1}{4}(\pi 2^2) = -\pi$

(d) $g'(x) = f(x) = 0 \Rightarrow x = -3, 1, 3$ and the sign chart for $g'(x) = f(x)$ is $\underset{-3}{|}+++\underset{1}{|}---\underset{3}{|}+++$. So g has a relative maximum at $x = 1$.

(e) $g'(-1) = f(-1) = 2$ is the slope and $g(-1) = \int_1^{-1} f(t)\,dt = -\pi$, by (c). Thus the equation is $y + \pi = 2(x+1)$ $y = 2x + 2 - \pi$.

(f) $g''(x) = f'(x) = 0$ at $x = -1$ and $g''(x) = f'(x)$ is negative on $(-3, -1)$ and positive on $(-1, 1)$ so there is an inflection point for g at $x = -1$. We notice that $g''(x) = f'(x) < 0$ for x on $(-1, 2)$ and $g''(x) = f'(x) > 0$ for x on $(2, 4)$, even though $g''(2)$ does not exist, g has a tangent line at $x = 2$, so there is an inflection point at $x = 2$.

(g) g is continuous on $[-3, 4]$ and so it attains its absolute maximum and minimum values on this interval. We saw in (d) that $g'(x) = 0 \Rightarrow x = -3, 1, 3$. We have that

$g(-3) = \int_1^{-3} f(t)\,dt = -\int_{-3}^1 f(t)\,dt = -\frac{\pi 2^2}{2} = -2\pi$

$g(1) = \int_1^1 f(t)\,dt = 0$

$g(3) = \int_1^3 f(t)\,dt = -1$

$g(4) = \int_1^4 f(t)\,dt = -1 + \frac{1}{2}\cdot 1\cdot 1 = -\frac{1}{2}$

Thus, the absolute minimum is -2π and the absolute maximum is 0. Thus, the range is $[-2\pi, 0]$.

27. $f(x) = \int_{1/x}^{x} \frac{1}{t}\,dt \Rightarrow f'(x) = \frac{1}{x}\left(\frac{dx}{dx}\right) - \left(\frac{1}{\frac{1}{x}}\right)\left(\frac{d}{dx}\left(\frac{1}{x}\right)\right) = \frac{1}{x} - x\left(-\frac{1}{x^2}\right) = \frac{1}{x} + \frac{1}{x} = \frac{2}{x}$

29. $g(y) = \int_{\sqrt{y}}^{2\sqrt{y}} \sin t^2\,dt \Rightarrow g'(y) = \left(\sin\left(2\sqrt{y}\right)^2\right)\left(\frac{d}{dy}\left(2\sqrt{y}\right)\right) - \left(\sin\left(\sqrt{y}\right)^2\right)\left(\frac{d}{dy}\left(\sqrt{y}\right)\right) = \frac{\sin 4y}{\sqrt{y}} - \frac{\sin y}{2\sqrt{y}}$

CHAPTER 6 APPLICATIONS OF DEFINITE INTEGRALS

6.1 VOLUMES USING CROSS-SECTIONS

1. $A(x) = \frac{(\text{diagonal})^2}{2} = \frac{(\sqrt{x}-(-\sqrt{x}))^2}{2} = 2x$; $a = 0, b = 4$;
$V = \int_a^b A(x)\,dx = \int_0^4 2x\,dx = [x^2]_0^4 = 16$

3. $A(x) = (\text{edge})^2 = \left[\sqrt{1-x^2} - \left(-\sqrt{1-x^2}\right)\right]^2 = \left(2\sqrt{1-x^2}\right)^2 = 4\,(1-x^2)$; $a = -1, b = 1$;
$V = \int_a^b A(x)\,dx = \int_{-1}^1 4(1-x^2)\,dx = 4\left[x - \frac{x^3}{3}\right]_{-1}^1 = 8\left(1 - \frac{1}{3}\right) = \frac{16}{3}$

5. (a) STEP 1) $A(x) = \frac{1}{2}(\text{side})\cdot(\text{side})\cdot\left(\sin\frac{\pi}{3}\right) = \frac{1}{2}\cdot\left(2\sqrt{\sin x}\right)\cdot\left(2\sqrt{\sin x}\right)\left(\sin\frac{\pi}{3}\right) = \sqrt{3}\sin x$
STEP 2) $a = 0, b = \pi$
STEP 3) $V = \int_a^b A(x)\,dx = \sqrt{3}\int_0^\pi \sin x\,dx = \left[-\sqrt{3}\cos x\right]_0^\pi = \sqrt{3}(1+1) = 2\sqrt{3}$
(b) STEP 1) $A(x) = (\text{side})^2 = \left(2\sqrt{\sin x}\right)\left(2\sqrt{\sin x}\right) = 4\sin x$
STEP 2) $a = 0, b = \pi$
STEP 3) $V = \int_a^b A(x)\,dx = \int_0^\pi 4\sin x\,dx = [-4\cos x]_0^\pi = 8$

7. (a) STEP 1) $A(x) = (\text{length})\cdot(\text{height}) = (6-3x)\cdot(10) = 60 - 30x$
STEP 2) $a = 0, b = 2$
STEP 3) $V = \int_a^b A(x)\,dx = \int_0^2 (60-30x)\,dx = [60x - 15x^2]_0^2 = (120-60) - 0 = 60$
(b) STEP 1) $A(x) = (\text{length})\cdot(\text{height}) = (6-3x)\cdot\left(\frac{20-2(6-3x)}{2}\right) = (6-3x)(4+3x) = 24 + 6x - 9x^2$
STEP 2) $a = 0, b = 2$
STEP 3) $V = \int_a^b A(x)\,dx = \int_0^2 (24+6x-9x^2)\,dx = [24x + 3x^2 - 3x^3]_0^2 = (48+12-24) - 0 = 36$

9. $A(y) = \frac{\pi}{4}(\text{diameter})^2 = \frac{\pi}{4}\left(\sqrt{5}y^2 - 0\right)^2 = \frac{5\pi}{4}y^4$;
$c = 0, d = 2$; $V = \int_c^d A(y)\,dy = \int_0^2 \frac{5\pi}{4}y^4\,dy$
$= \left[\left(\frac{5\pi}{4}\right)\left(\frac{y^5}{5}\right)\right]_0^2 = \frac{\pi}{4}(2^5 - 0) = 8\pi$

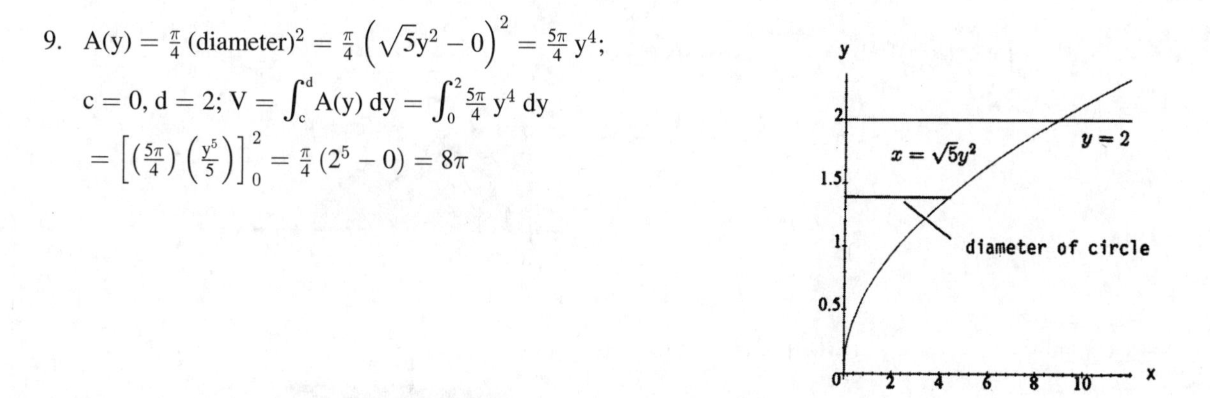

11. The slices perpendicular to the edge labeled 5 are triangles, and by similar triangles we have $\frac{b}{h} = \frac{4}{3} \Rightarrow h = \frac{3}{4}b$. The equation of the line through (5, 0) and (0, 4) is $y = -\frac{4}{5}x + 4$, thus the length of the base $= -\frac{4}{5}x + 4$ and the height $= \frac{3}{4}\left(-\frac{4}{5}x + 4\right) = -\frac{3}{5}x + 3$.Thus $A(x) = \frac{1}{2}(\text{base})\cdot(\text{height}) = \frac{1}{2}\left(-\frac{4}{5}x + 4\right)\cdot\left(-\frac{3}{5}x + 3\right) = \frac{6}{25}x^2 - \frac{12}{5}x + 6$ and $V = \int_a^b A(x)\,dx = \int_0^5 \left(\frac{6}{25}x^2 - \frac{12}{5}x + 6\right)dx = \left[\frac{2}{25}x^3 - \frac{6}{5}x^2 + 6x\right]_0^5 = (10 - 30 + 30) - 0 = 10$

13. (a) It follows from Cavalieri's Principle that the volume of a column is the same as the volume of a right prism with a square base of side length s and altitude h. Thus, STEP 1) $A(x) = (\text{side length})^2 = s^2$; STEP 2) $a = 0, b = h$; STEP 3) $V = \int_a^b A(x)\,dx = \int_0^h s^2\,dx = s^2h$

(b) From Cavalieri's Principle we conclude that the volume of the column is the same as the volume of the prism described above, regardless of the number of turns $\Rightarrow V = s^2h$

15. $R(x) = y = 1 - \frac{x}{2} \Rightarrow V = \int_0^2 \pi[R(x)]^2\,dx = \pi\int_0^2 \left(1 - \frac{x}{2}\right)^2 dx = \pi\int_0^2 \left(1 - x + \frac{x^2}{4}\right) dx = \pi\left[x - \frac{x^2}{2} + \frac{x^3}{12}\right]_0^2$
$= \pi\left(2 - \frac{4}{2} + \frac{8}{12}\right) = \frac{2\pi}{3}$

17. $R(y) = \tan\left(\frac{\pi}{4}y\right)$; $u = \frac{\pi}{4}y \Rightarrow du = \frac{\pi}{4}dy \Rightarrow 4\,du = \pi\,dy$; $y = 0 \Rightarrow u = 0$, $y = 1 \Rightarrow u = \frac{\pi}{4}$;
$V = \int_0^1 \pi[R(y)]^2\,dy = \pi\int_0^1 \left[\tan\left(\frac{\pi}{4}y\right)\right]^2 dy = 4\int_0^{\pi/4} \tan^2 u\,du = 4\int_0^{\pi/4}(-1 + \sec^2 u)\,du = 4[-u + \tan u]_0^{\pi/4}$
$= 4\left(-\frac{\pi}{4} + 1 - 0\right) = 4 - \pi$

19. $R(x) = x^2 \Rightarrow V = \int_0^2 \pi[R(x)]^2\,dx = \pi\int_0^2 (x^2)^2\,dx$
$= \pi\int_0^2 x^4\,dx = \pi\left[\frac{x^5}{5}\right]_0^2 = \frac{32\pi}{5}$

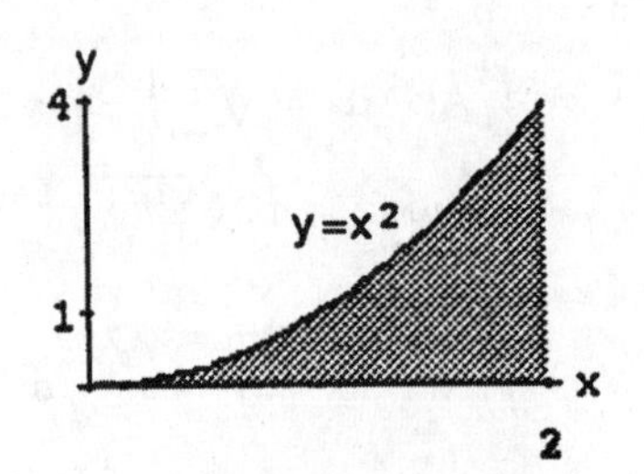

21. $R(x) = \sqrt{9 - x^2} \Rightarrow V = \int_{-3}^3 \pi[R(x)]^2\,dx = \pi\int_{-3}^3 (9 - x^2)\,dx$
$= \pi\left[9x - \frac{x^3}{3}\right]_{-3}^3 = 2\pi\left[9(3) - \frac{27}{3}\right] = 2\cdot\pi\cdot 18 = 36\pi$

23. $R(x) = \sqrt{\cos x} \Rightarrow V = \int_0^{\pi/2} \pi[R(x)]^2\,dx = \pi\int_0^{\pi/2} \cos x\,dx$
$= \pi[\sin x]_0^{\pi/2} = \pi(1 - 0) = \pi$

25. $R(x) = \sqrt{2} - \sec x \tan x \Rightarrow V = \int_0^{\pi/4} \pi[R(x)]^2\,dx$
$= \pi\int_0^{\pi/4}\left(\sqrt{2} - \sec x \tan x\right)^2 dx$
$= \pi\int_0^{\pi/4}\left(2 - 2\sqrt{2}\sec x \tan x + \sec^2 x \tan^2 x\right) dx$
$= \pi\left(\int_0^{\pi/4} 2\,dx - 2\sqrt{2}\int_0^{\pi/4} \sec x \tan x\,dx + \int_0^{\pi/4} (\tan x)^2 \sec^2 x\,dx\right)$
$= \pi\left([2x]_0^{\pi/4} - 2\sqrt{2}\,[\sec x]_0^{\pi/4} + \left[\frac{\tan^3 x}{3}\right]_0^{\pi/4}\right)$

$= \pi\left[\left(\frac{\pi}{2} - 0\right) - 2\sqrt{2}\left(\sqrt{2} - 1\right) + \frac{1}{3}\left(1^3 - 0\right)\right] = \pi\left(\frac{\pi}{2} + 2\sqrt{2} - \frac{11}{3}\right)$

27. $R(y) = \sqrt{5}\,y^2 \Rightarrow V = \int_{-1}^{1} \pi[R(y)]^2\,dy = \pi \int_{-1}^{1} 5y^4\,dy$
$= \pi\left[y^5\right]_{-1}^{1} = \pi[1 - (-1)] = 2\pi$

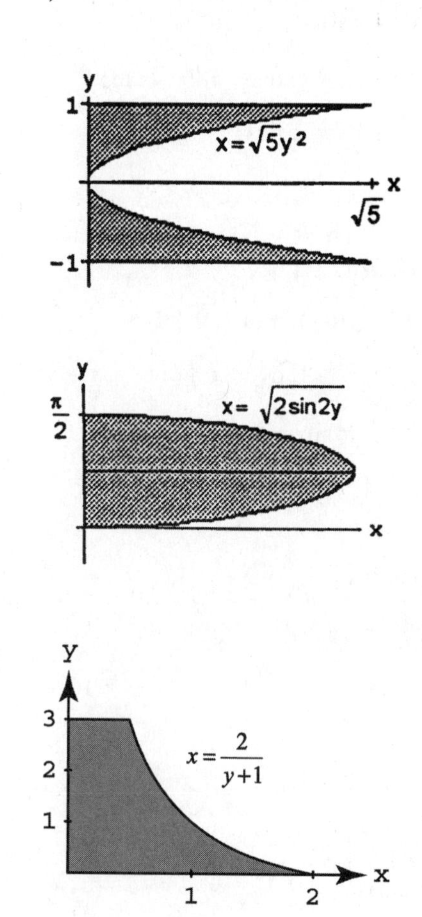

29. $R(y) = \sqrt{2\sin 2y} \Rightarrow V = \int_{0}^{\pi/2} \pi[R(y)]^2\,dy$
$= \pi\int_{0}^{\pi/2} 2\sin 2y\,dy = \pi\left[-\cos 2y\right]_{0}^{\pi/2}$
$= \pi[1 - (-1)] = 2\pi$

31. $R(y) = \frac{2}{y+1} \Rightarrow V = \int_{0}^{3} \pi[R(y)]^2\,dy = 4\pi \int_{0}^{3} \frac{1}{(y+1)^2}\,dy$
$= 4\pi\left[-\frac{1}{y+1}\right]_{0}^{3} = 4\pi\left[-\frac{1}{4} - (-1)\right] = 3\pi$

33. For the sketch given, $a = -\frac{\pi}{2}$, $b = \frac{\pi}{2}$; $R(x) = 1$, $r(x) = \sqrt{\cos x}$; $V = \int_{a}^{b} \pi\left([R(x)]^2 - [r(x)]^2\right) dx$
$= \int_{-\pi/2}^{\pi/2} \pi(1 - \cos x)\,dx = 2\pi\int_{0}^{\pi/2}(1 - \cos x)\,dx = 2\pi[x - \sin x]_{0}^{\pi/2} = 2\pi\left(\frac{\pi}{2} - 1\right) = \pi^2 - 2\pi$

35. $r(x) = x$ and $R(x) = 1 \Rightarrow V = \int_{0}^{1} \pi\left([R(x)]^2 - [r(x)]^2\right) dx$
$= \int_{0}^{1} \pi\left(1 - x^2\right) dx = \pi\left[x - \frac{x^3}{3}\right]_{0}^{1} = \pi\left[\left(1 - \frac{1}{3}\right) - 0\right] = \frac{2\pi}{3}$

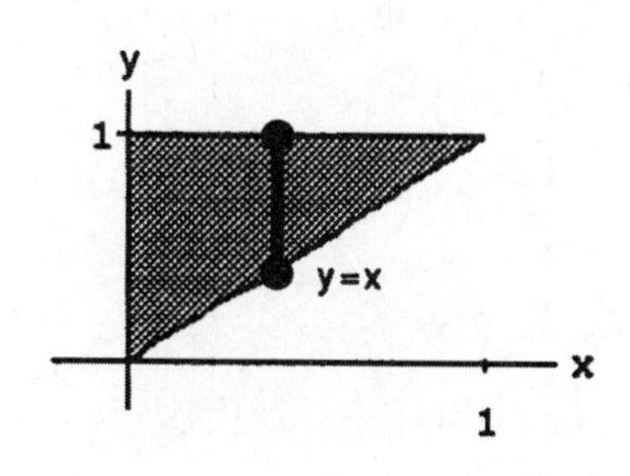

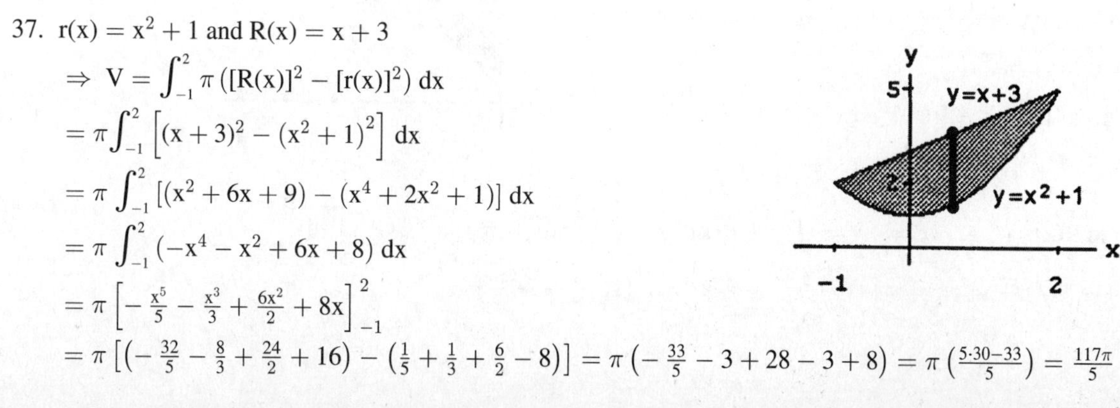

37. $r(x) = x^2 + 1$ and $R(x) = x + 3$
$\Rightarrow V = \int_{-1}^{2} \pi\left([R(x)]^2 - [r(x)]^2\right) dx$
$= \pi\int_{-1}^{2}\left[(x+3)^2 - \left(x^2+1\right)^2\right] dx$
$= \pi\int_{-1}^{2}\left[\left(x^2 + 6x + 9\right) - \left(x^4 + 2x^2 + 1\right)\right] dx$
$= \pi\int_{-1}^{2}\left(-x^4 - x^2 + 6x + 8\right) dx$
$= \pi\left[-\frac{x^5}{5} - \frac{x^3}{3} + \frac{6x^2}{2} + 8x\right]_{-1}^{2}$
$= \pi\left[\left(-\frac{32}{5} - \frac{8}{3} + \frac{24}{2} + 16\right) - \left(\frac{1}{5} + \frac{1}{3} + \frac{6}{2} - 8\right)\right] = \pi\left(-\frac{33}{5} - 3 + 28 - 3 + 8\right) = \pi\left(\frac{5\cdot 30 - 33}{5}\right) = \frac{117\pi}{5}$

39. $r(x) = \sec x$ and $R(x) = \sqrt{2}$

$$\Rightarrow V = \int_{-\pi/4}^{\pi/4} \pi\left([R(x)]^2 - [r(x)]^2\right) dx$$
$$= \pi \int_{-\pi/4}^{\pi/4} (2 - \sec^2 x)\, dx = \pi[2x - \tan x]_{-\pi/4}^{\pi/4}$$
$$= \pi\left[\left(\tfrac{\pi}{2} - 1\right) - \left(-\tfrac{\pi}{2} + 1\right)\right] = \pi(\pi - 2)$$

41. $r(y) = 1$ and $R(y) = 1 + y$

$$\Rightarrow V = \int_0^1 \pi\left([R(y)]^2 - [r(y)]^2\right) dy$$
$$= \pi \int_0^1 \left[(1 + y)^2 - 1\right] dy = \pi \int_0^1 (1 + 2y + y^2 - 1)\, dy$$
$$= \pi \int_0^1 (2y + y^2)\, dy = \pi\left[y^2 + \tfrac{y^3}{3}\right]_0^1 = \pi\left(1 + \tfrac{1}{3}\right) = \tfrac{4\pi}{3}$$

43. $R(y) = 2$ and $r(y) = \sqrt{y}$

$$\Rightarrow V = \int_0^4 \pi\left([R(y)]^2 - [r(y)]^2\right) dy$$
$$= \pi \int_0^4 (4 - y)\, dy = \pi\left[4y - \tfrac{y^2}{2}\right]_0^4 = \pi(16 - 8) = 8\pi$$

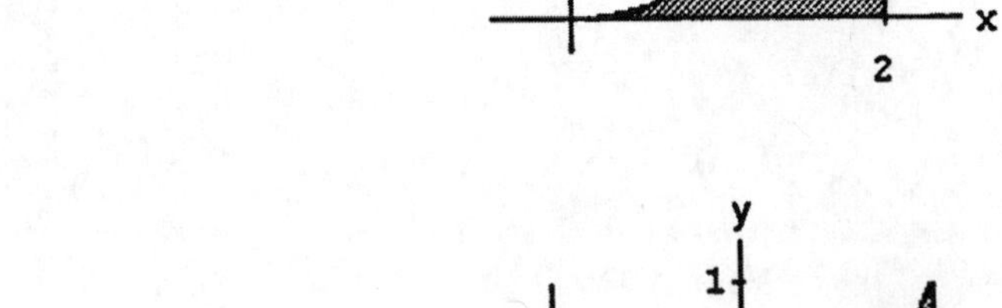

45. $R(y) = 2$ and $r(y) = 1 + \sqrt{y}$

$$\Rightarrow V = \int_0^1 \pi\left([R(y)]^2 - [r(y)]^2\right) dy$$
$$= \pi \int_0^1 \left[4 - \left(1 + \sqrt{y}\right)^2\right] dy$$
$$= \pi \int_0^1 \left(4 - 1 - 2\sqrt{y} - y\right) dy$$
$$= \pi \int_0^1 \left(3 - 2\sqrt{y} - y\right) dy$$
$$= \pi\left[3y - \tfrac{4}{3}y^{3/2} - \tfrac{y^2}{2}\right]_0^1$$
$$= \pi\left(3 - \tfrac{4}{3} - \tfrac{1}{2}\right) = \pi\left(\tfrac{18-8-3}{6}\right) = \tfrac{7\pi}{6}$$

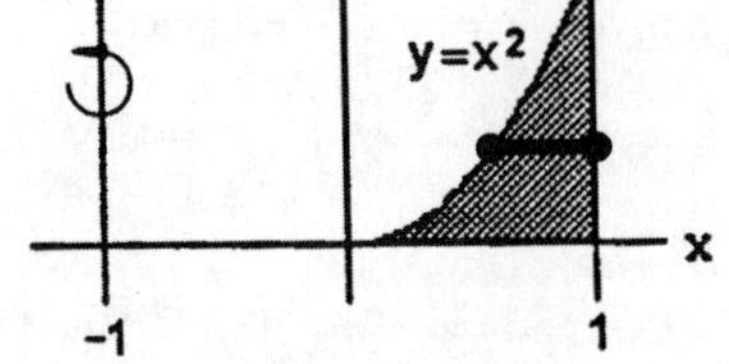

47. (a) $r(x) = \sqrt{x}$ and $R(x) = 2$

$$\Rightarrow V = \int_0^4 \pi\left([R(x)]^2 - [r(x)]^2\right) dx$$
$$= \pi \int_0^4 (4 - x)\, dx = \pi\left[4x - \tfrac{x^2}{2}\right]_0^4 = \pi(16 - 8) = 8\pi$$

(b) $r(y) = 0$ and $R(y) = y^2$

$$\Rightarrow V = \int_0^2 \pi\left([R(y)]^2 - [r(y)]^2\right) dy$$
$$= \pi \int_0^2 y^4\, dy = \pi\left[\tfrac{y^5}{5}\right]_0^2 = \tfrac{32\pi}{5}$$

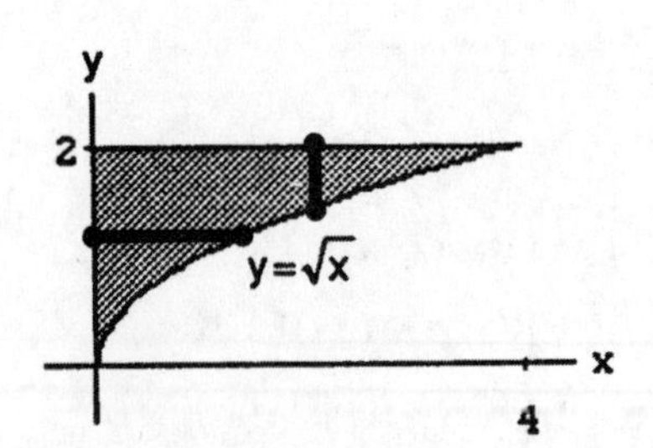

(c) $r(x) = 0$ and $R(x) = 2 - \sqrt{x} \Rightarrow V = \int_0^4 \pi\left([R(x)]^2 - [r(x)]^2\right) dx = \pi \int_0^4 \left(2 - \sqrt{x}\right)^2 dx$

$$= \pi \int_0^4 \left(4 - 4\sqrt{x} + x\right) dx = \pi\left[4x - \tfrac{8x^{3/2}}{3} + \tfrac{x^2}{2}\right]_0^4 = \pi\left(16 - \tfrac{64}{3} + \tfrac{16}{2}\right) = \tfrac{8\pi}{3}$$

(d) $r(y) = 4 - y^2$ and $R(y) = 4 \Rightarrow V = \int_0^2 \pi\left([R(y)]^2 - [r(y)]^2\right) dy = \pi \int_0^2 \left[16 - \left(4 - y^2\right)^2\right] dy$
$= \pi \int_0^2 \left(16 - 16 + 8y^2 - y^4\right) dy = \pi \int_0^2 \left(8y^2 - y^4\right) dy = \pi \left[\frac{8}{3}y^3 - \frac{y^5}{5}\right]_0^2 = \pi\left(\frac{64}{3} - \frac{32}{5}\right) = \frac{224\pi}{15}$

49. (a) $r(x) = 0$ and $R(x) = 1 - x^2$
$\Rightarrow V = \int_{-1}^{1} \pi\left([R(x)]^2 - [r(x)]^2\right) dx$
$= \pi \int_{-1}^{1} \left(1 - x^2\right)^2 dx = \pi \int_{-1}^{1} \left(1 - 2x^2 + x^4\right) dx$
$= \pi \left[x - \frac{2x^3}{3} + \frac{x^5}{5}\right]_{-1}^{1} = 2\pi\left(1 - \frac{2}{3} + \frac{1}{5}\right)$
$= 2\pi\left(\frac{15-10+3}{15}\right) = \frac{16\pi}{15}$

(b) $r(x) = 1$ and $R(x) = 2 - x^2 \Rightarrow V = \int_{-1}^{1} \pi\left([R(x)]^2 - [r(x)]^2\right) dx = \pi \int_{-1}^{1} \left[\left(2 - x^2\right)^2 - 1\right] dx$
$= \pi \int_{-1}^{1} \left(4 - 4x^2 + x^4 - 1\right) dx = \pi \int_{-1}^{1} \left(3 - 4x^2 + x^4\right) dx = \pi \left[3x - \frac{4}{3}x^3 + \frac{x^5}{5}\right]_{-1}^{1} = 2\pi\left(3 - \frac{4}{3} + \frac{1}{5}\right)$
$= \frac{2\pi}{15}(45 - 20 + 3) = \frac{56\pi}{15}$

(c) $r(x) = 1 + x^2$ and $R(x) = 2 \Rightarrow V = \int_{-1}^{1} \pi\left([R(x)]^2 - [r(x)]^2\right) dx = \pi \int_{-1}^{1} \left[4 - \left(1 + x^2\right)^2\right] dx$
$= \pi \int_{-1}^{1} \left(4 - 1 - 2x^2 - x^4\right) dx = \pi \int_{-1}^{1} \left(3 - 2x^2 - x^4\right) dx = \pi \left[3x - \frac{2}{3}x^3 - \frac{x^5}{5}\right]_{-1}^{1} = 2\pi\left(3 - \frac{2}{3} - \frac{1}{5}\right)$
$= \frac{2\pi}{15}(45 - 10 - 3) = \frac{64\pi}{15}$

51. $R(y) = b + \sqrt{a^2 - y^2}$ and $r(y) = b - \sqrt{a^2 - y^2}$
$\Rightarrow V = \int_{-a}^{a} \pi\left([R(y)]^2 - [r(y)]^2\right) dy$
$= \pi \int_{-a}^{a} \left[\left(b + \sqrt{a^2 - y^2}\right)^2 - \left(b - \sqrt{a^2 - y^2}\right)^2\right] dy$
$= \pi \int_{-a}^{a} 4b\sqrt{a^2 - y^2}\, dy = 4b\pi \int_{-a}^{a} \sqrt{a^2 - y^2}\, dy$
$= 4b\pi \cdot$ area of semicircle of radius $a = 4b\pi \cdot \frac{\pi a^2}{2} = 2a^2b\pi^2$

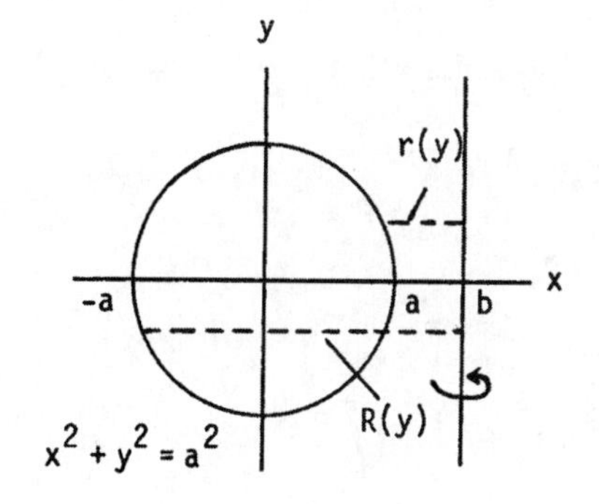

53. (a) $R(y) = \sqrt{a^2 - y^2} \Rightarrow V = \pi \int_{-a}^{h-a} \left(a^2 - y^2\right) dy = \pi \left[a^2y - \frac{y^3}{3}\right]_{-a}^{h-a} = \pi \left[a^2h - a^3 - \frac{(h-a)^3}{3} - \left(-a^3 + \frac{a^3}{3}\right)\right]$
$= \pi \left[a^2h - \frac{1}{3}\left(h^3 - 3h^2a + 3ha^2 - a^3\right) - \frac{a^3}{3}\right] = \pi\left(a^2h - \frac{h^3}{3} + h^2a - ha^2\right) = \frac{\pi h^2(3a - h)}{3}$

(b) Given $\frac{dV}{dt} = 0.2\ \text{m}^3/\text{sec}$ and $a = 5$ m, find $\left.\frac{dh}{dt}\right|_{h=4}$. From part (a), $V(h) = \frac{\pi h^2(15 - h)}{3} = 5\pi h^2 - \frac{\pi h^3}{3}$
$\Rightarrow \frac{dV}{dh} = 10\pi h - \pi h^2 \Rightarrow \frac{dV}{dt} = \frac{dV}{dh} \cdot \frac{dh}{dt} = \pi h(10 - h)\frac{dh}{dt} \Rightarrow \left.\frac{dh}{dt}\right|_{h=4} = \frac{0.2}{4\pi(10-4)} = \frac{1}{(20\pi)(6)} = \frac{1}{120\pi}$ m/sec.

55. The cross section of a solid right circular cylinder with a cone removed is a disk with radius R from which a disk of radius h has been removed. Thus its area is $A_1 = \pi R^2 - \pi h^2 = \pi\left(R^2 - h^2\right)$. The cross section of the hemisphere is a disk of radius $\sqrt{R^2 - h^2}$. Therefore its area is $A_2 = \pi\left(\sqrt{R^2 - h^2}\right)^2 = \pi\left(R^2 - h^2\right)$. We can see that $A_1 = A_2$. The altitudes of both solids are R. Applying Cavalieri's Principle we find
Volume of Hemisphere = (Volume of Cylinder) − (Volume of Cone) $= \left(\pi R^2\right)R - \frac{1}{3}\pi\left(R^2\right)R = \frac{2}{3}\pi R^3$.

57. $R(y) = \sqrt{256 - y^2} \Rightarrow V = \int_{-16}^{-7} \pi[R(y)]^2 dy = \pi \int_{-16}^{-7} \left(256 - y^2\right) dy = \pi \left[256y - \frac{y^3}{3}\right]_{-16}^{-7}$
$= \pi \left[(256)(-7) + \frac{7^3}{3} - \left((256)(-16) + \frac{16^3}{3}\right)\right] = \pi\left(\frac{7^3}{3} + 256(16 - 7) - \frac{16^3}{3}\right) = 1053\pi\ \text{cm}^3 \approx 3308\ \text{cm}^3$

59. Volume of the solid generated by rotating the region bounded by the x-axis and $y = f(x)$ from $x = a$ to $x = b$ about the x-axis is $V = \int_a^b \pi[f(x)]^2\,dx = 4\pi$, and the volume of the solid generated by rotating the same region about the line $y = -1$ is $V = \int_a^b \pi[f(x)+1]^2\,dx = 8\pi$. Thus $\int_a^b \pi[f(x)+1]^2\,dx - \int_a^b \pi[f(x)]^2\,dx = 8\pi - 4\pi$
$\Rightarrow \pi\int_a^b \left([f(x)]^2 + 2f(x) + 1 - [f(x)]^2\right)dx = 4\pi \Rightarrow \int_a^b (2f(x)+1)\,dx = 4 \Rightarrow 2\int_a^b f(x)\,dx + \int_a^b dx = 4$
$\Rightarrow \int_a^b f(x)\,dx + \frac{1}{2}(b-a) = 2 \Rightarrow \int_a^b f(x)\,dx = \frac{4-b+a}{2}$

6.2 VOLUME USING CYLINDRICAL SHELLS

1. For the sketch given, $a = 0$, $b = 2$;
$V = \int_a^b 2\pi\left(\begin{smallmatrix}\text{shell}\\ \text{radius}\end{smallmatrix}\right)\left(\begin{smallmatrix}\text{shell}\\ \text{height}\end{smallmatrix}\right)dx = \int_0^2 2\pi x\left(1 + \frac{x^2}{4}\right)dx = 2\pi\int_0^2\left(x + \frac{x^3}{4}\right)dx = 2\pi\left[\frac{x^2}{2} + \frac{x^4}{16}\right]_0^2 = 2\pi\left(\frac{4}{2} + \frac{16}{16}\right)$
$= 2\pi \cdot 3 = 6\pi$

3. For the sketch given, $c = 0$, $d = \sqrt{2}$;
$V = \int_c^d 2\pi\left(\begin{smallmatrix}\text{shell}\\ \text{radius}\end{smallmatrix}\right)\left(\begin{smallmatrix}\text{shell}\\ \text{height}\end{smallmatrix}\right)dy = \int_0^{\sqrt{2}} 2\pi y \cdot (y^2)\,dy = 2\pi\int_0^{\sqrt{2}} y^3\,dy = 2\pi\left[\frac{y^4}{4}\right]_0^{\sqrt{2}} = 2\pi$

5. For the sketch given, $a = 0$, $b = \sqrt{3}$;
$V = \int_a^b 2\pi\left(\begin{smallmatrix}\text{shell}\\ \text{radius}\end{smallmatrix}\right)\left(\begin{smallmatrix}\text{shell}\\ \text{height}\end{smallmatrix}\right)dx = \int_0^{\sqrt{3}} 2\pi x \cdot \left(\sqrt{x^2+1}\right)dx;$
$\left[u = x^2 + 1 \Rightarrow du = 2x\,dx;\ x = 0 \Rightarrow u = 1,\ x = \sqrt{3} \Rightarrow u = 4\right]$
$\rightarrow V = \pi\int_1^4 u^{1/2}\,du = \pi\left[\frac{2}{3}u^{3/2}\right]_1^4 = \frac{2\pi}{3}\left(4^{3/2} - 1\right) = \left(\frac{2\pi}{3}\right)(8-1) = \frac{14\pi}{3}$

7. $a = 0$, $b = 2$;
$V = \int_a^b 2\pi\left(\begin{smallmatrix}\text{shell}\\ \text{radius}\end{smallmatrix}\right)\left(\begin{smallmatrix}\text{shell}\\ \text{height}\end{smallmatrix}\right)dx = \int_0^2 2\pi x\left[x - \left(-\frac{x}{2}\right)\right]dx$
$= \int_0^2 2\pi x^2 \cdot \frac{3}{2}\,dx = \pi\int_0^2 3x^2\,dx = \pi\left[x^3\right]_0^2 = 8\pi$

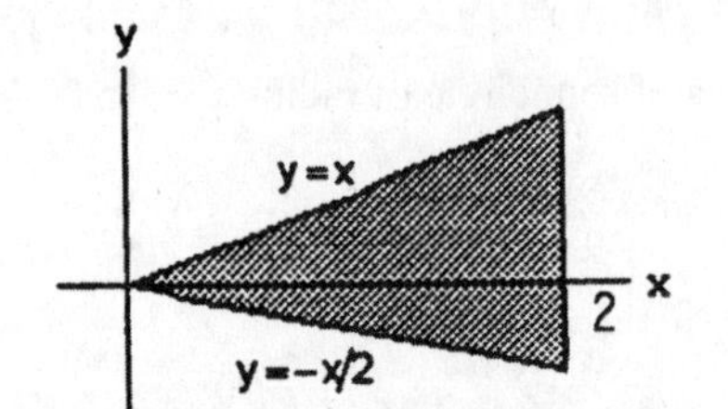

9. $a = 0$, $b = 1$;
$V = \int_a^b 2\pi\left(\begin{smallmatrix}\text{shell}\\ \text{radius}\end{smallmatrix}\right)\left(\begin{smallmatrix}\text{shell}\\ \text{height}\end{smallmatrix}\right)dx = \int_0^1 2\pi x\left[(2-x) - x^2\right]dx$
$= 2\pi\int_0^1 \left(2x - x^2 - x^3\right)dx = 2\pi\left[x^2 - \frac{x^3}{3} - \frac{x^4}{4}\right]_0^1$
$= 2\pi\left(1 - \frac{1}{3} - \frac{1}{4}\right) = 2\pi\left(\frac{12-4-3}{12}\right) = \frac{10\pi}{12} = \frac{5\pi}{6}$

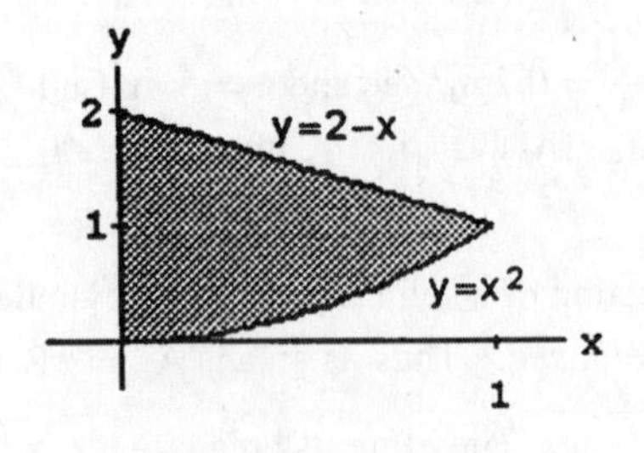

11. $a = 0, b = 1;$

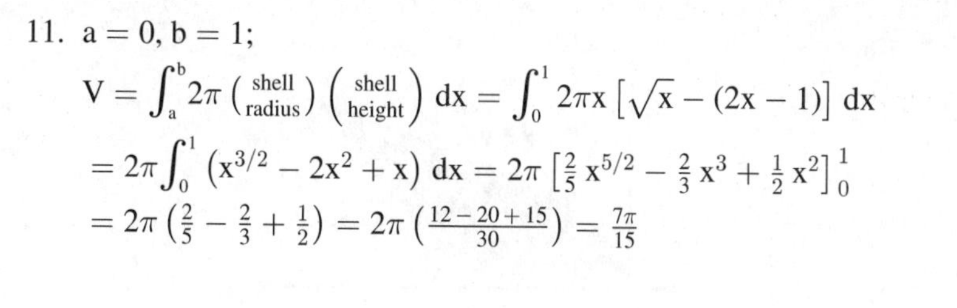

$$V = \int_a^b 2\pi \left(\begin{smallmatrix}\text{shell}\\ \text{radius}\end{smallmatrix}\right)\left(\begin{smallmatrix}\text{shell}\\ \text{height}\end{smallmatrix}\right) dx = \int_0^1 2\pi x\left[\sqrt{x} - (2x-1)\right] dx$$
$$= 2\pi \int_0^1 \left(x^{3/2} - 2x^2 + x\right) dx = 2\pi\left[\tfrac{2}{5}x^{5/2} - \tfrac{2}{3}x^3 + \tfrac{1}{2}x^2\right]_0^1$$
$$= 2\pi\left(\tfrac{2}{5} - \tfrac{2}{3} + \tfrac{1}{2}\right) = 2\pi\left(\tfrac{12-20+15}{30}\right) = \tfrac{7\pi}{15}$$

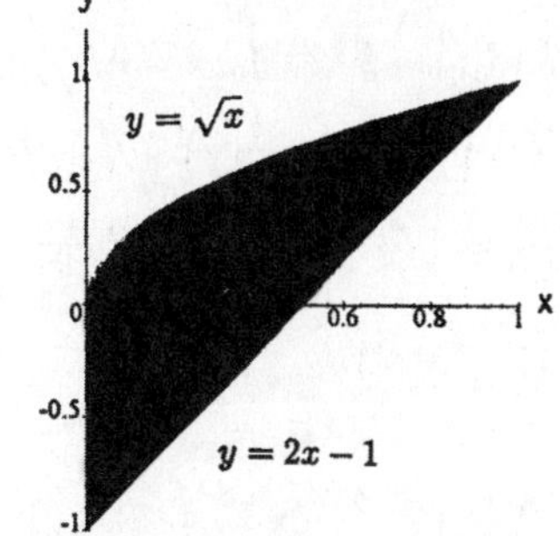

13. (a) $xf(x) = \begin{cases} x \cdot \frac{\sin x}{x}, & 0 < x \le \pi \\ x, & x = 0 \end{cases} \Rightarrow xf(x) = \begin{cases} \sin x, & 0 < x \le \pi \\ 0, & x = 0 \end{cases}$; since $\sin 0 = 0$ we have

$xf(x) = \begin{cases} \sin x, & 0 < x \le \pi \\ \sin x, & x = 0 \end{cases} \Rightarrow xf(x) = \sin x,\ 0 \le x \le \pi$

(b) $V = \int_a^b 2\pi \left(\begin{smallmatrix}\text{shell}\\ \text{radius}\end{smallmatrix}\right)\left(\begin{smallmatrix}\text{shell}\\ \text{height}\end{smallmatrix}\right) dx = \int_0^\pi 2\pi x \cdot f(x)\,dx$ and $x \cdot f(x) = \sin x,\ 0 \le x \le \pi$ by part (a)

$\Rightarrow V = 2\pi \int_0^\pi \sin x\,dx = 2\pi[-\cos x]_0^\pi = 2\pi(-\cos \pi + \cos 0) = 4\pi$

15. $c = 0, d = 2;$

$$V = \int_c^d 2\pi \left(\begin{smallmatrix}\text{shell}\\ \text{radius}\end{smallmatrix}\right)\left(\begin{smallmatrix}\text{shell}\\ \text{height}\end{smallmatrix}\right) dy = \int_0^2 2\pi y\left[\sqrt{y} - (-y)\right] dy$$
$$= 2\pi \int_0^2 \left(y^{3/2} + y^2\right) dy = 2\pi\left[\tfrac{2y^{5/2}}{5} + \tfrac{y^3}{3}\right]_0^2$$
$$= 2\pi\left[\tfrac{2}{5}\left(\sqrt{2}\right)^5 + \tfrac{2^3}{3}\right] = 2\pi\left(\tfrac{8\sqrt{2}}{5} + \tfrac{8}{3}\right) = 16\pi\left(\tfrac{\sqrt{2}}{5} + \tfrac{1}{3}\right)$$
$$= \tfrac{16\pi}{15}\left(3\sqrt{2} + 5\right)$$

17. $c = 0, d = 2;$

$$V = \int_c^d 2\pi \left(\begin{smallmatrix}\text{shell}\\ \text{radius}\end{smallmatrix}\right)\left(\begin{smallmatrix}\text{shell}\\ \text{height}\end{smallmatrix}\right) dy = \int_0^2 2\pi y\,(2y - y^2)\,dy$$
$$= 2\pi \int_0^2 \left(2y^2 - y^3\right) dy = 2\pi\left[\tfrac{2y^3}{3} - \tfrac{y^4}{4}\right]_0^2 = 2\pi\left(\tfrac{16}{3} - \tfrac{16}{4}\right)$$
$$= 32\pi\left(\tfrac{1}{3} - \tfrac{1}{4}\right) = \tfrac{32\pi}{12} = \tfrac{8\pi}{3}$$

19. $c = 0, d = 1;$

$$V = \int_c^d 2\pi \left(\begin{smallmatrix}\text{shell}\\ \text{radius}\end{smallmatrix}\right)\left(\begin{smallmatrix}\text{shell}\\ \text{height}\end{smallmatrix}\right) dy = 2\pi \int_0^1 y[y - (-y)]dy$$
$$= 2\pi \int_0^1 2y^2\,dy = \tfrac{4\pi}{3}\left[y^3\right]_0^1 = \tfrac{4\pi}{3}$$

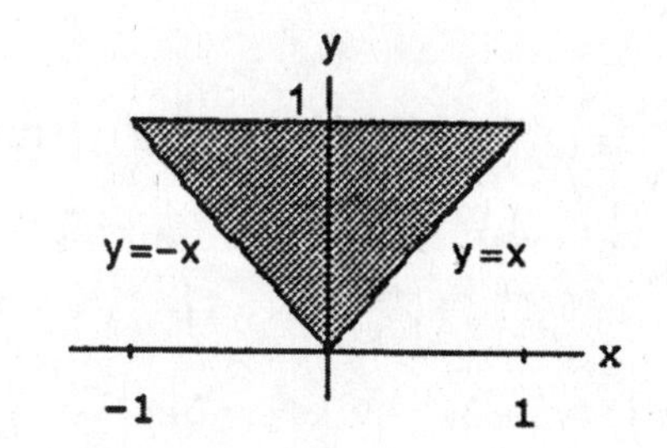

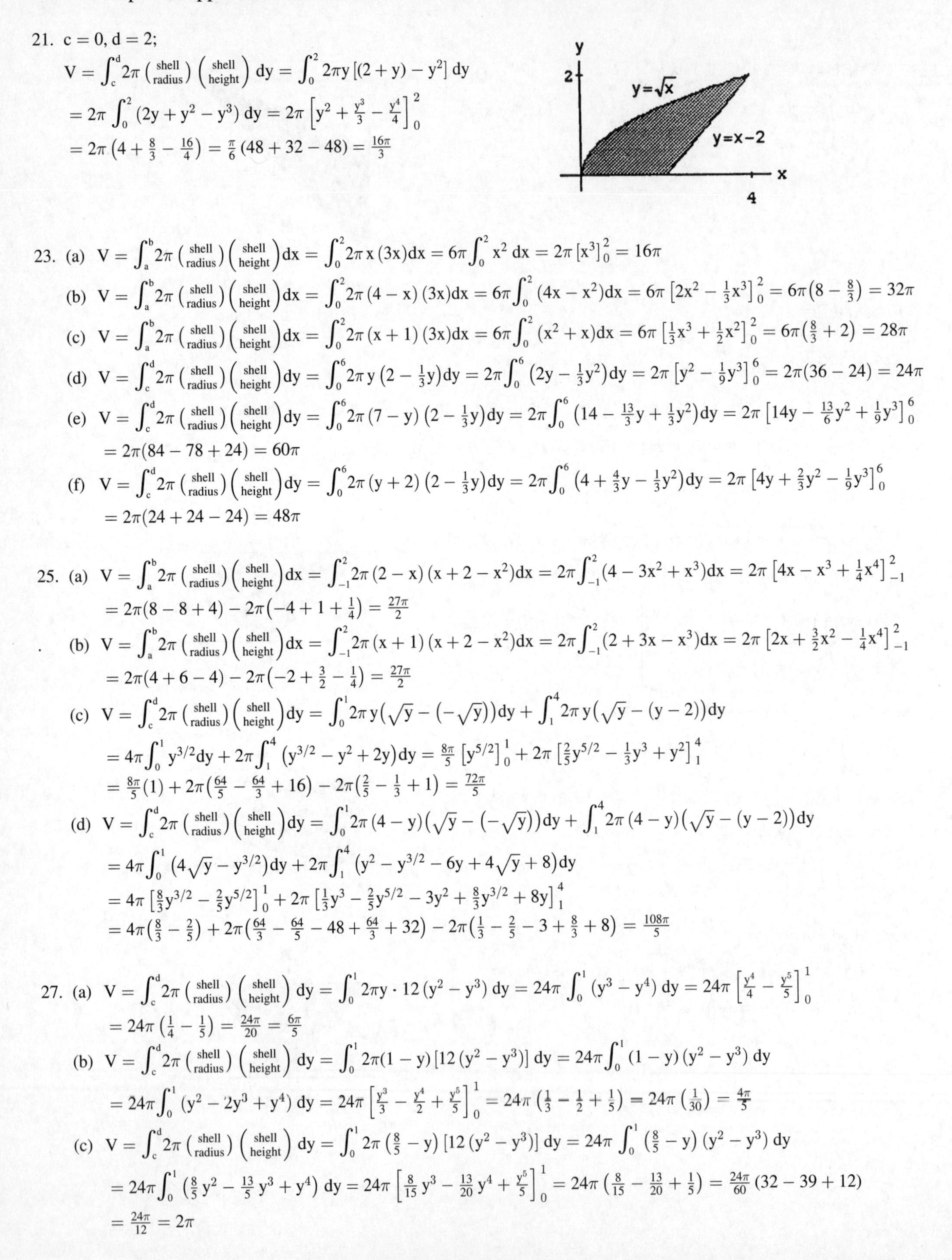

21. $c = 0, d = 2;$

$$V = \int_c^d 2\pi \left(\begin{smallmatrix}\text{shell}\\ \text{radius}\end{smallmatrix}\right)\left(\begin{smallmatrix}\text{shell}\\ \text{height}\end{smallmatrix}\right) dy = \int_0^2 2\pi y\,[(2+y) - y^2]\,dy$$
$$= 2\pi \int_0^2 (2y + y^2 - y^3)\,dy = 2\pi \left[y^2 + \frac{y^3}{3} - \frac{y^4}{4}\right]_0^2$$
$$= 2\pi \left(4 + \frac{8}{3} - \frac{16}{4}\right) = \frac{\pi}{6}(48 + 32 - 48) = \frac{16\pi}{3}$$

23. (a) $V = \int_a^b 2\pi \left(\begin{smallmatrix}\text{shell}\\ \text{radius}\end{smallmatrix}\right)\left(\begin{smallmatrix}\text{shell}\\ \text{height}\end{smallmatrix}\right) dx = \int_0^2 2\pi x\,(3x)\,dx = 6\pi \int_0^2 x^2\,dx = 2\pi\left[x^3\right]_0^2 = 16\pi$

(b) $V = \int_a^b 2\pi \left(\begin{smallmatrix}\text{shell}\\ \text{radius}\end{smallmatrix}\right)\left(\begin{smallmatrix}\text{shell}\\ \text{height}\end{smallmatrix}\right) dx = \int_0^2 2\pi (4 - x)(3x)\,dx = 6\pi \int_0^2 (4x - x^2)\,dx = 6\pi\left[2x^2 - \frac{1}{3}x^3\right]_0^2 = 6\pi\left(8 - \frac{8}{3}\right) = 32\pi$

(c) $V = \int_a^b 2\pi \left(\begin{smallmatrix}\text{shell}\\ \text{radius}\end{smallmatrix}\right)\left(\begin{smallmatrix}\text{shell}\\ \text{height}\end{smallmatrix}\right) dx = \int_0^2 2\pi (x + 1)(3x)\,dx = 6\pi \int_0^2 (x^2 + x)\,dx = 6\pi\left[\frac{1}{3}x^3 + \frac{1}{2}x^2\right]_0^2 = 6\pi\left(\frac{8}{3} + 2\right) = 28\pi$

(d) $V = \int_c^d 2\pi \left(\begin{smallmatrix}\text{shell}\\ \text{radius}\end{smallmatrix}\right)\left(\begin{smallmatrix}\text{shell}\\ \text{height}\end{smallmatrix}\right) dy = \int_0^6 2\pi y\left(2 - \frac{1}{3}y\right)dy = 2\pi \int_0^6 \left(2y - \frac{1}{3}y^2\right)dy = 2\pi\left[y^2 - \frac{1}{9}y^3\right]_0^6 = 2\pi(36 - 24) = 24\pi$

(e) $V = \int_c^d 2\pi \left(\begin{smallmatrix}\text{shell}\\ \text{radius}\end{smallmatrix}\right)\left(\begin{smallmatrix}\text{shell}\\ \text{height}\end{smallmatrix}\right) dy = \int_0^6 2\pi (7 - y)\left(2 - \frac{1}{3}y\right)dy = 2\pi \int_0^6 \left(14 - \frac{13}{3}y + \frac{1}{3}y^2\right)dy = 2\pi\left[14y - \frac{13}{6}y^2 + \frac{1}{9}y^3\right]_0^6$
$= 2\pi(84 - 78 + 24) = 60\pi$

(f) $V = \int_c^d 2\pi \left(\begin{smallmatrix}\text{shell}\\ \text{radius}\end{smallmatrix}\right)\left(\begin{smallmatrix}\text{shell}\\ \text{height}\end{smallmatrix}\right) dy = \int_0^6 2\pi (y + 2)\left(2 - \frac{1}{3}y\right)dy = 2\pi \int_0^6 \left(4 + \frac{4}{3}y - \frac{1}{3}y^2\right)dy = 2\pi\left[4y + \frac{2}{3}y^2 - \frac{1}{9}y^3\right]_0^6$
$= 2\pi(24 + 24 - 24) = 48\pi$

25. (a) $V = \int_a^b 2\pi \left(\begin{smallmatrix}\text{shell}\\ \text{radius}\end{smallmatrix}\right)\left(\begin{smallmatrix}\text{shell}\\ \text{height}\end{smallmatrix}\right) dx = \int_{-1}^2 2\pi (2 - x)(x + 2 - x^2)\,dx = 2\pi \int_{-1}^2 (4 - 3x^2 + x^3)\,dx = 2\pi\left[4x - x^3 + \frac{1}{4}x^4\right]_{-1}^2$
$= 2\pi(8 - 8 + 4) - 2\pi\left(-4 + 1 + \frac{1}{4}\right) = \frac{27\pi}{2}$

(b) $V = \int_a^b 2\pi \left(\begin{smallmatrix}\text{shell}\\ \text{radius}\end{smallmatrix}\right)\left(\begin{smallmatrix}\text{shell}\\ \text{height}\end{smallmatrix}\right) dx = \int_{-1}^2 2\pi (x + 1)(x + 2 - x^2)\,dx = 2\pi \int_{-1}^2 (2 + 3x - x^3)\,dx = 2\pi\left[2x + \frac{3}{2}x^2 - \frac{1}{4}x^4\right]_{-1}^2$
$= 2\pi(4 + 6 - 4) - 2\pi\left(-2 + \frac{3}{2} - \frac{1}{4}\right) = \frac{27\pi}{2}$

(c) $V = \int_c^d 2\pi \left(\begin{smallmatrix}\text{shell}\\ \text{radius}\end{smallmatrix}\right)\left(\begin{smallmatrix}\text{shell}\\ \text{height}\end{smallmatrix}\right) dy = \int_0^1 2\pi y\left(\sqrt{y} - \left(-\sqrt{y}\right)\right)dy + \int_1^4 2\pi y\left(\sqrt{y} - (y - 2)\right)dy$
$= 4\pi \int_0^1 y^{3/2}\,dy + 2\pi \int_1^4 \left(y^{3/2} - y^2 + 2y\right)dy = \frac{8\pi}{5}\left[y^{5/2}\right]_0^1 + 2\pi\left[\frac{2}{5}y^{5/2} - \frac{1}{3}y^3 + y^2\right]_1^4$
$= \frac{8\pi}{5}(1) + 2\pi\left(\frac{64}{5} - \frac{64}{3} + 16\right) - 2\pi\left(\frac{2}{5} - \frac{1}{3} + 1\right) = \frac{72\pi}{5}$

(d) $V = \int_c^d 2\pi \left(\begin{smallmatrix}\text{shell}\\ \text{radius}\end{smallmatrix}\right)\left(\begin{smallmatrix}\text{shell}\\ \text{height}\end{smallmatrix}\right) dy = \int_0^1 2\pi (4 - y)\left(\sqrt{y} - \left(-\sqrt{y}\right)\right)dy + \int_1^4 2\pi (4 - y)\left(\sqrt{y} - (y - 2)\right)dy$
$= 4\pi \int_0^1 \left(4\sqrt{y} - y^{3/2}\right)dy + 2\pi \int_1^4 \left(y^2 - y^{3/2} - 6y + 4\sqrt{y} + 8\right)dy$
$= 4\pi\left[\frac{8}{3}y^{3/2} - \frac{2}{5}y^{5/2}\right]_0^1 + 2\pi\left[\frac{1}{3}y^3 - \frac{2}{5}y^{5/2} - 3y^2 + \frac{8}{3}y^{3/2} + 8y\right]_1^4$
$= 4\pi\left(\frac{8}{3} - \frac{2}{5}\right) + 2\pi\left(\frac{64}{3} - \frac{64}{5} - 48 + \frac{64}{3} + 32\right) - 2\pi\left(\frac{1}{3} - \frac{2}{5} - 3 + \frac{8}{3} + 8\right) = \frac{108\pi}{5}$

27. (a) $V = \int_c^d 2\pi \left(\begin{smallmatrix}\text{shell}\\ \text{radius}\end{smallmatrix}\right)\left(\begin{smallmatrix}\text{shell}\\ \text{height}\end{smallmatrix}\right) dy = \int_0^1 2\pi y \cdot 12\,(y^2 - y^3)\,dy = 24\pi \int_0^1 (y^3 - y^4)\,dy = 24\pi\left[\frac{y^4}{4} - \frac{y^5}{5}\right]_0^1$
$= 24\pi\left(\frac{1}{4} - \frac{1}{5}\right) = \frac{24\pi}{20} = \frac{6\pi}{5}$

(b) $V = \int_c^d 2\pi \left(\begin{smallmatrix}\text{shell}\\ \text{radius}\end{smallmatrix}\right)\left(\begin{smallmatrix}\text{shell}\\ \text{height}\end{smallmatrix}\right) dy = \int_0^1 2\pi(1 - y)\,[12\,(y^2 - y^3)]\,dy = 24\pi \int_0^1 (1 - y)(y^2 - y^3)\,dy$
$= 24\pi \int_0^1 (y^2 - 2y^3 + y^4)\,dy = 24\pi\left[\frac{y^3}{3} - \frac{y^4}{2} + \frac{y^5}{5}\right]_0^1 = 24\pi\left(\frac{1}{3} - \frac{1}{2} + \frac{1}{5}\right) = 24\pi\left(\frac{1}{30}\right) = \frac{4\pi}{5}$

(c) $V = \int_c^d 2\pi \left(\begin{smallmatrix}\text{shell}\\ \text{radius}\end{smallmatrix}\right)\left(\begin{smallmatrix}\text{shell}\\ \text{height}\end{smallmatrix}\right) dy = \int_0^1 2\pi\left(\frac{8}{5} - y\right)[12\,(y^2 - y^3)]\,dy = 24\pi \int_0^1 \left(\frac{8}{5} - y\right)(y^2 - y^3)\,dy$
$= 24\pi \int_0^1 \left(\frac{8}{5}y^2 - \frac{13}{5}y^3 + y^4\right)dy = 24\pi\left[\frac{8}{15}y^3 - \frac{13}{20}y^4 + \frac{y^5}{5}\right]_0^1 = 24\pi\left(\frac{8}{15} - \frac{13}{20} + \frac{1}{5}\right) = \frac{24\pi}{60}(32 - 39 + 12)$
$= \frac{24\pi}{12} = 2\pi$

(d) $V = \int_c^d 2\pi \left(\begin{smallmatrix}\text{shell}\\ \text{radius}\end{smallmatrix}\right)\left(\begin{smallmatrix}\text{shell}\\ \text{height}\end{smallmatrix}\right) dy = \int_0^1 2\pi\left(y + \frac{2}{5}\right)\left[12\left(y^2 - y^3\right)\right] dy = 24\pi \int_0^1 \left(y + \frac{2}{5}\right)\left(y^2 - y^3\right) dy$
$= 24\pi \int_0^1 \left(y^3 - y^4 + \frac{2}{5}y^2 - \frac{2}{5}y^3\right) dy = 24\pi \int_0^1 \left(\frac{2}{5}y^2 + \frac{3}{5}y^3 - y^4\right) dy = 24\pi \left[\frac{2}{15}y^3 + \frac{3}{20}y^4 - \frac{y^5}{5}\right]_0^1$
$= 24\pi\left(\frac{2}{15} + \frac{3}{20} - \frac{1}{5}\right) = \frac{24\pi}{60}(8 + 9 - 12) = \frac{24\pi}{12} = 2\pi$

29. (a) About x-axis: $V = \int_c^d 2\pi \left(\begin{smallmatrix}\text{shell}\\ \text{radius}\end{smallmatrix}\right)\left(\begin{smallmatrix}\text{shell}\\ \text{height}\end{smallmatrix}\right) dy$
$= \int_0^1 2\pi y\left(\sqrt{y} - y\right) dy = 2\pi \int_0^1 \left(y^{3/2} - y^2\right) dy$
$= 2\pi\left[\frac{2}{5}y^{5/2} - \frac{1}{3}y^3\right]_0^1 = 2\pi\left(\frac{2}{5} - \frac{1}{3}\right) = \frac{2\pi}{15}$
About y-axis: $V = \int_a^b 2\pi \left(\begin{smallmatrix}\text{shell}\\ \text{radius}\end{smallmatrix}\right)\left(\begin{smallmatrix}\text{shell}\\ \text{height}\end{smallmatrix}\right) dx$
$= \int_0^1 2\pi x\left(x - x^2\right) dx = 2\pi \int_0^1 \left(x^2 - x^3\right) dx$
$= 2\pi\left[\frac{x^3}{3} - \frac{x^4}{4}\right]_0^1 = 2\pi\left(\frac{1}{3} - \frac{1}{4}\right) = \frac{\pi}{6}$

(b) About x-axis: $R(x) = x$ and $r(x) = x^2 \Rightarrow V = \int_a^b \pi\left[R(x)^2 - r(x)^2\right] dx = \int_0^1 \pi\left[x^2 - x^4\right] dx$
$= \pi\left[\frac{x^3}{3} - \frac{x^5}{5}\right]_0^1 = \pi\left(\frac{1}{3} - \frac{1}{5}\right) = \frac{2\pi}{15}$
About y-axis: $R(y) = \sqrt{y}$ and $r(y) = y \Rightarrow V = \int_c^d \pi\left[R(y)^2 - r(y)^2\right] dy = \int_0^1 \pi\left[y - y^2\right] dy$
$= \pi\left[\frac{y^2}{2} - \frac{y^3}{3}\right]_0^1 = \pi\left(\frac{1}{2} - \frac{1}{3}\right) = \frac{\pi}{6}$

31. (a) $V = \int_c^d 2\pi \left(\begin{smallmatrix}\text{shell}\\ \text{radius}\end{smallmatrix}\right)\left(\begin{smallmatrix}\text{shell}\\ \text{height}\end{smallmatrix}\right) dy = \int_1^2 2\pi y(y - 1)\, dy$
$= 2\pi \int_1^2 \left(y^2 - y\right) dy = 2\pi\left[\frac{y^3}{3} - \frac{y^2}{2}\right]_1^2$
$= 2\pi\left[\left(\frac{8}{3} - \frac{4}{2}\right) - \left(\frac{1}{3} - \frac{1}{2}\right)\right]$
$= 2\pi\left(\frac{7}{3} - 2 + \frac{1}{2}\right) = \frac{\pi}{3}(14 - 12 + 3) = \frac{5\pi}{3}$

(b) $V = \int_a^b 2\pi \left(\begin{smallmatrix}\text{shell}\\ \text{radius}\end{smallmatrix}\right)\left(\begin{smallmatrix}\text{shell}\\ \text{height}\end{smallmatrix}\right) dx = \int_1^2 2\pi x(2 - x)\, dx = 2\pi \int_1^2 \left(2x - x^2\right) dx = 2\pi\left[x^2 - \frac{x^3}{3}\right]_1^2$
$= 2\pi\left[\left(4 - \frac{8}{3}\right) - \left(1 - \frac{1}{3}\right)\right] = 2\pi\left[\left(\frac{12-8}{3}\right) - \left(\frac{3-1}{3}\right)\right] = 2\pi\left(\frac{4}{3} - \frac{2}{3}\right) = \frac{4\pi}{3}$

(c) $V = \int_a^b 2\pi \left(\begin{smallmatrix}\text{shell}\\ \text{radius}\end{smallmatrix}\right)\left(\begin{smallmatrix}\text{shell}\\ \text{height}\end{smallmatrix}\right) dx = \int_1^2 2\pi\left(\frac{10}{3} - x\right)(2 - x)\, dx = 2\pi \int_1^2 \left(\frac{20}{3} - \frac{16}{3}x + x^2\right) dx$
$= 2\pi\left[\frac{20}{3}x - \frac{8}{3}x^2 + \frac{1}{3}x^3\right]_1^2 = 2\pi\left[\left(\frac{40}{3} - \frac{32}{3} + \frac{8}{3}\right) - \left(\frac{20}{3} - \frac{8}{3} + \frac{1}{3}\right)\right] = 2\pi\left(\frac{3}{3}\right) = 2\pi$

(d) $V = \int_c^d 2\pi \left(\begin{smallmatrix}\text{shell}\\ \text{radius}\end{smallmatrix}\right)\left(\begin{smallmatrix}\text{shell}\\ \text{height}\end{smallmatrix}\right) dy = \int_1^2 2\pi(y - 1)(y - 1)\, dy = 2\pi \int_1^2 (y - 1)^2 = 2\pi\left[\frac{(y-1)^3}{3}\right]_1^2 = \frac{2\pi}{3}$

33. (a) $V = \int_c^d 2\pi \left(\begin{smallmatrix}\text{shell}\\ \text{radius}\end{smallmatrix}\right)\left(\begin{smallmatrix}\text{shell}\\ \text{height}\end{smallmatrix}\right) dy = \int_0^1 2\pi y\left(y - y^3\right) dy$
$= \int_0^1 2\pi\left(y^2 - y^4\right) dy = 2\pi\left[\frac{y^3}{3} - \frac{y^5}{5}\right]_0^1 = 2\pi\left(\frac{1}{3} - \frac{1}{5}\right)$
$= \frac{4\pi}{15}$

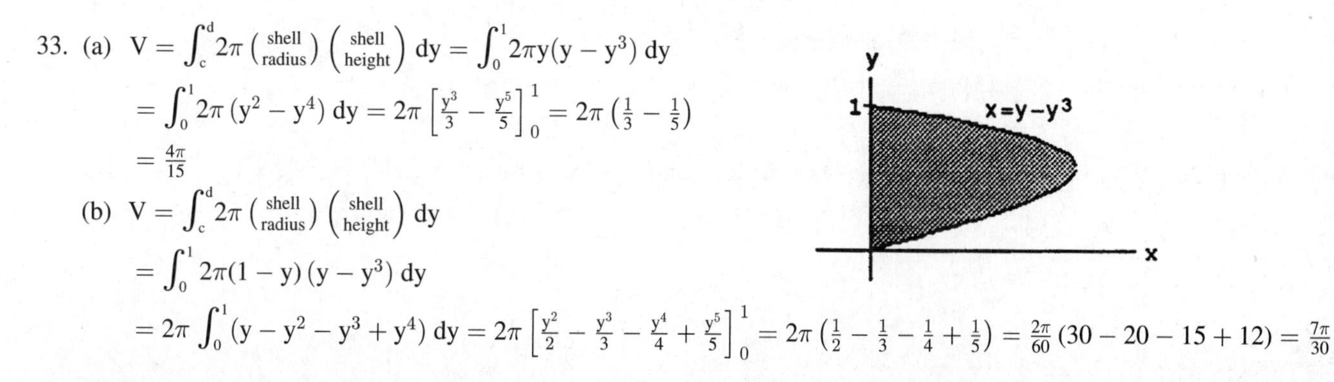

(b) $V = \int_c^d 2\pi \left(\begin{smallmatrix}\text{shell}\\ \text{radius}\end{smallmatrix}\right)\left(\begin{smallmatrix}\text{shell}\\ \text{height}\end{smallmatrix}\right) dy$
$= \int_0^1 2\pi(1 - y)\left(y - y^3\right) dy$
$= 2\pi \int_0^1 \left(y - y^2 - y^3 + y^4\right) dy = 2\pi\left[\frac{y^2}{2} - \frac{y^3}{3} - \frac{y^4}{4} + \frac{y^5}{5}\right]_0^1 = 2\pi\left(\frac{1}{2} - \frac{1}{3} - \frac{1}{4} + \frac{1}{5}\right) = \frac{2\pi}{60}(30 - 20 - 15 + 12) = \frac{7\pi}{30}$

35. (a) $V = \int_c^d 2\pi \left(\frac{\text{shell}}{\text{radius}}\right)\left(\frac{\text{shell}}{\text{height}}\right) dy = \int_0^2 2\pi y \left(\sqrt{8y} - y^2\right) dy$
$= 2\pi \int_0^2 \left(2\sqrt{2}\,y^{3/2} - y^3\right) dy = 2\pi \left[\frac{4\sqrt{2}}{5} y^{5/2} - \frac{y^4}{4}\right]_0^2$
$= 2\pi \left(\frac{4\sqrt{2}\cdot\left(\sqrt{2}\right)^5}{5} - \frac{2^4}{4}\right) = 2\pi \left(\frac{4\cdot 2^3}{5} - \frac{4\cdot 4}{4}\right)$
$= 2\pi \cdot 4\left(\frac{8}{5} - 1\right) = \frac{8\pi}{5}(8-5) = \frac{24\pi}{5}$

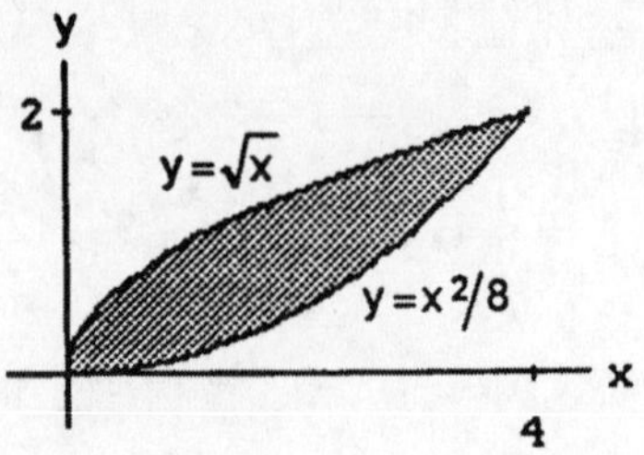

(b) $V = \int_a^b 2\pi \left(\frac{\text{shell}}{\text{radius}}\right)\left(\frac{\text{shell}}{\text{height}}\right) dx = \int_0^4 2\pi x \left(\sqrt{x} - \frac{x^2}{8}\right) dx = 2\pi \int_0^4 \left(x^{3/2} - \frac{x^3}{8}\right) dx = 2\pi \left[\frac{2}{5} x^{5/2} - \frac{x^4}{32}\right]_0^4$
$= 2\pi \left(\frac{2\cdot 2^5}{5} - \frac{4^4}{32}\right) = 2\pi \left(\frac{2^6}{5} - \frac{2^8}{32}\right) = \frac{\pi\cdot 2^7}{160}(32-20) = \frac{\pi\cdot 2^9\cdot 3}{160} = \frac{\pi\cdot 2^4\cdot 3}{5} = \frac{48\pi}{5}$

37. (a) $V = \int_a^b \pi\left[R^2(x) - r^2(x)\right] dx = \pi \int_{1/16}^1 \left(x^{-1/2} - 1\right) dx$
$= \pi\left[2x^{1/2} - x\right]_{1/16}^1 = \pi\left[(2-1) - \left(2\cdot\frac{1}{4} - \frac{1}{16}\right)\right]$
$= \pi\left(1 - \frac{7}{16}\right) = \frac{9\pi}{16}$

(b) $V = \int_a^b 2\pi \left(\frac{\text{shell}}{\text{radius}}\right)\left(\frac{\text{shell}}{\text{height}}\right) dy = \int_1^2 2\pi y \left(\frac{1}{y^4} - \frac{1}{16}\right) dy$
$= 2\pi \int_1^2 \left(y^{-3} - \frac{y}{16}\right) dy = 2\pi \left[-\frac{1}{2} y^{-2} - \frac{y^2}{32}\right]_1^2$
$= 2\pi\left[\left(-\frac{1}{8} - \frac{1}{8}\right) - \left(-\frac{1}{2} - \frac{1}{32}\right)\right] = 2\pi\left(\frac{1}{4} + \frac{1}{32}\right)$
$= \frac{2\pi}{32}(8+1) = \frac{9\pi}{16}$

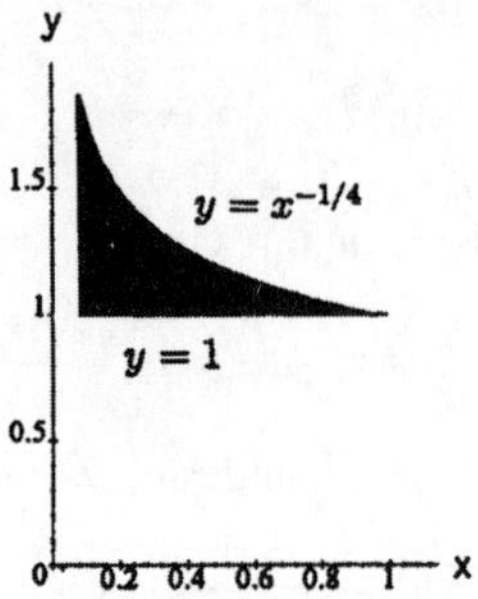

39. (a) *Disk*: $V = V_1 - V_2$
$V_1 = \int_{a_1}^{b_1} \pi[R_1(x)]^2\, dx$ and $V_2 = \int_{a_2}^{b_2} \pi[R_2(x)]^2$ with $R_1(x) = \sqrt{\frac{x+2}{3}}$ and $R_2(x) = \sqrt{x}$,
$a_1 = -2,\ b_1 = 1;\ a_2 = 0,\ b_2 = 1 \Rightarrow$ two integrals are required

(b) *Washer*: $V = V_1 + V_2$
$V_1 = \int_{a_1}^{b_1} \pi\left([R_1(x)]^2 - [r_1(x)]^2\right) dx$ with $R_1(x) = \sqrt{\frac{x+2}{3}}$ and $r_1(x) = 0$; $a_1 = -2$ and $b_1 = 0$;
$V_2 = \int_{a_2}^{b_2} \pi\left([R_2(x)]^2 - [r_2(x)]^2\right) dx$ with $R_2(x) = \sqrt{\frac{x+2}{3}}$ and $r_2(x) = \sqrt{x}$; $a_2 = 0$ and $b_2 = 1$
$\Rightarrow$ two integrals are required

(c) *Shell*: $V = \int_c^d 2\pi \left(\frac{\text{shell}}{\text{radius}}\right)\left(\frac{\text{shell}}{\text{height}}\right) dy = \int_c^d 2\pi y \left(\frac{\text{shell}}{\text{height}}\right) dy$ where shell height $= y^2 - (3y^2 - 2) = 2 - 2y^2$;
$c = 0$ and $d = 1$. Only *one* integral is required. It is, therefore preferable to use the *shell* method. However, whichever method you use, you will get $V = \pi$.

41. (a) $V = \int_a^b \pi\left[R^2(x) - r^2(x)\right] dx = \int_{-4}^4 \pi\left[\left(\sqrt{25 - x^2}\right)^2 - (3)^2\right] dx = \pi \int_{-4}^4 \left[25 - x^2 - 9\right] dx = \pi \int_{-4}^4 \left(16 - x^2\right) dx$
$= \pi\left[16x - \frac{1}{3}x^3\right]_{-4}^4 = \pi\left(64 - \frac{64}{3}\right) - \pi\left(-64 + \frac{64}{3}\right) = \frac{256\pi}{3}$

(b) Volume of sphere $= \frac{4}{3}\pi(5)^3 = \frac{500\pi}{3} \Rightarrow$ Volume of portion removed $= \frac{500\pi}{3} - \frac{256\pi}{3} = \frac{244\pi}{3}$

43. $V = \int_a^b 2\pi \left(\frac{\text{shell}}{\text{radius}}\right)\left(\frac{\text{shell}}{\text{height}}\right) dx = \int_0^r 2\pi x\left(-\frac{h}{r}x + h\right) dx = 2\pi \int_0^r \left(-\frac{h}{r}x^2 + hx\right) dx = 2\pi\left[-\frac{h}{3r}x^3 + \frac{h}{2}x^2\right]_0^r$
$= 2\pi\left(-\frac{r^2 h}{3} + \frac{r^2 h}{2}\right) = \frac{1}{3}\pi r^2 h$

6.3 ARC LENGTHS

1. $\frac{dy}{dx} = \frac{1}{3} \cdot \frac{3}{2}(x^2+2)^{1/2} \cdot 2x = \sqrt{(x^2+2)} \cdot x$
$\Rightarrow L = \int_0^3 \sqrt{1+(x^2+2)x^2}\,dx = \int_0^3 \sqrt{1+2x^2+x^4}\,dx$
$= \int_0^3 \sqrt{(1+x^2)^2}\,dx = \int_0^3 (1+x^2)\,dx = \left[x + \frac{x^3}{3}\right]_0^3$
$= 3 + \frac{27}{3} = 12$

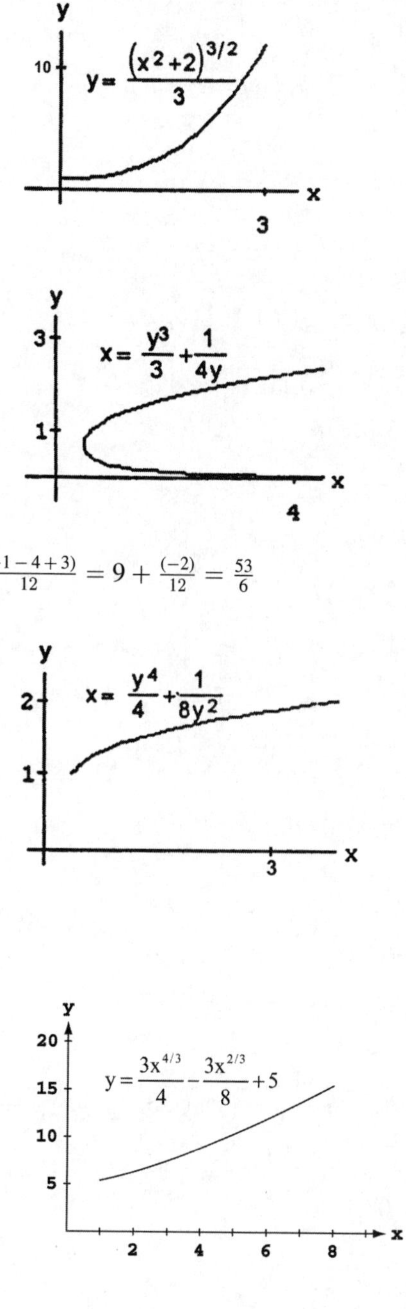

3. $\frac{dx}{dy} = y^2 - \frac{1}{4y^2} \Rightarrow \left(\frac{dx}{dy}\right)^2 = y^4 - \frac{1}{2} + \frac{1}{16y^4}$
$\Rightarrow L = \int_1^3 \sqrt{1 + y^4 - \frac{1}{2} + \frac{1}{16y^4}}\,dy$
$= \int_1^3 \sqrt{y^4 + \frac{1}{2} + \frac{1}{16y^4}}\,dy$
$= \int_1^3 \sqrt{\left(y^2 + \frac{1}{4y^2}\right)^2}\,dy = \int_1^3 \left(y^2 + \frac{1}{4y^2}\right)dy$
$= \left[\frac{y^3}{3} - \frac{y^{-1}}{4}\right]_1^3 = \left(\frac{27}{3} - \frac{1}{12}\right) - \left(\frac{1}{3} - \frac{1}{4}\right) = 9 - \frac{1}{12} - \frac{1}{3} + \frac{1}{4} = 9 + \frac{(-1-4+3)}{12} = 9 + \frac{(-2)}{12} = \frac{53}{6}$

5. $\frac{dx}{dy} = y^3 - \frac{1}{4y^3} \Rightarrow \left(\frac{dx}{dy}\right)^2 = y^6 - \frac{1}{2} + \frac{1}{16y^6}$
$\Rightarrow L = \int_1^2 \sqrt{1 + y^6 - \frac{1}{2} + \frac{1}{16y^6}}\,dy$
$= \int_1^2 \sqrt{y^6 + \frac{1}{2} + \frac{1}{16y^6}}\,dy = \int_1^2 \sqrt{\left(y^3 + \frac{y^{-3}}{4}\right)^2}\,dy$
$= \int_1^2 \left(y^3 + \frac{y^{-3}}{4}\right)dy = \left[\frac{y^4}{4} - \frac{y^{-2}}{8}\right]_1^2$
$= \left(\frac{16}{4} - \frac{1}{(16)(2)}\right) - \left(\frac{1}{4} - \frac{1}{8}\right) = 4 - \frac{1}{32} - \frac{1}{4} + \frac{1}{8} = \frac{128-1-8+4}{32} = \frac{123}{32}$

7. $\frac{dy}{dx} = x^{1/3} - \frac{1}{4}x^{-1/3} \Rightarrow \left(\frac{dy}{dx}\right)^2 = x^{2/3} - \frac{1}{2} + \frac{x^{-2/3}}{16}$
$\Rightarrow L = \int_1^8 \sqrt{1 + x^{2/3} - \frac{1}{2} + \frac{x^{-2/3}}{16}}\,dx$
$= \int_1^8 \sqrt{x^{2/3} + \frac{1}{2} + \frac{x^{-2/3}}{16}}\,dx$
$= \int_1^8 \sqrt{\left(x^{1/3} + \frac{1}{4}x^{-1/3}\right)^2}\,dx = \int_1^8 \left(x^{1/3} + \frac{1}{4}x^{-1/3}\right)dx$
$= \left[\frac{3}{4}x^{4/3} + \frac{3}{8}x^{2/3}\right]_1^8 = \frac{3}{8}\left[2x^{4/3} + x^{2/3}\right]_1^8$
$= \frac{3}{8}\left[(2 \cdot 2^4 + 2^2) - (2+1)\right] = \frac{3}{8}(32 + 4 - 3) = \frac{99}{8}$

9. $\frac{dx}{dy} = \sqrt{\sec^4 y - 1} \Rightarrow \left(\frac{dx}{dy}\right)^2 = \sec^4 y - 1$
$\Rightarrow L = \int_{-\pi/4}^{\pi/4} \sqrt{1 + (\sec^4 y - 1)}\,dy = \int_{-\pi/4}^{\pi/4} \sec^2 y\,dy$
$= [\tan y]_{-\pi/4}^{\pi/4} = 1 - (-1) = 2$

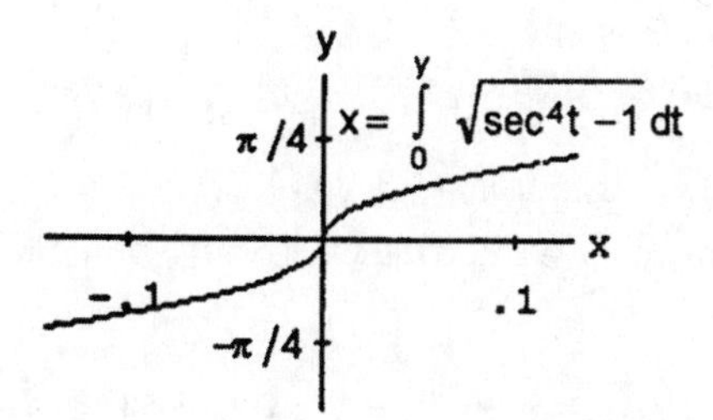

11. (a) $\frac{dy}{dx} = 2x \Rightarrow \left(\frac{dy}{dx}\right)^2 = 4x^2$
$\Rightarrow L = \int_{-1}^{2} \sqrt{1 + \left(\frac{dy}{dx}\right)^2}\, dx$
$= \int_{-1}^{2} \sqrt{1 + 4x^2}\, dx$
(b) 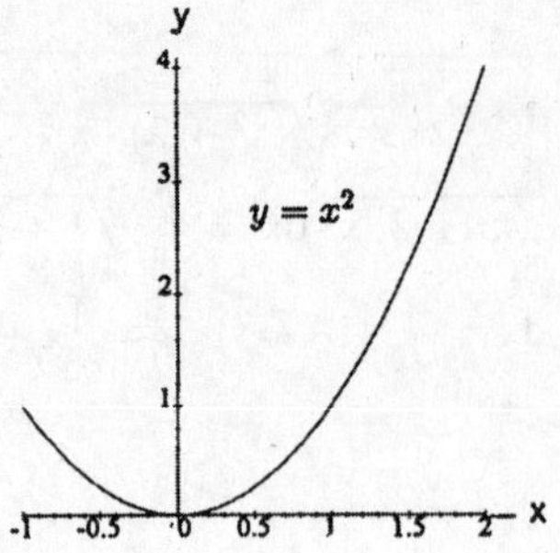
(c) $L \approx 6.13$

13. (a) $\frac{dx}{dy} = \cos y \Rightarrow \left(\frac{dx}{dy}\right)^2 = \cos^2 y$
$\Rightarrow L = \int_0^{\pi} \sqrt{1 + \cos^2 y}\, dy$
(b) 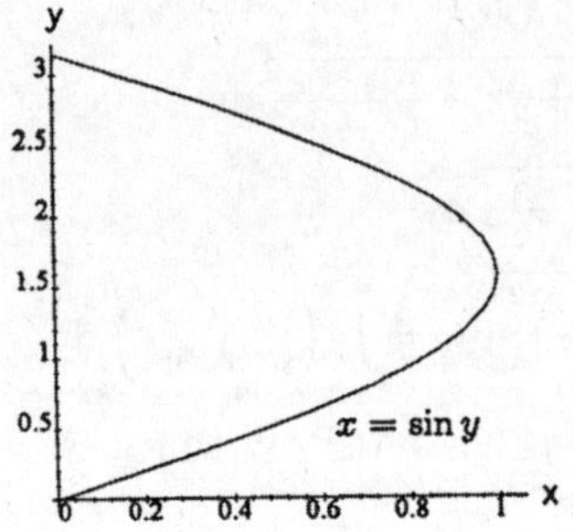
(c) $L \approx 3.82$

15. (a) $2y + 2 = 2\,\frac{dx}{dy} \Rightarrow \left(\frac{dx}{dy}\right)^2 = (y+1)^2$
$\Rightarrow L = \int_{-1}^{3} \sqrt{1 + (y+1)^2}\, dy$
(b) 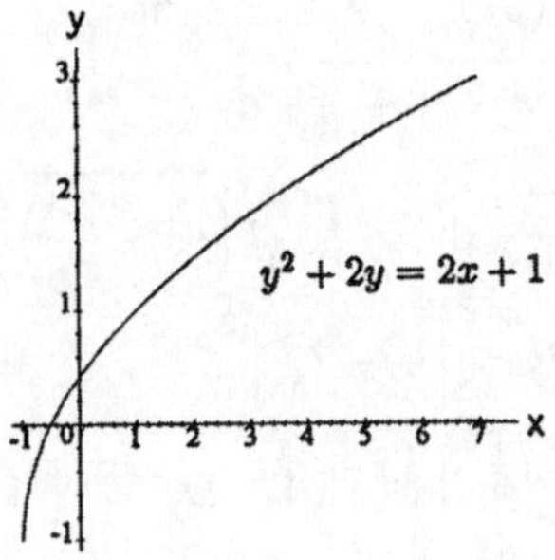
(c) $L \approx 9.29$

17. (a) $\frac{dy}{dx} = \tan x \Rightarrow \left(\frac{dy}{dx}\right)^2 = \tan^2 x$
$\Rightarrow L = \int_0^{\pi/6} \sqrt{1 + \tan^2 x}\, dx = \int_0^{\pi/6} \sqrt{\frac{\sin^2 x + \cos^2 x}{\cos^2 x}}\, dx$
$= \int_0^{\pi/6} \frac{dx}{\cos x} = \int_0^{\pi/6} \sec x\, dx$
(b) 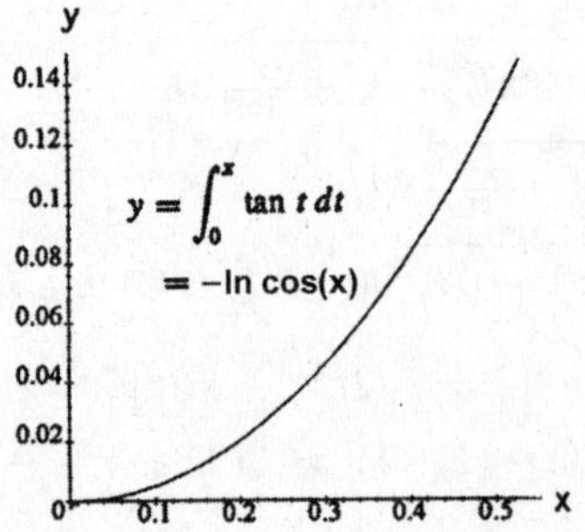
(c) $L \approx 0.55$

19. (a) $\left(\frac{dy}{dx}\right)^2$ corresponds to $\frac{1}{4x}$ here, so take $\frac{dy}{dx}$ as $\frac{1}{2\sqrt{x}}$. Then $y = \sqrt{x} + C$ and since $(1, 1)$ lies on the curve, $C = 0$. So $y = \sqrt{x}$ from $(1, 1)$ to $(4, 2)$.
(b) Only one. We know the derivative of the function and the value of the function at one value of x.

21. $y = \int_0^x \sqrt{\cos 2t}\, dt \Rightarrow \frac{dy}{dx} = \sqrt{\cos 2x} \Rightarrow L = \int_0^{\pi/4} \sqrt{1 + \left[\sqrt{\cos 2x}\right]^2}\, dx = \int_0^{\pi/4} \sqrt{1 + \cos 2x}\, dx = \int_0^{\pi/4} \sqrt{2\cos^2 x}\, dx$
$= \int_0^{\pi/4} \sqrt{2} \cos x\, dx = \sqrt{2}\left[\sin x\right]_0^{\pi/4} = \sqrt{2}\sin\left(\frac{\pi}{4}\right) - \sqrt{2}\sin(0) = 1$

23. $y = 3 - 2x,\ 0 \le x \le 2 \Rightarrow \frac{dy}{dx} = -2 \Rightarrow L = \int_0^2 \sqrt{1 + (-2)^2}\,dx = \int_0^2 \sqrt{5}\,dx = \left[\sqrt{5}\,x\right]_0^2 = 2\sqrt{5}.$

$d = \sqrt{(2-0)^2 + (3-(-1))^2} = 2\sqrt{5}$

25. $9x^2 = y(y-3)^2 \Rightarrow \frac{d}{dy}\left[9x^2\right] = \frac{d}{dy}\left[y(y-3)^2\right] \Rightarrow 18x\frac{dx}{dy} = 2y(y-3) + (y-3)^2 = 3(y-3)(y-1) \Rightarrow \frac{dx}{dy} = \frac{(y-3)(y-1)}{6x}$

$\Rightarrow dx = \frac{(y-3)(y-1)}{6x}dy;\ ds^2 = dx^2 + dy^2 = \left[\frac{(y-3)(y-1)}{6x}dy\right]^2 + dy^2 = \frac{(y-3)^2(y-1)^2}{36x^2}dy^2 + dy^2 = \frac{(y-3)^2(y-1)^2}{4y(y-3)^2}dy^2 + dy^2$

$= \left[\frac{(y-1)^2}{4y} + 1\right]dy^2 = \frac{y^2 - 2y + 1 + 4y}{4y}dy^2 = \frac{(y+1)^2}{4y}dy^2$

27. $\sqrt{2}\,x = \int_0^x \sqrt{1 + \left(\frac{dy}{dt}\right)^2}\,dt,\ x \ge 0 \Rightarrow \sqrt{2} = \sqrt{1 + \left(\frac{dy}{dx}\right)^2} \Rightarrow \frac{dy}{dx} = \pm 1 \Rightarrow y = f(x) = \pm x + C$ where C is any real number.

29. $x^2 + y^2 = 1 \Rightarrow y = \sqrt{1 - x^2};\ P = \{0, \frac{1}{4}, \frac{1}{2}, \frac{3}{4}, 1\} \Rightarrow L \approx \sum_{k=1}^{4} \sqrt{(x_i - x_{i-1})^2 + (y_i - y_{i-1})^2} = \sqrt{\left(\frac{1}{4} - 0\right)^2 + \left(\frac{\sqrt{15}}{4} - 1\right)^2}$

$+ \sqrt{\left(\frac{1}{2} - \frac{1}{4}\right)^2 + \left(\frac{\sqrt{3}}{2} - \frac{\sqrt{15}}{4}\right)^2} + \sqrt{\left(\frac{3}{4} - \frac{1}{2}\right)^2 + \left(\frac{\sqrt{7}}{4} - \frac{\sqrt{3}}{2}\right)^2} + \sqrt{\left(1 - \frac{3}{4}\right)^2 + \left(0 - \frac{\sqrt{7}}{4}\right)^2} \approx 1.55225$

31. $y = 2x^{3/2} \Rightarrow \frac{dy}{dx} = 3x^{1/2};\ L(x) = \int_0^x \sqrt{1 + \left[3t^{1/2}\right]^2}\,dt = \int_0^x \sqrt{1 + 9t}\,dt;\ [u = 1 + 9t \Rightarrow du = 9dt,\ t = 0 \Rightarrow u = 1,$

$t = x \Rightarrow u = 1 + 9x] \rightarrow \frac{1}{9}\int_1^{1+9x} \sqrt{u}\,du = \frac{2}{27}\left[u^{3/2}\right]_1^{1+9x} = \frac{2}{27}(1+9x)^{3/2} - \frac{2}{27};\ L(1) = \frac{2}{27}(10)^{3/2} - \frac{2}{27} = \frac{2\left(10\sqrt{10} - 1\right)}{27}$

6.4 AREAS OF SURFACES OF REVOLUTION

1. (a) $\frac{dy}{dx} = \sec^2 x \Rightarrow \left(\frac{dy}{dx}\right)^2 = \sec^4 x$

$\Rightarrow S = 2\pi\int_0^{\pi/4} (\tan x)\sqrt{1 + \sec^4 x}\,dx$

(c) $S \approx 3.84$

(b)

3. (a) $xy = 1 \Rightarrow x = \frac{1}{y} \Rightarrow \frac{dx}{dy} = -\frac{1}{y^2} \Rightarrow \left(\frac{dx}{dy}\right)^2 = \frac{1}{y^4}$

$\Rightarrow S = 2\pi\int_1^2 \frac{1}{y}\sqrt{1 + y^{-4}}\,dy$

(c) $S \approx 5.02$

(b)

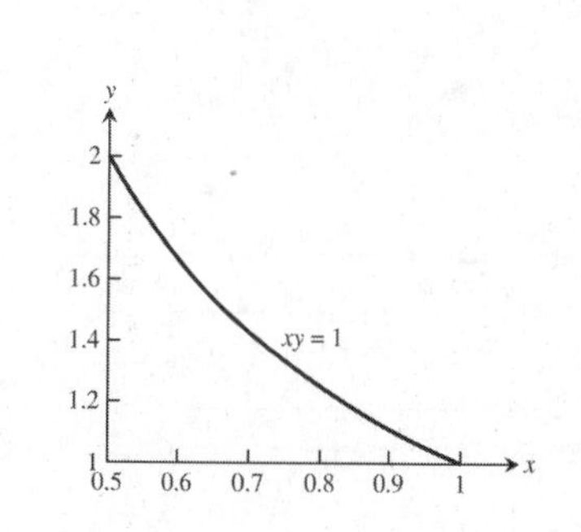

5. (a) $x^{1/2} + y^{1/2} = 3 \Rightarrow y = \left(3 - x^{1/2}\right)^2$
$\Rightarrow \frac{dy}{dx} = 2\left(3 - x^{1/2}\right)\left(-\frac{1}{2}x^{-1/2}\right)$
$\Rightarrow \left(\frac{dy}{dx}\right)^2 = \left(1 - 3x^{-1/2}\right)^2$
$\Rightarrow S = 2\pi \int_1^4 \left(3 - x^{1/2}\right)^2 \sqrt{1 + \left(1 - 3x^{-1/2}\right)^2}\, dx$

(b)

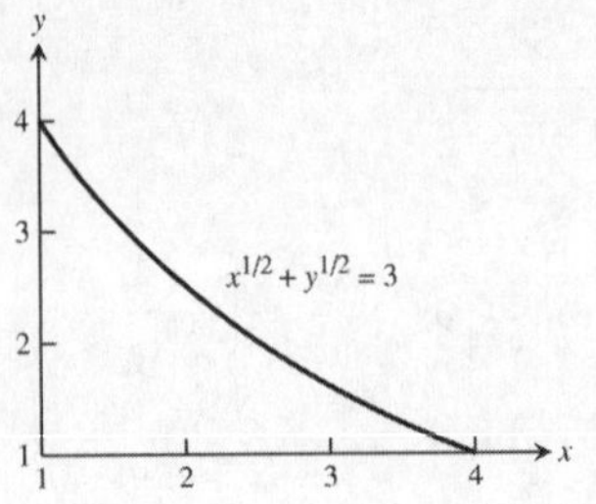

(c) $S \approx 63.37$

7. (a) $\frac{dx}{dy} = \tan y \Rightarrow \left(\frac{dx}{dy}\right)^2 = \tan^2 y$
$\Rightarrow S = 2\pi \int_0^{\pi/3} \left(\int_0^y \tan t\, dt\right) \sqrt{1 + \tan^2 y}\, dy$
$= 2\pi \int_0^{\pi/3} \left(\int_0^y \tan t\, dt\right) \sec y\, dy$

(b)

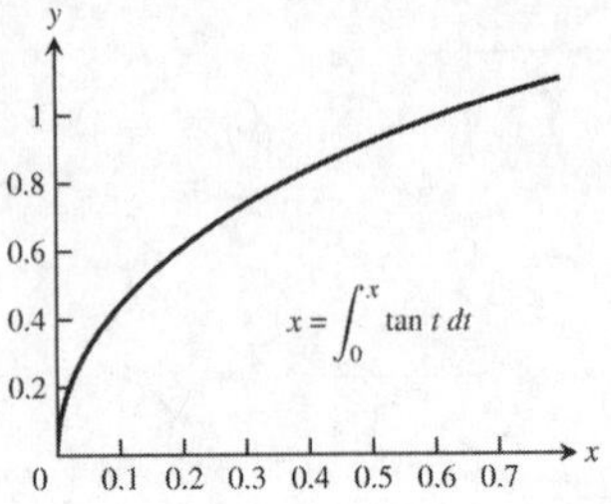

(c) $S \approx 2.08$

9. $y = \frac{x}{2} \Rightarrow \frac{dy}{dx} = \frac{1}{2}$; $S = \int_a^b 2\pi y \sqrt{1 + \left(\frac{dy}{dx}\right)^2}\, dx \Rightarrow S = \int_0^4 2\pi \left(\frac{x}{2}\right) \sqrt{1 + \frac{1}{4}}\, dx = \frac{\pi\sqrt{5}}{2} \int_0^4 x\, dx$
$= \frac{\pi\sqrt{5}}{2} \left[\frac{x^2}{2}\right]_0^4 = 4\pi\sqrt{5}$; Geometry formula: base circumference $= 2\pi(2)$, slant height $= \sqrt{4^2 + 2^2} = 2\sqrt{5}$
$\Rightarrow$ Lateral surface area $= \frac{1}{2}(4\pi)\left(2\sqrt{5}\right) = 4\pi\sqrt{5}$ in agreement with the integral value

11. $\frac{dy}{dx} = \frac{1}{2}$; $S = \int_a^b 2\pi y \sqrt{1 + \left(\frac{dy}{dx}\right)^2}\, dx = \int_1^3 2\pi \frac{(x+1)}{2} \sqrt{1 + \left(\frac{1}{2}\right)^2}\, dx = \frac{\pi\sqrt{5}}{2} \int_1^3 (x+1)\, dx = \frac{\pi\sqrt{5}}{2} \left[\frac{x^2}{2} + x\right]_1^3$
$= \frac{\pi\sqrt{5}}{2}\left[\left(\frac{9}{2} + 3\right) - \left(\frac{1}{2} + 1\right)\right] = \frac{\pi\sqrt{5}}{2}(4 + 2) = 3\pi\sqrt{5}$; Geometry formula: $r_1 = \frac{1}{2} + \frac{1}{2} = 1$, $r_2 = \frac{3}{2} + \frac{1}{2} = 2$,
slant height $= \sqrt{(2-1)^2 + (3-1)^2} = \sqrt{5} \Rightarrow$ Frustum surface area $= \pi(r_1 + r_2) \times$ slant height $= \pi(1+2)\sqrt{5}$
$= 3\pi\sqrt{5}$ in agreement with the integral value

13. $\frac{dy}{dx} = \frac{x^2}{3} \Rightarrow \left(\frac{dy}{dx}\right)^2 = \frac{x^4}{9} \Rightarrow S = \int_0^2 \frac{2\pi x^3}{9} \sqrt{1 + \frac{x^4}{9}}\, dx$;
$\left[u = 1 + \frac{x^4}{9} \Rightarrow du = \frac{4}{9}x^3\, dx \Rightarrow \frac{1}{4}\, du = \frac{x^3}{9}\, dx;\right.$
$\left. x = 0 \Rightarrow u = 1, x = 2 \Rightarrow u = \frac{25}{9}\right]$
$\rightarrow S = 2\pi \int_1^{25/9} u^{1/2} \cdot \frac{1}{4}\, du = \frac{\pi}{2}\left[\frac{2}{3}u^{3/2}\right]_1^{25/9}$
$= \frac{\pi}{3}\left(\frac{125}{27} - 1\right) = \frac{\pi}{3}\left(\frac{125 - 27}{27}\right) = \frac{98\pi}{81}$

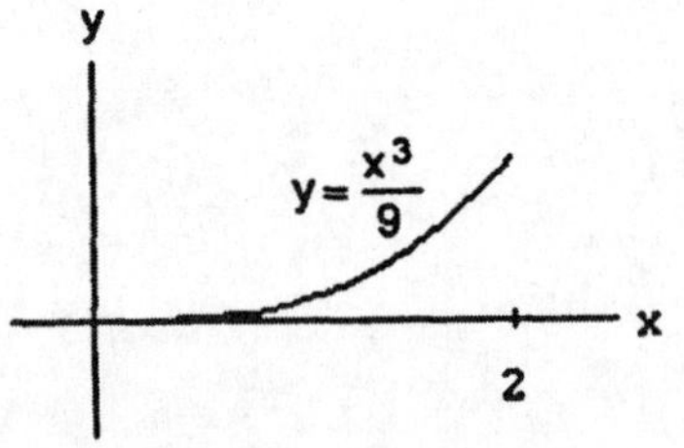

15. $\frac{dy}{dx} = \frac{1}{2}\frac{(2 - 2x)}{\sqrt{2x - x^2}} = \frac{1 - x}{\sqrt{2x - x^2}} \Rightarrow \left(\frac{dy}{dx}\right)^2 = \frac{(1-x)^2}{2x - x^2}$
$\Rightarrow S = \int_{0.5}^{1.5} 2\pi\sqrt{2x - x^2} \sqrt{1 + \frac{(1-x)^2}{2x - x^2}}\, dx$
$= 2\pi \int_{0.5}^{1.5} \sqrt{2x - x^2}\, \frac{\sqrt{2x - x^2 + 1 - 2x + x^2}}{\sqrt{2x - x^2}}\, dx$
$= 2\pi \int_{0.5}^{1.5} dx = 2\pi[x]_{0.5}^{1.5} = 2\pi$

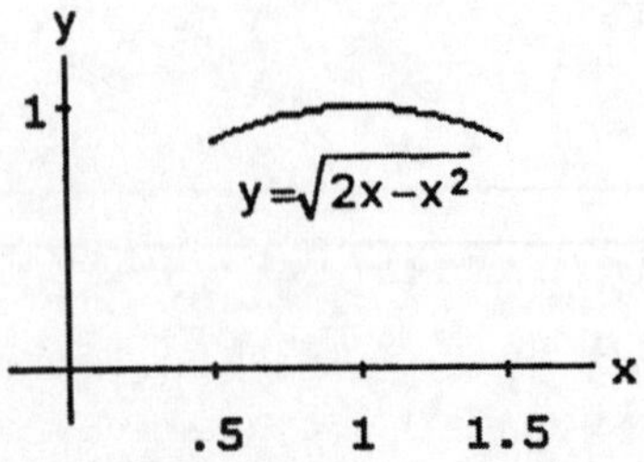

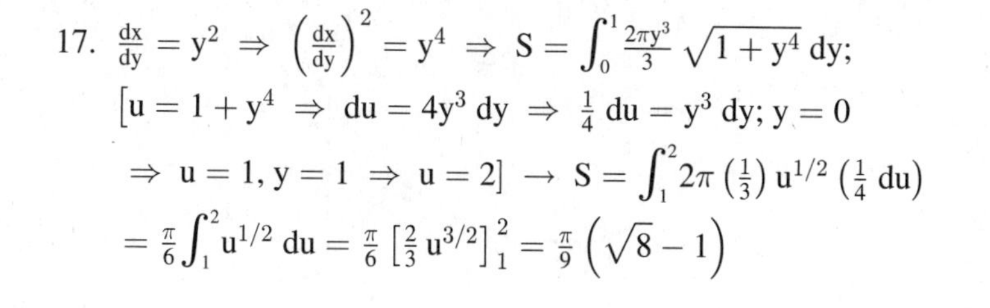

17. $\frac{dx}{dy} = y^2 \Rightarrow \left(\frac{dx}{dy}\right)^2 = y^4 \Rightarrow S = \int_0^1 \frac{2\pi y^3}{3}\sqrt{1+y^4}\,dy;$
$\left[u = 1 + y^4 \Rightarrow du = 4y^3\,dy \Rightarrow \frac{1}{4}\,du = y^3\,dy;\ y = 0\right.$
$\left.\Rightarrow u = 1,\ y = 1 \Rightarrow u = 2\right] \to S = \int_1^2 2\pi\left(\frac{1}{3}\right)u^{1/2}\left(\frac{1}{4}\,du\right)$
$= \frac{\pi}{6}\int_1^2 u^{1/2}\,du = \frac{\pi}{6}\left[\frac{2}{3}u^{3/2}\right]_1^2 = \frac{\pi}{9}\left(\sqrt{8}-1\right)$

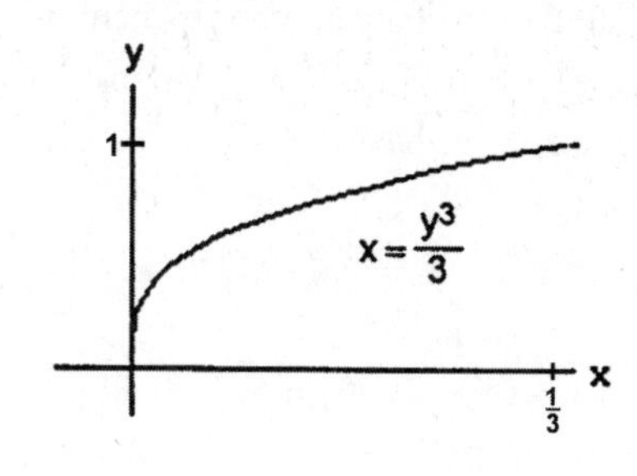

19. $\frac{dx}{dy} = \frac{-1}{\sqrt{4-y}} \Rightarrow \left(\frac{dx}{dy}\right)^2 = \frac{1}{4-y} \Rightarrow S = \int_0^{15/4} 2\pi\cdot 2\sqrt{4-y}\,\sqrt{1+\frac{1}{4-y}}\,dy = 4\pi\int_0^{15/4}\sqrt{(4-y)+1}\,dy$
$= 4\pi\int_0^{15/4}\sqrt{5-y}\,dy = -4\pi\left[\frac{2}{3}(5-y)^{3/2}\right]_0^{15/4} = -\frac{8\pi}{3}\left[\left(5-\frac{15}{4}\right)^{3/2} - 5^{3/2}\right] = -\frac{8\pi}{3}\left[\left(\frac{5}{4}\right)^{3/2} - 5^{3/2}\right]$
$= \frac{8\pi}{3}\left(5\sqrt{5} - \frac{5\sqrt{5}}{8}\right) = \frac{8\pi}{3}\left(\frac{40\sqrt{5}-5\sqrt{5}}{8}\right) = \frac{35\pi\sqrt{5}}{3}$

21. $S = 2\pi\int_{1/2}^1 \sqrt{2y-1}\,\sqrt{1+\left(\frac{1}{\sqrt{2y-1}}\right)^2}\,dy = 2\pi\int_{1/2}^1\sqrt{2y-1}\,\sqrt{1+\frac{1}{2y-1}}\,dy = 2\pi\int_{1/2}^1\sqrt{2y-1}\,\sqrt{\frac{2y}{2y-1}}\,dy$
$= 2\pi\int_{1/2}^1\sqrt{2y}\,dy = 2\sqrt{2}\pi\int_{1/2}^1\sqrt{y}\,dy = 2\sqrt{2}\pi\left[\frac{2}{3}y^{3/2}\right]_{1/2}^1 = 2\sqrt{2}\pi\left[\left(\frac{2}{3}\sqrt{1^3}\right) - \left(\frac{2}{3}\sqrt{\left(\frac{1}{2}\right)^3}\right)\right] = 2\sqrt{2}\pi\left(\frac{2}{3} - \frac{1}{3\sqrt{2}}\right)$
$= 2\sqrt{2}\pi\left(\frac{2\sqrt{2}-1}{3\sqrt{2}}\right) = \frac{2\pi}{3}\left(2\sqrt{2}-1\right)$

23. $ds = \sqrt{dx^2+dy^2} = \sqrt{\left(y^3 - \frac{1}{4y^3}\right)^2 + 1}\,dy = \sqrt{\left(y^6 - \frac{1}{2} + \frac{1}{16y^6}\right) + 1}\,dy = \sqrt{\left(y^6 + \frac{1}{2} + \frac{1}{16y^6}\right)}\,dy$
$= \sqrt{\left(y^3 + \frac{1}{4y^3}\right)^2}\,dy = \left(y^3 + \frac{1}{4y^3}\right)dy;\ S = \int_1^2 2\pi y\,ds = 2\pi\int_1^2 y\left(y^3 + \frac{1}{4y^3}\right)dy = 2\pi\int_1^2\left(y^4 + \frac{1}{4}y^{-2}\right)dy$
$= 2\pi\left[\frac{y^5}{5} - \frac{1}{4}y^{-1}\right]_1^2 = 2\pi\left[\left(\frac{32}{5} - \frac{1}{8}\right) - \left(\frac{1}{5} - \frac{1}{4}\right)\right] = 2\pi\left(\frac{31}{5} + \frac{1}{8}\right) = \frac{2\pi}{40}(8\cdot 31 + 5) = \frac{253\pi}{20}$

25. $y = \sqrt{a^2-x^2} \Rightarrow \frac{dy}{dx} = \frac{1}{2}\left(a^2-x^2\right)^{-1/2}(-2x) = \frac{-x}{\sqrt{a^2-x^2}} \Rightarrow \left(\frac{dy}{dx}\right)^2 = \frac{x^2}{(a^2-x^2)}$
$\Rightarrow S = 2\pi\int_{-a}^a\sqrt{a^2-x^2}\,\sqrt{1+\frac{x^2}{(a^2-x^2)}}\,dx = 2\pi\int_{-a}^a\sqrt{(a^2-x^2)+x^2}\,dx = 2\pi\int_{-a}^a a\,dx = 2\pi a[x]_{-a}^a$
$= 2\pi a[a-(-a)] = (2\pi a)(2a) = 4\pi a^2$

27. The area of the surface of one wok is $S = \int_c^d 2\pi x\sqrt{1+\left(\frac{dx}{dy}\right)^2}\,dy$. Now, $x^2 + y^2 = 16^2 \Rightarrow x = \sqrt{16^2 - y^2}$
$\Rightarrow \frac{dx}{dy} = \frac{-y}{\sqrt{16^2-y^2}} \Rightarrow \left(\frac{dx}{dy}\right)^2 = \frac{y^2}{16^2-y^2};\ S = \int_{-16}^{-7} 2\pi\sqrt{16^2-y^2}\,\sqrt{1+\frac{y^2}{16^2-y^2}}\,dy = 2\pi\int_{-16}^{-7}\sqrt{(16^2-y^2)+y^2}\,dy$
$= 2\pi\int_{-16}^{-7} 16\,dy = 32\pi\cdot 9 = 288\pi \approx 904.78\ \text{cm}^2$. The enamel needed to cover one surface of one wok is
$V = S\cdot 0.5\ \text{mm} = S\cdot 0.05\ \text{cm} = (904.78)(0.05)\ \text{cm}^3 = 45.24\ \text{cm}^3$. For 5000 woks, we need
$5000\cdot V = 5000\cdot 45.24\ \text{cm}^3 = (5)(45.24)\text{L} = 226.2\text{L} \Rightarrow$ 226.2 liters of each color are needed.

29. $y = \sqrt{R^2-x^2} \Rightarrow \frac{dy}{dx} = -\frac{1}{2}\,\frac{2x}{\sqrt{R^2-x^2}} = \frac{-x}{\sqrt{R^2-x^2}} \Rightarrow \left(\frac{dx}{dy}\right)^2 = \frac{x^2}{R^2-x^2};\ S = 2\pi\int_a^{a+h}\sqrt{R^2-x^2}\,\sqrt{1+\frac{x^2}{R^2-x^2}}\,dx$
$= 2\pi\int_a^{a+h}\sqrt{(R^2-x^2)+x^2}\,dx = 2\pi R\int_a^{a+h}dx = 2\pi Rh$

31. (a) An equation of the tangent line segment is (see figure) $y = f(m_k) + f'(m_k)(x - m_k)$. When $x = x_{k-1}$ we have $r_1 = f(m_k) + f'(m_k)(x_{k-1} - m_k)$ $= f(m_k) + f'(m_k)\left(-\frac{\Delta x_k}{2}\right) = f(m_k) - f'(m_k)\frac{\Delta x_k}{2}$; when $x = x_k$ we have $r_2 = f(m_k) + f'(m_k)(x_k - m_k)$ $= f(m_k) + f'(m_k)\frac{\Delta x_k}{2}$;

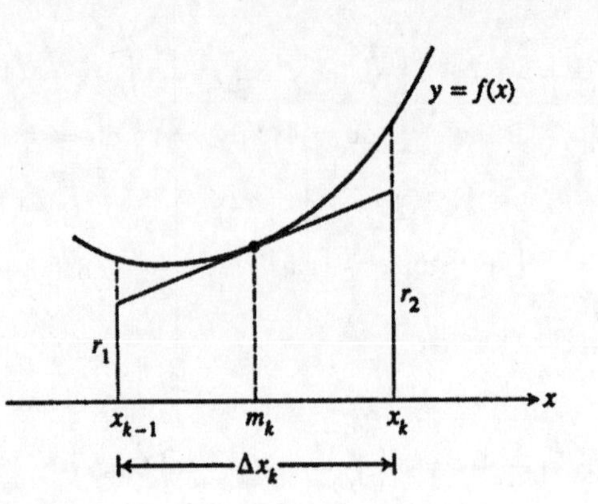

(b) $L_k^2 = (\Delta x_k)^2 + (r_2 - r_1)^2$
$= (\Delta x_k)^2 + \left[f'(m_k)\frac{\Delta x_k}{2} - \left(-f'(m_k)\frac{\Delta x_k}{2}\right)\right]^2$
$= (\Delta x_k)^2 + [f'(m_k)\Delta x_k]^2 \Rightarrow L_k = \sqrt{(\Delta x_k)^2 + [f'(m_k)\Delta x_k]^2}$, as claimed

(c) From geometry it is a fact that the lateral surface area of the frustum obtained by revolving the tangent line segment about the x-axis is given by $\Delta S_k = \pi(r_1 + r_2)L_k = \pi[2f(m_k)]\sqrt{(\Delta x_k)^2 + [f'(m_k)\Delta x_k]^2}$ using parts (a) and (b) above. Thus, $\Delta S_k = 2\pi f(m_k)\sqrt{1 + [f'(m_k)]^2}\,\Delta x_k$.

(d) $S = \lim_{n\to\infty}\sum_{k=1}^{n}\Delta S_k = \lim_{n\to\infty}\sum_{k=1}^{n} 2\pi f(m_k)\sqrt{1 + [f'(m_k)]^2}\,\Delta x_k = \int_a^b 2\pi f(x)\sqrt{1 + [f'(x)]^2}\,dx$

6.5 WORK AND FLUID FORCES

1. The force required to stretch the spring from its natural length of 2 m to a length of 5 m is $F(x) = kx$. The work done by F is $W = \int_0^3 F(x)\,dx = k\int_0^3 x\,dx = \frac{k}{2}\left[x^2\right]_0^3 = \frac{9k}{2}$. This work is equal to 1800 J $\Rightarrow \frac{9}{2}k = 1800 \Rightarrow k = 400$ N/m

3. We find the force constant from Hooke's law: $F = kx$. A force of 2 N stretches the spring to 0.02 m $\Rightarrow 2 = k \cdot (0.02)$ $\Rightarrow k = 100\,\frac{N}{m}$. The force of 4 N will stretch the rubber band y m, where $F = ky \Rightarrow y = \frac{F}{k} \Rightarrow y = \frac{4N}{100\,\frac{N}{m}} \Rightarrow y = 0.04$ m $= 4$ cm. The work done to stretch the rubber band 0.04 m is $W = \int_0^{0.04} kx\,dx = 100\int_0^{0.04} x\,dx = 100\left[\frac{x^2}{2}\right]_0^{0.04}$ $= \frac{(100)(0.04)^2}{2} = 0.08$ J

5. (a) We find the spring's constant from Hooke's law: $F = kx \Rightarrow k = \frac{F}{x} = \frac{21{,}714}{8-5} = \frac{21{,}714}{3} \Rightarrow k = 7238\,\frac{lb}{in}$

(b) The work done to compress the assembly the first half inch is $W = \int_0^{0.5} kx\,dx = 7238\int_0^{0.5} x\,dx = 7238\left[\frac{x^2}{2}\right]_0^{0.5}$ $= (7238)\frac{(0.5)^2}{2} = \frac{(7238)(0.25)}{2} \approx 905$ in · lb. The work done to compress the assembly the second half inch is: $W = \int_{0.5}^{1.0} kx\,dx = 7238\int_{0.5}^{1.0} x\,dx = 7238\left[\frac{x^2}{2}\right]_{0.5}^{1.0} = \frac{7238}{2}\left[1 - (0.5)^2\right] = \frac{(7238)(0.75)}{2} \approx 2714$ in · lb

7. The force required to haul up the rope is equal to the rope's weight, which varies steadily and is proportional to x, the length of the rope still hanging: $F(x) = 0.624x$. The work done is: $W = \int_0^{50} F(x)\,dx = \int_0^{50} 0.624x\,dx = 0.624\left[\frac{x^2}{2}\right]_0^{50}$ $= 780$ J

9. The force required to lift the cable is equal to the weight of the cable paid out: $F(x) = (4.5)(180 - x)$ where x is the position of the car off the first floor. The work done is: $W = \int_0^{180} F(x)\,dx = 4.5\int_0^{180}(180 - x)\,dx$ $= 4.5\left[180x - \frac{x^2}{2}\right]_0^{180} = 4.5\left(180^2 - \frac{180^2}{2}\right) = \frac{4.5 \cdot 180^2}{2} = 72{,}900$ ft · lb

11. Let r = the constant rate of leakage. Since the bucket is leaking at a constant rate and the bucket is rising at a constant rate, the amount of water in the bucket is proportional to $(20 - x)$, the distance the bucket is being raised. The leakage rate of

the water is 0.8 lb/ft raised and the weight of the water in the bucket is F = 0.8(20 − x). So:
$W = \int_0^{20} 0.8(20-x)\,dx = 0.8\left[20x - \frac{x^2}{2}\right]_0^{20} = 160$ ft · lb.

13. We will use the coordinate system given.

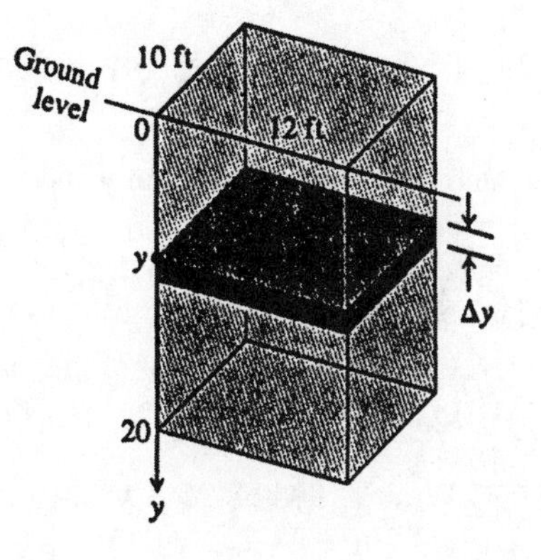

(a) The typical slab between the planes at y and y + Δy has a volume of $\Delta V = (10)(12)\,\Delta y = 120\,\Delta y$ ft³. The force F required to lift the slab is equal to its weight: $F = 62.4\,\Delta V = 62.4 \cdot 120\,\Delta y$ lb. The distance through which F must act is about y ft, so the work done lifting the slab is about ΔW = force × distance $= 62.4 \cdot 120 \cdot y \cdot \Delta y$ ft · lb. The work it takes to lift all the water is approximately $W \approx \sum_0^{20} \Delta W$ $= \sum_0^{20} 62.4 \cdot 120y \cdot \Delta y$ ft · lb. This is a Riemann sum for the function $62.4 \cdot 120y$ over the interval $0 \le y \le 20$. The work of pumping the tank empty is the limit of these sums:
$W = \int_0^{20} 62.4 \cdot 120y\,dy = (62.4)(120)\left[\frac{y^2}{2}\right]_0^{20} = (62.4)(120)\left(\frac{400}{2}\right) = (62.4)(120)(200) = 1{,}497{,}600$ ft · lb

(b) The time t it takes to empty the full tank with $\left(\frac{5}{11}\right)$–hp motor is $t = \frac{W}{250\,\frac{\text{ft}\cdot\text{lb}}{\text{sec}}} = \frac{1{,}497{,}600\ \text{ft}\cdot\text{lb}}{250\,\frac{\text{ft}\cdot\text{lb}}{\text{sec}}} = 5990.4$ sec = 1.664 hr
$\Rightarrow t \approx 1$ hr and 40 min

(c) Following all the steps of part (a), we find that the work it takes to lower the water level 10 ft is
$W = \int_0^{10} 62.4 \cdot 120y\,dy = (62.4)(120)\left[\frac{y^2}{2}\right]_0^{10} = (62.4)(120)\left(\frac{100}{2}\right) = 374{,}400$ ft · lb and the time is $t = \frac{W}{250\,\frac{\text{ft}\cdot\text{lb}}{\text{sec}}}$
= 1497.6 sec = 0.416 hr ≈ 25 min

(d) In a location where water weighs 62.26 $\frac{\text{lb}}{\text{ft}^3}$:
a) W = (62.26)(24,000) = 1,494,240 ft · lb.
b) $t = \frac{1{,}494{,}240}{250} = 5976.96$ sec ≈ 1.660 hr ⇒ t ≈ 1 hr and 40 min
In a location where water weighs 62.59 $\frac{\text{lb}}{\text{ft}^3}$
a) W = (62.59)(24,000) = 1,502,160 ft · lb
b) $t = \frac{1{,}502{,}160}{250} = 6008.64$ sec ≈ 1.669 hr ⇒ t ≈ 1 hr and 40.1 min

15. The slab is a disk of area $\pi x^2 = \pi\left(\frac{y}{2}\right)^2$, thickness Δy, and height below the top of the tank (10 − y). So the work to pump the oil in this slab, ΔW, is $57(10-y)\pi\left(\frac{y}{2}\right)^2$. The work to pump all the oil to the top of the tank is
$W = \int_0^{10} \frac{57\pi}{4}(10y^2 - y^3)dy = \frac{57\pi}{4}\left[\frac{10y^3}{3} - \frac{y^4}{4}\right]_0^{10} = 11{,}875\pi$ ft · lb ≈ 37,306 ft · lb.

17. The typical slab between the planes at y and and y + Δy has a volume of $\Delta V = \pi(\text{radius})^2(\text{thickness}) = \pi\left(\frac{20}{2}\right)^2\Delta y$ $= \pi \cdot 100\,\Delta y$ ft³. The force F required to lift the slab is equal to its weight: $F = 51.2\,\Delta V = 51.2 \cdot 100\pi\,\Delta y$ lb $\Rightarrow F = 5120\pi\,\Delta y$ lb. The distance through which F must act is about (30 − y) ft. The work it takes to lift all the kerosene is approximately $W \approx \sum_0^{30} \Delta W = \sum_0^{30} 5120\pi(30-y)\,\Delta y$ ft · lb which is a Riemann sum. The work to pump the tank dry is the limit of these sums: $W = \int_0^{30} 5120\pi(30-y)\,dy = 5120\pi\left[30y - \frac{y^2}{2}\right]_0^{30} = 5120\pi\left(\frac{900}{2}\right) = (5120)(450\pi)$
≈ 7,238,229.48 ft · lb

19. The typical slab between the planes at y and y+Δy has a volume of about $\Delta V = \pi(\text{radius})^2(\text{thickness}) = \pi\left(\sqrt{y}\right)^2\Delta y$ ft³. The force F(y) required to lift this slab is equal to its weight: $F(y) = 73 \cdot \Delta V = 73\pi\left(\sqrt{y}\right)^2\Delta y = 73\pi y\,\Delta y$ lb. The

distance through which F(y) must act to lift the slab to the top of the reservoir is about $(4-y)$ ft, so the work done is approximately $\Delta W \approx 73\pi\, y\,(4-y)\Delta y$ ft · lb. The work done lifting all the slabs from $y=0$ ft to $y=4$ ft is approximately $W \approx \sum_{k=0}^{n} 73\pi\, y_k\,(4-y_k)\Delta y$ ft · lb. Taking the limit of these Riemann sums as $n \to \infty$, we get

$W = \int_0^4 73\pi\, y\,(4-y)dy = 73\pi \int_0^4 (4y-y^2)dy = 73\pi\left[2y^2 - \frac{1}{3}y^3\right]_0^4 = 73\pi\left(32-\frac{64}{3}\right) = \frac{2336\pi}{3}$ ft · lb.

21. The typical slab between the planes at y and $y+\Delta y$ has a volume of about $\Delta V = \pi(\text{radius})^2(\text{thickness})$ $= \pi\left(\sqrt{25-y^2}\right)^2 \Delta y\ \text{m}^3$. The force F(y) required to lift this slab is equal to its weight: $F(y) = 9800 \cdot \Delta V$ $= 9800\pi\left(\sqrt{25-y^2}\right)^2 \Delta y = 9800\pi\,(25-y^2)\,\Delta y$ N. The distance through which F(y) must act to lift the slab to the level of 4 m above the top of the reservoir is about $(4-y)$ m, so the work done is approximately $\Delta W \approx 9800\pi\,(25-y^2)\,(4-y)\,\Delta y$ N · m. The work done lifting all the slabs from $y=-5$ m to $y=0$ m is approximately $W \approx \sum_{-5}^{0} 9800\pi\,(25-y^2)\,(4-y)\,\Delta y$ N · m. Taking the limit of these Riemann sums, we get

$W = \int_{-5}^{0} 9800\pi\,(25-y^2)\,(4-y)\,dy = 9800\pi \int_{-5}^{0} (100-25y-4y^2+y^3)\,dy = 9800\pi\left[100y - \frac{25}{2}y^2 - \frac{4}{3}y^3 + \frac{y^4}{4}\right]_{-5}^{0}$
$= -9800\pi\left(-500 - \frac{25\cdot 25}{2} + \frac{4}{3}\cdot 125 + \frac{625}{4}\right) \approx 15{,}073{,}099.75$ J

23. $F = m\frac{dv}{dt} = mv\frac{dv}{dx}$ by the chain rule $\Rightarrow W = \int_{x_1}^{x_2} mv\frac{dv}{dx}\,dx = m\int_{x_1}^{x_2}\left(v\frac{dv}{dx}\right)dx = m\left[\frac{1}{2}v^2(x)\right]_{x_1}^{x_2}$
$= \frac{1}{2}m\left[v^2(x_2) - v^2(x_1)\right] = \frac{1}{2}mv_2^2 - \frac{1}{2}mv_1^2$, as claimed.

25. $90\text{ mph} = \frac{90\text{ mi}}{1\text{ hr}} \cdot \frac{1\text{ hr}}{60\text{ min}} \cdot \frac{1\text{ min}}{60\text{ sec}} \cdot \frac{5280\text{ ft}}{1\text{ mi}} = 132$ ft/sec; $m = \frac{0.3125\text{ lb}}{32\text{ ft/sec}^2} = \frac{0.3125}{32}$ slugs;
$W = \left(\frac{1}{2}\right)\left(\frac{0.3125\text{ lb}}{32\text{ ft/sec}^2}\right)(132\text{ ft/sec})^2 \approx 85.1$ ft · lb

27. $v_1 = 0\text{ mph} = 0\frac{\text{ft}}{\text{sec}}$, $v_2 = 153\text{ mph} = 224.4\frac{\text{ft}}{\text{sec}}$; $2\text{ oz} = 0.125\text{ lb} \Rightarrow m = \frac{0.125\text{ lb}}{32\text{ ft/sec}^2} = \frac{1}{256}$ slugs;
$W = \int_{x_1}^{x_2} F(x)\,dx = \frac{1}{2}mv_2^2 - \frac{1}{2}mv_1^2 = \frac{1}{2}\left(\frac{1}{256}\right)(224.4)^2 - \frac{1}{2}\left(\frac{1}{256}\right)(0)^2 = 98.35$ ft-lb.

29. We imagine the milkshake divided into thin slabs by planes perpendicular to the y-axis at the points of a partition of the interval $[0, 7]$. The typical slab between the planes at y and $y+\Delta y$ has a volume of about $\Delta V = \pi(\text{radius})^2(\text{thickness})$ $= \pi\left(\frac{y+17.5}{14}\right)^2 \Delta y\ \text{in}^3$. The force F(y) required to lift this slab is equal to its weight: $F(y) = \frac{4}{9}\,\Delta V = \frac{4\pi}{9}\left(\frac{y+17.5}{14}\right)^2 \Delta y$ oz. The distance through which F(y) must act to lift this slab to the level of 1 inch above the top is about $(8-y)$ in. The work done lifting the slab is about $\Delta W = \left(\frac{4\pi}{9}\right)\frac{(y+17.5)^2}{14^2}(8-y)\Delta y$ in · oz. The work done lifting all the slabs from $y=0$ to $y=7$ is approximately $W = \sum_{0}^{7} \frac{4\pi}{9\cdot 14^2}(y+17.5)^2(8-y)\,\Delta y$ in · oz which is a Riemann sum. The work is the limit of these sums as the norm of the partition goes to zero: $W = \int_0^7 \frac{4\pi}{9\cdot 14^2}(y+17.5)^2(8-y)dy$
$= \frac{4\pi}{9\cdot 14^2}\int_0^7 (2450 - 26.25y - 27y^2 - y^3)dy = \frac{4\pi}{9\cdot 14^2}\left[-\frac{y^4}{4} - 9y^3 - \frac{26.25}{2}y^2 + 2450y\right]_0^7$
$= \frac{4\pi}{9\cdot 14^2}\left[-\frac{7^4}{4} - 9\cdot 7^3 - \frac{26.25}{2}\cdot 7^2 + 2450\cdot 7\right] \approx 91.32$ in · oz

31. To find the width of the plate at a typical depth y, we first find an equation for the line of the plate's right-hand edge: $y = x - 5$. If we let x denote the width of the right-hand half of the triangle at depth y, then $x = 5 + y$ and the total width is $L(y) = 2x = 2(5+y)$. The depth of the strip is $(-y)$. The force exerted by the water against one side of the plate is therefore $F = \int_{-5}^{-2} w(-y)\cdot L(y)\,dy = \int_{-5}^{-2} 62.4\cdot(-y)\cdot 2(5+y)\,dy$
$= 124.8\int_{-5}^{-2}(-5y - y^2)\,dy = 124.8\left[-\frac{5}{2}y^2 - \frac{1}{3}y^3\right]_{-5}^{-2} = 124.8\left[\left(-\frac{5}{2}\cdot 4 + \frac{1}{3}\cdot 8\right) - \left(-\frac{5}{2}\cdot 25 + \frac{1}{3}\cdot 125\right)\right]$
$= (124.8)\left(\frac{105}{2} - \frac{117}{3}\right) = (124.8)\left(\frac{315-234}{6}\right) = 1684.8$ lb

33. (a) The width of the strip is $L(y) = 4$, the depth of the strip is $(10 - y) \Rightarrow F = \int_a^b w \cdot \left(\begin{smallmatrix}\text{strip}\\ \text{depth}\end{smallmatrix}\right) F(y)dy$

$= \int_0^3 62.4(10 - y)(4)dy = 249.6\int_0^3 (10 - y)dy = 249.6\left[10y - \frac{y^2}{2}\right]_0^3 = 249.6\left(30 - \frac{9}{2}\right) = 6364.8 \text{ lb}$

(b) The width of the strip is $L(y) = 3$, the depth of the strip is $(10 - y) \Rightarrow F = \int_a^b w \cdot \left(\begin{smallmatrix}\text{strip}\\ \text{depth}\end{smallmatrix}\right) F(y)dy$

$= \int_0^4 62.4(10 - y)(3)dy = 187.2\int_0^4 (10 - y)dy = 187.2\left[10y - \frac{y^2}{2}\right]_0^4 = 187.2(40 - 8) = 5990.4 \text{ lb}$

35. Using the coordinate system of Exercise 32, we find the equation for the line of the plate's right-hand edge to be $y = 2x - 4 \Rightarrow x = \frac{y+4}{2}$ and $L(y) = 2x = y + 4$. The depth of the strip is $(1 - y)$.

(a) $F = \int_{-4}^0 w(1 - y)L(y)\,dy = \int_{-4}^0 62.4 \cdot (1 - y)(y + 4)\,dy = 62.4\int_{-4}^0 (4 - 3y - y^2)\,dy = 62.4\left[4y - \frac{3y^2}{2} - \frac{y^3}{3}\right]_{-4}^0$

$= (-62.4)\left[(-4)(4) - \frac{(3)(16)}{2} + \frac{64}{3}\right] = (-62.4)\left(-16 - 24 + \frac{64}{3}\right) = \frac{(-62.4)(-120+64)}{3} = 1164.8 \text{ lb}$

(b) $F = (-64.0)\left[(-4)(4) - \frac{(3)(16)}{2} + \frac{64}{3}\right] = \frac{(-64.0)(-120+64)}{3} \approx 1194.7 \text{ lb}$

37. Using the coordinate system given in the accompanying figure, we see that the total width is $L(y) = 63$ and the depth of the strip is $(33.5 - y) \Rightarrow F = \int_0^{33} w(33.5 - y)L(y)\,dy$

$= \int_0^{33} \frac{64}{12^3} \cdot (33.5 - y) \cdot 63\,dy = \left(\frac{64}{12^3}\right)(63)\int_0^{33} (33.5 - y)\,dy$

$= \left(\frac{64}{12^3}\right)(63)\left[33.5y - \frac{y^2}{2}\right]_0^{33} = \left(\frac{64 \cdot 63}{12^3}\right)\left[(33.5)(33) - \frac{33^2}{2}\right]$

$= \frac{(64)(63)(33)(67 - 33)}{(2)(12^3)} = 1309 \text{ lb}$

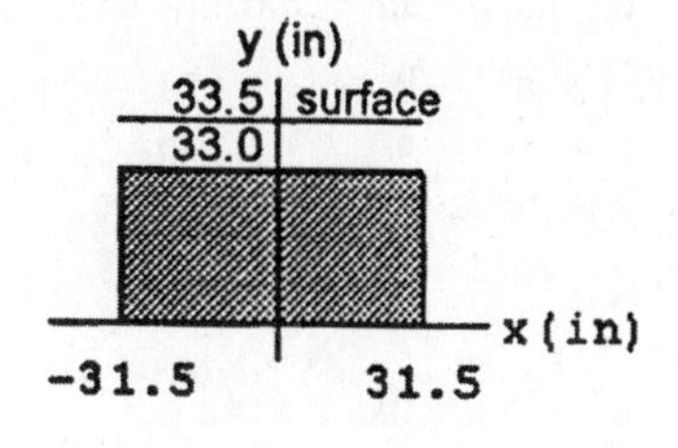

39. (a) $F = \left(62.4 \frac{\text{lb}}{\text{ft}^3}\right)(8 \text{ ft})(25 \text{ ft}^2) = 12480 \text{ lb}$

(b) The width of the strip is $L(y) = 5$, the depth of the strip is $(8 - y) \Rightarrow F = \int_a^b w \cdot \left(\begin{smallmatrix}\text{strip}\\ \text{depth}\end{smallmatrix}\right) F(y)dy$

$= \int_0^5 62.4(8 - y)(5)dy = 312\int_0^5 (8 - y)dy = 312\left[8y - \frac{y^2}{2}\right]_0^5 = 312\left(40 - \frac{25}{2}\right) = 8580 \text{ lb}$

(c) The width of the strip is $L(y) = 5$, the depth of the strip is $(8 - y)$, the height of the strip is $\sqrt{2}\,dy$

$\Rightarrow F = \int_a^b w \cdot \left(\begin{smallmatrix}\text{strip}\\ \text{depth}\end{smallmatrix}\right) F(y)dy = \int_0^{5/\sqrt{2}} 62.4(8 - y)(5)\sqrt{2}\,dy = 312\sqrt{2}\int_0^{5/\sqrt{2}} (8 - y)dy = 312\sqrt{2}\left[8y - \frac{y^2}{2}\right]_0^{5/\sqrt{2}}$

$= 312\sqrt{2}\left(\frac{40}{\sqrt{2}} - \frac{25}{4}\right) = 9722.3$

41. The coordinate system is given in the text. The right-hand edge is $x = \sqrt{y}$ and the total width is $L(y) = 2x = 2\sqrt{y}$.

(a) The depth of the strip is $(2 - y)$ so the force exerted by the liquid on the gate is $F = \int_0^1 w(2 - y)L(y)\,dy$

$= \int_0^1 50(2 - y) \cdot 2\sqrt{y}\,dy = 100\int_0^1 (2 - y)\sqrt{y}\,dy = 100\int_0^1 (2y^{1/2} - y^{3/2})\,dy = 100\left[\frac{4}{3}y^{3/2} - \frac{2}{5}y^{5/2}\right]_0^1$

$= 100\left(\frac{4}{3} - \frac{2}{5}\right) = \left(\frac{100}{15}\right)(20 - 6) = 93.33 \text{ lb}$

(b) We need to solve $160 = \int_0^1 w(H - y) \cdot 2\sqrt{y}\,dy$ for h. $160 = 100\left(\frac{2H}{3} - \frac{2}{5}\right) \Rightarrow H = 3$ ft.

43. The pressure at level y is $p(y) = w \cdot y \Rightarrow$ the average pressure is $\bar{p} = \frac{1}{b}\int_0^b p(y)\,dy = \frac{1}{b}\int_0^b w \cdot y\,dy = \frac{1}{b}w\left[\frac{y^2}{2}\right]_0^b$

$= \left(\frac{w}{b}\right)\left(\frac{b^2}{2}\right) = \frac{wb}{2}$. This is the pressure at level $\frac{b}{2}$, which is the pressure at the middle of the plate.

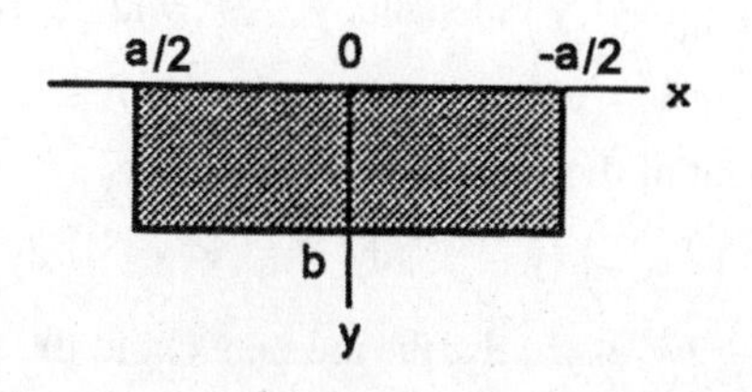

45. When the water reaches the top of the tank the force on the movable side is $\int_{-2}^{0} (62.4)\left(2\sqrt{4-y^2}\right)(-y)\,dy$ $= (62.4)\int_{-2}^{0} (4-y^2)^{1/2}(-2y)\,dy = (62.4)\left[\frac{2}{3}(4-y^2)^{3/2}\right]_{-2}^{0} = (62.4)\left(\frac{2}{3}\right)\left(4^{3/2}\right) = 332.8$ ft · lb. The force compressing the spring is $F = 100x$, so when the tank is full we have $332.8 = 100x \Rightarrow x \approx 3.33$ ft. Therefore the movable end does not reach the required 5 ft to allow drainage $\Rightarrow$ the tank will overflow.

6.6 MOMENTS AND CENTERS OF MASS

1. Since the plate is symmetric about the y-axis and its density is constant, the distribution of mass is symmetric about the y-axis and the center of mass lies on the y-axis. This means that $\bar{x} = 0$. It remains to find $\bar{y} = \frac{M_x}{M}$. We model the distribution of mass with *vertical* strips. The typical strip has center of mass: $(\tilde{x}, \tilde{y}) = \left(x, \frac{x^2+4}{2}\right)$, length: $4 - x^2$, width: dx, area:

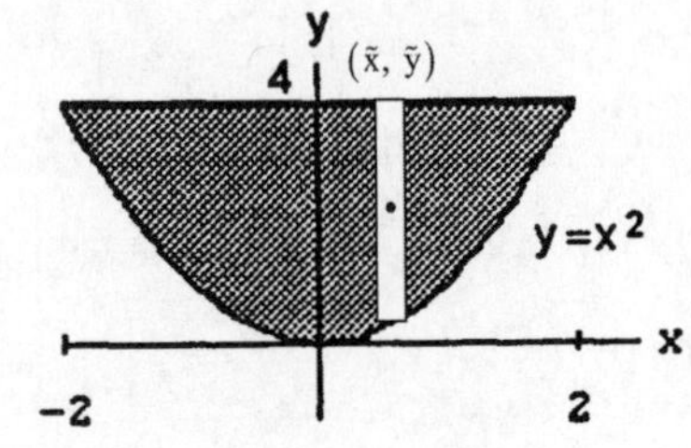

$dA = (4 - x^2)\,dx$, mass: $dm = \delta\,dA = \delta(4 - x^2)\,dx$. The moment of the strip about the x-axis is $\tilde{y}\,dm = \left(\frac{x^2+4}{2}\right)\delta(4 - x^2)\,dx = \frac{\delta}{2}(16 - x^4)\,dx$. The moment of the plate about the x-axis is $M_x = \int \tilde{y}\,dm$ $= \int_{-2}^{2} \frac{\delta}{2}(16 - x^4)\,dx = \frac{\delta}{2}\left[16x - \frac{x^5}{5}\right]_{-2}^{2} = \frac{\delta}{2}\left[\left(16 \cdot 2 - \frac{2^5}{5}\right) - \left(-16 \cdot 2 + \frac{2^5}{5}\right)\right] = \frac{\delta \cdot 2}{2}\left(32 - \frac{32}{5}\right) = \frac{128\delta}{5}$. The mass of the plate is $M = \int \delta(4 - x^2)\,dx = \delta\left[4x - \frac{x^3}{3}\right]_{-2}^{2} = 2\delta\left(8 - \frac{8}{3}\right) = \frac{32\delta}{3}$. Therefore $\bar{y} = \frac{M_x}{M} = \frac{\left(\frac{128\delta}{5}\right)}{\left(\frac{32\delta}{3}\right)} = \frac{12}{5}$. The plate's center of mass is the point $(\bar{x}, \bar{y}) = \left(0, \frac{12}{5}\right)$.

3. Intersection points: $x - x^2 = -x \Rightarrow 2x - x^2 = 0$ $\Rightarrow x(2 - x) = 0 \Rightarrow x = 0$ or $x = 2$. The typical *vertical* strip has center of mass: $(\tilde{x}, \tilde{y}) = \left(x, \frac{(x - x^2) + (-x)}{2}\right)$ $= \left(x, -\frac{x^2}{2}\right)$, length: $(x - x^2) - (-x) = 2x - x^2$, width: dx, area: $dA = (2x - x^2)\,dx$, mass: $dm = \delta\,dA = \delta(2x - x^2)\,dx$. The moment of the strip about the x-axis is $\tilde{y}\,dm = \left(-\frac{x^2}{2}\right)\delta(2x - x^2)\,dx$; about the y-axis it is $\tilde{x}\,dm = x \cdot \delta(2x - x^2)\,dx$. Thus, $M_x = \int \tilde{y}\,dm$ $= -\int_0^2 \left(\frac{\delta}{2}x^2\right)(2x - x^2)\,dx = -\frac{\delta}{2}\int_0^2 (2x^3 - x^4)\,dx = -\frac{\delta}{2}\left[\frac{x^4}{2} - \frac{x^5}{5}\right]_0^2 = -\frac{\delta}{2}\left(2^3 - \frac{2^5}{5}\right) = -\frac{\delta}{2} \cdot 2^3\left(1 - \frac{4}{5}\right)$ $= -\frac{4\delta}{5}$; $M_y = \int \tilde{x}\,dm = \int_0^2 x \cdot \delta(2x - x^2)\,dx = \delta\int_0^2 (2x^2 - x^3) = \delta\left[\frac{2}{3}x^3 - \frac{x^4}{4}\right]_0^2 = \delta\left(2 \cdot \frac{2^3}{3} - \frac{2^4}{4}\right) = \frac{\delta \cdot 2^4}{12} = \frac{4\delta}{3}$; $M = \int dm = \int_0^2 \delta(2x - x^2)\,dx = \delta\int_0^2 (2x - x^2)\,dx = \delta\left[x^2 - \frac{x^3}{3}\right]_0^2 = \delta\left(4 - \frac{8}{3}\right) = \frac{4\delta}{3}$. Therefore, $\bar{x} = \frac{M_y}{M}$ $= \left(\frac{4\delta}{3}\right)\left(\frac{3}{4\delta}\right) = 1$ and $\bar{y} = \frac{M_x}{M} = \left(-\frac{4\delta}{5}\right)\left(\frac{3}{4\delta}\right) = -\frac{3}{5} \Rightarrow (\bar{x}, \bar{y}) = \left(1, -\frac{3}{5}\right)$ is the center of mass.

5. The typical *horizontal* strip has center of mass: $(\tilde{x}, \tilde{y}) = \left(\frac{y - y^3}{2}, y\right)$, length: $y - y^3$, width: dy, area: $dA = (y - y^3)\,dy$, mass: $dm = \delta\,dA = \delta(y - y^3)\,dy$. The moment of the strip about the y-axis is $\tilde{x}\,dm = \delta\left(\frac{y - y^3}{2}\right)(y - y^3)\,dy = \frac{\delta}{2}(y - y^3)^2\,dy$ $= \frac{\delta}{2}(y^2 - 2y^4 + y^6)\,dy$; the moment about the x-axis is

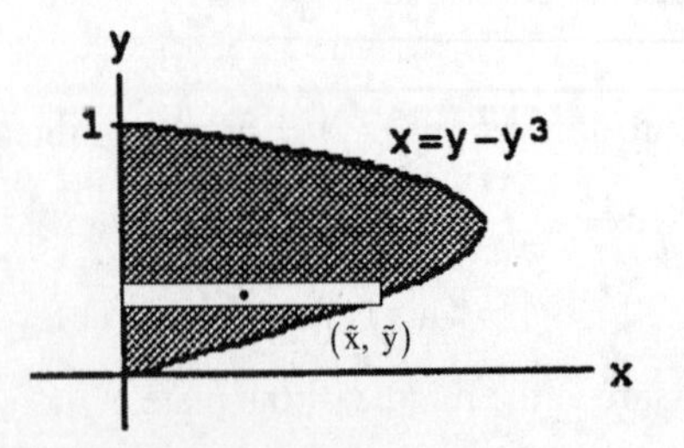

$\tilde{y}\,dm = \delta y\,(y - y^3)\,dy = \delta\,(y^2 - y^4)\,dy$. Thus, $M_x = \int \tilde{y}\,dm = \delta \int_0^1 (y^2 - y^4)\,dy = \delta\left[\frac{y^3}{3} - \frac{y^5}{5}\right]_0^1 = \delta\left(\frac{1}{3} - \frac{1}{5}\right) = \frac{2\delta}{15}$;
$M_y = \int \tilde{x}\,dm = \frac{\delta}{2}\int_0^1 (y^2 - 2y^4 + y^6)\,dy = \frac{\delta}{2}\left[\frac{y^3}{3} - \frac{2y^5}{5} + \frac{y^7}{7}\right]_0^1 = \frac{\delta}{2}\left(\frac{1}{3} - \frac{2}{5} + \frac{1}{7}\right) = \frac{\delta}{2}\left(\frac{35 - 42 + 15}{3\cdot 5\cdot 7}\right) = \frac{4\delta}{105}$; $M = \int dm$
$= \delta\int_0^1 (y - y^3)\,dy = \delta\left[\frac{y^2}{2} - \frac{y^4}{4}\right]_0^1 = \delta\left(\frac{1}{2} - \frac{1}{4}\right) = \frac{\delta}{4}$. Therefore, $\bar{x} = \frac{M_y}{M} = \left(\frac{4\delta}{105}\right)\left(\frac{4}{\delta}\right) = \frac{16}{105}$ and $\bar{y} = \frac{M_x}{M} = \left(\frac{2\delta}{15}\right)\left(\frac{4}{\delta}\right)$
$= \frac{8}{15} \Rightarrow (\bar{x}, \bar{y}) = \left(\frac{16}{105}, \frac{8}{15}\right)$ is the center of mass.

7. Applying the symmetry argument analogous to the onc used in Exercise 1, we find $\bar{x} = 0$. The typical *vertical* strip has center of mass: $(\tilde{x}, \tilde{y}) = \left(x, \frac{\cos x}{2}\right)$, length: $\cos x$, width: dx, area: $dA = \cos x\,dx$, mass: $dm = \delta\,dA = \delta \cos x\,dx$. The moment of the strip about the x-axis is $\tilde{y}\,dm = \delta \cdot \frac{\cos x}{2} \cdot \cos x\,dx$ $= \frac{\delta}{2}\cos^2 x\,dx = \frac{\delta}{2}\left(\frac{1 + \cos 2x}{2}\right)dx = \frac{\delta}{4}(1 + \cos 2x)\,dx$; thus,

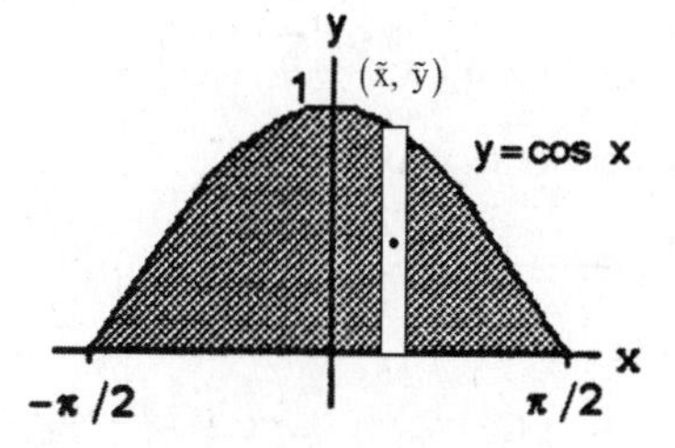

$M_x = \int \tilde{y}\,dm = \int_{-\pi/2}^{\pi/2} \frac{\delta}{4}(1 + \cos 2x)\,dx = \frac{\delta}{4}\left[x + \frac{\sin 2x}{2}\right]_{-\pi/2}^{\pi/2} = \frac{\delta}{4}\left[\left(\frac{\pi}{2} + 0\right) - \left(-\frac{\pi}{2}\right)\right] = \frac{\delta\pi}{4}$; $M = \int dm = \delta\int_{-\pi/2}^{\pi/2} \cos x\,dx$
$= \delta[\sin x]_{-\pi/2}^{\pi/2} = 2\delta$. Therefore, $\bar{y} = \frac{M_x}{M} = \frac{\delta\pi}{4\cdot 2\delta} = \frac{\pi}{8} \Rightarrow (\bar{x}, \bar{y}) = \left(0, \frac{\pi}{8}\right)$ is the center of mass.

9. Since the plate is symmetric about the line $x = 1$ and its density is constant, the distribution of mass is symmetric about this line and the center of mass lies on it. This means that $\bar{x} = 1$. The typical *vertical* strip has center of mass: $(\tilde{x}, \tilde{y}) = \left(x, \frac{(2x - x^2) + (2x^2 - 4x)}{2}\right) = \left(x, \frac{x^2 - 2x}{2}\right)$, length: $(2x - x^2) - (2x^2 - 4x) = -3x^2 + 6x = 3(2x - x^2)$, width: dx, area: $dA = 3(2x - x^2)\,dx$, mass: $dm = \delta\,dA$ $= 3\delta(2x - x^2)\,dx$. The moment about the x-axis is $\tilde{y}\,dm = \frac{3}{2}\delta(x^2 - 2x)(2x - x^2)\,dx = -\frac{3}{2}\delta(x^2 - 2x)^2\,dx$
$= -\frac{3}{2}\delta(x^4 - 4x^3 + 4x^2)\,dx$. Thus, $M_x = \int \tilde{y}\,dm = -\int_0^2 \frac{3}{2}\delta(x^4 - 4x^3 + 4x^2)\,dx = -\frac{3}{2}\delta\left[\frac{x^5}{5} - x^4 + \frac{4}{3}x^3\right]_0^2$
$= -\frac{3}{2}\delta\left(\frac{2^5}{5} - 2^4 + \frac{4}{3}\cdot 2^3\right) = -\frac{3}{2}\delta\cdot 2^4\left(\frac{2}{5} - 1 + \frac{2}{3}\right) = -\frac{3}{2}\delta\cdot 2^4\left(\frac{6 - 15 + 10}{15}\right) = -\frac{8\delta}{5}$; $M = \int dm$
$= \int_0^2 3\delta(2x - x^2)\,dx = 3\delta\left[x^2 - \frac{x^3}{3}\right]_0^2 = 3\delta\left(4 - \frac{8}{3}\right) = 4\delta$. Therefore, $\bar{y} = \frac{M_x}{M} = \left(-\frac{8\delta}{5}\right)\left(\frac{1}{4\delta}\right) = -\frac{2}{5}$
$\Rightarrow (\bar{x}, \bar{y}) = \left(1, -\frac{2}{5}\right)$ is the center of mass.

(b) Applying the symmetry argument analogous to the one used in Exercise 13, we find that $\bar{x} = 0$. The typical vertical strip has the same parameters as in part (a). Thus, $M_x = \int \tilde{y}\,dm = \int_{-3}^{3} \frac{\delta}{2}(9 - x^2)\,dx$
$= 2\int_0^3 \frac{\delta}{2}(9 - x^2)\,dx = 2(9\delta) = 18\delta$;
$M = \int dm = \int \delta\,dA = \delta\int dA$

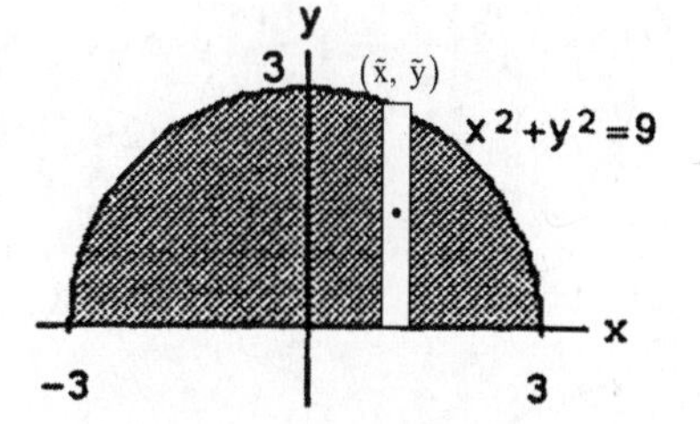

$= \delta$(Area of a semi-circle of radius 3) $= \delta\left(\frac{9\pi}{2}\right) = \frac{9\pi\delta}{2}$. Therefore, $\bar{y} = \frac{M_x}{M} = (18\delta)\left(\frac{2}{9\pi\delta}\right) = \frac{4}{\pi}$, the same $\bar{y}$ as in part (a) $\Rightarrow (\bar{x}, \bar{y}) = \left(0, \frac{4}{\pi}\right)$ is the center of mass.

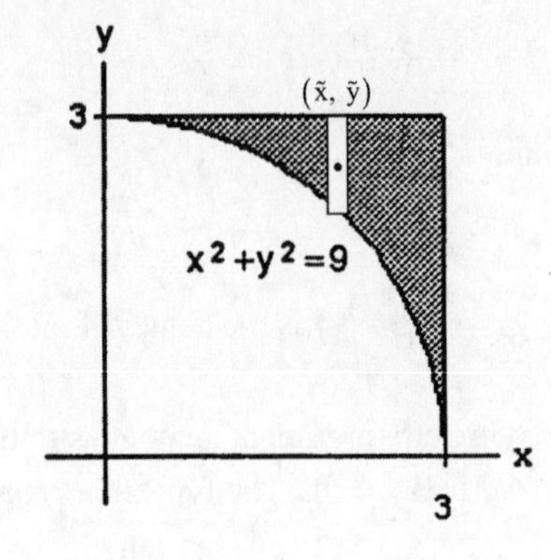

11. Since the plate is symmetric about the line $x = y$ and its density is constant, the distribution of mass is symmetric about this line. This means that $\bar{x} = \bar{y}$. The typical *vertical* strip has

center of mass: $(\tilde{x}, \tilde{y}) = \left(x, \frac{3+\sqrt{9-x^2}}{2}\right)$,

length: $3 - \sqrt{9-x^2}$, width: dx,

area: $dA = \left(3 - \sqrt{9-x^2}\right) dx$,

mass: $dm = \delta\, dA = \delta\left(3 - \sqrt{9-x^2}\right) dx$.

The moment about the x-axis is

$\tilde{y}\, dm = \delta \frac{\left(3+\sqrt{9-x^2}\right)\left(3-\sqrt{9-x^2}\right)}{2} dx = \frac{\delta}{2}\left[9 - \left(9 - x^2\right)\right] dx = \frac{\delta x^2}{2} dx$. Thus, $M_x = \int_0^3 \frac{\delta x^2}{2} dx = \frac{\delta}{6}\left[x^3\right]_0^3 = \frac{9\delta}{2}$. The area equals the area of a square with side length 3 minus one quarter the area of a disk with radius 3 $\Rightarrow A = 3^2 - \frac{\pi 9}{4}$ $= \frac{9}{4}(4 - \pi) \Rightarrow M = \delta A = \frac{9\delta}{4}(4 - \pi)$. Therefore, $\bar{y} = \frac{M_x}{M} = \left(\frac{9\delta}{2}\right)\left[\frac{4}{9\delta(4-\pi)}\right] = \frac{2}{4-\pi} \Rightarrow (\bar{x}, \bar{y}) = \left(\frac{2}{4-\pi}, \frac{2}{4-\pi}\right)$ is the center of mass.

13. $M_x = \int \tilde{y}\, dm = \int_1^2 \frac{\left(\frac{2}{x^2}\right)}{2} \cdot \delta \cdot \left(\frac{2}{x^2}\right) dx$
$= \int_1^2 \left(\frac{1}{x^2}\right)\left(x^2\right)\left(\frac{2}{x^2}\right) dx = \int_1^2 \frac{2}{x^2} dx = 2\int_1^2 x^{-2} dx$
$= 2\left[-x^{-1}\right]_1^2 = 2\left[\left(-\frac{1}{2}\right) - (-1)\right] = 2\left(\frac{1}{2}\right) = 1;$
$M_y = \int \tilde{x}\, dm = \int_1^2 x \cdot \delta \cdot \left(\frac{2}{x^2}\right) dx$
$= \int_1^2 x\left(x^2\right)\left(\frac{2}{x^2}\right) dx = 2\int_1^2 x\, dx = 2\left[\frac{x^2}{2}\right]_1^2$
$= 2\left(2 - \frac{1}{2}\right) = 4 - 1 = 3$; $M = \int dm = \int_1^2 \delta\left(\frac{2}{x^2}\right) dx = \int_1^2 x^2\left(\frac{2}{x^2}\right) dx = 2\int_1^2 dx = 2[x]_1^2 = 2(2-1) = 2$. So $\bar{x} = \frac{M_y}{M} = \frac{3}{2}$ and $\bar{y} = \frac{M_x}{M} = \frac{1}{2} \Rightarrow (\bar{x}, \bar{y}) = \left(\frac{3}{2}, \frac{1}{2}\right)$ is the center of mass.

15. (a) We use the shell method: $V = \int_a^b 2\pi \left(\begin{smallmatrix}\text{shell}\\ \text{radius}\end{smallmatrix}\right)\left(\begin{smallmatrix}\text{shell}\\ \text{height}\end{smallmatrix}\right) dx = \int_1^4 2\pi x\left[\frac{4}{\sqrt{x}} - \left(-\frac{4}{\sqrt{x}}\right)\right] dx = 16\pi \int_1^4 \frac{x}{\sqrt{x}} dx$
$= 16\pi \int_1^4 x^{1/2} dx = 16\pi\left[\frac{2}{3}x^{3/2}\right]_1^4 = 16\pi\left(\frac{2}{3}\cdot 8 - \frac{2}{3}\right) = \frac{32\pi}{3}(8-1) = \frac{224\pi}{3}$

(b) Since the plate is symmetric about the x-axis and its density $\delta(x) = \frac{1}{x}$ is a function of x alone, the distribution of its mass is symmetric about the x-axis. This means that $\bar{y} = 0$. We use the vertical strip approach to find $\bar{x}$:
$M_y = \int \tilde{x}\, dm = \int_1^4 x \cdot \left[\frac{4}{\sqrt{x}} - \left(-\frac{4}{\sqrt{x}}\right)\right] \cdot \delta\, dx = \int_1^4 x \cdot \frac{8}{\sqrt{x}} \cdot \frac{1}{x}\, dx = 8\int_1^4 x^{-1/2} dx = 8\left[2x^{1/2}\right]_1^4 = 8(2\cdot 2 - 2) = 16;$
$M = \int dm = \int_1^4 \left[\frac{4}{\sqrt{x}} - \left(\frac{-4}{\sqrt{x}}\right)\right] \cdot \delta\, dx = 8\int_1^4 \left(\frac{1}{\sqrt{x}}\right)\left(\frac{1}{x}\right) dx = 8\int_1^4 x^{-3/2} dx = 8\left[-2x^{-1/2}\right]_1^4 = 8[-1-(-2)] = 8.$
So $\bar{x} = \frac{M_y}{M} = \frac{16}{8} = 2 \Rightarrow (\bar{x}, \bar{y}) = (2, 0)$ is the center of mass.

(c)

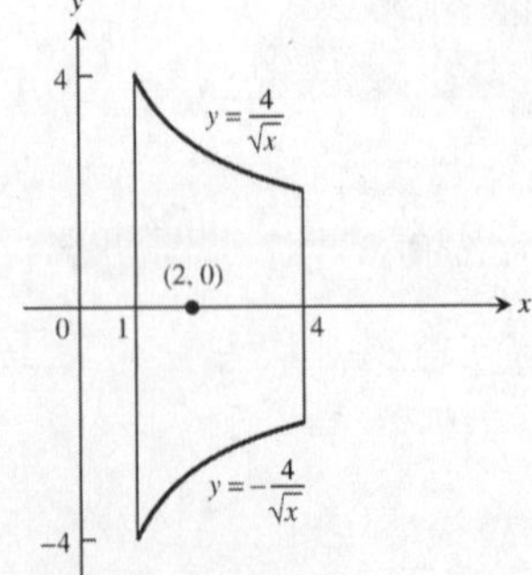

17. The mass of a horizontal strip is $dm = \delta\, dA = \delta L\, dy$, where L is the width of the triangle at a distance of y above its base on the x-axis as shown in the figure in the text. Also, by similar triangles we have $\frac{L}{b} = \frac{h-y}{h}$ $\Rightarrow L = \frac{b}{h}(h-y)$. Thus, $M_x = \int \tilde{y}\, dm = \int_0^h \delta y \left(\frac{b}{h}\right)(h-y)\, dy = \frac{\delta b}{h}\int_0^h (hy - y^2)\, dy = \frac{\delta b}{h}\left[\frac{hy^2}{2} - \frac{y^3}{3}\right]_0^h$ $= \frac{\delta b}{h}\left(\frac{h^3}{2} - \frac{h^3}{3}\right) = \delta b h^2\left(\frac{1}{2} - \frac{1}{3}\right) = \frac{\delta b h^2}{6}$; $M = \int dm = \int_0^h \delta\left(\frac{b}{h}\right)(h-y)\, dy = \frac{\delta b}{h}\int_0^h (h-y)\, dy = \frac{\delta b}{h}\left[hy - \frac{y^2}{2}\right]_0^h$ $= \frac{\delta b}{h}\left(h^2 - \frac{h^2}{2}\right) = \frac{\delta b h}{2}$. So $\bar{y} = \frac{M_x}{M} = \left(\frac{\delta b h^2}{6}\right)\left(\frac{2}{\delta b h}\right) = \frac{h}{3}$ $\Rightarrow$ the center of mass lies above the base of the triangle one-third of the way toward the opposite vertex. Similarly the other two sides of the triangle can be placed on the x-axis and the same results will occur. Therefore the centroid does lie at the intersection of the medians, as claimed.

19. From the symmetry about the line $x = y$ it follows that $\bar{x} = \bar{y}$. It also follows that the line through the points $(0,0)$ and $\left(\frac{1}{2}, \frac{1}{2}\right)$ is a median $\Rightarrow \bar{y} = \bar{x} = \frac{2}{3}\cdot\left(\frac{1}{2} - 0\right) = \frac{1}{3}$ $\Rightarrow (\bar{x}, \bar{y}) = \left(\frac{1}{3}, \frac{1}{3}\right)$.

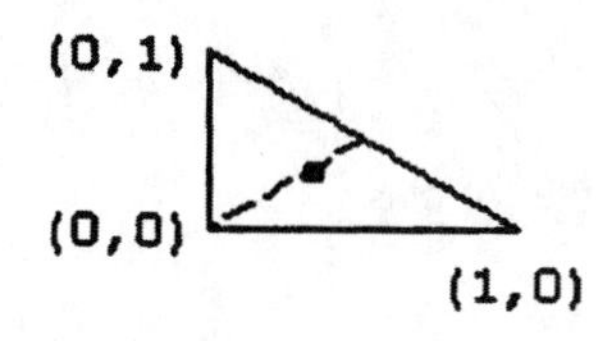

21. The point of intersection of the median from the vertex $(0, b)$ to the opposite side has coordinates $\left(0, \frac{a}{2}\right)$ $\Rightarrow \bar{y} = (b-0)\cdot\frac{1}{3} = \frac{b}{3}$ and $\bar{x} = \left(\frac{a}{2} - 0\right)\cdot\frac{2}{3} = \frac{a}{3}$ $\Rightarrow (\bar{x}, \bar{y}) = \left(\frac{a}{3}, \frac{b}{3}\right)$.

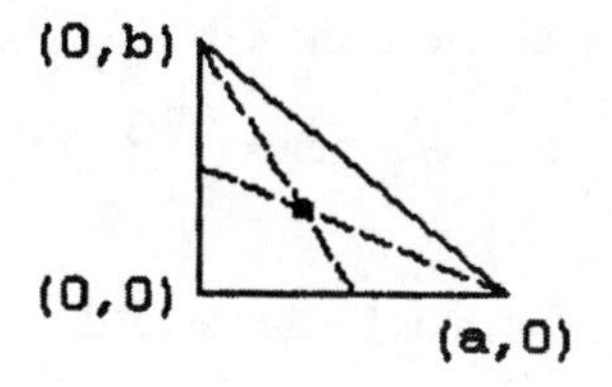

23. $y = x^{1/2} \Rightarrow dy = \frac{1}{2}x^{-1/2}\, dx$

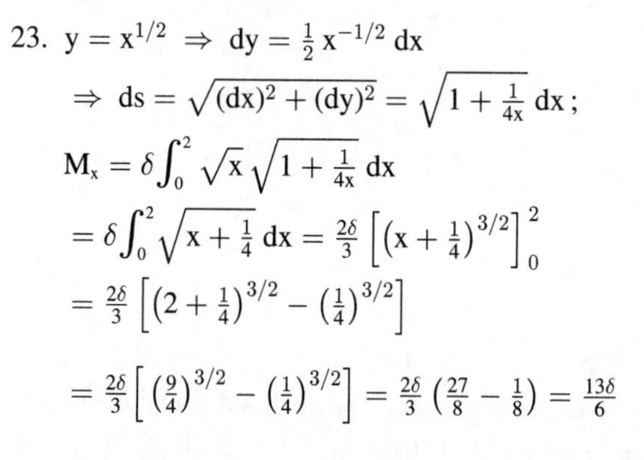

$\Rightarrow ds = \sqrt{(dx)^2 + (dy)^2} = \sqrt{1 + \frac{1}{4x}}\, dx$;

$M_x = \delta\int_0^2 \sqrt{x}\sqrt{1 + \frac{1}{4x}}\, dx$

$= \delta\int_0^2 \sqrt{x + \frac{1}{4}}\, dx = \frac{2\delta}{3}\left[\left(x + \frac{1}{4}\right)^{3/2}\right]_0^2$

$= \frac{2\delta}{3}\left[\left(2 + \frac{1}{4}\right)^{3/2} - \left(\frac{1}{4}\right)^{3/2}\right]$

$= \frac{2\delta}{3}\left[\left(\frac{9}{4}\right)^{3/2} - \left(\frac{1}{4}\right)^{3/2}\right] = \frac{2\delta}{3}\left(\frac{27}{8} - \frac{1}{8}\right) = \frac{13\delta}{6}$

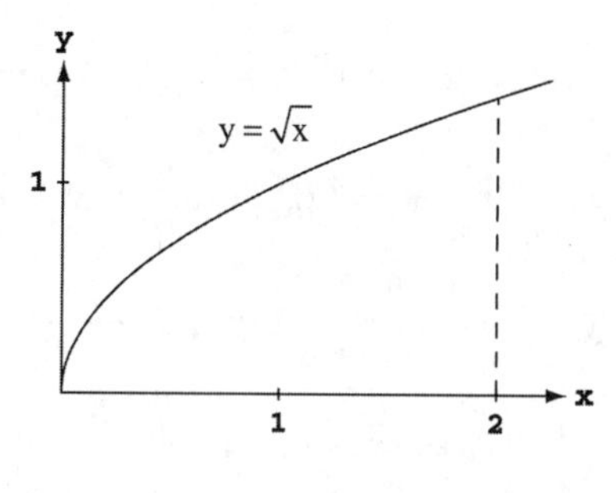

25. From Example 4 we have $M_x = \int_0^\pi a(a\sin\theta)(k\sin\theta)\, d\theta = a^2k\int_0^\pi \sin^2\theta\, d\theta = \frac{a^2k}{2}\int_0^\pi (1 - \cos 2\theta)\, d\theta = \frac{a^2k}{2}\left[\theta - \frac{\sin 2\theta}{2}\right]_0^\pi$ $= \frac{a^2k\pi}{2}$; $M_y = \int_0^\pi a(a\cos\theta)(k\sin\theta)\, d\theta = a^2k\int_0^\pi \sin\theta\cos\theta\, d\theta = \frac{a^2k}{2}\left[\sin^2\theta\right]_0^\pi = 0$; $M = \int_0^\pi ak\sin\theta\, d\theta = ak[-\cos\theta]_0^\pi$ $= 2ak$. Therefore, $\bar{x} = \frac{M_y}{M} = 0$ and $\bar{y} = \frac{M_x}{M} = \left(\frac{a^2k\pi}{2}\right)\left(\frac{1}{2ak}\right) = \frac{a\pi}{4} \Rightarrow \left(0, \frac{a\pi}{4}\right)$ is the center of mass.

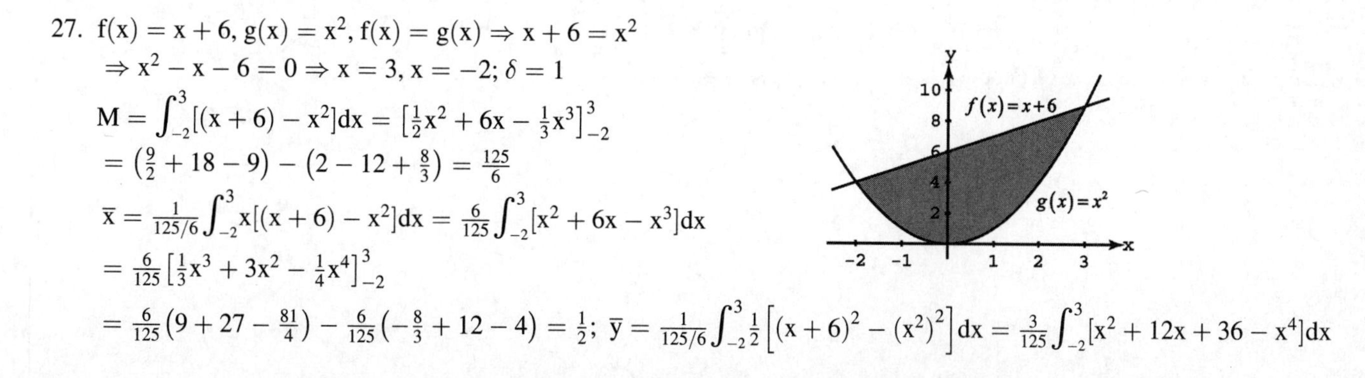

27. $f(x) = x + 6$, $g(x) = x^2$, $f(x) = g(x) \Rightarrow x + 6 = x^2$

$\Rightarrow x^2 - x - 6 = 0 \Rightarrow x = 3, x = -2; \delta = 1$

$M = \int_{-2}^3 [(x+6) - x^2]dx = \left[\frac{1}{2}x^2 + 6x - \frac{1}{3}x^3\right]_{-2}^3$

$= \left(\frac{9}{2} + 18 - 9\right) - \left(2 - 12 + \frac{8}{3}\right) = \frac{125}{6}$

$\bar{x} = \frac{1}{125/6}\int_{-2}^3 x[(x+6) - x^2]dx = \frac{6}{125}\int_{-2}^3 [x^2 + 6x - x^3]dx$

$= \frac{6}{125}\left[\frac{1}{3}x^3 + 3x^2 - \frac{1}{4}x^4\right]_{-2}^3$

$= \frac{6}{125}\left(9 + 27 - \frac{81}{4}\right) - \frac{6}{125}\left(-\frac{8}{3} + 12 - 4\right) = \frac{1}{2}$; $\bar{y} = \frac{1}{125/6}\int_{-2}^3 \frac{1}{2}\left[(x+6)^2 - (x^2)^2\right]dx = \frac{3}{125}\int_{-2}^3 [x^2 + 12x + 36 - x^4]dx$

$= \frac{3}{125}\left[\frac{1}{3}x^3 + 6x^2 + 36x - \frac{1}{5}x^5\right]_{-2}^{3} = \frac{3}{125}\left(9 + 54 + 108 - \frac{243}{5}\right) - \frac{3}{125}\left(-\frac{8}{3} + 24 - 72 + \frac{32}{5}\right) = 4$

$\Rightarrow \left(\frac{1}{2}, 4\right)$ is the center of mass.

29. $f(x) = x^2, g(x) = x^2(x-1), f(x) = g(x) \Rightarrow x^2 = x^2(x-1)$

$\Rightarrow x^3 - 2x^2 = 0 \Rightarrow x = 0, x = 2; \delta = 1$

$M = \int_0^2 [x^2 - x^2(x-1)]dx = \int_0^2 [2x^2 - x^3]dx$

$= \left[\frac{2}{3}x^3 - \frac{1}{4}x^4\right]_0^2 = \left(\frac{16}{3} - 4\right) - 0 = \frac{4}{3}$

$\overline{x} = \frac{1}{4/3}\int_0^2 x[x^2 - x^2(x-1)]dx = \frac{3}{4}\int_0^2 [2x^3 - x^4]dx$

$= \frac{3}{4}\left[\frac{1}{2}x^4 - \frac{1}{5}x^5\right]_0^2 = \frac{3}{4}\left(8 - \frac{32}{5}\right) - 0 = \frac{6}{5};$

$\overline{y} = \frac{1}{4/3}\int_0^2 \frac{1}{2}\left[(x^2)^2 - (x^2(x-1))^2\right]dx = \frac{3}{8}\int_0^2 [2x^5 - x^6]dx = \frac{3}{8}\left[\frac{1}{3}x^6 - \frac{1}{7}x^7\right]_0^2 = \frac{3}{8}\left(\frac{64}{3} - \frac{128}{7}\right) - 0 = \frac{8}{7}$

$\Rightarrow \left(\frac{6}{5}, \frac{8}{7}\right)$ is the center of mass.

31. Consider the curve as an infinite number of line segments joined together. From the derivation of arc length we have that the length of a particular segment is $ds = \sqrt{(dx)^2 + (dy)^2}$. This implies that $M_x = \int \delta y\, ds$, $M_y = \int \delta x\, ds$ and $M = \int \delta\, ds$. If δ is constant, then $\overline{x} = \frac{M_y}{M} = \frac{\int x\, ds}{\int ds} = \frac{\int x\, ds}{\text{length}}$ and $\overline{y} = \frac{M_x}{M} = \frac{\int y\, ds}{\int ds} = \frac{\int y\, ds}{\text{length}}$.

33. The centroid of the square is located at $(2, 2)$. The volume is $V = (2\pi)(\overline{y})(A) = (2\pi)(2)(8) = 32\pi$ and the surface area is $S = (2\pi)(\overline{y})(L) = (2\pi)(2)\left(4\sqrt{8}\right) = 32\sqrt{2}\pi$ (where $\sqrt{8}$ is the length of a side).

35. The centroid is located at $(2, 0) \Rightarrow V = (2\pi)(\overline{x})(A) = (2\pi)(2)(\pi) = 4\pi^2$

37. $S = 2\pi\,\overline{y}\,L \Rightarrow 4\pi a^2 = (2\pi\overline{y})(\pi a) \Rightarrow \overline{y} = \frac{2a}{\pi}$, and by symmetry $\overline{x} = 0$

39. $V = 2\pi\,\overline{y}A \Rightarrow \frac{4}{3}\pi ab^2 = (2\pi\overline{y})\left(\frac{\pi ab}{2}\right) \Rightarrow \overline{y} = \frac{4b}{3\pi}$ and by symmetry $\overline{x} = 0$

41. $V = 2\pi\rho\, A = (2\pi)$(area of the region) $\cdot$ (distance from the centroid to the line $y = x - a$). We must find the distance from $\left(0, \frac{4a}{3\pi}\right)$ to $y = x - a$. The line containing the centroid and perpendicular to $y = x - a$ has slope -1 and contains the point $\left(0, \frac{4a}{3\pi}\right)$. This line is $y = -x + \frac{4a}{3\pi}$. The intersection of $y = x - a$ and $y = -x + \frac{4a}{3\pi}$ is the point $\left(\frac{4a + 3a\pi}{6\pi}, \frac{4a - 3a\pi}{6\pi}\right)$. Thus, the distance from the centroid to the line $y = x - a$ is $\sqrt{\left(\frac{4a + 3a\pi}{6\pi}\right)^2 + \left(\frac{4a}{3\pi} - \frac{4a}{6\pi} + \frac{3a\pi}{6\pi}\right)^2} = \frac{\sqrt{2}(4a + 3a\pi)}{6\pi}$

$\Rightarrow V = (2\pi)\left(\frac{\sqrt{2}(4a + 3a\pi)}{6\pi}\right)\left(\frac{\pi a^2}{2}\right) = \frac{\sqrt{2}\pi a^3(4 + 3\pi)}{6}$

43. If we revolve the region about the y-axis: $r = a, h = b \Rightarrow A = \frac{1}{2}ab, V = \frac{1}{3}\pi a^2 b$, and $\rho = \overline{x}$. By the Theorem of Pappus: $\frac{1}{3}\pi a^2 b = 2\pi\,\overline{x}\left(\frac{1}{2}ab\right) \Rightarrow \overline{x} = \frac{a}{3}$; If we revolve the region about the x-axis: $r = b, h = a \Rightarrow A = \frac{1}{2}ab, V = \frac{1}{3}\pi b^2 a$, and $\rho = \overline{y}$. By the Theorem of Pappus: $\frac{1}{3}\pi\, b^2 a = 2\pi\,\overline{y}\left(\frac{1}{2}ab\right) \Rightarrow \overline{y} = \frac{b}{3} \Rightarrow \left(\frac{a}{3}, \frac{b}{3}\right)$ is the center of mass.

CHAPTER 6 PRACTICE EXERCISES

1. $A(x) = \frac{\pi}{4}(\text{diameter})^2 = \frac{\pi}{4}\left(\sqrt{x} - x^2\right)^2$
$= \frac{\pi}{4}\left(x - 2\sqrt{x}\cdot x^2 + x^4\right);\ a = 0,\ b = 1$
$\Rightarrow V = \int_a^b A(x)\,dx = \frac{\pi}{4}\int_0^1 \left(x - 2x^{5/2} + x^4\right) dx$
$= \frac{\pi}{4}\left[\frac{x^2}{2} - \frac{4}{7}x^{7/2} + \frac{x^5}{5}\right]_0^1 = \frac{\pi}{4}\left(\frac{1}{2} - \frac{4}{7} + \frac{1}{5}\right)$
$= \frac{\pi}{4\cdot 70}(35 - 40 + 14) = \frac{9\pi}{280}$

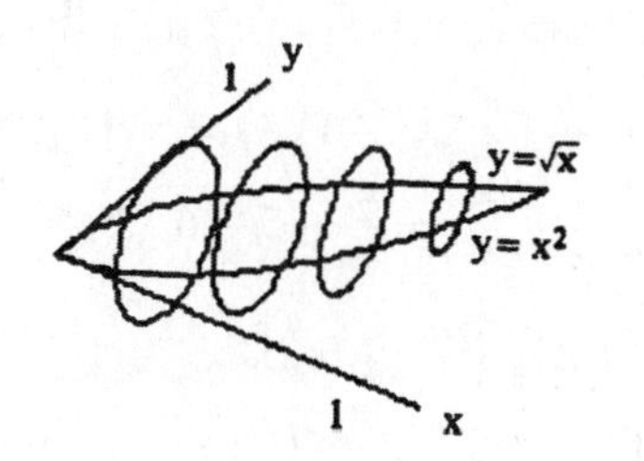

3. $A(x) = \frac{\pi}{4}(\text{diameter})^2 = \frac{\pi}{4}(2\sin x - 2\cos x)^2$
$= \frac{\pi}{4}\cdot 4\left(\sin^2 x - 2\sin x\cos x + \cos^2 x\right)$
$= \pi(1 - \sin 2x);\ a = \frac{\pi}{4},\ b = \frac{5\pi}{4}$
$\Rightarrow V = \int_a^b A(x)\,dx = \pi\int_{\pi/4}^{5\pi/4}(1 - \sin 2x)\,dx$
$= \pi\left[x + \frac{\cos 2x}{2}\right]_{\pi/4}^{5\pi/4}$
$= \pi\left[\left(\frac{5\pi}{4} + \frac{\cos\frac{5\pi}{2}}{2}\right) - \left(\frac{\pi}{4} - \frac{\cos\frac{\pi}{2}}{2}\right)\right] = \pi^2$

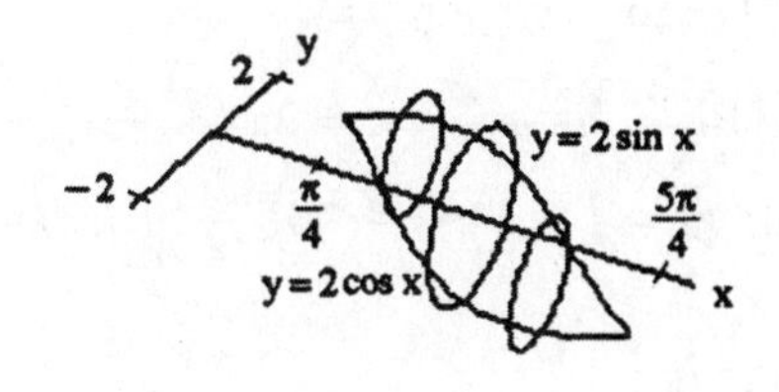

5. $A(x) = \frac{\pi}{4}(\text{diameter})^2 = \frac{\pi}{4}\left(2\sqrt{x} - \frac{x^2}{4}\right)^2 = \frac{\pi}{4}\left(4x - x^{5/2} + \frac{x^4}{16}\right);\ a = 0,\ b = 4 \Rightarrow V = \int_a^b A(x)\,dx$
$= \frac{\pi}{4}\int_0^4 \left(4x - x^{5/2} + \frac{x^4}{16}\right) dx = \frac{\pi}{4}\left[2x^2 - \frac{2}{7}x^{7/2} + \frac{x^5}{5\cdot 16}\right]_0^4 = \frac{\pi}{4}\left(32 - 32\cdot\frac{8}{7} + \frac{2}{5}\cdot 32\right)$
$= \frac{32\pi}{4}\left(1 - \frac{8}{7} + \frac{2}{5}\right) = \frac{8\pi}{35}(35 - 40 + 14) = \frac{72\pi}{35}$

7. (a) *disk method:*
$V = \int_a^b \pi R^2(x)\,dx = \int_{-1}^1 \pi\left(3x^4\right)^2 dx = \pi\int_{-1}^1 9x^8\,dx$
$= \pi\left[x^9\right]_{-1}^1 = 2\pi$

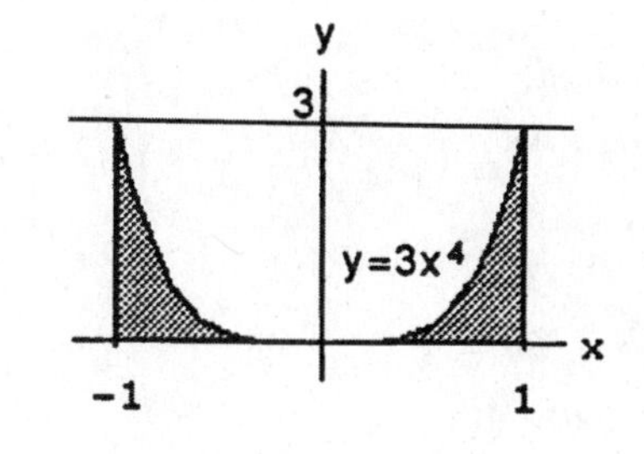

(b) *shell method:*
$V = \int_a^b 2\pi\left(\substack{\text{shell}\\ \text{radius}}\right)\left(\substack{\text{shell}\\ \text{height}}\right) dx = \int_0^1 2\pi x\left(3x^4\right) dx = 2\pi\cdot 3\int_0^1 x^5\,dx = 2\pi\cdot 3\left[\frac{x^6}{6}\right]_0^1 = \pi$
Note: The lower limit of integration is 0 rather than -1.

(c) *shell method:*
$V = \int_a^b 2\pi\left(\substack{\text{shell}\\ \text{radius}}\right)\left(\substack{\text{shell}\\ \text{height}}\right) dx = 2\pi\int_{-1}^1 (1 - x)\left(3x^4\right) dx = 2\pi\left[\frac{3x^5}{5} - \frac{x^6}{2}\right]_{-1}^1 = 2\pi\left[\left(\frac{3}{5} - \frac{1}{2}\right) - \left(-\frac{3}{5} - \frac{1}{2}\right)\right] = \frac{12\pi}{5}$

(d) *washer method:*
$R(x) = 3,\ r(x) = 3 - 3x^4 = 3\left(1 - x^4\right) \Rightarrow V = \int_a^b \pi\left[R^2(x) - r^2(x)\right] dx = \int_{-1}^1 \pi\left[9 - 9\left(1 - x^4\right)^2\right] dx$
$= 9\pi\int_{-1}^1 \left[1 - \left(1 - 2x^4 + x^8\right)\right] dx = 9\pi\int_{-1}^1 \left(2x^4 - x^8\right) dx = 9\pi\left[\frac{2x^5}{5} - \frac{x^9}{9}\right]_{-1}^1 = 18\pi\left[\frac{2}{5} - \frac{1}{9}\right] = \frac{2\pi\cdot 13}{5} = \frac{26\pi}{5}$

9. (a) *disk method:*
$V = \pi\int_1^5 \left(\sqrt{x - 1}\right)^2 dx = \pi\int_1^5 (x - 1)\,dx = \pi\left[\frac{x^2}{2} - x\right]_1^5$
$= \pi\left[\left(\frac{25}{2} - 5\right) - \left(\frac{1}{2} - 1\right)\right] = \pi\left(\frac{24}{2} - 4\right) = 8\pi$

(b) *washer method:*

$R(y) = 5,\ r(y) = y^2 + 1 \Rightarrow V = \int_c^d \pi\,[R^2(y) - r^2(y)]\,dy = \pi \int_{-2}^{2} \left[25 - (y^2+1)^2\right] dy$

$= \pi \int_{-2}^{2} (25 - y^4 - 2y^2 - 1)\,dy = \pi \int_{-2}^{2} (24 - y^4 - 2y^2)\,dy = \pi \left[24y - \frac{y^5}{5} - \frac{2}{3}y^3\right]_{-2}^{2} = 2\pi\left(24 \cdot 2 - \frac{32}{5} - \frac{2}{3} \cdot 8\right)$

$= 32\pi\left(3 - \frac{2}{5} - \frac{1}{3}\right) = \frac{32\pi}{15}(45 - 6 - 5) = \frac{1088\pi}{15}$

(c) *disk method:*

$R(y) = 5 - (y^2 + 1) = 4 - y^2$

$\Rightarrow V = \int_c^d \pi R^2(y)\,dy = \int_{-2}^{2} \pi\,(4 - y^2)^2\,dy$

$= \pi \int_{-2}^{2} (16 - 8y^2 + y^4)\,dy$

$= \pi \left[16y - \frac{8y^3}{3} + \frac{y^5}{5}\right]_{-2}^{2} = 2\pi\left(32 - \frac{64}{3} + \frac{32}{5}\right)$

$= 64\pi\left(1 - \frac{2}{3} + \frac{1}{5}\right) = \frac{64\pi}{15}(15 - 10 + 3) = \frac{512\pi}{15}$

11. *disk method:*

$R(x) = \tan x,\ a = 0,\ b = \frac{\pi}{3} \Rightarrow V = \pi \int_0^{\pi/3} \tan^2 x\,dx = \pi \int_0^{\pi/3} (\sec^2 x - 1)\,dx = \pi[\tan x - x]_0^{\pi/3} = \frac{\pi\left(3\sqrt{3} - \pi\right)}{3}$

13. (a) *disk method:*

$V = \pi \int_0^2 (x^2 - 2x)^2\,dx = \pi \int_0^2 (x^4 - 4x^3 + 4x^2)\,dx = \pi\left[\frac{x^5}{5} - x^4 + \frac{4}{3}x^3\right]_0^2 = \pi\left(\frac{32}{5} - 16 + \frac{32}{3}\right)$

$= \frac{16\pi}{15}(6 - 15 + 10) = \frac{16\pi}{15}$

(b) *washer method:*

$V = \int_0^2 \pi\left[1^2 - (x^2 - 2x + 1)^2\right] dx = \int_0^2 \pi\,dx - \int_0^2 \pi\,(x-1)^4\,dx = 2\pi - \left[\pi\,\frac{(x-1)^5}{5}\right]_0^2 = 2\pi - \pi \cdot \frac{2}{5} = \frac{8\pi}{5}$

(c) *shell method:*

$V = \int_a^b 2\pi \binom{\text{shell}}{\text{radius}} \binom{\text{shell}}{\text{height}} dx = 2\pi \int_0^2 (2 - x)\left[-(x^2 - 2x)\right] dx = 2\pi \int_0^2 (2 - x)(2x - x^2)\,dx$

$= 2\pi \int_0^2 (4x - 2x^2 - 2x^2 + x^3)\,dx = 2\pi \int_0^2 (x^3 - 4x^2 + 4x)\,dx = 2\pi\left[\frac{x^4}{4} - \frac{4}{3}x^3 + 2x^2\right]_0^2 = 2\pi\left(4 - \frac{32}{3} + 8\right)$

$= \frac{2\pi}{3}(36 - 32) = \frac{8\pi}{3}$

(d) *washer method:*

$V = \pi \int_0^2 [2 - (x^2 - 2x)]^2\,dx - \pi \int_0^2 2^2\,dx = \pi \int_0^2 \left[4 - 4(x^2 - 2x) + (x^2 - 2x)^2\right] dx - 8\pi$

$= \pi \int_0^2 (4 - 4x^2 + 8x + x^4 - 4x^3 + 4x^2)\,dx - 8\pi = \pi \int_0^2 (x^4 - 4x^3 + 8x + 4)\,dx - 8\pi$

$= \pi\left[\frac{x^5}{5} - x^4 + 4x^2 + 4x\right]_0^2 - 8\pi = \pi\left(\frac{32}{5} - 16 + 16 + 8\right) - 8\pi = \frac{\pi}{5}(32 + 40) - 8\pi = \frac{72\pi}{5} - \frac{40\pi}{5} = \frac{32\pi}{5}$

15. The material removed from the sphere consists of a cylinder and two "caps." From the diagram, the height of the cylinder is 2h, where $h^2 + \left(\sqrt{3}\right)^2 = 2^2$, i.e. $h = 1$. Thus $V_{cyl} = (2h)\pi\left(\sqrt{3}\right)^2 = 6\pi\ \text{ft}^3$. To get the volume of a cap, use the disk method and $x^2 + y^2 = 2^2$: $V_{cap} = \int_1^2 \pi x^2\,dy$

$= \int_1^2 \pi(4 - y^2)\,dy = \pi\left[4y - \frac{y^3}{3}\right]_1^2$

$= \pi\left[\left(8 - \frac{8}{3}\right) - \left(4 - \frac{1}{3}\right)\right] = \frac{5\pi}{3}\ \text{ft}^3$. Therefore,

$V_{removed} = V_{cyl} + 2V_{cap} = 6\pi + \frac{10\pi}{3} = \frac{28\pi}{3}\ \text{ft}^3$.

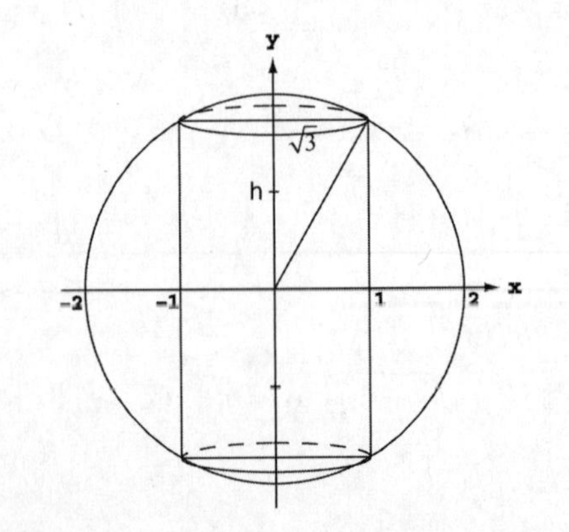

17. $y = x^{1/2} - \frac{x^{3/2}}{3} \Rightarrow \frac{dy}{dx} = \frac{1}{2}x^{-1/2} - \frac{1}{2}x^{1/2} \Rightarrow \left(\frac{dy}{dx}\right)^2 = \frac{1}{4}\left(\frac{1}{x} - 2 + x\right) \Rightarrow L = \int_1^4 \sqrt{1 + \frac{1}{4}\left(\frac{1}{x} - 2 + x\right)}\, dx$
$\Rightarrow L = \int_1^4 \sqrt{\frac{1}{4}\left(\frac{1}{x} + 2 + x\right)}\, dx = \int_1^4 \sqrt{\frac{1}{4}\left(x^{-1/2} + x^{1/2}\right)^2}\, dx = \int_1^4 \frac{1}{2}\left(x^{-1/2} + x^{1/2}\right) dx = \frac{1}{2}\left[2x^{1/2} + \frac{2}{3}x^{3/2}\right]_1^4$
$= \frac{1}{2}\left[\left(4 + \frac{2}{3}\cdot 8\right) - \left(2 + \frac{2}{3}\right)\right] = \frac{1}{2}\left(2 + \frac{14}{3}\right) = \frac{10}{3}$

19. $y = \frac{5}{12}x^{6/5} - \frac{5}{8}x^{4/5} \Rightarrow \frac{dy}{dx} = \frac{1}{2}x^{1/5} - \frac{1}{2}x^{-1/5} \Rightarrow \left(\frac{dy}{dx}\right)^2 = \frac{1}{4}\left(x^{2/5} - 2 + x^{-2/5}\right)$
$\Rightarrow L = \int_1^{32} \sqrt{1 + \frac{1}{4}\left(x^{2/5} - 2 + x^{-2/5}\right)}\, dx \Rightarrow L = \int_1^{32} \sqrt{\frac{1}{4}\left(x^{2/5} + 2 + x^{-2/5}\right)}\, dx = \int_1^{32} \sqrt{\frac{1}{4}\left(x^{1/5} + x^{-1/5}\right)^2}\, dx$
$= \int_1^{32} \frac{1}{2}\left(x^{1/5} + x^{-1/5}\right) dx = \frac{1}{2}\left[\frac{5}{6}x^{6/5} + \frac{5}{4}x^{4/5}\right]_1^{32} = \frac{1}{2}\left[\left(\frac{5}{6}\cdot 2^6 + \frac{5}{4}\cdot 2^4\right) - \left(\frac{5}{6} + \frac{5}{4}\right)\right] = \frac{1}{2}\left(\frac{315}{6} + \frac{75}{4}\right)$
$= \frac{1}{48}(1260 + 450) = \frac{1710}{48} = \frac{285}{8}$

21. $S = \int_a^b 2\pi y \sqrt{1 + \left(\frac{dy}{dx}\right)^2}\, dx;\ \frac{dy}{dx} = \frac{1}{\sqrt{2x+1}} \Rightarrow \left(\frac{dy}{dx}\right)^2 = \frac{1}{2x+1} \Rightarrow S = \int_0^3 2\pi\sqrt{2x+1}\sqrt{1 + \frac{1}{2x+1}}\, dx$
$= 2\pi \int_0^3 \sqrt{2x+1}\sqrt{\frac{2x+2}{2x+1}}\, dx = 2\sqrt{2}\pi \int_0^3 \sqrt{x+1}\, dx = 2\sqrt{2}\pi\left[\frac{2}{3}(x+1)^{3/2}\right]_0^3 = 2\sqrt{2}\pi\cdot\frac{2}{3}(8-1) = \frac{28\pi\sqrt{2}}{3}$

23. $S = \int_c^d 2\pi x \sqrt{1 + \left(\frac{dx}{dy}\right)^2}\, dy;\ \frac{dx}{dy} = \frac{\left(\frac{1}{2}\right)(4-2y)}{\sqrt{4y-y^2}} = \frac{2-y}{\sqrt{4y-y^2}} \Rightarrow 1 + \left(\frac{dx}{dy}\right)^2 = \frac{4y - y^2 + 4 - 4y + y^2}{4y - y^2} = \frac{4}{4y-y^2}$
$\Rightarrow S = \int_1^2 2\pi\sqrt{4y-y^2}\sqrt{\frac{4}{4y-y^2}}\, dy = 4\pi\int_1^2 dx = 4\pi$

25. The equipment alone: the force required to lift the equipment is equal to its weight $\Rightarrow F_1(x) = 100$ N. The work done is $W_1 = \int_a^b F_1(x)\, dx = \int_0^{40} 100\, dx = [100x]_0^{40} = 4000$ J; the rope alone: the force required to lift the rope is equal to the weight of the rope paid out at elevation x $\Rightarrow F_2(x) = 0.8(40 - x)$. The work done is $W_2 = \int_a^b F_2(x)\, dx = \int_0^{40} 0.8(40-x)\, dx = 0.8\left[40x - \frac{x^2}{2}\right]_0^{40} = 0.8\left(40^2 - \frac{40^2}{2}\right) = \frac{(0.8)(1600)}{2} = 640$ J; the total work is $W = W_1 + W_2 = 4000 + 640 = 4640$ J

27. Force constant: $F = kx \Rightarrow 20 = k\cdot 1 \Rightarrow k = 20$ lb/ft; the work to stretch the spring 1 ft is
$W = \int_0^1 kx\, dx = k\int_0^1 x\, dx = \left[20\,\frac{x^2}{2}\right]_0^1 = 10$ ft · lb; the work to stretch the spring an additional foot is
$W = \int_1^2 kx\, dx = k\int_1^2 x\, dx = 20\left[\frac{x^2}{2}\right]_1^2 = 20\left(\frac{4}{2} - \frac{1}{2}\right) = 20\left(\frac{3}{2}\right) = 30$ ft · lb

29. We imagine the water divided into thin slabs by planes perpendicular to the y-axis at the points of a partition of the interval [0, 8]. The typical slab between the planes at y and $y + \Delta y$ has a volume of about $\Delta V = \pi(\text{radius})^2(\text{thickness})$ $= \pi\left(\frac{5}{4}y\right)^2 \Delta y = \frac{25\pi}{16}y^2\,\Delta y$ ft³. The force F(y) required to lift this slab is equal to its weight: $F(y) = 62.4\,\Delta V$ $= \frac{(62.4)(25)}{16}\pi y^2\,\Delta y$ lb. The distance through which F(y) must act to lift this slab to the level 6 ft above the top is about $(6 + 8 - y)$ ft, so the work done lifting the slab is about $\Delta W = \frac{(62.4)(25)}{16}\pi y^2(14 - y)\,\Delta y$ ft · lb. The work done lifting all the slabs from y = 0 to y = 8 to the level 6 ft above the top is approximately

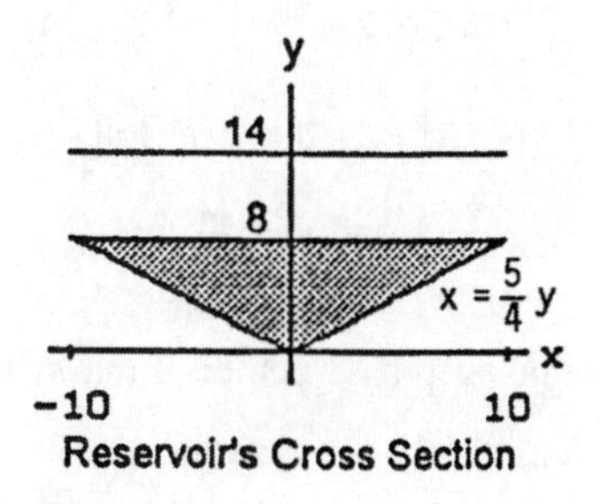

Reservoir's Cross Section

$W \approx \sum_0^8 \frac{(62.4)(25)}{16}\pi y^2(14 - y)\,\Delta y$ ft · lb so the work to pump the water is the limit of these Riemann sums as the norm of

the partition goes to zero: $W = \int_0^8 \frac{(62.4)(25)}{(16)} \pi y^2(14 - y)\,dy = \frac{(62.4)(25)\pi}{16} \int_0^8 (14y^2 - y^3)\,dy = (62.4)\left(\frac{25\pi}{16}\right)\left[\frac{14}{3}y^3 - \frac{y^4}{4}\right]_0^8$
$= (62.4)\left(\frac{25\pi}{16}\right)\left(\frac{14}{3} \cdot 8^3 - \frac{8^4}{4}\right) \approx 418{,}208.81 \text{ ft} \cdot \text{lb}$

31. The tank's cross section looks like the figure in Exercise 29 with right edge given by $x = \frac{5}{10}y = \frac{y}{2}$. A typical horizontal slab has volume $\Delta V = \pi(\text{radius})^2(\text{thickness}) = \pi\left(\frac{y}{2}\right)^2 \Delta y = \frac{\pi}{4} y^2 \Delta y$. The force required to lift thisslab is its weight: $F(y) = 60 \cdot \frac{\pi}{4} y^2 \Delta y$. The distance through which $F(y)$ must act is $(2 + 10 - y)$ ft, so the work to pump the liquid is $W = 60 \int_0^{10} \pi(12 - y)\left(\frac{y^2}{4}\right)dy = 15\pi\left[\frac{12y^3}{3} - \frac{y^4}{4}\right]_0^{10} = 22{,}500\pi \text{ ft} \cdot \text{lb}$; the time needed to empty the tank is $\frac{22{,}500\pi \text{ ft·lb}}{275 \text{ ft·lb/sec}} \approx 257 \text{ sec}$

33. Intersection points: $3 - x^2 = 2x^2 \Rightarrow 3x^2 - 3 = 0$ $\Rightarrow 3(x - 1)(x + 1) = 0 \Rightarrow x = -1$ or $x = 1$. Symmetry suggests that $\overline{x} = 0$. The typical *vertical* strip has center of mass: $(\tilde{x}, \tilde{y}) = \left(x, \frac{2x^2 + (3 - x^2)}{2}\right) = \left(x, \frac{x^2 + 3}{2}\right)$, length: $(3 - x^2) - 2x^2 = 3(1 - x^2)$, width: dx, area: $dA = 3(1 - x^2)\,dx$, and mass: $dm = \delta \cdot dA$ $= 3\delta(1 - x^2)\,dx \Rightarrow$ the moment about the x-axis is
$\tilde{y}\,dm = \frac{3}{2}\delta(x^2 + 3)(1 - x^2)\,dx = \frac{3}{2}\delta(-x^4 - 2x^2 + 3)\,dx \Rightarrow M_x = \int \tilde{y}\,dm = \frac{3}{2}\delta\int_{-1}^{1}(-x^4 - 2x^2 + 3)\,dx$
$= \frac{3}{2}\delta\left[-\frac{x^5}{5} - \frac{2x^3}{3} + 3x\right]_{-1}^{1} = 3\delta\left(-\frac{1}{5} - \frac{2}{3} + 3\right) = \frac{3\delta}{15}(-3 - 10 + 45) = \frac{32\delta}{5}$; $M = \int dm = 3\delta\int_{-1}^{1}(1 - x^2)\,dx$
$= 3\delta\left[x - \frac{x^3}{3}\right]_{-1}^{1} = 6\delta\left(1 - \frac{1}{3}\right) = 4\delta \Rightarrow \overline{y} = \frac{M_x}{M} = \frac{32\delta}{5 \cdot 4\delta} = \frac{8}{5}$. Therefore, the centroid is $(\overline{x}, \overline{y}) = \left(0, \frac{8}{5}\right)$.

35. The typical *vertical* strip has: center of mass: $(\tilde{x}, \tilde{y})$ $= \left(x, \frac{4 + \frac{x^2}{4}}{2}\right)$, length: $4 - \frac{x^2}{4}$, width: dx, area: $dA = \left(4 - \frac{x^2}{4}\right)dx$, mass: $dm = \delta \cdot dA$ $= \delta\left(4 - \frac{x^2}{4}\right)dx \Rightarrow$ the moment about the x-axis is $\tilde{y}\,dm = \delta \cdot \frac{\left(4 + \frac{x^2}{4}\right)}{2}\left(4 - \frac{x^2}{4}\right)dx = \frac{\delta}{2}\left(16 - \frac{x^4}{16}\right)dx$; the moment about the y-axis is $\tilde{x}\,dm = \delta\left(4 - \frac{x^2}{4}\right) \cdot x\,dx = \delta\left(4x - \frac{x^3}{4}\right)dx$. Thus, $M_x = \int \tilde{y}\,dm = \frac{\delta}{2}\int_0^4\left(16 - \frac{x^4}{16}\right)dx$
$= \frac{\delta}{2}\left[16x - \frac{x^5}{5 \cdot 16}\right]_0^4 = \frac{\delta}{2}\left[64 - \frac{64}{5}\right] = \frac{128\delta}{5}$; $M_y = \int \tilde{x}\,dm = \delta\int_0^4\left(4x - \frac{x^3}{4}\right)dx = \delta\left[2x^2 - \frac{x^4}{16}\right]_0^4$
$= \delta(32 - 16) = 16\delta$; $M = \int dm = \delta\int_0^4\left(4 - \frac{x^2}{4}\right)dx = \delta\left[4x - \frac{x^3}{12}\right]_0^4 = \delta\left(16 - \frac{64}{12}\right) = \frac{32\delta}{3}$
$\Rightarrow \overline{x} = \frac{M_y}{M} = \frac{16 \cdot \delta \cdot 3}{32 \cdot \delta} = \frac{3}{2}$ and $\overline{y} = \frac{M_x}{M} = \frac{128 \cdot \delta \cdot 3}{5 \cdot 32 \cdot \delta} = \frac{12}{5}$. Therefore, the centroid is $(\overline{x}, \overline{y}) = \left(\frac{3}{2}, \frac{12}{5}\right)$.

37. A typical horizontal strip has: center of mass: $(\tilde{x}, \tilde{y})$ $= \left(\frac{y^2 + 2y}{2}, y\right)$, length: $2y - y^2$, width: dy, area: $dA = (2y - y^2)\,dy$, mass: $dm = \delta \cdot dA$ $= (1 + y)(2y - y^2)\,dy \Rightarrow$ the moment about the x-axis is $\tilde{y}\,dm = y(1 + y)(2y - y^2)\,dy$ $= (2y^2 + 2y^3 - y^3 - y^4)\,dy$ $= (2y^2 + y^3 - y^4)\,dy$; the moment about the y-axis is

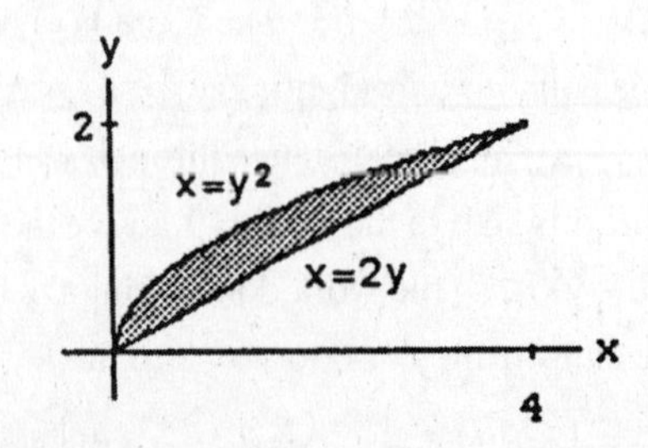

$\tilde{x}\,dm = \left(\frac{y^2+2y}{2}\right)(1+y)(2y-y^2)\,dy = \frac{1}{2}(4y^2-y^4)(1+y)\,dy = \frac{1}{2}(4y^2+4y^3-y^4-y^5)\,dy$

$\Rightarrow M_x = \int \tilde{y}\,dm = \int_0^2 (2y^2+y^3-y^4)\,dy = \left[\frac{2}{3}y^3+\frac{y^4}{4}-\frac{y^5}{5}\right]_0^2 = \left(\frac{16}{3}+\frac{16}{4}-\frac{32}{5}\right) = 16\left(\frac{1}{3}+\frac{1}{4}-\frac{2}{5}\right)$

$= \frac{16}{60}(20+15-24) = \frac{4}{15}(11) = \frac{44}{15}$; $M_y = \int \tilde{x}\,dm = \int_0^2 \frac{1}{2}(4y^2+4y^3-y^4-y^5)\,dy = \frac{1}{2}\left[\frac{4}{3}y^3+y^4-\frac{y^5}{5}-\frac{y^6}{6}\right]_0^2$

$= \frac{1}{2}\left(\frac{4\cdot 2^3}{3}+2^4-\frac{2^5}{5}-\frac{2^6}{6}\right) = 4\left(\frac{4}{3}+2-\frac{4}{5}-\frac{8}{6}\right) = 4\left(2-\frac{4}{5}\right) = \frac{24}{5}$; $M = \int dm = \int_0^2 (1+y)(2y-y^2)\,dy$

$= \int_0^2 (2y+y^2-y^3)\,dy = \left[y^2+\frac{y^3}{3}-\frac{y^4}{4}\right]_0^2 = \left(4+\frac{8}{3}-\frac{16}{4}\right) = \frac{8}{3} \Rightarrow \overline{x} = \frac{M_y}{M} = \left(\frac{24}{5}\right)\left(\frac{3}{8}\right) = \frac{9}{5}$ and $\overline{y} = \frac{M_x}{M}$

$= \left(\frac{44}{15}\right)\left(\frac{3}{8}\right) = \frac{44}{40} = \frac{11}{10}$. Therefore, the center of mass is $(\overline{x},\overline{y}) = \left(\frac{9}{5},\frac{11}{10}\right)$.

39. $F = \int_a^b W\cdot\left(\begin{smallmatrix}\text{strip}\\ \text{depth}\end{smallmatrix}\right)\cdot L(y)\,dy \Rightarrow F = 2\int_0^2 (62.4)(2-y)(2y)\,dy = 249.6\int_0^2 (2y-y^2)\,dy = 249.6\left[y^2-\frac{y^3}{3}\right]_0^2$
$= (249.6)\left(4-\frac{8}{3}\right) = (249.6)\left(\frac{4}{3}\right) = 332.8$ lb

41. $F = \int_a^b W\cdot\left(\begin{smallmatrix}\text{strip}\\ \text{depth}\end{smallmatrix}\right)\cdot L(y)\,dy \Rightarrow F = 62.4\int_0^4 (9-y)\left(2\cdot\frac{\sqrt{y}}{2}\right)dy = 62.4\int_0^4 \left(9y^{1/2}-3y^{3/2}\right)dy$
$= 62.4\left[6y^{3/2}-\frac{2}{5}y^{5/2}\right]_0^4 = (62.4)\left(6\cdot 8-\frac{2}{5}\cdot 32\right) = \left(\frac{62.4}{5}\right)(48\cdot 5-64) = \frac{(62.4)(176)}{5} = 2196.48$ lb

CHAPTER 6 ADDITIONAL AND ADVANCED EXERCISES

1. $V = \pi\int_a^b [f(x)]^2\,dx = b^2-ab \Rightarrow \pi\int_a^x [f(t)]^2\,dt = x^2-ax$ for all $x > a \Rightarrow \pi[f(x)]^2 = 2x-a \Rightarrow f(x) = \sqrt{\frac{2x-a}{\pi}}$

3. $s(x) = Cx \Rightarrow \int_0^x \sqrt{1+[f'(t)]^2}\,dt = Cx \Rightarrow \sqrt{1+[f'(x)]^2} = C \Rightarrow f'(x) = \sqrt{C^2-1}$ for $C \ge 1$
$\Rightarrow f(x) = \int_0^x \sqrt{C^2-1}\,dt + k$. Then $f(0) = a \Rightarrow a = 0+k \Rightarrow f(x) = \int_0^x \sqrt{C^2-1}\,dt + a \Rightarrow f(x) = x\sqrt{C^2-1}+a$,
where $C \ge 1$.

5. We can find the centroid and then use Pappus' Theorem to calculate the volume. $f(x) = x$, $g(x) = x^2$, $f(x) = g(x)$
$\Rightarrow x = x^2 \Rightarrow x^2-x = 0 \Rightarrow x = 0, x = 1$; $\delta = 1$; $M = \int_0^1 [x-x^2]dx = \left[\frac{1}{2}x^2-\frac{1}{3}x^3\right]_0^1 = \left(\frac{1}{2}-\frac{1}{3}\right)-0 = \frac{1}{6}$
$\overline{x} = \frac{1}{1/6}\int_0^1 x[x-x^2]dx = 6\int_0^1 [x^2-x^3]dx = 6\left[\frac{1}{3}x^3-\frac{1}{4}x^4\right]_0^1 = 6\left(\frac{1}{3}-\frac{1}{4}\right)-0 = \frac{1}{2}$
$\overline{y} = \frac{1}{1/6}\int_0^1 \frac{1}{2}\left[x^2-(x^2)^2\right]dx = 3\int_0^1 [x^2-x^4]dx = 3\left[\frac{1}{3}x^3-\frac{1}{5}x^5\right]_0^1 = 3\left(\frac{1}{3}-\frac{1}{5}\right)-0 = \frac{2}{5} \Rightarrow$ The centroid is $\left(\frac{1}{2},\frac{2}{5}\right)$.
ρ is the distance from $\left(\frac{1}{2},\frac{2}{5}\right)$ to the axis of rotation, $y = x$. To calculate this distance we must find the point on $y = x$ that also lies on the line perpendicular to $y = x$ that passes through $\left(\frac{1}{2},\frac{2}{5}\right)$. The equation of this line is $y-\frac{2}{5} = -1\left(x-\frac{1}{2}\right)$
$\Rightarrow x+y = \frac{9}{10}$. The point of intersection of the lines $x+y = \frac{9}{10}$ and $y = x$ is $\left(\frac{9}{20},\frac{9}{20}\right)$. Thus,
$\rho = \sqrt{\left(\frac{9}{20}-\frac{1}{2}\right)^2+\left(\frac{9}{20}-\frac{2}{5}\right)^2} = \frac{1}{10\sqrt{2}}$. Thus $V = 2\pi\left(\frac{1}{10\sqrt{2}}\right)\left(\frac{1}{6}\right) = \frac{\pi}{30\sqrt{2}}$.

7. $y = 2\sqrt{x} \Rightarrow ds = \sqrt{\frac{1}{x}+1}\,dx \Rightarrow A = \int_0^3 2\sqrt{x}\sqrt{\frac{1}{x}+1}\,dx = \frac{4}{3}\left[(1+x)^{3/2}\right]_0^3 = \frac{28}{3}$

9. $F = ma = t^2 \Rightarrow \frac{d^2x}{dt^2} = a = \frac{t^2}{m} \Rightarrow v = \frac{dx}{dt} = \frac{t^3}{3m}+C$; $v = 0$ when $t = 0 \Rightarrow C = 0 \Rightarrow \frac{dx}{dt} = \frac{t^3}{3m} \Rightarrow x = \frac{t^4}{12m}+C_1$;
$x = 0$ when $t = 0 \Rightarrow C_1 = 0 \Rightarrow x = \frac{t^4}{12m}$. Then $x = h \Rightarrow t = (12mh)^{1/4}$. The work done is
$W = \int F\,dx = \int_0^{(12mh)^{1/4}} F(t)\cdot\frac{dx}{dt}\,dt = \int_0^{(12mh)^{1/4}} t^2\cdot\frac{t^3}{3m}\,dt = \frac{1}{3m}\left[\frac{t^6}{6}\right]_0^{(12mh)^{1/4}} = \left(\frac{1}{18m}\right)(12mh)^{6/4}$
$= \frac{(12mh)^{3/2}}{18m} = \frac{12mh\cdot\sqrt{12mh}}{18m} = \frac{2h}{3}\cdot 2\sqrt{3mh} = \frac{4h}{3}\sqrt{3mh}$

11. From the symmetry of $y = 1 - x^n$, n even, about the y-axis for $-1 \le x \le 1$, we have $\bar{x} = 0$. To find $\bar{y} = \frac{M_x}{M}$, we use the vertical strips technique. The typical strip has center of mass: $(\tilde{x}, \tilde{y}) = \left(x, \frac{1-x^n}{2}\right)$, length: $1 - x^n$, width: dx, area: $dA = (1 - x^n)\,dx$, mass: $dm = 1 \cdot dA = (1 - x^n)\,dx$. The moment of the strip about the x-axis is $\tilde{y}\,dm = \frac{(1-x^n)^2}{2}\,dx \Rightarrow M_x = \int_{-1}^{1} \frac{(1-x^n)^2}{2}\,dx = 2\int_0^1 \frac{1}{2}(1 - 2x^n + x^{2n})\,dx = \left[x - \frac{2x^{n+1}}{n+1} + \frac{x^{2n+1}}{2n+1}\right]_0^1$ $= 1 - \frac{2}{n+1} + \frac{1}{2n+1} = \frac{(n+1)(2n+1) - 2(2n+1) + (n+1)}{(n+1)(2n+1)} = \frac{2n^2 + 3n + 1 - 4n - 2 + n + 1}{(n+1)(2n+1)} = \frac{2n^2}{(n+1)(2n+1)}$. Also, $M = \int_{-1}^{1} dA = \int_{-1}^{1} (1 - x^n)\,dx = 2\int_0^1 (1 - x^n)\,dx = 2\left[x - \frac{x^{n+1}}{n+1}\right]_0^1 = 2\left(1 - \frac{1}{n+1}\right) = \frac{2n}{n+1}$. Therefore, $\bar{y} = \frac{M_x}{M} = \frac{2n^2}{(n+1)(2n+1)} \cdot \frac{(n+1)}{2n} = \frac{n}{2n+1} \Rightarrow \left(0, \frac{n}{2n+1}\right)$ is the location of the centroid. As $n \to \infty$, $\bar{y} \to \frac{1}{2}$ so the limiting position of the centroid is $\left(0, \frac{1}{2}\right)$.

13. (a) Consider a single vertical strip with center of mass $(\tilde{x}, \tilde{y})$. If the plate lies to the right of the line, then the moment of this strip about the line $x = b$ is $(\tilde{x} - b)\,dm = (\tilde{x} - b)\,\delta\,dA \Rightarrow$ the plate's first moment about $x = b$ is the integral $\int (x - b)\delta\,dA = \int \delta x\,dA - \int \delta b\,dA = M_y - b\delta A$.

(b) If the plate lies to the left of the line, the moment of a vertical strip about the line $x = b$ is $(b - \tilde{x})\,dm = (b - \tilde{x})\,\delta\,dA \Rightarrow$ the plate's first moment about $x = b$ is $\int (b - x)\delta\,dA = \int b\delta\,dA - \int \delta x\,dA$ $= b\delta A - M_y$.

15. (a) On $[0, a]$ a typical *vertical* strip has center of mass: $(\tilde{x}, \tilde{y}) = \left(x, \frac{\sqrt{b^2 - x^2} + \sqrt{a^2 - x^2}}{2}\right)$, length: $\sqrt{b^2 - x^2} - \sqrt{a^2 - x^2}$, width: dx, area: $dA = \left(\sqrt{b^2 - x^2} - \sqrt{a^2 - x^2}\right)dx$, mass: $dm = \delta\,dA$ $= \delta\left(\sqrt{b^2 - x^2} - \sqrt{a^2 - x^2}\right)dx$. On $[a, b]$ a typical *vertical* strip has center of mass: $(\tilde{x}, \tilde{y}) = \left(x, \frac{\sqrt{b^2 - x^2}}{2}\right)$, length: $\sqrt{b^2 - x^2}$, width: dx, area: $dA = \sqrt{b^2 - x^2}\,dx$, mass: $dm = \delta\,dA = \delta\sqrt{b^2 - x^2}\,dx$. Thus, $M_x = \int \tilde{y}\,dm$

$$= \int_0^a \frac{1}{2}\left(\sqrt{b^2 - x^2} + \sqrt{a^2 - x^2}\right)\delta\left(\sqrt{b^2 - x^2} - \sqrt{a^2 - x^2}\right)dx + \int_a^b \frac{1}{2}\sqrt{b^2 - x^2}\,\delta\sqrt{b^2 - x^2}\,dx$$

$$= \frac{\delta}{2}\int_0^a [(b^2 - x^2) - (a^2 - x^2)]\,dx + \frac{\delta}{2}\int_a^b (b^2 - x^2)\,dx = \frac{\delta}{2}\int_0^a (b^2 - a^2)\,dx + \frac{\delta}{2}\int_a^b (b^2 - x^2)\,dx$$

$$= \frac{\delta}{2}\left[(b^2 - a^2)\,x\right]_0^a + \frac{\delta}{2}\left[b^2 x - \frac{x^3}{3}\right]_a^b = \frac{\delta}{2}[(b^2 - a^2)\,a] + \frac{\delta}{2}\left[\left(b^3 - \frac{b^3}{3}\right) - \left(b^2 a - \frac{a^3}{3}\right)\right]$$

$$= \frac{\delta}{2}(ab^2 - a^3) + \frac{\delta}{2}\left(\frac{2}{3}b^3 - ab^2 + \frac{a^3}{3}\right) = \frac{\delta b^3}{3} - \frac{\delta a^3}{3} = \delta\left(\frac{b^3 - a^3}{3}\right); \; M_y = \int \tilde{x}\,dm$$

$$= \int_0^a x\delta\left(\sqrt{b^2 - x^2} - \sqrt{a^2 - x^2}\right)dx + \int_a^b x\delta\sqrt{b^2 - x^2}\,dx$$

$$= \delta\int_0^a x\,(b^2 - x^2)^{1/2}\,dx - \delta\int_0^a x\,(a^2 - x^2)^{1/2}\,dx + \delta\int_a^b x\,(b^2 - x^2)^{1/2}\,dx$$

$$= \frac{-\delta}{2}\left[\frac{2(b^2 - x^2)^{3/2}}{3}\right]_0^a + \frac{\delta}{2}\left[\frac{2(a^2 - x^2)^{3/2}}{3}\right]_0^a - \frac{\delta}{2}\left[\frac{2(b^2 - x^2)^{3/2}}{3}\right]_a^b$$

$$= -\frac{\delta}{3}\left[(b^2 - a^2)^{3/2} - (b^2)^{3/2}\right] + \frac{\delta}{3}\left[0 - (a^2)^{3/2}\right] - \frac{\delta}{3}\left[0 - (b^2 - a^2)^{3/2}\right] = \frac{\delta b^3}{3} - \frac{\delta a^3}{3} = \frac{\delta(b^3 - a^3)}{3} = M_x;$$

We calculate the mass geometrically: $M = \delta A = \delta\left(\frac{\pi b^2}{4}\right) - \delta\left(\frac{\pi a^2}{4}\right) = \frac{\delta\pi}{4}(b^2 - a^2)$. Thus, $\bar{x} = \frac{M_y}{M}$ $= \frac{\delta(b^3 - a^3)}{3} \cdot \frac{4}{\delta\pi(b^2 - a^2)} = \frac{4}{3\pi}\left(\frac{b^3 - a^3}{b^2 - a^2}\right) = \frac{4}{3\pi}\,\frac{(b - a)(a^2 + ab + b^2)}{(b - a)(b + a)} = \frac{4(a^2 + ab + b^2)}{3\pi(a + b)}$; likewise $\bar{y} = \frac{M_x}{M} = \frac{4(a^2+ab+b^2)}{3\pi(a+b)}$.

(b) $\lim_{b \to a} \frac{4}{3\pi}\left(\frac{a^2 + ab + b^2}{a + b}\right) = \left(\frac{4}{3\pi}\right)\left(\frac{a^2 + a^2 + a^2}{a + a}\right) = \left(\frac{4}{3\pi}\right)\left(\frac{3a^2}{2a}\right) = \frac{2a}{\pi} \Rightarrow (\bar{x}, \bar{y}) = \left(\frac{2a}{\pi}, \frac{2a}{\pi}\right)$ is the limiting position of the centroid as $b \to a$. This is the centroid of a circle of radius a (and we note the two circles coincide when $b = a$).

17. The submerged triangular plate is depicted in the figure at the right. The hypotenuse of the triangle has slope -1 $\Rightarrow y - (-2) = -(x - 0) \Rightarrow x = -(y + 2)$ is an equation of the hypotenuse. Using a typical horizontal strip, the fluid pressure is $F = \int (62.4) \cdot \left(\begin{smallmatrix}\text{strip}\\ \text{depth}\end{smallmatrix}\right) \cdot \left(\begin{smallmatrix}\text{strip}\\ \text{length}\end{smallmatrix}\right) dy$
$= \int_{-6}^{-2} (62.4)(-y)[-(y+2)]\, dy = 62.4 \int_{-6}^{-2} (y^2 + 2y)\, dy$
$= 62.4 \left[\frac{y^3}{3} + y^2\right]_{-6}^{-2} = (62.4)\left[\left(-\frac{8}{3} + 4\right) - \left(-\frac{216}{3} + 36\right)\right]$
$= (62.4)\left(\frac{208}{3} - 32\right) = \frac{(62.4)(112)}{3} \approx 2329.6$ lb

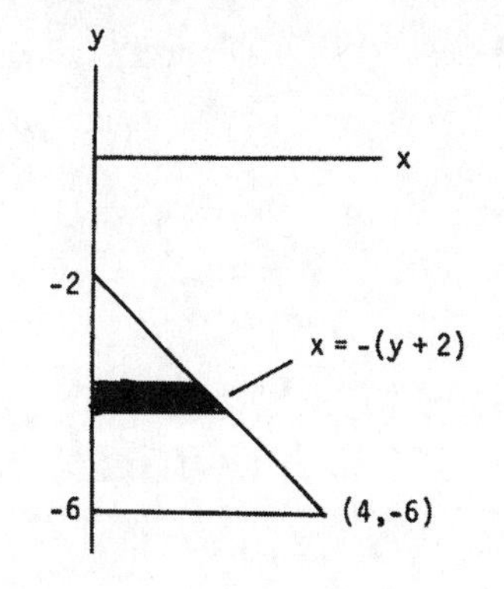

NOTES:

CHAPTER 7 TRANSCENDENTAL FUNCTIONS

7.1 INVERSE FUNCTIONS AND THEIR DERIVATIVES

1. Yes one-to-one, the graph passes the horizontal line test.

3. Not one-to-one since (for example) the horizontal line $y = 2$ intersects the graph twice.

5. Yes one-to-one, the graph passes the horizontal line test

7. Not one-to-one since the horizontal line $y = 3$ intersects the graph an infinite number of times.

9. Yes one-to-one, the graph passes the horizontal line test

11. Domain: $0 < x \le 1$, Range: $0 \le y$

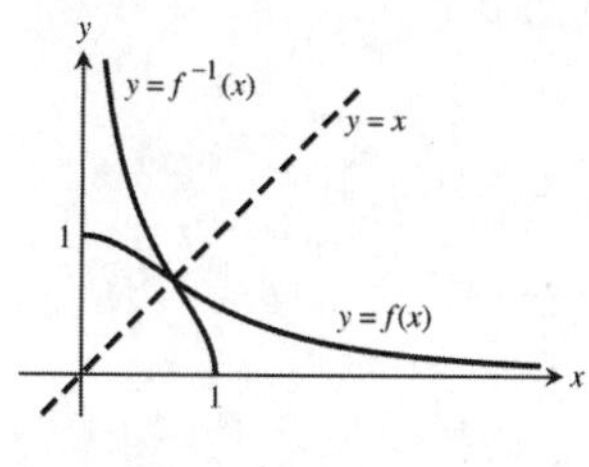

13. Domain: $-1 \le x \le 1$, Range: $-\frac{\pi}{2} \le y \le \frac{\pi}{2}$

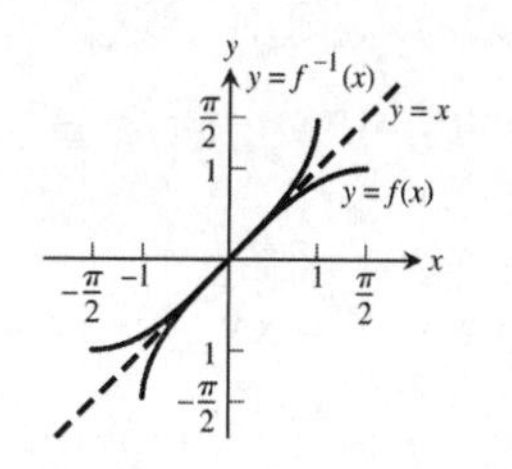

15. Domain: $0 \le x \le 6$, Range: $0 \le y \le 3$.

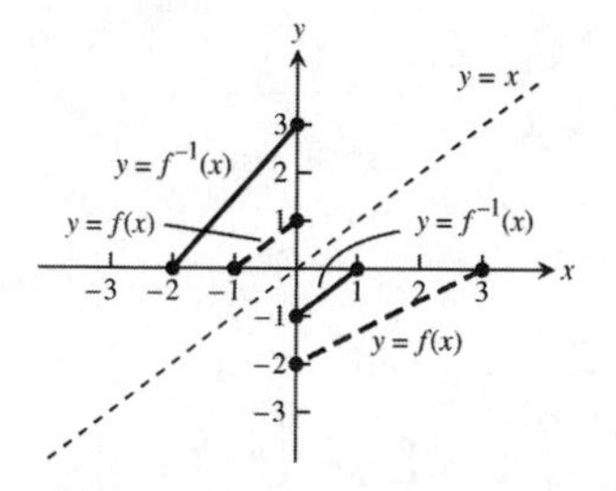

17. The graph is symmetric about $y = x$.

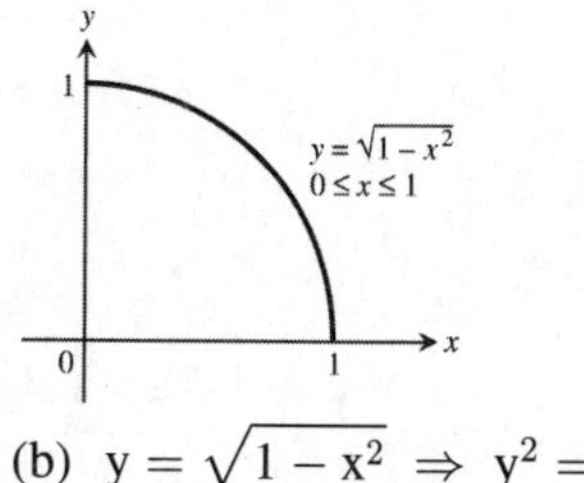

(b) $y = \sqrt{1 - x^2} \Rightarrow y^2 = 1 - x^2 \Rightarrow x^2 = 1 - y^2 \Rightarrow x = \sqrt{1 - y^2} \Rightarrow y = \sqrt{1 - x^2} = f^{-1}(x)$

19. Step 1: $y = x^2 + 1 \Rightarrow x^2 = y - 1 \Rightarrow x = \sqrt{y - 1}$
    Step 2: $y = \sqrt{x - 1} = f^{-1}(x)$

21. Step 1: $y = x^3 - 1 \Rightarrow x^3 = y + 1 \Rightarrow x = (y + 1)^{1/3}$
    Step 2: $y = \sqrt[3]{x + 1} = f^{-1}(x)$

23. Step 1: $y = (x+1)^2 \Rightarrow \sqrt{y} = x + 1$, since $x \ge -1 \Rightarrow x = \sqrt{y} - 1$
Step 2: $y = \sqrt{x} - 1 = f^{-1}(x)$

25. Step 1: $y = x^5 \Rightarrow x = y^{1/5}$
Step 2: $y = \sqrt[5]{x} = f^{-1}(x)$;
Domain and Range of f^{-1}: all reals;
$f(f^{-1}(x)) = \left(x^{1/5}\right)^5 = x$ and $f^{-1}(f(x)) = \left(x^5\right)^{1/5} = x$

27. Step 1: $y = x^3 + 1 \Rightarrow x^3 = y - 1 \Rightarrow x = (y-1)^{1/3}$
Step 2: $y = \sqrt[3]{x-1} = f^{-1}(x)$;
Domain and Range of f^{-1}: all reals;
$f(f^{-1}(x)) = \left((x-1)^{1/3}\right)^3 + 1 = (x-1) + 1 = x$ and $f^{-1}(f(x)) = \left((x^3+1) - 1\right)^{1/3} = \left(x^3\right)^{1/3} = x$

29. Step 1: $y = \frac{1}{x^2} \Rightarrow x^2 = \frac{1}{y} \Rightarrow x = \frac{1}{\sqrt{y}}$
Step 2: $y = \frac{1}{\sqrt{x}} = f^{-1}(x)$
Domain of f^{-1}: $x > 0$, Range of f^{-1}: $y > 0$;
$f(f^{-1}(x)) = \frac{1}{\left(\frac{1}{\sqrt{x}}\right)^2} = \frac{1}{\left(\frac{1}{x}\right)} = x$ and $f^{-1}(f(x)) = \frac{1}{\sqrt{\frac{1}{x^2}}} = \frac{1}{\left(\frac{1}{x}\right)} = x$ since $x > 0$

31. Step 1: $y = \frac{x+3}{x-2} \Rightarrow y(x-2) = x + 3 \Rightarrow xy - 2y = x + 3 \Rightarrow xy - x = 2y + 3 \Rightarrow x = \frac{2y+3}{y-1}$
Step 2: $y = \frac{2x+3}{x-1} = f^{-1}(x)$;
Domain of f^{-1}: $x \ne 1$, Range of f^{-1}: $y \ne 2$;
$f(f^{-1}(x)) = \frac{\left(\frac{2x+3}{x-1}\right)+3}{\left(\frac{2x+3}{x-1}\right)-2} = \frac{(2x+3)+3(x-1)}{(2x+3)-2(x-1)} = \frac{5x}{5} = x$ and $f^{-1}(f(x)) = \frac{2\left(\frac{x+3}{x-2}\right)+3}{\left(\frac{x+3}{x-2}\right)-1} = \frac{2(x+3)+3(x-2)}{(x+3)-(x-2)} = \frac{5x}{5} = x$

33. Step 1: $y = x^2 - 2x, x \le 1 \Rightarrow y + 1 = (x-1)^2, x \le 1 \Rightarrow -\sqrt{y+1} = x - 1, x \le 1 \Rightarrow x = 1 - \sqrt{y+1}$
Step 2: $y = 1 - \sqrt{x+1} = f^{-1}(x)$;
Domain of f^{-1}: $[-1, \infty)$, Range of f^{-1}: $(-\infty, 1]$;
$f(f^{-1}(x)) = \left(1 - \sqrt{x+1}\right)^2 - 2\left(1 - \sqrt{x+1}\right) = 1 - 2\sqrt{x+1} + x + 1 - 2 + 2\sqrt{x+1} = x$ and
$f^{-1}(f(x)) = 1 - \sqrt{(x^2 - 2x) + 1}, x \le 1 = 1 - \sqrt{(x-1)^2}, x \le 1 = 1 - |x - 1| = 1 - (1 - x) = x$

35. (a) $y = 2x + 3 \Rightarrow 2x = y - 3$
$\Rightarrow x = \frac{y}{2} - \frac{3}{2} \Rightarrow f^{-1}(x) = \frac{x}{2} - \frac{3}{2}$

(b)

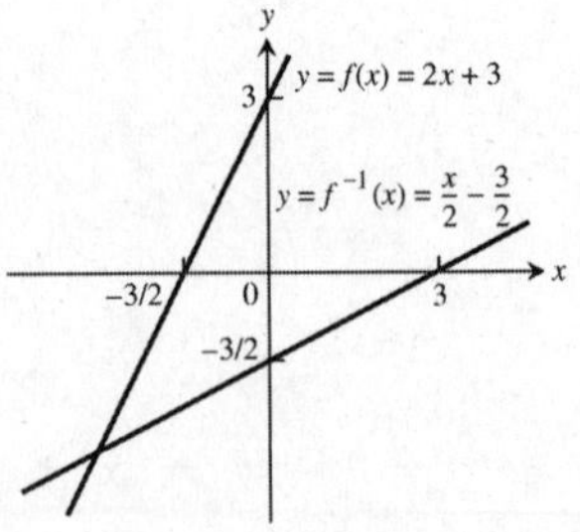

(c) $\left.\frac{df}{dx}\right|_{x=-1} = 2, \left.\frac{df^{-1}}{dx}\right|_{x=1} = \frac{1}{2}$

37. (a) $y = 5 - 4x \Rightarrow 4x = 5 - y$
$\Rightarrow x = \frac{5}{4} - \frac{y}{4} \Rightarrow f^{-1}(x) = \frac{5}{4} - \frac{x}{4}$
(b)

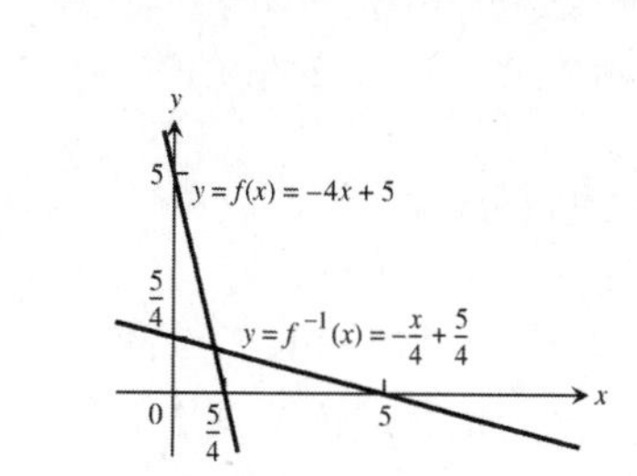

(c) $\left.\frac{df}{dx}\right|_{x=1/2} = -4, \left.\frac{df^{-1}}{dx}\right|_{x=3} = -\frac{1}{4}$

39. (a) $f(g(x)) = \left(\sqrt[3]{x}\right)^3 = x,\ g(f(x)) = \sqrt[3]{x^3} = x$
(b)

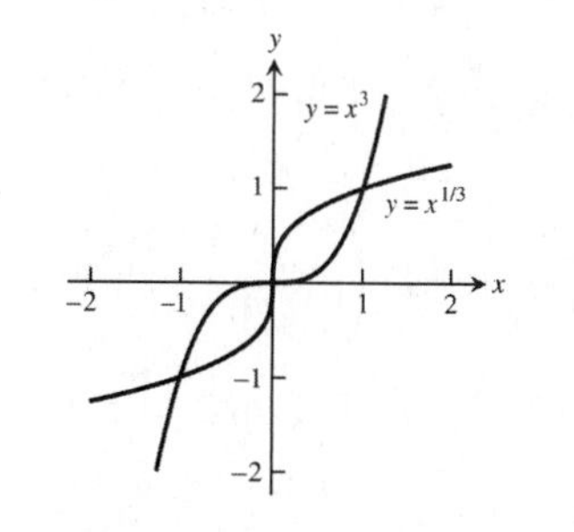

(c) $f'(x) = 3x^2 \Rightarrow f'(1) = 3, f'(-1) = 3;$
$g'(x) = \frac{1}{3}x^{-2/3} \Rightarrow g'(1) = \frac{1}{3}, g'(-1) = \frac{1}{3}$
(d) The line $y = 0$ is tangent to $f(x) = x^3$ at $(0, 0)$;
the line $x = 0$ is tangent to $g(x) = \sqrt[3]{x}$ at $(0, 0)$

41. $\frac{df}{dx} = 3x^2 - 6x \Rightarrow \left.\frac{df^{-1}}{dx}\right|_{x=f(3)} = \left.\frac{1}{\frac{df}{dx}}\right|_{x=3} = \frac{1}{9}$

43. $\left.\frac{df^{-1}}{dx}\right|_{x=4} = \left.\frac{df^{-1}}{dx}\right|_{x=f(2)} = \left.\frac{1}{\frac{df}{dx}}\right|_{x=2} = \frac{1}{\left(\frac{1}{3}\right)} = 3$

45. (a) $y = mx \Rightarrow x = \frac{1}{m}y \Rightarrow f^{-1}(x) = \frac{1}{m}x$
(b) The graph of $y = f^{-1}(x)$ is a line through the origin with slope $\frac{1}{m}$.

47. (a) $y = x + 1 \Rightarrow x = y - 1 \Rightarrow f^{-1}(x) = x - 1$
(b) $y = x + b \Rightarrow x = y - b \Rightarrow f^{-1}(x) = x - b$
(c) Their graphs will be parallel to one another and lie on opposite sides of the line $y = x$ equidistant from that line.

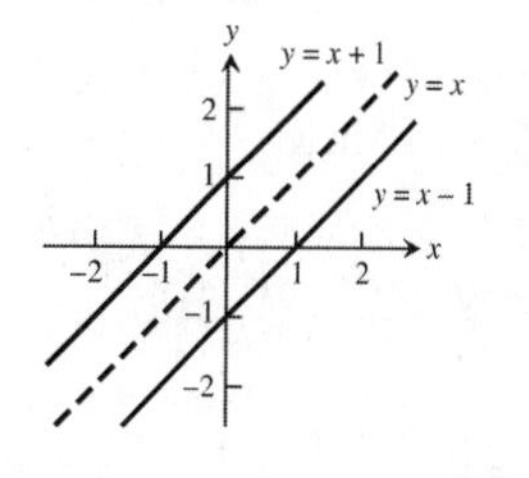

49. Let $x_1 \neq x_2$ be two numbers in the domain of an increasing function f. Then, either $x_1 < x_2$ or $x_1 > x_2$ which implies $f(x_1) < f(x_2)$ or $f(x_1) > f(x_2)$, since f(x) is increasing. In either case, $f(x_1) \neq f(x_2)$ and f is one-to-one. Similar arguments hold if f is decreasing.

51. f(x) is increasing since $x_2 > x_1 \Rightarrow 27x_2^3 > 27x_1^3$; $y = 27x^3 \Rightarrow x = \frac{1}{3}y^{1/3} \Rightarrow f^{-1}(x) = \frac{1}{3}x^{1/3}$;
$\frac{df}{dx} = 81x^2 \Rightarrow \frac{df^{-1}}{dx} = \left.\frac{1}{81x^2}\right|_{\frac{1}{3}x^{1/3}} = \frac{1}{9x^{2/3}} = \frac{1}{9}x^{-2/3}$

53. f(x) is decreasing since $x_2 > x_1 \Rightarrow (1 - x_2)^3 < (1 - x_1)^3$; $y = (1 - x)^3 \Rightarrow x = 1 - y^{1/3} \Rightarrow f^{-1}(x) = 1 - x^{1/3}$;
$\frac{df}{dx} = -3(1 - x)^2 \Rightarrow \frac{df^{-1}}{dx} = \left.\frac{1}{-3(1-x)^2}\right|_{1-x^{1/3}} = \frac{-1}{3x^{2/3}} = -\frac{1}{3}x^{-2/3}$

55. The function g(x) is also one-to-one. The reasoning: f(x) is one-to-one means that if $x_1 \neq x_2$ then $f(x_1) \neq f(x_2)$, so $-f(x_1) \neq -f(x_2)$ and therefore $g(x_1) \neq g(x_2)$. Therefore g(x) is one-to-one as well.

57. The composite is one-to-one also. The reasoning: If $x_1 \neq x_2$ then $g(x_1) \neq g(x_2)$ because g is one-to-one. Since $g(x_1) \neq g(x_2)$, we also have $f(g(x_1)) \neq f(g(x_2))$ because f is one-to-one. Thus, $f \circ g$ is one-to-one because $x_1 \neq x_2 \Rightarrow f(g(x_1)) \neq f(g(x_2))$.

59. $(g \circ f)(x) = x \Rightarrow g(f(x)) = x \Rightarrow g'(f(x))f'(x) = 1$

7.2 NATURAL LOGARITHMS

1. (a) $\ln 0.75 = \ln \frac{3}{4} = \ln 3 - \ln 4 = \ln 3 - \ln 2^2 = \ln 3 - 2\ln 2$
   (b) $\ln \frac{4}{9} = \ln 4 - \ln 9 = \ln 2^2 - \ln 3^2 = 2\ln 2 - 2\ln 3$
   (c) $\ln \frac{1}{2} = \ln 1 - \ln 2 = -\ln 2$
   (d) $\ln \sqrt[3]{9} = \frac{1}{3}\ln 9 = \frac{1}{3}\ln 3^2 = \frac{2}{3}\ln 3$
   (e) $\ln 3\sqrt{2} = \ln 3 + \ln 2^{1/2} = \ln 3 + \frac{1}{2}\ln 2$
   (f) $\ln \sqrt{13.5} = \frac{1}{2}\ln 13.5 = \frac{1}{2}\ln \frac{27}{2} = \frac{1}{2}(\ln 3^3 - \ln 2) = \frac{1}{2}(3\ln 3 - \ln 2)$

3. (a) $\ln \sin\theta - \ln\left(\frac{\sin\theta}{5}\right) = \ln\left(\frac{\sin\theta}{\left(\frac{\sin\theta}{5}\right)}\right) = \ln 5$
   (b) $\ln(3x^2 - 9x) + \ln\left(\frac{1}{3x}\right) = \ln\left(\frac{3x^2 - 9x}{3x}\right) = \ln(x - 3)$
   (c) $\frac{1}{2}\ln(4t^4) - \ln 2 = \ln\sqrt{4t^4} - \ln 2 = \ln 2t^2 - \ln 2 = \ln\left(\frac{2t^2}{2}\right) = \ln(t^2)$

5. $y = \ln 3x \Rightarrow y' = \left(\frac{1}{3x}\right)(3) = \frac{1}{x}$

7. $y = \ln(t^2) \Rightarrow \frac{dy}{dt} = \left(\frac{1}{t^2}\right)(2t) = \frac{2}{t}$

9. $y = \ln\frac{3}{x} = \ln 3x^{-1} \Rightarrow \frac{dy}{dx} = \left(\frac{1}{3x^{-1}}\right)(-3x^{-2}) = -\frac{1}{x}$

11. $y = \ln(\theta + 1) \Rightarrow \frac{dy}{d\theta} = \left(\frac{1}{\theta+1}\right)(1) = \frac{1}{\theta+1}$

13. $y = \ln x^3 \Rightarrow \frac{dy}{dx} = \left(\frac{1}{x^3}\right)(3x^2) = \frac{3}{x}$

15. $y = t(\ln t)^2 \Rightarrow \frac{dy}{dt} = (\ln t)^2 + 2t(\ln t)\cdot\frac{d}{dt}(\ln t) = (\ln t)^2 + \frac{2t\ln t}{t} = (\ln t)^2 + 2\ln t$

17. $y = \frac{x^4}{4}\ln x - \frac{x^4}{16} \Rightarrow \frac{dy}{dx} = x^3\ln x + \frac{x^4}{4}\cdot\frac{1}{x} - \frac{4x^3}{16} = x^3\ln x$

19. $y = \frac{\ln t}{t} \Rightarrow \frac{dy}{dt} = \frac{t\left(\frac{1}{t}\right) - (\ln t)(1)}{t^2} = \frac{1 - \ln t}{t^2}$

21. $y = \frac{\ln x}{1 + \ln x} \Rightarrow y' = \frac{(1 + \ln x)\left(\frac{1}{x}\right) - (\ln x)\left(\frac{1}{x}\right)}{(1 + \ln x)^2} = \frac{\frac{1}{x} + \frac{\ln x}{x} - \frac{\ln x}{x}}{(1 + \ln x)^2} = \frac{1}{x(1 + \ln x)^2}$

23. $y = \ln(\ln x) \Rightarrow y' = \left(\frac{1}{\ln x}\right)\left(\frac{1}{x}\right) = \frac{1}{x\ln x}$

25. $y = \theta[\sin(\ln\theta) + \cos(\ln\theta)] \Rightarrow \frac{dy}{d\theta} = [\sin(\ln\theta) + \cos(\ln\theta)] + \theta\left[\cos(\ln\theta)\cdot\frac{1}{\theta} - \sin(\ln\theta)\cdot\frac{1}{\theta}\right]$
   $= \sin(\ln\theta) + \cos(\ln\theta) + \cos(\ln\theta) - \sin(\ln\theta) = 2\cos(\ln\theta)$

27. $y = \ln\frac{1}{x\sqrt{x+1}} = -\ln x - \frac{1}{2}\ln(x+1) \Rightarrow y' = -\frac{1}{x} - \frac{1}{2}\left(\frac{1}{x+1}\right) = -\frac{2(x+1)+x}{2x(x+1)} = -\frac{3x+2}{2x(x+1)}$

29. $y = \frac{1 + \ln t}{1 - \ln t} \Rightarrow \frac{dy}{dt} = \frac{(1 - \ln t)\left(\frac{1}{t}\right) - (1 + \ln t)\left(\frac{-1}{t}\right)}{(1 - \ln t)^2} = \frac{\frac{1}{t} - \frac{\ln t}{t} + \frac{1}{t} + \frac{\ln t}{t}}{(1 - \ln t)^2} = \frac{2}{t(1 - \ln t)^2}$

31. $y = \ln(\sec(\ln\theta)) \Rightarrow \frac{dy}{d\theta} = \frac{1}{\sec(\ln\theta)}\cdot\frac{d}{d\theta}(\sec(\ln\theta)) = \frac{\sec(\ln\theta)\tan(\ln\theta)}{\sec(\ln\theta)}\cdot\frac{d}{d\theta}(\ln\theta) = \frac{\tan(\ln\theta)}{\theta}$

33. $y = \ln\left(\frac{(x^2+1)^5}{\sqrt{1-x}}\right) = 5\ln(x^2+1) - \frac{1}{2}\ln(1-x) \Rightarrow y' = \frac{5\cdot 2x}{x^2+1} - \frac{1}{2}\left(\frac{1}{1-x}\right)(-1) = \frac{10x}{x^2+1} + \frac{1}{2(1-x)}$

35. $y = \int_{x^2/2}^{x^2} \ln\sqrt{t}\,dt \Rightarrow \frac{dy}{dx} = \left(\ln\sqrt{x^2}\right)\cdot\frac{d}{dx}(x^2) - \left(\ln\sqrt{\frac{x^2}{2}}\right)\cdot\frac{d}{dx}\left(\frac{x^2}{2}\right) = 2x\ln|x| - x\ln\frac{|x|}{\sqrt{2}}$

37. $\int_{-3}^{-2} \frac{1}{x}\,dx = [\ln|x|]_{-3}^{-2} = \ln 2 - \ln 3 = \ln \frac{2}{3}$

39. $\int \frac{2y}{y^2-25}\,dy = \ln|y^2-25| + C$

41. $\int_0^{\pi} \frac{\sin t}{2-\cos t}\,dt = [\ln|2-\cos t|]_0^{\pi} = \ln 3 - \ln 1 = \ln 3$; or let $u = 2 - \cos t \Rightarrow du = \sin t\,dt$ with $t = 0$ $\Rightarrow u = 1$ and $t = \pi \Rightarrow u = 3 \Rightarrow \int_0^{\pi} \frac{\sin t}{2-\cos t}\,dt = \int_1^3 \frac{1}{u}\,du = [\ln|u|]_1^3 = \ln 3 - \ln 1 = \ln 3$

43. Let $u = \ln x \Rightarrow du = \frac{1}{x}\,dx$; $x = 1 \Rightarrow u = 0$ and $x = 2 \Rightarrow u = \ln 2$;
$\int_1^2 \frac{2\ln x}{x}\,dx = \int_0^{\ln 2} 2u\,du = [u^2]_0^{\ln 2} = (\ln 2)^2$

45. Let $u = \ln x \Rightarrow du = \frac{1}{x}\,dx$; $x = 2 \Rightarrow u = \ln 2$ and $x = 4 \Rightarrow u = \ln 4$;
$\int_2^4 \frac{dx}{x(\ln x)^2} = \int_{\ln 2}^{\ln 4} u^{-2}\,du = \left[-\frac{1}{u}\right]_{\ln 2}^{\ln 4} = -\frac{1}{\ln 4} + \frac{1}{\ln 2} = -\frac{1}{\ln 2^2} + \frac{1}{\ln 2} = -\frac{1}{2\ln 2} + \frac{1}{\ln 2} = \frac{1}{2\ln 2} = \frac{1}{\ln 4}$

47. Let $u = 6 + 3\tan t \Rightarrow du = 3\sec^2 t\,dt$;
$\int \frac{3\sec^2 t}{6+3\tan t}\,dt = \int \frac{du}{u} = \ln|u| + C = \ln|6+3\tan t| + C$

49. Let $u = \cos\frac{x}{2} \Rightarrow du = -\frac{1}{2}\sin\frac{x}{2}\,dx \Rightarrow -2\,du = \sin\frac{x}{2}\,dx$; $x = 0 \Rightarrow u = 1$ and $x = \frac{\pi}{2} \Rightarrow u = \frac{1}{\sqrt{2}}$;
$\int_0^{\pi/2} \tan\frac{x}{2}\,dx = \int_0^{\pi/2} \frac{\sin\frac{x}{2}}{\cos\frac{x}{2}}\,dx = -2\int_1^{1/\sqrt{2}} \frac{du}{u} = [-2\ln|u|]_1^{1/\sqrt{2}} = -2\ln\frac{1}{\sqrt{2}} = 2\ln\sqrt{2} = \ln 2$

51. Let $u = \sin\frac{\theta}{3} \Rightarrow du = \frac{1}{3}\cos\frac{\theta}{3}\,d\theta \Rightarrow 6\,du = 2\cos\frac{\theta}{3}\,d\theta$; $\theta = \frac{\pi}{2} \Rightarrow u = \frac{1}{2}$ and $\theta = \pi \Rightarrow u = \frac{\sqrt{3}}{2}$;
$\int_{\pi/2}^{\pi} 2\cot\frac{\theta}{3}\,d\theta = \int_{\pi/2}^{\pi} \frac{2\cos\frac{\theta}{3}}{\sin\frac{\theta}{3}}\,d\theta = 6\int_{1/2}^{\sqrt{3}/2} \frac{du}{u} = 6\,[\ln|u|]_{1/2}^{\sqrt{3}/2} = 6\left(\ln\frac{\sqrt{3}}{2} - \ln\frac{1}{2}\right) = 6\ln\sqrt{3} = \ln 27$

53. $\int \frac{dx}{2\sqrt{x}+2x} = \int \frac{dx}{2\sqrt{x}\,(1+\sqrt{x})}$; let $u = 1 + \sqrt{x} \Rightarrow du = \frac{1}{2\sqrt{x}}\,dx$; $\int \frac{dx}{2\sqrt{x}\,(1+\sqrt{x})} = \int \frac{du}{u} = \ln|u| + C$
$= \ln\left|1+\sqrt{x}\right| + C = \ln\left(1+\sqrt{x}\right) + C$

55. $y = \sqrt{x(x+1)} = (x(x+1))^{1/2} \Rightarrow \ln y = \frac{1}{2}\ln(x(x+1)) \Rightarrow 2\ln y = \ln(x) + \ln(x+1) \Rightarrow \frac{2y'}{y} = \frac{1}{x} + \frac{1}{x+1}$
$\Rightarrow y' = \left(\frac{1}{2}\right)\sqrt{x(x+1)}\left(\frac{1}{x} + \frac{1}{x+1}\right) = \frac{\sqrt{x(x+1)}\,(2x+1)}{2x(x+1)} = \frac{2x+1}{2\sqrt{x(x+1)}}$

57. $y = \sqrt{\frac{t}{t+1}} = \left(\frac{t}{t+1}\right)^{1/2} \Rightarrow \ln y = \frac{1}{2}[\ln t - \ln(t+1)] \Rightarrow \frac{1}{y}\frac{dy}{dt} = \frac{1}{2}\left(\frac{1}{t} - \frac{1}{t+1}\right)$
$\Rightarrow \frac{dy}{dt} = \frac{1}{2}\sqrt{\frac{t}{t+1}}\left(\frac{1}{t} - \frac{1}{t+1}\right) = \frac{1}{2}\sqrt{\frac{t}{t+1}}\left[\frac{1}{t(t+1)}\right] = \frac{1}{2\sqrt{t}\,(t+1)^{3/2}}$

59. $y = \sqrt{\theta+3}\,(\sin\theta) = (\theta+3)^{1/2}\sin\theta \Rightarrow \ln y = \frac{1}{2}\ln(\theta+3) + \ln(\sin\theta) \Rightarrow \frac{1}{y}\frac{dy}{d\theta} = \frac{1}{2(\theta+3)} + \frac{\cos\theta}{\sin\theta}$
$\Rightarrow \frac{dy}{d\theta} = \sqrt{\theta+3}\,(\sin\theta)\left[\frac{1}{2(\theta+3)} + \cot\theta\right]$

61. $y = t(t+1)(t+2) \Rightarrow \ln y = \ln t + \ln(t+1) + \ln(t+2) \Rightarrow \frac{1}{y}\frac{dy}{dt} = \frac{1}{t} + \frac{1}{t+1} + \frac{1}{t+2}$
$\Rightarrow \frac{dy}{dt} = t(t+1)(t+2)\left(\frac{1}{t} + \frac{1}{t+1} + \frac{1}{t+2}\right) = t(t+1)(t+2)\left[\frac{(t+1)(t+2)+t(t+2)+t(t+1)}{t(t+1)(t+2)}\right] = 3t^2 + 6t + 2$

63. $y = \frac{\theta+5}{\theta\cos\theta} \Rightarrow \ln y = \ln(\theta+5) - \ln\theta - \ln(\cos\theta) \Rightarrow \frac{1}{y}\frac{dy}{d\theta} = \frac{1}{\theta+5} - \frac{1}{\theta} + \frac{\sin\theta}{\cos\theta} \Rightarrow \frac{dy}{d\theta} = \left(\frac{\theta+5}{\theta\cos\theta}\right)\left(\frac{1}{\theta+5} - \frac{1}{\theta} + \tan\theta\right)$

65. $y = \frac{x\sqrt{x^2+1}}{(x+1)^{2/3}} \Rightarrow \ln y = \ln x + \frac{1}{2}\ln(x^2+1) - \frac{2}{3}\ln(x+1) \Rightarrow \frac{y'}{y} = \frac{1}{x} + \frac{x}{x^2+1} - \frac{2}{3(x+1)}$
$\Rightarrow y' = \frac{x\sqrt{x^2+1}}{(x+1)^{2/3}}\left[\frac{1}{x} + \frac{x}{x^2+1} - \frac{2}{3(x+1)}\right]$

67. $y = \sqrt[3]{\frac{x(x-2)}{x^2+1}} \Rightarrow \ln y = \frac{1}{3}[\ln x + \ln(x-2) - \ln(x^2+1)] \Rightarrow \frac{y'}{y} = \frac{1}{3}\left(\frac{1}{x} + \frac{1}{x-2} - \frac{2x}{x^2+1}\right)$
$\Rightarrow y' = \frac{1}{3}\sqrt[3]{\frac{x(x-2)}{x^2+1}}\left(\frac{1}{x} + \frac{1}{x-2} - \frac{2x}{x^2+1}\right)$

69. (a) $f(x) = \ln(\cos x) \Rightarrow f'(x) = -\frac{\sin x}{\cos x} = -\tan x = 0 \Rightarrow x = 0$; $f'(x) > 0$ for $-\frac{\pi}{4} \le x < 0$ and $f'(x) < 0$ for $0 < x \le \frac{\pi}{3} \Rightarrow$ there is a relative maximum at $x = 0$ with $f(0) = \ln(\cos 0) = \ln 1 = 0$; $f\left(-\frac{\pi}{4}\right) = \ln\left(\cos\left(-\frac{\pi}{4}\right)\right) = \ln\left(\frac{1}{\sqrt{2}}\right) = -\frac{1}{2}\ln 2$ and $f\left(\frac{\pi}{3}\right) = \ln\left(\cos\left(\frac{\pi}{3}\right)\right) = \ln\frac{1}{2} = -\ln 2$. Therefore, the absolute minimum occurs at $x = \frac{\pi}{3}$ with $f\left(\frac{\pi}{3}\right) = -\ln 2$ and the absolute maximum occurs at $x = 0$ with $f(0) = 0$.

(b) $f(x) = \cos(\ln x) \Rightarrow f'(x) = \frac{-\sin(\ln x)}{x} = 0 \Rightarrow x = 1$; $f'(x) > 0$ for $\frac{1}{2} \le x < 1$ and $f'(x) < 0$ for $1 < x \le 2$ $\Rightarrow$ there is a relative maximum at $x = 1$ with $f(1) = \cos(\ln 1) = \cos 0 = 1$; $f\left(\frac{1}{2}\right) = \cos\left(\ln\left(\frac{1}{2}\right)\right) = \cos(-\ln 2) = \cos(\ln 2)$ and $f(2) = \cos(\ln 2)$. Therefore, the absolute minimum occurs at $x = \frac{1}{2}$ and $x = 2$ with $f\left(\frac{1}{2}\right) = f(2) = \cos(\ln 2)$, and the absolute maximum occurs at $x = 1$ with $f(1) = 1$.

71. $\int_1^5 (\ln 2x - \ln x)\,dx = \int_1^5 (-\ln x + \ln 2 + \ln x)\,dx = (\ln 2)\int_1^5 dx = (\ln 2)(5-1) = \ln 2^4 = \ln 16$

73. $V = \pi\int_0^3 \left(\frac{2}{\sqrt{y+1}}\right)^2 dy = 4\pi\int_0^3 \frac{1}{y+1}\,dy = 4\pi\left[\ln|y+1|\right]_0^3 = 4\pi(\ln 4 - \ln 1) = 4\pi\ln 4$

75. $V = 2\pi\int_{1/2}^2 x\left(\frac{1}{x^2}\right)dx = 2\pi\int_{1/2}^2 \frac{1}{x}\,dx = 2\pi\left[\ln|x|\right]_{1/2}^2 = 2\pi\left(\ln 2 - \ln\frac{1}{2}\right) = 2\pi(2\ln 2) = \pi\ln 2^4 = \pi\ln 16$

77. (a) $y = \frac{x^2}{8} - \ln x \Rightarrow 1 + (y')^2 = 1 + \left(\frac{x}{4} - \frac{1}{x}\right)^2 = 1 + \left(\frac{x^2-4}{4x}\right)^2 = \left(\frac{x^2+4}{4x}\right)^2 \Rightarrow L = \int_4^8 \sqrt{1+(y')^2}\,dx$
$= \int_4^8 \frac{x^2+4}{4x}\,dx = \int_4^8 \left(\frac{x}{4} + \frac{1}{x}\right)dx = \left[\frac{x^2}{8} + \ln|x|\right]_4^8 = (8 + \ln 8) - (2 + \ln 4) = 6 + \ln 2$

(b) $x = \left(\frac{y}{4}\right)^2 - 2\ln\left(\frac{y}{4}\right) \Rightarrow \frac{dx}{dy} = \frac{y}{8} - \frac{2}{y} \Rightarrow 1 + \left(\frac{dx}{dy}\right)^2 = 1 + \left(\frac{y}{8} - \frac{2}{y}\right)^2 = 1 + \left(\frac{y^2-16}{8y}\right)^2 = \left(\frac{y^2+16}{8y}\right)^2$
$\Rightarrow L = \int_4^{12} \sqrt{1 + \left(\frac{dx}{dy}\right)^2}\,dy = \int_4^{12} \frac{y^2+16}{8y}\,dy = \int_4^{12}\left(\frac{y}{8} + \frac{2}{y}\right)dy = \left[\frac{y^2}{16} + 2\ln y\right]_4^{12} = (9 + 2\ln 12) - (1 + 2\ln 4)$
$= 8 + 2\ln 3 = 8 + \ln 9$

79. (a) $M_y = \int_1^2 x\left(\frac{1}{x}\right)dx = 1$, $M_x = \int_1^2 \left(\frac{1}{2x}\right)\left(\frac{1}{x}\right)dx = \frac{1}{2}\int_1^2 \frac{1}{x^2}\,dx = \left[-\frac{1}{2x}\right]_1^2 = \frac{1}{4}$, $M = \int_1^2 \frac{1}{x}\,dx = \left[\ln|x|\right]_1^2 = \ln 2$
$\Rightarrow \bar{x} = \frac{M_y}{M} = \frac{1}{\ln 2} \approx 1.44$ and $\bar{y} = \frac{M_x}{M} = \frac{\left(\frac{1}{4}\right)}{\ln 2} \approx 0.36$

(b)

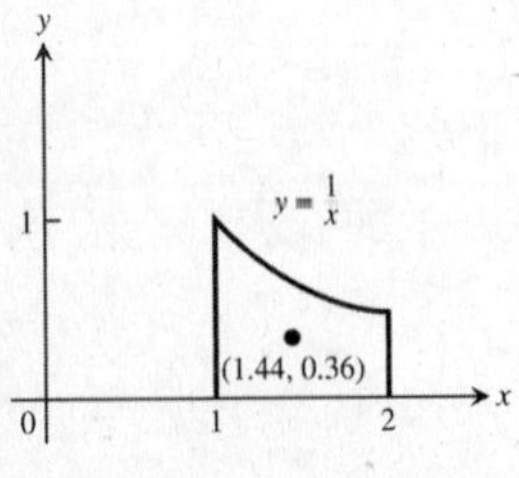

81. $f(x) = \ln(x^3 - 1)$, domain of f: $(1, \infty) \Rightarrow f'(x) = \frac{3x^2}{x^3 - 1}$; $f'(x) = 0 \Rightarrow 3x^2 = 0 \Rightarrow x = 0$, not in the domain; $f'(x) =$ undefined $\Rightarrow x^3 - 1 = 0 \Rightarrow x = 1$, not in domain. On $(1, \infty)$, $f'(x) > 0 \Rightarrow$ f is increasing on $(1, \infty)$ $\Rightarrow$ f is one-to-one

83. $\frac{dy}{dx} = 1 + \frac{1}{x}$ at $(1, 3) \Rightarrow y = x + \ln|x| + C$; $y = 3$ at $x = 1 \Rightarrow C = 2 \Rightarrow y = x + \ln|x| + 2$

85. (a) $L(x) = f(0) + f'(0) \cdot x$, and $f(x) = \ln(1 + x) \Rightarrow f'(x)\big|_{x=0} = \frac{1}{1+x}\Big|_{x=0} = 1 \Rightarrow L(x) = \ln 1 + 1 \cdot x \Rightarrow L(x) = x$

(b) Let $f(x) = \ln(x + 1)$. Since $f''(x) = -\frac{1}{(x+1)^2} < 0$ on $[0, 0.1]$, the graph of f is concave down on this interval and the largest error in the linear approximation will occur when $x = 0.1$. This error is $0.1 - \ln(1.1) \approx 0.00469$ to five decimal places.

(c) The approximation $y = x$ for $\ln(1 + x)$ is best for smaller positive values of x; in particular for $0 \le x \le 0.1$ in the graph. As x increases, so does the error $x - \ln(1 + x)$. From the graph an upper bound for the error is $0.5 - \ln(1 + 0.5) \approx 0.095$; i.e., $|E(x)| \le 0.095$ for $0 \le x \le 0.5$. Note from the graph that $0.1 - \ln(1 + 0.1) \approx 0.00469$ estimates the error in replacing $\ln(1 + x)$ by x over $0 \le x \le 0.1$. This is consistent with the estimate given in part (b) above.

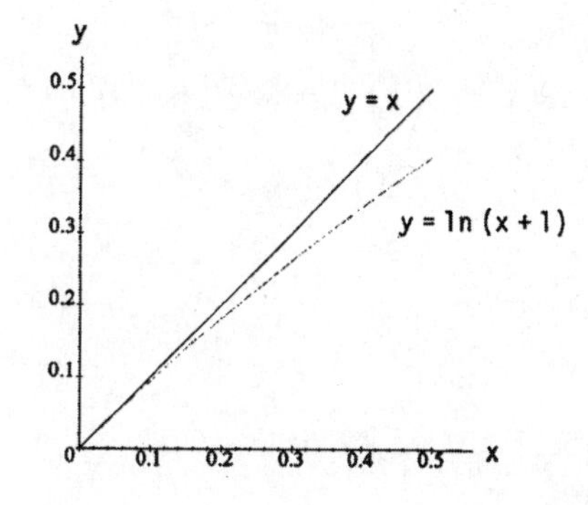

87. (a)

y = ln (a + sin x)

(b) $y' = \frac{\cos x}{a + \sin x}$. Since $|\sin x|$ and $|\cos x|$ are less than or equal to 1, we have for $a > 1$

$\frac{-1}{a-1} \le y' \le \frac{1}{a-1}$ for all x.

Thus, $\lim_{a \to +\infty} y' = 0$ for all $x \Rightarrow$ the graph of y looks more and more horizontal as $a \to +\infty$.

7.3 EXPONENTIAL FUNCTIONS

1. (a) $e^{-0.3t} = 27 \Rightarrow \ln e^{-0.3t} = \ln 3^3 \Rightarrow (-0.3t)\ln e = 3\ln 3 \Rightarrow -0.3t = 3\ln 3 \Rightarrow t = -10\ln 3$

(b) $e^{kt} = \frac{1}{2} \Rightarrow \ln e^{kt} = \ln 2^{-1} = kt\ln e = -\ln 2 \Rightarrow t = -\frac{\ln 2}{k}$

(c) $e^{(\ln 0.2)t} = 0.4 \Rightarrow \left(e^{\ln 0.2}\right)^t = 0.4 \Rightarrow 0.2^t = 0.4 \Rightarrow \ln 0.2^t = \ln 0.4 \Rightarrow t\ln 0.2 = \ln 0.4 \Rightarrow t = \frac{\ln 0.4}{\ln 0.2}$

3. $e^{\sqrt{t}} = x^2 \Rightarrow \ln e^{\sqrt{t}} = \ln x^2 \Rightarrow \sqrt{t} = 2\ln x \Rightarrow t = 4(\ln x)^2$

5. $y = e^{-5x} \Rightarrow y' = e^{-5x}\frac{d}{dx}(-5x) \Rightarrow y' = -5e^{-5x}$

7. $y = e^{5-7x} \Rightarrow y' = e^{5-7x}\frac{d}{dx}(5 - 7x) \Rightarrow y' = -7e^{5-7x}$

9. $y = xe^x - e^x \Rightarrow y' = (e^x + xe^x) - e^x = xe^x$

11. $y = (x^2 - 2x + 2)e^x \Rightarrow y' = (2x - 2)e^x + (x^2 - 2x + 2)e^x = x^2e^x$

13. $y = e^\theta(\sin\theta + \cos\theta) \Rightarrow y' = e^\theta(\sin\theta + \cos\theta) + e^\theta(\cos\theta - \sin\theta) = 2e^\theta\cos\theta$

15. $y = \cos\left(e^{-\theta^2}\right) \Rightarrow \frac{dy}{d\theta} = -\sin\left(e^{-\theta^2}\right)\frac{d}{d\theta}\left(e^{-\theta^2}\right) = \left(-\sin\left(e^{-\theta^2}\right)\right)\left(e^{-\theta^2}\right)\frac{d}{d\theta}(-\theta^2) = 2\theta e^{-\theta^2}\sin\left(e^{-\theta^2}\right)$

17. $y = \ln(3te^{-t}) = \ln 3 + \ln t + \ln e^{-t} = \ln 3 + \ln t - t \Rightarrow \frac{dy}{dt} = \frac{1}{t} - 1 = \frac{1-t}{t}$

19. $y = \ln\frac{e^\theta}{1+e^\theta} = \ln e^\theta - \ln(1+e^\theta) = \theta - \ln(1+e^\theta) \Rightarrow \frac{dy}{d\theta} = 1 - \left(\frac{1}{1+e^\theta}\right)\frac{d}{d\theta}(1+e^\theta) = 1 - \frac{e^\theta}{1+e^\theta} = \frac{1}{1+e^\theta}$

21. $y = e^{(\cos t + \ln t)} = e^{\cos t}e^{\ln t} = te^{\cos t} \Rightarrow \frac{dy}{dt} = e^{\cos t} + te^{\cos t}\frac{d}{dt}(\cos t) = (1 - t\sin t)e^{\cos t}$

23. $\int_0^{\ln x}\sin e^t\,dt \Rightarrow y' = \left(\sin e^{\ln x}\right)\cdot\frac{d}{dx}(\ln x) = \frac{\sin x}{x}$

25. $\ln y = e^y\sin x \Rightarrow \left(\frac{1}{y}\right)y' = (y'e^y)(\sin x) + e^y\cos x \Rightarrow y'\left(\frac{1}{y} - e^y\sin x\right) = e^y\cos x$
$\Rightarrow y'\left(\frac{1-ye^y\sin x}{y}\right) = e^y\cos x \Rightarrow y' = \frac{ye^y\cos x}{1-ye^y\sin x}$

27. $e^{2x} = \sin(x+3y) \Rightarrow 2e^{2x} = (1+3y')\cos(x+3y) \Rightarrow 1+3y' = \frac{2e^{2x}}{\cos(x+3y)} \Rightarrow 3y' = \frac{2e^{2x}}{\cos(x+3y)} - 1 \Rightarrow y' = \frac{2e^{2x} - \cos(x+3y)}{3\cos(x+3y)}$

29. $\int(e^{3x} + 5e^{-x})\,dx = \frac{e^{3x}}{3} - 5e^{-x} + C$

31. $\int_{\ln 2}^{\ln 3} e^x\,dx = [e^x]_{\ln 2}^{\ln 3} = e^{\ln 3} - e^{\ln 2} = 3 - 2 = 1$

33. $\int 8e^{(x+1)}\,dx = 8e^{(x+1)} + C$

35. $\int_{\ln 4}^{\ln 9} e^{x/2}\,dx = \left[2e^{x/2}\right]_{\ln 4}^{\ln 9} = 2\left[e^{(\ln 9)/2} - e^{(\ln 4)/2}\right] = 2\left(e^{\ln 3} - e^{\ln 2}\right) = 2(3-2) = 2$

37. Let $u = r^{1/2} \Rightarrow du = \frac{1}{2}r^{-1/2}\,dr \Rightarrow 2\,du = r^{-1/2}\,dr;$
$\int\frac{e^{\sqrt{r}}}{\sqrt{r}}\,dr = \int e^{r^{1/2}}\cdot r^{-1/2}\,dr = 2\int e^u\,du = 2e^u + C = 2e^{r^{1/2}} + C = 2e^{\sqrt{r}} + C$

39. Let $u = -t^2 \Rightarrow du = -2t\,dt \Rightarrow -du = 2t\,dt;$
$\int 2te^{-t^2}\,dt = -\int e^u\,du = -e^u + C = -e^{-t^2} + C$

41. Let $u = \frac{1}{x} \Rightarrow du = -\frac{1}{x^2}\,dx \Rightarrow -du = \frac{1}{x^2}\,dx;$
$\int\frac{e^{1/x}}{x^2}\,dx = \int -e^u\,du = -e^u + C = -e^{1/x} + C$

43. Let $u = \tan\theta \Rightarrow du = \sec^2\theta\,d\theta;\ \theta = 0 \Rightarrow u = 0,\ \theta = \frac{\pi}{4} \Rightarrow u = 1;$
$\int_0^{\pi/4}\left(1 + e^{\tan\theta}\right)\sec^2\theta\,d\theta = \int_0^{\pi/4}\sec^2\theta\,d\theta + \int_0^1 e^u\,du = [\tan\theta]_0^{\pi/4} + [e^u]_0^1 = \left[\tan\left(\frac{\pi}{4}\right) - \tan(0)\right] + (e^1 - e^0)$
$= (1-0) + (e-1) = e$

45. Let $u = \sec\pi t \Rightarrow du = \pi\sec\pi t\tan\pi t\,dt \Rightarrow \frac{du}{\pi} = \sec\pi t\tan\pi t\,dt;$
$\int e^{\sec(\pi t)}\sec(\pi t)\tan(\pi t)\,dt = \frac{1}{\pi}\int e^u\,du = \frac{e^u}{\pi} + C = \frac{e^{\sec(\pi t)}}{\pi} + C$

47. Let $u = e^v \Rightarrow du = e^v\,dv \Rightarrow 2\,du = 2e^v\,dv;\ v = \ln\frac{\pi}{6} \Rightarrow u = \frac{\pi}{6},\ v = \ln\frac{\pi}{2} \Rightarrow u = \frac{\pi}{2};$
$\int_{\ln(\pi/6)}^{\ln(\pi/2)} 2e^v\cos e^v\,dv = 2\int_{\pi/6}^{\pi/2}\cos u\,du = [2\sin u]_{\pi/6}^{\pi/2} = 2\left[\sin\left(\frac{\pi}{2}\right) - \sin\left(\frac{\pi}{6}\right)\right] = 2\left(1 - \frac{1}{2}\right) = 1$

49. Let $u = 1 + e^r \Rightarrow du = e^r\,dr$;
$\int \frac{e^r}{1+e^r}\,dr = \int \frac{1}{u}\,du = \ln|u| + C = \ln(1+e^r) + C$

51. $\frac{dy}{dt} = e^t \sin(e^t - 2) \Rightarrow y = \int e^t \sin(e^t - 2)\,dt$;
let $u = e^t - 2 \Rightarrow du = e^t\,dt \Rightarrow y = \int \sin u\,du = -\cos u + C = -\cos(e^t - 2) + C$; $y(\ln 2) = 0$
$\Rightarrow -\cos\left(e^{\ln 2} - 2\right) + C = 0 \Rightarrow -\cos(2-2) + C = 0 \Rightarrow C = \cos 0 = 1$; thus, $y = 1 - \cos(e^t - 2)$

53. $\frac{d^2y}{dx^2} = 2e^{-x} \Rightarrow \frac{dy}{dx} = -2e^{-x} + C$; $x = 0$ and $\frac{dy}{dx} = 0 \Rightarrow 0 = -2e^0 + C \Rightarrow C = 2$; thus $\frac{dy}{dx} = -2e^{-x} + 2$
$\Rightarrow y = 2e^{-x} + 2x + C_1$; $x = 0$ and $y = 1 \Rightarrow 1 = 2e^0 + C_1 \Rightarrow C_1 = -1 \Rightarrow y = 2e^{-x} + 2x - 1 = 2(e^{-x} + x) - 1$

55. $y = 2^x \Rightarrow y' = 2^x \ln 2$

57. $y = 5^{\sqrt{s}} \Rightarrow \frac{dy}{ds} = 5^{\sqrt{s}}(\ln 5)\left(\frac{1}{2}s^{-1/2}\right) = \left(\frac{\ln 5}{2\sqrt{s}}\right)5^{\sqrt{s}}$

59. $y = x^\pi \Rightarrow y' = \pi x^{(\pi - 1)}$

61. $y = (\cos\theta)^{\sqrt{2}} \Rightarrow \frac{dy}{d\theta} = -\sqrt{2}(\cos\theta)^{(\sqrt{2}-1)}(\sin\theta)$

63. $y = 7^{\sec\theta}\ln 7 \Rightarrow \frac{dy}{d\theta} = (7^{\sec\theta}\ln 7)(\ln 7)(\sec\theta\tan\theta) = 7^{\sec\theta}(\ln 7)^2(\sec\theta\tan\theta)$

65. $y = 2^{\sin 3t} \Rightarrow \frac{dy}{dt} = (2^{\sin 3t}\ln 2)(\cos 3t)(3) = (3\cos 3t)(2^{\sin 3t})(\ln 2)$

67. $y = \log_2 5\theta = \frac{\ln 5\theta}{\ln 2} \Rightarrow \frac{dy}{d\theta} = \left(\frac{1}{\ln 2}\right)\left(\frac{1}{5\theta}\right)(5) = \frac{1}{\theta\ln 2}$

69. $y = \frac{\ln x}{\ln 4} + \frac{\ln x^2}{\ln 4} = \frac{\ln x}{\ln 4} + 2\frac{\ln x}{\ln 4} = 3\frac{\ln x}{\ln 4} \Rightarrow y' = \frac{3}{x\ln 4}$

71. $y = x^3\log_{10}x = x^3\left(\frac{\ln x}{\ln 10}\right) = \frac{1}{\ln 10}x^3\ln x \Rightarrow y' = \frac{1}{\ln 10}\left(x^3\cdot\frac{1}{x} + 3x^2\ln x\right) = \frac{1}{\ln 10}x^2 + 3x^2\frac{\ln x}{\ln 10} = \frac{1}{\ln 10}x^2 + 3x^2\log_{10}x$

73. $y = \log_3\left(\left(\frac{x+1}{x-1}\right)^{\ln 3}\right) = \frac{\ln\left(\frac{x+1}{x-1}\right)^{\ln 3}}{\ln 3} = \frac{(\ln 3)\ln\left(\frac{x+1}{x-1}\right)}{\ln 3} = \ln\left(\frac{x+1}{x-1}\right) = \ln(x+1) - \ln(x-1)$
$\Rightarrow \frac{dy}{dx} = \frac{1}{x+1} - \frac{1}{x-1} = \frac{-2}{(x+1)(x-1)}$

75. $y = \theta\sin(\log_7\theta) = \theta\sin\left(\frac{\ln\theta}{\ln 7}\right) \Rightarrow \frac{dy}{d\theta} = \sin\left(\frac{\ln\theta}{\ln 7}\right) + \theta\left[\cos\left(\frac{\ln\theta}{\ln 7}\right)\right]\left(\frac{1}{\theta\ln 7}\right) = \sin(\log_7\theta) + \frac{1}{\ln 7}\cos(\log_7\theta)$

77. $y = \log_{10}e^x = \frac{\ln e^x}{\ln 10} = \frac{x}{\ln 10} \Rightarrow y' = \frac{1}{\ln 10}$

79. $y = 3^{\log_2 t} = 3^{(\ln t)/(\ln 2)} \Rightarrow \frac{dy}{dt} = [3^{(\ln t)/(\ln 2)}(\ln 3)]\left(\frac{1}{t\ln 2}\right) = \frac{1}{t}(\log_2 3)\,3^{\log_2 t}$

81. $y = \log_2(8t^{\ln 2}) = \frac{\ln 8 + \ln(t^{\ln 2})}{\ln 2} = \frac{3\ln 2 + (\ln 2)(\ln t)}{\ln 2} = 3 + \ln t \Rightarrow \frac{dy}{dt} = \frac{1}{t}$

83. $\int 5^x\,dx = \frac{5^x}{\ln 5} + C$

85. $\int_0^1 2^{-\theta}\,d\theta = \int_0^1 \left(\frac{1}{2}\right)^\theta d\theta = \left[\frac{\left(\frac{1}{2}\right)^\theta}{\ln\left(\frac{1}{2}\right)}\right]_0^1 = \frac{\frac{1}{2}}{\ln\left(\frac{1}{2}\right)} - \frac{1}{\ln\left(\frac{1}{2}\right)} = -\frac{\frac{1}{2}}{\ln\left(\frac{1}{2}\right)} = \frac{-1}{2(\ln 1 - \ln 2)} = \frac{1}{2\ln 2}$

87. Let $u = x^2 \Rightarrow du = 2x\,dx \Rightarrow \frac{1}{2}\,du = x\,dx$; $x = 1 \Rightarrow u = 1$, $x = \sqrt{2} \Rightarrow u = 2$;
$\int_1^{\sqrt{2}} x2^{(x^2)}\,dx = \int_1^2 \left(\frac{1}{2}\right) 2^u\,du = \frac{1}{2}\left[\frac{2^u}{\ln 2}\right]_1^2 = \left(\frac{1}{2\ln 2}\right)(2^2 - 2^1) = \frac{1}{\ln 2}$

89. Let $u = \cos t \Rightarrow du = -\sin t\,dt \Rightarrow -du = \sin t\,dt$; $t = 0 \Rightarrow u = 1$, $t = \frac{\pi}{2} \Rightarrow u = 0$;
$\int_0^{\pi/2} 7^{\cos t}\sin t\,dt = -\int_1^0 7^u\,du = \left[-\frac{7^u}{\ln 7}\right]_1^0 = \left(\frac{-1}{\ln 7}\right)(7^0 - 7) = \frac{6}{\ln 7}$

91. Let $u = x^{2x} \Rightarrow \ln u = 2x\ln x \Rightarrow \frac{1}{u}\frac{du}{dx} = 2\ln x + (2x)\left(\frac{1}{x}\right) \Rightarrow \frac{du}{dx} = 2u(\ln x + 1) \Rightarrow \frac{1}{2}\,du = x^{2x}(1 + \ln x)\,dx$;
$x = 2 \Rightarrow u = 2^4 = 16$, $x = 4 \Rightarrow u = 4^8 = 65{,}536$;
$\int_2^4 x^{2x}(1 + \ln x)\,dx = \frac{1}{2}\int_{16}^{65{,}536} du = \frac{1}{2}[u]_{16}^{65{,}536} = \frac{1}{2}(65{,}536 - 16) = \frac{65{,}520}{2} = 32{,}760$

93. $\int 3x^{\sqrt{3}}\,dx = \frac{3x^{(\sqrt{3}+1)}}{\sqrt{3}+1} + C$

95. $\int_0^3 \left(\sqrt{2}+1\right) x^{\sqrt{2}}\,dx = \left[x^{(\sqrt{2}+1)}\right]_0^3 = 3^{(\sqrt{2}+1)}$

97. $\int \frac{\log_{10} x}{x}\,dx = \int \left(\frac{\ln x}{\ln 10}\right)\left(\frac{1}{x}\right)dx$; $\left[u = \ln x \Rightarrow du = \frac{1}{x}\,dx\right]$
$\to \int \left(\frac{\ln x}{\ln 10}\right)\left(\frac{1}{x}\right)dx = \frac{1}{\ln 10}\int u\,du = \left(\frac{1}{\ln 10}\right)\left(\frac{1}{2}u^2\right) + C = \frac{(\ln x)^2}{2\ln 10} + C$

99. $\int_1^4 \frac{\ln 2\log_2 x}{x}\,dx = \int_1^4 \left(\frac{\ln 2}{x}\right)\left(\frac{\ln x}{\ln 2}\right)dx = \int_1^4 \frac{\ln x}{x}\,dx = \left[\frac{1}{2}(\ln x)^2\right]_1^4 = \frac{1}{2}\left[(\ln 4)^2 - (\ln 1)^2\right] = \frac{1}{2}(\ln 4)^2 = \frac{1}{2}(2\ln 2)^2 = 2(\ln 2)^2$

101. $\int_0^2 \frac{\log_2(x+2)}{x+2}\,dx = \frac{1}{\ln 2}\int_0^2 [\ln(x+2)]\left(\frac{1}{x+2}\right)dx = \left(\frac{1}{\ln 2}\right)\left[\frac{(\ln(x+2))^2}{2}\right]_0^2 = \left(\frac{1}{\ln 2}\right)\left[\frac{(\ln 4)^2}{2} - \frac{(\ln 2)^2}{2}\right]$
$= \left(\frac{1}{\ln 2}\right)\left[\frac{4(\ln 2)^2}{2} - \frac{(\ln 2)^2}{2}\right] = \frac{3}{2}\ln 2$

103. $\int_0^9 \frac{2\log_{10}(x+1)}{x+1}\,dx = \frac{2}{\ln 10}\int_0^9 \ln(x+1)\left(\frac{1}{x+1}\right)dx = \left(\frac{2}{\ln 10}\right)\left[\frac{(\ln(x+1))^2}{2}\right]_0^9 = \left(\frac{2}{\ln 10}\right)\left[\frac{(\ln 10)^2}{2} - \frac{(\ln 1)^2}{2}\right] = \ln 10$

105. $\int \frac{dx}{x\log_{10} x} = \int \left(\frac{\ln 10}{\ln x}\right)\left(\frac{1}{x}\right)dx = (\ln 10)\int \left(\frac{1}{\ln x}\right)\left(\frac{1}{x}\right)dx$; $\left[u = \ln x \Rightarrow du = \frac{1}{x}\,dx\right]$
$\to (\ln 10)\int \left(\frac{1}{\ln x}\right)\left(\frac{1}{x}\right)dx = (\ln 10)\int \frac{1}{u}\,du = (\ln 10)\ln|u| + C = (\ln 10)\ln|\ln x| + C$

107. $\int_1^{\ln x} \frac{1}{t}\,dt = [\ln|t|]_1^{\ln x} = \ln|\ln x| - \ln 1 = \ln(\ln x)$, $x > 1$

109. $\int_1^{1/x} \frac{1}{t}\,dt = [\ln|t|]_1^{1/x} = \ln\left|\frac{1}{x}\right| - \ln 1 = (\ln 1 - \ln|x|) - \ln 1 = -\ln x$, $x > 0$

111. $y = (x+1)^x \Rightarrow \ln y = \ln(x+1)^x = x\ln(x+1) \Rightarrow \frac{y'}{y} = \ln(x+1) + x\cdot\frac{1}{(x+1)} \Rightarrow y' = (x+1)^x\left[\frac{x}{x+1} + \ln(x+1)\right]$

113. $y = \left(\sqrt{t}\right)^t = \left(t^{1/2}\right)^t = t^{t/2} \Rightarrow \ln y = \ln t^{t/2} = \left(\frac{t}{2}\right)\ln t \Rightarrow \frac{1}{y}\frac{dy}{dt} = \left(\frac{1}{2}\right)(\ln t) + \left(\frac{t}{2}\right)\left(\frac{1}{t}\right) = \frac{\ln t}{2} + \frac{1}{2}$
$\Rightarrow \frac{dy}{dt} = \left(\sqrt{t}\right)^t\left(\frac{\ln t}{2} + \frac{1}{2}\right)$

115. $y = (\sin x)^x \Rightarrow \ln y = \ln(\sin x)^x = x\ln(\sin x) \Rightarrow \frac{y'}{y} = \ln(\sin x) + x\left(\frac{\cos x}{\sin x}\right) \Rightarrow y' = (\sin x)^x[\ln(\sin x) + x\cot x]$

117. $y = \sin x^x \Rightarrow y' = \cos x^x\,\frac{d}{dx}(x^x)$; if $u = x^x \Rightarrow \ln u = \ln x^x = x\ln x \Rightarrow \frac{u'}{u} = x\cdot\frac{1}{x} + 1\cdot\ln x = 1 + \ln x$
$\Rightarrow u' = x^x(1 + \ln x) \Rightarrow y' = \cos x^x\cdot x^x(1 + \ln x) = x^x\cos x^x(1 + \ln x)$

119. $f(x) = e^x - 2x \Rightarrow f'(x) = e^x - 2$; $f'(x) = 0 \Rightarrow e^x = 2 \Rightarrow x = \ln 2$; $f(0) = 1$, the absolute maximum; $f(\ln 2) = 2 - 2\ln 2 \approx 0.613706$, the absolute minimum; $f(1) = e - 2 \approx 0.71828$, a relative or local maximum since $f''(x) = e^x$ is always positive.

121. $f(x) = x e^{-x} \Rightarrow f'(x) = x e^{-x}(-1) + e^{-x} = e^{-x} - x e^{-x} \Rightarrow f''(x) = -e^{-x} - (x e^{-x}(-1) + e^{-x}) = x e^{-x} - 2e^{-x}$

(a) $f'(x) = 0 \Rightarrow e^{-x} - x e^{-x} = e^{-x}(1 - x) = 0 \Rightarrow e^{-x} = 0$ or $1 - x = 0 \Rightarrow x = 1$, $f(1) = (1)e^{-1} = \frac{1}{e}$; using second derivative test, $f''(1) = (1)e^{-1} - 2e^{-1} = -\frac{1}{e} < 0 \Rightarrow$ absolute maximum at $\left(1, \frac{1}{e}\right)$

(b) $f''(x) = 0 \Rightarrow x e^{-x} - 2e^{-x} = e^{-x}(x - 2) = 0 \Rightarrow e^{-x} = 0$ or $x - 2 = 0 \Rightarrow x = 2$, $f(2) = (2)e^{-2} = \frac{2}{e^2}$; since $f''(1) < 0$ and $f''(3) = e^{-3}(3 - 2) = \frac{1}{e^3} > 0 \Rightarrow$ point of inflection at $\left(2, \frac{2}{e^2}\right)$

123. $f(x) = x^2 \ln \frac{1}{x} \Rightarrow f'(x) = 2x \ln \frac{1}{x} + x^2 \left(\frac{1}{\frac{1}{x}}\right)(-x^{-2}) = 2x \ln \frac{1}{x} - x = -x(2\ln x + 1)$; $f'(x) = 0 \Rightarrow x = 0$ or $\ln x = -\frac{1}{2}$. Since $x = 0$ is not in the domain of f, $x = e^{-1/2} = \frac{1}{\sqrt{e}}$. Also, $f'(x) > 0$ for $0 < x < \frac{1}{\sqrt{e}}$ and $f'(x) < 0$ for $x > \frac{1}{\sqrt{e}}$. Therefore, $f\left(\frac{1}{\sqrt{e}}\right) = \frac{1}{e} \ln \sqrt{e} = \frac{1}{e} \ln e^{1/2} = \frac{1}{2e} \ln e = \frac{1}{2e}$ is the absolute maximum value of f assumed at $x = \frac{1}{\sqrt{e}}$.

125. $\int_0^{\ln 3} (e^{2x} - e^x)\,dx = \left[\frac{e^{2x}}{2} - e^x\right]_0^{\ln 3} = \left(\frac{e^{2\ln 3}}{2} - e^{\ln 3}\right) - \left(\frac{e^0}{2} - e^0\right) = \left(\frac{9}{2} - 3\right) - \left(\frac{1}{2} - 1\right) = \frac{8}{2} - 2 = 2$

127. $L = \int_0^1 \sqrt{1 + \frac{e^x}{4}}\,dx \Rightarrow \frac{dy}{dx} = \frac{e^{x/2}}{2} \Rightarrow y = e^{x/2} + C$; $y(0) = 0 \Rightarrow 0 = e^0 + C \Rightarrow C = -1 \Rightarrow y = e^{x/2} - 1$

129. $y = \frac{1}{2}(e^x + e^{-x}) \Rightarrow \frac{dy}{dx} = \frac{1}{2}(e^x - e^{-x})$; $L = \int_0^1 \sqrt{1 + \left(\frac{1}{2}(e^x - e^{-x})\right)^2}\,dx = \int_0^1 \sqrt{1 + \frac{e^{2x}}{4} - \frac{1}{2} + \frac{e^{-2x}}{4}}\,dx$
$= \int_0^1 \sqrt{\frac{e^{2x}}{4} + \frac{1}{2} + \frac{e^{-2x}}{4}}\,dx = \int_0^1 \sqrt{\left(\frac{1}{2}(e^x + e^{-x})\right)^2}\,dx = \int_0^1 \frac{1}{2}(e^x + e^{-x})\,dx = \frac{1}{2}\left[e^x - e^{-x}\right]_0^1 = \frac{1}{2}\left(e - \frac{1}{e}\right) - 0 = \frac{e^2 - 1}{2e}$

131. $y = \ln \cos x \Rightarrow \frac{dy}{dx} = \frac{-\sin x}{\cos x} = -\tan x$; $L = \int_0^{\pi/4} \sqrt{1 + (-\tan x)^2}\,dx = \int_0^{\pi/4} \sqrt{1 + \tan^2 x}\,dx = \int_0^{\pi/4} \sqrt{\sec^2 x}\,dx$
$= \int_0^{\pi/4} \sec x\,dx = \left[\ln|\sec x + \tan x|\right]_0^{\pi/4} = \left(\ln\left|\sec\left(\frac{\pi}{4}\right) + \tan\left(\frac{\pi}{4}\right)\right|\right) - (0) = \ln\left(\sqrt{2} + 1\right)$

133. (a) $\frac{d}{dx}(x \ln x - x + C) = x \cdot \frac{1}{x} + \ln x - 1 + 0 = \ln x$

(b) average value $= \frac{1}{e - 1}\int_1^e \ln x\,dx = \frac{1}{e - 1}\left[x \ln x - x\right]_1^e = \frac{1}{e - 1}\left[(e \ln e - e) - (1 \ln 1 - 1)\right] = \frac{1}{e - 1}(e - e + 1) = \frac{1}{e - 1}$

135. (a) $f(x) = e^x \Rightarrow f'(x) = e^x$; $L(x) = f(0) + f'(0)(x - 0) \Rightarrow L(x) = 1 + x$

(b) $f(0) = 1$ and $L(0) = 1 \Rightarrow$ error $= 0$; $f(0.2) = e^{0.2} \approx 1.22140$ and $L(0.2) = 1.2 \Rightarrow$ error ≈ 0.02140

(c) Since $y'' = e^x > 0$, the tangent line approximation always lies below the curve $y = e^x$. Thus $L(x) = x + 1$ never overestimates e^x.

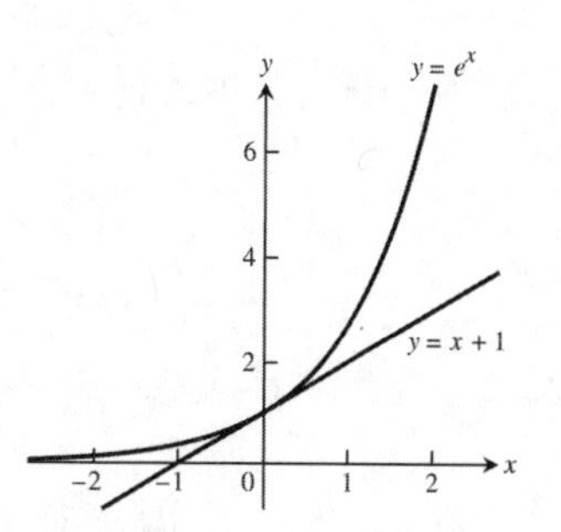

137. $A = \int_{-2}^{2} \frac{2x}{1 + x^2}\,dx = 2\int_0^2 \frac{2x}{1 + x^2}\,dx$; $[u = 1 + x^2 \Rightarrow du = 2x\,dx;\ x = 0 \Rightarrow u = 1,\ x = 2 \Rightarrow u = 5]$
$\rightarrow A = 2\int_1^5 \frac{1}{u}\,du = 2\left[\ln|u|\right]_1^5 = 2(\ln 5 - \ln 1) = 2\ln 5$

139. From zooming in on the graph at the right, we estimate the third root to be $x \approx -0.76666$

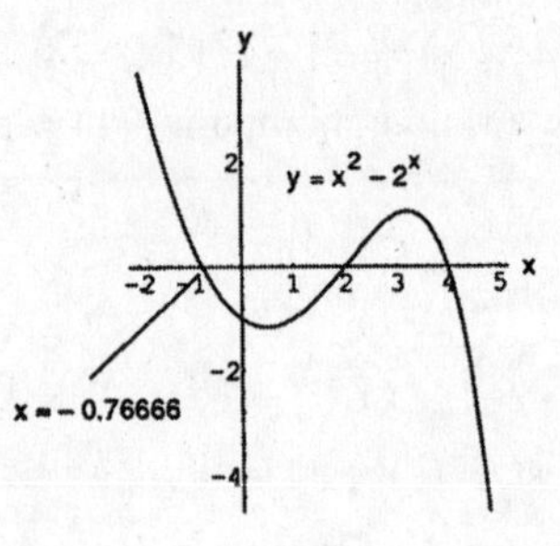

141. (a) $f(x) = 2^x \Rightarrow f'(x) = 2^x \ln 2$; $L(x) = (2^0 \ln 2)\,x + 2^0 = x \ln 2 + 1 \approx 0.69x + 1$

(b)

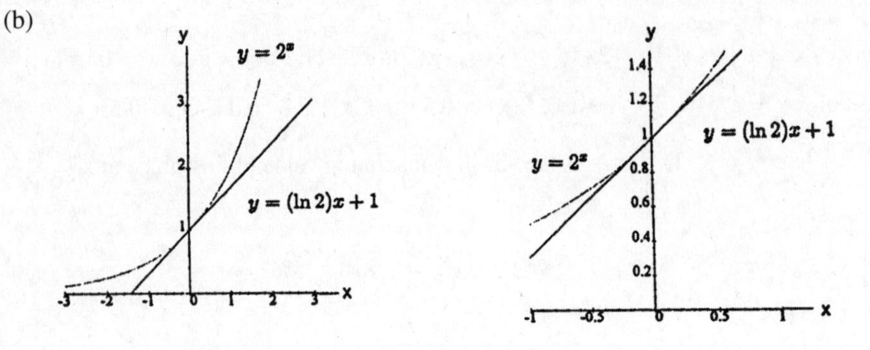

143. (a) The point of tangency is $(p, \ln p)$ and $m_{tangent} = \frac{1}{p}$ since $\frac{dy}{dx} = \frac{1}{x}$. The tangent line passes through $(0, 0) \Rightarrow$ the equation of the tangent line is $y = \frac{1}{p}x$. The tangent line also passes through $(p, \ln p) \Rightarrow \ln p = \frac{1}{p}p = 1 \Rightarrow p = e$, and the tangent line equation is $y = \frac{1}{e}x$.

(b) $\frac{d^2y}{dx^2} = -\frac{1}{x^2}$ for $x \neq 0 \Rightarrow y = \ln x$ is concave downward over its domain. Therefore, $y = \ln x$ lies below the graph of $y = \frac{1}{e}x$ for all $x > 0$, $x \neq e$, and $\ln x < \frac{x}{e}$ for $x > 0$, $x \neq e$.

(c) Multiplying by e, $e \ln x < x$ or $\ln x^e < x$.

(d) Exponentiating both sides of $\ln x^e < x$, we have $e^{\ln x^e} < e^x$, or $x^e < e^x$ for all positive $x \neq e$.

(e) Let $x = \pi$ to see that $\pi^e < e^\pi$. Therefore, e^π is bigger.

7.4 EXPONENTIAL CHANGE AND SEPARABLE DIFFERENTIAL EQUATIONS

1. (a) $y = e^{-x} \Rightarrow y' = -e^{-x} \Rightarrow 2y' + 3y = 2\left(-e^{-x}\right) + 3e^{-x} = e^{-x}$

(b) $y = e^{-x} + e^{-3x/2} \Rightarrow y' = -e^{-x} - \frac{3}{2}e^{-3x/2} \Rightarrow 2y' + 3y = 2\left(-e^{-x} - \frac{3}{2}e^{-3x/2}\right) + 3\left(e^{-x} + e^{-3x/2}\right) = e^{-x}$

(c) $y = e^{-x} + Ce^{-3x/2} \Rightarrow y' = -e^{-x} - \frac{3}{2}Ce^{-3x/2} \Rightarrow 2y' + 3y = 2\left(-e^{-x} - \frac{3}{2}Ce^{-3x/2}\right) + 3\left(e^{-x} + Ce^{-3x/2}\right) = e^{-x}$

3. $y = \frac{1}{x}\int_1^x \frac{e^t}{t}\,dt \Rightarrow y' = -\frac{1}{x^2}\int_1^x \frac{e^t}{t}\,dt + \left(\frac{1}{x}\right)\left(\frac{e^x}{x}\right) \Rightarrow x^2y' = -\int_1^x \frac{e^t}{t}\,dt + e^x = -x\left(\frac{1}{x}\int_1^x \frac{e^t}{t}\,dt\right) + e^x = -xy + e^x$
$\Rightarrow x^2y' + xy = e^x$

5. $y = e^{-x}\tan^{-1}(2e^x) \Rightarrow y' = -e^{-x}\tan^{-1}(2e^x) + e^{-x}\left[\frac{1}{1+(2e^x)^2}\right](2e^x) = -e^{-x}\tan^{-1}(2e^x) + \frac{2}{1+4e^{2x}}$
$\Rightarrow y' = -y + \frac{2}{1+4e^{2x}} \Rightarrow y' + y = \frac{2}{1+4e^{2x}}$; $y(-\ln 2) = e^{-(-\ln 2)}\tan^{-1}(2e^{-\ln 2}) = 2\tan^{-1} 1 = 2\left(\frac{\pi}{4}\right) = \frac{\pi}{2}$

7. $y = \frac{\cos x}{x} \Rightarrow y' = \frac{-x\sin x - \cos x}{x^2} \Rightarrow y' = -\frac{\sin x}{x} - \frac{1}{x}\left(\frac{\cos x}{x}\right) \Rightarrow y' = -\frac{\sin x}{x} - \frac{y}{x} \Rightarrow xy' = -\sin x - y \Rightarrow xy' + y = -\sin x$;
$y\left(\frac{\pi}{2}\right) = \frac{\cos(\pi/2)}{(\pi/2)} = 0$

9. $2\sqrt{xy}\,\frac{dy}{dx} = 1 \Rightarrow 2x^{1/2}y^{1/2}\,dy = dx \Rightarrow 2y^{1/2}\,dy = x^{-1/2}\,dx \Rightarrow \int 2y^{1/2}\,dy = \int x^{-1/2}\,dx \Rightarrow 2\left(\frac{2}{3}y^{3/2}\right) = 2x^{1/2} + C_1$
$\Rightarrow \frac{2}{3}y^{3/2} - x^{1/2} = C$, where $C = \frac{1}{2}C_1$

11. $\frac{dy}{dx} = e^{x-y} \Rightarrow dy = e^x e^{-y}\,dx \Rightarrow e^y\,dy = e^x\,dx \Rightarrow \int e^y\,dy = \int e^x\,dx \Rightarrow e^y = e^x + C \Rightarrow e^y - e^x = C$

13. $\frac{dy}{dx} = \sqrt{y}\cos^2\sqrt{y} \Rightarrow dy = \left(\sqrt{y}\cos^2\sqrt{y}\right)dx \Rightarrow \frac{\sec^2\sqrt{y}}{\sqrt{y}}dy = dx \Rightarrow \int \frac{\sec^2\sqrt{y}}{\sqrt{y}}dy = \int dx$. In the integral on the left-hand side, substitute $u = \sqrt{y} \Rightarrow du = \frac{1}{2\sqrt{y}}dy \Rightarrow 2\,du = \frac{1}{\sqrt{y}}dy$, and we have $\int \sec^2 u\,du = \int dx \Rightarrow 2\tan u = x + C$
$\Rightarrow -x + 2\tan\sqrt{y} = C$

15. $\sqrt{x}\,\frac{dy}{dx} = e^{y+\sqrt{x}} \Rightarrow \frac{dy}{dx} = \frac{e^y e^{\sqrt{x}}}{\sqrt{x}} \Rightarrow dy = \frac{e^y e^{\sqrt{x}}}{\sqrt{x}}dx \Rightarrow e^{-y}\,dy = \frac{e^{\sqrt{x}}}{\sqrt{x}}\,dx \Rightarrow \int e^{-y}\,dy = \int \frac{e^{\sqrt{x}}}{\sqrt{x}}\,dx$. In the integral on the right-hand side, substitute $u = \sqrt{x} \Rightarrow du = \frac{1}{2\sqrt{x}}\,dx \Rightarrow 2\,du = \frac{1}{\sqrt{x}}\,dx$, and we have $\int e^{-y}\,dy = 2\int e^u\,du \Rightarrow -e^{-y} = 2e^u + C_1$
$\Rightarrow -e^{-y} = 2e^{\sqrt{x}} + C$, where $C = -C_1$

17. $\frac{dy}{dx} = 2x\sqrt{1-y^2} \Rightarrow dy = 2x\sqrt{1-y^2}dx \Rightarrow \frac{dy}{\sqrt{1-y^2}} = 2x\,dx \Rightarrow \int \frac{dy}{\sqrt{1-y^2}} = \int 2x\,dx \Rightarrow \sin^{-1}y = x^2 + C$ since $|y| < 1$
$\Rightarrow y = \sin(x^2 + C)$

19. $y^2\frac{dy}{dx} = 3x^2y^3 - 6x^2 \Rightarrow y^2\,dy = 3x^2(y^3 - 2)dx \Rightarrow \frac{y^2}{y^3-2}\,dy = 3x^2dx \Rightarrow \int \frac{y^2}{y^3-2}\,dy = \int 3x^2dx \Rightarrow \frac{1}{3}\ln|y^3 - 2| = x^3 + C$

21. $\frac{1}{x}\frac{dy}{dx} = ye^{x^2} + 2\sqrt{y}\,e^{x^2} = e^{x^2}\left(y + 2\sqrt{y}\right) \Rightarrow \frac{1}{y+2\sqrt{y}}dy = x\,e^{x^2}dx \Rightarrow \int \frac{1}{y+2\sqrt{y}}dy = \int x\,e^{x^2}dx$
$\Rightarrow \int \frac{1}{\sqrt{y}(\sqrt{y}+2)}dy = \int x\,e^{x^2}dx \Rightarrow 2\ln|\sqrt{y}+2| = \frac{1}{2}e^{x^2} + C \Rightarrow 4\ln|\sqrt{y}+2| = e^{x^2} + C \Rightarrow 4\ln\left(\sqrt{y}+2\right) = e^{x^2} + C$

23. (a) $y = y_0e^{kt} \Rightarrow 0.99y_0 = y_0e^{1000k} \Rightarrow k = \frac{\ln 0.99}{1000} \approx -0.00001$
(b) $0.9 = e^{(-0.00001)t} \Rightarrow (-0.00001)t = \ln(0.9) \Rightarrow t = \frac{\ln(0.9)}{-0.00001} \approx 10{,}536$ years
(c) $y = y_0e^{(20{,}000)k} \approx y_0e^{-0.2} = y_0(0.82) \Rightarrow 82\%$

25. $\frac{dy}{dt} = -0.6y \Rightarrow y = y_0e^{-0.6t}$; $y_0 = 100 \Rightarrow y = 100e^{-0.6t} \Rightarrow y = 100e^{-0.6} \approx 54.88$ grams when $t = 1$ hr

27. $L(x) = L_0e^{-kx} \Rightarrow \frac{L_0}{2} = L_0e^{-18k} \Rightarrow \ln\frac{1}{2} = -18k \Rightarrow k = \frac{\ln 2}{18} \approx 0.0385 \Rightarrow L(x) = L_0e^{-0.0385x}$; when the intensity is one-tenth of the surface value, $\frac{L_0}{10} = L_0e^{-0.0385x} \Rightarrow \ln 10 = 0.0385x \Rightarrow x \approx 59.8$ ft

29. $y = y_0e^{kt}$ and $y_0 = 1 \Rightarrow y = e^{kt} \Rightarrow$ at $y = 2$ and $t = 0.5$ we have $2 = e^{0.5k} \Rightarrow \ln 2 = 0.5k \Rightarrow k = \frac{\ln 2}{0.5} = \ln 4$.
Therefore, $y = e^{(\ln 4)t} \Rightarrow y = e^{24\ln 4} = 4^{24} = 2.81474978 \times 10^{14}$ at the end of 24 hrs

31. (a) $10{,}000e^{k(1)} = 7500 \Rightarrow e^k = 0.75 \Rightarrow k = \ln 0.75$ and $y = 10{,}000e^{(\ln 0.75)t}$. Now $1000 = 10{,}000e^{(\ln 0.75)t}$
$\Rightarrow \ln 0.1 = (\ln 0.75)t \Rightarrow t = \frac{\ln 0.1}{\ln 0.75} \approx 8.00$ years (to the nearest hundredth of a year)
(b) $1 = 10{,}000e^{(\ln 0.75)t} \Rightarrow \ln 0.0001 = (\ln 0.75)t \Rightarrow t = \frac{\ln 0.0001}{\ln 0.75} \approx 32.02$ years (to the nearest hundredth of a year)

33. $0.9P_0 = P_0e^k \Rightarrow k = \ln 0.9$; when the well's output falls to one-fifth of its present value $P = 0.2P_0$
$\Rightarrow 0.2P_0 = P_0e^{(\ln 0.9)t} \Rightarrow 0.2 = e^{(\ln 0.9)t} \Rightarrow \ln(0.2) = (\ln 0.9)t \Rightarrow t = \frac{\ln 0.2}{\ln 0.9} \approx 15.28$ yr

35. $A = A_0e^{kt}$ and $A_0 = 10 \Rightarrow A = 10\,e^{kt}$, $5 = 10\,e^{k(24360)} \Rightarrow k = \frac{\ln(0.5)}{24360} \approx -0.000028454 \Rightarrow A = 10\,e^{-0.000028454t}$,
then $0.2(10) = 10\,e^{-0.000028254t} \Rightarrow t = \frac{\ln 0.2}{-0.000028454} \approx 56563$ years

37. $y = y_0e^{-kt} = y_0e^{-(k)(3/k)} = y_0e^{-3} = \frac{y_0}{e^3} < \frac{y_0}{20} = (0.05)(y_0) \Rightarrow$ after three mean lifetimes less than 5% remains

39. $T - T_s = (T_0 - T_s)e^{-kt}$, $T_0 = 90°C$, $T_s = 20°C$, $T = 60°C \Rightarrow 60 - 20 = 70e^{-10k} \Rightarrow \frac{4}{7} = e^{-10k} \Rightarrow k = \frac{\ln\left(\frac{7}{4}\right)}{10} \approx 0.05596$
(a) $35 - 20 = 70e^{-0.05596t} \Rightarrow t \approx 27.5$ min is the total time $\Rightarrow$ it will take $27.5 - 10 = 17.5$ minutes longer to reach 35°C
(b) $T - T_s = (T_0 - T_s)e^{-kt}$, $T_0 = 90°C$, $T_s = -15°C \Rightarrow 35 + 15 = 105e^{-0.05596t} \Rightarrow t \approx 13.26$ min

41. $T - T_s = (T_0 - T_s)e^{-kt} \Rightarrow 39 - T_s = (46 - T_s)e^{-10k}$ and $33 - T_s = (46 - T_s)e^{-20k} \Rightarrow \frac{39-T_s}{46-T_s} = e^{-10k}$ and $\frac{33-T_s}{46-T_s} = e^{-20k} = \left(e^{-10k}\right)^2 \Rightarrow \frac{33-T_s}{46-T_s} = \left(\frac{39-T_s}{46-T_s}\right)^2 \Rightarrow (33 - T_s)(46 - T_s) = (39 - T_s)^2 \Rightarrow 1518 - 79T_s + T_s^2$
$= 1521 - 78T_s + T_s^2 \Rightarrow -T_s = 3 \Rightarrow T_s = -3°C$

43. From Example 4, the half-life of carbon-14 is 5700 yr $\Rightarrow \frac{1}{2}c_0 = c_0e^{-k(5700)} \Rightarrow k = \frac{\ln 2}{5700} \approx 0.0001216 \Rightarrow c = c_0e^{-0.0001216t}$
$\Rightarrow (0.445)c_0 = c_0e^{-0.0001216t} \Rightarrow t = \frac{\ln(0.445)}{-0.0001216} \approx 6659$ years

45. From Exercise 43, $k \approx 0.0001216$ for carbon-14 $\Rightarrow y = y_0e^{-0.0001216t}$. When $t = 5000$
$\Rightarrow y = y_0e^{-0.0001216(5000)} \approx 0.5444y_0 \Rightarrow \frac{y}{y_0} \approx 0.5444 \Rightarrow$ approximately 54.44% remains

7.5 INDETERMINATE FORMS AND L'HÔPITAL'S RULE

1. l'Hôpital: $\lim_{x \to 2} \frac{x-2}{x^2-4} = \frac{1}{2x}\Big|_{x=2} = \frac{1}{4}$ or $\lim_{x \to 2} \frac{x-2}{x^2-4} = \lim_{x \to 2} \frac{x-2}{(x-2)(x+2)} = \lim_{x \to 2} \frac{1}{x+2} = \frac{1}{4}$

3. l'Hôpital: $\lim_{x \to \infty} \frac{5x^2-3x}{7x^2+1} = \lim_{x \to \infty} \frac{10x-3}{14x} = \lim_{x \to \infty} \frac{10}{14} = \frac{5}{7}$ or $\lim_{x \to \infty} \frac{5x^2-3x}{7x^2+1} = \lim_{x \to \infty} \frac{5-\frac{3}{x}}{7+\frac{1}{x^2}} = \frac{5}{7}$

5. l'Hôpital: $\lim_{x \to 0} \frac{1-\cos x}{x^2} = \lim_{x \to 0} \frac{\sin x}{2x} = \lim_{x \to 0} \frac{\cos x}{2} = \frac{1}{2}$ or $\lim_{x \to 0} \frac{1-\cos x}{x^2} = \lim_{x \to 0} \left[\frac{(1-\cos x)}{x^2}\left(\frac{1+\cos x}{1+\cos x}\right)\right]$
$= \lim_{x \to 0} \frac{\sin^2 x}{x^2(1+\cos x)} = \lim_{x \to 0} \left[\left(\frac{\sin x}{x}\right)\left(\frac{\sin x}{x}\right)\left(\frac{1}{1+\cos x}\right)\right] = \frac{1}{2}$

7. $\lim_{x \to 2} \frac{x-2}{x^2-4} = \lim_{x \to 2} \frac{1}{2x} = \frac{1}{4}$

9. $\lim_{t \to -3} \frac{t^3-4t+15}{t^2-t-12} = \lim_{t \to -3} \frac{3t^2-4}{2t-1} = \frac{3(-3)^2-4}{2(-3)-1} = -\frac{23}{7}$

11. $\lim_{x \to \infty} \frac{5x^3-2x}{7x^3+3} = \lim_{x \to \infty} \frac{15x^2-2}{21x^2} = \lim_{x \to \infty} \frac{30x}{42x} = \lim_{x \to \infty} \frac{30}{42} = \frac{5}{7}$

13. $\lim_{t \to 0} \frac{\sin t^2}{t} = \lim_{t \to 0} \frac{(\cos t^2)(2t)}{1} = 0$

15. $\lim_{x \to 0} \frac{8x^2}{\cos x - 1} = \lim_{x \to 0} \frac{16x}{-\sin x} = \lim_{x \to 0} \frac{16}{-\cos x} = \frac{16}{-1} = -16$

17. $\lim_{\theta \to \pi/2} \frac{2\theta-\pi}{\cos(2\pi-\theta)} = \lim_{\theta \to \pi/2} \frac{2}{\sin(2\pi-\theta)} = \frac{2}{\sin\left(\frac{3\pi}{2}\right)} = -2$

19. $\lim_{\theta \to \pi/2} \frac{1-\sin\theta}{1+\cos 2\theta} = \lim_{\theta \to \pi/2} \frac{-\cos\theta}{-2\sin 2\theta} = \lim_{\theta \to \pi/2} \frac{\sin\theta}{-4\cos 2\theta} = \frac{1}{(-4)(-1)} = \frac{1}{4}$

21. $\lim_{x \to 0} \frac{x^2}{\ln(\sec x)} = \lim_{x \to 0} \frac{2x}{\left(\frac{\sec x \tan x}{\sec x}\right)} = \lim_{x \to 0} \frac{2x}{\tan x} = \lim_{x \to 0} \frac{2}{\sec^2 x} = \frac{2}{1^2} = 2$

23. $\lim_{t \to 0} \frac{t(1-\cos t)}{t-\sin t} = \lim_{t \to 0} \frac{(1-\cos t)+t(\sin t)}{1-\cos t} = \lim_{t \to 0} \frac{\sin t+(\sin t+t\cos t)}{\sin t} = \lim_{t \to 0} \frac{\cos t+\cos t+\cos t-t\sin t}{\cos t} = \frac{1+1+1-0}{1} = 3$

25. $\lim_{x \to (\pi/2)^-} \left(x - \frac{\pi}{2}\right) \sec x = \lim_{x \to (\pi/2)^-} \frac{\left(x - \frac{\pi}{2}\right)}{\cos x} = \lim_{x \to (\pi/2)^-} \left(\frac{1}{-\sin x}\right) = \frac{1}{-1} = -1$

27. $\lim_{\theta \to 0} \frac{3^{\sin \theta} - 1}{\theta} = \lim_{\theta \to 0} \frac{3^{\sin \theta}(\ln 3)(\cos \theta)}{1} = \frac{(3^0)(\ln 3)(1)}{1} = \ln 3$

29. $\lim_{x \to 0} \frac{x\, 2^x}{2^x - 1} = \lim_{x \to 0} \frac{(1)(2^x) + (x)(\ln 2)(2^x)}{(\ln 2)(2^x)} = \frac{1 \cdot 2^0 + 0}{(\ln 2) \cdot 2^0} = \frac{1}{\ln 2}$

31. $\lim_{x \to \infty} \frac{\ln(x+1)}{\log_2 x} = \lim_{x \to \infty} \frac{\ln(x+1)}{\left(\frac{\ln x}{\ln 2}\right)} = (\ln 2) \lim_{x \to \infty} \frac{\left(\frac{1}{x+1}\right)}{\left(\frac{1}{x}\right)} = (\ln 2) \lim_{x \to \infty} \frac{x}{x+1} = (\ln 2) \lim_{x \to \infty} \frac{1}{1} = \ln 2$

33. $\lim_{x \to 0^+} \frac{\ln(x^2+2x)}{\ln x} = \lim_{x \to 0^+} \frac{\left(\frac{2x+2}{x^2+2x}\right)}{\left(\frac{1}{x}\right)} = \lim_{x \to 0^+} \frac{2x^2+2x}{x^2+2x} = \lim_{x \to 0^+} \frac{4x+2}{2x+2} = \lim_{x \to 0^+} \frac{2}{2} = 1$

35. $\lim_{y \to 0} \frac{\sqrt{5y+25} - 5}{y} = \lim_{y \to 0} \frac{(5y+25)^{1/2} - 5}{y} = \lim_{y \to 0} \frac{\left(\frac{1}{2}\right)(5y+25)^{-1/2}(5)}{1} = \lim_{y \to 0} \frac{5}{2\sqrt{5y+25}} = \frac{1}{2}$

37. $\lim_{x \to \infty} [\ln 2x - \ln(x+1)] = \lim_{x \to \infty} \ln\left(\frac{2x}{x+1}\right) = \ln\left(\lim_{x \to \infty} \frac{2x}{x+1}\right) = \ln\left(\lim_{x \to \infty} \frac{2}{1}\right) = \ln 2$

39. $\lim_{x \to 0^+} \frac{(\ln x)^2}{\ln(\sin x)} = \lim_{x \to 0^+} \frac{2(\ln x)\left(\frac{1}{x}\right)}{\frac{\cos x}{\sin x}} = \lim_{x \to 0^+} \frac{2(\ln x)(\sin x)}{x \cos x} = \lim_{x \to 0^+} \left[\frac{2(\ln x)}{\cos x} \cdot \frac{\sin x}{x}\right] = -\infty \cdot 1 = -\infty$

41. $\lim_{x \to 1^+} \left(\frac{1}{x-1} - \frac{1}{\ln x}\right) = \lim_{x \to 1^+} \left(\frac{\ln x - (x-1)}{(x-1)(\ln x)}\right) = \lim_{x \to 1^+} \left(\frac{\frac{1}{x} - 1}{(\ln x) + (x-1)\left(\frac{1}{x}\right)}\right) = \lim_{x \to 1^+} \left(\frac{1-x}{(x \ln x) + x - 1}\right)$
$= \lim_{x \to 1^+} \left(\frac{-1}{(\ln x + 1) + 1}\right) = \frac{-1}{(0+1)+1} = -\frac{1}{2}$

43. $\lim_{\theta \to 0} \frac{\cos \theta - 1}{e^\theta - \theta - 1} = \lim_{\theta \to 0} \frac{-\sin \theta}{e^\theta - 1} = \lim_{\theta \to 0} \frac{-\cos \theta}{e^\theta} = -1$

45. $\lim_{t \to \infty} \frac{e^t + t^2}{e^t - 1} = \lim_{t \to \infty} \frac{e^t + 2t}{e^t} = \lim_{t \to \infty} \frac{e^t + 2}{e^t} = \lim_{t \to \infty} \frac{e^t}{e^t} = 1$

47. $\lim_{x \to 0} \frac{x - \sin x}{x \tan x} = \lim_{x \to 0} \frac{1 - \cos x}{x \sec^2 x + \tan x} = \lim_{x \to 0} \frac{\sin x}{2x \sec^2 x \tan x + 2\sec^2 x} = \frac{0}{2} = 0$

49. $\lim_{\theta \to 0} \frac{\theta - \sin \theta \cos \theta}{\tan \theta - \theta} = \lim_{\theta \to 0} \frac{1 + \sin^2 \theta - \cos^2 \theta}{\sec^2 \theta - 1} = \lim_{\theta \to 0} \frac{2\sin^2 \theta}{\tan^2 \theta} = \lim_{\theta \to 0} 2\cos^2 \theta = 2$

51. The limit leads to the indeterminate form 1^∞. Let $f(x) = x^{1/(1-x)} \Rightarrow \ln f(x) = \ln\left(x^{1/(1-x)}\right) = \frac{\ln x}{1-x}$. Now $\lim_{x \to 1^+} \ln f(x) = \lim_{x \to 1^+} \frac{\ln x}{1-x} = \lim_{x \to 1^+} \frac{\left(\frac{1}{x}\right)}{-1} = -1$. Therefore $\lim_{x \to 1^+} x^{1/(1-x)} = \lim_{x \to 1^+} f(x) = \lim_{x \to 1^+} e^{\ln f(x)} = e^{-1} = \frac{1}{e}$

53. The limit leads to the indeterminate form ∞^0. Let $f(x) = (\ln x)^{1/x} \Rightarrow \ln f(x) = \ln(\ln x)^{1/x} = \frac{\ln(\ln x)}{x}$. Now $\lim_{x \to \infty} \ln f(x) = \lim_{x \to \infty} \frac{\ln(\ln x)}{x} = \lim_{x \to \infty} \frac{\left(\frac{1}{x \ln x}\right)}{1} = 0$. Therefore $\lim_{x \to \infty} (\ln x)^{1/x} = \lim_{x \to \infty} f(x) = \lim_{x \to \infty} e^{\ln f(x)} = e^0 = 1$

55. The limit leads to the indeterminate form 0^0. Let $f(x) = x^{-1/\ln x} \Rightarrow \ln f(x) = -\frac{\ln x}{\ln x} = -1$. Therefore $\lim_{x \to 0^+} x^{-1/\ln x} = \lim_{x \to 0^+} f(x) = \lim_{x \to 0^+} e^{\ln f(x)} = e^{-1} = \frac{1}{e}$

57. The limit leads to the indeterminate form ∞^0. Let $f(x) = (1+2x)^{1/(2\ln x)} \Rightarrow \ln f(x) = \frac{\ln(1+2x)}{2\ln x}$
$\Rightarrow \lim_{x \to \infty} \ln f(x) = \lim_{x \to \infty} \frac{\ln(1+2x)}{2\ln x} = \lim_{x \to \infty} \frac{x}{1+2x} = \lim_{x \to \infty} \frac{1}{2} = \frac{1}{2}$. Therefore $\lim_{x \to \infty} (1+2x)^{1/(2\ln x)}$
$= \lim_{x \to \infty} f(x) = \lim_{x \to \infty} e^{\ln f(x)} = e^{1/2}$

59. The limit leads to the indeterminate form 0^0. Let $f(x) = x^x \Rightarrow \ln f(x) = x \ln x \Rightarrow \ln f(x) = \frac{\ln x}{\left(\frac{1}{x}\right)}$

$= \lim_{x \to 0^+} \ln f(x) = \lim_{x \to 0^+} \frac{\ln x}{\left(\frac{1}{x}\right)} = \lim_{x \to 0^+} \frac{\left(\frac{1}{x}\right)}{\left(-\frac{1}{x^2}\right)} = \lim_{x \to 0^+} (-x) = 0$. Therefore $\lim_{x \to 0^+} x^x = \lim_{x \to 0^+} f(x)$

$= \lim_{x \to 0^+} e^{\ln f(x)} = e^0 = 1$

61. The limit leads to the indeterminate form 1^∞. Let $f(x) = \left(\frac{x+2}{x-1}\right)^x \Rightarrow \ln f(x) = \ln \left(\frac{x+2}{x-1}\right)^x = x \ln \left(\frac{x+2}{x-1}\right) \Rightarrow \lim_{x\to\infty} \ln f(x)$

$= \lim_{x\to\infty} x \ln \left(\frac{x+2}{x-1}\right) = \lim_{x\to\infty} \left(\frac{\ln \left(\frac{x+2}{x-1}\right)}{\frac{1}{x}}\right) = \lim_{x\to\infty} \left(\frac{\ln (x+2) - \ln (x-1)}{\frac{1}{x}}\right) = \lim_{x\to\infty} \left(\frac{\frac{1}{x+2} - \frac{1}{x-1}}{-\frac{1}{x^2}}\right) = \lim_{x\to\infty} \left(\frac{\frac{-3}{(x+2)(x-1)}}{-\frac{1}{x^2}}\right)$

$= \lim_{x\to\infty} \left(\frac{3x^2}{(x+2)(x-1)}\right) = \lim_{x\to\infty} \left(\frac{6x}{2x+1}\right) = \lim_{x\to\infty} \left(\frac{6}{2}\right) = 3$. Therefore, $\lim_{x\to\infty} \left(\frac{x+2}{x-1}\right)^x = \lim_{x\to\infty} f(x) = \lim_{x\to\infty} e^{\ln f(x)} = e^3$

63. $\lim_{x\to 0^+} x^2 \ln x = \lim_{x\to 0^+} \left(\frac{\ln x}{\frac{1}{x^2}}\right) = \lim_{x\to 0^+} \left(\frac{\frac{1}{x}}{-\frac{2}{x^3}}\right) = \lim_{x\to 0^+} \left(-\frac{x^3}{2x}\right) = \lim_{x\to 0^+} \left(-\frac{3x^2}{2}\right) = 0$

65. $\lim_{x\to 0^+} x \tan\left(\frac{\pi}{2} - x\right) = \lim_{x\to 0^+} \left(\frac{x}{\cot\left(\frac{\pi}{2} - x\right)}\right) = \lim_{x\to 0^+} \left(\frac{1}{\csc^2\left(\frac{\pi}{2} - x\right)}\right) = \frac{1}{1} = 1$

67. $\lim_{x \to \infty} \frac{\sqrt{9x+1}}{\sqrt{x+1}} = \sqrt{\lim_{x \to \infty} \frac{9x+1}{x+1}} = \sqrt{\lim_{x \to \infty} \frac{9}{1}} = \sqrt{9} = 3$

69. $\lim_{x \to \pi/2^-} \frac{\sec x}{\tan x} = \lim_{x \to \pi/2^-} \left(\frac{1}{\cos x}\right)\left(\frac{\cos x}{\sin x}\right) = \lim_{x \to \pi/2^-} \frac{1}{\sin x} = 1$

71. $\lim_{x \to \infty} \frac{2^x - 3^x}{3^x + 4^x} = \lim_{x \to \infty} \frac{\left(\frac{2}{3}\right)^x - 1}{1 + \left(\frac{4}{3}\right)^x} = 0$

73. $\lim_{x \to \infty} \frac{e^{x^2}}{x e^x} = \lim_{x \to \infty} \frac{e^{x^2 - x}}{x} = \lim_{x \to \infty} \frac{e^{x(x-1)}}{x} = \lim_{x \to \infty} \frac{e^{x(x-1)}(2x-1)}{1} = \infty$

75. Part (b) is correct because part (a) is neither in the $\frac{0}{0}$ nor $\frac{\infty}{\infty}$ form and so l'Hôpital's rule may not be used.

77. Part (d) is correct, the other parts are indeterminate forms and cannot be calculated by the incorrect arithmetic

79. If f(x) is to be continuous at $x = 0$, then $\lim_{x \to 0} f(x) = f(0) \Rightarrow c = f(0) = \lim_{x \to 0} \frac{9x - 3\sin 3x}{5x^3} = \lim_{x \to 0} \frac{9 - 9\cos 3x}{15x^2}$

$= \lim_{x \to 0} \frac{27 \sin 3x}{30x} = \lim_{x \to 0} \frac{81 \cos 3x}{30} = \frac{27}{10}$.

81. (a)

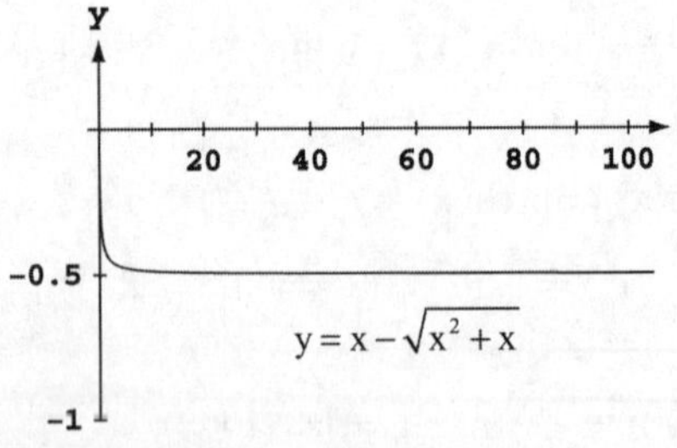

$y = x - \sqrt{x^2 + x}$

(b) The limit leads to the indeterminate form $\infty - \infty$:

$\lim_{x \to \infty} \left(x - \sqrt{x^2+x}\right) = \lim_{x \to \infty} \left(x - \sqrt{x^2+x}\right)\left(\frac{x+\sqrt{x^2+x}}{x+\sqrt{x^2+x}}\right) = \lim_{x \to \infty} \left(\frac{x^2 - (x^2+x)}{x+\sqrt{x^2+x}}\right) = \lim_{x \to \infty} \frac{-x}{x+\sqrt{x^2+x}}$

$= \lim_{x \to \infty} \frac{-1}{1+\sqrt{1+\frac{1}{x}}} = \frac{-1}{1+\sqrt{1+0}} = -\frac{1}{2}$

83. The graph indicates a limit near -1. The limit leads to the indeterminate form $\frac{0}{0}$: $\lim\limits_{x\to 1} \frac{2x^2-(3x+1)\sqrt{x}+2}{x-1}$

$= \lim\limits_{x\to 1} \frac{2x^2-3x^{3/2}-x^{1/2}+2}{x-1} = \lim\limits_{x\to 1} \frac{4x-\frac{9}{2}x^{1/2}-\frac{1}{2}x^{-1/2}}{1}$

$= \frac{4-\frac{9}{2}-\frac{1}{2}}{1} = \frac{4-5}{1} = -1$

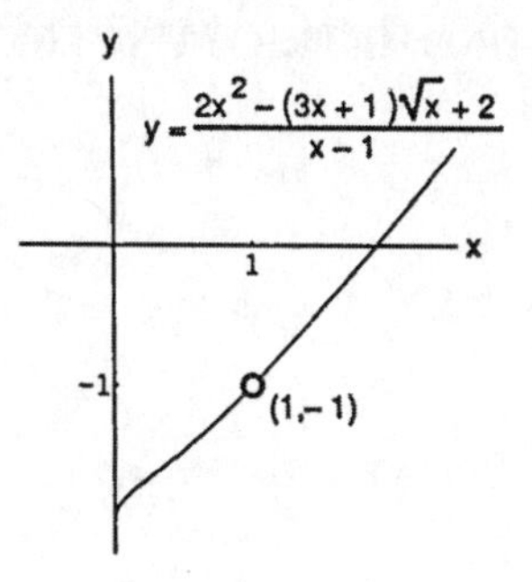

85. Let $f(k) = \left(1+\frac{r}{k}\right)^k \Rightarrow \ln f(k) = \frac{\ln(1+rk^{-1})}{k^{-1}} \Rightarrow \lim\limits_{k\to\infty} \frac{\ln(1+rk^{-1})}{k^{-1}} = \lim\limits_{k\to\infty} \frac{\left(\frac{-rk^{-2}}{1+rk^{-1}}\right)}{-k^{-2}} = \lim\limits_{k\to\infty} \frac{r}{1+rk^{-1}}$

$= \lim\limits_{k\to\infty} \frac{rk}{k+r} = \lim\limits_{k\to\infty} \frac{r}{1} = r$. Therefore $\lim\limits_{k\to\infty} \left(1+\frac{r}{k}\right)^k = \lim\limits_{k\to\infty} f(k) = \lim\limits_{k\to\infty} e^{\ln f(k)} = e^r$.

87. (a) $y = x\tan\left(\frac{1}{x}\right)$, $\lim\limits_{x\to\infty}\left(x\tan\left(\frac{1}{x}\right)\right) = \lim\limits_{x\to\infty}\left(\frac{\tan\left(\frac{1}{x}\right)}{\frac{1}{x}}\right) = \lim\limits_{x\to\infty}\left(\frac{\sec^2\left(\frac{1}{x}\right)\left(-\frac{1}{x^2}\right)}{\left(-\frac{1}{x^2}\right)}\right) = \lim\limits_{x\to\infty}\sec^2\left(\frac{1}{x}\right) = 1$; $\lim\limits_{x\to-\infty}\left(x\tan\left(\frac{1}{x}\right)\right)$

$= \lim\limits_{x\to-\infty}\left(\frac{\tan\left(\frac{1}{x}\right)}{\frac{1}{x}}\right) = \lim\limits_{x\to-\infty}\left(\frac{\sec^2\left(\frac{1}{x}\right)\left(-\frac{1}{x^2}\right)}{\left(-\frac{1}{x^2}\right)}\right) = \lim\limits_{x\to-\infty}\sec^2\left(\frac{1}{x}\right) = 1 \Rightarrow$ the horizontal asymptote is $y = 1$ as $x\to\infty$ and as $x\to-\infty$.

(b) $y = \frac{3x+e^{2x}}{2x+e^{3x}}$, $\lim\limits_{x\to\infty}\left(\frac{3x+e^{2x}}{2x+e^{3x}}\right) = \lim\limits_{x\to\infty}\left(\frac{3+2e^{2x}}{2+3e^{3x}}\right) = \lim\limits_{x\to\infty}\left(\frac{4e^{2x}}{9e^{3x}}\right) = \lim\limits_{x\to\infty}\left(\frac{4}{9e^x}\right) = 0$; $\lim\limits_{x\to-\infty}\left(\frac{3x+e^{2x}}{2x+e^{3x}}\right)$

$= \lim\limits_{x\to-\infty}\left(\frac{3+2e^{2x}}{2+3e^{3x}}\right) = \frac{3}{2} \Rightarrow$ the horizontal asymptotes are $y = 0$ as $x\to\infty$ and $y = \frac{3}{2}$ as $x\to-\infty$.

89. (a) We should assign the value 1 to $f(x) = (\sin x)^x$ to make it continuous at $x = 0$.

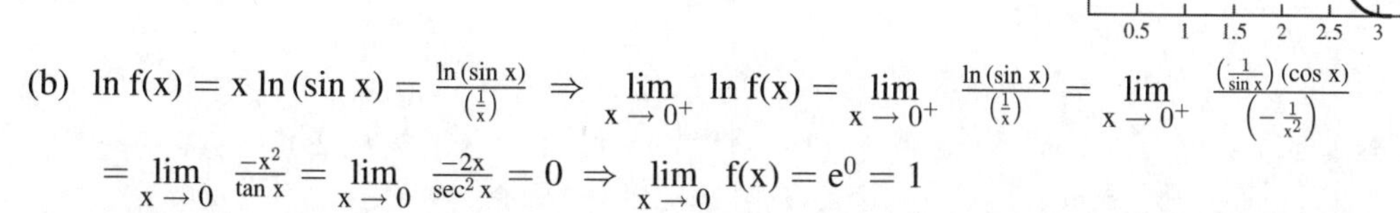

(b) $\ln f(x) = x\ln(\sin x) = \frac{\ln(\sin x)}{\left(\frac{1}{x}\right)} \Rightarrow \lim\limits_{x\to 0^+} \ln f(x) = \lim\limits_{x\to 0^+} \frac{\ln(\sin x)}{\left(\frac{1}{x}\right)} = \lim\limits_{x\to 0^+} \frac{\left(\frac{1}{\sin x}\right)(\cos x)}{\left(-\frac{1}{x^2}\right)}$

$= \lim\limits_{x\to 0} \frac{-x^2}{\tan x} = \lim\limits_{x\to 0} \frac{-2x}{\sec^2 x} = 0 \Rightarrow \lim\limits_{x\to 0} f(x) = e^0 = 1$

(c) The maximum value of f(x) is close to 1 near the point $x \approx 1.55$ (see the graph in part (a)).

(d) The root in question is near 1.57.

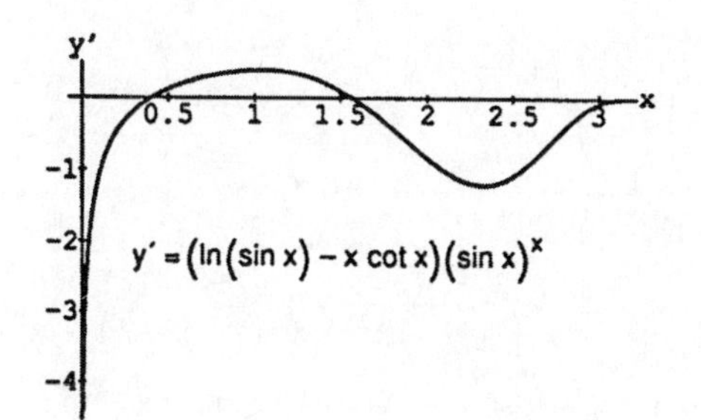

7.6 INVERSE TRIGONOMETRIC FUNCTIONS

1. (a) $\frac{\pi}{4}$ (b) $-\frac{\pi}{3}$ (c) $\frac{\pi}{6}$

3. (a) $-\frac{\pi}{6}$ (b) $\frac{\pi}{4}$ (c) $-\frac{\pi}{3}$

5. (a) $\frac{\pi}{3}$ (b) $\frac{3\pi}{4}$ (c) $\frac{\pi}{6}$

7. (a) $\frac{3\pi}{4}$ (b) $\frac{\pi}{6}$ (c) $\frac{2\pi}{3}$

9. $\sin\left(\cos^{-1}\frac{\sqrt{2}}{2}\right) = \sin\left(\frac{\pi}{4}\right) = \frac{1}{\sqrt{2}}$

11. $\tan\left(\sin^{-1}\left(-\frac{1}{2}\right)\right) = \tan\left(-\frac{\pi}{6}\right) = -\frac{1}{\sqrt{3}}$

13. $\lim\limits_{x \to 1^-} \sin^{-1} x = \frac{\pi}{2}$

15. $\lim\limits_{x \to \infty} \tan^{-1} x = \frac{\pi}{2}$

17. $\lim\limits_{x \to \infty} \sec^{-1} x = \frac{\pi}{2}$

19. $\lim\limits_{x \to \infty} \csc^{-1} x = \lim\limits_{x \to \infty} \sin^{-1}\left(\frac{1}{x}\right) = 0$

21. $y = \cos^{-1}(x^2) \Rightarrow \frac{dy}{dx} = -\frac{2x}{\sqrt{1-(x^2)^2}} = \frac{-2x}{\sqrt{1-x^4}}$

23. $y = \sin^{-1}\sqrt{2}t \Rightarrow \frac{dy}{dt} = \frac{\sqrt{2}}{\sqrt{1-\left(\sqrt{2}t\right)^2}} = \frac{\sqrt{2}}{\sqrt{1-2t^2}}$

25. $y = \sec^{-1}(2s+1) \Rightarrow \frac{dy}{ds} = \frac{2}{|2s+1|\sqrt{(2s+1)^2-1}} = \frac{2}{|2s+1|\sqrt{4s^2+4s}} = \frac{1}{|2s+1|\sqrt{s^2+s}}$

27. $y = \csc^{-1}(x^2+1) \Rightarrow \frac{dy}{dx} = -\frac{2x}{|x^2+1|\sqrt{(x^2+1)^2-1}} = \frac{-2x}{(x^2+1)\sqrt{x^4+2x^2}}$

29. $y = \sec^{-1}\left(\frac{1}{t}\right) = \cos^{-1} t \Rightarrow \frac{dy}{dt} = \frac{-1}{\sqrt{1-t^2}}$

31. $y = \cot^{-1}\sqrt{t} = \cot^{-1} t^{1/2} \Rightarrow \frac{dy}{dt} = -\frac{\left(\frac{1}{2}\right)t^{-1/2}}{1+\left(t^{1/2}\right)^2} = \frac{-1}{2\sqrt{t}(1+t)}$

33. $y = \ln(\tan^{-1} x) \Rightarrow \frac{dy}{dx} = \frac{\left(\frac{1}{1+x^2}\right)}{\tan^{-1} x} = \frac{1}{(\tan^{-1} x)(1+x^2)}$

35. $y = \csc^{-1}(e^t) \Rightarrow \frac{dy}{dt} = -\frac{e^t}{|e^t|\sqrt{(e^t)^2-1}} = \frac{-1}{\sqrt{e^{2t}-1}}$

37. $y = s\sqrt{1-s^2} + \cos^{-1} s = s(1-s^2)^{1/2} + \cos^{-1} s \Rightarrow \frac{dy}{ds} = (1-s^2)^{1/2} + s\left(\frac{1}{2}\right)(1-s^2)^{-1/2}(-2s) - \frac{1}{\sqrt{1-s^2}}$
$= \sqrt{1-s^2} - \frac{s^2}{\sqrt{1-s^2}} - \frac{1}{\sqrt{1-s^2}} = \sqrt{1-s^2} - \frac{s^2+1}{\sqrt{1-s^2}} = \frac{1-s^2-s^2-1}{\sqrt{1-s^2}} = \frac{-2s^2}{\sqrt{1-s^2}}$

39. $y = \tan^{-1}\sqrt{x^2-1} + \csc^{-1} x = \tan^{-1}(x^2-1)^{1/2} + \csc^{-1} x \Rightarrow \frac{dy}{dx} = \frac{\left(\frac{1}{2}\right)(x^2-1)^{-1/2}(2x)}{1+\left[(x^2-1)^{1/2}\right]^2} - \frac{1}{|x|\sqrt{x^2-1}}$
$= \frac{1}{x\sqrt{x^2-1}} - \frac{1}{|x|\sqrt{x^2-1}} = 0$, for $x > 1$

41. $y = x\sin^{-1} x + \sqrt{1-x^2} = x\sin^{-1} x + (1-x^2)^{1/2} \Rightarrow \frac{dy}{dx} = \sin^{-1} x + x\left(\frac{1}{\sqrt{1-x^2}}\right) + \left(\frac{1}{2}\right)(1-x^2)^{-1/2}(-2x)$
$= \sin^{-1} x + \frac{x}{\sqrt{1-x^2}} - \frac{x}{\sqrt{1-x^2}} = \sin^{-1} x$

43. $\int \frac{1}{\sqrt{9-x^2}}\,dx = \sin^{-1}\left(\frac{x}{3}\right) + C$

45. $\int \frac{1}{17+x^2}\,dx = \int \frac{1}{\left(\sqrt{17}\right)^2+x^2}\,dx = \frac{1}{\sqrt{17}}\tan^{-1}\frac{x}{\sqrt{17}} + C$

47. $\int \frac{dx}{x\sqrt{25x^2-2}} = \int \frac{du}{u\sqrt{u^2-2}}$, where $u = 5x$ and $du = 5\,dx$
$= \frac{1}{\sqrt{2}} \sec^{-1}\left|\frac{u}{\sqrt{2}}\right| + C = \frac{1}{\sqrt{2}} \sec^{-1}\left|\frac{5x}{\sqrt{2}}\right| + C$

49. $\int_0^1 \frac{4\,ds}{\sqrt{4-s^2}} = \left[4 \sin^{-1} \frac{s}{2}\right]_0^1 = 4\left(\sin^{-1} \frac{1}{2} - \sin^{-1} 0\right) = 4\left(\frac{\pi}{6} - 0\right) = \frac{2\pi}{3}$

51. $\int_0^2 \frac{dt}{8+2t^2} = \frac{1}{\sqrt{2}} \int_0^{2\sqrt{2}} \frac{du}{8+u^2}$, where $u = \sqrt{2}t$ and $du = \sqrt{2}\,dt$; $t = 0 \Rightarrow u = 0, t = 2 \Rightarrow u = 2\sqrt{2}$
$= \left[\frac{1}{\sqrt{2}} \cdot \frac{1}{\sqrt{8}} \tan^{-1} \frac{u}{\sqrt{8}}\right]_0^{2\sqrt{2}} = \frac{1}{4}\left(\tan^{-1} \frac{2\sqrt{2}}{\sqrt{8}} - \tan^{-1} 0\right) = \frac{1}{4}\left(\tan^{-1} 1 - \tan^{-1} 0\right) = \frac{1}{4}\left(\frac{\pi}{4} - 0\right) = \frac{\pi}{16}$

53. $\int_{-1}^{-\sqrt{2}/2} \frac{dy}{y\sqrt{4y^2-1}} = \int_{-2}^{-\sqrt{2}} \frac{du}{u\sqrt{u^2-1}}$, where $u = 2y$ and $du = 2\,dy$; $y = -1 \Rightarrow u = -2, y = -\frac{\sqrt{2}}{2} \Rightarrow u = -\sqrt{2}$
$= \left[\sec^{-1} |u|\right]_{-2}^{-\sqrt{2}} = \sec^{-1}\left|-\sqrt{2}\right| - \sec^{-1} |-2| = \frac{\pi}{4} - \frac{\pi}{3} = -\frac{\pi}{12}$

55. $\int \frac{3\,dr}{\sqrt{1-4(r-1)^2}} = \frac{3}{2} \int \frac{du}{\sqrt{1-u^2}}$, where $u = 2(r-1)$ and $du = 2\,dr$
$= \frac{3}{2} \sin^{-1} u + C = \frac{3}{2} \sin^{-1} 2(r-1) + C$

57. $\int \frac{dx}{2+(x-1)^2} = \int \frac{du}{2+u^2}$, where $u = x - 1$ and $du = dx$
$= \frac{1}{\sqrt{2}} \tan^{-1} \frac{u}{\sqrt{2}} + C = \frac{1}{\sqrt{2}} \tan^{-1}\left(\frac{x-1}{\sqrt{2}}\right) + C$

59. $\int \frac{dx}{(2x-1)\sqrt{(2x-1)^2-4}} = \frac{1}{2} \int \frac{du}{u\sqrt{u^2-4}}$, where $u = 2x - 1$ and $du = 2\,dx$
$= \frac{1}{2} \cdot \frac{1}{2} \sec^{-1}\left|\frac{u}{2}\right| + C = \frac{1}{4} \sec^{-1}\left|\frac{2x-1}{2}\right| + C$

61. $\int_{-\pi/2}^{\pi/2} \frac{2\cos\theta\,d\theta}{1+(\sin\theta)^2} = 2\int_{-1}^{1} \frac{du}{1+u^2}$, where $u = \sin\theta$ and $du = \cos\theta\,d\theta$; $\theta = -\frac{\pi}{2} \Rightarrow u = -1, \theta = \frac{\pi}{2} \Rightarrow u = 1$
$= \left[2 \tan^{-1} u\right]_{-1}^{1} = 2\left(\tan^{-1} 1 - \tan^{-1}(-1)\right) = 2\left[\frac{\pi}{4} - \left(-\frac{\pi}{4}\right)\right] = \pi$

63. $\int_0^{\ln\sqrt{3}} \frac{e^x\,dx}{1+e^{2x}} = \int_1^{\sqrt{3}} \frac{du}{1+u^2}$, where $u = e^x$ and $du = e^x\,dx$; $x = 0 \Rightarrow u = 1, x = \ln\sqrt{3} \Rightarrow u = \sqrt{3}$
$= \left[\tan^{-1} u\right]_1^{\sqrt{3}} = \tan^{-1}\sqrt{3} - \tan^{-1} 1 = \frac{\pi}{3} - \frac{\pi}{4} = \frac{\pi}{12}$

65. $\int \frac{y\,dy}{\sqrt{1-y^4}} = \frac{1}{2} \int \frac{du}{\sqrt{1-u^2}}$, where $u = y^2$ and $du = 2y\,dy$
$= \frac{1}{2} \sin^{-1} u + C = \frac{1}{2} \sin^{-1} y^2 + C$

67. $\int \frac{dx}{\sqrt{-x^2+4x-3}} = \int \frac{dx}{\sqrt{1-(x^2-4x+4)}} = \int \frac{dx}{\sqrt{1-(x-2)^2}} = \sin^{-1}(x-2) + C$

69. $\int_{-1}^{0} \frac{6\,dt}{\sqrt{3-2t-t^2}} = 6\int_{-1}^{0} \frac{dt}{\sqrt{4-(t^2+2t+1)}} = 6\int_{-1}^{0} \frac{dt}{\sqrt{2^2-(t+1)^2}} = 6\left[\sin^{-1}\left(\frac{t+1}{2}\right)\right]_{-1}^{0}$
$= 6\left[\sin^{-1}\left(\frac{1}{2}\right) - \sin^{-1} 0\right] = 6\left(\frac{\pi}{6} - 0\right) = \pi$

71. $\int \frac{dy}{y^2-2y+5} = \int \frac{dy}{4+y^2-2y+1} = \int \frac{dy}{2^2+(y-1)^2} = \frac{1}{2} \tan^{-1}\left(\frac{y-1}{2}\right) + C$

73. $\int_1^2 \frac{8\,dx}{x^2-2x+2} = 8\int_1^2 \frac{dx}{1+(x^2-2x+1)} = 8\int_1^2 \frac{dx}{1+(x-1)^2} = 8\left[\tan^{-1}(x-1)\right]_1^2 = 8\left(\tan^{-1} 1 - \tan^{-1} 0\right) = 8\left(\frac{\pi}{4} - 0\right) = 2\pi$

75. $\int \frac{x+4}{x^2+4}dx = \int \frac{x}{x^2+4}dx + \int \frac{4}{x^2+4}dx$; $\int \frac{x}{x^2+4}dx = \frac{1}{2}\int \frac{1}{u}du$ where $u = x^2+4 \Rightarrow du = 2x\,dx \Rightarrow \frac{1}{2}du = x\,dx$
$\Rightarrow \int \frac{x+4}{x^2+4}dx = \frac{1}{2}\ln(x^2+4) + 2\tan^{-1}\left(\frac{x}{2}\right) + C$

77. $\int \frac{x^2+2x-1}{x^2+9}dx = \int \left(1 + \frac{2x-10}{x^2+9}\right)dx = \int dx + \int \frac{2x}{x^2+9}dx - 10\int \frac{1}{x^2+9}dx$; $\int \frac{2x}{x^2+9}dx = \int \frac{1}{u}du$ where $u = x^2+9$
$\Rightarrow du = 2x\,dx \Rightarrow \int dx + \int \frac{2x}{x^2+9}dx - 10\int \frac{1}{x^2+9}dx = x + \ln(x^2+9) - \frac{10}{3}\tan^{-1}\left(\frac{x}{3}\right) + C$

79. $\int \frac{dx}{(x+1)\sqrt{x^2+2x}} = \int \frac{dx}{(x+1)\sqrt{x^2+2x+1-1}} = \int \frac{dx}{(x+1)\sqrt{(x+1)^2-1}} = \int \frac{du}{u\sqrt{u^2-1}}$, where $u = x+1$ and $du = dx$
$= \sec^{-1}|u| + C = \sec^{-1}|x+1| + C$

81. $\int \frac{e^{\sin^{-1}x}}{\sqrt{1-x^2}}\,dx = \int e^u\,du$, where $u = \sin^{-1}x$ and $du = \frac{dx}{\sqrt{1-x^2}}$
$= e^u + C = e^{\sin^{-1}x} + C$

83. $\int \frac{(\sin^{-1}x)^2}{\sqrt{1-x^2}}\,dx = \int u^2\,du$, where $u = \sin^{-1}x$ and $du = \frac{dx}{\sqrt{1-x^2}}$
$= \frac{u^3}{3} + C = \frac{(\sin^{-1}x)^3}{3} + C$

85. $\int \frac{1}{(\tan^{-1}y)(1+y^2)}\,dy = \int \frac{\left(\frac{1}{1+y^2}\right)}{\tan^{-1}y}\,dy = \int \frac{1}{u}\,du$, where $u = \tan^{-1}y$ and $du = \frac{dy}{1+y^2}$
$= \ln|u| + C = \ln|\tan^{-1}y| + C$

87. $\int_{\sqrt{2}}^{2} \frac{\sec^2(\sec^{-1}x)}{x\sqrt{x^2-1}}\,dx = \int_{\pi/4}^{\pi/3} \sec^2 u\,du$, where $u = \sec^{-1}x$ and $du = \frac{dx}{x\sqrt{x^2-1}}$; $x = \sqrt{2} \Rightarrow u = \frac{\pi}{4}$, $x = 2 \Rightarrow u = \frac{\pi}{3}$
$= [\tan u]_{\pi/4}^{\pi/3} = \tan\frac{\pi}{3} - \tan\frac{\pi}{4} = \sqrt{3} - 1$

89. $\int \frac{1}{\sqrt{x}(x+1)\left[\left(\tan^{-1}\sqrt{x}\right)^2+9\right]}\,dx = 2\int \frac{1}{u^2+9}du$ where $u = \tan^{-1}\sqrt{x} \Rightarrow du = \frac{1}{1+\left(\sqrt{x}\right)^2}\frac{1}{2\sqrt{x}}dx \Rightarrow 2du = \frac{1}{(1+x)\sqrt{x}}dx$
$= \frac{2}{3}\tan^{-1}\left(\frac{\tan^{-1}\sqrt{x}}{3}\right) + C$

91. $\lim\limits_{x\to 0} \frac{\sin^{-1}5x}{x} = \lim\limits_{x\to 0} \frac{\left(\frac{5}{\sqrt{1-25x^2}}\right)}{1} = 5$

93. $\lim\limits_{x\to\infty} x\tan^{-1}\left(\frac{2}{x}\right) = \lim\limits_{x\to\infty} \frac{\tan^{-1}(2x^{-1})}{x^{-1}} = \lim\limits_{x\to\infty} \frac{\left(\frac{-2x^{-2}}{1+4x^{-2}}\right)}{-x^{-2}} = \lim\limits_{x\to\infty} \frac{2}{1+4x^{-2}} = 2$

95. $\lim\limits_{x\to 0} \frac{\tan^{-1}x^2}{x\sin^{-1}x} = \lim\limits_{x\to 0} \left(\frac{\frac{2x}{1+x^4}}{x\frac{1}{\sqrt{1-x^2}} + \sin^{-1}x}\right) = \lim\limits_{x\to 0} \left(\frac{\frac{-2\left(3x^4-1\right)}{\left(1+x^4\right)^2}}{\frac{-x^2+2}{\left(1-x^2\right)^{3/2}}}\right) = \frac{\frac{-2(0-1)}{1^2}}{\frac{-0+2}{(1-0)^{3/2}}} = \frac{2}{2} = 1$

97. $\lim\limits_{x\to 0^+} \frac{\left[\tan^{-1}\left(\sqrt{x}\right)\right]^2}{x\sqrt{x+1}} = \lim\limits_{x\to 0^+} \frac{\tan^{-1}\left(\sqrt{x}\right)\frac{1}{\sqrt{x}(1+x)}}{\frac{x}{2\sqrt{x+1}} + \sqrt{x+1}} = \lim\limits_{x\to 0^+} \frac{\frac{\tan^{-1}\left(\sqrt{x}\right)}{\sqrt{x}(1+x)}}{\frac{3x+2}{2\sqrt{x+1}}} = \lim\limits_{x\to 0^+} \left(\frac{2\tan^{-1}\left(\sqrt{x}\right)}{(3x+2)\sqrt{x}\sqrt{x+1}}\right) = \lim\limits_{x\to 0^+} \left(\frac{\frac{1}{\sqrt{x}(1+x)}}{\frac{12x^2+13x+2}{2\sqrt{x}\sqrt{x+1}}}\right)$
$= \lim\limits_{x\to 0^+} \left(\frac{2}{(12x^2+13x+2)\sqrt{x+1}}\right) = \frac{2}{2} = 1$

99. If $y = \ln x - \frac{1}{2}\ln(1+x^2) - \frac{\tan^{-1}x}{x} + C$, then $dy = \left[\frac{1}{x} - \frac{x}{1+x^2} - \frac{\left(\frac{x}{1+x^2}\right) - \tan^{-1}x}{x^2}\right] dx$

$= \left(\frac{1}{x} - \frac{x}{1+x^2} - \frac{1}{x(1+x^2)} + \frac{\tan^{-1}x}{x^2}\right) dx = \frac{x(1+x^2) - x^3 - x + (\tan^{-1}x)(1+x^2)}{x^2(1+x^2)} dx = \frac{\tan^{-1}x}{x^2} dx$,

which verifies the formula

101. If $y = x(\sin^{-1}x)^2 - 2x + 2\sqrt{1-x^2}\sin^{-1}x + C$, then

$dy = \left[(\sin^{-1}x)^2 + \frac{2x(\sin^{-1}x)}{\sqrt{1-x^2}} - 2 + \frac{-2x}{\sqrt{1-x^2}}\sin^{-1}x + 2\sqrt{1-x^2}\left(\frac{1}{\sqrt{1-x^2}}\right)\right] dx = (\sin^{-1}x)^2 dx$, which verifies the formula

103. $\frac{dy}{dx} = \frac{1}{\sqrt{1-x^2}} \Rightarrow dy = \frac{dx}{\sqrt{1-x^2}} \Rightarrow y = \sin^{-1}x + C$; $x = 0$ and $y = 0 \Rightarrow 0 = \sin^{-1}0 + C \Rightarrow C = 0 \Rightarrow y = \sin^{-1}x$

105. $\frac{dy}{dx} = \frac{1}{x\sqrt{x^2-1}} \Rightarrow dy = \frac{dx}{x\sqrt{x^2-1}} \Rightarrow y = \sec^{-1}|x| + C$; $x = 2$ and $y = \pi \Rightarrow \pi = \sec^{-1}2 + C \Rightarrow C = \pi - \sec^{-1}2$

$= \pi - \frac{\pi}{3} = \frac{2\pi}{3} \Rightarrow y = \sec^{-1}(x) + \frac{2\pi}{3}, x > 1$

107. (a) The angle α is the large angle between the wall and the right end of the blackboard minus the small angle between the left end of the blackboard and the wall $\Rightarrow \alpha = \cot^{-1}\left(\frac{x}{15}\right) - \cot^{-1}\left(\frac{x}{3}\right)$.

(b) $\frac{d\alpha}{dt} = -\frac{\frac{1}{15}}{1+\left(\frac{x}{15}\right)^2} + \frac{\frac{1}{3}}{1+\left(\frac{x}{3}\right)^2} = -\frac{15}{225+x^2} + \frac{3}{9+x^2} = \frac{540-12x^2}{(225+x^2)(9+x^2)}$; $\frac{d\alpha}{dt} = 0 \Rightarrow 540 - 12x^2 = 0 \Rightarrow x = \pm 3\sqrt{5}$

Since $x > 0$, consider only $x = 3\sqrt{5} \Rightarrow \alpha\left(3\sqrt{5}\right) = \cot^{-1}\left(\frac{3\sqrt{5}}{15}\right) - \cot^{-1}\left(\frac{3\sqrt{5}}{3}\right) \approx 0.729728 \approx 41.8103°$. Using the first derivative test, $\left.\frac{d\alpha}{dt}\right|_{x=1} = \frac{132}{565} > 0$ and $\left.\frac{d\alpha}{dt}\right|_{x=10} = -\frac{132}{7085} < 0 \Rightarrow$ local maximum of $41.8103°$ when $x = 3\sqrt{5} \approx 6.7082$ ft.

109. $V = \left(\frac{1}{3}\right)\pi r^2 h = \left(\frac{1}{3}\right)\pi(3\sin\theta)^2(3\cos\theta) = 9\pi(\cos\theta - \cos^3\theta)$, where $0 \le \theta \le \frac{\pi}{2}$

$\Rightarrow \frac{dV}{d\theta} = -9\pi(\sin\theta)(1 - 3\cos^2\theta) = 0 \Rightarrow \sin\theta = 0$ or $\cos\theta = \pm\frac{1}{\sqrt{3}} \Rightarrow$ the critical points are: 0, $\cos^{-1}\left(\frac{1}{\sqrt{3}}\right)$, and $\cos^{-1}\left(-\frac{1}{\sqrt{3}}\right)$; but $\cos^{-1}\left(-\frac{1}{\sqrt{3}}\right)$ is not in the domain. When $\theta = 0$, we have a minimum and when $\theta = \cos^{-1}\left(\frac{1}{\sqrt{3}}\right)$ $\approx 54.7°$, we have a maximum volume.

111. Take each square as a unit square. From the diagram we have the following: the smallest angle α has a tangent of 1 $\Rightarrow \alpha = \tan^{-1}1$; the middle angle β has a tangent of 2 $\Rightarrow \beta = \tan^{-1}2$; and the largest angle γ has a tangent of 3 $\Rightarrow \gamma = \tan^{-1}3$. The sum of these three angles is $\pi \Rightarrow \alpha + \beta + \gamma = \pi$

$\Rightarrow \tan^{-1}1 + \tan^{-1}2 + \tan^{-1}3 = \pi$.

113. $\sin^{-1}(1) + \cos^{-1}(1) = \frac{\pi}{2} + 0 = \frac{\pi}{2}$; $\sin^{-1}(0) + \cos^{-1}(0) = 0 + \frac{\pi}{2} = \frac{\pi}{2}$; and $\sin^{-1}(-1) + \cos^{-1}(-1) = -\frac{\pi}{2} + \pi = \frac{\pi}{2}$.

If $x \in (-1, 0)$ and $x = -a$, then $\sin^{-1}(x) + \cos^{-1}(x) = \sin^{-1}(-a) + \cos^{-1}(-a) = -\sin^{-1}a + (\pi - \cos^{-1}a)$

$= \pi - (\sin^{-1}a + \cos^{-1}a) = \pi - \frac{\pi}{2} = \frac{\pi}{2}$ from Equations (3) and (4) in the text.

115. $\csc^{-1}u = \frac{\pi}{2} - \sec^{-1}u \Rightarrow \frac{d}{dx}(\csc^{-1}u) = \frac{d}{dx}\left(\frac{\pi}{2} - \sec^{-1}u\right) = 0 - \frac{\frac{du}{dx}}{|u|\sqrt{u^2-1}} = -\frac{\frac{du}{dx}}{|u|\sqrt{u^2-1}}, |u| > 1$

117. $f(x) = \sec x \Rightarrow f'(x) = \sec x \tan x \Rightarrow \left.\frac{df^{-1}}{dx}\right|_{x=b} = \frac{1}{\left.\frac{df}{dx}\right|_{x=f^{-1}(b)}} = \frac{1}{\sec(\sec^{-1}b)\tan(\sec^{-1}b)} = \frac{1}{b\left(\pm\sqrt{b^2-1}\right)}$.

Since the slope of $\sec^{-1}x$ is always positive, we the right sign by writing $\frac{d}{dx}\sec^{-1}x = \frac{1}{|x|\sqrt{x^2-1}}$.

119. The functions f and g have the same derivative (for $x \geq 0$), namely $\frac{1}{\sqrt{x}(x+1)}$. The functions therefore differ by a constant. To identify the constant we can set x equal to 0 in the equation $f(x) = g(x) + C$, obtaining $\sin^{-1}(-1) = 2\tan^{-1}(0) + C \Rightarrow -\frac{\pi}{2} = 0 + C \Rightarrow C = -\frac{\pi}{2}$. For $x \geq 0$, we have $\sin^{-1}\left(\frac{x-1}{x+1}\right) = 2\tan^{-1}\sqrt{x} - \frac{\pi}{2}$.

121. $V = \pi \int_{-\sqrt{3}/3}^{\sqrt{3}} \left(\frac{1}{\sqrt{1+x^2}}\right)^2 dx = \pi \int_{-\sqrt{3}/3}^{\sqrt{3}} \frac{1}{1+x^2}\,dx = \pi\left[\tan^{-1} x\right]_{-\sqrt{3}/3}^{\sqrt{3}} = \pi\left[\tan^{-1}\sqrt{3} - \tan^{-1}\left(-\frac{\sqrt{3}}{3}\right)\right]$
$= \pi\left[\frac{\pi}{3} - \left(-\frac{\pi}{6}\right)\right] = \frac{\pi^2}{2}$

123. (a) $A(x) = \frac{\pi}{4}(\text{diameter})^2 = \frac{\pi}{4}\left[\frac{1}{\sqrt{1+x^2}} - \left(-\frac{1}{\sqrt{1+x^2}}\right)\right]^2 = \frac{\pi}{1+x^2} \Rightarrow V = \int_a^b A(x)\,dx = \int_{-1}^{1} \frac{\pi\,dx}{1+x^2}$
$= \pi\left[\tan^{-1} x\right]_{-1}^{1} = (\pi)(2)\left(\frac{\pi}{4}\right) = \frac{\pi^2}{2}$

(b) $A(x) = (\text{edge})^2 = \left[\frac{1}{\sqrt{1+x^2}} - \left(-\frac{1}{\sqrt{1+x^2}}\right)\right]^2 = \frac{4}{1+x^2} \Rightarrow V = \int_a^b A(x)\,dx = \int_{-1}^{1} \frac{4\,dx}{1+x^2}$
$= 4\left[\tan^{-1} x\right]_{-1}^{1} = 4\left[\tan^{-1}(1) - \tan^{-1}(-1)\right] = 4\left[\frac{\pi}{4} - \left(-\frac{\pi}{4}\right)\right] = 2\pi$

125. (a) $\sec^{-1} 1.5 = \cos^{-1}\frac{1}{1.5} \approx 0.84107$ (b) $\csc^{-1}(-1.5) = \sin^{-1}\left(-\frac{1}{1.5}\right) \approx -0.72973$
(c) $\cot^{-1} 2 = \frac{\pi}{2} - \tan^{-1} 2 \approx 0.46365$

127. (a) Domain: all real numbers except those having the form $\frac{\pi}{2} + k\pi$ where k is an integer.
Range: $-\frac{\pi}{2} < y < \frac{\pi}{2}$

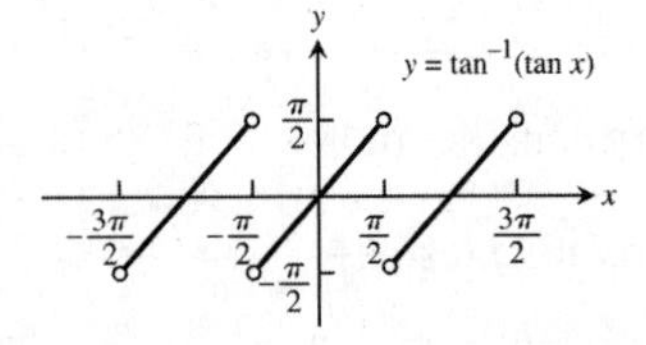

(b) Domain: $-\infty < x < \infty$; Range: $-\infty < y < \infty$
The graph of $y = \tan^{-1}(\tan x)$ is periodic, the graph of $y = \tan(\tan^{-1} x) = x$ for $-\infty \leq x < \infty$.

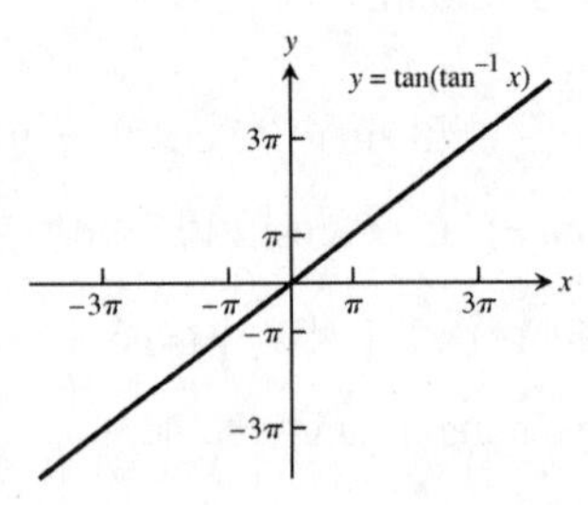

129. (a) Domain: $-\infty < x < \infty$; Range: $0 \leq y \leq \pi$

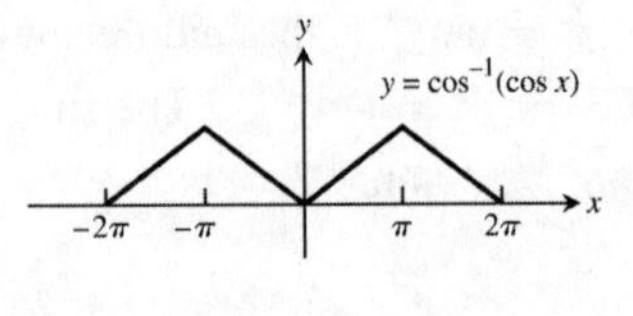

(b) Domain: $-1 \leq x \leq 1$; Range: $-1 \leq y \leq 1$
The graph of $y = \cos^{-1}(\cos x)$ is periodic; the graph of $y = \cos(\cos^{-1} x) = x$ for $-1 \leq x \leq 1$.

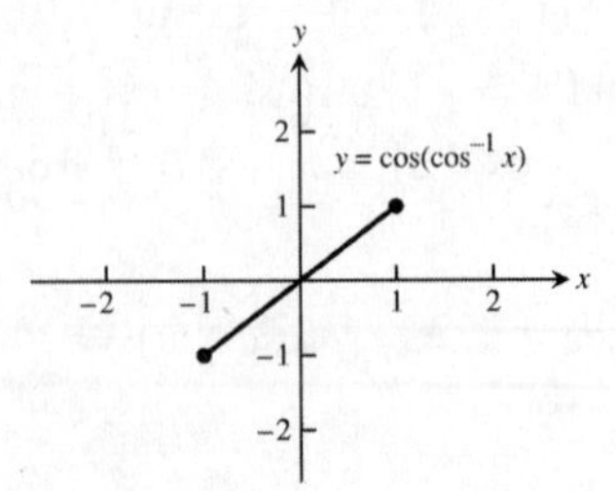

131. The graphs are identical for $y = 2\sin(2\tan^{-1} x)$

$= 4[\sin(\tan^{-1} x)][\cos(\tan^{-1} x)] = 4\left(\frac{x}{\sqrt{x^2+1}}\right)\left(\frac{1}{\sqrt{x^2+1}}\right)$

$= \frac{4x}{x^2+1}$ from the triangle

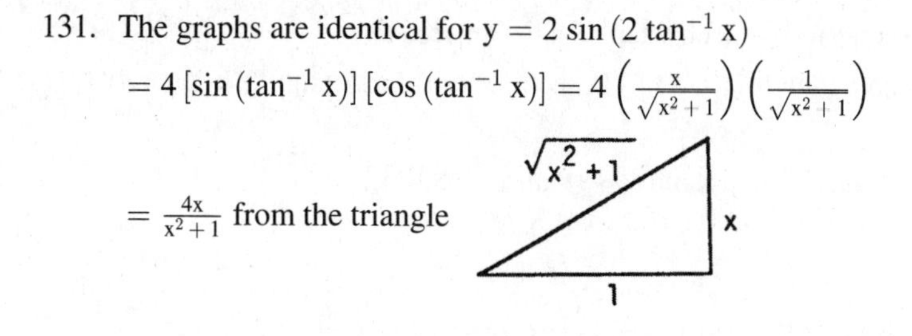

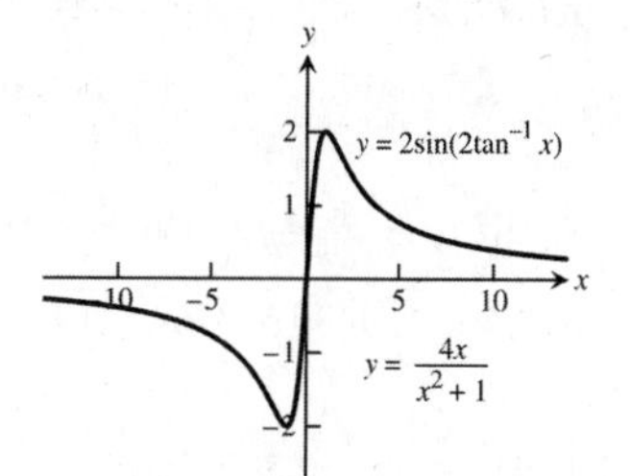

133. The values of f increase over the interval $[-1, 1]$ because $f' > 0$, and the graph of f steepens as the values of f' increase towards the ends of the interval. The graph of f is concave down to the left of the origin where $f'' < 0$, and concave up to the right of the origin where $f'' > 0$. There is an inflection point at $x = 0$ where $f'' = 0$ and f' has a local minimum value.

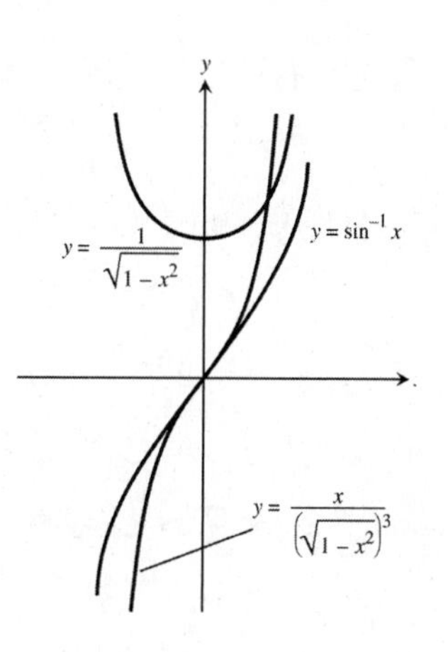

7.7 HYPERBOLIC FUNCTIONS

1. $\sinh x = -\frac{3}{4} \Rightarrow \cosh x = \sqrt{1+\sinh^2 x} = \sqrt{1+\left(-\frac{3}{4}\right)^2} = \sqrt{1+\frac{9}{16}} = \sqrt{\frac{25}{16}} = \frac{5}{4}$, $\tanh x = \frac{\sinh x}{\cosh x} = \frac{\left(-\frac{3}{4}\right)}{\left(\frac{5}{4}\right)} = -\frac{3}{5}$,
$\coth x = \frac{1}{\tanh x} = -\frac{5}{3}$, $\operatorname{sech} x = \frac{1}{\cosh x} = \frac{4}{5}$, and $\operatorname{csch} x = \frac{1}{\sin x} = -\frac{4}{3}$

3. $\cosh x = \frac{17}{15}$, $x > 0 \Rightarrow \sinh x = \sqrt{\cosh^2 x - 1} = \sqrt{\left(\frac{17}{15}\right)^2 - 1} = \sqrt{\frac{289}{225} - 1} = \sqrt{\frac{64}{225}} = \frac{8}{15}$, $\tanh x = \frac{\sinh x}{\cosh x} = \frac{\left(\frac{8}{15}\right)}{\left(\frac{17}{15}\right)}$
$= \frac{8}{17}$, $\coth x = \frac{1}{\tanh x} = \frac{17}{8}$, $\operatorname{sech} x = \frac{1}{\cosh x} = \frac{15}{17}$, and $\operatorname{csch} x = \frac{1}{\sinh x} = \frac{15}{8}$

5. $2\cosh(\ln x) = 2\left(\frac{e^{\ln x} + e^{-\ln x}}{2}\right) = e^{\ln x} + \frac{1}{e^{\ln x}} = x + \frac{1}{x}$

7. $\cosh 5x + \sinh 5x = \frac{e^{5x} + e^{-5x}}{2} + \frac{e^{5x} - e^{-5x}}{2} = e^{5x}$

9. $(\sinh x + \cosh x)^4 = \left(\frac{e^x - e^{-x}}{2} + \frac{e^x + e^{-x}}{2}\right)^4 = (e^x)^4 = e^{4x}$

11. (a) $\sinh 2x = \sinh(x + x) = \sinh x \cosh x + \cosh x \sinh x = 2 \sinh x \cosh x$
(b) $\cosh 2x = \cosh(x + x) = \cosh x \cosh x + \sinh x \sin x = \cosh^2 x + \sinh^2 x$

13. $y = 6\sinh\frac{x}{3} \Rightarrow \frac{dy}{dx} = 6\left(\cosh\frac{x}{3}\right)\left(\frac{1}{3}\right) = 2\cosh\frac{x}{3}$

15. $y = 2\sqrt{t}\tanh\sqrt{t} = 2t^{1/2}\tanh t^{1/2} \Rightarrow \frac{dy}{dt} = \left[\operatorname{sech}^2\left(t^{1/2}\right)\right]\left(\frac{1}{2}t^{-1/2}\right)\left(2t^{1/2}\right) + \left(\tanh t^{1/2}\right)\left(t^{-1/2}\right) = \operatorname{sech}^2\sqrt{t} + \frac{\tanh\sqrt{t}}{\sqrt{t}}$

17. $y = \ln(\sinh z) \Rightarrow \frac{dy}{dz} = \frac{\cosh z}{\sinh z} = \coth z$

19. $y = (\text{sech}\,\theta)(1 - \ln \text{sech}\,\theta) \Rightarrow \frac{dy}{d\theta} = \left(-\frac{-\text{sech}\,\theta \tanh\theta}{\text{sech}\,\theta}\right)(\text{sech}\,\theta) + (-\text{sech}\,\theta \tanh\theta)(1 - \ln \text{sech}\,\theta)$
$= \text{sech}\,\theta \tanh\theta - (\text{sech}\,\theta \tanh\theta)(1 - \ln \text{sech}\,\theta) = (\text{sech}\,\theta \tanh\theta)[1 - (1 - \ln \text{sech}\,\theta)] = (\text{sech}\,\theta \tanh\theta)(\ln \text{sech}\,\theta)$

21. $y = \ln \cosh v - \frac{1}{2}\tanh^2 v \Rightarrow \frac{dy}{dv} = \frac{\sinh v}{\cosh v} - \left(\frac{1}{2}\right)(2\tanh v)(\text{sech}^2 v) = \tanh v - (\tanh v)(\text{sech}^2 v)$
$= (\tanh v)(1 - \text{sech}^2 v) = (\tanh v)(\tanh^2 v) = \tanh^3 v$

23. $y = (x^2+1)\,\text{sech}(\ln x) = (x^2+1)\left(\frac{2}{e^{\ln x} + e^{-\ln x}}\right) = (x^2+1)\left(\frac{2}{x + x^{-1}}\right) = (x^2+1)\left(\frac{2x}{x^2+1}\right) = 2x \Rightarrow \frac{dy}{dx} = 2$

25. $y = \sinh^{-1}\sqrt{x} = \sinh^{-1}(x^{1/2}) \Rightarrow \frac{dy}{dx} = \frac{\left(\frac{1}{2}\right)x^{-1/2}}{\sqrt{1+(x^{1/2})^2}} = \frac{1}{2\sqrt{x}\sqrt{1+x}} = \frac{1}{2\sqrt{x(1+x)}}$

27. $y = (1-\theta)\tanh^{-1}\theta \Rightarrow \frac{dy}{d\theta} = (1-\theta)\left(\frac{1}{1-\theta^2}\right) + (-1)\tanh^{-1}\theta = \frac{1}{1+\theta} - \tanh^{-1}\theta$

29. $y = (1-t)\coth^{-1}\sqrt{t} = (1-t)\coth^{-1}(t^{1/2}) \Rightarrow \frac{dy}{dt} = (1-t)\left[\frac{\left(\frac{1}{2}\right)t^{-1/2}}{1-(t^{1/2})^2}\right] + (-1)\coth^{-1}(t^{1/2}) = \frac{1}{2\sqrt{t}} - \coth^{-1}\sqrt{t}$

31. $y = \cos^{-1}x - x\,\text{sech}^{-1}x \Rightarrow \frac{dy}{dx} = \frac{-1}{\sqrt{1-x^2}} - \left[x\left(\frac{-1}{x\sqrt{1-x^2}}\right) + (1)\,\text{sech}^{-1}x\right] = \frac{-1}{\sqrt{1-x^2}} + \frac{1}{\sqrt{1-x^2}} - \text{sech}^{-1}x = -\text{sech}^{-1}x$

33. $y = \text{csch}^{-1}\left(\frac{1}{2}\right)^\theta \Rightarrow \frac{dy}{d\theta} = -\frac{\left[\ln\left(\frac{1}{2}\right)\right]\left(\frac{1}{2}\right)^\theta}{\left(\frac{1}{2}\right)^\theta\sqrt{1+\left[\left(\frac{1}{2}\right)^\theta\right]^2}} = -\frac{\ln(1) - \ln(2)}{\sqrt{1+\left(\frac{1}{2}\right)^{2\theta}}} = \frac{\ln 2}{\sqrt{1+\left(\frac{1}{2}\right)^{2\theta}}}$

35. $y = \sinh^{-1}(\tan x) \Rightarrow \frac{dy}{dx} = \frac{\sec^2 x}{\sqrt{1+(\tan x)^2}} = \frac{\sec^2 x}{\sqrt{\sec^2 x}} = \frac{\sec^2 x}{|\sec x|} = \frac{|\sec x|\,|\sec x|}{|\sec x|} = |\sec x|$

37. (a) If $y = \tan^{-1}(\sinh x) + C$, then $\frac{dy}{dx} = \frac{\cosh x}{1+\sinh^2 x} = \frac{\cosh x}{\cosh^2 x} = \text{sech}\,x$, which verifies the formula
(b) If $y = \sin^{-1}(\tanh x) + C$, then $\frac{dy}{dx} = \frac{\text{sech}^2 x}{\sqrt{1-\tanh^2 x}} = \frac{\text{sech}^2 x}{\text{sech}\,x} = \text{sech}\,x$, which verifies the formula

39. If $y = \frac{x^2-1}{2}\coth^{-1}x + \frac{x}{2} + C$, then $\frac{dy}{dx} = x\coth^{-1}x + \left(\frac{x^2-1}{2}\right)\left(\frac{1}{1-x^2}\right) + \frac{1}{2} = x\coth^{-1}x$, which verifies the formula

41. $\int \sinh 2x\,dx = \frac{1}{2}\int \sinh u\,du$, where $u = 2x$ and $du = 2\,dx$
$= \frac{\cosh u}{2} + C = \frac{\cosh 2x}{2} + C$

43. $\int 6\cosh\left(\frac{x}{2} - \ln 3\right)dx = 12\int \cosh u\,du$, where $u = \frac{x}{2} - \ln 3$ and $du = \frac{1}{2}\,dx$
$= 12\sinh u + C = 12\sinh\left(\frac{x}{2} - \ln 3\right) + C$

45. $\int \tanh\frac{x}{7}\,dx = 7\int \frac{\sinh u}{\cosh u}\,du$, where $u = \frac{x}{7}$ and $du = \frac{1}{7}\,dx$
$= 7\ln|\cosh u| + C_1 = 7\ln\left|\cosh\frac{x}{7}\right| + C_1 = 7\ln\left|\frac{e^{x/7}+e^{-x/7}}{2}\right| + C_1 = 7\ln\left|e^{x/7}+e^{-x/7}\right| - 7\ln 2 + C_1$
$= 7\ln\left|e^{x/7}+e^{-x/7}\right| + C$

47. $\int \text{sech}^2\left(x - \frac{1}{2}\right)dx = \int \text{sech}^2 u\,du$, where $u = \left(x - \frac{1}{2}\right)$ and $du = dx$
$= \tanh u + C = \tanh\left(x - \frac{1}{2}\right) + C$

49. $\int \frac{\operatorname{sech}\sqrt{t}\tanh\sqrt{t}}{\sqrt{t}}\,dt = 2\int \operatorname{sech} u \tanh u\,du$, where $u = \sqrt{t} = t^{1/2}$ and $du = \frac{dt}{2\sqrt{t}}$
$= 2(-\operatorname{sech} u) + C = -2\operatorname{sech}\sqrt{t} + C$

51. $\int_{\ln 2}^{\ln 4} \coth x\,dx = \int_{\ln 2}^{\ln 4} \frac{\cosh x}{\sinh x}\,dx = \int_{3/4}^{15/8} \frac{1}{u}\,du = [\ln|u|]_{3/4}^{15/8} = \ln\left|\frac{15}{8}\right| - \ln\left|\frac{3}{4}\right| = \ln\left|\frac{15}{8}\cdot\frac{4}{3}\right| = \ln\frac{5}{2}$,

where $u = \sinh x$, $du = \cosh x\,dx$, the lower limit is $\sinh(\ln 2) = \frac{e^{\ln 2} - e^{-\ln 2}}{2} = \frac{2 - \left(\frac{1}{2}\right)}{2} = \frac{3}{4}$ and the upper limit is $\sinh(\ln 4) = \frac{e^{\ln 4} - e^{-\ln 4}}{2} = \frac{4 - \left(\frac{1}{4}\right)}{2} = \frac{15}{8}$

53. $\int_{-\ln 4}^{-\ln 2} 2e^{\theta}\cosh\theta\,d\theta = \int_{-\ln 4}^{-\ln 2} 2e^{\theta}\left(\frac{e^{\theta}+e^{-\theta}}{2}\right)d\theta = \int_{-\ln 4}^{-\ln 2} (e^{2\theta} + 1)\,d\theta = \left[\frac{e^{2\theta}}{2} + \theta\right]_{-\ln 4}^{-\ln 2}$
$= \left(\frac{e^{-2\ln 2}}{2} - \ln 2\right) - \left(\frac{e^{-2\ln 4}}{2} - \ln 4\right) = \left(\frac{1}{8} - \ln 2\right) - \left(\frac{1}{32} - \ln 4\right) = \frac{3}{32} - \ln 2 + 2\ln 2 = \frac{3}{32} + \ln 2$

55. $\int_{-\pi/4}^{\pi/4} \cosh(\tan\theta)\sec^2\theta\,d\theta = \int_{-1}^{1} \cosh u\,du = [\sinh u]_{-1}^{1} = \sinh(1) - \sinh(-1) = \left(\frac{e^1 - e^{-1}}{2}\right) - \left(\frac{e^{-1} - e^1}{2}\right)$
$= \frac{e - e^{-1} - e^{-1} + e}{2} = e - e^{-1}$, where $u = \tan\theta$, $du = \sec^2\theta\,d\theta$, the lower limit is $\tan\left(-\frac{\pi}{4}\right) = -1$ and the upper limit is $\tan\left(\frac{\pi}{4}\right) = 1$

57. $\int_{1}^{2} \frac{\cosh(\ln t)}{t}\,dt = \int_{0}^{\ln 2} \cosh u\,du = [\sinh u]_0^{\ln 2} = \sinh(\ln 2) - \sinh(0) = \frac{e^{\ln 2} - e^{-\ln 2}}{2} - 0 = \frac{2 - \frac{1}{2}}{2} = \frac{3}{4}$, where $u = \ln t$, $du = \frac{1}{t}\,dt$, the lower limit is $\ln 1 = 0$ and the upper limit is $\ln 2$

59. $\int_{-\ln 2}^{0} \cosh^2\left(\frac{x}{2}\right)dx = \int_{-\ln 2}^{0} \frac{\cosh x + 1}{2}\,dx = \frac{1}{2}\int_{-\ln 2}^{0} (\cosh x + 1)\,dx = \frac{1}{2}[\sinh x + x]_{-\ln 2}^{0}$
$= \frac{1}{2}[(\sinh 0 + 0) - (\sinh(-\ln 2) - \ln 2)] = \frac{1}{2}\left[(0 + 0) - \left(\frac{e^{-\ln 2} - e^{\ln 2}}{2} - \ln 2\right)\right] = \frac{1}{2}\left[-\frac{\left(\frac{1}{2}\right) - 2}{2} + \ln 2\right]$
$= \frac{1}{2}\left(1 - \frac{1}{4} + \ln 2\right) = \frac{3}{8} + \frac{1}{2}\ln 2 = \frac{3}{8} + \ln\sqrt{2}$

61. $\sinh^{-1}\left(\frac{-5}{12}\right) = \ln\left(-\frac{5}{12} + \sqrt{\frac{25}{144} + 1}\right) = \ln\left(\frac{2}{3}\right)$

63. $\tanh^{-1}\left(-\frac{1}{2}\right) = \frac{1}{2}\ln\left(\frac{1 - (1/2)}{1 + (1/2)}\right) = -\frac{\ln 3}{2}$

65. $\operatorname{sech}^{-1}\left(\frac{3}{5}\right) = \ln\left(\frac{1 + \sqrt{1 - (9/25)}}{(3/5)}\right) = \ln 3$

67. (a) $\int_0^{2\sqrt{3}} \frac{dx}{\sqrt{4 + x^2}} = \left[\sinh^{-1}\frac{x}{2}\right]_0^{2\sqrt{3}} = \sinh^{-1}\sqrt{3} - \sinh 0 = \sinh^{-1}\sqrt{3}$
(b) $\sinh^{-1}\sqrt{3} = \ln\left(\sqrt{3} + \sqrt{3 + 1}\right) = \ln\left(\sqrt{3} + 2\right)$

69. (a) $\int_{5/4}^{2} \frac{1}{1 - x^2}\,dx = \left[\coth^{-1} x\right]_{5/4}^{2} = \coth^{-1} 2 - \coth^{-1}\frac{5}{4}$
(b) $\coth^{-1} 2 - \coth^{-1}\frac{5}{4} = \frac{1}{2}\left[\ln 3 - \ln\left(\frac{9/4}{1/4}\right)\right] = \frac{1}{2}\ln\frac{1}{3}$

71. (a) $\int_{1/5}^{3/13} \frac{dx}{x\sqrt{1 - 16x^2}} = \int_{4/5}^{12/13} \frac{du}{u\sqrt{a^2 - u^2}}$, where $u = 4x$, $du = 4\,dx$, $a = 1$
$= \left[-\operatorname{sech}^{-1} u\right]_{4/5}^{12/13} = -\operatorname{sech}^{-1}\frac{12}{13} + \operatorname{sech}^{-1}\frac{4}{5}$
(b) $-\operatorname{sech}^{-1}\frac{12}{13} + \operatorname{sech}^{-1}\frac{4}{5} = -\ln\left(\frac{1 + \sqrt{1 - (12/13)^2}}{(12/13)}\right) + \ln\left(\frac{1 + \sqrt{1 - (4/5)^2}}{(4/5)}\right)$
$= -\ln\left(\frac{13 + \sqrt{169 - 144}}{12}\right) + \ln\left(\frac{5 + \sqrt{25 - 16}}{4}\right) = \ln\left(\frac{5+3}{4}\right) - \ln\left(\frac{13+5}{12}\right) = \ln 2 - \ln\frac{3}{2} = \ln\left(2\cdot\frac{2}{3}\right) = \ln\frac{4}{3}$

73. (a) $\int_0^{\pi} \frac{\cos x}{\sqrt{1+\sin^2 x}}\,dx = \int_0^0 \frac{1}{\sqrt{1+u^2}}\,du = \left[\sinh^{-1} u\right]_0^0 = \sinh^{-1} 0 - \sinh^{-1} 0 = 0$, where $u = \sin x$, $du = \cos x\,dx$

(b) $\sinh^{-1} 0 - \sinh^{-1} 0 = \ln\left(0 + \sqrt{0+1}\right) - \ln\left(0 + \sqrt{0+1}\right) = 0$

75. Let $E(x) = \frac{f(x)+f(-x)}{2}$ and $O(x) = \frac{f(x)-f(-x)}{2}$. Then $E(x) + O(x) = \frac{f(x)+f(-x)}{2} + \frac{f(x)-f(-x)}{2} = \frac{2f(x)}{2} = f(x)$. Also, $E(-x) = \frac{f(-x)+f(-(-x))}{2} = \frac{f(x)+f(-x)}{2} = E(x) \Rightarrow E(x)$ is even, and $O(-x) = \frac{f(-x)-f(-(-x))}{2} = -\frac{f(x)-f(-x)}{2} = -O(x)$ $\Rightarrow O(x)$ is odd. Consequently, $f(x)$ can be written as a sum of an even and an odd function.
$f(x) = \frac{f(x)+f(-x)}{2}$ because $\frac{f(x)-f(-x)}{2} = 0$ if f is even and $f(x) = \frac{f(x)-f(-x)}{2}$ because $\frac{f(x)+f(-x)}{2} = 0$ if f is odd.
Thus, if f is even $f(x) = \frac{2f(x)}{2} + 0$ and if f is odd, $f(x) = 0 + \frac{2f(x)}{2}$

77. (a) $v = \sqrt{\frac{mg}{k}} \tanh\left(\sqrt{\frac{gk}{m}}\,t\right) \Rightarrow \frac{dv}{dt} = \sqrt{\frac{mg}{k}}\left[\operatorname{sech}^2\left(\sqrt{\frac{gk}{m}}\,t\right)\right]\left(\sqrt{\frac{gk}{m}}\right) = g\operatorname{sech}^2\left(\sqrt{\frac{gk}{m}}\,t\right)$.

Thus $m\frac{dv}{dt} = mg\operatorname{sech}^2\left(\sqrt{\frac{gk}{m}}\,t\right) = mg\left(1 - \tanh^2\left(\sqrt{\frac{gk}{m}}\,t\right)\right) = mg - kv^2$. Also, since $\tanh x = 0$ when $x = 0$, $v = 0$ when $t = 0$.

(b) $\lim_{t\to\infty} v = \lim_{t\to\infty} \sqrt{\frac{mg}{k}} \tanh\left(\sqrt{\frac{kg}{m}}\,t\right) = \sqrt{\frac{mg}{k}} \lim_{t\to\infty} \tanh\left(\sqrt{\frac{kg}{m}}\,t\right) = \sqrt{\frac{mg}{k}}\,(1) = \sqrt{\frac{mg}{k}}$

(c) $\sqrt{\frac{160}{0.005}} = \sqrt{\frac{160{,}000}{5}} = \frac{400}{\sqrt{5}} = 80\sqrt{5} \approx 178.89$ ft/sec

79. $V = \pi\int_0^2 (\cosh^2 x - \sinh^2 x)\,dx = \pi\int_0^2 1\,dx = 2\pi$

81. $y = \frac{1}{2}\cosh 2x \Rightarrow y' = \sinh 2x \Rightarrow L = \int_0^{\ln\sqrt{5}} \sqrt{1 + (\sinh 2x)^2}\,dx = \int_0^{\ln\sqrt{5}} \cosh 2x\,dx = \left[\frac{1}{2}\sinh 2x\right]_0^{\ln\sqrt{5}}$
$= \left[\frac{1}{2}\left(\frac{e^{2x}-e^{-2x}}{2}\right)\right]_0^{\ln\sqrt{5}} = \frac{1}{4}\left(5 - \frac{1}{5}\right) = \frac{6}{5}$

83. (a) $y = \frac{H}{w}\cosh\left(\frac{w}{H}x\right) \Rightarrow \tan\phi = \frac{dy}{dx} = \left(\frac{H}{w}\right)\left[\frac{w}{H}\sinh\left(\frac{w}{H}x\right)\right] = \sinh\left(\frac{w}{H}x\right)$

(b) The tension at P is given by $T\cos\phi = H \Rightarrow T = H\sec\phi = H\sqrt{1+\tan^2\phi} = H\sqrt{1 + \left(\sinh\frac{w}{H}x\right)^2}$
$= H\cosh\left(\frac{w}{H}x\right) = w\left(\frac{H}{w}\right)\cosh\left(\frac{w}{H}x\right) = wy$

85. To find the length of the curve: $y = \frac{1}{a}\cosh ax \Rightarrow y' = \sinh ax \Rightarrow L = \int_0^b \sqrt{1 + (\sinh ax)^2}\,dx$
$\Rightarrow L = \int_0^b \cosh ax\,dx = \left[\frac{1}{a}\sinh ax\right]_0^b = \frac{1}{a}\sinh ab$. The area under the curve is $A = \int_0^b \frac{1}{a}\cosh ax\,dx$
$= \left[\frac{1}{a^2}\sinh ax\right]_0^b = \frac{1}{a^2}\sinh ab = \left(\frac{1}{a}\right)\left(\frac{1}{a}\sinh ab\right)$ which is the area of the rectangle of height $\frac{1}{a}$ and length L as claimed, and which is illustrated below.

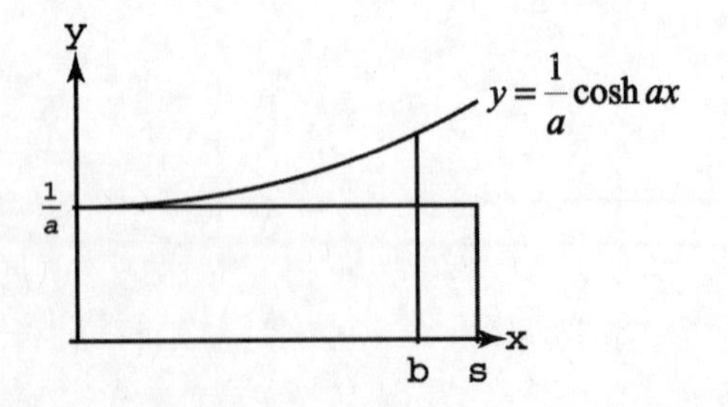

7.8 RELATIVE RATES OF GROWTH

1. (a) slower, $\lim_{x\to\infty} \frac{x+3}{e^x} = \lim_{x\to\infty} \frac{1}{e^x} = 0$

   (b) slower, $\lim_{x\to\infty} \frac{x^3+\sin^2 x}{e^x} = \lim_{x\to\infty} \frac{3x^2+2\sin x\cos x}{e^x} = \lim_{x\to\infty} \frac{6x+2\cos 2x}{e^x} = \lim_{x\to\infty} \frac{6-4\sin 2x}{e^x} = 0$ by the Sandwich Theorem because $\frac{2}{e^x} \le \frac{6-4\sin 2x}{e^x} \le \frac{10}{e^x}$ for all reals and $\lim_{x\to\infty} \frac{2}{e^x} = 0 = \lim_{x\to\infty} \frac{10}{e^x}$

   (c) slower, $\lim_{x\to\infty} \frac{\sqrt{x}}{e^x} = \lim_{x\to\infty} \frac{x^{1/2}}{e^x} = \lim_{x\to\infty} \frac{\left(\frac{1}{2}\right)x^{-1/2}}{e^x} = \lim_{x\to\infty} \frac{1}{2\sqrt{x}\,e^x} = 0$

   (d) faster, $\lim_{x\to\infty} \frac{4^x}{e^x} = \lim_{x\to\infty} \left(\frac{4}{e}\right)^x = \infty$ since $\frac{4}{e} > 1$

   (e) slower, $\lim_{x\to\infty} \frac{\left(\frac{3}{2}\right)^x}{e^x} = \lim_{x\to\infty} \left(\frac{3}{2e}\right)^x = 0$ since $\frac{3}{2e} < 1$

   (f) slower, $\lim_{x\to\infty} \frac{e^{x/2}}{e^x} = \lim_{x\to\infty} \frac{1}{e^{x/2}} = 0$

   (g) same, $\lim_{x\to\infty} \frac{\left(\frac{e^x}{2}\right)}{e^x} = \lim_{x\to\infty} \frac{1}{2} = \frac{1}{2}$

   (h) slower, $\lim_{x\to\infty} \frac{\log_{10} x}{e^x} = \lim_{x\to\infty} \frac{\ln x}{(\ln 10)\,e^x} = \lim_{x\to\infty} \frac{\frac{1}{x}}{(\ln 10)\,e^x} = \lim_{x\to\infty} \frac{1}{(\ln 10)xe^x} = 0$

3. (a) same, $\lim_{x\to\infty} \frac{x^2+4x}{x^2} = \lim_{x\to\infty} \frac{2x+4}{2x} = \lim_{x\to\infty} \frac{2}{2} = 1$

   (b) faster, $\lim_{x\to\infty} \frac{x^5-x^2}{x^2} = \lim_{x\to\infty} (x^3-1) = \infty$

   (c) same, $\lim_{x\to\infty} \frac{\sqrt{x^4+x^3}}{x^2} = \sqrt{\lim_{x\to\infty} \frac{x^4+x^3}{x^4}} = \sqrt{\lim_{x\to\infty} \left(1+\frac{1}{x}\right)} = \sqrt{1} = 1$

   (d) same, $\lim_{x\to\infty} \frac{(x+3)^2}{x^2} = \lim_{x\to\infty} \frac{2(x+3)}{2x} = \lim_{x\to\infty} \frac{2}{2} = 1$

   (e) slower, $\lim_{x\to\infty} \frac{x\ln x}{x^2} = \lim_{x\to\infty} \frac{\ln x}{x} = \lim_{x\to\infty} \frac{\left(\frac{1}{x}\right)}{1} = 0$

   (f) faster, $\lim_{x\to\infty} \frac{2^x}{x^2} = \lim_{x\to\infty} \frac{(\ln 2)\,2^x}{2x} = \lim_{x\to\infty} \frac{(\ln 2)^2\,2^x}{2} = \infty$

   (g) slower, $\lim_{x\to\infty} \frac{x^3e^{-x}}{x^2} = \lim_{x\to\infty} \frac{x}{e^x} = \lim_{x\to\infty} \frac{1}{e^x} = 0$

   (h) same, $\lim_{x\to\infty} \frac{8x^2}{x^2} = \lim_{x\to\infty} 8 = 8$

5. (a) same, $\lim_{x\to\infty} \frac{\log_3 x}{\ln x} = \lim_{x\to\infty} \frac{\left(\frac{\ln x}{\ln 3}\right)}{\ln x} = \lim_{x\to\infty} \frac{1}{\ln 3} = \frac{1}{\ln 3}$

   (b) same, $\lim_{x\to\infty} \frac{\ln 2x}{\ln x} = \lim_{x\to\infty} \frac{\left(\frac{2}{2x}\right)}{\left(\frac{1}{x}\right)} = 1$

   (c) same, $\lim_{x\to\infty} \frac{\ln\sqrt{x}}{\ln x} = \lim_{x\to\infty} \frac{\left(\frac{1}{2}\right)\ln x}{\ln x} = \lim_{x\to\infty} \frac{1}{2} = \frac{1}{2}$

   (d) faster, $\lim_{x\to\infty} \frac{\sqrt{x}}{\ln x} = \lim_{x\to\infty} \frac{x^{1/2}}{\ln x} = \lim_{x\to\infty} \frac{\left(\frac{1}{2}\right)x^{-1/2}}{\left(\frac{1}{x}\right)} = \lim_{x\to\infty} \frac{x}{2\sqrt{x}} = \lim_{x\to\infty} \frac{\sqrt{x}}{2} = \infty$

   (e) faster, $\lim_{x\to\infty} \frac{x}{\ln x} = \lim_{x\to\infty} \frac{1}{\left(\frac{1}{x}\right)} = \lim_{x\to\infty} x = \infty$

   (f) same, $\lim_{x\to\infty} \frac{5\ln x}{\ln x} = \lim_{x\to\infty} 5 = 5$

   (g) slower, $\lim_{x\to\infty} \frac{\left(\frac{1}{x}\right)}{\ln x} = \lim_{x\to\infty} \frac{1}{x\ln x} = 0$

   (h) faster, $\lim_{x\to\infty} \frac{e^x}{\ln x} = \lim_{x\to\infty} \frac{e^x}{\left(\frac{1}{x}\right)} = \lim_{x\to\infty} xe^x = \infty$

7. $\lim_{x\to\infty} \frac{e^x}{e^{x/2}} = \lim_{x\to\infty} e^{x/2} = \infty \Rightarrow e^x$ grows faster than $e^{x/2}$; since for $x > e^e$ we have $\ln x > e$ and $\lim_{x\to\infty} \frac{(\ln x)^x}{e^x} = \lim_{x\to\infty} \left(\frac{\ln x}{e}\right)^x = \infty \Rightarrow (\ln x)^x$ grows faster than e^x; since $x > \ln x$ for all $x > 0$ and $\lim_{x\to\infty} \frac{x^x}{(\ln x)^x} = \lim_{x\to\infty} \left(\frac{x}{\ln x}\right)^x = \infty \Rightarrow x^x$ grows faster than $(\ln x)^x$. Therefore, slowest to fastest are: $e^{x/2}$, e^x, $(\ln x)^x$, x^x so the order is d, a, c, b

9. (a) false; $\lim\limits_{x \to \infty} \frac{x}{x} = 1$

(b) false; $\lim\limits_{x \to \infty} \frac{x}{x+5} = \frac{1}{1} = 1$

(c) true; $x < x + 5 \Rightarrow \frac{x}{x+5} < 1$ if $x > 1$ (or sufficiently large)

(d) true; $x < 2x \Rightarrow \frac{x}{2x} < 1$ if $x > 1$ (or sufficiently large)

(e) true; $\lim\limits_{x \to \infty} \frac{e^x}{e^{2x}} = \lim\limits_{x \to 0} \frac{1}{e^x} = 0$

(f) true; $\frac{x + \ln x}{x} = 1 + \frac{\ln x}{x} < 1 + \frac{\sqrt{x}}{x} = 1 + \frac{1}{\sqrt{x}} < 2$ if $x > 1$ (or sufficiently large)

(g) false; $\lim\limits_{x \to \infty} \frac{\ln x}{\ln 2x} = \lim\limits_{x \to \infty} \frac{\left(\frac{1}{x}\right)}{\left(\frac{2}{2x}\right)} = \lim\limits_{x \to \infty} 1 = 1$

(h) true; $\frac{\sqrt{x^2+5}}{x} < \frac{\sqrt{(x+5)^2}}{x} < \frac{x+5}{x} = 1 + \frac{5}{x} < 6$ if $x > 1$ (or sufficiently large)

11. If f(x) and g(x) grow at the same rate, then $\lim\limits_{x \to \infty} \frac{f(x)}{g(x)} = L \neq 0 \Rightarrow \lim\limits_{x \to \infty} \frac{g(x)}{f(x)} = \frac{1}{L} \neq 0$. Then $\left|\frac{f(x)}{g(x)} - L\right| < 1$ if x is sufficiently large $\Rightarrow L - 1 < \frac{f(x)}{g(x)} < L + 1 \Rightarrow \frac{f(x)}{g(x)} \leq |L| + 1$ if x is sufficiently large $\Rightarrow f = O(g)$. Similarly, $\frac{g(x)}{f(x)} \leq \left|\frac{1}{L}\right| + 1 \Rightarrow g = O(f)$.

13. When the degree of f is less than or equal to the degree of g since $\lim\limits_{x \to \infty} \frac{f(x)}{g(x)} = 0$ when the degree of f is smaller than the degree of g, and $\lim\limits_{x \to \infty} \frac{f(x)}{g(x)} = \frac{a}{b}$ (the ratio of the leading coefficients) when the degrees are the same.

15. $\lim\limits_{x \to \infty} \frac{\ln(x+1)}{\ln x} = \lim\limits_{x \to \infty} \frac{\left(\frac{1}{x+1}\right)}{\left(\frac{1}{x}\right)} = \lim\limits_{x \to \infty} \frac{x}{x+1} = \lim\limits_{x \to \infty} \frac{1}{1} = 1$ and $\lim\limits_{x \to \infty} \frac{\ln(x+999)}{\ln x} = \lim\limits_{x \to \infty} \frac{\left(\frac{1}{x+999}\right)}{\left(\frac{1}{x}\right)}$ $= \lim\limits_{x \to \infty} \frac{x}{x+999} = 1$

17. $\lim\limits_{x \to \infty} \frac{\sqrt{10x+1}}{\sqrt{x}} = \sqrt{\lim\limits_{x \to \infty} \frac{10x+1}{x}} = \sqrt{10}$ and $\lim\limits_{x \to \infty} \frac{\sqrt{x+1}}{\sqrt{x}} = \sqrt{\lim\limits_{x \to \infty} \frac{x+1}{x}} = \sqrt{1} = 1$. Since the growth rate is transitive, we conclude that $\sqrt{10x+1}$ and $\sqrt{x+1}$ have the same growth rate $\left(\text{that of } \sqrt{x}\right)$.

19. $\lim\limits_{x \to \infty} \frac{x^n}{e^x} = \lim\limits_{x \to \infty} \frac{nx^{n-1}}{e^x} = \ldots = \lim\limits_{x \to \infty} \frac{n!}{e^x} = 0 \Rightarrow x^n = o(e^x)$ for any non-negative integer n

21. (a) $\lim\limits_{x \to \infty} \frac{x^{1/n}}{\ln x} = \lim\limits_{x \to \infty} \frac{x^{(1-n)/n}}{n\left(\frac{1}{x}\right)} = \left(\frac{1}{n}\right) \lim\limits_{x \to \infty} x^{1/n} = \infty \Rightarrow \ln x = o\left(x^{1/n}\right)$ for any positive integer n

(b) $\ln\left(e^{17{,}000{,}000}\right) = 17{,}000{,}000 < \left(e^{17 \times 10^6}\right)^{1/10^6} = e^{17} \approx 24{,}154{,}952.75$

(c) $x \approx 3.430631121 \times 10^{15}$

(d) In the interval $\left[3.41 \times 10^{15}, 3.45 \times 10^{15}\right]$ we have $\ln x = 10 \ln(\ln x)$. The graphs cross at about 3.4306311×10^{15}.

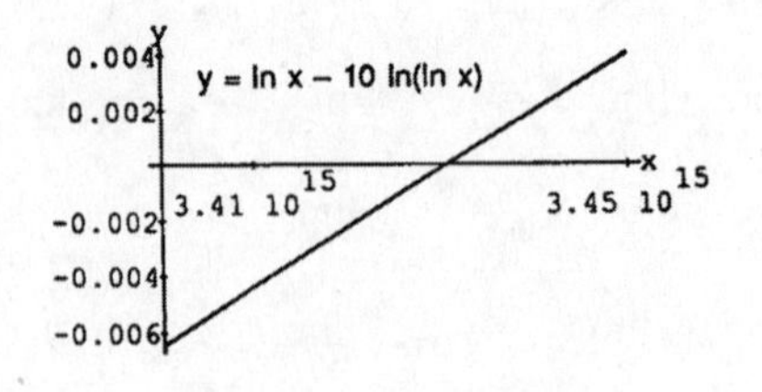

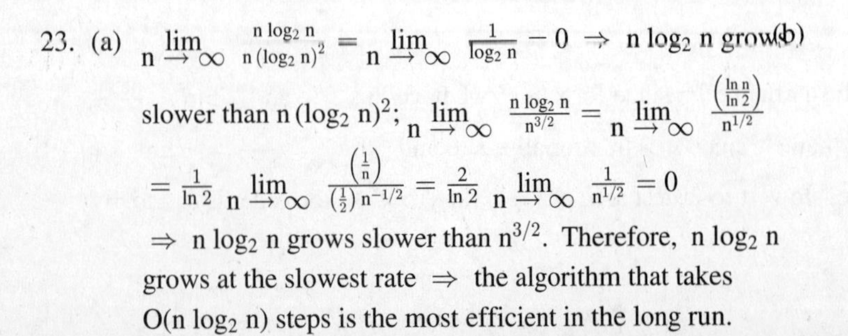

23. (a) $\lim\limits_{n \to \infty} \frac{n \log_2 n}{n (\log_2 n)^2} = \lim\limits_{n \to \infty} \frac{1}{\log_2 n} - 0 \Rightarrow n \log_2 n$ grows slower than $n(\log_2 n)^2$; $\lim\limits_{n \to \infty} \frac{n \log_2 n}{n^{3/2}} = \lim\limits_{n \to \infty} \frac{\left(\frac{\ln n}{\ln 2}\right)}{n^{1/2}}$

$= \frac{1}{\ln 2} \lim\limits_{n \to \infty} \frac{\left(\frac{1}{n}\right)}{\left(\frac{1}{2}\right) n^{-1/2}} = \frac{2}{\ln 2} \lim\limits_{n \to \infty} \frac{1}{n^{1/2}} = 0$

$\Rightarrow n \log_2 n$ grows slower than $n^{3/2}$. Therefore, $n \log_2 n$ grows at the slowest rate $\Rightarrow$ the algorithm that takes $O(n \log_2 n)$ steps is the most efficient in the long run.

(b)

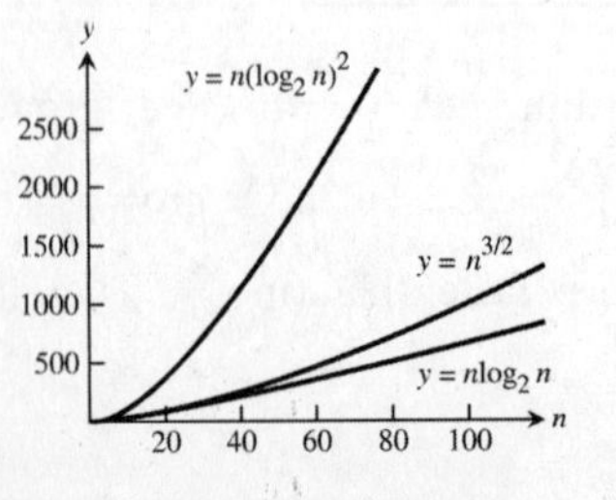

25. It could take one million steps for a sequential search, but at most 20 steps for a binary search because $2^{19} = 524{,}288 < 1{,}000{,}000 < 1{,}048{,}576 = 2^{20}$.

CHAPTER 7 PRACTICE EXERCISES

1. $y = 10e^{-x/5} \Rightarrow \frac{dy}{dx} = (10)\left(-\frac{1}{5}\right)e^{-x/5} = -2e^{-x/5}$

3. $y = \frac{1}{4}xe^{4x} - \frac{1}{16}e^{4x} \Rightarrow \frac{dy}{dx} = \frac{1}{4}\left[x\left(4e^{4x}\right) + e^{4x}(1)\right] - \frac{1}{16}\left(4e^{4x}\right) = xe^{4x} + \frac{1}{4}e^{4x} - \frac{1}{4}e^{4x} = xe^{4x}$

5. $y = \ln\left(\sin^2\theta\right) \Rightarrow \frac{dy}{d\theta} = \frac{2(\sin\theta)(\cos\theta)}{\sin^2\theta} = \frac{2\cos\theta}{\sin\theta} = 2\cot\theta$

7. $y = \log_2\left(\frac{x^2}{2}\right) = \frac{\ln\left(\frac{x^2}{2}\right)}{\ln 2} \Rightarrow \frac{dy}{dx} = \frac{1}{\ln 2}\left(\frac{x}{\left(\frac{x^2}{2}\right)}\right) = \frac{2}{(\ln 2)x}$

9. $y = 8^{-t} \Rightarrow \frac{dy}{dt} = 8^{-t}(\ln 8)(-1) = -8^{-t}(\ln 8)$

11. $y = 5x^{3.6} \Rightarrow \frac{dy}{dx} = 5(3.6)x^{2.6} = 18x^{2.6}$

13. $y = (x+2)^{x+2} \Rightarrow \ln y = \ln(x+2)^{x+2} = (x+2)\ln(x+2) \Rightarrow \frac{y'}{y} = (x+2)\left(\frac{1}{x+2}\right) + (1)\ln(x+2)$
$\Rightarrow \frac{dy}{dx} = (x+2)^{x+2}\left[\ln(x+2) + 1\right]$

15. $y = \sin^{-1}\sqrt{1-u^2} = \sin^{-1}\left(1-u^2\right)^{1/2} \Rightarrow \frac{dy}{du} = \frac{\frac{1}{2}\left(1-u^2\right)^{-1/2}(-2u)}{\sqrt{1-\left[\left(1-u^2\right)^{1/2}\right]^2}} = \frac{-u}{\sqrt{1-u^2}\sqrt{1-\left(1-u^2\right)}} = \frac{-u}{|u|\sqrt{1-u^2}}$
$= \frac{-u}{u\sqrt{1-u^2}} = \frac{-1}{\sqrt{1-u^2}},\ 0 < u < 1$

17. $y = \ln\left(\cos^{-1}x\right) \Rightarrow y' = \frac{\left(\frac{-1}{\sqrt{1-x^2}}\right)}{\cos^{-1}x} = \frac{-1}{\sqrt{1-x^2}\cos^{-1}x}$

19. $y = t\tan^{-1}t - \left(\frac{1}{2}\right)\ln t \Rightarrow \frac{dy}{dt} = \tan^{-1}t + t\left(\frac{1}{1+t^2}\right) - \left(\frac{1}{2}\right)\left(\frac{1}{t}\right) = \tan^{-1}t + \frac{t}{1+t^2} - \frac{1}{2t}$

21. $y = z\sec^{-1}z - \sqrt{z^2-1} = z\sec^{-1}z - \left(z^2-1\right)^{1/2} \Rightarrow \frac{dy}{dz} = z\left(\frac{1}{|z|\sqrt{z^2-1}}\right) + \left(\sec^{-1}z\right)(1) - \frac{1}{2}\left(z^2-1\right)^{-1/2}(2z)$
$= \frac{z}{|z|\sqrt{z^2-1}} - \frac{z}{\sqrt{z^2-1}} + \sec^{-1}z = \frac{1-z}{\sqrt{z^2-1}} + \sec^{-1}z,\ z > 1$

23. $y = \csc^{-1}(\sec\theta) \Rightarrow \frac{dy}{d\theta} = \frac{-\sec\theta\tan\theta}{|\sec\theta|\sqrt{\sec^2\theta - 1}} = -\frac{\tan\theta}{|\tan\theta|} = -1,\ 0 < \theta < \frac{\pi}{2}$

25. $y = \frac{2\left(x^2+1\right)}{\sqrt{\cos 2x}} \Rightarrow \ln y = \ln\left(\frac{2\left(x^2+1\right)}{\sqrt{\cos 2x}}\right) = \ln(2) + \ln\left(x^2+1\right) - \frac{1}{2}\ln(\cos 2x) \Rightarrow \frac{y'}{y} = 0 + \frac{2x}{x^2+1} - \left(\frac{1}{2}\right)\frac{(-2\sin 2x)}{\cos 2x}$
$\Rightarrow y' = \left(\frac{2x}{x^2+1} + \tan 2x\right)y = \frac{2\left(x^2+1\right)}{\sqrt{\cos 2x}}\left(\frac{2x}{x^2+1} + \tan 2x\right)$

27. $y = \left[\frac{(t+1)(t-1)}{(t-2)(t+3)}\right]^5 \Rightarrow \ln y = 5\left[\ln(t+1) + \ln(t-1) - \ln(t-2) - \ln(t+3)\right] \Rightarrow \left(\frac{1}{y}\right)\left(\frac{dy}{dt}\right)$
$= 5\left(\frac{1}{t+1} + \frac{1}{t-1} - \frac{1}{t-2} - \frac{1}{t+3}\right) \Rightarrow \frac{dy}{dt} = 5\left[\frac{(t+1)(t-1)}{(t-2)(t+3)}\right]^5\left(\frac{1}{t+1} + \frac{1}{t-1} - \frac{1}{t-2} - \frac{1}{t+3}\right)$

29. $y = (\sin\theta)^{\sqrt{\theta}} \Rightarrow \ln y = \sqrt{\theta}\ln(\sin\theta) \Rightarrow \left(\frac{1}{y}\right)\left(\frac{dy}{d\theta}\right) = \sqrt{\theta}\left(\frac{\cos\theta}{\sin\theta}\right) + \frac{1}{2}\theta^{-1/2}\ln(\sin\theta)$
$\Rightarrow \frac{dy}{d\theta} = (\sin\theta)^{\sqrt{\theta}}\left(\sqrt{\theta}\cot\theta + \frac{\ln(\sin\theta)}{2\sqrt{\theta}}\right)$

31. $\int e^x \sin(e^x)\,dx = \int \sin u\,du$, where $u = e^x$ and $du = e^x\,dx$
$= -\cos u + C = -\cos(e^x) + C$

33. $\int e^x \sec^2(e^x - 7)\,dx = \int \sec^2 u\,du$, where $u = e^x - 7$ and $du = e^x\,dx$
$= \tan u + C = \tan(e^x - 7) + C$

35. $\int (\sec^2 x)\,e^{\tan x}\,dx = \int e^u\,du$, where $u = \tan x$ and $du = \sec^2 x\,dx$
$= e^u + C = e^{\tan x} + C$

37. $\int_{-1}^{1} \frac{1}{3x-4}\,dx = \frac{1}{3}\int_{-7}^{-1} \frac{1}{u}\,du$, where $u = 3x - 4$, $du = 3\,dx$; $x = -1 \Rightarrow u = -7$, $x = 1 \Rightarrow u = -1$
$= \frac{1}{3}\left[\ln|u|\right]_{-7}^{-1} = \frac{1}{3}\left[\ln|-1| - \ln|-7|\right] = \frac{1}{3}\left[0 - \ln 7\right] = -\frac{\ln 7}{3}$

39. $\int_0^{\pi} \tan\left(\frac{x}{3}\right)dx = \int_0^{\pi} \frac{\sin\left(\frac{x}{3}\right)}{\cos\left(\frac{x}{3}\right)}\,dx = -3\int_1^{1/2} \frac{1}{u}\,du$, where $u = \cos\left(\frac{x}{3}\right)$, $du = -\frac{1}{3}\sin\left(\frac{x}{3}\right)dx$; $x = 0 \Rightarrow u = 1$, $x = \pi$
$\Rightarrow u = \frac{1}{2}$
$= -3\left[\ln|u|\right]_1^{1/2} = -3\left[\ln\left|\frac{1}{2}\right| - \ln|1|\right] = -3\ln\frac{1}{2} = \ln 2^3 = \ln 8$

41. $\int_0^4 \frac{2t}{t^2-25}\,dt = \int_{-25}^{-9} \frac{1}{u}\,du$, where $u = t^2 - 25$, $du = 2t\,dt$; $t = 0 \Rightarrow u = -25$, $t = 4 \Rightarrow u = -9$
$= \left[\ln|u|\right]_{-25}^{-9} = \ln|-9| - \ln|-25| = \ln 9 - \ln 25 = \ln\frac{9}{25}$

43. $\int \frac{\tan(\ln v)}{v}\,dv = \int \tan u\,du = \int \frac{\sin u}{\cos u}\,du$, where $u = \ln v$ and $du = \frac{1}{v}\,dv$
$= -\ln|\cos u| + C = -\ln|\cos(\ln v)| + C$

45. $\int \frac{(\ln x)^{-3}}{x}\,dx = \int u^{-3}\,du$, where $u = \ln x$ and $du = \frac{1}{x}\,dx$
$= \frac{u^{-2}}{-2} + C = -\frac{1}{2}(\ln x)^{-2} + C$

47. $\int \frac{1}{r}\csc^2(1 + \ln r)\,dr = \int \csc^2 u\,du$, where $u = 1 + \ln r$ and $du = \frac{1}{r}\,dr$
$= -\cot u + C = -\cot(1 + \ln r) + C$

49. $\int x3^{x^2}\,dx = \frac{1}{2}\int 3^u\,du$, where $u = x^2$ and $du = 2x\,dx$
$= \frac{1}{2\ln 3}(3^u) + C = \frac{1}{2\ln 3}\left(3^{x^2}\right) + C$

51. $\int_1^7 \frac{3}{x}\,dx = 3\int_1^7 \frac{1}{x}\,dx = 3\left[\ln|x|\right]_1^7 = 3(\ln 7 - \ln 1) = 3\ln 7$

53. $\int_1^4 \left(\frac{x}{8} + \frac{1}{2x}\right)dx = \frac{1}{2}\int_1^4 \left(\frac{1}{4}x + \frac{1}{x}\right)dx = \frac{1}{2}\left[\frac{1}{8}x^2 + \ln|x|\right]_1^4 = \frac{1}{2}\left[\left(\frac{16}{8} + \ln 4\right) - \left(\frac{1}{8} + \ln 1\right)\right] = \frac{15}{16} + \frac{1}{2}\ln 4$
$= \frac{15}{16} + \ln\sqrt{4} = \frac{15}{16} + \ln 2$

55. $\int_{-2}^{-1} e^{-(x+1)}\,dx = -\int_1^0 e^u\,du$, where $u = -(x+1)$, $du = -dx$; $x = -2 \Rightarrow u = 1$, $x = -1 \Rightarrow u = 0$
$= -\left[e^u\right]_1^0 = -(e^0 - e^1) = e - 1$

57. $\int_{1}^{\ln 5} e^{r}(3e^{r}+1)^{-3/2}\,dr = \frac{1}{3}\int_{4}^{16} u^{-3/2}\,du$, where $u = 3e^{r}+1$, $du = 3e^{r}dr$; $r = 0 \Rightarrow u = 4$, $r = \ln 5 \Rightarrow u = 16$

$= -\frac{2}{3}\left[u^{-1/2}\right]_{4}^{16} = -\frac{2}{3}\left(16^{-1/2} - 4^{-1/2}\right) = \left(-\frac{2}{3}\right)\left(\frac{1}{4}-\frac{1}{2}\right) = \left(-\frac{2}{3}\right)\left(-\frac{1}{4}\right) = \frac{1}{6}$

59. $\int_{1}^{e} \frac{1}{x}(1+7\ln x)^{-1/3}\,dx = \frac{1}{7}\int_{1}^{8} u^{-1/3}\,du$, where $u = 1+7\ln x$, $du = \frac{7}{x}\,dx$, $x = 1 \Rightarrow u = 1$, $x = e \Rightarrow u = 8$

$= \frac{3}{14}\left[u^{2/3}\right]_{1}^{8} = \frac{3}{14}\left(8^{2/3} - 1^{2/3}\right) = \left(\frac{3}{14}\right)(4-1) = \frac{9}{14}$

61. $\int_{1}^{3} \frac{[\ln(v+1)]^2}{v+1}\,dv = \int_{1}^{3} [\ln(v+1)]^2 \frac{1}{v+1}\,dv = \int_{\ln 2}^{\ln 4} u^2\,du$, where $u = \ln(v+1)$, $du = \frac{1}{v+1}\,dv$;

$v = 1 \Rightarrow u = \ln 2$, $v = 3 \Rightarrow u = \ln 4$;

$= \frac{1}{3}\left[u^3\right]_{\ln 2}^{\ln 4} = \frac{1}{3}\left[(\ln 4)^3 - (\ln 2)^3\right] = \frac{1}{3}\left[(2\ln 2)^3 - (\ln 2)^3\right] = \frac{(\ln 2)^3}{3}(8-1) = \frac{7}{3}(\ln 2)^3$

63. $\int_{1}^{8} \frac{\log_4 \theta}{\theta}\,d\theta = \frac{1}{\ln 4}\int_{1}^{8} (\ln\theta)\left(\frac{1}{\theta}\right)d\theta = \frac{1}{\ln 4}\int_{0}^{\ln 8} u\,du$, where $u = \ln\theta$, $du = \frac{1}{\theta}\,d\theta$, $\theta = 1 \Rightarrow u = 0$, $\theta = 8 \Rightarrow u = \ln 8$

$= \frac{1}{2\ln 4}\left[u^2\right]_{0}^{\ln 8} = \frac{1}{\ln 16}\left[(\ln 8)^2 - 0^2\right] = \frac{(3\ln 2)^2}{4\ln 2} = \frac{9\ln 2}{4}$

65. $\int_{-3/4}^{3/4} \frac{6}{\sqrt{9-4x^2}}\,dx = 3\int_{-3/4}^{3/4} \frac{2}{\sqrt{3^2-(2x)^2}}\,dx = 3\int_{-3/2}^{3/2} \frac{1}{\sqrt{3^2-u^2}}\,du$, where $u = 2x$, $du = 2\,dx$;

$x = -\frac{3}{4} \Rightarrow u = -\frac{3}{2}$, $x = \frac{3}{4} \Rightarrow u = \frac{3}{2}$

$= 3\left[\sin^{-1}\left(\frac{u}{3}\right)\right]_{-3/2}^{3/2} = 3\left[\sin^{-1}\left(\frac{1}{2}\right) - \sin^{-1}\left(-\frac{1}{2}\right)\right] = 3\left[\frac{\pi}{6} - \left(-\frac{\pi}{6}\right)\right] = 3\left(\frac{\pi}{3}\right) = \pi$

67. $\int_{-2}^{2} \frac{3}{4+3t^2}\,dt = \sqrt{3}\int_{-2}^{2} \frac{\sqrt{3}}{2^2+\left(\sqrt{3}t\right)^2}\,dt = \sqrt{3}\int_{-2\sqrt{3}}^{2\sqrt{3}} \frac{1}{2^2+u^2}\,du$, where $u = \sqrt{3}t$, $du = \sqrt{3}\,dt$;

$t = -2 \Rightarrow u = -2\sqrt{3}$, $t = 2 \Rightarrow u = 2\sqrt{3}$

$= \sqrt{3}\left[\frac{1}{2}\tan^{-1}\left(\frac{u}{2}\right)\right]_{-2\sqrt{3}}^{2\sqrt{3}} = \frac{\sqrt{3}}{2}\left[\tan^{-1}\left(\sqrt{3}\right) - \tan^{-1}\left(-\sqrt{3}\right)\right] = \frac{\sqrt{3}}{2}\left[\frac{\pi}{3} - \left(-\frac{\pi}{3}\right)\right] = \frac{\pi}{\sqrt{3}}$

69. $\int \frac{1}{y\sqrt{4y^2-1}}\,dy = \int \frac{2}{(2y)\sqrt{(2y)^2-1}}\,dy = \int \frac{1}{u\sqrt{u^2-1}}\,du$, where $u = 2y$ and $du = 2\,dy$

$= \sec^{-1}|u| + C = \sec^{-1}|2y| + C$

71. $\int_{\sqrt{2}/3}^{2/3} \frac{1}{|y|\sqrt{9y^2-1}}\,dy = \int_{\sqrt{2}/3}^{2/3} \frac{3}{|3y|\sqrt{(3y)^2-1}}\,dy = \int_{\sqrt{2}}^{2} \frac{1}{|u|\sqrt{u^2-1}}\,du$, where $u = 3y$, $du = 3\,dy$;

$y = \frac{\sqrt{2}}{3} \Rightarrow u = \sqrt{2}$, $y = \frac{2}{3} \Rightarrow u = 2$

$= \left[\sec^{-1} u\right]_{\sqrt{2}}^{2} = \left[\sec^{-1} 2 - \sec^{-1}\sqrt{2}\right] = \frac{\pi}{3} - \frac{\pi}{4} = \frac{\pi}{12}$

73. $\int \frac{1}{\sqrt{-2x-x^2}}\,dx = \int \frac{1}{\sqrt{1-(x^2+2x+1)}}\,dx = \int \frac{1}{\sqrt{1-(x+1)^2}}\,dx = \int \frac{1}{\sqrt{1-u^2}}\,du$, where $u = x+1$ and

$du = dx$

$= \sin^{-1} u + C = \sin^{-1}(x+1) + C$

75. $\int_{-2}^{-1} \frac{2}{v^2+4v+5}\,dv = 2\int_{-2}^{-1} \frac{1}{1+(v^2+4v+4)}\,dv = 2\int_{-2}^{-1} \frac{1}{1+(v+2)^2}\,dv = 2\int_{0}^{1} \frac{1}{1+u^2}\,du$,

where $u = v+2$, $du = dv$; $v = -2 \Rightarrow u = 0$, $v = -1 \Rightarrow u = 1$

$= 2\left[\tan^{-1} u\right]_{0}^{1} = 2\left(\tan^{-1} 1 - \tan^{-1} 0\right) = 2\left(\frac{\pi}{4} - 0\right) = \frac{\pi}{2}$

77. $\int \frac{1}{(t+1)\sqrt{t^2+2t-8}}\,dt = \int \frac{1}{(t+1)\sqrt{(t^2+2t+1)-9}}\,dt = \int \frac{1}{(t+1)\sqrt{(t+1)^2-3^2}}\,dt = \int \frac{1}{u\sqrt{u^2-3^2}}\,du$

where $u = t+1$ and $du = dt$

$= \frac{1}{3}\sec^{-1}\left|\frac{u}{3}\right| + C = \frac{1}{3}\sec^{-1}\left|\frac{t+1}{3}\right| + C$

79. $3^y = 2^{y+1} \Rightarrow \ln 3^y = \ln 2^{y+1} \Rightarrow y(\ln 3) = (y+1)\ln 2 \Rightarrow (\ln 3 - \ln 2)y = \ln 2 \Rightarrow \left(\ln \frac{3}{2}\right) y = \ln 2 \Rightarrow y = \frac{\ln 2}{\ln\left(\frac{3}{2}\right)}$

81. $9e^{2y} = x^2 \Rightarrow e^{2y} = \frac{x^2}{9} \Rightarrow \ln e^{2y} = \ln\left(\frac{x^2}{9}\right) \Rightarrow 2y(\ln e) = \ln\left(\frac{x^2}{9}\right) \Rightarrow y = \frac{1}{2}\ln\left(\frac{x^2}{9}\right) = \ln\sqrt{\frac{x^2}{9}} = \ln\left|\frac{x}{3}\right| = \ln|x| - \ln 3$

83. $\ln(y-1) = x + \ln y \Rightarrow e^{\ln(y-1)} = e^{(x+\ln y)} = e^x e^{\ln y} \Rightarrow y - 1 = ye^x \Rightarrow y - ye^x = 1 \Rightarrow y(1-e^x) = 1 \Rightarrow y = \frac{1}{1-e^x}$

85. $\lim\limits_{x\to 1} \frac{x^2+3x-4}{x-1} = \lim\limits_{x\to 1} \frac{2x+3}{1} = 5$

87. $\lim\limits_{x\to \pi} \frac{\tan x}{x} = \frac{\tan \pi}{\pi} = 0$

89. $\lim\limits_{x\to 0} \frac{\sin^2 x}{\tan(x^2)} = \lim\limits_{x\to 0} \frac{2\sin x\cdot\cos x}{2x\sec^2(x^2)} = \lim\limits_{x\to 0} \frac{\sin(2x)}{2x\sec^2(x^2)} = \lim\limits_{x\to 0} \frac{2\cos(2x)}{2x\left(2\sec^2(x^2)\tan(x^2)\cdot 2x\right) + 2\sec^2(x^2)} = \frac{2}{0+2\cdot 1} = 1$

91. $\lim\limits_{x\to \pi/2^-} \sec(7x)\cos(3x) = \lim\limits_{x\to \pi/2^-} \frac{\cos(3x)}{\cos(7x)} = \lim\limits_{x\to \pi/2^-} \frac{-3\sin(3x)}{-7\sin(7x)} = \frac{3}{7}$

93. $\lim\limits_{x\to 0} (\csc x - \cot x) = \lim\limits_{x\to 0} \frac{1-\cos x}{\sin x} = \lim\limits_{x\to 0} \frac{\sin x}{\cos x} = \frac{0}{1} = 0$

95. $\lim\limits_{x\to \infty} \left(\sqrt{x^2+x+1} - \sqrt{x^2-x}\right) = \lim\limits_{x\to \infty} \left(\sqrt{x^2+x+1} - \sqrt{x^2-x}\right) \cdot \frac{\sqrt{x^2+x+1}+\sqrt{x^2-x}}{\sqrt{x^2+x+1}+\sqrt{x^2-x}}$

$= \lim\limits_{x\to \infty} \frac{2x+1}{\sqrt{x^2+x+1}+\sqrt{x^2-x}}$

Notice that $x = \sqrt{x^2}$ for $x > 0$ so this is equivalent to

$= \lim\limits_{x\to \infty} \frac{\frac{2x+1}{x}}{\sqrt{\frac{x^2+x+1}{x^2}}+\sqrt{\frac{x^2-x}{x^2}}} = \lim\limits_{x\to \infty} \frac{2+\frac{1}{x}}{\sqrt{1+\frac{1}{x}+\frac{1}{x^2}}+\sqrt{1-\frac{1}{x}}} = \frac{2}{\sqrt{1}+\sqrt{1}} = 1$

97. The limit leads to the indeterminate form $\frac{0}{0}$: $\lim\limits_{x\to 0} \frac{10^x - 1}{x} = \lim\limits_{x\to 0} \frac{(\ln 10)10^x}{1} = \ln 10$

99. The limit leads to the indeterminate form $\frac{0}{0}$: $\lim\limits_{x\to 0} \frac{2^{\sin x}-1}{e^x-1} = \lim\limits_{x\to 0} \frac{2^{\sin x}(\ln 2)(\cos x)}{e^x} = \ln 2$

101. The limit leads to the indeterminate form $\frac{0}{0}$: $\lim\limits_{x\to 0} \frac{5-5\cos x}{e^x-x-1} = \lim\limits_{x\to 0} \frac{5\sin x}{e^x-1} = \lim\limits_{x\to 0} \frac{5\cos x}{e^x} = 5$

103. The limit leads to the indeterminate form $\frac{0}{0}$: $\lim\limits_{t\to 0^+} \frac{t-\ln(1+2t)}{t^2} = \lim\limits_{t\to 0^+} \frac{\left(1-\frac{2}{1+2t}\right)}{2t} = -\infty$

105. The limit leads to the indeterminate form $\frac{0}{0}$: $\lim\limits_{t\to 0^+} \left(\frac{e^t}{t} - \frac{1}{t}\right) = \lim\limits_{t\to 0^+} \left(\frac{e^t-1}{t}\right) = \lim\limits_{t\to 0^+} \frac{e^t}{1} = 1$

107. Let $f(x) = \left(\frac{e^x+1}{e^x-1}\right)^{\ln x} \Rightarrow \ln f(x) = \ln x \ln\left(\frac{e^x+1}{e^x-1}\right) \Rightarrow \lim\limits_{x\to \infty} \ln f(x) = \lim\limits_{x\to \infty} \ln x \ln\left(\frac{e^x+1}{e^x-1}\right)$; this is limit is currently of the form $0\cdot\infty$. Before we put in one of the indeterminate forms, we rewrite $\frac{e^x+1}{e^x-1} = \frac{e^{x/2}+e^{-x/2}}{e^{x/2}-e^{-x/2}} = \coth\left(\frac{x}{2}\right)$; the limit is $\lim\limits_{x\to \infty} \ln x \ln\coth\left(\frac{x}{2}\right) = \lim\limits_{x\to \infty} \frac{\ln\coth\left(\frac{x}{2}\right)}{\frac{1}{\ln x}}$; the limit leads to the indeterminate form $\frac{0}{0}$: $\lim\limits_{x\to \infty} \frac{\ln\coth\left(\frac{x}{2}\right)}{\frac{1}{\ln x}}$

$= \lim\limits_{x\to \infty} \left(\frac{\frac{\operatorname{csch}^2\left(\frac{x}{2}\right)}{\coth\left(\frac{x}{2}\right)}\left(-\frac{1}{2}\right)}{-\frac{1}{(\ln x)^2}\left(\frac{1}{x}\right)}\right) = \lim\limits_{x\to \infty} \left(\frac{x(\ln x)^2}{2\sinh\left(\frac{x}{2}\right)\cosh\left(\frac{x}{2}\right)}\right) = \lim\limits_{x\to \infty} \left(\frac{x(\ln x)^2}{\sinh x}\right) = \lim\limits_{x\to \infty} \left(\frac{2x(\ln x)\left(\frac{1}{x}\right)+(\ln x)^2}{\cosh x}\right)$

$= \lim_{x\to\infty} \left(\frac{2\ln x + (\ln x)^2}{\cosh x}\right) = \lim_{x\to\infty} \left(\frac{2\left(\frac{1}{x}\right) + 2(\ln x)\left(\frac{1}{x}\right)}{\sinh x}\right) = \lim_{x\to\infty} \left(\frac{2+2\ln x}{x\sinh x}\right) = \lim_{x\to\infty} \left(\frac{\frac{2}{x}}{x\cosh x + \sinh x}\right)$

$= \lim_{x\to\infty} \left(\frac{2}{x^2\cosh x + x\sinh x}\right) = 0 \Rightarrow \lim_{x\to\infty} \left(\frac{e^x+1}{e^x-1}\right)^{\ln x} = \lim_{x\to\infty} e^{\ln f(x)} = e^0 = 1$

109. (a) $\lim_{x\to\infty} \frac{\log_2 x}{\log_3 x} = \lim_{x\to\infty} \frac{\left(\frac{\ln x}{\ln 2}\right)}{\left(\frac{\ln x}{\ln 3}\right)} = \lim_{x\to\infty} \frac{\ln 3}{\ln 2} = \frac{\ln 3}{\ln 2} \Rightarrow$ same rate

(b) $\lim_{x\to\infty} \frac{x}{x+\left(\frac{1}{x}\right)} = \lim_{x\to\infty} \frac{x^2}{x^2+1} = \lim_{x\to\infty} \frac{2x}{2x} = \lim_{x\to\infty} 1 = 1 \Rightarrow$ same rate

(c) $\lim_{x\to\infty} \frac{\left(\frac{x}{100}\right)}{xe^{-x}} = \lim_{x\to\infty} \frac{xe^x}{100x} = \lim_{x\to\infty} \frac{e^x}{100} = \infty \Rightarrow$ faster

(d) $\lim_{x\to\infty} \frac{x}{\tan^{-1} x} = \infty \Rightarrow$ faster

(e) $\lim_{x\to\infty} \frac{\csc^{-1} x}{\left(\frac{1}{x}\right)} = \lim_{x\to\infty} \frac{\sin^{-1}(x^{-1})}{x^{-1}} = \lim_{x\to\infty} \frac{\left(\frac{-x^{-2}}{\sqrt{1-(x^{-1})^2}}\right)}{-x^{-2}} = \lim_{x\to\infty} \frac{1}{\sqrt{1-\left(\frac{1}{x^2}\right)}} = 1 \Rightarrow$ same rate

(f) $\lim_{x\to\infty} \frac{\sinh x}{e^x} = \lim_{x\to\infty} \frac{(e^x-e^{-x})}{2e^x} = \lim_{x\to\infty} \frac{1-e^{-2x}}{2} = \frac{1}{2} \Rightarrow$ same rate

111. (a) $\frac{\left(\frac{1}{x^2}+\frac{1}{x^4}\right)}{\left(\frac{1}{x^2}\right)} = 1 + \frac{1}{x^2} \le 2$ for x sufficiently large $\Rightarrow$ true

(b) $\frac{\left(\frac{1}{x^2}+\frac{1}{x^4}\right)}{\left(\frac{1}{x^4}\right)} = x^2 + 1 > M$ for any positive integer M whenever $x > \sqrt{M} \Rightarrow$ false

(c) $\lim_{x\to\infty} \frac{x}{x+\ln x} = \lim_{x\to\infty} \frac{1}{1+\frac{1}{x}} = 1 \Rightarrow$ the same growth rate $\Rightarrow$ false

(d) $\lim_{x\to\infty} \frac{\ln(\ln x)}{\ln x} = \lim_{x\to\infty} \frac{\left[\frac{\left(\frac{1}{x}\right)}{\ln x}\right]}{\left(\frac{1}{x}\right)} = \lim_{x\to\infty} \frac{1}{\ln x} = 0 \Rightarrow$ grows slower $\Rightarrow$ true

(e) $\frac{\tan^{-1} x}{1} \le \frac{\pi}{2}$ for all x $\Rightarrow$ true

(f) $\frac{\cosh x}{e^x} = \frac{1}{2}\left(1+e^{-2x}\right) \le \frac{1}{2}(1+1) = 1$ if $x > 0 \Rightarrow$ true

113. $\frac{df}{dx} = e^x + 1 \Rightarrow \left(\frac{df^{-1}}{dx}\right)_{x=f(\ln 2)} = \frac{1}{\left(\frac{df}{dx}\right)_{x=\ln 2}} \Rightarrow \left(\frac{df^{-1}}{dx}\right)_{x=f(\ln 2)} = \frac{1}{(e^x+1)_{x=\ln 2}} = \frac{1}{2+1} = \frac{1}{3}$

115. $y = x\ln 2x - x \Rightarrow y' = x\left(\frac{2}{2x}\right) + \ln(2x) - 1 = \ln 2x$; solving $y' = 0 \Rightarrow x = \frac{1}{2}$; $y' > 0$ for $x > \frac{1}{2}$ and $y' < 0$ for $x < \frac{1}{2} \Rightarrow$ relative minimum of $-\frac{1}{2}$ at $x = \frac{1}{2}$; $f\left(\frac{1}{2e}\right) = -\frac{1}{e}$ and $f\left(\frac{e}{2}\right) = 0 \Rightarrow$ absolute minimum is $-\frac{1}{2}$ at $x = \frac{1}{2}$ and the absolute maximum is 0 at $x = \frac{e}{2}$

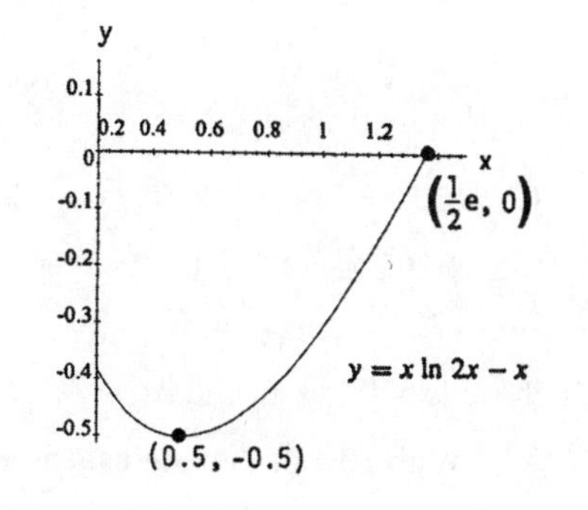

117. $A = \int_1^e \frac{2\ln x}{x}\,dx = \int_0^1 2u\,du = \left[u^2\right]_0^1 = 1$, where $u = \ln x$ and $du = \frac{1}{x}\,dx$; $x = 1 \Rightarrow u = 0$, $x = e \Rightarrow u = 1$

119. $y = \ln x \Rightarrow \frac{dy}{dx} = \frac{1}{x}$; $\frac{dy}{dt} = \frac{dy}{dx}\frac{dx}{dt} \Rightarrow \frac{dy}{dt} = \left(\frac{1}{x}\right)\sqrt{x} = \frac{1}{\sqrt{x}} \Rightarrow \left.\frac{dy}{dt}\right|_{e^2} = \frac{1}{e}$ m/sec

121. $A = xy = xe^{-x^2} \Rightarrow \frac{dA}{dx} = e^{-x^2} + (x)(-2x)e^{-x^2} = e^{-x^2}\left(1-2x^2\right)$. Solving $\frac{dA}{dx} = 0 \Rightarrow 1 - 2x^2 = 0$ $\Rightarrow x = \frac{1}{\sqrt{2}}$; $\frac{dA}{dx} < 0$ for $x > \frac{1}{\sqrt{2}}$ and $\frac{dA}{dx} > 0$ for $0 < x < \frac{1}{\sqrt{2}} \Rightarrow$ absolute maximum of $\frac{1}{\sqrt{2}}e^{-1/2} = \frac{1}{\sqrt{2e}}$ at $x = \frac{1}{\sqrt{2}}$ units long by $y = e^{-1/2} = \frac{1}{\sqrt{e}}$ units high.

123. (a) $y = \frac{\ln x}{\sqrt{x}} \Rightarrow y' = \frac{1}{x\sqrt{x}} - \frac{\ln x}{2x^{3/2}} = \frac{2-\ln x}{2x\sqrt{x}}$
$\Rightarrow y'' = -\frac{3}{4}x^{-5/2}(2-\ln x) - \frac{1}{2}x^{-5/2} = x^{-5/2}\left(\frac{3}{4}\ln x - 2\right)$; solving $y' = 0 \Rightarrow \ln x = 2 \Rightarrow x = e^2$; $y' < 0$ for $x > e^2$ and and $y' > 0$ for $x < e^2 \Rightarrow$ a maximum of $\frac{2}{e}$; $y'' = 0$ $\Rightarrow \ln x = \frac{8}{3} \Rightarrow x = e^{8/3}$; the curve is concave down on $\left(0, e^{8/3}\right)$ and concave up on $\left(e^{8/3}, \infty\right)$; so there is an inflection point at $\left(e^{8/3}, \frac{8}{3e^{4/3}}\right)$.

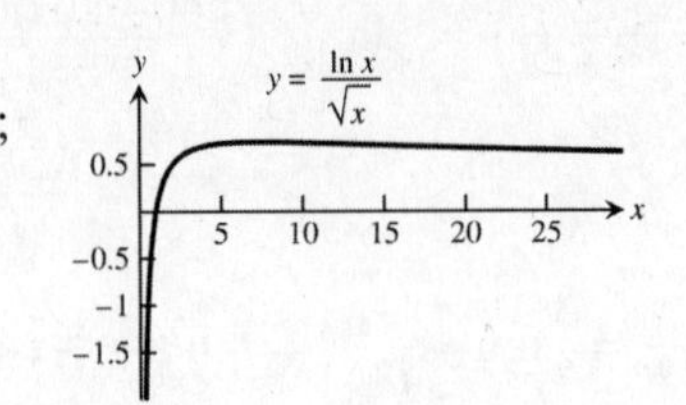

(b) $y = e^{-x^2} \Rightarrow y' = -2xe^{-x^2} \Rightarrow y'' = -2e^{-x^2} + 4x^2e^{-x^2}$ $= (4x^2-2)e^{-x^2}$; solving $y' = 0 \Rightarrow x = 0$; $y' < 0$ for $x > 0$ and $y' > 0$ for $x < 0 \Rightarrow$ a maximum at $x = 0$ of $e^0 = 1$; there are points of inflection at $x = \pm\frac{1}{\sqrt{2}}$; the curve is concave down for $-\frac{1}{\sqrt{2}} < x < \frac{1}{\sqrt{2}}$ and concave up otherwise.

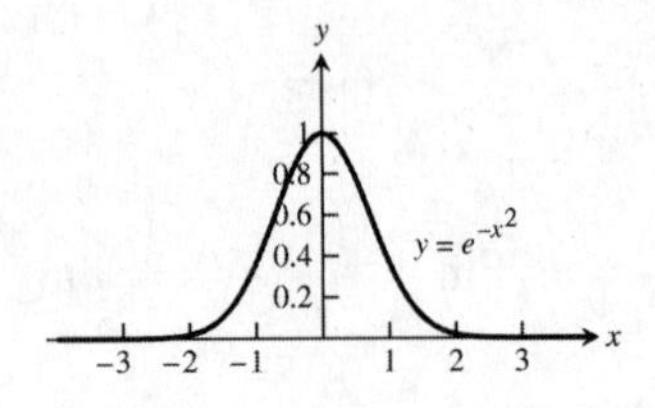

(c) $y = (1+x)e^{-x} \Rightarrow y' = e^{-x} - (1+x)e^{-x} = -xe^{-x}$ $\Rightarrow y'' = -e^{-x} + xe^{-x} = (x-1)e^{-x}$; solving $y' = 0$ $\Rightarrow -xe^{-x} = 0 \Rightarrow x = 0$; $y' < 0$ for $x > 0$ and $y' > 0$ for $x < 0 \Rightarrow$ a maximum at $x = 0$ of $(1+0)e^0 = 1$; there is a point of inflection at $x = 1$ and the curve is concave up for $x > 1$ and concave down for $x < 1$.

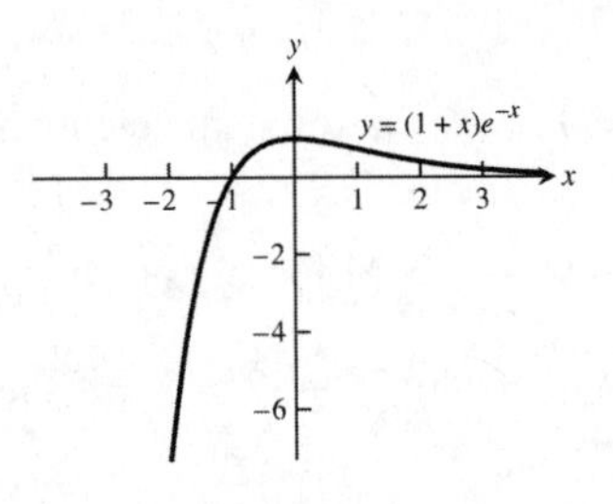

125. $\frac{dy}{dx} = \sqrt{y}\cos^2\sqrt{y} \Rightarrow \frac{dy}{\sqrt{y}\cos^2\sqrt{y}} = dx \Rightarrow 2\tan\sqrt{y} = x + C \Rightarrow y = \left(\tan^{-1}\left(\frac{x+C}{2}\right)\right)^2$

127. $yy' = \sec(y^2)\sec^2 x \Rightarrow \frac{y\,dy}{\sec(y^2)} = \sec^2 x\,dx \Rightarrow \frac{\sin(y^2)}{2} = \tan x + C \Rightarrow \sin(y^2) = 2\tan x + C_1$

129. $\frac{dy}{dx} = e^{-x-y-2} \Rightarrow e^y dy = e^{-(x+2)}dx \Rightarrow e^y = -e^{-(x+2)} + C$. We have $y(0) = -2$, so $e^{-2} = -e^{-2} + C \Rightarrow C = 2e^{-2}$ and $e^y = -e^{-(x+2)} + 2e^{-2} \Rightarrow y = \ln\left(-e^{-(x+2)} + 2e^{-2}\right)$

131. $x\,dy - (y + \sqrt{y})dx = 0 \Rightarrow \frac{dy}{(y+\sqrt{y})} = \frac{dx}{x} \Rightarrow 2\ln(\sqrt{y}+1) = \ln x + C$. We have $y(1) = 1 \Rightarrow 2\ln\left(\sqrt{1}+1\right) = \ln 1 + C$ $\Rightarrow 2\ln 2 = C = \ln 2^2 = \ln 4$. So $2\ln(\sqrt{y}+1) = \ln x + \ln 4 = \ln(4x) \Rightarrow \ln(\sqrt{y}+1) = \frac{1}{2}\ln(4x) = \ln(4x)^{1/2}$ $\Rightarrow e^{\ln(\sqrt{y}+1)} = e^{\ln(4x)^{1/2}} \Rightarrow \sqrt{y}+1 = 2\sqrt{x} \Rightarrow y = \left(2\sqrt{x}-1\right)^2$

133. Since the half life is 5700 years and $A(t) = A_0e^{kt}$ we have $\frac{A_0}{2} = A_0e^{5700k} \Rightarrow \frac{1}{2} = e^{5700k} \Rightarrow \ln(0.5) = 5700k$ $\Rightarrow k = \frac{\ln(0.5)}{5700}$. With 10% of the original carbon-14 remaining we have $0.1A_0 = A_0e^{\frac{\ln(0.5)}{5700}t} \Rightarrow 0.1 = e^{\frac{\ln(0.5)}{5700}t}$ $\Rightarrow \ln(0.1) = \frac{\ln(0.5)}{5700}t \Rightarrow t = \frac{(5700)\ln(0.1)}{\ln(0.5)} \approx 18{,}935$ years (rounded to the nearest year).

135. $\theta = \pi - \cot^{-1}\left(\frac{x}{60}\right) - \cot^{-1}\left(\frac{5}{3} - \frac{x}{30}\right), 0 < x < 50 \Rightarrow \frac{d\theta}{dx} = \frac{\left(\frac{1}{60}\right)}{1+\left(\frac{x}{60}\right)^2} + \frac{\left(-\frac{1}{30}\right)}{1+\left(\frac{50-x}{30}\right)^2}$

$= 30\left[\frac{2}{60^2+x^2} - \frac{1}{30^2+(50-x)^2}\right]$; solving $\frac{d\theta}{dx} = 0 \Rightarrow x^2 - 200x + 3200 = 0 \Rightarrow x = 100 \pm 20\sqrt{17}$, but $100 + 20\sqrt{17}$ is not in the domain; $\frac{d\theta}{dx} > 0$ for $x < 20\left(5-\sqrt{17}\right)$ and $\frac{d\theta}{dx} < 0$ for $20\left(5-\sqrt{17}\right) < x < 50$ $\Rightarrow x = 20\left(5-\sqrt{17}\right) \approx 17.54$ m maximizes θ

CHAPTER 7 ADDITIONAL AND ADVANCED EXERCISES

1. $\lim\limits_{b \to 1^-} \int_0^b \frac{1}{\sqrt{1-x^2}}\,dx = \lim\limits_{b \to 1^-} \left[\sin^{-1} x\right]_0^b = \lim\limits_{b \to 1^-} \left(\sin^{-1} b - \sin^{-1} 0\right) = \lim\limits_{b \to 1^-} \left(\sin^{-1} b - 0\right) = \lim\limits_{b \to 1^-} \sin^{-1} b = \frac{\pi}{2}$

3. $y = \left(\cos\sqrt{x}\right)^{1/x} \Rightarrow \ln y = \frac{1}{x}\ln\left(\cos\sqrt{x}\right)$ and $\lim\limits_{x \to 0^+} \frac{\ln\left(\cos\sqrt{x}\right)}{x} = \lim\limits_{x \to 0^+} \frac{-\sin\sqrt{x}}{2\sqrt{x}\cos\sqrt{x}} = \frac{-1}{2}\lim\limits_{x \to 0^+} \frac{\tan\sqrt{x}}{\sqrt{x}}$
$= -\frac{1}{2}\lim\limits_{x \to 0^+} \frac{\frac{1}{2}x^{-1/2}\sec^2\sqrt{x}}{\frac{1}{2}x^{-1/2}} = -\frac{1}{2} \Rightarrow \lim\limits_{x \to 0^+} \left(\cos\sqrt{x}\right)^{1/x} = e^{-1/2} = \frac{1}{\sqrt{e}}$

5. $\lim\limits_{x \to \infty} \left(\frac{1}{n+1} + \frac{1}{n+2} + \ldots + \frac{1}{2n}\right) = \lim\limits_{x \to \infty} \left(\left(\frac{1}{n}\right)\left[\frac{1}{1+\left(\frac{1}{n}\right)}\right] + \left(\frac{1}{n}\right)\left[\frac{1}{1+2\left(\frac{1}{n}\right)}\right] + \ldots + \left(\frac{1}{n}\right)\left[\frac{1}{1+n\left(\frac{1}{n}\right)}\right]\right)$
which can be interpreted as a Riemann sum with partitioning $\Delta x = \frac{1}{n} \Rightarrow \lim\limits_{x \to \infty} \left(\frac{1}{n+1} + \frac{1}{n+2} + \ldots + \frac{1}{2n}\right)$
$= \int_0^1 \frac{1}{1+x}\,dx = \left[\ln(1+x)\right]_0^1 = \ln 2$

7. $A(t) = \int_0^t e^{-x}\,dx = \left[-e^{-x}\right]_0^t = 1 - e^{-t}$, $V(t) = \pi\int_0^t e^{-2x}\,dx = \left[-\frac{\pi}{2}e^{-2x}\right]_0^t = \frac{\pi}{2}\left(1 - e^{-2t}\right)$
   (a) $\lim\limits_{t \to \infty} A(t) = \lim\limits_{t \to \infty} \left(1 - e^{-t}\right) = 1$
   (b) $\lim\limits_{t \to \infty} \frac{V(t)}{A(t)} = \lim\limits_{t \to \infty} \frac{\frac{\pi}{2}\left(1-e^{-2t}\right)}{1-e^{-t}} = \frac{\pi}{2}$
   (c) $\lim\limits_{t \to 0^+} \frac{V(t)}{A(t)} = \lim\limits_{t \to 0^+} \frac{\frac{\pi}{2}\left(1-e^{-2t}\right)}{1-e^{-t}} = \lim\limits_{t \to 0^+} \frac{\frac{\pi}{2}\left(1-e^{-t}\right)\left(1+e^{-t}\right)}{\left(1-e^{-t}\right)} = \lim\limits_{t \to 0^+} \frac{\pi}{2}\left(1 + e^{-t}\right) = \pi$

9. $A_1 = \int_1^e \frac{2\log_2 x}{x}\,dx = \frac{2}{\ln 2}\int_1^e \frac{\ln x}{x}\,dx = \left[\frac{(\ln x)^2}{\ln 2}\right]_1^e = \frac{1}{\ln 2}$; $A_2 = \int_1^e \frac{2\log_4 x}{4}\,dx = \frac{2}{\ln 4}\int_1^e \frac{\ln x}{x}\,dx$
$= \left[\frac{(\ln x)^2}{2\ln 2}\right]_1^e = \frac{1}{2\ln 2} \Rightarrow A_1 : A_2 = 2:1$

11. $\ln x^{(x^x)} = x^x \ln x$ and $\ln\left(x^x\right)^x = x\ln x^x = x^2\ln x$; then, $x^x\ln x = x^2\ln x \Rightarrow \left(x^x - x^2\right)\ln x = 0 \Rightarrow x^x = x^2$ or $\ln x = 0$.
$\ln x = 0 \Rightarrow x = 1$; $x^x = x^2 \Rightarrow x\ln x = 2\ln x \Rightarrow x = 2$. Therefore, $x^{(x^x)} = \left(x^x\right)^x$ when $x = 2$ or $x = 1$.

13. $f(x) = e^{g(x)} \Rightarrow f'(x) = e^{g(x)}\,g'(x)$, where $g'(x) = \frac{x}{1+x^4} \Rightarrow f'(2) = e^0\left(\frac{2}{1+16}\right) = \frac{2}{17}$

15. (a) $g(x) + h(x) = 0 \Rightarrow g(x) = -h(x)$; also $g(x) + h(x) = 0 \Rightarrow g(-x) + h(-x) = 0 \Rightarrow g(x) - h(x) = 0$
$\Rightarrow g(x) = h(x)$; therefore $-h(x) = h(x) \Rightarrow h(x) = 0 \Rightarrow g(x) = 0$
   (b) $\frac{f(x)+f(-x)}{2} = \frac{\left[f_E(x)+f_O(x)\right]+\left[f_E(-x)+f_O(-x)\right]}{2} = \frac{f_E(x)+f_O(x)+f_E(x)-f_O(x)}{2} = f_E(x)$;
$\frac{f(x)-f(-x)}{2} = \frac{\left[f_E(x)+f_O(x)\right]-\left[f_E(-x)+f_O(-x)\right]}{2} = \frac{f_E(x)+f_O(x)-f_E(x)+f_O(x)}{2} = f_O(x)$
   (c) Part b $\Rightarrow$ such a decomposition is unique.

17. $M = \int_0^1 \frac{2}{1+x^2}\,dx = 2\left[\tan^{-1} x\right]_0^1 = \frac{\pi}{2}$ and $M_y = \int_0^1 \frac{2x}{1+x^2}\,dx = \left[\ln\left(1+x^2\right)\right]_0^1 = \ln 2 \Rightarrow \overline{x} = \frac{M_y}{M}$
$= \frac{\ln 2}{\left(\frac{\pi}{2}\right)} = \frac{\ln 4}{\pi}$; $\overline{y} = 0$ by symmetry

19. (a) $L = k\left(\frac{a - b\cot\theta}{R^4} + \frac{b\csc\theta}{r^4}\right) \Rightarrow \frac{dL}{d\theta} = k\left(\frac{b\csc^2\theta}{R^4} - \frac{b\csc\theta\cot\theta}{r^4}\right)$; solving $\frac{dL}{d\theta} = 0$
$\Rightarrow r^4 b\csc^2\theta - bR^4\csc\theta\cot\theta = 0 \Rightarrow (b\csc\theta)\left(r^4\csc\theta - R^4\cot\theta\right) = 0$; but $b\csc\theta \neq 0$ since
$\theta \neq \frac{\pi}{2} \Rightarrow r^4\csc\theta - R^4\cot\theta = 0 \Rightarrow \cos\theta = \frac{r^4}{R^4} \Rightarrow \theta = \cos^{-1}\left(\frac{r^4}{R^4}\right)$, the critical value of θ
   (b) $\theta = \cos^{-1}\left(\frac{5}{6}\right)^4 \approx \cos^{-1}(0.48225) \approx 61°$

NOTES:

CHAPTER 8 TECHNIQUES OF INTEGRATION

8.1 INTEGRATION BY PARTS

1. $u = x,\ du = dx;\ dv = \sin\frac{x}{2}\,dx,\ v = -2\cos\frac{x}{2};$

$\int x\sin\frac{x}{2}\,dx = -2x\cos\frac{x}{2} - \int\left(-2\cos\frac{x}{2}\right)dx = -2x\cos\left(\frac{x}{2}\right) + 4\sin\left(\frac{x}{2}\right) + C$

3.

		$\cos t$
t^2	$\xrightarrow{(+)}$	$\sin t$
$2t$	$\xrightarrow{(-)}$	$-\cos t$
2	$\xrightarrow{(+)}$	$-\sin t$
0		

$\int t^2\cos t\,dt = t^2\sin t + 2t\cos t - 2\sin t + C$

5. $u = \ln x,\ du = \frac{dx}{x};\ dv = x\,dx,\ v = \frac{x^2}{2};$

$\int_1^2 x\ln x\,dx = \left[\frac{x^2}{2}\ln x\right]_1^2 - \int_1^2 \frac{x^2}{2}\frac{dx}{x} = 2\ln 2 - \left[\frac{x^2}{4}\right]_1^2 = 2\ln 2 - \frac{3}{4} = \ln 4 - \frac{3}{4}$

7. $u = x,\ du = dx;\ dv = e^x dx,\ v = e^x;$

$\int x e^x dx = x e^x - \int e^x dx = x e^x - e^x + C$

9.

		e^{-x}
x^2	$\xrightarrow{(+)}$	$-e^{-x}$
$2x$	$\xrightarrow{(-)}$	e^{-x}
2	$\xrightarrow{(+)}$	$-e^{-x}$
0		

$\int x^2 e^{-x}\,dx = -x^2 e^{-x} - 2x e^{-x} - 2e^{-x} + C$

11. $u = \tan^{-1} y,\ du = \frac{dy}{1+y^2};\ dv = dy,\ v = y;$

$\int \tan^{-1} y\,dy = y\tan^{-1} y - \int\frac{y\,dy}{(1+y^2)} = y\tan^{-1} y - \frac{1}{2}\ln(1+y^2) + C = y\tan^{-1} y - \ln\sqrt{1+y^2} + C$

13. $u = x,\ du = dx;\ dv = \sec^2 x\,dx,\ v = \tan x;$

$\int x\sec^2 x\,dx = x\tan x - \int\tan x\,dx = x\tan x + \ln|\cos x| + C$

15.

		e^x
x^3	$\xrightarrow{(+)}$	e^x
$3x^2$	$\xrightarrow{(-)}$	e^x
$6x$	$\xrightarrow{(+)}$	e^x
6	$\xrightarrow{(-)}$	e^x
0		

$\int x^3 e^x\,dx = x^3 e^x - 3x^2 e^x + 6x e^x - 6e^x + C = (x^3 - 3x^2 + 6x - 6)e^x + C$

17.

		e^x
$x^2 - 5x$	$\xrightarrow{(+)}$	e^x
$2x - 5$	$\xrightarrow{(-)}$	e^x
2	$\xrightarrow{(+)}$	e^x
0		

$$\int (x^2 - 5x)\, e^x\, dx = (x^2 - 5x)\, e^x - (2x - 5)e^x + 2e^x + C = x^2e^x - 7xe^x + 7e^x + C$$
$$= (x^2 - 7x + 7)\, e^x + C$$

19.

		e^x
x^5	$\xrightarrow{(+)}$	e^x
$5x^4$	$\xrightarrow{(-)}$	e^x
$20x^3$	$\xrightarrow{(+)}$	e^x
$60x^2$	$\xrightarrow{(-)}$	e^x
$120x$	$\xrightarrow{(+)}$	e^x
120	$\xrightarrow{(-)}$	e^x
0		

$$\int x^5e^x\, dx = x^5e^x - 5x^4e^x + 20x^3e^x - 60x^2e^x + 120xe^x - 120e^x + C$$
$$= (x^5 - 5x^4 + 20x^3 - 60x^2 + 120x - 120)\, e^x + C$$

21. $I = \int e^\theta \sin\theta\, d\theta$; $[u = \sin\theta,\ du = \cos\theta\, d\theta;\ dv = e^\theta\, d\theta,\ v = e^\theta] \Rightarrow I = e^\theta \sin\theta - \int e^\theta \cos\theta\, d\theta$;
$[u = \cos\theta,\ du = -\sin\theta\, d\theta;\ dv = e^\theta\, d\theta,\ v = e^\theta] \Rightarrow I = e^\theta \sin\theta - \left(e^\theta \cos\theta + \int e^\theta \sin\theta\, d\theta\right)$
$= e^\theta \sin\theta - e^\theta \cos\theta - I + C' \Rightarrow 2I = (e^\theta \sin\theta - e^\theta \cos\theta) + C' \Rightarrow I = \frac{1}{2}(e^\theta \sin\theta - e^\theta \cos\theta) + C$, where $C = \frac{C'}{2}$ is another arbitrary constant

23. $I = \int e^{2x} \cos 3x\, dx$; $\left[u = \cos 3x;\ du = -3 \sin 3x\, dx,\ dv = e^{2x}\, dx;\ v = \frac{1}{2} e^{2x}\right]$
$\Rightarrow I = \frac{1}{2} e^{2x} \cos 3x + \frac{3}{2} \int e^{2x} \sin 3x\, dx$; $\left[u = \sin 3x,\ du = 3 \cos 3x,\ dv = e^{2x}\, dx;\ v = \frac{1}{2} e^{2x}\right]$
$\Rightarrow I = \frac{1}{2} e^{2x} \cos 3x + \frac{3}{2}\left(\frac{1}{2} e^{2x} \sin 3x - \frac{3}{2} \int e^{2x} \cos 3x\, dx\right) = \frac{1}{2} e^{2x} \cos 3x + \frac{3}{4} e^{2x} \sin 3x - \frac{9}{4} I + C'$
$\Rightarrow \frac{13}{4} I = \frac{1}{2} e^{2x} \cos 3x + \frac{3}{4} e^{2x} \sin 3x + C' \Rightarrow \frac{e^{2x}}{13}(3 \sin 3x + 2 \cos 3x) + C$, where $C = \frac{4}{13} C'$

25. $\int e^{\sqrt{3s+9}}\, ds$; $\begin{bmatrix} 3s + 9 = x^2 \\ ds = \frac{2}{3} x\, dx \end{bmatrix} \to \int e^x \cdot \frac{2}{3} x\, dx = \frac{2}{3} \int xe^x\, dx$; $[u = x,\ du = dx;\ dv = e^x\, dx,\ v = e^x]$;
$\frac{2}{3} \int xe^x\, dx = \frac{2}{3}\left(xe^x - \int e^x\, dx\right) = \frac{2}{3}(xe^x - e^x) + C = \frac{2}{3}\left(\sqrt{3s+9}\, e^{\sqrt{3s+9}} - e^{\sqrt{3s+9}}\right) + C$

27. $u = x,\ du = dx;\ dv = \tan^2 x\, dx,\ v = \int \tan^2 x\, dx = \int \frac{\sin^2 x}{\cos^2 x}\, dx = \int \frac{1 - \cos^2 x}{\cos^2 x}\, dx = \int \frac{dx}{\cos^2 x} - \int dx$
$= \tan x - x$; $\int_0^{\pi/3} x \tan^2 x\, dx = [x(\tan x - x)]_0^{\pi/3} - \int_0^{\pi/3} (\tan x - x)\, dx = \frac{\pi}{3}\left(\sqrt{3} - \frac{\pi}{3}\right) + \left[\ln|\cos x| + \frac{x^2}{2}\right]_0^{\pi/3}$
$= \frac{\pi}{3}\left(\sqrt{3} - \frac{\pi}{3}\right) + \ln\frac{1}{2} + \frac{\pi^2}{18} = \frac{\pi\sqrt{3}}{3} - \ln 2 - \frac{\pi^2}{18}$

29. $\int \sin(\ln x)\,dx$; $\begin{bmatrix} u = \ln x \\ du = \frac{1}{x}\,dx \\ dx = e^u\,du \end{bmatrix} \to \int (\sin u)\,e^u\,du$. From Exercise 21, $\int (\sin u)\,e^u\,du = e^u\left(\frac{\sin u - \cos u}{2}\right) + C$
$= \frac{1}{2}\left[-x\cos(\ln x) + x\sin(\ln x)\right] + C$

31. $\int x \sec x^2\,dx$ $\left[\text{Let } u = x^2,\ du = 2x\,dx \Rightarrow \frac{1}{2}du = x\,dx\right] \to \int x\sec x^2\,dx = \frac{1}{2}\int \sec u\,du = \frac{1}{2}\ln|\sec u + \tan u| + C$
$= \frac{1}{2}\ln|\sec x^2 + \tan x^2| + C$

33. $\int x(\ln x)^2\,dx$; $\begin{bmatrix} u = \ln x \\ du = \frac{1}{x}\,dx \\ dx = e^u\,du \end{bmatrix} \to \int e^u \cdot u^2 \cdot e^u\,du = \int e^{2u} \cdot u^2\,du$;

e^{2u}

$u^2 \xrightarrow{(+)} \frac{1}{2}e^{2u}$

$2u \xrightarrow{(-)} \frac{1}{4}e^{2u}$

$2 \xrightarrow{(+)} \frac{1}{8}e^{2u}$

0

$\int u^2 e^{2u}\,du = \frac{u^2}{2}e^{2u} - \frac{u}{2}e^{2u} + \frac{1}{4}e^{2u} + C = \frac{e^{2u}}{4}\left[2u^2 - 2u + 1\right] + C$
$= \frac{x^2}{4}\left[2(\ln x)^2 - 2\ln x + 1\right] + C = \frac{x^2}{2}(\ln x)^2 - \frac{x^2}{2}\ln x + \frac{x^2}{4} + C$

35. $u = \ln x,\ du = \frac{1}{x}\,dx;\ dv = \frac{1}{x^2}\,dx,\ v = -\frac{1}{x}$;
$\int \frac{\ln x}{x^2}\,dx = -\frac{\ln x}{x} + \int \frac{1}{x^2}\,dx = -\frac{\ln x}{x} - \frac{1}{x} + C$

37. $\int x^3 e^{x^4}\,dx$ $\left[\text{Let } u = x^4,\ du = 4x^3\,dx \Rightarrow \frac{1}{4}du = x^3\,dx\right] \to \int x^3 e^{x^4}\,dx = \frac{1}{4}\int e^u\,du = \frac{1}{4}e^u + C = \frac{1}{4}e^{x^4} + C$

39. $u = x^2,\ du = 2x\,dx;\ dv = \sqrt{x^2+1}\,x\,dx,\ v = \frac{1}{3}\left(x^2+1\right)^{3/2}$;
$\int x^3\sqrt{x^2+1}\,dx = \frac{1}{3}x^2\left(x^2+1\right)^{3/2} - \frac{1}{3}\int\left(x^2+1\right)^{3/2}2x\,dx = \frac{1}{3}x^2\left(x^2+1\right)^{3/2} - \frac{2}{15}\left(x^2+1\right)^{5/2} + C$

41. $u = \sin 3x,\ du = 3\cos 3x\,dx;\ dv = \cos 2x\,dx,\ v = \frac{1}{2}\sin 2x$;
$\int \sin 3x\cos 2x\,dx = \frac{1}{2}\sin 3x\sin 2x - \frac{3}{2}\int \cos 3x\sin 2x\,dx$
$u = \cos 3x,\ du = -3\sin 3x\,dx;\ dv = \sin 2x\,dx,\ v = -\frac{1}{2}\cos 2x$;
$\int \sin 3x\cos 2x\,dx = \frac{1}{2}\sin 3x\sin 2x - \frac{3}{2}\left[-\frac{1}{2}\cos 3x\cos 2x - \frac{3}{2}\int \sin 3x\cos 2x\,dx\right]$
$= \frac{1}{2}\sin 3x\sin 2x + \frac{3}{4}\cos 3x\cos 2x + \frac{9}{4}\int \sin 3x\cos 2x\,dx \Rightarrow -\frac{5}{4}\int \sin 3x\cos 2x\,dx = \frac{1}{2}\sin 3x\sin 2x + \frac{3}{4}\cos 3x\cos 2x$
$\Rightarrow \int \sin 3x\cos 2x\,dx = -\frac{2}{5}\sin 3x\sin 2x - \frac{3}{5}\cos 3x\cos 2x + C$

43. $\int e^x \sin e^x\,dx$ $\left[\text{Let } u = e^x,\ du = e^x\,dx\right] \to \int e^x\sin e^x\,dx = \int \sin u\,du = -\cos u + C = -\cos e^x + C$

45. $\int \cos\sqrt{x}\,dx$; $\begin{bmatrix} y = \sqrt{x} \\ dy = \frac{1}{2\sqrt{x}}\,dx \\ dx = 2y\,dy \end{bmatrix} \to \int \cos y\,2y\,dy = \int 2y\cos y\,dy$;
$u = 2y,\ du = 2\,dy;\ dv = \cos y\,dy,\ v = \sin y$;
$\int 2y\cos y\,dy = 2y\sin y - \int 2\sin y\,dy = 2y\sin y + 2\cos y + C = 2\sqrt{x}\sin\sqrt{x} + 2\cos\sqrt{x} + C$

47.

		$\sin 2\theta$
θ^2	$\xrightarrow{(+)}$	$-\frac{1}{2}\cos 2\theta$
2θ	$\xrightarrow{(-)}$	$-\frac{1}{4}\sin 2\theta$
2	$\xrightarrow{(+)}$	$\frac{1}{8}\cos 2\theta$
0		

$$\int_0^{\pi/2} \theta^2 \sin 2\theta\, d\theta = \left[-\tfrac{\theta^2}{2}\cos 2\theta + \tfrac{\theta}{2}\sin 2\theta + \tfrac{1}{4}\cos 2\theta\right]_0^{\pi/2}$$
$$= \left[-\tfrac{\pi^2}{8}\cdot(-1) + \tfrac{\pi}{4}\cdot 0 + \tfrac{1}{4}\cdot(-1)\right] - \left[0+0+\tfrac{1}{4}\cdot 1\right] = \tfrac{\pi^2}{8} - \tfrac{1}{2} = \tfrac{\pi^2-4}{8}$$

49. $u = \sec^{-1} t,\ du = \frac{dt}{t\sqrt{t^2-1}}$; $dv = t\,dt,\ v = \frac{t^2}{2}$;

$$\int_{2/\sqrt{3}}^{2} t\sec^{-1} t\, dt = \left[\tfrac{t^2}{2}\sec^{-1} t\right]_{2/\sqrt{3}}^{2} - \int_{2/\sqrt{3}}^{2}\left(\tfrac{t^2}{2}\right)\frac{dt}{t\sqrt{t^2-1}} = \left(2\cdot\tfrac{\pi}{3} - \tfrac{2}{3}\cdot\tfrac{\pi}{6}\right) - \int_{2/\sqrt{3}}^{2}\frac{t\,dt}{2\sqrt{t^2-1}}$$
$$= \tfrac{5\pi}{9} - \left[\tfrac{1}{2}\sqrt{t^2-1}\right]_{2/\sqrt{3}}^{2} = \tfrac{5\pi}{9} - \tfrac{1}{2}\left(\sqrt{3} - \sqrt{\tfrac{4}{3}-1}\right) = \tfrac{5\pi}{9} - \tfrac{1}{2}\left(\sqrt{3} - \tfrac{\sqrt{3}}{3}\right) = \tfrac{5\pi}{9} - \tfrac{\sqrt{3}}{3} = \tfrac{5\pi - 3\sqrt{3}}{9}$$

51. (a) $u = x,\ du = dx;\ dv = \sin x\, dx,\ v = -\cos x$;

$$S_1 = \int_0^{\pi} x\sin x\, dx = [-x\cos x]_0^{\pi} + \int_0^{\pi}\cos x\, dx = \pi + [\sin x]_0^{\pi} = \pi$$

(b) $S_2 = -\int_{\pi}^{2\pi} x\sin x\, dx = -\left[[-x\cos x]_{\pi}^{2\pi} + \int_{\pi}^{2\pi}\cos x\, dx\right] = -\left[-3\pi + [\sin x]_{\pi}^{2\pi}\right] = 3\pi$

(c) $S_3 = \int_{2\pi}^{3\pi} x\sin x\, dx = [-x\cos x]_{2\pi}^{3\pi} + \int_{2\pi}^{3\pi}\cos x\, dx = 5\pi + [\sin x]_{2\pi}^{3\pi} = 5\pi$

(d) $S_{n+1} = (-1)^{n+1}\int_{n\pi}^{(n+1)\pi} x\sin x\, dx = (-1)^{n+1}\left[[-x\cos x]_{n\pi}^{(n+1)\pi} + [\sin x]_{n\pi}^{(n+1)\pi}\right]$
$= (-1)^{n+1}\left[-(n+1)\pi(-1)^n + n\pi(-1)^{n+1}\right] + 0 = (2n+1)\pi$

53. $V = \int_0^{\ln 2} 2\pi(\ln 2 - x)e^x\, dx = 2\pi\ln 2\int_0^{\ln 2} e^x\, dx - 2\pi\int_0^{\ln 2} xe^x\, dx$
$= (2\pi\ln 2)\left[e^x\right]_0^{\ln 2} - 2\pi\left(\left[xe^x\right]_0^{\ln 2} - \int_0^{\ln 2} e^x\, dx\right)$
$= 2\pi\ln 2 - 2\pi\left(2\ln 2 - \left[e^x\right]_0^{\ln 2}\right) = -2\pi\ln 2 + 2\pi = 2\pi(1 - \ln 2)$

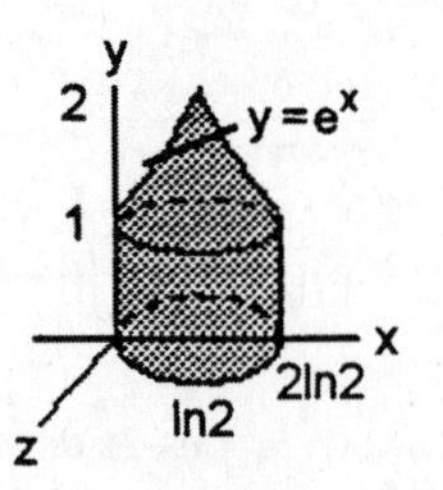

55. (a) $V = \int_0^{\pi/2} 2\pi x\cos x\, dx = 2\pi\left([x\sin x]_0^{\pi/2} - \int_0^{\pi/2}\sin x\, dx\right)$
$= 2\pi\left(\tfrac{\pi}{2} + [\cos x]_0^{\pi/2}\right) = 2\pi\left(\tfrac{\pi}{2} + 0 - 1\right) = \pi(\pi - 2)$

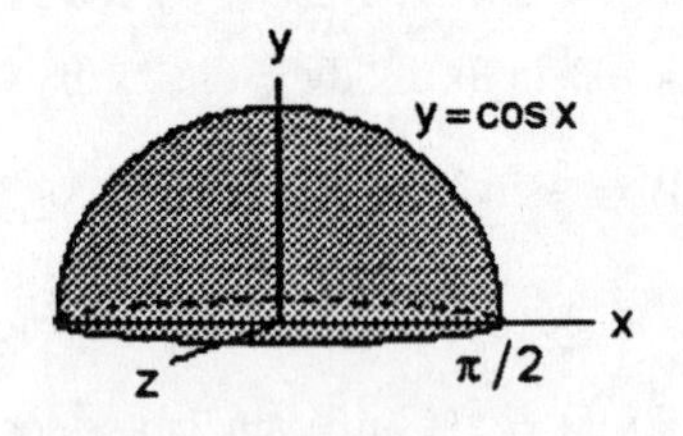

(b) $V = \int_0^{\pi/2} 2\pi\left(\tfrac{\pi}{2} - x\right)\cos x\, dx$; $u = \tfrac{\pi}{2} - x,\ du = -dx;\ dv = \cos x\, dx,\ v = \sin x$;
$V = 2\pi\left[\left(\tfrac{\pi}{2} - x\right)\sin x\right]_0^{\pi/2} + 2\pi\int_0^{\pi/2}\sin x\, dx = 0 + 2\pi[-\cos x]_0^{\pi/2} = 2\pi(0+1) = 2\pi$

57. (a) $A = \int_1^e \ln x\,dx = \left[x \ln x\right]_1^e - \int_1^e dx$

$= (e \ln e - 1 \ln 1) - \left[x\right]_1^e = e - (e - 1) = 1$

(b) $V = \int_1^e \pi (\ln x)^2\,dx = \pi\left(\left[x(\ln x)^2\right]_1^e - \int_1^e 2 \ln x\,dx\right)$

$= \pi\left[\left(e(\ln e)^2 - 1(\ln 1)^2\right) - \left(\left[2x \ln x\right]_1^e - \int_1^e 2\,dx\right)\right]$

$= \pi\left[e - \left((2e \ln e - 2(1)\ln 1) - \left[2x\right]_1^e\right)\right]$

$= \pi\left[e - (2e - (2e - 2))\right] = \pi(e - 2)$

(c) $V = \int_1^e 2\pi(x+2)\ln x\,dx = 2\pi \int_1^e (x+2)\ln x\,dx = 2\pi\left(\left[\left(\tfrac{1}{2}x^2 + 2x\right)\ln x\right]_1^e - \int_1^e \left(\tfrac{1}{2}x + 2\right)dx\right)$

$= 2\pi\left(\left(\tfrac{1}{2}e^2 + 2e\right)\ln e - \left(\tfrac{1}{2} + 2\right)\ln 1 - \left[\left(\tfrac{1}{4}x^2 + 2x\right)\right]_1^e\right) = 2\pi\left(\left(\tfrac{1}{2}e^2 + 2e\right) - \left(\left(\tfrac{1}{4}e^2 + 2e\right) - \tfrac{9}{4}\right)\right) = \tfrac{\pi}{2}(e^2 + 9)$

(d) $M = \int_1^e \ln x\,dx = 1$ (from part (a)); $\overline{x} = \tfrac{1}{1}\int_1^e x \ln x\,dx = \left[\tfrac{1}{2}x^2 \ln x\right]_1^e - \int_1^e \tfrac{1}{2}x\,dx = \left(\tfrac{1}{2}e^2 \ln e - \tfrac{1}{2}(1)^2 \ln 1\right) - \left[\tfrac{1}{4}x^2\right]_1^e$

$= \tfrac{1}{2}e^2 - \left(\tfrac{1}{4}e^2 - \tfrac{1}{4}(1)^2\right) = \tfrac{1}{4}(e^2 + 1)$; $\overline{y} = \tfrac{1}{1}\int_1^e \tfrac{1}{2}(\ln x)^2\,dx = \tfrac{1}{2}\left(\left[x(\ln x)^2\right]_1^e - \int_1^e 2 \ln x\,dx\right)$

$= \tfrac{1}{2}\left(\left(e(\ln e)^2 - 1 \cdot (\ln 1)^2\right) - \left(\left[2x \ln x\right]_1^e - \int_1^e 2\,dx\right)\right) = \tfrac{1}{2}\left(e - \left((2e \ln e - 2(1)\ln 1) - \left[2x\right]_1^e\right)\right)$

$= \tfrac{1}{2}(e - 2e + 2e - 2) = \tfrac{1}{2}(e - 2) \Rightarrow (\overline{x}, \overline{y}) = \left(\tfrac{e^2+1}{4}, \tfrac{e-2}{2}\right)$ is the centroid.

59. $\text{av}(y) = \tfrac{1}{2\pi}\int_0^{2\pi} 2e^{-t}\cos t\,dt$

$= \tfrac{1}{\pi}\left[e^{-t}\left(\tfrac{\sin t - \cos t}{2}\right)\right]_0^{2\pi}$

(see Exercise 22) $\Rightarrow \text{av}(y) = \tfrac{1}{2\pi}\left(1 - e^{-2\pi}\right)$

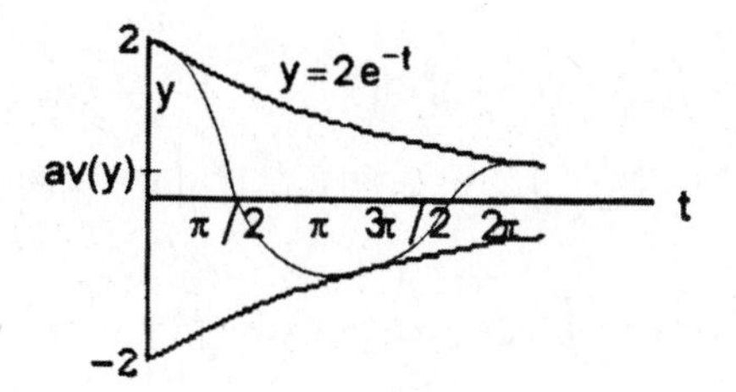

61. $I = \int x^n \cos x\,dx$; $[u = x^n,\ du = nx^{n-1}\,dx;\ dv = \cos x\,dx,\ v = \sin x]$

$\Rightarrow I = x^n \sin x - \int nx^{n-1}\sin x\,dx$

63. $I = \int x^n e^{ax}\,dx$; $\left[u = x^n,\ du = nx^{n-1}\,dx;\ dv = e^{ax}\,dx,\ v = \tfrac{1}{a}e^{ax}\right]$

$\Rightarrow I = \tfrac{x^n e^{ax}}{a}e^{ax} - \tfrac{n}{a}\int x^{n-1}e^{ax}\,dx,\ a \neq 0$

65. $\int_a^b (x - a) f(x)\,dx$; $\left[u = x - a,\ du = dx;\ dv = f(x)\,dx,\ v = \int_b^x f(t)\,dt = -\int_x^b f(t)\,dt\right]$

$= \left[(x - a)\int_b^x f(t)\,dt\right]_a^b - \int_a^b \left(\int_b^x f(t)\,dt\right)dx = \left((b - a)\int_b^b f(t)\,dt - (a - a)\int_b^a f(t)\,dt\right) - \int_a^b \left(-\int_x^b f(t)\,dt\right)dx$

$= 0 + \int_a^b \left(\int_x^b f(t)\,dt\right)dx = \int_a^b \left(\int_x^b f(t)\,dt\right)dx$

67. $\int \sin^{-1} x\,dx = x \sin^{-1} x - \int \sin y\,dy = x \sin^{-1} x + \cos y + C = x \sin^{-1} x + \cos(\sin^{-1} x) + C$

69. $\int \sec^{-1} x\, dx = x \sec^{-1} x - \int \sec y\, dy = x \sec^{-1} x - \ln|\sec y + \tan y| + C$
$= x \sec^{-1} x - \ln|\sec(\sec^{-1} x) + \tan(\sec^{-1} x)| + C = x \sec^{-1} x - \ln\left|x + \sqrt{x^2 - 1}\right| + C$

71. Yes, $\cos^{-1} x$ is the angle whose cosine is x which implies $\sin(\cos^{-1} x) = \sqrt{1 - x^2}$.

73. (a) $\int \sinh^{-1} x\, dx = x \sinh^{-1} x - \int \sinh y\, dy = x \sinh^{-1} x - \cosh y + C = x \sinh^{-1} x - \cosh(\sinh^{-1} x) + C$;
check: $d\,[x \sinh^{-1} x - \cosh(\sinh^{-1} x) + C] = \left[\sinh^{-1} x + \frac{x}{\sqrt{1+x^2}} - \sinh(\sinh^{-1} x)\frac{1}{\sqrt{1+x^2}}\right] dx$
$= \sinh^{-1} x\, dx$
(b) $\int \sinh^{-1} x\, dx = x \sinh^{-1} x - \int x\left(\frac{1}{\sqrt{1+x^2}}\right) dx = x \sinh^{-1} x - \frac{1}{2}\int (1 + x^2)^{-1/2} 2x\, dx$
$= x \sinh^{-1} x - (1 + x^2)^{1/2} + C$
check: $d\left[x \sinh^{-1} x - (1 + x^2)^{1/2} + C\right] = \left[\sinh^{-1} x + \frac{x}{\sqrt{1+x^2}} - \frac{x}{\sqrt{1+x^2}}\right] dx = \sinh^{-1} x\, dx$

8.2 TRIGONOMETRIC INTEGRALS

1. $\int \cos 2x\, dx = \frac{1}{2}\int \cos 2x \cdot 2dx = \frac{1}{2}\sin 2x + C$

3. $\int \cos^3 x \sin x\, dx = -\int \cos^3 x(-\sin x)dx = -\frac{1}{4}\cos^4 x + C$

5. $\int \sin^3 x\, dx = \int \sin^2 x \sin x\, dx = \int (1 - \cos^2 x) \sin x\, dx = \int \sin x\, dx - \int \cos^2 x \sin x\, dx = -\cos x + \frac{1}{3}\cos^3 x + C$

7. $\int \sin^5 x\, dx = \int (\sin^2 x)^2 \sin x\, dx = \int (1 - \cos^2 x)^2 \sin x\, dx = \int (1 - 2\cos^2 x + \cos^4 x)\sin x\, dx$
$= \int \sin x\, dx - \int 2\cos^2 x \sin x\, dx + \int \cos^4 x \sin x\, dx = -\cos x + \frac{2}{3}\cos^3 x - \frac{1}{5}\cos^5 x + C$

9. $\int \cos^3 x\, dx = \int (\cos^2 x)\cos x\, dx = \int (1 - \sin^2 x)\cos x\, dx = \int \cos x\, dx - \int \sin^2 x \cos x\, dx = \sin x - \frac{1}{3}\sin^3 x + C$

11. $\int \sin^3 x \cos^3 x\, dx = \int \sin^3 x \cos^2 x \cos x dx = \int \sin^3 x(1 - \sin^2 x)\cos x\, dx = \int \sin^3 x \cos x\, dx - \int \sin^5 x \cos x\, dx$
$= \frac{1}{4}\sin^4 x - \frac{1}{6}\sin^6 x + C$

13. $\int \cos^2 x\, dx = \int \frac{1+\cos 2x}{2}\, dx = \frac{1}{2}\int (1 + \cos 2x)dx = \frac{1}{2}\int dx + \frac{1}{2}\int \cos 2x\, dx = \frac{1}{2}\int dx + \frac{1}{4}\int \cos 2x \cdot 2dx$
$= \frac{1}{2}x + \frac{1}{4}\sin 2x + C$

15. $\int_0^{\pi/2} \sin^7 y\, dy = \int_0^{\pi/2} \sin^6 y \sin y\, dy = \int_0^{\pi/2} (1 - \cos^2 y)^3 \sin y\, dy = \int_0^{\pi/2} \sin y\, dy - 3\int_0^{\pi/2} \cos^2 y \sin y\, dy$
$+ 3\int_0^{\pi/2} \cos^4 y \sin y\, dy - \int_0^{\pi/2} \cos^6 y \sin y\, dy = \left[-\cos y + 3\frac{\cos^3 y}{3} - 3\frac{\cos^5 y}{5} + \frac{\cos^7 y}{7}\right]_0^{\pi/2} = (0) - \left(-1 + 1 - \frac{3}{5} + \frac{1}{7}\right) = \frac{16}{35}$

17. $\int_0^{\pi} 8\sin^4 x\, dx = 8\int_0^{\pi} \left(\frac{1-\cos 2x}{2}\right)^2 dx = 2\int_0^{\pi} (1 - 2\cos 2x + \cos^2 2x)dx = 2\int_0^{\pi} dx \quad 2\int_0^{\pi} \cos 2x \cdot 2dx + 2\int_0^{\pi} \frac{1+\cos 4x}{2}\, dx$
$= [2x - 2\sin 2x]_0^{\pi} + \int_0^{\pi} dx + \int_0^{\pi} \cos 4x\, dx = 2\pi + \left[x + \frac{1}{2}\sin 4x\right]_0^{\pi} = 2\pi + \pi = 3\pi$

19. $\int 16 \sin^2 x \cos^2 x\, dx = 16\int \left(\frac{1-\cos 2x}{2}\right)\left(\frac{1+\cos 2x}{2}\right) dx = 4\int (1-\cos^2 2x) dx = 4\int dx - 4\int \left(\frac{1+\cos 4x}{2}\right) dx$
$= 4x - 2\int dx - 2\int \cos 4x\, dx = 4x - 2x - \frac{1}{2}\sin 4x + C = 2x - \frac{1}{2}\sin 4x + C = 2x - \sin 2x \cos 2x + C$
$= 2x - 2\sin x \cos x\,(2\cos^2 x - 1) + C = 2x - 4\sin x \cos^3 x + 2\sin x \cos x\ + C$

21. $\int 8\cos^3 2\theta \sin 2\theta\, d\theta = 8\left(-\frac{1}{2}\right)\frac{\cos^4 2\theta}{4} + C = -\cos^4 2\theta + C$

23. $\int_0^{2\pi} \sqrt{\frac{1-\cos x}{2}}\, dx = \int_0^{2\pi} \left|\sin \frac{x}{2}\right| dx = \int_0^{2\pi} \sin \frac{x}{2}\, dx = \left[-2\cos \frac{x}{2}\right]_0^{2\pi} = 2 + 2 = 4$

25. $\int_0^{\pi} \sqrt{1-\sin^2 t}\, dt = \int_0^{\pi} |\cos t|\, dt = \int_0^{\pi/2} \cos t\, dt - \int_{\pi/2}^{\pi} \cos t\, dt = [\sin t]_0^{\pi/2} - [\sin t]_{\pi/2}^{\pi} = 1 - 0 - 0 + 1 = 2$

27. $\int_{\pi/3}^{\pi/2} \frac{\sin^2 x}{\sqrt{1-\cos x}}\, dx = \int_{\pi/3}^{\pi/2} \frac{\sin^2 x}{\sqrt{1-\cos x}} \frac{\sqrt{1+\cos x}}{\sqrt{1+\cos x}}\, dx = \int_{\pi/3}^{\pi/2} \frac{\sin^2 x\sqrt{1+\cos x}}{\sqrt{1-\cos^2 x}}\, dx = \int_{\pi/3}^{\pi/2} \frac{\sin^2 x\sqrt{1+\cos x}}{\sqrt{\sin^2 x}}\, dx$
$= \int_{\pi/3}^{\pi/2} \sin x\sqrt{1+\cos x}\, dx = \left[-\frac{2}{3}(1+\cos x)^{3/2}\right]_{\pi/3}^{\pi/2} = -\frac{2}{3}\left(1+\cos\left(\frac{\pi}{2}\right)\right)^{3/2} + \frac{2}{3}\left(1+\cos\left(\frac{\pi}{3}\right)\right)^{3/2} = -\frac{2}{3} + \frac{2}{3}\left(\frac{3}{2}\right)^{3/2}$
$= \sqrt{\frac{3}{2}} - \frac{2}{3}$

29. $\int_{5\pi/6}^{\pi} \frac{\cos^4 x}{\sqrt{1-\sin x}}\, dx = \int_{5\pi/6}^{\pi} \frac{\cos^4 x}{\sqrt{1-\sin x}} \frac{\sqrt{1+\sin x}}{\sqrt{1+\sin x}}\, dx = \int_{5\pi/6}^{\pi} \frac{\cos^4 x\sqrt{1+\sin x}}{\sqrt{1-\sin^2 x}}\, dx = \int_{5\pi/6}^{\pi} \frac{\cos^4 x\sqrt{1+\sin x}}{\sqrt{\cos^2 x}}\, dx$
$= \int_{5\pi/6}^{\pi} \frac{\cos^4 x\sqrt{1+\sin x}}{-\cos x}\, dx = -\int_{5\pi/6}^{\pi} \cos^3 x\sqrt{1+\sin x}\, dx = -\int_{5\pi/6}^{\pi} \cos x(1-\sin^2 x)\sqrt{1+\sin x}\, dx$
$= -\int_{5\pi/6}^{\pi} \cos x\sqrt{1+\sin x}\, dx + \int_{5\pi/6}^{\pi} \cos x \sin^2 x\sqrt{1+\sin x}\, dx;\ u^2\sqrt{u}\, du$
$\left[\text{Let } u = 1+\sin x \Rightarrow u - 1 = \sin x \Rightarrow du = \cos x\, dx,\ x = \frac{5\pi}{6} \Rightarrow u = 1+\sin\left(\frac{5\pi}{6}\right) = \frac{3}{2},\ x = \pi \Rightarrow u = 1+\sin\pi = 1\right]$
$= \left[-\frac{2}{3}(1+\sin x)^{3/2}\right]_{5\pi/6}^{\pi} + \int_{3/2}^{1} (u-1)^2\sqrt{u}\, du = \left[-\frac{2}{3}(1+\sin x)^{3/2}\right]_{5\pi/6}^{\pi} + \int_{3/2}^{1} \left(u^{5/2} - 2u^{3/2} + \sqrt{u}\right) du$
$= \left(-\frac{2}{3}(1+\sin\pi)^{3/2} + \frac{2}{3}\left(1+\sin\left(\frac{5\pi}{6}\right)\right)^{3/2}\right) + \left[\frac{2}{7}u^{7/2} - \frac{4}{5}u^{5/2} + \frac{2}{3}u^{3/2}\right]_{3/2}^{1}$
$= \left(-\frac{2}{3} + \frac{2}{3}\left(\frac{3}{2}\right)^{3/2}\right) + \left(\frac{2}{7} - \frac{4}{5} + \frac{2}{3}\right) - \left(\frac{2}{7}\left(\frac{3}{2}\right)^{7/2} - \frac{4}{5}\left(\frac{3}{2}\right)^{5/2} + \frac{2}{3}\left(\frac{3}{2}\right)^{3/2}\right) = \frac{4}{5}\left(\frac{3}{2}\right)^{5/2} - \frac{2}{7}\left(\frac{3}{2}\right)^{7/2} - \frac{18}{35}$

31. $\int_0^{\pi/2} \theta\sqrt{1-\cos 2\theta}\, d\theta = \int_0^{\pi/2} \theta\sqrt{2}\,|\sin\theta|\, d\theta = \sqrt{2}\int_0^{\pi/2} \theta \sin\theta\, d\theta = \sqrt{2}\,[-\theta\cos\theta + \sin\theta]_0^{\pi/2} = \sqrt{2}(1) = \sqrt{2}$

33. $\int \sec^2 x \tan x\, dx = \int \tan x \sec^2 x\, dx = \frac{1}{2}\tan^2 x + C$

35. $\int \sec^3 x \tan x\, dx = \int \sec^2 x \sec x \tan x\, dx = \frac{1}{3}\sec^3 x + C$

37. $\int \sec^2 x \tan^2 x\, dx = \int \tan^2 x \sec^2 x\, dx = \frac{1}{3}\tan^3 x + C$

39. $\int_{-\pi/3}^{0} 2\sec^3 x\, dx;\ u = \sec x,\ du = \sec x \tan x\, dx,\ dv = \sec^2 x\, dx,\ v = \tan x;$
$\int_{-\pi/3}^{0} 2\sec^3 x\, dx = [2\sec x \tan x]_{-\pi/3}^{0} - 2\int_{-\pi/3}^{0} \sec x \tan^2 x\, dx = 2\cdot 1\cdot 0 - 2\cdot 2\cdot\sqrt{3} - 2\int_{-\pi/3}^{0} \sec x\,(\sec^2 x - 1) dx$
$= 4\sqrt{3} - 2\int_{-\pi/3}^{0} \sec^3 x\, dx + 2\int_{-\pi/3}^{0} \sec x\, dx;\ 2\int_{-\pi/3}^{0} 2\sec^3 x\, dx = 4\sqrt{3} + [2\ln|\sec x + \tan x|]_{-\pi/3}^{0}$
$2\int_{-\pi/3}^{0} 2\sec^3 x\, dx = 4\sqrt{3} + 2\ln|1+0| - 2\ln|2-\sqrt{3}| = 4\sqrt{3} - 2\ln\left(2-\sqrt{3}\right)$
$\int_{-\pi/3}^{0} 2\sec^3 x\, dx = 2\sqrt{3} - \ln\left(2-\sqrt{3}\right)$

41. $\int \sec^4\theta\, d\theta = \int (1+\tan^2\theta)\sec^2\theta\, d\theta = \int \sec^2\theta\, d\theta + \int \tan^2\theta \sec^2\theta\, d\theta = \tan\theta + \frac{1}{3}\tan^3\theta + C$
$= \tan\theta + \frac{1}{3}\tan\theta(\sec^2\theta - 1) + C = \frac{1}{3}\tan\theta\sec^2\theta + \frac{2}{3}\tan\theta + C$

43. $\int_{\pi/4}^{\pi/2} \csc^4\theta\, d\theta = \int_{\pi/4}^{\pi/2} (1+\cot^2\theta)\csc^2\theta\, d\theta = \int_{\pi/4}^{\pi/2} \csc^2\theta\, d\theta + \int_{\pi/4}^{\pi/2} \cot^2\theta\csc^2\theta\, d\theta = \left[-\cot\theta - \frac{\cot^3\theta}{3}\right]_{\pi/4}^{\pi/2}$
$= (0) - \left(-1 - \frac{1}{3}\right) = \frac{4}{3}$

45. $\int 4\tan^3 x\, dx = 4\int (\sec^2 x - 1)\tan x\, dx = 4\int \sec^2 x \tan x\, dx - 4\int \tan x\, dx = 4\frac{\tan^2 x}{2} - 4\ln|\sec x| + C$
$= 2\tan^2 x - 4\ln|\sec x| + C = 2\tan^2 x - 2\ln|\sec^2 x| + C = 2\tan^2 x - 2\ln(1+\tan^2 x) + C$

47. $\int \tan^5 x\, dx = \int \tan^4 x \tan x\, dx = \int \left(\sec^2 x - 1\right)^2 \tan x\, dx = \int \left(\sec^4 x - 2\sec^2 x + 1\right)\tan x\, dx$
$= \int \sec^4 x \tan x\, dx - 2\int \sec^2 x \tan x\, dx + \int \tan x\, dx = \int \sec^3 x \sec x \tan x\, dx - 2\int \sec x \sec x \tan x\, dx + \int \tan x\, dx$
$= \frac{1}{4}\sec^4 x - \sec^2 x + \ln|\sec x| + C = \frac{1}{4}\left(\tan^2 x + 1\right)^2 - \left(\tan^2 x + 1\right) + \ln|\sec x| + C = \frac{1}{4}\tan^4 x - \frac{1}{2}\tan^2 x + \ln|\sec x| + C$

49. $\int_{\pi/6}^{\pi/3} \cot^3 x\, dx = \int_{\pi/6}^{\pi/3} (\csc^2 x - 1)\cot x\, dx = \int_{\pi/6}^{\pi/3} \csc^2 x \cot x\, dx - \int_{\pi/6}^{\pi/3} \cot x\, dx = \left[-\frac{\cot^2 x}{2} + \ln|\csc x|\right]_{\pi/6}^{\pi/3}$
$= -\frac{1}{2}\left(\frac{1}{3} - 3\right) + \left(\ln\frac{2}{\sqrt{3}} - \ln 2\right) = \frac{4}{3} - \ln\sqrt{3}$

51. $\int \sin 3x \cos 2x\, dx = \frac{1}{2}\int (\sin x + \sin 5x)\, dx = -\frac{1}{2}\cos x - \frac{1}{10}\cos 5x + C$

53. $\int_{-\pi}^{\pi} \sin 3x \sin 3x\, dx = \frac{1}{2}\int_{-\pi}^{\pi} (\cos 0 - \cos 6x)\, dx = \frac{1}{2}\int_{-\pi}^{\pi} dx - \frac{1}{2}\int_{-\pi}^{\pi} \cos 6x\, dx = \frac{1}{2}\left[x - \frac{1}{12}\sin 6x\right]_{-\pi}^{\pi} = \frac{\pi}{2} + \frac{\pi}{2} - 0 = \pi$

55. $\int \cos 3x \cos 4x\, dx = \frac{1}{2}\int (\cos(-x) + \cos 7x)\, dx = \frac{1}{2}\int (\cos x + \cos 7x)\, dx = \frac{1}{2}\sin x + \frac{1}{14}\sin 7x + C$

57. $\int \sin^2\theta \cos 3\theta\, d\theta = \int \frac{1-\cos 2\theta}{2}\cos 3\theta\, d\theta = \frac{1}{2}\int \cos 3\theta\, d\theta - \frac{1}{2}\int \cos 2\theta \cos 3\theta\, d\theta$
$= \frac{1}{2}\int \cos 3\theta\, d\theta - \frac{1}{2}\int \frac{1}{2}(\cos(2-3)\theta + \cos(2+3)\theta)\, d\theta = \frac{1}{2}\int \cos 3\theta\, d\theta - \frac{1}{4}\int (\cos(-\theta) + \cos 5\theta)\, d\theta$
$= \frac{1}{2}\int \cos 3\theta\, d\theta - \frac{1}{4}\int \cos\theta\, d\theta - \frac{1}{4}\int \cos 5\theta\, d\theta = \frac{1}{6}\sin 3\theta - \frac{1}{4}\sin\theta - \frac{1}{20}\sin 5\theta + C$

59. $\int \cos^3\theta \sin 2\theta\, d\theta = \int \cos^3\theta\,(2\sin\theta\cos\theta)\, d\theta = 2\int \cos^4\theta \sin\theta\, d\theta = -\frac{2}{5}\cos^5\theta + C$

61. $\int \sin\theta\cos\theta\cos 3\theta\, d\theta = \frac{1}{2}\int 2\sin\theta\cos\theta\cos 3\theta\, d\theta = \frac{1}{2}\int \sin 2\theta\cos 3\theta\, d\theta = \frac{1}{2}\int \frac{1}{2}(\sin(2-3)\theta + \sin(2+3)\theta)\, d\theta$
$= \frac{1}{4}\int (\sin(-\theta) + \sin 5\theta)\, d\theta = \frac{1}{4}\int (-\sin\theta + \sin 5\theta)\, d\theta = \frac{1}{4}\cos\theta - \frac{1}{20}\cos 5\theta + C$

63. $\int \frac{\sec^3 x}{\tan x}\, dx = \int \frac{\sec^2 x \sec x}{\tan x}\, dx = \int \frac{\left(\tan^2 x + 1\right)\sec x}{\tan x}\, dx = \int \frac{\tan^2 x \sec x}{\tan x}\, dx + \int \frac{\sec x}{\tan x}\, dx = \int \tan x \sec x\, dx + \int \csc x\, dx$
$= \sec x - \ln|\csc x + \cot x| + C$

65. $\int \frac{\tan^2 x}{\csc x}\, dx = \int \frac{\sin^2 x}{\cos^2 x}\sin x\, dx = \int \frac{\left(1-\cos^2 x\right)}{\cos^2 x}\sin x\, dx = \int \frac{1}{\cos^2 x}\sin x\, dx - \int \frac{\cos^2 x}{\cos^2 x}\sin x\, dx = \int \sec x \tan x\, dx - \int \sin x\, dx$
$= \sec x + \cos x + C$

67. $\int x \sin^2 x\, dx = \int x \frac{1-\cos 2x}{2}\, dx = \frac{1}{2}\int x\, dx - \frac{1}{2}\int x \cos 2x\, dx \left[u = x,\ du = dx,\ dv = \cos 2x\, dx,\ v = \frac{1}{2}\sin 2x\right]$
$= \frac{1}{4}x^2 - \frac{1}{2}\left[\frac{1}{2}x \sin 2x - \int \frac{1}{2}\sin 2x\, dx\right] = \frac{1}{4}x^2 - \frac{1}{4}x \sin 2x - \frac{1}{8}\cos 2x + C$

69. $y = \ln(\sec x);\ y' = \frac{\sec x \tan x}{\sec x} = \tan x; (y')^2 = \tan^2 x;\ \int_0^{\pi/4} \sqrt{1+\tan^2 x}\, dx = \int_0^{\pi/4} |\sec x|\, dx = [\ln|\sec x + \tan x|]_0^{\pi/4}$
$= \ln\left(\sqrt{2}+1\right) - \ln(0+1) = \ln\left(\sqrt{2}+1\right)$

71. $V = \pi\int_0^\pi \sin^2 x\, dx = \pi\int_0^\pi \frac{1-\cos 2x}{2}\, dx = \frac{\pi}{2}\int_0^\pi dx - \frac{\pi}{2}\int_0^\pi \cos 2x\, dx = \frac{\pi}{2}[x]_0^\pi - \frac{\pi}{4}[\sin 2x]_0^\pi = \frac{\pi}{2}(\pi - 0) - \frac{\pi}{4}(0-0) = \frac{\pi^2}{2}$

73. $M = \int_0^{2\pi} (x + \cos x) dx = \left[\frac{1}{2}x^2 + \sin x\right]_0^{2\pi} = \left(\frac{1}{2}(2\pi)^2 + \sin(2\pi)\right) - \left(\frac{1}{2}(0)^2 + \sin(0)\right) = 2\pi^2;$
$\overline{x} = \frac{1}{2\pi^2}\int_0^{2\pi} x(x+\cos x)dx = \frac{1}{2\pi^2}\int_0^{2\pi} (x^2 + x\cos x)dx = \frac{1}{2\pi^2}\int_0^{2\pi} x^2 dx + \frac{1}{2\pi^2}\int_0^{2\pi} x \cos x dx$
$\left[u = x,\ du = dx,\ dv = \cos x\, dx,\ v = \sin x\right]$
$= \frac{1}{6\pi^2}\left[x^3\right]_0^{2\pi} + \frac{1}{2\pi^2}\left(\left[x \sin x\right]_0^{2\pi} - \int_0^{2\pi} \sin x dx\right) = \frac{1}{6\pi^2}(8\pi^3 - 0) + \frac{1}{2\pi^2}\left(2\pi \sin 2\pi - 0 - \int_0^{2\pi} \sin x dx\right)$
$= \frac{4\pi}{3} + \frac{1}{2\pi^2}\left[\cos x\right]_0^{2\pi} = \frac{4\pi}{3} + \frac{1}{2\pi^2}(\cos 2\pi - \cos 0) = \frac{4\pi}{3} + 0 = \frac{4\pi}{3};\ \overline{y} = \frac{1}{2\pi^2}\int_0^{2\pi} \frac{1}{2}(x + \cos x)^2 dx$
$= \frac{1}{4\pi^2}\int_0^{2\pi} (x^2 + 2x\cos x + \cos^2 x)dx = \frac{1}{4\pi^2}\int_0^{2\pi} x^2 dx + \frac{1}{2\pi^2}\int_0^{2\pi} x \cos x dx + \frac{1}{4\pi^2}\int_0^{2\pi} \cos^2 x\, dx$
$= \frac{1}{12\pi^2}\left[x^3\right]_0^{2\pi} + \frac{1}{2\pi^2}\left[x \sin x + \cos x\right]_0^{2\pi} + \frac{1}{4\pi^2}\int_0^{2\pi} \frac{\cos 2x + 1}{2}\, dx = \frac{2\pi}{3} + 0 + \frac{1}{8\pi^2}\int_0^{2\pi} \cos 2x\, dx + \frac{1}{8\pi^2}\int_0^{2\pi} dx$
$= \frac{2\pi}{3} + \frac{1}{16\pi^2}\left[\sin 2x\right]_0^{2\pi} + \frac{1}{8\pi^2}\left[x\right]_0^{2\pi} = \frac{2\pi}{3} + 0 + \frac{1}{4\pi} = \frac{8\pi^2+3}{12\pi} \Rightarrow$ The centroid is $\left(\frac{4\pi}{3}, \frac{8\pi^2+3}{12\pi}\right)$.

8.3 TRIGONOMETRIC SUBSTITUTIONS

1. $x = 3\tan\theta,\ -\frac{\pi}{2} < \theta < \frac{\pi}{2},\ dx = \frac{3\, d\theta}{\cos^2\theta},\ 9 + x^2 = 9(1 + \tan^2\theta) = 9\sec^2\theta \Rightarrow \frac{1}{\sqrt{9+x^2}} = \frac{1}{3|\sec\theta|} = \frac{|\cos\theta|}{3} = \frac{\cos\theta}{3};$
(because $\cos\theta > 0$ when $-\frac{\pi}{2} < \theta < \frac{\pi}{2}$);
$\int \frac{dx}{\sqrt{9+x^2}} = 3\int \frac{\cos\theta\, d\theta}{3\cos^2\theta} = \int \frac{d\theta}{\cos\theta} = \ln|\sec\theta + \tan\theta| + C' = \ln\left|\frac{\sqrt{9+x^2}}{3} + \frac{x}{3}\right| + C' = \ln\left|\sqrt{9+x^2} + x\right| + C$

3. $\int_{-2}^{2} \frac{dx}{4+x^2} = \left[\frac{1}{2}\tan^{-1}\frac{x}{2}\right]_{-2}^{2} = \frac{1}{2}\tan^{-1}1 - \frac{1}{2}\tan^{-1}(-1) = \left(\frac{1}{2}\right)\left(\frac{\pi}{4}\right) - \left(\frac{1}{2}\right)\left(-\frac{\pi}{4}\right) = \frac{\pi}{4}$

5. $\int_0^{3/2} \frac{dx}{\sqrt{9-x^2}} = \left[\sin^{-1}\frac{x}{3}\right]_0^{3/2} = \sin^{-1}\frac{1}{2} - \sin^{-1}0 = \frac{\pi}{6} - 0 = \frac{\pi}{6}$

7. $t = 5\sin\theta,\ -\frac{\pi}{2} < \theta < \frac{\pi}{2},\ dt = 5\cos\theta\, d\theta,\ \sqrt{25-t^2} = 5\cos\theta;$
$\int \sqrt{25-t^2}\, dt = \int (5\cos\theta)(5\cos\theta)\, d\theta = 25\int \cos^2\theta\, d\theta = 25\int \frac{1+\cos 2\theta}{2}\, d\theta = 25\left(\frac{\theta}{2} + \frac{\sin 2\theta}{4}\right) + C$
$= \frac{25}{2}(\theta + \sin\theta\cos\theta) + C = \frac{25}{2}\left[\sin^{-1}\left(\frac{t}{5}\right) + \left(\frac{t}{5}\right)\left(\frac{\sqrt{25-t^2}}{5}\right)\right] + C = \frac{25}{2}\sin^{-1}\left(\frac{t}{5}\right) + \frac{t\sqrt{25-t^2}}{2} + C$

9. $x = \frac{7}{2}\sec\theta,\ 0 < \theta < \frac{\pi}{2},\ dx = \frac{7}{2}\sec\theta\tan\theta\, d\theta,\ \sqrt{4x^2-49} = \sqrt{49\sec^2\theta - 49} = 7\tan\theta;$
$\int \frac{dx}{\sqrt{4x^2-49}} = \int \frac{\left(\frac{7}{2}\sec\theta\tan\theta\right) d\theta}{7\tan\theta} = \frac{1}{2}\int \sec\theta\, d\theta = \frac{1}{2}\ln|\sec\theta + \tan\theta| + C = \frac{1}{2}\ln\left|\frac{2x}{7} + \frac{\sqrt{4x^2-49}}{7}\right| + C$

11. $y = 7\sec\theta,\ 0 < \theta < \frac{\pi}{2},\ dy = 7\sec\theta\tan\theta\,d\theta,\ \sqrt{y^2-49} = 7\tan\theta;$
$\int \frac{\sqrt{y^2-49}}{y}\,dy = \int \frac{(7\tan\theta)(7\sec\theta\tan\theta)\,d\theta}{7\sec\theta} = 7\int \tan^2\theta\,d\theta = 7\int (\sec^2\theta - 1)\,d\theta = 7(\tan\theta - \theta) + C$
$= 7\left[\frac{\sqrt{y^2-49}}{7} - \sec^{-1}\left(\frac{y}{7}\right)\right] + C$

13. $x = \sec\theta,\ 0 < \theta < \frac{\pi}{2},\ dx = \sec\theta\tan\theta\,d\theta,\ \sqrt{x^2-1} = \tan\theta;$
$\int \frac{dx}{x^2\sqrt{x^2-1}} = \int \frac{\sec\theta\tan\theta\,d\theta}{\sec^2\theta\tan\theta} = \int \frac{d\theta}{\sec\theta} = \sin\theta + C = \frac{\sqrt{x^2-1}}{x} + C$

15. $u = 9 - x^2 \Rightarrow du = -2x\,dx \Rightarrow -\frac{1}{2}du = x\,dx;$
$\int \frac{x\,dx}{\sqrt{9-x^2}} = -\frac{1}{2}\int \frac{1}{\sqrt{u}}du = -\sqrt{u} + C = -\sqrt{9-x^2} + C$

17. $x = 2\tan\theta,\ -\frac{\pi}{2} < \theta < \frac{\pi}{2},\ dx = \frac{2\,d\theta}{\cos^2\theta},\ \sqrt{x^2+4} = \frac{2}{\cos\theta};$
$\int \frac{x^3\,dx}{\sqrt{x^2+4}} = \int \frac{(8\tan^3\theta)(\cos\theta)\,d\theta}{\cos^2\theta} = 8\int \frac{\sin^3\theta\,d\theta}{\cos^4\theta} = 8\int \frac{(\cos^2\theta - 1)(-\sin\theta)\,d\theta}{\cos^4\theta};$
$[t = \cos\theta] \rightarrow 8\int \frac{t^2-1}{t^4}\,dt = 8\int \left(\frac{1}{t^2} - \frac{1}{t^4}\right)dt = 8\left(-\frac{1}{t} + \frac{1}{3t^3}\right) + C = 8\left(-\sec\theta + \frac{\sec^3\theta}{3}\right) + C$
$= 8\left(-\frac{\sqrt{x^2+4}}{2} + \frac{(x^2+4)^{3/2}}{8\cdot 3}\right) + C = \frac{1}{3}(x^2+4)^{3/2} - 4\sqrt{x^2+4} + C = \frac{1}{3}(x^2-8)\sqrt{x^2+4} + C$

19. $w = 2\sin\theta,\ -\frac{\pi}{2} < \theta < \frac{\pi}{2},\ dw = 2\cos\theta\,d\theta,\ \sqrt{4-w^2} = 2\cos\theta;$
$\int \frac{8\,dw}{w^2\sqrt{4-w^2}} = \int \frac{8\cdot 2\cos\theta\,d\theta}{4\sin^2\theta\cdot 2\cos\theta} = 2\int \frac{d\theta}{\sin^2\theta} = -2\cot\theta + C = \frac{-2\sqrt{4-w^2}}{w} + C$

21. $u = 5x \Rightarrow du = 5dx,\ a = 6$
$\int \frac{100}{36+25x^2}dx = 20\int \frac{1}{(6)^2+(5x)^2}5dx = 20\int \frac{1}{a^2+u^2}du = 20\cdot\frac{1}{6}\tan^{-1}\left(\frac{u}{6}\right) + C = \frac{10}{3}\tan^{-1}\left(\frac{5x}{6}\right) + C$

23. $x = \sin\theta,\ 0 \le \theta \le \frac{\pi}{3},\ dx = \cos\theta\,d\theta,\ (1-x^2)^{3/2} = \cos^3\theta;$
$\int_0^{\sqrt{3}/2} \frac{4x^2\,dx}{(1-x^2)^{3/2}} = \int_0^{\pi/3} \frac{4\sin^2\theta\cos\theta\,d\theta}{\cos^3\theta} = 4\int_0^{\pi/3}\left(\frac{1-\cos^2\theta}{\cos^2\theta}\right)d\theta = 4\int_0^{\pi/3}(\sec^2\theta - 1)\,d\theta$
$= 4\left[\tan\theta - \theta\right]_0^{\pi/3} = 4\sqrt{3} - \frac{4\pi}{3}$

25. $x = \sec\theta,\ 0 < \theta < \frac{\pi}{2},\ dx = \sec\theta\tan\theta\,d\theta,\ (x^2-1)^{3/2} = \tan^3\theta;$
$\int \frac{dx}{(x^2-1)^{3/2}} = \int \frac{\sec\theta\tan\theta\,d\theta}{\tan^3\theta} = \int \frac{\cos\theta\,d\theta}{\sin^2\theta} = -\frac{1}{\sin\theta} + C = -\frac{x}{\sqrt{x^2-1}} + C$

27. $x = \sin\theta,\ -\frac{\pi}{2} < \theta < \frac{\pi}{2},\ dx = \cos\theta\,d\theta,\ (1-x^2)^{3/2} = \cos^3\theta;$
$\int \frac{(1-x^2)^{3/2}\,dx}{x^6} = \int \frac{\cos^3\theta\cdot\cos\theta\,d\theta}{\sin^6\theta} = \int \cot^4\theta\csc^2\theta\,d\theta = -\frac{\cot^5\theta}{5} + C = -\frac{1}{5}\left(\frac{\sqrt{1-x^2}}{x}\right)^5 + C$

29. $x = \frac{1}{2}\tan\theta,\ -\frac{\pi}{2} < \theta < \frac{\pi}{2},\ dx = \frac{1}{2}\sec^2\theta\,d\theta,\ (4x^2+1)^2 = \sec^4\theta;$
$\int \frac{8\,dx}{(4x^2+1)^2} = \int \frac{8\left(\frac{1}{2}\sec^2\theta\right)d\theta}{\sec^4\theta} = 4\int \cos^2\theta\,d\theta = 2(\theta + \sin\theta\cos\theta) + C = 2\tan^{-1}2x + \frac{4x}{(4x^2+1)} + C$

31. $u = x^2 - 1 \Rightarrow du = 2x\,dx \Rightarrow \frac{1}{2}du = x\,dx$
$\int \frac{x^3}{x^2-1}dx = \int \left(x + \frac{x}{x^2-1}\right)dx = \int x\,dx + \int \frac{x}{x^2-1}dx = \frac{1}{2}x^2 + \frac{1}{2}\int \frac{1}{u}du = \frac{1}{2}x^2 + \frac{1}{2}\ln|u| + C = \frac{1}{2}x^2 + \frac{1}{2}\ln|x^2-1| + C$

33. $v = \sin\theta,\ -\frac{\pi}{2} < \theta < \frac{\pi}{2},\ dv = \cos\theta\, d\theta,\ (1-v^2)^{5/2} = \cos^5\theta;$

$\int \frac{v^2\, dv}{(1-v^2)^{5/2}} = \int \frac{\sin^2\theta\cos\theta\, d\theta}{\cos^5\theta} = \int \tan^2\theta\sec^2\theta\, d\theta = \frac{\tan^3\theta}{3} + C = \frac{1}{3}\left(\frac{v}{\sqrt{1-v^2}}\right)^3 + C$

35. Let $e^t = 3\tan\theta,\ t = \ln(3\tan\theta),\ \tan^{-1}\left(\frac{1}{3}\right) \le \theta \le \tan^{-1}\left(\frac{4}{3}\right),\ dt = \frac{\sec^2\theta}{\tan\theta}\, d\theta,\ \sqrt{e^{2t}+9} = \sqrt{9\tan^2\theta+9} = 3\sec\theta;$

$\int_0^{\ln 4} \frac{e^t\, dt}{\sqrt{e^{2t}+9}} = \int_{\tan^{-1}(1/3)}^{\tan^{-1}(4/3)} \frac{3\tan\theta\cdot\sec^2\theta\, d\theta}{\tan\theta\cdot 3\sec\theta} = \int_{\tan^{-1}(1/3)}^{\tan^{-1}(4/3)} \sec\theta\, d\theta = [\ln|\sec\theta + \tan\theta|]_{\tan^{-1}(1/3)}^{\tan^{-1}(4/3)}$

$= \ln\left(\frac{5}{3} + \frac{4}{3}\right) - \ln\left(\frac{\sqrt{10}}{3} + \frac{1}{3}\right) = \ln 9 - \ln\left(1 + \sqrt{10}\right)$

37. $\int_{1/12}^{1/4} \frac{2\, dt}{\sqrt{t} + 4t\sqrt{t}};\ \left[u = 2\sqrt{t},\ du = \frac{1}{\sqrt{t}}\, dt\right] \to \int_{1/\sqrt{3}}^{1} \frac{2\, du}{1+u^2};\ u = \tan\theta,\ \frac{\pi}{6} \le \theta \le \frac{\pi}{4},\ du = \sec^2\theta\, d\theta,\ 1 + u^2 = \sec^2\theta;$

$\int_{1/\sqrt{3}}^{1} \frac{2\, du}{1+u^2} = \int_{\pi/6}^{\pi/4} \frac{2\sec^2\theta\, d\theta}{\sec^2\theta} = [2\theta]_{\pi/6}^{\pi/4} = 2\left(\frac{\pi}{4} - \frac{\pi}{6}\right) = \frac{\pi}{6}$

39. $x = \sec\theta,\ 0 < \theta < \frac{\pi}{2},\ dx = \sec\theta\tan\theta\, d\theta,\ \sqrt{x^2-1} = \sqrt{\sec^2\theta - 1} = \tan\theta;$

$\int \frac{dx}{x\sqrt{x^2-1}} = \int \frac{\sec\theta\tan\theta\, d\theta}{\sec\theta\tan\theta} = \theta + C = \sec^{-1} x + C$

41. $x = \sec\theta,\ dx = \sec\theta\tan\theta\, d\theta,\ \sqrt{x^2-1} = \sqrt{\sec^2\theta - 1} = \tan\theta;$

$\int \frac{x\, dx}{\sqrt{x^2-1}} = \int \frac{\sec\theta\cdot\sec\theta\tan\theta\, d\theta}{\tan\theta} = \int \sec^2\theta\, d\theta = \tan\theta + C = \sqrt{x^2-1} + C$

43. Let $x^2 = \tan\theta,\ 0 \le \theta < \frac{\pi}{2},\ 2x\, dx = \sec^2\theta\, d\theta \Rightarrow x\, dx = \frac{1}{2}\sec^2\theta\, d\theta;\ \sqrt{1+x^4} = \sqrt{1+\tan^2\theta} = \sec\theta$

$\int \frac{x}{\sqrt{1+x^4}} dx = \frac{1}{2}\int \frac{\sec^2\theta}{\sec\theta} d\theta = \frac{1}{2}\int \sec\theta d\theta = \frac{1}{2}\ln|\sec\theta + \tan\theta| + C = \frac{1}{2}\ln|\sqrt{1+x^4} + x^2| + C$

45. Let $u = \sqrt{x} \Rightarrow x = u^2 \Rightarrow dx = 2u\, du \Rightarrow \int \sqrt{\frac{4-x}{x}}\, dx = \int \sqrt{\frac{4-u^2}{u^2}}\, 2u\, du = 2\int \sqrt{4-u^2}\, du;$

$u = 2\sin\theta,\ du = 2\cos\theta\, d\theta,\ 0 < \theta \le \frac{\pi}{2},\ \sqrt{4-u^2} = 2\cos\theta$

$2\int \sqrt{4-u^2}\, du = 2\int (2\cos\theta)(2\cos\theta)\, d\theta = 8\int \cos^2\theta\, d\theta = 8\int \frac{1+\cos 2\theta}{2}\, d\theta = 4\int d\theta + 4\int \cos 2\theta\, d\theta$

$= 4\theta + 2\sin 2\theta + C = 4\theta + 4\sin\theta\cos\theta + C = 4\sin^{-1}\left(\frac{u}{2}\right) + 4\left(\frac{u}{2}\right)\left(\frac{\sqrt{4-u^2}}{2}\right) + C = 4\sin^{-1}\left(\frac{\sqrt{x}}{2}\right) + \sqrt{x}\sqrt{4-x} + C$

$= 4\sin^{-1}\left(\frac{\sqrt{x}}{2}\right) + \sqrt{4x - x^2} + C$

47. Let $u = \sqrt{x} \Rightarrow x = u^2 \Rightarrow dx = 2u\, du \Rightarrow \int \sqrt{x}\sqrt{1-x}\, dx = \int u\sqrt{1-u^2}\, 2u\, du = 2\int u^2\sqrt{1-u^2}\, du;$

$u = \sin\theta,\ du = \cos\theta\, d\theta,\ -\frac{\pi}{2} < \theta \le \frac{\pi}{2},\ \sqrt{1-u^2} = \cos\theta$

$2\int u^2\sqrt{1-u^2}\, du = 2\int \sin^2\theta\cos\theta\cos\theta\, d\theta = 2\int \sin^2\theta\cos^2\theta\, d\theta = \frac{1}{2}\int \sin^2 2\theta\, d\theta = \frac{1}{2}\int \frac{1-\cos 4\theta}{2}\, d\theta$

$= \frac{1}{4}\int d\theta - \frac{1}{4}\int \cos 4\theta\, d\theta = \frac{1}{4}\theta - \frac{1}{16}\sin 4\theta + C = \frac{1}{4}\theta - \frac{1}{8}\sin 2\theta\cos 2\theta + C = \frac{1}{4}\theta - \frac{1}{4}\sin\theta\cos\theta\,(2\cos^2\theta - 1) + C$

$= \frac{1}{4}\theta - \frac{1}{2}\sin\theta\cos^3\theta + \frac{1}{4}\sin\theta\cos\theta + C = \frac{1}{4}\sin^{-1}u - \frac{1}{2}u\,(1-u^2)^{3/2} - \frac{1}{4}u\sqrt{1-u^2} + C$

$= \frac{1}{4}\sin^{-1}\sqrt{x} - \frac{1}{2}\sqrt{x}\,(1-x)^{3/2} - \frac{1}{4}\sqrt{x}\sqrt{1-x} + C$

49. $x\frac{dy}{dx} = \sqrt{x^2-4};\ dy = \sqrt{x^2-4}\,\frac{dx}{x};\ y = \int \frac{\sqrt{x^2-4}}{x}\, dx;\ \begin{bmatrix} x = 2\sec\theta,\ 0 < \theta < \frac{\pi}{2} \\ dx = 2\sec\theta\tan\theta\, d\theta \\ \sqrt{x^2-4} = 2\tan\theta \end{bmatrix}$

$\to y = \int \frac{(2\tan\theta)(2\sec\theta\tan\theta)\, d\theta}{2\sec\theta} = 2\int \tan^2\theta\, d\theta = 2\int (\sec^2\theta - 1)\, d\theta = 2(\tan\theta - \theta) + C$

$= 2\left[\frac{\sqrt{x^2-4}}{2} - \sec^{-1}\left(\frac{x}{2}\right)\right] + C;\ x = 2$ and $y = 0 \Rightarrow 0 = 0 + C \Rightarrow C = 0 \Rightarrow y = 2\left[\frac{\sqrt{x^2-4}}{2} - \sec^{-1}\frac{x}{2}\right]$

51. $(x^2+4)\frac{dy}{dx}=3$, $dy=\frac{3\,dx}{x^2+4}$; $y=3\int\frac{dx}{x^2+4}=\frac{3}{2}\tan^{-1}\frac{x}{2}+C$; $x=2$ and $y=0 \Rightarrow 0=\frac{3}{2}\tan^{-1}1+C$
$\Rightarrow C=-\frac{3\pi}{8} \Rightarrow y=\frac{3}{2}\tan^{-1}\left(\frac{x}{2}\right)-\frac{3\pi}{8}$

53. $A=\int_0^3\frac{\sqrt{9-x^2}}{3}\,dx$; $x=3\sin\theta$, $0\le\theta\le\frac{\pi}{2}$, $dx=3\cos\theta\,d\theta$, $\sqrt{9-x^2}=\sqrt{9-9\sin^2\theta}=3\cos\theta$;
$A=\int_0^{\pi/2}\frac{3\cos\theta\cdot 3\cos\theta\,d\theta}{3}=3\int_0^{\pi/2}\cos^2\theta\,d\theta=\frac{3}{2}\left[\theta+\sin\theta\cos\theta\right]_0^{\pi/2}=\frac{3\pi}{4}$

55. (a) $A=\int_0^{1/2}\sin^{-1}x\,dx\ \left[u=\sin^{-1}x,\ du=\frac{1}{\sqrt{1-x^2}}dx,\ dv=dx,\ v=x\right]$
$=\left[x\sin^{-1}x\right]_0^{1/2}-\int_0^{1/2}\frac{x}{\sqrt{1-x^2}}\,dx==\left(\frac{1}{2}\sin^{-1}\frac{1}{2}-0\right)+\left[\sqrt{1-x^2}\right]_0^{1/2}=\frac{\pi+6\sqrt{3}-12}{12}$

(b) $M=\int_0^{1/2}\sin^{-1}x\,dx=\frac{\pi+6\sqrt{3}-12}{12}$; $\bar{x}=\frac{1}{\frac{\pi+6\sqrt{3}-12}{12}}\int_0^{1/2}x\sin^{-1}x\,dx=\frac{12}{\pi+6\sqrt{3}-12}\int_0^{1/2}x\sin^{-1}x\,dx$
$\left[u=\sin^{-1}x,\ du=\frac{1}{\sqrt{1-x^2}}dx,\ dv=x\,dx,\ v=\frac{1}{2}x^2\right]$
$=\frac{12}{\pi+6\sqrt{3}-12}\left(\left[\frac{1}{2}x^2\sin^{-1}x\right]_0^{1/2}-\frac{1}{2}\int_0^{1/2}\frac{x^2}{\sqrt{1-x^2}}\,dx\right)$
$\left[x=\sin\theta,\ -\frac{\pi}{2}<\theta<\frac{\pi}{2},\ dx=\cos\theta\,d\theta,\ \sqrt{1-x^2}=\cos\theta,\ x=0=\sin\theta\Rightarrow\theta=0,\ x=\frac{1}{2}=\sin\theta\Rightarrow\theta=\frac{\pi}{6}\right]$
$=\frac{12}{\pi+6\sqrt{3}-12}\left(\left(\frac{1}{2}\left(\frac{1}{2}\right)^2\sin^{-1}\left(\frac{1}{2}\right)-0\right)-\frac{1}{2}\int_0^{\pi/6}\frac{\sin^2\theta}{\cos\theta}\cos\theta\,d\theta\right)=\frac{12}{\pi+6\sqrt{3}-12}\left(\frac{\pi}{48}-\frac{1}{2}\int_0^{\pi/6}\sin^2\theta\,d\theta\right)$
$=\frac{12}{\pi+6\sqrt{3}-12}\left(\frac{\pi}{48}-\frac{1}{2}\int_0^{\pi/6}\frac{1-\cos 2\theta}{2}\,d\theta\right)=\frac{12}{\pi+6\sqrt{3}-12}\left(\frac{\pi}{48}-\frac{1}{4}\int_0^{\pi/6}d\theta+\frac{1}{4}\int_0^{\pi/6}\cos 2\theta\,d\theta\right)$
$=\frac{12}{\pi+6\sqrt{3}-12}\left(\frac{\pi}{48}+\left[-\frac{\theta}{4}+\frac{1}{8}\sin 2\theta\right]_0^{\pi/6}\right)=\frac{3\sqrt{3}-\pi}{4\left(\pi+6\sqrt{3}-12\right)}$; $\bar{y}=\frac{1}{\frac{\pi+6\sqrt{3}-12}{12}}\int_0^{1/2}\frac{1}{2}\left(\sin^{-1}x\right)^2dx$
$\left[u=\left(\sin^{-1}x\right)^2,\ du=\frac{2\sin^{-1}x}{\sqrt{1-x^2}}dx,\ dv=dx,\ v=x\right]$
$=\frac{6}{\pi+6\sqrt{3}-12}\left(\left[x\left(\sin^{-1}x\,dx\right)^2\right]_0^{1/2}-\int_0^{1/2}\frac{2x\sin^{-1}x}{\sqrt{1-x^2}}dx\right)$
$\left[u=\sin^{-1}x,\ du=\frac{1}{\sqrt{1-x^2}}dx,\ dv=\frac{2x}{\sqrt{1-x^2}}dx,\ v=-2\sqrt{1-x^2}\right]$
$=\frac{6}{\pi+6\sqrt{3}-12}\left(\left(\frac{1}{2}\left(\sin^{-1}\left(\frac{1}{2}\right)\right)^2-0\right)+\left[2\sqrt{1-x^2}\sin^{-1}x\right]_0^{1/2}-\int_0^{1/2}\frac{2\sqrt{1-x^2}}{\sqrt{1-x^2}}dx\right)$
$=\frac{6}{\pi+6\sqrt{3}-12}\left(\frac{\pi^2}{72}+\left(2\sqrt{1-\left(\frac{1}{2}\right)^2}\sin^{-1}\left(\frac{1}{2}\right)-0\right)-\left[2x\right]_0^{1/2}\right)=\frac{6}{\pi+6\sqrt{3}-12}\left(\frac{\pi^2}{72}+\frac{\pi\sqrt{3}}{6}-1\right)=\frac{\pi^2+12\pi\sqrt{3}-72}{12\left(\pi+6\sqrt{3}-12\right)}$

57. (a) Integration by parts: $u=x^2$, $du=2x\,dx$, $dv=x\sqrt{1-x^2}\,dx$, $v=-\frac{1}{3}\left(1-x^2\right)^{3/2}$
$\int x^3\sqrt{1-x^2}\,dx=-\frac{1}{3}x^2\left(1-x^2\right)^{3/2}+\frac{1}{3}\int\left(1-x^2\right)^{3/2}2x\,dx=-\frac{1}{3}x^2\left(1-x^2\right)^{3/2}-\frac{2}{15}\left(1-x^2\right)^{5/2}+C$

(b) Substitution: $u=1-x^2\Rightarrow x^2=1-u\Rightarrow du=-2x\,dx\Rightarrow-\frac{1}{2}du=x\,dx$
$\int x^3\sqrt{1-x^2}\,dx=\int x^2\sqrt{1-x^2}\,x\,dx=-\frac{1}{2}\int(1-u)\sqrt{u}\,du=-\frac{1}{2}\int\left(\sqrt{u}-u^{3/2}\right)du=-\frac{1}{3}u^{3/2}+\frac{1}{5}u^{5/2}+C$
$=-\frac{1}{3}\left(1-x^2\right)^{3/2}+\frac{1}{5}\left(1-x^2\right)^{5/2}+C$

(c) Trig substitution: $x=\sin\theta$, $\frac{\pi}{2}\le\theta\le\frac{\pi}{2}$, $dx=\cos\theta\,d\theta$, $\sqrt{1-x^2}=\cos\theta$
$\int x^3\sqrt{1-x^2}\,dx=\int\sin^3\theta\cos\theta\cos\theta\,d\theta=\int\sin^2\theta\cos^2\theta\sin\theta\,d\theta=\int\left(1-\cos^2\theta\right)\cos^2\theta\sin\theta\,d\theta$
$=\int\cos^2\theta\sin\theta\,d\theta-\int\cos^4\theta\sin\theta\,d\theta=-\frac{1}{3}\cos^3\theta+\frac{1}{5}\cos^5\theta+C=-\frac{1}{3}\left(1-x^2\right)^{3/2}+\frac{1}{5}\left(1-x^2\right)^{5/2}+C$

8.4 INTEGRATION OF RATIONAL FUNCTIONS BY PARTIAL FRACTIONS

1. $\frac{5x-13}{(x-3)(x-2)} = \frac{A}{x-3} + \frac{B}{x-2} \Rightarrow 5x - 13 = A(x-2) + B(x-3) = (A+B)x - (2A+3B)$
$\Rightarrow \left.\begin{matrix} A+B=5 \\ 2A+3B=13 \end{matrix}\right\} \Rightarrow -B = (10-13) \Rightarrow B = 3 \Rightarrow A = 2$; thus, $\frac{5x-13}{(x-3)(x-2)} = \frac{2}{x-3} + \frac{3}{x-2}$

3. $\frac{x+4}{(x+1)^2} = \frac{A}{x+1} + \frac{B}{(x+1)^2} \Rightarrow x+4 = A(x+1) + B = Ax + (A+B) \Rightarrow \left.\begin{matrix} A=1 \\ A+B=4 \end{matrix}\right\} \Rightarrow A = 1$ and $B = 3$;
thus, $\frac{x+4}{(x+1)^2} = \frac{1}{x+1} + \frac{3}{(x+1)^2}$

5. $\frac{z+1}{z^2(z-1)} = \frac{A}{z} + \frac{B}{z^2} + \frac{C}{z-1} \Rightarrow z+1 = Az(z-1) + B(z-1) + Cz^2 \Rightarrow z+1 = (A+C)z^2 + (-A+B)z - B$
$\Rightarrow \left.\begin{matrix} A+C=0 \\ -A+B=1 \\ -B=1 \end{matrix}\right\} \Rightarrow B = -1 \Rightarrow A = -2 \Rightarrow C = 2$; thus, $\frac{z+1}{z^2(z-1)} = \frac{-2}{z} + \frac{-1}{z^2} + \frac{2}{z-1}$

7. $\frac{t^2+8}{t^2-5t+6} = 1 + \frac{5t+2}{t^2-5t+6}$ (after long division); $\frac{5t+2}{t^2-5t+6} = \frac{5t+2}{(t-3)(t-2)} = \frac{A}{t-3} + \frac{B}{t-2}$
$\Rightarrow 5t+2 = A(t-2) + B(t-3) = (A+B)t + (-2A-3B) \Rightarrow \left.\begin{matrix} A+B=5 \\ -2A-3B=2 \end{matrix}\right\} \Rightarrow -B = (10+2) = 12$
$\Rightarrow B = -12 \Rightarrow A = 17$; thus, $\frac{t^2+8}{t^2-5t+6} = 1 + \frac{17}{t-3} + \frac{-12}{t-2}$

9. $\frac{1}{1-x^2} = \frac{A}{1-x} + \frac{B}{1+x} \Rightarrow 1 = A(1+x) + B(1-x)$; $x = 1 \Rightarrow A = \frac{1}{2}$; $x = -1 \Rightarrow B = \frac{1}{2}$;
$\int \frac{dx}{1-x^2} = \frac{1}{2}\int \frac{dx}{1-x} + \frac{1}{2}\int \frac{dx}{1+x} = \frac{1}{2}[\ln|1+x| - \ln|1-x|] + C$

11. $\frac{x+4}{x^2+5x-6} = \frac{A}{x+6} + \frac{B}{x-1} \Rightarrow x+4 = A(x-1) + B(x+6)$; $x = 1 \Rightarrow B = \frac{5}{7}$; $x = -6 \Rightarrow A = \frac{-2}{-7} = \frac{2}{7}$;
$\int \frac{x+4}{x^2+5x-6}\,dx = \frac{2}{7}\int \frac{dx}{x+6} + \frac{5}{7}\int \frac{dx}{x-1} = \frac{2}{7}\ln|x+6| + \frac{5}{7}\ln|x-1| + C = \frac{1}{7}\ln\left|(x+6)^2(x-1)^5\right| + C$

13. $\frac{y}{y^2-2y-3} = \frac{A}{y-3} + \frac{B}{y+1} \Rightarrow y = A(y+1) + B(y-3)$; $y = -1 \Rightarrow B = \frac{-1}{-4} = \frac{1}{4}$; $y = 3 \Rightarrow A = \frac{3}{4}$;
$\int_4^8 \frac{y\,dy}{y^2-2y-3} = \frac{3}{4}\int_4^8 \frac{dy}{y-3} + \frac{1}{4}\int_4^8 \frac{dy}{y+1} = \left[\frac{3}{4}\ln|y-3| + \frac{1}{4}\ln|y+1|\right]_4^8 = \left(\frac{3}{4}\ln 5 + \frac{1}{4}\ln 9\right) - \left(\frac{3}{4}\ln 1 + \frac{1}{4}\ln 5\right)$
$= \frac{1}{2}\ln 5 + \frac{1}{2}\ln 3 = \frac{\ln 15}{2}$

15. $\frac{1}{t^3+t^2-2t} = \frac{A}{t} + \frac{B}{t+2} + \frac{C}{t-1} \Rightarrow 1 = A(t+2)(t-1) + Bt(t-1) + Ct(t+2)$; $t = 0 \Rightarrow A = -\frac{1}{2}$; $t = -2$
$\Rightarrow B = \frac{1}{6}$; $t = 1 \Rightarrow C = \frac{1}{3}$; $\int \frac{dt}{t^3+t^2-2t} = -\frac{1}{2}\int \frac{dt}{t} + \frac{1}{6}\int \frac{dt}{t+2} + \frac{1}{3}\int \frac{dt}{t-1}$
$= -\frac{1}{2}\ln|t| + \frac{1}{6}\ln|t+2| + \frac{1}{3}\ln|t-1| + C$

17. $\frac{x^3}{x^2+2x+1} = (x-2) + \frac{3x+2}{(x+1)^2}$ (after long division); $\frac{3x+2}{(x+1)^2} = \frac{A}{x+1} + \frac{B}{(x+1)^2} \Rightarrow 3x+2 = A(x+1) + B$
$= Ax + (A+B) \Rightarrow A = 3, A+B = 2 \Rightarrow A = 3, B = -1$; $\int_0^1 \frac{x^3\,dx}{x^2+2x+1}$
$= \int_0^1 (x-2)\,dx + 3\int_0^1 \frac{dx}{x+1} - \int_0^1 \frac{dx}{(x+1)^2} = \left[\frac{x^2}{2} - 2x + 3\ln|x+1| + \frac{1}{x+1}\right]_0^1$
$= \left(\frac{1}{2} - 2 + 3\ln 2 + \frac{1}{2}\right) - (1) = 3\ln 2 - 2$

19. $\frac{1}{(x^2-1)^2} = \frac{A}{x+1} + \frac{B}{x-1} + \frac{C}{(x+1)^2} + \frac{D}{(x-1)^2} \Rightarrow 1 = A(x+1)(x-1)^2 + B(x-1)(x+1)^2 + C(x-1)^2 + D(x+1)^2$;
$x = -1 \Rightarrow C = \frac{1}{4}$; $x = 1 \Rightarrow D = \frac{1}{4}$; coefficient of $x^3 = A + B \Rightarrow A + B = 0$; constant $= A - B + C + D$

$\Rightarrow A - B + C + D = 1 \Rightarrow A - B = \frac{1}{2}$; thus, $A = \frac{1}{4} \Rightarrow B = -\frac{1}{4}$; $\int \frac{dx}{(x^2-1)^2}$
$= \frac{1}{4}\int \frac{dx}{x+1} - \frac{1}{4}\int \frac{dx}{x-1} + \frac{1}{4}\int \frac{dx}{(x+1)^2} + \frac{1}{4}\int \frac{dx}{(x-1)^2} = \frac{1}{4}\ln\left|\frac{x+1}{x-1}\right| - \frac{x}{2(x^2-1)} + C$

21. $\frac{1}{(x+1)(x^2+1)} = \frac{A}{x+1} + \frac{Bx+C}{x^2+1} \Rightarrow 1 = A(x^2+1) + (Bx+C)(x+1)$; $x = -1 \Rightarrow A = \frac{1}{2}$; coefficient of x^2
$= A + B \Rightarrow A + B = 0 \Rightarrow B = -\frac{1}{2}$; constant $= A + C \Rightarrow A + C = 1 \Rightarrow C = \frac{1}{2}$; $\int_0^1 \frac{dx}{(x+1)(x^2+1)}$
$= \frac{1}{2}\int_0^1 \frac{dx}{x+1} + \frac{1}{2}\int_0^1 \frac{(-x+1)}{x^2+1}\,dx = \left[\frac{1}{2}\ln|x+1| - \frac{1}{4}\ln(x^2+1) + \frac{1}{2}\tan^{-1}x\right]_0^1$
$= \left(\frac{1}{2}\ln 2 - \frac{1}{4}\ln 2 + \frac{1}{2}\tan^{-1}1\right) - \left(\frac{1}{2}\ln 1 - \frac{1}{4}\ln 1 + \frac{1}{2}\tan^{-1}0\right) = \frac{1}{4}\ln 2 + \frac{1}{2}\left(\frac{\pi}{4}\right) = \frac{(\pi + 2\ln 2)}{8}$

23. $\frac{y^2+2y+1}{(y^2+1)^2} = \frac{Ay+B}{y^2+1} + \frac{Cy+D}{(y^2+1)^2} \Rightarrow y^2 + 2y + 1 = (Ay+B)(y^2+1) + Cy + D$
$= Ay^3 + By^2 + (A+C)y + (B+D) \Rightarrow A = 0, B = 1; A + C = 2 \Rightarrow C = 2; B + D = 1 \Rightarrow D = 0;$
$\int \frac{y^2+2y+1}{(y^2+1)^2}\,dy = \int \frac{1}{y^2+1}\,dy + 2\int \frac{y}{(y^2+1)^2}\,dy = \tan^{-1}y - \frac{1}{y^2+1} + C$

25. $\frac{2s+2}{(s^2+1)(s-1)^3} = \frac{As+B}{s^2+1} + \frac{C}{s-1} + \frac{D}{(s-1)^2} + \frac{E}{(s-1)^3} \Rightarrow 2s + 2$
$= (As+B)(s-1)^3 + C(s^2+1)(s-1)^2 + D(s^2+1)(s-1) + E(s^2+1)$
$= [As^4 + (-3A+B)s^3 + (3A-3B)s^2 + (-A+3B)s - B] + C(s^4 - 2s^3 + 2s^2 - 2s + 1) + D(s^3 - s^2 + s - 1)$
$+ E(s^2+1)$
$= (A+C)s^4 + (-3A+B-2C+D)s^3 + (3A-3B+2C-D+E)s^2 + (-A+3B-2C+D)s + (-B+C-D+E)$

$$\Rightarrow \left.\begin{array}{r} A + C = 0 \\ -3A + B - 2C + D = 0 \\ 3A - 3B + 2C - D + E = 0 \\ -A + 3B - 2C + D = 2 \\ -B + C - D + E = 2 \end{array}\right\} \text{summing all equations} \Rightarrow 2E = 4 \Rightarrow E = 2;$$

summing eqs (2) and (3) $\Rightarrow -2B + 2 = 0 \Rightarrow B = 1$; summing eqs (3) and (4) $\Rightarrow 2A + 2 = 2 \Rightarrow A = 0; C = 0$ from eq (1); then $-1 + 0 - D + 2 = 2$ from eq (5) $\Rightarrow D = -1$;
$\int \frac{2s+2}{(s^2+1)(s-1)^3}\,ds = \int \frac{ds}{s^2+1} - \int \frac{ds}{(s-1)^2} + 2\int \frac{ds}{(s-1)^3} = -(s-1)^{-2} + (s-1)^{-1} + \tan^{-1}s + C$

27. $\frac{x^2-x+2}{x^3-1} = \frac{A}{x-1} + \frac{Bx+C}{x^2+x+1} \Rightarrow x^2 - x + 2 = A(x^2+x+1) + (Bx+C)(x-1) = (A+B)x^2 + (A-B+C)x + (A-C)$
$\Rightarrow A + B = 1, A - B + C = -1, A - C = 2 \Rightarrow$ adding eq(2) and eq(3) $\Rightarrow 2A - B = 1$, add this equation to eq(1)
$\Rightarrow 3A = 2 \Rightarrow A = \frac{2}{3} \Rightarrow B = 1 - A = \frac{1}{3} \Rightarrow C = -1 - A + B = -\frac{4}{3}$; $\int \frac{x^2-x+2}{x^3-1}dx = \int\left(\frac{2/3}{x-1} + \frac{(1/3)x - 4/3}{x^2+x+1}\right)dx$
$= \frac{2}{3}\int \frac{1}{x-1}dx + \frac{1}{3}\int \frac{x-4}{\left(x+\frac{1}{2}\right)^2 + \frac{3}{4}}dx \quad \left[u = x + \frac{1}{2} \Rightarrow u - \frac{1}{2} = x \Rightarrow du = dx\right]$
$= \frac{2}{3}\int \frac{1}{x-1}dx + \frac{1}{3}\int \frac{u - \frac{9}{2}}{u^2 + \frac{3}{4}}du = \frac{2}{3}\int \frac{1}{x-1}dx + \frac{1}{3}\int \frac{u}{u^2+\frac{3}{4}}du - \frac{3}{2}\int \frac{1}{u^2+\frac{3}{4}}du$
$= \frac{2}{3}\ln|x-1| + \frac{1}{6}\ln\left|\left(x+\frac{1}{2}\right)^2 + \frac{3}{4}\right| - \frac{3}{\sqrt{3}}\tan^{-1}\left(\frac{x+\frac{1}{2}}{\sqrt{3}/2}\right) + C = \frac{2}{3}\ln|x-1| + \frac{1}{6}\ln|x^2+x+1| - \sqrt{3}\tan^{-1}\left(\frac{2x+1}{\sqrt{3}}\right) + C$

29. $\frac{x^2}{x^4-1} = \frac{A}{x+1} + \frac{B}{x-1} + \frac{Cx+D}{x^2+1} \Rightarrow x^2 = A(x-1)(x^2+1) + B(x+1)(x^2+1) + (Cx+D)(x-1)(x+1)$
$= (A+B+C)x^3 + (-A+B+D)x^2 + (A+B-C)x - A + B - D \Rightarrow A + B + C = 0, -A + B + D = 1,$
$A + B - C = 0, -A + B - D = 0 \Rightarrow$ adding eq(1) to eq (3) gives $2A + 2B = 0$, adding eq(2) to eq(4) gives
$-2A + 2B = 1$, adding these two equations gives $4B = 1 \Rightarrow B = \frac{1}{4}$, using $2A + 2B = 0 \Rightarrow A = -\frac{1}{4}$, using
$-A + B - D = 0 \Rightarrow D = \frac{1}{2}$, and using $A + B - C = 0 \Rightarrow C = 0$; $\int \frac{x^2}{x^4-1}dx = \int\left(\frac{-1/4}{x+1} + \frac{1/4}{x-1} + \frac{1/2}{x^2+1}\right)dx$
$= -\frac{1}{4}\int \frac{1}{x+1}dx + \frac{1}{4}\int \frac{1}{x-1}dx + \frac{1}{2}\int \frac{1}{x^2+1}dx = -\frac{1}{4}\ln|x+1| + \frac{1}{4}\ln|x-1| + \frac{1}{2}\tan^{-1}x + C = \frac{1}{4}\ln\left|\frac{x-1}{x+1}\right| + \frac{1}{2}\tan^{-1}x + C$

31. $\frac{2\theta^3+5\theta^2+8\theta+4}{(\theta^2+2\theta+2)^2} = \frac{A\theta+B}{\theta^2+2\theta+2} + \frac{C\theta+D}{(\theta^2+2\theta+2)^2} \Rightarrow 2\theta^3+5\theta^2+8\theta+4 = (A\theta+B)(\theta^2+2\theta+2)+C\theta+D$
$= A\theta^3+(2A+B)\theta^2+(2A+2B+C)\theta+(2B+D) \Rightarrow A=2;\ 2A+B=5 \Rightarrow B=1;\ 2A+2B+C=8 \Rightarrow C=2;$
$2B+D=4 \Rightarrow D=2;\ \int \frac{2\theta^3+5\theta^2+8\theta+4}{(\theta^2+2\theta+2)^2}\,d\theta = \int \frac{2\theta+1}{(\theta^2+2\theta+2)}\,d\theta + \int \frac{2\theta+2}{(\theta^2+2\theta+2)^2}\,d\theta$
$= \int \frac{2\theta+2}{\theta^2+2\theta+2}\,d\theta - \int \frac{d\theta}{\theta^2+2\theta+2} + \int \frac{d(\theta^2+2\theta+2)}{(\theta^2+2\theta+2)^2} = \int \frac{d(\theta^2+2\theta+2)}{\theta^2+2\theta+2} - \int \frac{d\theta}{(\theta+1)^2+1} - \frac{1}{\theta^2+2\theta+2}$
$= \frac{-1}{\theta^2+2\theta+2} + \ln(\theta^2+2\theta+2) - \tan^{-1}(\theta+1) + C$

33. $\frac{2x^3-2x^2+1}{x^2-x} = 2x + \frac{1}{x^2-x} = 2x + \frac{1}{x(x-1)};\ \frac{1}{x(x-1)} = \frac{A}{x} + \frac{B}{x-1} \Rightarrow 1 = A(x-1)+Bx;\ x=0 \Rightarrow A=-1;$
$x=1 \Rightarrow B=1;\ \int \frac{2x^3-2x^2+1}{x^2-x} = \int 2x\,dx - \int \frac{dx}{x} + \int \frac{dx}{x-1} = x^2 - \ln|x| + \ln|x-1| + C = x^2 + \ln\left|\frac{x-1}{x}\right| + C$

35. $\frac{9x^3-3x+1}{x^3-x^2} = 9 + \frac{9x^2-3x+1}{x^2(x-1)}$ (after long division); $\frac{9x^2-3x+1}{x^2(x-1)} = \frac{A}{x} + \frac{B}{x^2} + \frac{C}{x-1}$
$\Rightarrow 9x^2-3x+1 = Ax(x-1)+B(x-1)+Cx^2;\ x=1 \Rightarrow C=7;\ x=0 \Rightarrow B=-1;\ A+C=9 \Rightarrow A=2;$
$\int \frac{9x^3-3x+1}{x^3-x^2}\,dx = \int 9\,dx + 2\int \frac{dx}{x} - \int \frac{dx}{x^2} + 7\int \frac{dx}{x-1} = 9x + 2\ln|x| + \frac{1}{x} + 7\ln|x-1| + C$

37. $\frac{y^4+y^2-1}{y^3+y} = y - \frac{1}{y(y^2+1)};\ \frac{1}{y(y^2+1)} = \frac{A}{y} + \frac{By+C}{y^2+1} \Rightarrow 1 = A(y^2+1)+(By+C)y = (A+B)y^2+Cy+A$
$7 \Rightarrow A=1;\ A+B=0 \Rightarrow B=-1;\ C=0;\ \int \frac{y^4+y^2-1}{y^3+y}\,dy = \int y\,dy - \int \frac{dy}{y} + \int \frac{y\,dy}{y^2+1}$
$= \frac{y^2}{2} - \ln|y| + \frac{1}{2}\ln(1+y^2) + C$

39. $\int \frac{e^t\,dt}{e^{2t}+3e^t+2} = [e^t = y] \int \frac{dy}{y^2+3y+2} = \int \frac{dy}{y+1} - \int \frac{dy}{y+2} = \ln\left|\frac{y+1}{y+2}\right| + C = \ln\left(\frac{e^t+1}{e^t+2}\right) + C$

41. $\int \frac{\cos y\,dy}{\sin^2 y + \sin y - 6}$; $[\sin y = t,\ \cos y\,dy = dt] \rightarrow \int \frac{dy}{t^2+t-6} = \frac{1}{5}\int\left(\frac{1}{t-2} - \frac{1}{t+3}\right)dt = \frac{1}{5}\ln\left|\frac{t-2}{t+3}\right| + C$
$= \frac{1}{5}\ln\left|\frac{\sin y - 2}{\sin y + 3}\right| + C$

43. $\int \frac{(x-2)^2\tan^{-1}(2x) - 12x^3 - 3x}{(4x^2+1)(x-2)^2}\,dx = \int \frac{\tan^{-1}(2x)}{4x^2+1}\,dx - 3\int \frac{x}{(x-2)^2}\,dx$
$= \frac{1}{2}\int \tan^{-1}(2x)\,d(\tan^{-1}(2x)) - 3\int \frac{dx}{x-2} - 6\int \frac{dx}{(x-2)^2} = \frac{(\tan^{-1}2x)^2}{4} - 3\ln|x-2| + \frac{6}{x-2} + C$

45. $\int \frac{1}{x^{3/2}-\sqrt{x}}dx = \int \frac{1}{\sqrt{x}(x-1)}dx\ \left[\text{Let } u = \sqrt{x} \Rightarrow du = \frac{1}{2\sqrt{x}}dx \Rightarrow 2\,du = \frac{1}{\sqrt{x}}dx\right] \rightarrow \int \frac{2}{u^2-1}du;$
$\frac{2}{u^2-1} = \frac{A}{u+1} + \frac{B}{u-1} \Rightarrow 2 = A(u-1)+B(u+1) = (A+B)u - A + B \Rightarrow A+B=0,\ -A+B=2$

$\Rightarrow B=1 \Rightarrow A=-1;\ \int \frac{2}{u^2-1}du = \int\left(\frac{-1}{u+1} + \frac{1}{u-1}\right)du = -\int \frac{1}{u+1}du + \int \frac{1}{u-1}du = -\ln|u+1| + \ln|u-1| + C$
$= \ln\left|\frac{\sqrt{x}-1}{\sqrt{x}+1}\right| + C$

47. $\int \frac{\sqrt{x+1}}{x}\,dx\ \left[\text{Let } x+1 = u^2 \Rightarrow dx = 2u\,du\right] \rightarrow \int \frac{u}{u^2-1}\,2u\,du = \int \frac{2u^2}{u^2-1}\,du = \int\left(2 + \frac{2}{u^2-1}\right)du$
$= 2\int du + \int \frac{2}{u^2-1}du;\ \frac{2}{u^2-1} = \frac{A}{u+1} + \frac{B}{u-1} \Rightarrow 2 = A(u-1)+B(u+1) = (A+B)u - A + B \Rightarrow A+B=0,$
$-A+B=2 \Rightarrow B=1 \Rightarrow A=-1;\ 2\int du + \int \frac{2}{u^2-1}du = 2u + \int\left(\frac{-1}{u+1} + \frac{1}{u-1}\right)du = 2u - \int \frac{1}{u+1}du + \int \frac{1}{u-1}du$
$= 2u - \ln|u+1| + \ln|u-1| + C = 2\sqrt{x+1} + \ln\left|\frac{\sqrt{x+1}-1}{\sqrt{x+1}+1}\right| + C$

49. $\int \frac{1}{x(x^4+1)}\,dx = \int \frac{x^3}{x^4(x^4+1)}\,dx \left[\text{Let } u = x^4 \Rightarrow du = 4x^3\,dx\right] \to \frac{1}{4}\int \frac{1}{u(u+1)}\,du;\ \frac{1}{u(u+1)} = \frac{A}{u} + \frac{B}{u+1}$
$\Rightarrow 1 = A(u+1) + Bu = (A+B)u + A \Rightarrow A = 1 \Rightarrow B = -1;\ \frac{1}{4}\int \frac{1}{u(u+1)}\,du = \frac{1}{4}\int \left(\frac{1}{u} - \frac{1}{u+1}\right)du$
$= \frac{1}{4}\int \frac{1}{u}\,du - \frac{1}{4}\int \frac{1}{u+1}\,du = \frac{1}{4}\ln|u| - \frac{1}{4}\ln|u+1| + C = \frac{1}{4}\ln\left(\frac{x^4}{x^4+1}\right) + C$

51. $(t^2 - 3t + 2)\frac{dx}{dt} = 1;\ x = \int \frac{dt}{t^2-3t+2} = \int \frac{dt}{t-2} - \int \frac{dt}{t-1} = \ln\left|\frac{t-2}{t-1}\right| + C;\ \frac{t-2}{t-1} = Ce^x;\ t = 3 \text{ and } x = 0$
$\Rightarrow \frac{1}{2} = C \Rightarrow \frac{t-2}{t-1} = \frac{1}{2}e^x \Rightarrow x = \ln\left|2\left(\frac{t-2}{t-1}\right)\right| = \ln|t-2| - \ln|t-1| + \ln 2$

53. $(t^2 + 2t)\frac{dx}{dt} = 2x + 2;\ \frac{1}{2}\int \frac{dx}{x+1} = \int \frac{dt}{t^2+2t} \Rightarrow \frac{1}{2}\ln|x+1| = \frac{1}{2}\int \frac{dt}{t} - \frac{1}{2}\int \frac{dt}{t+2} \Rightarrow \ln|x+1| = \ln\left|\frac{t}{t+2}\right| + C;$
$t = 1 \text{ and } x = 1 \Rightarrow \ln 2 = \ln\frac{1}{3} + C \Rightarrow C = \ln 2 + \ln 3 = \ln 6 \Rightarrow \ln|x+1| = \ln 6\left|\frac{t}{t+2}\right| \Rightarrow x + 1 = \frac{6t}{t+2}$
$\Rightarrow x = \frac{6t}{t+2} - 1,\ t > 0$

55. $V = \pi\int_{0.5}^{2.5} y^2\,dx = \pi\int_{0.5}^{2.5} \frac{9}{3x-x^2}\,dx = 3\pi\left(\int_{0.5}^{2.5}\left(-\frac{1}{x-3} + \frac{1}{x}\right)\right)dx = \left[3\pi\ln\left|\frac{x}{x-3}\right|\right]_{0.5}^{2.5} = 3\pi\ln 25$

57. $A = \int_0^{\sqrt{3}} \tan^{-1}x\,dx = \left[x\tan^{-1}x\right]_0^{\sqrt{3}} - \int_0^{\sqrt{3}} \frac{x}{1+x^2}\,dx$
$= \frac{\pi\sqrt{3}}{3} - \left[\frac{1}{2}\ln(x^2+1)\right]_0^{\sqrt{3}} = \frac{\pi\sqrt{3}}{3} - \ln 2;$
$\overline{x} = \frac{1}{A}\int_0^{\sqrt{3}} x\tan^{-1}x\,dx$
$= \frac{1}{A}\left(\left[\frac{1}{2}x^2\tan^{-1}x\right]_0^{\sqrt{3}} - \frac{1}{2}\int_0^{\sqrt{3}} \frac{x^2}{1+x^2}\,dx\right)$
$= \frac{1}{A}\left[\frac{\pi}{2} - \left[\frac{1}{2}(x - \tan^{-1}x)\right]_0^{\sqrt{3}}\right]$
$= \frac{1}{A}\left(\frac{\pi}{2} - \frac{\sqrt{3}}{2} + \frac{\pi}{6}\right) = \frac{1}{A}\left(\frac{2\pi}{3} - \frac{\sqrt{3}}{2}\right) \cong 1.10$

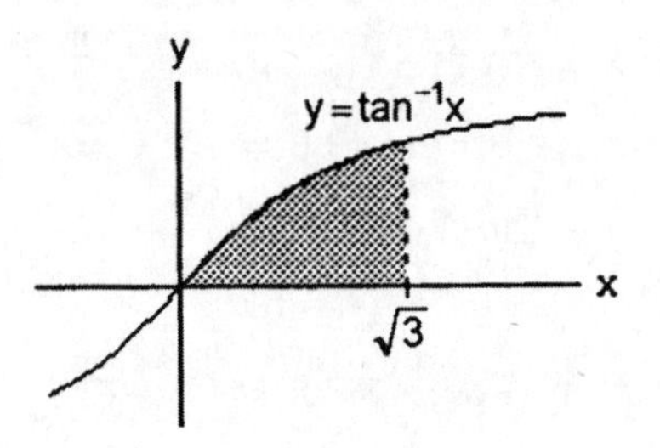

59. (a) $\frac{dx}{dt} = kx(N-x) \Rightarrow \int \frac{dx}{x(N-x)} = \int k\,dt \Rightarrow \frac{1}{N}\int \frac{dx}{x} + \frac{1}{N}\int \frac{dx}{N-x} = \int k\,dt \Rightarrow \frac{1}{N}\ln\left|\frac{x}{N-x}\right| = kt + C;$
$k = \frac{1}{250},\ N = 1000,\ t = 0 \text{ and } x = 2 \Rightarrow \frac{1}{1000}\ln\left|\frac{2}{998}\right| = C \Rightarrow \frac{1}{1000}\ln\left|\frac{x}{1000-x}\right| = \frac{t}{250} + \frac{1}{1000}\ln\left(\frac{1}{499}\right)$
$\Rightarrow \ln\left|\frac{499x}{1000-x}\right| = 4t \Rightarrow \frac{499x}{1000-x} = e^{4t} \Rightarrow 499x = e^{4t}(1000-x) \Rightarrow (499 + e^{4t})\,x = 1000e^{4t} \Rightarrow x = \frac{1000e^{4t}}{499+e^{4t}}$
(b) $x = \frac{1}{2}N = 500 \Rightarrow 500 = \frac{1000e^{4t}}{499+e^{4t}} \Rightarrow 500 \cdot 499 + 500e^{4t} = 1000e^{4t} \Rightarrow e^{4t} = 499 \Rightarrow t = \frac{1}{4}\ln 499 \approx 1.55$ days

8.5 INTEGRAL TABLES AND COMPUTER ALGEBRA SYSTEMS

1. $\int \frac{dx}{x\sqrt{x-3}} = \frac{2}{\sqrt{3}}\tan^{-1}\sqrt{\frac{x-3}{3}} + C$
(We used FORMULA 13(a) with a = 1, b = 3)

3. $\int \frac{x\,dx}{\sqrt{x-2}} = \int \frac{(x-2)\,dx}{\sqrt{x-2}} + 2\int \frac{dx}{\sqrt{x-2}} = \int \left(\sqrt{x-2}\right)^1 dx + 2\int \left(\sqrt{x-2}\right)^{-1} dx$
$= \left(\frac{2}{1}\right)\frac{\left(\sqrt{x-2}\right)^3}{3} + 2\left(\frac{2}{1}\right)\frac{\left(\sqrt{x-2}\right)^1}{1} = \sqrt{x-2}\left[\frac{2(x-2)}{3} + 4\right] + C$
(We used FORMULA 11 with a = 1, b = −2, n = 1 and a = 1, b = −2, n = −1)

5. $\int x\sqrt{2x-3}\,dx = \frac{1}{2}\int (2x-3)\sqrt{2x-3}\,dx + \frac{3}{2}\int \sqrt{2x-3}\,dx = \frac{1}{2}\int \left(\sqrt{2x-3}\right)^3 dx + \frac{3}{2}\int \left(\sqrt{2x-3}\right)^1 dx$
$= \left(\frac{1}{2}\right)\left(\frac{2}{2}\right)\frac{\left(\sqrt{2x-3}\right)^5}{5} + \left(\frac{3}{2}\right)\left(\frac{2}{2}\right)\frac{\left(\sqrt{2x-3}\right)^3}{3} + C = \frac{(2x-3)^{3/2}}{2}\left[\frac{2x-3}{5} + 1\right] + C = \frac{(2x-3)^{3/2}(x+1)}{5} + C$
(We used FORMULA 11 with a = 2, b = −3, n = 3 and a = 2, b = −3, n = 1)

7. $\int \frac{\sqrt{9-4x}}{x^2}\,dx = -\frac{\sqrt{9-4x}}{x} + \frac{(-4)}{2}\int \frac{dx}{x\sqrt{9-4x}} + C$
(We used FORMULA 14 with a = −4, b = 9)
$= -\frac{\sqrt{9-4x}}{x} - 2\left(\frac{1}{\sqrt{9}}\right)\ln\left|\frac{\sqrt{9-4x}-\sqrt{9}}{\sqrt{9-4x}+\sqrt{9}}\right| + C$
(We used FORMULA 13(b) with a = −4, b = 9)
$= \frac{-\sqrt{9-4x}}{x} - \frac{2}{3}\ln\left|\frac{\sqrt{9-4x}-3}{\sqrt{9-4x}+3}\right| + C$

9. $\int x\sqrt{4x-x^2}\,dx = \int x\sqrt{2\cdot 2x - x^2}\,dx = \frac{(x+2)(2x-3\cdot 2)\sqrt{2\cdot 2\cdot x - x^2}}{6} + \frac{2^3}{2}\sin^{-1}\left(\frac{x-2}{2}\right) + C$
$= \frac{(x+2)(2x-6)\sqrt{4x-x^2}}{6} + 4\sin^{-1}\left(\frac{x-2}{2}\right) + C = \frac{(x+2)(x-3)\sqrt{4x-x^2}}{3} + 4\sin^{-1}\left(\frac{x-2}{2}\right) + C$
(We used FORMULA 51 with a = 2)

11. $\int \frac{dx}{x\sqrt{7+x^2}} = \int \frac{dx}{x\sqrt{\left(\sqrt{7}\right)^2 + x^2}} = -\frac{1}{\sqrt{7}}\ln\left|\frac{\sqrt{7}+\sqrt{\left(\sqrt{7}\right)^2+x^2}}{x}\right| + C = -\frac{1}{\sqrt{7}}\ln\left|\frac{\sqrt{7}+\sqrt{7+x^2}}{x}\right| + C$
(We used FORMULA 26 with $a = \sqrt{7}$)

13. $\int \frac{\sqrt{4-x^2}}{x}\,dx = \int \frac{\sqrt{2^2-x^2}}{x}\,dx = \sqrt{2^2-x^2} - 2\ln\left|\frac{2+\sqrt{2^2-x^2}}{x}\right| + C = \sqrt{4-x^2} - 2\ln\left|\frac{2+\sqrt{4-x^2}}{x}\right| + C$
(We used FORMULA 31 with a = 2)

15. $\int e^{2t}\cos 3t\,dt = \frac{e^{2t}}{2^2+3^2}(2\cos 3t + 3\sin 3t) + C = \frac{e^{2t}}{13}(2\cos 3t + 3\sin 3t) + C$
(We used FORMULA 108 with a = 2, b = 3)

17. $\int x\cos^{-1}x\,dx = \int x^1\cos^{-1}x\,dx = \frac{x^{1+1}}{1+1}\cos^{-1}x + \frac{1}{1+1}\int \frac{x^{1+1}\,dx}{\sqrt{1-x^2}} = \frac{x^2}{2}\cos^{-1}x + \frac{1}{2}\int \frac{x^2\,dx}{\sqrt{1-x^2}}$
(We used FORMULA 100 with a = 1, n = 1)
$= \frac{x^2}{2}\cos^{-1}x + \frac{1}{2}\left(\frac{1}{2}\sin^{-1}x\right) - \frac{1}{2}\left(\frac{1}{2}x\sqrt{1-x^2}\right) + C = \frac{x^2}{2}\cos^{-1}x + \frac{1}{4}\sin^{-1}x - \frac{1}{4}x\sqrt{1-x^2} + C$
(We used FORMULA 33 with a = 1)

19. $\int x^2\tan^{-1}x\,dx = \frac{x^{2+1}}{2+1}\tan^{-1}x - \frac{1}{2+1}\int \frac{x^{2+1}}{1+x^2}\,dx = \frac{x^3}{3}\tan^{-1}x - \frac{1}{3}\int \frac{x^3}{1+x^2}\,dx$
(We used FORMULA 101 with a = 1, n = 2);
$\int \frac{x^3}{1+x^2}\,dx = \int x\,dx - \int \frac{x\,dx}{1+x^2} = \frac{x^2}{2} - \frac{1}{2}\ln(1+x^2) + C \Rightarrow \int x^2\tan^{-1}x\,dx$
$= \frac{x^3}{3}\tan^{-1}x - \frac{x^2}{6} + \frac{1}{6}\ln(1+x^2) + C$

21. $\int \sin 3x\cos 2x\,dx = -\frac{\cos 5x}{10} - \frac{\cos x}{2} + C$
(We used FORMULA 62(a) with a = 3, b = 2)

23. $\int 8\sin 4t\sin\frac{t}{2}\,dx = \frac{8}{7}\sin\left(\frac{7t}{2}\right) - \frac{8}{9}\sin\left(\frac{9t}{2}\right) + C = 8\left[\frac{\sin\left(\frac{7t}{2}\right)}{7} - \frac{\sin\left(\frac{9t}{2}\right)}{9}\right] + C$
(We used FORMULA 62(b) with a = 4, $b = \frac{1}{2}$)

25. $\int \cos\frac{\theta}{3}\cos\frac{\theta}{4}\,d\theta = 6\sin\left(\frac{\theta}{12}\right) + \frac{6}{7}\sin\left(\frac{7\theta}{12}\right) + C$
(We used FORMULA 62(c) with $a = \frac{1}{3}$, $b = \frac{1}{4}$)

27. $\int \frac{x^3+x+1}{(x^2+1)^2}\,dx = \int \frac{x\,dx}{x^2+1} + \int \frac{dx}{(x^2+1)^2} = \frac{1}{2}\int \frac{d(x^2+1)}{x^2+1} + \int \frac{dx}{(x^2+1)^2}$
$= \frac{1}{2}\ln(x^2+1) + \frac{x}{2(1+x^2)} + \frac{1}{2}\tan^{-1}x + C$
(For the second integral we used FORMULA 17 with $a = 1$)

29. $\int \sin^{-1}\sqrt{x}\,dx;\ \begin{bmatrix} u = \sqrt{x} \\ x = u^2 \\ dx = 2u\,du \end{bmatrix} \to 2\int u^1 \sin^{-1}u\,du = 2\left(\frac{u^{1+1}}{1+1}\sin^{-1}u - \frac{1}{1+1}\int \frac{u^{1+1}}{\sqrt{1-u^2}}\,du\right)$
$= u^2\sin^{-1}u - \int \frac{u^2\,du}{\sqrt{1-u^2}}$
(We used FORMULA 99 with $a = 1, n = 1$)
$= u^2\sin^{-1}u - \left(\frac{1}{2}\sin^{-1}u - \frac{1}{2}u\sqrt{1-u^2}\right) + C = \left(u^2 - \frac{1}{2}\right)\sin^{-1}u + \frac{1}{2}u\sqrt{1-u^2} + C$
(We used FORMULA 33 with $a = 1$)
$= \left(x - \frac{1}{2}\right)\sin^{-1}\sqrt{x} + \frac{1}{2}\sqrt{x - x^2} + C$

31. $\int \frac{\sqrt{x}}{\sqrt{1-x}}\,dx;\ \begin{bmatrix} u = \sqrt{x} \\ x = u^2 \\ dx = 2u\,du \end{bmatrix} \to \int \frac{u \cdot 2u}{\sqrt{1-u^2}}\,du = 2\int \frac{u^2}{\sqrt{1-u^2}}\,du = 2\left(\frac{1}{2}\sin^{-1}u - \frac{1}{2}u\sqrt{1-u^2}\right) + C$
$= \sin^{-1}u - u\sqrt{1-u^2} + C$
(We used FORMULA 33 with $a = 1$)
$= \sin^{-1}\sqrt{x} - \sqrt{x}\sqrt{1-x} + C = \sin^{-1}\sqrt{x} - \sqrt{x - x^2} + C$

33. $\int (\cot t)\sqrt{1 - \sin^2 t}\,dt = \int \frac{\sqrt{1-\sin^2 t}\,(\cos t)\,dt}{\sin t};\ \begin{bmatrix} u = \sin t \\ du = \cos t\,dt \end{bmatrix} \to \int \frac{\sqrt{1-u^2}\,du}{u}$
$= \sqrt{1-u^2} - \ln\left|\frac{1+\sqrt{1-u^2}}{u}\right| + C$
(We used FORMULA 31 with $a = 1$)
$= \sqrt{1-\sin^2 t} - \ln\left|\frac{1+\sqrt{1-\sin^2 t}}{\sin t}\right| + C$

35. $\int \frac{dy}{y\sqrt{3+(\ln y)^2}};\ \begin{bmatrix} u = \ln y \\ y = e^u \\ dy = e^u\,du \end{bmatrix} \to \int \frac{e^u\,du}{e^u\sqrt{3+u^2}} = \int \frac{du}{\sqrt{3+u^2}} = \ln\left|u + \sqrt{3+u^2}\right| + C$
$= \ln\left|\ln y + \sqrt{3+(\ln y)^2}\right| + C$
(We used FORMULA 20 with $a = \sqrt{3}$)

37. $\int \frac{1}{\sqrt{x^2+2x+5}}dx = \int \frac{1}{\sqrt{(x+1)^2+4}}dx;\ \begin{bmatrix} t = x+1 \\ dt = dx \end{bmatrix} \to \int \frac{1}{\sqrt{t^2+4}}dt$
(We used FORMULA 20 with $a = 2$)
$= \ln\left|t + \sqrt{t^2+4}\right| + C = \ln\left|(x+1) + \sqrt{(x+1)^2+4} + C = \ln\right|(x+1) + \sqrt{x^2+2x+5}\right| + C$

39. $\int \sqrt{5\ \ 4x\ \ x^2}dx = \int \sqrt{9-(x+2)^2}dx;\ \begin{bmatrix} t = x+2 \\ dt = dx \end{bmatrix} \to \int \sqrt{9-t^2}dt;$
(We used FORMULA 29 with $a = 3$)
$= \frac{t}{2}\sqrt{9-t^2} + \frac{3^2}{2}\sin^{-1}\left(\frac{t}{3}\right) + C = \frac{x+2}{2}\sqrt{9-(x+2)^2} + \frac{9}{2}\sin^{-1}\left(\frac{x+2}{3}\right) + C = \frac{x+2}{2}\sqrt{5-4x-x^2} + \frac{9}{2}\sin^{-1}\left(\frac{x+2}{3}\right) + C$

41. $\int \sin^5 2x\, dx = -\frac{\sin^4 2x \cos 2x}{5\cdot 2} + \frac{5-1}{5}\int \sin^3 2x\, dx = -\frac{\sin^4 2x \cos 2x}{10} + \frac{4}{5}\left[-\frac{\sin^2 2x \cos 2x}{3\cdot 2} + \frac{3-1}{3}\int \sin 2x\, dx\right]$

(We used FORMULA 60 with a = 2, n = 5 and a = 2, n = 3)

$= -\frac{\sin^4 2x \cos 2x}{10} - \frac{2}{15}\sin^2 2x \cos 2x + \frac{8}{15}\left(-\frac{1}{2}\right)\cos 2x + C = -\frac{\sin^4 2x \cos 2x}{10} - \frac{2\sin^2 2x \cos 2x}{15} - \frac{4\cos 2x}{15} + C$

43. $\int \sin^2 2\theta \cos^3 2\theta\, d\theta = \frac{\sin^3 2\theta \cos^2 2\theta}{2(2+3)} + \frac{3-1}{3+2}\int \sin^2 2\theta \cos 2\theta\, d\theta$

(We used FORMULA 69 with a = 2, m = 3, n = 2)

$= \frac{\sin^3 2\theta \cos^2 2\theta}{10} + \frac{2}{5}\int \sin^2 2\theta \cos 2\theta\, d\theta = \frac{\sin^3 2\theta \cos^2 2\theta}{10} + \frac{2}{5}\left[\frac{1}{2}\int \sin^2 2\theta\, d(\sin 2\theta)\right] = \frac{\sin^3 2\theta \cos^2 2\theta}{10} + \frac{\sin^3 2\theta}{15} + C$

45. $\int 4\tan^3 2x\, dx = 4\left(\frac{\tan^2 2x}{2\cdot 2} - \int \tan 2x\, dx\right) = \tan^2 2x - 4\int \tan 2x\, dx$

(We used FORMULA 86 with n = 3, a = 2)

$= \tan^2 2x - \frac{4}{2}\ln|\sec 2x| + C = \tan^2 2x - 2\ln|\sec 2x| + C$

47. $\int 2\sec^3 \pi x\, dx = 2\left[\frac{\sec \pi x \tan \pi x}{\pi(3-1)} + \frac{3-2}{3-1}\int \sec \pi x\, dx\right]$

(We used FORMULA 92 with n = 3, a = π)

$= \frac{1}{\pi}\sec \pi x \tan \pi x + \frac{1}{\pi}\ln|\sec \pi x + \tan \pi x| + C$

(We used FORMULA 88 with a = π)

49. $\int \csc^5 x\, dx = -\frac{\csc^3 x \cot x}{5-1} + \frac{5-2}{5-1}\int \csc^3 x\, dx = -\frac{\csc^3 x \cot x}{4} + \frac{3}{4}\left(-\frac{\csc x \cot x}{3-1} + \frac{3-2}{3-1}\int \csc x\, dx\right)$

(We used FORMULA 93 with n = 5, a = 1 and n = 3, a = 1)

$= -\frac{1}{4}\csc^3 x \cot x - \frac{3}{8}\csc x \cot x - \frac{3}{8}\ln|\csc x + \cot x| + C$

(We used FORMULA 89 with a = 1)

51. $\int e^t \sec^3(e^t - 1)\, dt;\ \begin{bmatrix} x = e^t - 1 \\ dx = e^t\, dt \end{bmatrix} \rightarrow \int \sec^3 x\, dx = \frac{\sec x \tan x}{3-1} + \frac{3-2}{3-1}\int \sec x\, dx$

(We used FORMULA 92 with a = 1, n = 3)

$= \frac{\sec x \tan x}{2} + \frac{1}{2}\ln|\sec x + \tan x| + C = \frac{1}{2}\left[\sec(e^t - 1)\tan(e^t - 1) + \ln|\sec(e^t - 1) + \tan(e^t - 1)|\right] + C$

53. $\int_0^1 2\sqrt{x^2+1}\, dx;\ [x = \tan t] \rightarrow 2\int_0^{\pi/4} \sec t \cdot \sec^2 t\, dt = 2\int_0^{\pi/4} \sec^3 t\, dt = 2\left[\left[\frac{\sec t \cdot \tan t}{3-1}\right]_0^{\pi/4} + \frac{3-2}{3-1}\int_0^{\pi/4} \sec t\, dt\right]$

(We used FORMULA 92 with n = 3, a = 1)

$= [\sec t \cdot \tan t + \ln|\sec t + \tan t|]_0^{\pi/4} = \sqrt{2} + \ln\left(\sqrt{2} + 1\right)$

55. $\int_1^2 \frac{(r^2-1)^{3/2}}{r}\, dr;\ [r = \sec\theta] \rightarrow \int_0^{\pi/3} \frac{\tan^3\theta}{\sec\theta}(\sec\theta \tan\theta)\, d\theta = \int_0^{\pi/3} \tan^4\theta\, d\theta = \left[\frac{\tan^3\theta}{4-1}\right]_0^{\pi/3} - \int_0^{\pi/3} \tan^2\theta\, d\theta$

$= \left[\frac{\tan^3\theta}{3} - \tan\theta + \theta\right]_0^{\pi/3} = \frac{3\sqrt{3}}{3} - \sqrt{3} + \frac{\pi}{3} = \frac{\pi}{3}$

(We used FORMULA 86 with a = 1, n = 4 and FORMULA 84 with a = 1)

57. $S = \int_0^{\sqrt{2}} 2\pi y\sqrt{1 + (y')^2}\, dx$

$= 2\pi \int_0^{\sqrt{2}} \sqrt{x^2+2}\sqrt{1 + \frac{x^2}{x^2+2}}\, dx$

$= 2\sqrt{2}\pi \int_0^{\sqrt{2}} \sqrt{x^2+1}\, dx$

$= 2\sqrt{2}\pi \left[\frac{x\sqrt{x^2+1}}{2} + \frac{1}{2}\ln\left|x + \sqrt{x^2+1}\right|\right]_0^{\sqrt{2}}$

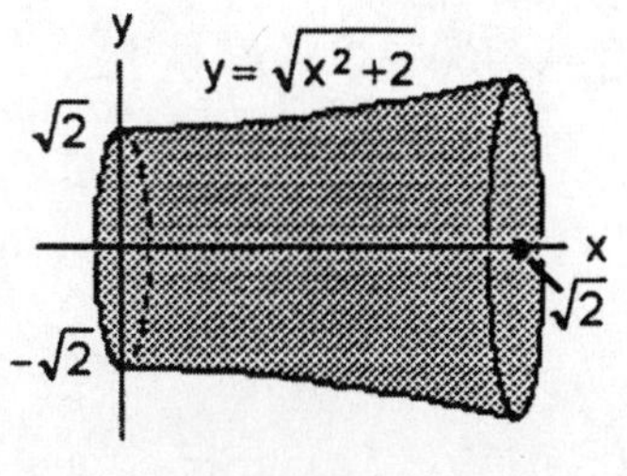

(We used FORMULA 21 with a = 1)

$= \sqrt{2}\pi\left[\sqrt{6} + \ln\left(\sqrt{2} + \sqrt{3}\right)\right] = 2\pi\sqrt{3} + \pi\sqrt{2}\ln\left(\sqrt{2} + \sqrt{3}\right)$

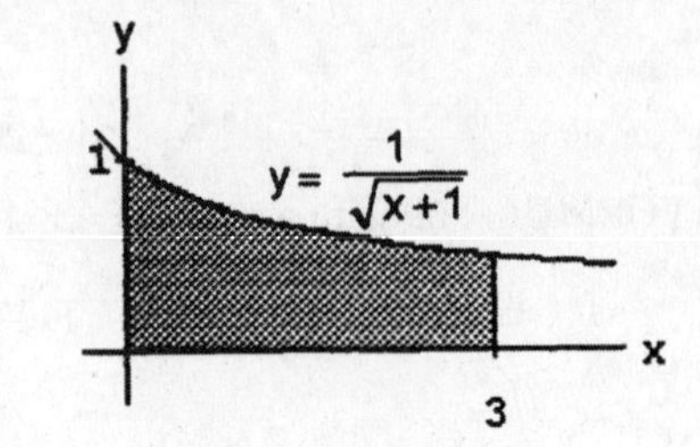

59. $A = \int_0^3 \frac{dx}{\sqrt{x+1}} = \left[2\sqrt{x+1}\right]_0^3 = 2;\ \bar{x} = \frac{1}{A}\int_0^3 \frac{x\,dx}{\sqrt{x+1}}$

$= \frac{1}{A}\int_0^3 \sqrt{x+1}\,dx - \frac{1}{A}\int_0^3 \frac{dx}{\sqrt{x+1}}$

$= \frac{1}{2}\cdot\frac{2}{3}\left[(x+1)^{3/2}\right]_0^3 - 1 = \frac{4}{3};$

(We used FORMULA 11 with a = 1, b = 1, n = 1 and a = 1, b = 1, n = −1)

$\bar{y} = \frac{1}{2A}\int_0^3 \frac{dx}{x+1} = \frac{1}{4}\left[\ln(x+1)\right]_0^3 = \frac{1}{4}\ln 4 = \frac{1}{2}\ln 2 = \ln\sqrt{2}$

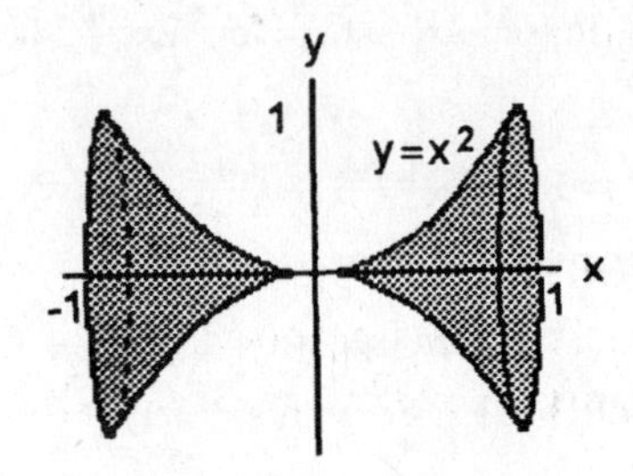

61. $S = 2\pi\int_{-1}^{1} x^2\sqrt{1+4x^2}\,dx;$

$\begin{bmatrix} u = 2x \\ du = 2\,dx \end{bmatrix} \rightarrow \frac{\pi}{4}\int_{-2}^{2} u^2\sqrt{1+u^2}\,du$

$= \frac{\pi}{4}\left[\frac{u}{8}(1+2u^2)\sqrt{1+u^2} - \frac{1}{8}\ln\left(u + \sqrt{1+u^2}\right)\right]_{-2}^{2}$

(We used FORMULA 22 with a = 1)

$= \frac{\pi}{4}\left[\frac{2}{8}(1+2\cdot 4)\sqrt{1+4} - \frac{1}{8}\ln\left(2+\sqrt{1+4}\right) + \frac{2}{8}(1+2\cdot 4)\sqrt{1+4} + \frac{1}{8}\ln\left(-2+\sqrt{1+4}\right)\right]$

$= \frac{\pi}{4}\left[\frac{9}{2}\sqrt{5} - \frac{1}{8}\ln\left(\frac{2+\sqrt{5}}{-2+\sqrt{5}}\right)\right] \approx 7.62$

63. The integrand $f(x) = \sqrt{x - x^2}$ is nonnegative, so the integral is maximized by integrating over the function's entire domain, which runs from x = 0 to x = 1

$\Rightarrow \int_0^1 \sqrt{x - x^2}\,dx = \int_0^1 \sqrt{2\cdot\frac{1}{2}x - x^2}\,dx = \left[\frac{\left(x-\frac{1}{2}\right)}{2}\sqrt{2\cdot\frac{1}{2}x - x^2} + \frac{\left(\frac{1}{2}\right)^2}{2}\sin^{-1}\left(\frac{x-\frac{1}{2}}{\frac{1}{2}}\right)\right]_0^1$

(We used FORMULA 48 with $a = \frac{1}{2}$)

$= \left[\frac{\left(x-\frac{1}{2}\right)}{2}\sqrt{x - x^2} + \frac{1}{8}\sin^{-1}(2x-1)\right]_0^1 = \frac{1}{8}\cdot\frac{\pi}{2} - \frac{1}{8}\left(-\frac{\pi}{2}\right) = \frac{\pi}{8}$

8.6 NUMERICAL INTEGRATION

1. $\int_1^2 x\,dx$

I. (a) For n = 4, $\Delta x = \frac{b-a}{n} = \frac{2-1}{4} = \frac{1}{4} \Rightarrow \frac{\Delta x}{2} = \frac{1}{8};$

$\sum mf(x_i) = 12 \Rightarrow T = \frac{1}{8}(12) = \frac{3}{2};$

$f(x) = x \Rightarrow f'(x) = 1 \Rightarrow f'' = 0 \Rightarrow M = 0$

$\Rightarrow |E_T| = 0$

	x_i	$f(x_i)$	m	$mf(x_i)$
x_0	1	1	1	1
x_1	5/4	5/4	2	5/2
x_2	3/2	3/2	2	3
x_3	7/4	7/4	2	7/2
x_4	2	2	1	2

(b) $\int_1^2 x\,dx = \left[\frac{x^2}{2}\right]_1^2 = 2 - \frac{1}{2} = \frac{3}{2} \Rightarrow |E_T| = \int_1^2 x\,dx - T = 0$

(c) $\frac{|E_T|}{\text{True Value}} \times 100 - 0\%$

II. (a) For $n = 4$, $\Delta x = \frac{b-a}{n} = \frac{2-1}{4} = \frac{1}{4} \Rightarrow \frac{\Delta x}{3} = \frac{1}{12}$;
$\sum mf(x_i) = 18 \Rightarrow S = \frac{1}{12}(18) = \frac{3}{2}$;
$f^{(4)}(x) = 0 \Rightarrow M = 0 \Rightarrow |E_S| = 0$

(b) $\int_1^2 x\,dx = \frac{3}{2} \Rightarrow |E_S| = \int_1^2 x\,dx - S = \frac{3}{2} - \frac{3}{2} = 0$

(c) $\frac{|E_S|}{\text{True Value}} \times 100 = 0\%$

	x_i	$f(x_i)$	m	$mf(x_i)$
x_0	1	1	1	1
x_1	5/4	5/4	4	5
x_2	3/2	3/2	2	3
x_3	7/4	7/4	4	7
x_4	2	2	1	2

3. $\int_{-1}^{1} (x^2 + 1)\,dx$

I. (a) For $n = 4$, $\Delta x = \frac{b-a}{n} = \frac{1-(-1)}{4} = \frac{2}{4} = \frac{1}{2} \Rightarrow \frac{\Delta x}{2} = \frac{1}{4}$;
$\sum mf(x_i) = 11 \Rightarrow T = \frac{1}{4}(11) = 2.75$;
$f(x) = x^2 + 1 \Rightarrow f'(x) = 2x \Rightarrow f''(x) = 2 \Rightarrow M = 2$
$\Rightarrow |E_T| \le \frac{1-(-1)}{12}\left(\frac{1}{2}\right)^2(2) = \frac{1}{12}$ or 0.08333

	x_i	$f(x_i)$	m	$mf(x_i)$
x_0	-1	2	1	2
x_1	$-1/2$	5/4	2	5/2
x_2	0	1	2	2
x_3	1/2	5/4	2	5/2
x_4	1	2	1	2

(b) $\int_{-1}^{1} (x^2 + 1)\,dx = \left[\frac{x^3}{3} + x\right]_{-1}^{1} = \left(\frac{1}{3} + 1\right) - \left(-\frac{1}{3} - 1\right) = \frac{8}{3} \Rightarrow E_T = \int_{-1}^{1} (x^2 + 1)\,dx - T = \frac{8}{3} - \frac{11}{4} = -\frac{1}{12}$
$\Rightarrow |E_T| = \left|-\frac{1}{12}\right| \approx 0.08333$

(c) $\frac{|E_T|}{\text{True Value}} \times 100 = \left(\frac{\frac{1}{12}}{\frac{8}{3}}\right) \times 100 \approx 3\%$

II. (a) For $n = 4$, $\Delta x = \frac{b-a}{n} = \frac{1-(-1)}{4} = \frac{2}{4} = \frac{1}{2} \Rightarrow \frac{\Delta x}{3} = \frac{1}{6}$;
$\sum mf(x_i) = 16 \Rightarrow S = \frac{1}{6}(16) = \frac{8}{3} = 2.66667$;
$f^{(3)}(x) = 0 \Rightarrow f^{(4)}(x) = 0 \Rightarrow M = 0 \Rightarrow |E_S| = 0$

(b) $\int_{-1}^{1} (x^2 + 1)\,dx = \left[\frac{x^3}{3} + x\right]_{-1}^{1} = \frac{8}{3}$
$\Rightarrow |E_S| = \int_{-1}^{1} (x^2 + 1)\,dx - S = \frac{8}{3} - \frac{8}{3} = 0$

(c) $\frac{|E_S|}{\text{True Value}} \times 100 = 0\%$

	x_i	$f(x_i)$	m	$mf(x_i)$
x_0	-1	2	1	2
x_1	$-1/2$	5/4	4	5
x_2	0	1	2	2
x_3	1/2	5/4	4	5
x_4	1	2	1	2

5. $\int_0^2 (t^3 + t)\,dt$

I. (a) For $n = 4$, $\Delta x = \frac{b-a}{n} = \frac{2-0}{4} = \frac{2}{4} = \frac{1}{2}$
$\Rightarrow \frac{\Delta x}{2} = \frac{1}{4}$; $\sum mf(t_i) = 25 \Rightarrow T = \frac{1}{4}(25) = \frac{25}{4}$;
$f(t) = t^3 + t \Rightarrow f'(t) = 3t^2 + 1 \Rightarrow f''(t) = 6t$
$\Rightarrow M = 12 = f''(2) \Rightarrow |E_T| \le \frac{2-0}{12}\left(\frac{1}{2}\right)^2(12) = \frac{1}{2}$

	t_i	$f(t_i)$	m	$mf(t_i)$
t_0	0	0	1	0
t_1	1/2	5/8	2	5/4
t_2	1	2	2	4
t_3	3/2	39/8	2	39/4
t_4	2	10	1	10

(b) $\int_0^2 (t^3 + t)\,dt = \left[\frac{t^4}{4} + \frac{t^2}{2}\right]_0^2 = \left(\frac{2^4}{4} + \frac{2^2}{2}\right) - 0 = 6 \Rightarrow |E_T| = \int_0^2 (t^3 + t)\,dt - T = 6 - \frac{25}{4} = -\frac{1}{4} \Rightarrow |E_T| = \frac{1}{4}$

(c) $\frac{|E_T|}{\text{True Value}} \times 100 = \frac{\left|-\frac{1}{4}\right|}{6} \times 100 \approx 4\%$

II. (a) For $n = 4$, $\Delta x = \frac{b-a}{n} = \frac{2-0}{4} = \frac{2}{4} = \frac{1}{2} \Rightarrow \frac{\Delta x}{3} = \frac{1}{6}$;
$\sum mf(t_i) = 36 \Rightarrow S = \frac{1}{6}(36) = 6$;
$f^{(3)}(t) = 6 \Rightarrow f^{(4)}(t) = 0 \Rightarrow M = 0 \Rightarrow |E_S| = 0$

(b) $\int_0^2 (t^3 + t)\,dt = 6 \Rightarrow |E_S| = \int_0^2 (t^3 + t)\,dt - S$
$= 6 - 6 = 0$

(c) $\frac{|E_S|}{\text{True Value}} \times 100 = 0\%$

	t_i	$f(t_i)$	m	$mf(t_i)$
t_0	0	0	1	0
t_1	1/2	5/8	4	5/2
t_2	1	2	2	4
t_3	3/2	39/8	4	39/2
t_4	2	10	1	10

7. $\int_1^2 \frac{1}{s^2}\,ds$

	s_i	$f(s_i)$	m	$mf(s_i)$
s_0	1	1	1	1
s_1	5/4	16/25	2	32/25
s_2	3/2	4/9	2	8/9
s_3	7/4	16/49	2	32/49
s_4	2	1/4	1	1/4

I. (a) For $n = 4$, $\Delta x = \frac{b-a}{n} = \frac{2-1}{4} = \frac{1}{4} \Rightarrow \frac{\Delta x}{2} = \frac{1}{8}$;
$\sum mf(s_i) = \frac{179{,}573}{44{,}100} \Rightarrow T = \frac{1}{8}\left(\frac{179{,}573}{44{,}100}\right) = \frac{179{,}573}{352{,}800}$
≈ 0.50899; $f(s) = \frac{1}{s^2} \Rightarrow f'(s) = -\frac{2}{s^3}$
$\Rightarrow f''(s) = \frac{6}{s^4} \Rightarrow M = 6 = f''(1)$
$\Rightarrow |E_T| \le \frac{2-1}{12}\left(\frac{1}{4}\right)^2(6) = \frac{1}{32} = 0.03125$

(b) $\int_1^2 \frac{1}{s^2}\,ds = \int_1^2 s^{-2}\,ds = \left[-\frac{1}{s}\right]_1^2 = -\frac{1}{2} - \left(-\frac{1}{1}\right) = \frac{1}{2} \Rightarrow E_T = \int_1^2 \frac{1}{s^2}\,ds - T = \frac{1}{2} - 0.50899 = -0.00899$
$\Rightarrow |E_T| = 0.00899$

(c) $\frac{|E_T|}{\text{True Value}} \times 100 = \frac{0.00899}{0.5} \times 100 \approx 2\%$

	s_i	$f(s_i)$	m	$mf(s_i)$
s_0	1	1	1	1
s_1	5/4	16/25	4	64/25
s_2	3/2	4/9	2	8/9
s_3	7/4	16/49	4	64/49
s_4	2	1/4	1	1/4

II. (a) For $n = 4$, $\Delta x = \frac{b-a}{n} = \frac{2-1}{4} = \frac{1}{4} \Rightarrow \frac{\Delta x}{3} = \frac{1}{12}$;
$\sum mf(s_i) = \frac{264{,}821}{44{,}100} \Rightarrow S = \frac{1}{12}\left(\frac{264{,}821}{44{,}100}\right) = \frac{264{,}821}{529{,}200}$
≈ 0.50042; $f^{(3)}(s) = -\frac{24}{s^5} \Rightarrow f^{(4)}(s) = \frac{120}{s^6}$
$\Rightarrow M = 120 \Rightarrow |E_S| \le \left|\frac{2-1}{180}\right|\left(\frac{1}{4}\right)^4(120)$
$= \frac{1}{384} \approx 0.00260$

(b) $\int_1^2 \frac{1}{s^2}\,ds = \frac{1}{2} \Rightarrow E_S = \int_1^2 \frac{1}{s^2}\,ds - S = \frac{1}{2} - 0.50042 = -0.00042 \Rightarrow |E_S| = 0.00042$

(c) $\frac{|E_S|}{\text{True Value}} \times 100 = \frac{0.0004}{0.5} \times 100 \approx 0.08\%$

9. $\int_0^\pi \sin t\,dt$

	t_i	$f(t_i)$	m	$mf(t_i)$
t_0	0	0	1	0
t_1	$\pi/4$	$\sqrt{2}/2$	2	$\sqrt{2}$
t_2	$\pi/2$	1	2	2
t_3	$3\pi/4$	$\sqrt{2}/2$	2	$\sqrt{2}$
t_4	π	0	1	0

I. (a) For $n = 4$, $\Delta x = \frac{b-a}{n} = \frac{\pi-0}{4} = \frac{\pi}{4} \Rightarrow \frac{\Delta x}{2} = \frac{\pi}{8}$;
$\sum mf(t_i) = 2 + 2\sqrt{2} \approx 4.8284$
$\Rightarrow T = \frac{\pi}{8}\left(2 + 2\sqrt{2}\right) \approx 1.89612$;
$f(t) = \sin t \Rightarrow f'(t) = \cos t \Rightarrow f''(t) = -\sin t$
$\Rightarrow M = 1 \Rightarrow |E_T| \le \frac{\pi-0}{12}\left(\frac{\pi}{4}\right)^2(1) = \frac{\pi^3}{192}$
≈ 0.16149

(b) $\int_0^\pi \sin t\,dt = [-\cos t]_0^\pi = (-\cos \pi) - (-\cos 0) = 2 \Rightarrow |E_T| = \int_0^\pi \sin t\,dt - T \approx 2 - 1.89612 = 0.10388$

(c) $\frac{|E_T|}{\text{True Value}} \times 100 = \frac{0.10388}{2} \times 100 \approx 5\%$

	t_i	$f(t_i)$	m	$mf(t_i)$
t_0	0	0	1	0
t_1	$\pi/4$	$\sqrt{2}/2$	4	$2\sqrt{2}$
t_2	$\pi/2$	1	2	2
t_3	$3\pi/4$	$\sqrt{2}/2$	4	$2\sqrt{2}$
t_4	π	0	1	0

II. (a) For $n = 4$, $\Delta x = \frac{b-a}{n} = \frac{\pi-0}{4} = \frac{\pi}{4} \Rightarrow \frac{\Delta x}{3} = \frac{\pi}{12}$;
$\sum mf(t_i) = 2 + 4\sqrt{2} \approx 7.6569$
$\Rightarrow S = \frac{\pi}{12}\left(2 + 4\sqrt{2}\right) \approx 2.00456$;
$f^{(3)}(t) = -\cos t \Rightarrow f^{(4)}(t) = \sin t$
$\Rightarrow M = 1 \Rightarrow |E_S| \le \frac{\pi-0}{180}\left(\frac{\pi}{4}\right)^4(1) \approx 0.00664$

(b) $\int_0^\pi \sin t\,dt = 2 \Rightarrow E_S = \int_0^\pi \sin t\,dt - S \approx 2 - 2.00456 = -0.00456 \Rightarrow |E_S| \approx 0.00456$

(c) $\frac{|E_S|}{\text{True Value}} \times 100 = \frac{0.00456}{2} \times 100 \approx 0\%$

11. (a) $M = 0$ (see Exercise 1): Then $n = 1 \Rightarrow \Delta x = 1 \Rightarrow |E_T| = \frac{1}{12}(1)^2(0) = 0 < 10^{-4}$

(b) $M = 0$ (see Exercise 1): Then $n = 2$ (n must be even) $\Rightarrow \Delta x = \frac{1}{2} \Rightarrow |E_S| = \frac{1}{180}\left(\frac{1}{2}\right)^4(0) = 0 < 10^{-4}$

13. (a) $M = 2$ (see Exercise 3): Then $\Delta x = \frac{2}{n} \Rightarrow |E_T| \le \frac{2}{12}\left(\frac{2}{n}\right)^2(2) = \frac{4}{3n^2} < 10^{-4} \Rightarrow n^2 > \frac{4}{3}(10^4) \Rightarrow n > \sqrt{\frac{4}{3}(10^4)}$
$\Rightarrow n > 115.4$, so let $n = 116$

(b) $M = 0$ (see Exercise 3): Then $n = 2$ (n must be even) $\Rightarrow \Delta x = 1 \Rightarrow |E_S| = \frac{2}{180}(1)^4(0) = 0 < 10^{-4}$

15. (a) $M = 12$ (see Exercise 5): Then $\Delta x = \frac{2}{n} \Rightarrow |E_T| \le \frac{2}{12}\left(\frac{2}{n}\right)^2(12) = \frac{8}{n^2} < 10^{-4} \Rightarrow n^2 > 8\left(10^4\right) \Rightarrow n > \sqrt{8\left(10^4\right)}$
$\Rightarrow n > 282.8$, so let $n = 283$
(b) $M = 0$ (see Exercise 5): Then $n = 2$ (n must be even) $\Rightarrow \Delta x = 1 \Rightarrow |E_S| = \frac{2}{180}(1)^4(0) = 0 < 10^{-4}$

17. (a) $M = 6$ (see Exercise 7): Then $\Delta x = \frac{1}{n} \Rightarrow |E_T| \le \frac{1}{12}\left(\frac{1}{n}\right)^2(6) = \frac{1}{2n^2} < 10^{-4} \Rightarrow n^2 > \frac{1}{2}\left(10^4\right) \Rightarrow n > \sqrt{\frac{1}{2}\left(10^4\right)}$
$\Rightarrow n > 70.7$, so let $n = 71$
(b) $M = 120$ (see Exercise 7): Then $\Delta x = \frac{1}{n} \Rightarrow |E_S| = \frac{1}{180}\left(\frac{1}{n}\right)^4(120) = \frac{2}{3n^4} < 10^{-4} \Rightarrow n^4 > \frac{2}{3}\left(10^4\right)$
$\Rightarrow n > \sqrt[4]{\frac{2}{3}\left(10^4\right)} \Rightarrow n > 9.04$, so let $n = 10$ (n must be even)

19. (a) $f(x) = \sqrt{x+1} \Rightarrow f'(x) = \frac{1}{2}(x+1)^{-1/2} \Rightarrow f''(x) = -\frac{1}{4}(x+1)^{-3/2} = -\frac{1}{4\left(\sqrt{x+1}\right)^3} \Rightarrow M = \frac{1}{4\left(\sqrt{1}\right)^3} = \frac{1}{4}$.
Then $\Delta x = \frac{3}{n} \Rightarrow |E_T| \le \frac{3}{12}\left(\frac{3}{n}\right)^2\left(\frac{1}{4}\right) = \frac{9}{16n^2} < 10^{-4} \Rightarrow n^2 > \frac{9}{16}\left(10^4\right) \Rightarrow n > \sqrt{\frac{9}{16}\left(10^4\right)} \Rightarrow n > 75$,
so let $n = 76$
(b) $f^{(3)}(x) = \frac{3}{8}(x+1)^{-5/2} \Rightarrow f^{(4)}(x) = -\frac{15}{16}(x+1)^{-7/2} = -\frac{15}{16\left(\sqrt{x+1}\right)^7} \Rightarrow M = \frac{15}{16\left(\sqrt{1}\right)^7} = \frac{15}{16}$. Then $\Delta x = \frac{3}{n}$
$\Rightarrow |E_S| \le \frac{3}{180}\left(\frac{3}{n}\right)^4\left(\frac{15}{16}\right) = \frac{3^5(15)}{16(180)n^4} < 10^{-4} \Rightarrow n^4 > \frac{3^5(15)\left(10^4\right)}{16(180)} \Rightarrow n > \sqrt[4]{\frac{3^5(15)\left(10^4\right)}{16(180)}} \Rightarrow n > 10.6$, so let
$n = 12$ (n must be even)

21. (a) $f(x) = \sin(x+1) \Rightarrow f'(x) = \cos(x+1) \Rightarrow f''(x) = -\sin(x+1) \Rightarrow M = 1$. Then $\Delta x = \frac{2}{n} \Rightarrow |E_T| \le \frac{2}{12}\left(\frac{2}{n}\right)^2(1)$
$= \frac{8}{12n^2} < 10^{-4} \Rightarrow n^2 > \frac{8\left(10^4\right)}{12} \Rightarrow n > \sqrt{\frac{8\left(10^4\right)}{12}} \Rightarrow n > 81.6$, so let $n = 82$
(b) $f^{(3)}(x) = -\cos(x+1) \Rightarrow f^{(4)}(x) = \sin(x+1) \Rightarrow M = 1$. Then $\Delta x = \frac{2}{n} \Rightarrow |E_S| \le \frac{2}{180}\left(\frac{2}{n}\right)^4(1) = \frac{32}{180n^4} < 10^{-4}$
$\Rightarrow n^4 > \frac{32\left(10^4\right)}{180} \Rightarrow n > \sqrt[4]{\frac{32\left(10^4\right)}{180}} \Rightarrow n > 6.49$, so let $n = 8$ (n must be even)

23. $\frac{5}{2}(6.0 + 2(8.2) + 2(9.1)\ldots + 2(12.7) + 13.0)(30) = 15{,}990\ \text{ft}^3$.

25. Using Simpson's Rule, $\Delta x = 1 \Rightarrow \frac{\Delta x}{3} = \frac{1}{3}$;
$\sum my_i = 33.6 \Rightarrow$ Cross Section Area $\approx \frac{1}{3}(33.6)$
$= 11.2\ \text{ft}^2$. Let x be the length of the tank. Then the Volume $V =$ (Cross Sectional Area) $x = 11.2x$.
Now 5000 lb of gasoline at $42\ \text{lb/ft}^3$
$\Rightarrow V = \frac{5000}{42} = 119.05\ \text{ft}^3$
$\Rightarrow 119.05 = 11.2x \Rightarrow x \approx 10.63$ ft

	x_i	y_i	m	my_i
x_0	0	1.5	1	1.5
x_1	1	1.6	4	6.4
x_2	2	1.8	2	3.6
x_3	3	1.9	4	7.6
x_4	4	2.0	2	4.0
x_5	5	2.1	4	8.4
x_6	6	2.1	1	2.1

27. (a) $|E_S| \le \frac{b-a}{180}\left(\Delta x^4\right)M$; $n = 4 \Rightarrow \Delta x = \frac{\frac{\pi}{2} - 0}{4} = \frac{\pi}{8}$; $\left|f^{(4)}\right| \le 1 \Rightarrow M = 1 \Rightarrow |E_S| \le \frac{\left(\frac{\pi}{2} - 0\right)}{180}\left(\frac{\pi}{8}\right)^4(1) \approx 0.00021$
(b) $\Delta x = \frac{\pi}{8} \Rightarrow \frac{\Delta x}{3} = \frac{\pi}{24}$;
$\sum mf(x_i) = 10.47208705$
$\Rightarrow S = \frac{\pi}{24}(10.47208705) \approx 1.37079$

	x_i	$f(x_i)$	m	$mf(x_{1i})$
x_0	0	1	1	1
x_1	$\pi/8$	0.974495358	4	3.897981432
x_2	$\pi/4$	0.900316316	2	1.800632632
x_3	$3\pi/8$	0.784213303	4	3.136853212
x_4	$\pi/2$	0.636619772	1	0.636619772

(c) $\approx \left(\frac{0.00021}{1.37079}\right) \times 100 \approx 0.015\%$

29. $T = \frac{\Delta x}{2}(y_0 + 2y_1 + 2y_2 + 2y_3 + \ldots + 2y_{n-1} + y_n)$ where $\Delta x = \frac{b-a}{n}$ and f is continuous on [a, b]. So
$T = \frac{b-a}{n}\frac{(y_0 + y_1 + y_1 + y_2 + y_2 + \ldots + y_{n-1} + y_{n-1} + y_n)}{2} = \frac{b-a}{n}\left(\frac{f(x_0)+f(x_1)}{2} + \frac{f(x_1)+f(x_2)}{2} + \ldots + \frac{f(x_{n-1})+f(x_n)}{2}\right)$.
Since f is continuous on each interval $[x_{k-1}, x_k]$, and $\frac{f(x_{k-1})+f(x_k)}{2}$ is always between $f(x_{k-1})$ and $f(x_k)$, there is a point c_k in $[x_{k-1}, x_k]$ with $f(c_k) = \frac{f(x_{k-1})+f(x_k)}{2}$; this is a consequence of the Intermediate Value Theorem. Thus our sum is $\sum_{k=1}^{n}\left(\frac{b-a}{n}\right)f(c_k)$ which has the form $\sum_{k=1}^{n}\Delta x_k f(c_k)$ with $\Delta x_k = \frac{b-a}{n}$ for all k. This is a Riemann Sum for f on [a, b].

31. (a) $a = 1, e = \frac{1}{2} \Rightarrow \text{Length} = 4\int_0^{\pi/2}\sqrt{1 - \frac{1}{4}\cos^2 t}\,dt$
$= 2\int_0^{\pi/2}\sqrt{4 - \cos^2 t}\,dt = \int_0^{\pi/2} f(t)\,dt$; use the Trapezoid Rule with $n = 10 \Rightarrow \Delta t = \frac{b-a}{n} = \frac{\left(\frac{\pi}{2}\right)-0}{10}$
$= \frac{\pi}{20}.\ \int_0^{\pi/2}\sqrt{4-\cos^2 t}\,dt \approx \sum_{n=0}^{10} mf(x_n) = 37.3686183$
$\Rightarrow T = \frac{\Delta t}{2}(37.3686183) = \frac{\pi}{40}(37.3686183)$
$= 2.934924419 \Rightarrow \text{Length} = 2(2.934924419)$
≈ 5.870

	x_i	$f(x_i)$	m	$mf(x_i)$
x_0	0	1.732050808	1	1.732050808
x_1	$\pi/20$	1.739100843	2	3.478201686
x_2	$\pi/10$	1.759400893	2	3.518801786
x_3	$3\pi/20$	1.790560631	2	3.581121262
x_4	$\pi/5$	1.82906848	1	3.658136959
x_5	$\pi/4$	1.870828693	1	3.741657387
x_6	$3\pi/10$	1.911676881	2	3.823353762
x_7	$7\pi/20$	1.947791731	2	3.895583461
x_8	$2\pi/5$	1.975982919	2	3.951965839
x_9	$9\pi/20$	1.993872679	2	3.987745357
x_{10}	$\pi/2$	2	1	2

(b) $|f''(t)| < 1 \Rightarrow M = 1$
$\Rightarrow |E_T| \le \frac{b-a}{12}(\Delta t^2 M) \le \frac{\left(\frac{\pi}{2}\right)-0}{12}\left(\frac{\pi}{20}\right)^2 1 \le 0.0032$

33. The length of the curve $y = \sin\left(\frac{3\pi}{20}x\right)$ from 0 to 20 is: $L = \int_0^{20}\sqrt{1+\left(\frac{dy}{dx}\right)^2}\,dx$; $\frac{dy}{dx} = \frac{3\pi}{20}\cos\left(\frac{3\pi}{20}x\right) \Rightarrow \left(\frac{dy}{dx}\right)^2 = \frac{9\pi^2}{400}\cos^2\left(\frac{3\pi}{20}x\right) \Rightarrow L = \int_0^{20}\sqrt{1+\frac{9\pi^2}{400}\cos^2\left(\frac{3\pi}{20}x\right)}\,dx$. Using numerical integration we find $L \approx 21.07$ in

35. $y = \sin x \Rightarrow \frac{dy}{dx} = \cos x \Rightarrow \left(\frac{dy}{dx}\right)^2 = \cos^2 x \Rightarrow S = \int_0^{\pi} 2\pi(\sin x)\sqrt{1+\cos^2 x}\,dx$; a numerical integration gives $S \approx 14.4$

37. A calculator or computer numerical integrator yields $\sin^{-1} 0.6 \approx 0.643501109$.

8.7 IMPROPER INTEGRALS

1. $\int_0^{\infty}\frac{dx}{x^2+1} = \lim_{b\to\infty}\int_0^b \frac{dx}{x^2+1} = \lim_{b\to\infty}\left[\tan^{-1}x\right]_0^b = \lim_{b\to\infty}\left(\tan^{-1}b - \tan^{-1}0\right) = \frac{\pi}{2} - 0 = \frac{\pi}{2}$

3. $\int_0^1 \frac{dx}{\sqrt{x}} = \lim_{b\to 0^+}\int_b^1 x^{-1/2}\,dx = \lim_{b\to 0^+}\left[2x^{1/2}\right]_b^1 = \lim_{b\to 0^+}\left(2 - 2\sqrt{b}\right) = 2 - 0 = 2$

5. $\int_{-1}^1 \frac{dx}{x^{2/3}} = \int_{-1}^0 \frac{dx}{x^{2/3}} + \int_0^1 \frac{dx}{x^{2/3}} = \lim_{b\to 0^-}\left[3x^{1/3}\right]_{-1}^b + \lim_{c\to 0^+}\left[3x^{1/3}\right]_c^1$
$= \lim_{b\to 0^-}\left[3b^{1/3} - 3(-1)^{1/3}\right] + \lim_{c\to 0^+}\left[3(1)^{1/3} - 3c^{1/3}\right] = (0+3) + (3-0) = 6$

7. $\int_0^1 \frac{dx}{\sqrt{1-x^2}} = \lim_{b\to 1^-}\left[\sin^{-1}x\right]_0^b = \lim_{b\to 1^-}\left(\sin^{-1}b - \sin^{-1}0\right) = \frac{\pi}{2} - 0 = \frac{\pi}{2}$

9. $\int_{-\infty}^{-2}\frac{2\,dx}{x^2-1} = \int_{-\infty}^{-2}\frac{dx}{x-1} - \int_{-\infty}^{-2}\frac{dx}{x+1} = \lim_{b\to-\infty}\left[\ln|x-1|\right]_b^{-2} - \lim_{b\to-\infty}\left[\ln|x+1|\right]_b^{-2} = \lim_{b\to-\infty}\left[\ln\left|\frac{x-1}{x+1}\right|\right]_b^{-2}$
$= \lim_{b\to-\infty}\left(\ln\left|\frac{-3}{-1}\right| - \ln\left|\frac{b-1}{b+1}\right|\right) = \ln 3 - \ln\left(\lim_{b\to-\infty}\frac{b-1}{b+1}\right) = \ln 3 - \ln 1 = \ln 3$

11. $\int_2^\infty \frac{2\,dv}{v^2-v} = \lim_{b\to\infty} \left[2\ln\left|\frac{v-1}{v}\right|\right]_2^b = \lim_{b\to\infty} \left(2\ln\left|\frac{b-1}{b}\right| - 2\ln\left|\frac{2-1}{2}\right|\right) = 2\ln(1) - 2\ln\left(\frac{1}{2}\right) = 0 + 2\ln 2 = \ln 4$

13. $\int_{-\infty}^\infty \frac{2x\,dx}{(x^2+1)^2} = \int_{-\infty}^0 \frac{2x\,dx}{(x^2+1)^2} + \int_0^\infty \frac{2x\,dx}{(x^2+1)^2}$; $\begin{bmatrix} u = x^2+1 \\ du = 2x\,dx \end{bmatrix} \to \int_\infty^1 \frac{du}{u^2} + \int_1^\infty \frac{du}{u^2} = \lim_{b\to\infty} \left[-\frac{1}{u}\right]_b^1 + \lim_{c\to\infty} \left[-\frac{1}{u}\right]_1^c$
$= \lim_{b\to\infty} \left(-1 + \frac{1}{b}\right) + \lim_{c\to\infty} \left[-\frac{1}{c} - (-1)\right] = (-1+0) + (0+1) = 0$

15. $\int_0^1 \frac{\theta+1}{\sqrt{\theta^2+2\theta}}\,d\theta$; $\begin{bmatrix} u = \theta^2+2\theta \\ du = 2(\theta+1)\,d\theta \end{bmatrix} \to \int_0^3 \frac{du}{2\sqrt{u}} = \lim_{b\to 0^+} \int_b^3 \frac{du}{2\sqrt{u}} = \lim_{b\to 0^+} \left[\sqrt{u}\right]_b^3 = \lim_{b\to 0^+} \left(\sqrt{3} - \sqrt{b}\right) = \sqrt{3} - 0$
$= \sqrt{3}$

17. $\int_0^\infty \frac{dx}{(1+x)\sqrt{x}}$; $\begin{bmatrix} u = \sqrt{x} \\ du = \frac{dx}{2\sqrt{x}} \end{bmatrix} \to \int_0^\infty \frac{2\,du}{u^2+1} = \lim_{b\to\infty} \int_0^b \frac{2\,du}{u^2+1} = \lim_{b\to\infty} \left[2\tan^{-1} u\right]_0^b$
$= \lim_{b\to\infty} \left(2\tan^{-1} b - 2\tan^{-1} 0\right) = 2\left(\frac{\pi}{2}\right) - 2(0) = \pi$

19. $\int_0^\infty \frac{dv}{(1+v^2)(1+\tan^{-1} v)} = \lim_{b\to\infty} \left[\ln|1+\tan^{-1} v|\right]_0^b = \lim_{b\to\infty} \left[\ln|1+\tan^{-1} b|\right] - \ln|1+\tan^{-1} 0|$
$= \ln\left(1+\frac{\pi}{2}\right) - \ln(1+0) = \ln\left(1+\frac{\pi}{2}\right)$

21. $\int_{-\infty}^0 \theta e^\theta\,d\theta = \lim_{b\to-\infty} \left[\theta e^\theta - e^\theta\right]_b^0 = (0\cdot e^0 - e^0) - \lim_{b\to-\infty} \left[be^b - e^b\right] = -1 - \lim_{b\to-\infty} \left(\frac{b-1}{e^{-b}}\right)$
$= -1 - \lim_{b\to-\infty} \left(\frac{1}{-e^{-b}}\right)$ (l'Hôpital's rule for $\frac{\infty}{\infty}$ form)
$= -1 - 0 = -1$

23. $\int_{-\infty}^0 e^{-|x|}\,dx = \int_{-\infty}^0 e^x\,dx = \lim_{b\to-\infty} \left[e^x\right]_b^0 = \lim_{b\to-\infty} \left(1 - e^b\right) = (1-0) = 1$

25. $\int_0^1 x\ln x\,dx = \lim_{b\to 0^+} \left[\frac{x^2}{2}\ln x - \frac{x^2}{4}\right]_b^1 = \left(\frac{1}{2}\ln 1 - \frac{1}{4}\right) - \lim_{b\to 0^+} \left(\frac{b^2}{2}\ln b - \frac{b^2}{4}\right) = -\frac{1}{4} - \lim_{b\to 0^+} \frac{\ln b}{\left(\frac{2}{b^2}\right)} + 0$
$= -\frac{1}{4} - \lim_{b\to 0^+} \frac{\left(\frac{1}{b}\right)}{\left(-\frac{4}{b^3}\right)} = -\frac{1}{4} + \lim_{b\to 0^+} \left(\frac{b^2}{4}\right) = -\frac{1}{4} + 0 = -\frac{1}{4}$

27. $\int_0^2 \frac{ds}{\sqrt{4-s^2}} = \lim_{b\to 2^-} \left[\sin^{-1}\frac{s}{2}\right]_0^b = \lim_{b\to 2^-} \left(\sin^{-1}\frac{b}{2}\right) - \sin^{-1} 0 = \frac{\pi}{2} - 0 = \frac{\pi}{2}$

29. $\int_1^2 \frac{ds}{s\sqrt{s^2-1}} = \lim_{b\to 1^+} \left[\sec^{-1} s\right]_b^2 = \sec^{-1} 2 - \lim_{b\to 1^+} \sec^{-1} b = \frac{\pi}{3} - 0 = \frac{\pi}{3}$

31. $\int_{-1}^4 \frac{dx}{\sqrt{|x|}} = \lim_{b\to 0^-} \int_{-1}^b \frac{dx}{\sqrt{-x}} + \lim_{c\to 0^+} \int_c^4 \frac{dx}{\sqrt{x}} = \lim_{b\to 0^-} \left[-2\sqrt{-x}\right]_{-1}^b + \lim_{c\to 0^+} \left[2\sqrt{x}\right]_c^4$
$= \lim_{b\to 0^-} \left(-2\sqrt{-b}\right) - \left(-2\sqrt{-(-1)}\right) + 2\sqrt{4} - \lim_{c\to 0^+} 2\sqrt{c} = 0 + 2 + 2\cdot 2 - 0 = 6$

33. $\int_{-1}^\infty \frac{d\theta}{\theta^2+5\theta+6} = \lim_{b\to\infty} \left[\ln\left|\frac{\theta+2}{\theta+3}\right|\right]_{-1}^b = \lim_{b\to\infty} \left[\ln\left|\frac{b+2}{b+3}\right|\right] - \ln\left|\frac{-1+2}{-1+3}\right| = 0 - \ln\left(\frac{1}{2}\right) = \ln 2$

35. $\int_0^{\pi/2} \tan\theta\,d\theta = \lim_{b\to\frac{\pi}{2}^-} \left[-\ln|\cos\theta|\right]_0^b = \lim_{b\to\frac{\pi}{2}^-} \left[-\ln|\cos b|\right] + \ln 1 = \lim_{b\to\frac{\pi}{2}^-} \left[-\ln|\cos b|\right] = +\infty$, the integral diverges

37. $\int_0^{\pi} \frac{\sin\theta\, d\theta}{\sqrt{\pi-\theta}}$; $[\pi - \theta = x] \rightarrow -\int_{\pi}^{0} \frac{\sin x\, dx}{\sqrt{x}} = \int_0^{\pi} \frac{\sin x\, dx}{\sqrt{x}}$. Since $0 \le \frac{\sin x}{\sqrt{x}} \le \frac{1}{\sqrt{x}}$ for all $0 \le x \le \pi$ and $\int_0^{\pi} \frac{dx}{\sqrt{x}}$ converges, then $\int_0^{\pi} \frac{\sin x}{\sqrt{x}}\, dx$ converges by the Direct Comparison Test.

39. $\int_0^{\ln 2} x^{-2} e^{-1/x}\, dx$; $\left[\frac{1}{x} = y\right] \rightarrow \int_{\infty}^{1/\ln 2} \frac{y^2 e^{-y}\, dy}{-y^2} = \int_{1/\ln 2}^{\infty} e^{-y}\, dy = \lim\limits_{b \to \infty} \left[-e^{-y}\right]_{1/\ln 2}^{b} = \lim\limits_{b \to \infty} \left[-e^{-b}\right] - \left[-e^{-1/\ln 2}\right]$ $= 0 + e^{-1/\ln 2} = e^{-1/\ln 2}$, so the integral converges.

41. $\int_0^{\pi} \frac{dt}{\sqrt{t} + \sin t}$. Since for $0 \le t \le \pi$, $0 \le \frac{1}{\sqrt{t} + \sin t} \le \frac{1}{\sqrt{t}}$ and $\int_0^{\pi} \frac{dt}{\sqrt{t}}$ converges, then the original integral converges as well by the Direct Comparison Test.

43. $\int_0^2 \frac{dx}{1-x^2} = \int_0^1 \frac{dx}{1-x^2} + \int_1^2 \frac{dx}{1-x^2}$ and $\int_0^1 \frac{dx}{1-x^2} = \lim\limits_{b \to 1^-} \left[\frac{1}{2} \ln\left|\frac{1+x}{1-x}\right|\right]_0^b = \lim\limits_{b \to 1^-} \left[\frac{1}{2} \ln\left|\frac{1+b}{1-b}\right|\right] - 0 = \infty$, which diverges $\Rightarrow \int_0^2 \frac{dx}{1-x^2}$ diverges as well.

45. $\int_{-1}^{1} \ln|x|\, dx = \int_{-1}^{0} \ln(-x)\, dx + \int_0^1 \ln x\, dx$; $\int_0^1 \ln x\, dx = \lim\limits_{b \to 0^+} \left[x \ln x - x\right]_b^1 = [1 \cdot 0 - 1] - \lim\limits_{b \to 0^+} [b \ln b - b]$ $= -1 - 0 = -1$; $\int_{-1}^{0} \ln(-x)\, dx = -1 \Rightarrow \int_{-1}^{1} \ln|x|\, dx = -2$ converges.

47. $\int_1^{\infty} \frac{dx}{1+x^3}$; $0 \le \frac{1}{x^3+1} \le \frac{1}{x^3}$ for $1 \le x < \infty$ and $\int_1^{\infty} \frac{dx}{x^3}$ converges $\Rightarrow \int_1^{\infty} \frac{dx}{1+x^3}$ converges by the Direct Comparison Test.

49. $\int_2^{\infty} \frac{dv}{\sqrt{v-1}}$; $\lim\limits_{v \to \infty} \frac{\left(\frac{1}{\sqrt{v-1}}\right)}{\left(\frac{1}{\sqrt{v}}\right)} = \lim\limits_{v \to \infty} \frac{\sqrt{v}}{\sqrt{v-1}} = \lim\limits_{v \to \infty} \frac{1}{\sqrt{1-\frac{1}{v}}} = \frac{1}{\sqrt{1-0}} = 1$ and $\int_2^{\infty} \frac{dv}{\sqrt{v}} = \lim\limits_{b \to \infty} \left[2\sqrt{v}\right]_2^b = \infty$, which diverges $\Rightarrow \int_2^{\infty} \frac{dv}{\sqrt{v-1}}$ diverges by the Limit Comparison Test.

51. $\int_0^{\infty} \frac{dx}{\sqrt{x^6+1}} = \int_0^1 \frac{dx}{\sqrt{x^6+1}} + \int_1^{\infty} \frac{dx}{\sqrt{x^6+1}} < \int_0^1 \frac{dx}{\sqrt{x^6+1}} + \int_1^{\infty} \frac{dx}{x^3}$ and $\int_1^{\infty} \frac{dx}{x^3} = \lim\limits_{b \to \infty} \left[-\frac{1}{2x^2}\right]_1^b$ $= \lim\limits_{b \to \infty} \left(-\frac{1}{2b^2} + \frac{1}{2}\right) = \frac{1}{2} \Rightarrow \int_0^{\infty} \frac{dx}{\sqrt{x^6+1}}$ converges by the Direct Comparison Test.

53. $\int_1^{\infty} \frac{\sqrt{x+1}}{x^2}\, dx$; $\lim\limits_{x \to \infty} \frac{\left(\frac{\sqrt{x}}{x^2}\right)}{\left(\frac{\sqrt{x+1}}{x^2}\right)} = \lim\limits_{x \to \infty} \frac{\sqrt{x}}{\sqrt{x+1}} = \lim\limits_{x \to \infty} \frac{1}{\sqrt{1+\frac{1}{x}}} = 1$; $\int_1^{\infty} \frac{\sqrt{x}}{x^2}\, dx = \int_1^{\infty} \frac{dx}{x^{3/2}}$ $= \lim\limits_{b \to \infty} \left[-2x^{-1/2}\right]_1^b = \lim\limits_{b \to \infty} \left(\frac{-2}{\sqrt{b}} + 2\right) = 2 \Rightarrow \int_1^{\infty} \frac{\sqrt{x+1}}{x^2}\, dx$ converges by the Limit Comparison Test.

55. $\int_{\pi}^{\infty} \frac{2+\cos x}{x}\, dx$; $0 < \frac{1}{x} \le \frac{2+\cos x}{x}$ for $x \ge \pi$ and $\int_{\pi}^{\infty} \frac{dx}{x} = \lim\limits_{b \to \infty} [\ln x]_{\pi}^{b} = \infty$, which diverges $\Rightarrow \int_{\pi}^{\infty} \frac{2+\cos x}{x}\, dx$ diverges by the Direct Comparison Test.

57. $\int_4^{\infty} \frac{2\, dt}{t^{3/2}-1}$; $\lim\limits_{t \to \infty} \frac{t^{3/2}}{t^{3/2}-1} = 1$ and $\int_4^{\infty} \frac{2\, dt}{t^{3/2}} = \lim\limits_{b \to \infty} \left[-4t^{-1/2}\right]_4^b = \lim\limits_{b \to \infty} \left(\frac{-4}{\sqrt{b}} + 2\right) = 2 \Rightarrow \int_4^{\infty} \frac{2\, dt}{t^{3/2}}$ converges $\Rightarrow \int_4^{\infty} \frac{2\, dt}{t^{3/2}+1}$ converges by the Limit Comparison Test.

59. $\int_1^{\infty} \frac{e^x}{x}\, dx$; $0 < \frac{1}{x} < \frac{e^x}{x}$ for $x > 1$ and $\int_1^{\infty} \frac{dx}{x}$ diverges $\Rightarrow \int_1^{\infty} \frac{e^x\, dx}{x}$ diverges by the Direct Comparison Test.

61. $\int_1^\infty \frac{dx}{\sqrt{e^x - x}}$; $\lim_{x\to\infty} \frac{\left(\frac{1}{\sqrt{e^x-x}}\right)}{\left(\frac{1}{\sqrt{e^x}}\right)} = \lim_{x\to\infty} \frac{\sqrt{e^x}}{\sqrt{e^x-x}} = \lim_{x\to\infty} \frac{1}{\sqrt{1-\frac{x}{e^x}}} = \frac{1}{\sqrt{1-0}} = 1$; $\int_1^\infty \frac{dx}{\sqrt{e^x}} = \int_1^\infty e^{-x/2}\,dx$

$= \lim_{b\to\infty} \left[-2e^{-x/2}\right]_1^b = \lim_{b\to\infty} \left(-2e^{-b/2} + 2e^{-1/2}\right) = \frac{2}{\sqrt{e}} \Rightarrow \int_1^\infty e^{-x/2}\,dx$ converges $\Rightarrow \int_1^\infty \frac{dx}{\sqrt{e^x-x}}$ converges by the Limit Comparison Test.

63. $\int_{-\infty}^\infty \frac{dx}{\sqrt{x^4+1}} = 2\int_0^\infty \frac{dx}{\sqrt{x^4+1}}$; $\int_0^\infty \frac{dx}{\sqrt{x^4+1}} = \int_0^1 \frac{dx}{\sqrt{x^4+1}} + \int_1^\infty \frac{dx}{\sqrt{x^4+1}} < \int_0^1 \frac{dx}{\sqrt{x^4+1}} + \int_1^\infty \frac{dx}{x^2}$ and

$\int_1^\infty \frac{dx}{x^2} = \lim_{b\to\infty} \left[-\frac{1}{x}\right]_1^b = \lim_{b\to\infty} \left(-\frac{1}{b}+1\right) = 1 \Rightarrow \int_{-\infty}^\infty \frac{dx}{\sqrt{x^4+1}}$ converges by the Direct Comparison Test.

65. (a) $\int_1^2 \frac{dx}{x(\ln x)^p}$; $[t = \ln x] \to \int_0^{\ln 2} \frac{dt}{t^p} = \lim_{b\to 0^+} \left[\frac{1}{-p+1} t^{1-p}\right]_b^{\ln 2} = \lim_{b\to 0^+} \frac{b^{1-p}}{p-1} + \frac{1}{1-p}(\ln 2)^{1-p}$

$\Rightarrow$ the integral converges for $p < 1$ and diverges for $p \ge 1$

(b) $\int_2^\infty \frac{dx}{x(\ln x)^p}$; $[t = \ln x] \to \int_{\ln 2}^\infty \frac{dt}{t^p}$ and this integral is essentially the same as in Exercise 65(a): it converges for $p > 1$ and diverges for $p \le 1$

67. $A = \int_0^\infty e^{-x}\,dx = \lim_{b\to\infty} \left[-e^{-x}\right]_0^b = \lim_{b\to\infty} \left(-e^{-b}\right) - \left(-e^{-0}\right)$

$= 0 + 1 = 1$

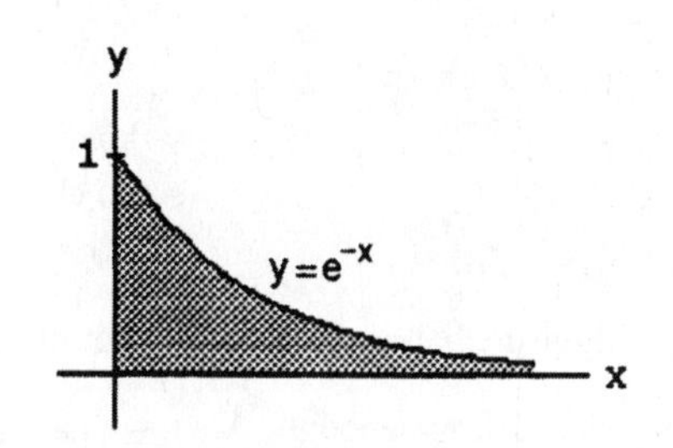

69. $V = \int_0^\infty 2\pi x e^{-x}\,dx = 2\pi \int_0^\infty x e^{-x}\,dx = 2\pi \lim_{b\to\infty} \left[-xe^{-x} - e^{-x}\right]_0^b = 2\pi \left[\lim_{b\to\infty} \left(-be^{-b} - e^{-b}\right) - 1\right] = 2\pi$

71. $A = \int_0^{\pi/2} (\sec x - \tan x)\,dx = \lim_{b\to\frac{\pi}{2}^-} \left[\ln|\sec x + \tan x| - \ln|\sec x|\right]_0^b = \lim_{b\to\frac{\pi}{2}^-} \left(\ln\left|1 + \frac{\tan b}{\sec b}\right| - \ln|1+0|\right)$

$= \lim_{b\to\frac{\pi}{2}^-} \ln|1 + \sin b| = \ln 2$

73. (a) $\int_3^\infty e^{-3x}\,dx = \lim_{b\to\infty} \left[-\frac{1}{3}e^{-3x}\right]_3^b = \lim_{b\to\infty} \left(-\frac{1}{3}e^{-3b}\right) - \left(-\frac{1}{3}e^{-3\cdot 3}\right) = 0 + \frac{1}{3}\cdot e^{-9} = \frac{1}{3}e^{-9}$

$\approx 0.0000411 < 0.000042$. Since $e^{-x^2} \le e^{-3x}$ for $x > 3$, then $\int_3^\infty e^{-x^2}\,dx < 0.000042$ and therefore $\int_0^\infty e^{-x^2}\,dx$ can be replaced by $\int_0^3 e^{-x^2}\,dx$ without introducing an error greater than 0.000042.

(b) $\int_0^3 e^{-x^2}\,dx \cong 0.88621$

75. (a)

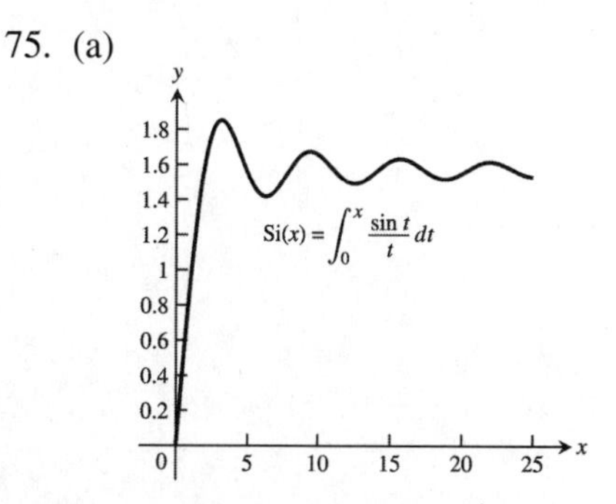

(b) > int((sin(t))/t, t=0..infinity); (answer is $\frac{\pi}{2}$)

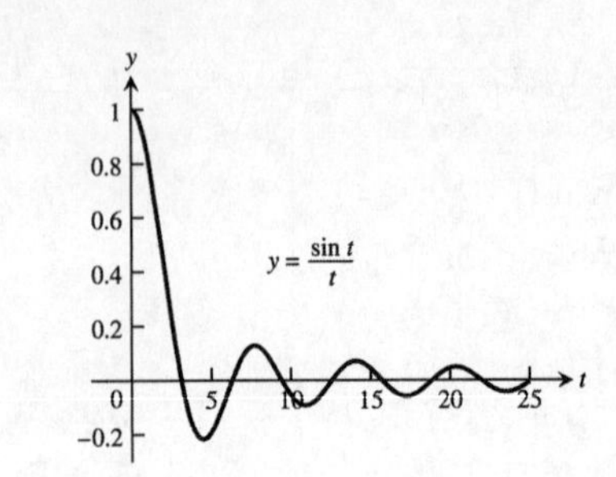

77. (a) $f(x) = \frac{1}{\sqrt{2\pi}} e^{-x^2/2}$

f is increasing on $(-\infty, 0]$. f is decreasing on $[0, \infty)$.

f has a local maximum at $(0, f(0)) = \left(0, \frac{1}{\sqrt{2\pi}}\right)$

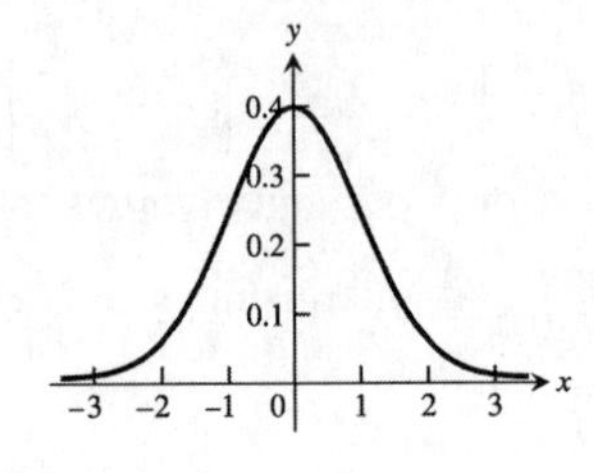

(b) Maple commands:

```
>f: = exp(−x^2/2)(sqrt(2*pi);
>int(f, x = −1..1);        ≈ 0.683
>int(f, x = −2..2);        ≈ 0.954
>int(f, x = −3..3);        ≈ 0.997
```

(c) Part (b) suggests that as n increases, the integral approaches 1. We can take $\int_{-n}^{n} f(x)\, dx$ as close to 1 as we want by choosing $n > 1$ large enough. Also, we can make $\int_{n}^{\infty} f(x)\, dx$ and $\int_{-\infty}^{-n} f(x)\, dx$ as small as we want by choosing n large enough. This is because $0 < f(x) < e^{-x/2}$ for $x > 1$. (Likewise, $0 < f(x) < e^{x/2}$ for $x < -1$.)

Thus, $\int_{n}^{\infty} f(x)\, dx < \int_{n}^{\infty} e^{-x/2} dx$.

$\int_{n}^{\infty} e^{-x/2} dx = \lim_{c\to\infty} \int_{n}^{c} e^{-x/2} dx = \lim_{c\to\infty} \left[-2e^{-x/2}\right]_{n}^{c} = \lim_{c\to\infty} \left[-2e^{-c/2} + 2e^{-n/2}\right] = 2e^{-n/2}$

As $n \to \infty$, $2e^{-n/2} \to 0$, for large enough n, $\int_{n}^{\infty} f(x)\, dx$ is as small as we want. Likewise for large enough n, $\int_{-\infty}^{-n} f(x)\, dx$ is as small as we want.

CHAPTER 8 PRACTICE EXERCISES

1. $u = \ln(x+1)$, $du = \frac{dx}{x+1}$; $dv = dx$, $v = x$;

$\int \ln(x+1)\, dx = x \ln(x+1) - \int \frac{x}{x+1}\, dx = x \ln(x+1) - \int dx + \int \frac{dx}{x+1} = x \ln(x+1) - x + \ln(x+1) + C_1$
$= (x+1) \ln(x+1) - x + C_1 = (x+1) \ln(x+1) - (x+1) + C$, where $C = C_1 + 1$

3. $u = \tan^{-1} 3x$, $du = \frac{3\, dx}{1+9x^2}$; $dv = dx$, $v = x$;

$\int \tan^{-1} 3x\, dx = x \tan^{-1} 3x - \int \frac{3x\, dx}{1+9x^2}$; $\begin{bmatrix} y = 1 + 9x^2 \\ dy = 18x\, dx \end{bmatrix} \to x \tan^{-1} 3x - \frac{1}{6} \int \frac{dy}{y}$
$= x \tan^{-1}(3x) - \frac{1}{6} \ln(1 + 9x^2) + C$

5.

		e^x
$(x+1)^2$	$\xrightarrow{(+)}$	e^x
$2(x+1)$	$\xrightarrow{(-)}$	e^x
2	$\xrightarrow{(+)}$	e^x
0		

$\Rightarrow \int (x+1)^2 e^x\,dx = \left[(x+1)^2 - 2(x+1) + 2\right] e^x + C$

7. $u = \cos 2x,\ du = -2\sin 2x\,dx;\ dv = e^x\,dx,\ v = e^x;$
$I = \int e^x \cos 2x\,dx = e^x \cos 2x + 2\int e^x \sin 2x\,dx;$
$u = \sin 2x,\ du = 2\cos 2x\,dx;\ dv = e^x\,dx,\ v = e^x;$
$I = e^x \cos 2x + 2\left[e^x \sin 2x - 2\int e^x \cos 2x\,dx\right] = e^x \cos 2x + 2e^x \sin 2x - 4I \Rightarrow I = \frac{e^x \cos 2x}{5} + \frac{2e^x \sin 2x}{5} + C$

9. $\int \frac{x\,dx}{x^2-3x+2} = \int \frac{2\,dx}{x-2} - \int \frac{dx}{x-1} = 2\ln|x-2| - \ln|x-1| + C$

11. $\int \frac{dx}{x(x+1)^2} = \int \left(\frac{1}{x} - \frac{1}{x+1} + \frac{-1}{(x+1)^2}\right) dx = \ln|x| - \ln|x+1| + \frac{1}{x+1} + C$

13. $\int \frac{\sin\theta\,d\theta}{\cos^2\theta + \cos\theta - 2};\ [\cos\theta = y] \rightarrow -\int \frac{dy}{y^2+y-2} = -\frac{1}{3}\int \frac{dy}{y-1} + \frac{1}{3}\int \frac{dy}{y+2} = \frac{1}{3}\ln\left|\frac{y+2}{y-1}\right| + C$
$= \frac{1}{3}\ln\left|\frac{\cos\theta+2}{\cos\theta-1}\right| + C = -\frac{1}{3}\ln\left|\frac{\cos\theta-1}{\cos\theta+2}\right| + C$

15. $\int \frac{3x^2+4x+4}{x^3+x}\,dx = \int \frac{4}{x}\,dx - \int \frac{x-4}{x^2+1}\,dx = 4\ln|x| - \frac{1}{2}\ln(x^2+1) + 4\tan^{-1}x + C$

17. $\int \frac{(v+3)\,dv}{2v^3-8v} = \frac{1}{2}\int \left(-\frac{3}{4v} + \frac{5}{8(v-2)} + \frac{1}{8(v+2)}\right) dv = -\frac{3}{8}\ln|v| + \frac{5}{16}\ln|v-2| + \frac{1}{16}\ln|v+2| + C$
$= \frac{1}{16}\ln\left|\frac{(v-2)^5(v+2)}{v^6}\right| + C$

19. $\int \frac{dt}{t^4+4t^2+3} = \frac{1}{2}\int \frac{dt}{t^2+1} - \frac{1}{2}\int \frac{dt}{t^2+3} = \frac{1}{2}\tan^{-1}t - \frac{1}{2\sqrt{3}}\tan^{-1}\left(\frac{t}{\sqrt{3}}\right) + C = \frac{1}{2}\tan^{-1}t - \frac{\sqrt{3}}{6}\tan^{-1}\frac{t}{\sqrt{3}} + C$

21. $\int \frac{x^3+x^2}{x^2+x-2}\,dx = \int \left(x + \frac{2x}{x^2+x-2}\right) dx = \int x\,dx + \frac{2}{3}\int \frac{dx}{x-1} + \frac{4}{3}\int \frac{dx}{x+2} = \frac{x^2}{2} + \frac{4}{3}\ln|x+2| + \frac{2}{3}\ln|x-1| + C$

23. $\int \frac{x^3+4x^2}{x^2+4x+3}\,dx = \int \left(x - \frac{3x}{x^2+4x+3}\right) dx = \int x\,dx + \frac{3}{2}\int \frac{dx}{x+1} - \frac{9}{2}\int \frac{dx}{x+3} = \frac{x^2}{2} - \frac{9}{2}\ln|x+3| + \frac{3}{2}\ln|x+1| + C$

25. $\int \frac{dx}{x(3\sqrt{x+1})};\ \begin{bmatrix} u = \sqrt{x+1} \\ du = \frac{dx}{2\sqrt{x+1}} \\ dx = 2u\,du \end{bmatrix} \rightarrow \frac{2}{3}\int \frac{u\,du}{(u^2-1)u} = \frac{1}{3}\int \frac{du}{u-1} - \frac{1}{3}\int \frac{du}{u+1} = \frac{1}{3}\ln|u-1| - \frac{1}{3}\ln|u+1| + C$
$= \frac{1}{3}\ln\left|\frac{\sqrt{x+1}-1}{\sqrt{x+1}+1}\right| + C$

27. $\int \frac{ds}{e^s-1};\ \begin{bmatrix} u = e^s - 1 \\ du = e^s\,ds \\ ds = \frac{du}{u+1} \end{bmatrix} \rightarrow \int \frac{du}{u(u+1)} = -\int \frac{du}{u+1} + \int \frac{du}{u} = \ln\left|\frac{u}{u+1}\right| + C = \ln\left|\frac{e^s-1}{e^s}\right| + C = \ln|1-e^{-s}| + C$

29. (a) $\int \frac{y\,dy}{\sqrt{16-y^2}} = -\frac{1}{2}\int \frac{d(16-y^2)}{\sqrt{16-y^2}} = -\sqrt{16-y^2} + C$
(b) $\int \frac{y\,dy}{\sqrt{16-y^2}};\ [y = 4\sin x] \rightarrow 4\int \frac{\sin x \cos x\,dx}{\cos x} = -4\cos x + C = -\frac{4\sqrt{16-y^2}}{4} + C = -\sqrt{16-y^2} + C$

31. (a) $\int \frac{x\,dx}{4-x^2} = -\frac{1}{2}\int \frac{d(4-x^2)}{4-x^2} = -\frac{1}{2}\ln|4-x^2| + C$

(b) $\int \frac{x\,dx}{4-x^2}$; $[x = 2\sin\theta] \rightarrow \int \frac{2\sin\theta \cdot 2\cos\theta\,d\theta}{4\cos^2\theta} = \int \tan\theta\,d\theta = -\ln|\cos\theta| + C = -\ln\left(\frac{\sqrt{4-x^2}}{2}\right) + C$
$= -\frac{1}{2}\ln|4-x^2| + C$

33. $\int \frac{x\,dx}{9-x^2}$; $\begin{bmatrix} u = 9-x^2 \\ du = -2x\,dx \end{bmatrix} \rightarrow -\frac{1}{2}\int \frac{du}{u} = -\frac{1}{2}\ln|u| + C = \ln\frac{1}{\sqrt{u}} + C = \ln\frac{1}{\sqrt{9-x^2}} + C$

35. $\int \frac{dx}{9-x^2} = \frac{1}{6}\int \frac{dx}{3-x} + \frac{1}{6}\int \frac{dx}{3+x} = -\frac{1}{6}\ln|3-x| + \frac{1}{6}\ln|3+x| + C = \frac{1}{6}\ln\left|\frac{x+3}{x-3}\right| + C$

37. $\int \sin^3 x\cos^4 x\,dx = \int \cos^4 x(1-\cos^2 x)\sin x\,dx = \int \cos^4 x\sin x\,dx - \int \cos^6 x\sin x\,dx = -\frac{\cos^5 x}{5} + \frac{\cos^7 x}{7} + C$

39. $\int \tan^4 x\sec^2 x\,dx = \frac{\tan^5 x}{5} + C$

41. $\int \sin 5\theta\cos 6\theta\,d\theta = \frac{1}{2}\int (\sin(-\theta) + \sin(11\theta))\,d\theta = \frac{1}{2}\int \sin(-\theta)\,d\theta + \frac{1}{2}\int \sin(11\theta)\,d\theta = \frac{1}{2}\cos(-\theta) - \frac{1}{22}\cos 11\theta + C$
$= \frac{1}{2}\cos\theta - \frac{1}{22}\cos 11\theta + C$

43. $\int \sqrt{1+\cos\left(\frac{t}{2}\right)}\,dt = \int \sqrt{2}\left|\cos\frac{t}{4}\right|\,dt = 4\sqrt{2}\left|\sin\frac{t}{4}\right| + C$

45. $|E_s| \le \frac{3-1}{180}(\Delta x)^4 M$ where $\Delta x = \frac{3-1}{n} = \frac{2}{n}$; $f(x) = \frac{1}{x} = x^{-1} \Rightarrow f'(x) = -x^{-2} \Rightarrow f''(x) = 2x^{-3} \Rightarrow f'''(x) = -6x^{-4}$
$\Rightarrow f^{(4)}(x) = 24x^{-5}$ which is decreasing on $[1, 3] \Rightarrow$ maximum of $f^{(4)}(x)$ on $[1, 3]$ is $f^{(4)}(1) = 24 \Rightarrow M = 24$. Then
$|E_s| \le 0.0001 \Rightarrow \left(\frac{3-1}{180}\right)\left(\frac{2}{n}\right)^4(24) \le 0.0001 \Rightarrow \left(\frac{768}{180}\right)\left(\frac{1}{n^4}\right) \le 0.0001 \Rightarrow \frac{1}{n^4} \le (0.0001)\left(\frac{180}{768}\right) \Rightarrow n^4 \ge 10{,}000\left(\frac{768}{180}\right)$
$\Rightarrow n \ge 14.37 \Rightarrow n \ge 16$ (n must be even)

47. $\Delta x = \frac{b-a}{n} = \frac{\pi-0}{6} = \frac{\pi}{6} \Rightarrow \frac{\Delta x}{2} = \frac{\pi}{12}$;

$\sum_{i=0}^{6} mf(x_i) = 12 \Rightarrow T = \left(\frac{\pi}{12}\right)(12) = \pi$;

	x_i	$f(x_i)$	m	$mf(x_i)$
x_0	0	0	1	0
x_1	$\pi/6$	1/2	2	1
x_2	$\pi/3$	3/2	2	3
x_3	$\pi/2$	2	2	4
x_4	$2\pi/3$	3/2	2	3
x_5	$5\pi/6$	1/2	2	1
x_6	π	0	1	0

$\sum_{i=0}^{6} mf(x_i) = 18$ and $\frac{\Delta x}{3} = \frac{\pi}{18} \Rightarrow$

$S = \left(\frac{\pi}{18}\right)(18) = \pi$.

	x_i	$f(x_i)$	m	$mf(x_i)$
x_0	0	0	1	0
x_1	$\pi/6$	1/2	4	2
x_2	$\pi/3$	3/2	2	3
x_3	$\pi/2$	2	4	8
x_4	$2\pi/3$	3/2	2	3
x_5	$5\pi/6$	1/2	4	2
x_6	π	0	1	0

49. $y_{av} = \frac{1}{365-0}\int_0^{365}\left[37\sin\left(\frac{2\pi}{365}(x-101)\right) + 25\right]dx = \frac{1}{365}\left[-37\left(\frac{365}{2\pi}\cos\left(\frac{2\pi}{365}(x-101)\right) + 25x\right)\right]_0^{365}$
$= \frac{1}{365}\left[\left(-37\left(\frac{365}{2\pi}\right)\cos\left[\frac{2\pi}{365}(365-101)\right] + 25(365)\right) - \left(-37\left(\frac{365}{2\pi}\right)\cos\left[\frac{2\pi}{365}(0-101)\right] + 25(0)\right)\right]$
$= -\frac{37}{2\pi}\cos\left(\frac{2\pi}{365}(264)\right) + 25 + \frac{37}{2\pi}\cos\left(\frac{2\pi}{365}(-101)\right) = -\frac{37}{2\pi}\left(\cos\left(\frac{2\pi}{365}(264)\right) - \cos\left(\frac{2\pi}{365}(-101)\right)\right) + 25$
$\approx -\frac{37}{2\pi}(0.16705 - 0.16705) + 25 = 25°$ F

51. (a) Each interval is 5 min $= \frac{1}{12}$ hour.
$\frac{1}{24}[2.5 + 2(2.4) + 2(2.3) + \ldots + 2(2.4) + 2.3] = \frac{29}{12} \approx 2.42$ gal
(b) $(60 \text{ mph})\left(\frac{12}{29} \text{ hours/gal}\right) \approx 24.83$ mi/gal

53. $\int_0^3 \frac{dx}{\sqrt{9-x^2}} = \lim_{b \to 3^-} \int_0^b \frac{dx}{\sqrt{9-x^2}} = \lim_{b \to 3^-} \left[\sin^{-1}\left(\frac{x}{3}\right)\right]_0^b = \lim_{b \to 3^-} \sin^{-1}\left(\frac{b}{3}\right) - \sin^{-1}\left(\frac{0}{3}\right) = \frac{\pi}{2} - 0 = \frac{\pi}{2}$

55. $\int_{-1}^1 \frac{dy}{y^{2/3}} = \int_{-1}^0 \frac{dy}{y^{2/3}} + \int_0^1 \frac{dy}{y^{2/3}} = 2\int_0^1 \frac{dy}{y^{2/3}} = 2 \cdot 3 \lim_{b \to 0^+} \left[y^{1/3}\right]_b^1 = 6\left(1 - \lim_{b \to 0^+} b^{1/3}\right) = 6$

57. $\int_3^\infty \frac{2\,du}{u^2 - 2u} = \int_3^\infty \frac{du}{u-2} - \int_3^\infty \frac{du}{u} = \lim_{b \to \infty} \left[\ln\left|\frac{u-2}{u}\right|\right]_3^b = \lim_{b \to \infty} \left[\ln\left|\frac{b-2}{b}\right|\right] - \ln\left|\frac{3-2}{3}\right| = 0 - \ln\left(\frac{1}{3}\right) = \ln 3$

59. $\int_0^\infty x^2 e^{-x}\,dx = \lim_{b \to \infty} \left[-x^2e^{-x} - 2xe^{-x} - 2e^{-x}\right]_0^b = \lim_{b \to \infty} \left(-b^2e^{-b} - 2be^{-b} - 2e^{-b}\right) - (-2) = 0 + 2 = 2$

61. $\int_{-\infty}^\infty \frac{dx}{4x^2+9} = 2\int_0^\infty \frac{dx}{4x^2+9} = \frac{1}{2}\int_0^\infty \frac{dx}{x^2+\frac{9}{4}} = \frac{1}{2}\lim_{b \to \infty} \left[\frac{2}{3}\tan^{-1}\left(\frac{2x}{3}\right)\right]_0^b = \frac{1}{2}\lim_{b \to \infty} \left[\frac{2}{3}\tan^{-1}\left(\frac{2b}{3}\right)\right] - \frac{1}{3}\tan^{-1}(0)$
$= \frac{1}{2}\left(\frac{2}{3} \cdot \frac{\pi}{2}\right) - 0 = \frac{\pi}{6}$

63. $\lim_{\theta \to \infty} \frac{\theta}{\sqrt{\theta^2+1}} = 1$ and $\int_6^\infty \frac{d\theta}{\theta}$ diverges $\Rightarrow \int_6^\infty \frac{d\theta}{\sqrt{\theta^2+1}}$ diverges

65. $\int_1^\infty \frac{\ln z}{z}\,dz = \int_1^e \frac{\ln z}{z}\,dz + \int_e^\infty \frac{\ln z}{z}\,dz = \left[\frac{(\ln z)^2}{2}\right]_1^e + \lim_{b \to \infty} \left[\frac{(\ln z)^2}{2}\right]_e^b = \left(\frac{1^2}{2} - 0\right) + \lim_{b \to \infty} \left[\frac{(\ln b)^2}{2} - \frac{1}{2}\right] = \infty$
$\Rightarrow$ diverges

67. $\int_{-\infty}^\infty \frac{2\,dx}{e^x + e^{-x}} = 2\int_0^\infty \frac{2\,dx}{e^x + e^{-x}} < \int_0^\infty \frac{4\,dx}{e^x}$ converges $\Rightarrow \int_{-\infty}^\infty \frac{2\,dx}{e^x + e^{-x}}$ converges

69. $\int \frac{x\,dx}{1+\sqrt{x}}$; $\begin{bmatrix} u = \sqrt{x} \\ du = \frac{dx}{2\sqrt{x}} \end{bmatrix} \to \int \frac{u^2 \cdot 2u\,du}{1+u} = \int \left(2u^2 - 2u + 2 - \frac{2}{1+u}\right) du = \frac{2}{3}u^3 - u^2 + 2u - 2\ln|1+u| + C$
$= \frac{2x^{3/2}}{3} - x + 2\sqrt{x} - 2\ln\left(1 + \sqrt{x}\right) + C$

71. $\int \frac{dx}{x\left(x^2+1\right)^2}$; $\begin{bmatrix} x = \tan\theta \\ dx = \sec^2\theta\,d\theta \end{bmatrix} \to \int \frac{\sec^2\theta\,d\theta}{\tan\theta\sec^4\theta} = \int \frac{\cos^3\theta\,d\theta}{\sin\theta} = \int \left(\frac{1-\sin^2\theta}{\sin\theta}\right) d(\sin\theta)$
$= \ln|\sin\theta| - \frac{1}{2}\sin^2\theta + C = \ln\left|\frac{x}{\sqrt{x^2+1}}\right| - \frac{1}{2}\left(\frac{x}{\sqrt{x^2+1}}\right)^2 + C$

73. $\int \frac{2 - \cos x + \sin x}{\sin^2 x}\,dx = \int 2\csc^2 x\,dx - \int \frac{\cos x\,dx}{\sin^2 x} + \int \csc x\,dx = -2\cot x + \frac{1}{\sin x} - \ln|\csc x + \cot x| + C$
$= -2\cot x + \csc x - \ln|\csc x + \cot x| + C$

75. $\int \frac{9\,dv}{81 - v^4} = \frac{1}{2}\int \frac{dv}{v^2+9} + \frac{1}{12}\int \frac{dv}{3-v} + \frac{1}{12}\int \frac{dv}{3+v} = \frac{1}{12}\ln\left|\frac{3+v}{3-v}\right| + \frac{1}{6}\tan^{-1}\frac{v}{3} + C$

77.

		$\cos(2\theta + 1)$
θ	$\xrightarrow{(+)}$	$\frac{1}{2}\sin(2\theta + 1)$
1	$\xrightarrow{(-)}$	$-\frac{1}{4}\cos(2\theta + 1)$
0		

$\Rightarrow \int \theta\cos(2\theta + 1)\,d\theta = \frac{\theta}{2}\sin(2\theta + 1) + \frac{1}{4}\cos(2\theta + 1) + C$

79. $\int \frac{\sin 2\theta \, d\theta}{(1+\cos 2\theta)^2} = -\frac{1}{2}\int \frac{d(1+\cos 2\theta)}{(1+\cos 2\theta)^2} = \frac{1}{2(1+\cos 2\theta)} + C = \frac{1}{4}\sec^2\theta + C$

81. $\int \frac{x\,dx}{\sqrt{2-x}}$; $\left[\begin{array}{l} y = 2-x \\ dy = -dx \end{array}\right] \to -\int \frac{(2-y)\,dy}{\sqrt{y}} = \frac{2}{3}y^{3/2} - 4y^{1/2} + C = \frac{2}{3}(2-x)^{3/2} - 4(2-x)^{1/2} + C$
$= 2\left[\frac{\left(\sqrt{2-x}\right)^3}{3} - 2\sqrt{2-x}\right] + C$

83. $\int \frac{dy}{y^2-2y+2} = \int \frac{d(y-1)}{(y-1)^2+1} = \tan^{-1}(y-1) + C$

85. $\int \frac{z+1}{z^2(z^2+4)}\,dz = \frac{1}{4}\int\left(\frac{1}{z} + \frac{1}{z^2} - \frac{z+1}{z^2+4}\right)dz = \frac{1}{4}\ln|z| - \frac{1}{4z} - \frac{1}{8}\ln(z^2+4) - \frac{1}{8}\tan^{-1}\frac{z}{2} + C$

87. $\int \frac{t\,dt}{\sqrt{9-4t^2}} = -\frac{1}{8}\int \frac{d(9-4t^2)}{\sqrt{9-4t^2}} = -\frac{1}{4}\sqrt{9-4t^2} + C$

89. $\int \frac{e^t\,dt}{e^{2t}+3e^t+2}$; $[e^t = x] \to \int \frac{dx}{(x+1)(x+2)} = \int \frac{dx}{x+1} - \int \frac{dx}{x+2} = \ln|x+1| - \ln|x+2| + C = \ln\left|\frac{x+1}{x+2}\right| + C$
$= \ln\left(\frac{e^t+1}{e^t+2}\right) + C$

91. $\int_1^\infty \frac{\ln y\,dy}{y^3}$; $\left[\begin{array}{c} x = \ln y \\ dx = \frac{dy}{y} \\ dy = e^x\,dx \end{array}\right] \to \int_0^\infty \frac{x \cdot e^x}{e^{3x}}\,dx = \int_0^\infty xe^{-2x}\,dx = \lim_{b\to\infty}\left[-\frac{x}{2}e^{-2x} - \frac{1}{4}e^{-2x}\right]_0^b$
$= \lim_{b\to\infty}\left(\frac{-b}{2e^{2b}} - \frac{1}{4e^{2b}}\right) - \left(0 - \frac{1}{4}\right) = \frac{1}{4}$

93. $\int e^{\ln\sqrt{x}}\,dx = \int \sqrt{x}\,dx = \frac{2}{3}x^{3/2} + C$

95. $\int \frac{\sin 5t\,dt}{1+(\cos 5t)^2}$; $\left[\begin{array}{c} u = \cos 5t \\ du = -5\sin 5t\,dt \end{array}\right] \to -\frac{1}{5}\int \frac{du}{1+u^2} = -\frac{1}{5}\tan^{-1}u + C = -\frac{1}{5}\tan^{-1}(\cos 5t) + C$

97. $\int \frac{dr}{1+\sqrt{r}}$; $\left[\begin{array}{c} u = \sqrt{r} \\ du = \frac{dr}{2\sqrt{r}} \end{array}\right] \to \int \frac{2u\,du}{1+u} = \int\left(2 - \frac{2}{1+u}\right)du = 2u - 2\ln|1+u| + C = 2\sqrt{r} - 2\ln\left(1+\sqrt{r}\right) + C$

99. $\int \frac{x^3}{1+x^2}\,dx = \int\left(x - \frac{x}{1+x^2}\right)dx = \int x\,dx - \frac{1}{2}\int \frac{2x}{1+x^2}\,dx = \frac{1}{2}x^2 - \frac{1}{2}\ln(1+x^2) + C$

101. $\int \frac{1+x^2}{1+x^3}\,dx$; $\frac{1+x^2}{1+x^3} = \frac{A}{1+x} + \frac{Bx+C}{1-x+x^2} \Rightarrow 1+x^2 = A(1-x+x^2) + (Bx+C)(1+x)$
$= (A+B)x^2 + (-A+B+C)x + (A+C) \Rightarrow A+B = 1,\ -A+B+C = 0,\ A+C = 1 \Rightarrow A = \frac{2}{3}, B = \frac{1}{3}, C = \frac{1}{3}$;
$\int \frac{1+x^2}{1+x^3}\,dx = \int\left(\frac{2/3}{1+x} + \frac{(1/3)x+1/3}{1-x+x^2}\right)dx = \frac{2}{3}\int \frac{1}{1+x}\,dx + \frac{1}{3}\int \frac{x+1}{1-x+x^2}\,dx = \frac{2}{3}\int \frac{1}{1+x}\,dx + \frac{1}{3}\int \frac{x+1}{\frac{3}{4}+\left(x-\frac{1}{2}\right)^2}\,dx$;
$\left[\begin{array}{c} u = x - \frac{1}{2} \\ du = dx \end{array}\right] \to \frac{1}{3}\int \frac{u+\frac{3}{2}}{\frac{3}{4}+u^2}\,du = \frac{1}{3}\int \frac{u}{\frac{3}{4}+u^2}\,du + \frac{1}{2}\int \frac{1}{\frac{3}{4}+u^2}\,du = \frac{1}{6}\ln\left|\frac{3}{4} + u^2\right| + \frac{1}{\sqrt{3}}\tan^{-1}\left(\frac{u}{\sqrt{3}/2}\right)$
$= \frac{1}{6}\ln\left|\frac{3}{4} + \left(x - \frac{1}{2}\right)^2\right| + \frac{1}{\sqrt{3}}\tan^{-1}\left(\frac{x-\frac{1}{2}}{\sqrt{3}/2}\right) - \frac{1}{6}\ln|1 - x + x^2| + \frac{1}{\sqrt{3}}\tan^{-1}\left(\frac{2x-1}{\sqrt{3}}\right)$
$\Rightarrow \frac{2}{3}\int \frac{1}{1+x}\,dx + \frac{1}{3}\int \frac{x+1}{1-x+x^2}\,dx = \frac{2}{3}\ln|1+x| + \frac{1}{6}\ln|1 - x + x^2| + \frac{1}{\sqrt{3}}\tan^{-1}\left(\frac{2x-1}{\sqrt{3}}\right) + C$

103. $\int \sqrt{x}\sqrt{1+\sqrt{x}}\,dx;\ \begin{bmatrix} w=\sqrt{x} \Rightarrow w^2 = x \\ 2w\,dw = dx \end{bmatrix} \to \int 2w^2\sqrt{1+w}\,dw$

$$\begin{array}{lcl} & & \sqrt{1+w} \\ 2w^2 & \xrightarrow{(+)} & \frac{2}{3}(1+w)^{3/2} \\ 4w & \xrightarrow{(-)} & \frac{4}{15}(1+w)^{5/2} \\ 4 & \xrightarrow{(+)} & \frac{8}{105}(1+w)^{7/2} \\ 0 & & \end{array}$$

$\Rightarrow \int 2w^2\sqrt{1+w}\,dw = \frac{4}{3}w^2(1+w)^{3/2} - \frac{16}{15}w(1+w)^{5/2} + \frac{32}{105}(1+w)^{7/2} + C$

$= \frac{4}{3}x\left(1+\sqrt{x}\right)^{3/2} - \frac{16}{15}\sqrt{x}\left(1+\sqrt{x}\right)^{5/2} + \frac{32}{105}\left(1+\sqrt{x}\right)^{7/2} + C$

105. $\int \frac{1}{\sqrt{x}\sqrt{1+x}}\,dx;\ \begin{bmatrix} u=\sqrt{x} \Rightarrow u^2 = x \\ 2u\,du = dx \end{bmatrix} \to \int \frac{2}{\sqrt{1+u^2}}\,du;\ \left[u = \tan\theta,\ -\frac{\pi}{2} < \theta < \frac{\pi}{2},\ du = \sec^2\theta\,d\theta,\ \sqrt{1+u^2} = \sec\theta\right]$

$\int \frac{2}{\sqrt{1+u^2}}\,du = \int \frac{2\sec^2\theta}{\sec\theta}\,d\theta = \int 2\sec\theta\,d\theta = 2\ln|\sec\theta + \tan\theta| + C = 2\ln\left|\sqrt{1+u^2}+u\right| + C$

$= 2\ln\left|\sqrt{1+x}+\sqrt{x}\right| + C$

107. $\int \frac{\ln x}{x + x\ln x}\,dx = \int \frac{\ln x}{x(1+\ln x)}\,dx;\ \begin{bmatrix} u = 1+\ln x \\ du = \frac{1}{x}dx \end{bmatrix} \to \int \frac{u-1}{u}\,du = \int du - \int \frac{1}{u}\,du = u - \ln|u| + C$

$= (1+\ln x) - \ln|1+\ln x| + C = \ln x - \ln|1+\ln x| + C$

109. $\int \frac{x^{\ln x}\ln x}{x}dx;\ \left[u = x^{\ln x} \Rightarrow \ln u = \ln x^{\ln x} = (\ln x)^2 \Rightarrow \frac{1}{u}du = \frac{2\ln x}{x}\,dx \Rightarrow du = \frac{2u\ln x}{x}\,dx = \frac{2x^{\ln x}\ln x}{x}\,dx\right] \to \frac{1}{2}\int du$

$= \frac{1}{2}u + C = \frac{1}{2}x^{\ln x} + C$

111. $\int \frac{1}{x\sqrt{1-x^4}}\,dx = \int \frac{x}{x^2\sqrt{1-x^4}}\,dx;\ \left[x^2 = \sin\theta,\ 0 \le \theta < \frac{\pi}{2},\ 2x\,dx = \cos\theta\,d\theta,\ \sqrt{1-x^4} = \cos\theta\right] \to \frac{1}{2}\int \frac{\cos\theta}{\sin\theta\cos\theta}\,d\theta$

$= \frac{1}{2}\int \csc\theta\,d\theta = -\frac{1}{2}\ln|\csc\theta + \cot\theta| + C = -\frac{1}{2}\ln\left|\frac{1}{x^2} + \frac{\sqrt{1-x^4}}{x^2}\right| + C = -\frac{1}{2}\ln\left|\frac{1+\sqrt{1-x^4}}{x^2}\right| + C$

113. (a) $\int_0^a f(a-x)\,dx;\ \left[u = a - x \Rightarrow du = -dx,\ x = 0 \Rightarrow u = a,\ x = a \Rightarrow u = 0\right] \to -\int_a^0 f(u)\,du = \int_0^a f(u)\,du$, which is the same integral as $\int_0^a f(x)\,dx$.

(b) $\int_0^{\pi/2} \frac{\sin x}{\sin x + \cos x}\,dx = \int_0^{\pi/2} \frac{\sin\left(\frac{\pi}{2}-x\right)}{\sin\left(\frac{\pi}{2}-x\right) + \cos\left(\frac{\pi}{2}-x\right)}\,dx = \int_0^{\pi/2} \frac{\sin\left(\frac{\pi}{2}\right)\cos x - \cos\left(\frac{\pi}{2}\right)\sin x}{\sin\left(\frac{\pi}{2}\right)\cos x - \cos\left(\frac{\pi}{2}\right)\sin x + \cos\left(\frac{\pi}{2}\right)\cos x + \sin\left(\frac{\pi}{2}\right)\sin x}\,dx$

$= \int_0^{\pi/2} \frac{\cos x}{\cos x + \sin x}\,dx \Rightarrow 2\int_0^{\pi/2} \frac{\sin x}{\sin x + \cos x}\,dx = \int_0^{\pi/2} \frac{\sin x}{\sin x + \cos x}\,dx + \int_0^{\pi/2} \frac{\cos x}{\cos x + \sin x}\,dx = \int_0^{\pi/2} \frac{\sin x + \cos x}{\sin x + \cos x}\,dx = \int_0^{\pi/2} dx$

$= \left[x\right]_0^{\pi/2} = \frac{\pi}{2} \Rightarrow 2\int_0^{\pi/2} \frac{\sin x}{\sin x + \cos x}\,dx = \frac{\pi}{2} \Rightarrow \int_0^{\pi/2} \frac{\sin x}{\sin x + \cos x}\,dx = \frac{\pi}{4}$

115. $\int \frac{\sin^2 x}{1+\sin^2 x}\,dx = \int \frac{\frac{\sin^2 x}{\cos^2 x}}{\frac{1}{\cos^2 x} + \frac{\sin^2 x}{\cos^2 x}}\,dx = \int \frac{\tan^2 x}{\sec^2 x + \tan^2 x}\,dx = \int \frac{\tan^2 x + \sec^2 x - \sec^2 x}{\sec^2 x + \tan^2 x}\,dx = \int \frac{\tan^2 x + \sec^2 x}{\sec^2 x + \tan^2 x}\,dx - \int \frac{\sec^2 x}{\sec^2 x + \tan^2 x}\,dx$

$= \int dx - \int \frac{\sec^2 x}{1 + 2\tan^2 x}\,dx = x - \frac{1}{\sqrt{2}}\tan^{-1}\left(\sqrt{2}\tan x\right) + C$

CHAPTER 8 ADDITIONAL AND ADVANCED EXERCISES

1. $u = (\sin^{-1} x)^2,\ du = \frac{2\sin^{-1} x\,dx}{\sqrt{1-x^2}}$; $dv = dx,\ v = x$;
$\int (\sin^{-1} x)^2\,dx = x(\sin^{-1} x)^2 - \int \frac{2x\sin^{-1} x\,dx}{\sqrt{1-x^2}}$;
$u = \sin^{-1} x,\ du = \frac{dx}{\sqrt{1-x^2}}$; $dv = -\frac{2x\,dx}{\sqrt{1-x^2}},\ v = 2\sqrt{1-x^2}$;
$-\int \frac{2x\sin^{-1} x\,dx}{\sqrt{1-x^2}} = 2(\sin^{-1} x)\sqrt{1-x^2} - \int 2\,dx = 2(\sin^{-1} x)\sqrt{1-x^2} - 2x + C$; therefore
$\int (\sin^{-1} x)^2\,dx = x(\sin^{-1} x)^2 + 2(\sin^{-1} x)\sqrt{1-x^2} - 2x + C$

3. $u = \sin^{-1} x,\ du = \frac{dx}{\sqrt{1-x^2}}$; $dv = x\,dx,\ v = \frac{x^2}{2}$;
$\int x\sin^{-1} x\,dx = \frac{x^2}{2}\sin^{-1} x - \int \frac{x^2\,dx}{2\sqrt{1-x^2}}$; $\begin{bmatrix} x = \sin\theta \\ dx = \cos\theta\,d\theta \end{bmatrix} \to \int x\sin^{-1} x\,dx = \frac{x^2}{2}\sin^{-1} x - \int \frac{\sin^2\theta\cos\theta\,d\theta}{2\cos\theta}$
$= \frac{x^2}{2}\sin^{-1} x - \frac{1}{2}\int \sin^2\theta\,d\theta = \frac{x^2}{2}\sin^{-1} x - \frac{1}{2}\left(\frac{\theta}{2} - \frac{\sin 2\theta}{4}\right) + C = \frac{x^2}{2}\sin^{-1} x + \frac{\sin\theta\cos\theta - \theta}{4} + C$
$= \frac{x^2}{2}\sin^{-1} x + \frac{x\sqrt{1-x^2} - \sin^{-1} x}{4} + C$

5. $\int \frac{dt}{t-\sqrt{1-t^2}}$; $\begin{bmatrix} t = \sin\theta \\ dt = \cos\theta\,d\theta \end{bmatrix} \to \int \frac{\cos\theta\,d\theta}{\sin\theta - \cos\theta} = \int \frac{d\theta}{\tan\theta - 1}$; $\begin{bmatrix} u = \tan\theta \\ du = \sec^2\theta\,d\theta \\ d\theta = \frac{du}{u^2+1} \end{bmatrix} \to \int \frac{du}{(u-1)(u^2+1)}$
$= \frac{1}{2}\int \frac{du}{u-1} - \frac{1}{2}\int \frac{du}{u^2+1} - \frac{1}{2}\int \frac{u\,du}{u^2+1} = \frac{1}{2}\ln\left|\frac{u-1}{\sqrt{u^2+1}}\right| - \frac{1}{2}\tan^{-1} u + C = \frac{1}{2}\ln\left|\frac{\tan\theta - 1}{\sec\theta}\right| - \frac{1}{2}\theta + C$
$= \frac{1}{2}\ln\left(t - \sqrt{1-t^2}\right) - \frac{1}{2}\sin^{-1} t + C$

7. $\lim_{x\to\infty} \int_{-x}^{x} \sin t\,dt = \lim_{x\to\infty} [-\cos t]_{-x}^{x} = \lim_{x\to\infty} [-\cos x + \cos(-x)] = \lim_{x\to\infty} (-\cos x + \cos x) = \lim_{x\to\infty} 0 = 0$

9. $\lim_{n\to\infty} \sum_{k=1}^{n} \ln \sqrt[n]{1 + \frac{k}{n}} = \lim_{n\to\infty} \sum_{k=1}^{n} \ln\left(1 + k\left(\frac{1}{n}\right)\right)\left(\frac{1}{n}\right) = \int_0^1 \ln(1+x)\,dx$; $\begin{bmatrix} u = 1+x,\ du = dx \\ x = 0 \Rightarrow u = 1,\ x = 1 \Rightarrow u = 2 \end{bmatrix}$
$\to \int_1^2 \ln u\,du = [u\ln u - u]_1^2 = (2\ln 2 - 2) - (\ln 1 - 1) = 2\ln 2 - 1 = \ln 4 - 1$

11. $\frac{dy}{dx} = \sqrt{\cos 2x} \Rightarrow 1 + \left(\frac{dy}{dx}\right)^2 = 1 + \cos 2x = 2\cos^2 x$; $L = \int_0^{\pi/4} \sqrt{1 + \left(\sqrt{\cos 2t}\right)^2}\,dt = \sqrt{2}\int_0^{\pi/4} \sqrt{\cos^2 t}\,dt$
$= \sqrt{2}\,[\sin t]_0^{\pi/4} = 1$

13. $V = \int_a^b 2\pi\left(\begin{smallmatrix}\text{shell}\\ \text{radius}\end{smallmatrix}\right)\left(\begin{smallmatrix}\text{shell}\\ \text{height}\end{smallmatrix}\right)dx = \int_0^1 2\pi xy\,dx$
$= 6\pi\int_0^1 x^2\sqrt{1-x}\,dx$; $\begin{bmatrix} u = 1-x \\ du = -dx \\ x^2 = (1-u)^2 \end{bmatrix}$
$\to -6\pi\int_1^0 (1-u)^2\sqrt{u}\,du$
$= -6\pi\int_1^0 \left(u^{1/2} - 2u^{3/2} + u^{5/2}\right)du$
$= -6\pi\left[\frac{2}{3}u^{3/2} - \frac{4}{5}u^{5/2} + \frac{2}{7}u^{7/2}\right]_1^0 = 6\pi\left(\frac{2}{3} - \frac{4}{5} + \frac{2}{7}\right)$
$= 6\pi\left(\frac{70 - 84 + 30}{105}\right) = 6\pi\left(\frac{16}{105}\right) = \frac{32\pi}{35}$

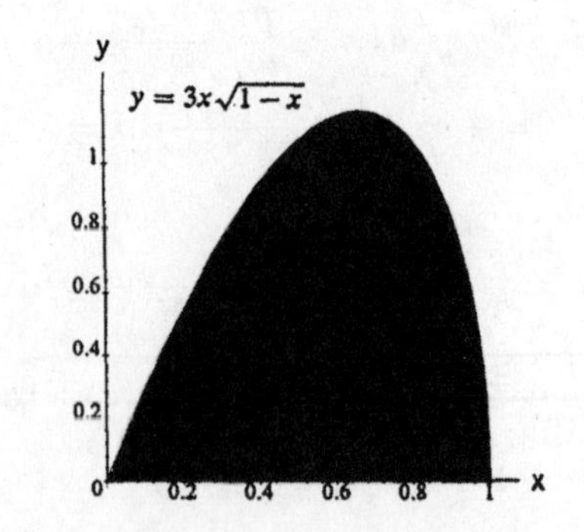

15. $V = \int_a^b 2\pi \left(\begin{smallmatrix}\text{shell}\\ \text{radius}\end{smallmatrix}\right)\left(\begin{smallmatrix}\text{shell}\\ \text{height}\end{smallmatrix}\right) dx = \int_0^1 2\pi x e^x\, dx$

$= 2\pi \left[xe^x - e^x\right]_0^1 = 2\pi$

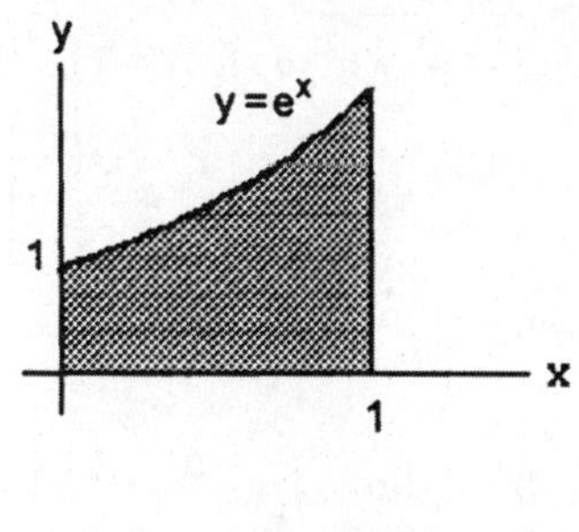

17. (a) $V = \int_1^e \pi\left[1 - (\ln x)^2\right] dx$

$= \pi\left[x - x(\ln x)^2\right]_1^e + 2\pi \int_1^e \ln x\, dx$

(FORMULA 110)

$= \pi\left[x - x(\ln x)^2 + 2(x \ln x - x)\right]_1^e$

$= \pi\left[-x - x(\ln x)^2 + 2x \ln x\right]_1^e$

$= \pi\left[-e - e + 2e - (-1)\right] = \pi$

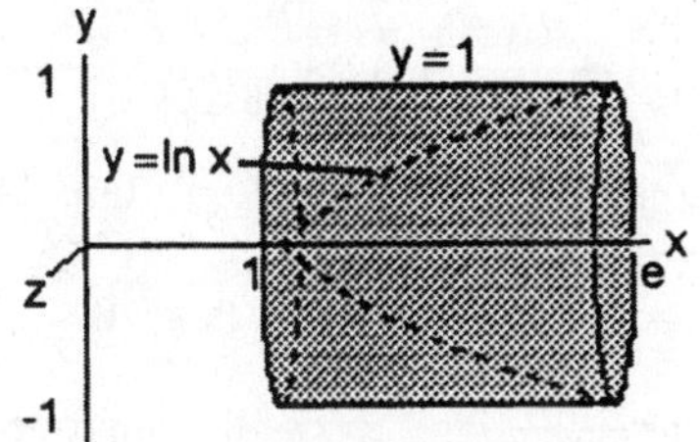

(b) $V = \int_1^e \pi(1 - \ln x)^2\, dx = \pi \int_1^e \left[1 - 2\ln x + (\ln x)^2\right] dx$

$= \pi\left[x - 2(x \ln x - x) + x(\ln x)^2\right]_1^e - 2\pi \int_1^e \ln x\, dx$

$= \pi\left[x - 2(x \ln x - x) + x(\ln x)^2 - 2(x \ln x - x)\right]_1^e$

$= \pi\left[5x - 4x \ln x + x(\ln x)^2\right]_1^e$

$= \pi\left[(5e - 4e + e) - (5)\right] = \pi(2e - 5)$

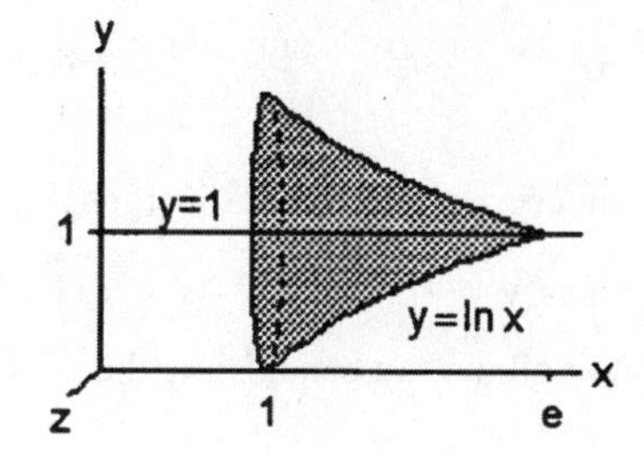

19. (a) $\lim_{x \to 0^+} x \ln x = 0 \Rightarrow \lim_{x \to 0^+} f(x) = 0 = f(0) \Rightarrow$ f is continuous

(b) $V = \int_0^2 \pi x^2 (\ln x)^2\, dx; \left[\begin{array}{c} u = (\ln x)^2 \\ du = (2 \ln x)\frac{dx}{x} \\ dv = x^2 dx \\ v = \frac{x^3}{3} \end{array}\right] \to \pi\left(\lim_{b \to 0^+} \left[\frac{x^3}{3}(\ln x)^2\right]_b^2 - \int_0^2 \left(\frac{x^3}{3}\right)(2 \ln x)\frac{dx}{x}\right)$

$= \pi\left[\left(\frac{8}{3}\right)(\ln 2)^2 - \left(\frac{2}{3}\right)\lim_{b \to 0^+}\left[\frac{x^3}{3}\ln x - \frac{x^3}{9}\right]_b^2\right] = \pi\left[\frac{8(\ln 2)^2}{3} - \frac{16(\ln 2)}{9} + \frac{16}{27}\right]$

21. $M = \int_1^e \ln x\, dx = \left[x \ln x - x\right]_1^e = (e - e) - (0 - 1) = 1;$

$M_x = \int_1^e (\ln x)\left(\frac{\ln x}{2}\right) dx = \frac{1}{2}\int_1^e (\ln x)^2\, dx$

$= \frac{1}{2}\left(\left[x(\ln x)^2\right]_1^e - 2\int_1^e \ln x\, dx\right) = \frac{1}{2}(e - 2);$

$M_y = \int_1^e x \ln x\, dx = \left[\frac{x^2 \ln x}{2}\right]_1^e - \frac{1}{2}\int_1^e x\, dx$

$= \frac{1}{2}\left[x^2 \ln x - \frac{x^2}{2}\right]_1^e = \frac{1}{2}\left[\left(e^2 - \frac{e^2}{2}\right) + \frac{1}{2}\right] = \frac{1}{4}(e^2 + 1);$

therefore, $\bar{x} = \frac{M_y}{M} = \frac{e^2+1}{4}$ and $\bar{y} = \frac{M_x}{M} = \frac{e-2}{2}$

23. $L = \int_1^e \sqrt{1 + \frac{1}{x^2}}\, dx = \int_1^e \frac{\sqrt{x^2+1}}{x}\, dx; \left[\begin{array}{c} x = \tan\theta \\ dx = \sec^2\theta\, d\theta \end{array}\right] \to L = \int_{\pi/4}^{\tan^{-1} e} \frac{\sec\theta \cdot \sec^2\theta\, d\theta}{\tan\theta}$

$= \int_{\pi/4}^{\tan^{-1} e} \frac{(\sec\theta)(\tan^2\theta + 1)}{\tan\theta}\, d\theta = \int_{\pi/4}^{\tan^{-1} e} (\tan\theta \sec\theta + \csc\theta)\, d\theta = \left[\sec\theta - \ln\left|\csc\theta + \cot\theta\right|\right]_{\pi/4}^{\tan^{-1} e}$

$= \left(\sqrt{1+e^2} - \ln\left|\frac{\sqrt{1+e^2}}{e} + \frac{1}{e}\right|\right) - \left[\sqrt{2} - \ln\left(1 + \sqrt{2}\right)\right] = \sqrt{1+e^2} - \ln\left(\frac{\sqrt{1+e^2}}{e} + \frac{1}{e}\right) - \sqrt{2} + \ln\left(1 + \sqrt{2}\right)$

25. $S = 2\pi \int_{-1}^{1} f(x)\sqrt{1 + [f'(x)]^2}\, dx$; $f(x) = \left(1 - x^{2/3}\right)^{3/2} \Rightarrow [f'(x)]^2 + 1 = \frac{1}{x^{2/3}} \Rightarrow S = 2\pi \int_{-1}^{1} \left(1 - x^{2/3}\right)^{3/2} \cdot \frac{dx}{\sqrt{x^{2/3}}}$

$= 4\pi \int_0^1 \left(1 - x^{2/3}\right)^{3/2}\left(\frac{1}{x^{1/3}}\right) dx$; $\left[\begin{array}{c} u = x^{2/3} \\ du = \frac{2}{3}\frac{dx}{x^{1/3}} \end{array}\right] \to 4 \cdot \frac{3}{2}\pi \int_0^1 (1 - u)^{3/2}\, du = -6\pi \int_0^1 (1 - u)^{3/2}\, d(1 - u)$

$= -6\pi \cdot \frac{2}{5}\left[(1 - u)^{5/2}\right]_0^1 = \frac{12\pi}{5}$

27. $\int_1^{\infty} \left(\frac{ax}{x^2 + 1} - \frac{1}{2x}\right) dx = \lim_{b \to \infty} \int_1^b \left(\frac{ax}{x^2 + 1} - \frac{1}{2x}\right) dx = \lim_{b \to \infty} \left[\frac{a}{2} \ln(x^2 + 1) - \frac{1}{2} \ln x\right]_1^b = \lim_{b \to \infty} \left[\frac{1}{2} \ln \frac{(x^2 + 1)^a}{x}\right]_1^b$

$= \lim_{b \to \infty} \frac{1}{2}\left[\ln \frac{(b^2 + 1)^a}{b} - \ln 2^a\right]$; $\lim_{b \to \infty} \frac{(b^2 + 1)^a}{b} > \lim_{b \to \infty} \frac{b^{2a}}{b} = \lim_{b \to \infty} b^{2\left(a - \frac{1}{2}\right)} = \infty$ if $a > \frac{1}{2} \Rightarrow$ the improper integral diverges if $a > \frac{1}{2}$; for $a = \frac{1}{2}$: $\lim_{b \to \infty} \frac{\sqrt{b^2 + 1}}{b} = \lim_{b \to \infty} \sqrt{1 + \frac{1}{b^2}} = 1 \Rightarrow \lim_{b \to \infty} \frac{1}{2}\left[\ln \frac{(b^2+1)^{1/2}}{b} - \ln 2^{1/2}\right]$

$= \frac{1}{2}\left(\ln 1 - \frac{1}{2} \ln 2\right) = -\frac{\ln 2}{4}$; if $a < \frac{1}{2}$: $0 \le \lim_{b \to \infty} \frac{(b^2+1)^a}{b} < \lim_{b \to \infty} \frac{(b+1)^{2a}}{b+1} = \lim_{b \to \infty} (b + 1)^{2a-1} = 0$

$\Rightarrow \lim_{b \to \infty} \ln \frac{(b^2 + 1)^a}{b} = -\infty \Rightarrow$ the improper integral diverges if $a < \frac{1}{2}$; in summary, the improper integral $\int_1^{\infty} \left(\frac{ax}{x^2 + 1} - \frac{1}{2x}\right) dx$ converges only when $a = \frac{1}{2}$ and has the value $-\frac{\ln 2}{4}$

29. $A = \int_1^{\infty} \frac{dx}{x^p}$ converges if $p > 1$ and diverges if $p \le 1$. Thus, $p \le 1$ for infinite area. The volume of the solid of revolution about the x-axis is $V = \int_1^{\infty} \pi \left(\frac{1}{x^p}\right)^2 dx = \pi \int_1^{\infty} \frac{dx}{x^{2p}}$ which converges if $2p > 1$ and diverges if $2p \le 1$. Thus we want $p > \frac{1}{2}$ for finite volume. In conclusion, the curve $y = x^{-p}$ gives infinite area and finite volume for values of p satisfying $\frac{1}{2} < p \le 1$.

31. (a) $\Gamma(1) = \int_0^{\infty} e^{-t}\, dt = \lim_{b \to \infty} \int_0^b e^{-t}\, dt = \lim_{b \to \infty} \left[-e^{-t}\right]_0^b = \lim_{b \to \infty} \left[-\frac{1}{e^b} - (-1)\right] = 0 + 1 = 1$

(b) $u = t^x$, $du = xt^{x-1}\, dt$; $dv = e^{-t}\, dt$, $v = -e^{-t}$; x = fixed positive real

$\Rightarrow \Gamma(x + 1) = \int_0^{\infty} t^x e^{-t}\, dt = \lim_{b \to \infty} \left[-t^x e^{-t}\right]_0^b + x \int_0^{\infty} t^{x-1} e^{-t}\, dt = \lim_{b \to \infty} \left(-\frac{b^x}{e^b} + 0^x e^0\right) + x\Gamma(x) = x\Gamma(x)$

(c) $\Gamma(n + 1) = n\Gamma(n) = n!$:

$n = 0$: $\Gamma(0 + 1) = \Gamma(1) = 0!$;	
$n = k$: Assume $\Gamma(k + 1) = k!$	for some $k > 0$;
$n = k + 1$: $\Gamma(k + 1 + 1) = (k + 1)\,\Gamma(k + 1)$	from part (b)
$= (k + 1)k!$	induction hypothesis
$= (k + 1)!$	definition of factorial

Thus, $\Gamma(n + 1) = n\Gamma(n) = n!$ for every positive integer n.

33.

e^{2x}	(+)	$\cos 3x$
$2e^{2x}$	(−)	$\frac{1}{3}\sin 3x$
$4e^{2x}$	(+)	$-\frac{1}{9}\cos 3x$

$I = \frac{e^{2x}}{3} \sin 3x + \frac{2e^{2x}}{9} \cos 3x - \frac{4}{9} I \Rightarrow \frac{13}{9} I = \frac{e^{2x}}{9}(3 \sin 3x + 2 \cos 3x) \Rightarrow I = \frac{e^{2x}}{13}(3 \sin 3x + 2 \cos 3x) + C$

35.

$\sin 3x$	(+)	$\sin x$
$3 \cos 3x$	(−)	$-\cos x$
$-9 \sin 3x$	(+)	$-\sin x$

$I = -\sin 3x \cos x + 3 \cos 3x \sin x + 9I \Rightarrow -8I = -\sin 3x \cos x + 3 \cos 3x \sin x$

$\Rightarrow I = \frac{\sin 3x \cos x - 3 \cos 3x \sin x}{8} + C$

37. e^{ax} (+) $\sin bx$

ae^{ax} (−) $-\frac{1}{b}\cos bx$

a^2e^{ax} (+) $-\frac{1}{b^2}\sin bx$

$I = -\frac{e^{ax}}{b}\cos bx + \frac{ae^{ax}}{b^2}\sin bx - \frac{a^2}{b^2}I \Rightarrow \left(\frac{a^2+b^2}{b^2}\right)I = \frac{e^{ax}}{b^2}(a\sin bx - b\cos bx)$

$\Rightarrow I = \frac{e^{ax}}{a^2+b^2}(a\sin bx - b\cos bx) + C$

39. $\ln(ax)$ (+) 1

$\frac{1}{x}$ (−) x

$I = x\ln(ax) - \int\left(\frac{1}{x}\right)x\,dx = x\ln(ax) - x + C$

41. $\int \frac{dx}{1-\sin x} = \int \frac{\left(\frac{2\,dz}{1+z^2}\right)}{1-\left(\frac{2z}{1+z^2}\right)} = \int \frac{2\,dz}{(1-z)^2} = \frac{2}{1-z} + C = \frac{2}{1-\tan\left(\frac{x}{2}\right)} + C$

43. $\int_0^{\pi/2} \frac{dx}{1+\sin x} = \int_0^1 \frac{\left(\frac{2\,dz}{1+z^2}\right)}{1+\left(\frac{2z}{1+z^2}\right)} = \int_0^1 \frac{2\,dz}{(1+z)^2} = -\left[\frac{2}{1+z}\right]_0^1 = -(1-2) = 1$

45. $\int_0^{\pi/2} \frac{d\theta}{2+\cos\theta} = \int_0^1 \frac{\left(\frac{2\,dz}{1+z^2}\right)}{2+\left(\frac{1-z^2}{1+z^2}\right)} = \int_0^1 \frac{2\,dz}{2+2z^2+1-z^2} = \int_0^1 \frac{2\,dz}{z^2+3} = \frac{2}{\sqrt{3}}\left[\tan^{-1}\frac{z}{\sqrt{3}}\right]_0^1 = \frac{2}{\sqrt{3}}\tan^{-1}\frac{1}{\sqrt{3}}$

$= \frac{\pi}{3\sqrt{3}} = \frac{\sqrt{3}\pi}{9}$

47. $\int \frac{dt}{\sin t - \cos t} = \int \frac{\left(\frac{2\,dz}{1+z^2}\right)}{\left(\frac{2z}{1+z^2} - \frac{1-z^2}{1+z^2}\right)} = \int \frac{2\,dz}{2z-1+z^2} = \int \frac{2\,dz}{(z+1)^2-2} = \frac{1}{\sqrt{2}}\ln\left|\frac{z+1-\sqrt{2}}{z+1+\sqrt{2}}\right| + C$

$= \frac{1}{\sqrt{2}}\ln\left|\frac{\tan\left(\frac{t}{2}\right)+1-\sqrt{2}}{\tan\left(\frac{t}{2}\right)+1+\sqrt{2}}\right| + C$

49. $\int \sec\theta\,d\theta = \int \frac{d\theta}{\cos\theta} = \int \frac{\left(\frac{2\,dz}{1+z^2}\right)}{\left(\frac{1-z^2}{1+z^2}\right)} = \int \frac{2\,dz}{1-z^2} = \int \frac{2\,dz}{(1+z)(1-z)} = \int \frac{dz}{1+z} + \int \frac{dz}{1-z}$

$= \ln|1+z| - \ln|1-z| + C = \ln\left|\frac{1+\tan\left(\frac{\theta}{2}\right)}{1-\tan\left(\frac{\theta}{2}\right)}\right| + C$

NOTES:

CHAPTER 9 FIRST-ORDER DIFFERENTIAL EQUATIONS

9.1 SOLUTIONS, SLOPE FIELDS AND EULER'S METHOD

1. $y' = x + y \Rightarrow$ slope of 0 for the line $y = -x$.
   For $x, y > 0$, $y' = x + y \Rightarrow$ slope > 0 in Quadrant I.
   For $x, y < 0$, $y' = x + y \Rightarrow$ slope < 0 in Quadrant III.
   For $|y| > |x|$, $y > 0$, $x < 0$, $y' = x + y \Rightarrow$ slope > 0 in Quadrant II above $y = -x$.
   For $|y| < |x|$, $y > 0$, $x < 0$, $y' = x + y \Rightarrow$ slope < 0 in Quadrant II below $y = -x$.
   For $|y| < |x|$, $x > 0$, $y < 0$, $y' = x + y \Rightarrow$ slope > 0 in Quadrant IV above $y = -x$.
   For $|y| > |x|$, $x > 0$, $y < 0$, $y' = x + y \Rightarrow$ slope < 0 in Quadrant IV below $y = -x$.
   All of the conditions are seen in slope field (d).

3. $y' = -\frac{x}{y} \Rightarrow$ slope $= 1$ on $y = -x$ and -1 on $y = x$.
   $y' = -\frac{x}{y} \Rightarrow$ slope $= 0$ on the y-axis, excluding $(0, 0)$, and is undefined on the x-axis. Slopes are positive for $x > 0$, $y < 0$ and $x < 0$, $y > 0$ (Quadrants II and IV), otherwise negative. Field (a) is consistent with these conditions.

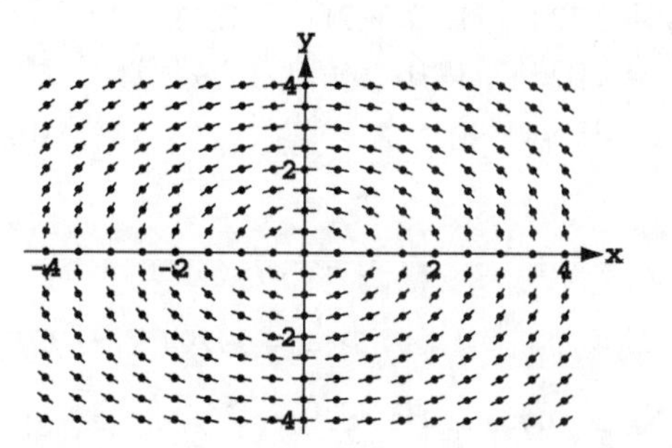

5.

7. $y = -1 + \int_1^x (t - y(t))dt \Rightarrow \frac{dy}{dx} = x - y(x)$; $y(1) = -1 + \int_1^1 (t - y(t))dt = -1$; $\frac{dy}{dx} = x - y$, $y(1) = -1$

9. $y = 2 - \int_0^x (1 + y(t))\sin t\,dt \Rightarrow \frac{dy}{dx} = -(1 + y(x))\sin x$; $y(0) = 2 - \int_0^0 (1 + y(t))\sin t\,dt = 2$; $\frac{dy}{dx} = -(1 + y)\sin x$, $y(0) = 2$

11. $y_1 = y_0 + \left(1 - \frac{y_0}{x_0}\right) dx = -1 + \left(1 - \frac{-1}{2}\right)(.5) = -0.25$,
    $y_2 = y_1 + \left(1 - \frac{y_1}{x_1}\right) dx = -0.25 + \left(1 - \frac{-0.25}{2.5}\right)(.5) = 0.3$,
    $y_3 = y_2 + \left(1 - \frac{y_2}{x_2}\right) dx = 0.3 + \left(1 - \frac{0.3}{3}\right)(.5) = 0.75$;
    $\frac{dy}{dx} + \left(\frac{1}{x}\right) y = 1 \Rightarrow P(x) = \frac{1}{x}$, $Q(x) = 1 \Rightarrow \int P(x)\,dx = \int \frac{1}{x}\,dx = \ln|x| = \ln x,\ x > 0 \Rightarrow v(x) = e^{\ln x} = x$

$\Rightarrow y = \frac{1}{x}\int x \cdot 1\, dx = \frac{1}{x}\left(\frac{x^2}{2} + C\right)$; $x = 2, y = -1 \Rightarrow -1 = 1 + \frac{C}{2} \Rightarrow C = -4 \Rightarrow y = \frac{x}{2} - \frac{4}{x}$
$\Rightarrow y(3.5) = \frac{3.5}{2} - \frac{4}{3.5} = \frac{4.25}{7} \approx 0.6071$

13. $y_1 = y_0 + (2x_0y_0 + 2y_0)\, dx = 3 + [2(0)(3) + 2(3)](.2) = 4.2$,
$y_2 = y_1 + (2x_1y_1 + 2y_1)\, dx = 4.2 + [2(.2)(4.2) + 2(4.2)](.2) = 6.216$,
$y_3 = y_2 + (2x_2y_2 + 2y_2)\, dx = 6.216 + [2(.4)(6.216) + 2(6.216)](.2) = 9.6969$;
$\frac{dy}{dx} = 2y(x+1) \Rightarrow \frac{dy}{y} = 2(x+1)\, dx \Rightarrow \ln|y| = (x+1)^2 + C$; $x = 0, y = 3 \Rightarrow \ln 3 = 1 + C \Rightarrow C = \ln 3 - 1$
$\Rightarrow \ln y = (x+1)^2 + \ln 3 - 1 \Rightarrow y = e^{(x+1)^2 + \ln 3 - 1} = e^{\ln 3} e^{x^2+2x} = 3e^{x(x+2)} \Rightarrow y(.6) \approx 14.2765$

15. $y_1 = y_0 + 2x_0e^{x_0^2}\, dx = 2 + 2(0)(.1) = 2$,
$y_2 = y_1 + 2x_1e^{x_1^2}\, dx = 2 + 2(.1)\, e^{.1^2}(.1) = 2.0202$,
$y_3 = y_2 + 2x_2e^{x_2^2}\, dx = 2.0202 + 2(.2)\, e^{.2^2}(.1) = 2.0618$,
$dy = 2xe^{x^2}\, dx \Rightarrow y = e^{x^2} + C$; $y(0) = 2 \Rightarrow 2 = 1 + C \Rightarrow C = 1 \Rightarrow y = e^{x^2} + 1 \Rightarrow y(.3) = e^{.3^2} + 1 \approx 2.0942$

17. $y_1 = 1 + 1(.2) = 1.2$,
$y_2 = 1.2 + (1.2)(.2) = 1.44$,
$y_3 = 1.44 + (1.44)(.2) = 1.728$,
$y_4 = 1.728 + (1.728)(.2) = 2.0736$,
$y_5 = 2.0736 + (2.0736)(.2) = 2.48832$;
$\frac{dy}{y} = dx \Rightarrow \ln y = x + C_1 \Rightarrow y = Ce^x$; $y(0) = 1 \Rightarrow 1 = Ce^0 \Rightarrow C = 1 \Rightarrow y = e^x \Rightarrow y(1) = e \approx 2.7183$

19. $y_1 = -1 + \left[\frac{(-1)^2}{\sqrt{1}}\right](.5) = -.5$,
$y_2 = -.5 + \left[\frac{(-.5)^2}{\sqrt{1.5}}\right](.5) = -.39794$,
$y_3 = -.39794 + \left[\frac{(-.39794)^2}{\sqrt{2}}\right](.5) = -.34195$,
$y_4 = -.34195 + \left[\frac{(-.34195)^2}{\sqrt{2.5}}\right](.5) = -.30497$,
$y_5 = -.27812$, $y_6 = -.25745$, $y_7 = -.24088$, $y_8 = -.2272$;
$\frac{dy}{y^2} = \frac{dx}{\sqrt{x}} \Rightarrow -\frac{1}{y} = 2\sqrt{x} + C$; $y(1) = -1 \Rightarrow 1 = 2 + C \Rightarrow C = -1 \Rightarrow y = \frac{1}{1 - 2\sqrt{x}} \Rightarrow y(5) = \frac{1}{1 - 2\sqrt{5}} \approx -.2880$

21. $y = -1 - x + (1 + x_0 + y_0)e^{x - x_0} \Rightarrow y(x_0) = -1 - x_0 + (1 + x_0 + y_0)e^{x_0 - x_0} = -1 - x_0 + (1 + x_0 + y_0)(1) = y_0$
$\frac{dy}{dx} = -1 + (1 + x_0 + y_0)e^{x - x_0} \Rightarrow y = -1 - x + (1 + x_0 + y_0)e^{x - x_0} = \frac{dy}{dx} - x \Rightarrow \frac{dy}{dx} = x + y$

9.2 FIRST-ORDER LINEAR DIFFERENTIAL EQUATIONS

1. $x\frac{dy}{dx} + y = e^x \Rightarrow \frac{dy}{dx} + \left(\frac{1}{x}\right)y = \frac{e^x}{x}$, $P(x) = \frac{1}{x}$, $Q(x) = \frac{e^x}{x}$
$\int P(x)\, dx = \int \frac{1}{x}\, dx = \ln|x| = \ln x, x > 0 \Rightarrow v(x) = e^{\int P(x)\, dx} = e^{\ln x} = x$
$y = \frac{1}{v(x)}\int v(x)\, Q(x)\, dx = \frac{1}{x}\int x\left(\frac{e^x}{x}\right) dx = \frac{1}{x}(e^x + C) = \frac{e^x + C}{x}, x > 0$

3. $xy' + 3y = \frac{\sin x}{x^2}, x > 0 \Rightarrow \frac{dy}{dx} + \left(\frac{3}{x}\right)y = \frac{\sin x}{x^3}$, $P(x) = \frac{3}{x}$, $Q(x) = \frac{\sin x}{x^3}$
$\int \frac{3}{x}\, dx = 3\ln|x| = \ln x^3, x > 0 \Rightarrow v(x) = e^{\ln x^3} = x^3$
$y = \frac{1}{x^3}\int x^3\left(\frac{\sin x}{x^3}\right) dx = \frac{1}{x^3}\int \sin x\, dx = \frac{1}{x^3}(-\cos x + C) = \frac{C - \cos x}{x^3}, x > 0$

5. $x\,\frac{dy}{dx} + 2y = 1 - \frac{1}{x}$, $x > 0 \Rightarrow \frac{dy}{dx} + \left(\frac{2}{x}\right) y = \frac{1}{x} - \frac{1}{x^2}$, $P(x) = \frac{2}{x}$, $Q(x) = \frac{1}{x} - \frac{1}{x^2}$
$\int \frac{2}{x}\,dx = 2\ln|x| = \ln x^2$, $x > 0 \Rightarrow v(x) = e^{\ln x^2} = x^2$
$y = \frac{1}{x^2}\int x^2\left(\frac{1}{x} - \frac{1}{x^2}\right)dx = \frac{1}{x^2}\int (x-1)\,dx = \frac{1}{x^2}\left(\frac{x^2}{2} - x + C\right) = \frac{1}{2} - \frac{1}{x} + \frac{C}{x^2}$, $x > 0$

7. $\frac{dy}{dx} - \frac{1}{2}y = \frac{1}{2}e^{x/2} \Rightarrow P(x) = -\frac{1}{2}$, $Q(x) = \frac{1}{2}e^{x/2} \Rightarrow \int P(x)\,dx = -\frac{1}{2}x \Rightarrow v(x) = e^{-x/2}$
$\Rightarrow y = \frac{1}{e^{-x/2}}\int e^{-x/2}\left(\frac{1}{2}e^{x/2}\right)dx = e^{x/2}\int \frac{1}{2}\,dx = e^{x/2}\left(\frac{1}{2}x + C\right) = \frac{1}{2}xe^{x/2} + Ce^{x/2}$

9. $\frac{dy}{dx} - \left(\frac{1}{x}\right)y = 2\ln x \Rightarrow P(x) = -\frac{1}{x}$, $Q(x) = 2\ln x \Rightarrow \int P(x)\,dx = -\int \frac{1}{x}\,dx = -\ln x$, $x > 0$
$\Rightarrow v(x) = e^{-\ln x} = \frac{1}{x} \Rightarrow y = x\int \left(\frac{1}{x}\right)(2\ln x)\,dx = x\left[(\ln x)^2 + C\right] = x(\ln x)^2 + Cx$

11. $\frac{ds}{dt} + \left(\frac{4}{t-1}\right)s = \frac{t+1}{(t-1)^3} \Rightarrow P(t) = \frac{4}{t-1}$, $Q(t) = \frac{t+1}{(t-1)^3} \Rightarrow \int P(t)\,dt = \int \frac{4}{t-1}\,dt = 4\ln|t-1| = \ln(t-1)^4$
$\Rightarrow v(t) = e^{\ln(t-1)^4} = (t-1)^4 \Rightarrow s = \frac{1}{(t-1)^4}\int (t-1)^4\left[\frac{t+1}{(t-1)^3}\right]dt = \frac{1}{(t-1)^4}\int (t^2-1)\,dt$
$= \frac{1}{(t-1)^4}\left(\frac{t^3}{3} - t + C\right) = \frac{t^3}{3(t-1)^4} - \frac{t}{(t-1)^4} + \frac{C}{(t-1)^4}$

13. $\frac{dr}{d\theta} + (\cot\theta)r = \sec\theta \Rightarrow P(\theta) = \cot\theta$, $Q(\theta) = \sec\theta \Rightarrow \int P(\theta)\,d\theta = \int \cot\theta\,d\theta = \ln|\sin\theta| \Rightarrow v(\theta) = e^{\ln|\sin\theta|}$
$= \sin\theta$ because $0 < \theta < \frac{\pi}{2} \Rightarrow r = \frac{1}{\sin\theta}\int (\sin\theta)(\sec\theta)\,d\theta = \frac{1}{\sin\theta}\int \tan\theta\,d\theta = \frac{1}{\sin\theta}(\ln|\sec\theta| + C)$
$= (\csc\theta)(\ln|\sec\theta| + C)$

15. $\frac{dy}{dt} + 2y = 3 \Rightarrow P(t) = 2$, $Q(t) = 3 \Rightarrow \int P(t)\,dt = \int 2\,dt = 2t \Rightarrow v(t) = e^{2t} \Rightarrow y = \frac{1}{e^{2t}}\int 3e^{2t}\,dt$
$= \frac{1}{e^{2t}}\left(\frac{3}{2}e^{2t} + C\right)$; $y(0) = 1 \Rightarrow \frac{3}{2} + C = 1 \Rightarrow C = -\frac{1}{2} \Rightarrow y = \frac{3}{2} - \frac{1}{2}e^{-2t}$

17. $\frac{dy}{d\theta} + \left(\frac{1}{\theta}\right)y = \frac{\sin\theta}{\theta} \Rightarrow P(\theta) = \frac{1}{\theta}$, $Q(\theta) = \frac{\sin\theta}{\theta} \Rightarrow \int P(\theta)\,d\theta = \ln|\theta| \Rightarrow v(\theta) = e^{\ln|\theta|} = |\theta|$
$\Rightarrow y = \frac{1}{|\theta|}\int |\theta|\left(\frac{\sin\theta}{\theta}\right)d\theta = \frac{1}{\theta}\int \theta\left(\frac{\sin\theta}{\theta}\right)d\theta$ for $\theta \neq 0 \Rightarrow y = \frac{1}{\theta}\int \sin\theta\,d\theta = \frac{1}{\theta}(-\cos\theta + C)$
$= -\frac{1}{\theta}\cos\theta + \frac{C}{\theta}$; $y\left(\frac{\pi}{2}\right) = 1 \Rightarrow C = \frac{\pi}{2} \Rightarrow y = -\frac{1}{\theta}\cos\theta + \frac{\pi}{2\theta}$

19. $(x+1)\frac{dy}{dx} - 2(x^2 + x)y = \frac{e^{x^2}}{x+1} \Rightarrow \frac{dy}{dx} - 2\left[\frac{x(x+1)}{x+1}\right]y = \frac{e^{x^2}}{(x+1)^2} \Rightarrow \frac{dy}{dx} - 2xy = \frac{e^{x^2}}{(x+1)^2} \Rightarrow P(x) = -2x$,
$Q(x) = \frac{e^{x^2}}{(x+1)^2} \Rightarrow \int P(x)\,dx = \int -2x\,dx = -x^2 \Rightarrow v(x) = e^{-x^2} \Rightarrow y = \frac{1}{e^{-x^2}}\int e^{-x^2}\left[\frac{e^{x^2}}{(x+1)^2}\right]dx$
$= e^{x^2}\int \frac{1}{(x+1)^2}\,dx = e^{x^2}\left[\frac{(x+1)^{-1}}{-1} + C\right] = -\frac{e^{x^2}}{x+1} + Ce^{x^2}$; $y(0) = 5 \Rightarrow -\frac{1}{0+1} + C = 5 \Rightarrow -1 + C = 5$
$\Rightarrow C = 6 \Rightarrow y = 6e^{x^2} - \frac{e^{x^2}}{x+1}$

21. $\frac{dy}{dt} - ky = 0 \Rightarrow P(t) = -k$, $Q(t) = 0 \Rightarrow \int P(t)\,dt = \int -k\,dt = -kt \Rightarrow v(t) = e^{-kt}$
$\Rightarrow y = \frac{1}{e^{-kt}}\int (e^{-kt})(0)\,dt = e^{kt}(0 + C) = Ce^{kt}$; $y(0) = y_0 \Rightarrow C = y_0 \Rightarrow y = y_0e^{kt}$

23. $x\int \frac{1}{x}\,dx = x(\ln|x| + C) = x\ln|x| + Cx \Rightarrow$ (b) is correct

25. Steady State $= \frac{V}{R}$ and we want $i = \frac{1}{2}\left(\frac{V}{R}\right) \Rightarrow \frac{1}{2}\left(\frac{V}{R}\right) = \frac{V}{R}\left(1 - e^{-Rt/L}\right) \Rightarrow \frac{1}{2} = 1 - e^{-Rt/L} \Rightarrow -\frac{1}{2} = -e^{-Rt/L}$
$\Rightarrow \ln\frac{1}{2} = -\frac{Rt}{L} \Rightarrow -\frac{L}{R}\ln\frac{1}{2} = t \Rightarrow t = \frac{L}{R}\ln 2$ sec

27. (a) $t = \frac{3L}{R} \Rightarrow i = \frac{V}{R}\left(1 - e^{(-R/L)(3L/R)}\right) = \frac{V}{R}\left(1 - e^{-3}\right) \approx 0.9502\,\frac{V}{R}$ amp, or about 95% of the steady state value

(b) $t = \frac{2L}{R} \Rightarrow i = \frac{V}{R}\left(1 - e^{(-R/L)(2L/R)}\right) = \frac{V}{R}\left(1 - e^{-2}\right) \approx 0.8647\, \frac{V}{R}$ amp, or about 86% of the steady state value

29. $y' - y = -y^2$; we have $n = 2$, so let $u = y^{1-2} = y^{-1}$. Then $y = u^{-1}$ and $\frac{du}{dx} = -1y^{-2}\frac{dy}{dx} \Rightarrow \frac{dy}{dx} = -y^2\frac{du}{dx}$
$\Rightarrow -u^{-2}\frac{du}{dx} - u^{-1} = -u^{-2} \Rightarrow \frac{du}{dx} + u = 1$. With $e^{\int dx} = e^x$ as the integrating factor, we have
$e^x\left(\frac{du}{dx} + u\right) = \frac{d}{dx}(e^x u) = e^x$. Integrating, we get $e^x u = e^x + C \Rightarrow u = 1 + \frac{C}{e^x} = \frac{1}{y} \Rightarrow y = \frac{1}{1+\frac{C}{e^x}} = \frac{e^x}{e^x + C}$

31. $xy' + y = y^{-2} \Rightarrow y' + \left(\frac{1}{x}\right)y = \left(\frac{1}{x}\right)y^{-2}$. Let $u = y^{1-(-2)} = y^3 \Rightarrow y = u^{1/3}$ and $y^{-2} = u^{-2/3}$.
$\frac{du}{dx} = 3y^2\frac{dy}{dx} \Rightarrow y' = \frac{dy}{dx} = \left(\frac{1}{3}\right)\left(\frac{du}{dx}\right)\left(y^{-2}\right) = \left(\frac{1}{3}\right)\left(\frac{du}{dx}\right)\left(u^{-2/3}\right)$. Thus we have

$\left(\frac{1}{3}\right)\left(\frac{du}{dx}\right)\left(u^{-2/3}\right) + \left(\frac{1}{x}\right)u^{1/3} = \left(\frac{1}{x}\right)u^{-2/3} \Rightarrow \frac{du}{dx} + \left(\frac{3}{x}\right)u = \left(\frac{3}{x}\right)1$. The integrating factor, $v(x)$, is
$e^{\int \frac{3}{x}dx} = e^{3\ln x} = e^{\ln x^3} = x^3$. Thus $\frac{d}{dx}(x^3 u) = \left(\frac{3}{x}\right)x^3 = 3x^2 \Rightarrow x^3 u = x^3 + C \Rightarrow u = 1 + \frac{C}{x^3} = y^3$
$\Rightarrow y = \left(1 + \frac{C}{x^3}\right)^{1/3}$

9.3 APPLICATIONS

1. Note that the total mass is $66 + 7 = 73$ kg, therefore, $v = v_0 e^{-(k/m)t} \Rightarrow v = 9e^{-3.9t/73}$
(a) $s(t) = \int 9e^{-3.9t/73}dt = -\frac{2190}{13}e^{-3.9t/73} + C$
Since $s(0) = 0$ we have $C = \frac{2190}{13}$ and $\lim_{t\to\infty} s(t) = \lim_{t\to\infty} \frac{2190}{13}\left(1 - e^{-3.9t/73}\right) = \frac{2190}{13} \approx 168.5$
The cyclist will coast about 168.5 meters.
(b) $1 = 9e^{-3.9t/73} \Rightarrow \frac{3.9t}{73} = \ln 9 \Rightarrow t = \frac{73\ln 9}{3.9} \approx 41.13$ sec
It will take about 41.13 seconds.

3. The total distance traveled $= \frac{v_0 m}{k} \Rightarrow \frac{(2.75)(39.92)}{k} = 4.91 \Rightarrow k = 22.36$. Therefore, the distance traveled is given by the function $s(t) = 4.91\left(1 - e^{-(22.36/39.92)t}\right)$. The graph shows $s(t)$ and the data points.

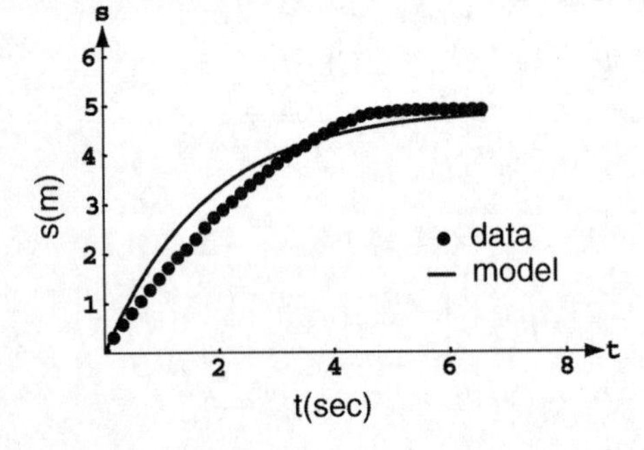

5. $y = mx \Rightarrow \frac{y}{x} = m \Rightarrow \frac{xy' - y}{x^2} = 0 \Rightarrow y' = \frac{y}{x}$. So for orthogonals: $\frac{dy}{dx} = -\frac{x}{y} \Rightarrow y\,dy = -x\,dx \Rightarrow \frac{y^2}{2} + \frac{x^2}{2} = C$
$\Rightarrow x^2 + y^2 = C_1$

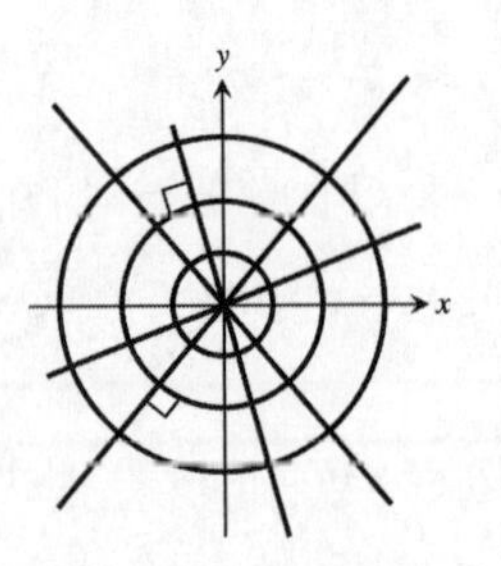

7. $kx^2 + y^2 = 1 \Rightarrow 1 - y^2 = kx^2 \Rightarrow \frac{1-y^2}{x^2} = k$

$\Rightarrow \frac{x^2(2y)y' - (1-y^2)2x}{x^4} = 0 \Rightarrow -2yx^2y' = (1-y^2)(2x)$

$\Rightarrow y' = \frac{(1-y^2)(2x)}{-2xy^2} = \frac{(1-y^2)}{-xy}$. So for the orthogonals:

$\frac{dy}{dx} = \frac{xy}{1-y^2} \Rightarrow \frac{(1-y^2)}{y} dy = x\,dx \Rightarrow \ln y - \frac{y^2}{2} = \frac{x^2}{2} + C$

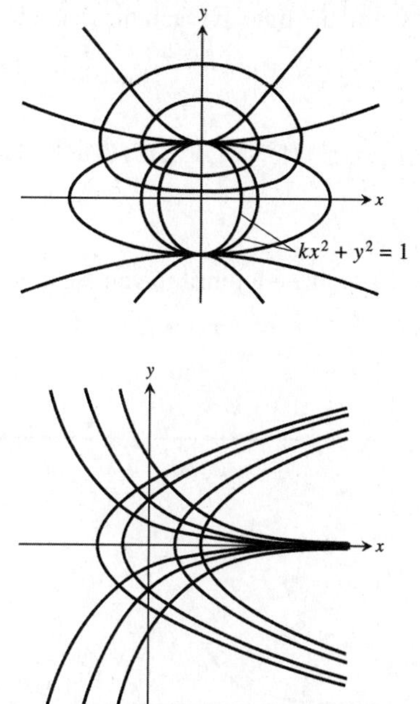

9. $y = ce^{-x} \Rightarrow \frac{y}{e^{-x}} = c \Rightarrow \frac{e^{-x}y' - y(e^{-x})(-1)}{(e^{-x})^2} = 0$

$\Rightarrow e^{-x}y' = -ye^{-x} \Rightarrow y' = -y$. So for the orthogonals:

$\frac{dy}{dx} = \frac{1}{y} \Rightarrow y\,dy = dx \Rightarrow \frac{y^2}{2} = x + C$

$\Rightarrow y^2 = 2x + C_1 \Rightarrow y = \pm\sqrt{2x + C_1}$

11. $2x^2 + 3y^2 = 5$ and $y^2 = x^3$ intersect at $(1, 1)$. Also, $2x^2 + 3y^2 = 5 \Rightarrow 4x + 6y\,y' = 0 \Rightarrow y' = -\frac{4x}{6y} \Rightarrow y'(1, 1) = -\frac{2}{3}$

$y_1^2 = x^3 \Rightarrow 2y_1y_1' = 3x^2 \Rightarrow y_1' = \frac{3x^2}{2y_1} \Rightarrow y_1'(1, 1) = \frac{3}{2}$. Since $y' \cdot y_1' = \left(-\frac{2}{3}\right)\left(\frac{3}{2}\right) = -1$, the curves are orthogonal.

13. Let $y(t)$ = the amount of salt in the container and $V(t)$ = the total volume of liquid in the tank at time t. Then, the departure rate is $\frac{y(t)}{V(t)}$ (the outflow rate).

(a) Rate entering $= \frac{2 \text{ lb}}{\text{gal}} \cdot \frac{5 \text{ gal}}{\text{min}} = 10$ lb/min

(b) Volume $= V(t) = 100$ gal $+ (5t$ gal $- 4t$ gal$) = (100 + t)$ gal

(c) The volume at time t is $(100 + t)$ gal. The amount of salt in the tank at time t is y lbs. So the concentration at any time t is $\frac{y}{100+t}$ lbs/gal. Then, the rate leaving $= \frac{y}{100+t}$ (lbs/gal) $\cdot$ 4 (gal/min)

$= \frac{4y}{100+t}$ lbs/min

(d) $\frac{dy}{dt} = 10 - \frac{4y}{100+t} \Rightarrow \frac{dy}{dt} + \left(\frac{4}{100+t}\right) y = 10 \Rightarrow P(t) = \frac{4}{100+t}, Q(t) = 10 \Rightarrow \int P(t)\,dt = \int \frac{4}{100+t}\,dt$

$= 4\ln(100+t) \Rightarrow v(t) = e^{4\ln(100+t)} = (100+t)^4 \Rightarrow y = \frac{1}{(100+t)^4} \int (100+t)^4 (10\,dt)$

$= \frac{10}{(100+t)^4}\left(\frac{(100+t)^5}{5} + C\right) = 2(100+t) + \frac{C}{(100+t)^4}$; $y(0) = 50 \Rightarrow 2(100+0) + \frac{C}{(100+0)^4} = 50$

$\Rightarrow C = -(150)(100)^4 \Rightarrow y = 2(100+t) - \frac{(150)(100)^4}{(100+t)^4} \Rightarrow y = 2(100+t) - \frac{150}{\left(1+\frac{t}{100}\right)^4}$

(e) $y(25) = 2(100+25) - \frac{(150)(100)^4}{(100+25)^4} \approx 188.56$ lbs $\Rightarrow$ concentration $= \frac{y(25)}{\text{volume}} \approx \frac{188.6}{125} \approx 1.5$ lb/gal

15. Let y be the amount of fertilizer in the tank at time t. Then rate entering $= 1\,\frac{\text{lb}}{\text{gal}} \cdot 1\,\frac{\text{gal}}{\text{min}} = 1\,\frac{\text{lb}}{\text{min}}$ and the volume in the tank at time t is $V(t) = 100$ (gal) + [1 (gal/min) − 3 (gal/min)]t min = (100 − 2t) gal. Hence rate out $= \left(\frac{y}{100-2t}\right)3 = \frac{3y}{100-2t}$ lbs/min $\Rightarrow \frac{dy}{dt} = \left(1 - \frac{3y}{100-2t}\right)$ lbs/min $\Rightarrow \frac{dy}{dt} + \left(\frac{3}{100-2t}\right)y = 1$

$\Rightarrow P(t) = \frac{3}{100-2t}, Q(t) = 1 \Rightarrow \int P(t)\,dt = \int \frac{3}{100-2t}\,dt = \frac{3\ln(100-2t)}{-2} \Rightarrow v(t) = e^{(-3\ln(100-2t))2}$

$= (100-2t)^{-3/2} \Rightarrow y = \frac{1}{(100-2t)^{-3/2}} \int (100-2t)^{-3/2}\,dt = (100-2t)^{-3/2}\left[\frac{-2(100-2t)^{-1/2}}{-2} + C\right]$

$= (100-2t) + C(100-2t)^{3/2}$; $y(0) = 0 \Rightarrow [100 - 2(0)] + C[100 - 2(0)]^{3/2} \Rightarrow C(100)^{3/2} = -100$

$\Rightarrow C = -(100)^{-1/2} = -\frac{1}{10} \Rightarrow y = (100-2t) - \frac{(100-2t)^{3/2}}{10}$. Let $\frac{dy}{dt} = 0 \Rightarrow \frac{dy}{dt} = -2 - \frac{\left(\frac{3}{2}\right)(100-2t)^{1/2}(-2)}{10}$

$= -2 + \frac{3\sqrt{100-2t}}{10} = 0 \Rightarrow 20 = 3\sqrt{100-2t} \Rightarrow 400 = 9(100-2t) \Rightarrow 400 = 900 - 18t \Rightarrow -500 = -18t$

$\Rightarrow t \approx 27.8$ min, the time to reach the maximum. The maximum amount is then $y(27.8) = [100 - 2(27.8)] - \frac{[100 - 2(27.8)]^{3/2}}{10} \approx 14.8$ lb

9.4 GRAPHICAL SOUTIONS OF AUTONOMOUS EQUATIONS

1. $y' = (y+2)(y-3)$
   (a) $y = -2$ is a stable equilibrium value and $y = 3$ is an unstable equilibrium.
   (b) $y'' = (2y-1)y' = 2(y+2)\left(y - \frac{1}{2}\right)(y-3)$

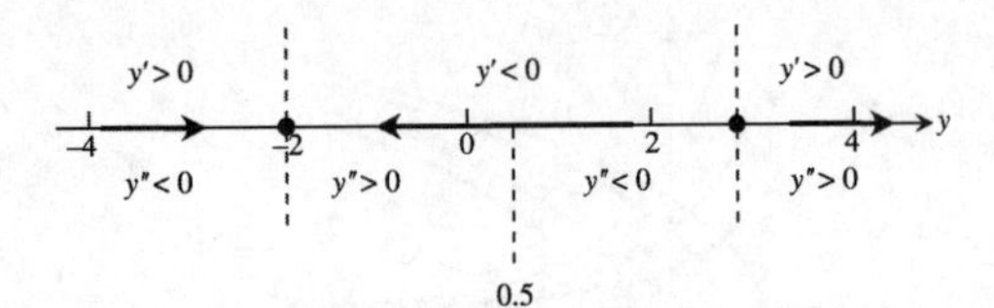

(c)

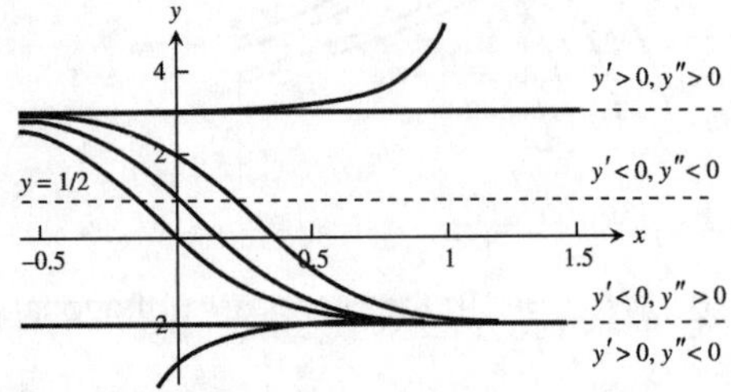

3. $y' = y^3 - y = (y+1)y(y-1)$
   (a) $y = -1$ and $y = 1$ is an unstable equilibrium and $y = 0$ is a stable equilibrium value.
   (b) $y'' = (3y^2 - 1)y' = 3(y+1)\left(y + \frac{1}{\sqrt{3}}\right)y\left(y - \frac{1}{\sqrt{3}}\right)(y-1)$

$y' < 0$ | $y' > 0$ | $y' < 0$ | $y' > 0$

-1.5, -1, -0.5, 0, 0.5, 1, 1.5, y

$y'' < 0$ | $y'' > 0$ | $y'' < 0$ | $y'' > 0$ | $y'' < 0$ | $y'' > 0$

$-\frac{1}{\sqrt{3}}$ $\quad \frac{1}{\sqrt{3}}$

(c)

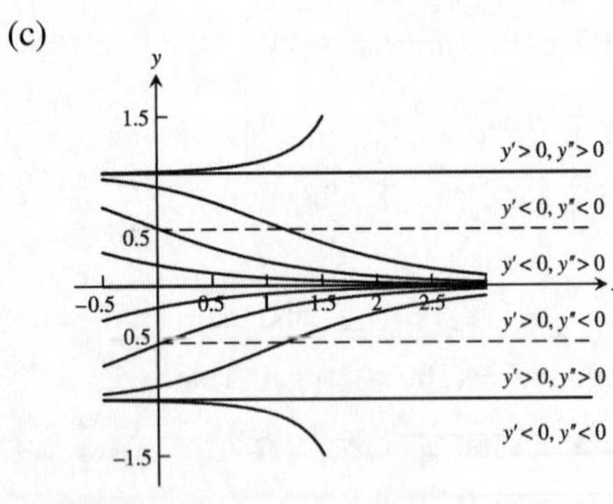

5. $y' = \sqrt{y}, y > 0$
   (a) There are no equilibrium values.
   (b) $y'' = \frac{1}{2\sqrt{y}}\, y' = \frac{1}{2\sqrt{y}}\sqrt{y} = \frac{1}{2}$

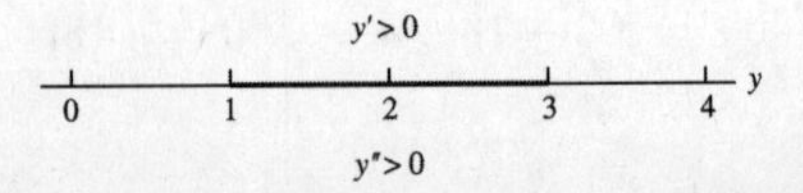

(c)

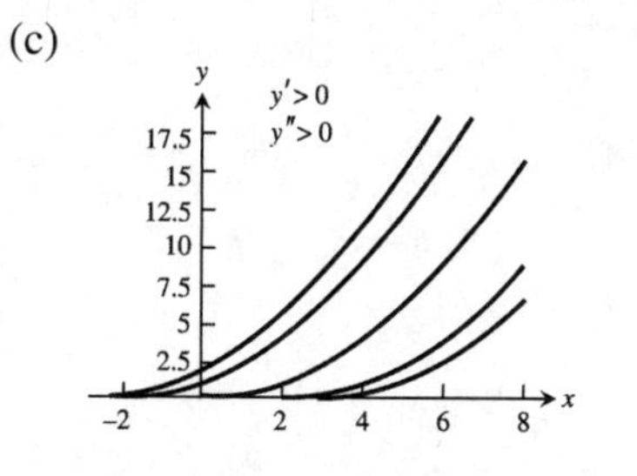

7. $y' = (y-1)(y-2)(y-3)$

(a) $y = 1$ and $y = 3$ is an unstable equilibrium and $y = 2$ is a stable equilibrium value.

(b) $y'' = (3y^2 - 12y + 11)(y-1)(y-2)(y-3) = 3(y-1)\left(y - \frac{6-\sqrt{3}}{3}\right)(y-2)\left(y - \frac{6+\sqrt{3}}{3}\right)(y-3)$

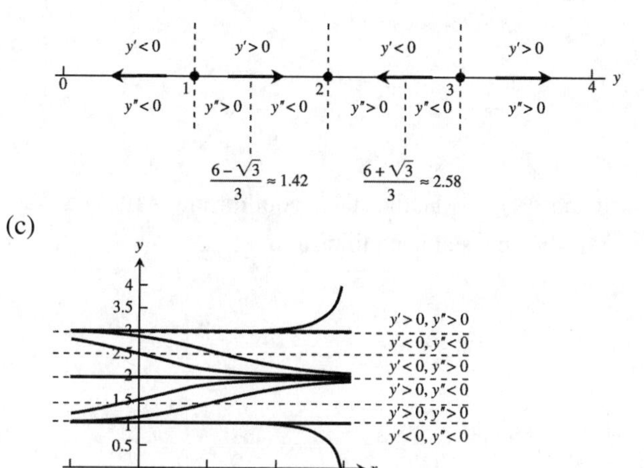

9. $\frac{dP}{dt} = 1 - 2P$ has a stable equilibrium at $P = \frac{1}{2}$. $\frac{d^2P}{dt^2} = -2\frac{dP}{dt} = -2(1 - 2P)$

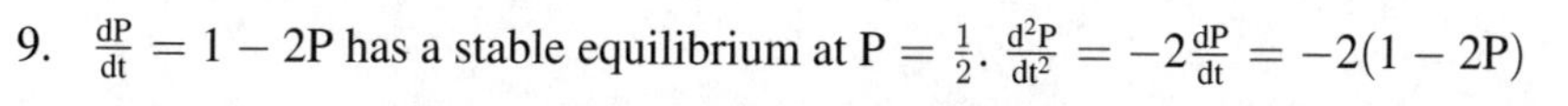

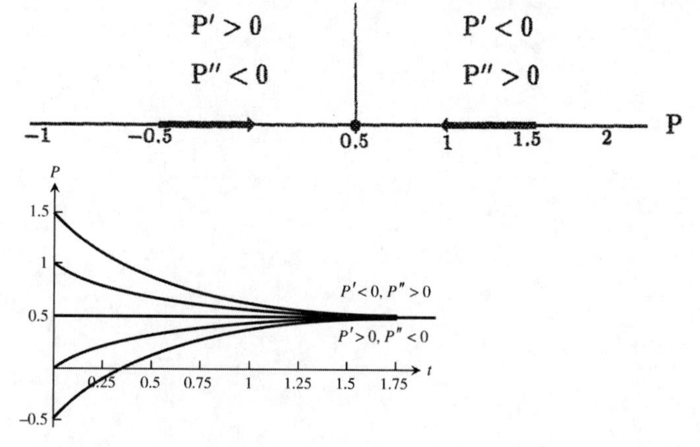

11. $\frac{dP}{dt} = 2P(P-3)$ has a stable equilibrium at $P = 0$ and an unstable equilibrium at $P = 3$.
$\frac{d^2P}{dt^2} = 2(2P-3)\frac{dP}{dt} = 4P(2P-3)(P-3)$

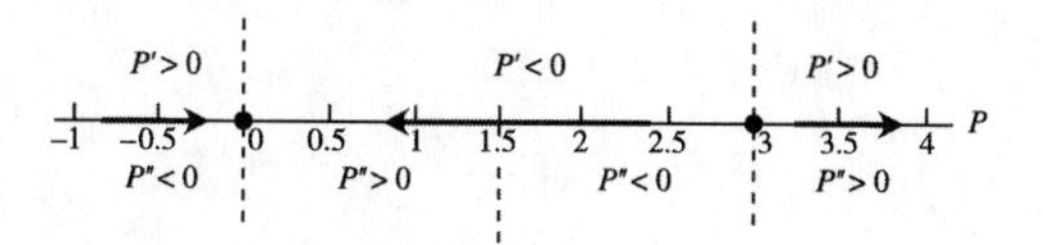

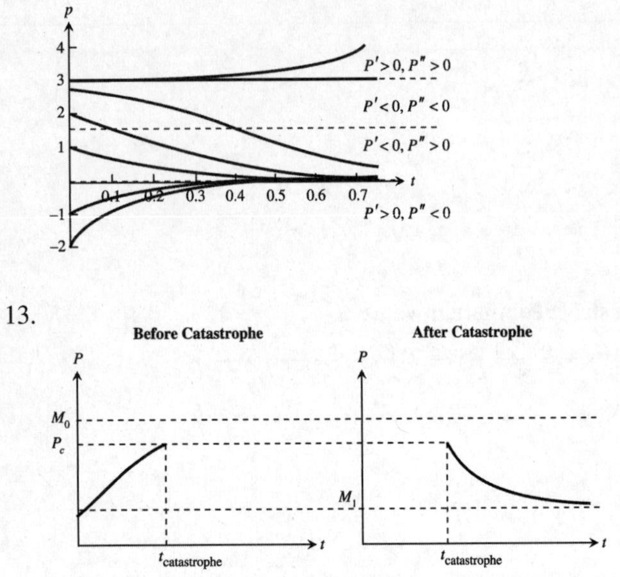

13. Before the catastrophe, the population exhibits logistic growth and $P(t) \to M_0$, the stable equilibrium. After the catastrophe, the population declines logistically and $P(t) \to M_1$, the new stable equilibrium.

15. $\frac{dv}{dt} = g - \frac{k}{m}v^2$, g, k, m > 0 and $v(t) \geq 0$

Equilibrium: $\frac{dv}{dt} = g - \frac{k}{m}v^2 = 0 \Rightarrow v = \sqrt{\frac{mg}{k}}$

Concavity: $\frac{d^2v}{dt^2} = -2\left(\frac{k}{m}v\right)\frac{dv}{dt} = -2\left(\frac{k}{m}v\right)\left(g - \frac{k}{m}v^2\right)$

(a)

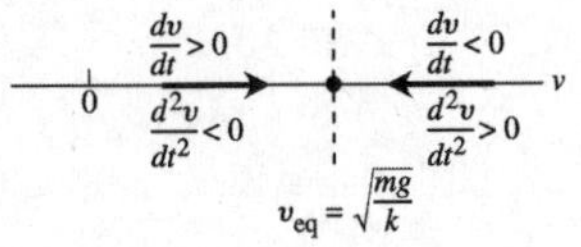

(b)

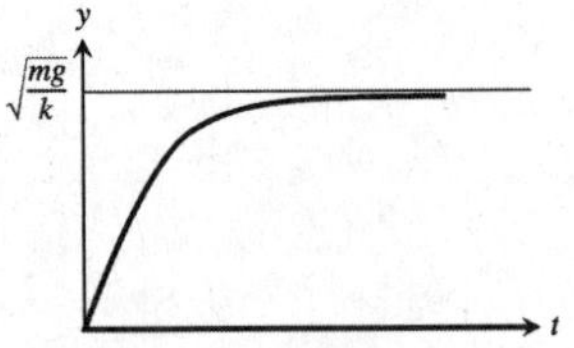

(c) $v_{terminal} = \sqrt{\frac{160}{0.005}} = 178.9\ \frac{ft}{s} = 122$ mph

17. $F = F_p - F_r$

$ma = 50 - 5|v|$

$\frac{dv}{dt} = \frac{1}{m}(50 - 5|v|)$

The maximum velocity occurs when $\frac{dv}{dt} = 0$ or $v = 10\ \frac{ft}{sec}$.

19. $L\frac{di}{dt} + Ri = V \Rightarrow \frac{di}{dt} = \frac{V}{L} - \frac{R}{L}i = \frac{R}{L}\left(\frac{V}{R} - i\right)$, V, L, R > 0

Equilibrium: $\frac{di}{dt} = \frac{R}{L}\left(\frac{V}{R} - i\right) = 0 \Rightarrow i = \frac{V}{R}$

Concavity: $\frac{d^2i}{dt^2} = -\left(\frac{R}{L}\right)\frac{di}{dt} = -\left(\frac{R}{L}\right)^2\left(\frac{V}{R} - i\right)$

Phase Line:

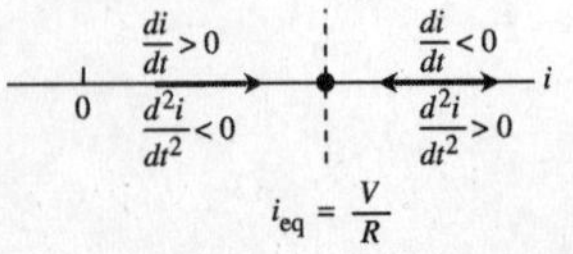

If the switch is closed at $t = 0$, then $i(0) = 0$, and the graph of the solution looks like this:

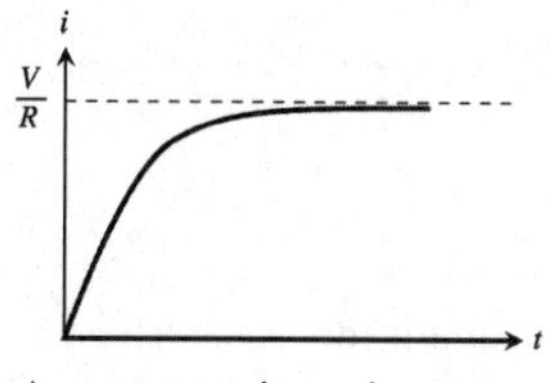

As $t \to \infty$, it $\to i_{\text{steady state}} = \frac{V}{R}$. (In the steady state condition, the self-inductance acts like a simple wire connector and, as a result, the current throught the resistor can be calculated using the familiar version of Ohm's Law.)

9.5 SYSTEMS OF EQUATIONS AND PHASE PLANES

1. Seasonal variations, nonconformity of the environments, effects of other interactions, unexpected disasters, etc.

3. This model assumes that the number of interactions is porportional to the product of x and y:
$\frac{dx}{dt} = (a - by)x$, $a < 0$, $\frac{dy}{dt} = m\left(1 - \frac{y}{M}\right)y - nxy = y\left(m - \frac{m}{M}y - nx\right)$.
To find the equilibrium points:
$\frac{dx}{dt} = 0 \Rightarrow (a - by)x = 0 \Rightarrow x = 0$ or $y = \frac{a}{b}$ (remember $\frac{a}{b} < 0$);
$\frac{dy}{dt} = 0 \Rightarrow y\left(m - \frac{m}{M}y - nx\right) \Rightarrow y = 0$ or $y = -\frac{Mn}{m}x + M$;
Thus there are two equlibrium points, both occur when $x = 0$, $(0, 0)$ and $(0, M)$.

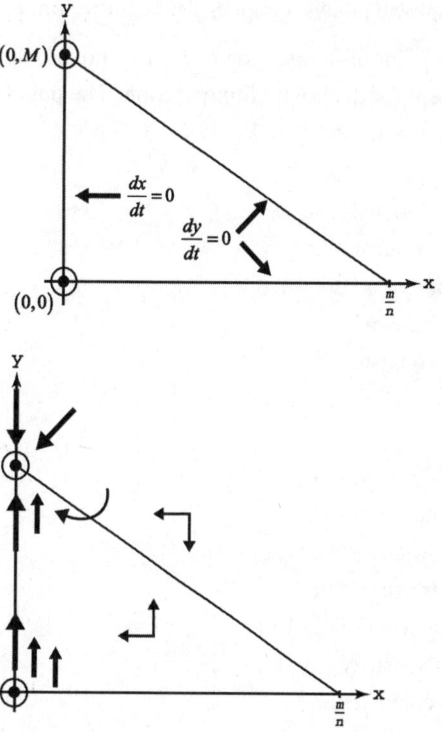

Implies coexistence is not possible because eventually trout die out and bass reach their population limit.

5. (a) Logistic growth occurs in the absence of the competitor, and simple interaction of the species: growth dominates the competition when either population is small so it is difficult to drive either species to extinction.
   (b) $a =$ per capita growth rate for trout
   $m =$ per capita growth rate for bass
   $b =$ intensity of competition to the trout
   $n =$ intensity of competition to the bass
   $k_1 =$ environmental carrying capacity for the trout
   $k_2 =$ environmental carrying capacity for the bass

(c) $\frac{dx}{dt} = 0 \Rightarrow a\left(1 - \frac{x}{k_1}\right)x - bxy = \left[a\left(1 - \frac{x}{k_1}\right) - by\right]x = 0 \Rightarrow x = 0$ or $a\left(1 - \frac{x}{k_1}\right) - by = 0 \Rightarrow x = 0$ or $y = \frac{a}{b} - \frac{a}{bk_1}x$; $\frac{dy}{dt} = 0 \Rightarrow m\left(1 - \frac{y}{k_2}\right)y - nxy = \left[m\left(1 - \frac{y}{k_2}\right) - nx\right]y = 0 \Rightarrow y = 0$ or $m\left(1 - \frac{y}{k_2}\right) - nx = 0 \Rightarrow y = 0$ or $y = k_2 - \frac{nk_2}{m}x$. There are five cases to consider.

Case I: $\frac{a}{b} > k_2$ and $\frac{m}{n} > k_1$.

By picking $\frac{a}{b} > k_2$ and $\frac{m}{n} > k_1$ we ensure an equilibrium point exists inside the first quadrant.

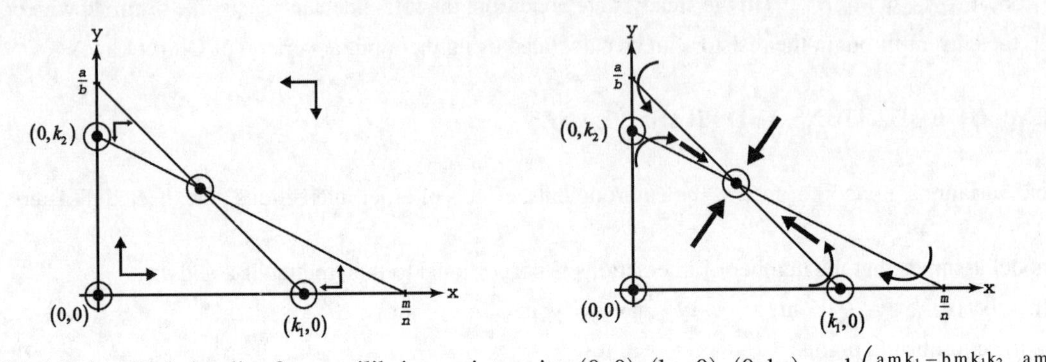

Graphical analysis implies four equilibrium points exist: (0, 0), $(k_1, 0)$, $(0, k_2)$, and $\left(\frac{amk_1 - bmk_1k_2}{am - bnk_1k_2}, \frac{amk_2 - ank_1k_2}{am - bnk_1k_2}\right)$ (the point of intersection of the two boundaries in the first quadrant). All of these equilibrium points are unstable except for the point of intersection. The possibility of coexistence is predicted by this model.

Case II: $\frac{a}{b} > k_2$ and $\frac{m}{n} < k_1$.

$(0, k_2)$: unstable
$(k_1, 0)$: stable
(0, 0): unstable
Trout wins: $(k_1, 0)$
Not sensitive
No coexistence

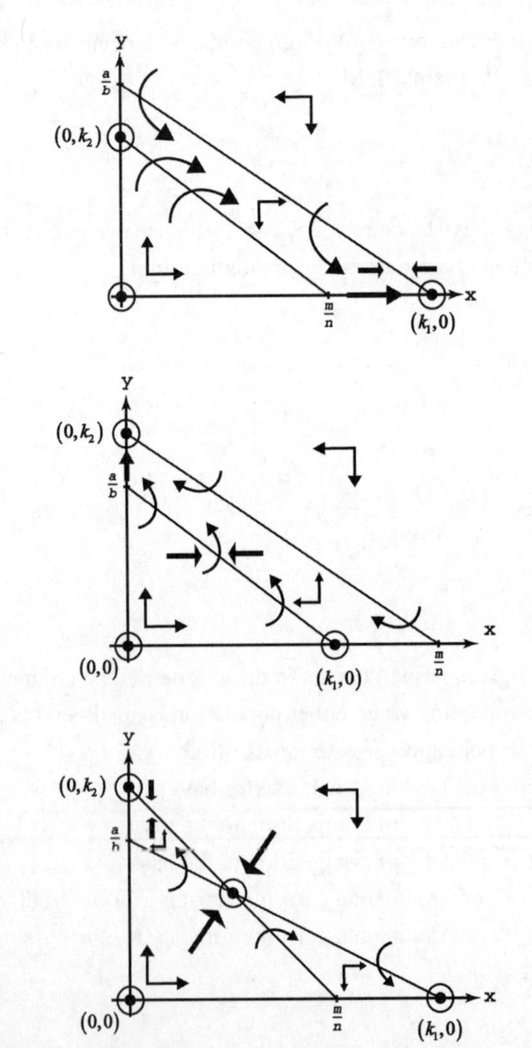

Case III: $\frac{a}{b} < k_2$ and $\frac{m}{n} > k_1$.

$(0, k_2)$: stable
$(k_1, 0)$: unstable
(0, 0): unstable
Bass wins: $(0, k_2)$
Not sensitive
No coexistence

Case IV: $\frac{a}{b} < k_2$ and $\frac{m}{n} < k_1$.

$(0, k_2)$: stable
$(k_1, 0)$: stable
(0, 0): unstable
$\left(\frac{amk_1 - bmk_1k_2}{am - bnk_1k_2}, \frac{amk_2 - ank_1k_2}{am - bnk_1k_2}\right)$: unstable
Bass or trout: $(0, k_2)$ or $(k_1, 0)$
Very sensitive
Coexistence is possible but not predicted

If we assume $\frac{a}{b} < k_2$ and $\frac{m}{n} < k_1$ then graphical analysis implies four equilibrium poins exist: $(0, k_2)$, $(k_1, 0)$, $(0, 0)$, and $\left(\frac{amk_1 - bmk_1k_2}{am - bnk_1k_2}, \frac{amk_2 - ank_1k_2}{am - bnk_1k_2}\right)$ (the point of intersection of the two boundaries in the first quadrant).

Case V: $\frac{a}{b} = k_2$ and $\frac{a}{bk_1} = \frac{nk_2}{m}$ (lines coincide).

$(0, k_2)$: stable
$(k_1, 0)$: stable
$(0, 0)$: unstable
Line segment joining $(0, k_2)$ and $(k_1, 0)$: stable
Bass wins: $(0, k_2)$
Not sensitive
Coexistence is likely outcome

Note that all points on the line segment joining $(0, k_2)$ and $(k_1, 0)$ are rest points.

7. (a) $\frac{dx}{dt} = ax - bxy = (a - by)x$ and $\frac{dy}{dt} = my - nxy = (m - nx)y \Rightarrow \frac{dy}{dt} = \frac{dy}{dx}\frac{dx}{dt} \Rightarrow \frac{dy}{dx} = \frac{dy/dt}{dx/dt} = \frac{(m - nx)y}{(a - by)x}$

(b) $\frac{dy}{dx} = \frac{(m - nx)y}{(a - by)x} \Rightarrow \left(\frac{a}{y} - b\right)dy = \left(\frac{m}{x} - n\right)dx \Rightarrow \int\left(\frac{a}{y} - b\right)dy = \int\left(\frac{m}{x} - n\right)dx \Rightarrow a\ln|y| - by = m\ln|x| - nx + C$
$\Rightarrow \ln|y^a| + \ln e^{-by} = \ln|x^m| + \ln e^{-nx} + \ln e^C \Rightarrow \ln|y^a e^{-by}| = \ln|x^m e^{-nx} e^C| \Rightarrow y^a e^{-by} = x^m e^{-nx} e^C$, let $K = e^C$
$\Rightarrow y^a e^{-by} = Kx^m e^{-nx}$

(c) $f(y) = y^a e^{-by} \Rightarrow f'(y) = ay^{a-1}e^{-by} - by^a e^{-by} = y^{a-1}e^{-by}(a - by)$ and $f'(y) = 0 \Rightarrow y = 0$ or $y = \frac{a}{b}$;
$f''\left(\frac{a}{b}\right) = -b\left(\frac{a}{b}\right)^{a-1}e^{-a} < 0 \Rightarrow f(y)$ has a unique max of $M_y = \left(\frac{a}{eb}\right)^a$ when $y = \frac{a}{b}$. $g(x) = x^m e^{-nx}$
$\Rightarrow g'(x) = mx^{m-1}e^{-nx} - nx^m e^{-nx} = x^{m-1}e^{-nx}(m - nx)$ and $g'(x) = 0 \Rightarrow x = 0$ or $x = \frac{m}{n}$;
$g''\left(\frac{m}{n}\right) = -n\left(\frac{m}{n}\right)^{m-1}e^{-m} < 0 \Rightarrow g(x)$ has a unique max of $M_x = \left(\frac{m}{en}\right)^m$ when $x = \frac{m}{n}$.

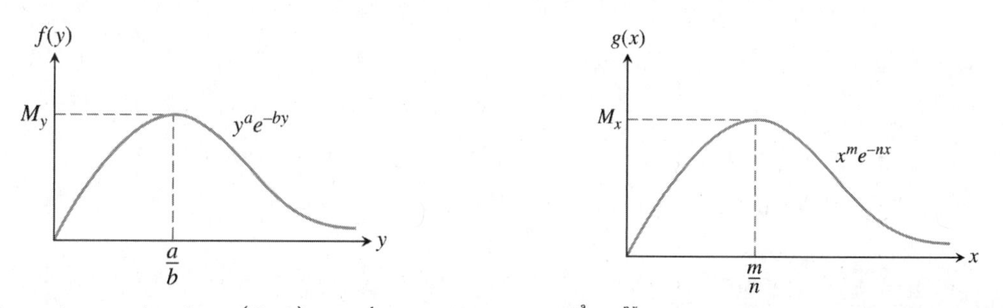

(d) Consider trajectory $(x, y) \to \left(\frac{m}{n}, \frac{a}{b}\right)$. $y^a e^{-by} = Kx^m e^{-nx} \Rightarrow \frac{y^a}{e^{by}} \cdot \frac{e^{nx}}{x^m} = K$, taking the limit of both sides
$\Rightarrow \lim_{\substack{x \to m/n \\ y \to a/b}} \left(\frac{y^a}{e^{by}} \cdot \frac{e^{nx}}{x^m}\right) = \lim_{\substack{x \to m/n \\ y \to a/b}} K \Rightarrow \frac{M_y}{M_x} = K$. Thus, $\frac{y^a}{e^{by}} = \frac{M_y}{M_x}\frac{x^m}{e^{nx}}$ represents the equation any solution trajectory must satisfy if the trajectory approaches the rest point asymptotically.

(e) Pick initial condition $y_0 < \frac{a}{b}$. Then, from the figure at right, $f(y_0) < M_y$ implies $\frac{M_y}{M_x}\frac{x^m}{e^{nx}} = \frac{y_0^a}{e^{by_0}} < M_y$ and thus $\frac{x^m}{e^{nx}} < M_x$. From the figure for $g(x)$, there exists a unique $x_0 < \frac{m}{n}$ satisfying $\frac{x^m}{e^{nx}} < M_x$. That is, for each $y < \frac{a}{b}$ there is a unique x satisfying $\frac{y^a}{e^{by}} = \frac{M_y}{M_x}\frac{x^m}{e^{nx}}$. Thus, there can exist only one trajectory solution approaching $\left(\frac{m}{n}, \frac{a}{b}\right)$. (You can think of the point (x_0, y_0) as the initial condition for that trajectory.)

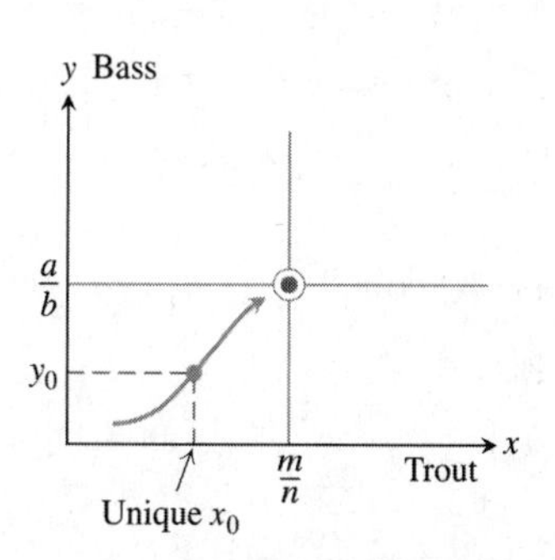

(f) Likewise there exists a unique trajectory when $y_0 > \frac{a}{b}$. Again, $f(y_0) < M_y$ implies $\frac{M_y}{M_x}\frac{x^m}{e^{nx}} = \frac{y_0^a}{e^{by_0}} < M_y$ and thus $\frac{x^m}{e^{nx}} < M_x$. From the figure for $g(x)$, there exists a unique $x_0 > \frac{m}{n}$ satisfying $\frac{x^m}{e^{nx}} < M_x$. That is, for each $y > \frac{a}{b}$ there is a unique x satisfying $\frac{y^a}{e^{by}} = \frac{M_y}{M_x}\frac{x^m}{e^{nx}}$. Thus, there can exist only one trajectory solution approaching $\left(\frac{m}{n}, \frac{a}{b}\right)$.

9. In the absence of foxes $\Rightarrow b = 0 \Rightarrow \frac{dx}{dt} = ax$ and the population of rabbits grows at a rate proportional to the number of rabbits.

11. $\frac{dx}{dt} = (a - by)x = 0 \Rightarrow y = \frac{a}{b}$ or $x = 0$; $\frac{dy}{dt} = (-c + dx)y = 0 \Rightarrow x = \frac{c}{d}$ or $y = 0 \Rightarrow$ equilibrium points at $(0, 0)$ or $\left(\frac{c}{d}, \frac{a}{b}\right)$. For the point $(0, 0)$, there are no rabbits and no foxes. It is an unstable equilibrium point, if there are no foxes, but a few rabbits are introduced, then $\frac{dx}{dt} = a \Rightarrow$ the rabbit population will grow exponentially away from $(0, 0)$

13. Consider a particular trajectory and suppose that (x_0, y_0) is such that $x_0 < \frac{c}{d}$ and $y_0 < \frac{a}{b}$, then $\frac{dx}{dt} > 0$ and $\frac{dy}{dt} < 0 \Rightarrow$ the rabbit population is increasing while the fox population is decreasing, points on the trajectory are moving down and to the right; if $x_0 > \frac{c}{d}$ and $y_0 < \frac{a}{b}$, then $\frac{dx}{dt} > 0$ and $\frac{dy}{dt} > 0 \Rightarrow$ both the rabbit and fox populations are increasing, points on the trajectory are moving up and to the right; if $x_0 > \frac{c}{d}$ and $y_0 > \frac{a}{b}$, then $\frac{dx}{dt} < 0$ and $\frac{dy}{dt} > 0 \Rightarrow$ the rabbit population is decreasing while the fox population is increasing, points on the trajectory are moving up and to the left; and finally if $x_0 < \frac{c}{d}$ and $y_0 > \frac{a}{b}$, then $\frac{dx}{dt} < 0$ and $\frac{dy}{dt} < 0 \Rightarrow$ both the rabbit and fox populations are decreasing, points on the trajectory are moving down and to the left. Thus, points travel around the trajectory in a counterclockwise direction. Note that we will follow the same trajectory if (x_0, y_0) starts at a different point on the trajectory.

CHAPTER 9 PRACTICE EXERCISES

1. $y' = xe^y\sqrt{x-2} \Rightarrow e^{-y}dy = x\sqrt{x-2}\,dx \Rightarrow -e^{-y} = \frac{2(x-2)^{3/2}(3x+4)}{15} + C \Rightarrow e^{-y} = \frac{-2(x-2)^{3/2}(3x+4)}{15} - C$
$\Rightarrow -y = \ln\left[\frac{-2(x-2)^{3/2}(3x+4)}{15} - C\right] \Rightarrow y = -\ln\left[\frac{-2(x-2)^{3/2}(3x+4)}{15} - C\right]$

3. $\sec x\,dy + x\cos^2 y\,dx = 0 \Rightarrow \frac{dy}{\cos^2 y} = -\frac{x\,dx}{\sec x} \Rightarrow \tan y = -\cos x - x\sin x + C$

5. $y' = \frac{e^y}{xy} \Rightarrow ye^{-y}dy = \frac{dx}{x} \Rightarrow (y+1)e^{-y} = -\ln|x| + C$

7. $x(x-1)dy - y\,dx = 0 \Rightarrow x(x-1)dy = y\,dx \Rightarrow \frac{dy}{y} = \frac{dx}{x(x-1)} \Rightarrow \ln y = \ln(x-1) - \ln(x) + C$
$\Rightarrow \ln y = \ln(x-1) - \ln(x) + \ln C_1 \Rightarrow \ln y = \ln\left(\frac{C_1(x-1)}{x}\right) \Rightarrow y = \frac{C_1(x-1)}{x}$

9. $2y' - y = xe^{x/2} \Rightarrow y' - \frac{1}{2}y = \frac{x}{2}e^{x/2}$.
$p(x) = -\frac{1}{2}$, $v(x) = e^{\int\left(-\frac{1}{2}\right)dx} = e^{-x/2}$.
$e^{-x/2}y' - \frac{1}{2}e^{-x/2}y = \left(e^{-x/2}\right)\left(\frac{x}{2}\right)\left(e^{x/2}\right) = \frac{x}{2} \Rightarrow \frac{d}{dx}\left(e^{-x/2}y\right) = \frac{x}{2} \Rightarrow e^{-x/2}y = \frac{x^2}{4} + C \Rightarrow y = e^{x/2}\left(\frac{x^2}{4} + C\right)$

11. $xy' + 2y = 1 - x^{-1} \Rightarrow y' + \left(\frac{2}{x}\right)y = \frac{1}{x} - \frac{1}{x^2}$.
$v(x) = e^{2\int\frac{dx}{x}} = e^{2\ln x} = e^{\ln x^2} = x^2$.
$x^2y' + 2xy = x - 1 \Rightarrow \frac{d}{dx}(x^2y) = x - 1 \Rightarrow x^2y = \frac{x^2}{2} - x + C \Rightarrow y = \frac{1}{2} - \frac{1}{x} + \frac{C}{x^2}$

13. $(1+e^x)dy + (ye^x + e^{-x})dx = 0 \Rightarrow (1+e^x)y' + e^xy = -e^{-x} \Rightarrow y' = \frac{e^x}{1+e^x}y = \frac{-e^{-x}}{(1+e^x)}$.
$v(x) = e^{\int\frac{e^x dx}{(1+e^x)}} = e^{\ln(e^x+1)} = e^x + 1$.
$(e^x+1)y' + (e^x+1)\left(\frac{e^x}{1+e^x}\right)y = \frac{-e^{-x}}{(1+e^x)}(e^x+1) \Rightarrow \frac{d}{dx}[(e^x+1)y] = -e^{-x} \Rightarrow (e^x+1)y = e^{-x} + C$
$\Rightarrow y = \frac{e^{-x}+C}{e^x+1} = \frac{e^{-x}+C}{1+e^x}$

15. $(x+3y^2)\,dy + y\,dx = 0 \Rightarrow x\,dy + y\,dx = -3y^2dy \Rightarrow \frac{d}{dx}(xy) = -3y^2dy \Rightarrow xy = -y^3 + C$

17. $(x+1)\frac{dy}{dx}+2y=x \Rightarrow y'+\left(\frac{2}{x+1}\right)y=\frac{x}{x+1}$. Let $v(x)=e^{\int \frac{2}{x+1}dx}=e^{2\ln(x+1)}=e^{\ln(x+1)^2}=(x+1)^2$.
So $y'(x+1)^2+\frac{2}{(x+1)}(x+1)^2y=\frac{x}{(x+1)}(x+1)^2 \Rightarrow \frac{d}{dx}\left[y(x+1)^2\right]=x(x+1) \Rightarrow y(x+1)^2=\int x(x+1)dx$
$\Rightarrow y(x+1)^2=\frac{x^3}{3}+\frac{x^2}{2}+C \Rightarrow y=(x+1)^{-2}\left(\frac{x^3}{3}+\frac{x^2}{2}+C\right)$. We have $y(0)=1 \Rightarrow 1=C$. So
$y=(x+1)^{-2}\left(\frac{x^3}{3}+\frac{x^2}{2}+1\right)$

19. $\frac{dy}{dx}+3x^2y=x^2$. Let $v(x)=e^{\int 3x^2dx}=e^{x^3}$. So $e^{x^3}y'+3x^2e^{x^3}y=x^2e^{x^3} \Rightarrow \frac{d}{dx}\left(e^{x^3}y\right)=x^2e^{x^3} \Rightarrow e^{x^3}y=\frac{1}{3}e^{x^3}+C$.
We have $y(0)=-1 \Rightarrow e^{0^3}(-1)=\frac{1}{3}e^{0^3}+C \Rightarrow -1=\frac{1}{3}+C \Rightarrow C=-\frac{4}{3}$ and $e^{x^3}y=\frac{1}{3}e^{x^3}-\frac{4}{3} \Rightarrow y=\frac{1}{3}-\frac{4}{3}e^{-x^3}$

21. $xy'+(x-2)y=3x^3e^{-x} \Rightarrow y'+\left(\frac{x-2}{x}\right)y=3x^2e^{-x}$. Let $v(x)=e^{\int\left(\frac{x-2}{x}\right)dx}=e^{x-2\ln x}=\frac{e^x}{x^2}$. So
$\frac{e^x}{x^2}y'+\frac{e^x}{x^2}\left(\frac{x-2}{x}\right)y=3 \Rightarrow \frac{d}{dx}\left(y\cdot\frac{e^x}{x^2}\right)=3 \Rightarrow y\cdot\frac{e^x}{x^2}=3x+C$. We have $y(1)=0 \Rightarrow 0=3(1)+C \Rightarrow C=-3$
$\Rightarrow y\cdot\frac{e^x}{x^2}=3x-3 \Rightarrow y=x^2e^{-x}(3x-3)$

23. To find the approximate values let $y_n=y_{n-1}+(y_{n-1}+\cos x_{n-1})(0.1)$ with $x_0=0$, $y_0=0$, and 20 steps. Use a spreadsheet, graphing calculator, or CAS to obtain the values in the following table.

x	y
0	0
0.1	0.1000
0.2	0.2095
0.3	0.3285
0.4	0.4568
0.5	0.5946
0.6	0.7418
0.7	0.8986
0.8	1.0649
0.9	1.2411
1.0	1.4273

x	y
1.1	1.6241
1.2	1.8319
1.3	2.0513
1.4	2.2832
1.5	2.5285
1.6	2.7884
1.7	3.0643
1.8	3.3579
1.9	3.6709
2.0	4.0057

25. To estimate $y(3)$, let $y=y_{n-1}+\left(\frac{x_{n-1}-2y_{n-1}}{x_{n-1}+1}\right)(0.05)$ with initial values $x_0=0$, $y_0=1$, and 60 steps. Use a spreadsheet, graphing calculator, or CAS to obtain $y(3)\approx 0.8981$.

27. Let $y_n=y_{n-1}+\left(\frac{1}{e^{x_{n-1}+y_{n-1}+2}}\right)(dx)$ with starting values $x_0=0$ and $y_0=2$, and steps of 0.1 and -0.1. Use a spreadsheet, programmable calculator, or CAS to generate the following graphs.

(a)

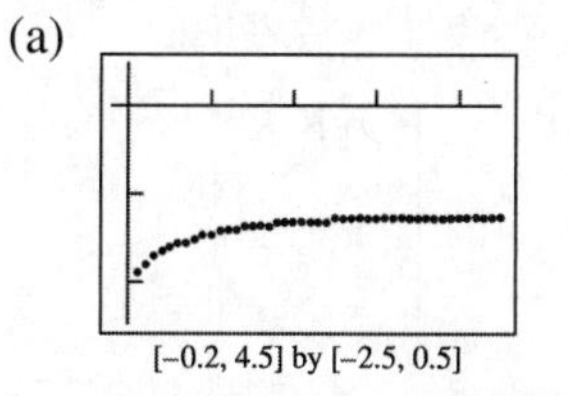

[−0.2, 4.5] by [−2.5, 0.5]

(b) Note that we choose a small interval of x-values because the y-values decrease very rapidly and our calculator cannot handle the calculations for $x\le -1$. (This occurs because the analytic solution is $y=-2+\ln(2-e^{-x})$, which has an asymptote at $x=-\ln 2\approx 0.69$. Obviously, the Euler approximations are misleading for $x\le -0.7$.)

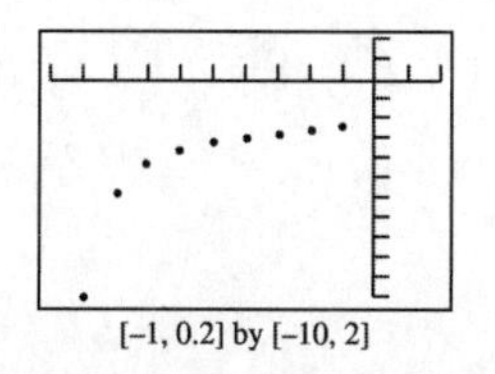

[−1, 0.2] by [−10, 2]

29.

x	1	1.2	1.4	1.6	1.8	2.0
y	−1	−0.8	−0.56	−0.28	0.04	0.4

$\frac{dy}{dx} = x \Rightarrow dy = x\,dx \Rightarrow y = \frac{x^2}{2} + C$; $x = 1$ and $y = -1$
$\Rightarrow -1 = \frac{1}{2} + C \Rightarrow C = -\frac{3}{2} \Rightarrow y(\text{exact}) = \frac{x^2}{2} - \frac{3}{2}$
$\Rightarrow y(2) = \frac{2^2}{2} - \frac{3}{2} = \frac{1}{2}$ is the exact value.

31.

x	1	1.2	1.4	1.6	1.8	2.0
y	−1	−1.2	−0.488	−1.9046	−2.5141	−3.4192

$\frac{dy}{dx} = xy \Rightarrow \frac{dy}{y} = x\,dx \Rightarrow \ln|y| = \frac{x^2}{2} + C$
$\Rightarrow y = e^{\frac{x^2}{2}+C} = e^{\frac{x^2}{2}} \cdot e^C = C_1 e^{\frac{x^2}{2}}$; $x = 1$ and $y = -1$
$\Rightarrow -1 = C_1 e^{1/2} \Rightarrow C_1 = -e^{1/2} y(\text{exact}) = -e^{1/2} \cdot e^{\frac{x^2}{2}}$
$= -e^{(x^2-1)/2} \Rightarrow y(2) = -e^{3/2} \approx -4.4817$ is the exact value.

33. $\frac{dy}{dx} = y^2 - 1 \Rightarrow y' = (y+1)(y-1)$. We have $y' = 0 \Rightarrow (y+1) = 0, (y-1) = 0 \Rightarrow y = -1, 1$.

(a) Equilibrium points are -1 (stable) and 1 (unstable)

(b) $y' = y^2 - 1 \Rightarrow y'' = 2yy' \Rightarrow y'' = 2y(y^2-1) = 2y(y+1)(y-1)$. So $y'' = 0 \Rightarrow y = 0, y = -1, y = 1$.

(c)

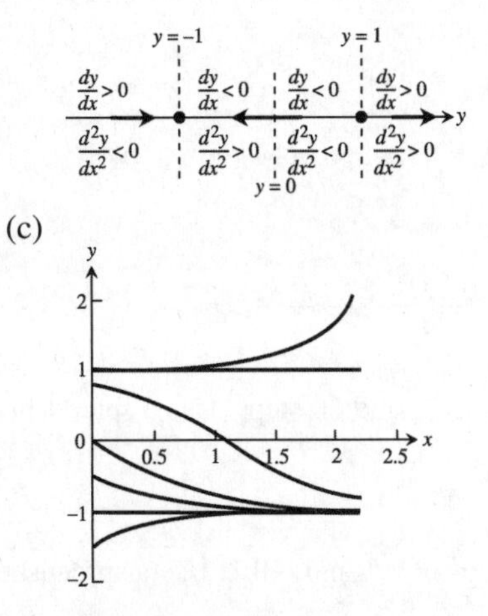

35. (a) Force $=$ Mass times Acceleration (Newton's Second Law) or $F = ma$. Let $a = \frac{dv}{dt} = \frac{dv}{ds} \cdot \frac{ds}{dt} = v\frac{dv}{ds}$. Then
$ma = -mgR^2s^{-2} \Rightarrow a = -gR^2s^{-2} \Rightarrow v\frac{dv}{ds} = -gR^2s^{-2} \Rightarrow v\,dv = -gR^2s^{-2}ds \Rightarrow \int v\,dv = \int -gR^2s^{-2}ds$
$\Rightarrow \frac{v^2}{2} = \frac{gR^2}{s} + C_1 \Rightarrow v^2 = \frac{2gR^2}{s} + 2C_1 = \frac{2gR^2}{s} + C$. When $t = 0$, $v = v_0$ and $s = R \Rightarrow v_0^2 = \frac{2gR^2}{R} + C$
$\Rightarrow C = v_0^2 - 2gR \Rightarrow v^2 = \frac{2gR^2}{s} + v_0^2 - 2gR$

(b) If $v_0 = \sqrt{2gR}$, then $v^2 = \frac{2gR^2}{s} \Rightarrow v = \sqrt{\frac{2gR^2}{s}}$, since $v \ge 0$ if $v_0 \ge \sqrt{2gR}$. Then $\frac{ds}{dt} = \frac{\sqrt{2gR^2}}{\sqrt{s}} \Rightarrow \sqrt{s}\,ds = \sqrt{2gR^2}\,dt$
$\Rightarrow \int s^{1/2}ds = \int \sqrt{2gR^2}\,dt \Rightarrow \frac{2}{3}s^{3/2} = \sqrt{2gR^2}t + C_1 \Rightarrow s^{3/2} = \left(\frac{3}{2}\sqrt{2gR^2}\right)t + C$; $t = 0$ and $s = R$
$\Rightarrow R^{3/2} = \left(\frac{3}{2}\sqrt{2gR^2}\right)(0) + C \Rightarrow C = R^{3/2} \Rightarrow s^{3/2} = \left(\frac{3}{2}\sqrt{2gR^2}\right)t + R^{3/2} = \left(\frac{3}{2}R\sqrt{2g}\right)t + R^{3/2}$
$= R^{3/2}\left[\left(\frac{3}{2}R^{-1/2}\sqrt{2g}\right)t + 1\right] = R^{3/2}\left[\left(\frac{3\sqrt{2gR}}{2R}\right)t + 1\right] = R^{3/2}\left[\left(\frac{3v_0}{2R}\right)t + 1\right] \Rightarrow s = R\left[1 + \left(\frac{3v_0}{2R}\right)t\right]^{2/3}$

CHAPTER 9 ADDITIONAL AND ADVANCED EXERCISES

1. (a) $\frac{dy}{dt} = k\frac{A}{V}(c - y) \Rightarrow dy = -k\frac{A}{V}(y - c)dt \Rightarrow \frac{dy}{y-c} = -k\frac{A}{V}dt \Rightarrow \int \frac{dy}{y-c} = -\int k\frac{A}{V}dt \Rightarrow \ln|y - c| = -k\frac{A}{V}t + C_1$
$\Rightarrow y - c = \pm e^{C_1}e^{-k\frac{A}{V}t}$. Apply the initial condition, $y(0) = y_0 \Rightarrow y_0 = c + C \Rightarrow C = y_0 - c$
$\Rightarrow y = c + (y_0 - c)e^{-k\frac{A}{V}t}$.
   (b) Steady state solution: $y_\infty = \lim_{t\to\infty} y(t) = \lim_{t\to\infty}\left[c + (y_0 - c)e^{-k\frac{A}{V}t}\right] = c + (y_0 - c)(0) = c$

3. (a) Let y be any function such that $v(x)y = \int v(x)Q(x)\,dx + C$, $v(x) = e^{\int P(x)\,dx}$. Then
$\frac{d}{dx}(v(x)\cdot y) = v(x)\cdot y' + y\cdot v'(x) = v(x)Q(x)$. We have $v(x) = e^{\int P(x)\,dx} \Rightarrow v'(x) = \ = e^{\int P(x)\,dx}P(x) = v(x)P(x)$.
Thus $v(x)\cdot y' + y\cdot v(x)\,P(x) = v(x)Q(x) \Rightarrow y' + y\,P(x) = Q(x) \Rightarrow$ the given y is a solution.
   (b) If v and Q are continuous on $[a, b]$ and $x \in (a, b)$, then $\frac{d}{dx}\left[\int_{x_0}^{x} v(t)Q(t)\,dt\right] = v(x)Q(x)$
$\Rightarrow \int_{x_0}^{x} v(t)Q(t)\,dt = \int v(x)Q(x)\,dx$. So $C = y_0v(x_0) - \int v(x)Q(x)\,dx$. From part (a), $v(x)y = \int v(x)Q(x)\,dx + C$.
Substituting for C: $v(x)y = \int v(x)Q(x)\,dx + y_0v(x_0) - \int v(x)Q(x)\,dx \Rightarrow v(x)y = y_0v(x_0)$ when $x = x_0$.

5. $(x^2 + y^2)dx + xy\,dy = 0 \Rightarrow \frac{dy}{dx} = \frac{-(x^2+y^2)}{xy} = -\frac{x}{y} - \frac{y}{x} = -\frac{1}{y/x} - \frac{y}{x} = F\left(\frac{y}{x}\right) \Rightarrow F(v) = -\frac{1}{v} - v \Rightarrow \frac{dx}{x} + \frac{dv}{v - F(v)} = 0$
$\Rightarrow \frac{dx}{x} + \frac{dv}{v - \left(-\frac{1}{v} - v\right)} = 0 \Rightarrow \int \frac{dx}{x} + \int \frac{v\,dv}{2v^2+1} = C \Rightarrow \ln|x| + \frac{1}{4}\ln|2v^2 + 1| = C \Rightarrow 4\ln|x| + \ln\left|2\left(\frac{y}{x}\right)^2 + 1\right| = C$
$\Rightarrow \ln|x^4| + \ln\left|\frac{2y^2+x^2}{x^2}\right| = C \Rightarrow \ln\left|x^2(2y^2 + x^2)\right| = C \Rightarrow x^2(2y^2 + x^2) = e^C \Rightarrow x^2(2y^2 + x^2) = C$

7. $(x e^{y/x} + y)dx - x\,dy = 0 \Rightarrow \frac{dy}{dx} = \frac{x e^{y/x} + y}{x} = e^{y/x} + \frac{y}{x} = F\left(\frac{y}{x}\right) \Rightarrow F(v) = e^v + v \Rightarrow \frac{dx}{x} + \frac{dv}{v - (e^v + v)} = 0$
$\Rightarrow \int \frac{dx}{x} - \int \frac{dv}{e^v} = C \Rightarrow \ln|x| + e^{-v} = C \Rightarrow \ln|x| + e^{-y/x} = C$

9. $y' = \frac{y}{x} + \cos\left(\frac{y-x}{x}\right) = \frac{y}{x} + \cos\left(\frac{y}{x} - 1\right) = F\left(\frac{y}{x}\right) \Rightarrow F(v) = v + \cos(v - 1) \Rightarrow \frac{dx}{x} + \frac{dv}{v - (v + \cos(v-1))} = 0$
$\Rightarrow \int \frac{dx}{x} - \int \sec(v - 1)\,dv = 0 \Rightarrow \ln|x| - \ln|\sec(v - 1) + \tan(v - 1)| = C \Rightarrow \ln|x| - \ln\left|\sec\left(\frac{y}{x} - 1\right) + \tan\left(\frac{y}{x} - 1\right)\right| = C$

NOTES:

CHAPTER 10 INFINITE SEQUENCES AND SERIES

10.1 SEQUENCES

1. $a_1 = \frac{1-1}{1^2} = 0, a_2 = \frac{1-2}{2^2} = -\frac{1}{4}, a_3 = \frac{1-3}{3^2} = -\frac{2}{9}, a_4 = \frac{1-4}{4^2} = -\frac{3}{16}$

3. $a_1 = \frac{(-1)^2}{2-1} = 1, a_2 = \frac{(-1)^3}{4-1} = -\frac{1}{3}, a_3 = \frac{(-1)^4}{6-1} = \frac{1}{5}, a_4 = \frac{(-1)^5}{8-1} = -\frac{1}{7}$

5. $a_1 = \frac{2}{2^2} = \frac{1}{2}, a_2 = \frac{2^2}{2^3} = \frac{1}{2}, a_3 = \frac{2^3}{2^4} = \frac{1}{2}, a_4 = \frac{2^4}{2^5} = \frac{1}{2}$

7. $a_1 = 1, a_2 = 1 + \frac{1}{2} = \frac{3}{2}, a_3 = \frac{3}{2} + \frac{1}{2^2} = \frac{7}{4}, a_4 = \frac{7}{4} + \frac{1}{2^3} = \frac{15}{8}, a_5 = \frac{15}{8} + \frac{1}{2^4} = \frac{31}{16}, a_6 = \frac{63}{32},$
$a_7 = \frac{127}{64}, a_8 = \frac{255}{128}, a_9 = \frac{511}{256}, a_{10} = \frac{1023}{512}$

9. $a_1 = 2, a_2 = \frac{(-1)^2(2)}{2} = 1, a_3 = \frac{(-1)^3(1)}{2} = -\frac{1}{2}, a_4 = \frac{(-1)^4\left(-\frac{1}{2}\right)}{2} = -\frac{1}{4}, a_5 = \frac{(-1)^5\left(-\frac{1}{4}\right)}{2} = \frac{1}{8},$
$a_6 = \frac{1}{16}, a_7 = -\frac{1}{32}, a_8 = -\frac{1}{64}, a_9 = \frac{1}{128}, a_{10} = \frac{1}{256}$

11. $a_1 = 1, a_2 = 1, a_3 = 1 + 1 = 2, a_4 = 2 + 1 = 3, a_5 = 3 + 2 = 5, a_6 = 8, a_7 = 13, a_8 = 21, a_9 = 34, a_{10} = 55$

13. $a_n = (-1)^{n+1}, n = 1, 2, \ldots$

15. $a_n = (-1)^{n+1}n^2, n = 1, 2, \ldots$

17. $a_n = \frac{2^{n-1}}{3(n+2)}, n = 1, 2, \ldots$

19. $a_n = n^2 - 1, n = 1, 2, \ldots$

21. $a_n = 4n - 3, n = 1, 2, \ldots$

23. $a_n = \frac{3n+2}{n!}, n = 1, 2, \ldots$

25. $a_n = \frac{1+(-1)^{n+1}}{2}, n = 1, 2, \ldots$

27. $\lim_{n\to\infty} 2 + (0.1)^n = 2 \Rightarrow$ converges (Theorem 5, #4)

29. $\lim_{n\to\infty} \frac{1-2n}{1+2n} = \lim_{n\to\infty} \frac{\left(\frac{1}{n}\right)-2}{\left(\frac{1}{n}\right)+2} = \lim_{n\to\infty} \frac{-2}{2} = -1 \Rightarrow$ converges

31. $\lim_{n\to\infty} \frac{1-5n^4}{n^4+8n^3} = \lim_{n\to\infty} \frac{\left(\frac{1}{n^4}\right)-5}{1+\left(\frac{8}{n}\right)} = -5 \Rightarrow$ converges

33. $\lim_{n\to\infty} \frac{n^2-2n+1}{n-1} = \lim_{n\to\infty} \frac{(n-1)(n-1)}{n-1} = \lim_{n\to\infty} (n-1) = \infty \Rightarrow$ diverges

35. $\lim_{n\to\infty} (1 + (-1)^n)$ does not exist $\Rightarrow$ diverges

37. $\lim_{n\to\infty} \left(\frac{n+1}{2n}\right)\left(1 - \frac{1}{n}\right) = \lim_{n\to\infty} \left(\frac{1}{2} + \frac{1}{2n}\right)\left(1 - \frac{1}{n}\right) = \frac{1}{2} \Rightarrow$ converges

39. $\lim_{n\to\infty} \frac{(-1)^{n+1}}{2n-1} = 0 \Rightarrow$ converges

41. $\lim_{n\to\infty} \sqrt{\frac{2n}{n+1}} = \sqrt{\lim_{n\to\infty} \frac{2n}{n+1}} = \sqrt{\lim_{n\to\infty} \left(\frac{2}{1+\frac{1}{n}}\right)} = \sqrt{2} \Rightarrow$ converges

43. $\lim_{n\to\infty} \sin\left(\frac{\pi}{2}+\frac{1}{n}\right) = \sin\left(\lim_{n\to\infty}\left(\frac{\pi}{2}+\frac{1}{n}\right)\right) = \sin\frac{\pi}{2} = 1 \Rightarrow$ converges

45. $\lim_{n\to\infty} \frac{\sin n}{n} = 0$ because $-\frac{1}{n} \le \frac{\sin n}{n} \le \frac{1}{n} \Rightarrow$ converges by the Sandwich Theorem for sequences

47. $\lim_{n\to\infty} \frac{n}{2^n} = \lim_{n\to\infty} \frac{1}{2^n \ln 2} = 0 \Rightarrow$ converges (using l'Hôpital's rule)

49. $\lim_{n\to\infty} \frac{\ln(n+1)}{\sqrt{n}} = \lim_{n\to\infty} \frac{\left(\frac{1}{n+1}\right)}{\left(\frac{1}{2\sqrt{n}}\right)} = \lim_{n\to\infty} \frac{2\sqrt{n}}{n+1} = \lim_{n\to\infty} \frac{\left(\frac{2}{\sqrt{n}}\right)}{1+\left(\frac{1}{n}\right)} = 0 \Rightarrow$ converges

51. $\lim_{n\to\infty} 8^{1/n} = 1 \Rightarrow$ converges (Theorem 5, #3)

53. $\lim_{n\to\infty} \left(1+\frac{7}{n}\right)^n = e^7 \Rightarrow$ converges (Theorem 5, #5)

55. $\lim_{n\to\infty} \sqrt[n]{10n} = \lim_{n\to\infty} 10^{1/n} \cdot n^{1/n} = 1 \cdot 1 = 1 \Rightarrow$ converges (Theorem 5, #3 and #2)

57. $\lim_{n\to\infty} \left(\frac{3}{n}\right)^{1/n} = \frac{\lim_{n\to\infty} 3^{1/n}}{\lim_{n\to\infty} n^{1/n}} = \frac{1}{1} = 1 \Rightarrow$ converges (Theorem 5, #3 and #2)

59. $\lim_{n\to\infty} \frac{\ln n}{n^{1/n}} = \frac{\lim_{n\to\infty} \ln n}{\lim_{n\to\infty} n^{1/n}} = \frac{\infty}{1} = \infty \Rightarrow$ diverges (Theorem 5, #2)

61. $\lim_{n\to\infty} \sqrt[n]{4^n n} = \lim_{n\to\infty} 4\sqrt[n]{n} = 4 \cdot 1 = 4 \Rightarrow$ converges (Theorem 5, #2)

63. $\lim_{n\to\infty} \frac{n!}{n^n} = \lim_{n\to\infty} \frac{1\cdot 2\cdot 3\cdots(n-1)(n)}{n\cdot n\cdot n\cdots n\cdot n} \le \lim_{n\to\infty} \left(\frac{1}{n}\right) = 0$ and $\frac{n!}{n^n} \ge 0 \Rightarrow \lim_{n\to\infty} \frac{n!}{n^n} = 0 \Rightarrow$ converges

65. $\lim_{n\to\infty} \frac{n!}{10^{6n}} = \lim_{n\to\infty} \frac{1}{\left(\frac{(10^6)^n}{n!}\right)} = \infty \Rightarrow$ diverges (Theorem 5, #6)

67. $\lim_{n\to\infty} \left(\frac{1}{n}\right)^{1/(\ln n)} = \lim_{n\to\infty} \exp\left(\frac{1}{\ln n}\ln\left(\frac{1}{n}\right)\right) = \lim_{n\to\infty} \exp\left(\frac{\ln 1 - \ln n}{\ln n}\right) = e^{-1} \Rightarrow$ converges

69. $\lim_{n\to\infty} \left(\frac{3n+1}{3n-1}\right)^n = \lim_{n\to\infty} \exp\left(n \ln\left(\frac{3n+1}{3n-1}\right)\right) = \lim_{n\to\infty} \exp\left(\frac{\ln(3n+1) - \ln(3n-1)}{\frac{1}{n}}\right)$

$= \lim_{n\to\infty} \exp\left(\frac{\frac{3}{3n+1} - \frac{3}{3n-1}}{\left(-\frac{1}{n^2}\right)}\right) = \lim_{n\to\infty} \exp\left(\frac{6n^2}{(3n+1)(3n-1)}\right) = \exp\left(\frac{6}{9}\right) = e^{2/3} \Rightarrow$ converges

71. $\lim_{n\to\infty} \left(\frac{x^n}{2n+1}\right)^{1/n} = \lim_{n\to\infty} x\left(\frac{1}{2n+1}\right)^{1/n} = x \lim_{n\to\infty} \exp\left(\frac{1}{n}\ln\left(\frac{1}{2n+1}\right)\right) = x \lim_{n\to\infty} \exp\left(\frac{-\ln(2n+1)}{n}\right)$

$= x \lim_{n\to\infty} \exp\left(\frac{-2}{2n+1}\right) = xe^0 = x,\ x > 0 \Rightarrow$ converges

73. $\lim_{n\to\infty} \frac{3^n \cdot 6^n}{2^{-n} \cdot n!} = \lim_{n\to\infty} \frac{36^n}{n!} = 0 \Rightarrow$ converges (Theorem 5, #6)

75. $\lim_{n\to\infty} \tanh n = \lim_{n\to\infty} \frac{e^n - e^{-n}}{e^n + e^{-n}} = \lim_{n\to\infty} \frac{e^{2n} - 1}{e^{2n} + 1} = \lim_{n\to\infty} \frac{2e^{2n}}{2e^{2n}} = \lim_{n\to\infty} 1 = 1 \Rightarrow$ converges

77. $\lim_{n\to\infty} \frac{n^2 \sin\left(\frac{1}{n}\right)}{2n-1} = \lim_{n\to\infty} \frac{\sin\left(\frac{1}{n}\right)}{\left(\frac{2}{n} - \frac{1}{n^2}\right)} = \lim_{n\to\infty} \frac{-\left(\cos\left(\frac{1}{n}\right)\right)\left(\frac{1}{n^2}\right)}{\left(-\frac{2}{n^2} + \frac{2}{n^3}\right)} = \lim_{n\to\infty} \frac{-\cos\left(\frac{1}{n}\right)}{-2 + \left(\frac{2}{n}\right)} = \frac{1}{2} \Rightarrow$ converges

79. $\lim_{n \to \infty} \sqrt{n} \sin\left(\frac{1}{\sqrt{n}}\right) = \lim_{n \to \infty} \frac{\sin\left(\frac{1}{\sqrt{n}}\right)}{\frac{1}{\sqrt{n}}} = \lim_{n \to \infty} \frac{\cos\left(\frac{1}{\sqrt{n}}\right)\left(-\frac{1}{2n^{3/2}}\right)}{-\frac{1}{2n^{3/2}}} = \lim_{n \to \infty} \cos\left(\frac{1}{\sqrt{n}}\right) = \cos 0 = 1 \Rightarrow$ converges

81. $\lim_{n \to \infty} \tan^{-1} n = \frac{\pi}{2} \Rightarrow$ converges

83. $\lim_{n \to \infty} \left(\frac{1}{3}\right)^n + \frac{1}{\sqrt{2^n}} = \lim_{n \to \infty} \left(\left(\frac{1}{3}\right)^n + \left(\frac{1}{\sqrt{2}}\right)^n\right) = 0 \Rightarrow$ converges (Theorem 5, #4)

85. $\lim_{n \to \infty} \frac{(\ln n)^{200}}{n} = \lim_{n \to \infty} \frac{200 (\ln n)^{199}}{n} = \lim_{n \to \infty} \frac{200 \cdot 199 (\ln n)^{198}}{n} = \ldots = \lim_{n \to \infty} \frac{200!}{n} = 0 \Rightarrow$ converges

87. $\lim_{n \to \infty} \left(n - \sqrt{n^2 - n}\right) = \lim_{n \to \infty} \left(n - \sqrt{n^2 - n}\right)\left(\frac{n + \sqrt{n^2 - n}}{n + \sqrt{n^2 - n}}\right) = \lim_{n \to \infty} \frac{n}{n + \sqrt{n^2 - n}} = \lim_{n \to \infty} \frac{1}{1 + \sqrt{1 - \frac{1}{n}}}$
$= \frac{1}{2} \Rightarrow$ converges

89. $\lim_{n \to \infty} \frac{1}{n} \int_1^n \frac{1}{x}\, dx = \lim_{n \to \infty} \frac{\ln n}{n} = \lim_{n \to \infty} \frac{1}{n} = 0 \Rightarrow$ converges (Theorem 5, #1)

91. Since a_n converges $\Rightarrow \lim_{n \to \infty} a_n = L \Rightarrow \lim_{n \to \infty} a_{n+1} = \lim_{n \to \infty} \frac{72}{1 + a_n} \Rightarrow L = \frac{72}{1 + L} \Rightarrow L(1 + L) = 72 \Rightarrow L^2 + L - 72 = 0$
$\Rightarrow L = -9$ or $L = 8$; since $a_n > 0$ for $n \ge 1 \Rightarrow L = 8$

93. Since a_n converges $\Rightarrow \lim_{n \to \infty} a_n = L \Rightarrow \lim_{n \to \infty} a_{n+1} = \lim_{n \to \infty} \sqrt{8 + 2a_n} \Rightarrow L = \sqrt{8 + 2L} \Rightarrow L^2 - 2L - 8 = 0 \Rightarrow L = -2$
or $L = 4$; since $a_n > 0$ for $n \ge 3 \Rightarrow L = 4$

95. Since a_n converges $\Rightarrow \lim_{n \to \infty} a_n = L \Rightarrow \lim_{n \to \infty} a_{n+1} = \lim_{n \to \infty} \sqrt{5a_n} \Rightarrow L = \sqrt{5L} \Rightarrow L^2 - 5L = 0 \Rightarrow L = 0$ or $L = 5$; since
$a_n > 0$ for $n \ge 1 \Rightarrow L = 5$

97. $a_{n+1} = 2 + \frac{1}{a_n}$, $n \ge 1$, $a_1 = 2$. Since a_n converges $\Rightarrow \lim_{n \to \infty} a_n = L \Rightarrow \lim_{n \to \infty} a_{n+1} = \lim_{n \to \infty} \left(2 + \frac{1}{a_n}\right) \Rightarrow L = 2 + \frac{1}{L}$
$\Rightarrow L^2 - 2L - 1 = 0 \Rightarrow L = 1 \pm \sqrt{2}$; since $a_n > 0$ for $n \ge 1 \Rightarrow L = 1 + \sqrt{2}$

99. $1, 1, 2, 4, 8, 16, 32, \ldots = 1, 2^0, 2^1, 2^2, 2^3, 2^4, 2^5, \ldots \Rightarrow x_1 = 1$ and $x_n = 2^{n-2}$ for $n \ge 2$

101. (a) $f(x) = x^2 - 2$; the sequence converges to $1.414213562 \approx \sqrt{2}$
(b) $f(x) = \tan(x) - 1$; the sequence converges to $0.7853981635 \approx \frac{\pi}{4}$
(c) $f(x) = e^x$; the sequence $1, 0, -1, -2, -3, -4, -5, \ldots$ diverges

103. (a) If $a = 2n + 1$, then $b = \lfloor \frac{a^2}{2} \rfloor = \lfloor \frac{4n^2 + 4n + 1}{2} \rfloor = \lfloor 2n^2 + 2n + \frac{1}{2} \rfloor = 2n^2 + 2n$, $c = \lceil \frac{a^2}{2} \rceil = \lceil 2n^2 + 2n + \frac{1}{2} \rceil$
$= 2n^2 + 2n + 1$ and $a^2 + b^2 = (2n + 1)^2 + \left(2n^2 + 2n\right)^2 = 4n^2 + 4n + 1 + 4n^4 + 8n^3 + 4n^2$
$= 4n^4 + 8n^3 + 8n^2 + 4n + 1 = \left(2n^2 + 2n + 1\right)^2 = c^2$.
(b) $\lim_{a \to \infty} \frac{\lfloor \frac{a^2}{2} \rfloor}{\lceil \frac{a^2}{2} \rceil} = \lim_{a \to \infty} \frac{2n^2 + 2n}{2n^2 + 2n + 1} = 1$ or $\lim_{a \to \infty} \frac{\lfloor \frac{a^2}{2} \rfloor}{\lceil \frac{a^2}{2} \rceil} = \lim_{a \to \infty} \sin\theta = \lim_{\theta \to \pi/2} \sin\theta = 1$

105. (a) $\lim_{n \to \infty} \frac{\ln n}{n^c} = \lim_{n \to \infty} \frac{\left(\frac{1}{n}\right)}{cn^{c-1}} = \lim_{n \to \infty} \frac{1}{cn^c} = 0$
(b) For all $\epsilon > 0$, there exists an $N = e^{-(\ln \epsilon)/c}$ such that $n > e^{-(\ln \epsilon)/c} \Rightarrow \ln n > -\frac{\ln \epsilon}{c} \Rightarrow \ln n^c > \ln\left(\frac{1}{\epsilon}\right)$
$\Rightarrow n^c > \frac{1}{\epsilon} \Rightarrow \frac{1}{n^c} < \epsilon \Rightarrow \left|\frac{1}{n^c} - 0\right| < \epsilon \Rightarrow \lim_{n \to \infty} \frac{1}{n^c} = 0$

107. $\lim_{n \to \infty} n^{1/n} = \lim_{n \to \infty} \exp\left(\frac{1}{n} \ln n\right) = \lim_{n \to \infty} \exp\left(\frac{1}{n}\right) = e^0 = 1$

109. Assume the hypotheses of the theorem and let ϵ be a positive number. For all ϵ there exists a N_1 such that when $n > N_1$ then $|a_n - L| < \epsilon \Rightarrow -\epsilon < a_n - L < \epsilon \Rightarrow L - \epsilon < a_n$, and there exists a N_2 such that when $n > N_2$ then $|c_n - L| < \epsilon \Rightarrow -\epsilon < c_n - L < \epsilon \Rightarrow c_n < L + \epsilon$. If $n > \max\{N_1, N_2\}$, then $L - \epsilon < a_n \le b_n \le c_n < L + \epsilon \Rightarrow |b_n - L| < \epsilon \Rightarrow \lim_{n \to \infty} b_n = L$.

111. $a_{n+1} \ge a_n \Rightarrow \frac{3(n+1)+1}{(n+1)+1} > \frac{3n+1}{n+1} \Rightarrow \frac{3n+4}{n+2} > \frac{3n+1}{n+1} \Rightarrow 3n^2 + 3n + 4n + 4 > 3n^2 + 6n + n + 2$ $\Rightarrow 4 > 2$; the steps are reversible so the sequence is nondecreasing; $\frac{3n+1}{n+1} < 3 \Rightarrow 3n + 1 < 3n + 3$ $\Rightarrow 1 < 3$; the steps are reversible so the sequence is bounded above by 3

113. $a_{n+1} \le a_n \Rightarrow \frac{2^{n+1}3^{n+1}}{(n+1)!} \le \frac{2^n 3^n}{n!} \Rightarrow \frac{2^{n+1}3^{n+1}}{2^n 3^n} \le \frac{(n+1)!}{n!} \Rightarrow 2 \cdot 3 \le n + 1$ which is true for $n \ge 5$; the steps are reversible so the sequence is decreasing after a_5, but it is not nondecreasing for all its terms; $a_1 = 6$, $a_2 = 18$, $a_3 = 36$, $a_4 = 54$, $a_5 = \frac{324}{5} = 64.8 \Rightarrow$ the sequence is bounded from above by 64.8

115. $a_n = 1 - \frac{1}{n}$ converges because $\frac{1}{n} \to 0$ by Example 1; also it is a nondecreasing sequence bounded above by 1

117. $a_n = \frac{2^n - 1}{2^n} = 1 - \frac{1}{2^n}$ and $0 < \frac{1}{2^n} < \frac{1}{n}$; since $\frac{1}{n} \to 0$ (by Example 1) $\Rightarrow \frac{1}{2^n} \to 0$, the sequence converges; also it is a nondecreasing sequence bounded above by 1

119. $a_n = ((-1)^n + 1)\left(\frac{n+1}{n}\right)$ diverges because $a_n = 0$ for n odd, while for n even $a_n = 2\left(1 + \frac{1}{n}\right)$ converges to 2; it diverges by definition of divergence

121. $a_n \ge a_{n+1} \Leftrightarrow \frac{1+\sqrt{2n}}{\sqrt{n}} \ge \frac{1+\sqrt{2(n+1)}}{\sqrt{n+1}} \Leftrightarrow \sqrt{n+1} + \sqrt{2n^2+2n} \ge \sqrt{n} + \sqrt{2n^2+2n} \Leftrightarrow \sqrt{n+1} \ge \sqrt{n}$ and $\frac{1+\sqrt{2n}}{\sqrt{n}} \ge \sqrt{2}$; thus the sequence is nonincreasing and bounded below by $\sqrt{2} \Rightarrow$ it converges

123. $\frac{4^{n+1}+3^n}{4^n} = 4 + \left(\frac{3}{4}\right)^n$ so $a_n \ge a_{n+1} \Leftrightarrow 4 + \left(\frac{3}{4}\right)^n \ge 4 + \left(\frac{3}{4}\right)^{n+1} \Leftrightarrow \left(\frac{3}{4}\right)^n \ge \left(\frac{3}{4}\right)^{n+1} \Leftrightarrow 1 \ge \frac{3}{4}$ and $4 + \left(\frac{3}{4}\right)^n \ge 4$; thus the sequence is nonincreasing and bounded below by 4 $\Rightarrow$ it converges

125. Let $0 < M < 1$ and let N be an integer greater than $\frac{M}{1-M}$. Then $n > N \Rightarrow n > \frac{M}{1-M} \Rightarrow n - nM > M$ $\Rightarrow n > M + nM \Rightarrow n > M(n+1) \Rightarrow \frac{n}{n+1} > M$.

127. The sequence $a_n = 1 + \frac{(-1)^n}{2}$ is the sequence $\frac{1}{2}, \frac{3}{2}, \frac{1}{2}, \frac{3}{2}, \ldots$. This sequence is bounded above by $\frac{3}{2}$, but it clearly does not converge, by definition of convergence.

129. Given an $\epsilon > 0$, by definition of convergence there corresponds an N such that for all $n > N$, $|L_1 - a_n| < \epsilon$ and $|L_2 - a_n| < \epsilon$. Now $|L_2 - L_1| = |L_2 - a_n + a_n - L_1| \le |L_2 - a_n| + |a_n - L_1| < \epsilon + \epsilon = 2\epsilon$. $|L_2 - L_1| < 2\epsilon$ says that the difference between two fixed values is smaller than any positive number 2ϵ. The only nonnegative number smaller than every positive number is 0, so $|L_1 - L_2| = 0$ or $L_1 = L_2$.

131. $a_{2k} \to L \Leftrightarrow$ given an $\epsilon > 0$ there corresponds an N_1 such that $[2k > N_1 \Rightarrow |a_{2k} - L| < \epsilon]$. Similarly, $a_{2k+1} \to L \Leftrightarrow [2k + 1 > N_2 \Rightarrow |a_{2k+1} - L| < \epsilon]$. Let $N = \max\{N_1, N_2\}$. Then $n > N \Rightarrow |a_n - L| < \epsilon$ whether n is even or odd, and hence $a_n \to L$.

133. (a) $f(x) = x^2 - a \Rightarrow f'(x) = 2x \Rightarrow x_{n+1} = x_n - \frac{x_n^2 - a}{2x_n} \Rightarrow x_{n+1} = \frac{2x_n^2 - (x_n^2 - a)}{2x_n} = \frac{x_n^2 + a}{2x_n} = \frac{\left(x_n + \frac{a}{x_n}\right)}{2}$

(b) $x_1 = 2$, $x_2 = 1.75$, $x_3 = 1.732142857$, $x_4 = 1.73205081$, $x_5 = 1.732050808$; we are finding the positive number where $x^2 - 3 = 0$; that is, where $x^2 = 3$, $x > 0$, or where $x = \sqrt{3}$.

10.2 INFINITE SERIES

1. $s_n = \frac{a(1-r^n)}{(1-r)} = \frac{2\left(1-\left(\frac{1}{3}\right)^n\right)}{1-\left(\frac{1}{3}\right)} \Rightarrow \lim\limits_{n\to\infty} s_n = \frac{2}{1-\left(\frac{1}{3}\right)} = 3$

3. $s_n = \frac{a(1-r^n)}{(1-r)} = \frac{1-\left(-\frac{1}{2}\right)^n}{1-\left(-\frac{1}{2}\right)} \Rightarrow \lim\limits_{n\to\infty} s_n = \frac{1}{\left(\frac{3}{2}\right)} = \frac{2}{3}$

5. $\frac{1}{(n+1)(n+2)} = \frac{1}{n+1} - \frac{1}{n+2} \Rightarrow s_n = \left(\frac{1}{2}-\frac{1}{3}\right) + \left(\frac{1}{3}-\frac{1}{4}\right) + \ldots + \left(\frac{1}{n+1}-\frac{1}{n+2}\right) = \frac{1}{2} - \frac{1}{n+2} \Rightarrow \lim\limits_{n\to\infty} s_n = \frac{1}{2}$

7. $1 - \frac{1}{4} + \frac{1}{16} - \frac{1}{64} + \ldots$, the sum of this geometric series is $\frac{1}{1-\left(-\frac{1}{4}\right)} = \frac{1}{1+\left(\frac{1}{4}\right)} = \frac{4}{5}$

9. $\frac{7}{4} + \frac{7}{16} + \frac{7}{64} + \ldots$, the sum of this geometric series is $\frac{\left(\frac{7}{4}\right)}{1-\left(\frac{1}{4}\right)} = \frac{7}{3}$

11. $(5+1) + \left(\frac{5}{2}+\frac{1}{3}\right) + \left(\frac{5}{4}+\frac{1}{9}\right) + \left(\frac{5}{8}+\frac{1}{27}\right) + \ldots$, is the sum of two geometric series; the sum is $\frac{5}{1-\left(\frac{1}{2}\right)} + \frac{1}{1-\left(\frac{1}{3}\right)} = 10 + \frac{3}{2} = \frac{23}{2}$

13. $(1+1) + \left(\frac{1}{2}-\frac{1}{5}\right) + \left(\frac{1}{4}+\frac{1}{25}\right) + \left(\frac{1}{8}-\frac{1}{125}\right) + \ldots$, is the sum of two geometric series; the sum is $\frac{1}{1-\left(\frac{1}{2}\right)} + \frac{1}{1+\left(\frac{1}{5}\right)} = 2 + \frac{5}{6} = \frac{17}{6}$

15. Series is geometric with $r = \frac{2}{5} \Rightarrow \left|\frac{2}{5}\right| < 1 \Rightarrow$ Converges to $\frac{1}{1-\frac{2}{5}} = \frac{5}{3}$

17. Series is geometric with $r = \frac{1}{8} \Rightarrow \left|\frac{1}{8}\right| < 1 \Rightarrow$ Converges to $\frac{\frac{1}{8}}{1-\frac{1}{8}} = \frac{1}{7}$

19. $0.\overline{23} = \sum\limits_{n=0}^{\infty} \frac{23}{100}\left(\frac{1}{10^2}\right)^n = \frac{\left(\frac{23}{100}\right)}{1-\left(\frac{1}{100}\right)} = \frac{23}{99}$

21. $0.\overline{7} = \sum\limits_{n=0}^{\infty} \frac{7}{10}\left(\frac{1}{10}\right)^n = \frac{\left(\frac{7}{10}\right)}{1-\left(\frac{1}{10}\right)} = \frac{7}{9}$

23. $0.0\overline{6} = \sum\limits_{n=0}^{\infty} \left(\frac{1}{10}\right)\left(\frac{6}{10}\right)\left(\frac{1}{10}\right)^n = \frac{\left(\frac{6}{100}\right)}{1-\left(\frac{1}{10}\right)} = \frac{6}{90} = \frac{1}{15}$

25. $1.24\overline{123} = \frac{124}{100} + \sum\limits_{n=0}^{\infty} \frac{123}{10^5}\left(\frac{1}{10^3}\right)^n = \frac{124}{100} + \frac{\left(\frac{123}{10^5}\right)}{1-\left(\frac{1}{10^3}\right)} = \frac{124}{100} + \frac{123}{10^5-10^2} = \frac{124}{100} + \frac{123}{99{,}900} = \frac{123{,}999}{99{,}900} = \frac{41{,}333}{33{,}300}$

27. $\lim\limits_{n\to\infty} \frac{n}{n+10} = \lim\limits_{n\to\infty} \frac{1}{1} = 1 \neq 0 \Rightarrow$ diverges

29. $\lim\limits_{n\to\infty} \frac{1}{n+4} = 0 \Rightarrow$ test inconclusive

31. $\lim\limits_{n\to\infty} \cos\frac{1}{n} = \cos 0 = 1 \neq 0 \Rightarrow$ diverges

33. $\lim\limits_{n\to\infty} \ln\frac{1}{n} = -\infty \neq 0 \Rightarrow$ diverges

35. $s_k = \left(1-\frac{1}{2}\right) + \left(\frac{1}{2}-\frac{1}{3}\right) + \left(\frac{1}{3}-\frac{1}{4}\right) + \ldots + \left(\frac{1}{k-1}-\frac{1}{k}\right) + \left(\frac{1}{k}-\frac{1}{k+1}\right) = 1 - \frac{1}{k+1} \Rightarrow \lim\limits_{k\to\infty} s_k$ $= \lim\limits_{k\to\infty}\left(1-\frac{1}{k+1}\right) = 1$, series converges to 1

37. $s_k = \left(\ln\sqrt{2} - \ln\sqrt{1}\right) + \left(\ln\sqrt{3} - \ln\sqrt{2}\right) + \left(\ln\sqrt{4} - \ln\sqrt{3}\right) + \ldots + \left(\ln\sqrt{k} - \ln\sqrt{k-1}\right) + \left(\ln\sqrt{k+1} - \ln\sqrt{k}\right)$
$= \ln\sqrt{k+1} - \ln\sqrt{1} = \ln\sqrt{k+1} \Rightarrow \lim\limits_{k\to\infty} s_k = \lim\limits_{k\to\infty} \ln\sqrt{k+1} = \infty$; series diverges

39. $s_k = \left(\cos^{-1}\left(\frac{1}{2}\right) - \cos^{-1}\left(\frac{1}{3}\right)\right) + \left(\cos^{-1}\left(\frac{1}{3}\right) - \cos^{-1}\left(\frac{1}{4}\right)\right) + \left(\cos^{-1}\left(\frac{1}{4}\right) - \cos^{-1}\left(\frac{1}{5}\right)\right) + \ldots$
$+ \left(\cos^{-1}\left(\frac{1}{k}\right) - \cos^{-1}\left(\frac{1}{k+1}\right)\right) + \left(\cos^{-1}\left(\frac{1}{k+1}\right) - \cos^{-1}\left(\frac{1}{k+2}\right)\right) = \frac{\pi}{3} - \cos^{-1}\left(\frac{1}{k+2}\right)$
$\Rightarrow \lim\limits_{k\to\infty} s_k = \lim\limits_{k\to\infty} \left[\frac{\pi}{3} - \cos^{-1}\left(\frac{1}{k+2}\right)\right] = \frac{\pi}{3} - \frac{\pi}{2} = \frac{\pi}{6}$, series converges to $\frac{\pi}{6}$

41. $\frac{4}{(4n-3)(4n+1)} = \frac{1}{4n-3} - \frac{1}{4n+1} \Rightarrow s_k = \left(1 - \frac{1}{5}\right) + \left(\frac{1}{5} - \frac{1}{9}\right) + \left(\frac{1}{9} - \frac{1}{13}\right) + \ldots + \left(\frac{1}{4k-7} - \frac{1}{4k-3}\right)$
$+ \left(\frac{1}{4k-3} - \frac{1}{4k+1}\right) = 1 - \frac{1}{4k+1} \Rightarrow \lim\limits_{k\to\infty} s_k = \lim\limits_{k\to\infty} \left(1 - \frac{1}{4k+1}\right) = 1$

43. $\frac{40n}{(2n-1)^2(2n+1)^2} = \frac{A}{(2n-1)} + \frac{B}{(2n-1)^2} + \frac{C}{(2n+1)} + \frac{D}{(2n+1)^2} = \frac{A(2n-1)(2n+1)^2 + B(2n+1)^2 + C(2n+1)(2n-1)^2 + D(2n-1)^2}{(2n-1)^2(2n+1)^2}$
$\Rightarrow A(2n-1)(2n+1)^2 + B(2n+1)^2 + C(2n+1)(2n-1)^2 + D(2n-1)^2 = 40n$
$\Rightarrow A\left(8n^3 + 4n^2 - 2n - 1\right) + B\left(4n^2 + 4n + 1\right) + C\left(8n^3 - 4n^2 - 2n + 1\right) = D\left(4n^2 - 4n + 1\right) = 40n$
$\Rightarrow (8A + 8C)n^3 + (4A + 4B - 4C + 4D)n^2 + (-2A + 4B - 2C - 4D)n + (-A + B + C + D) = 40n$

$$\Rightarrow \begin{cases} 8A + 8C = 0 \\ 4A + 4B - 4C + 4D = 0 \\ -2A + 4B - 2C - 4D = 40 \\ -A + B + C + D = 0 \end{cases} \Rightarrow \begin{cases} 8A + 8C = 0 \\ A + B - C + D = 0 \\ -A + 2B - C - 2D = 20 \\ -A + B + C + D = 0 \end{cases} \Rightarrow \begin{cases} B + D = 0 \\ 2B - 2D = 20 \end{cases} \Rightarrow 4B = 20 \Rightarrow B = 5$$

and $D = -5 \Rightarrow \begin{cases} A + C = 0 \\ -A + 5 + C - 5 = 0 \end{cases} \Rightarrow C = 0$ and $A = 0$. Hence, $\sum\limits_{n=1}^{k} \left[\frac{40n}{(2n-1)^2(2n+1)^2}\right]$
$= 5\sum\limits_{n=1}^{k} \left[\frac{1}{(2n-1)^2} - \frac{1}{(2n+1)^2}\right] = 5\left(\frac{1}{1} - \frac{1}{9} + \frac{1}{9} - \frac{1}{25} + \frac{1}{25} - \ldots - \frac{1}{(2(k-1)+1)^2} + \frac{1}{(2k-1)^2} - \frac{1}{(2k+1)^2}\right)$
$= 5\left(1 - \frac{1}{(2k+1)^2}\right) \Rightarrow$ the sum is $\lim\limits_{n\to\infty} 5\left(1 - \frac{1}{(2k+1)^2}\right) = 5$

45. $s_k = \left(1 - \frac{1}{\sqrt{2}}\right) + \left(\frac{1}{\sqrt{2}} - \frac{1}{\sqrt{3}}\right) + \left(\frac{1}{\sqrt{3}} - \frac{1}{\sqrt{4}}\right) + \ldots + \left(\frac{1}{\sqrt{k-1}} + \frac{1}{\sqrt{k}}\right) + \left(\frac{1}{\sqrt{k}} - \frac{1}{\sqrt{k+1}}\right) = 1 - \frac{1}{\sqrt{k+1}}$
$\Rightarrow \lim\limits_{k\to\infty} s_k = \lim\limits_{k\to\infty} \left(1 - \frac{1}{\sqrt{k+1}}\right) = 1$

47. $s_k = \left(\frac{1}{\ln 3} - \frac{1}{\ln 2}\right) + \left(\frac{1}{\ln 4} - \frac{1}{\ln 3}\right) + \left(\frac{1}{\ln 5} - \frac{1}{\ln 4}\right) + \ldots + \left(\frac{1}{\ln(k+1)} - \frac{1}{\ln k}\right) + \left(\frac{1}{\ln(k+2)} - \frac{1}{\ln(k+1)}\right)$
$= -\frac{1}{\ln 2} + \frac{1}{\ln(k+2)} \Rightarrow \lim\limits_{k\to\infty} s_k = -\frac{1}{\ln 2}$

49. convergent geometric series with sum $\frac{1}{1-\left(\frac{1}{\sqrt{2}}\right)} = \frac{\sqrt{2}}{\sqrt{2}-1} = 2 + \sqrt{2}$

51. convergent geometric series with sum $\frac{\left(\frac{3}{2}\right)}{1-\left(-\frac{1}{2}\right)} = 1$

53. $\lim\limits_{n\to\infty} \cos(n\pi) = \lim\limits_{n\to\infty} (-1)^n \neq 0 \Rightarrow$ diverges

55. convergent geometric series with sum $\frac{1}{1-\left(\frac{1}{e^2}\right)} = \frac{e^2}{e^2-1}$

57. convergent geometric series with sum $\frac{2}{1-\left(\frac{1}{10}\right)} - 2 = \frac{20}{9} - \frac{18}{9} = \frac{2}{9}$

59. difference of two geometric series with sum $\frac{1}{1-\left(\frac{2}{3}\right)} - \frac{1}{1-\left(\frac{1}{3}\right)} = 3 - \frac{3}{2} = \frac{3}{2}$

61. $\lim_{n\to\infty} \frac{n!}{1000^n} = \infty \neq 0 \Rightarrow$ diverges

63. $\sum_{n=1}^{\infty} \frac{2^n+3^n}{4^n} = \sum_{n=1}^{\infty} \frac{2^n}{4^n} + \sum_{n=1}^{\infty} \frac{3^n}{4^n} = \sum_{n=1}^{\infty} \left(\frac{1}{2}\right)^n + \sum_{n=1}^{\infty} \left(\frac{3}{4}\right)^n$; both $= \sum_{n=1}^{\infty} \left(\frac{1}{2}\right)^n$ and $\sum_{n=1}^{\infty} \left(\frac{3}{4}\right)^n$ are geometric series, and both converge since $r = \frac{1}{2} \Rightarrow \left|\frac{1}{2}\right| < 1$ and $r = \frac{3}{4} \Rightarrow \left|\frac{3}{4}\right| < 1$, respectivley $\Rightarrow \sum_{n=1}^{\infty} \left(\frac{1}{2}\right)^n = \frac{\frac{1}{2}}{1-\frac{1}{2}} = 1$ and $\sum_{n=1}^{\infty} \left(\frac{3}{4}\right)^n = \frac{\frac{3}{4}}{1-\frac{3}{4}} = 3 \Rightarrow$ $\sum_{n=1}^{\infty} \frac{2^n+3^n}{4^n} = 1 + 3 = 4$ by Theorem 8, part (1)

65. $\sum_{n=1}^{\infty} \ln\left(\frac{n}{n+1}\right) = \sum_{n=1}^{\infty} [\ln(n) - \ln(n+1)] \Rightarrow s_k = [\ln(1) - \ln(2)] + [\ln(2) - \ln(3)] + [\ln(3) - \ln(4)] + \ldots$ $+ [\ln(k-1) - \ln(k)] + [\ln(k) - \ln(k+1)] = -\ln(k+1) \Rightarrow \lim_{k\to\infty} s_k = -\infty, \Rightarrow$ diverges

67. convergent geometric series with sum $\frac{1}{1-\left(\frac{e}{\pi}\right)} = \frac{\pi}{\pi - e}$

69. $\sum_{n=0}^{\infty} (-1)^n x^n = \sum_{n=0}^{\infty} (-x)^n$; $a = 1$, $r = -x$; converges to $\frac{1}{1-(-x)} = \frac{1}{1+x}$ for $|x| < 1$

71. $a = 3$, $r = \frac{x-1}{2}$; converges to $\frac{3}{1-\left(\frac{x-1}{2}\right)} = \frac{6}{3-x}$ for $-1 < \frac{x-1}{2} < 1$ or $-1 < x < 3$

73. $a = 1$, $r = 2x$; converges to $\frac{1}{1-2x}$ for $|2x| < 1$ or $|x| < \frac{1}{2}$

75. $a = 1$, $r = -(x+1)^n$; converges to $\frac{1}{1+(x+1)} = \frac{1}{2+x}$ for $|x+1| < 1$ or $-2 < x < 0$

77. $a = 1$, $r = \sin x$; converges to $\frac{1}{1-\sin x}$ for $x \neq (2k+1)\frac{\pi}{2}$, k an integer

79. (a) $\sum_{n=-2}^{\infty} \frac{1}{(n+4)(n+5)}$ (b) $\sum_{n=0}^{\infty} \frac{1}{(n+2)(n+3)}$ (c) $\sum_{n=5}^{\infty} \frac{1}{(n-3)(n-2)}$

81. (a) one example is $\frac{1}{2} + \frac{1}{4} + \frac{1}{8} + \frac{1}{16} + \ldots = \frac{\left(\frac{1}{2}\right)}{1-\left(\frac{1}{2}\right)} = 1$

(b) one example is $-\frac{3}{2} - \frac{3}{4} - \frac{3}{8} - \frac{3}{16} - \ldots = \frac{\left(-\frac{3}{2}\right)}{1-\left(\frac{1}{2}\right)} = -3$

(c) one example is $1 - \frac{1}{2} - \frac{1}{4} - \frac{1}{8} - \frac{1}{16} - \ldots = 1 - \frac{\left(\frac{1}{2}\right)}{1-\left(\frac{1}{2}\right)} = 0.$

83. Let $a_n = b_n = \left(\frac{1}{2}\right)^n$. Then $\sum_{n=1}^{\infty} a_n = \sum_{n=1}^{\infty} b_n = \sum_{n=1}^{\infty} \left(\frac{1}{2}\right)^n = 1$, while $\sum_{n=1}^{\infty} \left(\frac{a_n}{b_n}\right) = \sum_{n=1}^{\infty} (1)$ diverges.

85. Let $a_n = \left(\frac{1}{4}\right)^n$ and $b_n = \left(\frac{1}{2}\right)^n$. Then $A = \sum_{n=1}^{\infty} a_n = \frac{1}{3}$, $B = \sum_{n=1}^{\infty} b_n = 1$ and $\sum_{n=1}^{\infty} \left(\frac{a_n}{b_n}\right) = \sum_{n=1}^{\infty} \left(\frac{1}{2}\right)^n = 1 \neq \frac{A}{B}$.

87. Since the sum of a finite number of terms is finite, adding or subtracting a finite number of terms from a series that diverges does not change the divergence of the series.

89. (a) $\frac{2}{1-r} = 5 \Rightarrow \frac{2}{5} = 1 - r \Rightarrow r = \frac{3}{5}$; $2 + 2\left(\frac{3}{5}\right) + 2\left(\frac{3}{5}\right)^2 + \ldots$

(b) $\frac{\left(\frac{13}{2}\right)}{1-r} = 5 \Rightarrow \frac{13}{10} = 1 - r \Rightarrow r = -\frac{3}{10}$; $\frac{13}{2} - \frac{13}{2}\left(\frac{3}{10}\right) + \frac{13}{2}\left(\frac{3}{10}\right)^2 - \frac{13}{2}\left(\frac{3}{10}\right)^3 + \ldots$

91. $s_n = 1 + 2r + r^2 + 2r^3 + r^4 + 2r^5 + \ldots + r^{2n} + 2r^{2n+1}$, $n = 0, 1, \ldots$
$\Rightarrow s_n = (1 + r^2 + r^4 + \ldots + r^{2n}) + (2r + 2r^3 + 2r^5 + \ldots + 2r^{2n+1}) \Rightarrow \lim_{n\to\infty} s_n = \frac{1}{1-r^2} + \frac{2r}{1-r^2}$
$= \frac{1+2r}{1-r^2}$, if $|r^2| < 1$ or $|r| < 1$

93. $\text{area} = 2^2 + \left(\sqrt{2}\right)^2 + (1)^2 + \left(\frac{1}{\sqrt{2}}\right)^2 + \ldots = 4 + 2 + 1 + \frac{1}{2} + \ldots = \frac{4}{1-\frac{1}{2}} = 8 \text{ m}^2$

10.3 THE INTEGRAL TEST

1. $f(x) = \frac{1}{x^2}$ is positive, continuous, and decreasing for $x \ge 1$; $\int_1^\infty \frac{1}{x^2}\,dx = \lim_{b\to\infty} \int_1^b \frac{1}{x^2}\,dx = \lim_{b\to\infty} \left[-\frac{1}{x}\right]_1^b$
$= \lim_{b\to\infty} \left(-\frac{1}{b} + 1\right) = 1 \Rightarrow \int_1^\infty \frac{1}{x^2}\,dx$ converges $\Rightarrow \sum_{n=1}^{\infty} \frac{1}{n^2}$ converges

3. $f(x) = \frac{1}{x^2+4}$ is positive, continuous, and decreasing for $x \ge 1$; $\int_1^\infty \frac{1}{x^2+4}\,dx = \lim_{b\to\infty} \int_1^b \frac{1}{x^2+4}\,dx = \lim_{b\to\infty} \left[\frac{1}{2}\tan^{-1}\frac{x}{2}\right]_1^b$
$= \lim_{b\to\infty} \left(\frac{1}{2}\tan^{-1}\frac{b}{2} - \frac{1}{2}\tan^{-1}\frac{1}{2}\right) = \frac{\pi}{4} - \frac{1}{2}\tan^{-1}\frac{1}{2} \Rightarrow \int_1^\infty \frac{1}{x^2+4}\,dx$ converges $\Rightarrow \sum_{n=1}^{\infty} \frac{1}{n^2+4}$ converges

5. $f(x) = e^{-2x}$ is positive, continuous, and decreasing for $x \ge 1$; $\int_1^\infty e^{-2x}\,dx = \lim_{b\to\infty} \int_1^b e^{-2x}\,dx = \lim_{b\to\infty} \left[-\frac{1}{2}e^{-2x}\right]_1^b$
$= \lim_{b\to\infty} \left(-\frac{1}{2e^{2b}} + \frac{1}{2e^2}\right) = \frac{1}{2e^2} \Rightarrow \int_1^\infty e^{-2x}\,dx$ converges $\Rightarrow \sum_{n=1}^{\infty} e^{-2n}$ converges

7. $f(x) = \frac{x}{x^2+4}$ is positive and continuous for $x \ge 1$, $f'(x) = \frac{4-x^2}{(x^2+4)^2} < 0$ for $x > 2$, thus f is decreasing for $x \ge 3$;
$\int_3^\infty \frac{x}{x^2+4}\,dx = \lim_{b\to\infty} \int_3^b \frac{x}{x^2+4}\,dx = \lim_{b\to\infty} \left[\frac{1}{2}\ln(x^2+4)\right]_3^b = \lim_{b\to\infty} \left(\frac{1}{2}\ln(b^2+4) - \frac{1}{2}\ln(13)\right) = \infty \Rightarrow \int_3^\infty \frac{x}{x^2+4}\,dx$
diverges $\Rightarrow \sum_{n=3}^{\infty} \frac{n}{n^2+4}$ diverges $\Rightarrow \sum_{n=1}^{\infty} \frac{n}{n^2+4} = \frac{1}{5} + \frac{2}{8} + \sum_{n=3}^{\infty} \frac{n}{n^2+4}$ diverges

9. $f(x) = \frac{x^2}{e^{x/3}}$ is positive and continuous for $x \ge 1$, $f'(x) = \frac{-x(x-6)}{3e^{x/3}} < 0$ for $x > 6$, thus f is decreasing for $x \ge 7$;
$\int_7^\infty \frac{x^2}{e^{x/3}}\,dx = \lim_{b\to\infty} \int_7^b \frac{x^2}{e^{x/3}}\,dx = \lim_{b\to\infty} \left[-\frac{3x^2}{e^{x/3}} - \frac{18x}{e^{x/3}} - \frac{54}{e^{x/3}}\right]_7^b = \lim_{b\to\infty} \left(\frac{-3b^2 - 18b - 54}{e^{b/3}} + \frac{327}{e^{7/3}}\right) =$
$= \lim_{b\to\infty} \left(\frac{3(-6b-18)}{e^{b/3}}\right) + \frac{327}{e^{7/3}} = \lim_{b\to\infty} \left(\frac{-54}{e^{b/3}}\right) + \frac{327}{e^{7/3}} = \frac{327}{e^{7/3}} \Rightarrow \int_7^\infty \frac{x^2}{e^{x/3}}\,dx$ converges $\Rightarrow \sum_{n=7}^{\infty} \frac{n^2}{e^{n/3}}$ converges
$\Rightarrow \sum_{n=1}^{\infty} \frac{n^2}{e^{n/3}} = \frac{1}{e^{1/3}} + \frac{4}{e^{2/3}} + \frac{9}{e^1} + \frac{16}{e^{4/3}} + \frac{25}{e^{5/3}} + \frac{36}{e^2} + \sum_{n=7}^{\infty} \frac{n^2}{e^{n/3}}$ converges

11. converges; a geometric series with $r = \frac{1}{10} < 1$

13. diverges; by the nth-Term Test for Divergence, $\lim_{n\to\infty} \frac{n}{n+1} = 1 \ne 0$

15. diverges; $\sum_{n=1}^{\infty} \frac{3}{\sqrt{n}} = 3 \sum_{n=1}^{\infty} \frac{1}{\sqrt{n}}$, which is a divergent p-series $\left(p = \frac{1}{2}\right)$

17. converges; a geometric series with $r = \frac{1}{8} < 1$

19. diverges by the Integral Test: $\int_2^n \frac{\ln x}{x}\,dx = \frac{1}{2}(\ln^2 n - \ln 2) \Rightarrow \int_2^\infty \frac{\ln x}{x}\,dx \to \infty$

21. converges; a geometric series with $r = \frac{2}{3} < 1$

23. diverges; $\sum_{n=0}^{\infty} \frac{-2}{n+1} = -2 \sum_{n=0}^{\infty} \frac{1}{n+1}$, which diverges by the Integral Test

25. diverges; $\lim_{n \to \infty} a_n = \lim_{n \to \infty} \frac{2^n}{n+1} = \lim_{n \to \infty} \frac{2^n \ln 2}{1} = \infty \neq 0$

27. diverges; $\lim_{n \to \infty} \frac{\sqrt{n}}{\ln n} = \lim_{n \to \infty} \frac{\left(\frac{1}{2\sqrt{n}}\right)}{\left(\frac{1}{n}\right)} = \lim_{n \to \infty} \frac{\sqrt{n}}{2} = \infty \neq 0$

29. diverges; a geometric series with $r = \frac{1}{\ln 2} \approx 1.44 > 1$

31. converges by the Integral Test: $\int_3^{\infty} \frac{\left(\frac{1}{x}\right)}{(\ln x)\sqrt{(\ln x)^2 - 1}}\, dx$; $\left[\begin{array}{c} u = \ln x \\ du = \frac{1}{x}\, dx \end{array}\right] \rightarrow \int_{\ln 3}^{\infty} \frac{1}{u\sqrt{u^2-1}}\, du$
$= \lim_{b \to \infty} \left[\sec^{-1} |u|\right]_{\ln 3}^{b} = \lim_{b \to \infty} \left[\sec^{-1} b - \sec^{-1} (\ln 3)\right] = \lim_{b \to \infty} \left[\cos^{-1}\left(\frac{1}{b}\right) - \sec^{-1} (\ln 3)\right]$
$= \cos^{-1}(0) - \sec^{-1}(\ln 3) = \frac{\pi}{2} - \sec^{-1}(\ln 3) \approx 1.1439$

33. diverges by the nth-Term Test for divergence; $\lim_{n \to \infty} n \sin\left(\frac{1}{n}\right) = \lim_{n \to \infty} \frac{\sin\left(\frac{1}{n}\right)}{\left(\frac{1}{n}\right)} = \lim_{x \to 0} \frac{\sin x}{x} = 1 \neq 0$

35. converges by the Integral Test: $\int_1^{\infty} \frac{e^x}{1+e^{2x}}\, dx$; $\left[\begin{array}{c} u = e^x \\ du = e^x\, dx \end{array}\right] \rightarrow \int_e^{\infty} \frac{1}{1+u^2}\, du = \lim_{n \to \infty} \left[\tan^{-1} u\right]_e^b$
$= \lim_{b \to \infty} \left(\tan^{-1} b - \tan^{-1} e\right) = \frac{\pi}{2} - \tan^{-1} e \approx 0.35$

37. converges by the Integral Test: $\int_1^{\infty} \frac{8 \tan^{-1} x}{1+x^2}\, dx$; $\left[\begin{array}{c} u = \tan^{-1} x \\ du = \frac{dx}{1+x^2} \end{array}\right] \rightarrow \int_{\pi/4}^{\pi/2} 8u\, du = \left[4u^2\right]_{\pi/4}^{\pi/2} = 4\left(\frac{\pi^2}{4} - \frac{\pi^2}{16}\right) = \frac{3\pi^2}{4}$

39. converges by the Integral Test: $\int_1^{\infty} \operatorname{sech} x\, dx = 2 \lim_{b \to \infty} \int_1^b \frac{e^x}{1+(e^x)^2}\, dx = 2 \lim_{b \to \infty} \left[\tan^{-1} e^x\right]_1^b$
$= 2 \lim_{b \to \infty} \left(\tan^{-1} e^b - \tan^{-1} e\right) = \pi - 2 \tan^{-1} e \approx 0.71$

41. $\int_1^{\infty} \left(\frac{a}{x+2} - \frac{1}{x+4}\right) dx = \lim_{b \to \infty} \left[a \ln|x+2| - \ln|x+4|\right]_1^b = \lim_{b \to \infty} \ln \frac{(b+2)^a}{b+4} - \ln\left(\frac{3^a}{5}\right)$;
$\lim_{b \to \infty} \frac{(b+2)^a}{b+4} = a \lim_{b \to \infty} (b+2)^{a-1} = \begin{cases} \infty, & a > 1 \\ 1, & a = 1 \end{cases}$ $\Rightarrow$ the series converges to $\ln\left(\frac{5}{3}\right)$ if $a = 1$ and diverges to ∞ if $a > 1$. If $a < 1$, the terms of the series eventually become negative and the Integral Test does not apply. From that point on, however, the series behaves like a negative multiple of the harmonic series, and so it diverges.

43. (a)

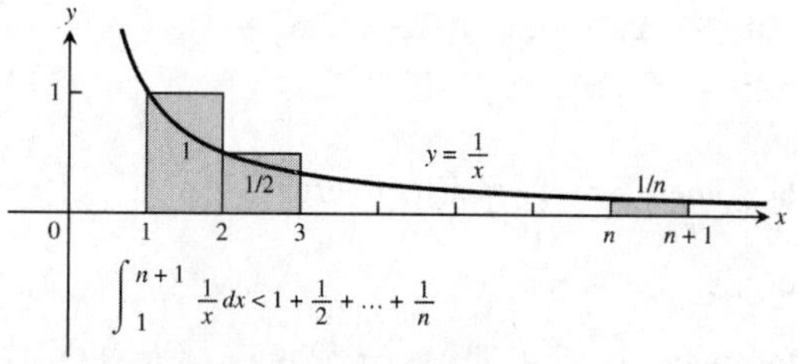
$y = \frac{1}{x}$
1
1/2
1/n
0
1
2
3
n
n + 1
$\int_1^{n+1} \frac{1}{x}\, dx < 1 + \frac{1}{2} + \ldots + \frac{1}{n}$

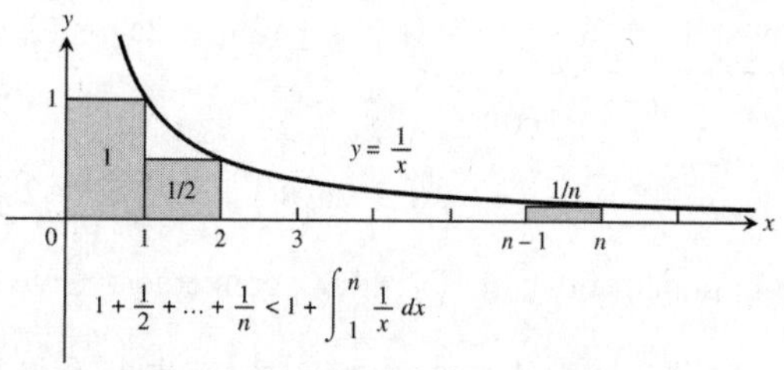
$y = \frac{1}{x}$
1
1/2
1/n
0
1
2
3
n − 1
n
$1 + \frac{1}{2} + \ldots + \frac{1}{n} < 1 + \int_1^n \frac{1}{x}\, dx$

(b) There are $(13)(365)(24)(60)(60)\,(10^9)$ seconds in 13 billion years; by part (a) $s_n \le 1 + \ln n$ where $n = (13)(365)(24)(60)(60)\,(10^9) \Rightarrow s_n \le 1 + \ln\left((13)(365)(24)(60)(60)\,(10^9)\right)$
$= 1 + \ln(13) + \ln(365) + \ln(24) + 2\ln(60) + 9\ln(10) \approx 41.55$

45. Yes. If $\sum_{n=1}^{\infty} a_n$ is a divergent series of positive numbers, then $\left(\frac{1}{2}\right)\sum_{n=1}^{\infty} a_n = \sum_{n=1}^{\infty}\left(\frac{a_n}{2}\right)$ also diverges and $\frac{a_n}{2} < a_n$. There is no "smallest" divergent series of positive numbers: for any divergent series $\sum_{n=1}^{\infty} a_n$ of positive numbers $\sum_{n=1}^{\infty}\left(\frac{a_n}{2}\right)$ has smaller terms and still diverges.

47. (a) Both integrals can represent the area under the curve $f(x) = \frac{1}{\sqrt{x+1}}$, and the sum s_{50} can be considered an approximation of either integral using rectangles with $\Delta x = 1$. The sum $s_{50} = \sum_{n=1}^{50} \frac{1}{\sqrt{n+1}}$ is an overestimate of the integral $\int_1^{51} \frac{1}{\sqrt{x+1}}dx$. The sum s_{50} represents a left-hand sum (that is, the we are choosing the left-hand endpoint of each subinterval for c_i) and because f is a decreasing function, the value of f is a maximum at the left-hand endpoint of each sub interval. The area of each rectangle overestimates the true area, thus $\int_1^{51} \frac{1}{\sqrt{x+1}}dx < \sum_{n=1}^{50} \frac{1}{\sqrt{n+1}}$. In a similar manner, s_{50} underestimates the integral $\int_0^{50} \frac{1}{\sqrt{x+1}}dx$. In this case, the sum s_{50} represents a right-hand sum and because f is a decreasing function, the value of f is aminimum at the right-hand endpoint of each subinterval. The area of each rectangle underestimates the true area, thus $\sum_{n=1}^{50} \frac{1}{\sqrt{n+1}} < \int_0^{50} \frac{1}{\sqrt{x+1}}dx$. Evaluating the integrals we find $\int_1^{51} \frac{1}{\sqrt{x+1}}dx = \left[2\sqrt{x+1}\right]_1^{51} = 2\sqrt{52} - 2\sqrt{2} \approx 11.6$ and $\int_0^{50} \frac{1}{\sqrt{x+1}}dx = \left[2\sqrt{x+1}\right]_0^{50} = 2\sqrt{51} - 2\sqrt{1} \approx 12.3$. Thus, $11.6 < \sum_{n=1}^{50} \frac{1}{\sqrt{n+1}} < 12.3$.

(b) $s_n > 1000 \Rightarrow \int_1^{n+1} \frac{1}{\sqrt{x+1}}dx = \left[2\sqrt{x+1}\right]_1^{n+1} = 2\sqrt{n+1} - 2\sqrt{2} > 1000 \Rightarrow n > \left(500 + 2\sqrt{2}\right)^2 - \approx 251414.2$ $\Rightarrow n \geq 251415$.

49. We want $S - s_n < 0.01 \Rightarrow \int_n^{\infty} \frac{1}{x^3}dx < 0.01 \Rightarrow \int_n^{\infty} \frac{1}{x^3}dx = \lim_{b\to\infty} \int_n^b \frac{1}{x^3}dx = \lim_{b\to\infty}\left[-\frac{1}{2x^2}\right]_n^b = \lim_{b\to\infty}\left(-\frac{1}{2b^2} + \frac{1}{2n^2}\right)$ $= \frac{1}{2n^2} < 0.01 \Rightarrow n > \sqrt{50} \approx 7.071 \Rightarrow n \geq 8 \Rightarrow S \approx s_8 = \sum_{n=1}^{8} \frac{1}{n^3} \approx 1.195$

51. $S - s_n < 0.00001 \Rightarrow \int_n^{\infty} \frac{1}{x^{1.1}}dx < 0.00001 \Rightarrow \int_n^{\infty} \frac{1}{x^{1.1}}dx = \lim_{b\to\infty} \int_n^b \frac{1}{x^{1.1}}dx = \lim_{b\to\infty}\left[-\frac{10}{x^{0.1}}\right]_n^b = \lim_{b\to\infty}\left(-\frac{10}{b^{0.1}} + \frac{10}{n^{0.1}}\right)$ $= \frac{10}{n^{0.1}} < 0.00001 \Rightarrow n > 1000000^{10} \Rightarrow n > 10^{60}$

53. Let $A_n = \sum_{k=1}^{n} a_k$ and $B_n = \sum_{k=1}^{n} 2^k a_{(2^k)}$, where $\{a_k\}$ is a nonincreasing sequence of positive terms converging to 0. Note that $\{A_n\}$ and $\{B_n\}$ are nondecreasing sequences of positive terms. Now,

$$B_n = 2a_2 + 4a_4 + 8a_8 + \ldots + 2^n a_{(2^n)} = 2a_2 + (2a_4 + 2a_4) + (2a_8 + 2a_8 + 2a_8 + 2a_8) + \ldots$$
$$+ \underbrace{\left(2a_{(2^n)} + 2a_{(2^n)} + \ldots + 2a_{(2^n)}\right)}_{2^{n-1}\text{ terms}} \leq 2a_1 + 2a_2 + (2a_3 + 2a_4) + (2a_5 + 2a_6 + 2a_7 + 2a_8) + \ldots$$
$$+ \left(2a_{(2^{n-1})} + 2a_{(2^{n-1}+1)} + \ldots + 2a_{(2^n)}\right) - 2A_{(2^n)} \leq 2\sum_{k=1}^{\infty} a_k.$$

Therefore if $\sum a_k$ converges, then $\{B_n\}$ is bounded above $\Rightarrow \sum 2^k a_{(2^k)}$ converges. Conversely,

$$A_n = a_1 + (a_2 + a_3) + (a_4 + a_5 + a_6 + a_7) + \ldots + a_n < a_1 + 2a_2 + 4a_4 + \ldots + 2^n a_{(2^n)} = a_1 + B_n < a_1 + \sum_{k=1}^{\infty} 2^k a_{(2^k)}.$$

Therefore, if $\sum_{k=1}^{\infty} 2^k a_{(2^k)}$ converges, then $\{A_n\}$ is bounded above and hence converges.

55. (a) $\int_2^\infty \frac{dx}{x(\ln x)^p}$; $\begin{bmatrix} u = \ln x \\ du = \frac{dx}{x} \end{bmatrix} \to \int_{\ln 2}^\infty u^{-p}\, du = \lim_{b \to \infty} \left[\frac{u^{-p+1}}{-p+1}\right]_{\ln 2}^b = \lim_{b \to \infty} \left(\frac{1}{1-p}\right)\left[b^{-p+1} - (\ln 2)^{-p+1}\right]$

$= \begin{cases} \frac{1}{p-1}(\ln 2)^{-p+1}, & p > 1 \\ \infty, & p < 1 \end{cases} \Rightarrow$ the improper integral converges if $p > 1$ and diverges if $p < 1$.

For $p = 1$: $\int_2^\infty \frac{dx}{x \ln x} = \lim_{b \to \infty} [\ln(\ln x)]_2^b = \lim_{b \to \infty} [\ln(\ln b) - \ln(\ln 2)] = \infty$, so the improper integral diverges if $p = 1$.

(b) Since the series and the integral converge or diverge together, $\sum_{n=2}^{\infty} \frac{1}{n(\ln n)^p}$ converges if and only if $p > 1$.

57. (a) From Fig. 10.11(a) in the text with $f(x) = \frac{1}{x}$ and $a_k = \frac{1}{k}$, we have $\int_1^{n+1} \frac{1}{x}\, dx \le 1 + \frac{1}{2} + \frac{1}{3} + \ldots + \frac{1}{n}$ $\le 1 + \int_1^n f(x)\, dx \Rightarrow \ln(n+1) \le 1 + \frac{1}{2} + \frac{1}{3} + \ldots + \frac{1}{n} \le 1 + \ln n \Rightarrow 0 \le \ln(n+1) - \ln n$ $\le \left(1 + \frac{1}{2} + \frac{1}{3} + \ldots + \frac{1}{n}\right) - \ln n \le 1$. Therefore the sequence $\left\{\left(1 + \frac{1}{2} + \frac{1}{3} + \ldots + \frac{1}{n}\right) - \ln n\right\}$ is bounded above by 1 and below by 0.

(b) From the graph in Fig. 10.11(b) with $f(x) = \frac{1}{x}$, $\frac{1}{n+1} < \int_n^{n+1} \frac{1}{x}\, dx = \ln(n+1) - \ln n$ $\Rightarrow 0 > \frac{1}{n+1} - [\ln(n+1) - \ln n] = \left(1 + \frac{1}{2} + \frac{1}{3} + \ldots + \frac{1}{n+1} - \ln(n+1)\right) - \left(1 + \frac{1}{2} + \frac{1}{3} + \ldots + \frac{1}{n} - \ln n\right)$.
If we define $a_n = 1 + \frac{1}{2} = \frac{1}{3} + \frac{1}{n} - \ln n$, then $0 > a_{n+1} - a_n \Rightarrow a_{n+1} < a_n \Rightarrow \{a_n\}$ is a decreasing sequence of nonnegative terms.

59. (a) $s_{10} = \sum_{n=1}^{10} \frac{1}{n^3} = 1.97531986$; $\int_{11}^\infty \frac{1}{x^3}\, dx = \lim_{b \to \infty} \int_{11}^b x^{-3}\, dx = \lim_{b \to \infty} \left[-\frac{x^{-2}}{2}\right]_{11}^b = \lim_{b \to \infty} \left(-\frac{1}{2b^2} + \frac{1}{242}\right) = \frac{1}{242}$ and $\int_{10}^\infty \frac{1}{x^3}\, dx = \lim_{b \to \infty} \int_{10}^b x^{-3}\, dx = \lim_{b \to \infty} \left[-\frac{x^{-2}}{2}\right]_{10}^b = \lim_{b \to \infty} \left(-\frac{1}{2b^2} + \frac{1}{200}\right) = \frac{1}{200}$
$\Rightarrow 1.97531986 + \frac{1}{242} < s < 1.97531986 + \frac{1}{200} \Rightarrow 1.20166 < s < 1.20253$

(b) $s = \sum_{n=1}^{\infty} \frac{1}{n^3} \approx \frac{1.20166 + 1.20253}{2} = 1.202095$; error $\le \frac{1.20253 - 1.20166}{2} = 0.000435$

10.4 COMPARISON TESTS

1. Compare with $\sum_{n=1}^{\infty} \frac{1}{n^2}$, which is a convergent p-series, since $p = 2 > 1$. Both series have nonnegative terms for $n \ge 1$. For $n \ge 1$, we have $n^2 \le n^2 + 30 \Rightarrow \frac{1}{n^2} \ge \frac{1}{n^2 + 30}$. Then by Comparison Test, $\sum_{n=1}^{\infty} \frac{1}{n^2 + 30}$ converges.

3. Compare with $\sum_{n=2}^{\infty} \frac{1}{\sqrt{n}}$, which is a divergent p-series, since $p = \frac{1}{2} \le 1$. Both series have nonnegative terms for $n \ge 2$. For $n \ge 2$, we have $\sqrt{n} - 1 \le \sqrt{n} \Rightarrow \frac{1}{\sqrt{n} - 1} \ge \frac{1}{\sqrt{n}}$. Then by Comparison Test, $\sum_{n=2}^{\infty} \frac{1}{\sqrt{n} - 1}$ diverges.

5. Compare with $\sum_{n=1}^{\infty} \frac{1}{n^{3/2}}$, which is a convergent p-series, since $p = \frac{3}{2} > 1$. Both series have nonnegative terms for $n \ge 1$. For $n \ge 1$, we have $0 \le \cos^2 n \le 1 \Rightarrow \frac{\cos^2 n}{n^{3/2}} \le \frac{1}{n^{3/2}}$. Then by Comparison Test, $\sum_{n=1}^{\infty} \frac{\cos^2 n}{n^{3/2}}$ converges.

7. Compare with $\sum_{n=1}^{\infty} \frac{\sqrt{5}}{n^{3/2}}$. The series $\sum_{n=1}^{\infty} \frac{1}{n^{3/2}}$ is a convergent p-series, since $p = \frac{3}{2} > 1$, and the series $\sum_{n=1}^{\infty} \frac{\sqrt{5}}{n^{3/2}}$ $= \sqrt{5} \sum_{n=1}^{\infty} \frac{1}{n^{3/2}}$ converges by Theorem 8 part 3. Both series have nonnegative terms for $n \ge 1$. For $n \ge 1$, we have

$n^3 \le n^4 \Rightarrow 4n^3 \le 4n^4 \Rightarrow n^4 + 4n^3 \le n^4 + 4n^4 = 5n^4 \Rightarrow n^4 + 4n^3 \le 5n^4 + 20 = 5(n^4+4) \Rightarrow \frac{n^4+4n^3}{n^4+4} \le 5.$

$\Rightarrow \frac{n^3(n+4)}{n^4+4} \le 5 \Rightarrow \frac{n+4}{n^4+4} \le \frac{5}{n^3} \Rightarrow \sqrt{\frac{n+4}{n^4+4}} \le \sqrt{\frac{5}{n^3}} = \frac{\sqrt{5}}{n^{3/2}}$ Then by Comparison Test, $\sum\limits_{n=1}^{\infty} \sqrt{\frac{n+4}{n^4+4}}$ converges.

9. Compare with $\sum\limits_{n=1}^{\infty} \frac{1}{n^2}$, which is a convergent p-series, since $p = 2 > 1$. Both series have positive terms for $n \ge 1$. $\lim\limits_{n\to\infty} \frac{a_n}{b_n}$ $= \lim\limits_{n\to\infty} \frac{\frac{n-2}{n^3-n^2+3}}{1/n^2} = \lim\limits_{n\to\infty} \frac{n^3-2n^2}{n^3-n^2+3} = \lim\limits_{n\to\infty} \frac{3n^2-4n}{3n^2-2n} = \lim\limits_{n\to\infty} \frac{6n-4}{6n-2} = \lim\limits_{n\to\infty} \frac{6}{6} = 1 > 0$. Then by Limit Comparison Test, $\sum\limits_{n=1}^{\infty} \frac{n-2}{n^3-n^2+3}$ converges.

11. Compare with $\sum\limits_{n=2}^{\infty} \frac{1}{n}$, which is a divergent p-series, since $p = 1 \le 1$. Both series have positive terms for $n \ge 2$. $\lim\limits_{n\to\infty} \frac{a_n}{b_n}$ $= \lim\limits_{n\to\infty} \frac{\frac{n(n+1)}{(n^2+1)(n-1)}}{1/n} = \lim\limits_{n\to\infty} \frac{n^3+n^2}{n^3-n^2+n-1} = \lim\limits_{n\to\infty} \frac{3n^2+2n}{3n^2-2n+1} = \lim\limits_{n\to\infty} \frac{6n+2}{6n-2} = \lim\limits_{n\to\infty} \frac{6}{6} = 1 > 0$. Then by Limit Comparison Test, $\sum\limits_{n=2}^{\infty} \frac{n(n+1)}{(n^2+1)(n-1)}$ diverges.

13. Compare with $\sum\limits_{n=1}^{\infty} \frac{1}{\sqrt{n}}$, which is a divergent p-series, since $p = \frac{1}{2} \le 1$. Both series have positive terms for $n \ge 1$. $\lim\limits_{n\to\infty} \frac{a_n}{b_n}$ $= \lim\limits_{n\to\infty} \frac{\frac{5^n}{\sqrt{n}\cdot 4^n}}{1/\sqrt{n}} = \lim\limits_{n\to\infty} \frac{5^n}{4^n} = \lim\limits_{n\to\infty} \left(\frac{5}{4}\right)^n = \infty$. Then by Limit Comparison Test, $\sum\limits_{n=1}^{\infty} \frac{5^n}{\sqrt{n}\cdot 4^n}$ diverges.

15. Compare with $\sum\limits_{n=2}^{\infty} \frac{1}{n}$, which is a divergent p-series, since $p = 1 \le 1$. Both series have positive terms for $n \ge 2$. $\lim\limits_{n\to\infty} \frac{a_n}{b_n}$ $= \lim\limits_{n\to\infty} \frac{\frac{1}{\ln n}}{1/n} = \lim\limits_{n\to\infty} \frac{n}{\ln n} = \lim\limits_{n\to\infty} \frac{1}{1/n} = \lim\limits_{n\to\infty} n = \infty$. Then by Limit Comparison Test, $\sum\limits_{n=2}^{\infty} \frac{1}{\ln n}$ diverges.

17. diverges by the Limit Comparison Test (part 1) when compared with $\sum\limits_{n=1}^{\infty} \frac{1}{\sqrt{n}}$, a divergent p-series:

$$\lim_{n\to\infty} \frac{\left(\frac{1}{2\sqrt{n}+\sqrt[3]{n}}\right)}{\left(\frac{1}{\sqrt{n}}\right)} = \lim_{n\to\infty} \frac{\sqrt{n}}{2\sqrt{n}+\sqrt[3]{n}} = \lim_{n\to\infty} \left(\frac{1}{2+n^{-1/6}}\right) = \frac{1}{2}$$

19. converges by the Direct Comparison Test; $\frac{\sin^2 n}{2^n} \le \frac{1}{2^n}$, which is the nth term of a convergent geometric series

21. diverges since $\lim\limits_{n\to\infty} \frac{2n}{3n-1} = \frac{2}{3} \ne 0$

23. converges by the Limit Comparison Test (part 1) with $\frac{1}{n^2}$, the nth term of a convergent p-series:

$$\lim_{n\to\infty} \frac{\left(\frac{10n+1}{n(n+1)(n+2)}\right)}{\left(\frac{1}{n^2}\right)} = \lim_{n\to\infty} \frac{10n^2+n}{n^2+3n+2} = \lim_{n\to\infty} \frac{20n+1}{2n+3} = \lim_{n\to\infty} \frac{20}{2} = 10$$

25. converges by the Direct Comparison Test; $\left(\frac{n}{3n+1}\right)^n < \left(\frac{n}{3n}\right)^n = \left(\frac{1}{3}\right)^n$, the nth term of a convergent geometric series

27. diverges by the Direct Comparison Test; $n > \ln n \Rightarrow \ln n > \ln \ln n \Rightarrow \frac{1}{n} < \frac{1}{\ln n} < \frac{1}{\ln(\ln n)}$ and $\sum\limits_{n=3}^{\infty} \frac{1}{n}$ diverges

29. diverges by the Limit Comparison Test (part 3) with $\frac{1}{n}$, the nth term of the divergent harmonic series:
$\lim\limits_{n \to \infty} \frac{\left[\frac{1}{\sqrt{n}\ln n}\right]}{\left(\frac{1}{n}\right)} = \lim\limits_{n \to \infty} \frac{\sqrt{n}}{\ln n} = \lim\limits_{n \to \infty} \frac{\left(\frac{1}{2\sqrt{n}}\right)}{\left(\frac{1}{n}\right)} = \lim\limits_{n \to \infty} \frac{\sqrt{n}}{2} = \infty$

31. diverges by the Limit Comparison Test (part 3) with $\frac{1}{n}$, the nth term of the divergent harmonic series:
$\lim\limits_{n \to \infty} \frac{\left(\frac{1}{1+\ln n}\right)}{\left(\frac{1}{n}\right)} = \lim\limits_{n \to \infty} \frac{n}{1+\ln n} = \lim\limits_{n \to \infty} \frac{1}{\left(\frac{1}{n}\right)} = \lim\limits_{n \to \infty} n = \infty$

33. converges by the Direct Comparison Test with $\frac{1}{n^{3/2}}$, the nth term of a convergent p-series: $n^2 - 1 > n$ for $n \ge 2 \Rightarrow n^2(n^2 - 1) > n^3 \Rightarrow n\sqrt{n^2 - 1} > n^{3/2} \Rightarrow \frac{1}{n^{3/2}} > \frac{1}{n\sqrt{n^2 - 1}}$ or use Limit Comparison Test with $\frac{1}{n^2}$.

35. converges because $\sum\limits_{n=1}^{\infty} \frac{1-n}{n2^n} = \sum\limits_{n=1}^{\infty} \frac{1}{n2^n} + \sum\limits_{n=1}^{\infty} \frac{-1}{2^n}$ which is the sum of two convergent series:
$\sum\limits_{n=1}^{\infty} \frac{1}{n2^n}$ converges by the Direct Comparison Test since $\frac{1}{n2^n} < \frac{1}{2^n}$, and $\sum\limits_{n=1}^{\infty} \frac{-1}{2^n}$ is a convergent geometric series

37. converges by the Direct Comparison Test: $\frac{1}{3^{n-1}+1} < \frac{1}{3^{n-1}}$, which is the nth term of a convergent geometric series

39. converges by Limit Comparison Test: compare with $\sum\limits_{n=1}^{\infty} \left(\frac{1}{5}\right)^n$, which is a convergent geometric series with $|r| = \frac{1}{5} < 1$,
$\lim\limits_{n \to \infty} \frac{\left(\frac{n+1}{n^2+3n} \cdot \frac{1}{5^n}\right)}{(1/5)^n} = \lim\limits_{n \to \infty} \frac{n+1}{n^2+3n} = \lim\limits_{n \to \infty} \frac{1}{2n+3} = 0.$

41. diverges by Limit Comparison Test: compare with $\sum\limits_{n=1}^{\infty} \frac{1}{n}$, which is a divergent p-series, $\lim\limits_{n \to \infty} \frac{\left(\frac{2^n - n}{n \cdot 2^n}\right)}{1/n} = \lim\limits_{n \to \infty} \frac{2^n - n}{2^n}$
$= \lim\limits_{n \to \infty} \frac{2^n \ln 2 - 1}{2^n \ln 2} = \lim\limits_{n \to \infty} \frac{2^n (\ln 2)^2}{2^n (\ln 2)^2} = 1 > 0.$

43. converges by Comparison Test with $\sum\limits_{n=2}^{\infty} \frac{1}{n(n-1)}$ which converges since $\sum\limits_{n=2}^{\infty} \frac{1}{n(n-1)} = \sum\limits_{n=2}^{\infty} \left[\frac{1}{n-1} - \frac{1}{n}\right]$, and
$s_k = \left(1 - \frac{1}{2}\right) + \left(\frac{1}{2} - \frac{1}{3}\right) + \ldots + \left(\frac{1}{k-2} - \frac{1}{k-1}\right) + \left(\frac{1}{k-1} - \frac{1}{k}\right) = 1 - \frac{1}{k} \Rightarrow \lim\limits_{k \to \infty} s_k = 1$; for $n \ge 2$, $(n-2)! \ge 1$
$\Rightarrow n(n-1)(n-2)! \ge n(n-1) \Rightarrow n! \ge n(n-1) \Rightarrow \frac{1}{n!} \le \frac{1}{n(n-1)}$

45. diverges by the Limit Comparison Test (part 1) with $\frac{1}{n}$, the nth term of the divergent harmonic series:
$\lim\limits_{n \to \infty} \frac{\left(\sin \frac{1}{n}\right)}{\left(\frac{1}{n}\right)} = \lim\limits_{x \to 0} \frac{\sin x}{x} = 1$

47. converges by the Direct Comparison Test: $\frac{\tan^{-1} n}{n^{1.1}} < \frac{\frac{\pi}{2}}{n^{1.1}}$ and $\sum\limits_{n=1}^{\infty} \frac{\frac{\pi}{2}}{n^{1.1}} = \frac{\pi}{2} \sum\limits_{n=1}^{\infty} \frac{1}{n^{1.1}}$ is the product of a convergent p-series and a nonzero constant

49. converges by the Limit Comparison Test (part 1) with $\frac{1}{n^2}$: $\lim\limits_{n \to \infty} \frac{\left(\frac{\coth n}{n^2}\right)}{\left(\frac{1}{n^2}\right)} = \lim\limits_{n \to \infty} \coth n = \lim\limits_{n \to \infty} \frac{e^n + e^{-n}}{e^n - e^{-n}}$
$= \lim\limits_{n \to \infty} \frac{1 + e^{-2n}}{1 - e^{-2n}} = 1$

51. diverges by the Limit Comparison Test (part 1) with $\frac{1}{n}$: $\lim\limits_{n \to \infty} \frac{\left(\frac{1}{n\sqrt[n]{n}}\right)}{\left(\frac{1}{n}\right)} = \lim\limits_{n \to \infty} \frac{1}{\sqrt[n]{n}} = 1.$

53. $\frac{1}{1+2+3+\ldots+n} = \frac{1}{\left(\frac{n(n+1)}{2}\right)} = \frac{2}{n(n+1)}$. The series converges by the Limit Comparison Test (part 1) with $\frac{1}{n^2}$:

$$\lim_{n\to\infty} \frac{\left(\frac{2}{n(n+1)}\right)}{\left(\frac{1}{n^2}\right)} = \lim_{n\to\infty} \frac{2n^2}{n^2+n} = \lim_{n\to\infty} \frac{4n}{2n+1} = \lim_{n\to\infty} \frac{4}{2} = 2.$$

55. (a) If $\lim_{n\to\infty} \frac{a_n}{b_n} = 0$, then there exists an integer N such that for all $n > N$, $\left|\frac{a_n}{b_n} - 0\right| < 1 \Rightarrow -1 < \frac{a_n}{b_n} < 1$ $\Rightarrow a_n < b_n$. Thus, if $\sum b_n$ converges, then $\sum a_n$ converges by the Direct Comparison Test.

(b) If $\lim_{n\to\infty} \frac{a_n}{b_n} = \infty$, then there exists an integer N such that for all $n > N$, $\frac{a_n}{b_n} > 1 \Rightarrow a_n > b_n$. Thus, if $\sum b_n$ diverges, then $\sum a_n$ diverges by the Direct Comparison Test.

57. $\lim_{n\to\infty} \frac{a_n}{b_n} = \infty \Rightarrow$ there exists an integer N such that for all $n > N$, $\frac{a_n}{b_n} > 1 \Rightarrow a_n > b_n$. If $\sum a_n$ converges, then $\sum b_n$ converges by the Direct Comparison Test

59. Since $a_n > 0$ and $\lim_{n\to\infty} a_n = \infty \neq 0$, by n^{th} term test for divergence, $\sum a_n$ diverges.

61. Let $-\infty < q < \infty$ and $p > 1$. If $q = 0$, then $\sum_{n=2}^{\infty} \frac{(\ln n)^q}{n^p} = \sum_{n=2}^{\infty} \frac{1}{n^p}$, which is a convergent p-series. If $q \neq 0$, compare with $\sum_{n=2}^{\infty} \frac{1}{n^r}$ where $1 < r < p$, then $\lim_{n\to\infty} \frac{\frac{(\ln n)^q}{n^p}}{1/n^r} = \lim_{n\to\infty} \frac{(\ln n)^q}{n^{p-r}}$, and $p - r > 0$. If $q < 0 \Rightarrow -q > 0$ and $\lim_{n\to\infty} \frac{(\ln n)^q}{n^{p-r}}$ $= \lim_{n\to\infty} \frac{1}{(\ln n)^{-q} n^{p-r}} = 0$. If $q > 0$, $\lim_{n\to\infty} \frac{(\ln n)^q}{n^{p-r}} = \lim_{n\to\infty} \frac{q(\ln n)^{q-1}\left(\frac{1}{n}\right)}{(p-r)n^{p-r-1}} = \lim_{n\to\infty} \frac{q(\ln n)^{q-1}}{(p-r)n^{p-r}}$. If $q - 1 \le 0 \Rightarrow 1 - q \ge 0$ and $\lim_{n\to\infty} \frac{q(\ln n)^{q-1}}{(p-r)n^{p-r}} = \lim_{n\to\infty} \frac{q}{(p-r)n^{p-r}(\ln n)^{1-q}} = 0$, otherwise, we apply L'Hopital's Rule again. $\lim_{n\to\infty} \frac{q(q-1)(\ln n)^{q-2}\left(\frac{1}{n}\right)}{(p-r)^2 n^{p-r-1}}$ $= \lim_{n\to\infty} \frac{q(q-1)(\ln n)^{q-2}}{(p-r)^2 n^{p-r}}$. If $q - 2 \le 0 \Rightarrow 2 - q \ge 0$ and $\lim_{n\to\infty} \frac{q(q-1)(\ln n)^{q-2}}{(p-r)^2 n^{p-r}} = \lim_{n\to\infty} \frac{q(q-1)}{(p-r)^2 n^{p-r}(\ln n)^{2-q}} = 0$; otherwise, we apply L'Hopital's Rule again. Since q is finite, there is a positive integer k such that $q - k \le 0 \Rightarrow k - q \ge 0$. Thus, after k applications of L'Hopital's Rule we obtain $\lim_{n\to\infty} \frac{q(q-1)\cdots(q-k+1)(\ln n)^{q-k}}{(p-r)^k n^{p-r}} = \lim_{n\to\infty} \frac{q(q-1)\cdots(q-k+1)}{(p-r)^k n^{p-r}(\ln n)^{k-q}} = 0$. Since the limit is 0 in every case, by Limit Comparison Test, the series $\sum_{n=1}^{\infty} \frac{(\ln n)^q}{n^p}$ converges.

63. Converges by Exercise 61 with $q = 3$ and $p = 4$.

65. Converges by Exercise 61 with $q = 1000$ and $p = 1.001$.

67. Converges by Exercise 61 with $q = -3$ and $p = 1.1$.

10.5 THE RATIO AND ROOT TESTS

1. $\frac{2^n}{n!} > 0$ for all $n \ge 1$; $\lim_{n\to\infty} \left(\frac{\frac{2^{n+1}}{(n+1)!}}{\frac{2^n}{n!}}\right) = \lim_{n\to\infty} \left(\frac{2^n \cdot 2}{(n+1)\cdot n!} \cdot \frac{n!}{2^n}\right) = \lim_{n\to\infty} \left(\frac{2}{n+1}\right) = 0 < 1 \Rightarrow \sum_{n=1}^{\infty} \frac{2^n}{n!}$ converges

3. $\frac{(n-1)!}{(n+1)^2} > 0$ for all $n \ge 1$; $\lim_{n\to\infty} \left(\frac{\frac{((n+1)-1)!}{((n+1)+1)^2}}{\frac{(n-1)!}{(n+1)^2}}\right) = \lim_{n\to\infty} \left(\frac{n\cdot(n-1)!}{(n+2)^2} \cdot \frac{(n+1)^2}{(n-1)!}\right) = \lim_{n\to\infty} \left(\frac{n^3+2n^2+n}{n^2+4n+4}\right) = \lim_{n\to\infty} \left(\frac{3n^2+4n+1}{2n+4}\right)$

$= \lim_{n\to\infty} \left(\frac{6n+4}{2}\right) = \infty > 1 \Rightarrow \sum_{n=1}^{\infty} \frac{(n-1)!}{(n+1)^2}$ diverges

5. $\frac{n^4}{4^n} > 0$ for all $n \geq 1$; $\lim_{n\to\infty}\left(\frac{\frac{(n+1)^4}{4^{n+1}}}{\frac{n^4}{4^n}}\right) = \lim_{n\to\infty}\left(\frac{(n+1)^4}{4^n\cdot 4}\cdot\frac{4^n}{n^4}\right) = \lim_{n\to\infty}\left(\frac{n^4+4n^3+6n^2+4n+1}{4n^4}\right)$

$= \lim_{n\to\infty}\left(\frac{1}{4}+\frac{1}{n}+\frac{3}{2n^2}+\frac{1}{n^3}+\frac{1}{4n^4}\right) = \frac{1}{4} < 1 \Rightarrow \sum_{n=1}^{\infty}\frac{n^4}{4^n}$ converges

7. $\frac{n^2(n+2)!}{n!3^{2n}} > 0$ for all $n \geq 1$; $\lim_{n\to\infty}\left(\frac{\frac{(n+1)^2((n+1)+2)!}{(n+1)!3^{2(n+1)}}}{\frac{n^2(n+2)!}{n!3^{2n}}}\right) = \lim_{n\to\infty}\left(\frac{(n+1)^2(n+3)(n+2)!}{(n+1)\cdot n!3^{2n}\cdot 3^2}\cdot\frac{n!3^{2n}}{n^2(n+2)!}\right) = \lim_{n\to\infty}\left(\frac{n^3+5n^2+7n+3}{9n^3+9n^2}\right)$

$= \lim_{n\to\infty}\left(\frac{3n^2+15n+7}{27n^2+18n}\right) = \lim_{n\to\infty}\left(\frac{6n+15}{54n+18}\right) = \lim_{n\to\infty}\left(\frac{6}{54}\right) = \frac{1}{9} < 1 \Rightarrow \sum_{n=1}^{\infty}\frac{n^2(n+2)!}{n!3^{2n}}$ converges

9. $\frac{7}{(2n+5)^n} \geq 0$ for all $n \geq 1$; $\lim_{n\to\infty}\sqrt[n]{\frac{7}{(2n+5)^n}} = \lim_{n\to\infty}\left(\frac{\sqrt[n]{7}}{2n+5}\right) = 0 < 1 \Rightarrow \sum_{n=1}^{\infty}\frac{7}{(2n+5)^n}$ converges

11. $\left(\frac{4n+3}{3n-5}\right)^n \geq 0$ for all $n \geq 2$; $\lim_{n\to\infty}\sqrt[n]{\left(\frac{4n+3}{3n-5}\right)^n} = \lim_{n\to\infty}\left(\frac{4n+3}{3n-5}\right) = \lim_{n\to\infty}\left(\frac{4}{3}\right) = \frac{4}{3} > 1 \Rightarrow \sum_{n=1}^{\infty}\left(\frac{4n+3}{3n-5}\right)^n$ diverges

13. $\frac{8}{\left(3+\frac{1}{n}\right)^{2n}} \geq 0$ for all $n \geq 1$; $\lim_{n\to\infty}\sqrt[n]{\frac{8}{\left(3+\frac{1}{n}\right)^{2n}}} = \lim_{n\to\infty}\left(\frac{\sqrt[n]{8}}{\left(3+\frac{1}{n}\right)^2}\right) = \frac{1}{9} < 1 \Rightarrow \sum_{n=1}^{\infty}\frac{8}{\left(3+\frac{1}{n}\right)^{2n}}$ converges

15. $\left(1-\frac{1}{n}\right)^{n^2} \geq 0$ for all $n \geq 1$; $\lim_{n\to\infty}\sqrt[n]{\left(1-\frac{1}{n}\right)^{n^2}} = \lim_{n\to\infty}\left(1-\frac{1}{n}\right)^n = e^{-1} < 1 \Rightarrow \sum_{n=1}^{\infty}\left(1-\frac{1}{n}\right)^{n^2}$ converges

17. converges by the Ratio Test: $\lim_{n\to\infty}\frac{a_{n+1}}{a_n} = \lim_{n\to\infty}\frac{\left[\frac{(n+1)^{\sqrt{2}}}{2^{n+1}}\right]}{\left[\frac{n^{\sqrt{2}}}{2^n}\right]} = \lim_{n\to\infty}\frac{(n+1)^{\sqrt{2}}}{2^{n+1}}\cdot\frac{2^n}{n^{\sqrt{2}}} = \lim_{n\to\infty}\left(1+\frac{1}{n}\right)^{\sqrt{2}}\left(\frac{1}{2}\right) = \frac{1}{2} < 1$

19. diverges by the Ratio Test: $\lim_{n\to\infty}\frac{a_{n+1}}{a_n} = \lim_{n\to\infty}\frac{\left(\frac{(n+1)!}{e^{n+1}}\right)}{\left(\frac{n!}{e^n}\right)} = \lim_{n\to\infty}\frac{(n+1)!}{e^{n+1}}\cdot\frac{e^n}{n!} = \lim_{n\to\infty}\frac{n+1}{e} = \infty$

21. converges by the Ratio Test: $\lim_{n\to\infty}\frac{a_{n+1}}{a_n} = \lim_{n\to\infty}\frac{\left(\frac{(n+1)^{10}}{10^{n+1}}\right)}{\left(\frac{n^{10}}{10^n}\right)} = \lim_{n\to\infty}\frac{(n+1)^{10}}{10^{n+1}}\cdot\frac{10^n}{n^{10}} = \lim_{n\to\infty}\left(1+\frac{1}{n}\right)^{10}\left(\frac{1}{10}\right) = \frac{1}{10} < 1$

23. converges by the Direct Comparison Test: $\frac{2+(-1)^n}{(1.25)^n} = \left(\frac{4}{5}\right)^n\left[2+(-1)^n\right] \leq \left(\frac{4}{5}\right)^n(3)$ which is the nth term of a convergent geometric series

25. diverges; $\lim_{n\to\infty} a_n = \lim_{n\to\infty}\left(1-\frac{3}{n}\right)^n = \lim_{n\to\infty}\left(1+\frac{-3}{n}\right)^n = e^{-3} \approx 0.05 \neq 0$

27. converges by the Direct Comparison Test: $\frac{\ln n}{n^3} < \frac{n}{n^3} = \frac{1}{n^2}$ for $n \geq 2$, the nth term of a convergent p-series.

29. diverges by the Direct Comparison Test: $\frac{1}{n} - \frac{1}{n^2} = \frac{n-1}{n^2} > \frac{1}{2}\left(\frac{1}{n}\right)$ for $n > 2$ or by the Limit Comparison Test (part 1) with $\frac{1}{n}$.

31. diverges by the Direct Comparison Test: $\frac{\ln n}{n} > \frac{1}{n}$ for $n \geq 3$

33. converges by the Ratio Test: $\lim_{n\to\infty}\frac{a_{n+1}}{a_n} = \lim_{n\to\infty}\frac{(n+2)(n+3)}{(n+1)!}\cdot\frac{n!}{(n+1)(n+2)} = 0 < 1$

35. converges by the Ratio Test: $\lim_{n\to\infty}\frac{a_{n+1}}{a_n} = \lim_{n\to\infty}\frac{(n+4)!}{3!(n+1)!3^{n+1}}\cdot\frac{3!n!3^n}{(n+3)!} = \lim_{n\to\infty}\frac{n+4}{3(n+1)} = \frac{1}{3} < 1$

37. converges by the Ratio Test: $\lim\limits_{n \to \infty} \frac{a_{n+1}}{a_n} = \lim\limits_{n \to \infty} \frac{(n+1)!}{(2n+3)!} \cdot \frac{(2n+1)!}{n!} = \lim\limits_{n \to \infty} \frac{n+1}{(2n+3)(2n+2)} = 0 < 1$

39. converges by the Root Test: $\lim\limits_{n \to \infty} \sqrt[n]{a_n} = \lim\limits_{n \to \infty} \sqrt[n]{\frac{n}{(\ln n)^n}} = \lim\limits_{n \to \infty} \frac{\sqrt[n]{n}}{\ln n} = \lim\limits_{n \to \infty} \frac{1}{\ln n} = 0 < 1$

41. converges by the Direct Comparison Test: $\frac{n! \ln n}{n(n+2)!} = \frac{\ln n}{n(n+1)(n+2)} < \frac{n}{n(n+1)(n+2)} = \frac{1}{(n+1)(n+2)} < \frac{1}{n^2}$ which is the nth-term of a convergent p-series

43. converges by the Ratio Test: $\lim\limits_{n \to \infty} \frac{a_{n+1}}{a_n} = \lim\limits_{n \to \infty} \frac{[(n+1)!]^2}{[2(n+1)]!} \cdot \frac{(2n)!}{[n!]^2} = \lim\limits_{n \to \infty} \frac{(n+1)^2}{(2n+2)(2n+1)} = \lim\limits_{n \to \infty} \frac{n^2+2n+1}{4n^2+6n+2} = \frac{1}{4} < 1$

45. converges by the Ratio Test: $\lim\limits_{n \to \infty} \frac{a_{n+1}}{a_n} = \lim\limits_{n \to \infty} \frac{\left(\frac{1+\sin n}{n}\right) a_n}{a_n} = 0 < 1$

47. diverges by the Ratio Test: $\lim\limits_{n \to \infty} \frac{a_{n+1}}{a_n} = \lim\limits_{n \to \infty} \frac{\left(\frac{3n-1}{2n+5}\right) a_n}{a_n} = \lim\limits_{n \to \infty} \frac{3n-1}{2n+5} = \frac{3}{2} > 1$

49. converges by the Ratio Test: $\lim\limits_{n \to \infty} \frac{a_{n+1}}{a_n} = \lim\limits_{n \to \infty} \frac{\left(\frac{2}{n}\right) a_n}{a_n} = \lim\limits_{n \to \infty} \frac{2}{n} = 0 < 1$

51. converges by the Ratio Test: $\lim\limits_{n \to \infty} \frac{a_{n+1}}{a_n} = \lim\limits_{n \to \infty} \frac{\left(\frac{1+\ln n}{n}\right) a_n}{a_n} = \lim\limits_{n \to \infty} \frac{1+\ln n}{n} = \lim\limits_{n \to \infty} \frac{1}{n} = 0 < 1$

53. diverges by the nth-Term Test: $a_1 = \frac{1}{3}$, $a_2 = \sqrt[2]{\frac{1}{3}}$, $a_3 = \sqrt[3]{\sqrt[2]{\frac{1}{3}}} = \sqrt[6]{\frac{1}{3}}$, $a_4 = \sqrt[4]{\sqrt[3]{\sqrt[2]{\frac{1}{3}}}} = \sqrt[4!]{\frac{1}{3}}, \ldots,$

$a_n = \sqrt[n!]{\frac{1}{3}} \Rightarrow \lim\limits_{n \to \infty} a_n = 1$ because $\left\{\sqrt[n!]{\frac{1}{3}}\right\}$ is a subsequence of $\left\{\sqrt[n]{\frac{1}{3}}\right\}$ whose limit is 1 by Table 8.1

55. converges by the Ratio Test: $\lim\limits_{n \to \infty} \frac{a_{n+1}}{a_n} = \lim\limits_{n \to \infty} \frac{2^{n+1}(n+1)!\,(n+1)!}{(2n+2)!} \cdot \frac{(2n)!}{2^n n!\, n!} = \lim\limits_{n \to \infty} \frac{2(n+1)(n+1)}{(2n+2)(2n+1)}$

$= \lim\limits_{n \to \infty} \frac{n+1}{2n+1} = \frac{1}{2} < 1$

57. diverges by the Root Test: $\lim\limits_{n \to \infty} \sqrt[n]{a_n} \equiv \lim\limits_{n \to \infty} \sqrt[n]{\frac{(n!)^n}{(n^n)^2}} = \lim\limits_{n \to \infty} \frac{n!}{n^2} = \infty > 1$

59. converges by the Root Test: $\lim\limits_{n \to \infty} \sqrt[n]{a_n} = \lim\limits_{n \to \infty} \sqrt[n]{\frac{n^n}{2^{n^2}}} = \lim\limits_{n \to \infty} \frac{n}{2^n} = \lim\limits_{n \to \infty} \frac{1}{2^n \ln 2} = 0 < 1$

61. converges by the Ratio Test: $\lim\limits_{n \to \infty} \frac{a_{n+1}}{a_n} = \lim\limits_{n \to \infty} \frac{1 \cdot 3 \cdot \cdots \cdot (2n-1)(2n+1)}{4^{n+1} 2^{n+1} (n+1)!} \cdot \frac{4^n 2^n n!}{1 \cdot 3 \cdot \cdots \cdot (2n-1)} = \lim\limits_{n \to \infty} \frac{2n+1}{(4 \cdot 2)(n+1)} = \frac{1}{4} < 1$

63. Ratio: $\lim\limits_{n \to \infty} \frac{a_{n+1}}{a_n} = \lim\limits_{n \to \infty} \frac{1}{(n+1)^p} \cdot \frac{n^p}{1} = \lim\limits_{n \to \infty} \left(\frac{n}{n+1}\right)^p = 1^p = 1 \Rightarrow$ no conclusion

Root: $\lim\limits_{n \to \infty} \sqrt[n]{a_n} = \lim\limits_{n \to \infty} \sqrt[n]{\frac{1}{n^p}} = \lim\limits_{n \to \infty} \frac{1}{\left(\sqrt[n]{n}\right)^p} = \frac{1}{(1)^p} = 1 \Rightarrow$ no conclusion

65. $a_n \le \frac{n}{2^n}$ for every n and the series $\sum\limits_{n=1}^{\infty} \frac{n}{2^n}$ converges by the Ratio Test since $\lim\limits_{n \to \infty} \frac{(n+1)}{2^{n+1}} \cdot \frac{2^n}{n} = \frac{1}{2} < 1$

$\Rightarrow \sum\limits_{n=1}^{\infty} a_n$ converges by the Direct Comparison Test

10.6 ALTERNATING SERIES, ABSOLUTE AND CONDITIONAL CONVERGENCE

1. converges by the Alternating Convergence Test since: $u_n = \frac{1}{\sqrt{n}} > 0$ for all $n \ge 1$; $n \ge 1 \Rightarrow n+1 \ge n \Rightarrow \sqrt{n+1} \ge \sqrt{n}$ $\Rightarrow \frac{1}{\sqrt{n+1}} \le \frac{1}{\sqrt{n}} \Rightarrow u_{n+1} \le u_n$; $\lim\limits_{n\to\infty} u_n = \lim\limits_{n\to\infty} \frac{1}{\sqrt{n}} = 0.$

3. converges $\Rightarrow$ converges by Alternating Series Test since: $u_n = \frac{1}{n3^n} > 0$ for all $n \ge 1$; $n \ge 1 \Rightarrow n+1 \ge n \Rightarrow 3^{n+1} \ge 3^n$ $\Rightarrow (n+1)3^{n+1} \ge n\,3^n \Rightarrow \frac{1}{(n+1)3^{n+1}} \le \frac{1}{n\,3^n} \Rightarrow u_{n+1} \le u_n$; $\lim\limits_{n\to\infty} u_n = \lim\limits_{n\to\infty} \frac{1}{n\,3^n} = 0.$

5. converges $\Rightarrow$ converges by Alternating Series Test since: $u_n = \frac{n}{n^2+1} > 0$ for all $n \ge 1$; $n \ge 1 \Rightarrow 2n^2 + 2n \ge n^2 + n + 1$ $\Rightarrow n^3 + 2n^2 + 2n \ge n^3 + n^2 + n + 1 \Rightarrow n(n^2 + 2n + 2) \ge n^3 + n^2 + n + 1 \Rightarrow n\left((n+1)^2 + 1\right) \ge (n^2+1)(n+1)$ $\Rightarrow \frac{n}{n^2+1} \ge \frac{n+1}{(n+1)^2+1} \Rightarrow u_{n+1} \le u_n$; $\lim\limits_{n\to\infty} u_n = \lim\limits_{n\to\infty} \frac{n}{n^2+1} = 0.$

7. diverges $\Rightarrow$ diverges by n^{th} Term Test for Divergence since: $\lim\limits_{n\to\infty} \frac{2^n}{n^2} = \infty \Rightarrow \lim\limits_{n\to\infty} (-1)^{n+1} \frac{2^n}{n^2} =$ does not exist

9. diverges by the nth-Term Test since for $n > 10 \Rightarrow \frac{n}{10} > 1 \Rightarrow \lim\limits_{n\to\infty} \left(\frac{n}{10}\right)^n \ne 0 \Rightarrow \sum\limits_{n=1}^{\infty} (-1)^{n+1} \left(\frac{n}{10}\right)^n$ diverges

11. converges by the Alternating Series Test since $f(x) = \frac{\ln x}{x} \Rightarrow f'(x) = \frac{1 - \ln x}{x^2} < 0$ when $x > e \Rightarrow f(x)$ is decreasing $\Rightarrow u_n \ge u_{n+1}$; also $u_n \ge 0$ for $n \ge 1$ and $\lim\limits_{n\to\infty} u_n = \lim\limits_{n\to\infty} \frac{\ln n}{n} = \lim\limits_{n\to\infty} \frac{\left(\frac{1}{n}\right)}{1} = 0$

13. converges by the Alternating Series Test since $f(x) = \frac{\sqrt{x}+1}{x+1} \Rightarrow f'(x) = \frac{1 - x - 2\sqrt{x}}{2\sqrt{x}(x+1)^2} < 0 \Rightarrow f(x)$ is decreasing $\Rightarrow u_n \ge u_{n+1}$; also $u_n \ge 0$ for $n \ge 1$ and $\lim\limits_{n\to\infty} u_n = \lim\limits_{n\to\infty} \frac{\sqrt{n}+1}{n+1} = 0$

15. converges absolutely since $\sum\limits_{n=1}^{\infty} |a_n| = \sum\limits_{n=1}^{\infty} \left(\frac{1}{10}\right)^n$ a convergent geometric series

17. converges conditionally since $\frac{1}{\sqrt{n}} > \frac{1}{\sqrt{n+1}} > 0$ and $\lim\limits_{n\to\infty} \frac{1}{\sqrt{n}} = 0 \Rightarrow$ convergence; but $\sum\limits_{n=1}^{\infty} |a_n| = \sum\limits_{n=1}^{\infty} \frac{1}{n^{1/2}}$ is a divergent p-series

19. converges absolutely since $\sum\limits_{n=1}^{\infty} |a_n| = \sum\limits_{n=1}^{\infty} \frac{n}{n^3+1}$ and $\frac{n}{n^3+1} < \frac{1}{n^2}$ which is the nth-term of a converging p-series

21. converges conditionally since $\frac{1}{n+3} > \frac{1}{(n+1)+3} > 0$ and $\lim\limits_{n\to\infty} \frac{1}{n+3} = 0 \Rightarrow$ convergence; but $\sum\limits_{n=1}^{\infty} |a_n|$ $= \sum\limits_{n=1}^{\infty} \frac{1}{n+3}$ diverges because $\frac{1}{n+3} \ge \frac{1}{4n}$ and $\sum\limits_{n=1}^{\infty} \frac{1}{n}$ is a divergent series

23. diverges by the nth-Term Test since $\lim\limits_{n\to\infty} \frac{3+n}{5+n} = 1 \ne 0$

25. converges conditionally since $f(x) = \frac{1}{x^2} + \frac{1}{x} \Rightarrow f'(x) = -\left(\frac{2}{x^3} + \frac{1}{x^2}\right) < 0 \Rightarrow f(x)$ is decreasing and hence $u_n > u_{n+1} > 0$ for $n \ge 1$ and $\lim\limits_{n\to\infty} \left(\frac{1}{n^2} + \frac{1}{n}\right) = 0 \Rightarrow$ convergence; but $\sum\limits_{n=1}^{\infty} |a_n| = \sum\limits_{n=1}^{\infty} \frac{1+n}{n^2}$ $= \sum\limits_{n=1}^{\infty} \frac{1}{n^2} + \sum\limits_{n=1}^{\infty} \frac{1}{n}$ is the sum of a convergent and divergent series, and hence diverges

27. converges absolutely by the Ratio Test: $\lim_{n\to\infty}\left(\frac{u_{n+1}}{u_n}\right) = \lim_{n\to\infty}\left[\frac{(n+1)^2\left(\frac{2}{3}\right)^{n+1}}{n^2\left(\frac{2}{3}\right)^n}\right] = \frac{2}{3} < 1$

29. converges absolutely by the Integral Test since $\int_1^{\infty}(\tan^{-1}x)\left(\frac{1}{1+x^2}\right)dx = \lim_{b\to\infty}\left[\frac{(\tan^{-1}x)^2}{2}\right]_1^b$ $= \lim_{b\to\infty}\left[(\tan^{-1}b)^2 - (\tan^{-1}1)^2\right] = \frac{1}{2}\left[\left(\frac{\pi}{2}\right)^2 - \left(\frac{\pi}{4}\right)^2\right] = \frac{3\pi^2}{32}$

31. diverges by the nth-Term Test since $\lim_{n\to\infty}\frac{n}{n+1} = 1 \neq 0$

33. converges absolutely by the Ratio Test: $\lim_{n\to\infty}\left(\frac{u_{n+1}}{u_n}\right) = \lim_{n\to\infty}\frac{(100)^{n+1}}{(n+1)!}\cdot\frac{n!}{(100)^n} = \lim_{n\to\infty}\frac{100}{n+1} = 0 < 1$

35. converges absolutely since $\sum_{n=1}^{\infty}|a_n| = \sum_{n=1}^{\infty}\left|\frac{(-1)^n}{n\sqrt{n}}\right| = \sum_{n=1}^{\infty}\frac{1}{n^{3/2}}$ is a convergent p-series

37. converges absolutely by the Root Test: $\lim_{n\to\infty}\sqrt[n]{|a_n|} = \lim_{n\to\infty}\left(\frac{(n+1)^n}{(2n)^n}\right)^{1/n} = \lim_{n\to\infty}\frac{n+1}{2n} = \frac{1}{2} < 1$

39. diverges by the nth-Term Test since $\lim_{n\to\infty}|a_n| = \lim_{n\to\infty}\frac{(2n)!}{2^n n!\,n} = \lim_{n\to\infty}\frac{(n+1)(n+2)\cdots(2n)}{2^n n}$ $= \lim_{n\to\infty}\frac{(n+1)(n+2)\cdots(n+(n-1))}{2^{n-1}} > \lim_{n\to\infty}\left(\frac{n+1}{2}\right)^{n-1} = \infty \neq 0$

41. converges conditionally since $\frac{\sqrt{n+1}-\sqrt{n}}{1}\cdot\frac{\sqrt{n+1}+\sqrt{n}}{\sqrt{n+1}+\sqrt{n}} = \frac{1}{\sqrt{n+1}+\sqrt{n}}$ and $\left\{\frac{1}{\sqrt{n+1}+\sqrt{n}}\right\}$ is a decreasing sequence of positive terms which converges to 0 $\Rightarrow \sum_{n=1}^{\infty}\frac{(-1)^n}{\sqrt{n+1}+\sqrt{n}}$ converges; but $\sum_{n=1}^{\infty}|a_n| = \sum_{n=1}^{\infty}\frac{1}{\sqrt{n+1}+\sqrt{n}}$ diverges by the Limit Comparison Test (part 1) with $\frac{1}{\sqrt{n}}$; a divergent p-series:

$$\lim_{n\to\infty}\left(\frac{\frac{1}{\sqrt{n+1}+\sqrt{n}}}{\frac{1}{\sqrt{n}}}\right) = \lim_{n\to\infty}\frac{\sqrt{n}}{\sqrt{n+1}+\sqrt{n}} = \lim_{n\to\infty}\frac{1}{\sqrt{1+\frac{1}{n}}+1} = \frac{1}{2}$$

43. diverges by the nth-Term Test since $\lim_{n\to\infty}\left(\sqrt{n+\sqrt{n}}-\sqrt{n}\right) = \lim_{n\to\infty}\left[\left(\sqrt{n+\sqrt{n}}-\sqrt{n}\right)\left(\frac{\sqrt{n+\sqrt{n}}+\sqrt{n}}{\sqrt{n+\sqrt{n}}+\sqrt{n}}\right)\right]$ $= \lim_{n\to\infty}\frac{\sqrt{n}}{\sqrt{n+\sqrt{n}}+\sqrt{n}} = \lim_{n\to\infty}\frac{1}{\sqrt{1+\frac{1}{\sqrt{n}}}+1} = \frac{1}{2} \neq 0$

45. converges absolutely by the Direct Comparison Test since $\operatorname{sech}(n) = \frac{2}{e^n+e^{-n}} = \frac{2e^n}{e^{2n}+1} < \frac{2e^n}{e^{2n}} = \frac{2}{e^n}$ which is the nth term of a convergent geometric series

47. $\frac{1}{4}-\frac{1}{6}+\frac{1}{8}-\frac{1}{10}+\frac{1}{12}-\frac{1}{14}+\ldots = \sum_{n=1}^{\infty}\frac{(-1)^{n+1}}{2(n+1)}$; converges by Alternating Series Test since: $u_n = \frac{1}{2(n+1)} > 0$ for all $n \geq 1$; $n+2 \geq n+1 \Rightarrow 2(n+2) \geq 2(n+1) \Rightarrow \frac{1}{2((n+1)+1)} \leq \frac{1}{2(n+1)} \Rightarrow u_{n+1} \leq u_n$; $\lim_{n\to\infty}u_n = \lim_{n\to\infty}\frac{1}{2(n+1)} = 0$.

49. $|\text{error}| < \left|(-1)^6\left(\frac{1}{5}\right)\right| = 0.2$

51. $|\text{error}| < \left|(-1)^6\frac{(0.01)^5}{5}\right| = 2\times10^{-11}$

53. $|\text{error}| < 0.001 \Rightarrow u_{n+1} < 0.001 \Rightarrow \frac{1}{(n+1)^2+3} < 0.001 \Rightarrow (n+1)^2+3 > 1000 \Rightarrow n > -1+\sqrt{997} \approx 30.5753 \Rightarrow n \geq 31$

55. $|\text{error}| < 0.001 \Rightarrow u_{n+1} < 0.001 \Rightarrow \frac{1}{\left((n+1)+3\sqrt{n+1}\right)^3} < 0.001 \Rightarrow \left((n+1)+3\sqrt{n+1}\right)^3 > 1000$

$\Rightarrow \left(\sqrt{n+1}\right)^2 + 3\sqrt{n+1} - 10 > 0 \Rightarrow \sqrt{n+1} = -\frac{3+\sqrt{9+40}}{2} = 2 \Rightarrow n = 3 \Rightarrow n \geq 4$

57. $\frac{1}{(2n)!} < \frac{5}{10^6} \Rightarrow (2n)! > \frac{10^6}{5} = 200{,}000 \Rightarrow n \geq 5 \Rightarrow 1 - \frac{1}{2!} + \frac{1}{4!} - \frac{1}{6!} + \frac{1}{8!} \approx 0.54030$

59. (a) $a_n \geq a_{n+1}$ fails since $\frac{1}{3} < \frac{1}{2}$

(b) Since $\sum_{n=1}^{\infty} |a_n| = \sum_{n=1}^{\infty} \left[\left(\frac{1}{3}\right)^n + \left(\frac{1}{2}\right)^n\right] = \sum_{n=1}^{\infty} \left(\frac{1}{3}\right)^n + \sum_{n=1}^{\infty} \left(\frac{1}{2}\right)^n$ is the sum of two absolutely convergent series, we can rearrange the terms of the original series to find its sum:

$\left(\frac{1}{3} + \frac{1}{9} + \frac{1}{27} + \dots\right) - \left(\frac{1}{2} + \frac{1}{4} + \frac{1}{8} + \dots\right) = \frac{\left(\frac{1}{3}\right)}{1-\left(\frac{1}{3}\right)} - \frac{\left(\frac{1}{2}\right)}{1-\left(\frac{1}{2}\right)} = \frac{1}{2} - 1 = -\frac{1}{2}$

61. The unused terms are $\sum_{j=n+1}^{\infty} (-1)^{j+1} a_j = (-1)^{n+1}(a_{n+1} - a_{n+2}) + (-1)^{n+3}(a_{n+3} - a_{n+4}) + \dots$

$= (-1)^{n+1}\left[(a_{n+1} - a_{n+2}) + (a_{n+3} - a_{n+4}) + \dots\right]$. Each grouped term is positive, so the remainder has the same sign as $(-1)^{n+1}$, which is the sign of the first unused term.

63. Theorem 16 states that $\sum_{n=1}^{\infty} |a_n|$ converges $\Rightarrow \sum_{n=1}^{\infty} a_n$ converges. But this is equivalent to $\sum_{n=1}^{\infty} a_n$ diverges $\Rightarrow \sum_{n=1}^{\infty} |a_n|$ diverges

65. (a) $\sum_{n=1}^{\infty} |a_n + b_n|$ converges by the Direct Comparison Test since $|a_n + b_n| \leq |a_n| + |b_n|$ and hence $\sum_{n=1}^{\infty} (a_n + b_n)$ converges absolutely

(b) $\sum_{n=1}^{\infty} |b_n|$ converges $\Rightarrow \sum_{n=1}^{\infty} -b_n$ converges absolutely; since $\sum_{n=1}^{\infty} a_n$ converges absolutely and $\sum_{n=1}^{\infty} -b_n$ converges absolutely, we have $\sum_{n=1}^{\infty} [a_n + (-b_n)] = \sum_{n=1}^{\infty} (a_n - b_n)$ converges absolutely by part (a)

(c) $\sum_{n=1}^{\infty} |a_n|$ converges $\Rightarrow |k| \sum_{n=1}^{\infty} |a_n| = \sum_{n=1}^{\infty} |ka_n|$ converges $\Rightarrow \sum_{n=1}^{\infty} ka_n$ converges absolutely

67. $s_1 = -\frac{1}{2}$, $s_2 = -\frac{1}{2} + 1 = \frac{1}{2}$,

$s_3 = -\frac{1}{2} + 1 - \frac{1}{4} - \frac{1}{6} - \frac{1}{8} - \frac{1}{10} - \frac{1}{12} - \frac{1}{14} - \frac{1}{16} - \frac{1}{18} - \frac{1}{20} - \frac{1}{22} \approx -0.5099$,

$s_4 = s_3 + \frac{1}{3} \approx -0.1766$,

$s_5 = s_4 - \frac{1}{24} - \frac{1}{26} - \frac{1}{28} - \frac{1}{30} - \frac{1}{32} - \frac{1}{34} - \frac{1}{36} - \frac{1}{38} - \frac{1}{40} - \frac{1}{42} - \frac{1}{44} \approx -0.512$,

$s_6 = s_5 + \frac{1}{5} \approx -0.312$,

$s_7 = s_6 - \frac{1}{46} - \frac{1}{48} - \frac{1}{50} - \frac{1}{52} - \frac{1}{54} - \frac{1}{56} - \frac{1}{58} - \frac{1}{60} - \frac{1}{62} - \frac{1}{64} - \frac{1}{66} \approx -0.51106$

0.4
0.2
2 4 6 8
-0.2
-0.4
y = 1/2

10.7 POWER SERIES

1. $\lim_{n\to\infty}\left|\frac{u_{n+1}}{u_n}\right| < 1 \Rightarrow \lim_{n\to\infty}\left|\frac{x^{n+1}}{x^n}\right| < 1 \Rightarrow |x| < 1 \Rightarrow -1 < x < 1$; when $x = -1$ we have $\sum_{n=1}^{\infty}(-1)^n$, a divergent series; when $x = 1$ we have $\sum_{n=1}^{\infty} 1$, a divergent series
   (a) the radius is 1; the interval of convergence is $-1 < x < 1$
   (b) the interval of absolute convergence is $-1 < x < 1$
   (c) there are no values for which the series converges conditionally

3. $\lim_{n\to\infty}\left|\frac{u_{n+1}}{u_n}\right| < 1 \Rightarrow \lim_{n\to\infty}\left|\frac{(4x+1)^{n+1}}{(4x+1)^n}\right| < 1 \Rightarrow |4x+1| < 1 \Rightarrow -1 < 4x+1 < 1 \Rightarrow -\frac{1}{2} < x < 0$; when $x = -\frac{1}{2}$ we have $\sum_{n=1}^{\infty}(-1)^n(-1)^n = \sum_{n=1}^{\infty}(-1)^{2n} = \sum_{n=1}^{\infty} 1^n$, a divergent series; when $x = 0$ we have $\sum_{n=1}^{\infty}(-1)^n(1)^n = \sum_{n=1}^{\infty}(-1)^n$, a divergent series
   (a) the radius is $\frac{1}{4}$; the interval of convergence is $-\frac{1}{2} < x < 0$
   (b) the interval of absolute convergence is $-\frac{1}{2} < x < 0$
   (c) there are no values for which the series converges conditionally

5. $\lim_{n\to\infty}\left|\frac{u_{n+1}}{u_n}\right| < 1 \Rightarrow \lim_{n\to\infty}\left|\frac{(x-2)^{n+1}}{10^{n+1}} \cdot \frac{10^n}{(x-2)^n}\right| < 1 \Rightarrow \frac{|x-2|}{10} < 1 \Rightarrow |x-2| < 10 \Rightarrow -10 < x-2 < 10$ $\Rightarrow -8 < x < 12$; when $x = -8$ we have $\sum_{n=1}^{\infty}(-1)^n$, a divergent series; when $x = 12$ we have $\sum_{n=1}^{\infty} 1$, a divergent series
   (a) the radius is 10; the interval of convergence is $-8 < x < 12$
   (b) the interval of absolute convergence is $-8 < x < 12$
   (c) there are no values for which the series converges conditionally

7. $\lim_{n\to\infty}\left|\frac{u_{n+1}}{u_n}\right| < 1 \Rightarrow \lim_{n\to\infty}\left|\frac{(n+1)x^{n+1}}{(n+3)} \cdot \frac{(n+2)}{nx^n}\right| < 1 \Rightarrow |x| \lim_{n\to\infty}\frac{(n+1)(n+2)}{(n+3)(n)} < 1 \Rightarrow |x| < 1$ $\Rightarrow -1 < x < 1$; when $x = -1$ we have $\sum_{n=1}^{\infty}(-1)^n\frac{n}{n+2}$, a divergent series by the nth-term Test; when $x = 1$ we have $\sum_{n=1}^{\infty}\frac{n}{n+2}$, a divergent series
   (a) the radius is 1; the interval of convergence is $-1 < x < 1$
   (b) the interval of absolute convergence is $-1 < x < 1$
   (c) there are no values for which the series converges conditionally

9. $\lim_{n\to\infty}\left|\frac{u_{n+1}}{u_n}\right| < 1 \Rightarrow \lim_{n\to\infty}\left|\frac{x^{n+1}}{(n+1)\sqrt{n+1}\,3^{n+1}} \cdot \frac{n\sqrt{n}\,3^n}{x^n}\right| < 1 \Rightarrow \frac{|x|}{3}\left(\lim_{n\to\infty}\frac{n}{n+1}\right)\left(\sqrt{\lim_{n\to\infty}\frac{n}{n+1}}\right) < 1$ $\Rightarrow \frac{|x|}{3}(1)(1) < 1 \Rightarrow |x| < 3 \Rightarrow -3 < x < 3$; when $x = -3$ we have $\sum_{n=1}^{\infty}\frac{(-1)^n}{n^{3/2}}$, an absolutely convergent series; when $x = 3$ we have $\sum_{n=1}^{\infty}\frac{1}{n^{3/2}}$, a convergent p-series
   (a) the radius is 3; the interval of convergence is $-3 \le x \le 3$
   (b) the interval of absolute convergence is $-3 \le x \le 3$
   (c) there are no values for which the series converges conditionally

11. $\lim_{n\to\infty}\left|\frac{u_{n+1}}{u_n}\right| < 1 \Rightarrow \lim_{n\to\infty}\left|\frac{x^{n+1}}{(n+1)!} \cdot \frac{n!}{x^n}\right| < 1 \Rightarrow |x| \lim_{n\to\infty}\left(\frac{1}{n+1}\right) < 1$ for all x
   (a) the radius is ∞; the series converges for all x
   (b) the series converges absolutely for all x
   (c) there are no values for which the series converges conditionally

13. $\lim_{n\to\infty}\left|\frac{u_{n+1}}{u_n}\right|<1 \Rightarrow \lim_{n\to\infty}\left|\frac{4^{n+1}x^{2n+2}}{n+1}\cdot\frac{n}{4^n x^{2n}}\right|<1 \Rightarrow x^2\lim_{n\to\infty}\left(\frac{4n}{n+1}\right)=4x^2<1 \Rightarrow x^2<\frac{1}{4}$

$\Rightarrow -\frac{1}{2}<x<\frac{1}{2}$; when $x=-\frac{1}{2}$ we have $\sum_{n=1}^{\infty}\frac{4^n}{n}\left(-\frac{1}{2}\right)^{2n}=\sum_{n=1}^{\infty}\frac{1}{n}$, a divergent p-series; when $x=\frac{1}{2}$ we have

$\sum_{n=1}^{\infty}\frac{4^n}{n}\left(\frac{1}{2}\right)^{2n}=\sum_{n=1}^{\infty}\frac{1}{n}$, a divergent p-series

(a) the radius is $\frac{1}{2}$; the interval of convergence is $-\frac{1}{2}<x<\frac{1}{2}$

(b) the interval of absolute convergence is $-\frac{1}{2}<x<\frac{1}{2}$

(c) there are no values for which the series converges conditionally

15. $\lim_{n\to\infty}\left|\frac{u_{n+1}}{u_n}\right|<1 \Rightarrow \lim_{n\to\infty}\left|\frac{x^{n+1}}{\sqrt{(n+1)^2+3}}\cdot\frac{\sqrt{n^2+3}}{x^n}\right|<1 \Rightarrow |x|\sqrt{\lim_{n\to\infty}\frac{n^2+3}{n^2+2n+4}}<1 \Rightarrow |x|<1$

$\Rightarrow -1<x<1$; when $x=-1$ we have $\sum_{n=1}^{\infty}\frac{(-1)^n}{\sqrt{n^2+3}}$, a conditionally convergent series; when $x=1$ we have

$\sum_{n=1}^{\infty}\frac{1}{\sqrt{n^2+3}}$, a divergent series

(a) the radius is 1; the interval of convergence is $-1\le x<1$

(b) the interval of absolute convergence is $-1<x<1$

(c) the series converges conditionally at $x=-1$

17. $\lim_{n\to\infty}\left|\frac{u_{n+1}}{u_n}\right|<1 \Rightarrow \lim_{n\to\infty}\left|\frac{(n+1)(x+3)^{n+1}}{5^{n+1}}\cdot\frac{5^n}{n(x+3)^n}\right|<1 \Rightarrow \frac{|x+3|}{5}\lim_{n\to\infty}\left(\frac{n+1}{n}\right)<1 \Rightarrow \frac{|x+3|}{5}<1$

$\Rightarrow |x+3|<5 \Rightarrow -5<x+3<5 \Rightarrow -8<x<2$; when $x=-8$ we have $\sum_{n=1}^{\infty}\frac{n(-5)^n}{5^n}=\sum_{n=1}^{\infty}(-1)^n n$, a divergent

series; when $x=2$ we have $\sum_{n=1}^{\infty}\frac{n5^n}{5^n}=\sum_{n=1}^{\infty}n$, a divergent series

(a) the radius is 5; the interval of convergence is $-8<x<2$

(b) the interval of absolute convergence is $-8<x<2$

(c) there are no values for which the series converges conditionally

19. $\lim_{n\to\infty}\left|\frac{u_{n+1}}{u_n}\right|<1 \Rightarrow \lim_{n\to\infty}\left|\frac{\sqrt{n+1}\,x^{n+1}}{3^{n+1}}\cdot\frac{3^n}{\sqrt{n}\,x^n}\right|<1 \Rightarrow \frac{|x|}{3}\sqrt{\lim_{n\to\infty}\left(\frac{n+1}{n}\right)}<1 \Rightarrow \frac{|x|}{3}<1 \Rightarrow |x|<3$

$\Rightarrow -3<x<3$; when $x=-3$ we have $\sum_{n=1}^{\infty}(-1)^n\sqrt{n}$, a divergent series; when $x=3$ we have $\sum_{n=1}^{\infty}\sqrt{n}$, a divergent series

(a) the radius is 3; the interval of convergence is $-3<x<3$

(b) the interval of absolute convergence is $-3<x<3$

(c) there are no values for which the series converges conditionally

21. First, rewrite the series as $\sum_{n=1}^{\infty}(2+(-1)^n)(x+1)^{n-1}=\sum_{n=1}^{\infty}2(x+1)^{n-1}+\sum_{n=1}^{\infty}(-1)^n(x+1)^{n-1}$. For the series

$\sum_{n=1}^{\infty}2(x+1)^{n-1}$: $\lim_{n\to\infty}\left|\frac{u_{n+1}}{u_n}\right|<1 \Rightarrow \lim_{n\to\infty}\left|\frac{2(x+1)^n}{2(x+1)^{n-1}}\right|<1 \Rightarrow |x+1|\lim_{n\to\infty}1=|x+1|<1 \Rightarrow -2<x<0$; For the

series $\sum_{n=1}^{\infty}(-1)^n(x+1)^{n-1}$: $\lim_{n\to\infty}\left|\frac{u_{n+1}}{u_n}\right|<1 \Rightarrow \lim_{n\to\infty}\left|\frac{(-1)^{n+1}(x+1)^n}{(-1)^n(x+1)^{n-1}}\right|<1 \Rightarrow |x+1|\lim_{n\to\infty}1=|x+1|<1$

$\Rightarrow -2<x<0$; when $x=-2$ we have $\sum_{n=1}^{\infty}(2+(-1)^n)(-1)^{n-1}$, a divergent series; when $x=0$ we have

$\sum_{n=1}^{\infty}(2+(-1)^n)$, a divergent series

(a) the radius is 1; the interval of convergence is $-2<x<0$

(b) the interval of absolute convergence is $-2<x<0$

(c) there are no values for which the series converges conditionally

23. $\lim_{n\to\infty}\left|\frac{u_{n+1}}{u_n}\right|<1 \Rightarrow \lim_{n\to\infty}\left|\frac{\left(1+\frac{1}{n+1}\right)^{n+1}x^{n+1}}{\left(1+\frac{1}{n}\right)^{n}x^{n}}\right|<1 \Rightarrow |x|\left(\frac{\lim_{t\to\infty}\left(1+\frac{1}{t}\right)^{t}}{\lim_{n\to\infty}\left(1+\frac{1}{n}\right)^{n}}\right)<1 \Rightarrow |x|\left(\frac{e}{e}\right)<1 \Rightarrow |x|<1$

$\Rightarrow -1<x<1$; when $x=-1$ we have $\sum_{n=1}^{\infty}(-1)^n\left(1+\frac{1}{n}\right)^n$, a divergent series by the nth-Term Test since

$\lim_{n\to\infty}\left(1+\frac{1}{n}\right)^n=e\neq 0$; when $x=1$ we have $\sum_{n=1}^{\infty}\left(1+\frac{1}{n}\right)^n$, a divergent series

(a) the radius is 1; the interval of convergence is $-1<x<1$
(b) the interval of absolute convergence is $-1<x<1$
(c) there are no values for which the series converges conditionally

25. $\lim_{n\to\infty}\left|\frac{u_{n+1}}{u_n}\right|<1 \Rightarrow \lim_{n\to\infty}\left|\frac{(n+1)^{n+1}x^{n+1}}{n^n x^n}\right|<1 \Rightarrow |x|\left(\lim_{n\to\infty}\left(1+\frac{1}{n}\right)^n\right)\left(\lim_{n\to\infty}(n+1)\right)<1$

$\Rightarrow e|x|\lim_{n\to\infty}(n+1)<1 \Rightarrow$ only $x=0$ satisfies this inequality

(a) the radius is 0; the series converges only for $x=0$
(b) the series converges absolutely only for $x=0$
(c) there are no values for which the series converges conditionally

27. $\lim_{n\to\infty}\left|\frac{u_{n+1}}{u_n}\right|<1 \Rightarrow \lim_{n\to\infty}\left|\frac{(x+2)^{n+1}}{(n+1)2^{n+1}}\cdot\frac{n2^n}{(x+2)^n}\right|<1 \Rightarrow \frac{|x+2|}{2}\lim_{n\to\infty}\left(\frac{n}{n+1}\right)<1 \Rightarrow \frac{|x+2|}{2}<1 \Rightarrow |x+2|<2$

$\Rightarrow -2<x+2<2 \Rightarrow -4<x<0$; when $x=-4$ we have $\sum_{n=1}^{\infty}\frac{-1}{n}$, a divergent series; when $x=0$ we have $\sum_{n=1}^{\infty}\frac{(-1)^{n+1}}{n}$,

the alternating harmonic series which converges conditionally

(a) the radius is 2; the interval of convergence is $-4<x\le 0$
(b) the interval of absolute convergence is $-4<x<0$
(c) the series converges conditionally at $x=0$

29. $\lim_{n\to\infty}\left|\frac{u_{n+1}}{u_n}\right|<1 \Rightarrow \lim_{n\to\infty}\left|\frac{x^{n+1}}{(n+1)(\ln(n+1))^2}\cdot\frac{n(\ln n)^2}{x^n}\right|<1 \Rightarrow |x|\left(\lim_{n\to\infty}\frac{n}{n+1}\right)\left(\lim_{n\to\infty}\frac{\ln n}{\ln(n+1)}\right)^2<1$

$\Rightarrow |x|(1)\left(\lim_{n\to\infty}\frac{\left(\frac{1}{n}\right)}{\left(\frac{1}{n+1}\right)}\right)^2<1 \Rightarrow |x|\left(\lim_{n\to\infty}\frac{n+1}{n}\right)^2<1 \Rightarrow |x|<1 \Rightarrow -1<x<1$; when $x=-1$ we have

$\sum_{n=1}^{\infty}\frac{(-1)^n}{n(\ln n)^2}$ which converges absolutely; when $x=1$ we have $\sum_{n=1}^{\infty}\frac{1}{n(\ln n)^2}$ which converges

(a) the radius is 1; the interval of convergence is $-1\le x\le 1$
(b) the interval of absolute convergence is $-1\le x\le 1$
(c) there are no values for which the series converges conditionally

31. $\lim_{n\to\infty}\left|\frac{u_{n+1}}{u_n}\right|<1 \Rightarrow \lim_{n\to\infty}\left|\frac{(4x-5)^{2n+3}}{(n+1)^{3/2}}\cdot\frac{n^{3/2}}{(4x-5)^{2n+1}}\right|<1 \Rightarrow (4x-5)^2\left(\lim_{n\to\infty}\frac{n}{n+1}\right)^{3/2}<1 \Rightarrow (4x-5)^2<1$

$\Rightarrow |4x-5|<1 \Rightarrow -1<4x-5<1 \Rightarrow 1<x<\frac{3}{2}$; when $x=1$ we have $\sum_{n=1}^{\infty}\frac{(-1)^{2n+1}}{n^{3/2}}=\sum_{n=1}^{\infty}\frac{-1}{n^{3/2}}$ which is

absolutely convergent; when $x=\frac{3}{2}$ we have $\sum_{n=1}^{\infty}\frac{(1)^{2n+1}}{n^{3/2}}$, a convergent p-series

(a) the radius is $\frac{1}{4}$; the interval of convergence is $1\le x\le\frac{3}{2}$
(b) the interval of absolute convergence is $1\le x\le\frac{3}{2}$
(c) there are no values for which the series converges conditionally

33. $\lim_{n\to\infty}\left|\frac{u_{n+1}}{u_n}\right|<1 \Rightarrow \lim_{n\to\infty}\left|\frac{x^{n+1}}{2\cdot 4\cdot 6\cdots(2n)(2(n+1))}\cdot\frac{2\cdot 4\cdot 6\cdots(2n)}{x^n}\right|<1 \Rightarrow |x|\lim_{n\to\infty}\left(\frac{1}{2n+2}\right)<1$ for all x

(a) the radius is ∞; the series converges for all x
(b) the series converges absolutely for all x
(c) there are no values for which the series converges conditionally

35. For the series $\sum_{n=1}^{\infty} \frac{1+2+\cdots+n}{1^2+2^2+\cdots+n^2} x^n$, recall $1+2+\cdots+n = \frac{n(n+1)}{2}$ and $1^2+2^2+\cdots+n^2 = \frac{n(n+1)(2n+1)}{6}$ so that we can rewrite the series as $\sum_{n=1}^{\infty} \left(\frac{\frac{n(n+1)}{2}}{\frac{n(n+1)(2n+1)}{6}} \right) x^n = \sum_{n=1}^{\infty} \left(\frac{3}{2n+1}\right) x^n$; then $\lim_{n\to\infty} \left|\frac{u_{n+1}}{u_n}\right| < 1 \Rightarrow \lim_{n\to\infty} \left| \frac{3x^{n+1}}{(2(n+1)+1)} \cdot \frac{(2n+1)}{3x^n} \right| < 1$ $\Rightarrow |x| \lim_{n\to\infty} \left|\frac{(2n+1)}{(2n+3)}\right| < 1 \Rightarrow |x| < 1 \Rightarrow -1 < x < 1$; when $x = -1$ we have $\sum_{n=1}^{\infty} \left(\frac{3}{2n+1}\right)(-1)^n$, a conditionally convergent series; when $x = 1$ we have $\sum_{n=1}^{\infty} \left(\frac{3}{2n+1}\right)$, a divergent series.

(a) the radius is 1; the interval of convergence is $-1 \le x < 1$

(b) the interval of absolute convergence is $-1 < x < 1$

(c) the series converges conditionally at $x = -1$

37. $\lim_{n\to\infty} \left|\frac{u_{n+1}}{u_n}\right| < 1 \Rightarrow \lim_{n\to\infty} \left| \frac{(n+1)!x^{n+1}}{3\cdot6\cdot9\cdots(3n)(3(n+1))} \cdot \frac{3\cdot6\cdot9\cdots(3n)}{n!\,x^n} \right| < 1 \Rightarrow |x| \lim_{n\to\infty} \left|\frac{(n+1)}{3(n+1)}\right| < 1 \Rightarrow \frac{|x|}{3} < 1 \Rightarrow |x| < 3 \Rightarrow R = 3$

39. $\lim_{n\to\infty} \left|\frac{u_{n+1}}{u_n}\right| < 1 \Rightarrow \lim_{n\to\infty} \left| \frac{((n+1)!)^2 x^{n+1}}{2^{n+1}(2(n+1))!} \cdot \frac{2^n(2n)!}{(n!)^2 x^n} \right| < 1 \Rightarrow |x| \lim_{n\to\infty} \left|\frac{(n+1)^2}{2(2n+2)(2n+1)}\right| < 1 \Rightarrow \frac{|x|}{8} < 1 \Rightarrow |x| < 8 \Rightarrow R = 8$

41. $\lim_{n\to\infty} \left|\frac{u_{n+1}}{u_n}\right| < 1 \Rightarrow \lim_{n\to\infty} \left|\frac{3^{n+1}x^{n+1}}{3^n x^n}\right| < 1 \Rightarrow |x| \lim_{n\to\infty} 3 < 1 \Rightarrow |x| < \frac{1}{3} \Rightarrow -\frac{1}{3} < x < \frac{1}{3}$; at $x = -\frac{1}{3}$ we have $\sum_{n=0}^{\infty} 3^n \left(-\frac{1}{3}\right)^n = \sum_{n=0}^{\infty} (-1)^n$, which diverges; at $x = \frac{1}{3}$ we have $\sum_{n=0}^{\infty} 3^n \left(\frac{1}{3}\right)^n = \sum_{n=0}^{\infty} 1$, which diverges. The series $\sum_{n=0}^{\infty} 3^n x^n$ $= \sum_{n=0}^{\infty} (3x)^n$ is a convergent geometric series when $-\frac{1}{3} < x < \frac{1}{3}$ and the sum is $\frac{1}{1-3x}$.

43. $\lim_{n\to\infty} \left|\frac{u_{n+1}}{u_n}\right| < 1 \Rightarrow \lim_{n\to\infty} \left| \frac{(x-1)^{2n+2}}{4^{n+1}} \cdot \frac{4^n}{(x-1)^{2n}} \right| < 1 \Rightarrow \frac{(x-1)^2}{4} \lim_{n\to\infty} |1| < 1 \Rightarrow (x-1)^2 < 4 \Rightarrow |x-1| < 2$ $\Rightarrow -2 < x-1 < 2 \Rightarrow -1 < x < 3$; at $x = -1$ we have $\sum_{n=0}^{\infty} \frac{(-2)^{2n}}{4^n} = \sum_{n=0}^{\infty} \frac{4^n}{4^n} = \sum_{n=0}^{\infty} 1$, which diverges; at $x = 3$ we have $\sum_{n=0}^{\infty} \frac{2^{2n}}{4^n} = \sum_{n=0}^{\infty} \frac{4^n}{4^n} = \sum_{n=0}^{\infty} 1$, a divergent series; the interval of convergence is $-1 < x < 3$; the series $\sum_{n=0}^{\infty} \frac{(x-1)^{2n}}{4^n} = \sum_{n=0}^{\infty} \left(\left(\frac{x-1}{2}\right)^2\right)^n$ is a convergent geometric series when $-1 < x < 3$ and the sum is $\frac{1}{1-\left(\frac{x-1}{2}\right)^2} = \frac{1}{\left[\frac{4-(x-1)^2}{4}\right]} = \frac{4}{4-x^2+2x-1} = \frac{4}{3+2x-x^2}$

45. $\lim_{n\to\infty} \left|\frac{u_{n+1}}{u_n}\right| < 1 \Rightarrow \lim_{n\to\infty} \left| \frac{\left(\sqrt{x}-2\right)^{n+1}}{2^{n+1}} \cdot \frac{2^n}{\left(\sqrt{x}-2\right)^n} \right| < 1 \Rightarrow \left|\sqrt{x}-2\right| < 2 \Rightarrow -2 < \sqrt{x}-2 < 2 \Rightarrow 0 < \sqrt{x} < 4$ $\Rightarrow 0 < x < 16$; when $x = 0$ we have $\sum_{n=0}^{\infty} (-1)^n$, a divergent series; when $x = 16$ we have $\sum_{n=0}^{\infty} (1)^n$, a divergent series; the interval of convergence is $0 < x < 16$; the series $\sum_{n=0}^{\infty} \left(\frac{\sqrt{x}-2}{2}\right)^n$ is a convergent geometric series when $0 < x < 16$ and its sum is $\frac{1}{1-\left(\frac{\sqrt{x}-2}{2}\right)} = \frac{1}{\left(\frac{2-\sqrt{x}+2}{2}\right)} = \frac{2}{4-\sqrt{x}}$

47. $\lim_{n\to\infty} \left|\frac{u_{n+1}}{u_n}\right| < 1 \Rightarrow \lim_{n\to\infty} \left| \left(\frac{x^2+1}{3}\right)^{n+1} \cdot \left(\frac{3}{x^2+1}\right)^n \right| < 1 \Rightarrow \frac{(x^2+1)}{3} \lim_{n\to\infty} |1| < 1 \Rightarrow \frac{x^2+1}{3} < 1 \Rightarrow x^2 < 2$ $\Rightarrow |x| < \sqrt{2} \Rightarrow -\sqrt{2} < x < \sqrt{2}$; at $x = \pm\sqrt{2}$ we have $\sum_{n=0}^{\infty} (1)^n$ which diverges; the interval of convergence is $-\sqrt{2} < x < \sqrt{2}$; the series $\sum_{n=0}^{\infty} \left(\frac{x^2+1}{3}\right)^n$ is a convergent geometric series when $-\sqrt{2} < x < \sqrt{2}$ and its sum is $\frac{1}{1-\left(\frac{x^2+1}{3}\right)} = \frac{1}{\left(\frac{3-x^2-1}{3}\right)} = \frac{3}{2-x^2}$

49. $\lim_{n\to\infty}\left|\frac{(x-3)^{n+1}}{2^{n+1}}\cdot\frac{2^n}{(x-3)^n}\right|<1 \Rightarrow |x-3|<2 \Rightarrow 1<x<5$; when $x=1$ we have $\sum_{n=1}^{\infty}(1)^n$ which diverges; when $x=5$ we have $\sum_{n=1}^{\infty}(-1)^n$ which also diverges; the interval of convergence is $1<x<5$; the sum of this convergent geometric series is $\frac{1}{1+\left(\frac{x-3}{2}\right)}=\frac{2}{x-1}$. If $f(x)=1-\frac{1}{2}(x-3)+\frac{1}{4}(x-3)^2+\ldots+\left(-\frac{1}{2}\right)^n(x-3)^n+\ldots$ $=\frac{2}{x-1}$ then $f'(x)=-\frac{1}{2}+\frac{1}{2}(x-3)+\ldots+\left(-\frac{1}{2}\right)^n n(x-3)^{n-1}+\ldots$ is convergent when $1<x<5$, and diverges when $x=1$ or 5. The sum for $f'(x)$ is $\frac{-2}{(x-1)^2}$, the derivative of $\frac{2}{x-1}$.

51. (a) Differentiate the series for sin x to get $\cos x = 1-\frac{3x^2}{3!}+\frac{5x^4}{5!}-\frac{7x^6}{7!}+\frac{9x^8}{9!}-\frac{11x^{10}}{11!}+\ldots$ $=1-\frac{x^2}{2!}+\frac{x^4}{4!}-\frac{x^6}{6!}+\frac{x^8}{8!}-\frac{x^{10}}{10!}+\ldots$. The series converges for all values of x since $\lim_{n\to\infty}\left|\frac{x^{2n+2}}{(2n+2)!}\cdot\frac{(2n)!}{x^{2n}}\right|=x^2\lim_{n\to\infty}\left(\frac{1}{(2n+1)(2n+2)}\right)=0<1$ for all x.

(b) $\sin 2x = 2x-\frac{2^3x^3}{3!}+\frac{2^5x^5}{5!}-\frac{2^7x^7}{7!}+\frac{2^9x^9}{9!}-\frac{2^{11}x^{11}}{11!}+\ldots=2x-\frac{8x^3}{3!}+\frac{32x^5}{5!}-\frac{128x^7}{7!}+\frac{512x^9}{9!}-\frac{2048x^{11}}{11!}+\ldots$

(c) $2\sin x\cos x = 2\left[(0\cdot 1)+(0\cdot 0+1\cdot 1)x+\left(0\cdot\frac{-1}{2}+1\cdot 0+0\cdot 1\right)x^2+\left(0\cdot 0-1\cdot\frac{1}{2}+0\cdot 0-1\cdot\frac{1}{3!}\right)x^3\right.$ $+\left(0\cdot\frac{1}{4!}+1\cdot 0-0\cdot\frac{1}{2}-0\cdot\frac{1}{3!}+0\cdot 1\right)x^4+\left(0\cdot 0+1\cdot\frac{1}{4!}+0\cdot 0+\frac{1}{2}\cdot\frac{1}{3!}+0\cdot 0+1\cdot\frac{1}{5!}\right)x^5$ $\left.+\left(0\cdot\frac{1}{6!}+1\cdot 0+0\cdot\frac{1}{4!}+0\cdot\frac{1}{3!}+0\cdot\frac{1}{2}+0\cdot\frac{1}{5!}+0\cdot 1\right)x^6+\ldots\right]=2\left[x-\frac{4x^3}{3!}+\frac{16x^5}{5!}-\ldots\right]$ $=2x-\frac{2^3x^3}{3!}+\frac{2^5x^5}{5!}-\frac{2^7x^7}{7!}+\frac{2^9x^9}{9!}-\frac{2^{11}x^{11}}{11!}+\ldots$

53. (a) $\ln|\sec x|+C=\int \tan x\,dx=\int\left(x+\frac{x^3}{3}+\frac{2x^5}{15}+\frac{17x^7}{315}+\frac{62x^9}{2835}+\ldots\right)dx$ $=\frac{x^2}{2}+\frac{x^4}{12}+\frac{x^6}{45}+\frac{17x^8}{2520}+\frac{31x^{10}}{14{,}175}+\ldots+C$; $x=0\Rightarrow C=0\Rightarrow \ln|\sec x|=\frac{x^2}{2}+\frac{x^4}{12}+\frac{x^6}{45}+\frac{17x^8}{2520}+\frac{31x^{10}}{14{,}175}+\ldots$, converges when $-\frac{\pi}{2}<x<\frac{\pi}{2}$

(b) $\sec^2 x=\frac{d(\tan x)}{dx}=\frac{d}{dx}\left(x+\frac{x^3}{3}+\frac{2x^5}{15}+\frac{17x^7}{315}+\frac{62x^9}{2835}+\ldots\right)=1+x^2+\frac{2x^4}{3}+\frac{17x^6}{45}+\frac{62x^8}{315}+\ldots$, converges when $-\frac{\pi}{2}<x<\frac{\pi}{2}$

(c) $\sec^2 x=(\sec x)(\sec x)=\left(1+\frac{x^2}{2}+\frac{5x^4}{24}+\frac{61x^6}{720}+\ldots\right)\left(1+\frac{x^2}{2}+\frac{5x^4}{24}+\frac{61x^6}{720}+\ldots\right)$ $=1+\left(\frac{1}{2}+\frac{1}{2}\right)x^2+\left(\frac{5}{24}+\frac{1}{4}+\frac{5}{24}\right)x^4+\left(\frac{61}{720}+\frac{5}{48}+\frac{5}{48}+\frac{61}{720}\right)x^6+\ldots$ $=1+x^2+\frac{2x^4}{3}+\frac{17x^6}{45}+\frac{62x^8}{315}+\ldots,\ -\frac{\pi}{2}<x<\frac{\pi}{2}$

55. (a) If $f(x)=\sum_{n=0}^{\infty}a_nx^n$, then $f^{(k)}(x)=\sum_{n=k}^{\infty}n(n-1)(n-2)\cdots(n-(k-1))\,a_nx^{n-k}$ and $f^{(k)}(0)=k!a_k$ $\Rightarrow a_k=\frac{f^{(k)}(0)}{k!}$; likewise if $f(x)=\sum_{n=0}^{\infty}b_nx^n$, then $b_k=\frac{f^{(k)}(0)}{k!}\Rightarrow a_k=b_k$ for every nonnegative integer k

(b) If $f(x)=\sum_{n=0}^{\infty}a_nx^n=0$ for all x, then $f^{(k)}(x)=0$ for all x $\Rightarrow$ from part (a) that $a_k=0$ for every nonnegative integer k

10.8 TAYLOR AND MACLAURIN SERIES

1. $f(x)=e^{2x}$, $f'(x)=2e^{2x}$, $f''(x)=4e^{2x}$, $f'''(x)=8e^{2x}$; $f(0)=e^{2(0)}=1$, $f'(0)=2$, $f''(0)=4$, $f'''(0)=8\Rightarrow P_0(x)=1$, $P_1(x)=1+2x$, $P_2(x)=1+x+2x^2$, $P_3(x)=1+x+2x^2+\frac{4}{3}x^3$

3. $f(x)=\ln x$, $f'(x)=\frac{1}{x}$, $f''(x)=-\frac{1}{x^2}$, $f'''(x)=\frac{2}{x^3}$; $f(1)=\ln 1=0$, $f'(1)=1$, $f''(1)=-1$, $f'''(1)=2\Rightarrow P_0(x)=0$, $P_1(x)=(x-1)$, $P_2(x)=(x-1)-\frac{1}{2}(x-1)^2$, $P_3(x)=(x-1)-\frac{1}{2}(x-1)^2+\frac{1}{3}(x-1)^3$

5. $f(x) = \frac{1}{x} = x^{-1}$, $f'(x) = -x^{-2}$, $f''(x) = 2x^{-3}$, $f'''(x) = -6x^{-4}$; $f(2) = \frac{1}{2}$, $f'(2) = -\frac{1}{4}$, $f''(2) = \frac{1}{4}$, $f'''(x) = -\frac{3}{8}$
$\Rightarrow P_0(x) = \frac{1}{2}$, $P_1(x) = \frac{1}{2} - \frac{1}{4}(x-2)$, $P_2(x) = \frac{1}{2} - \frac{1}{4}(x-2) + \frac{1}{8}(x-2)^2$,
$P_3(x) = \frac{1}{2} - \frac{1}{4}(x-2) + \frac{1}{8}(x-2)^2 - \frac{1}{16}(x-2)^3$

7. $f(x) = \sin x$, $f'(x) = \cos x$, $f''(x) = -\sin x$, $f'''(x) = -\cos x$; $f\left(\frac{\pi}{4}\right) = \sin\frac{\pi}{4} = \frac{\sqrt{2}}{2}$, $f'\left(\frac{\pi}{4}\right) = \cos\frac{\pi}{4} = \frac{\sqrt{2}}{2}$,
$f''\left(\frac{\pi}{4}\right) = -\sin\frac{\pi}{4} = -\frac{\sqrt{2}}{2}$, $f'''\left(\frac{\pi}{4}\right) = -\cos\frac{\pi}{4} = -\frac{\sqrt{2}}{2} \Rightarrow P_0 = \frac{\sqrt{2}}{2}$, $P_1(x) = \frac{\sqrt{2}}{2} + \frac{\sqrt{2}}{2}\left(x - \frac{\pi}{4}\right)$,
$P_2(x) = \frac{\sqrt{2}}{2} + \frac{\sqrt{2}}{2}\left(x - \frac{\pi}{4}\right) - \frac{\sqrt{2}}{4}\left(x - \frac{\pi}{4}\right)^2$, $P_3(x) = \frac{\sqrt{2}}{2} + \frac{\sqrt{2}}{2}\left(x - \frac{\pi}{4}\right) - \frac{\sqrt{2}}{4}\left(x - \frac{\pi}{4}\right)^2 - \frac{\sqrt{2}}{12}\left(x - \frac{\pi}{4}\right)^3$

9. $f(x) = \sqrt{x} = x^{1/2}$, $f'(x) = \left(\frac{1}{2}\right)x^{-1/2}$, $f''(x) = \left(-\frac{1}{4}\right)x^{-3/2}$, $f'''(x) = \left(\frac{3}{8}\right)x^{-5/2}$; $f(4) = \sqrt{4} = 2$,
$f'(4) = \left(\frac{1}{2}\right)4^{-1/2} = \frac{1}{4}$, $f''(4) = \left(-\frac{1}{4}\right)4^{-3/2} = -\frac{1}{32}$, $f'''(4) = \left(\frac{3}{8}\right)4^{-5/2} = \frac{3}{256} \Rightarrow P_0(x) = 2$, $P_1(x) = 2 + \frac{1}{4}(x-4)$,
$P_2(x) = 2 + \frac{1}{4}(x-4) - \frac{1}{64}(x-4)^2$, $P_3(x) = 2 + \frac{1}{4}(x-4) - \frac{1}{64}(x-4)^2 + \frac{1}{512}(x-4)^3$

11. $f(x) = e^{-x}$, $f'(x) = -e^{-x}$, $f''(x) = e^{-x}$, $f'''(x) = -e^{-x} \Rightarrow \ldots f^{(k)}(x) = (-1)^k e^{-x}$; $f(0) = e^{-(0)} = 1$, $f'(0) = -1$,
$f''(0) = 1$, $f'''(0) = -1, \ldots, f^{(k)}(0) = (-1)^k \Rightarrow e^{-x} = 1 - x + \frac{1}{2}x^2 - \frac{1}{6}x^3 + \ldots = \sum_{n=0}^{\infty} \frac{(-1)^n}{n!}x^n$

13. $f(x) = (1+x)^{-1} \Rightarrow f'(x) = -(1+x)^{-2}$, $f''(x) = 2(1+x)^{-3}$, $f'''(x) = -3!(1+x)^{-4} \Rightarrow \ldots f^{(k)}(x)$
$= (-1)^k k!(1+x)^{-k-1}$; $f(0) = 1$, $f'(0) = -1$, $f''(0) = 2$, $f'''(0) = -3!, \ldots, f^{(k)}(0) = (-1)^k k!$
$\Rightarrow 1 - x + x^2 - x^3 + \ldots = \sum_{n=0}^{\infty}(-x)^n = \sum_{n=0}^{\infty}(-1)^n x^n$

15. $\sin x = \sum_{n=0}^{\infty} \frac{(-1)^n x^{2n+1}}{(2n+1)!} \Rightarrow \sin 3x = \sum_{n=0}^{\infty} \frac{(-1)^n (3x)^{2n+1}}{(2n+1)!} = \sum_{n=0}^{\infty} \frac{(-1)^n 3^{2n+1} x^{2n+1}}{(2n+1)!} = 3x - \frac{3^3x^3}{3!} + \frac{3^5x^5}{5!} - \ldots$

17. $7\cos(-x) = 7\cos x = 7\sum_{n=0}^{\infty} \frac{(-1)^n x^{2n}}{(2n)!} = 7 - \frac{7x^2}{2!} + \frac{7x^4}{4!} - \frac{7x^6}{6!} + \ldots$, since the cosine is an even function

19. $\cosh x = \frac{e^x + e^{-x}}{2} = \frac{1}{2}\left[\left(1 + x^2 + \frac{x^2}{2!} + \frac{x^3}{3!} + \frac{x^4}{4!} + \ldots\right) + \left(1 - x + \frac{x^2}{2!} - \frac{x^3}{3!} + \frac{x^4}{4!} - \ldots\right)\right] = 1 + \frac{x^2}{2!} + \frac{x^4}{4!} + \frac{x^6}{6!} + \ldots$
$= \sum_{n=0}^{\infty} \frac{x^{2n}}{(2n)!}$

21. $f(x) = x^4 - 2x^3 - 5x + 4 \Rightarrow f'(x) = 4x^3 - 6x^2 - 5$, $f''(x) = 12x^2 - 12x$, $f'''(x) = 24x - 12$, $f^{(4)}(x) = 24$
$\Rightarrow f^{(n)}(x) = 0$ if $n \ge 5$; $f(0) = 4$, $f'(0) = -5$, $f''(0) = 0$, $f'''(0) = -12$, $f^{(4)}(0) = 24$, $f^{(n)}(0) = 0$ if $n \ge 5$
$\Rightarrow x^4 - 2x^3 - 5x + 4 = 4 - 5x - \frac{12}{3!}x^3 + \frac{24}{4!}x^4 = x^4 - 2x^3 - 5x + 4$

23. $f(x) = x^3 - 2x + 4 \Rightarrow f'(x) = 3x^2 - 2$, $f''(x) = 6x$, $f'''(x) = 6 \Rightarrow f^{(n)}(x) = 0$ if $n \ge 4$; $f(2) = 8$, $f'(2) = 10$,
$f''(2) = 12$, $f'''(2) = 6$, $f^{(n)}(2) = 0$ if $n \ge 4 \Rightarrow x^3 - 2x + 4 = 8 + 10(x-2) + \frac{12}{2!}(x-2)^2 + \frac{6}{3!}(x-2)^3$
$= 8 + 10(x-2) + 6(x-2)^2 + (x-2)^3$

25. $f(x) = x^4 + x^2 + 1 \Rightarrow f'(x) = 4x^3 + 2x$, $f''(x) = 12x^2 + 2$, $f'''(x) = 24x$, $f^{(4)}(x) = 24$, $f^{(n)}(x) = 0$ if $n \ge 5$;
$f(-2) = 21$, $f'(-2) = -36$, $f''(-2) = 50$, $f'''(-2) = -48$, $f^{(4)}(-2) = 24$, $f^{(n)}(-2) = 0$ if $n \ge 5 \Rightarrow x^4 + x^2 + 1$
$= 21 - 36(x+2) + \frac{50}{2!}(x+2)^2 - \frac{48}{3!}(x+2)^3 + \frac{24}{4!}(x+2)^4 = 21 - 36(x+2) + 25(x+2)^2 - 8(x+2)^3 + (x+2)^4$

27. $f(x) = x^{-2} \Rightarrow f'(x) = -2x^{-3}, f''(x) = 3!\,x^{-4}, f'''(x) = -4!\,x^{-5} \Rightarrow f^{(n)}(x) = (-1)^n(n+1)!\,x^{-n-2}$;
$f(1) = 1, f'(1) = -2, f''(1) = 3!, f'''(1) = -4!, f^{(n)}(1) = (-1)^n(n+1)! \Rightarrow \frac{1}{x^2}$
$= 1 - 2(x-1) + 3(x-1)^2 - 4(x-1)^3 + \ldots = \sum_{n=0}^{\infty} (-1)^n(n+1)(x-1)^n$

29. $f(x) = e^x \Rightarrow f'(x) = e^x, f''(x) = e^x \Rightarrow f^{(n)}(x) = e^x; f(2) = e^2, f'(2) = e^2, \ldots f^{(n)}(2) = e^2$
$\Rightarrow e^x = e^2 + e^2(x-2) + \frac{e^2}{2}(x-2)^2 + \frac{e^3}{3!}(x-2)^3 + \ldots = \sum_{n=0}^{\infty} \frac{e^2}{n!}(x-2)^n$

31. $f(x) = \cos\left(2x + \frac{\pi}{2}\right), f'(x) = -2\sin\left(2x + \frac{\pi}{2}\right), f''(x) = -4\cos\left(2x + \frac{\pi}{2}\right), f'''(x) = 8\sin\left(2x + \frac{\pi}{2}\right)$,
$f^{(4)}(x) = 2^4\cos\left(2x + \frac{\pi}{2}\right), f^{(5)}(x) = -2^5\sin\left(2x + \frac{\pi}{2}\right), \ldots; f\left(\frac{\pi}{4}\right) = -1, f'\left(\frac{\pi}{4}\right) = 0, f''\left(\frac{\pi}{4}\right) = 4, f'''\left(\frac{\pi}{4}\right) = 0, f^{(4)}\left(\frac{\pi}{4}\right) = 2^4$,
$f^{(5)}\left(\frac{\pi}{4}\right) = 0, \ldots, f^{(2n)}\left(\frac{\pi}{4}\right) = (-1)^n 2^{2n} \Rightarrow \cos\left(2x + \frac{\pi}{2}\right) = -1 + 2\left(x - \frac{\pi}{4}\right)^2 - \frac{2}{3}\left(x - \frac{\pi}{4}\right)^4 + \ldots$
$= \sum_{n=0}^{\infty} \frac{(-1)^n 2^{2n}}{(2n)!}\left(x - \frac{\pi}{4}\right)^{2n}$

33. The Maclaurin series generated by $\cos x$ is $\sum_{n=0}^{\infty} \frac{(-1)^n}{(2n)!}x^{2n}$ which converges on $(-\infty, \infty)$ and the Maclaurin series generated by $\frac{2}{1-x}$ is $2\sum_{n=0}^{\infty} x^n$ which converges on $(-1, 1)$. Thus the Maclaurin series generated by $f(x) = \cos x - \frac{2}{1-x}$ is given by $\sum_{n=0}^{\infty} \frac{(-1)^n}{(2n)!}x^{2n} - 2\sum_{n=0}^{\infty} x^n = -1 - 2x - \frac{5}{2}x^2 - \ldots$. which converges on the intersection of $(-\infty, \infty)$ and $(-1, 1)$, so the interval of convergence is $(-1, 1)$.

35. The Maclaurin series generated by $\sin x$ is $\sum_{n=0}^{\infty} \frac{(-1)^n}{(2n+1)!}x^{2n+1}$ which converges on $(-\infty, \infty)$ and the Maclaurin series generated by $\ln(1+x)$ is $\sum_{n=1}^{\infty} \frac{(-1)^{n-1}}{n}x^n$ which converges on $(-1, 1)$. Thus the Maclaurin series genereated by $f(x) = \sin x \cdot \ln(1+x)$ is given by $\left(\sum_{n=0}^{\infty} \frac{(-1)^n}{(2n+1)!}x^{2n+1}\right)\left(\sum_{n=1}^{\infty} \frac{(-1)^{n-1}}{n}x^n\right) = x^2 - \frac{1}{2}x^3 + \frac{1}{6}x^4 - \ldots$. which converges on the intersection of $(-\infty, \infty)$ and $(-1, 1)$, so the interval of convergence is $(-1, 1)$.

37. If $e^x = \sum_{n=0}^{\infty} \frac{f^{(n)}(a)}{n!}(x-a)^n$ and $f(x) = e^x$, we have $f^{(n)}(a) = e^a$ f or all $n = 0, 1, 2, 3, \ldots$
$\Rightarrow e^x = e^a\left[\frac{(x-a)^0}{0!} + \frac{(x-a)^1}{1!} + \frac{(x-a)^2}{2!} + \ldots\right] = e^a\left[1 + (x-a) + \frac{(x-a)^2}{2!} + \ldots\right]$ at $x = a$

39. $f(x) = f(a) + f'(a)(x-a) + \frac{f''(a)}{2}(x-a)^2 + \frac{f'''(a)}{3!}(x-a)^3 + \ldots \Rightarrow f'(x)$
$= f'(a) + f''(a)(x-a) + \frac{f'''(a)}{3!}3(x-a)^2 + \ldots \Rightarrow f''(x) = f''(a) + f'''(a)(x-a) + \frac{f^{(4)}(a)}{4!}4 \cdot 3(x-a)^2 + \ldots$
$\Rightarrow f^{(n)}(x) = f^{(n)}(a) + f^{(n+1)}(a)(x-a) + \frac{f^{(n+2)}(a)}{2}(x-a)^2 + \ldots$
$\Rightarrow f(a) = f(a) + 0, f'(a) = f'(a) + 0, \ldots, f^{(n)}(a) = f^{(n)}(a) + 0$

41. $f(x) = \ln(\cos x) \Rightarrow f'(x) = -\tan x$ and $f''(x) = -\sec^2 x$; $f(0) = 0, f'(0) = 0, f''(0) = -1 \Rightarrow L(x) = 0$ and $Q(x) = -\frac{x^2}{2}$

43. $f(x) = (1-x^2)^{-1/2} \Rightarrow f'(x) = x(1-x^2)^{-3/2}$ and $f''(x) = (1-x^2)^{-3/2} + 3x^2(1-x^2)^{-5/2}$; $f(0) = 1, f'(0) = 0$,
$f''(0) = 1 \Rightarrow L(x) = 1$ and $Q(x) = 1 + \frac{x^2}{2}$

45. $f(x) = \sin x \Rightarrow f'(x) = \cos x$ and $f''(x) = -\sin x$; $f(0) = 0, f'(0) = 1, f''(0) = 0 \Rightarrow L(x) = x$ and $Q(x) = x$

10.9 CONVERGENCE OF TAYLOR SERIES

1. $e^x = 1 + x + \frac{x^2}{2!} + \dots = \sum_{n=0}^{\infty} \frac{x^n}{n!} \Rightarrow e^{-5x} = 1 + (-5x) + \frac{(-5x)^2}{2!} + \dots = 1 - 5x + \frac{5^2x^2}{2!} - \frac{5^3x^3}{3!} + \dots = \sum_{n=0}^{\infty} \frac{(-1)^n 5^n x^n}{n!}$

3. $\sin x = x - \frac{x^3}{3!} + \frac{x^5}{5!} - \dots = \sum_{n=0}^{\infty} \frac{(-1)^n x^{2n+1}}{(2n+1)!} \Rightarrow 5\sin(-x) = 5\left[(-x) - \frac{(-x)^3}{3!} + \frac{(-x)^5}{5!} - \dots\right] = \sum_{n=0}^{\infty} \frac{5(-1)^{n+1} x^{2n+1}}{(2n+1)!}$

5. $\cos x = \sum_{n=0}^{\infty} \frac{(-1)^n x^{2n}}{(2n)!} \Rightarrow \cos 5x^2 = \sum_{n=0}^{\infty} \frac{(-1)^n \left[5x^2\right]^{2n}}{(2n)!} = \sum_{n=0}^{\infty} \frac{(-1)^n 5^{2n} x^{4n}}{(2n)!} = 1 - \frac{25x^4}{2!} + \frac{625x^8}{4!} - \frac{15625x^{12}}{6!} + \dots$

7. $\ln(1+x) = \sum_{n=1}^{\infty} \frac{(-1)^{n-1} x^n}{n} \Rightarrow \ln(1+x^2) = \sum_{n=1}^{\infty} \frac{(-1)^{n-1}(x^2)^n}{n} = \sum_{n=1}^{\infty} \frac{(-1)^{n-1} x^{2n}}{n} = x^2 - \frac{x^4}{2} + \frac{x^6}{3} - \frac{x^8}{4} + \dots$

9. $\frac{1}{1+x} = \sum_{n=0}^{\infty} (-1)^n x^n \Rightarrow \frac{1}{1+\frac{3}{4}x^3} = \sum_{n=0}^{\infty} (-1)^n \left(\frac{3}{4}x^3\right)^n = \sum_{n=0}^{\infty} (-1)^n \left(\frac{3}{4}\right)^n x^{3n} = 1 - \frac{3}{4}x^3 + \frac{9}{16}x^6 - \frac{27}{64}x^9 + \dots$

11. $e^x = \sum_{n=0}^{\infty} \frac{x^n}{n!} \Rightarrow xe^x = x\left(\sum_{n=0}^{\infty} \frac{x^n}{n!}\right) = \sum_{n=0}^{\infty} \frac{x^{n+1}}{n!} = x + x^2 + \frac{x^3}{2!} + \frac{x^4}{3!} + \frac{x^5}{4!} + \dots$

13. $\cos x = \sum_{n=0}^{\infty} \frac{(-1)^n x^{2n}}{(2n)!} \Rightarrow \frac{x^2}{2} - 1 + \cos x = \frac{x^2}{2} - 1 + \sum_{n=0}^{\infty} \frac{(-1)^n x^{2n}}{(2n)!} = \frac{x^2}{2} - 1 + 1 - \frac{x^2}{2} + \frac{x^4}{4!} - \frac{x^6}{6!} + \frac{x^8}{8!} - \frac{x^{10}}{10!} + \dots$
$= \frac{x^4}{4!} - \frac{x^6}{6!} + \frac{x^8}{8!} - \frac{x^{10}}{10!} + \dots = \sum_{n=2}^{\infty} \frac{(-1)^n x^{2n}}{(2n)!}$

15. $\cos x = \sum_{n=0}^{\infty} \frac{(-1)^n x^{2n}}{(2n)!} \Rightarrow x\cos \pi x = x\sum_{n=0}^{\infty} \frac{(-1)^n (\pi x)^{2n}}{(2n)!} = \sum_{n=0}^{\infty} \frac{(-1)^n \pi^{2n} x^{2n+1}}{(2n)!} = x - \frac{\pi^2 x^3}{2!} + \frac{\pi^4 x^5}{4!} - \frac{\pi^6 x^7}{6!} + \dots$

17. $\cos^2 x = \frac{1}{2} + \frac{\cos 2x}{2} = \frac{1}{2} + \frac{1}{2}\sum_{n=0}^{\infty} \frac{(-1)^n (2x)^{2n}}{(2n)!} = \frac{1}{2} + \frac{1}{2}\left[1 - \frac{(2x)^2}{2!} + \frac{(2x)^4}{4!} - \frac{(2x)^6}{6!} + \frac{(2x)^8}{8!} - \dots\right]$
$= 1 - \frac{(2x)^2}{2 \cdot 2!} + \frac{(2x)^4}{2 \cdot 4!} - \frac{(2x)^6}{2 \cdot 6!} + \frac{(2x)^8}{2 \cdot 8!} - \dots = 1 + \sum_{n=1}^{\infty} \frac{(-1)^n (2x)^{2n}}{2 \cdot (2n)!} = 1 + \sum_{n=1}^{\infty} \frac{(-1)^n 2^{2n-1} x^{2n}}{(2n)!}$

19. $\frac{x^2}{1-2x} = x^2\left(\frac{1}{1-2x}\right) = x^2 \sum_{n=0}^{\infty} (2x)^n = \sum_{n=0}^{\infty} 2^n x^{n+2} = x^2 + 2x^3 + 2^2x^4 + 2^3x^5 + \dots$

21. $\frac{1}{1-x} = \sum_{n=0}^{\infty} x^n = 1 + x + x^2 + x^3 + \dots \Rightarrow \frac{d}{dx}\left(\frac{1}{1-x}\right) = \frac{1}{(1-x)^2} = 1 + 2x + 3x^2 + \dots = \sum_{n=1}^{\infty} nx^{n-1} = \sum_{n=0}^{\infty} (n+1)x^n$

23. $\tan^{-1}x = x - \frac{1}{3}x^3 + \frac{1}{5}x^5 - \frac{1}{7}x^7 + \dots \Rightarrow x\tan^{-1}x^2 = x\left(x^2 - \frac{1}{3}(x^2)^3 + \frac{1}{5}(x^2)^5 - \frac{1}{7}(x^2)^7 + \dots\right)$
$= x^3 - \frac{1}{3}x^7 + \frac{1}{5}x^{11} - \frac{1}{7}x^{15} + \dots = \sum_{n=1}^{\infty} \frac{(-1)^n x^{4n-1}}{2n-1}$

25. $e^x = 1 + x + \frac{x^2}{2!} + \frac{x^3}{3!} + \dots$ and $\frac{1}{1+x} = 1 - x + x^2 - x^3 + \dots \Rightarrow e^x + \frac{1}{1+x}$
$= \left(1 + x + \frac{x^2}{2!} + \frac{x^3}{3!} + \dots\right) + (1 - x + x^2 - x^3 + \dots) = 2 + \frac{3}{2}x^2 - \frac{5}{6}x^3 + \frac{25}{24}x^4 + \dots = \sum_{n=0}^{\infty}\left(\frac{1}{n!} + (-1)^n\right)x^n$

27. $\ln(1+x) = x - \frac{1}{2}x^2 + \frac{1}{3}x^3 - \frac{1}{4}x^4 + \ldots \Rightarrow \frac{x}{3}\ln(1+x^2) = \frac{x}{3}\left(x^2 - \frac{1}{2}(x^2)^2 + \frac{1}{3}(x^2)^3 - \frac{1}{4}(x^2)^4 + \ldots\right)$
$= \frac{1}{3}x^3 - \frac{1}{6}x^5 + \frac{1}{9}x^7 - \frac{1}{12}x^9 + \ldots = \sum_{n=1}^{\infty} \frac{(-1)^{n-1}}{3n} x^{2n+1}$

29. $e^x = 1 + x + \frac{x^2}{2!} + \frac{x^3}{3!} + \ldots$ and $\sin x = x - \frac{x^3}{3!} + \frac{x^5}{5!} - \frac{x^7}{7!} + \ldots \Rightarrow e^x \cdot \sin x$
$= \left(1 + x + \frac{x^2}{2!} + \frac{x^3}{3!} + \ldots\right)\left(x - \frac{x^3}{3!} + \frac{x^5}{5!} - \frac{x^7}{7!} + \ldots\right) = x + x^2 + \frac{1}{3}x^3 - \frac{1}{30}x^5 - \ldots.$

31. $\tan^{-1}x = x - \frac{1}{3}x^3 + \frac{1}{5}x^5 - \frac{1}{7}x^7 + \ldots \Rightarrow (\tan^{-1}x)^2 = (\tan^{-1}x)(\tan^{-1}x)$
$= \left(x - \frac{1}{3}x^3 + \frac{1}{5}x^5 - \frac{1}{7}x^7 + \ldots\right)\left(x - \frac{1}{3}x^3 + \frac{1}{5}x^5 - \frac{1}{7}x^7 + \ldots\right) = x^2 - \frac{2}{3}x^4 - \frac{23}{45}x^6 - \frac{44}{105}x^8 + \ldots.$

33. $\sin x = x - \frac{x^3}{3!} + \frac{x^5}{5!} - \frac{x^7}{7!} + \ldots$ and $e^x = 1 + x + \frac{x^2}{2!} + \frac{x^3}{3!} + \ldots$
$\Rightarrow e^{\sin x} = 1 + \left(x - \frac{x^3}{3!} + \frac{x^5}{5!} - \frac{x^7}{7!} + \ldots\right) + \frac{1}{2}\left(x - \frac{x^3}{3!} + \frac{x^5}{5!} - \frac{x^7}{7!} + \ldots\right)^2 + \frac{1}{6}\left(x - \frac{x^3}{3!} + \frac{x^5}{5!} - \frac{x^7}{7!} + \ldots\right)^3 + \ldots$
$= 1 + x + \frac{1}{2}x^2 - \frac{1}{8}x^4 + \ldots.$

35. Since $n = 3$, then $f^{(4)}(x) = \sin x$, $|f^{(4)}(x)| \le M$ on $[0, 0.1] \Rightarrow |\sin x| \le 1$ on $[0, 0.1] \Rightarrow M = 1$. Then $|R_3(0.1)| \le 1\frac{|0.1-0|^4}{4!}$
$= 4.2 \times 10^{-6} \Rightarrow$ error $\le 4.2 \times 10^{-6}$

37. By the Alternating Series Estimation Theorem, the error is less than $\frac{|x|^5}{5!} \Rightarrow |x|^5 < (5!)(5 \times 10^{-4}) \Rightarrow |x|^5 < 600 \times 10^{-4}$
$\Rightarrow |x| < \sqrt[5]{6 \times 10^{-2}} \approx 0.56968$

39. If $\sin x = x$ and $|x| < 10^{-3}$, then the error is less than $\frac{(10^{-3})^3}{3!} \approx 1.67 \times 10^{-10}$, by Alternating Series Estimation Theorem; The Alternating Series Estimation Theorem says $R_2(x)$ has the same sign as $-\frac{x^3}{3!}$. Moreover, $x < \sin x$
$\Rightarrow 0 < \sin x - x = R_2(x) \Rightarrow x < 0 \Rightarrow -10^{-3} < x < 0.$

41. $|R_2(x)| = \left|\frac{e^c x^3}{3!}\right| < \frac{3^{(0.1)}(0.1)^3}{3!} < 1.87 \times 10^{-4}$, where c is between 0 and x

43. $\sin^2 x = \left(\frac{1-\cos 2x}{2}\right) = \frac{1}{2} - \frac{1}{2}\cos 2x = \frac{1}{2} - \frac{1}{2}\left(1 - \frac{(2x)^2}{2!} + \frac{(2x)^4}{4!} - \frac{(2x)^6}{6!} + \ldots\right) = \frac{2x^2}{2!} - \frac{2^3x^4}{4!} + \frac{2^5x^6}{6!} - \ldots$
$\Rightarrow \frac{d}{dx}(\sin^2 x) = \frac{d}{dx}\left(\frac{2x^2}{2!} - \frac{2^3x^4}{4!} + \frac{2^5x^6}{6!} - \ldots\right) = 2x - \frac{(2x)^3}{3!} + \frac{(2x)^5}{5!} - \frac{(2x)^7}{7!} + \ldots \Rightarrow 2\sin x\cos x$
$= 2x - \frac{(2x)^3}{3!} + \frac{(2x)^5}{5!} - \frac{(2x)^7}{7!} + \ldots = \sin 2x$, which checks

45. A special case of Taylor's Theorem is $f(b) = f(a) + f'(c)(b-a)$, where c is between a and b $\Rightarrow f(b) - f(a) = f'(c)(b-a)$, the Mean Value Theorem.

47. (a) $f'' \le 0$, $f'(a) = 0$ and $x = a$ interior to the interval I $\Rightarrow f(x) - f(a) = \frac{f''(c_2)}{2}(x-a)^2 \le 0$ throughout I
$\Rightarrow f(x) \le f(a)$ throughout I $\Rightarrow$ f has a local maximum at $x = a$
(b) similar reasoning gives $f(x) - f(a) = \frac{f''(c_2)}{2}(x-a)^2 \ge 0$ throughout I $\Rightarrow f(x) \ge f(a)$ throughout I $\Rightarrow$ f has a local minimum at $x = a$

49. (a) $f(x) = (1+x)^k \Rightarrow f'(x) = k(1+x)^{k-1} \Rightarrow f''(x) = k(k-1)(1+x)^{k-2}$; $f(0) = 1$, $f'(0) = k$, and $f''(0) = k(k-1)$
$\Rightarrow Q(x) = 1 + kx + \frac{k(k-1)}{2}x^2$
(b) $|R_2(x)| = \left|\frac{3 \cdot 2 \cdot 1}{3!}x^3\right| < \frac{1}{100} \Rightarrow |x^3| < \frac{1}{100} \Rightarrow 0 < x < \frac{1}{100^{1/3}}$ or $0 < x < .21544$

51. If $f(x) = \sum_{n=0}^{\infty} a_n x^n$, then $f^{(k)}(x) = \sum_{n=k}^{\infty} n(n-1)(n-2)\cdots(n-k+1)a_n x^{n-k}$ and $f^{(k)}(0) = k!\, a_k$ $\Rightarrow a_k = \frac{f^{(k)}(0)}{k!}$ for k a nonnegative integer. Therefore, the coefficients of f(x) are identical with the corresponding coefficients in the Maclaurin series of f(x) and the statement follows.

10.10 THE BINOMIAL SERIES

1. $(1+x)^{1/2} = 1 + \frac{1}{2}x + \frac{\left(\frac{1}{2}\right)\left(-\frac{1}{2}\right)x^2}{2!} + \frac{\left(\frac{1}{2}\right)\left(-\frac{1}{2}\right)\left(-\frac{3}{2}\right)x^3}{3!} + \ldots = 1 + \frac{1}{2}x - \frac{1}{8}x^2 + \frac{1}{16}x^3 - \ldots$

3. $(1-x)^{-1/2} = 1 - \frac{1}{2}(-x) + \frac{\left(-\frac{1}{2}\right)\left(-\frac{3}{2}\right)(-x)^2}{2!} + \frac{\left(-\frac{1}{2}\right)\left(-\frac{3}{2}\right)\left(-\frac{5}{2}\right)(-x)^3}{3!} + \ldots = 1 + \frac{1}{2}x + \frac{3}{8}x^2 + \frac{5}{16}x^3 + \ldots$

5. $\left(1+\frac{x}{2}\right)^{-2} = 1 - 2\left(\frac{x}{2}\right) + \frac{(-2)(-3)\left(\frac{x}{2}\right)^2}{2!} + \frac{(-2)(-3)(-4)\left(\frac{x}{2}\right)^3}{3!} + \ldots = 1 - x + \frac{3}{4}x^2 - \frac{1}{2}x^3$

7. $\left(1+x^3\right)^{-1/2} = 1 - \frac{1}{2}x^3 + \frac{\left(-\frac{1}{2}\right)\left(-\frac{3}{2}\right)\left(x^3\right)^2}{2!} + \frac{\left(-\frac{1}{2}\right)\left(-\frac{3}{2}\right)\left(-\frac{5}{2}\right)\left(x^3\right)^3}{3!} + \ldots = 1 - \frac{1}{2}x^3 + \frac{3}{8}x^6 - \frac{5}{16}x^9 + \ldots$

9. $\left(1+\frac{1}{x}\right)^{1/2} = 1 + \frac{1}{2}\left(\frac{1}{x}\right) + \frac{\left(\frac{1}{2}\right)\left(-\frac{1}{2}\right)\left(\frac{1}{x}\right)^2}{2!} + \frac{\left(\frac{1}{2}\right)\left(-\frac{1}{2}\right)\left(-\frac{3}{2}\right)\left(\frac{1}{x}\right)^3}{3!} + \ldots = 1 + \frac{1}{2x} - \frac{1}{8x^2} + \frac{1}{16x^3} + \ldots$

11. $(1+x)^4 = 1 + 4x + \frac{(4)(3)x^2}{2!} + \frac{(4)(3)(2)x^3}{3!} + \frac{(4)(3)(2)x^4}{4!} = 1 + 4x + 6x^2 + 4x^3 + x^4$

13. $(1-2x)^3 = 1 + 3(-2x) + \frac{(3)(2)(-2x)^2}{2!} + \frac{(3)(2)(1)(-2x)^3}{3!} = 1 - 6x + 12x^2 - 8x^3$

15. $\int_0^{0.2} \sin x^2\, dx = \int_0^{0.2}\left(x^2 - \frac{x^6}{3!} + \frac{x^{10}}{5!} - \ldots\right) dx = \left[\frac{x^3}{3} - \frac{x^7}{7\cdot 3!} + \ldots\right]_0^{0.2} \approx \left[\frac{x^3}{3}\right]_0^{0.2} \approx 0.00267$ with error $|E| \le \frac{(.2)^7}{7\cdot 3!} \approx 0.0000003$

17. $\int_0^{0.1} \frac{1}{\sqrt{1+x^4}}\, dx = \int_0^{0.1}\left(1 - \frac{x^4}{2} + \frac{3x^8}{8} - \ldots\right) dx = \left[x - \frac{x^5}{10} + \ldots\right]_0^{0.1} \approx [x]_0^{0.1} \approx 0.1$ with error $|E| \le \frac{(0.1)^5}{10} = 0.000001$

19. $\int_0^{0.1} \frac{\sin x}{x}\, dx = \int_0^{0.1}\left(1 - \frac{x^2}{3!} + \frac{x^4}{5!} - \frac{x^6}{7!} + \ldots\right) dx = \left[x - \frac{x^3}{3\cdot 3!} + \frac{x^5}{5\cdot 5!} - \frac{x^7}{7\cdot 7!} + \ldots\right]_0^{0.1} \approx \left[x - \frac{x^3}{3\cdot 3!} + \frac{x^5}{5\cdot 5!}\right]_0^{0.1}$ ≈ 0.0999444611, $|E| \le \frac{(0.1)^7}{7\cdot 7!} \approx 2.8 \times 10^{-12}$

21. $\left(1+x^4\right)^{1/2} = (1)^{1/2} + \frac{\left(\frac{1}{2}\right)}{1}(1)^{-1/2}\left(x^4\right) + \frac{\left(\frac{1}{2}\right)\left(-\frac{1}{2}\right)}{2!}(1)^{-3/2}\left(x^4\right)^2 + \frac{\left(\frac{1}{2}\right)\left(-\frac{1}{2}\right)\left(-\frac{3}{2}\right)}{3!}(1)^{-5/2}\left(x^4\right)^3$ $+ \frac{\left(\frac{1}{2}\right)\left(-\frac{1}{2}\right)\left(-\frac{3}{2}\right)\left(-\frac{5}{2}\right)}{4!}(1)^{-7/2}\left(x^4\right)^4 + \ldots = 1 + \frac{x^4}{2} - \frac{x^8}{8} + \frac{x^{12}}{16} - \frac{5x^{16}}{128} + \ldots$
$\Rightarrow \int_0^{0.1}\left(1 + \frac{x^4}{2} - \frac{x^8}{8} + \frac{x^{12}}{16} - \frac{5x^{16}}{128} + \ldots\right) dx \approx \left[x + \frac{x^5}{10}\right]_0^{0.1} \approx 0.100001$, $|E| \le \frac{(0.1)^9}{72} \approx 1.39 \times 10^{-11}$

23. $\int_0^1 \cos t^2\, dt = \int_0^1\left(1 - \frac{t^4}{2} + \frac{t^8}{4!} - \frac{t^{12}}{6!} + \ldots\right) dt = \left[t - \frac{t^5}{10} + \frac{t^9}{9\cdot 4!} - \frac{t^{13}}{13\cdot 6!} + \ldots\right]_0^1 \Rightarrow |\text{error}| < \frac{1}{13\cdot 6!} \approx .00011$

25. $F(x) = \int_0^x\left(t^2 - \frac{t^6}{3!} + \frac{t^{10}}{5!} - \frac{t^{14}}{7!} + \ldots\right) dt = \left[\frac{t^3}{3} - \frac{t^7}{7\cdot 3!} + \frac{t^{11}}{11\cdot 5!} - \frac{t^{15}}{15\cdot 7!} + \ldots\right]_0^x \approx \frac{x^3}{3} - \frac{x^7}{7\cdot 3!} + \frac{x^{11}}{11\cdot 5!}$ $\Rightarrow |\text{error}| < \frac{1}{15\cdot 7!} \approx 0.000013$

27. (a) $F(x) = \int_0^x \left(t - \frac{t^3}{3} + \frac{t^5}{5} - \frac{t^7}{7} + \ldots\right) dt = \left[\frac{t^2}{2} - \frac{t^4}{12} + \frac{t^6}{30} - \ldots\right]_0^x \approx \frac{x^2}{2} - \frac{x^4}{12} \Rightarrow |\text{error}| < \frac{(0.5)^6}{30} \approx .00052$

(b) $|\text{error}| < \frac{1}{33 \cdot 34} \approx .00089$ when $F(x) \approx \frac{x^2}{2} - \frac{x^4}{3 \cdot 4} + \frac{x^6}{5 \cdot 6} - \frac{x^8}{7 \cdot 8} + \ldots + (-1)^{15} \frac{x^{32}}{31 \cdot 32}$

29. $\frac{1}{x^2}(e^x - (1+x)) = \frac{1}{x^2}\left(\left(1 + x + \frac{x^2}{2} + \frac{x^3}{3!} + \ldots\right) - 1 - x\right) = \frac{1}{2} + \frac{x}{3!} + \frac{x^2}{4!} + \ldots \Rightarrow \lim_{x \to 0} \frac{e^x - (1+x)}{x^2}$
$= \lim_{x \to 0} \left(\frac{1}{2} + \frac{x}{3!} + \frac{x^2}{4!} + \ldots\right) = \frac{1}{2}$

31. $\frac{1}{t^4}\left(1 - \cos t - \frac{t^2}{2}\right) = \frac{1}{t^4}\left[1 - \frac{t^2}{2} - \left(1 - \frac{t^2}{2} + \frac{t^4}{4!} - \frac{t^6}{6!} + \ldots\right)\right] = -\frac{1}{4!} + \frac{t^2}{6!} - \frac{t^4}{8!} + \ldots \Rightarrow \lim_{t \to 0} \frac{1 - \cos t - \left(\frac{t^2}{2}\right)}{t^4}$
$= \lim_{t \to 0} \left(-\frac{1}{4!} + \frac{t^2}{6!} - \frac{t^4}{8!} + \ldots\right) = -\frac{1}{24}$

33. $\frac{1}{y^3}(y - \tan^{-1} y) = \frac{1}{y^3}\left[y - \left(y - \frac{y^3}{3} + \frac{y^5}{5} - \ldots\right)\right] = \frac{1}{3} - \frac{y^2}{5} + \frac{y^4}{7} - \ldots \Rightarrow \lim_{y \to 0} \frac{y - \tan^{-1} y}{y^3} = \lim_{y \to 0} \left(\frac{1}{3} - \frac{y^2}{5} + \frac{y^4}{7} - \ldots\right)$
$= \frac{1}{3}$

35. $x^2\left(-1 + e^{-1/x^2}\right) = x^2\left(-1 + 1 - \frac{1}{x^2} + \frac{1}{2x^4} - \frac{1}{6x^6} + \ldots\right) = -1 + \frac{1}{2x^2} - \frac{1}{6x^4} + \ldots \Rightarrow \lim_{x \to \infty} x^2\left(e^{-1/x^2} - 1\right)$
$= \lim_{x \to \infty} \left(-1 + \frac{1}{2x^2} - \frac{1}{6x^4} + \ldots\right) = -1$

37. $\frac{\ln(1+x^2)}{1 - \cos x} = \frac{\left(x^2 - \frac{x^4}{2} + \frac{x^6}{3} - \ldots\right)}{1 - \left(1 - \frac{x^2}{2!} + \frac{x^4}{4!} - \ldots\right)} = \frac{\left(1 - \frac{x^2}{2} + \frac{x^4}{3} - \ldots\right)}{\left(\frac{1}{2!} - \frac{x^2}{4!} + \ldots\right)} \Rightarrow \lim_{x \to 0} \frac{\ln(1+x^2)}{1 - \cos x} = \lim_{x \to 0} \frac{\left(1 - \frac{x^2}{2} + \frac{x^4}{3} - \ldots\right)}{\left(\frac{1}{2!} - \frac{x^2}{4!} + \ldots\right)} = 2! = 2$

39. $\sin 3x^2 = 3x^2 - \frac{9}{2}x^6 + \frac{81}{40}x^{10} - \ldots$ and $1 - \cos 2x = 2x^2 - \frac{2}{3}x^4 + \frac{4}{45}x^6 - \ldots \Rightarrow \lim_{x \to 0} \frac{\sin 3x^2}{1 - \cos 2x}$
$= \lim_{x \to 0} \frac{3x^2 - \frac{9}{2}x^6 + \frac{81}{40}x^{10} - \ldots}{2x^2 - \frac{2}{3}x^4 + \frac{4}{45}x^6 - \ldots} = \lim_{x \to 0} \frac{3 - \frac{9}{2}x^4 + \frac{81}{40}x^8 - \ldots}{2 - \frac{2}{3}x^2 + \frac{4}{45}x^4 - \ldots} = \frac{3}{2}$

41. $1 + 1 + \frac{1}{2!} + \frac{1}{3!} + \frac{1}{4!} + \ldots = e^1 = e$

43. $1 - \frac{3^2}{4^2 2!} + \frac{3^4}{4^4 4!} - \frac{3^6}{4^6 6!} + \ldots = 1 - \frac{1}{2!}\left(\frac{3}{4}\right)^2 + \frac{1}{4!}\left(\frac{3}{4}\right)^4 - \frac{1}{6!}\left(\frac{3}{4}\right)^6 + \ldots = \cos\left(\frac{3}{4}\right)$

45. $\frac{\pi}{3} - \frac{\pi^3}{3^3 3!} + \frac{\pi^5}{3^5 5!} - \frac{\pi^7}{3^7 7!} + \ldots = \frac{\pi}{3} - \frac{1}{3!}\left(\frac{\pi}{3}\right)^3 + \frac{1}{5!}\left(\frac{\pi}{3}\right)^5 - \frac{1}{7!}\left(\frac{\pi}{3}\right)^7 + \ldots = \sin\left(\frac{\pi}{3}\right) = \frac{\sqrt{3}}{2}$

47. $x^3 + x^4 + x^5 + x^6 + \ldots = x^3(1 + x + x^2 + x^3 + \ldots) = x^3\left(\frac{1}{1-x}\right) = \frac{x^3}{1-x}$

49. $x^3 - x^5 + x^7 - x^9 + \ldots = x^3\left(1 - x^2 + (x^2)^2 - (x^2)^3 + \ldots\right) = x^3\left(\frac{1}{1+x^2}\right) = \frac{x^3}{1+x^2}$

51. $-1 + 2x - 3x^2 + 4x^3 - 5x^4 + \ldots = \frac{d}{dx}(1 - x + x^2 - x^3 + x^4 - x^5 + \ldots) = \frac{d}{dx}\left(\frac{1}{1+x}\right) = \frac{-1}{(1+x)^2}$

53. $\ln\left(\frac{1+x}{1-x}\right) = \ln(1+x) - \ln(1-x) = \left(x - \frac{x^2}{2} + \frac{x^3}{3} - \frac{x^4}{4} + \ldots\right) - \left(-x - \frac{x^2}{2} - \frac{x^3}{3} - \frac{x^4}{4} - \ldots\right) = 2\left(x + \frac{x^3}{3} + \frac{x^5}{5} + \ldots\right)$

55. $\tan^{-1} x = x - \frac{x^3}{3} + \frac{x^5}{5} - \frac{x^7}{7} + \frac{x^9}{9} - \ldots + \frac{(-1)^{n-1}x^{2n-1}}{2n-1} + \ldots \Rightarrow |\text{error}| = \left|\frac{(-1)^{n-1}x^{2n-1}}{2n-1}\right| = \frac{1}{2n-1}$ when $x = 1$;
$\frac{1}{2n-1} < \frac{1}{10^3} \Rightarrow n > \frac{1001}{2} = 500.5 \Rightarrow$ the first term not used is the 501st $\Rightarrow$ we must use 500 terms

57. $\tan^{-1} x = x - \frac{x^3}{3} + \frac{x^5}{5} - \frac{x^7}{7} + \frac{x^9}{9} - \ldots + \frac{(-1)^{n-1}x^{2n-1}}{2n-1} + \ldots$ and when the series representing $48 \tan^{-1}\left(\frac{1}{18}\right)$ has an error less than $\frac{1}{3} \cdot 10^{-6}$, then the series representing the sum
$48 \tan^{-1}\left(\frac{1}{18}\right) + 32 \tan^{-1}\left(\frac{1}{57}\right) - 20 \tan^{-1}\left(\frac{1}{239}\right)$ also has an error of magnitude less than 10^{-6}; thus
$|\text{error}| = 48 \frac{\left(\frac{1}{18}\right)^{2n-1}}{2n-1} < \frac{1}{3 \cdot 10^6} \Rightarrow n \geq 4$ using a calculator $\Rightarrow$ 4 terms

59. (a) $(1 - x^2)^{-1/2} \approx 1 + \frac{x^2}{2} + \frac{3x^4}{8} + \frac{5x^6}{16} \Rightarrow \sin^{-1} x \approx x + \frac{x^3}{6} + \frac{3x^5}{40} + \frac{5x^7}{112}$; Using the Ratio Test:
$\lim_{n \to \infty} \left| \frac{1 \cdot 3 \cdot 5 \cdots (2n-1)(2n+1)x^{2n+3}}{2 \cdot 4 \cdot 6 \cdots (2n)(2n+2)(2n+3)} \cdot \frac{2 \cdot 4 \cdot 6 \cdots (2n)(2n+1)}{1 \cdot 3 \cdot 5 \cdots (2n-1)x^{2n+1}} \right| < 1 \Rightarrow x^2 \lim_{n \to \infty} \left| \frac{(2n+1)(2n+1)}{(2n+2)(2n+3)} \right| < 1$
$\Rightarrow |x| < 1 \Rightarrow$ the radius of convergence is 1. See Exercise 69.

(b) $\frac{d}{dx}(\cos^{-1} x) = -(1 - x^2)^{-1/2} \Rightarrow \cos^{-1} x = \frac{\pi}{2} - \sin^{-1} x \approx \frac{\pi}{2} - \left(x + \frac{x^3}{6} + \frac{3x^5}{40} + \frac{5x^7}{112}\right) \approx \frac{\pi}{2} - x - \frac{x^3}{6} - \frac{3x^5}{40} - \frac{5x^7}{112}$

61. $\frac{-1}{1+x} = -\frac{1}{1-(-x)} = -1 + x - x^2 + x^3 - \ldots \Rightarrow \frac{d}{dx}\left(\frac{-1}{1+x}\right) = \frac{1}{1+x^2} = \frac{d}{dx}(-1 + x - x^2 + x^3 - \ldots)$
$= 1 - 2x + 3x^2 - 4x^3 + \ldots$

63. Wallis' formula gives the approximation $\pi \approx 4\left[\frac{2 \cdot 4 \cdot 4 \cdot 6 \cdot 6 \cdot 8 \cdots (2n-2) \cdot (2n)}{3 \cdot 3 \cdot 5 \cdot 5 \cdot 7 \cdot 7 \cdots (2n-1) \cdot (2n-1)}\right]$ to produce the table

n	$\sim \pi$
10	3.221088998
20	3.181104886
30	3.167880758
80	3.151425420
90	3.150331383
93	3.150049112
94	3.149959030
95	3.149870848
100	3.149456425

At $n = 1929$ we obtain the first approximation accurate to 3 decimals: 3.141999845. At $n = 30{,}000$ we still do not obtain accuracy to 4 decimals: 3.141617732, so the convergence to π is very slow. Here is a Maple CAS procedure to produce these approximations:

```
pie :=
  proc(n)
  local i,j;
     a(2) := evalf(8/9);
     for i from 3 to n do a(i) := evalf(2*(2*i−2)*i/(2*i−1)^2*a(i−1)) od;
     [[j,4*a(j)] $ (j = n−5 .. n)]
  end
```

65. $(1 - x^2)^{-1/2} = (1 + (-x^2))^{-1/2} = (1)^{-1/2} + \left(-\frac{1}{2}\right)(1)^{-3/2}(-x^2) + \frac{\left(-\frac{1}{2}\right)\left(-\frac{3}{2}\right)(1)^{-5/2}(-x^2)^2}{2!}$
$+ \frac{\left(-\frac{1}{2}\right)\left(-\frac{3}{2}\right)\left(-\frac{5}{2}\right)(1)^{-7/2}(-x^2)^3}{3!} + \ldots = 1 + \frac{x^2}{2} + \frac{1 \cdot 3x^4}{2^2 \cdot 2!} + \frac{1 \cdot 3 \cdot 5x^6}{2^3 \cdot 3!} + \ldots = 1 + \sum_{n=1}^{\infty} \frac{1 \cdot 3 \cdot 5 \cdots (2n-1)x^{2n}}{2^n \cdot n!}$
$\Rightarrow \sin^{-1} x = \int_0^x (1 - t^2)^{-1/2}\, dt = \int_0^x \left(1 + \sum_{n=1}^{\infty} \frac{1 \cdot 3 \cdot 5 \cdots (2n-1)x^{2n}}{2^n \cdot n!}\right) dt = x + \sum_{n=1}^{\infty} \frac{1 \cdot 3 \cdot 5 \cdots (2n-1)x^{2n+1}}{2 \cdot 4 \cdots (2n)(2n+1)}$,
where $|x| < 1$

67. (a) $e^{-i\pi} = \cos(-\pi) + i \sin(-\pi) = -1 + i(0) = -1$
(b) $e^{i\pi/4} = \cos\left(\frac{\pi}{4}\right) + i \sin\left(\frac{\pi}{4}\right) = \frac{1}{\sqrt{2}} + \frac{i}{\sqrt{2}} = \left(\frac{1}{\sqrt{2}}\right)(1 + i)$

(c) $e^{-i\pi/2} = \cos\left(-\frac{\pi}{2}\right) + i\sin\left(-\frac{\pi}{2}\right) = 0 + i(-1) = -i$

69. $e^x = 1 + x + \frac{x^2}{2!} + \frac{x^3}{3!} + \frac{x^4}{4!} + \ldots \Rightarrow e^{i\theta} = 1 + i\theta + \frac{(i\theta)^2}{2!} + \frac{(i\theta)^3}{3!} + \frac{(i\theta)^4}{4!} + \ldots$ and
$e^{-i\theta} = 1 - i\theta + \frac{(-i\theta)^2}{2!} + \frac{(-i\theta)^3}{3!} + \frac{(-i\theta)^4}{4!} + \ldots = 1 - i\theta + \frac{(i\theta)^2}{2!} - \frac{(i\theta)^3}{3!} + \frac{(i\theta)^4}{4!} - \ldots$
$\Rightarrow \frac{e^{i\theta}+e^{-i\theta}}{2} = \frac{\left(1+i\theta+\frac{(i\theta)^2}{2!}+\frac{(i\theta)^3}{3!}+\frac{(i\theta)^4}{4!}+\ldots\right)+\left(1-i\theta+\frac{(i\theta)^2}{2!}-\frac{(i\theta)^3}{3!}+\frac{(i\theta)^4}{4!}-\ldots\right)}{2}$
$= 1 - \frac{\theta^2}{2!} + \frac{\theta^4}{4!} - \frac{\theta^6}{6!} + \ldots = \cos\theta;$
$\frac{e^{i\theta}-e^{-i\theta}}{2i} = \frac{\left(1+i\theta+\frac{(i\theta)^2}{2!}+\frac{(i\theta)^3}{3!}+\frac{(i\theta)^4}{4!}+\ldots\right)-\left(1-i\theta+\frac{(i\theta)^2}{2!}-\frac{(i\theta)^3}{3!}+\frac{(i\theta)^4}{4!}-\ldots\right)}{2i}$
$= \theta - \frac{\theta^3}{3!} + \frac{\theta^5}{5!} - \frac{\theta^7}{7!} + \ldots = \sin\theta$

71. $e^x \sin x = \left(1 + x + \frac{x^2}{2!} + \frac{x^3}{3!} + \frac{x^4}{4!} + \ldots\right)\left(x - \frac{x^3}{3!} + \frac{x^5}{5!} - \frac{x^7}{7!} + \ldots\right)$
$= (1)x + (1)x^2 + \left(-\frac{1}{6} + \frac{1}{2}\right)x^3 + \left(-\frac{1}{6} + \frac{1}{6}\right)x^4 + \left(\frac{1}{120} - \frac{1}{12} + \frac{1}{24}\right)x^5 + \ldots = x + x^2 + \frac{1}{3}x^3 - \frac{1}{30}x^5 + \ldots;$
$e^x \cdot e^{ix} = e^{(1+i)x} = e^x(\cos x + i\sin x) = e^x\cos x + i(e^x \sin x) \Rightarrow e^x \sin x$ is the series of the imaginary part of $e^{(1+i)x}$ which we calculate next; $e^{(1+i)x} = \sum_{n=0}^{\infty} \frac{(x+ix)^n}{n!} = 1 + (x + ix) + \frac{(x+ix)^2}{2!} + \frac{(x+ix)^3}{3!} + \frac{(x+ix)^4}{4!} + \ldots$
$= 1 + x + ix + \frac{1}{2!}(2ix^2) + \frac{1}{3!}(2ix^3 - 2x^3) + \frac{1}{4!}(-4x^4) + \frac{1}{5!}(-4x^5 - 4ix^5) + \frac{1}{6!}(-8ix^6) + \ldots \Rightarrow$ the imaginary part of $e^{(1+i)x}$ is $x + \frac{2}{2!}x^2 + \frac{2}{3!}x^3 - \frac{4}{5!}x^5 - \frac{8}{6!}x^6 + \ldots = x + x^2 + \frac{1}{3}x^3 - \frac{1}{30}x^5 - \frac{1}{90}x^6 + \ldots$ in agreement with our product calculation. The series for $e^x \sin x$ converges for all values of x.

73. (a) $e^{i\theta_1}e^{i\theta_2} = (\cos\theta_1 + i\sin\theta_1)(\cos\theta_2 + i\sin\theta_2) = (\cos\theta_1\cos\theta_2 - \sin\theta_1\sin\theta_2) + i(\sin\theta_1\cos\theta_2 + \sin\theta_2\cos\theta_1)$
$= \cos(\theta_1 + \theta_2) + i\sin(\theta_1 + \theta_2) = e^{i(\theta_1+\theta_2)}$
(b) $e^{-i\theta} = \cos(-\theta) + i\sin(-\theta) = \cos\theta - i\sin\theta = (\cos\theta - i\sin\theta)\left(\frac{\cos\theta + i\sin\theta}{\cos\theta + i\sin\theta}\right) = \frac{1}{\cos\theta + i\sin\theta} = \frac{1}{e^{i\theta}}$

CHAPTER 10 PRACTICE EXERCISES

1. converges to 1, since $\lim_{n\to\infty} a_n = \lim_{n\to\infty}\left(1 + \frac{(-1)^n}{n}\right) = 1$

3. converges to -1, since $\lim_{n\to\infty} a_n = \lim_{n\to\infty}\left(\frac{1-2^n}{2^n}\right) = \lim_{n\to\infty}\left(\frac{1}{2^n} - 1\right) = -1$

5. diverges, since $\left\{\sin\frac{n\pi}{2}\right\} = \{0, 1, 0, -1, 0, 1, \ldots\}$

7. converges to 0, since $\lim_{n\to\infty} a_n = \lim_{n\to\infty}\frac{\ln n^2}{n} = 2\lim_{n\to\infty}\frac{\left(\frac{1}{n}\right)}{1} = 0$

9. converges to 1, since $\lim_{n\to\infty} a_n = \lim_{n\to\infty}\left(\frac{n+\ln n}{n}\right) = \lim_{n\to\infty}\frac{1+\left(\frac{1}{n}\right)}{1} = 1$

11. converges to e^{-5}, since $\lim_{n\to\infty} a_n = \lim_{n\to\infty}\left(\frac{n-5}{n}\right)^n = \lim_{n\to\infty}\left(1 + \frac{(-5)}{n}\right)^n = e^{-5}$ by Theorem 5

13. converges to 3, since $\lim_{n\to\infty} a_n = \lim_{n\to\infty}\left(\frac{3^n}{n}\right)^{1/n} = \lim_{n\to\infty}\frac{3}{n^{1/n}} = \frac{3}{1} = 3$ by Theorem 5

15. converges to ln 2, since $\lim_{n\to\infty} a_n = \lim_{n\to\infty} n\left(2^{1/n} - 1\right) = \lim_{n\to\infty}\frac{2^{1/n}-1}{\left(\frac{1}{n}\right)} = \lim_{n\to\infty}\frac{\left[\frac{\left(-2^{1/n}\ln 2\right)}{n^2}\right]}{\left(\frac{-1}{n^2}\right)} = \lim_{n\to\infty} 2^{1/n}\ln 2$
$= 2^0 \cdot \ln 2 = \ln 2$

17. diverges, since $\lim_{n\to\infty} a_n = \lim_{n\to\infty} \frac{(n+1)!}{n!} = \lim_{n\to\infty} (n+1) = \infty$

19. $\frac{1}{(2n-3)(2n-1)} = \frac{\left(\frac{1}{2}\right)}{2n-3} - \frac{\left(\frac{1}{2}\right)}{2n-1} \Rightarrow s_n = \left[\frac{\left(\frac{1}{2}\right)}{3} - \frac{\left(\frac{1}{2}\right)}{5}\right] + \left[\frac{\left(\frac{1}{2}\right)}{5} - \frac{\left(\frac{1}{2}\right)}{7}\right] + \ldots + \left[\frac{\left(\frac{1}{2}\right)}{2n-3} - \frac{\left(\frac{1}{2}\right)}{2n-1}\right] = \frac{\left(\frac{1}{2}\right)}{3} - \frac{\left(\frac{1}{2}\right)}{2n-1}$

$\Rightarrow \lim_{n\to\infty} s_n = \lim_{n\to\infty} \left[\frac{1}{6} - \frac{\left(\frac{1}{2}\right)}{2n-1}\right] = \frac{1}{6}$

21. $\frac{9}{(3n-1)(3n+2)} = \frac{3}{3n-1} - \frac{3}{3n+2} \Rightarrow s_n = \left(\frac{3}{2} - \frac{3}{5}\right) + \left(\frac{3}{5} - \frac{3}{8}\right) + \left(\frac{3}{8} - \frac{3}{11}\right) + \ldots + \left(\frac{3}{3n-1} - \frac{3}{3n+2}\right)$
$= \frac{3}{2} - \frac{3}{3n+2} \Rightarrow \lim_{n\to\infty} s_n = \lim_{n\to\infty} \left(\frac{3}{2} - \frac{3}{3n+2}\right) = \frac{3}{2}$

23. $\sum_{n=0}^{\infty} e^{-n} = \sum_{n=0}^{\infty} \frac{1}{e^n}$, a convergent geometric series with $r = \frac{1}{e}$ and $a = 1 \Rightarrow$ the sum is $\frac{1}{1-\left(\frac{1}{e}\right)} = \frac{e}{e-1}$

25. diverges, a p-series with $p = \frac{1}{2}$

27. Since $f(x) = \frac{1}{x^{1/2}} \Rightarrow f'(x) = -\frac{1}{2x^{3/2}} < 0 \Rightarrow f(x)$ is decreasing $\Rightarrow a_{n+1} < a_n$, and $\lim_{n\to\infty} a_n = \lim_{n\to\infty} \frac{1}{\sqrt{n}} = 0$, the series $\sum_{n=1}^{\infty} \frac{(-1)^n}{\sqrt{n}}$ converges by the Alternating Series Test. Since $\sum_{n=1}^{\infty} \frac{1}{\sqrt{n}}$ diverges, the given series converges conditionally.

29. The given series does not converge absolutely by the Direct Comparison Test since $\frac{1}{\ln(n+1)} > \frac{1}{n+1}$, which is the nth term of a divergent series. Since $f(x) = \frac{1}{\ln(x+1)} \Rightarrow f'(x) = -\frac{1}{(\ln(x+1))^2(x+1)} < 0 \Rightarrow f(x)$ is decreasing $\Rightarrow a_{n+1} < a_n$, and $\lim_{n\to\infty} a_n = \lim_{n\to\infty} \frac{1}{\ln(n+1)} = 0$, the given series converges conditionally by the Alternating Series Test.

31. converges absolutely by the Direct Comparison Test since $\frac{\ln n}{n^3} < \frac{n}{n^3} = \frac{1}{n^2}$, the nth term of a convergent p-series

33. $\lim_{n\to\infty} \frac{\left(\frac{1}{n\sqrt{n^2+1}}\right)}{\left(\frac{1}{n^2}\right)} = \sqrt{\lim_{n\to\infty} \frac{n^2}{n^2+1}} = \sqrt{1} = 1 \Rightarrow$ converges absolutely by the Limit Comparison Test

35. converges absolutely by the Ratio Test since $\lim_{n\to\infty} \left[\frac{n+2}{(n+1)!} \cdot \frac{n!}{n+1}\right] = \lim_{n\to\infty} \frac{n+2}{(n+1)^2} = 0 < 1$

37. converges absolutely by the Ratio Test since $\lim_{n\to\infty} \left[\frac{3^{n+1}}{(n+1)!} \cdot \frac{n!}{3^n}\right] = \lim_{n\to\infty} \frac{3}{n+1} = 0 < 1$

39. converges absolutely by the Limit Comparison Test since $\lim_{n\to\infty} \frac{\left(\frac{1}{n^{3/2}}\right)}{\left(\frac{1}{\sqrt{n(n+1)(n+2)}}\right)} = \sqrt{\lim_{n\to\infty} \frac{n(n+1)(n+2)}{n^3}} = 1$

41. $\lim_{n\to\infty} \left|\frac{u_{n+1}}{u_n}\right| < 1 \Rightarrow \lim_{n\to\infty} \left|\frac{(x+4)^{n+1}}{(n+1)3^{n+1}} \cdot \frac{n3^n}{(x+4)^n}\right| < 1 \Rightarrow \frac{|x+4|}{3} \lim_{n\to\infty} \left(\frac{n}{n+1}\right) < 1 \Rightarrow \frac{|x+4|}{3} < 1$
$\Rightarrow |x+4| < 3 \Rightarrow -3 < x+4 < 3 \Rightarrow -7 < x < -1$; at $x = -7$ we have $\sum_{n=1}^{\infty} \frac{(-1)^n 3^n}{n3^n} = \sum_{n=1}^{\infty} \frac{(-1)^n}{n}$, the alternating harmonic series, which converges conditionally; at $x = -1$ we have $\sum_{n=1}^{\infty} \frac{3^n}{n3^n} = \sum_{n=1}^{\infty} \frac{1}{n}$, the divergent harmonic series

(a) the radius is 3; the interval of convergence is $-7 \le x < -1$
(b) the interval of absolute convergence is $-7 < x < -1$
(c) the series converges conditionally at $x = -7$

43. $\lim_{n\to\infty}\left|\frac{u_{n+1}}{u_n}\right| < 1 \Rightarrow \lim_{n\to\infty}\left|\frac{(3x-1)^{n+1}}{(n+1)^2}\cdot\frac{n^2}{(3x-1)^n}\right| < 1 \Rightarrow |3x-1|\lim_{n\to\infty}\frac{n^2}{(n+1)^2} < 1 \Rightarrow |3x-1| < 1$
$\Rightarrow -1 < 3x-1 < 1 \Rightarrow 0 < 3x < 2 \Rightarrow 0 < x < \frac{2}{3}$; at $x = 0$ we have $\sum_{n=1}^{\infty}\frac{(-1)^{n-1}(-1)^n}{n^2} = \sum_{n=1}^{\infty}\frac{(-1)^{2n-1}}{n^2}$
$= -\sum_{n=1}^{\infty}\frac{1}{n^2}$, a nonzero constant multiple of a convergent p-series, which is absolutely convergent; at $x = \frac{2}{3}$ we have $\sum_{n=1}^{\infty}\frac{(-1)^{n-1}(1)^n}{n^2} = \sum_{n=1}^{\infty}\frac{(-1)^{n-1}}{n^2}$, which converges absolutely
(a) the radius is $\frac{1}{3}$; the interval of convergence is $0 \le x \le \frac{2}{3}$
(b) the interval of absolute convergence is $0 \le x \le \frac{2}{3}$
(c) there are no values for which the series converges conditionally

45. $\lim_{n\to\infty}\left|\frac{u_{n+1}}{u_n}\right| < 1 \Rightarrow \lim_{n\to\infty}\left|\frac{x^{n+1}}{(n+1)^{n+1}}\cdot\frac{n^n}{x^n}\right| < 1 \Rightarrow |x|\lim_{n\to\infty}\left|\left(\frac{n}{n+1}\right)^n\left(\frac{1}{n+1}\right)\right| < 1 \Rightarrow \frac{|x|}{e}\lim_{n\to\infty}\left(\frac{1}{n+1}\right) < 1$
$\Rightarrow \frac{|x|}{e}\cdot 0 < 1$, which holds for all x
(a) the radius is ∞; the series converges for all x
(b) the series converges absolutely for all x
(c) there are no values for which the series converges conditionally

47. $\lim_{n\to\infty}\left|\frac{u_{n+1}}{u_n}\right| < 1 \Rightarrow \lim_{n\to\infty}\left|\frac{(n+2)x^{2n+1}}{3^{n+1}}\cdot\frac{3^n}{(n+1)x^{2n-1}}\right| < 1 \Rightarrow \frac{x^2}{3}\lim_{n\to\infty}\left(\frac{n+2}{n+1}\right) < 1 \Rightarrow -\sqrt{3} < x < \sqrt{3}$;
the series $\sum_{n=1}^{\infty} -\frac{n+1}{\sqrt{3}}$ and $\sum_{n=1}^{\infty}\frac{n+1}{\sqrt{3}}$, obtained with $x = \pm\sqrt{3}$, both diverge
(a) the radius is $\sqrt{3}$; the interval of convergence is $-\sqrt{3} < x < \sqrt{3}$
(b) the interval of absolute convergence is $-\sqrt{3} < x < \sqrt{3}$
(c) there are no values for which the series converges conditionally

49. $\lim_{n\to\infty}\left|\frac{u_{n+1}}{u_n}\right| < 1 \Rightarrow \lim_{n\to\infty}\left|\frac{\operatorname{csch}(n+1)x^{n+1}}{\operatorname{csch}(n)x^n}\right| < 1 \Rightarrow |x|\lim_{n\to\infty}\left|\frac{\left(\frac{2}{e^{n+1}-e^{-n-1}}\right)}{\left(\frac{2}{e^n-e^{-n}}\right)}\right| < 1$
$\Rightarrow |x|\lim_{n\to\infty}\left|\frac{e^{-1}-e^{-2n-1}}{1-e^{-2n-2}}\right| < 1 \Rightarrow \frac{|x|}{e} < 1 \Rightarrow -e < x < e$; the series $\sum_{n=1}^{\infty}(\pm e)^n \operatorname{csch} n$, obtained with $x = \pm e$, both diverge since $\lim_{n\to\infty}(\pm e)^n \operatorname{csch} n \ne 0$
(a) the radius is e; the interval of convergence is $-e < x < e$
(b) the interval of absolute convergence is $-e < x < e$
(c) there are no values for which the series converges conditionally

51. The given series has the form $1 - x + x^2 - x^3 + \ldots + (-x)^n + \ldots = \frac{1}{1+x}$, where $x = \frac{1}{4}$; the sum is $\frac{1}{1+\left(\frac{1}{4}\right)} = \frac{4}{5}$

53. The given series has the form $x - \frac{x^3}{3!} + \frac{x^5}{5!} - \ldots + (-1)^n\frac{x^{2n+1}}{(2n+1)!} + \ldots = \sin x$, where $x = \pi$; the sum is $\sin\pi = 0$

55. The given series has the form $1 + x + \frac{x^2}{2!} + \frac{x^2}{3!} + \ldots + \frac{x^n}{n!} + \ldots = e^x$, where $x = \ln 2$; the sum is $e^{\ln(2)} = 2$

57. Consider $\frac{1}{1-2x}$ as the sum of a convergent geometric series with $a = 1$ and $r = 2x \Rightarrow \frac{1}{1-2x}$
$= 1 + (2x) + (2x)^2 + (2x)^3 + \ldots = \sum_{n=0}^{\infty}(2x)^n = \sum_{n=0}^{\infty}2^n x^n$ where $|2x| < 1 \Rightarrow |x| < \frac{1}{2}$

59. $\sin x = \sum_{n=0}^{\infty}\frac{(-1)^n x^{2n+1}}{(2n+1)!} \Rightarrow \sin\pi x = \sum_{n=0}^{\infty}\frac{(-1)^n(\pi x)^{2n+1}}{(2n+1)!} = \sum_{n=0}^{\infty}\frac{(-1)^n\pi^{2n+1}x^{2n+1}}{(2n+1)!}$

61. $\cos x = \sum_{n=0}^{\infty} \frac{(-1)^n x^{2n}}{(2n)!} \Rightarrow \cos\left(x^{5/3}\right) = \sum_{n=0}^{\infty} \frac{(-1)^n \left(x^{5/3}\right)^{2n}}{(2n)!} = \sum_{n=0}^{\infty} \frac{(-1)^n x^{10n/3}}{(2n)!}$

63. $e^x = \sum_{n=0}^{\infty} \frac{x^n}{n!} \Rightarrow e^{(\pi x/2)} = \sum_{n=0}^{\infty} \frac{\left(\frac{\pi x}{2}\right)^n}{n!} = \sum_{n=0}^{\infty} \frac{\pi^n x^n}{2^n n!}$

65. $f(x) = \sqrt{3+x^2} = (3+x^2)^{1/2} \Rightarrow f'(x) = x\,(3+x^2)^{-1/2} \Rightarrow f''(x) = -x^2\,(3+x^2)^{-3/2} + (3+x^2)^{-1/2}$
$\Rightarrow f'''(x) = 3x^3\,(3+x^2)^{-5/2} - 3x\,(3+x^2)^{-3/2}$; $f(-1) = 2$, $f'(-1) = -\frac{1}{2}$, $f''(-1) = -\frac{1}{8} + \frac{1}{2} = \frac{3}{8}$,
$f'''(-1) = -\frac{3}{32} + \frac{3}{8} = \frac{9}{32} \Rightarrow \sqrt{3+x^2} = 2 - \frac{(x+1)}{2 \cdot 1!} + \frac{3(x+1)^2}{2^3 \cdot 2!} + \frac{9(x+1)^3}{2^5 \cdot 3!} + \ldots$

67. $f(x) = \frac{1}{x+1} = (x+1)^{-1} \Rightarrow f'(x) = -(x+1)^{-2} \Rightarrow f''(x) = 2(x+1)^{-3} \Rightarrow f'''(x) = -6(x+1)^{-4}$; $f(3) = \frac{1}{4}$,
$f'(3) = -\frac{1}{4^2}$, $f''(3) = \frac{2}{4^3}$, $f'''(2) = \frac{-6}{4^4} \Rightarrow \frac{1}{x+1} = \frac{1}{4} - \frac{1}{4^2}(x-3) + \frac{1}{4^3}(x-3)^2 - \frac{1}{4^4}(x-3)^3 + \ldots$

69. $\int_0^{1/2} \exp(-x^3)\,dx = \int_0^{1/2} \left(1 - x^3 + \frac{x^6}{2!} - \frac{x^9}{3!} + \frac{x^{12}}{4!} + \ldots\right) dx = \left[x - \frac{x^4}{4} + \frac{x^7}{7 \cdot 2!} - \frac{x^{10}}{10 \cdot 3!} + \frac{x^{13}}{13 \cdot 4!} - \ldots\right]_0^{1/2}$
$\approx \frac{1}{2} - \frac{1}{2^4 \cdot 4} + \frac{1}{2^7 \cdot 7 \cdot 2!} - \frac{1}{2^{10} \cdot 10 \cdot 3!} + \frac{1}{2^{13} \cdot 13 \cdot 4!} - \frac{1}{2^{16} \cdot 16 \cdot 5!} \approx 0.484917143$

71. $\int_1^{1/2} \frac{\tan^{-1} x}{x}\,dx = \int_1^{1/2} \left(1 - \frac{x^2}{3} + \frac{x^4}{5} - \frac{x^6}{7} + \frac{x^8}{9} - \frac{x^{10}}{11} + \ldots\right) dx = \left[x - \frac{x^3}{9} + \frac{x^5}{25} - \frac{x^7}{49} + \frac{x^9}{81} - \frac{x^{11}}{121} + \ldots\right]_0^{1/2}$
$\approx \frac{1}{2} - \frac{1}{9 \cdot 2^3} + \frac{1}{5^2 \cdot 2^5} - \frac{1}{7^2 \cdot 2^7} + \frac{1}{9^2 \cdot 2^9} - \frac{1}{11^2 \cdot 2^{11}} + \frac{1}{13^2 \cdot 2^{13}} - \frac{1}{15^2 \cdot 2^{15}} + \frac{1}{17^2 \cdot 2^{17}} - \frac{1}{19^2 \cdot 2^{19}} + \frac{1}{21^2 \cdot 2^{21}} \approx 0.4872223583$

73. $\lim_{x \to 0} \frac{7 \sin x}{e^{2x} - 1} = \lim_{x \to 0} \frac{7\left(x - \frac{x^3}{3!} + \frac{x^5}{5!} - \ldots\right)}{\left(2x + \frac{2^2 x^2}{2!} + \frac{2^3 x^3}{3!} + \ldots\right)} = \lim_{x \to 0} \frac{7\left(1 - \frac{x^2}{3!} + \frac{x^4}{5!} - \ldots\right)}{\left(2 + \frac{2^2 x}{2!} + \frac{2^3 x^2}{3!} + \ldots\right)} = \frac{7}{2}$

75. $\lim_{t \to 0} \left(\frac{1}{2 - 2\cos t} - \frac{1}{t^2}\right) = \lim_{t \to 0} \frac{t^2 - 2 + 2\cos t}{2t^2(1 - \cos t)} = \lim_{t \to 0} \frac{t^2 - 2 + 2\left(1 - \frac{t^2}{2} + \frac{t^4}{4!} - \ldots\right)}{2t^2\left(1 - 1 + \frac{t^2}{2} - \frac{t^4}{4!} + \ldots\right)} = \lim_{t \to 0} \frac{2\left(\frac{t^4}{4!} - \frac{t^6}{6!} + \ldots\right)}{\left(t^4 - \frac{2t^6}{4!} + \ldots\right)}$
$= \lim_{t \to 0} \frac{2\left(\frac{1}{4!} - \frac{t^2}{6!} + \ldots\right)}{\left(1 - \frac{2t^2}{4!} + \ldots\right)} = \frac{1}{12}$

77. $\lim_{z \to 0} \frac{1 - \cos^2 z}{\ln(1 - z) + \sin z} = \lim_{z \to 0} \frac{1 - \left(1 - z^2 + \frac{z^4}{3} - \ldots\right)}{\left(-z - \frac{z^2}{2} - \frac{z^3}{3} - \ldots\right) + \left(z - \frac{z^3}{3!} + \frac{z^5}{5!} - \ldots\right)} = \lim_{z \to 0} \frac{\left(z^2 - \frac{z^4}{3} + \ldots\right)}{\left(-\frac{z^2}{2} - \frac{2z^3}{3} - \frac{z^4}{4} - \ldots\right)}$
$= \lim_{z \to 0} \frac{\left(1 - \frac{z^2}{3} + \ldots\right)}{\left(-\frac{1}{2} - \frac{2z}{3} - \frac{z^2}{4} - \ldots\right)} = -2$

79. $\lim_{x \to 0} \left(\frac{\sin 3x}{x^3} + \frac{r}{x^2} + s\right) = \lim_{x \to 0} \left[\frac{\left(3x - \frac{(3x)^3}{6} + \frac{(3x)^5}{120} - \ldots\right)}{x^3} + \frac{r}{x^2} + s\right] = \lim_{x \to 0} \left(\frac{3}{x^2} - \frac{9}{2} + \frac{81x^2}{40} + \ldots + \frac{r}{x^2} + s\right) = 0$
$\Rightarrow \frac{r}{x^2} + \frac{3}{x^2} = 0$ and $s - \frac{9}{2} = 0 \Rightarrow r = -3$ and $s = \frac{9}{2}$

81. $\lim_{n \to \infty} \left|\frac{2 \cdot 5 \cdot 8 \cdots (3n-1)(3n+2)x^{n+1}}{2 \cdot 4 \cdot 6 \cdots (2n)(2n+2)} \cdot \frac{2 \cdot 4 \cdot 6 \cdots (2n)}{2 \cdot 5 \cdot 8 \cdots (3n-1)x^n}\right| < 1 \Rightarrow |x| \lim_{n \to \infty} \left|\frac{3n+2}{2n+2}\right| < 1 \Rightarrow |x| < \frac{2}{3}$
$\Rightarrow$ the radius of convergence is $\frac{2}{3}$

83. $\sum_{k=2}^{n} \ln\left(1 - \frac{1}{k^2}\right) = \sum_{k=2}^{n} \left[\ln\left(1 + \frac{1}{k}\right) + \ln\left(1 - \frac{1}{k}\right)\right] = \sum_{k=2}^{n} \left[\ln(k+1) - \ln k + \ln(k-1) - \ln k\right]$
$= [\ln 3 - \ln 2 + \ln 1 - \ln 2] + [\ln 4 - \ln 3 + \ln 2 - \ln 3] + [\ln 5 - \ln 4 + \ln 3 - \ln 4] + [\ln 6 - \ln 5 + \ln 4 - \ln 5]$

$+ \ldots + [\ln(n+1) - \ln n + \ln(n-1) - \ln n] = [\ln 1 - \ln 2] + [\ln(n+1) - \ln n]$ after cancellation

$\Rightarrow \sum_{k=2}^{n} \ln\left(1 - \frac{1}{k^2}\right) = \ln\left(\frac{n+1}{2n}\right) \Rightarrow \sum_{k=2}^{\infty} \ln\left(1 - \frac{1}{k^2}\right) = \lim_{n \to \infty} \ln\left(\frac{n+1}{2n}\right) = \ln \frac{1}{2}$ is the sum

85. (a) $\lim_{n \to \infty} \left| \frac{1 \cdot 4 \cdot 7 \cdots (3n-2)(3n+1)x^{3n+3}}{(3n+3)!} \cdot \frac{(3n)!}{1 \cdot 4 \cdot 7 \cdots (3n-2)x^{3n}} \right| < 1 \Rightarrow |x^3| \lim_{n \to \infty} \frac{(3n+1)}{(3n+1)(3n+2)(3n+3)}$
$= |x^3| \cdot 0 < 1 \Rightarrow$ the radius of convergence is ∞

(b) $y = 1 + \sum_{n=1}^{\infty} \frac{1 \cdot 4 \cdot 7 \cdots (3n-2)}{(3n)!} x^{3n} \Rightarrow \frac{dy}{dx} = \sum_{n=1}^{\infty} \frac{1 \cdot 4 \cdot 7 \cdots (3n-2)}{(3n-1)!} x^{3n-1}$

$\Rightarrow \frac{d^2y}{dx^2} = \sum_{n=1}^{\infty} \frac{1 \cdot 4 \cdot 7 \cdots (3n-2)}{(3n-2)!} x^{3n-2} = x + \sum_{n=2}^{\infty} \frac{1 \cdot 4 \cdot 7 \cdots (3n-5)}{(3n-3)!} x^{3n-2}$

$= x\left(1 + \sum_{n=1}^{\infty} \frac{1 \cdot 4 \cdot 7 \cdots (3n-2)}{(3n)!} x^{3n}\right) = xy + 0 \Rightarrow a = 1$ and $b = 0$

87. Yes, the series $\sum_{n=1}^{\infty} a_n b_n$ converges as we now show. Since $\sum_{n=1}^{\infty} a_n$ converges it follows that $a_n \to 0 \Rightarrow a_n < 1$ for $n >$ some index $N \Rightarrow a_n b_n < b_n$ for $n > N \Rightarrow \sum_{n=1}^{\infty} a_n b_n$ converges by the Direct Comparison Test with $\sum_{n=1}^{\infty} b_n$

89. $\sum_{n=1}^{\infty} (x_{n+1} - x_n) = \lim_{n \to \infty} \sum_{k=1}^{n} (x_{k+1} - x_k) = \lim_{n \to \infty} (x_{n+1} - x_1) = \lim_{n \to \infty} (x_{n+1}) - x_1 \Rightarrow$ both the series and sequence must either converge or diverge.

91. $\sum_{n=1}^{\infty} \frac{a_n}{n} = a_1 + \frac{a_2}{2} + \frac{a_3}{3} + \frac{a_4}{4} + \ldots \geq a_1 + \left(\frac{1}{2}\right) a_2 + \left(\frac{1}{3} + \frac{1}{4}\right) a_4 + \left(\frac{1}{5} + \frac{1}{6} + \frac{1}{7} + \frac{1}{8}\right) a_8$
$+ \left(\frac{1}{9} + \frac{1}{10} + \frac{1}{11} + \ldots + \frac{1}{16}\right) a_{16} + \ldots \geq \frac{1}{2}(a_2 + a_4 + a_8 + a_{16} + \ldots)$ which is a divergent series

CHAPTER 10 ADDITIONAL AND ADVANCED EXERCISES

1. converges since $\frac{1}{(3n-2)^{(2n+1)/2}} < \frac{1}{(3n-2)^{3/2}}$ and $\sum_{n=1}^{\infty} \frac{1}{(3n-2)^{3/2}}$ converges by the Limit Comparison Test:

$$\lim_{n \to \infty} \frac{\left(\frac{1}{n^{3/2}}\right)}{\left(\frac{1}{(3n-2)^{3/2}}\right)} = \lim_{n \to \infty} \left(\frac{3n-2}{n}\right)^{3/2} = 3^{3/2}$$

3. diverges by the nth-Term Test since $\lim_{n \to \infty} a_n = \lim_{n \to \infty} (-1)^n \tanh n = \lim_{b \to \infty} (-1)^n \left(\frac{1-e^{-2n}}{1+e^{-2n}}\right) = \lim_{n \to \infty} (-1)^n$ does not exist

5. converges by the Direct Comparison Test: $a_1 = 1 = \frac{12}{(1)(3)(2)^2}$, $a_2 = \frac{1 \cdot 2}{3 \cdot 4} = \frac{12}{(2)(4)(3)^2}$, $a_3 = \left(\frac{2 \cdot 3}{4 \cdot 5}\right)\left(\frac{1 \cdot 2}{3 \cdot 4}\right)$
$= \frac{12}{(3)(5)(4)^2}$, $a_4 = \left(\frac{3 \cdot 4}{5 \cdot 6}\right)\left(\frac{2 \cdot 3}{4 \cdot 5}\right)\left(\frac{1 \cdot 2}{3 \cdot 4}\right) = \frac{12}{(4)(6)(5)^2}, \ldots \Rightarrow 1 + \sum_{n=1}^{\infty} \frac{12}{(n+1)(n+3)(n+2)^2}$ represents the given series and $\frac{12}{(n+1)(n+3)(n+2)^2} < \frac{12}{n^4}$, which is the nth-term of a convergent p-series

7. diverges by the nth-Term Test since if $a_n \to L$ as $n \to \infty$, then $L = \frac{1}{1+L} \Rightarrow L^2 + L - 1 = 0 \Rightarrow L = \frac{-1 \pm \sqrt{5}}{2} \neq 0$

9. $f(x) = \cos x$ with $a = \frac{\pi}{3} \Rightarrow f\left(\frac{\pi}{3}\right) = 0.5$, $f'\left(\frac{\pi}{3}\right) = -\frac{\sqrt{3}}{2}$, $f''\left(\frac{\pi}{3}\right) = -0.5$, $f'''\left(\frac{\pi}{3}\right) = \frac{\sqrt{3}}{2}$, $f^{(4)}\left(\frac{\pi}{3}\right) = 0.5$;
$\cos x = \frac{1}{2} - \frac{\sqrt{3}}{2}\left(x - \frac{\pi}{3}\right) - \frac{1}{4}\left(x - \frac{\pi}{3}\right)^2 + \frac{\sqrt{3}}{12}\left(x - \frac{\pi}{3}\right)^3 + \ldots$

11. $e^x = 1 + x + \frac{x^2}{2!} + \frac{x^3}{3!} + \ldots$ with $a = 0$

13. $f(x) = \cos x$ with $a = 22\pi \Rightarrow f(22\pi) = 1, f'(22\pi) = 0, f''(22\pi) = -1, f'''(22\pi) = 0, f^{(4)}(22\pi) = 1,$ $f^{(5)}(22\pi) = 0, f^{(6)}(22\pi) = -1; \cos x = 1 - \frac{1}{2}(x - 22\pi)^2 + \frac{1}{4!}(x - 22\pi)^4 - \frac{1}{6!}(x - 22\pi)^6 + \ldots$

15. Yes, the sequence converges: $c_n = (a^n + b^n)^{1/n} \Rightarrow c_n = b\left(\left(\frac{a}{b}\right)^n + 1\right)^{1/n} \Rightarrow \lim_{n\to\infty} c_n = \ln b + \lim_{n\to\infty} \frac{\ln\left(\left(\frac{a}{b}\right)^n + 1\right)}{n}$ $= \ln b + \lim_{n\to\infty} \frac{\left(\frac{a}{b}\right)^n \ln\left(\frac{a}{b}\right)}{\left(\frac{a}{b}\right)^n + 1} = \ln b + \frac{0 \cdot \ln\left(\frac{a}{b}\right)}{0+1} = \ln b$ since $0 < a < b$. Thus, $\lim_{n\to\infty} c_n = e^{\ln b} = b$.

17. $s_n = \sum_{k=0}^{n-1} \int_k^{k+1} \frac{dx}{1+x^2} \Rightarrow s_n = \int_0^1 \frac{dx}{1+x^2} + \int_1^2 \frac{dx}{1+x^2} + \ldots + \int_{n-1}^n \frac{dx}{1+x^2} \Rightarrow s_n = \int_0^n \frac{dx}{1+x^2}$ $\Rightarrow \lim_{n\to\infty} s_n = \lim_{n\to\infty} (\tan^{-1} n - \tan^{-1} 0) = \frac{\pi}{2}$

19. (a) No, the limit does not appear to depend on the value of the constant a
(b) Yes, the limit depends on the value of b
(c) $s = \left(1 - \frac{\cos\left(\frac{a}{n}\right)}{n}\right)^n \Rightarrow \ln s = \frac{\ln\left(1 - \frac{\cos\left(\frac{a}{n}\right)}{n}\right)}{\left(\frac{1}{n}\right)} \Rightarrow \lim_{n\to\infty} \ln s = \frac{\left(\frac{1}{1 - \frac{\cos\left(\frac{a}{n}\right)}{n}}\right)\left(\frac{-\frac{a}{n}\sin\left(\frac{a}{n}\right) + \cos\left(\frac{a}{n}\right)}{n^2}\right)}{\left(-\frac{1}{n^2}\right)}$ $= \lim_{n\to\infty} \frac{\frac{a}{n}\sin\left(\frac{a}{n}\right) - \cos\left(\frac{a}{n}\right)}{1 - \frac{\cos\left(\frac{a}{n}\right)}{n}} = \frac{0-1}{1-0} = -1 \Rightarrow \lim_{n\to\infty} s = e^{-1} \approx 0.3678794412$; similarly, $\lim_{n\to\infty} \left(1 - \frac{\cos\left(\frac{a}{n}\right)}{bn}\right)^n = e^{-1/b}$

21. $\lim_{n\to\infty} \left|\frac{u_{n+1}}{u_n}\right| < 1 \Rightarrow \lim_{n\to\infty} \left|\frac{b^{n+1}x^{n+1}}{\ln(n+1)} \cdot \frac{\ln n}{b^n x^n}\right| < 1 \Rightarrow |bx| < 1 \Rightarrow -\frac{1}{b} < x < \frac{1}{b} = 5 \Rightarrow b = \pm\frac{1}{5}$

23. $\lim_{x\to 0} \frac{\sin(ax) - \sin x - x}{x^3} = \lim_{x\to 0} \frac{\left(ax - \frac{a^3x^3}{3!} + \ldots\right) - \left(x - \frac{x^3}{3!} + \ldots\right) - x}{x^3} = \lim_{x\to 0} \left[\frac{a-2}{x^2} - \frac{a^3}{3!} + \frac{1}{3!} - \left(\frac{a^5}{5!} - \frac{1}{5!}\right)x^2 + \ldots\right]$ is finite if $a - 2 = 0 \Rightarrow a = 2$; $\lim_{x\to 0} \frac{\sin 2x - \sin x - x}{x^3} = -\frac{2^3}{3!} + \frac{1}{3!} = -\frac{7}{6}$

25. (a) $\frac{u_n}{u_{n+1}} = \frac{(n+1)^2}{n^2} = 1 + \frac{2}{n} + \frac{1}{n^2} \Rightarrow C = 2 > 1$ and $\sum_{n=1}^{\infty} \frac{1}{n^2}$ converges
(b) $\frac{u_n}{u_{n+1}} = \frac{n+1}{n} = 1 + \frac{1}{n} + \frac{0}{n^2} \Rightarrow C = 1 \le 1$ and $\sum_{n=1}^{\infty} \frac{1}{n}$ diverges

27. (a) $\sum_{n=1}^{\infty} a_n = L \Rightarrow a_n^2 \le a_n \sum_{n=1}^{\infty} a_n = a_n L \Rightarrow \sum_{n=1}^{\infty} a_n^2$ converges by the Direct Comparison Test
(b) converges by the Limit Comparison Test: $\lim_{n\to\infty} \frac{\left(\frac{a_n}{1-a_n}\right)}{a_n} = \lim_{n\to\infty} \frac{1}{1-a_n} = 1$ since $\sum_{n=1}^{\infty} a_n$ converges and therefore $\lim_{x\to\infty} a_n = 0$

29. $(1-x)^{-1} = 1 + \sum_{n=1}^{\infty} x^n$ where $|x| < 1 \Rightarrow \frac{1}{(1-x)^2} = \frac{d}{dx}(1-x)^{-1} = \sum_{n=1}^{\infty} nx^{n-1}$ and when $x = \frac{1}{2}$ we have $4 = 1 + 2\left(\frac{1}{2}\right) + 3\left(\frac{1}{2}\right)^2 + 4\left(\frac{1}{2}\right)^3 + \ldots + n\left(\frac{1}{2}\right)^{n-1} + \ldots$

31. (a) $\frac{1}{(1-x)^2} = \frac{d}{dx}\left(\frac{1}{1-x}\right) = \frac{d}{dx}(1 + x + x^2 + x^3 + \ldots) = 1 + 2x + 3x^2 + 4x^3 + \ldots = \sum_{n=1}^{\infty} nx^{n-1}$
(b) from part (a) we have $\sum_{n=1}^{\infty} n\left(\frac{5}{6}\right)^{n-1}\left(\frac{1}{6}\right) = \left(\frac{1}{6}\right)\left[\frac{1}{1-\left(\frac{5}{6}\right)}\right]^2 = 6$
(c) from part (a) we have $\sum_{n=1}^{\infty} np^{n-1}q = \frac{q}{(1-p)^2} = \frac{q}{q^2} = \frac{1}{q}$

33. (a) $R_n = C_0e^{-kt_0} + C_0e^{-2kt_0} + \ldots + C_0e^{-nkt_0} = \frac{C_0e^{-kt_0}\left(1-e^{-nkt_0}\right)}{1-e^{-kt_0}} \Rightarrow R = \lim_{n \to \infty} R_n = \frac{C_0e^{-kt_0}}{1-e^{-kt_0}} = \frac{C_0}{e^{kt_0}-1}$

(b) $R_n = \frac{e^{-1}\left(1-e^{-n}\right)}{1-e^{-1}} \Rightarrow R_1 = e^{-1} \approx 0.36787944$ and $R_{10} = \frac{e^{-1}\left(1-e^{-10}\right)}{1-e^{-1}} \approx 0.58195028$;
$R = \frac{1}{e-1} \approx 0.58197671$; $R - R_{10} \approx 0.00002643 \Rightarrow \frac{R-R_{10}}{R} < 0.0001$

(c) $R_n = \frac{e^{-.1}\left(1-e^{-.1n}\right)}{1-e^{-.1}}$, $\frac{R}{2} = \frac{1}{2}\left(\frac{1}{e^{.1}-1}\right) \approx 4.7541659$; $R_n > \frac{R}{2} \Rightarrow \frac{1-e^{-.1n}}{e^{.1}-1} > \left(\frac{1}{2}\right)\left(\frac{1}{e^{.1}-1}\right)$
$\Rightarrow 1 - e^{-n/10} > \frac{1}{2} \Rightarrow e^{-n/10} < \frac{1}{2} \Rightarrow -\frac{n}{10} < \ln\left(\frac{1}{2}\right) \Rightarrow \frac{n}{10} > -\ln\left(\frac{1}{2}\right) \Rightarrow n > 6.93 \Rightarrow n = 7$

CHAPTER 11 PARAMETRIC EQUATIONS AND POLAR COORDINATES

11.1 PARAMETRIZATIONS OF PLANE CURVES

1. $x = 3t,\ y = 9t^2,\ -\infty < t < \infty \Rightarrow y = x^2$

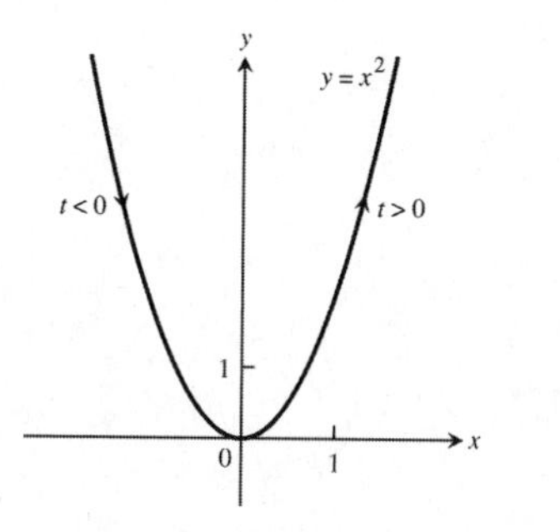

3. $x = 2t - 5,\ y = 4t - 7,\ -\infty < t < \infty$
$\Rightarrow x + 5 = 2t \Rightarrow 2(x + 5) = 4t$
$\Rightarrow y = 2(x + 5) - 7 \Rightarrow y = 2x + 3$

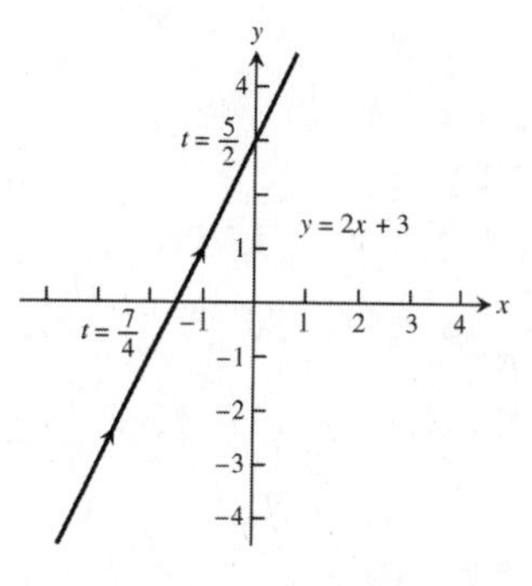

5. $x = \cos 2t,\ y = \sin 2t,\ 0 \le t \le \pi$
$\Rightarrow \cos^2 2t + \sin^2 2t = 1 \Rightarrow x^2 + y^2 = 1$

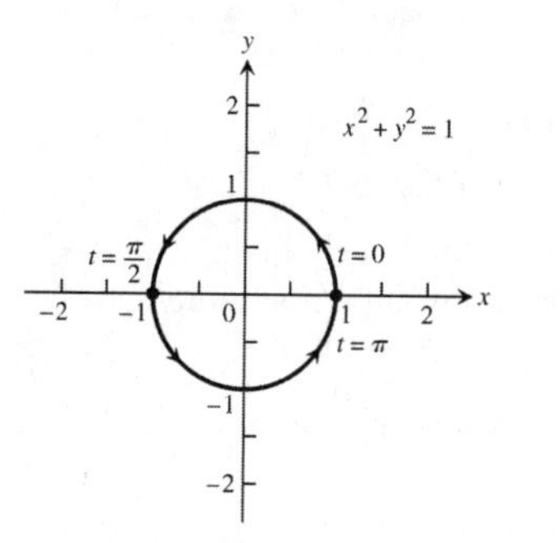

7. $x = 4 \cos t,\ y = 2 \sin t,\ 0 \le t \le 2\pi$
$\Rightarrow \frac{16 \cos^2 t}{16} + \frac{4 \sin^2 t}{4} = 1 \Rightarrow \frac{x^2}{16} + \frac{y^2}{4} = 1$

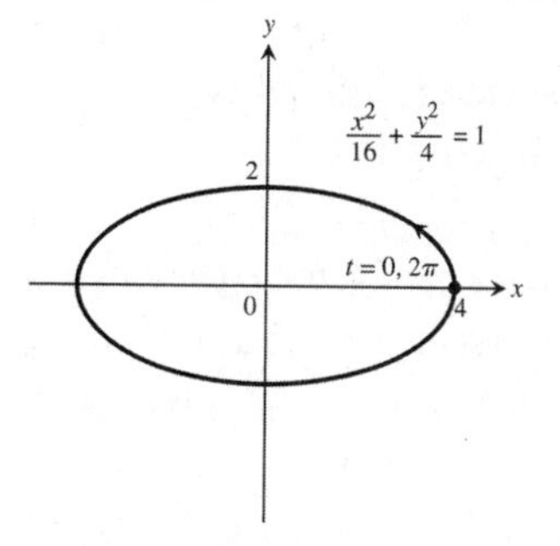

9. $x = \sin t,\ y = \cos 2t,\ -\frac{\pi}{2} \le t \le \frac{\pi}{2}$
$\Rightarrow y = \cos 2t = 1 - 2\sin^2 t \Rightarrow y = 1 - 2x^2$

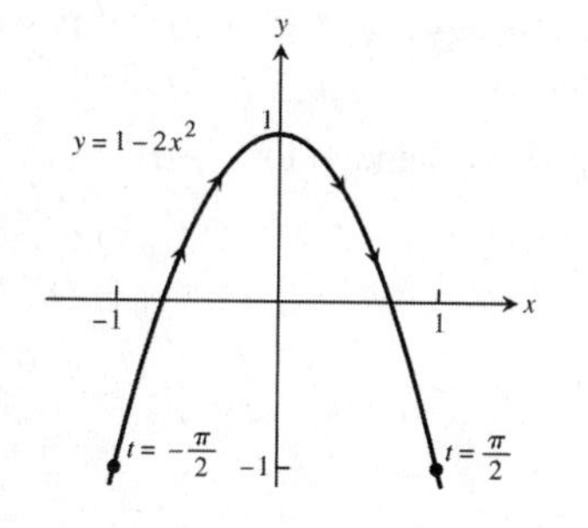

11. $x = t^2,\ y = t^6 - 2t^4,\ -\infty < t < \infty$
$\Rightarrow y = (t^2)^3 - 2(t^2)^2 \Rightarrow y = x^3 - 2x^2$

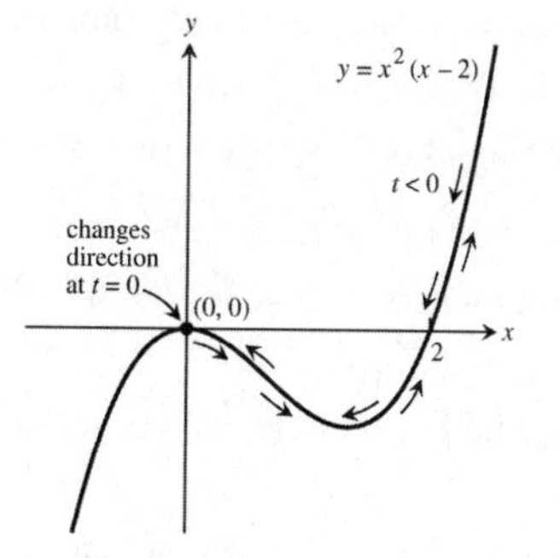

13. $x = t,\ y = \sqrt{1 - t^2},\ -1 \le t \le 0$
$\Rightarrow\ y = \sqrt{1 - x^2}$

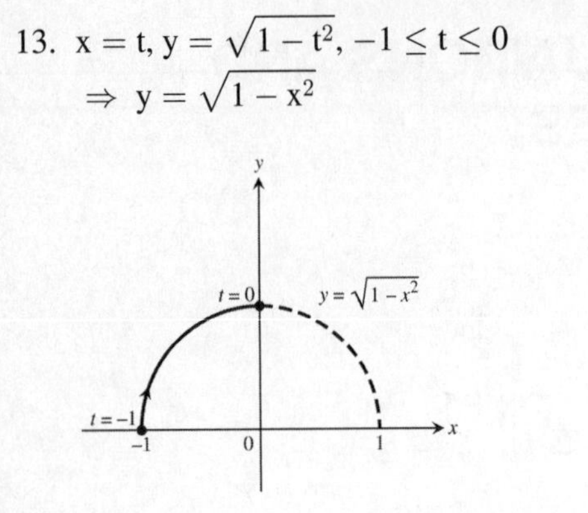

15. $x = \sec^2 t - 1,\ y = \tan t,\ -\frac{\pi}{2} < t < \frac{\pi}{2}$
$\Rightarrow\ \sec^2 t - 1 = \tan^2 t\ \Rightarrow\ x = y^2$

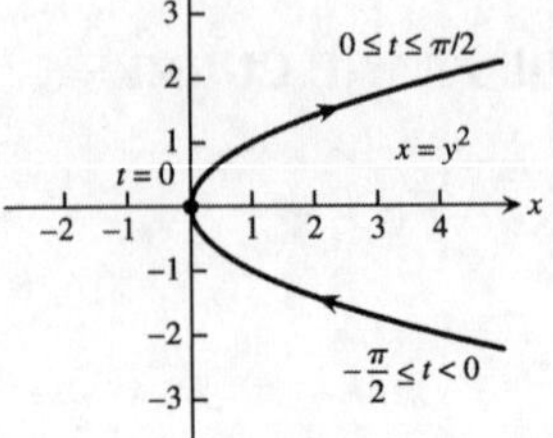

17. $x = -\cosh t,\ y = \sinh t,\ -\infty < 1 < \infty$
$\Rightarrow\ \cosh^2 t - \sinh^2 t = 1\ \Rightarrow\ x^2 - y^2 = 1$

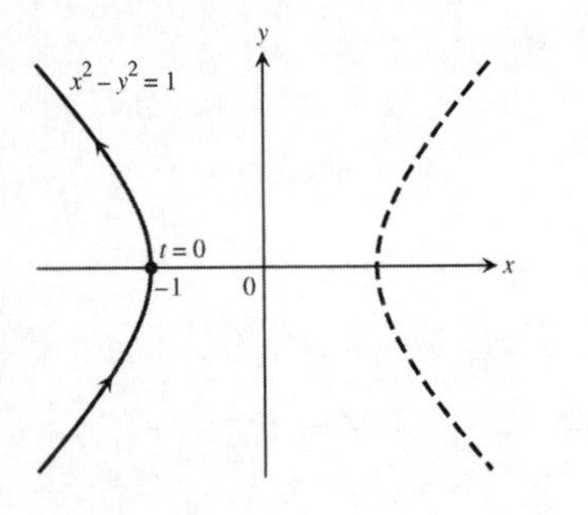

19. (a) $x = a \cos t,\ y = -a \sin t,\ 0 \le t \le 2\pi$
(b) $x = a \cos t,\ y = a \sin t,\ 0 \le t \le 2\pi$
(c) $x = a \cos t,\ y = -a \sin t,\ 0 \le t \le 4\pi$
(d) $x = a \cos t,\ y = a \sin t,\ 0 \le t \le 4\pi$

21. Using $(-1, -3)$ we create the parametric equations $x = -1 + at$ and $y = -3 + bt$, representing a line which goes through $(-1, -3)$ at $t = 0$. We determine a and b so that the line goes through $(4, 1)$ when $t = 1$. Since $4 = -1 + a \Rightarrow a = 5$. Since $1 = -3 + b \Rightarrow b = 4$. Therefore, one possible parameterization is $x = -1 + 5t$, $y = -3 + 4t,\ 0 \le t \le 1$.

23. The lower half of the parabola is given by $x = y^2 + 1$ for $y \le 0$. Substituting t for y, we obtain one possible parameterization $x = t^2 + 1,\ y = t,\ t \le 0$.

25. For simplicity, we assume that x and y are linear functions of t and that the point(x, y) starts at $(2, 3)$ for $t = 0$ and passes through $(-1, -1)$ at $t = 1$. Then $x = f(t)$, where $f(0) = 2$ and $f(1) = -1$.
Since slope $= \frac{\Delta x}{\Delta t} = \frac{-1-2}{1-0} = -3$, $x = f(t) = -3t + 2 = 2 - 3t$. Also, $y = g(t)$, where $g(0) = 3$ and $g(1) = -1$.
Since slope $= \frac{\Delta y}{\Delta t} = \frac{-1-3}{1-0} = -4$. $y = g(t) = -4t + 3 = 3 - 4t$.
One possible parameterization is: $x = 2 - 3t,\ y = 3 - 4t,\ t \ge 0$.

27. Since we only want the top half of a circle, $y \ge 0$, so let $x = 2\cos t,\ y = 2|\sin t|,\ 0 \le t \le 4\pi$

29. $x^2 + y^2 = a^2\ \Rightarrow\ 2x + 2y \frac{dy}{dx} = 0 \Rightarrow \frac{dy}{dx} = -\frac{x}{y}$; let $t = \frac{dy}{dx}\ \Rightarrow\ -\frac{x}{y} = t \Rightarrow x = -yt$. Substitution yields $y^2t^2 + y^2 = a^2\ \Rightarrow\ y = \frac{a}{\sqrt{1+t^2}}$ and $x = \frac{-at}{\sqrt{1+t}}$, $-\infty < t < \infty$

31. Drop a vertical line from the point (x, y) to the x-axis, then θ is an angle in a right triangle, and from trigonometry we know that $\tan\theta = \frac{y}{x} \Rightarrow y = x\tan\theta$. The equation of the line through $(0, 2)$ and $(4, 0)$ is given by $y = -\frac{1}{2}x + 2$. Thus $x\tan\theta = -\frac{1}{2}x + 2 \Rightarrow x = \frac{4}{2\tan\theta + 1}$ and $y = \frac{4\tan\theta}{2\tan\theta + 1}$ where $0 \le \theta < \frac{\pi}{2}$.

33. The equation of the circle is given by $(x-2)^2 + y^2 = 1$. Drop a vertical line from the point (x, y) on the circle to the x-axis, then θ is an angle in a right triangle. So that we can start at $(1, 0)$ and rotate in a clockwise direction, let $x = 2 - \cos\theta,\ y = \sin\theta,\ 0 \le \theta \le 2\pi$.

35. Extend the vertical line through A to the x-axis and let C be the point of intersection. Then $OC = AQ = x$ and $\tan t = \frac{2}{OC} = \frac{2}{x} \Rightarrow x = \frac{2}{\tan t} = 2\cot t$; $\sin t = \frac{2}{OA} \Rightarrow OA = \frac{2}{\sin t}$; and $(AB)(OA) = (AQ)^2 \Rightarrow AB\left(\frac{2}{\sin t}\right) = x^2$ $\Rightarrow AB\left(\frac{2}{\sin t}\right) = \left(\frac{2}{\tan t}\right)^2 \Rightarrow AB = \frac{2\sin t}{\tan^2 t}$. Next $y = 2 - AB\sin t \Rightarrow y = 2 - \left(\frac{2\sin t}{\tan^2 t}\right)\sin t =$ $2 - \frac{2\sin^2 t}{\tan^2 t} = 2 - 2\cos^2 t = 2\sin^2 t$. Therefore let $x = 2\cot t$ and $y = 2\sin^2 t,\ 0 < t < \pi$.

37. Draw line AM in the figure and note that $\angle AMO$ is a right angle since it is an inscribed angle which spans the diameter of a circle. Then $AN^2 = MN^2 + AM^2$. Now, $OA = a$, $\frac{AN}{a} = \tan t$, and $\frac{AM}{a} = \sin t$. Next $MN = OP$ $\Rightarrow OP^2 = AN^2 - AM^2 = a^2\tan^2 t - a^2\sin^2 t$ $\Rightarrow OP = \sqrt{a^2\tan^2 t - a^2\sin^2 t}$ $= (a\sin t)\sqrt{\sec^2 t - 1} = \frac{a\sin^2 t}{\cos t}$. In triangle BPO, $x = OP\sin t = \frac{a\sin^3 t}{\cos t} = a\sin^2 t\tan t$ and $y = OP\cos t = a\sin^2 t \Rightarrow x = a\sin^2 t\tan t$ and $y = a\sin^2 t$.

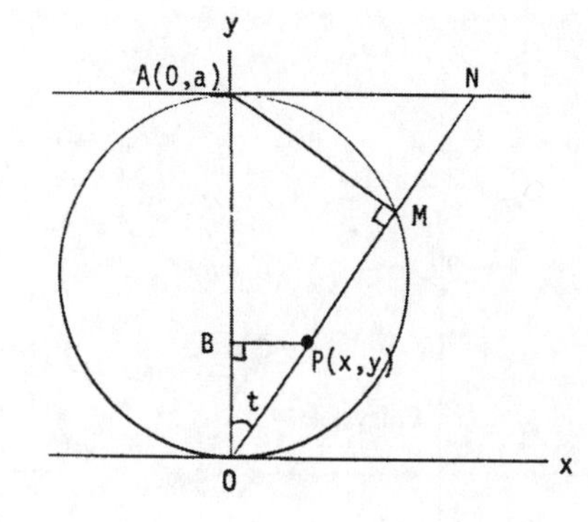

39. $D = \sqrt{(x-2)^2 + \left(y - \frac{1}{2}\right)^2} \Rightarrow D^2 = (x-2)^2 + \left(y-\frac{1}{2}\right)^2 = (t-2)^2 + \left(t^2 - \frac{1}{2}\right)^2 \Rightarrow D^2 = t^4 - 4t + \frac{17}{4}$ $\Rightarrow \frac{d(D^2)}{dt} = 4t^3 - 4 = 0 \Rightarrow t = 1$. The second derivative is always positive for $t \ne 0 \Rightarrow t = 1$ gives a local minimum for D^2 (and hence D) which is an absolute minimum since it is the only extremum $\Rightarrow$ the closest point on the parabola is $(1, 1)$.

41. (a)

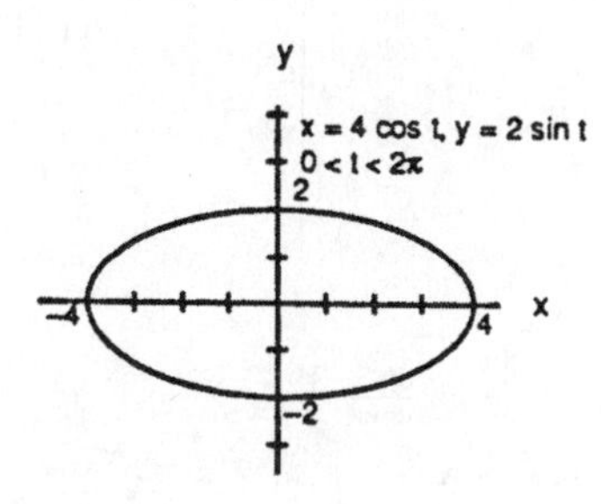

(b)

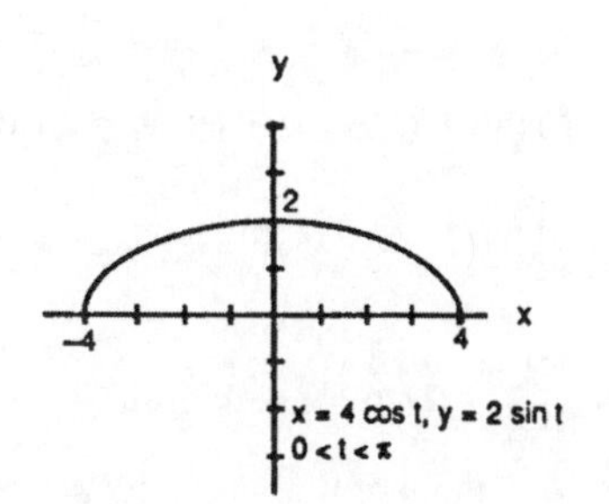

(c)

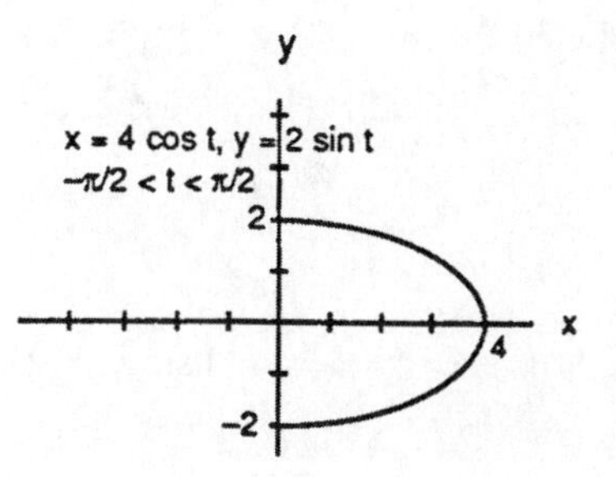

43.

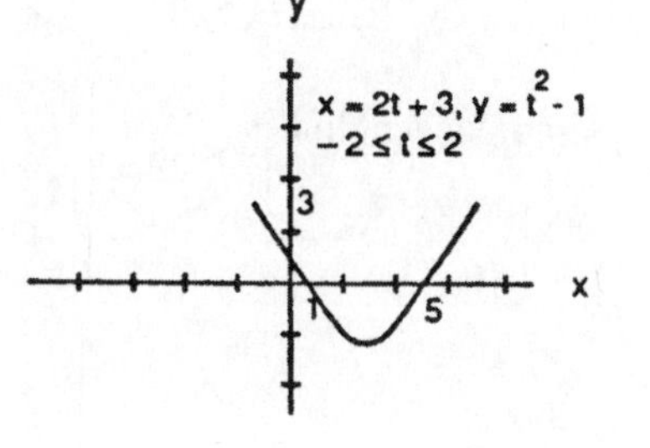

45.

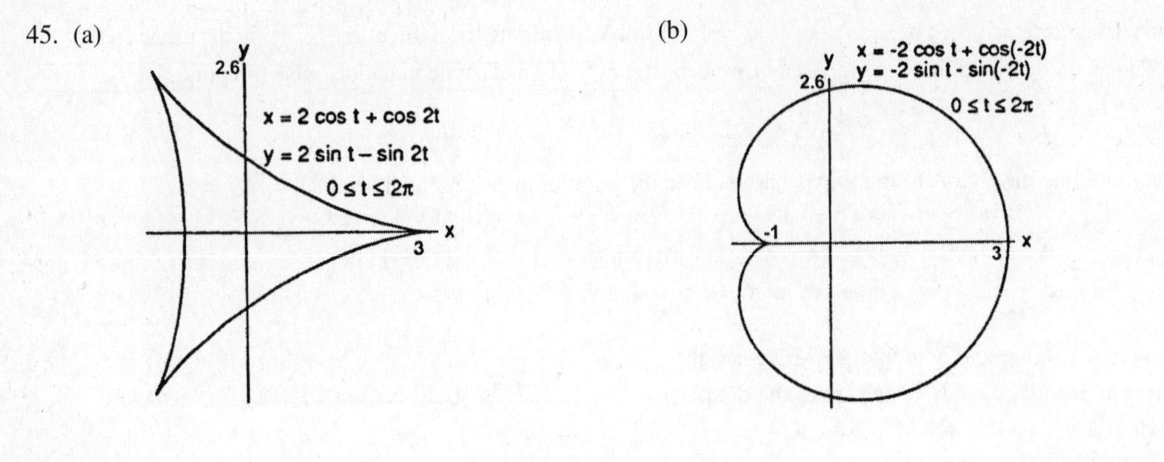

47.

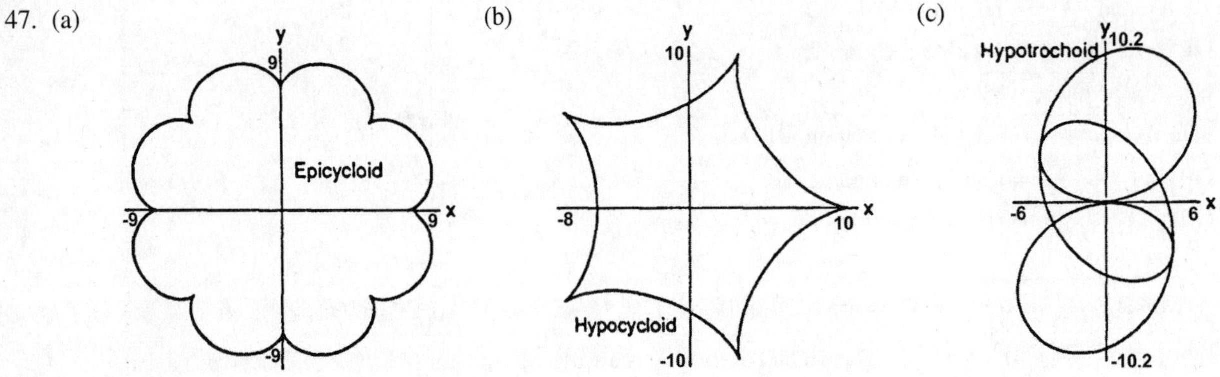

11.2 CALCULUS WITH PARAMETRIC CURVES

1. $t = \frac{\pi}{4} \Rightarrow x = 2\cos\frac{\pi}{4} = \sqrt{2},\ y = 2\sin\frac{\pi}{4} = \sqrt{2};\ \frac{dx}{dt} = -2\sin t,\ \frac{dy}{dt} = 2\cos t \Rightarrow \frac{dy}{dx} = \frac{dy/dt}{dx/dt} = \frac{2\cos t}{-2\sin t} = -\cot t$
$\Rightarrow \left.\frac{dy}{dx}\right|_{t=\frac{\pi}{4}} = -\cot\frac{\pi}{4} = -1$; tangent line is $y - \sqrt{2} = -1\left(x - \sqrt{2}\right)$ or $y = -x + 2\sqrt{2};\ \frac{dy'}{dt} = \csc^2 t$
$\Rightarrow \frac{d^2y}{dx^2} = \frac{dy'/dt}{dx/dt} = \frac{\csc^2 t}{-2\sin t} = -\frac{1}{2\sin^3 t} \Rightarrow \left.\frac{d^2y}{dx^2}\right|_{t=\frac{\pi}{4}} = -\sqrt{2}$

3. $t = \frac{\pi}{4} \Rightarrow x = 4\sin\frac{\pi}{4} = 2\sqrt{2},\ y = 2\cos\frac{\pi}{4} = \sqrt{2};\ \frac{dx}{dt} = 4\cos t,\ \frac{dy}{dt} = -2\sin t \Rightarrow \frac{dy}{dx} = \frac{dy/dt}{dx/dt} = \frac{-2\sin t}{4\cos t}$
$= -\frac{1}{2}\tan t \Rightarrow \left.\frac{dy}{dx}\right|_{t=\frac{\pi}{4}} = -\frac{1}{2}\tan\frac{\pi}{4} = -\frac{1}{2}$; tangent line is $y - \sqrt{2} = -\frac{1}{2}\left(x - 2\sqrt{2}\right)$ or $y = -\frac{1}{2}x + 2\sqrt{2}$;
$\frac{dy'}{dt} = -\frac{1}{2}\sec^2 t \Rightarrow \frac{d^2y}{dx^2} = \frac{dy'/dt}{dx/dt} = \frac{-\frac{1}{2}\sec^2 t}{4\cos t} = -\frac{1}{8\cos^3 t} \Rightarrow \left.\frac{d^2y}{dx^2}\right|_{t=\frac{\pi}{4}} = -\frac{\sqrt{2}}{4}$

5. $t = \frac{1}{4} \Rightarrow x = \frac{1}{4},\ y = \frac{1}{2};\ \frac{dx}{dt} = 1,\ \frac{dy}{dt} = \frac{1}{2\sqrt{t}} \Rightarrow \frac{dy}{dx} = \frac{dy/dt}{dx/dt} = \frac{1}{2\sqrt{t}} \Rightarrow \left.\frac{dy}{dx}\right|_{t=\frac{1}{4}} = \frac{1}{2\sqrt{\frac{1}{4}}} = 1$; tangent line is
$y - \frac{1}{2} = 1\cdot\left(x - \frac{1}{4}\right)$ or $y = x + \frac{1}{4};\ \frac{dy'}{dt} = -\frac{1}{4}t^{-3/2} \Rightarrow \frac{d^2y}{dx^2} = \frac{dy'/dt}{dx/dt} = -\frac{1}{4}t^{-3/2} \Rightarrow \left.\frac{d^2y}{dx^2}\right|_{t=\frac{1}{4}} = -2$

7. $t = \frac{\pi}{6} \Rightarrow x = \sec\frac{\pi}{6} = \frac{2}{\sqrt{3}},\ y = \tan\frac{\pi}{6} = \frac{1}{\sqrt{3}};\ \frac{dx}{dt} = \sec t\tan t,\ \frac{dy}{dt} = \sec^2 t \Rightarrow \frac{dy}{dx} = \frac{dy/dt}{dx/dt}$
$= \frac{\sec^2 t}{\sec t\tan t} = \csc t \Rightarrow \left.\frac{dy}{dx}\right|_{t=\frac{\pi}{6}} = \csc\frac{\pi}{6} = 2$; tangent line is $y - \frac{1}{\sqrt{3}} = 2\left(x - \frac{2}{\sqrt{3}}\right)$ or $y = 2x - \sqrt{3}$;
$\frac{dy'}{dt} = -\csc t\cot t \Rightarrow \frac{d^2y}{dx^2} = \frac{dy'/dt}{dx/dt} = \frac{-\csc t\cot t}{\sec t\tan t} = -\cot^3 t \Rightarrow \left.\frac{d^2y}{dx^2}\right|_{t=\frac{\pi}{6}} = -3\sqrt{3}$

9. $t = -1 \Rightarrow x = 5,\ y = 1;\ \frac{dx}{dt} = 4t,\ \frac{dy}{dt} = 4t^3 \Rightarrow \frac{dy}{dx} = \frac{dy/dt}{dx/dt} = \frac{4t^3}{4t} = t^2 \Rightarrow \left.\frac{dy}{dx}\right|_{t=-1} = (-1)^2 = 1$; tangent line is
$y - 1 = 1\cdot(x - 5)$ or $y = x - 4;\ \frac{dy'}{dt} = 2t \Rightarrow \frac{d^2y}{dx^2} = \frac{dy'/dt}{dx/dt} = \frac{2t}{4t} = \frac{1}{2} \Rightarrow \left.\frac{d^2y}{dx^2}\right|_{t=-1} = \frac{1}{2}$

11. $t = \frac{\pi}{3} \Rightarrow x = \frac{\pi}{3} - \sin\frac{\pi}{3} = \frac{\pi}{3} - \frac{\sqrt{3}}{2}$, $y = 1 - \cos\frac{\pi}{3} = 1 - \frac{1}{2} = \frac{1}{2}$; $\frac{dx}{dt} = 1 - \cos t$, $\frac{dy}{dt} = \sin t \Rightarrow \frac{dy}{dx} = \frac{dy/dt}{dx/dt}$
$= \frac{\sin t}{1-\cos t} \Rightarrow \left.\frac{dy}{dx}\right|_{t=\frac{\pi}{3}} = \frac{\sin\left(\frac{\pi}{3}\right)}{1-\cos\left(\frac{\pi}{3}\right)} = \frac{\left(\frac{\sqrt{3}}{2}\right)}{\left(\frac{1}{2}\right)} = \sqrt{3}$; tangent line is $y - \frac{1}{2} = \sqrt{3}\left(x - \frac{\pi}{3} + \frac{\sqrt{3}}{2}\right)$
$\Rightarrow y = \sqrt{3}x - \frac{\pi\sqrt{3}}{3} + 2$; $\frac{dy'}{dt} = \frac{(1-\cos t)(\cos t) - (\sin t)(\sin t)}{(1-\cos t)^2} = \frac{-1}{1-\cos t} \Rightarrow \frac{d^2y}{dx^2} = \frac{dy'/dt}{dx/dt} = \frac{\left(\frac{-1}{1-\cos t}\right)}{1-\cos t}$
$= \frac{-1}{(1-\cos t)^2} \Rightarrow \left.\frac{d^2y}{dx^2}\right|_{t=\frac{\pi}{3}} = -4$

13. $t = 2 \Rightarrow x = \frac{1}{2+1} = \frac{1}{3}$, $y = \frac{2}{2-1} = 2$; $\frac{dx}{dt} = \frac{-1}{(t+1)^2}$, $\frac{dy}{dt} = \frac{-1}{(t-1)^2} \Rightarrow \frac{dy}{dx} = \frac{(t+1)^2}{(t-1)^2} \Rightarrow \left.\frac{dy}{dx}\right|_{t=2} = \frac{(2+1)^2}{(2-1)^2} = 9$;
tangent line is $y = 9x - 1$; $\frac{dy'}{dt} = -\frac{4(t+1)}{(t-1)^3} \Rightarrow \frac{d^2y}{dx^2} = \frac{4(t+1)^3}{(t-1)^3} \Rightarrow \left.\frac{d^2y}{dx^2}\right|_{t=2} = \frac{4(2+1)^3}{(2-1)^3} = 108$

15. $x^3 + 2t^2 = 9 \Rightarrow 3x^2\frac{dx}{dt} + 4t = 0 \Rightarrow 3x^2\frac{dx}{dt} = -4t \Rightarrow \frac{dx}{dt} = \frac{-4t}{3x^2}$;
$2y^3 - 3t^2 = 4 \Rightarrow 6y^2\frac{dy}{dt} - 6t = 0 \Rightarrow \frac{dy}{dt} = \frac{6t}{6y^2} = \frac{t}{y^2}$; thus $\frac{dy}{dx} = \frac{dy/dt}{dx/dt} = \frac{\left(\frac{t}{y^2}\right)}{\left(\frac{-4t}{3x^2}\right)} = \frac{t(3x^2)}{y^2(-4t)} = \frac{3x^2}{-4y^2}$; $t = 2$
$\Rightarrow x^3 + 2(2)^2 = 9 \Rightarrow x^3 + 8 = 9 \Rightarrow x^3 = 1 \Rightarrow x = 1$; $t = 2 \Rightarrow 2y^3 - 3(2)^2 = 4$
$\Rightarrow 2y^3 = 16 \Rightarrow y^3 = 8 \Rightarrow y = 2$; therefore $\left.\frac{dy}{dx}\right|_{t=2} = \frac{3(1)^2}{-4(2)^2} = -\frac{3}{16}$

17. $x + 2x^{3/2} = t^2 + t \Rightarrow \frac{dx}{dt} + 3x^{1/2}\frac{dx}{dt} = 2t + 1 \Rightarrow \left(1 + 3x^{1/2}\right)\frac{dx}{dt} = 2t + 1 \Rightarrow \frac{dx}{dt} = \frac{2t+1}{1+3x^{1/2}}$; $y\sqrt{t+1} + 2t\sqrt{y} = 4$
$\Rightarrow \frac{dy}{dt}\sqrt{t+1} + y\left(\frac{1}{2}\right)(t+1)^{-1/2} + 2\sqrt{y} + 2t\left(\frac{1}{2}y^{-1/2}\right)\frac{dy}{dt} = 0 \Rightarrow \frac{dy}{dt}\sqrt{t+1} + \frac{y}{2\sqrt{t+1}} + 2\sqrt{y} + \left(\frac{t}{\sqrt{y}}\right)\frac{dy}{dt} = 0$
$\Rightarrow \left(\sqrt{t+1} + \frac{t}{\sqrt{y}}\right)\frac{dy}{dt} = \frac{-y}{2\sqrt{t+1}} - 2\sqrt{y} \Rightarrow \frac{dy}{dt} = \frac{\left(\frac{-y}{2\sqrt{t+1}} - 2\sqrt{y}\right)}{\left(\sqrt{t+1} + \frac{t}{\sqrt{y}}\right)} = \frac{-y\sqrt{y} - 4y\sqrt{t+1}}{2\sqrt{y}(t+1) + 2t\sqrt{t+1}}$; thus
$\frac{dy}{dx} = \frac{dy/dt}{dx/dt} = \frac{\left(\frac{-y\sqrt{y} - 4y\sqrt{t+1}}{2\sqrt{y}(t+1) + 2t\sqrt{t+1}}\right)}{\left(\frac{2t+1}{1+3x^{1/2}}\right)}$; $t = 0 \Rightarrow x + 2x^{3/2} = 0 \Rightarrow x\left(1 + 2x^{1/2}\right) = 0 \Rightarrow x = 0$; $t = 0$
$\Rightarrow y\sqrt{0+1} + 2(0)\sqrt{y} = 4 \Rightarrow y = 4$; therefore $\left.\frac{dy}{dx}\right|_{t=0} = \frac{\left(\frac{-4\sqrt{4} - 4(4)\sqrt{0+1}}{2\sqrt{4}(0+1) + 2(0)\sqrt{0+1}}\right)}{\left(\frac{2(0)+1}{1+3(0)^{1/2}}\right)} = -6$

19. $x = t^3 + t$, $y + 2t^3 = 2x + t^2 \Rightarrow \frac{dx}{dt} = 3t^2 + 1$, $\frac{dy}{dt} + 6t^2 = 2\frac{dx}{dt} + 2t \Rightarrow \frac{dy}{dt} = 2(3t^2 + 1) + 2t - 6t^2 = 2t + 2$
$\Rightarrow \frac{dy}{dx} = \frac{2t+2}{3t^2+1} \Rightarrow \left.\frac{dy}{dx}\right|_{t=1} = \frac{2(1)+2}{3(1)^2+1} = 1$

21. $A = \int_0^{2\pi} y\,dx = \int_0^{2\pi} a(1-\cos t)a(1-\cos t)dt = a^2\int_0^{2\pi}(1-\cos t)^2 dt = a^2\int_0^{2\pi}(1 - 2\cos t + \cos^2 t)dt$
$= a^2\int_0^{2\pi}\left(1 - 2\cos t + \frac{1+\cos 2t}{2}\right)dt = a^2\int_0^{2\pi}\left(\frac{3}{2} - 2\cos t + \frac{1}{2}\cos 2t\right)dt = a^2\left[\frac{3}{2}t - 2\sin t + \frac{1}{4}\sin 2t\right]_0^{2\pi}$
$= a^2(3\pi - 0 + 0) - 0 = 3\pi a^2$

23. $A = 2\int_\pi^0 y\,dx = 2\int_\pi^0 (b\sin t)(-a\sin t)dt = 2ab\int_0^\pi \sin^2 t\,dt = 2ab\int_0^\pi \frac{1-\cos 2t}{2}dt = ab\int_0^\pi (1 - \cos 2t)\,dt$
$= ab\left[t - \frac{1}{2}\sin 2t\right]_0^\pi = ab((\pi - 0) - 0) = \pi ab$

25. $\frac{dx}{dt} = -\sin t$ and $\frac{dy}{dt} = 1 + \cos t \Rightarrow \sqrt{\left(\frac{dx}{dt}\right)^2 + \left(\frac{dy}{dt}\right)^2} = \sqrt{(-\sin t)^2 + (1+\cos t)^2} = \sqrt{2 + 2\cos t}$
$\Rightarrow \text{Length} = \int_0^\pi \sqrt{2 + 2\cos t}\,dt = \sqrt{2}\int_0^\pi \sqrt{\left(\frac{1-\cos t}{1-\cos t}\right)(1+\cos t)}\,dt = \sqrt{2}\int_0^\pi \sqrt{\frac{\sin^2 t}{1-\cos t}}\,dt$

$= \sqrt{2} \int_0^{\pi} \frac{\sin t}{\sqrt{1-\cos t}}\, dt$ (since $\sin t \ge 0$ on $[0, \pi]$); $[u = 1 - \cos t \Rightarrow du = \sin t\, dt;\ t = 0 \Rightarrow u = 0,$
$t = \pi \Rightarrow u = 2] \rightarrow \sqrt{2} \int_0^2 u^{-1/2}\, du = \sqrt{2}\left[2u^{1/2}\right]_0^2 = 4$

27. $\frac{dx}{dt} = t$ and $\frac{dy}{dt} = (2t+1)^{1/2} \Rightarrow \sqrt{\left(\frac{dx}{dt}\right)^2 + \left(\frac{dy}{dt}\right)^2} = \sqrt{t^2 + (2t+1)} = \sqrt{(t+1)^2} = |t+1| = t+1$ since $0 \le t \le 4$
$\Rightarrow \text{Length} = \int_0^4 (t+1)\, dt = \left[\frac{t^2}{2} + t\right]_0^4 = (8+4) = 12$

29. $\frac{dx}{dt} = 8t \cos t$ and $\frac{dy}{dt} = 8t \sin t \Rightarrow \sqrt{\left(\frac{dx}{dt}\right)^2 + \left(\frac{dy}{dt}\right)^2} = \sqrt{(8t\cos t)^2 + (8t \sin t)^2} = \sqrt{64t^2\cos^2 t + 64t^2 \sin^2 t}$
$= |8t| = 8t$ since $0 \le t \le \frac{\pi}{2} \Rightarrow \text{Length} = \int_0^{\pi/2} 8t\, dt = \left[4t^2\right]_0^{\pi/2} = \pi^2$

31. $\frac{dx}{dt} = -\sin t$ and $\frac{dy}{dt} = \cos t \Rightarrow \sqrt{\left(\frac{dx}{dt}\right)^2 + \left(\frac{dy}{dt}\right)^2} = \sqrt{(-\sin t)^2 + (\cos t)^2} = 1 \Rightarrow \text{Area} = \int 2\pi y\, ds$
$= \int_0^{2\pi} 2\pi(2+\sin t)(1)dt = 2\pi\left[2t - \cos t\right]_0^{2\pi} = 2\pi[(4\pi - 1) - (0-1)] = 8\pi^2$

33. $\frac{dx}{dt} = 1$ and $\frac{dy}{dt} = t + \sqrt{2} \Rightarrow \sqrt{\left(\frac{dx}{dt}\right)^2 + \left(\frac{dy}{dt}\right)^2} = \sqrt{1^2 + \left(t+\sqrt{2}\right)^2} = \sqrt{t^2 + 2\sqrt{2}t + 3} \Rightarrow \text{Area} = \int 2\pi x\, ds$
$= \int_{-\sqrt{2}}^{\sqrt{2}} 2\pi\left(t + \sqrt{2}\right)\sqrt{t^2 + 2\sqrt{2}t + 3}\, dt;\ \left[u = t^2 + 2\sqrt{2}t + 3 \Rightarrow du = \left(2t + 2\sqrt{2}\right) dt;\ t = -\sqrt{2} \Rightarrow u = 1,\right.$
$\left.t = \sqrt{2} \Rightarrow u = 9\right] \rightarrow \int_1^9 \pi\sqrt{u}\, du = \left[\frac{2}{3}\pi u^{3/2}\right]_1^9 = \frac{2\pi}{3}(27-1) = \frac{52\pi}{3}$

35. $\frac{dx}{dt} = 2$ and $\frac{dy}{dt} = 1 \Rightarrow \sqrt{\left(\frac{dx}{dt}\right)^2 + \left(\frac{dy}{dt}\right)^2} = \sqrt{2^2 + 1^2} = \sqrt{5} \Rightarrow \text{Area} = \int 2\pi y\, ds = \int_0^1 2\pi(t+1)\sqrt{5}\, dt$
$= 2\pi\sqrt{5}\left[\frac{t^2}{2} + t\right]_0^1 = 3\pi\sqrt{5}$. Check: slant height is $\sqrt{5} \Rightarrow$ Area is $\pi(1+2)\sqrt{5} = 3\pi\sqrt{5}$.

37. Let the density be $\delta = 1$. Then $x = \cos t + t \sin t \Rightarrow \frac{dx}{dt} = t \cos t$, and $y = \sin t - t \cos t \Rightarrow \frac{dy}{dt} = t \sin t$
$\Rightarrow dm = 1 \cdot ds = \sqrt{\left(\frac{dx}{dt}\right)^2 + \left(\frac{dy}{dt}\right)^2}\, dt = \sqrt{(t\cos t)^2 + (t \sin t)^2} = |t|\, dt = t\, dt$ since $0 \le t \le \frac{\pi}{2}$. The curve's mass is
$M = \int dm = \int_0^{\pi/2} t\, dt = \frac{\pi^2}{8}$. Also $M_x = \int \tilde{y}\, dm = \int_0^{\pi/2} (\sin t - t\cos t)\, t\, dt = \int_0^{\pi/2} t \sin t\, dt - \int_0^{\pi/2} t^2 \cos t\, dt$
$= \left[\sin t - t\cos t\right]_0^{\pi/2} - \left[t^2 \sin t - 2\sin t + 2t\cos t\right]_0^{\pi/2} = 3 - \frac{\pi^2}{4}$, where we integrated by parts. Therefore,
$\overline{y} = \frac{M_x}{M} = \frac{\left(3 - \frac{\pi^2}{4}\right)}{\left(\frac{\pi^2}{8}\right)} = \frac{24}{\pi^2} - 2$. Next, $M_y = \int \tilde{x}\, dm = \int_0^{\pi/2} (\cos t + t \sin t)\, t\, dt = \int_0^{\pi/2} t\cos t\, dt + \int_0^{\pi/2} t^2 \sin t\, dt$
$= \left[\cos t + t\sin t\right]_0^{\pi/2} + \left[-t^2\cos t + 2\cos t + 2t \sin t\right]_0^{\pi/2} = \frac{3\pi}{2} - 3$, again integrating by parts. Hence
$\overline{x} = \frac{M_y}{M} = \frac{\left(\frac{3\pi}{2} - 3\right)}{\left(\frac{\pi^2}{8}\right)} = \frac{12}{\pi} - \frac{24}{\pi^2}$. Therefore $(\overline{x}, \overline{y}) = \left(\frac{12}{\pi} - \frac{24}{\pi^2}, \frac{24}{\pi^2} - 2\right)$.

39. Let the density be $\delta = 1$. Then $x = \cos t \Rightarrow \frac{dx}{dt} = -\sin t$, and $y = t + \sin t \Rightarrow \frac{dy}{dt} = 1 + \cos t$
$\Rightarrow dm = 1 \cdot ds = \sqrt{\left(\frac{dx}{dt}\right)^2 + \left(\frac{dy}{dt}\right)^2}\, dt = \sqrt{(-\sin t)^2 + (1+\cos t)^2}\, dt = \sqrt{2 + 2\cos t}\, dt$. The curve's mass
is $M = \int dm = \int_0^{\pi} \sqrt{2 + 2\cos t}\, dt = \sqrt{2}\int_0^{\pi}\sqrt{1+\cos t}\, dt = \sqrt{2}\int_0^{\pi}\sqrt{2\cos^2\left(\frac{t}{2}\right)}\, dt = 2\int_0^{\pi}\left|\cos\left(\frac{t}{2}\right)\right| dt$
$= 2\int_0^{\pi}\cos\left(\frac{t}{2}\right) dt$ (since $0 \le t \le \pi \Rightarrow 0 \le \frac{t}{2} \le \frac{\pi}{2}$) $= 2\left[2\sin\left(\frac{t}{2}\right)\right]_0^{\pi} = 4$. Also $M_x = \int \tilde{y}\, dm$
$= \int_0^{\pi} (t + \sin t)\left(2\cos\frac{t}{2}\right) dt = \int_0^{\pi} 2t\cos\left(\frac{t}{2}\right) dt + \int_0^{\pi} 2\sin t \cos\left(\frac{t}{2}\right) dt$

$= 2\left[4\cos\left(\frac{t}{2}\right) + 2t\sin\left(\frac{t}{2}\right)\right]_0^{\pi} + 2\left[-\frac{1}{3}\cos\left(\frac{3}{2}t\right) - \cos\left(\frac{1}{2}t\right)\right]_0^{\pi} = 4\pi - \frac{16}{3} \Rightarrow \bar{y} = \frac{M_x}{M} = \frac{\left(4\pi - \frac{16}{3}\right)}{4} = \pi - \frac{4}{3}.$
Next $M_y = \int \tilde{x}\, dm = \int_0^{\pi} (\cos t)\left(2\cos\frac{t}{2}\right) dt = \int_0^{\pi} \cos t \cos\left(\frac{t}{2}\right) dt = 2\left[\sin\left(\frac{t}{2}\right) + \frac{\sin\left(\frac{3}{2}t\right)}{3}\right]_0^{\pi} = 2 - \frac{2}{3}$
$= \frac{4}{3} \Rightarrow \bar{x} = \frac{M_y}{M} = \frac{\left(\frac{4}{3}\right)}{4} = \frac{1}{3}$. Therefore $(\bar{x}, \bar{y}) = \left(\frac{1}{3}, \pi - \frac{4}{3}\right)$.

41. (a) $\frac{dx}{dt} = -2\sin 2t$ and $\frac{dy}{dt} = 2\cos 2t \Rightarrow \sqrt{\left(\frac{dx}{dt}\right)^2 + \left(\frac{dy}{dt}\right)^2} = \sqrt{(-2\sin 2t)^2 + (2\cos 2t)^2} = 2$

$\Rightarrow \text{Length} = \int_0^{\pi/2} 2\, dt = [2t]_0^{\pi/2} = \pi$

(b) $\frac{dx}{dt} = \pi\cos\pi t$ and $\frac{dy}{dt} = -\pi\sin\pi t \Rightarrow \sqrt{\left(\frac{dx}{dt}\right)^2 + \left(\frac{dy}{dt}\right)^2} = \sqrt{(\pi\cos\pi t)^2 + (-\pi\sin\pi t)^2} = \pi$

$\Rightarrow \text{Length} = \int_{-1/2}^{1/2} \pi\, dt = [\pi t]_{-1/2}^{1/2} = \pi$

43. $x = (1 + 2\sin\theta)\cos\theta,\ y = (1 + 2\sin\theta)\sin\theta \Rightarrow \frac{dx}{d\theta} = 2\cos^2\theta - \sin\theta(1 + 2\sin\theta),\ \frac{dy}{d\theta} = 2\cos\theta\sin\theta + \cos\theta(1 + 2\sin\theta)$
$\Rightarrow \frac{dy}{dx} = \frac{2\cos\theta\sin\theta + \cos\theta(1+2\sin\theta)}{2\cos^2\theta - \sin\theta(1+2\sin\theta)} = \frac{4\cos\theta\sin\theta + \cos\theta}{2\cos^2\theta - 2\sin^2\theta - \sin\theta} = \frac{2\sin 2\theta + \cos\theta}{2\cos 2\theta - \sin\theta}$

(a) $x = (1 + 2\sin(0))\cos(0) = 1,\ y = (1 + 2\sin(0))\sin(0) = 0;\ \left.\frac{dy}{dx}\right|_{\theta=0} = \frac{2\sin(2(0)) + \cos(0)}{2\cos(2(0)) - \sin(0)} = \frac{0+1}{2-0} = \frac{1}{2}$

(b) $x = \left(1 + 2\sin\left(\frac{\pi}{2}\right)\right)\cos\left(\frac{\pi}{2}\right) = 0,\ y = \left(1 + 2\sin\left(\frac{\pi}{2}\right)\right)\sin\left(\frac{\pi}{2}\right) = 3;\ \left.\frac{dy}{dx}\right|_{\theta=\pi/2} = \frac{2\sin\left(2\left(\frac{\pi}{2}\right)\right) + \cos\left(\frac{\pi}{2}\right)}{2\cos\left(2\left(\frac{\pi}{2}\right)\right) - \sin\left(\frac{\pi}{2}\right)} = \frac{0+0}{-2-1} = 0$

(c) $x = \left(1 + 2\sin\left(\frac{4\pi}{3}\right)\right)\cos\left(\frac{4\pi}{3}\right) = \frac{\sqrt{3}-1}{2},\ y = \left(1 + 2\sin\left(\frac{4\pi}{3}\right)\right)\sin\left(\frac{4\pi}{3}\right) = \frac{3-\sqrt{3}}{2};\ \left.\frac{dy}{dx}\right|_{\theta=4\pi/3} = \frac{2\sin\left(2\left(\frac{4\pi}{3}\right)\right) + \cos\left(\frac{4\pi}{3}\right)}{2\cos\left(2\left(\frac{4\pi}{3}\right)\right) - \sin\left(\frac{4\pi}{3}\right)}$
$= \frac{\sqrt{3} - \frac{1}{2}}{-1 + \frac{\sqrt{3}}{2}} = \frac{2\sqrt{3} - 1}{\sqrt{3} - 2} = -\left(4 + 3\sqrt{3}\right)$

45. $\frac{dx}{dt} = \cos t$ and $\frac{dy}{dt} = 2\cos 2t \Rightarrow \frac{dy}{dx} = \frac{dy/dt}{dx/dt} = \frac{2\cos 2t}{\cos t} = \frac{2(2\cos^2 t - 1)}{\cos t}$; then $\frac{dy}{dx} = 0 \Rightarrow \frac{2(2\cos^2 t - 1)}{\cos t} = 0$
$\Rightarrow 2\cos^2 t - 1 = 0 \Rightarrow \cos t = \pm\frac{1}{\sqrt{2}} \Rightarrow t = \frac{\pi}{4}, \frac{3\pi}{4}, \frac{5\pi}{4}, \frac{7\pi}{4}$. In the 1st quadrant: $t = \frac{\pi}{4} \Rightarrow x = \sin\frac{\pi}{4} = \frac{\sqrt{2}}{2}$ and
$y = \sin 2\left(\frac{\pi}{4}\right) = 1 \Rightarrow \left(\frac{\sqrt{2}}{2}, 1\right)$ is the point where the tangent line is horizontal. At the origin: $x = 0$ and $y = 0$
$\Rightarrow \sin t = 0 \Rightarrow t = 0$ or $t = \pi$ and $\sin 2t = 0 \Rightarrow t = 0, \frac{\pi}{2}, \pi, \frac{3\pi}{2}$; thus $t = 0$ and $t = \pi$ give the tangent lines at the origin. Tangents at origin: $\left.\frac{dy}{dx}\right|_{t=0} = 2 \Rightarrow y = 2x$ and $\left.\frac{dy}{dx}\right|_{t=\pi} = -2 \Rightarrow y = -2x$

47. (a) $x = a(t - \sin t),\ y = a(1 - \cos t),\ 0 \le t \le 2\pi \Rightarrow \frac{dx}{dt} = a(1 - \cos t),\ \frac{dy}{dt} = a\sin t \Rightarrow$ Length
$= \int_0^{2\pi} \sqrt{(a(1-\cos t))^2 + (a\sin t)^2}\, dt = \int_0^{2\pi} \sqrt{a^2 - 2a^2\cos t + a^2\cos^2 t + a^2\sin^2 t}\, dt$
$= a\sqrt{2}\int_0^{2\pi} \sqrt{1 - \cos t}\, dt = a\sqrt{2}\int_0^{2\pi} \sqrt{2\sin^2\left(\frac{t}{2}\right)}\, dt = 2a\int_0^{2\pi} \sin\left(\frac{t}{2}\right) dt = \left[-4a\cos\left(\frac{t}{2}\right)\right]_0^{2\pi}$
$= -4a\cos\pi + 4a\cos(0) = 8a$

(b) $a = 1 \Rightarrow x = t - \sin t,\ y = 1 - \cos t,\ 0 \le t \le 2\pi \Rightarrow \frac{dx}{dt} = 1 - \cos t,\ \frac{dy}{dt} = \sin t \Rightarrow$ Surface area $=$
$= \int_0^{2\pi} 2\pi(1 - \cos t)\sqrt{(1-\cos t)^2 + (\sin t)^2}\, dt = \int_0^{2\pi} 2\pi(1 - \cos t)\sqrt{1 - 2\cos t + \cos^2 t + \sin^2 t}\, dt$
$= 2\pi\int_0^{2\pi} (1 - \cos t)\sqrt{2 - 2\cos t}\, dt = 2\sqrt{2}\pi\int_0^{2\pi} (1 - \cos t)^{3/2}\, dt = 2\sqrt{2}\pi\int_0^{2\pi} \left(1 - \cos\left(2 \cdot \frac{t}{2}\right)\right)^{3/2} dt$
$= 2\sqrt{2}\pi\int_0^{2\pi} \left(2\sin^2\left(\frac{t}{2}\right)\right)^{3/2} dt = 8\pi\int_0^{2\pi} \sin^3\left(\frac{t}{2}\right) dt$
$\left[u = \frac{t}{2} \Rightarrow du = \frac{1}{2}dt \Rightarrow dt = 2\, du;\ t = 0 \Rightarrow u = 0,\ t = 2\pi \Rightarrow u = \pi\right]$
$= 16\pi\int_0^{\pi} \sin^3 u\, du = 16\pi\int_0^{\pi} \sin^2 u \sin u\, du = 16\pi\int_0^{\pi} (1 - \cos^2 u)\sin u\, du = 16\pi\int_0^{\pi} \sin u\, du - 16\pi\int_0^{\pi} \cos^2 u \sin u\, du$
$= \left[-16\pi\cos u + \frac{16\pi}{3}\cos^3 u\right]_0^{\pi} = \left(16\pi - \frac{16\pi}{3}\right) - \left(-16\pi + \frac{16\pi}{3}\right) = \frac{64\pi}{3}$

11.3 POLAR COORDINATES

1. a, e; b, g; c, h; d, f

3. (a) $\left(2, \frac{\pi}{2} + 2n\pi\right)$ and $\left(-2, \frac{\pi}{2} + (2n+1)\pi\right)$, n an integer
   (b) $(2, 2n\pi)$ and $(-2, (2n+1)\pi)$, n an integer
   (c) $\left(2, \frac{3\pi}{2} + 2n\pi\right)$ and $\left(-2, \frac{3\pi}{2} + (2n+1)\pi\right)$, n an integer
   (d) $(2, (2n+1)\pi)$ and $(-2, 2n\pi)$, n an integer

y
$(2, \frac{\pi}{2})$
x
(−2, 0)
(2, 0)
$(-2, \frac{\pi}{2})$

5. (a) $x = r\cos\theta = 3\cos 0 = 3$, $y = r\sin\theta = 3\sin 0 = 0 \Rightarrow$ Cartesian coordinates are $(3, 0)$
   (b) $x = r\cos\theta = -3\cos 0 = -3$, $y = r\sin\theta = -3\sin 0 = 0 \Rightarrow$ Cartesian coordinates are $(-3, 0)$
   (c) $x = r\cos\theta = 2\cos\frac{2\pi}{3} = -1$, $y = r\sin\theta = 2\sin\frac{2\pi}{3} = \sqrt{3} \Rightarrow$ Cartesian coordinates are $\left(-1, \sqrt{3}\right)$
   (d) $x = r\cos\theta = 2\cos\frac{7\pi}{3} = 1$, $y = r\sin\theta = 2\sin\frac{7\pi}{3} = \sqrt{3} \Rightarrow$ Cartesian coordinates are $\left(1, \sqrt{3}\right)$
   (e) $x = r\cos\theta = -3\cos\pi = 3$, $y = r\sin\theta = -3\sin\pi = 0 \Rightarrow$ Cartesian coordinates are $(3, 0)$
   (f) $x = r\cos\theta = 2\cos\frac{\pi}{3} = 1$, $y = r\sin\theta = 2\sin\frac{\pi}{3} = \sqrt{3} \Rightarrow$ Cartesian coordinates are $\left(1, \sqrt{3}\right)$
   (g) $x = r\cos\theta = -3\cos 2\pi = -3$, $y = r\sin\theta = -3\sin 2\pi = 0 \Rightarrow$ Cartesian coordinates are $(-3, 0)$
   (h) $x = r\cos\theta = -2\cos\left(-\frac{\pi}{3}\right) = -1$, $y = r\sin\theta = -2\sin\left(-\frac{\pi}{3}\right) = \sqrt{3} \Rightarrow$ Cartesian coordinates are $\left(-1, \sqrt{3}\right)$

7. (a) $(1, 1) \Rightarrow r = \sqrt{1^2 + 1^2} = \sqrt{2}$, $\sin\theta = \frac{1}{\sqrt{2}}$ and $\cos\theta = \frac{1}{\sqrt{2}} \Rightarrow \theta = \frac{\pi}{4} \Rightarrow$ Polar coordinates are $\left(\sqrt{2}, \frac{\pi}{4}\right)$
   (b) $(-3, 0) \Rightarrow r = \sqrt{(-3)^2 + 0^2} = 3$, $\sin\theta = 0$ and $\cos\theta = -1 \Rightarrow \theta = \pi \Rightarrow$ Polar coordinates are $(3, \pi)$
   (c) $\left(\sqrt{3}, -1\right) \Rightarrow r = \sqrt{\left(\sqrt{3}\right)^2 + (-1)^2} = 2$, $\sin\theta = -\frac{1}{2}$ and $\cos\theta = \frac{\sqrt{3}}{2} \Rightarrow \theta = \frac{11\pi}{6} \Rightarrow$ Polar coordinates are $\left(2, \frac{11\pi}{6}\right)$
   (d) $(-3, 4) \Rightarrow r = \sqrt{(-3)^2 + 4^2} = 5$, $\sin\theta = \frac{4}{5}$ and $\cos\theta = -\frac{3}{5} \Rightarrow \theta = \pi - \arctan\left(\frac{4}{3}\right) \Rightarrow$ Polar coordinates are $\left(5, \pi - \arctan\left(\frac{4}{3}\right)\right)$

9. (a) $(3, 3) \Rightarrow r = -\sqrt{3^2 + 3^2} = -3\sqrt{2}$, $\sin\theta = -\frac{1}{\sqrt{2}}$ and $\cos\theta = -\frac{1}{\sqrt{2}} \Rightarrow \theta = \frac{5\pi}{4} \Rightarrow$ Polar coordinates are $\left(-3\sqrt{2}, \frac{5\pi}{4}\right)$
   (b) $(-1, 0) \Rightarrow r = -\sqrt{(-1)^2 + 0^2} = -1$, $\sin\theta = 0$ and $\cos\theta = 1 \Rightarrow \theta = 0 \Rightarrow$ Polar coordinates are $(-1, 0)$
   (c) $\left(-1, \sqrt{3}\right) \Rightarrow r = -\sqrt{(-1)^2 + \left(\sqrt{3}\right)^2} = -2$, $\sin\theta = -\frac{\sqrt{3}}{2}$ and $\cos\theta = \frac{1}{2} \Rightarrow \theta = \frac{5\pi}{3} \Rightarrow$ Polar coordinates are $\left(-2, \frac{5\pi}{3}\right)$
   (d) $(4, -3) \Rightarrow r = -\sqrt{4^2 + (-3)^2} = -5$, $\sin\theta = \frac{3}{5}$ and $\cos\theta = -\frac{4}{5} \Rightarrow \theta = \pi - \arctan\left(\frac{3}{4}\right) \Rightarrow$ Polar coordinates are $\left(-5, \pi - \arctan\left(\frac{4}{3}\right)\right)$

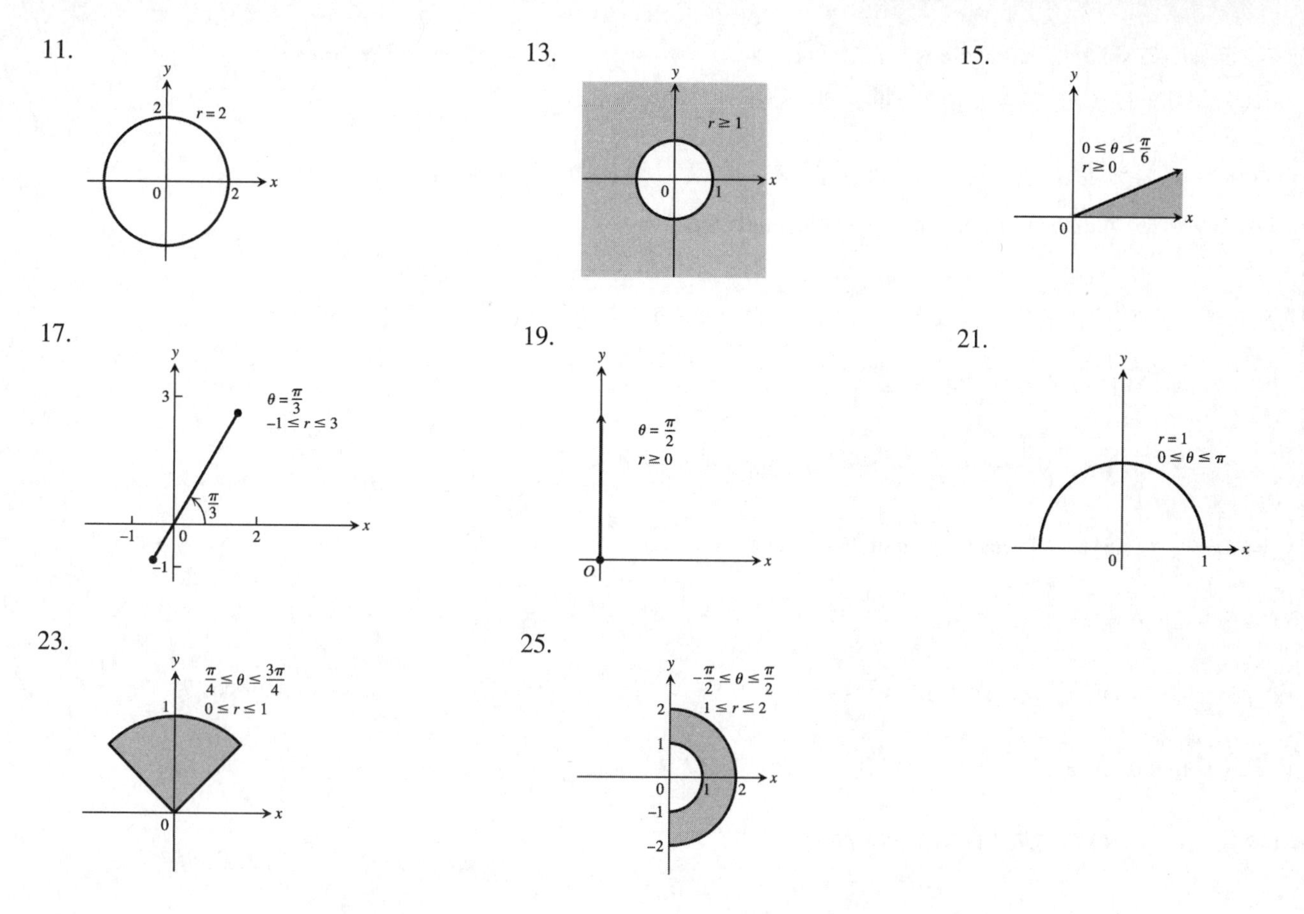

27. $r \cos \theta = 2 \Rightarrow x = 2$, vertical line through $(2, 0)$

29. $r \sin \theta = 0 \Rightarrow y = 0$, the x-axis

31. $r = 4 \csc \theta \Rightarrow r = \frac{4}{\sin \theta} \Rightarrow r \sin \theta = 4 \Rightarrow y = 4$, a horizontal line through $(0, 4)$

33. $r \cos \theta + r \sin \theta = 1 \Rightarrow x + y = 1$, line with slope $m = -1$ and intercept $b = 1$

35. $r^2 = 1 \Rightarrow x^2 + y^2 = 1$, circle with center $C = (0, 0)$ and radius 1

37. $r = \frac{5}{\sin \theta - 2 \cos \theta} \Rightarrow r \sin \theta - 2r \cos \theta = 5 \Rightarrow y - 2x = 5$, line with slope $m = 2$ and intercept $b = 5$

39. $r = \cot \theta \csc \theta = \left(\frac{\cos \theta}{\sin \theta}\right)\left(\frac{1}{\sin \theta}\right) \Rightarrow r \sin^2 \theta = \cos \theta \Rightarrow r^2 \sin^2 \theta = r \cos \theta \Rightarrow y^2 = x$, parabola with vertex $(0, 0)$ which opens to the right

41. $r = (\csc \theta) e^{r \cos \theta} \Rightarrow r \sin \theta = e^{r \cos \theta} \Rightarrow y = e^x$, graph of the natural exponential function

43. $r^2 + 2r^2 \cos \theta \sin \theta = 1 \Rightarrow x^2 + y^2 + 2xy = 1 \Rightarrow x^2 + 2xy + y^2 = 1 \Rightarrow (x + y)^2 = 1 \Rightarrow x + y = \pm 1$, two parallel straight lines of slope -1 and y-intercepts $b = \pm 1$

45. $r^2 = -4r \cos \theta \Rightarrow x^2 + y^2 = -4x \Rightarrow x^2 + 4x + y^2 = 0 \Rightarrow x^2 + 4x + 4 + y^2 = 4 \Rightarrow (x + 2)^2 + y^2 = 4$, a circle with center $C(-2, 0)$ and radius 2

47. $r = 8 \sin \theta \Rightarrow r^2 = 8r \sin \theta \Rightarrow x^2 + y^2 = 8y \Rightarrow x^2 + y^2 - 8y = 0 \Rightarrow x^2 + y^2 - 8y + 16 = 16 \Rightarrow x^2 + (y - 4)^2 = 16$, a circle with center $C(0, 4)$ and radius 4

49. $r = 2\cos\theta + 2\sin\theta \Rightarrow r^2 = 2r\cos\theta + 2r\sin\theta \Rightarrow x^2 + y^2 = 2x + 2y \Rightarrow x^2 - 2x + y^2 - 2y = 0$
$\Rightarrow (x-1)^2 + (y-1)^2 = 2$, a circle with center C(1, 1) and radius $\sqrt{2}$

51. $r\sin\left(\theta + \frac{\pi}{6}\right) = 2 \Rightarrow r\left(\sin\theta\cos\frac{\pi}{6} + \cos\theta\sin\frac{\pi}{6}\right) = 2 \Rightarrow \frac{\sqrt{3}}{2} r\sin\theta + \frac{1}{2} r\cos\theta = 2 \Rightarrow \frac{\sqrt{3}}{2} y + \frac{1}{2} x = 2$
$\Rightarrow \sqrt{3}\,y + x = 4$, line with slope $m = -\frac{1}{\sqrt{3}}$ and intercept $b = \frac{4}{\sqrt{3}}$

53. $x = 7 \Rightarrow r\cos\theta = 7$

55. $x = y \Rightarrow r\cos\theta = r\sin\theta \Rightarrow \theta = \frac{\pi}{4}$

57. $x^2 + y^2 = 4 \Rightarrow r^2 = 4 \Rightarrow r = 2$ or $r = -2$

59. $\frac{x^2}{9} + \frac{y^2}{4} = 1 \Rightarrow 4x^2 + 9y^2 = 36 \Rightarrow 4r^2\cos^2\theta + 9r^2\sin^2\theta = 36$

61. $y^2 = 4x \Rightarrow r^2\sin^2\theta = 4r\cos\theta \Rightarrow r\sin^2\theta = 4\cos\theta$

63. $x^2 + (y-2)^2 = 4 \Rightarrow x^2 + y^2 - 4y + 4 = 4 \Rightarrow x^2 + y^2 = 4y \Rightarrow r^2 = 4r\sin\theta \Rightarrow r = 4\sin\theta$

65. $(x-3)^2 + (y+1)^2 = 4 \Rightarrow x^2 - 6x + 9 + y^2 + 2y + 1 = 4 \Rightarrow x^2 + y^2 = 6x - 2y - 6 \Rightarrow r^2 = 6r\cos\theta - 2r\sin\theta - 6$

67. $(0, \theta)$ where θ is any angle

11.4 GRAPHING IN POLAR COORDINATES

1. $1 + \cos(-\theta) = 1 + \cos\theta = r \Rightarrow$ symmetric about the x-axis; $1 + \cos(-\theta) \neq -r$ and $1 + \cos(\pi - \theta) = 1 - \cos\theta \neq r \Rightarrow$ not symmetric about the y-axis; therefore not symmetric about the origin

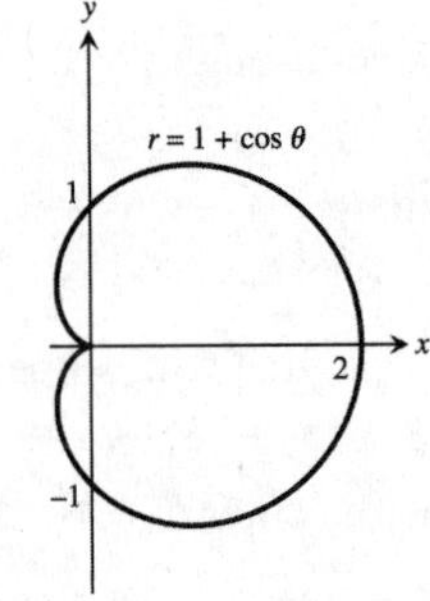

3. $1 - \sin(-\theta) = 1 + \sin\theta \neq r$ and $1 - \sin(\pi - \theta) = 1 - \sin\theta \neq -r \Rightarrow$ not symmetric about the x-axis; $1 - \sin(\pi - \theta) = 1 - \sin\theta = r \Rightarrow$ symmetric about the y-axis; therefore not symmetric about the origin

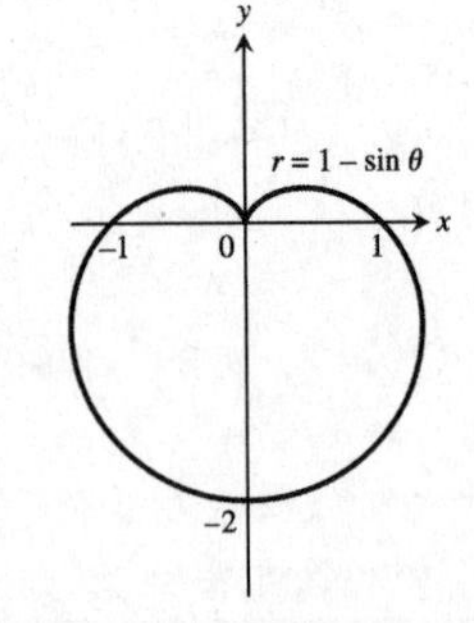

5. $2 + \sin(-\theta) = 2 - \sin\theta \neq r$ and $2 + \sin(\pi - \theta)$
$= 2 + \sin\theta \neq -r \Rightarrow$ not symmetric about the x-axis;
$2 + \sin(\pi - \theta) = 2 + \sin\theta = r \Rightarrow$ symmetric about the y-axis; therefore not symmetric about the origin

7. $\sin\left(-\frac{\theta}{2}\right) = -\sin\left(\frac{\theta}{2}\right) = -r \Rightarrow$ symmetric about the y-axis; $\sin\left(\frac{2\pi-\theta}{2}\right) = \sin\left(\frac{\theta}{2}\right)$, so the graph is symmetric about the x-axis, and hence the origin.

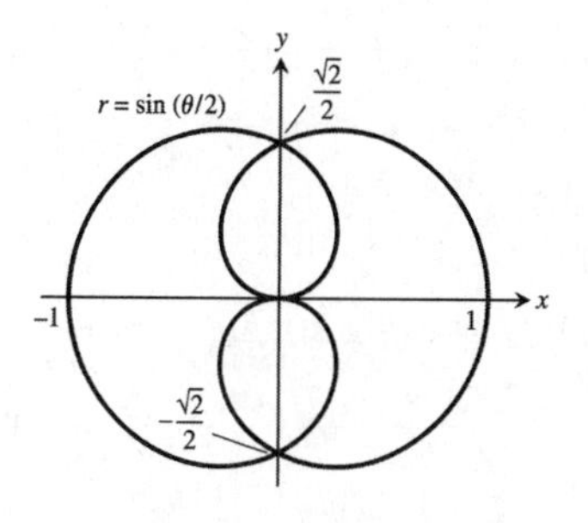

9. $\cos(-\theta) = \cos\theta = r^2 \Rightarrow (r, -\theta)$ and $(-r, -\theta)$ are on the graph when (r, θ) is on the graph $\Rightarrow$ symmetric about the x-axis and the y-axis; therefore symmetric about the origin

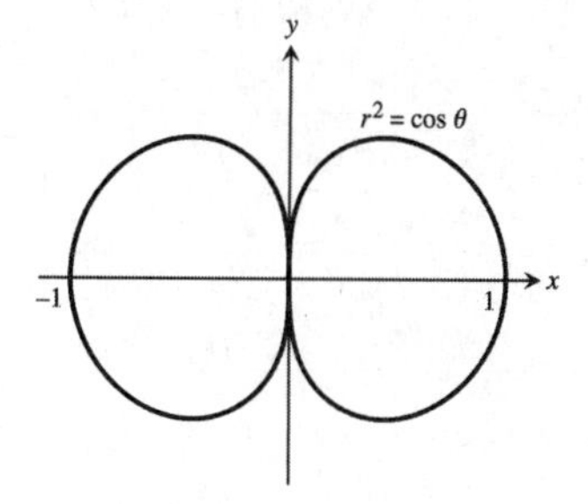

11. $-\sin(\pi - \theta) = -\sin\theta = r^2 \Rightarrow (r, \pi - \theta)$ and $(-r, \pi - \theta)$ are on the graph when (r, θ) is on the graph $\Rightarrow$ symmetric about the y-axis and the x-axis; therefore symmetric about the origin

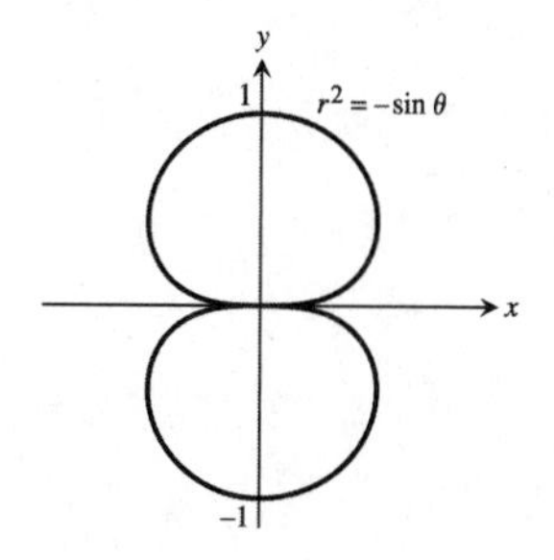

13. Since $(\pm r, -\theta)$ are on the graph when (r, θ) is on the graph $\left((\pm r)^2 = 4\cos 2(-\theta) \Rightarrow r^2 = 4\cos 2\theta\right)$, the graph is symmetric about the x-axis and the y-axis $\Rightarrow$ the graph is symmetric about the origin

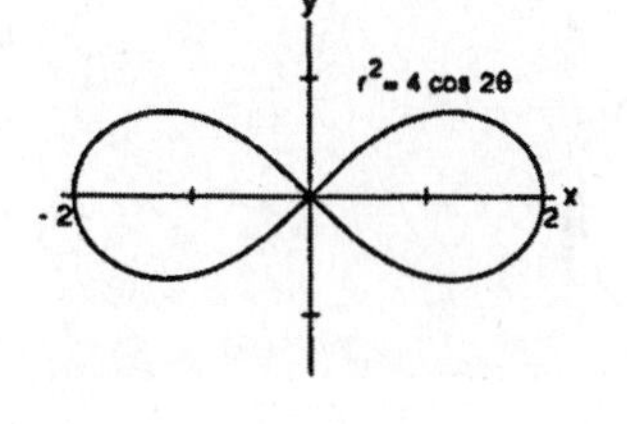

15. Since (r, θ) on the graph $\Rightarrow (-r, \theta)$ is on the graph $\left((\pm r)^2 = -\sin 2\theta \Rightarrow r^2 = -\sin 2\theta\right)$, the graph is symmetric about the origin. But $-\sin 2(-\theta) = -(-\sin 2\theta)$
$\sin 2\theta \neq r^2$ and $-\sin 2(\pi - \theta) = -\sin(2\pi - 2\theta)$
$= -\sin(-2\theta) = -(-\sin 2\theta) = \sin 2\theta \neq r^2 \Rightarrow$ the graph is not symmetric about the x-axis; therefore the graph is not symmetric about the y-axis

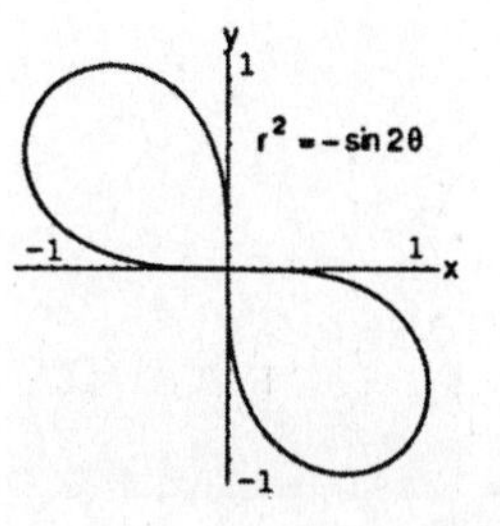

17. $\theta = \frac{\pi}{2} \Rightarrow r = -1 \Rightarrow \left(-1, \frac{\pi}{2}\right)$, and $\theta = -\frac{\pi}{2} \Rightarrow r = -1$
$\Rightarrow \left(-1, -\frac{\pi}{2}\right)$; $r' = \frac{dr}{d\theta} = -\sin\theta$; Slope $= \frac{r'\sin\theta + r\cos\theta}{r'\cos\theta - r\sin\theta}$
$= \frac{-\sin^2\theta + r\cos\theta}{-\sin\theta\cos\theta - r\sin\theta} \Rightarrow$ Slope at $\left(-1, \frac{\pi}{2}\right)$ is
$\frac{-\sin^2\left(\frac{\pi}{2}\right) + (-1)\cos\frac{\pi}{2}}{-\sin\frac{\pi}{2}\cos\frac{\pi}{2} - (-1)\sin\frac{\pi}{2}} = -1$; Slope at $\left(-1, -\frac{\pi}{2}\right)$ is
$\frac{-\sin^2\left(-\frac{\pi}{2}\right) + (-1)\cos\left(-\frac{\pi}{2}\right)}{-\sin\left(-\frac{\pi}{2}\right)\cos\left(-\frac{\pi}{2}\right) - (-1)\sin\left(-\frac{\pi}{2}\right)} = 1$

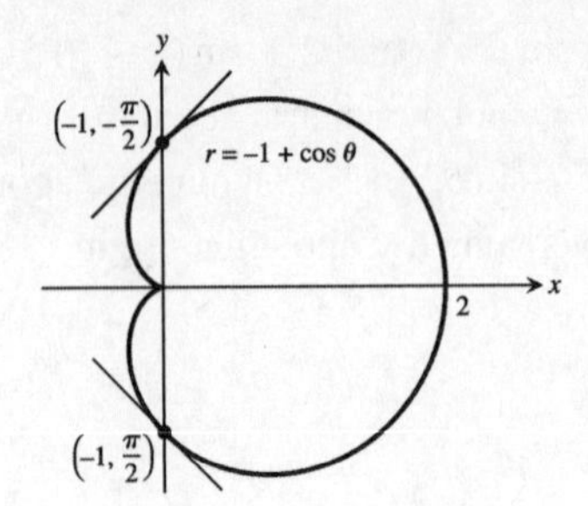

19. $\theta = \frac{\pi}{4} \Rightarrow r = 1 \Rightarrow \left(1, \frac{\pi}{4}\right)$; $\theta = -\frac{\pi}{4} \Rightarrow r = -1$
$\Rightarrow \left(-1, -\frac{\pi}{4}\right)$; $\theta = \frac{3\pi}{4} \Rightarrow r = -1 \Rightarrow \left(-1, \frac{3\pi}{4}\right)$;
$\theta = -\frac{3\pi}{4} \Rightarrow r = 1 \Rightarrow \left(1, -\frac{3\pi}{4}\right)$;
$r' = \frac{dr}{d\theta} = 2\cos 2\theta$;
Slope $= \frac{r'\sin\theta + r\cos\theta}{r'\cos\theta - r\sin\theta} = \frac{2\cos 2\theta\sin\theta + r\cos\theta}{2\cos 2\theta\cos\theta - r\sin\theta}$
$\Rightarrow$ Slope at $\left(1, \frac{\pi}{4}\right)$ is $\frac{2\cos\left(\frac{\pi}{2}\right)\sin\left(\frac{\pi}{4}\right) + (1)\cos\left(\frac{\pi}{4}\right)}{2\cos\left(\frac{\pi}{2}\right)\cos\left(\frac{\pi}{4}\right) - (1)\sin\left(\frac{\pi}{4}\right)} = -1$;

Slope at $\left(-1, -\frac{\pi}{4}\right)$ is $\frac{2\cos\left(-\frac{\pi}{2}\right)\sin\left(-\frac{\pi}{4}\right) + (-1)\cos\left(-\frac{\pi}{4}\right)}{2\cos\left(-\frac{\pi}{2}\right)\cos\left(-\frac{\pi}{4}\right) - (-1)\sin\left(-\frac{\pi}{4}\right)} = 1$;

Slope at $\left(-1, \frac{3\pi}{4}\right)$ is $\frac{2\cos\left(\frac{3\pi}{2}\right)\sin\left(\frac{3\pi}{4}\right) + (-1)\cos\left(\frac{3\pi}{4}\right)}{2\cos\left(\frac{3\pi}{2}\right)\cos\left(\frac{3\pi}{4}\right) - (-1)\sin\left(\frac{3\pi}{4}\right)} = 1$;

Slope at $\left(1, -\frac{3\pi}{4}\right)$ is $\frac{2\cos\left(-\frac{3\pi}{2}\right)\sin\left(-\frac{3\pi}{4}\right) + (1)\cos\left(-\frac{3\pi}{4}\right)}{2\cos\left(-\frac{3\pi}{2}\right)\cos\left(-\frac{3\pi}{4}\right) - (1)\sin\left(-\frac{3\pi}{4}\right)} = -1$

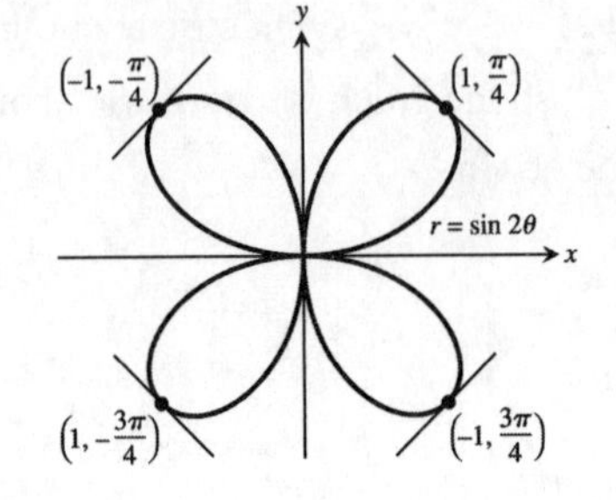

21. (a)

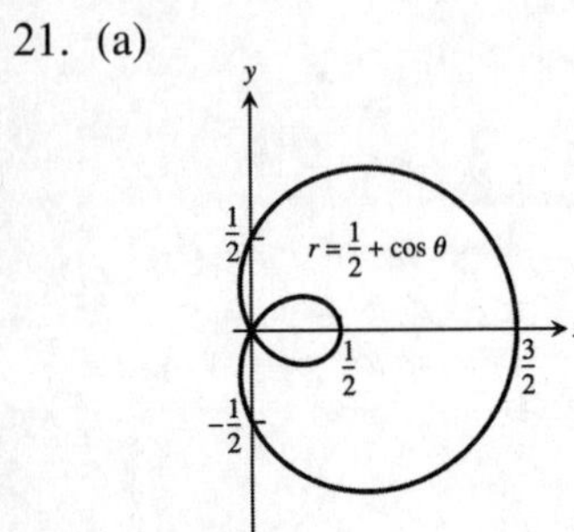

(b)

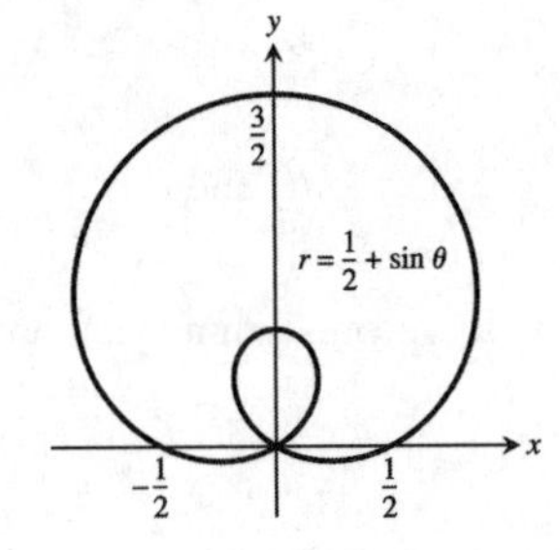

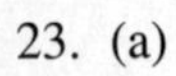

23. (a)

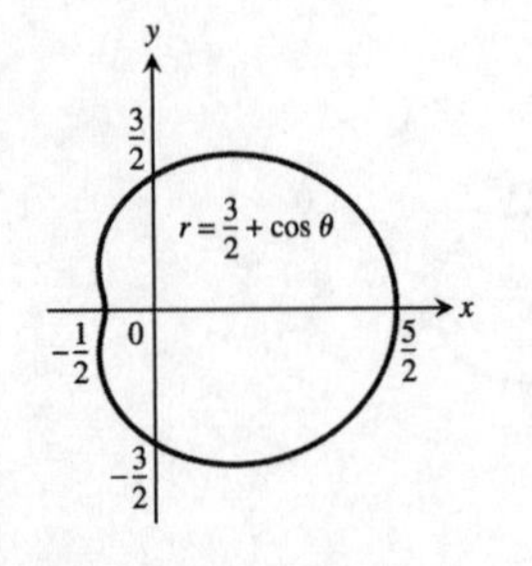

(b)

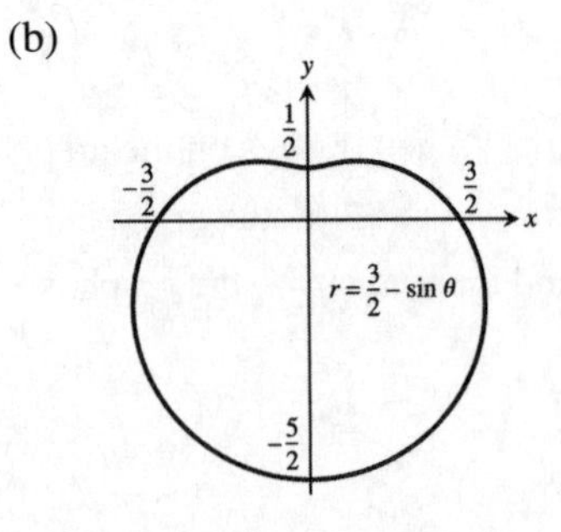

25.

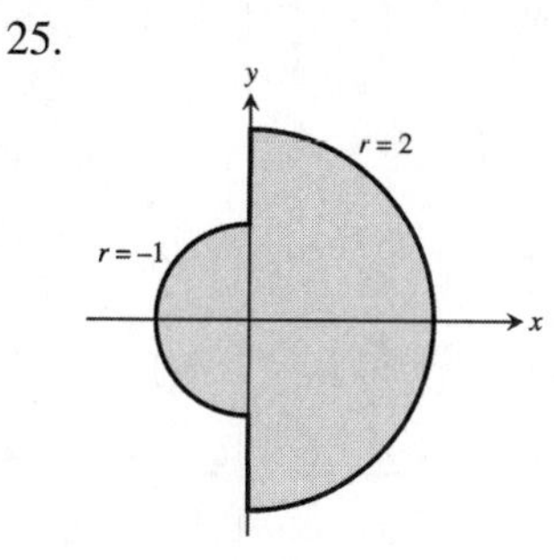

27.

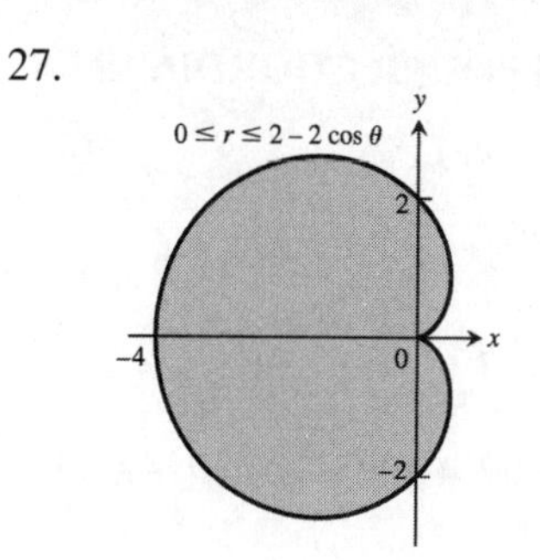

29. Note that (r, θ) and $(-r, \theta + \pi)$ describe the same point in the plane. Then $r = 1 - \cos\theta \Leftrightarrow -1 - \cos(\theta + \pi)$ $= -1 - (\cos\theta\cos\pi - \sin\theta\sin\pi) = -1 + \cos\theta = -(1 - \cos\theta) = -r$; therefore (r, θ) is on the graph of $r = 1 - \cos\theta \Leftrightarrow (-r, \theta + \pi)$ is on the graph of $r = -1 - \cos\theta \Rightarrow$ the answer is (a).

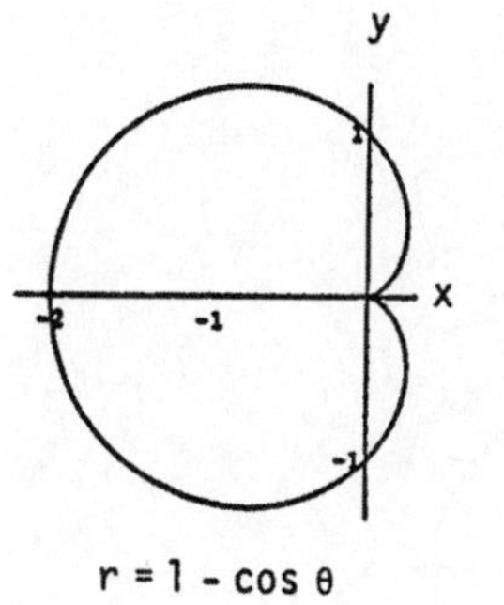

$r = 1 - \cos\theta$

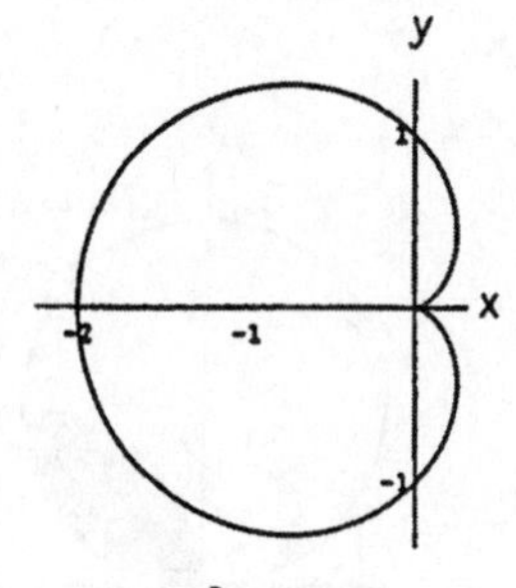

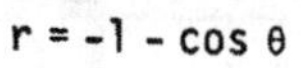

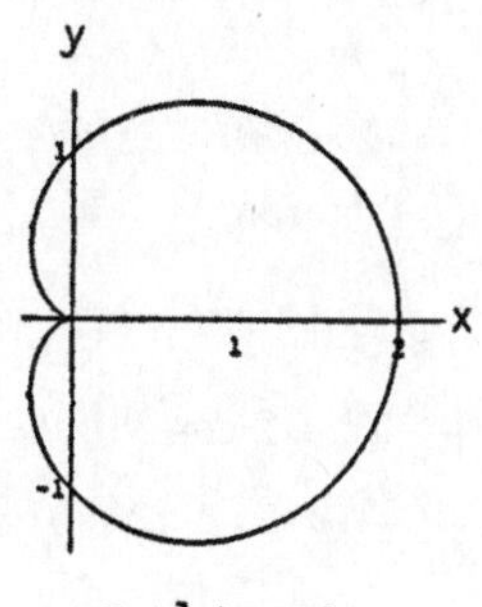

$r = 1 + \cos\theta$

31.

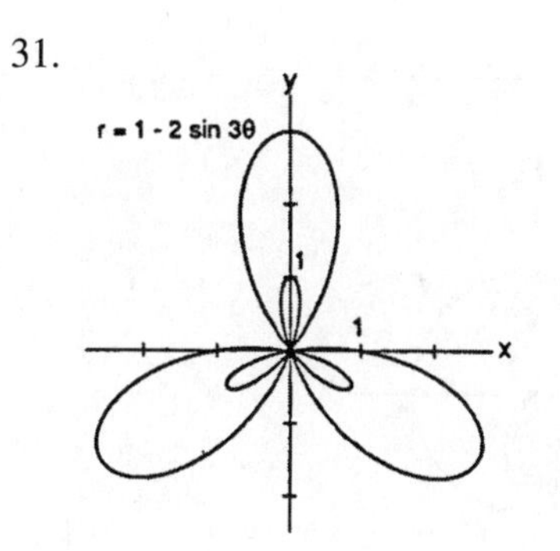

33. (a) (b) (c) (d)

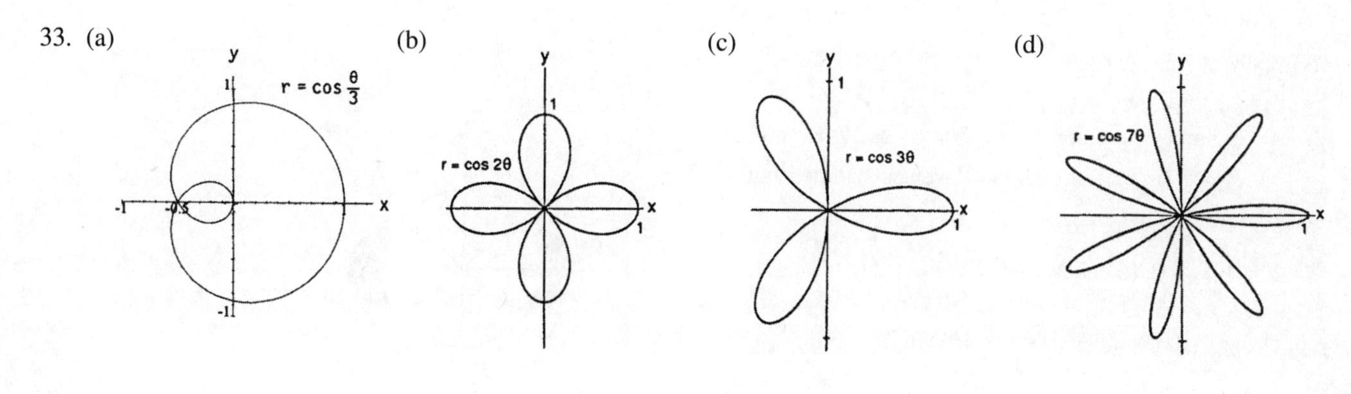

11.5 AREA AND LENGTHS IN POLAR COORDINATES

1. $A = \int_0^{\pi} \frac{1}{2}\theta^2\, d\theta = \left[\frac{1}{6}\theta^3\right]_0^{\pi} = \frac{\pi^3}{6}$

3. $A = \int_0^{2\pi} \frac{1}{2}(4 + 2\cos\theta)^2\, d\theta = \int_0^{2\pi} \frac{1}{2}(16 + 16\cos\theta + 4\cos^2\theta)\, d\theta = \int_0^{2\pi} \left[8 + 8\cos\theta + 2\left(\frac{1+\cos 2\theta}{2}\right)\right] d\theta$
$= \int_0^{2\pi} (9 + 8\cos\theta + \cos 2\theta)\, d\theta = \left[9\theta + 8\sin\theta + \frac{1}{2}\sin 2\theta\right]_0^{2\pi} = 18\pi$

5. $A = 2\int_0^{\pi/4} \frac{1}{2}\cos^2 2\theta\, d\theta = \int_0^{\pi/4} \frac{1+\cos 4\theta}{2}\, d\theta = \frac{1}{2}\left[\theta + \frac{\sin 4\theta}{4}\right]_0^{\pi/4} = \frac{\pi}{8}$

7. $A = \int_0^{\pi/2} \frac{1}{2}(4\sin 2\theta)\, d\theta = \int_0^{\pi/2} 2\sin 2\theta\, d\theta = [-\cos 2\theta]_0^{\pi/2} = 2$

9. $r = 2\cos\theta$ and $r = 2\sin\theta \Rightarrow 2\cos\theta = 2\sin\theta$
$\Rightarrow \cos\theta = \sin\theta \Rightarrow \theta = \frac{\pi}{4}$; therefore
$A = 2\int_0^{\pi/4} \frac{1}{2}(2\sin\theta)^2\, d\theta = \int_0^{\pi/4} 4\sin^2\theta\, d\theta$
$= \int_0^{\pi/4} 4\left(\frac{1-\cos 2\theta}{2}\right) d\theta = \int_0^{\pi/4} (2 - 2\cos 2\theta)\, d\theta$
$= [2\theta - \sin 2\theta]_0^{\pi/4} = \frac{\pi}{2} - 1$

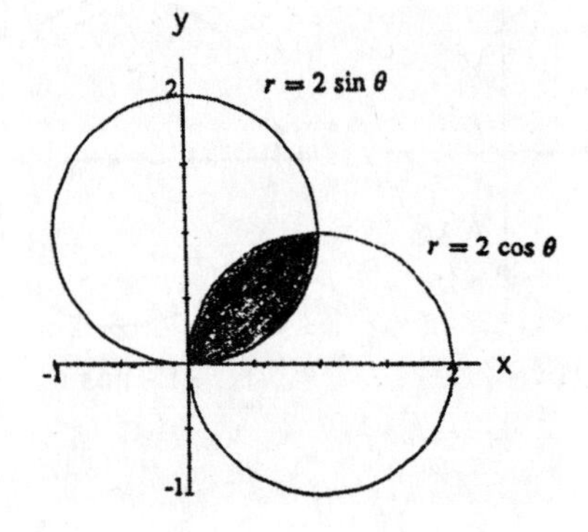

11. $r = 2$ and $r = 2(1 - \cos\theta) \Rightarrow 2 = 2(1 - \cos\theta)$
$\Rightarrow \cos\theta = 0 \Rightarrow \theta = \pm\frac{\pi}{2}$; therefore
$A = 2\int_0^{\pi/2} \frac{1}{2}[2(1 - \cos\theta)]^2\, d\theta + \frac{1}{2}$area of the circle
$= \int_0^{\pi/2} 4(1 - 2\cos\theta + \cos^2\theta)\, d\theta + \left(\frac{1}{2}\pi\right)(2)^2$
$= \int_0^{\pi/2} 4\left(1 - 2\cos\theta + \frac{1+\cos 2\theta}{2}\right) d\theta + 2\pi$
$= \int_0^{\pi/2} (4 - 8\cos\theta + 2 + 2\cos 2\theta)\, d\theta + 2\pi$
$= [6\theta - 8\sin\theta + \sin 2\theta]_0^{\pi/2} + 2\pi = 5\pi - 8$

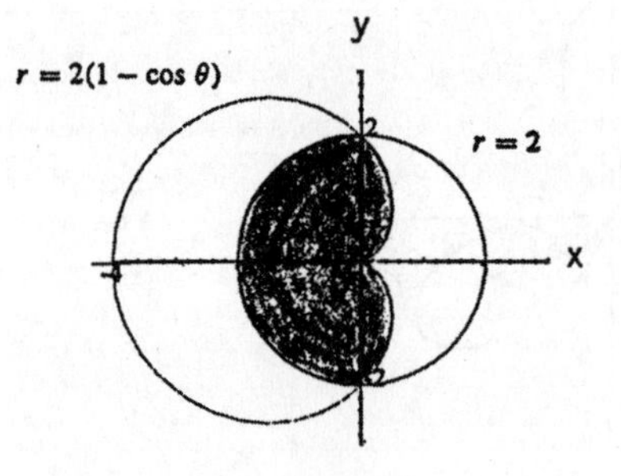

13. $r = \sqrt{3}$ and $r^2 = 6\cos 2\theta \Rightarrow 3 = 6\cos 2\theta \Rightarrow \cos 2\theta = \frac{1}{2}$
$\Rightarrow \theta = \frac{\pi}{6}$ (in the 1st quadrant); we use symmetry of the graph to find the area, so
$A = 4\int_0^{\pi/6} \left[\frac{1}{2}(6\cos 2\theta) - \frac{1}{2}\left(\sqrt{3}\right)^2\right] d\theta$
$= 2\int_0^{\pi/6} (6\cos 2\theta - 3)\, d\theta = 2[3\sin 2\theta - 3\theta]_0^{\pi/6}$
$= 3\sqrt{3} - \pi$

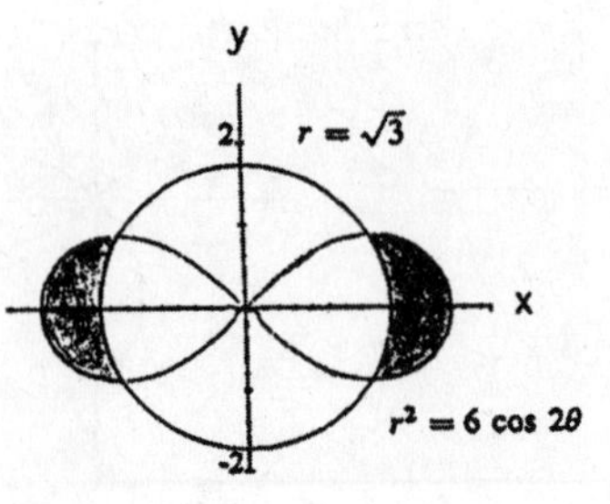

15. $r = 1$ and $r = -2\cos\theta \Rightarrow 1 = -2\cos\theta \Rightarrow \cos\theta = -\frac{1}{2}$
$\Rightarrow \theta = \frac{2\pi}{3}$ in quadrant II; therefore
$A = 2\int_{2\pi/3}^{\pi} \frac{1}{2}\left[(-2\cos\theta)^2 - 1^2\right] d\theta = \int_{2\pi/3}^{\pi} (4\cos^2\theta - 1)\, d\theta$
$= \int_{2\pi/3}^{\pi} [2(1+\cos 2\theta) - 1]\, d\theta = \int_{2\pi/3}^{\pi} (1 + 2\cos 2\theta)\, d\theta$
$= [\theta + \sin 2\theta]_{2\pi/3}^{\pi} = \frac{\pi}{3} + \frac{\sqrt{3}}{2}$

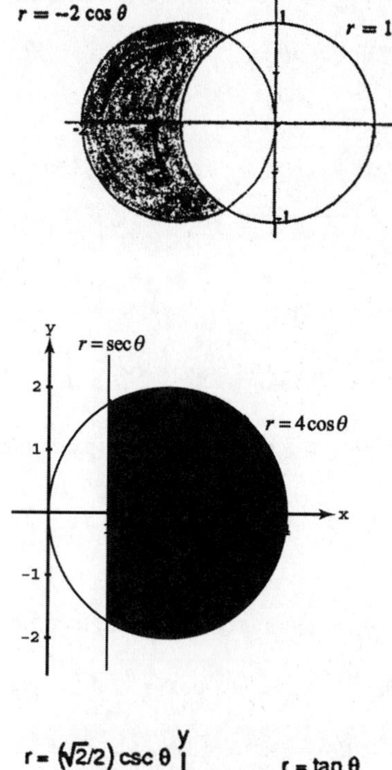

17. $r = \sec\theta$ and $r = 4\cos\theta \Rightarrow 4\cos\theta = \sec\theta \Rightarrow \cos^2\theta = \frac{1}{4}$
$\Rightarrow \theta = \frac{\pi}{3}, \frac{2\pi}{3}, \frac{4\pi}{3}$, or $\frac{5\pi}{3}$; therefore
$A = 2\int_0^{\pi/3} \frac{1}{2}\left(16\cos^2\theta - \sec^2\theta\right) d\theta$
$= \int_0^{\pi/3} (8 + 8\cos 2\theta - \sec^2\theta)\, d\theta$
$= [8\theta + 4\sin 2\theta - \tan\theta]_0^{\pi/3}$
$= \left(\frac{8\pi}{3} + 2\sqrt{3} - \sqrt{3}\right) - (0 + 0 - 0) = \frac{8\pi}{3} + \sqrt{3}$

19. (a) $r = \tan\theta$ and $r = \left(\frac{\sqrt{2}}{2}\right)\csc\theta \Rightarrow \tan\theta = \left(\frac{\sqrt{2}}{2}\right)\csc\theta$
$\Rightarrow \sin^2\theta = \left(\frac{\sqrt{2}}{2}\right)\cos\theta \Rightarrow 1 - \cos^2\theta = \left(\frac{\sqrt{2}}{2}\right)\cos\theta$
$\Rightarrow \cos^2\theta + \left(\frac{\sqrt{2}}{2}\right)\cos\theta - 1 = 0 \Rightarrow \cos\theta = -\sqrt{2}$ or
$\frac{\sqrt{2}}{2}$ (use the quadratic formula) $\Rightarrow \theta = \frac{\pi}{4}$ (the solution in the first quadrant); therefore the area of R_1 is
$A_1 = \int_0^{\pi/4} \frac{1}{2}\tan^2\theta\, d\theta = \frac{1}{2}\int_0^{\pi/4} (\sec^2\theta - 1)\, d\theta = \frac{1}{2}[\tan\theta - \theta]_0^{\pi/4} = \frac{1}{2}\left(\tan\frac{\pi}{4} - \frac{\pi}{4}\right) = \frac{1}{2} - \frac{\pi}{8}$; $AO = \left(\frac{\sqrt{2}}{2}\right)\csc\frac{\pi}{2}$
$= \frac{\sqrt{2}}{2}$ and $OB = \left(\frac{\sqrt{2}}{2}\right)\csc\frac{\pi}{4} = 1 \Rightarrow AB = \sqrt{1^2 - \left(\frac{\sqrt{2}}{2}\right)^2} = \frac{\sqrt{2}}{2} \Rightarrow$ the area of R_2 is $A_2 = \frac{1}{2}\left(\frac{\sqrt{2}}{2}\right)\left(\frac{\sqrt{2}}{2}\right) = \frac{1}{4}$;
therefore the area of the region shaded in the text is $2\left(\frac{1}{2} - \frac{\pi}{8} + \frac{1}{4}\right) = \frac{3}{2} - \frac{\pi}{4}$. Note: The area must be found this way since no common interval generates the region. For example, the interval $0 \le \theta \le \frac{\pi}{4}$ generates the arc OB of $r = \tan\theta$ but does not generate the segment AB of the liner $= \frac{\sqrt{2}}{2}\csc\theta$. Instead the interval generates the half-line from B to $+\infty$ on the line $r = \frac{\sqrt{2}}{2}\csc\theta$.

(b) $\lim_{\theta \to \pi/2^-} \tan\theta = \infty$ and the line $x = 1$ is $r = \sec\theta$ in polar coordinates; then $\lim_{\theta \to \pi/2^-} (\tan\theta - \sec\theta)$
$= \lim_{\theta \to \pi/2^-} \left(\frac{\sin\theta}{\cos\theta} - \frac{1}{\cos\theta}\right) = \lim_{\theta \to \pi/2^-} \left(\frac{\sin\theta - 1}{\cos\theta}\right) = \lim_{\theta \to \pi/2^-} \left(\frac{\cos\theta}{-\sin\theta}\right) = 0 \Rightarrow r = \tan\theta$ approaches

$r = \sec\theta$ as $\theta \to \frac{\pi^-}{2} \Rightarrow r = \sec\theta$ (or $x = 1$) is a vertical asymptote of $r = \tan\theta$. Similarly, $r = -\sec\theta$ (or $x = -1$) is a vertical asymptote of $r = \tan\theta$.

21. $r = \theta^2, 0 \le \theta \le \sqrt{5} \Rightarrow \frac{dr}{d\theta} = 2\theta$; therefore Length $= \int_0^{\sqrt{5}} \sqrt{(\theta^2)^2 + (2\theta)^2}\, d\theta = \int_0^{\sqrt{5}} \sqrt{\theta^4 + 4\theta^2}\, d\theta$
$= \int_0^{\sqrt{5}} |\theta|\sqrt{\theta^2 + 4}\, d\theta =$ (since $\theta \ge 0$) $\int_0^{\sqrt{5}} \theta\sqrt{\theta^2 + 4}\, d\theta$; $\left[u = \theta^2 + 4 \Rightarrow \frac{1}{2}\, du = \theta\, d\theta; \theta = 0 \Rightarrow u = 4,\right.$
$\left.\theta = \sqrt{5} \Rightarrow u = 9\right] \to \int_4^9 \frac{1}{2}\sqrt{u}\, du = \frac{1}{2}\left[\frac{2}{3}u^{3/2}\right]_4^9 = \frac{19}{3}$

23. $r = 1 + \cos\theta \Rightarrow \frac{dr}{d\theta} = -\sin\theta$; therefore Length $= \int_0^{2\pi} \sqrt{(1+\cos\theta)^2 + (-\sin\theta)^2}\, d\theta$
$= 2\int_0^{\pi} \sqrt{2 + 2\cos\theta}\, d\theta = 2\int_0^{\pi} \sqrt{\frac{4(1+\cos\theta)}{2}}\, d\theta = 4\int_0^{\pi} \sqrt{\frac{1+\cos\theta}{2}}\, d\theta = 4\int_0^{\pi} \cos\left(\frac{\theta}{2}\right) d\theta = 4\left[2\sin\frac{\theta}{2}\right]_0^{\pi} = 8$

25. $r = \frac{6}{1+\cos\theta}, 0 \le \theta \le \frac{\pi}{2} \Rightarrow \frac{dr}{d\theta} = \frac{6\sin\theta}{(1+\cos\theta)^2}$; therefore Length $= \int_0^{\pi/2} \sqrt{\left(\frac{6}{1+\cos\theta}\right)^2 + \left(\frac{6\sin\theta}{(1+\cos\theta)^2}\right)^2}\, d\theta$
$= \int_0^{\pi/2} \sqrt{\frac{36}{(1+\cos\theta)^2} + \frac{36\sin^2\theta}{(1+\cos\theta)^4}}\, d\theta = 6\int_0^{\pi/2} \left|\frac{1}{1+\cos\theta}\right| \sqrt{1 + \frac{\sin^2\theta}{(1+\cos\theta)^2}}\, d\theta$
$= \left(\text{since } \frac{1}{1+\cos\theta} > 0 \text{ on } 0 \le \theta \le \frac{\pi}{2}\right)\ 6\int_0^{\pi/2} \left(\frac{1}{1+\cos\theta}\right) \sqrt{\frac{1 + 2\cos\theta + \cos^2\theta + \sin^2\theta}{(1+\cos\theta)^2}}\, d\theta$
$= 6\int_0^{\pi/2} \left(\frac{1}{1+\cos\theta}\right) \sqrt{\frac{2+2\cos\theta}{(1+\cos\theta)^2}}\, d\theta = 6\sqrt{2}\int_0^{\pi/2} \frac{d\theta}{(1+\cos\theta)^{3/2}} = 6\sqrt{2}\int_0^{\pi/2} \frac{d\theta}{\left(2\cos^2\frac{\theta}{2}\right)^{3/2}} = 3\int_0^{\pi/2} \left|\sec^3\frac{\theta}{2}\right| d\theta$
$= 3\int_0^{\pi/2} \sec^3\frac{\theta}{2}\, d\theta = 6\int_0^{\pi/4} \sec^3 u\, du = \text{(use tables) } 6\left(\left[\frac{\sec u \tan u}{2}\right]_0^{\pi/4} + \frac{1}{2}\int_0^{\pi/4} \sec u\, du\right)$
$= 6\left(\frac{1}{\sqrt{2}} + \left[\frac{1}{2}\ln|\sec u + \tan u|\right]_0^{\pi/4}\right) = 3\left[\sqrt{2} + \ln\left(1+\sqrt{2}\right)\right]$

27. $r = \cos^3\frac{\theta}{3} \Rightarrow \frac{dr}{d\theta} = -\sin\frac{\theta}{3}\cos^2\frac{\theta}{3}$; therefore Length $= \int_0^{\pi/4} \sqrt{\left(\cos^3\frac{\theta}{3}\right)^2 + \left(-\sin\frac{\theta}{3}\cos^2\frac{\theta}{3}\right)^2}\, d\theta$
$= \int_0^{\pi/4} \sqrt{\cos^6\left(\frac{\theta}{3}\right) + \sin^2\left(\frac{\theta}{3}\right)\cos^4\left(\frac{\theta}{3}\right)}\, d\theta = \int_0^{\pi/4} \left(\cos^2\frac{\theta}{3}\right) \sqrt{\cos^2\left(\frac{\theta}{3}\right) + \sin^2\left(\frac{\theta}{3}\right)}\, d\theta = \int_0^{\pi/4} \cos^2\left(\frac{\theta}{3}\right) d\theta$
$= \int_0^{\pi/4} \frac{1+\cos\left(\frac{2\theta}{3}\right)}{2}\, d\theta = \frac{1}{2}\left[\theta + \frac{3}{2}\sin\frac{2\theta}{3}\right]_0^{\pi/4} = \frac{\pi}{8} + \frac{3}{8}$

29. Let $r = f(\theta)$. Then $x = f(\theta)\cos\theta \Rightarrow \frac{dx}{d\theta} = f'(\theta)\cos\theta - f(\theta)\sin\theta \Rightarrow \left(\frac{dx}{d\theta}\right)^2 = [f'(\theta)\cos\theta - f(\theta)\sin\theta]^2$
$= [f'(\theta)]^2\cos^2\theta - 2f'(\theta)\,f(\theta)\sin\theta\cos\theta + [f(\theta)]^2\sin^2\theta$; $y = f(\theta)\sin\theta \Rightarrow \frac{dy}{d\theta} = f'(\theta)\sin\theta + f(\theta)\cos\theta$
$\Rightarrow \left(\frac{dy}{d\theta}\right)^2 = [f'(\theta)\sin\theta + f(\theta)\cos\theta]^2 = [f'(\theta)]^2\sin^2\theta + 2f'(\theta)f(\theta)\sin\theta\cos\theta + [f(\theta)]^2\cos^2\theta$. Therefore
$\left(\frac{dx}{d\theta}\right)^2 + \left(\frac{dy}{d\theta}\right)^2 = [f'(\theta)]^2(\cos^2\theta + \sin^2\theta) + [f(\theta)]^2(\cos^2\theta + \sin^2\theta) = [f'(\theta)]^2 + [f(\theta)]^2 = r^2 + \left(\frac{dr}{d\theta}\right)^2$.
Thus, $L = \int_\alpha^\beta \sqrt{\left(\frac{dx}{d\theta}\right)^2 + \left(\frac{dy}{d\theta}\right)^2}\, d\theta = \int_\alpha^\beta \sqrt{r^2 + \left(\frac{dr}{d\theta}\right)^2}\, d\theta$.

31. (a) $r_{av} = \frac{1}{2\pi - 0}\int_0^{2\pi} a(1-\cos\theta)\, d\theta = \frac{a}{2\pi}[\theta - \sin\theta]_0^{2\pi} = a$
(b) $r_{av} = \frac{1}{2\pi - 0}\int_0^{2\pi} a\, d\theta = \frac{1}{2\pi}[a\theta]_0^{2\pi} = a$
(c) $r_{av} = \frac{1}{\left(\frac{\pi}{2}\right) - \left(-\frac{\pi}{2}\right)}\int_{-\pi/2}^{\pi/2} a\cos\theta\, d\theta = \frac{1}{\pi}[a\sin\theta]_{-\pi/2}^{\pi/2} = \frac{2a}{\pi}$

11.6 CONIC SECTIONS

1. $x = \frac{y^2}{8} \Rightarrow 4p = 8 \Rightarrow p = 2$; focus is $(2, 0)$, directrix is $x = -2$

3. $y = -\frac{x^2}{6} \Rightarrow 4p = 6 \Rightarrow p = \frac{3}{2}$; focus is $\left(0, -\frac{3}{2}\right)$, directrix is $y = \frac{3}{2}$

5. $\frac{x^2}{4} - \frac{y^2}{9} = 1 \Rightarrow c = \sqrt{4+9} = \sqrt{13} \Rightarrow$ foci are $\left(\pm\sqrt{13}, 0\right)$; vertices are $(\pm 2, 0)$; asymptotes are $y = \pm\frac{3}{2}x$

7. $\frac{x^2}{2} + y^2 = 1 \Rightarrow c = \sqrt{2-1} = 1 \Rightarrow$ foci are $(\pm 1, 0)$; vertices are $\left(\pm\sqrt{2}, 0\right)$

9. $y^2 = 12x \Rightarrow x = \frac{y^2}{12} \Rightarrow 4p = 12 \Rightarrow p = 3$;
focus is $(3, 0)$, directrix is $x = -3$

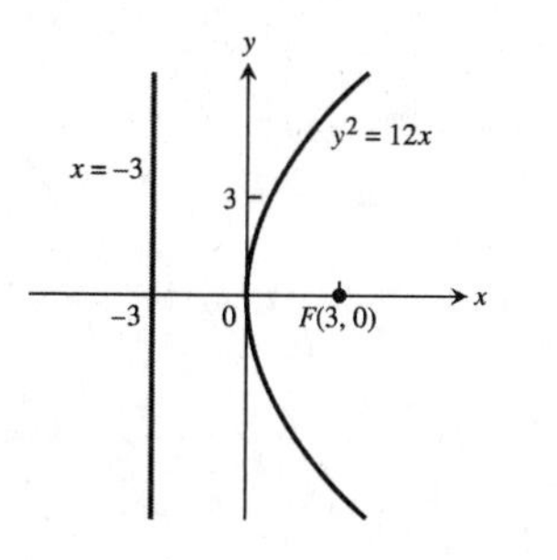

11. $x^2 = -8y \Rightarrow y = \frac{x^2}{-8} \Rightarrow 4p = 8 \Rightarrow p = 2$;
focus is $(0, -2)$, directrix is $y = 2$

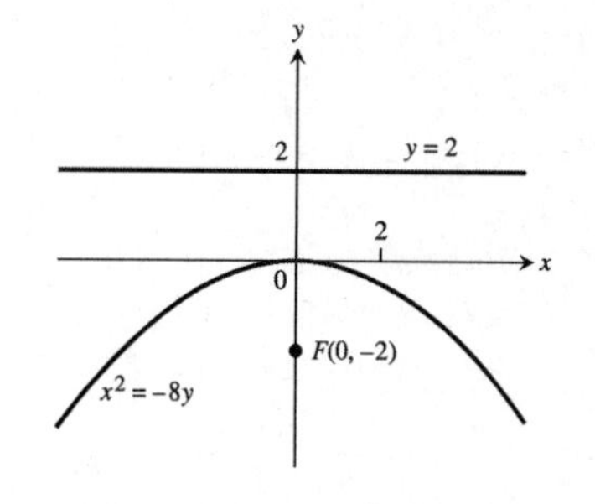

13. $y = 4x^2 \Rightarrow y = \frac{x^2}{\left(\frac{1}{4}\right)} \Rightarrow 4p = \frac{1}{4} \Rightarrow p = \frac{1}{16}$;
focus is $\left(0, \frac{1}{16}\right)$, directrix is $y = -\frac{1}{16}$

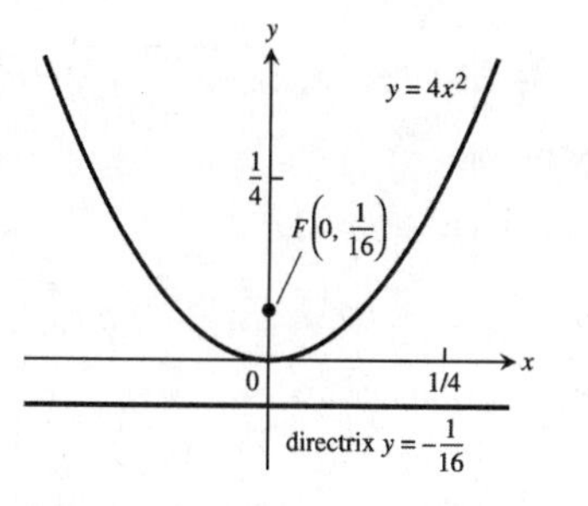

15. $x = -3y^2 \Rightarrow x = -\frac{y^2}{\left(\frac{1}{3}\right)} \Rightarrow 4p = \frac{1}{3} \Rightarrow p = \frac{1}{12}$;
focus is $\left(-\frac{1}{12}, 0\right)$, directrix is $x = \frac{1}{12}$

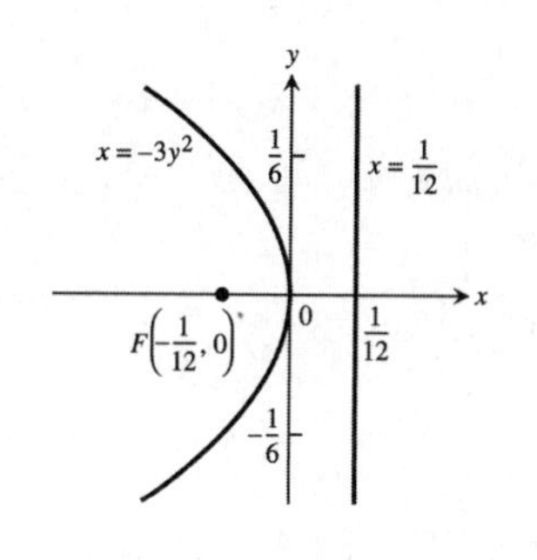

17. $16x^2 + 25y^2 = 400 \Rightarrow \frac{x^2}{25} + \frac{y^2}{16} = 1$
$\Rightarrow c = \sqrt{a^2 - b^2} = \sqrt{25 - 16} = 3$

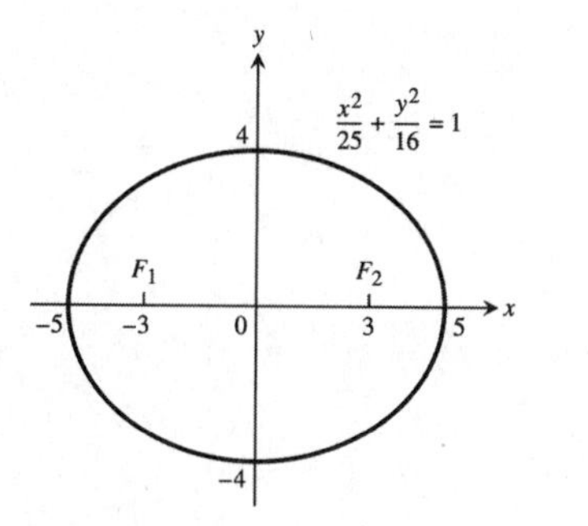

19. $2x^2 + y^2 = 2 \Rightarrow x^2 + \frac{y^2}{2} = 1$
$\Rightarrow c = \sqrt{a^2 - b^2} = \sqrt{2 - 1} = 1$

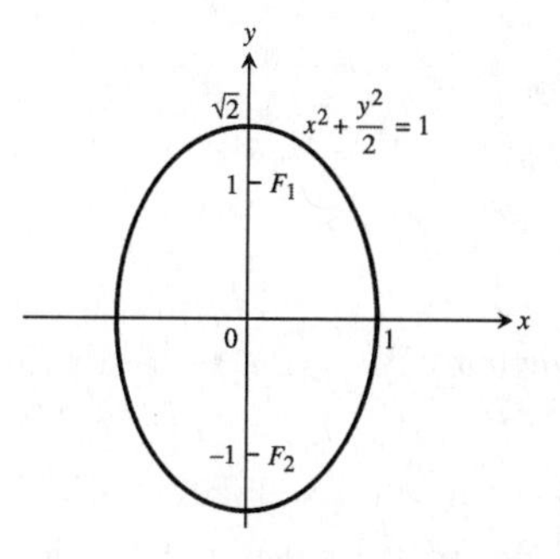

21. $3x^2 + 2y^2 = 6 \Rightarrow \frac{x^2}{2} + \frac{y^2}{3} = 1$
$\Rightarrow c = \sqrt{a^2 - b^2} = \sqrt{3 - 2} = 1$

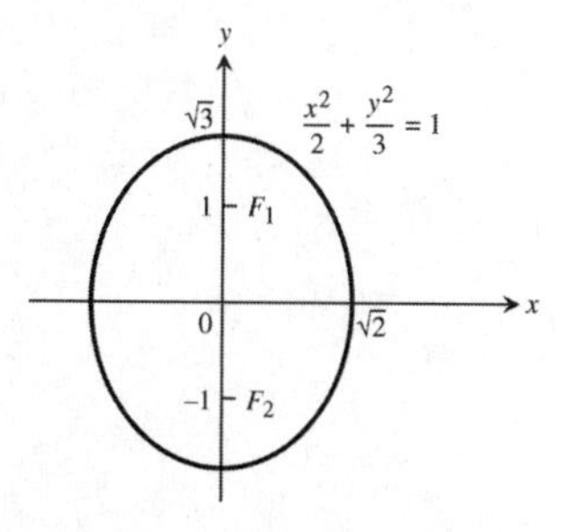

23. $6x^2 + 9y^2 = 54 \Rightarrow \frac{x^2}{9} + \frac{y^2}{6} = 1$
$\Rightarrow c = \sqrt{a^2 - b^2} = \sqrt{9 - 6} = \sqrt{3}$

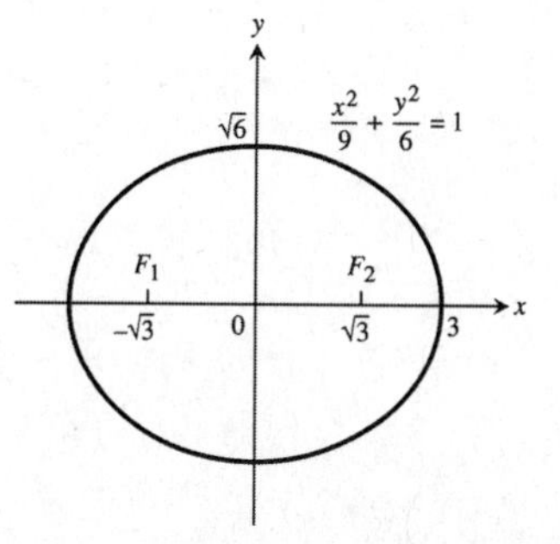

25. Foci: $\left(\pm\sqrt{2},0\right)$, Vertices: $(\pm 2,0) \Rightarrow a=2, c=\sqrt{2} \Rightarrow b^2=a^2-c^2=4-\left(\sqrt{2}\right)^2=2 \Rightarrow \frac{x^2}{4}+\frac{y^2}{2}=1$

27. $x^2-y^2=1 \Rightarrow c=\sqrt{a^2+b^2}=\sqrt{1+1}=\sqrt{2}$; asymptotes are $y=\pm x$

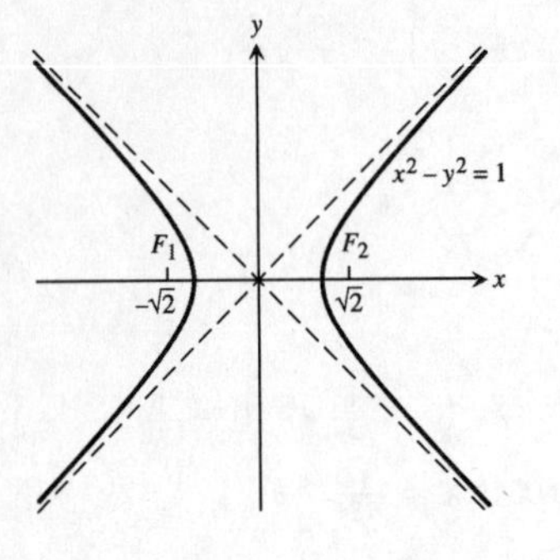

29. $y^2-x^2=8 \Rightarrow \frac{y^2}{8}-\frac{x^2}{8}=1 \Rightarrow c=\sqrt{a^2+b^2}$ $=\sqrt{8+8}=4$; asymptotes are $y=\pm x$

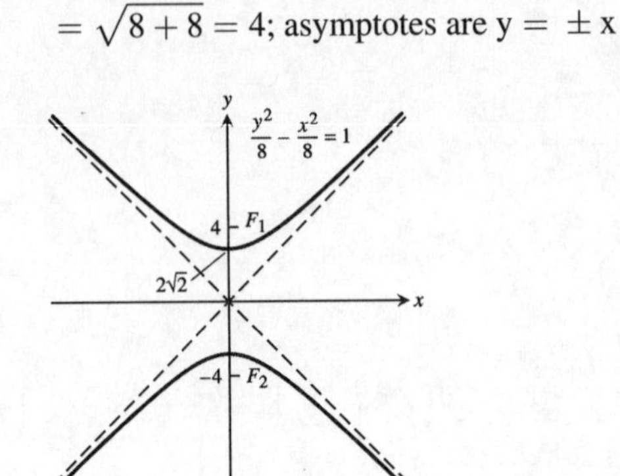

31. $8x^2-2y^2=16 \Rightarrow \frac{x^2}{2}-\frac{y^2}{8}=1 \Rightarrow c=\sqrt{a^2+b^2}$ $=\sqrt{2+8}=\sqrt{10}$; asymptotes are $y=\pm 2x$

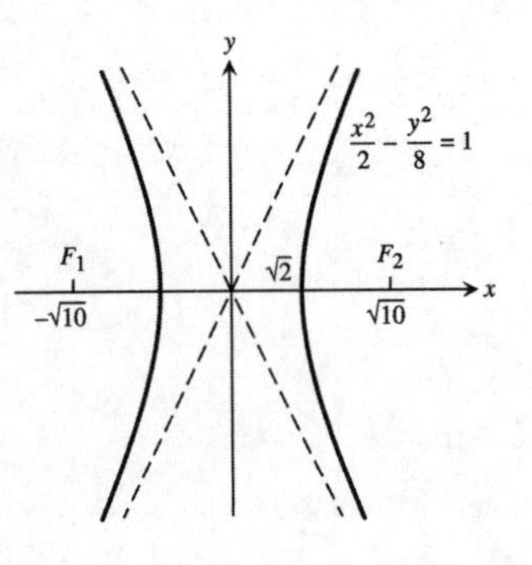

33. $8y^2-2x^2=16 \Rightarrow \frac{y^2}{2}-\frac{x^2}{8}=1 \Rightarrow c=\sqrt{a^2+b^2}$ $=\sqrt{2+8}=\sqrt{10}$; asymptotes are $y=\pm\frac{x}{2}$

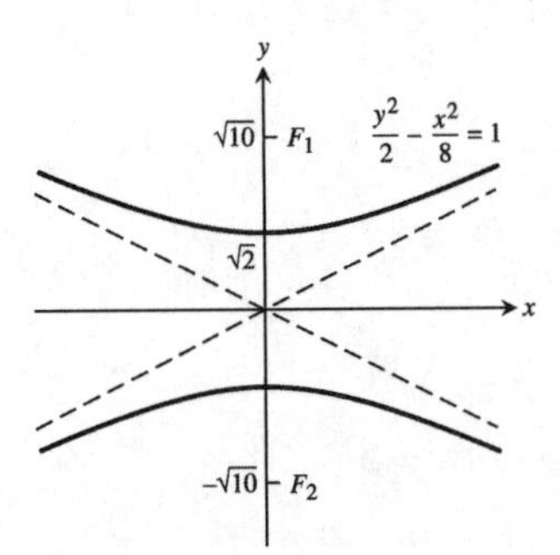

35. Foci: $\left(0,\pm\sqrt{2}\right)$, Asymptotes: $y=\pm x \Rightarrow c=\sqrt{2}$ and $\frac{a}{b}=1 \Rightarrow a=b \Rightarrow c^2=a^2+b^2=2a^2 \Rightarrow 2=2a^2$ $\Rightarrow a=1 \Rightarrow b=1 \Rightarrow y^2-x^2=1$

37. Vertices: $(\pm 3,0)$, Asymptotes: $y=\pm\frac{4}{3}x \Rightarrow a=3$ and $\frac{b}{a}=\frac{4}{3} \Rightarrow b=\frac{4}{3}(3)=4 \Rightarrow \frac{x^2}{9}-\frac{y^2}{16}=1$

39. (a) $y^2=8x \Rightarrow 4p=8 \Rightarrow p=2 \Rightarrow$ directrix is $x=-2$, focus is $(2,0)$, and vertex is $(0,0)$; therefore the new directrix is $x=-1$, the new focus is $(3,-2)$, and the new vertex is $(1,-2)$

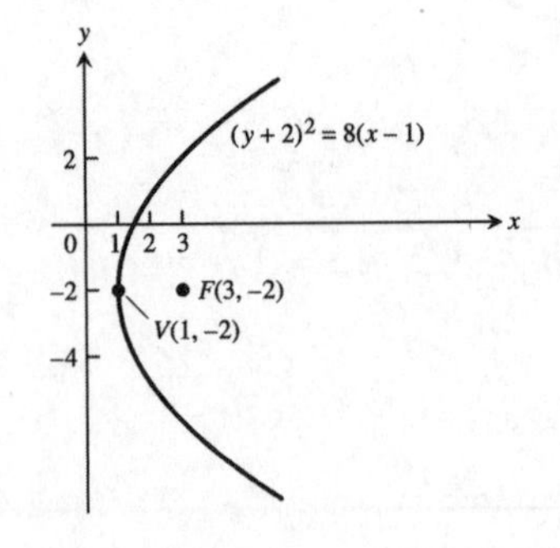

41. (a) $\frac{x^2}{16} + \frac{y^2}{9} = 1 \Rightarrow$ center is $(0, 0)$, vertices are $(-4, 0)$ and $(4, 0)$; $c = \sqrt{a^2 - b^2} = \sqrt{7} \Rightarrow$ foci are $\left(\sqrt{7}, 0\right)$ and $\left(-\sqrt{7}, 0\right)$; therefore the new center is $(4, 3)$, the new vertices are $(0, 3)$ and $(8, 3)$, and the new foci are $\left(4 \pm \sqrt{7}, 3\right)$

(b)

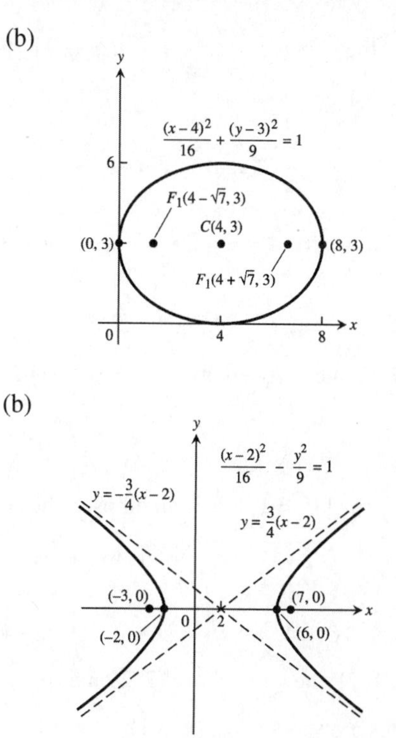

43. (a) $\frac{x^2}{16} - \frac{y^2}{9} = 1 \Rightarrow$ center is $(0, 0)$, vertices are $(-4, 0)$ and $(4, 0)$, and the asymptotes are $\frac{x}{4} = \pm \frac{y}{3}$ or $y = \pm \frac{3x}{4}$; $c = \sqrt{a^2 + b^2} = \sqrt{25} = 5 \Rightarrow$ foci are $(-5, 0)$ and $(5, 0)$; therefore the new center is $(2, 0)$, the new vertices are $(-2, 0)$ and $(6, 0)$, the new foci are $(-3, 0)$ and $(7, 0)$, and the new asymptotes are $y = \pm \frac{3(x-2)}{4}$

(b)

45. $y^2 = 4x \Rightarrow 4p = 4 \Rightarrow p = 1 \Rightarrow$ focus is $(1, 0)$, directrix is $x = -1$, and vertex is $(0, 0)$; therefore the new vertex is $(-2, -3)$, the new focus is $(-1, -3)$, and the new directrix is $x = -3$; the new equation is $(y + 3)^2 = 4(x + 2)$

47. $x^2 = 8y \Rightarrow 4p = 8 \Rightarrow p = 2 \Rightarrow$ focus is $(0, 2)$, directrix is $y = -2$, and vertex is $(0, 0)$; therefore the new vertex is $(1, -7)$, the new focus is $(1, -5)$, and the new directrix is $y = -9$; the new equation is $(x - 1)^2 = 8(y + 7)$

49. $\frac{x^2}{6} + \frac{y^2}{9} = 1 \Rightarrow$ center is $(0, 0)$, vertices are $(0, 3)$ and $(0, -3)$; $c = \sqrt{a^2 - b^2} = \sqrt{9 - 6} = \sqrt{3} \Rightarrow$ foci are $\left(0, \sqrt{3}\right)$ and $\left(0, -\sqrt{3}\right)$; therefore the new center is $(-2, -1)$, the new vertices are $(-2, 2)$ and $(-2, -4)$, and the new foci are $\left(-2, -1 \pm \sqrt{3}\right)$; the new equation is $\frac{(x+2)^2}{6} + \frac{(y+1)^2}{9} = 1$

51. $\frac{x^2}{3} + \frac{y^2}{2} = 1 \Rightarrow$ center is $(0, 0)$, vertices are $\left(\sqrt{3}, 0\right)$ and $\left(-\sqrt{3}, 0\right)$; $c = \sqrt{a^2 - b^2} = \sqrt{3 - 2} = 1 \Rightarrow$ foci are $(-1, 0)$ and $(1, 0)$; therefore the new center is $(2, 3)$, the new vertices are $\left(2 \pm \sqrt{3}, 3\right)$, and the new foci are $(1, 3)$ and $(3, 3)$; the new equation is $\frac{(x-2)^2}{3} + \frac{(y-3)^2}{2} = 1$

53. $\frac{x^2}{4} - \frac{y^2}{5} = 1 \Rightarrow$ center is $(0, 0)$, vertices are $(2, 0)$ and $(-2, 0)$; $c = \sqrt{a^2 + b^2} = \sqrt{4 + 5} = 3 \Rightarrow$ foci are $(3, 0)$ and $(-3, 0)$; the asymptotes are $\pm \frac{x}{2} = \frac{y}{\sqrt{5}} \Rightarrow y = \pm \frac{\sqrt{5}x}{2}$; therefore the new center is $(2, 2)$, the new vertices are $(4, 2)$ and $(0, 2)$, and the new foci are $(5, 2)$ and $(-1, 2)$; the new asymptotes are $y - 2 = \pm \frac{\sqrt{5}(x-2)}{2}$; the new equation is $\frac{(x-2)^2}{4} - \frac{(y-2)^2}{5} = 1$

55. $y^2 - x^2 = 1 \Rightarrow$ center is $(0, 0)$, vertices are $(0, 1)$ and $(0, -1)$; $c = \sqrt{a^2 + b^2} = \sqrt{1 + 1} = \sqrt{2} \Rightarrow$ foci are $\left(0, \pm\sqrt{2}\right)$; the asymptotes are $y = \pm x$; therefore the new center is $(-1, -1)$, the new vertices are $(-1, 0)$ and

$(-1,-2)$, and the new foci are $\left(-1,-1\pm\sqrt{2}\right)$; the new asymptotes are $y+1=\pm(x+1)$; the new equation is $(y+1)^2-(x+1)^2=1$

57. $x^2+4x+y^2=12 \Rightarrow x^2+4x+4+y^2=12+4 \Rightarrow (x+2)^2+y^2=16$; this is a circle: center at $C(-2,0)$, $a=4$

59. $x^2+2x+4y-3=0 \Rightarrow x^2+2x+1=-4y+3+1 \Rightarrow (x+1)^2=-4(y-1)$; this is a parabola: $V(-1,1)$, $F(-1,0)$

61. $x^2+5y^2+4x=1 \Rightarrow x^2+4x+4+5y^2=5 \Rightarrow (x+2)^2+5y^2=5 \Rightarrow \frac{(x+2)^2}{5}+y^2=1$; this is an ellipse: the center is $(-2,0)$, the vertices are $\left(-2\pm\sqrt{5},0\right)$; $c=\sqrt{a^2-b^2}=\sqrt{5-1}=2 \Rightarrow$ the foci are $(-4,0)$ and $(0,0)$

63. $x^2+2y^2-2x-4y=-1 \Rightarrow x^2-2x+1+2\left(y^2-2y+1\right)=2 \Rightarrow (x-1)^2+2(y-1)^2=2$
$\Rightarrow \frac{(x-1)^2}{2}+(y-1)^2=1$; this is an ellipse: the center is $(1,1)$, the vertices are $\left(1\pm\sqrt{2},1\right)$;
$c=\sqrt{a^2-b^2}=\sqrt{2-1}=1 \Rightarrow$ the foci are $(2,1)$ and $(0,1)$

65. $x^2-y^2-2x+4y=4 \Rightarrow x^2-2x+1-\left(y^2-4y+4\right)=1 \Rightarrow (x-1)^2-(y-2)^2=1$; this is a hyperbola: the center is $(1,2)$, the vertices are $(2,2)$ and $(0,2)$; $c=\sqrt{a^2+b^2}=\sqrt{1+1}=\sqrt{2} \Rightarrow$ the foci are $\left(1\pm\sqrt{2},2\right)$; the asymptotes are $y-2=\pm(x-1)$

67. $2x^2-y^2+6y=3 \Rightarrow 2x^2-\left(y^2-6y+9\right)=-6 \Rightarrow \frac{(y-3)^2}{6}-\frac{x^2}{3}=1$; this is a hyperbola: the center is $(0,3)$, the vertices are $\left(0,3\pm\sqrt{6}\right)$; $c=\sqrt{a^2+b^2}=\sqrt{6+3}=3 \Rightarrow$ the foci are $(0,6)$ and $(0,0)$; the asymptotes are $\frac{y-3}{\sqrt{6}}=\pm\frac{x}{\sqrt{3}} \Rightarrow y=\pm\sqrt{2}x+3$

69. (a) $y^2=kx \Rightarrow x=\frac{y^2}{k}$; the volume of the solid formed by revolving R_1 about the y-axis is $V_1=\int_0^{\sqrt{kx}}\pi\left(\frac{y^2}{k}\right)^2 dy$ $=\frac{\pi}{k^2}\int_0^{\sqrt{kx}} y^4\,dy=\frac{\pi x^2\sqrt{kx}}{5}$; the volume of the right circular cylinder formed by revolving PQ about the y-axis is $V_2=\pi x^2\sqrt{kx} \Rightarrow$ the volume of the solid formed by revolving R_2 about the y-axis is $V_3=V_2-V_1=\frac{4\pi x^2\sqrt{kx}}{5}$. Therefore we can see the ratio of V_3 to V_1 is 4:1.

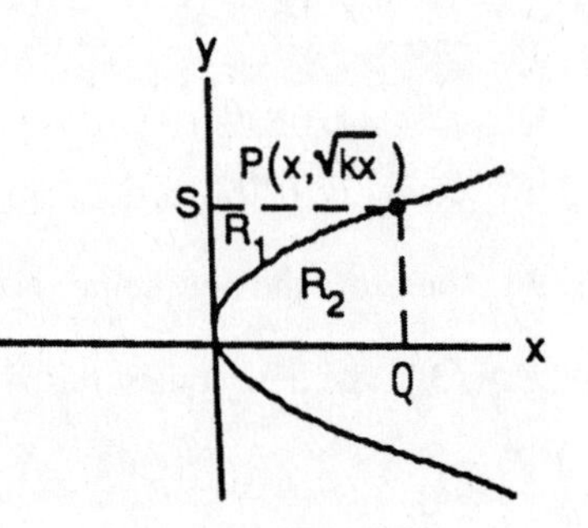

(b) The volume of the solid formed by revolving R_2 about the x-axis is $V_1=\int_0^x\pi\left(\sqrt{kt}\right)^2 dt=\pi k\int_0^x t\,dt$ $=\frac{\pi kx^2}{2}$. The volume of the right circular cylinder formed by revolving PS about the x-axis is $V_2=\pi\left(\sqrt{kx}\right)^2 x=\pi kx^2 \Rightarrow$ the volume of the solid formed by revolving R_1 about the x-axis is $V_3=V_2-V_1=\pi kx^2-\frac{\pi kx^2}{2}=\frac{\pi kx^2}{2}$. Therefore the ratio of V_3 to V_1 is 1:1.

71. $x^2=4py$ and $y=p \Rightarrow x^2=4p^2 \Rightarrow x=\pm 2p$. Therefore the line $y=p$ cuts the parabola at points $(-2p,p)$ and $(2p,p)$, and these points are $\sqrt{[2p-(-2p)]^2+(p-p)^2}=4p$ units apart.

73. Let $y=\sqrt{1-\frac{x^2}{4}}$ on the interval $0\le x\le 2$. The area of the inscribed rectangle is given by $A(x)=2x\left(2\sqrt{1-\frac{x^2}{4}}\right)=4x\sqrt{1-\frac{x^2}{4}}$ (since the length is 2x and the height is 2y)

$\Rightarrow A'(x) = 4\sqrt{1-\frac{x^2}{4}} - \frac{x^2}{\sqrt{1-\frac{x^2}{4}}}$. Thus $A'(x) = 0 \Rightarrow 4\sqrt{1-\frac{x^2}{4}} - \frac{x^2}{\sqrt{1-\frac{x^2}{4}}} = 0 \Rightarrow 4\left(1-\frac{x^2}{4}\right) - x^2 = 0 \Rightarrow x^2 = 2$

$\Rightarrow x = \sqrt{2}$ (only the positive square root lies in the interval). Since $A(0) = A(2) = 0$ we have that $A\left(\sqrt{2}\right) = 4$ is the maximum area when the length is $2\sqrt{2}$ and the height is $\sqrt{2}$.

75. $9x^2 - 4y^2 = 36 \Rightarrow y^2 = \frac{9x^2-36}{4} \Rightarrow y = \pm\frac{3}{2}\sqrt{x^2-4}$ on the interval $2 \le x \le 4 \Rightarrow V = \int_2^4 \pi\left(\frac{3}{2}\sqrt{x^2-4}\right)^2 dx$
$= \frac{9\pi}{4}\int_2^4 (x^2-4)\,dx = \frac{9\pi}{4}\left[\frac{x^3}{3} - 4x\right]_2^4 = \frac{9\pi}{4}\left[\left(\frac{64}{3}-16\right) - \left(\frac{8}{3}-8\right)\right] = \frac{9\pi}{4}\left(\frac{56}{3}-8\right) = \frac{3\pi}{4}(56-24) = 24\pi$

77. $(x-2)^2 + (y-1)^2 = 5 \Rightarrow 2(x-2) + 2(y-1)\frac{dy}{dx} = 0 \Rightarrow \frac{dy}{dx} = -\frac{x-2}{y-1}$; $y = 0 \Rightarrow (x-2)^2 + (0-1)^2 = 5$
$\Rightarrow (x-2)^2 = 4 \Rightarrow x = 4$ or $x = 0 \Rightarrow$ the circle crosses the x-axis at $(4, 0)$ and $(0, 0)$; $x = 0$
$\Rightarrow (0-2)^2 + (y-1)^2 = 5 \Rightarrow (y-1)^2 = 1 \Rightarrow y = 2$ or $y = 0 \Rightarrow$ the circle crosses the y-axis at $(0, 2)$ and $(0, 0)$.
At $(4, 0)$: $\frac{dy}{dx} = -\frac{4-2}{0-1} = 2 \Rightarrow$ the tangent line is $y = 2(x-4)$ or $y = 2x - 8$
At $(0, 0)$: $\frac{dy}{dx} = -\frac{0-2}{0-1} = -2 \Rightarrow$ the tangent line is $y = -2x$
At $(0, 2)$: $\frac{dy}{dx} = -\frac{0-2}{2-1} = 2 \Rightarrow$ the tangent line is $y - 2 = 2x$ or $y = 2x + 2$

79. Let $y = \sqrt{16 - \frac{16}{9}x^2}$ on the interval $-3 \le x \le 3$. Since the plate is symmetric about the y-axis, $\bar{x} = 0$. For a vertical strip: $(\tilde{x}, \tilde{y}) = \left(x, \frac{\sqrt{16-\frac{16}{9}x^2}}{2}\right)$, length $= \sqrt{16-\frac{16}{9}x^2}$, width $= dx \Rightarrow$ area $= dA = \sqrt{16-\frac{16}{9}x^2}\,dx$
$\Rightarrow$ mass $= dm = \delta\,dA = \delta\sqrt{16-\frac{16}{9}x^2}\,dx$. Moment of the strip about the x-axis:
$\tilde{y}\,dm = \frac{\sqrt{16-\frac{16}{9}x^2}}{2}\left(\delta\sqrt{16-\frac{16}{9}x^2}\right)dx = \delta\left(8 - \frac{8}{9}x^2\right)dx$ so the moment of the plate about the x-axis is
$M_x = \int \tilde{y}\,dm = \int_{-3}^{3} \delta\left(8-\frac{8}{9}x^2\right)dx = \delta\left[8x - \frac{8}{27}x^3\right]_{-3}^{3} = 32\delta$; also the mass of the plate is
$M = \int_{-3}^{3} \delta\sqrt{16-\frac{16}{9}x^2}\,dx = \int_{-3}^{3} 4\delta\sqrt{1-\left(\frac{1}{3}x\right)^2}\,dx = 4\delta\int_{-1}^{1} 3\sqrt{1-u^2}\,du$ where $u = \frac{x}{3} \Rightarrow 3\,du = dx$; $x = -3$
$\Rightarrow u = -1$ and $x = 3 \Rightarrow u = 1$. Hence, $4\delta\int_{-1}^{1} 3\sqrt{1-u^2}\,du = 12\delta\int_{-1}^{1}\sqrt{1-u^2}\,du$
$= 12\delta\left[\frac{1}{2}\left(u\sqrt{1-u^2} + \sin^{-1} u\right)\right]_{-1}^{1} = 6\pi\delta \Rightarrow \bar{y} = \frac{M_x}{M} = \frac{32\delta}{6\pi\delta} = \frac{16}{3\pi}$. Therefore the center of mass is $\left(0, \frac{16}{3\pi}\right)$.

81. (a) $\tan\beta = m_L \Rightarrow \tan\beta = f'(x_0)$ where $f(x) = \sqrt{4px}$;
$f'(x) = \frac{1}{2}(4px)^{-1/2}(4p) = \frac{2p}{\sqrt{4px}} \Rightarrow f'(x_0) = \frac{2p}{\sqrt{4px_0}}$
$= \frac{2p}{y_0} \Rightarrow \tan\beta = \frac{2p}{y_0}$.

(b) $\tan\phi = m_{FP} = \frac{y_0 - 0}{x_0 - p} = \frac{y_0}{x_0 - p}$

(c) $\tan\alpha = \frac{\tan\phi - \tan\beta}{1 + \tan\phi\tan\beta} = \frac{\left(\frac{y_0}{x_0-p} - \frac{2p}{y_0}\right)}{1 + \left(\frac{y_0}{x_0-p}\right)\left(\frac{2p}{y_0}\right)}$
$= \frac{y_0^2 - 2p(x_0-p)}{y_0(x_0-p+2p)} = \frac{4px_0 - 2px_0 + 2p^2}{y_0(x_0+p)} = \frac{2p(x_0+p)}{y_0(x_0+p)} = \frac{2p}{y_0}$

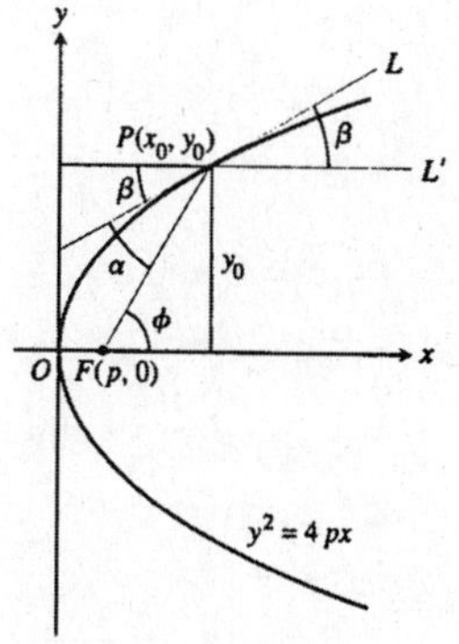

11.7 CONICS IN POLAR COORDINATES

1. $16x^2 + 25y^2 = 400 \Rightarrow \frac{x^2}{25} + \frac{y^2}{16} = 1 \Rightarrow c = \sqrt{a^2 - b^2}$
$= \sqrt{25 - 16} = 3 \Rightarrow e = \frac{c}{a} = \frac{3}{5}; F(\pm 3, 0);$
directrices are $x = 0 \pm \frac{a}{e} = \pm \frac{5}{\left(\frac{3}{5}\right)} = \pm \frac{25}{3}$

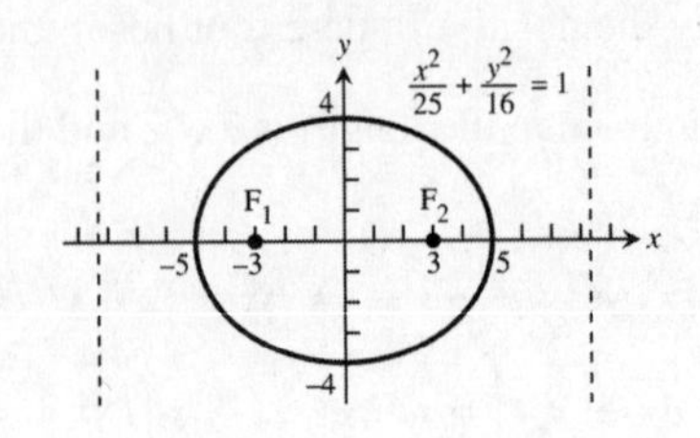

3. $2x^2 + y^2 = 2 \Rightarrow x^2 + \frac{y^2}{2} = 1 \Rightarrow c = \sqrt{a^2 - b^2}$
$= \sqrt{2 - 1} = 1 \Rightarrow e = \frac{c}{a} = \frac{1}{\sqrt{2}}; F(0, \pm 1);$
directrices are $y = 0 \pm \frac{a}{e} = \pm \frac{\sqrt{2}}{\left(\frac{1}{\sqrt{2}}\right)} = \pm 2$

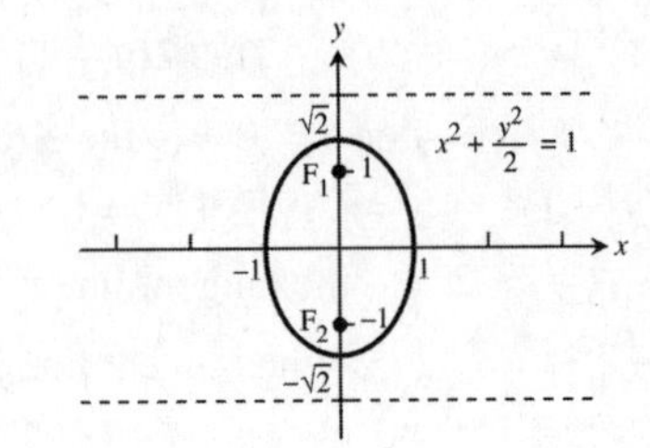

5. $3x^2 + 2y^2 = 6 \Rightarrow \frac{x^2}{2} + \frac{y^2}{3} = 1 \Rightarrow c = \sqrt{a^2 - b^2}$
$= \sqrt{3 - 2} = 1 \Rightarrow e = \frac{c}{a} = \frac{1}{\sqrt{3}}; F(0, \pm 1);$
directrices are $y = 0 \pm \frac{a}{e} = \pm \frac{\sqrt{3}}{\left(\frac{1}{\sqrt{3}}\right)} = \pm 3$

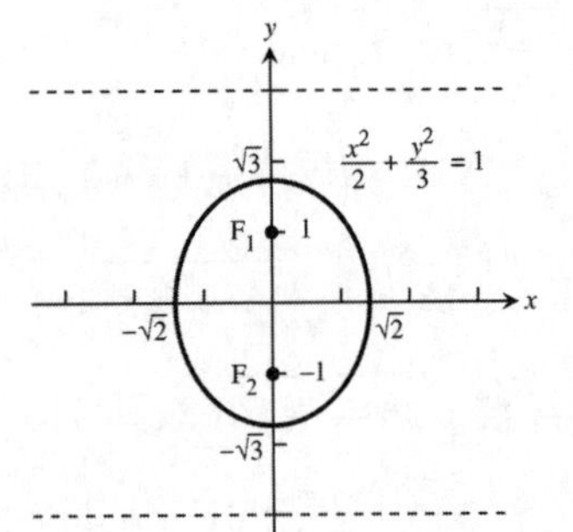

7. $6x^2 + 9y^2 = 54 \Rightarrow \frac{x^2}{9} + \frac{y^2}{6} = 1 \Rightarrow c = \sqrt{a^2 - b^2}$
$= \sqrt{9 - 6} = \sqrt{3} \Rightarrow e = \frac{c}{a} = \frac{\sqrt{3}}{3}; F\left(\pm \sqrt{3}, 0\right);$
directrices are $x = 0 \pm \frac{a}{e} = \pm \frac{3}{\left(\frac{\sqrt{3}}{3}\right)} = \pm 3\sqrt{3}$

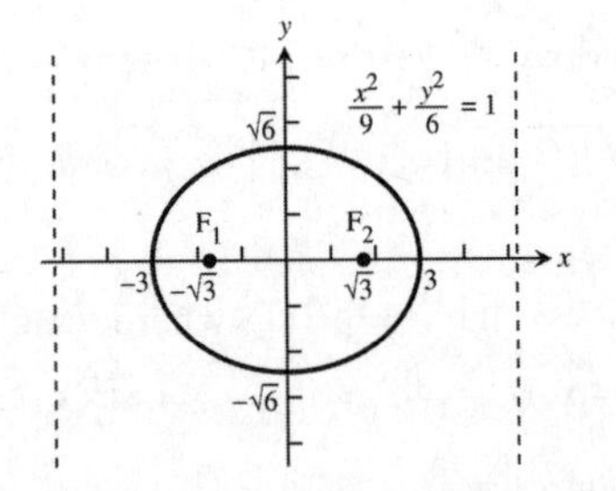

9. Foci: $(0, \pm 3), e = 0.5 \Rightarrow c = 3$ and $a = \frac{c}{e} = \frac{3}{0.5} = 6 \Rightarrow b^2 = 36 - 9 = 27 \Rightarrow \frac{x^2}{27} + \frac{y^2}{36} = 1$

11. Vertices: $(0, \pm 70), e = 0.1 \Rightarrow a = 70$ and $c = ae = 70(0.1) = 7 \Rightarrow b^2 = 4900 - 49 = 4851 \Rightarrow \frac{x^2}{4851} + \frac{y^2}{4900} = 1$

13. Focus: $\left(\sqrt{5}, 0\right)$, Directrix: $x = \frac{9}{\sqrt{5}} \Rightarrow c = ae = \sqrt{5}$ and $\frac{a}{e} = \frac{9}{\sqrt{5}} \Rightarrow \frac{ae}{e^2} = \frac{9}{\sqrt{5}} \Rightarrow \frac{\sqrt{5}}{e^2} = \frac{9}{\sqrt{5}} \Rightarrow e^2 = \frac{5}{9}$
$\Rightarrow e = \frac{\sqrt{5}}{3}$. Then $PF = \frac{\sqrt{5}}{3} PD \Rightarrow \sqrt{\left(x - \sqrt{5}\right)^2 + (y - 0)^2} = \frac{\sqrt{5}}{3} \left|x - \frac{9}{\sqrt{5}}\right| \Rightarrow \left(x - \sqrt{5}\right)^2 + y^2 = \frac{5}{9}\left(x - \frac{9}{\sqrt{5}}\right)^2$
$\Rightarrow x^2 - 2\sqrt{5}x + 5 + y^2 = \frac{5}{9}\left(x^2 - \frac{18}{\sqrt{5}}x + \frac{81}{5}\right) \Rightarrow \frac{4}{9}x^2 + y^2 = 4 \Rightarrow \frac{x^2}{9} + \frac{y^2}{4} = 1$

15. Focus: $(-4, 0)$, Directrix: $x = -16 \Rightarrow c = ae = 4$ and $\frac{a}{e} = 16 \Rightarrow \frac{ae}{e^2} = 16 \Rightarrow \frac{4}{e^2} = 16 \Rightarrow e^2 = \frac{1}{4} \Rightarrow e = \frac{1}{2}$. Then
$PF = \frac{1}{2} PD \Rightarrow \sqrt{(x + 4)^2 + (y - 0)^2} = \frac{1}{2}|x + 16| \Rightarrow (x + 4)^2 + y^2 = \frac{1}{4}(x + 16)^2 \Rightarrow x^2 + 8x + 16 + y^2$

$= \frac{1}{4}(x^2 + 32x + 256) \Rightarrow \frac{3}{4}x^2 + y^2 = 48 \Rightarrow \frac{x^2}{64} + \frac{y^2}{48} = 1$

17. $x^2 - y^2 = 1 \Rightarrow c = \sqrt{a^2 + b^2} = \sqrt{1+1} = \sqrt{2} \Rightarrow e = \frac{c}{a}$
$= \frac{\sqrt{2}}{1} = \sqrt{2}$; asymptotes are $y = \pm x$; $F\left(\pm\sqrt{2}, 0\right)$;
directrices are $x = 0 \pm \frac{a}{e} = \pm \frac{1}{\sqrt{2}}$

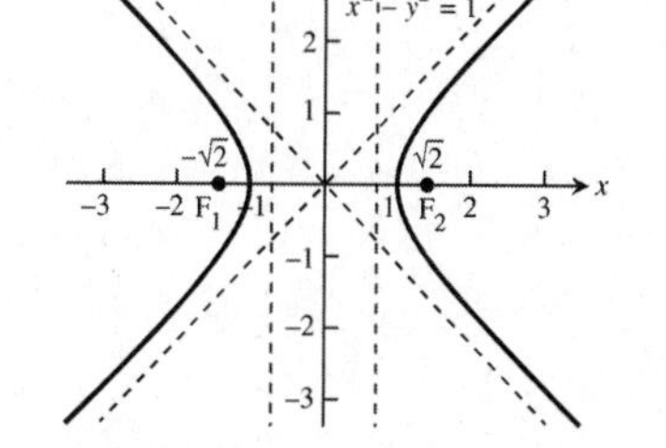

19. $y^2 - x^2 = 8 \Rightarrow \frac{y^2}{8} - \frac{x^2}{8} = 1 \Rightarrow c = \sqrt{a^2 + b^2}$
$= \sqrt{8+8} = 4 \Rightarrow e = \frac{c}{a} = \frac{4}{\sqrt{8}} = \sqrt{2}$; asymptotes are
$y = \pm x$; $F(0, \pm 4)$; directrices are $y = 0 \pm \frac{a}{e}$
$= \pm \frac{\sqrt{8}}{\sqrt{2}} = \pm 2$

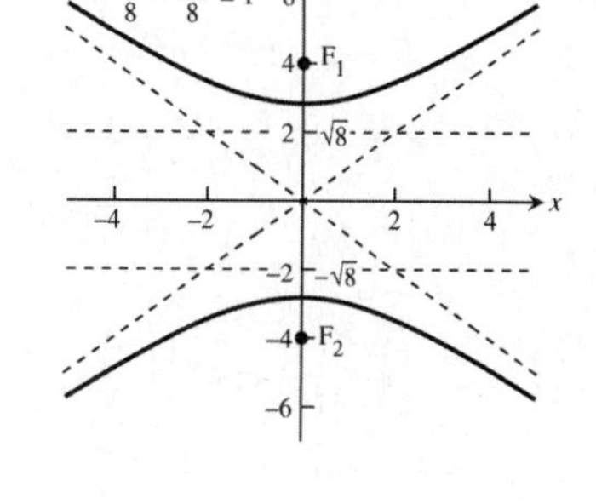

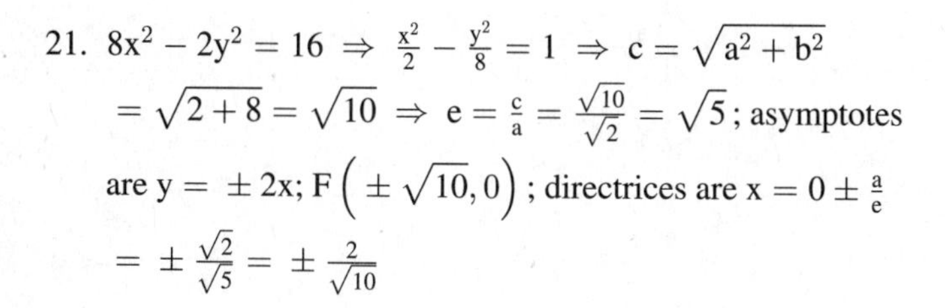

21. $8x^2 - 2y^2 = 16 \Rightarrow \frac{x^2}{2} - \frac{y^2}{8} = 1 \Rightarrow c = \sqrt{a^2 + b^2}$
$= \sqrt{2+8} = \sqrt{10} \Rightarrow e = \frac{c}{a} = \frac{\sqrt{10}}{\sqrt{2}} = \sqrt{5}$; asymptotes
are $y = \pm 2x$; $F\left(\pm\sqrt{10}, 0\right)$; directrices are $x = 0 \pm \frac{a}{e}$
$= \pm \frac{\sqrt{2}}{\sqrt{5}} = \pm \frac{2}{\sqrt{10}}$

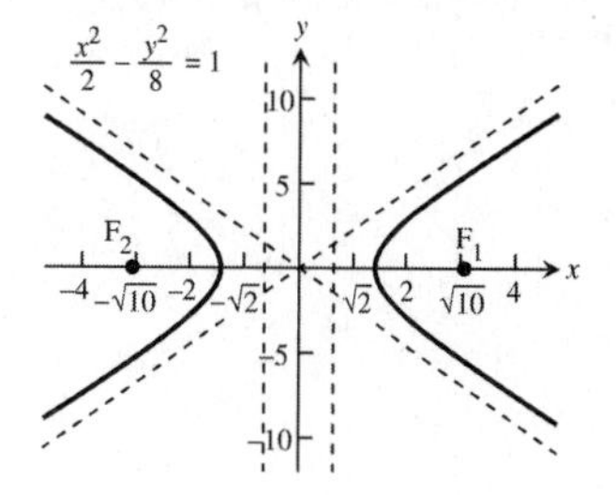

23. $8y^2 - 2x^2 = 16 \Rightarrow \frac{y^2}{2} - \frac{x^2}{8} = 1 \Rightarrow c = \sqrt{a^2 + b^2}$
$= \sqrt{2+8} = \sqrt{10} \Rightarrow e = \frac{c}{a} = \frac{\sqrt{10}}{\sqrt{2}} = \sqrt{5}$; asymptotes
are $y = \pm \frac{x}{2}$; $F\left(0, \pm\sqrt{10}\right)$; directrices are $y = 0 \pm \frac{a}{e}$
$= \pm \frac{\sqrt{2}}{\sqrt{5}} = \pm \frac{2}{\sqrt{10}}$

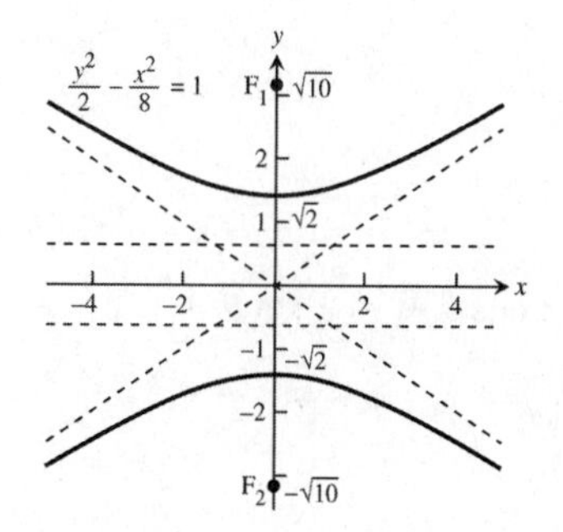

25. Vertices $(0, \pm 1)$ and $e = 3 \Rightarrow a = 1$ and $e = \frac{c}{a} = 3 \Rightarrow c = 3a = 3 \Rightarrow b^2 = c^2 - a^2 = 9 - 1 = 8 \Rightarrow y^2 - \frac{x^2}{8} = 1$

27. Foci $(\pm 3, 0)$ and $e = 3 \Rightarrow c = 3$ and $e = \frac{c}{a} = 3 \Rightarrow c = 3a \Rightarrow a = 1 \Rightarrow b^2 = c^2 - a^2 = 9 - 1 = 8 \Rightarrow x^2 - \frac{y^2}{8} = 1$

29. $e = 1, x = 2 \Rightarrow k = 2 \Rightarrow r = \frac{2(1)}{1 + (1)\cos\theta} = \frac{2}{1+\cos\theta}$

31. $e = 5, y = -6 \Rightarrow k = 6 \Rightarrow r = \frac{6(5)}{1 - 5\sin\theta} = \frac{30}{1-5\sin\theta}$

33. $e = \frac{1}{2}, x = 1 \Rightarrow k = 1 \Rightarrow r = \frac{\left(\frac{1}{2}\right)(1)}{1+\left(\frac{1}{2}\right)\cos\theta} = \frac{1}{2+\cos\theta}$

35. $e = \frac{1}{5}, y = -10 \Rightarrow k = 10 \Rightarrow r = \frac{\left(\frac{1}{5}\right)(10)}{1-\left(\frac{1}{5}\right)\sin\theta} = \frac{10}{5-\sin\theta}$

37. $r = \frac{1}{1+\cos\theta} \Rightarrow e = 1, k = 1 \Rightarrow x = 1$

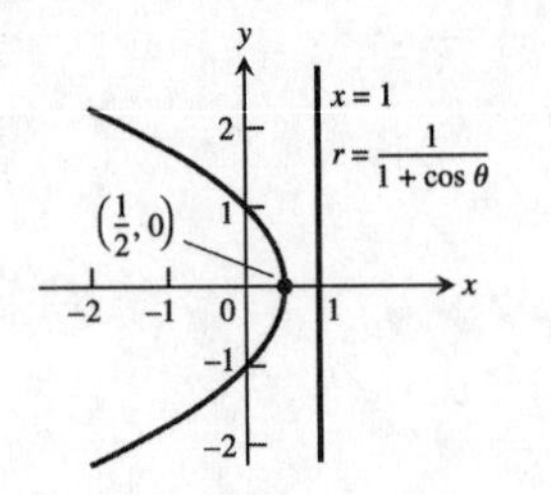

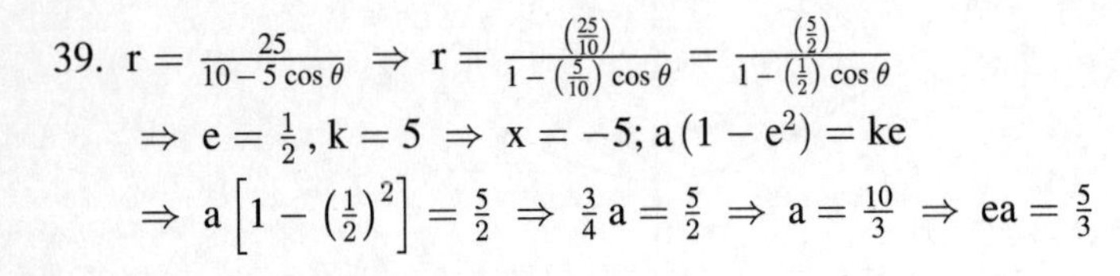

39. $r = \frac{25}{10-5\cos\theta} \Rightarrow r = \frac{\left(\frac{25}{10}\right)}{1-\left(\frac{5}{10}\right)\cos\theta} = \frac{\left(\frac{5}{2}\right)}{1-\left(\frac{1}{2}\right)\cos\theta}$

$\Rightarrow e = \frac{1}{2}, k = 5 \Rightarrow x = -5;\ a\left(1-e^2\right) = ke$

$\Rightarrow a\left[1-\left(\frac{1}{2}\right)^2\right] = \frac{5}{2} \Rightarrow \frac{3}{4}a = \frac{5}{2} \Rightarrow a = \frac{10}{3} \Rightarrow ea = \frac{5}{3}$

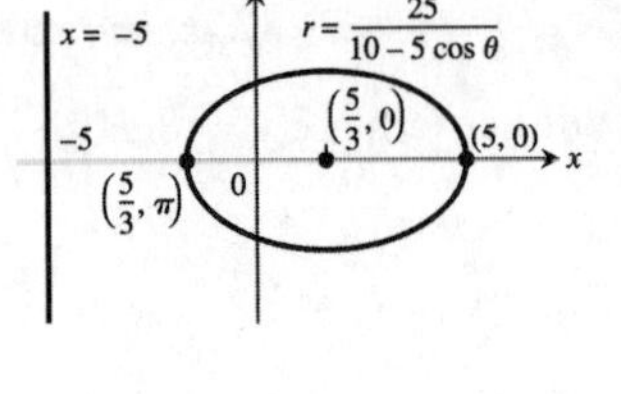

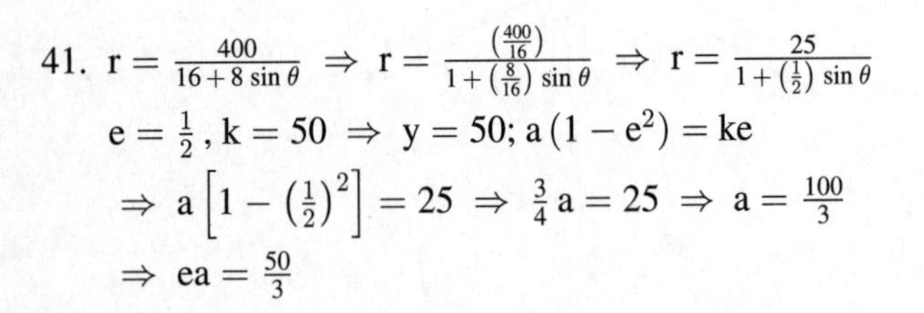

41. $r = \frac{400}{16+8\sin\theta} \Rightarrow r = \frac{\left(\frac{400}{16}\right)}{1+\left(\frac{8}{16}\right)\sin\theta} \Rightarrow r = \frac{25}{1+\left(\frac{1}{2}\right)\sin\theta}$

$e = \frac{1}{2}, k = 50 \Rightarrow y = 50;\ a\left(1-e^2\right) = ke$

$\Rightarrow a\left[1-\left(\frac{1}{2}\right)^2\right] = 25 \Rightarrow \frac{3}{4}a = 25 \Rightarrow a = \frac{100}{3}$

$\Rightarrow ea = \frac{50}{3}$

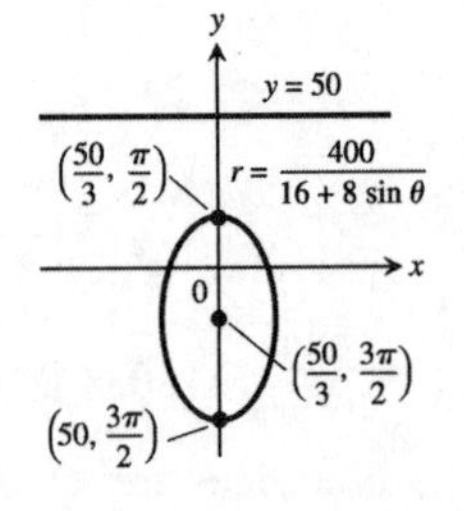

43. $r = \frac{8}{2-2\sin\theta} \Rightarrow r = \frac{4}{1-\sin\theta} \Rightarrow e = 1,$

$k = 4 \Rightarrow y = -4$

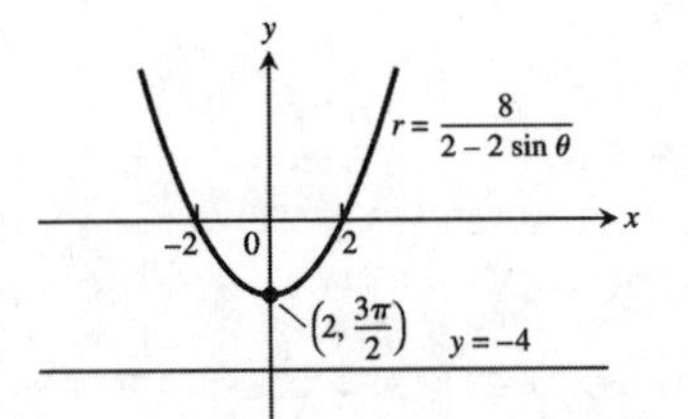

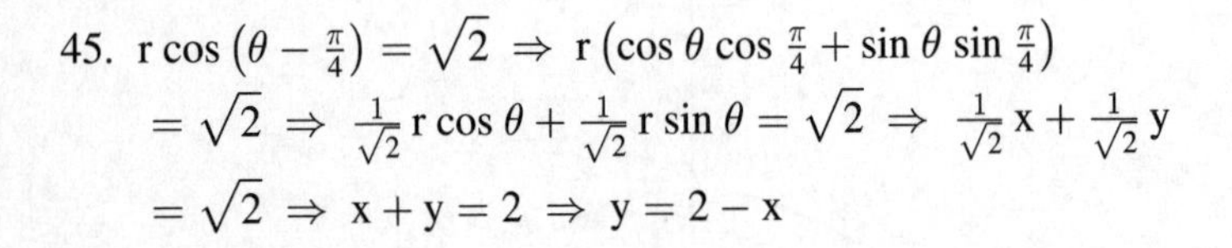

45. $r\cos\left(\theta-\frac{\pi}{4}\right) = \sqrt{2} \Rightarrow r\left(\cos\theta\cos\frac{\pi}{4}+\sin\theta\sin\frac{\pi}{4}\right)$

$= \sqrt{2} \Rightarrow \frac{1}{\sqrt{2}}r\cos\theta + \frac{1}{\sqrt{2}}r\sin\theta = \sqrt{2} \Rightarrow \frac{1}{\sqrt{2}}x + \frac{1}{\sqrt{2}}y$

$= \sqrt{2} \Rightarrow x + y = 2 \Rightarrow y = 2 - x$

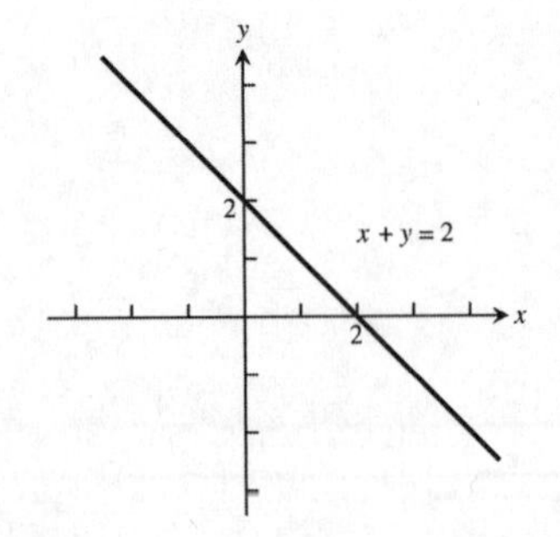

47. $r\cos\left(\theta - \frac{2\pi}{3}\right) = 3 \Rightarrow r\left(\cos\theta\cos\frac{2\pi}{3} + \sin\theta\sin\frac{2\pi}{3}\right) = 3$
$\Rightarrow -\frac{1}{2}r\cos\theta + \frac{\sqrt{3}}{2}r\sin\theta = 3 \Rightarrow -\frac{1}{2}x + \frac{\sqrt{3}}{2}y = 3$
$\Rightarrow -x + \sqrt{3}y = 6 \Rightarrow y = \frac{\sqrt{3}}{3}x + 2\sqrt{3}$

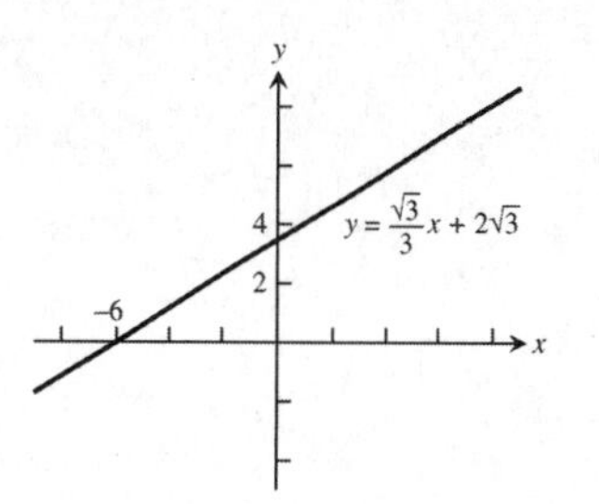

49. $\sqrt{2}x + \sqrt{2}y = 6 \Rightarrow \sqrt{2}r\cos\theta + \sqrt{2}r\sin\theta = 6 \Rightarrow r\left(\frac{\sqrt{2}}{2}\cos\theta + \frac{\sqrt{2}}{2}\sin\theta\right) = 3 \Rightarrow r\left(\cos\frac{\pi}{4}\cos\theta + \sin\frac{\pi}{4}\sin\theta\right)$
$= 3 \Rightarrow r\cos\left(\theta - \frac{\pi}{4}\right) = 3$

51. $y = -5 \Rightarrow r\sin\theta = -5 \Rightarrow -r\sin\theta = 5 \Rightarrow r\sin(-\theta) = 5 \Rightarrow r\cos\left(\frac{\pi}{2} - (-\theta)\right) = 5 \Rightarrow r\cos\left(\theta + \frac{\pi}{2}\right) = 5$

53.

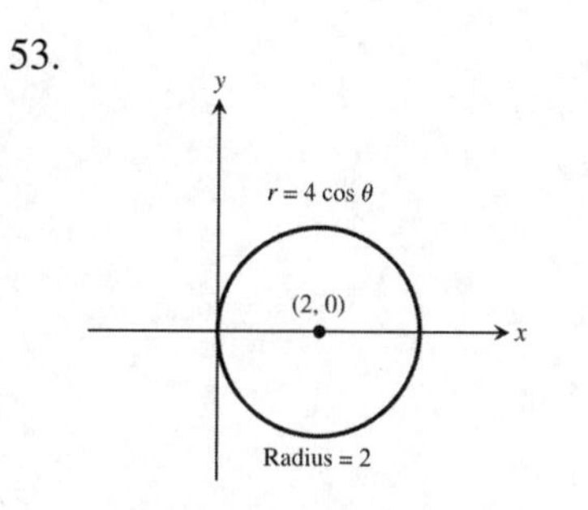

55.

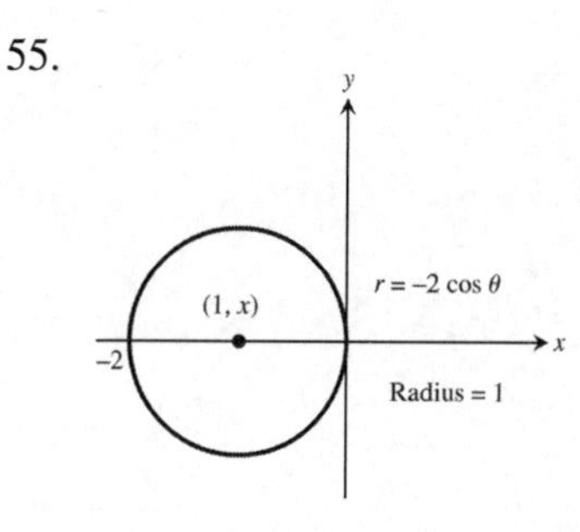

57. $(x - 6)^2 + y^2 = 36 \Rightarrow C = (6, 0),\ a = 6$
$\Rightarrow r = 12\cos\theta$ is the polar equation

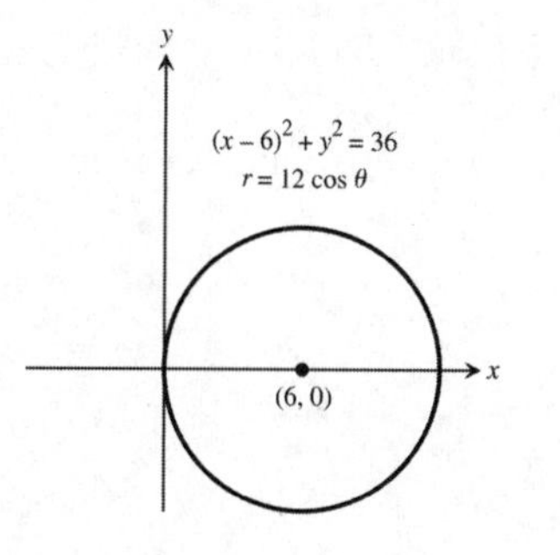

59. $x^2 + (y - 5)^2 = 25 \Rightarrow C = (0, 5),\ a = 5$
$\Rightarrow r = 10\sin\theta$ is the polar equation

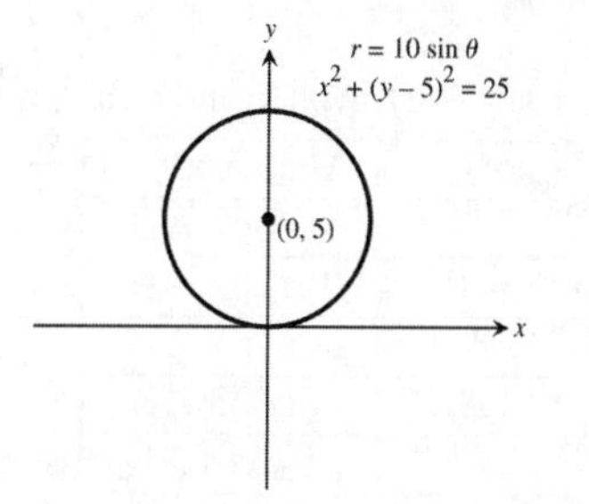

61. $x^2 + 2x + y^2 = 0 \Rightarrow (x + 1)^2 + y^2 = 1$
$\Rightarrow C = (-1, 0),\ a = 1 \Rightarrow r = -2\cos\theta$ is the polar equation

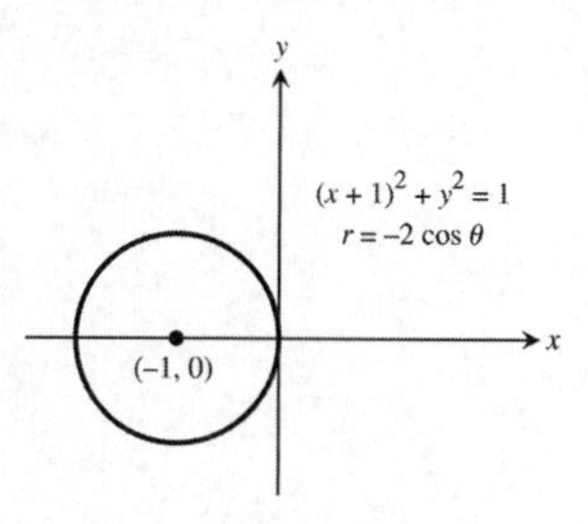

63. $x^2 + y^2 + y = 0 \Rightarrow x^2 + \left(y + \frac{1}{2}\right)^2 = \frac{1}{4}$
$\Rightarrow C = \left(0, -\frac{1}{2}\right),\ a = \frac{1}{2} \Rightarrow r = -\sin\theta$ is the polar equation

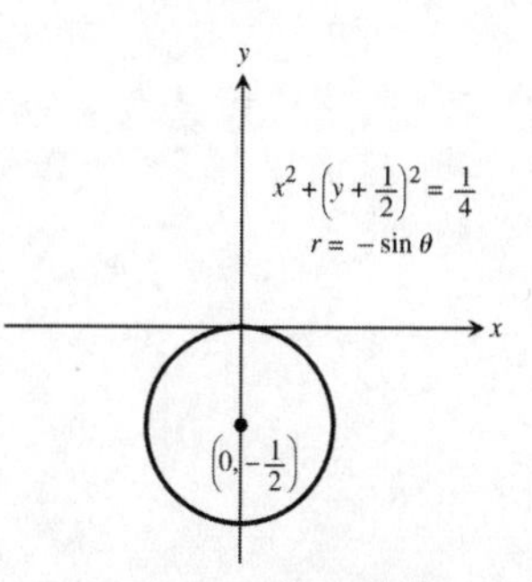

65.

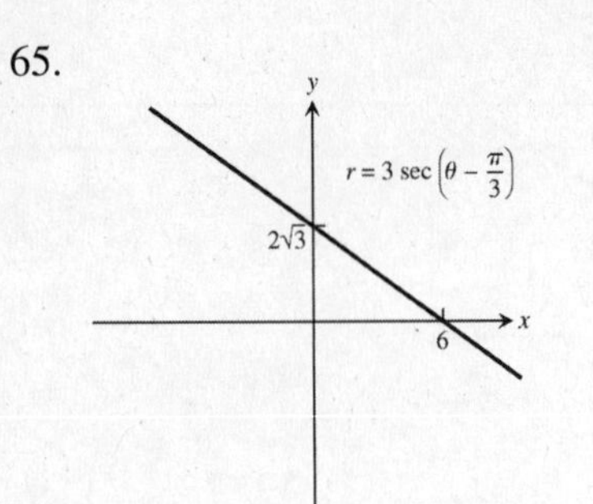

67.

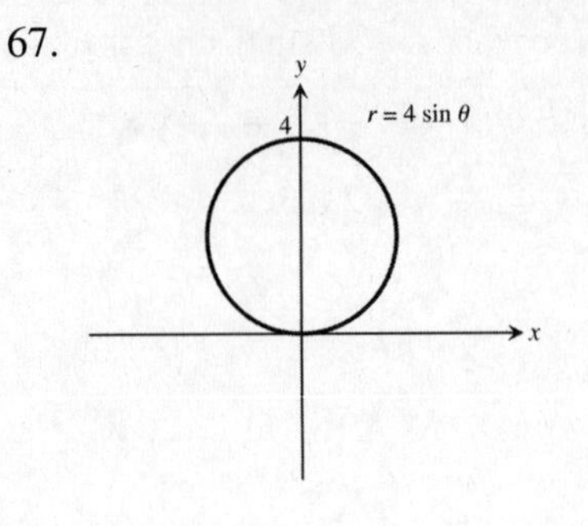

69.

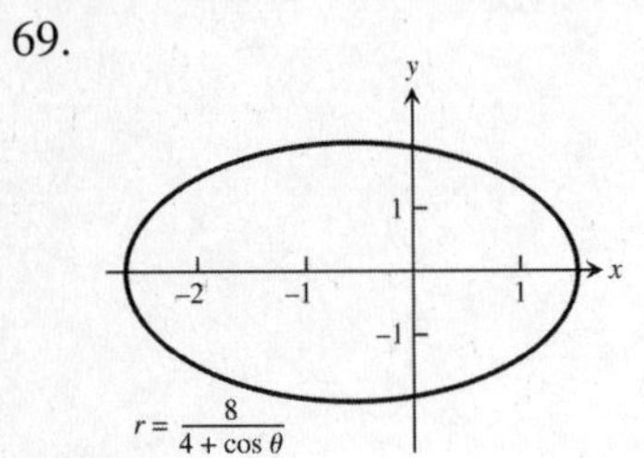

71.

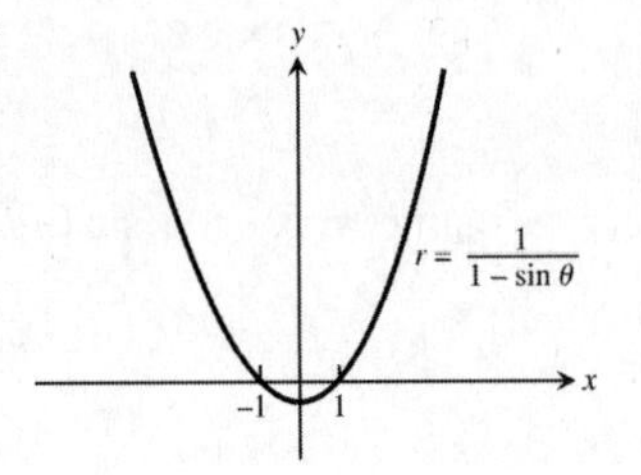

73.

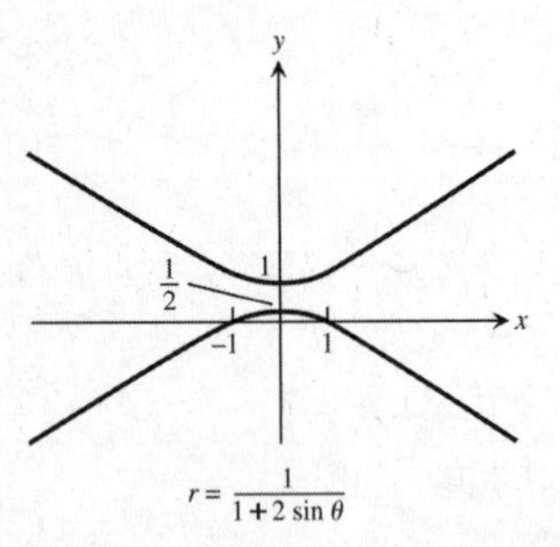

75. (a) Perihelion $= a - ae = a(1-e)$, Aphelion $= ea + a = a(1+e)$

(b)

Planet	Perihelion	Aphelion
Mercury	0.3075 AU	0.4667 AU
Venus	0.7184 AU	0.7282 AU
Earth	0.9833 AU	1.0167 AU
Mars	1.3817 AU	1.6663 AU
Jupiter	4.9512 AU	5.4548 AU
Saturn	9.0210 AU	10.0570 AU
Uranus	18.2977 AU	20.0623 AU
Neptune	29.8135 AU	30.3065 AU

CHAPTER 11 PRACTICE EXERCISES

1. $x = \frac{t}{2}$ and $y = t + 1 \Rightarrow 2x = t \Rightarrow y = 2x + 1$

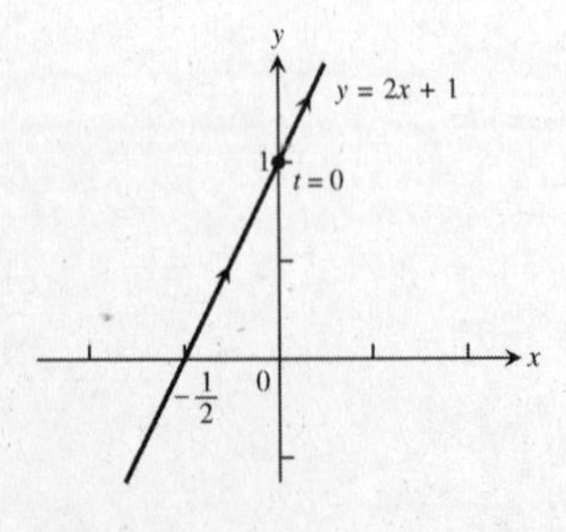

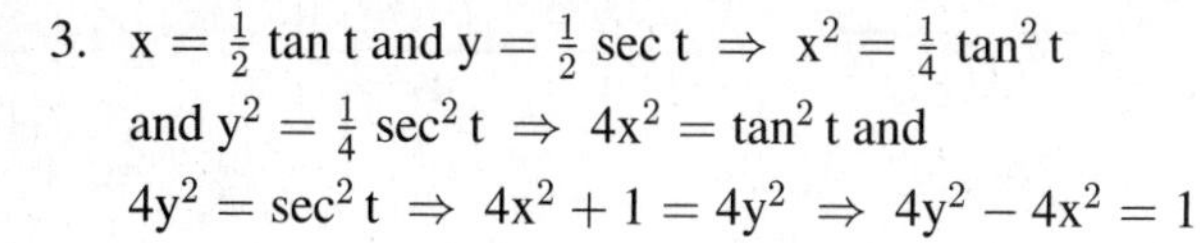

3. $x = \frac{1}{2}\tan t$ and $y = \frac{1}{2}\sec t \Rightarrow x^2 = \frac{1}{4}\tan^2 t$ and $y^2 = \frac{1}{4}\sec^2 t \Rightarrow 4x^2 = \tan^2 t$ and $4y^2 = \sec^2 t \Rightarrow 4x^2 + 1 = 4y^2 \Rightarrow 4y^2 - 4x^2 = 1$

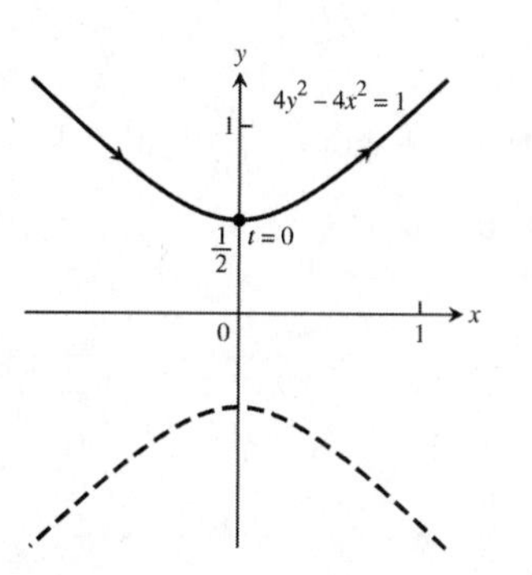

5. $x = -\cos t$ and $y = \cos^2 t \Rightarrow y = (-x)^2 = x^2$

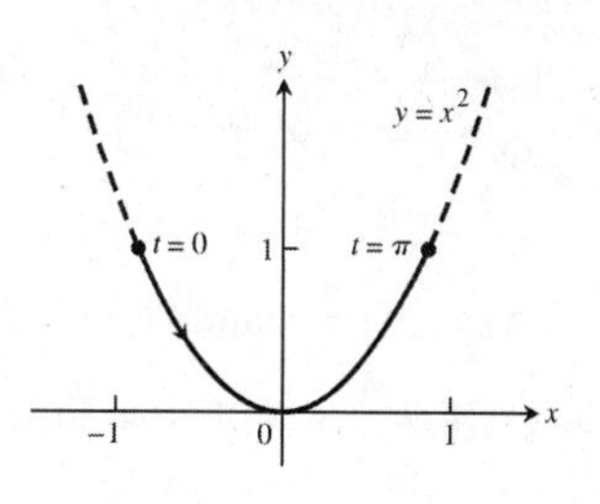

7. $16x^2 + 9y^2 = 144 \Rightarrow \frac{x^2}{9} + \frac{y^2}{16} = 1 \Rightarrow a = 3$ and $b = 4 \Rightarrow x = 3\cos t$ and $y = 4\sin t,\ 0 \le t \le 2\pi$

9. $x = \frac{1}{2}\tan t,\ y = \frac{1}{2}\sec t \Rightarrow \frac{dy}{dx} = \frac{dy/dt}{dx/dt} = \frac{\frac{1}{2}\sec t\tan t}{\frac{1}{2}\sec^2 t} = \frac{\tan t}{\sec t} = \sin t \Rightarrow \left.\frac{dy}{dx}\right|_{t=\pi/3} = \sin\frac{\pi}{3} = \frac{\sqrt{3}}{2};\ t = \frac{\pi}{3}$
$\Rightarrow x = \frac{1}{2}\tan\frac{\pi}{3} = \frac{\sqrt{3}}{2}$ and $y = \frac{1}{2}\sec\frac{\pi}{3} = 1 \Rightarrow y = \frac{\sqrt{3}}{2}x + \frac{1}{4};\ \frac{d^2y}{dx^2} = \frac{dy'/dt}{dx/dt} = \frac{\cos t}{\frac{1}{2}\sec^2 t} = 2\cos^3 t \Rightarrow \left.\frac{d^2y}{dx^2}\right|_{t=\pi/3}$
$= 2\cos^3\left(\frac{\pi}{3}\right) = \frac{1}{4}$

11. (a) $x = 4t^2,\ y = t^3 - 1 \Rightarrow t = \pm\frac{\sqrt{x}}{2} \Rightarrow y = \left(\pm\frac{\sqrt{x}}{2}\right)^3 - 1 = \pm\frac{x^{3/2}}{8} - 1$

(b) $x = \cos t,\ y = \tan t \Rightarrow \sec t = \frac{1}{x} \Rightarrow \tan^2 t + 1 = \sec^2 t \Rightarrow y^2 = \frac{1}{x^2} - 1 = \frac{1-x^2}{x^2} \Rightarrow y = \pm\frac{\sqrt{1-x^2}}{x}$

13. $y = x^{1/2} - \frac{x^{3/2}}{3} \Rightarrow \frac{dy}{dx} = \frac{1}{2}x^{-1/2} - \frac{1}{2}x^{1/2} \Rightarrow \left(\frac{dy}{dx}\right)^2 = \frac{1}{4}\left(\frac{1}{x} - 2 + x\right) \Rightarrow L = \int_1^4 \sqrt{1 + \frac{1}{4}\left(\frac{1}{x} - 2 + x\right)}\,dx$
$\Rightarrow L = \int_1^4 \sqrt{\frac{1}{4}\left(\frac{1}{x} + 2 + x\right)}\,dx = \int_1^4 \sqrt{\frac{1}{4}\left(x^{-1/2} + x^{1/2}\right)^2}\,dx = \int_1^4 \frac{1}{2}\left(x^{-1/2} + x^{1/2}\right)dx = \frac{1}{2}\left[2x^{1/2} + \frac{2}{3}x^{3/2}\right]_1^4$
$= \frac{1}{2}\left[\left(4 + \frac{2}{3}\cdot 8\right) - \left(2 + \frac{2}{3}\right)\right] = \frac{1}{2}\left(2 + \frac{14}{3}\right) = \frac{10}{3}$

15. $y = \frac{5}{12}x^{6/5} - \frac{5}{8}x^{4/5} \Rightarrow \frac{dy}{dx} = \frac{1}{2}x^{1/5} - \frac{1}{2}x^{-1/5} \Rightarrow \left(\frac{dy}{dx}\right)^2 = \frac{1}{4}\left(x^{2/5} - 2 + x^{-2/5}\right)$
$\Rightarrow L = \int_1^{32} \sqrt{1 + \frac{1}{4}\left(x^{2/5} - 2 + x^{-2/5}\right)}\,dx \Rightarrow L = \int_1^{32} \sqrt{\frac{1}{4}\left(x^{2/5} + 2 + x^{-2/5}\right)}\,dx = \int_1^{32} \sqrt{\frac{1}{4}\left(x^{1/5} + x^{-1/5}\right)^2}\,dx$
$= \int_1^{32} \frac{1}{2}\left(x^{1/5} + x^{-1/5}\right)dx = \frac{1}{2}\left[\frac{5}{6}x^{6/5} + \frac{5}{4}x^{4/5}\right]_1^{32} = \frac{1}{2}\left[\left(\frac{5}{6}\cdot 2^6 + \frac{5}{4}\cdot 2^4\right) - \left(\frac{5}{6} + \frac{5}{4}\right)\right] = \frac{1}{2}\left(\frac{315}{6} + \frac{75}{4}\right)$
$= \frac{1}{48}(1260 + 450) = \frac{1710}{48} = \frac{285}{8}$

17. $\frac{dx}{dt} = -5\sin t + 5\sin 5t$ and $\frac{dy}{dt} = 5\cos t - 5\cos 5t \Rightarrow \sqrt{\left(\frac{dx}{dt}\right)^2 + \left(\frac{dy}{dt}\right)^2}$
$= \sqrt{(-5\sin t + 5\sin 5t)^2 + (5\cos t - 5\cos 5t)^2}$
$= 5\sqrt{\sin^2 5t - 2\sin t\sin 5t + \sin^2 t + \cos^2 t - 2\cos t\cos 5t + \cos^2 5t} = 5\sqrt{2 - 2(\sin t\sin 5t + \cos t\cos 5t)}$
$= 5\sqrt{2(1 - \cos 4t)} = 5\sqrt{4\left(\frac{1}{2}\right)(1 - \cos 4t)} = 10\sqrt{\sin^2 2t} = 10|\sin 2t| = 10\sin 2t$ (since $0 \le t \le \frac{\pi}{2}$)
$\Rightarrow \text{Length} = \int_0^{\pi/2} 10\sin 2t\,dt = [-5\cos 2t]_0^{\pi/2} = (-5)(-1) - (-5)(1) = 10$

19. $\frac{dx}{d\theta} = -3\sin\theta$ and $\frac{dy}{d\theta} = 3\cos\theta \Rightarrow \sqrt{\left(\frac{dx}{d\theta}\right)^2 + \left(\frac{dy}{d\theta}\right)^2} = \sqrt{(-3\sin\theta)^2 + (3\cos\theta)^2} = \sqrt{3(\sin^2\theta + \cos^2\theta)} = 3$
$\Rightarrow \text{Length} = \int_0^{3\pi/2} 3\,d\theta = 3\int_0^{3\pi/2} d\theta = 3\left(\frac{3\pi}{2} - 0\right) = \frac{9\pi}{2}$

21. $x = \frac{t^2}{2}$ and $y = 2t$, $0 \le t \le \sqrt{5} \Rightarrow \frac{dx}{dt} = t$ and $\frac{dy}{dt} = 2 \Rightarrow$ Surface Area $= \int_0^{\sqrt{5}} 2\pi(2t)\sqrt{t^2+4}\,dt = \int_4^9 2\pi u^{1/2}\,du$
$= 2\pi\left[\frac{2}{3}u^{3/2}\right]_4^9 = \frac{76\pi}{3}$, where $u = t^2 + 4 \Rightarrow du = 2t\,dt$; $t = 0 \Rightarrow u = 4$, $t = \sqrt{5} \Rightarrow u = 9$

23. $r\cos\left(\theta + \frac{\pi}{3}\right) = 2\sqrt{3} \Rightarrow r\left(\cos\theta\cos\frac{\pi}{3} - \sin\theta\sin\frac{\pi}{3}\right)$
$= 2\sqrt{3} \Rightarrow \frac{1}{2}r\cos\theta - \frac{\sqrt{3}}{2}r\sin\theta = 2\sqrt{3}$
$\Rightarrow r\cos\theta - \sqrt{3}r\sin\theta = 4\sqrt{3} \Rightarrow x - \sqrt{3}y = 4\sqrt{3}$
$\Rightarrow y = \frac{\sqrt{3}}{3}x - 4$

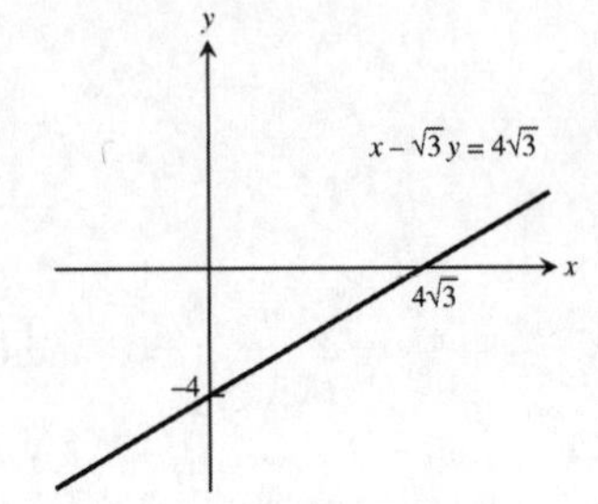

25. $r = 2\sec\theta \Rightarrow r = \frac{2}{\cos\theta} \Rightarrow r\cos\theta = 2 \Rightarrow x = 2$

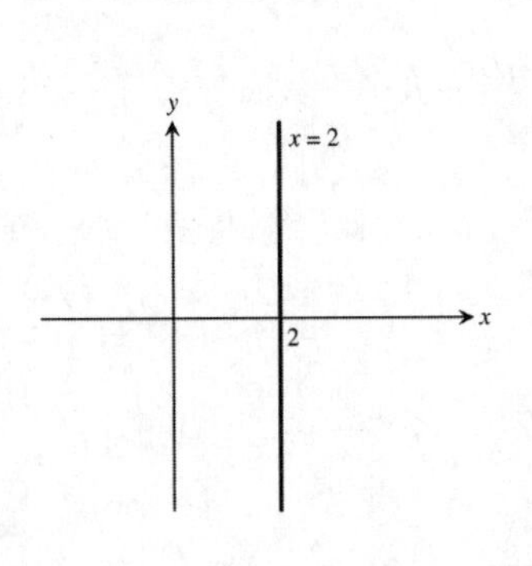

27. $r = -\frac{3}{2}\csc\theta \Rightarrow r\sin\theta = -\frac{3}{2} \Rightarrow y = -\frac{3}{2}$

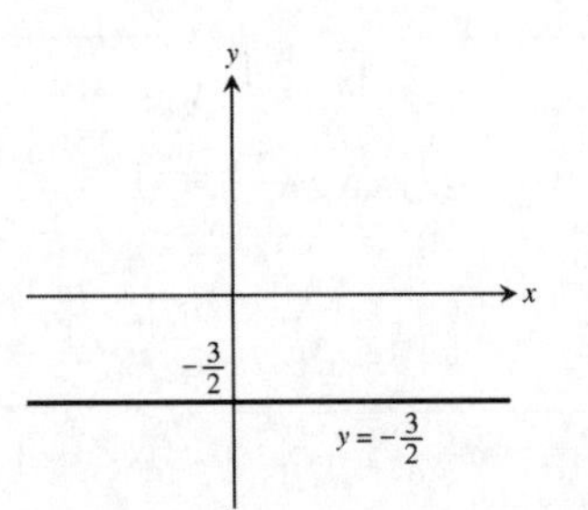

29. $r = -4\sin\theta \Rightarrow r^2 = -4r\sin\theta \Rightarrow x^2 + y^2 + 4y = 0$
$\Rightarrow x^2 + (y+2)^2 = 4$; circle with center $(0, -2)$ and radius 2.

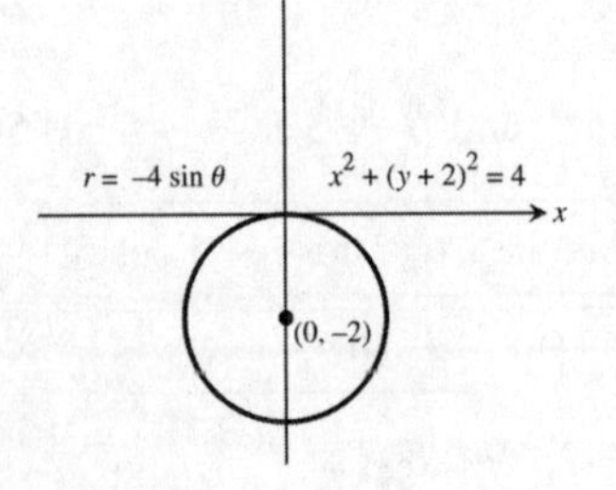

31. $r = 2\sqrt{2}\cos\theta \Rightarrow r^2 = 2\sqrt{2}\,r\cos\theta$
$\Rightarrow x^2 + y^2 - 2\sqrt{2}\,x = 0 \Rightarrow \left(x - \sqrt{2}\right)^2 + y^2 = 2;$
circle with center $\left(\sqrt{2}, 0\right)$ and radius $\sqrt{2}$

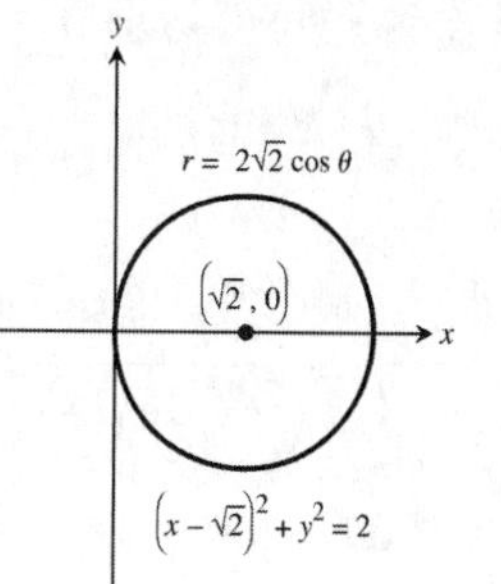

33. $x^2 + y^2 + 5y = 0 \Rightarrow x^2 + \left(y + \frac{5}{2}\right)^2 = \frac{25}{4} \Rightarrow C = \left(0, -\frac{5}{2}\right)$
and $a = \frac{5}{2}$; $r^2 + 5r\sin\theta = 0 \Rightarrow r = -5\sin\theta$

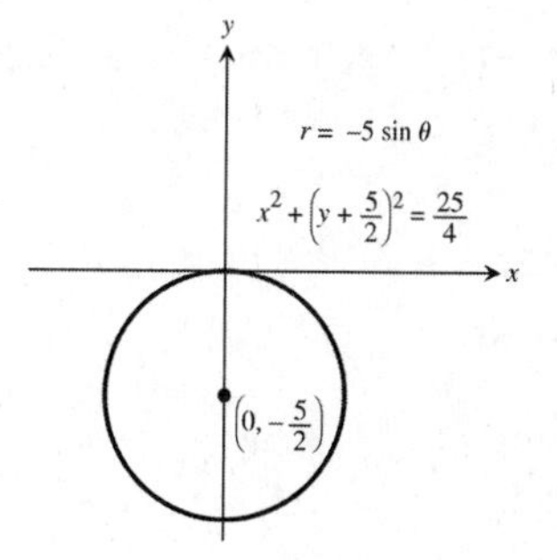

35. $x^2 + y^2 - 3x = 0 \Rightarrow \left(x - \frac{3}{2}\right)^2 + y^2 = \frac{9}{4} \Rightarrow C = \left(\frac{3}{2}, 0\right)$
and $a = \frac{3}{2}$; $r^2 - 3r\cos\theta = 0 \Rightarrow r = 3\cos\theta$

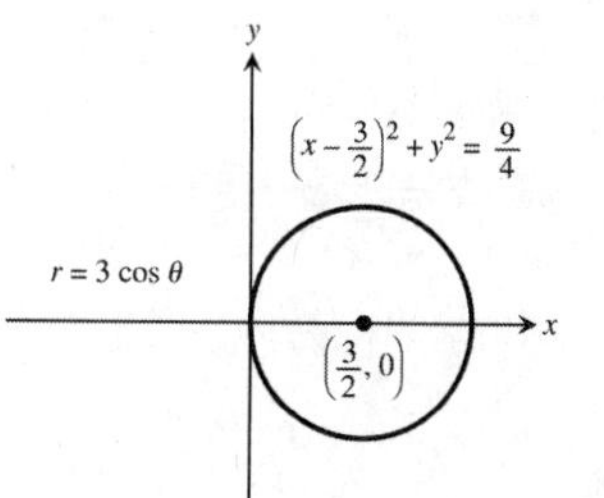

37.

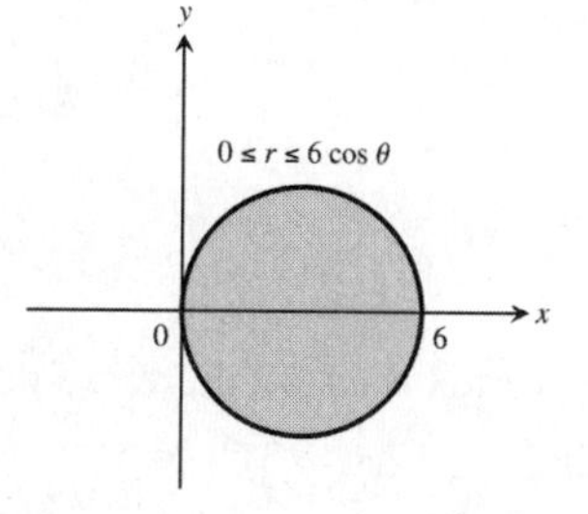

39. d 41. l 43. k 45. i

47. $A = 2\int_0^{\pi} \frac{1}{2} r^2\, d\theta = \int_0^{\pi} (2 - \cos\theta)^2\, d\theta = \int_0^{\pi} (4 - 4\cos\theta + \cos^2\theta)\, d\theta = \int_0^{\pi} \left(4 - 4\cos\theta + \frac{1 + \cos 2\theta}{2}\right) d\theta$
$= \int_0^{\pi} \left(\frac{9}{2} - 4\cos\theta + \frac{\cos 2\theta}{2}\right) d\theta = \left[\frac{9}{2}\theta - 4\sin\theta + \frac{\sin 2\theta}{4}\right]_0^{\pi} = \frac{9}{2}\pi$

49. $r = 1 + \cos 2\theta$ and $r = 1 \Rightarrow 1 = 1 + \cos 2\theta \Rightarrow 0 = \cos 2\theta \Rightarrow 2\theta = \frac{\pi}{2} \Rightarrow \theta = \frac{\pi}{4}$; therefore
$A = 4\int_0^{\pi/4} \frac{1}{2}\left[(1 + \cos 2\theta)^2 - 1^2\right] d\theta = 2\int_0^{\pi/4} (1 + 2\cos 2\theta + \cos^2 2\theta - 1)\, d\theta$
$= 2\int_0^{\pi/4} \left(2\cos 2\theta + \frac{1}{2} + \frac{\cos 4\theta}{2}\right) d\theta = 2\left[\sin 2\theta + \frac{1}{2}\theta + \frac{\sin 4\theta}{8}\right]_0^{\pi/4} = 2\left(1 + \frac{\pi}{8} + 0\right) = 2 + \frac{\pi}{4}$

51. $r = -1 + \cos\theta \Rightarrow \frac{dr}{d\theta} = -\sin\theta$; Length $= \int_0^{2\pi} \sqrt{(-1+\cos\theta)^2 + (-\sin\theta)^2}\, d\theta = \int_0^{2\pi} \sqrt{2 - 2\cos\theta}\, d\theta$
$= \int_0^{2\pi} \sqrt{\frac{4(1-\cos\theta)}{2}}\, d\theta = \int_0^{2\pi} 2\sin\frac{\theta}{2}\, d\theta = \left[-4\cos\frac{\theta}{2}\right]_0^{2\pi} = (-4)(-1) - (-4)(1) = 8$

53. $r = 8\sin^3\left(\frac{\theta}{3}\right), 0 \le \theta \le \frac{\pi}{4} \Rightarrow \frac{dr}{d\theta} = 8\sin^2\left(\frac{\theta}{3}\right)\cos\left(\frac{\theta}{3}\right); r^2 + \left(\frac{dr}{d\theta}\right)^2 = \left[8\sin^3\left(\frac{\theta}{3}\right)\right]^2 + \left[8\sin^2\left(\frac{\theta}{3}\right)\cos\left(\frac{\theta}{3}\right)\right]^2$
$= 64\sin^4\left(\frac{\theta}{3}\right) \Rightarrow L = \int_0^{\pi/4} \sqrt{64\sin^4\left(\frac{\theta}{3}\right)}\, d\theta = \int_0^{\pi/4} 8\sin^2\left(\frac{\theta}{3}\right) d\theta = \int_0^{\pi/4} 8\left[\frac{1-\cos\left(\frac{2\theta}{3}\right)}{2}\right] d\theta$
$= \int_0^{\pi/4} \left[4 - 4\cos\left(\frac{2\theta}{3}\right)\right] d\theta = \left[4\theta - 6\sin\left(\frac{2\theta}{3}\right)\right]_0^{\pi/4} = 4\left(\frac{\pi}{4}\right) - 6\sin\left(\frac{\pi}{6}\right) - 0 = \pi - 3$

55. $x^2 = -4y \Rightarrow y = -\frac{x^2}{4} \Rightarrow 4p = 4 \Rightarrow p = 1$;
therefore Focus is $(0, -1)$, Directrix is $y = 1$

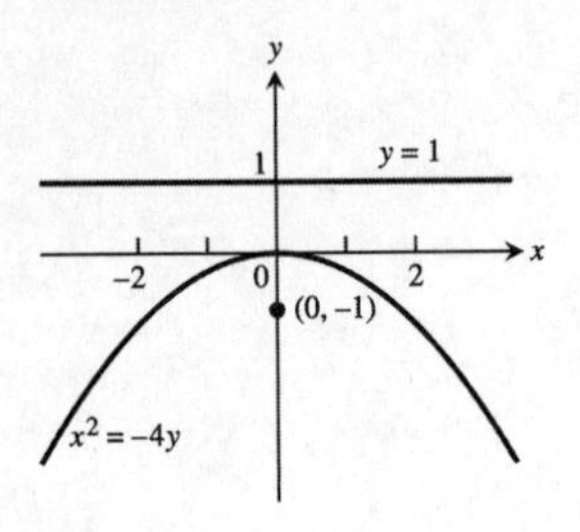

57. $y^2 = 3x \Rightarrow x = \frac{y^2}{3} \Rightarrow 4p = 3 \Rightarrow p = \frac{3}{4}$;
therefore Focus is $\left(\frac{3}{4}, 0\right)$, Directrix is $x = -\frac{3}{4}$

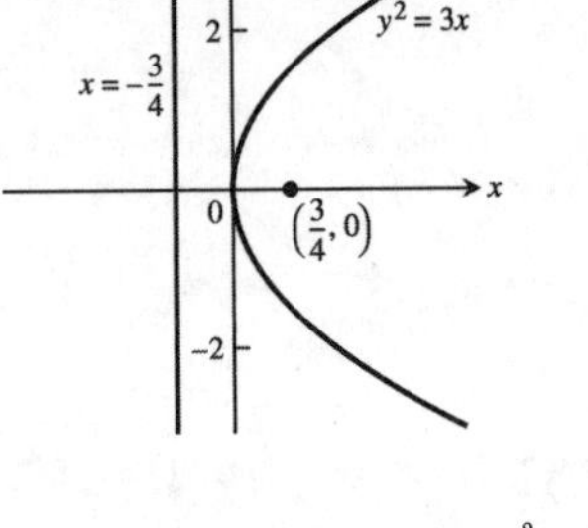

59. $16x^2 + 7y^2 = 112 \Rightarrow \frac{x^2}{7} + \frac{y^2}{16} = 1$
$\Rightarrow c^2 = 16 - 7 = 9 \Rightarrow c = 3; e = \frac{c}{a} = \frac{3}{4}$

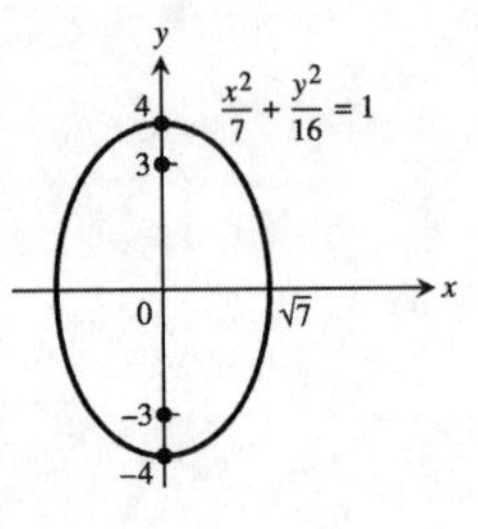

61. $3x^2 - y^2 = 3 \Rightarrow x^2 - \frac{y^2}{3} = 1 \Rightarrow c^2 = 1 + 3 = 4$
$\Rightarrow c = 2; e = \frac{c}{a} = \frac{2}{1} = 2$; the asymptotes are
$y = \pm\sqrt{3}x$

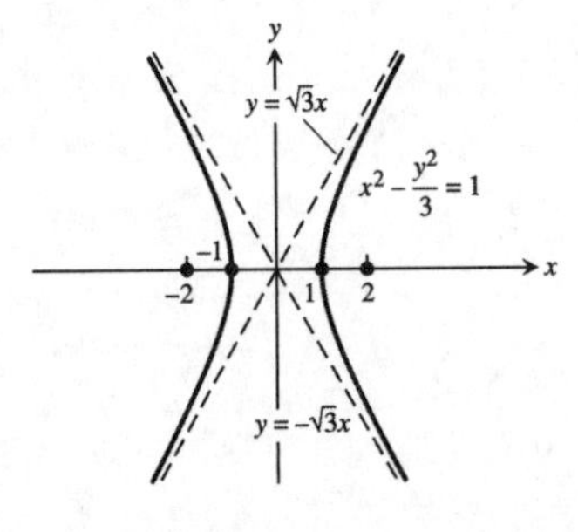

63. $x^2 = -12y \Rightarrow -\frac{x^2}{12} = y \Rightarrow 4p = 12 \Rightarrow p = 3 \Rightarrow$ focus is $(0, -3)$, directrix is $y = 3$, vertex is $(0, 0)$; therefore new vertex is $(2, 3)$, new focus is $(2, 0)$, new directrix is $y = 6$, and the new equation is $(x-2)^2 = -12(y-3)$

65. $\frac{x^2}{9} + \frac{y^2}{25} = 1 \Rightarrow a = 5$ and $b = 3 \Rightarrow c = \sqrt{25 - 9} = 4 \Rightarrow$ foci are $(0, \pm 4)$, vertices are $(0, \pm 5)$, center is $(0, 0)$; therefore the new center is $(-3, -5)$, new foci are $(-3, -1)$ and $(-3, -9)$, new vertices are $(-3, -10)$ and $(-3, 0)$, and the new equation is $\frac{(x+3)^2}{9} + \frac{(y+5)^2}{25} = 1$

67. $\frac{y^2}{8} - \frac{x^2}{2} = 1 \Rightarrow a = 2\sqrt{2}$ and $b = \sqrt{2} \Rightarrow c = \sqrt{8+2} = \sqrt{10} \Rightarrow$ foci are $\left(0, \pm\sqrt{10}\right)$, vertices are $\left(0, \pm 2\sqrt{2}\right)$, center is $(0, 0)$, and the asymptotes are $y = \pm 2x$; therefore the new center is $\left(2, 2\sqrt{2}\right)$, new foci are $\left(2, 2\sqrt{2} \pm \sqrt{10}\right)$, new vertices are $\left(2, 4\sqrt{2}\right)$ and $(2, 0)$, the new asymptotes are $y = 2x - 4 + 2\sqrt{2}$ and $y = -2x + 4 + 2\sqrt{2}$; the new equation is $\frac{\left(y - 2\sqrt{2}\right)^2}{8} - \frac{(x-2)^2}{2} = 1$

69. $x^2 - 4x - 4y^2 = 0 \Rightarrow x^2 - 4x + 4 - 4y^2 = 4 \Rightarrow (x-2)^2 - 4y^2 = 4 \Rightarrow \frac{(x-2)^2}{4} - y^2 = 1$, a hyperbola; $a = 2$ and $b = 1 \Rightarrow c = \sqrt{1+4} = \sqrt{5}$; the center is $(2, 0)$, the vertices are $(0, 0)$ and $(4, 0)$; the foci are $\left(2 \pm \sqrt{5}, 0\right)$ and the asymptotes are $y = \pm \frac{x-2}{2}$

71. $y^2 - 2y + 16x = -49 \Rightarrow y^2 - 2y + 1 = -16x - 48 \Rightarrow (y-1)^2 = -16(x+3)$, a parabola; the vertex is $(-3, 1)$; $4p = 16 \Rightarrow p = 4 \Rightarrow$ the focus is $(-7, 1)$ and the directrix is $x = 1$

73. $9x^2 + 16y^2 + 54x - 64y = -1 \Rightarrow 9(x^2 + 6x) + 16(y^2 - 4y) = -1 \Rightarrow 9(x^2 + 6x + 9) + 16(y^2 - 4y + 4) = 144$ $\Rightarrow 9(x+3)^2 + 16(y-2)^2 = 144 \Rightarrow \frac{(x+3)^2}{16} + \frac{(y-2)^2}{9} = 1$, an ellipse; the center is $(-3, 2)$; $a = 4$ and $b = 3$ $\Rightarrow c = \sqrt{16-9} = \sqrt{7}$; the foci are $\left(-3 \pm \sqrt{7}, 2\right)$; the vertices are $(1, 2)$ and $(-7, 2)$

75. $x^2 + y^2 - 2x - 2y = 0 \Rightarrow x^2 - 2x + 1 + y^2 - 2y + 1 = 2 \Rightarrow (x-1)^2 + (y-1)^2 = 2$, a circle with center $(1, 1)$ and radius $= \sqrt{2}$

77. $r = \frac{2}{1 + \cos\theta} \Rightarrow e = 1 \Rightarrow$ parabola with vertex at $(1, 0)$

79. $r = \frac{6}{1 - 2\cos\theta} \Rightarrow e = 2 \Rightarrow$ hyperbola; $ke = 6 \Rightarrow 2k = 6$ $\Rightarrow k = 3 \Rightarrow$ vertices are $(2, \pi)$ and $(6, \pi)$

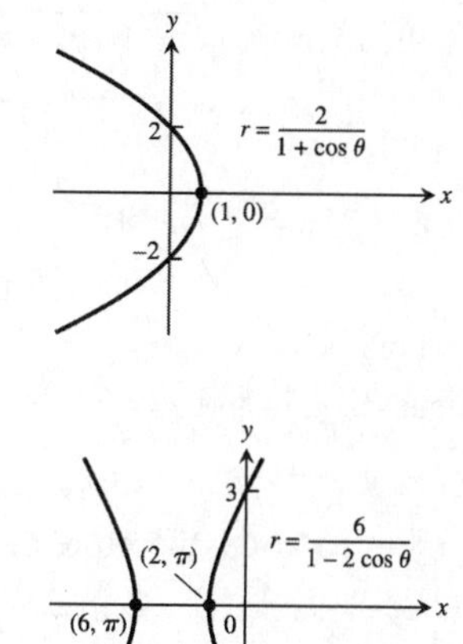

81. $e = 2$ and $r\cos\theta = 2 \Rightarrow x = 2$ is directrix $\Rightarrow k = 2$; the conic is a hyperbola; $r = \frac{ke}{1 + e\cos\theta} \Rightarrow r = \frac{(2)(2)}{1 + 2\cos\theta}$ $\Rightarrow r = \frac{4}{1 + 2\cos\theta}$

83. $e = \frac{1}{2}$ and $r\sin\theta = 2 \Rightarrow y = 2$ is directrix $\Rightarrow k = 2$; the conic is an ellipse; $r = \frac{ke}{1 + e\sin\theta} \Rightarrow r = \frac{(2)\left(\frac{1}{2}\right)}{1 + \left(\frac{1}{2}\right)\sin\theta}$ $\Rightarrow r = \frac{2}{2 + \sin\theta}$

85. (a) Around the x-axis: $9x^2 + 4y^2 = 36 \Rightarrow y^2 = 9 - \frac{9}{4}x^2 \Rightarrow y = \pm\sqrt{9 - \frac{9}{4}x^2}$ and we use the positive root:
$$V = 2\int_0^2 \pi \left(\sqrt{9 - \frac{9}{4}x^2}\right)^2 dx = 2\int_0^2 \pi \left(9 - \frac{9}{4}x^2\right) dx = 2\pi \left[9x - \frac{3}{4}x^3\right]_0^2 = 24\pi$$
(b) Around the y-axis: $9x^2 + 4y^2 = 36 \Rightarrow x^2 = 4 - \frac{4}{9}y^2 \Rightarrow x = \pm\sqrt{4 - \frac{4}{9}y^2}$ and we use the positive root:
$$V = 2\int_0^3 \pi \left(\sqrt{4 - \frac{4}{9}y^2}\right)^2 dy = 2\int_0^3 \pi \left(4 - \frac{4}{9}y^2\right) dy = 2\pi \left[4y - \frac{4}{27}y^3\right]_0^3 = 16\pi$$

87. (a) $r = \frac{k}{1 + e\cos\theta} \Rightarrow r + er\cos\theta = k \Rightarrow \sqrt{x^2 + y^2} + ex = k \Rightarrow \sqrt{x^2 + y^2} = k - ex \Rightarrow x^2 + y^2$ $= k^2 - 2kex + e^2x^2 \Rightarrow x^2 - e^2x^2 + y^2 + 2kex - k^2 = 0 \Rightarrow (1 - e^2)x^2 + y^2 + 2kex - k^2 = 0$

(b) $e = 0 \Rightarrow x^2 + y^2 - k^2 = 0 \Rightarrow x^2 + y^2 = k^2 \Rightarrow$ circle;
$0 < e < 1 \Rightarrow e^2 < 1 \Rightarrow e^2 - 1 < 0 \Rightarrow B^2 - 4AC = 0^2 - 4(1 - e^2)(1) = 4(e^2 - 1) < 0 \Rightarrow$ ellipse;
$e = 1 \Rightarrow B^2 - 4AC = 0^2 - 4(0)(1) = 0 \Rightarrow$ parabola;
$e > 1 \Rightarrow e^2 > 1 \Rightarrow B^2 - 4AC = 0^2 - 4(1 - e^2)(1) = 4e^2 - 4 > 0 \Rightarrow$ hyperbola

CHAPTER 11 ADDITIONAL AND ADVANCED EXERCISES

1. Directrix $x = 3$ and focus $(4, 0) \Rightarrow$ vertex is $\left(\frac{7}{2}, 0\right)$
$\Rightarrow p = \frac{1}{2} \Rightarrow$ the equation is $x - \frac{7}{2} = \frac{y^2}{2}$

3. $x^2 = 4y \Rightarrow$ vertex is $(0, 0)$ and $p = 1 \Rightarrow$ focus is $(0, 1)$; thus the distance from $P(x, y)$ to the vertex is $\sqrt{x^2 + y^2}$ and the distance from P to the focus is $\sqrt{x^2 + (y - 1)^2} \Rightarrow \sqrt{x^2 + y^2} = 2\sqrt{x^2 + (y - 1)^2}$
$\Rightarrow x^2 + y^2 = 4\left[x^2 + (y - 1)^2\right] \Rightarrow x^2 + y^2 = 4x^2 + 4y^2 - 8y + 4 \Rightarrow 3x^2 + 3y^2 - 8y + 4 = 0$, which is a circle

5. Vertices are $(0, \pm 2) \Rightarrow a = 2$; $e = \frac{c}{a} \Rightarrow 0.5 = \frac{c}{2} \Rightarrow c = 1 \Rightarrow$ foci are $(0, \pm 1)$

7. Let the center of the hyperbola be $(0, y)$.
(a) Directrix $y = -1$, focus $(0, -7)$ and $e = 2 \Rightarrow c - \frac{a}{e} = 6 \Rightarrow \frac{a}{e} = c - 6 \Rightarrow a = 2c - 12$. Also $c = ae = 2a$
$\Rightarrow a = 2(2a) - 12 \Rightarrow a = 4 \Rightarrow c = 8$; $y - (-1) = \frac{a}{e} = \frac{4}{2} = 2 \Rightarrow y = 1 \Rightarrow$ the center is $(0, 1)$; $c^2 = a^2 + b^2$
$\Rightarrow b^2 = c^2 - a^2 = 64 - 16 = 48$; therefore the equation is $\frac{(y-1)^2}{16} - \frac{x^2}{48} = 1$
(b) $e = 5 \Rightarrow c - \frac{a}{e} = 6 \Rightarrow \frac{a}{e} = c - 6 \Rightarrow a = 5c - 30$. Also, $c = ae = 5a \Rightarrow a = 5(5a) - 30 \Rightarrow 24a = 30 \Rightarrow a = \frac{5}{4}$
$\Rightarrow c = \frac{25}{4}$; $y - (-1) = \frac{a}{e} = \frac{\left(\frac{5}{4}\right)}{5} = \frac{1}{4} \Rightarrow y = -\frac{3}{4} \Rightarrow$ the center is $\left(0, -\frac{3}{4}\right)$; $c^2 = a^2 + b^2 \Rightarrow b^2 = c^2 - a^2$
$= \frac{625}{16} - \frac{25}{16} = \frac{75}{2}$; therefore the equation is $\frac{\left(y + \frac{3}{4}\right)^2}{\left(\frac{25}{16}\right)} - \frac{x^2}{\left(\frac{75}{2}\right)} = 1$ or $\frac{16\left(y + \frac{3}{4}\right)^2}{25} - \frac{2x^2}{75} = 1$

9. $b^2x^2 + a^2y^2 = a^2b^2 \Rightarrow \frac{dy}{dx} = -\frac{b^2x}{a^2y}$; at (x_1, y_1) the tangent line is $y - y_1 = \left(-\frac{b^2x_1}{a^2y_1}\right)(x - x_1)$
$\Rightarrow a^2yy_1 + b^2xx_1 = b^2x_1^2 + a^2y_1^2 = a^2b^2 \Rightarrow b^2xx_1 + a^2yy_1 - a^2b^2 = 0$

11.

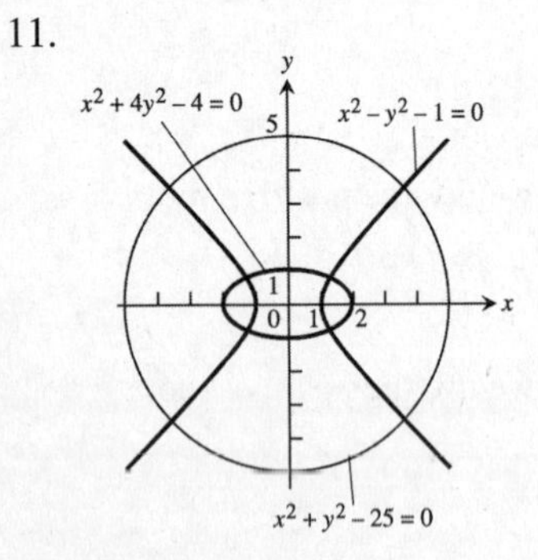

13.

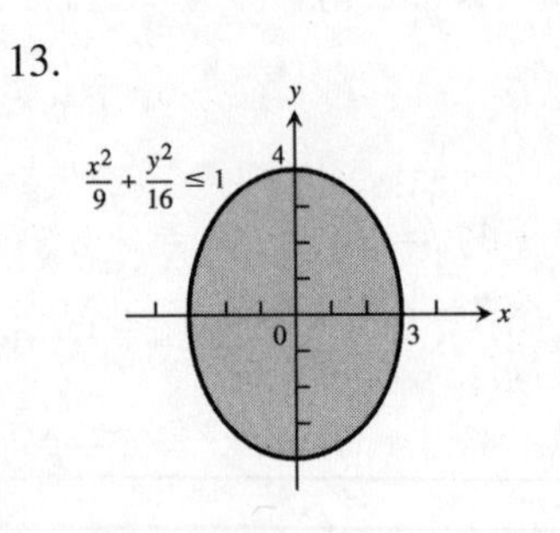

15. $(9x^2 + 4y^2 - 36)(4x^2 + 9y^2 - 16) \le 0$
$\Rightarrow 9x^2 + 4y^2 - 36 \le 0$ and $4x^2 + 9y^2 - 16 \ge 0$
or $9x^2 + 4y^2 - 36 \ge 0$ and $4x^2 + 9y^2 - 16 \le 0$

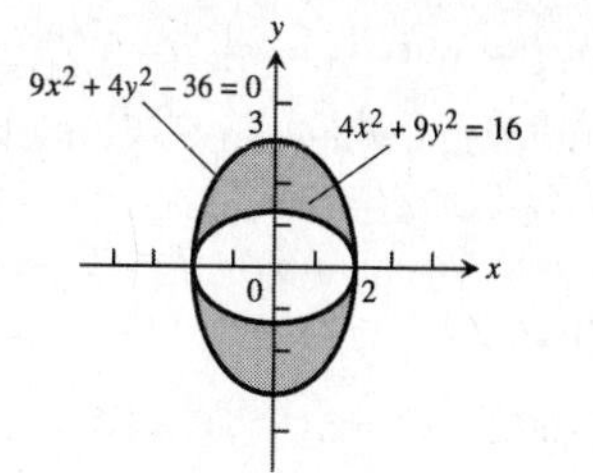

17. (a) $x = e^{2t}\cos t$ and $y = e^{2t}\sin t \Rightarrow x^2 + y^2 = e^{4t}\cos^2 t + e^{4t}\sin^2 t = e^{4t}$. Also $\frac{y}{x} = \frac{e^{2t}\sin t}{e^{2t}\cos t} = \tan t$
$\Rightarrow t = \tan^{-1}\left(\frac{y}{x}\right) \Rightarrow x^2 + y^2 = e^{4\tan^{-1}(y/x)}$ is the Cartesian equation. Since $r^2 = x^2 + y^2$ and
$\theta = \tan^{-1}\left(\frac{y}{x}\right)$, the polar equation is $r^2 = e^{4\theta}$ or $r = e^{2\theta}$ for $r > 0$

(b) $ds^2 = r^2\,d\theta^2 + dr^2;\ r = e^{2\theta} \Rightarrow dr = 2e^{2\theta}\,d\theta$
$\Rightarrow ds^2 = r^2\,d\theta^2 + \left(2e^{2\theta}\,d\theta\right)^2 = \left(e^{2\theta}\right)^2 d\theta^2 + 4e^{4\theta}\,d\theta^2$
$= 5e^{4\theta}\,d\theta^2 \Rightarrow ds = \sqrt{5}\,e^{2\theta}\,d\theta \Rightarrow L = \int_0^{2\pi} \sqrt{5}\,e^{2\theta}\,d\theta$
$= \left[\frac{\sqrt{5}\,e^{2\theta}}{2}\right]_0^{2\pi} = \frac{\sqrt{5}}{2}\left(e^{4\pi} - 1\right)$

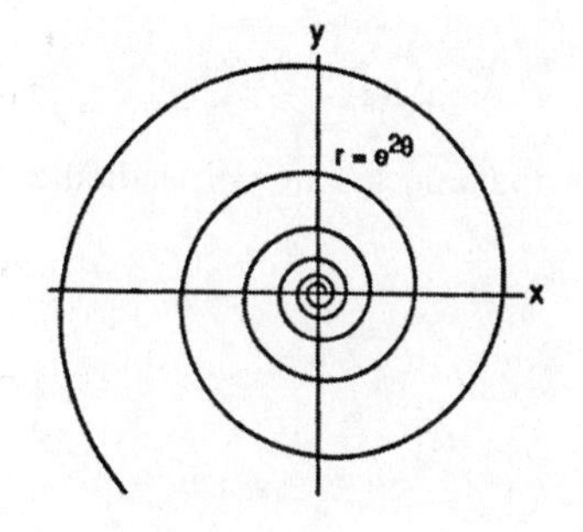

19. $e = 2$ and $r\cos\theta = 2 \Rightarrow x = 2$ is the directrix $\Rightarrow k = 2$; the conic is a hyperbola with $r = \frac{ke}{1 + e\cos\theta}$
$\Rightarrow r = \frac{(2)(2)}{1 + 2\cos\theta} = \frac{4}{1 + 2\cos\theta}$

21. $e = \frac{1}{2}$ and $r\sin\theta = 2 \Rightarrow y = 2$ is the directrix $\Rightarrow k = 2$; the conic is an ellipse with $r = \frac{ke}{1 + e\sin\theta}$
$\Rightarrow r = \frac{2\left(\frac{1}{2}\right)}{1 + \left(\frac{1}{2}\right)\sin\theta} = \frac{2}{2 + \sin\theta}$

23. Arc PF = Arc AF since each is the distance rolled;
$\angle PCF = \frac{\text{Arc PF}}{b} \Rightarrow \text{Arc PF} = b(\angle PCF);\ \theta = \frac{\text{Arc AF}}{a}$
$\Rightarrow \text{Arc AF} = a\theta \Rightarrow a\theta = b(\angle PCF) \Rightarrow \angle PCF = \left(\frac{a}{b}\right)\theta;$
$\angle OCB = \frac{\pi}{2} - \theta$ and $\angle OCB = \angle PCF - \angle PCE$
$= \angle PCF - \left(\frac{\pi}{2} - \alpha\right) = \left(\frac{a}{b}\right)\theta - \left(\frac{\pi}{2} - \alpha\right) \Rightarrow \frac{\pi}{2} - \theta$
$= \left(\frac{a}{b}\right)\theta - \left(\frac{\pi}{2} - \alpha\right) \Rightarrow \frac{\pi}{2} - \theta = \left(\frac{a}{b}\right)\theta - \frac{\pi}{2} + \alpha$
$\Rightarrow \alpha = \pi - \theta - \left(\frac{a}{b}\right)\theta \Rightarrow \alpha = \pi - \left(\frac{a+b}{b}\right)\theta.$

Now $x = OB + BD = OB + EP = (a + b)\cos\theta + b\cos\alpha = (a + b)\cos\theta + b\cos\left(\pi - \left(\frac{a+b}{b}\right)\theta\right)$
$= (a + b)\cos\theta + b\cos\pi\cos\left(\left(\frac{a+b}{b}\right)\theta\right) + b\sin\pi\sin\left(\left(\frac{a+b}{b}\right)\theta\right) = (a + b)\cos\theta - b\cos\left(\left(\frac{a+b}{b}\right)\theta\right)$ and
$y = PD = CB - CE = (a + b)\sin\theta - b\sin\alpha = (a + b)\sin\theta - b\sin\left(\left(\frac{a+b}{b}\right)\theta\right)$
$= (a + b)\sin\theta - b\sin\pi\cos\left(\left(\frac{a+b}{b}\right)\theta\right) + b\cos\pi\sin\left(\left(\frac{a+b}{b}\right)\theta\right) = (a + b)\sin\theta - b\sin\left(\left(\frac{a+b}{b}\right)\theta\right);$
therefore $x = (a + b)\cos\theta - b\cos\left(\left(\frac{a+b}{b}\right)\theta\right)$ and $y = (a + b)\sin\theta - b\sin\left(\left(\frac{a+b}{b}\right)\theta\right)$

25. $\beta = \psi_2 - \psi_1 \Rightarrow \tan\beta = \tan(\psi_2 - \psi_1) = \frac{\tan\psi_2 - \tan\psi_1}{1 + \tan\psi_2 \tan\psi_1}$; the curves will be orthogonal when $\tan\beta$ is undefined, or when $\tan\psi_2 = \frac{-1}{\tan\psi_1} \Rightarrow \frac{r}{g'(\theta)} = \frac{-1}{\left[\frac{r}{f'(\theta)}\right]}$ $\Rightarrow r^2 = -f'(\theta)\,g'(\theta)$

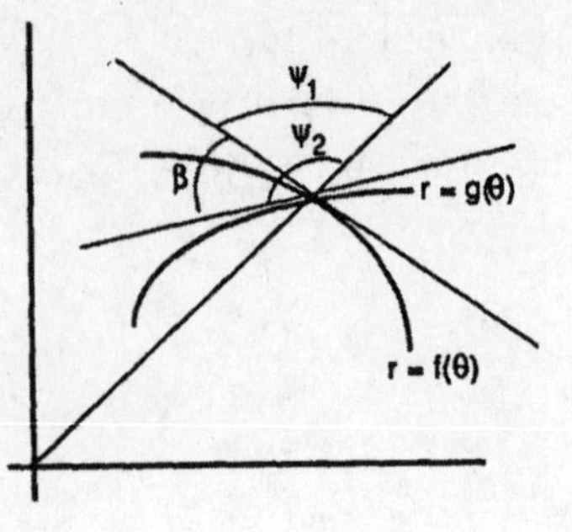

27. $r = 2a\sin 3\theta \Rightarrow \frac{dr}{d\theta} = 6a\cos 3\theta \Rightarrow \tan\psi = \frac{r}{\left(\frac{dr}{d\theta}\right)} = \frac{2a\sin 3\theta}{6a\cos 3\theta} = \frac{1}{3}\tan 3\theta$; when $\theta = \frac{\pi}{6}$, $\tan\psi = \frac{1}{3}\tan\frac{\pi}{2} \Rightarrow \psi = \frac{\pi}{2}$

29. $\tan\psi_1 = \frac{\sqrt{3}\cos\theta}{-\sqrt{3}\sin\theta} = -\cot\theta$ is $-\frac{1}{\sqrt{3}}$ at $\theta = \frac{\pi}{3}$; $\tan\psi_2 = \frac{\sin\theta}{\cos\theta} = \tan\theta$ is $\sqrt{3}$ at $\theta = \frac{\pi}{3}$; since the product of these slopes is -1, the tangents are perpendicular.